ISBN 978-1-5279-0050-9
PIBN 10928725

1 MONTH OF
FREE
READING

at

www.ForgottenBooks.com

By purchasing this book you are
eligible for one month membership to
ForgottenBooks.com, giving you
unlimited access to our entire
collection of over 1,000,000 titles via
our web site and mobile apps.

To claim your free month visit:
www.forgottenbooks.com/free928725

English
Français
Deutsche
Italiano
Español
Português

www.forgottenbooks.com

Mythology Photography **Fiction**
Fishing Christianity **Art** Cooking
Essays Buddhism Freemasonry
Medicine **Biology** Music **Ancient
Egypt** Evolution Carpentry Physics
Dance Geology **Mathematics** Fitness
Shakespeare **Folklore** Yoga Marketing
Confidence Immortality Biographies
Poetry **Psychology** Witchcraft
Electronics Chemistry History **Law**
Accounting **Philosophy** Anthropology
Alchemy Drama Quantum Mechanics
Atheism Sexual Health **Ancient History**
Entrepreneurship Languages Sport
Paleontology Needlework Islam
Metaphysics Investment Archaeology
Parenting Statistics Criminology
Motivational

THE

ENGINEERING MAGAZINE

AN INDUSTRIAL REVIEW

VOLUME XXXIV

October, 1907, to March, 1908

NEW YORK

THE ENGINEERING MAGAZINE

1908

INDEX TO VOLUME XXXIV.

October, 1907, to March, 1908.

LEADING ARTICLES.

REVIEWS, INDEX, AND EDITORIALS.

Vol. XXXIV.   OCTOBER, 1907.   No. 1

THE COPPER SITUATION IN THE UNITED STATES.

A STUDY OF THE STATISTICAL POSITION AND ECONOMIC OUTLOOK.

By James Douglas.

COPPER has of late occupied in the public eye a larger field of view than perhaps its importance as a national industry warrants. This has been due to the abnormal price to which it has rapidly risen—a price which has been in a certain sense fictitious, as probably none of the producing companies ever realized this price on much of their product. The producing companies, though not interested in the manufacturing companies, very properly contract ahead to supply the manufacturing companies with the raw material they have to deliver as a manufactured article. There is, so far, a community of interest between the two branches of the industry. The producing companies, therefore, have been selling ahead on a steadily rising market, and have consequently not realized by a very substantial figure the current prices published weekly. But, nevertheless, the price has been sufficiently remunerative to excite in the public mind such a feverish desire to own copper stocks that these have run up to a high figure. There is no reason to assign any other explanation for the high price than the perfectly legitimate one that the demand was for the time being in excess of the supply, that stocks have been exhausted, and metal was therefore scarce. Imagining that the scarcity would continue, and would be perennial, the price continued to soar until the demand fell off, and the delusion was dissipated.

There has been for the last ten years an average rate of 84 tons of iron consumed to 1 ton of copper, as per the following tables of production and consumption of the two metals:

Year.	Pig-Iron Production in United States.	Per Cent of Increase.	Pig-Iron Consumption in United States.	Per Cent of Increase.
1897	9,652,680	11.94	9,381,914	13.36
8	11,773,934	20.97	12,005,674	27.96
9	13,620,703	15.67	13,779,442	14.69
1900	13,789,242	1.23	13,177,409	*4.45
1	15,878,354	15.29	16,232,446	23.18
2	17,821,307	12.23	18,436,870	13.58
3	18,009,252	*1.61	18,039,909	*2.15
4	16,497,003	8.39	16,679,555	*7.54
5	23,010,625	39.48	23,155,624	38.82
6	25,307,191	9.98	26,603,202	14.88

Total 165,360,291 Average 12.00   Total 167,492,045 Average 13.23

The world's demand for the two metals was in the same ratio, and the world's production of copper also shows a notable percentage decrease. The production was in 1897,—405,350 tons, and in 1906, 718,368 tons, or an average increase for ten years of 4.3 per cent, whereas between the production of 1905, which was 699,504 tons, and that of 1906, there was a gain of only 15,754 tons, or 2.03 per cent. —hence the rise in price. That iron did not sympathize in price is due to the fact that the undeveloped resources of iron ore, and the metallurgical appliances for its reduction to metal, are much greater than those of copper. Consequently iron can bear a temporary strain better than copper.

Moreover, the standard price for iron has been fixed at a figure high enough to compensate for wide variations in the market. The average price of the two metals for ten years has been as follows:

Year.	Southern Pig Iron.	Lake Copper.
1897	10.25 per ton	11.29 per lb.
1898	10.45 " "	12.03 " "
1899	17.75 " "	17.61 " "
1900	18.35 "	16.52 '
1901	14.60 "	16.55 '
1902	20.10 "	11.89 '
1903	18.31 "	13.42 '
1904	13.92 "	12.99 "
1905	16.66 "	15.70 "
1906	19.44 "	19.62 '

* Decrease.

† We have accepted the statistics as published in the *Mineral Industry*. They do not include copper from the lead desilverizers, which, until 1901 was considerable; but on the other hand they include copper sulphate which the Government statistics omit. The consumption, which is of course largely a matter of speculation and doubt, as the stock on hand can never be determined with accuracy, is only calculated in the Government report since 1901.

Pig-Copper Production in United States.	Per Cent of Increase.	Pig-Copper Consumption in United States.	Per Cent of Increase.	Tons Pig Iron Consumed to 1 Ton Pig Copper.
221,958	6.32	122,500	20.	77
236,109	6.38	122,383	* .10	98
255,272	8.12	174,822	42.85	79
263,178	3.09	155,169	*11.24	85
266,716	1.34	196,837	26.85	82
284,284	6.59	209,241	6.30	88
316,239	11.24	221,222	5.73	82
365,051	15.43	208,082	* 5.94	80
390,733	7.04	273,652	31.51	85
409,652	4.84	298,472	9.07	89
Total 3,009,192	Average 7.39	Total 1,982,379	Average 12.50	Average 84

Between the price of rails early in 1897 of $16.00 per ton, and the present price of $28.00 per ton, there is, however, as wide a difference as between the price of copper in 1896, $10.88, and the price of copper in 1906, $19.62. But there can be a policy pursued in regulating the home prices of iron which cannot be followed in the case of copper. Iron is protected; copper is not; and, therefore, if desired by the large manufacturers of iron to maintain a staple price, the very large margin of $11.20 per ton of steel and of $4.00 per ton of pig iron, which the duty gives them, may be used. In case of a low foreign price, a high domestic price can be sustained; or in case of high foreign price, the domestic price may be maintained at a given standard up to the limit of the duty. The McKinley Bill struck off all duty on crude and refined copper, though retaining it on manufactured copper. Thus, the world's price regulates the American domestic price, and the copper producers have been saved from any temptation to form a copper trust, in order to derive the benefit of the duty. As to the regulation of price, considering our supreme position as producers it may be more proper to say that we fix the price of the world. For a short period this year, the English price ruled slightly lower than the price demanded in the United States, and copper immediately commenced to flow in from abroad, although over any considerable period of time, say the past ten years, 54.5 per cent of our production has been exported and we make 54.7 per cent of the world's total.

In spite of the active demand and the high price, last year's domestic production showed, as per the above table, that there were only two years, since copper became a prominent article of domestic production, when copper production showed as slight a percentage of gain.

A steadily progressive high percentage increase, however, cannot be maintained and should not be expected when the gross increase reaches certain limits. For instance, in 1896, when the percentage of increase over 1895 was 21 per cent, the actual increase was 81,168,640 pounds; but during the last decade, the production has increased at such a rate that the 4.8 per cent increase of 1906 over 1905 represents an actual increase of 42,378,560 pounds of copper—an increase of 137.4 per cent over the production of 1895, or of 451.1 per cent over that of 1885.

Still, it is significant that this notable falling off in percentage increase should occur when the high price of copper may be supposed to have served as a stimulus to the utmost activity. The world's increase, as well as our own, showed a diminished percentage of increase—being 2.16 per cent instead of 6.16 per cent of the previous years. To explain this anomaly, it must be borne in mind that as the average of the percent of the ore treated was probable lower, more ore in proportion to the output of copper was probably mined. When copper is 20 cents, money can be made out of ore which is valueless with copper standing at 12 cents; and, therefore, should the contents of the mines permit of selection, ore is utilized at one time which at another time is rejected. Moreover, the average percentage of ore of the great mines of the country is declining, and will continue to decline. Rich bodies of secondary deposits may be discovered in any deep mine, which may arrest the decline in its average yield, but the world's experience is not likely to be reversed in our favor—more especially as it coincides with our own—that the average yield of deep copper mines declines with depth; and as with declining yield generally occurs the increased cost of mining incident to greater depth, two factors concur to bring about the reduced production, which must occur sooner or later.

In comparing iron and copper mining, and the ease with which the varying demands of trade may be met by the producers, it must be remembered that 2 tons of average iron ore from the most prolific source now available—Lake Superior—will make 1 ton of metal; whereas from 33 to 75 tons (say an average of 50 tons) of copper ore must be mined to produce 1 ton of copper. A sudden increase in demand for the baser metal can be more readily met than when such a call is made for the rarer. The relative supply in nature may be gauged by the value on which the price of each is made, the value of iron being generally quoted by the ton, and of copper by the pound. Last year's production of iron was made from approximately 50,000,-

000 tons of ore. Our copper mines, to yield a production of 900,000,-000 pounds, must have handled approximately 22,000,000 tons of ore, or nearly half the quantity of ore raised from our iron mines. The ore of iron moreover occurs in very large deposits, worked in some cases by steam shovels, whereas our copper comes from comparatively narrow veins or irregular masses.

Referring to the relative growth and decline of the three principal sources of the metal since the Western States began to grow into prominence, say from 1882, we find that at that date the Lakes supplied from ores of metallic copper 25,000 long tons of copper; Arizona, from oxidized ores, 8,000 long tons of copper; and Montana, from sulphide ores, 4,000 long tons. After that, the Lake production relatively declined. The Montana production continued to increase rapidly until 1887, when Montana took the first place with 35,000 tons, Lake Superior second place with 34,000, and Arizona third, with 8,000. This order was maintained until 1905, when Arizona attained the second rank.

Meanwhile, however, notable contributions were made from neglected States which had again become active producers, or from entirely new territory. Before the Civil War broke out, Tennessee was the largest producer on the Appalachian Range. The active mines shipped during the first nine months of 1855, 14,291 tons of ore of 25 per cent—the Hiawassee alone shipping 4,156 tons. Smelting works were successfully started that year, and before the war broke out, the copper produced was being refined on the spot. After the war mining declined and for a time production ceased. But when the mines were reopened new metallurgical methods were very skilfully applied, and in 1896 Tennessee made 3,750,124 pounds of copper; but last year the State is credited with 18,821,000 pounds. Utah until recently has made an almost negligible quantity, but last year she added to the nation's stock 49,712,000 pounds. California in the sixties and seventies occupied a prominent position, but her production sank to 120,156 pounds in 1894. The revival commenced with the opening of the Mountain Copper Mine in 1896 and with a product of 1,120,000 pounds, which sprung to 14,000,000 in the following year. It now stands at 24,421,000 pounds. What Alaska will do, the future only will show. The large lenses of cupriferous pyrites on the coast in Hetta inlet sent into the market last year 8,700,000 pounds of copper, most of it treated at the Tacoma works. The largest sources of new supply, which can be counted on with tolerable certainty, will be found in the low-grade ores of Utah and Nevada, where works on a very large scale are being planned or erected for their treatment.

It is expected that within two or three years 15,000 to 20,000 tons of ore a day, containing not only copper but carrying notable quantities of the precious metals, will be concentrated and smelted in these two States, which should add to our total about 200,000,000 pounds a year.

The presence of small quantities of gold and silver in copper ores, now that the process of electrolytic copper refining permits these elements to be cheaply separated among the impurities, is putting within the range of profitable extraction ores of very low percentage in copper. When the Anaconda mine was first opened, and its rich ores (and subsequently the mattes, made from them) were shipped to Europe for refining, the margin of gold and silver deducted from the ton of copper, before any allowance was made to the shipper, was $60.00. As this deduction was substantially the whole of the precious metals in the Butte copper, the company derived no benefit. Whatever margin there was in the $60 over and above the heavy cost of separation went to the refiners. Most of the rich argentiferous copper was then converted into bluestone. But since electrolysis has been applied to copper refining, and the process has been carried out on a large scale, the commercial charges for refining the copper and separating the precious metals have been reduced from $60 to $16, and as the value of the electrolytically refined copper is ordinarily about $10 a ton above that of furnace-refined copper, copper bars containing $6 in precious metals can be economically converted into electrolytic, and their gold and silver added to the world's wealth. If the copper bars contain the concentrated gold and silver of 30 tons of ore, the ore need contain only 20 cents of gold and silver per ton to make it profitable to submit the resulting copper to electrolysis. The amount thus extracted from domestic and foreign ores in the refineries of the United States is approximately:—from Montana, $6,650,000; Arizona and Sonora, $2,105,000; Canada, $460,000; and other sources $2,500,-000—or over $11,000,000—most of which formerly entered the refined copper and was wasted.

While the presence of precious metals in high-grade ores redounds to the fortunate copper company which mines them, when ores of 2 per cent and under (like those of the Boundary District of British Columbia, or Bingham Cañon in Utah, or of Ely County in Nevada) are treated, the addition of $2 or $3 in gold and silver per ton may convert an unprofitable into a very profitable mineral. And from such ores will probably come much of the copper from our new mineral developments; for it is not probable that within the explored territory, many new large deposits will be discovered which are also of high grade, whereas lean ores in very large quantities are known to exist in many districts of the country. Apart from their contents in precious metals, better machinery and improved metallurgical methods permit of ores being treated economically today, which would not have been classed as ore a few years ago; and today ores are rejected in developed mines, which at no distant day may help us to maintain our position in the copper world.

But as already pointed out, we cannot expect to grow as vigorously as we did in the past. We are using up our natural resources of iron, copper, and lumber at headlong speed, and they cannot last. Nevertheless, there need be no immediate alarm as to a heavy decline in our copper supply. While even such large reserves as the Calumet and Hecla once had in the conglomerate beds are within measureable distance of exhaustion, others of the company's mines on the Osceola and Kearsarge amygdaloid promise to enable it to maintain its output, and the success of the Wolverine and Mohawk to the northeast of Portage Lake, and of the Baltic and other mines on the Copper Range to the southwest, are prophetic of what will gradually happen as the vast area of the amygdaloid and conglomerate beds which circle round from the Keweenaw promontory through Wisconsin into Minnesota are explored. They are so buried under forests and soil that their development will be slow, but the possibilities hidden within them are incalculable. Montana, which heretofore has meant Butte, has held her own simply by increasing her concentrating and smelting facilities to compensate for the gradual decline in her ores; but the discovery of large copper bodies at great depth below the argentiferous ores of the North Butte mine and of others within what was considered formerly as the silver zone, and also to the south and east of what was once supposed to be the limit of the copper zone, may extend the life of Butte indefinitely beyond the limits assigned to it even by its most enthusiastic supporters; and Butte will not be the only copper district of Montana, though probably none as productive will ever be discovered.

Arizona, which has made more rapid strides of late than its competitors, owes its progress mainly to the remarkable mining developments in the limestones of the Warren district in the neighborhood of Bisbee; and as yet, the limit of these large deposits, either laterally or in depth, has not been determined. In the Clifton district, the very large quantities of low-grade ores which support the Arizona Copper Company, the Detroit Copper Company, and the Shannon and other minor companies, show no signs of exhaustion. Other deposits of the same general character and of great magnitude, in Mineral Creek and elsewhere in Arizona, and in the Burro Mountains in New Mexico, assure us of a sure supply, if not a very rapid increase for some time to come from the Southwestern States. This assurance is confirmed by the rapid growth of the mines in the neighborhood of Globe, and from the production of the large number of small mines in the southwest, many of which are producing, and all of which are appealing to the public for support. The experience of the past year warns us, however, that the period of rapid expansion has passed, and that as the home demand will probably not fall off, we shall have less of our own copper for export than formerly. But this deficiency will probably be supplied from foreign ore or metallurgical products treated in our furnaces and refineries. Last year's statistics show an importation of foreign copper, partly in ores, partly in matte, and partly in unrefined bars, of 136,826,906 pounds. Most of this was from Mexico, some from Canada, some even from New Zealand. It is to the credit of our smelters and refineries that so large a quantity of copper should seek our shores for final treatment, some coming even from a British colony at the antipodes. This is due as much to their business methods as to the excellency and economy of their metallurgical work. The scale on which they work is liberal, and that is one reason why their scale of charges is attractive.

The *Mineral Industry* for 1906 gives the amount of electrolytic copper turned out by nine electrolytic refineries at 865,000,000 pounds, and their capacity, at 1,120,000,000 pounds. Distributing the actual amount refined proportionately among the refineries, if the capacity of each is correctly given, the production of each is as follows:

APPROXIMATE ANNUAL CAPACITY AND PRODUCTION OF ELECTROLYTIC COPPER, FOR YEAR ENDING DEC. 31, 1907.

	Capacity, lb.	Calculated production based on proportionate distribution of assumed total electrolytic copper output.
Nichols Copper Co........Laurel Hill, N. Y..	288,000,000	222,516,000
Raritan Works ....Perth Amboy, N. J.	288,000,000	222,516,000
American Sm. & Refg. Co...Perth Amboy, N. J.	144,000,000	111,168,000
U. S. Metals Refg. Co......Chrome, N. J......	144,000,000	111,168,000
Baltimore ....Baltimore, Md.	130,000,000	100,360,000
Balbach Sm. & Refg. Co....Newark, N. J......	75,000,000	57,900,000
Boston & Montana Cop. Co..Great Falls, Mont..	30,000,000	23,160,000
Tacoma Smelting Co........Tacoma, Wash. ...	18,000,000	13,896,000
Mountain Copper Co........Oakland, Cal.	3,000,000	2,316,000
	1,120,000,000	865,000,000

The very large quantity of copper carried in the vats of these electrolytic works introduces an element of uncertainty into the calculation of stocks on hand, and of consumption. It also influences the price, as the value of copper is undoubtedly influenced by the quantity in stock, actually available for sale and yet unsold. This is a matter of some moment at a period when production is on the increase. It must be remembered that any increase in production at the mines and at the mines' smelting works, does not enter the market for at least 90 days—as 30 days must be allowed for western copper in transit and 60 days for its electrolytic and furnace treatment at the refineries. If, therefore, there has been an increase in production at all the mines of the country of, say, 5,000,000 pounds, in any given month, this 5,000,000 becomes 15,000,000, before it reaches the market as salable copper. Moreover this addition to the copper in the vats of the electrolytic refineries, as long as the refineries continue to run up to a given capacity, should be no more counted as marketable stock than if it were still copper unmined and in the ground. At the present time there must be under treatment in the vats about 100,000,000 pounds of copper. In case of a great scarcity and a rise in price which would tempt the owners of this reserve stock to sacrifice its gold-silver contents, it would be speedily available. A reserve of this kind is a safeguard, not a menace to the market. Outside of this item it is very difficult to determine at any time what is in stock. In England a certain amount of copper is held under warrant and can be, and is, counted on as stock; but this is delusively small. In the United States the stock on hand is subject to guess work. That at the present moment it is increasing cannot be questioned; but it is small compared with the volume of business the metal has to support and the fluctuating needs of trade.

To forecast the future of copper is impossible. It is safe to predict that, if no substitute is discovered and the demand for the metal increases as it has done during the modern industrial era, the supply will fall short of the demand and the value of the metal will rise. The same is true of all the mineral products that are consumed in the arts. The experience of the last few years in the United States illustrates, however, the sensitive connection between demand and price. The table at the top of the following page shows that sudden excess of demand is generally followed by augmentation of price, because production does not as quickly respond.

AVERAGE PRICE OF ·LAKE COPPER IN NEW YORK.

Year.	Price.	Per Cent Increase in Production.	Per Cent Increase in Consumption.
1896	10.88	....	.....
1897	11.29	6.32	10.
1898	12.03	6.38	42.85
1899	17.61	8.12	11.24
1900	16.52	3.09	26.85
1901	16.55	1.34	6.30
1902	11.88	6.50	5.73
1903	13.41	11.24	5.94
1904	12.99	15.43	31.51
1905	15.69	7.04	9.07

While such spasmodic variations are transitory, there can be no
doubt that if the demand grows and no very important new discoveries are made, the metal will become scarcer and will command a
higher price than the average price of the past. One consequence,
which will also be a corrective, will be that the metal will cease to be
used for purposes to which it is turned today, and for which a
cheaper material can be discovered. For it is an invariable rule that
when the metal has run up above a figure which deters purchasers
from using it except for unavoidable purposes, the consumption drops
off sharply, and a more normal value is restored.

WORLD'S PRODUCTION BY QUINQUENNIAL PERIODS FROM 1881-1905.

	1881-1885. Tons.	1886-1890. Tons.	1891-1895. Tons.	1896-1900. Tons.	1901-1905. Tons.
United Kingdom	16,082	5,200	2,665	3,000	2,542
Spain and Portugal	215,306	267,592	273,497	263,800	244,996
Germany	74,064	79,551	83,350	104,165	107,735
Austria	2,775	5,061	6,200	5,175	5,535
Hungary	3,240	2,355	1,290	2,165	1,460
Sweden	3,962	3,985	2,790	2,495	2,300
Norway	13,126	7,987	9,092	17,110	25,575
Russia	20,248	23,445	24,980	31,335	46,395
Italy	6,640	8,562	12,300	15,765	15,755
Turkey	....	....	....	2,885	5,130
Chili	202,145	145,785	107,205	120,950	149,915
Venezuela	19,252	22,316	14,950	....	....
Japan	36,300	66,097	91,480	125,325	159,370
China	Unknown.	Unknown.	Unknown.	Unknown.	Unknown.
United States	261,649	468,998	760,004	1,179,932	1,643,810
Australasia	56,283	45,987	40,500	89,770	159,235
Argentine	1,792	840	950	565	770
Africa:—					
Algeria	2,310	590	155	50	....
Cape	25,608	34,915	31,520	35,160	31,180
Newfoundland	5,717	8,705	10,170	10,300	10,540
Bolivia	10,594	6,950	12,060	10,850	10,000
Peru	2,041	800	1,920	18,165	40,280
Canada	4,791	10,815	21,000	33,175	95,325
Mexico	1,889	13,771	44,385	82,340	227,660
	985,814	1,230,307	1,552,463	2,154,477	2,985,508

The all-important factors in discussing the future are the questions of sources of supply and the uses to which the metal is turned. The table on the opposite page is extracted from Brown and Turnbull's Century of Copper, and illustrates the decay and actual extinction of certain copper regions, and the growth of others. But the most vigorous of today will sooner or later be in the declining list.

The most notable features of the table are the virtual disappearance of Great Britain and Venezuela. Venezuela depended, however, upon only one great mine—the New Quebreda. In Great Britain, it has been a case of slow decline. The reduction in Spanish and Portuguese production is much more important. It is due to the approaching exhaustion of the large pyrites bodies worked by the Tharsis Co. and Mason & Barry, and points to the time, still happily distant, when even the vast resources of the Rio Tinto Company will have been worked out. To their huge lenses of sulphate of iron the chemical trade of the world looks for a cheap supply of the material for making sulphuric acid, and no similar source in such quantity or so advantageously situated exists elsewhere. The German production comes from one mine, the famous old Mansfeld. It has been worked since the thirteenth century, and has still comparatively a long life ahead. Turning to the American continent, we have already discussed the possibilities of our own future. Chili suffered a terrible decline after the exhaustion of her rich ores from her three great mining districts between 1870 and 1880, but she is now imitating us, and beginning to recognize the value of her leaner ores, and learning to utilize them. She, Bolivia, and Peru will show an increase in the future. So also will Mexico; and British Columbia will make progress with moderate speed. But upon the American continent, both South and North, there are wonderful possibilities of future discovery—in Alaska, and in British Columbia; in the Labrador peninsula; in the great belt of country in which cobalt and native silver deposits have so recently been discovered between the Lakes and Hudson Bay. Nor should the great Canadian Northwest even within the Arctic Circle be eliminated from view, for even Hearne's Copper River would not be shut off by ice from the world's markets by the closure of Hudson Bay for many more months than were the mines of Lake Superior till within very recent years, when their copper production could reach the east only after the opening of navigation.

The African Continent, as the table shows, has maintained a very stationary production. But in Africa there is in view the only large copper region said to be within reach of a future market. The Tanganyika Concessions, Ltd., claims to have discovered and sufficiently explored deposits of great extent in the Congo Free State, and the adjacent North-West Rhodesia, to make it probable (they consider certain) that this region will become one of the world's greatest sources of supply. But the deposits lie in the very heart of the continent, on the 12th degree of south latitude, and a thousand miles of railroad must be built before supplies and copper can be transported in and out between the West Coast and the mines.

Australia, the table shows, has made some progress, and will probably continue to move forward. So far as Russia in Asia has been explored, no mines of great promise have been exposed, but very little is really known of the possibilities of either Siberia or China.

Japan appears as a growing copper power, but this year her production will probably show no increase, owing to the labor riots and destruction of property at some of her copper mines. While she has undeveloped and partially developed resources which will increase her output, she has also created for herself in Corea, Manchuria, and Formosa, fields for industrial expansion which may absorb all her excess.

A review, therefore, of the world's mines would seem to indicate that the old mines are approaching exhaustion, that no new large deposits of rich ore, except possibly those of Tanganyika, are in sight; and therefore the trade must turn to deposits of lower grade than have heretofore been worked, if the demand continues as active as at present. And copper cannot be made as cheaply from mines of that class as from the mines we have been drawing upon so recklessly for a generation.

The next question bearing upon values is that of probable consumption. The three principal uses to which copper is now turned are:—in the manufacture of electrical machinery and the transmission of electrical current; in the composition of alloys, principally brass, for stationary engines, locomotives, railroad cars, automobiles, and for arms and ammunition, and in architecture, where it replaces iron and lead.

The Government has issued an advance sheet, giving its figures of production, importation, and exportation for 1906. The production differs from that of the *Mineral Industry* by 11,000,000 pounds, and makes the increase over 1905 only 2.03 per cent instead of 4.08 per cent. The statement makes the total of copper actually produced from domestic and foreign ores 1,097,961,969 pounds, and the exports

446,750,711, leaving a balance of 651,211,258 pounds. The Boston News Bureau assigns the following consumption to the five principal consumers in the United States:

American Brass Co..................................	250,000,000
Roebling & Sons Co................................	100,000,000
General Electric Co................................	75,000,000
American Steel & Wire Co..........................	50,000,000
National Conduit Co...............................	50,000,000
	525,000,000

This leaves for all other consumers and for margin of error, 126,-211,358 pounds, which is probably not far from the mark.

Glancing at the three sources of consumption, the draft on copper for telegraph wires, trolley lines, long-distance transmission and other electrical purposes, is of comparatively recent date, and is doubtless very large—probably about one-third of the world's supply goes into this new industry; but comparatively little of it is actually consumed, though a great deal of it is through carelessness wasted during the installation of plants. The copper remains in service with little waste, or goes back to the refineries as scrap. But the demand for this purpose will not fall off. The urban and interurban mileage of trolley lines in the United States is between 30,000 and 40,000 miles. They have been built within very recent years, and new lines are reticulating the country. No substitute for copper in this class of roads has been suggested, for though aluminum may indifferently replace copper for current transmission, it cannot bear the wear and tear of the trolley. Gradually, also, the motor will displace the locomotive on all our present steam roads where traffic is dense; and though the third rail instead of the trolley wire will be used, the amount of copper which goes into the transmission, the generators, the motors, and all subsidiary appliances, will increase the proportionate demand for the metal by electricians.

The amount of copper which enters brass and other alloys it is difficult to determine. From information derived from several of our large railroad systems I would infer that about 5 grains of copper are consumed and actually go out of existence per car mile. This would represent about 5,000 tons of copper consumed by our railroads per annum for this purpose alone. Automobiles must make a new and very heavy call on copper, and the demand for stationary engines and steamboats is increasing with the growing industrial activity of the country. The quantity used up in war material and ammunition—during peace and war—is very great. In spite of probable fluctuations in the brass trade, there will be a large and satisfactory average growth.

But when we come to the third principal use for copper, that is, in the building and house-furnishing trade, we find it playing the part of an article of luxury. Our forefathers were satisfied with wrought-iron railings in their city houses, beautiful examples of which still exist in all the west-side streets of the old residential portion of New York City. Now we must have bronze balustrades in our palatial offices, hotels and private buildings. Brass or bronze hinges and locks, gas fixtures, etc., appear in profusion in our houses and railroad cars, and brass and copper have driven out lead for roofing and plumbing. When copper is abundant and sells at a moderate price, it is the most desirable material for these purposes, but as it becomes rarer and dearer, substitutes for it will be used. As long, however, as money is forthcoming, the public taste for brass and bronze will be gratified, and probably more of the world's supply will go into architecture and house-furnishing than into any other use.

From the above review of the sources of supply and the sources of demand, it would seem that while there is no risk of copper becoming a rare metal in the near future, it will certainly have to be extracted from much leaner ores at an increased cost. Also while there is no reason to believe that its consumption in the arts will grow less, there are metals which can be substituted for it, should the popular taste or financial exigencies require.

PROSPERITY; ITS RELATION TO THE INCREAS-
ING PRODUCTION OF GOLD.

By Alex. Del Mar.

Mr. Del Mar was formerly Director of the United States Bureau of Commerce, Navigation and Statistics; official delegate to the International Statistical Congresses held in Florence, The Hague, and St. Petersburgh; a member of various governmental Monetary Commissions; and he is an authority of international reputation on the use of the precious metals
as money. Three years ago, in answer to a private inquiry by the editors of this Magazine,
he outlined the certain consequences of the vast additions which were and still are being
made to the world's store of gold, and predicted with most striking accuracy the phenomena
that have since appeared and are now apparent in economic and industrial affairs. Responding to our invitation to follow the analysis further, Mr. Del Mar has extended the range of
his observation over a wider field, with results even more keenly interesting in view of
the present state of the business and financial mind. His article will be read with close
attention and with profit, by those who demur as well as those who fully assent to his
reasoning.—THE EDITORS.

WHEN a period of commercial prosperity becomes prolonged, it
begins to invite the caution of the prudent and arouse the
fears of the timid. Such an interval occurred during the
Civil War and such another one is passing at the present time. It is
at periods of this character that every circumstance which tends to
heighten the prevailing distrust is brought forward by interest, prejudice or passion, and displayed as the true cause of the commotion.
The prosecution of certain railways for secret rebating; the condemnation of certain monopolies for repeated violations of law and good
faith; these and other pretexts are being sedulously put forward as
the causes of the existing financial uneasiness. Nor are such grounds
of alarm entirely destitute of plausibility; for alarm feeds as readily
upon belief as fact; upon sentiment, as upon reason. But after a
careful examination of the subject, it will be found that, aside from
a remediable constriction in the monetary field, there is nothing in
the situation to warrant any apprehension for the immediate future;
and as for such constriction, it is not pressing.

The circumstances which denote prosperity or adversity in the
United States embrace the outlook of Peace or War; Legislation; the
Crops; Mines; Forests and Fisheries; Immigration; Manufactures;
Foreign and Domestic Commerce; Railroads; Real Estate; Labor;
Interest; Prices; and Money. Of these in their turn.

The United States is at peace with all the world and itself; and appears likely so to remain. When it is considered that this country has never experienced a panic that could not be traced to its own legislative blunders, the importance of mature legislation by jurists thoroughly acquainted with economic laws can not be overestimated. The new conditions of this century demand the repeal of much of our lawyers' law and the enactment in its place of a little scientific law. These preliminaries bring us at once to the consideration of our principal source of prosperity—the agricultural crops.

THE CROPS of 1907 can happily be rapidly summarized. The report of the Agricultural Department for August indicates a crop of maize (corn), of about 2,650 million bushels; wheat, winter, 410, spring, 230, total 640 million bushels; together with crops of oats, barley, rye, buckwheat, tobacco, potatoes and hay equal to the average harvest of recent years. The cotton crop is still a matter of speculation. Of last year's crop there were "brought into sight," down to the first week in August, no less than 13,383,661 bales, against 11,052,063 bales at the same time the preceding year. The outlook for 1907 is for over 12 million bales, and the European demand so keen that the Texas planters are clamoring for a union to hold for 15 cents a pound, the prevailing price being about $13\frac{1}{2}$ cents for mid-upland.

When these colossal crop figures are multiplied by the enhanced prices which these staples now command, it will be found that the farmers and planters are likely to receive as much money for this year's produce as they got for last year's, and perhaps more. This means about six billion dollars.

The early fruit harvests have yielded, and the later ones, such as apples and grapes, promise, the usual abundance. From Maine to California and from the Lake region to the Gulf, Nature has poured forth such a profusion of agricultural wealth that nothing short of a *cadastre* can do justice to the subject. In almost every respect and in all important respects, it is a year of plenty.

The statistical returns from the mines, forests, and fisheries are not yet completed. Indications point to an increased product of gold, silver, copper, iron and coal, and a normal product of the other commodities included under these heads. The latest returns of the mineral product are those for the calendar year 1906, which Mr. Ingalls summarizes in his *Mineral Industry* as follows: The production of some two-score ores and minerals aggregated 819 millions, besides 177 millions of secondary minerals and chemicals. Metals yielded 873 millions, the gross total being 1,869 millions of dollars in value. The mines of the Transvaal yielded in July 54,438 fine ounces of gold, against 54,918 in June and 48,485 in July last year. The monthly average output this year amounts to nearly 50,000 ounces, much more than doubling the average of 1904, and making an increase of nearly 3,000 ounces per month over last year. The output of the last three months has been the greatest ever recorded.

Thus, in the production of the year 1907, there is to be observed no material falling off, in any general direction; and in most directions, a material gain. In a few words, the original material elements of prosperity produced this year should bring to the producers as much money as they did last year or more. The foundations of our welfare are thus seen to be solid; but what of the superstructure?

Immigration, whatever its intellectual or moral influences, has always been regarded as a source of wealth to the country, the immigrant bringing with him his money, goods, and knowledge. This year's immigration is the greatest in the history of the United States. The total number of immigrants in 1906 was 1,100,735, whilst in 1907 it was 1,285,349. As the immigrant brings with him on the average something over $100 in money and goods, we derive from this source alone an accession of material wealth to the country amounting to about 150 millions of dollars per annum.

The cry of the Far West for labor, which continues to fall as much short of demands as at any time last year, is one of the signs that prosperity remains at high pressure in that section of the country. The railroads, which were forced last Fall to advertise all over the East for men, offering inducements in transportation and wages, are still suffering; and not only are present extension projects in the West being held up for shortage of labor, but the fuel-supply problem for next winter is complicated by the same cause. In Utah mine labor is so scarce that Japanese are drawing as high as $170 a month.

DOMESTIC MANUFACTURES AND TRADE. With the object of ascertaining the true condition and prospects of manufactures and trade an enquiry was addressed in August to the leading trade papers in each principal line of industry, those best qualified to furnish reliable replies. The general result was summarized in these words:

> "With scarcely an exception each line of industrial activity shows an encouraging degree of progress, and in no case is there a forecast of the future colored by pessimistic forebodings. It is pointed out, as the one negative feature in the situation, that there is a scarcity of money in the markets; but this very lack of money is quoted as an evidence of business activity. Thus, on all sides America's business future is deemed to be built on a solid basis and there is no room given by these experts in trade-conditions to the belief that the threat of coming hard times is visible on the country's horizon."

The language in which the assurance is couched might have been improved; but the assurance itself was both ample and satisfactory.

Dun's Commercial Review reported that the industry and trade of the commercial world had outstripped its supply of monetary facilities. Wall Street had suffered from the fall in railway shares, yet the earnings of the roads were greater than ever. The mills and factories throughout this country have been kept busy and have found a ready sale for finished products at profitable prices.

The R. H. Macy department store stated that business this year had shown no curtailment and the prospects for the Fall trade were most encouraging.

The *Implement Age* reported an unusually heavy demand for agricultural implements. In all lines the demand has exceeded the supply. Manufacturers have but one difficulty—that of filling orders. Manufactories are being enlarged, more plants erected, and additional workmen employed. Europe cannot compete with America in this trade, and we are exporting 27 millions a year in harvesters, plows, cultivators, etc. The hardware-trade journals all report increased activity in manufactures and trade, largely due to the immense crops of agricultural produce and to the enhanced ability of the rural classes to buy. In machinery and builders' hardware the reports are to the same effect. Prospects for the Fall trade were never so good.

The *Motor,* a journal of the automobile trade, reported that the prospects of the trade were of the highest character and that after the market for pleasure carriages was supplied, the trade would find its extension in the power-wagon for trade purposes—a vehicle which was steadily supplanting the horse-drawn wagon.

The *Bicycling World* said that 1906 was a better year than 1905, and 1907 better than 1906. The trade had recovered from the craze of past years, and is now on a healthy basis. It doubted if there was more than one manufacturer in the trade whose books would not show both increase of production and profit.

The *American Pottery Gazette* reported depleted stocks in the hands of traders; an active demand in all branches of the china, glass and pottery trades; and well-filled order books of the manufacturers; the general outlook being "roseate."

The *American Silk Journal* said that the unusual depression in the first half of 1906, due to short crops of raw silk, had been overcome and that there is every prospect of a large business in the coming season. This year's crop reports from Japan, China, and Italy being favorable, additional mills are starting up in New England, New Jersey and Pennsylvania, for weaving ribbons and broad silk; and new throwing and dyeing plants in New York, New Jersey, and Pennsylvania. Nearly all the weaving mills have increased their capacity. Selling agents complain of not being able to obtain sufficient goods to fill their orders. Buyers in July and August have already placed enough orders to ensure a big business for the Fall.

The *House-Furnishing Review* reports that manufacturers have found it difficult to catch up with orders; many new retail concerns established; the construction of additional manufacturing plants is under way; and the prospect is of a lively Fall trade. The favorable condition of the agricultural crops has afforded an additional stimulus to an already busy trade. Prices are firm and markets steady; very little accumulation of stock in any line; and general condition of trade highly satisfactory.

The *Upholsterer* reports that the vast number of new dwellings being constructed throughout the country has occasioned an unusual demand for upholstery and house decorations. There are more American homes being built today than ever before. There is hardly a dry-goods store that has not found it necessary to put in furniture and drapery departments. Owing to the backwardness of the Spring weather, this Fall promises an enormous trade. Many buyers, anticipating difficulty in filling orders, have gone to complete their purchases in Europe. No indications of lower prices. The importers have never been busier. American furniture manufactories are being driven to their fullest capacity. All the mills are employing more hands and making provision to increase their trade.

The *Real-Estate Record,* a purely local journal, reports that, so far as New York City is concerned, the number of new buildings is slowly declining from previous high records—an indication that the local supply has in some measure caught up with the demand. The number of new buildings this year since January 1 is about 14,000, against 22,000 same time last year. The high price of land, the increased number of very large apartment-houses, and the removal of many families to the suburbs and to Western and Southern cities and villages, will probably account for this situation. The building news from the suburbs and from the countless towns and villages along the principal lines of railways and trolley routes, shows increased building operations, and the result is reflected in the increasing demand for lumber, builders' hardware, furniture, upholstery, and building and furnishing materials generally. Leaving New York city out of view, the condition of the house-building trade was never more prosperous than it is at present, and the prospects for Fall construction were never greater.

The *Wall Paper News* reports the construction of additional plants, larger buildings, and other evidences of an expected increase of trade in wall papers. The absence of Spring weather had occasioned a falling off, which it was believed would be more than made good during the coming Fall.

The *Lead and Zinc News* reports this year's production of lead in the Missouri-Kansas district to have been greater by 5,517 tons than for the corresponding period last year. Prices of lead have declined; while those of zinc have advanced, under large European orders. The mail-order houses continue to sell enormous quantities of prepared paint for house painting, an indication of increased means on the part of the rural population, the principal buyers.

The *Oil, Paint and Drug Reporter* says that business is exceptionally good in all lines, and 15 to 25 per cent greater than in 1906. The demand for drugs and pharmaceutical preparations is steady; chemical products are contracted for several years ahead. Petroleum and linseed oils continue in steady demand. Cotton-seed oil demand, falling off. The general outlook is optimistic.

The *Boot and Shoe Recorder* reports requests for early deliveries of Fall orders, indicating reduced stocks and large expected demand. The prices of raw leather have risen faster than those of manufactured goods; but it is believed that the latter will soon follow suit. An unusual number of buyers in the Boston market indicates a growing demand, and everything points to a large Fall trade.

The *International Studio,* a journal of the trade in works of art reports that the present is a period of unprecedented prosperity in all lines of the art trade.

The *Jewellers' Weekly* reports that the condition of the jewelry trade is eminently satisfactory, and that the prospects for the Fall trade are unusually bright. All lines have been steadily progressing. The prosperity of the jeweller has been practically uninterrupted. Over 41-million dollars' worth of diamonds had been imported during the year, at the port of New York alone. All departments of the trade showed increased activity, and in low-grade gold and plated jewelry, the demand far exceeded the supply. It is not the perturbations of the stock market, but the amplitude of the agricultural crops, that determines the prosperity of the principal lines of trade in jewelry.

Broomhall's July estimate of the wheat crop of the world is 2,992 million bushels, against 3,281 million bushels in 1906, Europe exhibiting a shortage of 173 million bushels. This condition of the world's principal breadstuffs bespeaks an active demand for American wheat, and continued prosperity for American millers, railways and shippers. And from the *Carpet and Upholstery Trade Review*, from *Men's Wear* and the *Apparel Gazette*, from *Fabrics, Fancy Goods and Notions*, from the *American Hatter*, from *Playthings*, from the *Publisher and Retailer*, representing the book trade, and from the *Daily Trade Record* representing men's wearing apparel, comes the same story; demand overtaxing the supply, outlook for Fall trade excellent, traders short of desirable goods.

RAILROADS. The statements by a number of roads showing their gross earnings for the month of July, 1907, exhibit the striking feature practically without exception of large increases of earnings over those for the corresponding month of last year. One scans in vain the railroad statements which have recently come to hand for any indication of a let-up in general business. The gross earnings for July during the last eight years, commencing with 1900, were respectively, 46, 46, 57, 62, 56, 47, 71, and 81 millions.

IRON AND STEEL. The *Iron Age* reports continued and increasing activity in the iron and steel markets. The August production of pig iron is expected to exceed that of July. The United States Steel Corporation, to fill its orders, has been obliged to purchase, since April 1, no less than 86,500 tons of billets from outside companies; its own immense capacity being inadequate to cope with demand.

COPPER. The production of the Calumet, Lake Superior, Butte, and British Columbia districts for July equalled 130 million pounds, valued at about 28 million dollars.*

In short, the reports of our domestic manufactures and trade exhibit an increase in all directions; provisions, metals, machinery, hardware, building materials, agricultural implements, carriages, dry goods (draperies), lumber, wooden-wares, skins and leather, oils, paper, publications, chemicals, drugs, dyes and medicines, soaps, spirits, tobacco, and a host of minor articles. Increased millions of men and machines are employed in their preparation and fabrication, and increased railways and ships in their transportation and distribution.

Our commerce, both foreign and domestic is greater than ever.

FOREIGN. The imports during the fiscal year 1905 were 1,118 millions; 1906, 1,227 millions; and 1907, 1,434 millions; the domestic exports were in 1905, 1,492; 1906, 1,718; and 1907, 1,855 millions; showing an immense gain this year over recent and indeed all previous years. The re-exports of foreign goods were about the same, 26 millions, in each of the years mentioned; so that they call for no

---

* The copper situation is discussed at length and with authority by Dr. Douglas, in the article immediately preceding this

special remark. The foreign commerce of 1907 alone affords us an apparent balance of trade amounting to not less than 447 million dollars. The imports of gold in 1906 were 92 millions; in 1907, 115 millions. The exports of gold in 1906 were 39 millions; in 1907, 51 mililons. Among the imports the most noticeable gains over 1906 were in the articles listed in the table just below; the sums attached showing the value in millions of dollars. As many of the articles imported, notably dry-goods, iron and steel wares, sugar, silk, spirits, wines, ales, tobacco, and woolens, had to pay a high rate of duty, it is evident that the rise in the cost of materials, rents, and wages in the United States enables the foreign articles, plus duty, to enter American markets at a price below their domestic rivals, when there is one; a fact of the highest importance in considering the workings of a tariff.

The principal articles of doméstic exports in 1907, all but four of which show a gain over the previous year, are summarized on the right-hand side of the table; values being given in millions of dollars:

PRINCIPAL IMPORTS AND EXPORTS DURING THE FISCAL YEAR ENDING JUNE 30, 1907.

IMPORTS.	MILLIONS OF DOLLARS.	EXPORTS.	MILLIONS OF DOLLARS.
Chemicals, Drugs and Dyes	83	Agricultural Implements	27
Cocoa	14	Cattle, Horses and Live Animals	41
Coffee	78	Cereals, Grains and Flour	184
Copper	48	Carriages and all other Vehicles	21
Cotton Goods	74	Chemicals, Drugs and Dyes	20
Diamonds and Precious Stones	43	Coal and Coke, excluding Bunker Coal for Foreign Trade	38
Flax, Hemp and Jute	42	Copper and Manufactures of	95
Manfs. of Hemp and Jute Flax.	67	Cotton, Raw	482
Fruits and Nuts	36	Cotton, Manufactures of	32
Furs and Manufactures of	30	Flax and other Seeds	10
Gums	15	Fruits and Nuts	18
Hides and Skins	83	Instruments (Scientific)	14
India Rubber	63	Iron and Steel, Machinery and Hardware	182
Iron and Steel and Manfs. of	41	Leather and Manufactures of	45
Leather and Manufactures of	20	Lumber	70
Oils	17	Meats and Dairy Products	202
Paper	11	Naval Stores	22
Raw silk	71	Oil Cake and Meal	26
Manufactures of Raw Silk	39	Mineral Oils	85
Spirits, Wines and Malt Liquors	22	Paper and Manufactures of	10
Sugars	93	Paraffine and Wax	9
Tin	38	Tobacco and Manufactures of	39
Tobacco and Manufactures of	30	Vegetable Oils	20
Woods and Manufactures of	43	Wood, Manufactures of	14
Wool	42		
Manufactures of Wool	23		
Total	1166	Total	1706

INTEREST. It is only when we enter the obscure ·domain of finance that the industrial problem becomes puzzling; but even this clears up before the lime-lights of those economical principles, the working of which, but ill understood, is the cause of the prevailing uncertainty in the commercial mind. One of these is the Rise in the Rate of Interest; a second is the Rise of Prices; and a third is the Supply of Money.

As any given rate of interest must consist of three elements, namely, profits in trade, the cost of the superintendence of loans, and insurance against loss, the first being variable and of the greatest importance, the others being usually constant and (in the United States) of less importance, it is easy to deduce that any general rise of profits will cause a rise in the rate of interest. As a matter of fact, the recent rise in the rate was foreseen and publicly predicted several years ago.

One of the inevitable consequences of a rise in interest is a fall in bonds, or in the value of anything yielding merely a fixed income. Hence the fall in New York municipal bonds and in numerous other securities with a fixed income, a fall which has puzzled our younger financiers and driven them into pessimism and alarm. So far is a high rate of interest, when accompanied by a normal cost of superintendance and insurance, removed from a bad symptom, that it is in fact the best of all symptoms; it is a certain evidence of high profits in production and trade. There is little risk in predicting that before many years the tide will turn the other way, when the now despised municipal bonds will rise to 130 or thereabouts.

PRICES. We now approach a subject which brings us close to the crux of the present situation—the Rise of Prices; for it is under this head that stocks, commodities, real-estate and labor are engaged in that struggle, which the unsophisticated view with doubt and distrust.

Two years ago and in the July number of this MAGAZINE, it was shown that the gold dredgers and other recently invented gold-producing and gold-saving devices were turning out nearly a million of dollars a day of gold; and it was predicted that "within the next ten years this output would increase to two millions a day." The world's product has already passed the million mark; and is steadily approaching the two-million mark.

Under the impulse of this great increase of the precious metal from which money is coined, and of the bank-note issues based upon it, and of the cheque, clearing-house, telegraphic, and telephone systems and other agencies and arrangements which have been established to expedite the velocity and efficiency of money, there has occurred a rise of prices, which everybody perceives, but nobody has yet resolved into order. From very careful observations which have been

made in the United States, it appears that following a given increase
of money, and taking no account of the time involved in geograph-
ically distributing the increase, it nevertheless requires a period of
some years for all prices to conform to such increase. During this
time, the prices of a certain few classes of commodities or services
will double; after which the prices of others will double; and so on
successively, until the doubling of all classes is completed. In other
words, the doubling of prices will not be simultaneous, but will take
the form of a precession, the order of which will be somewhat as
follows:

I, Stocks, or shares of incorporated companies. II, Staples, or
crude and imperishable commodities. III, Merchandise, such as perishable commodities, crude articles of subsistence, etc. IV, Fabrics,
such as machinery, manufactured food, dry-goods, etc. V, Landed
property, or real estate. VI, Skilled labor, or artisans' wages. VII,
Unskilled labor, or the wages of laborers, etc. VIII, Professional
services, or the emolument of authors, inventors, and other professional and clerical classes. The interval between the doubling of the
prices in these various classes of commodities or services, is not uniform; in other words, supposing eight years to be the time required
for the entire doubling of prices and the classes of commodities and
services to be eight in number, it would not follow that each successive year would add one to the classes with doubled prices. After
once commencing to feel the influence of the increase of money, some
classes will double in prices more quickly than others. Leaving this
irregularity out of view; assuming that population and trade and
therefore the demand of money remains unchanged; and supposing
that no further increase of money or devices for enhancing its efficiency are brought into use; the precession of prices will appear somewhat as is shown in the following diagram:

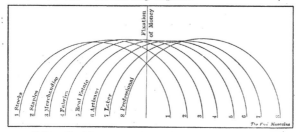

DIAGRAM OF PRECESSION OF PRICES DURING AN INCREASE OF MONEY.

At the point where money and credits become fixed in combined volume (both being reduced to the same ratio of velocity or efficiency) this diagram shows Stocks and Staples falling in price, while manufactured goods (Fabrics, and Labor) are still rising; both of these conditions being due to the same initial impulse.

Of course it will be understood that this diagram is not intended to represent or exemplify the actual condition of affairs at the present time, its sole design being to show that under the same influence, the prices of some classes of commodities or services may be falling, while others are rising: a phenomenon due to the precessional order in which they rise and fall under the influence of an increase of money which has ceased. The actual position of affairs is quite different. During the past eight years we have had an increase, chiefly of "National" bank notes, but far from a "doubling" of money. As nearly as can be made out from the very dubious "Circulation Statements" issued by the Treasury, which continue to overestimate the amount of gold "in circulation" by more than 500 millions, the per capita circulation of metallic and paper money is very little above what it was eight years ago; the effective increase of money being due more to activity than volume; more to the loaning of Treasury balances to the banks and to the use of improved devices for transmitting money and economizing its employment, than to any material increase in its amount.

On the other hand, the requirement for money to effect the purchases, sales, payments, and exchanges of the country has largely outstripped its population; a fact due to the universal employment of machinery in production. *Electricity has quickened industry; and industry needs more circulating money to effect its exchanges.* Capital wealth has increased from 88 to 115 billions; bank clearings from 88 to 175 billions; and products and commerce in like proportions. Last year the farmers received over five billions for their year's product; or more than twice as much as the entire currency of metal and paper. As for the requirement of the manufacturers, merchants, and carriers, it is many times as much. And yet all of this enormous series of exchanges has to be effected by means of a currency so ridiculously inadequate that had it not been eked out by the mechanical and financial devices of private persons, it would have plunged this rapidly growing country into very serious disorders. The events of 1837 to 1842 should not be entirely forgotten.

STOCKS. The fluctuations in stock quotations for railroad and industrial shares since January last are shown in the chart on page 27, taken from the New York *Times*, which added this comment:—

"The downward course of stocks since January 1 is well illustrated in the accompanying chart, the movement of sixty of the principal railroad stocks being averaged and indicated by the heavy line, that of ten industrials being averaged and indicated by the dotted line.

'It would be impossible, because of the confusion which would arise from the multiplicity of lines, to map the individual movements from day to day of each issue in which trading has been active on the Stock Exchange. The chart appended does not, therefore, show the violent fluctuations during the seven and a half months of such favorites of the professional trader as the Pacifics, Reading, Amalgamated, Great Northern, Steel, and Smelters. These and the score of other highly sensitive railroad and industrial stocks are toned down in the chart by association in their respective classes with the steadier stocks which go to make up the averages.

"It will be noticed how closely the general course of the ten industrials which have been averaged approximates the movement of the sixty railroad stocks."

Upon examining our classification of commodity prices it will be observed that the order of precession conforms to the marketability or ease of selling or exchanging, the various classes of commodities and services enumerated. The most marketable of all commodities are stocks; a fact which explains why, after having been the first to rise in response to increased money, they were the first to fall when the increase stopped.

MONEY. To increase money irregularly, is not merely to enhance all prices simultaneously; it is to enhance the price of some things in point of time before others; it is to benefit certain classes of the community at the expense of the remainder; it is to derange and throw into disorder all the varied and complicated interests of society. Contrariwise, to diminish its sum, or to permit it to diminish, or to neglect to increase it with the growing requirements of trade, is to depress the prices of certain commodities sooner than others, and to occasion a derangement of affairs even more perilous to the public welfare; for it so happens that, although theoretically labor benefits from a general fall of prices (it being the last in point of time to feel the effects of a paralysed currency) it practically suffers even more than during a general rise of prices; because a fall of prices hinders commerce and depresses production; and thus partially deprives labor of employment, or tangible existence.

The most cursory examination of the circulation will suffice to reveal its pinched condition and inadequate supply. As a matter of fact it consists entirely of paper notes and small change, whose combined sum does not exceed 2,000 millions. On August 1 the Treasury reported "Money in circulation, 2,781 millions; population 86 millions;

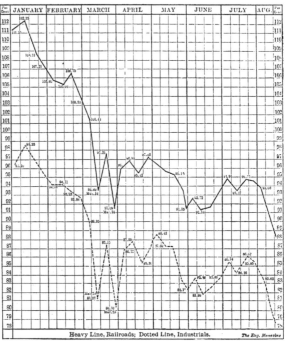

Heavy Line, Railroads; Dotted Line, Industrials.

The Eng. Magazine

FLUCTUATIONS OF STOCKS, JANUARY-AUGUST, 1907. FROM THE N. Y. TIMES.

circulation per capita, $32." No such statement is required by law;
it is not official; it is entirely voluntary; and it is false. There is no
such amount of money in circulation; nor even is there so much
money in the country. The statement includes 566 millions of gold
coin, which are wholly fictitious. Last month, in deference to the
oft-repeated objections to this item, the Treasury threw off 135
millions from its estimate of gold coin in circulation. It remains for
it to throw off 500 millions more; it being a liberal estimate to assume that there are 66 millions circulating in a country where, except
to a small extent in the Pacific States, gold coin is never seen at all.
The Treasury may deceive the people, but it cannot deceive the dealers
in bullion and foreign exchange; nor is there any necessity to deceive
anybody. The credit of this country needs no such crutch.

The way this estimate is made up is to the last degree pathetic. The Director of the Mint innocently adds together the reports of production from the Wells-Fargo Express Company and of importation from the customs "entries," and from the quotient he deducts the reported exportations. The remainder, barring an estimated amount melted in the arts, is assumed to be "gold coin in circulation."

But the estimates of the express company, as shown in the Report of the Monetary Commission, include many duplications, and are unreliable. There is no penalty for understating or concealing bullion exports. Banking houses which deal in foreign exchange are not anxious to publish their bullion exports. Moreover, a great deal of gold is carried out of the country by travelers. Hence the returns of exports are deficient and misleading. A similar remark applies to gold consumed in the arts. The returns are non-compulsory, non-official, and largely deficient. These and other sources of error have been accumulated in the Treasury Statement of Circulation for eighteen years, with the result of creating an imaginary mountain of gold, where there is in fact only a molehill. Substantially, the only gold in the country is in the Treasury and the banks, and is not in circulation at all.

The 204 millions of silver dollars and subsidiary silver "in circulation" is also a mis-statement. This is doubtless the net amount coined and paid out, but much of it has gone to Cuba, Porto Rico, the Philippines, and even to South America and other foreign countries, where it can be seen doing active duty; but not in the exchanges of the United States. As to the 150 millions of gold reserve; the 146 millions loaned to the banks; the 33 millions of currency certificates included in the estimate of greenbacks in circulation; and various other items of this illegal and misleading account, we leave their discussion to the partisan press. Avoiding all disputable points and refraining from unessential detail, it is regarded a fair estimate to put the entire circulation at 2,000 millions, or about $24, instead of $32, per capita; a sum which at present prices is regarded as entirely inadequate to effect the exchanges of a country so extensive and busy.

In short, prices have grown up to the limits of the circulation, plus the cheque system, and the other devices of commercial credit and economy; a phenomenon which is quite sufficient to explain existing conditions, without recourse to chimeras. The remedy is in the hands of the Treasury, which for some unexplained reason is heaping up an idle balance of nearly 500 millions. When this War Chest is permitted to enter the exchanges, there is every reason to believe that the passing financial cloud will disappear.

THE MECHANICAL ENGINEERING OF THE MINE.

By Chas. C. Christensen.

· The mine, as Mr. Christensen says, calls upon the mechanical engineer for co-operation throughout the entire range of his professional practice. Some of its leading demands—for power transmission, hoisting and haulage, electric service, and ventilation—have been discussed lately in this magazine. Another subject of prime importance in the same field —pumping—is treated elsewhere in this issue. Mr. Christensen therefore leaves it to the illustrations to suggest the interest of these applications of power, and devotes his text to a summary of methods and equipment employed in modern ore dressing and treatment.— THE EDITORS.

THE mechanical engineering of the mine is a broad subject and, if treated fairly, could fill volumes, but confined to the limited space of an article, one can only briefly depict the extent to which mining has come to depend upon machinery and the part which the mechanical engineer has played as expert associate with the mining engineer and mine owner in order to make mining a commercial success. The California miner of 1849 found only a shovel and a pick at his disposal, and with these ancient tools he started mining and was fairly successful. Soon, however, the mechanical, the civil, and the mining engineer came to his aid. The mechanical engineer brought him the air drill, the hoist, the deep mine pump, the stamp battery, and the ore crusher. The civil engineer gave him high explosives, and the mining engineer came and stayed with him, assayed his ore, and outlined better processes which in turn made the mechanical engineer design and invent newer and more suitable machinery.

Invention followed invention and when I arrived in the United States in 1879, a green graduate of an European college, I found as far east as Chicago all the work I could wish for, waiting in the mining-machinery branch.

In those old days machine shops devoted to the manufacture of mining machinery were few and small. The Pacific Coast shops were of course first in the field after the days of 1849. The Pelton Water Wheel Company had introduced its wonderful water wheel for high heads and small quantities of water. The Leffel Company in the East had brought to the mine the well designed Leffel water motor for low heads and large quantities of water. and the Rand Drill Company's air compressors and drills were opening up the mines both at home and abroad.

The way this estimate
The Director of the Mint i
duction from the Wells-Fa
from the customs "entries,'
ported exportations. The
melted in the arts, is assum

But the estimates of the
of the Monetary Commissi
reliable. There is no pen
exports. Banking houses
anxious to publish their b
gold is carried out of the c
exports are deficient and n
gold consumed in the arts.
official, and largely deficient.
been accumulated in the T
eighteen years, with the resu
gold, where there is in fact
gold in the country is in the
circulation at all.

The 204 millions of silve
also a mis-staten
d out, but mu
even to So
w een doin
of t . As
millio ba
include
other it
discussio
refrainin
the entire
per capita
inadequate

In sho
plus the che
and economy;
existing conditio
the hands of the
heaping up an idle b
Chest is permitted to e
lieve that the passing fi

is to gree pathe
lds to eports of p
s Con f importat
the q educts the
barr nated am
ld c tion."
npa in the R
han , and a
rst aling
n nge
great
el he re
rk a
pulse
f e

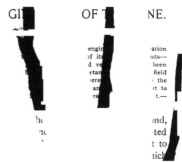

engi...
of its
d ve
rtan
ere
an
es

ation
nds—
been
field
the
t to
t.—

h...
n...

nd,
ted
t to
ich

The way this estimate is made up is to the last degree pathetic. The Director of the Mint innocently adds together the reports of production from the Wells-Fargo Express Company and of importation from the customs "entries," and from the quotient he deducts the reported exportations. The remainder, barring an estimated amount melted in the arts, is assumed to be "gold coin in circulation."

But the estimates of the express company, as shown in the Report of the Monetary Commission, include many duplications, and are unreliable. There is no penalty for understating or concealing bullion exports. Banking houses which deal in foreign exchange are not anxious to publish their bullion exports. Moreover, a great deal of gold is carried out of the country by travelers. Hence the returns of exports are deficient and misleading. A similar remark applies to gold consumed in the arts. The returns are non-compulsory, non-official, and largely deficient. These and other sources of error have been accumulated in the Treasury Statement of Circulation for eighteen years, with the result of creating an imaginary mountain of gold, where there is in fact only a molehill. Substantially, the only gold in the country is in the Treasury and the banks, and is not in circulation at all.

The 204 millions of silver dollars and subsidiary silver "in circulation" is also a mis-statement. This is doubtless the net amount coined and paid out, but much of it has gone to Cuba, Porto Rico, the Philippines, and even to South America and other foreign countries, where it can be seen doing active duty; but not in the exchanges of the United States. As to the 150 millions of gold reserve; the 146 millions loaned to the banks; the 33 millions of currency certificates included in the estimate of greenbacks in circulation; and various other items of this illegal and misleading account, we leave their discussion to the partisan press. Avoiding all disputable points and refraining from unessential detail, it is regarded a fair estimate to put the entire circulation at 2,000 millions, or about $24, instead of $32, per capita; a sum which at present prices is regarded as entirely inadequate to effect the exchanges of a country so extensive and busy.

In short, prices have grown up to the limits of the circulation, plus the cheque system, and the other devices of commercial credit and economy; a phenomenon which is quite sufficient to explain existing conditions, without recourse to chimeras. The remedy is in the hands of the Treasury, which for some unexplained reason is heaping up an idle balance of nearly 500 millions. When this War Chest is permitted to enter the exchanges, there is every reason to believe that the passing financial cloud will disappear.

THE MECHANICAL ENGINEERING OF THE MINE.

By Chas. C. Christensen.

The mine, as Mr. Christensen says, calls upon the mechanical engineer for co-operation throughout the entire range of his professional practice. Some of its leading demands—for power transmission, hoisting and haulage, electric service, and ventilation—have been discussed lately in this magazine. Another subject of prime importance in the same field —pumping—is treated elsewhere in this issue. Mr. Christensen therefore leaves it to the illustrations to suggest the interest of these applications of power, and devotes his text to a summary of methods and equipment employed in modern ore dressing and treatment.—THE EDITORS.

THE mechanical engineering of the mine is a broad subject and, if treated fairly, could fill volumes, but confined to the limited space of an article, one can only briefly depict the extent to which mining has come to depend upon machinery and the part which the mechanical engineer has played as expert associate with the mining engineer and mine owner in order to make mining a commercial success. The California miner of 1849 found only a shovel and a pick at his disposal, and with these ancient tools he started mining and was fairly successful. Soon, however, the mechanical, the civil, and the mining engineer came to his aid. The mechanical engineer brought him the air drill, the hoist, the deep mine pump, the stamp battery, and the ore crusher. The civil engineer gave him high explosives, and the mining engineer came and stayed with him, assayed his ore, and outlined better processes which in turn made the mechanical engineer design and invent newer and more suitable machinery.

Invention followed invention and when I arrived in the United States in 1879, a green graduate of an European college, I found as far east as Chicago all the work I could wish for, waiting in the mining-machinery branch.

In those old days machine shops devoted to the manufacture of mining machinery were few and small. The Pacific Coast shops were of course first in the field after the days of 1849. The Pelton Water Wheel Company had introduced its wonderful water wheel for high heads and small quantities of water. The Leffel Company in the East had brought to the mine the well designed Leffel water motor for low heads and large quantities of water, and the Rand Drill Company's air compressors and drills were opening up the mines both at home and abroad.

GROUP OF SULLIVAN 2¼-INCH PISTON DRILLS STOPING IN A COLORADO GOLD MINE.

The Gates Iron Works in Chicago started to make mining machinery in the seventies, and the famous Gates crusher, when it appeared,
marked a new era in the mechanical engineering of the mine. The
Edw. P. Allis Company in Milwaukee also early established a mining-
machinery department, and their mechanical engineers can be proud
of what they have contributed to the development of the mine. But
it was Fraser & Chalmers who, in the early seventies, started a machine shop on the corner of Fulton and Union Streets, Chicago, exclusively devoted to the manufacture of machinery for the mine and
smelter, and this Chicago corner became in the eighties the Mecca
for mining men from all the States in the Union, as well as from
many foreign lands. Inside the red walls of the plain building many
a mechanical-engineering problem of the mine has been solved and
many mining plants designed.

My first experience was to make the plans of the 120-stamp free-
milling gold mill erected in 1879 at Lead City, Dakota (Black Hills)
by the Highland Milling Co. This mill is what we may call a double
60-stamp mill, as it is arranged with two 60-stamp mills backing one
another. The ore bins are placed in the center, tapering down to
both rows of batteries and above the bins are located the crushers,
the grizzlies and the car tracks. The plans were ordered on the large
scale of ½ inch to 1 foot; consequently when making them, I had
to lie on the 6 by 16 foot drawing board and handle a T-square as

long as the table, but these drawings when completed, showed all details; and being made before the time of blue printing, they were colored and shaded in true European style. The next year (1880) a water-jacketed furnace was erected by the Hecla Consolidated Mining Company. It was a lead-silver furnace 36 by 60 inches in size at the tuyeres and was one of the first water-jacketed furnaces ever built. It had a 48-inch high cast-iron jacket made in sections suitable for transportation and for removal in case of repair. Compared with the furnace of today this first furnace was crude and clumsy, but I mention it here as it was, at that time, a great forward step in the mechanical engineering of the mine.

A number of furnaces were designed in the eighties with cast-iron jackets, because the first steel jackets built gave very little satisfaction; but it did not take many years before the steel jacket was perfected and today it reigns supreme. In the design of smelting furnaces, brick is at the present time used but little, and for copper furnaces it has been abolished altogether. Instead are used steel, cast-iron, and cast steel, from the bottom to the top, and a copper furnace of today is a marvel of mechanical genius. It is water-jacketed up to the charging floor, has water-jacketed aprons above at the charging openings, steel hood, stack and downtake, steel or cast-iron air or

DUPLEX DIRECT-ACTING DOUBLE-DRUM CORLISS HOISTING ENGINE, ALASKA-
TREADWELL GOLD MINING CO.
Allis-Chalmers Co., Milwaukee.

A 36-INCH SECTIONAL WATER-JACKETED COPPER FURNACE.
Allis-Chalmers Co., Chicago.

hydraulic-operated charging doors, heavy cast-iron crucible plate in sections, resting on short cast-iron columns, I-beam mantel frame supported by Z-bar columns carrying the cast-iron dumping plates and the whole superstructure.

The sizes of these modern furnaces run all the way from 36 by 84 inches to 56 by 612 inches, but though so perfect in design now, new ideas are springing up all the time and I am confronted with about a dozen new furnace designs every year.

Small, round water jackets are sometimes welded up in one piece or, if need be, made in sections for mule-back transportation. They are made in sizes from 20 inches diameter up to 48 inches diameter.

The old reliable stamp mill is perhaps among all apparatus the most standard piece of mining machinery in existence. It is as standard in the mining world as the high silk hat is in society circles, and like this headgear it has not changed much since it first appeared. Cast-iron anvil blocks, heavier stamps, shorter lifts, Blanton self-tightening cams, and sectional guides, are the improvements made so far. In the African mills stamps are used as heavy as 1,300 to 1,400 pounds and with 10 inches drop and about 100 drops per minute will crush 5 to 5½ tons per stamp per day of 24 hours. In Alaska at the Alaska Treadwell Gold Mining Company's Mills, stamps are 1,050 pounds and the capacity varies from 4.95 tons to 5.79 tons per

EXAMPLES OF RECENT AMERICAN MINE HOISTS.

By the Wellman-Seaver-Morgan Co. The upper is an electrically operated (a. c.) double-reel hoist for 2,000-lb. load on a 2,000-lb. skip from a depth of 1,200 ft. at 500 ft. per minute; friction clutches, post brakes, and automatic safety device against overwinding. The lower is the latest type of double-drum electric hoist for balanced cage hoisting, operated by continuous speed a. c. motor. The operation of the drum is reversed by two bevel gears controlled by band friction clutches.

stamp per day. In the Black Hills most mills have 850-pound stamps and crush 4½ tons per stamp per 24 hours at an average. This record, however, has been broken by the 2-stamp 4-screen opening mill, which at the Tenero Mining & Milling Company's plant crushes up to 6 tons per stamp per day through 24-mesh screens.

The Tremain steam stamp has been designed especially to meet the demand of those who own mining properties in the first stages of development, enabling them to establish a thoroughly good crushing, amalgamating, or concentrating plant of moderate capacity.

The Tremain steam stamp consists of two 300-pound stamps,

the stems of which terminate in pistons working in cylinders like a steam engine. The speed is 200 or more drops per minute and the capacity of the mill varies with the character of the ore and size mesh of screen from 8 tons to 20 tons per day of 24 hours, or 4 to 10 tons per stamp per day. The three-stamp prospecting mill for free-milling ore, consisting of a battery of three 250-pound stamps, is very popular. It has been designed to meet the demand for a light, compact plant, capable of being easily and quickly taken apart, transported on mule-back, and erected where desired. The mill is driven by a 3 horse-power combined vertical boiler and engine.

Heavy steam stamps, of which one will reduce as much ore as an ordinary 60 to 100-stamp mill, are in operation in the United States as well as abroad. The stamp is driven by a vertical steam cylinder with valve gearing designed to keep the top steam port fully open during the down stroke, adding the power of steam to the weight of the stamp, while for the work of the up stroke—merely lifting the stamp—a small steam admission suffices.

The Krause atmospheric stamp is a machine operated from any source of power by pulley and crank-shaft. This crank-shaft is connected by means of a connecting rod and crosshead to the upper piston in an air cylinder. The upper end of the stamp shaft forms the lower piston. As the upper piston starts to rise, a partial vacuum is created between the pistons which raises the lower piston and stamp shaft. A tappet on the stamp shaft coming in contact with a horizontal lever raises a vertical stem, which opens an inlet valve, thus admitting atmospheric pressure between the pistons. The lower piston and stamp shaft

BEVEL-GEAR ENDLESS-ROPE HAULAGE SET.
90 horse-power; driven by 550-volt enclosed multipolar motor.
J. H. Holmes & Co., Newcastle-on-Tyne.

continue their upward movement until arrested by a cushion of high initial pressure, caused by the down-coming upper piston, which drives the stamp down with a rapidly increasing movement until at the end of the stroke the stamp has attained a velocity of 1,200 feet per minute and delivers a very powerful crushing blow. The speed of the stamp is 190 blows per minute and it will crush hard 1½ to 2-inch size copper-bearing rock at the rate of 2½ tons per hour through a 16-mesh screen.

The Gates crusher has been an important factor in the development of the mine. It is a rock and ore-breaker of the gyratory form. The crushing is done between a cone placed on a gyratory shaft vertical through the center of a cylindrical shell. As it gyrates the crushing cone impinges against the sides of the shell in relation to which it is constantly approaching and receding. The top of the shaft carrying the crushing cone is held rigid while the bottom is gyrated with a certain amount of eccentricity by means of a single gearing. The crusher is built in about a dozen sizes and the capacity per hour, in tons of 2,000 pounds passing a 2½ inch ring, runs all the way from 2 to 200 and over.

The best-known and most successful crushers of the jaw type are the Blake stone and ore crusher and the Dodge crusher. Blake and Dodge crushers affect the reduction of the material between two jaws, protected from wear by removable plates. One of the jaws is stationary and the other movable. In comparing the relative merits of the two machines, it is obvious that as the width between the jaws at the point of discharge in the Dodge crusher remains more constant, a finer and more uniform product can be obtained than from the Blake, but it is inferior to the latter in capacity.

Dodge crushers are admirably adapted for the reduction of the product of a gyratory breaker adjusted for coarse work, the fine, even product thus obtained being very suitable for feeding rolls or any other kind of pulverizers. For the preliminary crushing of ore in stamp mills of small or medium capacity, Blake crushers have been extensively used and with very good results. The Blake and the Dodge crusher are built in respectively six and four sizes and their crushing capacity runs from 4 to 24 tons per hour, 1½-inch size, for the Blake type and from 1 to 8 tons nut size for the Dodge.

The several types of stamp mills, and other mining plants, now in successful operation, may be sketched as follows :—

A.—STAMP MILL FOR FREE-MILLING ORE. Free gold is caught on the inside copper plates of the mortar and the outside copper tables in front of the stamp batteries, by means of quicksilver, forming gold amalgam, which at intervals—about twice a month—is scraped off and retorted, vaporizing the quicksilver which is then condensed in water. The gold thus gained is melted and runs into moulds, while the quicksilver is saved and used over again.

The machinery for such a mill consists of: crusher, stamp feeders, which receive the ore from the bins and feed it into the mortars; stamps, lifted by cams; copper tables, and for a large mill, a clean-up pan in which the amalgam when dirty and impure is worked with additional quicksilver and the waste matter washed off before retorting. The camshaft of the stamp battery generally runs 45 to 50 revolutions per minute and by the double cams the stamps are lifted and drop by gravitation 90 to 100 times per minute.

B.—Stamp Mill with Concentrators or Frue Vanner Mill. Most gold ores will yield only a portion of their contents to the plain amalgamating process outlined above, because some of the precious metals is tied up in combination with sulphides and baser metals. Values in this form are generally saved by adding concentrators— the Frue vanner type of which have given excellent result in close competition perhaps with the Overstrom table and the Wilfley table. The Frue vanner consists of an endless flanged rubber belt 4 to 6 feet wide and 12 feet long. The belt travels up an incline and round a lower drum which dips into a water tank in which the mineral is collected. In addition to the travel of the belt the latter receives a steady shaking or settling motion from a crank shaft along one side of the machine, the shake being at right angles to the inclination and travel of the belt. The ore is fed upon the belt in a stream of water about 3 feet from the head, and flows slowly down the incline, subjected to the steady shaking motion which deposits the mineral on the belt. At the head of the belt is a row of water jets. The slow travel of the belt brings up the deposited mineral and the water jets wash back the lighter sand, letting only the heavy mineral pass and become deposited in the water tank below.

The machinery for a vanner mill is the same as for a free-milling ore mill, with the concentrators added.

C.—Wet-Crushing Silver Mills. The ore brought in by cars at the top of the mill is dumped over the inclined grizzly or screen and rolls onto the crusher floor. All the small pieces pass through the screen or grizzly into the ore bins underneath. The coarse rock is shoveled into the crusher from the floor, which is on a level with its charging hopper, and is crushed to the size of walnuts, falling into the ore bins, whence it passes into automatic stamp feeders through inclined chutes controlled by ore gates.

THE LARGEST AIR COMPRESSOR IN THE WORLD. HOMESTAKE GOLD MINING CO.,
LEAD, S. D.
Steam cylinders 32 and 60 in.; air cylinders 52¼ and 32¼; stroke 72 inches. One of seven
Ingersoll-Rand compressors used by the Homestake Co.

The finely stamped ore suspended in water, and known technically as pulp, flows into large settling tanks, where excess of water is drawn off. The thick pulp remaining is shoveled in regular charges into a row of amalgamating pans, in which it is worked several hours, first with salt, bluestone and other chemicals, then with addition of quicksilver. The contents of the pans are run into large settlers placed below and in front, in which the pulp is thinned by addition of water with gentle agitation and all the quicksilver, with precious metals in the form of amalgam, is settled to the bottom. The pulp is gradually run off from the settler and flows to waste. The amalgam is strained from excess of quicksilver, retorted to drive off the remaining quicksilver, and the silver and gold is melted into bars. Generally two 5-foot pans and one 8-foot settler are supplied with each 5-stamp battery. Most mills are also provided with a quicksilver system for elevating and distributing quicksilver throughout the mill automatically. The machinery for a wet-crushing mill would be:—crusher, stamp feeders, stamps, settling tanks placed in front of the stamp battery, pans, settlers, amalgam safes and strainers, retort, and bulletin furnace. When the Boss continuous process for amalgamation is used, the pulp flows direct from the battery through pipes to special grinding pans and is from there conveyed to the first amalgamating pan and flows continuously through the line of pans and settlers. The tailings are run off or led over concentrators.

The quicksilver is charged in the pans by means of pipes from the distributing tank and the amalgam flows through suitably arranged pipes to the strainer. The chemicals are supplied to the pans by two chemical feeders. The main line shaft runs directly under the pans and settlers, each of which is driven from it by a friction clutch. All the water from the battery must pass through the pans, so that all slimes are treated.

ELECTRICALLY OPERATED MAIN AND TAIL HAULAGE, WEST STANLEY COLLIERY, DURHAM.

Driving motor of slip-ring induction type, 500-volt, for three-phase 40-period circuit. Controller of reversing drum type. The British Westinghouse Co.

D.—DRY-CRUSHING MILLS. When ore is base and needs roasting and desulphurizing before amalgamation can be successful, a modification of the wet-crushing mill is necessary. After passing the rock crusher, the ore is dried by passing through a continuously revolving dryer, located beneath the crusher. The dried ore is then taken by a car, or run in chutes if height will admit, to the automatic feeders, which feed the stamps while the ore is still hot. The pulverized product from the stamps is conveyed by screw conveyors to an elevator, by which it is carried to an iron storage hopper above. From here another conveyor conducts the pulverized ore to the roasting furnace in the furnace room. The roasting furnace may be a Howell-White, reverberatory, Bruckner, or Brown, Wethy, Ropp, Jackling, or McDougal furnace. In the furnace, the ore, with the addition of common salt, is desulphurized and chloridized, thus preparing it for the pans and settlers. After roasting, the ore is allowed to remain in pits to continue the chloridizing and afterwards is spread on a cooling floor and is taken as required to the pans. Amalgamation follows then on the same plan as in the wet-crushing mill.

E.—Cyanide Plant. The ore upon arriving in cars is dumped into a bin, whence it is drawn to feed a crusher. This machine reduces the ore to a maximum size of 1 inch, and discharges it into a revolving dryer, which thoroughly dries the ore so that the screens may be worked to their full capacity. The ore passes from the dryer

ELECTRIC WINDING GEAR, WITH TWO DRUMS FOR TWO DIFFERENT LEVELS.
Large drum 8 ft. 8 in., smaller 5 ft. 8 in., the levels being 180 and 120 feet; 80 horse-power motor, three-phase alternating current. Ernest Scott & Mountain, Gateshead-on-Tyne.

into a special fine crusher and is reduced to ¼ inch maximum size. The fine crusher discharges into an elevator, which carries the ore to a double shaking screen arranged in such a manner that the stream of ore may be instantly diverted to either screen. A gentle shaking motion is imparted to them by simple mechanism. The rejections from the screen are returned to the fine crusher for recrushing; that portion passing through the meshes of the screen is spouted to the crushing rolls to be reduced to the size required. From the rolls the

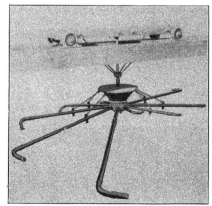

BUTTERS DISTRIBUTOR FOR CYANIDE TANKS.

ore is elevated and passed over a second double shaking s c r e e n and into the storage bins. The rejections from the screen are spouted back to the rolls.

T h e finished product, generally 30-mesh size, may now be d r a w n from the bins into small scoop cars standing on the p l a t f o r m scales and the contents of the cars weighed. A suspended track is constructed over the percolating tanks, by means of which the cars may be dumped directly into the tank. The ore is then leveled with a hoe, but a method producing more even percolation is to charge the tanks by means of the Butters distributor. This distributor consists of a hemispherical bowl, the top of which is covered with an iron sieve or grating. Into this bowl are inserted from eight to sixteen pieces of steel pipes of different lengths, the outer ends of which are bent at right angles horizontally and terminate in flattened nozzles, the whole being supported on an iron column in the center, or suspended from a light traveling crane running the whole length of the tank room. The centrifugal action of the discharging water causes the device to revolve slowly, thus equally distributing the charge.

The tailings from the percolating tanks are shoveled into cars and cast upon the dumps.

In a so-called cyanide tailings plant, the crushing machinery mentioned above is left out, the tailings being conveyed from the battery directly to the cyanide tanks. The tanks used in cyanide plants are generally made of steel plates and angle steel in sizes to suit the capacity of the plant. The percolating tanks are sometimes made as large as 40 feet in diameter by 5 to 8 feet deep. The time of percolation varies according to the character of the ore from 2 to 6 days.

F.—CHLORINATION WORKS. The ore, after being crushed, dried, screened, recrushed by rolls and roasted—see different types of roasters above—is spouted to the cooling floor. After the ore has become sufficiently cool it is conveyed to an elevator which delivers it to the bin situated over the chlorinating barrels. From this bin, it is drawn into a car and weighed, after which it is ready for charging the hopper over the barrels.

The chlorinating barrels are generally of five tons capacity each. They have two charging doors or manholes, and a filter secured opposite the charging doors; the filter should be level when the barrel is in such a position that the doors are up, or on top. These barrels have under favorable circumstances, handled as much as eight charges of 5 tons each per day of 24 hours.

The barrels are charged first with the requisite amount of water; then sulphuric acid which settles immediately on the bottom; after which the ore is added, and lastly chloride of lime. The charging doors are securely fastened and the barrel revolved. The mixture of the acid and chloride of lime which occurs when the barrel is revolved, causes a rapid evolution of chlorine gas which soon produces a pressure in the barrel. Chlorine gas under slight pressure combines readily with water and rapidly chloridizes the metals present. After chlorination is completed, the barrel is stopped with the charging doors up, so that the filter will be level. Connection by hose is then made between the water tank, (which is elevated enough to give 50-pounds pressure) and the top of the barrel; a connection is also made between the lower portion of the barrel and the settling tank.

ELECTRIC POWER HOUSE, MCKELL COAL & COKE CO., KILSYTH, W. VA.
Allis-Chalmers 26 by 42 engine and 500-kilowatt 25-cycle 3-phase 6,600-volt generator.

The fresh water under pressure, entering on top of the charge, absorbs all free chlorine gas remaining after the ore has been chlorinated and forces the solution through the filter, and thence through the pipe leading to the settling tanks.

ELECTRIC ROOM AND PILLAR UNDERCUTTING MACHINE IN AN OHIO MINE.
Adapts longwall principle to room and pillar work, making a continuous cut across the face
without withdrawal until room is finished. Sullivan Machinery Co., Chicago.

The solutions containing gold are allowed to settle for some time, after which the clear solution is drawn off into the storage tanks. From the storage tanks the solutions are pumped into the precipitating tanks which are situated about 25 to 30 feet above the filter press, thus generating the necessary pressure to force the solution properly through the press. The precipitation of gold from the solution is now to be accomplished. Gas generators are placed level with the top of the precipitating tanks. One is a closed vessel in which the sulphur is burned, while air under pressure is forced through it from an air compressor. Another generator is lined with lead, in which hydrogen sulphide is produced. The precipitating tank being full of chlorine solution, the SO_2 generator is put into commission and sulphurous acid gas forced through a hose, to the bottom of the tank. The H_2S generator is charged with iron matter and diluted sulphuric acid. When the compressed air is admitted the gas is conveyed to the solution. As soon as the precipitation is completed, the generator is stopped and the solution now settles for about an hour after which time it is passed through a filter press which collects the sulphides.

After the solutions have passed through the filter press, air under pressure is forced through and partially dries the sulphide cake, which is now placed in iron trays in large cast iron mufflers, roasted, fluxed, fused, refined, and cast into bars.

The machinery required for chlorination mill would be: crushers, dryers, revolving screens, rollers for recrushing and any type roaster; elevators, chlorination barrels, cars, platform scales, steel water tanks, precipitating tanks, solution pumps, gas generators, air compressor, filter presses, iron trays, and cast-iron muffler.

G.—CONCENTRATING PLANTS. Concentration is classed either as coarse or fine, terms relating primarily to the size of the material, whether granules if coarse, or slimes if fine.

Coarse concentration deals with the coarse or partial crushing of the ore, sizing in revolving screens or trommels and hydraulic classifiers, and treatment upon jigs (eccentric-driven plungers forcing water constantly through the ore placed on stationary screens or grates) which yield a product of clean or valuable concentrates.

Fine concentration, by concentrating tables, often follows the coarse, when it includes the treatment of the finer residual products of coarse concentration, or the ore may be crushed fine at the outset.

Concentrating ores with values chiefly of copper, zinc, lead, tin, or iron are generally treated by coarse concentration, but gold and silver-bearing ores, contain most of the precious metal in a fine state, and are therefore subject to fine crushing and concentration.

H.—HOISTING AND PUMPING WORKS. From the small horse-power hoist or whim, where, with one horse walking, 400 pounds of ore can be lifted 60 feet per minute, the mechanical engineer has co-operated with the miner in developing the modern powerful duplex direct-acting double-drum Corliss hoisting engine, with high-grade hoisting cages and safety appliances.

At the pumping works, the old Cornish pump, which in the eighties was so popular, has now, at most mines, been replaced by the modern, steam and electric-driven deep mine pump, of higher efficiency. The engineer in providing for the drainage of a mine, and after fixing upon the capacity and lift, has two things chiefly to bear in mind: First, the commercial efficiency of installation, considered with due reference to the mining risk and the length of time that the plant will probably be in use; second, the safety against "drowning out" that the plant should afford. To judge from the many modern installations of mine pumping plants, the points referred to above have been duly considered and successfully solved, and the mine pump of today is a fine piece of engineering.

I.—Smelting Plants. To melt or fuse ore for the purpose of separating the metal from its gangue is called smelting. It consists in subjecting ores and suitable fluxes to the action of intense heat, whereby the materials become fluid, the fluxes combining with the gangue of the ore to form a slag, while the valuable parts combine together to form a matte, the separation taking place while the materials are in a molten condition, by difference in their specific gravity.

ROOT BLOWER FOR BLAST FURNACE, ANACONDA MINING CO.

Blast Furnaces for Silver-Lead Ores.—The general form of the blast furnace for silver-lead ore is a rectangular brick shaft bound with iron, resting on an I-beam mantel frame supported at the four corners by columns, generally made up of four Z bars. Below this shaft are the water jackets made of flanged steel plates. The side jackets have openings for tuyeres, a short distance from the bottom. The jackets rest on a curb of fire-brick enclosed in a steel or cast-iron frame, and in the center of this curb is a large cavity, named the internal crucible, from which the lead is removed through the syphon tap. The smallest standard steel water-jacketed furnace designed is 20 inches inside diameter—a combined lead and copper furnace. The jacket, made in one piece by welding, is hung from the charging floor, which permit the curb being removed without disturbing any other part of the furnace. This furnace has an approximate capacity from 8 to 10 tons per day of 24 hours. Round silver-lead furnaces 36 and 48 inches are very popular sizes for smaller smelting plants, and these sizes are also designed and built in sections for mule-back transportation. The rectangular silver-lead furnaces are designed and built in many different sizes and to suit all kinds of local conditions; 33 by 72 inches, 36 by 100 inches, 36 by 120 inches, 36 by 160 inches and 40 by 96 inches—a good average 100-ton furnace; 42 by 120, 44 by 160, and 44 by 168 inches are the most common sizes.

BLAST FURNACES FOR COPPER ORES.—These furnaces differ considerably from those used for silver-lead ores. The steel jackets are generally made in two tiers and extend up to the charging floor. The internal crucible is dispensed with altogether and the jackets rest upon the heavy cast-iron crucible plate supported on jackscrews or columns.

A very popular size of the round type is the 36 inch inside by 6 feet 6 inches high, single water-jacketed furnace, having a capacity of about 30 tons per day of 24 hours. Such a furnace needs about 1,820 cubic feet of blast per minute, and the nearest blower of the iron-impeller—top-discharge—type would be one having a capacity of 9 cubic feet per revolution and running 220 to 225 revolutions per minute. Other items for such a small copper-smelting plant would be: Six slag pots and trucks, three bullion pots and trucks, two large settling pots and trucks, one 4-beam charging scale, one steel charging barrow, one coke barrow, six tubular steel wheelbarrows, one coke fork, one scoop shovel, six long-handled steel shovels, two sledges with handles, 300 pounds of steel furnace bars; a 12 horse-power engine and boiler, or electric motor attached to the blower to run it direct is all the power needed. The rectangular types are built in 36, 42, 44, 48, 54 and 56 inches width and in lengths up to 612 inches or 51 feet.

J.—COPPER CONVERTER PLANTS. To bessemerize copper matte, which consists mostly of copper, sulphur and iron, is to convert it into metallic copper by slagging off its iron contents and burning off its sulphur.

The process is performed in a steel vessel very similar to that used in the bessemer steel process, and said vessel is called a converter. It has a thick lining of crushed quartz in which is mixed a sufficient quantity of clay to hold the quartz together. On one side, near the bottom of the converter, is attached an air box, which supplies air to a number of horizontal tuyeres or holes pierced through the lining. Molten copper matte is charged into the converter and air from a blowing engine under a pressure of from 6 to 15 pounds is forced through the molten mass. The iron is oxidized and united with the quartz of the lining to form a slag, while the sulphur combines with the oxygen of the air, forming sulphurous acid gas, which

INTERIOR OF SMELTER POWER HOUSE, SHOWING COMPRESSORS, ANACONDA MINES.

Fifteen 4-stage Ingersoll-Rand compressors; steam cylinders 20 and 40 in., air cylinders 37¼, 20¾, 12½ and 6 in.; stroke 48 in. Capacity 2,000 ft. of free air per minute. Pressure 950 lb.

PORTIONS OF THE ANACONDA SMELTING PLANT.
Above, the converters, served by a 60-ton electric traveling crane. Below, the Straight Line casting machine for copper

escapes. The slag is skimmed off and returned to the blast furnace, and the metallic copper is poured into moulds.

In this operation no fuel is used, the heat being supplied by the combustion of the sulphur in the matte. The lining is rapidly eaten out—in fact, the operation is performed at the expense of the lining, and after a small number of operations, the converter shell is taken to the relining room to be relined.

The shells of the converter are made of heavy steel plates, heads

SULLIVAN CORLISS HOIST FOR CENTENNIAL COPPER CO., CALUMET, MICH.
Capacity 4,000 ft. depth; speed 3,500 ft. per minute; drum 15 ft. diameter. Engines 32 by 60. Sullivan Machinery Co.

and top of cast steel, supported on a cast-iron stand with flanged rollers; the converter is tilted either by hydraulic machinery or by gears and an electric motor attached to the stand. The converter shells are generally handled by an electric traveling crane and the air is furnished by a blowing engine of the Reynold or Nordberg type or by the Parsons steam-turbine blower. A Chilean mill is used in the relining room for grinding clay for the converter lining.

The converters are designed and built in different sizes 72 inches diameter by 100 long, and 84 inches diameter by 126 inches long being the most common. The "blows" per 24 hours run from 10 to 16.

The tuyeres are provided with the Dyblie tuyere valves and most converters nowadays have attached the Bennett pouring spoon. Trucks made of steel beams supported on four wheels carry the moulds when they are pushed under the converters to receive the blister copper.

THE MANUFACTURE OF STEEL AND WROUGHT IRON IN AMERICA.

By Bradley Stoughton.

The three metals most interesting from an economic point of view—always, but especially just now—are steel, copper and gold. Of these the first is the most closely interwoven with the development of engineering construction in every line. Professor Stoughton treats the American steel industry in its metallurgical rather than its economic aspects, and like the authors preceding him in this issue he brings to the discussion the equipment of a specialist with a record of creative professional work in his own department.—THE EDITORS.

IRON, as first obtained from its ores in the form of a commercial product today, appears almost wholly in the form of "pig."

The large amount of carbon in pig iron, however, makes it both weak and brittle, so that it is unfit for most engineering purposes. It is used for castings that are to be subjected only to compression, transverse, or very slight tensile strains, as, for example, supporting columns, engine bed-plates, railroad car wheels, water mains, etc; but the relatively increasing amount of steel used shows the preference of engineers for the stronger and more ductile material. To-day three quarters of the pig iron made in the United States is subsequently purified by either the Bessemer, open-hearth, or puddling process. Each of these will reduce the carbon to any desired point, while the silicon and manganese are eliminated as a necessary accompaniment of the reactions—indeed, we might almost say, as "a condition precedent" to carbon reduction. Phosphorus and sulphur are reduced by the puddling process, and by a special form of open-hearth process known as "the basic open-hearth process."* The complete scheme of American iron and steel manufacture is given on page 50.

The figure expresses graphically data which are interpreted at greater length as follows: Practically all the iron ore mined is smelted in about 325 blast furnaces, to produce annually 25,000,000 tons of pig iron. About 3 per cent of this pig iron is remelted and made into malleable cast iron; 20 per cent is made into gray cast iron; 52 per cent is purified in 62 Bessemer converters to Bessemer steel; 20 per cent is purified in 465 basic open-hearth furnaces; 2 per cent is purified in 195 acid open-hearth furnaces, while the remaining 3 per cent is purified in 3,000 puddling furnaces to make wrought iron. Only active furnaces are included, and the numbers are estimated

---

* The basic Bessemer process is not in operation in America.

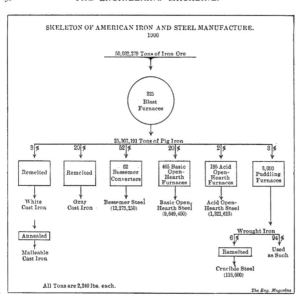

SKELETON OF AMERICAN IRON AND STEEL MANUFACTURE.
1900

50,032,279 Tons of Iron Ore

825
Blast
Furnaces

23,307,191 Tons of Pig Iron

3%	20%	52%	20%	2%	3%
Remelted	Remelted	62 Bessemer Converters	465 Basic Open-Hearth Furnaces	195 Acid Open-Hearth Furnaces	3,000 Puddling Furnaces
White Cast Iron	Gray Cast Iron	Bessemer Steel (12,275,250)	Basic Open-Hearth Steel (9,649,400)	Acid Open-Hearth Steel (1,321,613)	Wrought Iron

Annealed

Malleable Cast Iron

Wrought Iron
6% | 94%
Remelted | Used as Such

Crucible Steel
(118,000)

All Tons are 2,240 lbs. each.

The Eng. Magazine

as the recent statistics are not yet published.* The wrought iron may
be used as such for pipe, blacksmith work, small structural shapes,
etc, and 94 per cent of it is so used; the other 6 per cent is remelted
in crucibles to make crucible steel. To sum up, about 23 per cent of
the pig iron made is used without purification,† and 77 per cent is
purified and converted into another form. In all cases of purification the impurities are removed by oxidizing them, and we must
emphasize the rule of all smelting that unoxidized elements dissolve
in the metal, while those in the oxidized condition pass into the slag,
or, if there is no slag, form a slag for, and of, themselves. In considering the Bessemer, open-hearth, and puddling processes, then, we
have to do with oxidizing conditions, whereas the opposite is the case
in the blast furnace. The oxidization is effected by means of the

---

* The numbers and percentages given in this Figure will change slightly from time to
time, but this will convey an idea of the relative amounts of the different products made.
† It is true that the annealing process for malleable cast iron purifies the outer layers
of the castings from carbon, and, if the castings are very thin, this purification may extend
to the centre; but this is not primarily a purification process and will be treated at length
in another section.

oxygen of the air or that of iron ore, Fe_2O_3, or its equivalent, or of both air and oxide of iron.

There is not an exact relation between the amounts of pig iron used for the different purposes and the amounts of the resulting materials. In 1906 the following production was made:—

TABLE I.

	Cast Iron Used.	Made.
Malleable Cast Iron	600,000* tons	750,000* tons
Gray Cast Iron	5,100,000 "	6,000,000 "
Bessemer Steel	13,150,000 "	12,275,250 "
Basic Open-Hearth Steel	5,150,000 "	9,649,400 "
Acid Open-Hearth Steel	500,000 "	1,321,613 "
Wrought Iron	800,000 "	2,000,000* "
Crucible Steel	....	118,000 "

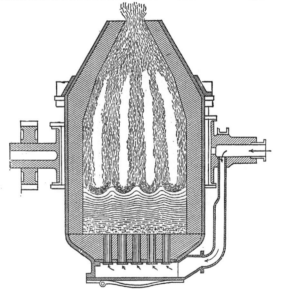

SECTION OF BESSEMER CONVERTER IN ACTION.

The cause of the descrepancy is the scrap iron or steel mixed with the pig iron in the manufacture of gray-iron castings and open-hearth steel. Perhaps an average of 25 per cent of old scrap will be mixed with 75 per cent of new pig iron for making iron castings, and 50 per cent or so of steel scrap will be mixed with 50 per cent of so of

* These are as close estimates as can be made in the absence of exact statistics.

SERIES OF FLAMES DURING A TYPICAL BESSEMER BLOW.

The series runs from left to right, and downward. At the beginning silicon and manganese are being oxidized; the second view is slightly later. In the third, carbon is
beginning to burn, with some spitting. The lower row shows the full carbon flame.

pig iron in the basic open-hearth process. Over one-half of the
wrought-iron production of the United States is made from wrought-
iron scrap by building it up into "piles," heating to a welding heat, and
re-rolling into the desired shapes and sizes.

BESSEMER PROCESS.—In the Bessemer process, perhaps 10 tons
of melted pig iron is poured into a hollow- pear-shaped converter lined
with silicious material (page 51). Through the molten material is
then forced 25,000 cubic feet of cold air per minute. In about four
minutes the silicon and manganese are all oxidized by the oxygen of
the air and have formed a slag. The carbon then begins to oxidize to

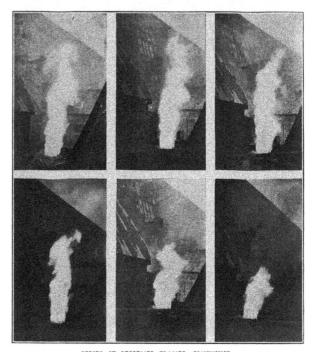

SERIES OF BESSEMER FLAMES, CONTINUED.

The flame is beginning to drop, until the end of the blow is reached. The views were taken at vessel No. 1, Edgar Thomson Works, with an 18-ton charge and a blow of about 9½ minutes.

carbon monoxide, CO, and this boils up through the metal and pours out the mouth of the vessel in a long brilliant flame. After another six minutes the flame shortens or "drops"; the operator knows that the carbon has been eliminated to the lowest practicable limit (say 0.04 per cent) and the operation is stopped. So great has been the heat evolved by the oxidation of the three impurities that the temperature is now higher than it was at the start, and we have a white-hot, liquid mass of relatively pure metal. To this is added a carefully calculated amount of carbon to produce the desired degree of strength or hardness or both; also about 1½ per cent of manganese and 0.2

A BESSEMER CONVERTER BLAZING.

per cent of silicon.* The manganese is added to remove from the
bath the oxygen with which it has become charged during the operation and which would render the steel unfit for use, and also to
neutralize partially the effect of sulphur. The silicon is added to get
rid of the gases which are contained in the bath and thus to prevent

* In the case of making rail steel. Only one-half to two-thirds of this silicon and manganese are found in the final steel.

the formation of blowholes. After adding these materials, or "recarburizing" as it is called, the metal is poured into ingots which are allowed to solidify and then rolled, while hot, into the desired size and form. The characteristics of the Bessemer process are: (1), great rapidity of purification, say ten minutes per "heat," (2), no extraneous fuel used; and (3), the metal is not melted in the furnace where the purification takes place.

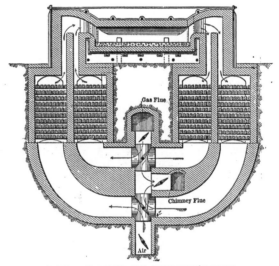

DIAGRAM OF REGENERATIVE OPEN-HEARTH FURNACE.

The four chambers below this furnace are filled with checkerwork of brick with horizontal and vertical channels through which the gas and air may pass. The gas enters the furnace through the inner regenerative chamber on one side and the air enters through the corresponding outer one. They meet and unite, passing through the furnace and thence dividing into proportional parts and passing to the chimney through the two regenerative chambers at the opposite end. In this way the brickwork in the chambers is heated up. The current of gas, air and products of combustion is changed every 20 minutes whereby all four regenerators are always kept hot. The gas and air enter in a highly preheated condition and thus give a greater temperature of combustion, while the products of combustion go out of the chimney at a relatively low heat and thus fuel economy is promoted.

ACID OPEN-HEARTH PROCESS.—The acid open-hearth furnace is heated by burning within it gas and air, each of which has been highly preheated before it enters the combustion chamber. A section of the furnace is shown just above. The metal lies in a shallow pool on

PUDDLING IRON, AND WITHDRAWING THE BALL FROM THE PUDDLING FURNACE.

the long hearth, composed of silicious material, and is heated by
radiation from the intense flame. The impurities are oxidized by an
excess of oxygen in the furnace gases over that necessary to burn the
gas. This oxidizes the slag and then the slag oxidizes the impurities.

This action is so slow, however, that the 3 per cent to 4 per cent of carbon in the pig iron would take a long time for combustion. The operation is therefore hastened in two ways: (1), iron ore is added to the slag to produce the following reaction:—

$$Fe_2O_3 + 3C = 3CO + 2Fe;$$

and (2), the carbon is diluted by adding varying amounts of cold steel scrap instead of all pig iron. The steel scrap and ore are added to the charge at the beginning of the process, and it takes about 6 to 10 hours to purify a charge, after which the metal is "recarburized" and cast into ingots. The characteristics of the open-hearth process are: (1), a long time occupied in purification; (2), large charges treated in the furnace (the modern practice is usually 30 to 75 tons to a furnace); (3), at least a part of the charge is melted in the purification furnace; and (4), the furnace is heated with preheated gas and air.

BASIC OPEN-HEARTH PROCESS.—The basic open-hearth operation is similar to the acid, with the difference that to the bath is added a sufficient amount of lime to form a very basic slag. This slag will dissolve all the phosphorus that is oxidized, which an acid slag will not do. We can oxidize the phosphorus in any of these processes, but in the acid Bessemer and the acid open-hearth furnaces the highly silicious slag rejects the phosphorus and it is immediately deoxidized again and returns to the iron. The characteristics of the basic open-hearth process are the same as those of the acid open-hearth with the addition of: (5), lime is added to produce a basic slag; (6), the hearth is lined with basic, instead of silicious, material in order that it shall not be eaten away by this slag; and (7), impure iron and scrap may be used, because phosphorus, and (to a limited extent) sulphur can be removed in the operation.

PUDDLING PROCESS.—Almost all the wrought iron is to-day made by the puddling process, invented by Henry Cort about 1780, with certain valuable improvements by Joseph Hall fifty years later. In this process the pig iron is melted on the hearth of a reverberatory furnace, lined with oxide of iron (page 58). During the melting there is an elimination of silicon, manganese, phosphorus, and sulphur and the formation of a slag, which automatically adjusts itself to a very high content of iron oxide by dissolving it from the lining. After melting the heat is reduced and a reaction set up between the iron oxide of the slag and the silicon, manganese, carbon, phosphorus, and sulphur of the bath, whereby the impurities are oxidized and all removed to a greater or less extent. The slag, because of its basicity (by iron oxide), will retain all the phosphorus oxidized, and therefore the greater part of this element is removed. The oxidation of all the impurities is produced chiefly by the iron oxide in the slag and the lining of the furnace, although it is probable that excess oxygen in the furnace gases assists, the slag acting as a carrier of oxygen from it to the impurities.

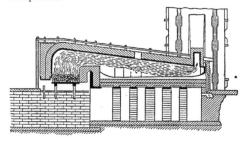

SECTIONAL VIEW OF PUDDLING FURNACE.

The purification finally reaches that stage at which the utmost heat of the furnace is not sufficient to keep the charge molten, because iron like almost every other metal melts at a higher temperature the purer it is. The metal therefore "comes to nature" as it is called; that is to say, it assumes a pasty state. The iron is then rolled up into several balls, weighing 125 to 180 pounds apiece (page 56), which are removed from the furnace dripping with slag and carried over to an apparatus in which they are squeezed into a much smaller size and a large amount of slag separated from them. The squeezed ball is then rolled between grooved rolls to a bar, whereby the slag is still further reduced so that the bar contains at the end usually about 1 per cent or 2 per cent. This puddled bar, or "muck bar," is cut into strips and piled up, as shown just below, into a bundle of bars which are bound

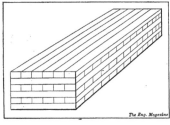

METHOD OF PILING MUCK BAR.

together by wire, raised to a welding heat, and again rolled into a smaller size. This rolled material is then known as "merchant bar" and all wrought iron, except that which is to be used for manufacture into crucible steel, is treated in this way before sale. The effect of the further rolling is to eject more slag and also to make a cross network of fibres, instead of a line of fibres all running in the same direction, *i.e.*, lengthwise of the bar. The fibres are produced by the action in rolling or drawing out the slag into strings and also producing long fibres of metal, each of which is surrounded by an envelope of slag.

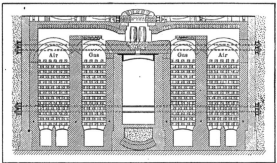

REGENERATIVE GAS CRUCIBLE-STEEL MELTING FURNACE.

CRUCIBLE PROCESS.—Wrought iron is converted into steel by the operation of carburizing, or adding carbon to it. This is today accomplished in two ways: (1), by the cementation or steel conversion process, in which carbon is allowed to soak into red-hot iron in a manner like in nature to the absorption of ink by blotting paper; and (2), by the crucible process in which wrought iron is melted in a crucible with carbon, or with iron containing carbon, *e.g.*, cast iron. The cementation process, on account of the length of time required and the very large amount of fuel used, has now been largely superseded by the crucible process, in which the wrought iron is cut up into small pieces and melted in covered crucibles, the desired amount of carbon being placed on top of the charge before the melting, together with any other alloying element desired, such as chromium, tungsten, manganese, etc. In Sheffield, England, the capital of the crucible-steel industry of the world, coke furnaces, or "melting holes," containing each two crucibles, are almost universally used; while in America regenerative gas furnaces containing each six crucibles are the common type. In the gas furnace it is necessary that the gas and air for combustion shall be preheated in order that we may obtain fuel economy and also reach the desired temperature for melting quickly. For this preheating the regenerative principle is used.

The characteristics of the crucible process are: (1), wrought iron is the raw material, not pig iron; (2), small units and hand labor are used; (3), labor and fuel costs are very high; and (4), during the operation the material is contained in entirely closed vessels.

COMPARISON OF PURIFICATION PROCESSES.

ACID WITH BASIC OPEN-HEARTH.—Acid open-hearth steel is believed by engineers to be better than basic, and is usually specified in all important parts of structures—although not so rigidly today as a few years ago. This is in spite of the fact that phosphorus and sulphur, two very harmful elements, are lower in the basic steel. The basic process is much less expensive than the acid, because high-phosphorus pig iron and scrap are cheap, the lower cost of materials used more than balancing the greater cost of the basic lining and the lime additions, and because also the acid furnace has a higher output, since the heats are shorter.

The reasons for the preference of acid steel are as follows: (1), a basic slag will dissolve silicon from the metal; we therefore recarburize in the basic process by adding the recarburizer to the steel after it has left the furnace, instead of in the furnace, as we do in the acid process. Should any basic slag be carried over with the metal, however, which is liable to happen, there is the danger that the ingots will be too low in silicon. They are then impregnated with gas bubbles, or "blow holes;" (2), moreover the recarburizer does not mix with the steel as well if it is not added in the furnace and this sometimes produce irregularities; (3), a basic slag is usually more highly oxidized than an acid one. Therefore the metal at the end of the operation is more highly charged with oxygen. For this reason we add a larger amount of manganese in the recarburizer, but the remedy is never quite as good as prevention; (4), since we cannot remove the phosphorus from the bath in the acid process, it is necessary to use only picked iron and scrap, whereas, in the basic process, good steel can be made from almost any quality of material. Many engineers believe, however, that a better grade of steel results from using the picked material; (5), it occasionally happens in the basic process that, after the phosphorus has all been oxidized in the slag and the operation is ended, some of it will get back into the metal again. This is especially liable to happen when basic slag is carried over into the ladle upon tapping the furnace. If this occurs, and if the bath is very hot, a reaction may take place between the basic slag and the acid lining of the ladle whereby the slag will be enriched in silica and phosphorus will be forced out of it.

BASIC OPEN-HEARTH WITH BESSEMER.—Basic open-hearth steel

is better than Bessemer steel. The reasons for this are believed to be: (1), the open-hearth process being slower, more attention and care can be given to each detail. This is particularly true of the ending of the process, for if the Bessemer process is continued only a second or so too long, the bath is highly charged with oxygen, to its detriment, and even under normal circumstances there is more oxygen in the metal at the end of the Bessemer process than at the end of the basic open-hearth, because there has been so intimate a mixture between metal and air; (2), for the same reason the Bessemer metal is believed to contain more nitrogen and hydrogen* which are thought to be deleterious; (3), the heat of the Bessemer process is dependent upon the impurities in the pig iron, and especially upon the amount of the silicon, and can be controlled only to a limited extent by methods that are not perfect in their operation. Furthermore, the heat is regulated according to the judgment of the operator and his skill in estimating the temperature of the flame. Irregularities therefore result at times and these produce an effect on the steel, because the temperature at which the ingots are cast should be neither too high nor too low. It is true that the temperature of the open-hearth steel is also regulated by the judgment of the operator, but more time is afforded for exercising this judgment and the heat of the operation is more easily controlled; (4), in the Bessemer process we must get rid of all the carbon first and then recarburize to the desired point. In the open-hearth process we may stop the operation at any desired amount of carbon and then recarburize only a small amount. Therefore the open-hearth has the advantage of greater homogeneity when making high-carbon steel, since a large amount of recarburizer may not distribute itself uniformly. In order to produce the best quality of steel it must be cast into ingot molds within a certain limited range of temperature which varies according to the amount of carbon, etc., that it contains. Therefore in casting the very large heats of the open-hearth process, the ingots must either be very large or else the first one will be too hot and the last one too cold for the best results. If the ingots are large, on the other hand, segregation is liable to be excessive. The large heats of the open-hearth process are therefore disadvantageous to the quality of the steel.

For nearly fifteen years the Bessemer process has been fighting a losing battle to maintain its supremacy against the inroads of the basic open-hearth, which have been possible because of the increasing cost of Bessemer pig iron, due to the exhaustion of the low-phosphorus ores. The pig iron for the Bessemer process must contain

* From moisture in the blast.

so little phosphorus that, after allowing 10 per cent loss of metal during the blow, the phosphorus in the steel shall be not over 0.100 per cent. Ores low enough in phosphorus to make this grade of metal have, therefore, come to be known as "Bessemer ores." The requirement of such an ore is that the percentage of iron in it must be at least 1,000 times the percentage of phosphorus. During the year 1906, the Bessemer process in the United States yielded very much to the basic open-hearth and it would seem as if there was no chance of its ever taking so important a position again unless new iron ores low in phosphorus are discovered.

On account of its ability to make low-carbon steel more readily than the basic open-hearth, the Bessemer process has a firm hold on the wire and welded steel pipe industry, although even here the open-hearth process has encroached. For rolling very thin for tin plate, etc, we want a metal relatively high in phosphorus, and therefore the Bessemer process is largely used here, although in some cases ferro-phosphorus is being added to basic open-hearth metal to accomplish the same result. The reason low phosphorus is desired is because the plates are rolled very thin by doubling them up and putting several thicknesses through the rolls at the same time. Low-phosphorus metal welds together too much under these circumstances.

The chief requisites of railroad rails are lack of brittleness and ability to withstand wear. The Bessemer process is able to provide such a material, and it works so well in conjunction with the rapid, continuous operation of the rail-rolling mill that it has a decided advantage. It produces a small tonnage of ingots at frequent intervals (say 15 tons every 7 minutes), while the open-hearth process provides a large tonnage of ingots, which may come at irregular intervals and thus alternately delay and over-crowd the rail-mill operations. But notwithstanding these advantages, an increasing tonnage of basic open-hearth rails is made every year in the United States. Lately this has attracted widespread interest, owing to reports of an alarming number of rail breakages and the action of some railroads in blaming the Bessemer process therefor. It is true that every year there becomes a greater scarcity of Bessemer ores, and therefore an increasing amount of phosphorus in the steel, so that it is no secret that many rails have been made within the past year containing more than the allowable 0.100 per cent phosphorus. Phosphorus makes the steel brittle, especially under shock and in cold weather. It also makes the steel hard to resist wear, but this hardness is better obtained by means of carbon, and low-phosphorus, high-carbon steel rails would undoubtedly break less in the track. It is to be remembered that heavier trains are being run every year, and this brings greater strains upon the rails in comparison with which they have not been correspondingly increased in size. At the present time so very large an amount of capital is tied up in Bessemer rail mills, and it would take so long to change them over into open-hearth mills, that there is no immediate prospect of a great replacement. The acid and Bessemer steel production of the principal countries of the world is shown in Tables II and III, while the recent history of open-hearth steel-rail manufacture is shown briefly in Table IV.

TABLE II. STEEL PRODUCTION OF PRINCIPAL COUNTRIES. 1906.

	United States.	Germany.	Great Britain.
Acid Converter	12,275,253	407,688	1,307,149
Basic Converter	....	6,772,804	600,189
Total Converter	12,275,253	7,180,492	1,907,338
Acid Open-hearth	1,321,613	230,668	3,378,691
Basic Open-hearth	9,649,385	3,534,612	1,176,245
Total Open-hearth	10,970,998	3,765,280	4,554,936
Crucible and Special	118,500	189,313	....?....
Total	23,364,751	11,135,085	6,462,274
Proportion Steel to Pig Iron.	92.3	89.2	....

TABLE III. MAKE OF ACID AND BASIC STEEL. 1906.

	Acid.		Basic.	
	Tons.	Per Cent.	Tons.	Per Cent.
United States	13,715,366	58.7	9,649,385	41.3
Germany	715,952	6.4	10,419,133	93.6
Great Britain	4,685,840	72.5	1,776,434	27.5
Total	19,117,158	46.7	21,844,952	53.3

TABLE IV. AMERICAN RAILROAD-RAIL MANUFACTURE.

	Bessemer.*	Gross Tons. Open Hearth.†	Wrought Iron.‡	Total.
1900	2,383,654	1,333	695	2,385,682
1901	2,870,816	2,093	1,730	2,874,639
1902	2,935,392	6,029	6,512	2,947,933
1903	2,946,756	45,054	667	2,992,477
1904	2,137,957	145,883	871	2,284,711
1905	3,188,675	183,264	318	3,372,257
1906	3,705,642	....	...	.....•..

CRUCIBLE STEEL WITH OTHERS.—Crucible steel is the most expensive of all, costing at least three times as much as the next in price —acid open-hearth steel. It is also the best quality of steel manufactured, and for very severe service, such as the points and edges of cutting tools, the highest grades of springs, armor-piercing projec-

* The first Bessemer rails were made commercially in 1867.
† The first open-hearth rails were made in 1878. In 1881, 22,515 gross tons of open-hearth rails were produced.
‡ The maximum production of iron rails was 808,866 gross tons in 1872.

tiles, etc., it should always be employed. The reason for its superiority is believed to be that it is manufactured in a vessel which excludes the air and furnace gases, and therefore is freer from oxygen, hydrogen, and nitrogen. Perhaps the fact that the process is in some ways under a little better control than any of the others, and the metal receives more care, on account of being manufactured in small units, assists in raising the grade. Crucible steels are usually higher in carbon than Bessemer and open-hearth steels, because the special service to which the crucible steels are adapted is usually one requiring steel that can be hardened and tempered—for example, cutting tools, springs, etc., and only the high-carbon steels are capable of this hardening and tempering.

Wrought Iron with Low-Carbon Steel.—Wrought iron costs 10 per cent to 20 per cent more than the cheapest steel. Its claims to superiority over dead-soft steel consist in its purity and the presence in it of slag. Just how much advantage the slag is, has never been proven; it gives the metal a fibrous structure which, perhaps, increases its toughness and its resistance to breaking under bending, or under a sudden blow, or shock. Some think that the slag also assists in the welding of the material, but this is doubtful and it is probable that the easy weldability of wrought iron is due alone to its being low in carbon. Some also believe that the slag assists the metal in resisting corrosion; hence one reason for the preference of engineers for wrought-iron pipe for boilers and other purposes. There are other qualities of wrought iron which may tend to make it corrode less than steel, chief among which are the absence of blowholes and possibly the absence of manganese, and the presence of phosphorus. It is now believed by many that manganese starts an electrolytic action which hastens corrosion. An advantage of wrought iron in this connection is its rough surface to which paint or other protective coatings will adhere more firmly than to the comparatively smooth surface of steel. Nevertheless the evidence goes to show that *properly made steel* corrodes very little more than wrought iron, especially in boilers, pipe, and other articles which cannot be coated.

The properties of wrought iron are the nearest to those of pure iron of any commercial material, notwithstanding its slag, which because it is mechanically mingled with the metal does not interfere with its chemical or physical behavior. Therefore wrought iron is greatly preferred for electrical-conductivity purposes, and as a metal with high magnetic power for use as armatures of electro-magnets, etc. The advantages I have mentioned—the conservatism of engineers and the capital previously invested in puddling furnaces—are the chief factors in keeping alive the manufacture of wrought iron. It was freely predicted that the invention of the Bessemer and open-hearth processes would bring about the extinction of the puddling process, but these prophecies have never been fulfilled, although the importance of wrought iron has waned very greatly in fifty years. When under strain greater than it can withstand wrought iron stretches more uniformly over its entire length than steel, as shown thus:

TABLE V.

	Elastic Limit. Lb. per Sq. In.	Ultimate Strength. Lb. per Sq. In.	Elongation, Per Cent. In 12 In.	In 18 Ft.	Reduction of Area. Per Cent.
Wrought Iron ...	31,550	48,810	23	15.22	28.30
Steel	33,150	59,260	39	14.40	51.50

This makes wrought iron a much more valuable material for boilers, for example.

SUMMARY.—In order of expense and of quality the different steels are arranged as follows: (1), Crucible; (2), Acid open-hearth; (3), Basic open-hearth; (4), Bessemer. The amounts of the different kinds made today and ten years ago are shown in Table VI. Though I have not made a direct comparison between certain of the classes, *e.g.*, acid open-hearth with Bessemer, their relations may be easily learned by collating the other comparisons given.

TABLE VI.

	Bessemer.		Open-Hearth.				Crucible, etc.	
1906..	12,275,253	52%	9,649,385	41%	1,321,613	6%	118,500	1%
1896..	4,909,128	78%	776,256	12%	522,444	9%	68,524	1%

Many engineers will be interested in the uses to which the annual steel and wrought iron production of the United States is put, which are shown below:

TABLE VII. USES OF STEEL AND WROUGHT IRON. Gross Tons. 1905.

	Steel.	Wrought Iron.
Railroad rails	3,375,611	318
Railroad splice bars.............................	179,900	17,422
Structural shapes	1,648,889	11,630
Merchant bars *	2,271,162	1,322,439
Rods for wire, and wire products.................	1,307,407	1,281
Rods for wire nails..............................	500,000	....
Plate for cut nails...............................	40,483	24,059
Plates †	2,031,184	10,022
Sheets †	1,428,890	62,134
Skelp ‡	983,198	452,797
Hoops, bands and cotton ties....................	442,664	2,863
Spike and chain rods, horseshoe bars, strips, etc.....	529,288	154,620
Blooms and billets...............................	41,349	405
Castings	560,767	....

* Merchant bars are bars of small sizes which are usually sold to be worked up into other forms.
† Plates are gauge No. 12 and thicker; sheets are gauge No. 13 and thinner.
‡ Skelp is flat strips which are welded into pipe.

DISTINGUISHING BETWEEN THE DIFFERENT PRODUCTS.

Low-carbon steel pipe, merchant bars, horse-shoe blanks, etc., sometimes masquerade under the name of wrought-iron; high-carbon open-hearth and Bessemer-steel merchant bars, tool blanks, etc, sometimes masquerade as "crucible steel," or perhaps, "cast steel," which is the trade name for crucible steel. Other deceptions are not unknown; indeed, even malleable cast iron is sold oftentimes as "steel castings." It is therefore important for engineers to understand the essential differences between these materials, although care in the wording of contracts and specifications should be the important consideration and should precede watchfulness over the products. The definitions of iron and steel materials are in such a confused and unsettled condition that it does not do to rely upon them at all, especially where a law suit may be involved, and contracts in clear, simple language, free from legal and metallurgical phraseology, are the best safeguards. But even where it is entirely plain what material is called for there is always a temptation to substitute steel for wrought iron, Bessemer for open-hearth, basic for acid, and Bessemer or open-hearth for crucible steel. In case any such substitution is suspected there are means by which the material may be tested, aside from its strength and ductility which may or may not be in the contract. The tests are somewhat delicate and usually require the judgment and experience of an expert and one who has standard samples of the different grades of material for comparison, because the details of manufacture vary from district to district, and still more so depending upon the purpose for which the products are to be used.

Wrought iron may be distinguished from low-carbon steel by the fact that it contains slag. Usually there is more than 1 per cent. of slag in iron and less than 0.20 per cent. slag (including metallic oxides) in steel. The slag may be determined either by chemical or microscopical analysis. Normal wrought iron is practically free from manganese, while normal Bessemer and open-hearth steel will contain 0.50 per cent. or more. Wrought iron generally contains more than 0.100 per cent. phosphorus, while good steel should never do so.

Crucible steel normally has less than 0.40 per cent. manganese and more than 0.20 per cent. silocon, while open-hearth and Bessemer steels normally have more than 0.50 per cent manganese and less than 0.20 per cent silicon. In the case of steel castings, however, this rule for silicon does not apply, as Bessemer and open-hearth steel castings are sometimes as high as 0.60 per cent. silicon. It is possible to make both Bessemer and open-hearth steels low in manganese, but they cannot be made low in both manganese and silicon without great danger from blowholes, while the difficulty is not met with to the same extent in crucible steel. When crucible steel is ordered low in carbon there is a much greater temptation to substitute another steel for it.

Acid open-hearth steel may be distinguished from basic open-hearth steel because it will normally be higher in silicon, and usually in phosphorus also, and lower in manganese. The same differences exist between acid and basic Bessemer steel.

Basic open-hearth steel may be distinguished from Bessemer steel by its lower manganese, silicon, phosphorus, and (generally) sulphur, as well as by the fact that it dissolves much more slowly in dilute hydrochloric acid.

It is possible to place such physical specifications in a contract as practically to insure obtaining the grade of material ordered. For example, such a high degree of ductility may be demanded, especially the percentage elongation in 10 or 20 feet, that nothing but wrought iron will give it; the strength and ductility may be put so high as to make it too dangerous to try to supply anything but crucible steel for the order; or they may be put a little lower so as to practically exclude Bessemer steel. The average physical differences between acid and basic open-hearth steel are not great enough to make this method of assurance so practicable, but it is again possible in the case of basic and acid Bessemer steel in England, where alone both these kinds of steel are made in important quantities.

POURING CRUCIBLE STEEL.

ASBESTOS; ITS MINING, PREPARATION, MARKETS, AND USES.

By E. Schaaf-Regelman.

ASBESTOS derives its name from the Greek, the term signifying "incombustible," "inconsumable," and while we can trace nearly to the dawn of history, the knowledge of the existence and application of this peculiar "stone for spinning purposes," it is but a decade or two since we have had a real and important asbestos industry. During the last four to six years this mineral has been applied to so many new uses that it is safe to say that asbestos goods are indispensable today for a great number of industrial purposes, and the recent unprecedented rise in prices has attracted attention so widespread that a more intimate knowledge of the conditions actually governing the industry should be beneficial to many.

The mineral asbestos, chiefly a silicate of magnesia, but containing also lime, iron, alumina, and water, is fibrous in structure, and in color varies from snowy white to green, blue, yellow, pink, and brownish hues. The mineralogist distinguishes two chief varieties, the chrysotile and amphibole asbestos, the principal difference in composition, aside from a slightly varying percentage of the above-named ingredients, being the amount and condition of the water present. Amphibole asbestos contains about 5 per cent of water, all of which is "combined," while chrysotile contains in the average about 15 per cent of water, of which only 12 to 13 per cent are chemically bound to the other constituents, while about 2 to 3 per cent are free or hygroscopic. For this reason chrysotile asbestos loses its strength, when heated, at about 660 degrees C., while amphibole asbestos can easily stand a temperature twice as high without disintegrating; from a purely technical point of view, therefore, the latter variety is a better material for protection against fire than chrysotile asbestos; but all deposits of amphibole asbestos known to this day exhibit a very brittle fibre, which cannot be used for spinning or weaving; its application is therefore limited, and this variety is unimportant from a commercial standpoint, while the chrysotile asbestos, occurring in much greater abundance in nature, has much stronger

BELL ASBESTOS MINES, THETFORD, P. Q. MAIN PIT.

fibres of great flexibility and sufficient tensile strength, qualities which give it great economic importance.

The historians of old Greek and Roman culture have mentioned repeatedly that the corpses of their kings and heroes, when prepared for cremation were wrapped in incombustible blankets in order to separate their ashes from those of the funeral pile, and Pausanias states that the wick of the eternal lamp in the temple of Pallas Athene in Athens was made of "crystallic flax." Charlemagne amused and astonished his guests by having a table cloth, made from "cotton stone," cleaned after a meal by throwing it into the fire and taking it out again unburnt and uninjured.

However, the practical use of asbestos was for a long time of a sporadic nature only, and the first experiments for using it on something like a commercial scale were made in the Alps in the early seventies; at about the same time the first specimens of a very fine asbestos from Canada, with silk-like fibres, were exhibited in London, but it was not until 1878 that mining operations of a tentative character were commenced in the Dominion. A lot of about 50 tons of selected crude asbestos was shipped to England, but great difficulty was experienced in marketing it, as no regular demand for asbestos was established as yet; the uncertainty of the supply, as well as the high prices due to very crude mining methods conducted on a small scale, were responsible for this lack of interest.

BELL ASBESTOS MINES. SHOWING THE CHARACTER OF THE SERPENTINE ROCK, AND
THE METHOD OF HOISTING BY CABLEWAYS.

The good quality of the Canadian asbestos and the ease with
which it can be spun and woven, however, attracted sufficient attention to induce some enterprising English and American capitalists to
investigate the deposits further, and as these investigations revealed
very encouraging features, busy mining camps soon sprang up around
Black Lake and Thetford in the Canadian province of Quebec, and a
steady betterment of the mines as well as an ever-increasing demand
for their products rewarded those who had ventured to create the
new industry. While the scope of operations increased steadily from
that time on, the methods of mining remained rather primitive, and
little or no machinery was used to replace hand labor until in the
nineties adverse industrial conditions in general caused a depreciation
in values, and it became obvious that cheaper production was necessary to leave the producers an adequate profit, inasmuch as other
countries, especially Italy and Russia, had also, in the meanwhile,
begun the exploitation of their asbestos deposits and had brought them
to a productive basis. In the middle of the nineties the Canadian
asbestos industry therefore experienced a crisis; many of the smaller
mines, and those producing but a limited amount of the better grades,
had to close down, and the adoption of cheaper working methods
and a better saving of values became imperative. The treatment of

the asbestos rock, requiring devices entirely different from those used in other concentrating or milling plants, demanded the construction of special machinery and (as Mr. Fritz Cirkel says, in speaking of this period in his monograph on "Asbestos, its Occurrence, Exploitation and Uses," one of the best and most exhaustive treatises on the subject) : "The asbestos industry is a striking example of what human ingenuity, if applied in the right direction, may accomplish. It demonstrates that in order to attain success it is necessary to strive, to seek to find, and not to yield." Many a set of experimental machinery was installed, discarded, and remodeled, before satisfactory results were reached. The difficulties to be overcome and the problems to be solved were manifold, as the aim was not only to replace hand labor by mechanical treatment, wherever possible, but also to liberate the asbestos from the adhering rock better than had hitherto been possible. Furthermore, it was necessary to turn the short-fibre material, large quantities of which were then lying on the waste dumps, into a marketable product, thus prolonging the life of a mine considerably; nothing but mechanical treatment could accomplish this.

It would lead too far to describe all the separate phases of evolution in this industry; be it sufficient therefore to say that there are more than a dozen plants in successful operation now with complete drying, milling, screening, fiberizing, and classifying devices, and while no two plants are exactly alike, the general principles fol-

CRUDE ASBESTOS, BEAVER ASBESTOS COMPANY'S MINES.

lowed are almost the same in all of them and may be shortly described as follows

The first step towards the production of a marketable asbestos is, of course, the mining of the asbestos-bearing rock, and this is done in quarry fashion in rather large open pits; the methods of working in them have experienced the well-known and identical changes which mining in general has undergone during the last two decades. Leaving the peculiarities of some of the mines out of consideration, it can be said that in most of them the broken rock is cobbed and assorted into the so-called crude asbestos, i. e., one with a fibre length of at least

MAIN PIT, BROUGHTON ASBESTOS FIBRE COMPANY, SHOWING FLOOR AND HOISTING
PLANT ON THE RIM.

⅜ inch, and into rock with shorter fibre, which is treated in the mill. The crude asbestos is then transported into cobbing sheds and by hammering and screening is further liberated from adhering rock and classified into two qualities, No. 1 crude measuring at least ¾ inch in length of fibre, and No. 2 crude the smaller sizes. This classification is done by rule-of-thumb methods, and considering the fact that it is nearly impossible to free the crude asbestos entirely from the rock, some stony matter adhering firmly to the end of a vein if the crude asbestos is in lumps, it can easily be seen that even the quality of

ENTRANCE TO TUNNEL, BELL ASBESTOS MINES. THE FIRST SUCCESSFUL ATTEMPT TO MINE ASBESTOS BY UNDERGROUND WORKINGS.

these standard grades is apt to vary and depends to a great degree upon the supervision over the laborers doing the cobbing, especially as this is done for the greater part by contract work. This method of preparing the crude asbestos for the market has not changed at all and it is not likely that mechanical treatment can be substituted.

All the rock mined, as well as the tailings from the cobbing sheds, goes to the fiberizing plant for further treatment and, as the material

contains much moisture, it must be dried first. This is done mostly
in rotary dryers, heated by fire from the outside, but in some cases
the drying is effected by spreading the material on a system of steam
pipes supplied with exhaust steam; while this latter method is economical, much floor space is required for a large drying capacity.
Before it is possible to liberate the fibre from the rock, the latter has
to pass through jaw crushers; it is then further reduced in rotating
crushers, and after having thereafter passed through rolls, the most

ASBESTOS QUARRY OF STANDARD MINE, BLACK LAKE. SHOWING THE ASBESTOS
VEINS RUNNING IN ALL DIRECTIONS.

important part of the work begins—that which it has been most
difficult to accomplish—namely, the fiberizing proper, and the complete separation of the asbestos fibres from rock particles. This is
now done by beaters and by cyclones, the first named consisting
chiefly of a large cylindrical trommel of boiler plates, in which strong
cutting knives, attached to a quickly rotating center axle, desintegrate
the material and (the whole apparatus being inclined) make it travel
from the feed hole to the discharge. Cyclones did not give much satisfaction during the early period of their application, but now they
have been perfected and no mill would be complete without them
today. They consist essentially of two beaters, shaped like screw
propellers, running in opposite directions and making as many as
2,500 revolutions per minute. They are encased in a cast-iron chamber, and all material reduced to about the size of a hazel nut and

smaller is taken off by air suction and falls onto a shaking screen, whence the liberated fibre is removed by an exhaust fan and the stony matter is shaken through; this is either further reduced in size and treated again in the same manner, or milled to an impalpable powder and used for plaster, cement, and similar purposes. The fiberized asbestos taken up from the screens by air suction is blown into settling chambers and must now be graded, this being done chiefly in rotating screens. In most cases the fibre passes over series of rotating and shaking screens and endless conveyors—so called picking belts—and from them the longer fibre, hidden in the rock and suitable for crude asbestos, can be taken up by hand; or barren rock can be picked out by hand to relieve the subsequent operations from useless material. The residue from the screens is either re-ground and re-fiberized or finely powdered according to size and character, while the fibre is always removed by air suction.

In no other country has the mining of asbestos, and especially its preparation for the market, attained such proportions as in Canada, and the excellent quality of the Canadian asbestos, the careful and clean separation of the fibre, and the comparative richness and accessibility of the deposits, will enable Canada to hold the lead in the production of asbestos for a great number of years to come. While the total production of all grades of asbestos in Canada has probably exceeded 75,000 tons during the last year, with prices ranging from $300 per ton for the best grade of crude down to $10 per ton for the shortest fibre, the United States has produced but 3,500 tons (approximately), very little of which is chrysotile asbestos, and the amphibole, as pointed out before, is of a much cheaper grade, giving the American asbestos an average value of about $15 per ton.

Of other producing countries the most important is Russia, whose output during the last year has probably exceeded 3,000 tons of good grade material, all suitable for spinning purposes, but no statistical figures are available for the production of inferior grades. The Russian fibre is not nearly as silky as the Canadian; it is harsher and of a yellowish to brownish hue. The chief drawbacks in Russia are the remoteness of the deposits from the consuming centres in Europe, and the shortness of the season, lasting in the average scarcely two months. As no machinery or mechanical means are employed, the work must be rushed feverishly to attain this comparatively large output in so short a time, and the largest mine is said to employ about 5,000 men when in full operation.

The first country in which asbestos was produced is Italy, and while the chemical composition of the Italian asbestos is practically

MILL AND FIBERIZING PLANT, BEAVER ASBESTOS COMPANY, THETFORD, P. Q.

the same as that of the Canadian product, its physical properties are quite different, and the asbestos fibres from the three principal Italian localities differ one from another; their common characteristics, however, are a grayish to brownish color, a touch somewhat similar to that of soapstone, and a comparative rarity of good, long fibre, the occurrence of which is pockety. The high altitudes at which workable asbestos deposits are found, the nature of the rock making mining by hand imperative, and the ever existing danger from landslides, have greatly hampered the development of this industry and made it but little remunerative, though the extent of asbestos-bearing ground seems practically unlimited.

A very promising country seems to be South Africa, and great attention has been paid of late to the asbestos shipped from there to various European countries. The African fibre has most of the good qualities required; it is long, soft, pliable, but harsher than the best Canadian; it can be separated most easily from the rock and its specific gravity is lighter than that of any other asbestos. Its most salient feature is the beautiful blue color, ranging from lavender to marine blue; even the finest fibre retains this color, while most of the other varieties of asbestos, being greenish, gray, or yellow when in masses, look white when fiberized. This blue color is due to a large proportion of protoxide of iron present, which averages 30 to 35 per cent and makes the blue asbestos less desirable for electrical insulations.

Asbestos has also been discovered in other countries, and Newfoundland, New Zealand, Queensland, and South Australia seem to possess workable asbestos deposits, but so far they are undeveloped and no quantities of commercial importance have been shipped. The newest discovery is that of a pink-colored asbestos in India, which looks very pretty, but, being of the amphibole variety, the low value forbids shipping to European or American markets and it is therefore of scarcely more than mineralogical interest.

As mentioned before, the uses of asbestos have increased enormously of late and it would require a special treatise to cover this phase of the subject. The application of asbestos in its various forms

BROUGHTON ASBESTOS FIBRE COMPANY'S PLANT, SHOWING FIBERIZING PLANT, TRESTLE FROM WHICH ASBESTOS ROCK IS FED TO THE MILL, AND CONVEYING CABLEWAY.

varies greatly in different countries, but if one be called upon to make
a general distinction between American and European practise, it can
be said that America uses the shorter fibres in the manufacture of pipe
and boiler coverings, as well as for plaster, cement, etc., in proportionately much larger quantities than European countries, while the
latter use more crude asbestos and the longest grades of fiberized
asbestos for spinning, weaving, and braiding, and the long and
medium fibres for the manufacture of insulating materials with admixtures of kieselguhr (infusorial earth), ground cork, felt, or magnesia,
or for stuffing and packing purposes as a rubber-asbestos-combina-
tion.

The spinning and weaving of asbestos has offered many difficulties, as the asbestos fibres have no rough surface like wool or cotton,
but are very smooth and thus have a tendency to slip by one another,
when twisted and subject to tension. An admixture of vegetable
or animal fibre was therefore often necessary, but while these facilitated the manufacturing operations, they impaired the fire resistance
of the fabric, and special machinery and ingenious devices had to be
invented to enable the successful spinning of a pure asbestos yarn; it
is, however, now possible to make a single asbestos thread which,
though weighing no more than one ounce per hundred yards, has a
fair strength, and braided material can be made much more resistant
to torsion and tension; while asbestos ropes, chiefly used by the fire
department, can be strengthened either by interwoven wires or by
having a wire-rope core. The use of asbestos for theatre curtains,
fireman's and electrician's gloves and garments, partition walls, and
for the isolation and insulation of certain parts of buildings or steamers from other parts, are well-known, and large quantities of asbestos
are annually used for these purposes. The packing and insulation
of steam pipes, boilers, refrigerators, etc., has assumed large proportions in America, especially in the form of magnesia sectional coverings in which a proper proportion of long asbestos fibre is used
as a binder, while European practise tends to mix asbestos with
a number of different ingredients. Large surfaces, such as may be
found on boilers, furnaces, or in ice plants are often covered with
asbestos cement, which in America is mostly but a mixture of asbestos
and clay and therefore cracks easily, especially if subject to vibration,
while in Europe infusorial earth, in itself a good insulator, replaces
the clay, and cow hair or soft felt refuse is added to give the mass a
better binding quality. Although the application of asbestos cement is
most simple, a paste being made by the addition of water and the
mixture smeared over the surface to be insulated, a number of rather

thin layers are necessary, each subsequent one being added after the preceding one has dried sufficiently.

In its spun state asbestos is used for insulations, especially in the form of wick for packing pistons and piston rods, valves, cocks, and glands. This application is bound to increase in America with the growing use of superheated steam, as asbestos packing can scarcely be excelled in economy and reliability by other packings, most of which owe the best of their properties to the asbestos contained in

ABOVE, VIEW ACROSS MAIN PIT OF THE BEAVER ASBESTOS COMPANY, SHOWING THE
DERRICKS. BELOW, KING'S ASBESTOS MINES, SHOWING DERRICKS AND
WIRE ROPE HOISTS.

them; but the addition of vulcanized rubber, metallic wires, and even thin metal sheets is very desirable for a number of special purposes.

Although the fire-resisting qualities of asbestos are suggestive for making paper for deeds and important documents, no satisfactory results have been obtained in this direction on account of the difficulties experienced in making the surface of the asbestos paper glossy and smooth, thus preventing the ink from blurring; but the manufacture of asbestos cardboard, or as it is generally called, millboard, is quite an important branch of the asbestos industry, and that of asbestos paper for structural and insulating uses is many times larger still, affording an important outlet for the fibre. This material adapting itself readily to uneven surfaces, is used extensively for insulating pipe joints, cylinder covers, etc.; it has a very wide range of usefulness in electrical machinery, in the construction of dynamos, in tubes and elbows for turning corners, as accumulator casings, in motors, switchboards and a great variety of other electrical appliances. In its corrugated form asbestos paper is known as air-cell covering and enjoys a lively demand on account of its cleanliness, the ease with which it can be handled, and its great economy in use, as it can be removed quickly and used over and over again if properly handled.

A few years ago patents of Austrian and German origin were taken out in every civilized country for new processes to make roofing shingles out of asbestos, for which purpose the fibre is mixed with other ingredients, making an adhesive mass which is then subject to a strong hydraulic pressure. These shingles can be made to resemble roofing slate closely, but they are of much lighter weight than these; or they can be colored, producing very nice effects on roofs as well as walls of buildings, making them at the same time fire-, water-, and vermin-proof and lessening the variations of temperature within the building to a remarkable degree. This new manufacture has assumed such proportions that many thousand tons of asbestos fibre are now annually consumed in the manufacture of asbestos shingles.

Asbestos paint, plaster, flooring, ceiling, and wall decorations. asbestos bricks, tiles, slabs, and even asbestos lumber, are now used largely in making buildings fireproof. Whole blocks of buildings made entirely of these asbestos materials demonstrate that hundreds of conflagrations do not impair their solidity, as the fire shows in Earl's Court, or on the Vogelwiese and elsewhere, have proved.

MACHINERY AND METHODS FOR THE EXCAVA-
TION OF SKY-SCRAPER FOUNDATIONS.

By T. Kennard Thomson.

Among the engineering developments of the last few years which have sprung from new conditions and created new methods and machinery, few if any have been more impressive than that of the tall building. It has been freely discussed in its architectural and structural aspects. Mr. Thomson presents it in a new phase—that of the mechanical plant and processes which it assembles.—THE EDITORS.

THE introduction of sky scrapers, only a few years ago, has not only revolutionized the amount of plant required to run these immense buildings but also brought into play much ingenious contractors' plant and a higher grade of contractors.

The permanent machinery for these buildings takes up much room, which, of course, can not be spared either from the ground floor or basement under the street floor, so that not only have sub-basements been put in, but also cellars and sub-cellars under them, making from two to four floors below the street level, the first of these being the addition to the Mutual Life Building on Cedar and Liberty Streets, New York, in 1900-1, which caused the words "Earth Scraper" to be coined.

The construction of these buildings also requires a very elaborate temporary plant—that is, derricks, cranes, pumps, electric-light plant, hoisting engines, compressors, air receivers, air coolers, rock drills, pile drivers, etc., all of which must be quickly brought to the site as needed and as quickly removed. The illustrations show typical examples of building sites completely covered with caissons and temporary plant, all of which disappear from sight before the erection of the steel work starts, the caissons being sunk until the tops are 15 to 20 feet below the street level and the plant all being removed.

This is especially true where the foundations are carried to bed rock by means of pneumatic caissons—as, indeed, all buildings in lower New York should be, on account of the bed of quicksand from 30 to 60 feet thick, which covers the whole of this part of the city. This quicksand makes a good foundation provided it be confined and no opportunity be afforded of escaping into some adjoining and deeper excavation for other buildings or tunnel. This danger is exceedingly apt to occur where the building is placed on this material, for it is so fine that it will run wherever water will and soon leave a big cavity

THE NEW YORK SKY-SCRAPER LINE, WITH THE NEW SINGER BUILDING INTRODUCED AS IT WILL APPEAR.

behind for the new or old structures to fall into. As we know, the subway tunnels are fairly close to the surface, but soon we will have tunnels from Brooklyn to Jersey which must, perforce, run below the present tunnels, and it is quite probable that we will, ere long, have other tunnels crossing under these again, so that no foundation above good hardpan or bed rock can be considered safe or economical for a sky scraper.

In Chicago, they put off using caisson foundations much longer than in New York, but the settlement has been so great (some of the big offices on what are called spread or floating foundations having settled two feet or more) that they are now coming around to rock foundations in Chicago, too. It is, however, only 14 years or so since that eminent architect, Mr. Francis H. Kimball, insisted upon using pneumatic caissons to bed rock to support the Manhattan Life building at 66 Broadway. At that time, he was very severely criticized for spending so much money to get a safe understanding; but within the last few years, the public has begun to see the wisdom of following the advice given in the Bible about building on rock.

One result of these high structures is the obliteration of old lot lines, for no one desires to erect a sky scraper down town on a single 25-foot lot, preferring, where possible, to absorb the whole block which, of course, gives the best chance for economical construction. When the investor has obtained two or more lots and selected his engineer and architect, he generally lets a contract for the removal of the old and generally ramshackle house—the reverse of an ordinary contract, for, in this case, the contractor usually pays the owner for the privilege of removing everything above the street or curb level, the price obviously depending on what the house-wrecker thinks he can sell the old material for. Generally the old brick work, set in lime mortar, can be removed by hand without using any force; but occasionally the wrecker runs up against a comparatively new and well built affair, like the Coal and Iron Exchange on the corner of Cortlandt and Church Streets in the removal of which dynamite is required, greatly increasing the expense and decreasing the value of the material taken out. These contractors often leave the old basement full of rubbish up to the street level for the foundation contractors to remove. It might be stated here that, before the advent of caissons, the old cellars were necessarily above the water level (which, in lower New York, is from 15 to 25 feet below the street level and generally a few feet above the high-tide level in the bay) and the old walls were seldom carried more than 2 feet below this level. Many of the older walls are supported by piles which, however, are unre-

TWO STAGES IN THE PROGRESS OF FOUNDATION WORK FOR A NEW YORK SKY SCRAPER.
In the earlier one, above, the erection of the air-compressors has begun—two Ingersoll-Rand machines, one 19 by 22 and one 18 by 20, with a total capacity of 660 cu. ft. of air per minute. In the later stage shown below the compressors are housed, a Smith concrete mixer is being lowered, and a four-masted derrick is being put up.

liable for they depend, of course, on the friction of the earth against the sides of the piles, and if the new operation should draw the material from around the piles, the frictional resistance would vanish and a collapse might result; in fact, this did happen in the case of a $400,000 wharf in Baltimore recently, where the superstructure was carried on an immense number of piles which had no frictional support for some 35 feet under water.

Where caissons have been decided on, the foundation contractor often lets a sub-contract for the removal of material from the street level to the water level, to men who confine themselves to this business. It might be stated here that in the country excavation can be done for 25 cents per cubic yard similar to that which, in New York, costs from $2.00 to $3.00 per cubic yard. At other times the head contractor will erect one or more stiff-leg derricks to handle iron buckets holding from 1 to 2 cubic yards each, which raise the material from the cellar and dump it into wagons which are, in turn, dumped onto scows and the load towed away, all bringing the cost up to the above figures. The next step (see the top of page 84, which shows the preparations of the Foundation Company at the junction of Maiden Lane and Liberty Street) is to erect the air-compressor plant. The air receivers here, shown in a horizontal position, are two, 41 inches in diameter and 16 feet long, and the air cooler shown between the compressors and receivers consists of a pipe 16 inches in diameter and 12 feet long, filled with gas pipes. A continual stream of cold water keeps the pipes cool and prevents the air from getting too hot. Sometimes the air is passed through a number of gas pipes which are kept bedded in and covered with ice—rather an expensive arrangement. Occasionally the cooling apparatus gets out of order, and the air in the caisson gets so hot (I have felt it over 106 degrees) that the men can not or will not work. The lower view, taken 5 days later, shows the compressors, etc., all housed in and a four-masted traveling derrick partially erected. This derrick takes the caissons off the wagon and lands them in place and handles all the material on the job. The first of these derricks was designed by me for the Commercial Cable building on Broad and New Streets in 1896, and the second for the Mutual Life Building in 1900. Since then many have been built, chiefly of wood, but the one shown in this photograph is the first and only one to be built of all steel and is the most up-to-date and best of its kind ever erected. It was first built by the Foundation Company last year for the Trust Company of America Building on Wall Street. On July 10, 1907, four days later, this derrick was entirely ready for use, and the picture on page 86

PROGRESS ON THE FOUNDATIONS ILLUSTRATED ON PAGE 84.

In this view, taken sixteen days after the upper one on page 84, and eleven days after the lower one there shown, the four-masted derrick is in full use and five caissons are in place.

86

taken on July 17, one week later, shows five caissons in place and the work in full blast, the compressors being boxed in, and offices, sheds, etc., built on top.

These four-masted derricks are rather too expensive to be built for a single job, and a steel derrick is obviously much better adapted to being taken apart and put up again than the old style of wood, for the members are naturally held together by bolts which are supposed to fit tight in the holes. Of course, if a bolt is repeatedly driven in the same hole, it enlarges the hole and spoils the tight fit, especially in wood.

UNWATERING A CITY BUILDING FOUNDATION.
Cameron steam pumps draining the excavation for new Edison Power House. It is one foot below low tide and 21 feet deep, and there was much water.

On a small job, one of these travelers is sufficient; but on a large site, like the City Investing building on Broadway and Cortlandt Street, two or more are used, and also a number of stiff-leg derricks, and, at the same time, the compressor and other mechanical plant also increases in some cases, until the capacity reaches 3,000 or 4,000 cubic feet of air per minute. The early traveler we built had an ordinary two-drum hoisting engine, about 7 by 10 or 8 by 10 inch cylinder, with a vertical 25 to 40 horse-power boiler for each boom; but now, wherever possible, the boiler is done away with, getting rid of the nuisance of coal, smoke, etc., and the engines are run direct by city steam which is also used to run the compressors, drills, etc., thereby saving not only much annoyance to the public, but also much valuable space.

STIFF-LEG DERRICK WITH HOISTING GEAR ATTACHED TO MAST.
Lidgerwood hoist, General Electric direct-current motor, capacity 4,000 lb. at 180 ft. per minute.

Each boom is capable of lifting from 10 to 20 tons. When hoisting the bucket out of the caisson, speed, of course, is the principal object, so a single steel cable of ⅝, ¾ or ⅞-inch wire rope is used, leading direct to the engine; but when heavy loads are to be lifted, it is "doubled up" or changed to four lines by means of blocks. Much ingenuity has been displayed in getting up schemes for making quick changes from single to double lines, for from 15 to 20 minutes a change is an expensive delay on a big rush job, with all the sand-hogs and other labor waiting for the bucket, etc. These booms, on a well designed derrick, are capable of swinging through an angle of about 270 degrees horizontally, and can be lowered to the ground or raised nearly vertical, giving a very wide range of action.

Of course, no caisson work would be carried out now without electric light, and an ample supply of fresh water; so for bridge caissons outside of the cities, a complete electric-light plant and pump-ing plant are at once installed, but fortunately in New York satis-factory terms can be made to tap the nearest pipe and wires. In fact, on two jobs this year, the City Investing building and the Cortlandt Street terminal, the O'Rourke Company had the entire plant equipped for electrical power, being the first foundation jobs to be so handled. The pictures on these facing pages show a stiff-leg

A COMPLETE VIEW OF THE STIFF-LEG DERRICK OF WHICH THE HOIST IS SHOWN OPPOSITE.

derrick with the hoisting gear attached to the mast. These are made for two- or three-phase current, 60 cycles, 110, 220, 440 and 550 volts. General Electric alternating-current induction-motor hoists, with friction drum and solenoid brakes, and in sizes from 15 to 150 horse power. This new type of electric-mast hoist as developed for these buildings is complete in itself, containing not only the electric motor, controller, and resistances, the two drums for operating the boom, and the fall lines of the derrick, but also the boom-swinging gear, giving the operator full control of his work which he is all the time facing. This machine is certainly compact and is also capable of high power and reliability.

The hoist shown in the view on page 88 was operated by a 250-volt direct-current General Electric motor, with a nominal capacity of 4,000 pounds lifted at a speed of 130 feet per minute; the drums, 12 inches diameter and 19-¾-inch face, were each complete with brakes, ratchets and pawl. In addition to the regular brakes, operated by foot levers, it was also equipped with electric solenoid brakes. This brake is applied to the shaft of the electric motor. When the operating current is turned onto the motor, it passes through the coils of the solenoid, lifts the armature, and automatically releases the brake. When the current is shut off, or in case it is cut off by accident, the armature falls, and its weight sets the brake with power to support the entire load of the hoist. The current to operate this hoist is claimed to be only one-half a kilowatt hour to lift a bucket weighing 2,200 pounds with its load out of the caisson and to return it empty to the air chamber.

Before the first contract for the foundations is let, and often before the old buildings are vacated, borings are made in the old cellars to ascertain where rock is and the nature of the intermediate material. These generally consist of wash borings made by forcing a pipe of about 3 inches (more or less) in diameter into the ground by means of a small pile driver, built for the purpose, and a free use of the water jet. These borings are only reliable as an indication of the depth of hard pan below the surface, for they very seldom penetrate the hard pan and reach bed rock, although this has been accomplished by occasionally dropping a small stick of dynamite and exploding it in the hole after the pipe has been raised a few feet.

The result of these insufficient borings usually is that a lump-sum contract is let on condition that the caissons are carried down to the depth shown by the borings, with an additional price per cubic yard for each additional cubic yard excavated. Occasionally diamond-drill or core borings are made and the core is taken out until the drill has bitten several feet into the rock. But while the top of hard pan is fairly level, and it is easy to arrive at a fair average level for it, New York bed rock, gneiss, is just the reverse, and it would take an immense number of diamond-drill borings to get a reliable average depth. Before commencing to sink any of the caissons, it is necessary to see that the adjoining buildings are safely shored up or under-

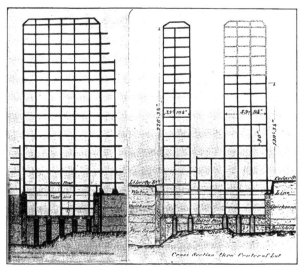

STEEL WORK, CAISSON FOUNDATIONS AND SUB-CELLARS OF A TYPICAL TALL BUILDING.
The Mutual Life Building, Cedar and Liberty Streets, N. Y.

pinned, sometimes a very serious matter, which has cost as much as
$75,000.

If the building is small, it is often sufficient to put a few inclined
shores or 12 by 12 inch timbers, say 20 to 30 feet long, up against
the side of the wall and rest them on sills placed on the ground.
When the building is heavier, one of the oldest methods is to cut a
hole through the wall and insert one or more wooden or steel beams,
called needles, supported by sills in the old cellars of the adjoining
buildings and the new lot. The centre of the beams being directly
under the wall to be supported, the weight of the wall is then transferred to the beams by means of wooden or steel wedges or hydraulic
or screw jacks or other means.

In some cases even this is not sufficient, and as it takes up a great
deal of space which can ill be spared, a patented method has been
developed of driving small caissons, say, 3 feet in diameter, under
the wall, one at a time. This has been done so successfully that
when the new foundation of an 18-story building had thus been
transferred to rock, 60 feet below the old foundation, there was not

FOUNDATION WORK IN FULL PROGRESS, CHURCH STREET TERMINAL BUILDING, NEW YORK CITY.

a settlement of 1/16 inch, causing the president of one of the safe-deposit companies to testify that a possible financial panic had been averted by the care with which the underpinning had been done; for, said he, had the building been allowed to settle even 1/32 inch, the doors of his vaults would have been thrown out of gear, Wall Street would have wakened up some morning and not been able to get at their securities, with the result of a panic.

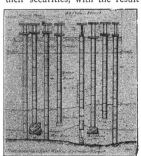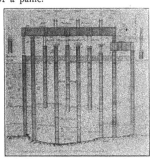

UNDERPINNING IN SKY-SCRAPER FOUNDATION WORK.

Like everything else in foundation work, the underpinning for each individual case is a study by itself. The left-hand cut shows a building under which the last of these small caissons is just being put in place. The outside diameter of these particular caissons was 36 inches and the inside 33 inches, leaving very little space for the men to work. Looking closely at the fourth column from the right, it will be seen what cramped quarters the sandhog has. He is shown chipping off the rock at the cutting edge. About 8 feet above the cutting edge is an air lock 8 or 10 feet high in which a lock-tender is waiting to pull the material in canvas bags up into the lock. When he has pulled four or five bags up that way, he closes the lower door and allows the air in the lock to escape, thus removing the pressure which holds the upper door shut. The air pressure having fallen to atmospheric, the bags are drawn up the shaft and emptied outside, and finally concrete is taken through the lock until the air chamber is sealed with concrete, when the air is taken off, and the rest of the cylinder filled with concrete. This is naturally a very expensive as well as the most reliable method, and can easily cost as much as $200 per cubic yard of material removed, including the cost of the cylinders, jacking, wedging, etc.

The adjacent walls having been safely secured or, at least, those

walls near where the first caisson is to be sunk, the contractor is
ready to start the most important part of the work—the caisson
proper. It might be stated here that a pneumatic caisson is built in
the shape of a box, having four sides and roof but no bottom, the bottom of the sides being called the cutting edges. The roof or deck
has one or more holes, 3 feet in diameter, and over these holes are
bolted 3-foot steel shafts. The drawings opposite show a large caisson
with two such shafts, one marked "excavating shaft" through which
the buckets of earth, boulders, etc., are removed, and concrete taken
in, and the other marked "main shaft," showing how men can enter or
leave the air chamber. The air chamber is usually 6 or 7 feet high
and, when the caisson reaches the site, it is usually 12 feet or more
in height. The width for New York caissons runs from 6 feet up,
and the lengths 6 to 30 feet or so. Circular caissons from 5 to 12
feet in diameter are used, but not often in the large sizes.

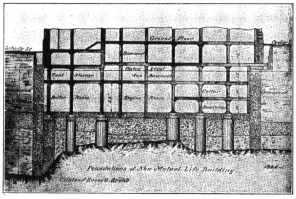

FOUNDATIONS AND LOWER STEEL WORK, MUTUAL LIFE BUILDING, NEW YORK.

When a caisson has been placed in position by means of derricks,
the caisson men (called "sandhogs") enter the air chamber to under-
mine the cutting edge, and shovel the material into the buckets, which
are about 29 inches in diameter and about 3 feet high, containing
about ½ cubic yard. While they are doing this, other men are
placing concrete on top of the roof or deck to form part of the permanent foundation and, at the same time, to add weight to the caisson
to force it to follow the excavation below the cutting edge. A great
weight is required for this purpose, for not only has the friction on

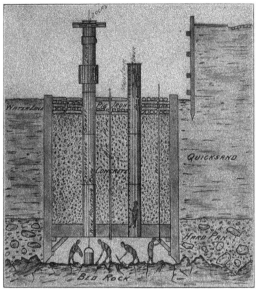

A LARGE CAISSON WITH ITS SHAFTS.

the sides of the caisson (which runs from 50 to 650 pounds per square foot of surface) to be overcome, but also the pressure of the compressed air against the roof of the caisson. It might be stated here that the only object of using compressed air is to prevent the water from flowing into the air chamber where the men are working, and therefore compressed air is used only where water is encountered; and the pressure of the air must always equal the pressure of the water, for, if the water pressure is greater, the water will flow in, and, if the air pressure is greater, the air will "blow out" with more or less serious results.

. Now, we know that a cubic foot of water weighs 62-½ pounds, 10 cubic feet, 625 pounds and so on, and this is equivalent to .434 pounds per square inch, or nearly ½ pound per square inch of surface for each vertical foot of water; so if the cutting edge is 1 foot below water, ½ a pound of air would be required and if the penetration were 100 feet then the pressure would be 43-½ pounds per square inch, in addition to the atmospheric pressure, which is about the limit

THE CITY INVESTING BUILDING AND THE NEW SINGER BUILDING WITH ITS 41-STORY TOWER, NEW YORK CITY. TWO OF THE LATEST AND GREATEST "SKY SCRAPERS," NOW UNDER CONSTRUCTION.

of human endurance. The illustration at the top of page 95 shows a large caisson in its final position with all the cofferdam on concrete on deck and additional weight, pig iron, required for sinking. As the cutting edge is undermined, the sides of the caisson (called cofferdam, when above the deck) are built up and filled with concrete. The cofferdam, necessarily, must be built up as fast as the caisson sinks, so as to keep its top always above the ground; otherwise the ground would fall into the cofferdam. The steel columns which rest on top of the caissons have broad bases to distribute the strain over the entire surface of the concrete, and these bases are generally kept below the lowest floor level to economize room, thus limiting the height to which the concrete can be built on the deck or roof; the result often is that the concrete does not weigh enough to force the caisson all the way down, and it is necessary to add more weight, which most often takes the shape of pig iron or heavy cast-iron blocks, frequently to the extent of hundreds of tons even on a small caisson.

WORKMEN ENTERING AIR LOCK, PREPARATORY TO DESCENT INTO A FOUNDATION CAISSON.

The deepest caisson put down in New York City is under the Mutual Life building, where we penetrated 100 feet below the street level, or 85 feet below water line with the foundations, while the cellar floor was stopped 35 feet under the standing water level. On

A FOUNDATION CAISSON TOP AND AIR LOCK.

the Cedar Street side, there are four stories underground (see page 94
while on the Liberty Street side there are only three, the bottom one being 22 feet high; two ordinary cellars, to make room
for the boilers, etc., the floor immediately above being used for coal.

But to come back to the caissons; they are built of wood, steel, or concrete or some combination of these three, and, of course, have pipes carried down through the concrete to keep up the supply of compressed air, as well as a small gas pipe with a whistle on top so that when the sand hogs want to signal to the man outside, they open the valve of the gas pipe in the air chamber, allowing the air to rush out and blow the whistle. One whistle means "hoist the bucket"; two, "hold the bucket where it is"; three, "lower"; five, "the gang want to go out, etc."

AIR COMPRESSOR PLANT FOR A CITY EXCAVATING JOB.
Ingersoll-Rand compressor, Class A, 30-inch stroke.

In big river caissons and tunnels, telephones are used, but they are very seldom used in city work. Sand hogs work in 8-hour shifts —that is, three gangs in 24 hours, each gang taking half an hour for lunch and 7-½ hours for actual work, for which they receive $3.50 a day, until the air pressure exceeds 20 pounds per square inch. As the pressure increases the pay increases, but the hours of labor decrease until at 45 pounds (in addition to atmospheric) the men only work 1½ hours a day, and even that is divided into two shifts of ¾ hour each and 4 hours apart; this is all the men can stand, and even then there is great danger of the bends, or worse, of being paralyzed. Many can not even stand the light pressure.

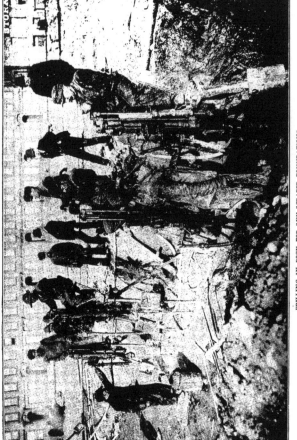

DRILLING IN ROCK FOR SKY-SCRAPER FOUNDATIONS.

The first effect generally noticed when the air is let into the lock is the pressure of the air on the ear drums, and if this pressure is not quickly equalized, the ear drum is ruptured, of course, for life. At other times, this plugged sensation results in blood vessels being ruptured in the head which danger is much greater if the person has a cold to start with. In fact, if one has a cold or anything wrong with his heart or lungs, he has no business to try to enter the lock at all. Even men in apparent perfect condition can not always stand it. A lock might be described as a small room with two doors like the vestibule of a house, so that if one enters and shuts the outside door before opening the inside door only the amount of air contained between the two doors is lost in passing in or out. It is necessary to have an air lock to prevent all the air escaping from the air chamber every time men or material pass through the lock, for. if all the compressed air were allowed to escape from the working chamber, even for a very short time, the chamber would quickly fill up with mud and water, with probably disastrous results to the adjoining buildings, to say nothing of the loss of human life. When the compressed air is allowed to enter the lock too quickly, and sometimes even when considerable time has been taken in entering and leaving the lock, it is supposed that the air bubbles force the blood away from the surface and then, when coming out of the compression, many of the bubbles remain in the system, which results in the "bends," a very painful experience. The attack is usually in the arm or legs. The longer one stays in compressed air and the more exertion taken in it, the greater the risk of the bends, which, however, often do not make themselves felt for several hours after coming out, although some claim that they can tell that they are getting the bends while still in the air chamber. The worst form of caisson disease is paralysis, from which some die at once, some recover, and others are afflicted for the rest of their lives ; and which it will be none can tell. Many of the old watchmen seen on these jobs humping around with a cane are such victims.

A school girl was once asked where could you walk the faster, on top of a high mountain, where the pressure is very much lighter than what we call atmospheric (nearly 15 pounds per square inch) or down in a caisson where one has to carry say 20 to 40 pounds on every square inch of the body in addition to the atmospheric. At first she thought on top of the mountain, but quickly corrected herself by saying: "Of course not, for more work could be done in the compressed air on account of the great supply of oxygen." The excess of oxygen not only gives the men great energy and appetites—you see seldom a

NORTH TRINITY AND U. S. REALTY BUILDING, WHERE WORLD'S TIME RECORD WAS MADE IN 1906.

thin sand hog—but it also makes candles, matches, cigars, etc., burn much faster; in fact, frequently, men have blown out a candle and put it in their pockets; only to find their coat on fire in a few minutes. One seldom sees an old sand hog; they must burn up their energies. It has often been a matter of comment that even the best of sand hogs are about useless if given an outside job; whether they lose their inclination or ability to do good work, I know not.

When the sand hogs have removed all the

INGERSOLL-RAND DRILL USED AS A PILE DRIVER.

material in the air chamber above good hard pan or bed rock, they fill the air chamber with good Portland-cement concrete, gradually backing up to the shaft until there is a solid mass of concrete, with the exception perhaps of the wooden or steel roof from the bottom of the foundation or bed rock up to the base of the columns. When the cellar is to be above the water line, the caissons are sunk to support the individual columns only, but when the excavation for the cellar has to be carried below the water line, it is necessary to put a continuous line of caissons around the entire lot and then put a water-tight joint between them, in which case the entire interior is sometimes excavated before the columns are placed and, in other cases, caissons are used for the interior columns also. While the pictures on pages 84 and 86 of the building between Maiden Lane and Liberty Street now under construction show a marvelous amount of work done in the first 19 days, the world's record for rapid caisson sinking is held by the Foundation Company, who, last year, sank and sealed 87 caissons for the North Trinity and United States Realty buildings on Broadway in 60 days, taking only 30 days for the last 57.

CONTRACTORS' PLANT AND MACHINERY, FOUNDATION OF THE 42-STORY SINGER BUILDING, N. Y. CITY.

There is a limit to the space allowed for this paper, but there is no limit to the subject, as a dozen or more branches of it which have been merely alluded to would require more space than this whole article to do them justice.

APPLIED ELECTRO-METALLURGY UP TO THE END OF 1906.

By John B. C. Kershaw.

Mr. Kershaw has condensed, into a remarkably close compass, an admirably clear view of the advance of electro-metallurgy up to the present year. It is significant of the extent of the development that even so concentrated a summary of its data exceeds the space of a single magazine article, and we have to reserve for next month another very interesting section, covering among other things the latest progress in the manufacture of ferro-alloys and of iron and steel direct from the ore—THE EDITORS.

THE electro-metallurgical industries are the growth of the last twenty years, but in that period very remarkable progress has been made. Only one industry existed prior to 1886, namely that of copper refining; this was carried on in a few works upon an extremely limited scale of operations. Today, the electrolytic copper-refining industry is second in importance only to that of copper smelting, and over one-half of the world's production of copper is submitted to the former process. The manufacture of aluminium, calcium carbide, carborundum, ferro-alloys, and sodium are other important and expanding electro-metallurgical industries, while the application of the electric furnace to steel refining is a new development which may lead to very important changes in the iron and steel industries—for in conjunction with gas engines and dynamos, it may serve as a means of utilizing the enormous power now lost in the waste gases from our blast-furnaces.

The following pages deal with the various electro-metallurgical industries in alphabetical order, describing briefly the processes or methods in use and the extent to which these methods have been applied upon an industrial scale.

ALUMINIUM.—The manufacture of aluminium by the electrolytic method was commenced at New Kensington in America in the year 1888, and at Neuhausen in Switzerland in the year 1889. The processes were worked out independently, by Hall in America, and by Heroult in France, but as now operated they are practically identical, and consist in the electrolysis, with carbon electrodes, of aluminium oxide held in solution in a fused bath of cryolite and fluorspar. Since the introduction of the electrolytic method of manufacture in 1889 the production of aluminium has increased from 85 tons to

POWER PLANT, LA PRAZ WORKS OF THE SOCIÉTÉ ELECTRO-METALLURGIQUE FRANÇAISE.

12,000 tons in 1906. The following tabular statement shews the gradual increase in output and fall in price which has marked the industrial development of the electrolytic process:—

YEAR.	PRODUCTION TONS.	PRICE PENCE PER LB.	YEAR.	PRODUCTION TONS.	PRICE PENCE PER LB.
1885...	3.12	600	1896	1,755	17.0
1886...	14.80	500	1897	3,327	17.5
1887...	22.50	400	1898	3,953	16.2
1888...	39.40	250	1899	5,459	16.2
1889...	85.20	125	1900	7,192	16.0
1890...	174.50	100	1901	7,420	15.5
1891...	345.50	125	1902	7,750	15.5
1892...	485.00	32	1903	8,102	15.5
1893...	713.00	...	1904	8,550	15.5
1894...	1,057.00	25 .	1905	9,000	16.0
1895...	1,129.00	22	1906	12,000	21.4

NOTE.—The production is given in tons of 2,240 lb. and the price in pence per lb. From 1902 onwards the production figures are estimated, and from 1897 to 1904 the figures for price are based on the American values.

The manufacture of aluminium is now carried on in a number of works, controlling over 84,000 horse power. Details of these so far as they are known, are given below:

NAME OF COMPANY.	LOCALITY OF WORKS.	HORSE POWER.
1. British Aluminium Co	Foyers, N. B.	5,000
	Norway	10,000*
	Switzerland, in course of erection	45,000*
2. Société Electro-Metallurgique Française	La Praz	7,500
	Gardannes	7,500
3. Compagnie des Produits Chimiques d'Alais	Calypso	10,000
	St. Felix	2,500
4. Aluminium Industrie Aktien-Gesellschaft	Neuhausen	5,000
	Rheinfelden	5,000
	Lend Gastein	15,000
5. Compagnia Italiana	Pescara	3,000*
6. Pittsburg Reduction Co	Niagara Falls	10,000
	Shawinigan Falls (Canada)	5,000
	Massena (U. S. A)	12,000

Assuming that 4 horse power are required for one year, to produce one ton of aluminium, the aggregate power available in these works would suffice to produce 35,000 tons of the metal per annum. Owing, however, to the diminished power available during the summer droughts and to other causes, the maximum total of power is not available for the manufacture all the year round, and my estimate of the 1906 production is about 12,000 tons.

* On the authority of the *Revue Industrielle,* 1907, page 222, as quoted in *L'Industria.* —ED. E. M.

The past year has been marked by the expiration of five of the United States patents granted to Hall in 1889. The Heroult patents lapsed in Europe in 1902, and the manufacture of aluminium by the electrolytic method can therefore now be carried on without the payment of patent royalties. The use of the electric current for keeping the bath in the molten state is however still covered in America by the Bradley electric-furnace patents which do not expire until 1909. In that country the Pittsburg Reduction Company therefore still possess the monopoly of the electrolytic reduction process.

As regards utilization, the demand for the metal in Europe during 1906 has been in excess of the output, and the reduction plants are being extended in several of the works, in order to benefit by the higher prices now obtainable for the metal. The British Aluminium Company, in addition to the development of a new water power in Switzerland, are carrying out a very large scheme on Loch Leven in Scotland, which when completed will add enormously to their power resources in Scotland. The Aluminium Industrie Aktien Gesellschaft, of Neuhausen, are likewise developing a large power scheme on the River Navisonce in Switzerland, from which it is expected that 25,000 horse power will be derived. A new aluminium works also has been erected by an Italian Company in the Valley of Pescara in Northern Italy, and is about to commence operations. In a few years, therefore, the productive capacities of the aluminium companies will be more than doubled, and it will be of interest to note whether the demand shews a similar expansion.

The metal is now being used in very large quantities for motor-car construction, and for general foundry work, while the "Thermit" and "Weldite" processes also consume large quantities of aluminium in the form of powder. In every direction in which the metal has been applied with success, its use has increased during 1906.

Mr. Schoop of Paris has worked out the details of a process for the autogenous welding of aluminium, which overcomes the difficulty of finding a suitable solder for the metal. By this process aluminium sheets, rods, or tubes, of any thickness, can be welded without any difficulty, and the joints are said to be as strong as the other parts of the metal. This method of welding will probably lead up to increased consumption of the metal in many industries, and to its use for larger articles and vessels than have yet been manufactured from it. Another direction in which the use of aluminium is extending is for the manufacture of pans, etc., for use in the wax-refining and jam-boiling industries, which have hitherto employed copper vessels for this purpose.

WESTINGHOUSE INSTALLATIONS IN AMERICAN ELECTROLYTIC AND ELECTROMETAL-
LURGICAL WORKS.
Above, power plant of the Pittsburg Reduction Co., Shawinigan Falls, P. Q., containing four
938-kilowatt 665-volt self-contained shunt generators. Below, plant of the Union
Carbide Co., Sault Ste Marie, Mich; twenty-one 375-kilowatt alternators.

PART OF THE ELECTRIC INSTALLATION OF THE PITTSBURG REDUCTION COMPANY,
NIAGARA FALLS, N. Y.
General Electric Company's multipolar 2100-kilowatt direct-current generators and rotary
converters.

BULLION REFINING.—Electrolytic methods have been applied with great success on both sides of the Atlantic in the refining of gold and silver bullion, the Moebius process being used for silver and the Wohlwill process for gold. In the Moebius process a dilute solution of silver nitrate containing free nitric acid, is employed as electrolyte, while in the Wohlwill process a solution of gold chloride is utilized. In America, the Philadelphia and Denver Mints are equipped with electrolytic parting apparatus, and a similar installation of electrolytic baths is now being erected at the Government mint in San Francisco. Many of the American copper refineries also have

VAT ROOM, AMERICAN METALS REFINING COMPANY.
Crocker-Wheeler electrical equipment.

an electrolytic plant for refining the silver obtained as a by-product in the copper-refining process. In Europe, electrolytic refining is carried on at Frankfort by the Deutsche Gold-und Silber-Scheide Anstalt, and by the Norddeutsche Affinerie at Hamburg, details of the Wohlwill gold-refining process having been worked out at the latter refinery. Electrolytic bullion-refining is also carried out in Great Britain and in France, but no details of the works are available for publication. A recent improvement of the Moebius process is the use of gelatine, which gives a smooth coherent, in place of a rough crystalline deposit at the cathode.

CALCIUM CARBIDE AND ACETYLENE.—Calcium carbide is obtained by heating lime and coke in an electric furnace, and it was first produced in a large scale by Willson at Spray in the United States in

the year 1893. The late Henri Moisson about the same time produced this compound in his laboratory in Paris, and the European
patents granted to Willson have not been upheld, owing to the earlier
publication of the results of Moisson's chemical researches upon the
electric-furnace and its products in the *"Comptes Rendus"* of 1894.

TRANSFORMER PLANT AT UNION CARBIDE WORKS, NIAGARA FALLS.
General Electric Company's 25-cycle 1,870-kilowatt 11,000-2,200 volt transformers.

The early history of the calcium-carbide and acetylene industries
is chiefly a record of reckless finance, worthless patents having been
used for company flotations upon a large scale, with serious results
for the investors and for the industry. The period culminated in
1899-1900, with a series of failures and financial "reconstructions."
·Since that year, the companies have been slowly recovering from the
effects of this unwise boom. Though acetylene gas has not displaced
other illuminants to the extent that was at one time expected, it is
now used for various purposes much more widely than is generally
recognized, and central acetylene-generating stations are found in
very many small village communities in Europe and America.

According to the most recent estimates there are now between
sixty and seventy works engaged in the production of calcium carbide, and the aggregate production amounts to between 90,000 and
100,000 tons per annum, valued at £1,000,000. The United States,
Italy, and France head the list of producing countries, and are also

GENERATING PLANT IN AMERICAN ELECTRO-METALLURGICAL WORKS.

Above, the Orford Copper Co., Bayonne, N. J.; four Crocker-Wheeler generators, 300-kilowatts, 125 volts, 150 revolutions. Mesta Machine Works engines. Below, the American Smelting & Refining Co., Maurer, N. J.; two Crocker-Wheeler generators, 520-kilowatt, 125-volts, 110 revolutions; Hooven-Owens-Rentschler engines.

the largest consumers of carbide for acetylene-generation purposes. During the period of inflated finance, several works for the manufacture of carbide were started in the United Kingdom. All of these have ceased operating, and only one small works is now active, at Askeaton in Ireland. The greater portion of the carbide consumed in the United Kingdom is therefore imported from Norway, and from other countries which produce in excess of their requirements. In France, the industry is controlled by a syndicate with headquarters in Paris, and this exercises a close watch over output and price. Eleven works are reported to be still operating, situated around the following centres of cheap water power:—Bellegarde, Grenoble, Nice and Toulouse. The estimated output of these works in 1906 was 24,000 tons; the annual consumption in France is about 15,000 tons.

Germany is dependent upon Switzerland, Austria, and Norway for two-thirds of its supply of carbide, only 8,000 tons being produced at home whilst 16,000 tons are imported. In the United States, the production of carbide is estimated to amount to 25,000 tons per annum, the Union Carbide Co., with works at Niagara Falls, being the chief producers. A large new factory designed for the utilization of 10,000 horse power is now being erected, however, in a new centre in the States.

Although calcium carbide is being employed chiefly for generating acetylene for illuminating purposes, its application for production of "calcium cyanamide" is likely to lead to developments of some importance. The use of acetylene gas in the oxy-acetylene blow pipe, for the autogenous welding of metals, is another application of considerable industrial importance, since temperatures can be obtained with this apparatus which approach those of the electric arc, and the size and shape of the flame are more suited for welding purposes.

CALCIUM.—Calcium in the metallic state is one of the latest electro-metallurgical products, the metal being produced by electrolysis of fused calcium chloride and fluoride with a rising cathode, which just touches the surface of the fused electrolyte. This method is adopted to prevent the re-solution in the molten electrolyte of the calcium deposited at the cathode. The temperature of the bath is kept at about 670 degrees C. and the process works most satisfactorily with fresh and neutral calcium chloride. The metal is obtained in the form of an irregular rod, made up of a series of buttons, fused together. The metal is dark grey in color, of specific gravity 1.51.

Calcium is now being manufactured upon a commercial scale by the Elektrochemische Werke at Bitterfeld in Germany, under the Rathenau patents, and is being placed upon the market by the same firm. The only difficulty in the development of the new manufacture lies in the lack of applications or uses for the metal. It has been suggested that it might be used in place of aluminium, for removing the oxides from steel, but at present aluminium is the cheaper metal. For many other reduction processes, calcium cannot replace sodium, since its affinity for oxygen is not so great. Attempts to form alloys of calcium with copper and other metals have also failed, as one would have expected.

A CARBORUNDUM FURNACE AFTER A RUN.

CARBORUNDUM.—Carborundum is the trade name given to a carbide of silicon—first made by E. G. Acheson at Niagara Falls, by heating coke, sand and sawdust, to a temperature of between 2,000 degrees and 3,000 degrees C. in an electric furnace of the resistance type. The product has the formula SiC, and the manufacture has grown into one of considerable importance on account of the excellent abrasive properties of the carbide. In 1892, 1,000 pounds of carborundum were produced at the Niagara Works, whereas in the last year for which complete figures are available (1906) the output had increased to 6,225,000 pounds. For many years the Niagara Falls works supplied all the demand for this compound.

Another artificial substitute for emery has also appeared, in the form of an electric-furnace product called *"alundum,"* obtained by heating bauxite to a high temperature. In order to meet the increased

INTERIOR OF FURNACE ROOM, THE CARBORUNDUM COMPANY, NIAGARA FALLS, N..Y.

competition, the Carborundum Company of the United States have arranged to carry on the subsidiary manufacture of grinding wheels, abrasive tools, and materials, in Germany, a new works for this purpose having been erected there in 1906.

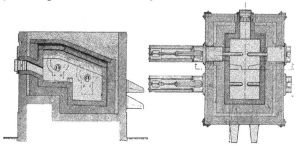

STASSANO FIXED-TYPE ELECTRIC FURNACE FOR COPPER.

Tucker & Lampen recently have carried out some laboratory experiments with carborundum, and have found that the temperature originally given by Acheson for its formation and dissociation are too high. According to Acheson these temperatures were over 2,500 degrees C., while Tucker and Lampen give 1,600 degrees to 1,900 degrees C., and 2,220 degrees C.

COPPER.—The electrolytic copper-refining industry is the oldest of the electro-metallurgical industries, having been started by James Elkington at Pembrey in South Wales in the year 1869. The process and methods used by Elkington in this small refinery were similar in all respects to those in use at the present day, copper sulphate being employed as the electrolyte with raw-copper anodes and thin sheets of pure copper as cathodes. The only change has been in the magnitude of the operations. At Pembrey the electrolyte was contained in small earthenware pots, and the output was 15 hundred weight per day, or 250 tons per annum. Today there is one refinery in America producing electrolytic copper at the rate of 350 tons per 24 hours, and the aggregate output of all the refineries is estimated at 400,000 tons, or 53 per cent of the total raw-copper production of the world. This enormous growth of the industry has occurred chiefly in recent years, the capacity and output of the American refineries which contribute over 85 per cent of the total having been doubled within the last seven years. The expansion is due partly to the great demand for a very pure copper for electrical purposes, and partly to the presence of silver and gold in the American raw copper, in sufficient amount to

INTERIOR OF FURNACE ROOM, INTERNATIONAL ACHESON GRAPHITE CO., NIAGARA FALLS, N. Y.

The nearer furnace has just been shut off and disconnected and the electrodes are lying in the foreground. The second furnace is in operation.

118

pay for their recovery from the slimes obtained in the electrolytic
process of copper refining. Thirty-four electrolytic refineries are now
operating in Europe and America.

The chief progress of recent years in this industry has been in
the substitution of machine for hand labour, the casting of the
raw-copper anodes, and the charging and discharging of the vats
by mechanical methods, now being carried out in all the large and
up-to-date refineries. The chief improvement on the chemical side
of the process has been the addition of a small amount of hydrochloric
acid to the electrolyte in the vats. This, according to Carlson, prevents the loss of silver which otherwise occurs, the insoluble silver
chloride being precipitated with the slimes.

DIAMANTINE.—This is a trade name given to a new product obtained by heating alumina with small quantities of silica to a high
temperature in the electric furnace. When finely powdered and
mixed with clay and water, the new material is said to form a useful
wash for the inside lining and walls of furnaces exposed to a high
temperature. The new product is being manufactured upon a commercial scale by the Diamantine Werke at Rheinfelden, Germany.

GRAPHITE.—The production of a hard variety of artificial graphite
has been carried on since 1892 by Acheson at Niagara Falls. The
method of manufacture is to form first a carbide in the electric furnace, and then to decompose it by increasing the heat up to a point
at which it dissociates and the second element is volatilized. Under
these conditions the carbon remains in the furnace in the form of
graphite. Acheson in his earlier work used coke mixed with silica
or sand, but he has since found that it is simply necessary to start
with ordinary anthracite coal; the impurities of this suffice to provide the second element of the carbide, and when raised to a definite
temperature, these elements volatilize and leave the carbon as
graphite. The manufacture has been a very successful one, and
the works of the International Acheson Graphite Company at Niagara Falls now utilize 2,000 horse power, and produce over 2,000
tons of artificial graphite per annum. The greater portion of this
output is used for electro-chemical and electro-metallurgical work,
the Acheson artificial graphite having been found specially suited for
electrodes. During 1906 Acheson discovered a process by which the
soft variety of graphite can be produced in the electric furnace, and
it is expected that this new artificial graphite will become a keen
competitor of the natural variety, especially as it shews more uniformity of composition. No details of the new process of manufacture have yet been published.

STEAM PRODUCTION FROM THE CHEAPER GRADES OF ANTHRACITE.

By *William D. Ennis.*

Mr. Ennis in the following pages pursues a discussion of strong practical interest to the coal miner, although it is primarily intended to appeal to the power user. The present subject follows logically as a development of "economy in burning fuel under the steam boiler," which he treated in our June, July and August issues. This particular phase—the burning of the small and cheap grades of anthracite—will be carried to its conclusion next month, and later we shall present another review by Mr. Ennis dealing with the burning of powdered fuels.—THE EDITORS.

A T a large market, say like New York, the price of coal is made up of two elements—the f. o. b. mine cost, and the freight. The latter is a large proportion—sometimes the larger proportion—of the whole. The freight rate is uniform for all qualities of coal. If the price of coal at the mine were fixed in strict proportion to the heating value, there would therefore be a loss to the consumer in buying the cheaper coal, because the reduction in price, proportionate to the reduction in quality, applies to a portion only of the elements making up the price. The logical outcome of resorting to cheap fuel under such conditions would be that the purchaser would eventually receive incombustible dirt, for which he would pay the miner nothing, but for which the railroad would receive its usual freight revenue. Obviously, therefore, for cheap fuel to be attractive, outside of the coal mining regions, either the freight rate must be reduced as the quality is reduced, or the price f. o. b. mines must be reduced more rapidly than the quality. The former outcome seems scarcely probable; the latter condition already exists, and for this reason the burning of the cheaper grades of anthracite is found highly profitable by steam producers who possess the necessary equipment for using such fuel.

Anthracite coal in sizes larger than pea is seldom used for steam production. Here again there enters a price variation which is out of alignment with quality variation. The actual heat units per pound are very little less (not over 5 per cent.) with good pea coal than with egg; but the domestic demand for the larger size, and its comparatively slight advantage to the steam producer, make the price of the latter grade of coal entirely out of line with that of pea.

The fine sizes—from pea down—are the coals to be considered in the present article. For the reason stated, they will not be compared with larger sizes of anthracite, but rather with their actual competitors—bituminous run-of-mine and slack. Run-of-mine coal has been stated to be about 30 per cent less economical than No. 2 buckwheat at present New York prices. The actual figures are: Run-of-mine, price $2.98 per ton. B. T. U. per pound = 14,100; whence for one dollar, the B. T. U. obtained = 9,450,000. A mixture of rice and barley coal containing 12,400 B. T. U. per pound cost $1.82½ per ton, giving 13,650,000 B. T. U. for one dollar. A sample of rice coal costing $2.00 per ton gave 12,457 B. T. U. per pound, or 12,457,000 for a dollar. Barley coal costing $1.65 tested 11,906 B. T. U. per pound, or 14,450,000 for one dollar. These prices are f. o. b. dock, and do not include lighterage or team delivery. The cost of lighterage and team delivery to city buildings is from 60 to 75 cents per ton. Like freight, these are fixed expenses, independent of the quality of fuel. The relative values of the various coals should therefore be based on the f. o. b. dock prices, as given. The following table gives the results in brief:

NEW YORK MARKET COAL PRICES—1907.

Grade of Coal.	Price per Ton.	B.T.U. per Dollar.	Saving as Compared with Run-of-Mine.	Saving per Ton, Based on Cost of Run-of-Mine.
Run-of-Mine	$2.98	9,450,000	....	.....
Rice (No. 2 Buck)..	2.00	12,457,000	32 per cent.	$0.95
Barley (No. 3 Buck).	1.65	14,450,000	53 " "	1.58
Rice and Barley.....	1.82½	13,650,000	44 " "	1.31

It is evident, from the above, that the estimated saving of 30 per cent by burning buckwheat as preferred to run-of-mine is sufficiently conservative, even when some deduction is made from the above figures to cover the cost of providing the necessary draft. The same condition applies in other markets. In Boston, for example, samples of hard and soft coal tested, respectively, hard (No. 2, buckwheat) 11,650, B. T. U.; soft (run-of-mine) 14,547 B. T. U. The prices were $2.34 and $3.37 f. o. b. cars, respectively, corresponding to 10,000,000 B. T. U. and 8,600,000 B. T. U. for one dollar; the saving by using the hard coal amounting to 16.3 per cent, or 55 cents per ton on the price of soft coal. This was with a hard coal of evidently inferior grade.

If it be objected that we should properly compare prices of the cheaper grades of hard coal with those of slack, rather than run-of-mine, it should be remembered that the slack ranges much lower in heating value than the run-of-mine; two recent samples of slack cor-

responding to the $2.98 run-of-mine above mentioned, analyzed 12,-
169 B. T. U. The cost was $2.45 per ton, so that the slack is far more
expensive, in proportion to heating value, than even the No. 2 buckwheat size of hard coal. Further, the slack is apt to produce much
more complaint on account of smoke, and as ordinarily fired will
generate at least 20 per cent less useful heat in proportion to its
heating value, than will the hard coal.

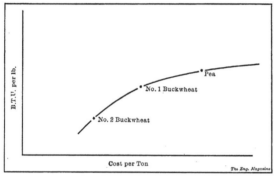

COMPARATIVE MONEY AND THERMAL VALUES OF VARIOUS GRADES OF COAL.

There is only one "size" of slack, and only one grade, excepting
as fate interposes. The manufacturer can either burn it or let it alone.
With the cheaper hard coals, there is quite a range to choose from; so
that it is quite possible that if one grade cannot be used, another can.
The following are the sizes below "nut" usually marketed, with the
common designation for each :—

SIZES OF HARD COAL—EASTERN MARKETS.

Sizes.	Limits.	Trade Name.
Pea....................	7/8 to 9/16 inch	....
No. 1 Buckwheat.......	9/16 to 3/8 inch	"Buckwheat."
No. 2 Buckwheat.......	3/8 to 3/16 inch	"Rice."
....	5/16 to 1/8 inch	"Birdseye."
No. 3 Buckwheat.......	3/16 to 2/32 inch	"Barley."
Culm.................	....	....

The point has already been made that the tidewater prices of coals
decrease with reduction of quality, but more rapidly than the quality decreases. The same rule applies, not only as between, say, egg
and pea coal, but also as between the various inferior coals listed
in the table. Thus, No. 2 buckwheat costs possibly 35 cents per ton
more than the No. 3; but No. 1 buckwheat costs 50 to 60 cents per

ton more than No. 2, and pea, 60 to 75 cents more than No. 1. There
is nothing like such a difference in the quality, which makes it all the
more noticeable that the slightly better grades not only each command a markedly higher price, but that the augmentation of price itself increases. A representation of this is shown in the chart opposite,
which might be interpreted, roughly, by saying, that the better grades
of coal command "fancy prices." Those who do not wish to pay
"fancy prices" will therefore burn the cheap coals, and in all cases,
the cheapest of all the sizes that can be burned will give the largest
useful amount of heat for dollar expended. The grade selected must,
however, be one that is sufficiently high in heating value and uniform
in quality to permit of keeping up the necessary pressure of steam.
Culm, it is safe to say, simply cannot be burned unless mixed with at
least an equal weight of buckwheat; and not then, unless fresh-mined.
It is only about 30 cents per ton cheaper than No. 3 buckwheat, and
little if any of it is ever shipped to tidewater *as culm*. Some of it ap-

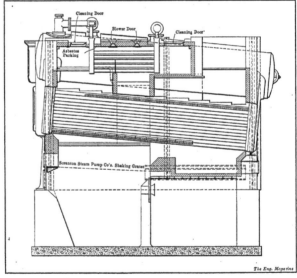

SETTING FOR 316 HORSE-POWER HEINE BOILER FOR BURNING ANTHRACITE-CULM,
SUBURBAN ELECTRIC LIGHT CO., SCRANTON, PA.

parently comes through at times, intermixed with the various grades of buckwheat and pea coal.

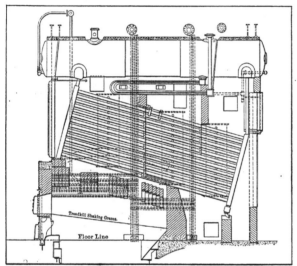

SECTIONAL ELEVATION OF SETTING FOR BABCOCK & WILCOX 650 HORSE-POWER BOILER FOR BURNING NO. 3 BUCKWHEAT COAL, WATERSIDE STATION OF N. Y. EDISON CO.

Equipped with Webster furnaces, the patents for which are owned by the Babcock & Wilcox Co.; with hand firing and moderate ash-pit pressure, 30 to 35 lb. buckwheat coal can be burned per hour per square foot of grate surface. The grates are 12 feet deep.

The finer sizes of anthracite differ to some extent, from the larger sizes in chemical composition. They carry a much larger percentage of sulphur, in the form of iron pyrites. This sulphur results in some loss of heating value, since the heat of combustion of sulphur burned to SO_2 is only about 4,000 B. T. U., as against 14,500 B. T. U. for carbon burned to CO_2. A far more serious disadvantage due to the larger sulphur content is that this substance exerts a detrimental influence on grates and even on boiler surfaces, producing rapid corrosion. It also greatly increases the tendency to clinker, always noticeable with low-grade fuels.

These cheaper coals also naturally run higher in percentage of ash. This is of detriment, not merely in proportion to the actual percentage contained, but in considerably higher proportion. That is, a coal containing 20 per cent of ash is considerably more than 10 per cent less valuable than one containing only 10 per cent of ash, other things being equal. A series of boiler tests demonstrating this point was reported by W. L. Abbott in the *Electrical Age,* November, 1906. These tests showed a gradually decreasing rate of evaporation from the boiler as the percentage of ash increased, until finally when the coal reached that point of inferiority where the ash amounted to 40 per cent, no evaporation whatever was obtained. This must have been due, of course, to the impossibility of keeping the fire in shape, as even 60 per cent of the total heating value of the coal would give more available heat than would be necessarily dissipated by radiation, incomplete combustion, etc. One way of looking at this therefore unexpected result is this: that the efficiency of the boiler, based on combustible, was 40 per cent; consequently, when the 40 per cent of maximum heat transmitted to the steam was just balanced by the 40 per cent less

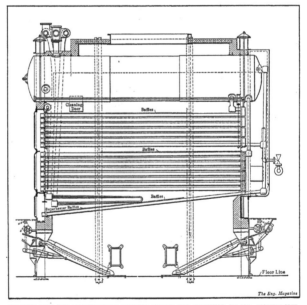

SETTING FOR 700 HORSE-POWER PARKER BOILER FOR BURNING SMALL ANTHRACITE,
WYOMING AVENUE POWER PLANT OF PHILADELPHIA RAPID TRANSIT CO.

heating value of the coal, due to that percentage of ash, the evaporation became *nil*.

The high percentage of ash necessitates carrying the fires thinner and cleaning them oftener. The frequent breaking up of the fire results in a high loss of fuel to the ash pit. This loss would be high in any case, on account of the fineness of the coal, which not only sifts through ordinary grate bars, like sand, but frequently fuses and runs like a liquid, with the fused ash, under forced draft. The Report of the Pennsylvania State Commission on "Waste of Coal Mining," 1883, mentions* analyses of ash from ash pits of boilers burning culm, in which as much as 58 per cent of carbon was found. If this culm contained (by analysis) 25 per cent of ash to start with, the loss of heating value due to the presence of unconsumed carbon in the ash was 14½ per cent. This loss, with ordinary fuels and firing, is from 1 to 2 per cent.

The percentage of ash in low-grade anthracite coals such as are used at tide-water points is seldom as

* Quoted by Professor Kent, "Steam Boiler Economy," First Edition, p. 149.

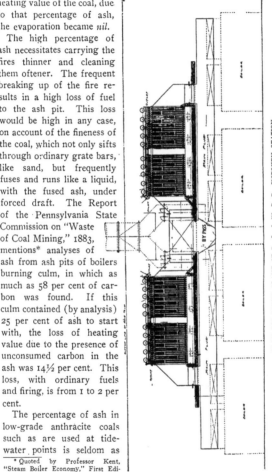

PLAN FOR POWER PLANT OF D., L. & W. R. R., HOBOKEN STATION.
Green fuel economizers and mechanical draft, for burning small anthracite.

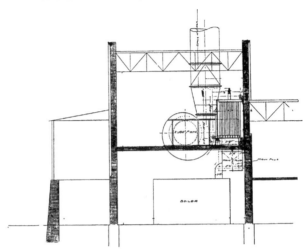

SECTIONAL ELEVATION OF GREEN FUEL ECONOMIZER INSTALLATION WITH
MECHANICAL DRAFT, D., L. & W. R. R., HOBOKEN STATION.

high as 25. The following list gives the ash content of various coals
sampled at New York during the past year:

PERCENTAGES OF ASH IN SCREENED COAL.

Kind of Coal.	Percentage of Ash.
Mixture of Rice and Barley	17.25
Rice Coal	20.50
Rice Coal	13.81
Barley Coal	16.96
Rice Coal	16.84
For Comparison.	
Mean of two samples of bituminous-slack	18.90
Mean of three samples of bituminous, run-of-mine	7.88
Average of Rice Coal, above figures	17.05

The rice coal seems, therefore, to contain more than twice the ash
that the run-of-mine does, and rather less ash than slack. The percentages vary widely, however, barley sometimes containing less ash
than rice, and occasionally less even than is found in samples of No. 1
buckwheat. The ash can of course be decreased by careful washing
and screening; but where the coal is billed to the customer at the
weight determined immediately after washing, he does not gain much.

The difficulty arising from high percentages of ash is purely
mechanical. The heat is there, in close proportion to the percentage
of combustible; the thing is to get it out. It is hard to do this, because

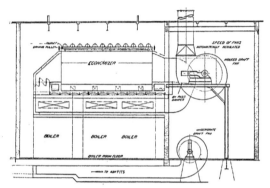

DIAGRAM OF TYPICAL MECHANICAL DRAFT SYSTEM WITH COMBINED FORCED AND
INDUCED DRAFT AND ECONOMIZERS.
The B. F. SturteVant Co., Boston, Mass

the ash slags into a hard cake that stops the air openings, because
the fuel falls unconsumed through the grate bars, and because the
fine coal packs so closely on the grate that ordinary draft intensities
are insufficient to give the air supply access to it. If by mechanical
means we remove these mechanical impediments to combustion, we
will have no difficulty in getting steam from fine coals. One method is
to pulverize the coal and blow it into the furnace. This gives a fuel
much finer than even the most powdery culm; but the combustion
secured is easily the most effective that is ever produced from coal.
The burning of pulverized coal is a subject for discussion in a later
article, but it should be referred to here, as demonstrating that neither
mere smallness of bulk nor high percentage of ash necessarily makes
a coal unfit for use under steam boilers. Mr. Eckley B. Coxe is
quoted as having in 1893 placed the following values upon fine sizes
of anthracite, f. o. b. mines:

Chestnut .... $2.75 per ton
Pea .... 1.25 "   "
No. 1 Buckwheat...................... .75 "   "
No. 2 Buckwheat...................... .25 "
No. 3 Buckwheat...................... .10 "

These differences, of course, are sufficient to pay the cost of pulverizing, but the interesting point is that effective combustion of the
lower grades can be secured *without* pulverizing. This has already
increased the value of the relatively cheaper fuels, and will no doubt
still further increase their values. The selling of No. 3 buckwheat

MECHANICAL-DRAFT INSTALLATION, LEHIGH VALLEY R. R. POWER HOUSE, SAYRE, PA.

Two Buffalo Forge Company's blowers with 11-ft. wheels, driven by Buffalo cross-compound enclosed engines. Each fan supplies air for burning 17,000 lb. of coal per hour in 14 boilers, each boiler having 88 sq. ft. of grate surface.

at $1.15 less per ton than pea—a differential of 92 per cent—will soon seem like flagrant economic waste.

Forced draft in some form is essential to burn fine sizes of anthracite. It is not necessary to go very deeply into mathematics to show that a chimney, considered as a machine, is about the most inefficient machine on earth. If the area of the chimney in square feet, be called A, and the height in feet H, then for a flue-gas temperature of 500 degrees and an air temperature of 62 degrees, the density inside is .0414 and outside .0761, pounds per foot. The difference in weight of gas inside and outside the chimney is then $A \times H \times (.00761 - .0414) = .0347$ AH. The pressure per square foot is .0347 H, and the height of air column necessary to produce this pressure $= .0347$ H $\div .0761 = .456$ H. The velocity of air corresponding to this height is $\sqrt{2g \times .456\,H} = 5.425\sqrt{H}$, feet per second. The weight passing through the chimney per second $= 5.425\sqrt{H} \times A \times .0761 = .413A\sqrt{H}$. The heat expended in warming this air from 62 to 500 degrees $= (500 - 62)\,.2375 \times .413A\sqrt{H} = 43.A\sqrt{H}$ B. T. U. $= 33,400\ A\sqrt{H}$ foot pounds. The work done in producing draft is equal to the difference in pressure, inside and outside, multiplied by

TYPICAL MECHANICAL DRAFT INSTALLATIONS BY THE AMERICAN BLOWER CO., DETROIT, MICH.

On the left, a full-housed bottom-discharge fan with direct connected two-cylinder double-acting self-oiling engine. On the right, a top horizontal-discharge fan with single-cylinder self-oiling engine.

the velocity of the air in feet per second, or $.0347$ AH $\times$ $5.425\sqrt{H} =$ $.188$ AH$^{\frac{3}{2}}$. The efficiency is equal to the work divided by the heat expended, or $.188$ AH$^{\frac{3}{2}}$ $\div$ $33,400$ A$\sqrt{H}$ $= .00000563$ H. When friction and radiation losses are considered, the efficiency is even lower than this. For any reasonable height of chimney not exceeding 300 feet, the efficiency cannot, therefore, exceed one-sixth of one per cent. The useful work obtained is of course the result of energy that is ordinarily wasted. It need not all be wasted, as fuel economizers may be made to absorb one-half of it; but under proper regulation of air supply, the loss of heat to the chimney is so small that it is rarely that any profit can be seen in economizers; and in ninety-nine per cent of all power plants in the United States, at least, the low effi-

ELECTRICALLY DRIVEN INDUCED-DRAFT APPARATUS, SHEFFIELD CORPORATION ELECTRIC POWER STATION.
B. F. Sturtevant Co., Boston, Mass.

ciency of the chimney for draft production is scarcely recognized as
a disadvantage.

The cost is, however, a serious disadvantage. Brick chimneys
used to be figured to cost $5 to $6 per boiler horse power equivalent
to the combustion of 5 pounds of coal per hour. This was, however, for reasonably good coals, such as would ordinarily burn when
subjected to a draft intensity of ⅜ to ⅝ inches of water. For such
coals as we are considering, draft of at least 1 inch should be provided, and much more draft than this will be necessary unless the rate
of combustion is to be very low. As the draft intensity from a chimney varies as the square root of the height of the chimney, it is easy
to see that when we double or triple the necessary amount of draft
we arrive at chimney heights beyond the limits of possible practice.

INDUCED-DRAFT APPARATUS AT ERIE RAILROAD SHOPS, SUSQUEHANNA, PA.
Buffalo Forge Company's 7-ft. fan for two 400 horse-power B. & W. boilers carrying 125-lb.
pressure. An example of an induced-draft installation in close quarters.

The use of reinforced concrete for chimneys has reduced their cost
to about $2 per boiler horse power, but has not made it any more
possible to build chimneys from 600 to 1,000 feet high instead of from
80 to 200 feet. Both on account of cost, and on account of physical
limitations, chimneys for draft are entirely inadequate, in burning
fine sizes of anthracite.

HEINE SAFETY BOILER INSTALLATIONS FOR BURNING RICE COAL.

Above, the Federal Sugar Co., Yonkers, N. Y.; twelve 316 horse-power boilers; Wilkinson stokers; economizers and induced-draft fan. Below, the Hotel St. Regis, N. Y.; four 362 horse-power boilers, with 62 sq. ft. dumping grates each. Wing's turbine fans.

A favorite method of providing the high draft required has been to use the steam jet. The only merit possessed by the steam jet, if in this case it can be considered a merit, is that of low cost of installation. In its simplest form, the device may consist of a few pieces of pipe, possibly installed complete at a cost of ten dollars or less, per boiler. With the steam jet, almost any desired intensity of draft can be produced. Clinkers can be kept broken up and smoke so diluted as to be unobjectionable in most instances. The steam consumed

PARKER BOILERS BURNING BUCKWHEAT COAL, POWER PLANT OF THE J. B. STETSON CO., PHILADELPHIA.
Single-ended boilers of 515 horse-power each. The Parker Boiler Co., Philadelphia.

BATTERY OF TWELVE 250 HORSE-POWER FITZGIBBONS BOILERS BURNING RICE COAL,
POWER PLANT OF THE INTERNATIONAL SALT CO., LUDLOWVILLE, N. Y.

AN INSTALLATION OF BABCOCK & WILCOX BOILERS BURNING NO. 3 ANTHRACITE COAL.
Webster patent furnaces, with grates 12-ft. deep; ordinary flush fire fronts.

COMBINED INDUCED-DRAFT AND FORCED-DRAFT APPARATUS AND FUEL ECONOMIZER,
B. F. STURTEVANT COMPANY'S POWER PLANT, HYDE PARK, MASS.

by the jet is enormous. The lowest reported tests put it at 8 per cent of the total steam produced; other reports make it as high as 40 per cent. The jet supplies, not air, which is the natural supporter of combustion, but water, which while containing more oxygen in proportion to its weight than air, contains this oxygen *in combination.* The same amount of heat is lost in decomposing the water as is afterwards evolved in burning the fuel with the oxygen thereby set free. The steam jet has been found to increase considerably the percentage of fuel lost in the ash, an always serious item in the case of these fine-sized coals; but what is more important still, even with the practically unlimited draft intensity obtained, the steam jet fails to produce a rate of combustion per square foot of grate per hour equal to that ordinarily obtained by a fan. The breaking up of clinker, which is quite effectively done by the jet, can be accomplished equally well by the use of *exhaust* steam under the grates. Exhaust steam is, in fact, commonly used in connection with many types of stokers, for this very purpose, when burning a clinkering soft coal; and it is beginning to be generally used in the same manner with fine sizes of anthracite, even where fans are employed to produce the necessary intensity of draft.

THE NEW JERSEY PORTAL OF THE PENNSYLVANIA TUNNELS UNDER THE HUDSON RIVER.

UNDERGROUND WORKINGS IN NEW YORK CITY.

By H. T. Hildage.

In a paper before the Institute of Mining Engineers, Mr. Hildage lately showed the strong analogy between the tunneling operations in and about New York, and practice already familiar to the mining engineer in his own work. This analogy is again suggested by some of the illustrations accompanying this article, but the author's text refers now more especially to work in soft ground and the interesting special methods for its accomplishment. He traces the origin and development of the practice which has found its most conspicuous applications in the several large undertakings now nearly finished under and through Manhattan Island. For the reproduction of the map and of some of the drawings of the shields we are indebted to the courtesy of the American Institute of Mining Engineers.—THE EDITORS.

THE individual undertakings in the construction of tunnels to "emancipate New York City from the limitations of its insular position" have been described in more or less detail, both as regards their design and the methods of construction used. The unparalleled magnitude of the work considered as a whole, the enormous difficulty with which it has been attended, the novelty of the problems in tunnel construction that have been solved, and the almost unbroken success that has attended its execution have been sufficiently demonstrated in various publications. It remains now and it is the purpose of this article to endeavor to discover how the

GENERAL VIEW OF THE OPEN-CUT WORK FOR THE PENNSYLVANIA TUNNEL TERMINAL SITE, 31ST TO 33RD STREETS AND NINTH AVENUE, NEW YORK CITY.

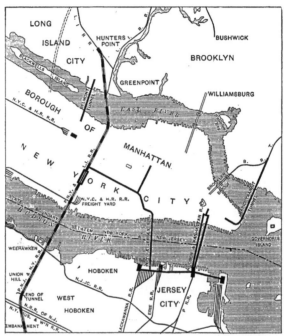

MAP SHOWING TUNNELS UNDER CONSTRUCTION.

work that has just been done is connected with, or is a sequel to, what had . been previously accomplished, with a view possibly to obtaining guidance for the future.

There are many steps between the old days of tunnelling, when the only method of temporarily holding the ground was by timbering, when the only means of preventing the flooding of the heading was by pumping, when the "muck" was removed by mule or pony haulage, when the only light was candle or oil light, and when the progress (which was won by dogged perseverance and sheer pluck) was counted in inches per week—and the modern iron-lined tunnel which is clean, well lighted by electricity, equipped with electric traction or mechanical haulage; which is constructed with the aid of a shield fitted with every conceivable device for advancing the work and main-

taining the advance, for supporting the ground and protecting the
men, and for handling the heavy plates that constitute the tunnel shell;
in which progress is steady and can be counted upon and be measured
in yards per day.

The first step was taken in the Thames tunnel that was commenced in 1825 by Brunel, who conceived the idea of a shield to
support the ground whilst the permanent lining of the tunnel was
being built. This shield consisted of a number of small cells with
friction rollers between them, each forced forward independently by
means of screws. The only thing about this shield that has persisted
is the idea. Almost all the completed tunnels in soft ground constructed since have been of circular cross section, and such a tunnel
could scarcely be constructed with a shield of Brunel's type.

ELECTRIC LOCOMOTIVE IN THE PENNSYLVANIA RAILROAD'S CROSSTOWN TUNNELS.

The great era of development of the art really began with the
works of Barlow and Greathead, for the construction of the Thames
tunnel must be regarded as a triumph of the splendid personality of
Brunel rather than as a success for the shield method.

The first important step in this development was the use of cast-
iron lining, and this was taken by Mr. Peter Barlow in the Tower
subway in 1869. This tunnel was driven through clay, with the aid

HEADWORKS AT THE WEEHAWKEN SHAFT, PENNSYLVANIA TUNNELS.

THE INTERIOR OF A CAISSON UNDER AIR, SHOWING AIR-TIGHT FLOOR AND CUTTING
EDGE.

of a cylindrical shield, and the maximum progress made was 9
feet per day of 24 hours. The shield consisted of a single cylinder
of wrought-iron plates, ½ inch thick, fitting loosely over the cast-
iron lining, stiffened by a cast-iron ring at the forward end to which
was bolted a diaphragm. Access to the face was obtained through a
rectangular opening in the diaphragm. The shield was advanced by
six screw jacks, each 2½ inches in diameter, attached to the shield
and abutting against the leading end of the cast-iron lining. The
cast-iron lining consisted of rings, 18 inches long and 7 feet 1 inch in
external diameter. Each was made up of three segments and a key.

In 1870, a short length of experimental tunnel was driven by
means of a shield under Broadway, New York. This tunnel was the
first one to be built with a shield that was provided with hydraulic
jacks to move it forward. The power for the jacks was supplied
by a hand pump attached to the shield, and could be admitted to all or
any of the jacks.

The use of compressed air was first suggested by Lord Cochrane
who, in 1830, took out a patent for air locks and other appliances
to permit its use in shaft-sinking and tunnelling. It was first used,
without a shield, in the old Hudson Tunnel, the construction of which
was undertaken to afford direct land communication between New

York and the railroad terminals in Jersey City. This tunnel was commenced in 1879, although the sinking of the first shaft in Jersey City was commenced in 1874 and stopped when a depth of 14 feet below mean high water had been reached. When the shaft had been sunk to its full depth, an air-lock was built into the side of the shaft, the excavation was commenced under compressed air, and a chamber was formed that was large enough to start two tunnels from. Each of the tunnels consisted of a very light wrought-iron shell in segments lined with from 2 feet to 2 feet 6 inches of brickwork, making the internal dimensions 16 feet wide and 18 feet high.

The method of working originally used was to excavate a place for a plate and bolt it in before the mud closed in, keeping the top well in advance. This method soon proved too slow and troublesome and the pilot-tunnel method was devised by Mr. J. F. Anderson. A smaller tunnel about 5 feet 6 inches in diameter was driven from 15 to 20 feet ahead of the working face. This tube, made up of wrought-iron plates, was erected one segment at a time, and broad flange plates were bolted in between the segments to give a footing for radial struts. The thin wrought-iron shell of the large tunnel was erected as before and each plate held in position by a radial strut from the pilot tunnel. The brickwork was built inside the wrought-iron shell in lengths of about 10 feet. By these methods about 2,000 feet of the north and 500 feet of the south tunnel had been constructed, and a shaft had been sunk on the New York side and the north tunnel started when the work was suspended in November, 1882, for lack of funds.

The next progressive step in the art of tunnelling was made in 1887, when Greathead devised the shield that bears his name and which was used for

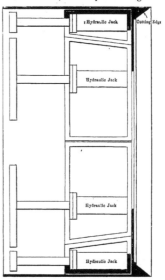

GREATHEAD'S SHIELD FOR CITY & SOUTH LONDON RAILWAY. LONGITUDINAL SECTION.

tunnelling through London clay, and with compressed air for tunnelling through Thames ballast. This shield (shewn in diagram on page 143) consisted of a steel cylinder stiffened at the forward end with a heavy cast-iron ring built in six segments, to which were attached hydraulic jacks and the cutting edge. To this also was bolted a diaphragm with an opening to the face; in the first shield, this opening was provided with iron doors, working upon rollers; these were subsequently removed and in their place timbers were used that were cut to drop in between channels attached to the diaphragm for the purpose. There were six hydraulic jacks and the rams were provided with large heads to distribute the thrust over the flanges of the cast-iron lining plates. Two hand pumps were provided to supply hydraulic power to these jacks. The rear end of the steel cylinder formed a "tail" under the shelter of which the cast-iron lining plates could be erected.

Tunnelling in Silt and Soft Clay.—Following the construction of the City & South London Railway tunnels, the development of the art has progressed much more rapidly. In the year 1889, the construction of the Hudson Tunnel that had been abandoned seven years previously was recommenced with a shield and cast-iron lining, and the St. Clair River tunnel was commenced and finished thirteen months later. Both of these undertakings marked a distinct advance in shield tunnelling, inasmuch as both tunnels were much larger than any iron-lined tunnel previously built with a cylindrical shield, and the St. Clair shield moreover was the first shield actually used that was provided with an erector, although Greathead designed and had one built in 1876 with a hydraulic erector, for use in the Woolwich subway.

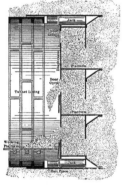

The second period of activity in the history of the Hudson tunnel just referred to as having commenced early in 1889, began when the work in the north tunnel from Jersey City was taken up by S. Pearson & Son of London, who built a shield at the end of the brickwork and commenced to drive an iron-lined tunnel. This shield (as shown here) consisted of a steel shell or skin, 19 feet 11 inches in external diameter, stiffened by the addition of an

SHIELD USED FOR CONTINUATION
OF OLD HUDSON TUNNEL.

inner shell connected to it by radial webs. The compartments of this ring were utilized to hold the sixteen hydraulic jacks for propelling the shield, using the iron lining as an abutment. The interior, shut off from the tunnel by a diaphragm, was divided into nine pockets by two vertical partitions and two horizontal platforms, and these pockets were open to the face and accessible from the tunnel through hinged doors. The method of working then used was to open some of the doors of the shield and to turn power on to the jacks. The shield moved slowly ahead, laying bare as it went the "tail" which had been lying over the outside of the lining, and ultimately giving, when the jacks were returned, space for the erection of another ring. The soft silt squeezed slowly in through the open doors, and falling upon the working platform, was loaded into cars and removed. When the shield had travelled the full length of a ring, the power was shut off, the doors closed, the invert cleaned and the ring erected. This was done by means of a hydraulic erector that trailed behind the shield on rails supported by the iron lining. This erector consisted of an arm revolving on a pivot by two hydraulic jacks and carrying a third hydraulic jack, the ram of which was fitted with a jaw capable of being attached to one of the tunnel segments. The whole arrangement was mounted on a carriage with four wheels. The inside diameter of the cast-iron rings was 19 feet 6 inches and the length about 20 inches. This method of tunnelling proved much more successful than the pilot-tunnel method, and with it Messrs. S. Pearson & Son completed about 1,900 feet of tunnel by July, 1891, when lack of funds again suspended operations.

As far as tunnelling in soft semi-fluid silt or soft clay is concerned, the evolution of the shield was completed when this shield was built. The essentials were an outer skin, the rear part of which covered the iron lining, and the forward part of which served to hold the hydraulic jacks to propel it, and was stiffened by an inner ring of cast iron or steel and cross bracing; some means of partially closing the face and at the same time permitting the flow of material into the tunnel, in the case of soft, flowing material, or of giving more or less free access to the face in the case of material that required to be dug out; in the case of large tunnels, erectors for the purpose of handling the iron lining; and platforms within the shield to enable the whole face to be reached. With the exception of details of more or less importance, no further changes have been made in the construction of shields used exclusively for silt or clay tunnelling. Sliding platforms actuated by hydraulic jacks have been used with profit, but their chief value has been in dealing with other conditions to be mentioned later. In some large shields such as the shields for station and cross-over tunnels on the City & South London Railway extension, the Waterloo & City Railway and the Baker Street and Waterloo Railway, two erectors have been used instead of one. These shields were really larger editions of Greathead's first shield, with erectors and sliding platforms added.

As shown in the previous description, the method of working varied with the stability of the material; in the London clay, a very stiff clay that stood without measurable flow, a box heading was often driven, but in every case a space was excavated for the shield to be shoved into. In the softer clay, through which the St. Clair tunnel passes, the men worked ahead of the shield to excavate, but apparently the flow was great, for they frequently excavated as much as 50 per cent more clay than was displaced by the length of tunnel built. In the fluid silt of the Hudson River, it was not necessary to excavate at all; the material simply flowed in through the open doors. At a later date, some time after the construction of the Hudson tunnel had been taken up by the present owners and the south tube was being completed, even this was changed; the shield doors were kept closed and very little silt was permitted to enter the tunnel; it was simply shoved out of the way. This method was applied to both the north and south tunnels, and very rapid progress was made in the silt.

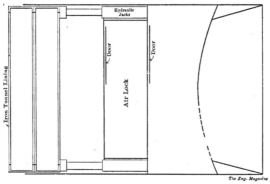

GREATHEAD'S SHIELD FOR THE NORTH WOOLWICH SUBWAY.

When the North River tunnels of the Pennsylvania Railroad Co. were being built, it was found impossible to use this method an l at the same time control the shield, maintaining line and grade. It was necessary to take in part of the silt through the doors in order to keep the shield on grade and to prevent the tunnel from rising considerably immediately after construction.

Tunnelling in Open Ground and Gravel.—Practically all the changes and additions that were made to shields since the St. Clair and the Hudson north tunnel shields were built, have been made for the purpose of permitting the ground to be excavated ahead

SHIELD FOR BLACKWALL TUNNEL. LONGITUDINAL SECTION.

of the shield, when it consisted either of (a) sand or gravel, or (b) rock, overlain by sand, gravel or silt. In both of these cases, for safe, rapid and efficient work it is almost always necessary to maintain a vertical face in front of the shield.

Probably the first shield especially designed to deal with sand and gravel, was designed and built by Greathead for the North Woolwich Subway already mentioned (page 146), but never used. This shield contained an air lock within its length, and also a siphon-trap arrangement that rendered it possible to prevent the flooding of the tunnel when the face was lost, by maintaining sufficient air pressure.

The shield that was used for the Blackwall tunnel (figured above) designed by Mr. E. W. Moir, (as also was the one used in the Hudson north tunnel), was required to meet a variety of conditions. The forward part of the shield was divided into twelve pockets or compartments by vertical diaphragms and horizontal platforms that gave considerable stiffness to the structure. The face was closed by two plate diaphragms, suitably stiffened, one of which was about the centre of the length of the shield and the other about 3 feet behind it. Four air-locks were provided, so that a higher air pressure could be

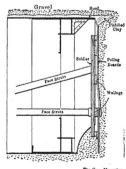

HOODED SHIELD FOR WATERLOO &
CITY RAILWAY. READY TO
ADVANCE.

used in the face than was in the tunnel, and each pocket was fitted with a chute, fitted with air-tight doors at each end, for the removal of the spoil. In each end of the compartments were hanging curtains or screens similar to the one in the Greathead shield last mentioned, for the protection of the men in case of a sudden inrush of water.

Two hydraulic erectors were attached to the back of the shield. To the sides of each compartment in the upper three floors were attached guides in which ran sliding shutters controlled by screws for poling the face. In the lower part of the shield, where shutters were not provided, horizontal poling boards were used to hold the face, and these were strutted against the shield with telescopic struts that would shorten under great pressure. When the shield was shoved, the screws controlling the shutters were released by the men as it advanced, and the struts holding the poling in the lower pockets shortened. The position of the face was thus to a certain extent independent of the position or movement of the shield.

Hooded Shield.—In the construction of that part of the Waterloo & City Railway (1894) that was in gravel, the method at first used was the same as that used by Greathead in the same kind of ground and consisted of excavating a timbered length ahead of the shield, into which it could be shoved. After a time this was abandoned and a hooded shield (see figure above), was built and used with much success. The hood in this case was a prolongation of the skin plates round the upper two-thirds of the circumference for some distance beyond the front of the shield. This prolongation forms in the front of the shield a working chamber, under

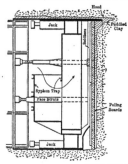

SHIELD FOR BAKER ST. & WATERLOO
RAILWAY.

cover of which the men poling the face can work in safety and need not take care of the roof. It reduced the amount of excavation to the net section of the shield, avoided leaving poling boards in the ground as was done in the City & South London tunnels, and escaped much of the disturbance of the unexcavated ground that necessarily occurred in the Blackwall tunnel, and rendered it possible to remove all the ballast from in front of the cutting edge in the invert.

This shield and the method of working it were designed by H. H. Dalrymple-Hay. Pockets were excavated in front of the cutting edge to a distance equal to the advance of the shield and filled with puddled clay. When the shield was shoved ahead, the cutting edge buried itself in the soft clay, and as the work progressed a layer of clay was left outside the skin of the shield and tunnel and reduced the losses of air considerably. This method of working was used in the Baker & Waterloo Railway tunnels in 1900 with some modifications, and in the Morton Street extension of the Hudson tunnel.

Tunnelling in Mixed Ground.—In all the cases so far mentioned, the face worked has consisted either entirely of clay, entirely of sand and gravel, or of clay overlain with sand and gravel. The latter case was probably no more difficult to handle than the second. In the tunnels under the North and East Rivers, however, it has frequently been necessary to work a face consisting partly of rock and partly of silt or quicksand, and special appliances and methods were necessary.

This difficulty was met for the first time in the Hudson north tunnel soon after the work was recommenced in 1901, when it was found that the tunnel would pass through a ridge of rock near the New York shore. A horizontal platform was built projecting about 2 feet in front of the shield and supported by raking struts from below, to carry the roof of the excavation, and all work was carried out under its shelter. That portion of the silt between the rock surface and the platform was held up by poling boards, stretched against the shield, whilst a path for the shield was blasted out below with light charges of dynamite. On account of the position of the shield (it had rolled on its axis about 45 degrees), it was not possible to hold the face whilst the shield was shoved and consequently the pockets became filled with silt each time and had to be cleaned out. Besides rendering the work tedious and slow, this resulted in the denuding of the tunnel of the cover of clay and contributed to the occurrence of blows. The evil was mitigated or avoided by laying a blanket of good clay on the river bed just above the shield, as had been done in the Blackwall tunnel.

The companion (south) tunnel was smaller in diameter, and being driven with its crown at the same level as the north tunnel, it was expected that a good deal of the rock work would be avoided. The shield (shown below) was different in some respects. It consisted of an outer skin stiffened by an inner shell. The space inside this shell was divided into compartments or pockets by vertical stiffening par-

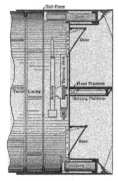

titions and a horizontal platform. The face was closed with a diaphragm which was provided with doors. A horizontal sliding platform in three parts was provided, each part being actuated by two hydraulic jacks. The hydraulic erector was attached to the shield and rotated about an axis coincident with the axis of the shield.

In the design of the shield for the North River Tunnels of the Pennsylvania Railroad Company, full advantage was taken of the experience in the old Hudson tunnel and in the tunnels through the Thames gravel. This shield was designed chiefly to deal with the difficult ground that it was expected to

SHIELD USED FOR HUDSON
SOUTH TUNNEL.

meet with on both sides of the river, before the shield was entirely in silt. It was expected that as the shield left the rock there would be a face partly rock and partly sand, with the rock line gradually lowering until it disappeared, and then a face partly sand and partly silt until a full face of silt was reached. These conditions were real-

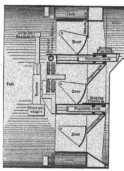
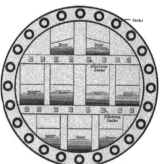

SHIELD USED FOR PENNSYLVANIA COMPANY'S NORTH RIVER TUNNEL.

ized on the New Jersey side, but on the Manhattan side there was practically no sand.

The shields (foot of page 150) had an external diameter of 23 feet 6¼ inches; the internal diameter was 23 feet 2 inches, allowing a clearance of 1 inch all around the tunnel lining, which was 23 feet in external diameter. A hood was attached which covered rather more than one-third of the circumference, and extended 2 feet 3⅝ inches in front of the face. The length of the tail was long enough to cover two and a half rings of the lining, which were 30 inches long. The forward part of the shield was divided into pockets by vertical partitions and horizontal platforms. Each of the platforms had a sliding extension made in four pieces, each piece being actuated by two hydraulic jacks.

The pockets were closed at the end nearest the face by inclined plates covering about half the opening. The remainder of the opening could be closed by means of a door. The doors were segments of a cylinder and swung on pivots attached to a·vertical diaghragm. A single erector was used and pivoted at the axis of the shield.

The method of working a face consisting partly of rock and partly of silt was to pole the silt with vertical poling boards and blast out the rock. When the shield was shoved, walings were put against the poling boards and held in place by the sliding platforms and by struts which passed through the tunnel lining against the shield. When the rock line fell entirely below the level of one of the platforms, that part of the silt above the platform was not poled.

In a face that consisted partly of rock and partly of sand, a similar method was used, except that the polings were horizontal instead of vertical. The top board was advanced first and strutted to the shield, and then the second and so on until the whole face was advanced, when the "soldiers" were put against it and held by the sliding platforms and by struts that passed through the shield to the tunnel lining. The ground being rather better than was expected, it was possible to advance the poling boards for a full ring length at one operation, and as the hood was not long enough to permit of this, roof and side polings were necessary.

The shield (see page 152) that is being used for the East River tunnels of the Pennsylvania Railroad Company is very similar to the one that was used in the Blackwall tunnel, the changes that have been made, having no doubt been suggested by the experience there obtained (the contractors for the East River tunnels, S. Pearson & Son, were the contractors for the Blackwall tunnel and Mr. Moir, the managing director, designed also this shield). The

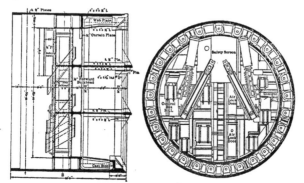

SHIELD FOR PENNSYLVANIA COMPANY'S EAST RIVER TUNNEL.

jacks were more closely spaced around the lower part of the shield than around the upper. The reason for this arrangement, as stated before, is that in a large tunnel it is not always possible to keep the air pressure high enough to keep out the sand and water at the invert without blowing the "cover" off at the roof (where the pressure would be excessive) ; hence the lower part of the shield is more or less full of sand, and additional power is necessary to displace it.

There is a heavy cast-steel cutting edge around the lower part of the circumference; the need for it has been shewn by many shields being damaged at this point and requiring repairs whilst the work is in progress. The shield is also fitted with a hood not shewn in the diagram, that can be advanced or retracted in segments by means of screws running in bearings attached to the skin and fitted with nuts. Air locks and face shutters are provided as in the Blackwell tunnel, but these have not, so far, been used for their intended purpose.

The method of working is generally similar to that used in other ·tunnels. The roof and sides are poled in front of the hood and the face is poled down to the level of the lower platform, the lower part of the shield being left practically full of sand; the polings are held up with struts that will telescope under the great pressure that comes upon them when the shield is advanced. With a face of part rock and part sand, the sand is held up by poling boards, whilst the rock is blasted out.

Shields of the Future.—Whilst the beds of the North and East Rivers present greater difficulties to the tunnel builder than the bed of the Thames, chiefly on account of the fact that it is not possible to build a tunnel under either of them that is wholly in silt and sand,- or wholly in rock, without approaching too near the bed of the river in the one case, or going to excessive depths and even then running the risk of meeting unsound rock in the other, and whilst these difficulties have been increased considerably by the fact that the tunnels that have been built in the neighborhood of New York are much larger than the tunnels that have been built under the Thames, with the exception of the Blackwall tunnel and the Rotherhithe tunnel recently constructed, it is evident that the methods that have been successfully used in New York are the same methods that were used in London. Moreover, there is plenty of evidence that these methods were not copied, but were arrived at as a result of experience and a careful study of the conditions met with. It therefore seems that these methods are pretty generally applicable and that it ought now to be possible to construct a shield which will deal efficiently and satisfactorily with any conditions that can be developed in tunnel work.

Several writers have divided shields into three classes according to the character of the material they are designed to tunnel through. Whilst the system of classification thus suggested may be excellent for shields that have been constructed in the past, of which it can probably be truly said that they have all been experimental, it is better to adopt a somewhat different classification for shields to be used in future work.

The first class, as before, would contain shields designed for work with an open face, which would not differ materially from the original Greathead shield for small tunnels and from the Central London Railway and Waterloo & City Railway station tunnel shields, which have movable tables to facilitate working at different levels. The second class would contain shields that could be satisfactorily used where it is known that it will not be necessary to work in front of the shield. Such cases are however so rare and the consequences of adopting an unsuitable design of shield so costly and serious, that this class might well be omitted.

The third class of shield would be used in all cases of soft or loose ground where there was any possibility that work would have to be done in front of the shield; in tunnels that pass through gravel or sands containing boulders or other obstructions; that include in the depth of their cross section the contact between sand or gravel and rock, or silt and rock; and generally in cases where there is any doubt about the nature of the ground to be dealt with. The shields in this class should be fitted with appliances for supporting the unexcavated ground in an undisturbed condition, such as an efficient hood, and in

- larger sizes of shields, sliding platforms, with struts for transferring the pressure of the face to the iron lining, or possibly shutters fitted with a satisfactory arrangement for release when the shield is advanced; with safety appliances to facilitate the escape of the men in case of blows or runs of material, such as siphon traps and screens; and if possible, with air locks to permit of a graduated air pressure being used in the face.

The value of the hood was amply demonstrated in the construction of the Waterloo & City and the Baker Street & Waterloo Railways. It has been emphasized in the experience in the North and East River tunnels. It should project so far in front of the shield that it will never be necessary to pole the roof and the sides of the excavation, and should be continued around at least two-thirds of the circumference.

The use of sliding platforms in supporting the face is exemplified only in the case of the Pennsylvania Railroad Company's North River tunnels. They have a further use. If they are advanced immediately after the shove, to their full stroke—that is, until they project as far as the hood—the miners can with care work safely under them as under a hood; in this way the work can be done at each level in the shield at the same time and the time necessary to pole the entire face will be proportionately shortened.

Shutters for supporting the face were used on the Blackwall tunnel, and were provided in the Pennsylvania Company's East River tunnels, but, apparently, they have not so far been used.

It is very desirable, if possible, that the arrangement for poling the face should be an integral part of the shield. There are two difficulties in the way, viz., the necessity for a satisfactory release arrangement that will hold the shutter up to the face with sufficient force to prevent disturbance, and at the same time will release it as the shield is advanced; and the fact that in order to avoid disturbing the ground, the area covered by the shutters should be as nearly as possible equal to the total area of the cross section of the shield.

The first difficulty might be overcome by holding the shutters up to the face with hydraulic or compressed-air cylinders suitably fitted with release valves, but in introducing such an arrangement, it is necessary to bear in mind that it will be subjected to very rough usage. Materials and tools used in the construction of tunnels are probably worked more nearly to their breaking point than in any other kind of work and it is undesirable to introduce intricate or delicate mechanism.

The second difficulty is a more serious one. The proper place

PUMPING STATION IN THE BELMONT TUNNEL. THREE CAMERON PUMPS HANDLING
1,000 GALLONS PER MINUTE WITH A LIFT OF 110 FEET.

for shutters is undoubtedly in the hood, but on the other hand, the
hood is often subjected to enormous forces and must be made very
strong and if necessary stiffened with brackets. Moreover the use of
shutters will render it very difficult, if not impossible, to use the clay
pocket system, which is of great value.

The siphon traps are very valuable safe-guards in tunnelling
through open ground, and can be very readily attached to the door
openings of large shields. In the event of a run of material in the
face, resulting in temporary loss of air pressure in the tunnel, the
water rushes into the shield until its level rises above the lower lip of
the upper screen, when a seal is established, the escape of air ceases
and the pressure rises, and further inflow of water is prevented. In
the absence of a screen or trap, the escape of air and the inflow of
water continues until the tunnel is full, and any attempts to check it
by increasing the input of air only aggravate the trouble. A run
in a face may be easily caused at any time by a trivial accident or
carelessness, but if checked by some such contrivance as the siphon
trap, is of no great importance. If, however, it is allowed to develop
to the extreme suggested above, it may cause serious delays and

COMPRESSOR PLANT FOR THE MANHATTAN END OF THE PENNSYLVANIA TUNNEL UNDER THE EAST RIVER.

The plant consists of eight Ingersoll-Rand compressors, six cross-compound Corliss and two straight-line, with a combined capacity of 22,000 cubic feet of air per minute. Four of the compressors supply air at a maximum pressure of 50 pounds, the others can deliver at 140 pounds. Steam is supplied by six 500 horse-power Stirling boilers at 150 pounds pressure, and the whole plant is run condensing so far as possible.

greatly increase the cost of tunnelling. It is probably more true in tunnelling than in any other kind of work, that as long as the forces of Nature are kept in check they can be readily dealt with, but once loosed, they are irresistible.

The use of air locks, as suggested in Greathead's shield (page 146) for the North Woolwich tunnel, was to restrict the high-pressure work to the men actually at work in the face. In the Blackwall and the East River Tunnels, it was probably designed to fill a far more important function. A great source of trouble in driving a large tunnel is the fact that whilst the hydraulic head is greater at the invert than at the crown by an amount that depends on the depth of the tunnel, the air pressure which is designed to balance it is the same everywhere. The consequence is that there is either an excess of pressure at the crown or a deficiency at the invert. The former condition results in excessive escape of air, and may cause blows unless the face is kept carefully poled and plastered tight with clay, and the latter allows a flow of material into the tunnel from under the shield and may cause settlement of the shield and other troubles.

In addition to the difficulties already mentioned, the tunnel builders in New York have had a condition to contend with that did not occur in London, viz., the scarcity of skilled labourers. In London there was a class—it might be almost called a race—of men, consisting largely of Welsh and Cornish miners that could be safely trusted to run the shield with comparatively little supervision and to take care of the details of the excavation and timbering with very little guidance. A pair of plumb lines and a pair of sighting rods hung up in the tunnel near the face to give line and level respectively, graduated rods at the sides of the shields and a line at right angles to the axis of the tunnel, together with some general instructions given to the ganger by the engineer once a day, was all that was required as far as keeping line and grade was concerned, with the addition of a table of distances and corresponding offsets if the tunnel was going around a curve. A little more supervision to prevent recklessness in opening too much of the face, etc., was necessary if the tunnel was passing through gravel. But it has been necessary in New York to train men to their work, and to produce from the rawest material, more or less skilled tunnel builders. This has been a task of considerable difficulty, and until it was accomplished, progress was very slow. When it had been done, the men often formed unions, manifested a desire to undertake the complete management of the work, and were in many ways a source of anxiety.

It is impossible to over-estimate the value of compressed air for tunnelling through soft ground. It was first used, as before stated, in the construction of the old Hudson tunnel by Haskins in 1879, and as far as tunnelling by its aid is concerned, there has been little change in the method of its application. The locks that are used now are little different from the locks that were used in that tunnel. In his paper on the City & South London Railway, published in 1893, Mr. Greathead said that he had proved that it was practicable to remove the material from the path of the shield by mechanical means, or by a current of water, or by the two combined; and that in such cases, men might work under reduced air pressure, or even under the normal air pressure, at depths below which they could work at all under pressure. There is some evidence to support this in the recent history of the Hudson tunnel, reduced air pressures having been used in silt, but no such method has been applied to sand or gravel, and no tunnel has recently been driven even through silt without compressed air.

Whilst there have been many deaths and cases of paralysis, especially in the early days, and innumerable temporary illnesses caused by compressed air under certain conditions, and whilst there is no generally accepted explanation of the manner in which these effects are produced, there is no doubt, if certain arrangements are made and precautions taken that have been dictated by experience, that men can work with safety in pressures up to about 35 pounds per square inch. On the other hand, it is generally believed, by engineers at any rate, that it is at least undesirable to use greater pressures than these.

These arrangements and precautions have been frequently stated and they cannot be too strongly insisted upon. Obviously, it is necessary that the men should be physically fit and free from heart and lung affections. After this, probably the most necessary condition is that at every point in the tunnel where men are working there shall be efficient ventilation, that the air shall be kept pure and proved so by analysis; this condition is necesssary in all pressures, but a higher standard of purity should be insisted upon in higher pressures. In no case should the proportion of carbon dioxide exceed 1 part in 1,000. Regulation of the time of the "locking out," provision of arrangements for giving the men hot coffee on coming out of the tunnel, for providing them with dry clothing, provision of medical air locks and medical supervision of their use, are precautions the observance of which will tend to eliminate cases of sickness. On the part of the men, abstinence from the use of alcohol, the avoidance of exposure to chills, and the careful observance of the ordinary rules of health are necessary.

TRIPLE-EXPANSION DUPLEX OUTSIDE-PACKED PLUNGER MINE PUMP.
Cylinders 11, 17 and 30 in., plunger 10 in., stroke 36 in. Built by Jeanesville Iron Works Co., Hazleton, Pa.

THE UNWATERING OF MINES IN THE ANTHRA-CITE REGION.

By R. V. Norris.

THE anthracite region of Pennsylvania, though confined to a relatively very small area in six counties, ships annually over 60,000,000 tons of coal, a far larger mineral production probably than is produced from any similar area in the world. The early mining was largely by water-level drifts which took care of the drainage by gravity, this mining in most cases being continued to the surface, permitting the entrance of surface water through crop falls; these, in the portions of the region where the measures are at a considerable inclination, form an almost continuous line along the outcrops of the beds, and make a very efficient device for conducting practically all the surface water into the mines, and this with the natural mine water makes the drainage problem one of the first importance.

From the figures reported to the State mine inspectors for 1905, it appears that the quantity pumped in the region reached in that year the enormous total of 633,000,000 gallons per day—enough water to supply several large cities, and sufficient to make a river 50 feet wide and 4 feet deep, flowing at the rate of nearly 4 miles per hour; or to fill daily a lake a mile long, one quarter of a mile wide and nearly 20 feet in depth.

When it is considered that all this water must be pumped from depths ranging from a few hundred feet to nearly two thousand, the magnitude of the work is apparent; in fact, fully 8 per cent of all the coal mined is used for steam purposes at the collieries, and of this a large proportion goes to supplying the pumps. In some collieries as much as 40 tons of water are pumped for every ton of coal hoisted, and that from great depths, exceeding in many instances 1,000 feet, and in one colliery reaching 1,800 feet vertical. Where such water conditions exist, instead of a mere 8 to 10 per cent of the output being used for fuel, as high as 40 per cent of all the coal mined is needed for this purpose, so that the operators often wish that they might sell the water and run the coal to waste.

One of the large operating companies figures for the year 1906, on 3,500,000 tons of coal shipped, that they pumped or hoisted 35,-500,000 tons of water, an average of over 10 tons of water per ton of coal.

Modern mining methods would leave water-level chain pillars to catch the surface water and prevent its entrance into the deeper workings; but unfortunatly capital was rarely available in the middle of the last century to permit this self-evident precaution, and the water has generally been permitted to pass into the deeper workings, either through direct connections or through cracks and fissures in the strata caused by the subsidence of the worked-out areas. So badly has the surface been caved and fissured in many parts of the region that it has been necessary to build miles of wooden flumes to carry the streams and prevent them from flowing directly into the mines. Such flumes are frequently necessary for mine water, as without this precaution it would often be pumped out at great expense only to find its way back again into the mines through crop falls or fissures. Where possible, too, the outcrops of the seams are protected by ditches arranged to intercept the surface water and conduct it on safe ground to the drainage channels; often a proper arrangement of ditches and their careful maintainance will very materially reduce the quantity of water to be pumped.

In the Wyoming region a large area of the workings lies under the valleys of the Susquehanna and Lackawanna Rivers, and the rock is in many places overlain by water-bearing gravels and quicksands to depths exceeding 250 feet. This condition makes imperative great care in mining to avoid fractures of the rock strata overlying the coal beds; even exploratory drill holes in such territory should be cemented solid to the rock surface, as the water which will flow through even a 2-inch hole under 200 feet head is by no means a negligible quantity.

Even with the most thorough surface precautions there still remains a vast amount of water to be raised.

The earlier pumping plants were of the simplest description, no attempt being made at steam economy, which indeed was unnecessary. All the coal below chestnut size was wasted on the culm banks, and the only cost of steam was the labor of firing and removing the ashes. The pumps were practically all simple, non-condensing, direct-acting single pumps, and frequently of very large size—as 38-inch steam, 72-inch stroke, and 16-inch plunger; these large pumps could be run at a very high plunger speed, 200 feet per minute being not unusual, and even higher speeds up to 400 feet were occasionally attained. Owing to the great weight of the moving parts a card somewhat resembling that of a throttling engine is obtained, as shown below. Such pumps require approximately 50 to 65 pounds of steam per horse-power hour; they are however simple and reliable and raise enormous quantities of water. An improvement in these long-stroke pumps (patented by Jas. Delaney, master mechanic Mineral Railroad & Mining Co.) consists in by-passing, through a check valve, sufficient steam from the pressure side of the piston to cushion the moving

parts; this not only improves the steam econ- omy but per- mits much higher speed.

A serious ob- jection to the long-stroke sin- gle-acting pump

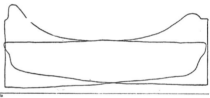

STEAM CARD FROM OLD STYLE LONG-STROKE PUMP.

is found in the excessive pressure in starting each stroke, shown by the steam card, this pressure being often twice and occasionally three or four times that due to the head. I have seen a 1,000-pound hydraulic gauge on the column pipe of a 600-foot lift pump broken by excessive pressure.

From the simple pumps the change was rapid to duplex. These, while not helping the steam economy (in fact often reducing it) have the advantage of giving a more regular flow and reducing the impact pressures in the column pipe. An objection to many duplex pumps is found in the fact that with the usual mine attendance they rarely make full stroke, resulting not only in a loss of water capacity but in a very serious decrease in economy. Owing to this the steam consumption of simple duplex pumps of good size is usually in the vicinity of 55 to 70 pounds of steam per horse-power hour under mine conditions.

Direct-acting multiple-expansion pumps have a great advantage as to steam economy over simple pumps; when it is remembered that in simple direct-acting pumps the steam must follow practically full stroke, it is evident that the economy in compounding such pumps, even with comparatively low-pressure steam, is very much greater than in engines, as the compound cylinders permit the expansive working of the steam besides reducing the cylinder condensation. The cylinder ratios of compound pumps in the anthracite region are usually for non-condensing about three to one, and for condensing four to one, and the steam economy ranges from 35 to 40 pounds per horse-power hour for non-condensing to 25 to 30 pounds for compound condensing pumps.

It would seem that with condensing water available, all mine pumps should be run condensing, but the effect of hot mine water

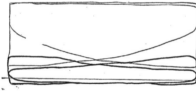

CARD FROM COMPOUND CONDENSING PUMP.

must always be considered. It must be remembered that practically all coal-mine water is acid, and that hot acid is much more active than cold; if the sumps are unduly heated the damage to the pumps is intensified, and a balance must be struck between steam economy with practically waste fuel and maintenance of pumps. The decision whether or not to condense must depend in each case on many factors, and the fact that any particular colliery is pumping non-condensing is by no means proof of poor engineering. Further, the deleterious effect on the roof and timbering of a mine of a hot, moist atmosphere must be considered, and this is often the governing factor in deciding on the type of pump.

The cost of pumping with an ordinary low-duty plant was determined for the years 1901 and 1902, at the Short Mountain colliery of the Pennsylvania Railroad* accurately and found to be for 1901. 567,113,616 gallons pumped an average lift of 1,141 feet. 4.95 cents per 1,000 gallons 1,000 feet vertical, or 0.6 cent per 1,000,000 foot

---

* Trans. A. I. M. E., February, 1903, Vol. XXXIV., page 127.

pounds of water; and for 1902, a flood year, 1,116,320,523 gallons pumped an average of 1,093 feet vertical, 3.9 cents per 1,000 gallons 1,000 feet vertical, or 0.47 cent per 1,000,000 foot pounds in water.

A few high-duty pumping plants have been installed in the mines. Notable and probably pre-eminent amongst these is the high-duty plant of the Hazleton shaft of the Lehigh Valley Coal Company. This consists of two compound-condensing flywheel pumping engines 32 and 60 by 48 inches, 13¾-inch plungers, which attained on test an average of 28.93 revolutions per minute, indicated 545.71 horse power, pumped each 5,141,247 gallons in 24 hours, and showed a duty of 103,805,456 foot pounds per 1,000 pounds steam. The delivery averaged 515.78 feet head, and the cost of pumping on an average of five years was 1.998 cents per 1,000 gallons, equal to 3½ cents per 1,000 gallons 1,000 feet vertical, showing that, with a very low cost of coal, steam economy is by no means the largest factor in pumping cost.

Single acting Cornish or "Bull" pumps of huge size were early used, and a very few still remain, as shown here. The steam cylinder is on the surface, connecting by wooden rods with the water plungers; the rods are raised by the steam, and the pumping stroke is made by the weight of the descending rods controlled and cushioned by throttling the exhaust steam

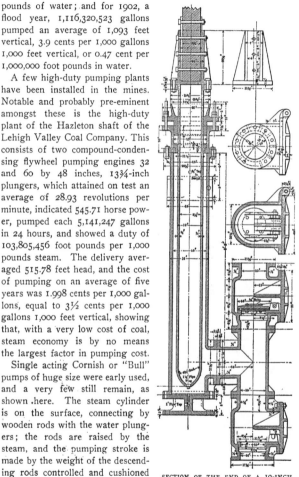

SECTION OF THE END OF A 10-INCH CORNISH PUMP.

CARDS FROM CORNISH OR "BULL" PUMP.

Above, from the lower end of cylinder, lifting stroke. Below, from return stroke, upper end of cylinder.

which is by-passed into the idle end of the cylinder to keep up its tempera-ture. Such pumps have the advantage of re-taining the steam end on the surface, and are not put out of business by flooding; the speed is very low owing to the weight of rods to be handled, about 50 feet useful plunger speed being near the maximum. One such pump tested by me in 1898 was 44 by 116 inch steam, 21½ inch water plungers, made a maximum of 4¾ double strokes per min-ute on 319 feet vertical lift, and showed a total steam consumption of 49.1 pounds per horse-power hour.

A TYPICAL SELF-CONTAINED COMPACT MINE PUMP.
It will work at almost any angle.

ANOTHER TYPICAL SINGLE-PISTON MINE PUMP.
Positive and certain in operation, under all conditions, on admission of steam. Epping-Carpenter Co., Pittsburg.

In many cases, as in sinking, it is important that the pumps be so designed as to require a minimum of attendance, and be as little as possible liable to damage from falling rock. In such cases short self-contained pumps are very largely used; these have the advantage of working in a very contracted space and not requiring to be placed perfectly level. In unwatering flooded slope workings where the pumps must be lowered down the pitch as the water recedes, it is of prime importance to select such types of pumps as will work in an inclined position; for this work pumps with a relatively low number of spring-closed valves are satisfactory, while those with a large number of guided metal valves closing by gravity are to be avoided; the valves are apt to cock sidewise and stick, causing constant losses of time in taking off the covers and freeing the valves.

In the work of unwatering it is of great advantage to keep counters on all pumps and have the revolutions of each pump reported daily by an independent observer. There are too many cases of attempted unwatering where the daily report "we are holding the water" seems to be considered satisfactory. There must for this work be ample pumping capacity and every pump must be made to count by being driven to its limit. As the pumps must be continuously moved as the water recedes, it is manifestly desirable to use pumps of moderate size and simple construction, and not to attempt steam economy.

COMPOUND CONDENSING MINE PUMP, PHOENIX PARK COLLIERY, P. & R. C. & I. CO.
Weight 135,000 lb.; capacity 2,500,000 gal. daily; 861 ft. lift. Steam cylinders 42 and 24 in., plunger 12 in., stroke 48 in. Note separate valve chambers and suction air column.
Goyne Steam Pump Co., Ashland, Pa.

The design of mine pumps involves considerations not found under ordinary conditions. The valve chambers should be so placed that the valves and seats are easily accessible for repairs and renewals; valve chambers should not be cast as integral parts of the working barrels; all packing should be readily accessible, and the metal of the water end should be very heavy to allow for considerable eating of the castings by acid water before renewals are necessary. Some duplex pumps are so arranged that either side may be run independently; such an arrangement may be of immense importance in case of accident, permitting of the repair of one side of a pump at a time without total stoppage of pumping.

All mine pumps should be provided with very ample air chambers both on the suction and on the column, the former so arranged as to take the suction water-hammer in a direct line from the suction pipes beyond the valves; suction air chambers placed between the suction and the valves are of very little service. Where height is limited, a Y in the column pipe close to the pump with a couple of pieces of column blanked off at the top on one branch makes a very satisfactory and efficient column air chamber. Usually the entrained air will be ample to keep the air chambers full, but it is well to arrange for means of accomplishing this mechanically when necessary, and to insist on the air chambers being kept constantly in working order. A "run around" arrangement of suction pipe on duplex pumps or a central suction pipe is also an advantage in reducing the suction impact and assisting in the rapid closing of the valves, thereby reducing slip. Where the mine water is very acid, the water ends of pumps require protection; this may be secured by using anti-acid bronze linings, or even bronze water ends, or by lead lining. Both of these methods are efficient but costly; a cheaper and equally satisfactory protection is found in wood-lining all parts exposed to water; to do this satisfactorily the water end must be designed for the purpose and completely lined, usually with dry pine wood driven into place and swelled with fresh water just before being put in use. Column pipes are also frequently wood-lined; this is best accomplished by making the lining of staves *sawed* to shape with a fine saw and the joints *not* planed; when swelled the rough sawn joints interlock and form a tighter joint than do planed strips. The wood lining should project slightly beyond the ends of the iron pipes so that the gaskets rest directly on the wood; of course if the acid water gets behind the wood lining its usefulness is gone, so that after swelling with fresh water the linings should never be allowed to dry out. Such linings carefully put in have protected for years the cast column pipes in collieries where the water is so acid as literally to eat all the iron work from a shovel left in the ditch for a single night.

TRIPLEX DOUBLE-ACTING ELECTRICALLY DRIVEN PLUNGER PUMP, AVONDALE COLLIERY
OF D. L. & W. R. R.
Capacity 800 gal. per min., 800 ft. lift. Jeanesville Iron Works.

Sumps for mine water are usually in the coal, generally in the form of a dip working; they should be at least of sufficient extent to hold one day maximum water, and are frequently much larger.

Important pumps should always be put in good pump houses, and too much weight cannot be put on the necessity of having the pump houses fireproof. Very many, if not most, mine fires start in timbered pump houses, where more or less grease is sure to be found; the best arrangement is either to wall and arch the pump houses with masonry and put in concrete floors, or in pitching seams to drive them in the rock and use exclusively steel and masonry supports for the roof. All waste should be kept in covered metal cans or metal closets, all oils carefully protected, and cleanliness enforced—not an easy contract. Open lights should never be allowed in pump houses or pumpways; where electric lights are not available closed lanterns should be absolutely insisted on. Further, all loafing or waiting in pump houses must be absolutely prohibited.

Pump columns should not be taken up the hoisting compartments in shafts, nor up working slopes, to avoid the double danger of injury to the pump column from accidents in hoisting and injury to the hoisting plant from breakages in the column.

TRIPLEX SINGLE-ACTING PUMP FOR 200 GALLONS PER MINUTE AGAINST 400 FFET
HEAD.
Driven by four-pole 35 horse-power motor. Automatically started and stopped by relay
switch connected to float. J. H. Holmes & Co., Newcastle-on-Tyne.

Occasionally it is practicable to place the pumping plant in an underlying or overlying seam from that mined, conducting the suction pipes through a rock tunnel so dammed that the water may raise to a considerable extent in the mine without flooding the pumps.

In some instances it is possible to drive drainage tunnels from adjacent valleys to the coal basins, placing a large portion of these above water-level and greatly reducing the head required for pumping from the lower levels. Notable amongst such tunnels are the "Centralia" tunnel, about 9,000 feet long, driven in 1889, draining a number of collieries near Ashland; the "Jeddo" tunnel, about 3½ miles

WORTHINGTON EIGHT-STAGE PUMP, 900 GALLONS PER MINUTE, 1,350 FEET HEAD.

long, driven by G. B. Markle & Co., to drain their collieries near Hazleton; the "Quakake" tunnel completed in 1906 by Coxe Bros. & Co., and the Lehigh Valley Coal Co., to drain collieries in the Beaver Meadow basin. All of these have been successful and have fully justified the expense of driving in the reduction of drainage costs. They have the further advantage of absolutely safeguarding the workings above the drainage level from flooding during cessations of work, such as the last two anthracite strikes, and making unnecessary the operation of the steam plants and pumps under armed guards at such times.

A very comprehensive drainage-tunnel proposition is now under construction by the Lehigh Coal & Navigation Company, to drain their collieries in the Panther Creek Basin. This company is driving a drainage tunnel from the Lehigh River near Mauch Chunk, starting at an elevation of 537 feet above sea level, and rising with a grade of 13 feet per mile to extend in tunnel and water-level gangways all the way to Tamaqua, over 12 miles away. The scheme when completed will involve about 24,000 feet of rock tunnel and 54,000 feet of water-

TWO-STAGE CENTRIFUGAL PUMP FOR WORKING AGAINST HEADS UP TO 200 FEET.
Dayton Hydraulic Machinery Co.

level gangway in the coal; the cost is estimated by Mr. Baird Snyder,
Jr., general superintendent, to be about $700,000 and the annual
saving to be in the vicinity of $150,000. Most of this company's col-
lieries will be placed above water level for a number of years, and the
pumping lift of the deeper ones reduced by about 400 feet vertical.

EIGHT-STAGE CENTRIFUGAL TURBINE PUMP INSTALLED IN A MINE.
For 300 gals. per minute. 1,400 ft. lift; 250 horse-power motor.   Buffalo Steam Pump Co.

While it is perfectly practicable to pump in single lifts to heights
of 1,000 feet or more with moderate sized pumps, the difficulties
with valves and column pipes in larger pumps are such that it is the
general practice in the anthracite region to confine single lifts for
pumps of 10-inch plunger and over to approximately 600 to 700
feet; a notable instance of this is in the new pumping plant of the
Lykens Valley Coal Company, where an old plant (described fully
in February, 1903, Trans. A. I. M. E. Vol. XXXIV, page 127) has
now been replaced by a modern compound condensing plant. The
total depth of about 1,650 feet vertical has been divided into three
lifts, the lower lift 665 feet vertical taken care of by two Jeanesville
compound condensing direct-acting duplex pumps, 38 inch and 25
inch steam, 10 inch plunger, 36 inch stroke, and the two upper lifts
of approximately 500 feet each, by eight similar pumps 16 inch and
36 inch steam 12 inch plunger, 48 inch stroke, wood-lined (four installed on each lift). The pump houses are all in rock in the
"Whites," a small vein overlying the Lykens Big Vein, and are fully

HIGH-LIFT TURBINE PUMP AND MOTOR.
For 700 gals. per minute at 750 ft. head. Three-phase motor of 375 horse power, 2,500 volts, 1,140 revolutions. Ernest Scott & Mountain, Gateshead-on-Tyne.

fireproof; the columns are also in the "Whites" vein and fully protected from danger from the other workings. Only a small portion of the water reaches the lowest level, and the pumps at this are protected from the main workings by dams so that it is possible to flood the lower mine workings to a depth of over 200 feet before putting these pumps out of business; this condition actually occurred during the last strike, when the Lykens workings were flooded to a very considerable depth above the level of the pumps, only such water as these could conveniently handle being admitted to them through valves in the dams. These lower pumps are notable for having worked some years ago for over a month under an average of 30 feet of water; the pumps were left running when the colliery was flooded and actually pumped themselves out.

While the practice of running pumps under water cannot be commended on economic principles, the simply constructed pumps

WORTHINGTON 4-STAGE TURBINE PUMP, 2,000 GALLONS PER MINUTE.

used in the region can generally be depended upon to run for considerable times under this condition, and that too without much injury to the mechanism. I am familiar with one small pump which has been under water most of the time for years on one of the Lehigh region collieries, and which could be started and stopped at will by merely opening the steam valve on a higher level.

In some instances it is necessary to run pumps by compressed air, usually in cases where the heat from steam pipes and from exhaust steam has a serious effect on the roof. The pumps used for this purpose are usually ordinary steam-type pumps, but a few have been built with special attention to clearance with a view to increasing the efficiency of the air. As the condition requiring air usually precludes the possibility of re-heating, the efficiency is low, probably in actual practice with the pumps as used rarely reaching 25 per cent; a large plant tested gave 21½ per cent efficiency between the indicated powers of the compressors and pumps.

Electrically driven reciprocating

TWO-STAGE VERTICAL-SHAFT CENTRIFUGAL PUMP.
For 150 ft. head; 1,440 revolutions. Buffalo Steam Pump Co.

FIVE-STAGE CENTRIFUGAL PUMP WITH CASING REMOVED.
Capacity 1,000 gals. per minute, 520 ft. lift. Jeanesville Iron Works, Hazleton, Pa.

pumps (pp. 167, 168) of relatively small size are largely used in collieries equipped with electrical plants. These are particularly convenient and efficient for dip workings below the general drainage level; and large centrifugal electrically driven pumps of the multiple-stage type (pp. 170-173) are coming into use; notable amongst these is the 10-inch six-stage electrically driven centrifugal pump, with a rated capacity of 5,000 gallons per minute, now successfully operating at the Hampton Shaft of the Delaware, Lackawanna & Western Railroad Company, in Scranton. Pumps of this type have not been in use long enough to determine fully their qualifications for the work, but

SIX-STAGE ELECTRICALLY DRIVEN WORTHINGTON CENTRIFUGAL PUMP, HAMPTON
WATER SHAFT OF D., L. & W. R. R.
Operating against a head of 485 ft.; capacity 5,000 gals. a minute. General Electric 3-phase
2,300-volt motor of 900 horse power. Photograph by courtesy of H. M. Warren.

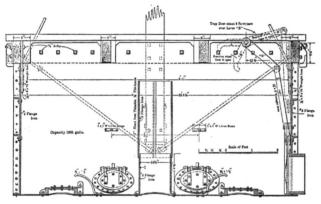

CAGE TANK, SUSQUEHANNA COAL COMPANY.

thus far are said to be giving excellent service; no accurate data as to their economical performance have yet become available.

Another largely used method of drainage consists in hoisting the water in vertical shafts. This was regularly done at the No. 1 shaft of the Susquehanna Coal Company (Pennsylvania Railroad), as early as 1880 by means of a tank (shown above) suspended under the regular hoisting cage; no other means of handling the water existing at this time. The use of a separate shaft with tanks especially designed for water hoisting was probably initiated at the Luke Fidler Colliery (Pennsylvania Railroad), Shamokin, in 1896, by Messrs. Irving A. Stearns, manager, and Morris Williams, superintendent of the Mineral Railroad & Mining Co. This practice is now very common for the deeper collieries; the early plants were described Trans. A. I. M. E., February 1903, Vol. XXXIV, page 106. In this paper the cost of hoisting was shown to be for three prominent water-hoisting plants, as follows:— ·

	PLANT.		
	LUKE FIDLER.	WM. PENN.	LYTLE COLL'Y.
Depth shaft	960 ft.	953 ft.	1,500 ft.
Record: gallons 24 hours.........	......	2,291,140	3,772,600
Capacity of tanks, gals..........	1,400	1,440	2,600
Cost per 1,000 gallons...........	$0.0306	$0.0234	$0.0219
Cost per 1,000 gallons 1,000 feet vertical	$0.032	$0.029	$0.028

The Fidler costs were from an average of three years' work, the

Lytle and William Penn collieries from unwatering after the flooding due to the 1902 strike.

The hoisting is done by very large first-motion engines, hoisting at very high speeds (page 176). The 1,500-foot hoist at the Lytle shaft has frequently been made at an average of 50 seconds per trip, including filling and dumping. The duty of the 36 by 60 inch first-motion water-hoisting engines at the Lytle shaft was found to be approximately 33,-250,000 foot pounds per 1,000 pounds of dry steam, and the cost of the installation, complete for 2,300,000 gallons daily capacity,

DISCHARGE OF A WATER-HOIST TANK.
L. C. & N. Company's No. 10 Colliery.
Tank capacity 3,000 gals. Engines,
first motion, 42 by 60.

including shaft sinking, machinery, and steam plant, was $80,777.96.

The tanks are now generally cylindrical, and are made either to dump (as shown on pp. 177-178) or to discharge from the bottom,

WATER HOIST OF THE HAMPTON SHAFT, D., L. & W. R. R.
Built by the Wellman-Seaver-Morgan Co.

AUTOMATIC ELECTRICALLY OPERATED WATER HOIST.
Built for the Hampton Shaft of the D., L. & W. R. R. by the Wellman-Seaver-Morgan Co.

FIRST MOTION WATER-HOIST ENGINES, LYTLE COAL CO., MINERSVILLE, PA.
Cylinders 36 by 60; cast coned drums 12 to 16 ft. diameter; two steam-actuated band brakes.
Shaft 1,500 ft. deep. Tanks 2,600 gals. capacity each; time of hoist 43 seconds.
Vulcan Iron Works, Wilkes-Barre.

as illustrated above in which case spout castings are necessary, and these of the double discharge type give the minimum of trouble.

A very interesting electrically operated water-hoisting plant has been installed at the Hampton Shaft of the Delaware, Lackawanna & Western Railroad Company; this plant, built by the Wellman-Seaver-Morgan Company, has been successfully operated for several years, but is now held in reserve, the water being pumped by the centrifugal pump above mentioned.

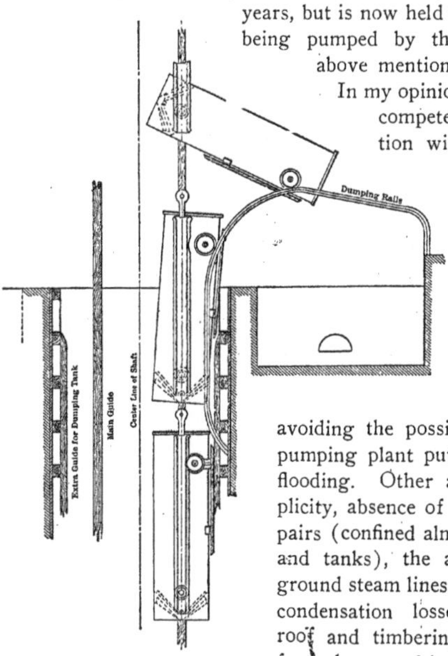

In my opinion hoisting plants can compete in cost of operation with modern pumping plants only at depths well above those practicable for pumping in single lifts; hoisting has, however, the inestimable advantage of having all the machinery on the surface and avoiding the possibility of having the pumping plant put out of business by flooding. Other advantages are simplicity, absence of slip, low cost of repairs (confined almost entirely to ropes and tanks), the avoidance of underground steam lines, with their attendant condensation losses and damage to roof and timbering, and the freedom from danger of interruption from falls or squeezes in the mines.

SHOWING GUIDES AND METHOD OF DUMPING.

It has been notable that where a water-hoisting has replaced a pumping plant the quantity of water previously figured on plunger displacement has invariably diminished to about 60 per cent of that previously reported. The difference is accounted for partly in slip, and partly because proper allowance is rarely made for delays and low speeds in mine pumps, which until recently were rarely provided with counters.

When conditions warrant the expenditure, the combination at the

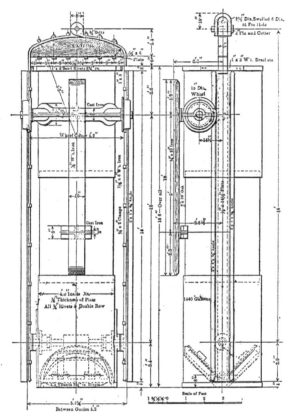

WATER HOIST DUMPING TANK OF THE WILLIAM PENN COLLIERY.

Delaware, Lackawanna & Western Railroad Co., Hampton Shaft, of pumps for regular work with a water hoist in reserve, would seem for low and moderate lifts to give a promise of the combination of maximum economy and absolute safety.

EDITORIAL COMMENT

SOME months ago, commenting in these columns upon the huge increase in the production of gold (which was then beginning to impress many observers as the most significant phenomenon of the times) we pointed out that it was attended by another, of possibly even higher potentiality—the gigantic demands on our finances imposed by the expansion of industry. We remarked that, while the world's store of primary money was being enlarged at a rate never before known, the production of other forms of wealth and the attendant demands for machinery, transportation, and tributary construction of every kind were advancing at a pace never before dreamed of.

Mr. Del Mar's noteworthy study in this issue demonstrates the conclusion that this second phenomenon has overtopped the first in magnitude and has really become the controlling influence in present economic conditions. And since his paper was in type, the impressive utterances of Sir Hugh Bell, president of the Iron and Steel Institute, have urged the same view, and it has been confirmed and repeated again and again by financiers at the National Convention of the American Bankers' Association, which has just closed.

The world's resources cannot be developed as rapidly as they are being produced, nor as largely as they are being demanded, simply for lack of money. An extraordinary situation to confront a whole great nation—a situation which may change the rate of our activities somewhat during the process of readjustment, but one in which "depression" is out of the question. Everything organic in our national life is sound; the functional disorder of a constricted currency system is within certain reach of cure.

THE confusion which befogs many students of phenomena of this class comes from economic nearsightedness. They exhaust themselves in survey and study of the meanderings of a small stream, when the problem involves the regimen of a whole watershed.

The larger point of view is ably taken in Dr. Douglas's analysis of the copper situation, which leads this number. He deals not with the flurries and dips of local and temporary oscillations, contenting himself with pointing out that the close margin between demand and supply makes such wide occasional fluctuations inevitable; but he draws the great curves of production and consumption so firmly that their trend and meaning are plain to all. And the result is wholly reassuring to those who depend upon the copper industry, either as buyers or as sellers. In brief, his article is not for the speculator; it is for the miner, the machinery builder, and the manufacturer. And by these it will be valued as its authority deserves.

* * *

THE greater the strain which passing conditions put upon the economic system, however, the larger the part which confidence has to play in maintaining equilibrium during the period of readjustment. At times, therefore, when the symptoms even of an ultimately wholesome change give rise to a certain general nervousness, it would be well for all who are active in public affairs to see that they move carefully. It is not necessary that offenders should go unpunished for fear their arrest might startle society, but it is well that the arrest should be made without wanton shouting and firing in the air. We shall never have a sound and safe civilization until corporate honesty and corporate, obedience to law have become a national habit; and it is to be feared that they will not become a habit until corporate dishonesty and disobedience have been severely punished. But the greatest effect and the greatest good will follow when the punishment descends with the greatest dignity and the least tumult.

And further, however upsetting to bystanders the clamorous castigation of offenders may be, the artificial hysterics of those who oppose all policing of our erring corporations are more harmful. We need sanity and straightforward justice on both sides.

* * *

It is quite possible that the anti-railroad legislation now so much indulged in finds popular favor for reasons more human and less moral than would first appear. No doubt abuse of franchise privilege and opportunity is a very serious civic offense, but its effect and gravity are borne in upon but few individuals. Secret rebates, however reprehensible, do not touch the mass of the people to the quick as they seem to have been touched. Something more than discontent with fares as they are popularizes the two-cents-a-mile legislation. It is quite possible, as hinted above, that another animus is at work, and that it is nothing nobler than revenge for the "petty tyrannies" of the transportation companies. Everyone who travels has tramped rebelliously the weary and devious distances imposed upon him by our badly designed terminals; he has fumed at the delays of slow-scheduled ferries—slow-scheduled, apparently, in order that they might be overcrowded; he has struggled, luggage-laden, through narrow and congested gateways where his ticket must be shown for the needless inconvenience of a preliminary punching, or, if not a traveler, he has been turned back from this same gate and denied the privilege of seeing even

an invalid or an aged relative safely down the long platform to the cars. He has been chivied hither and yon by employees in brass hats, more or less surly, but always inexorable in "enforcing their rules," at no matter what expense of annoyance or actual loss to the passenger. And he is sore. Plush and mahogany have not sufficed to soothe him. His sense of decency and justice has been trampled, and now that he sees what seems an opportunity, he is willing to trample in his turn. Vanderbilt's "the public be damned," like other curses, comes home to roost.

* * *

The Quebec bridge disaster, reviewed elsewhere in this issue, is appalling in its deliberateness. The Tay bridge went down in as sudden and utter ruin, but it had stood to its appointed work for years, and collapsed before a Titanic onslaught of Nature which had not been conceived possible by the mind of its builders. The Quebec bridge, before it had even attained the dimensions for which it was designed or endured its first service test, crumpled up on a summer afternoon.

The commission now investigating the wreck may well be trusted to answer finally and fully all the questionings of the engineering world. So far, the key to the physical aspects of the collapse seems to lie close to the now sadly well-known bottom chord A9L. It is noteworthy that Mr. A. W. Buel is disposed to attach much more importance than some other authorities do to the field injury and repair to this particular member. So far as we have seen, the fundamental re-study of the design has not yet been completed. And the thought will recur that this Quebec cantilever, while designed to be the longest in the world, had not yet exceeded the dimensions of the Forth bridge—and the Forth bridge is standing like a part of the natural Earth itself.

REVIEW OF
THE ENGINEERING PRESS.
"THE WORLD IS ITS FIELD"

THE QUEBEC BRIDGE DISASTER.

AN ANALYSIS OF THE AVAILABLE EVIDENCE AS TO THE CAUSES OF FAILURE OF THIS IMMENSE STRUCTURE.

Engineering News.

THE failure of the Quebec bridge may fairly be classed as the greatest disaster which has ever overtaken an engineering structure. The bridge was to be the largest of its kind in the world, but it was simple in design and all the problems involved were accurately determinate. All possible care was bestowed on its design by men who are leaders in the civil-engineering profession, and even the details of construction were carefully considered and specified before the work commenced. Both materials and workmanship were of the highest order of excellence and were subjected to the most rigid examination and inspection. It is evident, then, that none of the ordinary causes of bridge failure was responsible for this disaster, and that the real cause has to do with the fundamental principles of engineering construction, the theories as to the strength and behavior under stress of built-up members, which form the bases of bridge design. This is the ground taken by *Engineering News* in an editorial analysis of the available evidence as to the cause of failure of the Quebec bridge, of which the following is an abstract. The general details of the bridge and of the failure are too widely known to require recapitulation.

"It may seem to the reader that it should be an easy matter to promptly settle, beyond possibility of doubt, what caused the wreck by an examination made on the spot. The first sight of that gigantic heap of torn and twisted steel promptly negatives

such an idea. Only the most careful expert analysis can determine the hidden cause of the wreck. Elsewhere in this issue we have endeavored to record the chief superficially available facts regarding the disaster; such as are needed to enable an engineer to form in some degree his own opinion. Here we shall try to set down the deductions which the facts seem to permit. We make no pretense that the analysis is complete. Time has been insufficient and the data have been incomplete.

"At the time this is written, at Quebec, on the fourth day after the wreck, *the initial cause of the wreck appears to be the failure of some compression member in the anchor arm of the cantilever.*

"It is important to trace the analysis leading to this conclusion, an analysis independent of what we recount, in our descriptive article, about the curiously buckled ninth bottom chord of the west truss.

"(1) Neither the main pier nor the anchor pier show the slightest sign of settlement or failure. They are monumental examples of high-class masonry, and, except for a few coping stones displaced or broken by the falling superstructure, both piers are, to all appearance, absolutely uninjured. It is possible that the impact of the superincumbent mass may have loosened the bond of the cement in places, but this cannot yet be determined. But it is certain that the piers still stand in place, with the truss bars on the main pier and the anchorage in the anchor pier still there. The piers did not cause the fall.

"(2) *The initial failure was not in a tension member.* This is so important that we deem it well to show the proof in detail as follows:

"*a.* Had a tension member failed first it would have snapped with a loud, sharp report or a series of reports as successive eyebars in a panel parted, which would have impressed every eyewitness. All accounts agree that the very first yielding was silent. The first warning was when the men felt the floor sinking beneath them.

"*b.* Had a tension member failed, especially one of the top-chord members, the structure would have dropped instantly, like a falling body. The failure did not occur in this way. The hazy accounts given by eye-witnesses agree in the one fact that the collapse was rather gradual, at least in its first stages. A time of 10 to 15 seconds is indicated. The time-keeper, who was near the anchor end and had to run up-hill to safety, must have occupied at least this much before he turned around. This man, by the way, does not report jumping a gap at the anchorage, although a gap must have opened there at once if the anchor arm was the seat of the failure. But there are now hanging from the approaching span three long strings of track rail, about 75 ft. long, which pulled away from the rest of the track on the bridge; these doubtless carried their ties with them for a time and bridged a gap, if a gap did form. The absence of a gap, therefore, argues neither for nor against compression failure or anchor-arm failure. But the slowness of the fall points to a compression member as the one to give way. The gradual fall could not possibly occur in the case of sudden snapping of an eyebar, or, rather, a panel of eyebars.

"*c. The eyebars of the upper chords of the anchor arm are intact, unbroken and still joined in a continuous chain,* from the bottom of the anchor pier clear across the pile of wreckage, over the top of the main pier until they disappear in the waters of the St. Lawrence. In all that pile of torn and twisted and wrenched steel there is only one single broken eyebar thus far discovered, and that is but one bar out of 26 composing that particular top-chord member. It is true the top anchor chain fell on top of the pile of wreckage, and, naturally, suffered much less than the members at the bottom; yet there are places where eyebars received terrible punishment, were bent, twisted and distorted, and in spite of this, during a whole day's examination by two of the editors of this journal, not one eyebar, with the single exception noted, was found with even a crack across its edge. The importance of this fact to the art of bridge engineering we shall enlarge upon at another time. Its bearing on the question whether a tension or compression member caused the wreck is the point we would emphasize now.

"(3) The failure did not occur in a compression member of the river cantilever arm.

"If a strut had collapsed somewhere near the end of the river cantilever, the rest of the structure shoreward would have remained standing, or, at least, would have very slowly collapsed progressively and probably not far toward shore. All accounts agree with the appearance of the anchor-arm wreckage in showing that the failure did not occur in this way. There is, indeed, a bare possibility that a failing member on the river side of the main pier and near the pier might have buckled the towering main post over this pier and so caused the wreck of the structure; but everything indicates that this is not a correct record of the failure. As the wreck lies, the members, with the exception of some members of the bottom chord, have all been dragged over toward the river, due, it is believed, to the fall of the river arm as a whole, dragging the top chord and its attached members with it, when the failure in the anchor arm deprived the river arm of support. The main tower posts are bent and broken, it is true; and very badly broken. But this would have been caused both by the collapse of the web members attached to them on the shore side and by the riverward pull of the eyebar chain over their tops. The survivors agree, moreover, in the opinion that failure began in the anchor arm.

"(4) The failure was in a main truss member and not in the cross bracing.

"All the wind and sway bracing was in place and fully connected, and there was no wind of any account when the bridge fell. Further, the trusses fell quite generally in the plane of their original position. The evidence of witnesses, moreover, speaks of no sidewise swinging, but only of the downward motion..

"(5) The probabilities are against the failure beginning in a main post above the floor level. The time-keeper, who was facing the probable point of failure, had his first warning by feeling the floor yielding beneath him. A buckling upper post within a hundred feet or so of him would almost certainly have been seen.

"(6) No indication as to the probable point of failure is to be seen in the fact that the wreck of the anchor arm on the foreshore lies slightly to the east of its original position. The center of gravity may be 8 to 10 ft., more or less, east of its location when the span was in position on the piers. But when the great height of fall is considered, it is impossible to conclude anything but that the structure, or its anchor arm at least, went down in perfect verticality.

"Thus the probability is established, we think, that a compression member in the anchor arm, and that member not a post, but a section of the bottom chord, was the seat of initial failure.

"It may be thought that the next step, locating the actual responsible member, would be easy; that one need only go to the site, examine the wreck, and point to the weak link. The facts set forth in our descriptive article as to the buckling tendency noticed in the ninth left bottom-chord section may seem to the reader to put the matter practically on the basis of a certainty. This would be far from warrantable. Not what went before, but what happened at the time, is called for. This must rest wholly on the evidence of the wreckage. The task of constructing herefrom an explanation is not an easy one.

"But at present the explanation of greatest probability is the failure of this ninth left-hand bottom chord. This explanation rests on one most weighty fact: Of all the bottom-chord sections of the anchor arm (all of which have been fairly well traced) there is only one that exhibits character-

istic buckling distortion, and that is this ninth chord. All the others are bent and crushed, broken at the splices, cracked across, burst open, and most variously battered, but none has a well-defined buckle. The ninth of the left truss, however, is not merely buckled in indisputable manner, but it is doubly bent in a closely folded S-shape, and both its ends still lie practically in their original direction. Did this chord crush as a *result* of the fall? That event happened to all the posts, as our photographic views show strikingly, but not to the chords. The posts had the crushing endwise impact of the fall to withstand, but the chords only fell laterally. A chord member swung forward against the pier masonry might well be buckled. On the left side a tendency in this direction is indeed observable in the eighth chord. But the ninth chord lies far from the pier, in place, as it were, and can hardly be conceived to have returned after striking the stonework; especially is this obvious as chord 10 lies not far out of place as compared with its original position relative to the ninth, being on the pier side of chord 9.

"We believe that the most thorough study of all that relates to the present and past state of chord A 9 L must be among the first things to engage the attention of those charged with the investigation of the disaster.

"We have already alluded above to the lack of absolute knowledge as to the strength of steel columns of enormous size, such as were of necessity used here. These compression members were designed for a unit stress under full dead, live and wind loads of 24,000 lbs. per sq. in., about two-thirds of the elastic limit of the metal. They were carrying at the time of failure not more than two-thirds of this amount. Were these compression members able to safely carry this stress? Were the plates and angles of which they were built up so thoroughly braced and connected together as to make the whole member act as a unit? Their design was made and approved by the ablest engineers in the profession. No one has dreamed of doubting their strength; but now, with the testimony of that gigantic collapse, every engineer must long to know, by absolute trial, what such columns can safely bear.

"It is at exactly this point that the Quebec Bridge failure becomes of importance to the whole engineering profession. Until the cause is absolutely determined—if indeed it can ever be—or until the profession has actual results of tests of huge columns at its command, a cloud of doubt rests upon us as to the margin of safety in every great bridge structure; at any rate when the unit stresses are forced up to the point deemed safe by the designers of this bridge."

THE ICE PROBLEM IN ENGINEERING WORK.

A DISCUSSION OF THE IMPORTANCE OF THE ICE PROBLEM IN HYDRAULIC POWER DEVELOPMENT IN NORTHERN LATITUDES.

Howard T. Barnes—British Association for the Advancement of Science.

THE development of water powers is greatly complicated in northern latitudes by the ice problem, an element not considered in Mr. von Schon's recent series of articles in THE ENGINEERING MAGAZINE but which is occasionally of fully as great importance and worthy of as careful study and investigation as the variations in flow due to floods and drought. An interesting paper on the ice problem, referring especially to Canadian rivers, was presented by Dr. Howard T. Barnes before the recent meeting of the British Association for the Advancement of Science. Dr. Barnes is widely known as an authority on the formation of ice, his researches carried out in the St. Lawrence being of a most elaborate and careful character and extending over a long period of years.

"During the severe Canadian winter there is excellent opportunity for the physicist to study, on a grand scale, the operation of the natural laws governing the formation of ice in the many forms with which it is met in the large and often turbulent rivers. To the engineer the problem is more serious, for the development of the vast water-powers of the country must include means of combating the ice troubles which arise each winter. What presents itself during the summer months for consideration is small compared to what must be met during the winter months, when ice is forming rapidly, and ice-bridges, dams, and shoves, may change the whole character of the levels and channels in a single night. Rivers are thus known to have been turned entirely out of their course into new channels during a winter of unusual severity; and in some instances the reversal of a rapid is of yearly occurrence. No one set of conditions may be said to hold from year to year,

on account of the variation in the severity of the winters. Therefore, before an engineering scheme is carried out, a careful study is usually made of neighbouring conditions, previous summer and winter levels, and deductions made from a consideration of native traditions over an extended region round about.

"In general, three kinds of ice are distinguished, and present characteristics brought about by their method of production. Surface or sheet ice forms over the surface of quiet lakes or rivers, and is desirable or not, depending on the particular conditions. Spicular-ice, or, as it is called in Canada, frazil-ice, is formed by surface agitation in the more turbulent rivers and in water-falls, and accumulates in great quantities in the quieter waters. It is this form of ice which gives the most trouble in hydraulic work. It varies in size from thin plates to fine needle crystals, depending on the degree of agitation of the water. Anchor ice, or ground ice, is the most interesting, from the fact that it grows along the bed of a river not protected by surface ice, and often causes considerable inconvenience when it rises in great masses, carrying up with it boulders and stones of considerable size. Anchor ice is formed, in the first instance, by radiation of heat during a cold clear night, but increases to great depths by entangling and freezing large quantities of the frazil ice carried down by the shifting currents.

"A study of the temperature conditions in the water during the production of these forms of ice shows that this is accompanied by a small temperature depression in the water amounting to a few thousandths of a degree Centigrade. During the severe weather the water is thus thrown into a slightly supercooled state, during which time the ice crystals are growing rapidly by continued freezing, and giving rise to the agglomerating stage when they stick together into lumps and spongy masses, and adhere to the racks or to the machinery of the wheel-gates or turbines. So firmly does the ice freeze that it will interfere in a short time with the operation of the machinery, and cause a temporary cessation of operations. The rack-bars frequently become clogged with ice, and cut off entirely the supply of water. Fortunately, it is only a small temperature depression which brings about these conditions, and methods of artificial heat, applied about the affected spots, are found effectively to relieve the situation. The sun is the most powerful agent in preventing ice troubles, since the absorption of the heat rays prevents the state of supercooling. At night, however, it has been found most important to have available a system of steam injection, or electric heating, which can be readily applied about the machinery, in order to prevent it from becoming supercooled. It is not found necessary to warm the entire volume of water passing through, which would be very costly and difficult; but by applying the heat in the racks or wheel-cases, or blowing steam about the affected parts, the ice is prevented from gaining a foothold. The ice is as effective as so much water in producing a head; hence the necessity of passing it through, and not allowing it to freeze to the metal surfaces of the machinery.

"In places where the steam-injection system is installed, no trouble is experienced, even in the most severe weather; thus completely demonstrating the feasibility of coping with a situation which, for many years, has been regarded as involving inevitable interruption to the continuous operation of the plant.

"The hanging dams of frazil-ice accumulate in great quantity under the surface ice, and tend to produce a uniform flow of water throughout the spaces under the ice. Thus, in the quieter spots, larger masses accumulate. In this way the under surface of the masses is found to follow more or less completely the contour of the bed of the river. In some parts of the St. Lawrence, thicknesses 80 ft. deep have been

measured by a sounding-rod let down through the spongy accumulations through an opening in the surface ice.

"Anchor ice grows in large quantities during the severe weather, not only by radiation, but by the general adhesive properties of the frazil-ice during the time in which the water is in a super-cooled condition. Open portions of the St. Lawrence, are observed to rise, during cold weather, several feet, owing to the accumulation of ice on the bottom. This accumulation is not without benefit to power-users, while it lasts, for it is probably the most effective agent in clearing the water from frazil-ice that is known. Shooting up in long needle crystals through the water, taking arborescent forms, it attaches and filters a great deal of fine floating ice swept down by currents. It is a matter of comment amongst power-users that there is less fear from ice-troubles during prolonged cold weather than when the weather is intermittently warm and cold, a condition which keeps the bottom fairly clear of anchor ice.

"Once the ice gains a foothold on the metal surface of the machinery, it does not take long for it to accumulate, and cause a serious falling-off, or a complete stoppage, of the wheels. It has now been found possible to prevent this by passing a steam-pipe into the wheel-case, and blowing steam into the water whenever the ice shows a sign of sticking. It is not often that this is necessary, but chiefly at night or when the sun is not shining. A turbine, losing capacity, is found to pick up its load within five minutes of turning on the steam, and occasional injections of steam during a severely cold night are found to have beneficial results.

"Many of the Canadian rivers, such as the St. Lawrence, are subject to winter floods, occasioned by the accumulation of frazil and disintegrated anchor ice. Wherever a stretch of open water intervenes, an immense quantity of frazil ice is produced, and is carried down with the currents far under the surface ice covering the quieter portions. Large masses of ice are thus produced, which frequently so reduce the available water-way as to cause the waters to rise until sufficient head is produced to deliver an effective blow at the obstruction, and clear a passage-way.

"There is evidence to show, by the character of old river-channels, that some rivers have been completely turned from their courses to find new ones. That this might be accomplished during a winter of great severity, by an ice-bridge or jam, seems perfectly evident, when the tremendous upheavals are witnessed which take place at various points on a river like the St. Lawrence.

"All these conditions must be considered in the location of a power company, and great care must be exercised. When works are supplied by water from a river completely frozen over, little trouble is ever experienced. The most effective prevention to the formation of both frazil and anchor ice is the protection afforded by a surface sheet. When a power-house is located at the foot of rapids, or at the head of a rapid with open water above, means are taken to construct a head-race of sufficient magnitude to serve as a settling basin for the ice drawn in. Much of the ice is deflected at the head of such a channel by the construction of booms or crib-work. Even in this case a large staff of men may have to be employed to cut channels through the ice of the head-race to allow of sufficient water for the turbines.

"In cases where a channel to a power-house is fed from rapids, the growth of surface ice over the channel is often a disadvantage, and artificial means are employed to keep the channel open. The frazil-ice which passes under the booms is passed along as quickly as possible, and handled by artificial heat at the wheel-house.

"So many and varied are the problems to be met, not only from the varying conditions of locality, but also from the variation in the severity of the winters in any one locality, that it is only by a general study of the laws governing the formation of ice that means may be found to cope with the situation. It may safely be said, however, that the ice problem in Canada is no bar to the future development of her vast water-powers."

THE PROPULSION OF SHIPS.

A DISCUSSION OF THE UNSOLVED PROBLEMS OF SHIP PROPULSION AND THE PRESENT POSITION OF THE MARINE STEAM TURBINE AND INTERNAL-COMBUSTION ENGINE.

Francis Elgar—Institution of Civil Engineers.

FOR some years the James Forrest lectures before the Institution of Civil Engineers have discussed the unsolved problems of engineering, drawing attention to the direction in which the aid of the physicist is more immediately required by the engineer. This year's lecturer, Mr. Francis Elgar, following the established custom, presented an elaborate paper on "Unsolved Problems in the Design and Propulsion of Ships," with particular consideration of "the direction future progress may be expected to follow in the production of even finer and faster liners than those which now traverse the ocean for trading and commercial purposes." The greater part of Mr. Elgar's paper relates to the architectural side of ship design. The following extracts are taken from the section dealing with the mechanical problems of propulsion.

"The greatest problem of all with regard to the propulsion of ships is the form which propelling machinery is likely to take in the immediate future. Already an important change is in progress from the ordinary reciprocating marine engine to the steam-turbine; and the question is not only how far that change will extend, but also whether the whole of the cumbrous apparatus required for producing steam may not before very long be swept out of mercantile steamers, and the power be obtained from some form of internal-combustion engine. It would be rash to attempt to prophesy what will happen, but a short reference to what appears to be the present position with regard to these fundamental questions might be expected.

"The progress of the steam-turbine is remarkable. It has often been described by Mr. C. A. Parsons and others, and is pretty well known. The reasons why its progress has not been greater, and why it is not already more generally employed in ships of all classes, are not so well known, and it might be useful to consider them. The Parsons steam-turbine has practically superseded the reciprocating engine in the battleships, cruisers, and smaller very fast craft of our Navy; and it has had equal success in the very important class of Channel steamers and other boats of similar type. Turbine-steamers now employed on cross-Channel services are running at speeds that could not be reached with the best reciprocating engines. They are strictly limited in many cases to a very shallow draught of water; they carry very little dead-weight of coal and cargo; the weight of their machinery and boilers constitutes a large proportion of their gross weight, and the percentage of the latter that is saved by the use of fast-running turbines is sufficient to give a substantial advantage in speed. The coal consumption compares favourably with that of the best reciprocating engines of equal power that could be fitted in that type of ship; and still better, of course, with the paddle-engines or screw-propeller engines of old type which were in the boats they superseded.

"When these facts are stated, as they often have been, surprise is expressed by many that such apparent advantages are not secured more extensively for other classes of vessels. Very few ocean steamers have been fitted with turbine machinery or are being so fitted; and although this may not cause surprise in the case of cargo-boats and other vessels of low or even moderate speeds, it appears strange that liners of high speeds are still fitted with reciprocating engines, and that the bold lead given by the Cunard Company with their two fastest new boats and the Carmania should not be generally followed.

"The chief reason for hesitation to put turbine machinery into ocean liners is the doubt which exists as to coal consumption. The amounts at stake are so large in these costly vessels when experiments with novel propelling machinery are tried that everybody prefers to see some one else make them. The Cunard Company are making the crucial experiment upon the largest scale that is now possible, and every one interested in progress must wish those responsible for it all the success they hope for and deserve. But the result is to some

extent uncertain, and the immediate future of the turbine in fast liners depends greatly upon the issue.

"In warships the consumption of coal has been brought down to 1.7 lb. per equivalent indicated horse-power of reciprocating engines per hour, and to about the same figure in mercantile boats of cross-Channel type. That is as good as can be obtained with reciprocating engines in the same classes of vessels, as weight has to be kept down as much as possible in these by shortening the stroke, and using high mean pressures of steam in the cylinders, in order to get all the power that is practicable out of a moderate size and weight of machinery. It pays better, in these cases, to stop somewhat short of the maximum efficiency attainable than to carry the additional weight that the increase would involve. In ocean liners the conditions are different, and economy of consumption is the chief point aimed at. Their consumption with quadruple-expansion engines and a boiler pressure of 210 lb. to 220 lb. has been brought down to 1.3 lb. of coal per indicated horse-power per hour. The substitution of turbines for reciprocating engines in ocean vessels depends chiefly upon whether the consumption with turbines can be brought down to this figure, and there is no satisfactory evidence that this is now practicable. It is probable that the marine turbine will ultimately be so improved as to beat the best reciprocating engines in economy of consumption in ocean liners, but no proof is forthcoming that it can yet do it.

"In ocean liners the turbine has to compete with the reciprocating engine at its best, and the development of high efficiency with the turbine is counteracted by loss of propeller efficiency as speed of turning is increased. The problem is to secure a combination of turbine and propeller such as will give an efficient speed of turbine without reducing unduly the diameter of propeller, and experience with regard to this is much needed. The smaller propellers of turbine engines doubtless gain in efficiency on service by being immersed more deeply, and being thus less affected by the rising and falling of the sea at the stern. This is especially the case with some of the cross-Channel boats, in which the propellers have to be so large relatively to the shallow draught of vessel with reciprocating engines, even when these are run at the greatest possible number of revolutions per minute, that they are sensibly affected by the least disturbance of water-level. The difference in propeller efficiency soon becomes very marked as the weather gets bad. In the Newhaven and Dieppe service, for instance, it is found that whereas the turbine boat Brighton only gains an average of three minutes upon the time of the reciprocating-engine boat Arundel in fine weather, she gains fifteen minutes in anything like bad weather. The difference in favor of the turbine on account of extra immersion of propeller is not likely to be so great as that in ocean liners, because in them the immersion of propeller is not relatively so unfavourable with reciprocating engines, but it would doubtless be of appreciable value. There appears to be some other cause, however, which operates unfavourably in bad weather at sea, especially when running against a strong head wind, to turbine vessels with comparatively small propellers. In such circumstances they fall off in speed more than reciprocating-engine boats with their large propellers would do; and, altogether, notwithstanding the greater depth of immersion, the efficiency of the propellers appears to diminish more rapidly as they are reduced in diameter, when driving the vessel against a head wind and sea. . . . . . .

"The question of some form of internal-combustion marine engine which would be suitable for large ocean vessels is still about where it was when Mr. Milton's paper was read and discussed here last January, and I do not feel that I could now add anything with advantage to that paper and discussion. I will merely enumerate the conditions, most of which were mentioned by Mr. Milton, that must be satisfied by a successful marine engine of any type whatever. . . . .

1. The engine must be reversible.

2. It must be capable of being stopped quickly, and of being started quickly, either ahead or astern.

3. It must be capable of being promptly speeded to any desired number of revolutions between dead slow and full speed, and of being kept steadily at the required

speed for any length of time. 'Dead slow' ought not to be faster than one-quarter of full speed, and should be less in very fast vessels. . . .

4. It must be capable of running continuously for long distances, with but short intervals between the runs, without risk of stoppage or breakdown. . . .

5. It must be capable of working well, not only in smooth water, but also in heavy weather, in a seaway in which the varying immersion of the propeller causes rapidly-changing conditions of resistance.

6. All working parts must be readily accessible for overhauling, and all wearing surfaces must be capable of being promptly and easily adjusted.

7. The engine must be economical in fuel, especially at its ordinary working speed.

8. It must be compact, light in weight, and well balanced, so as not to cause vibration.

9. It must not involve any risk of accumulation of gas in the ship such as could form an explosive mixture.

10. It is a *sine quâ non* that it must be capable of using a fuel whose supply at moderate price is practically unlimited, and that could be readily obtained in whatever part of the world a ship might happen to be.

"We have since heard of gas and oil machinery for 16,000-horse-power battleships, but this only exists at present in imagination. It is impossible for anyone to judge by what has been achieved up to the present in this direction what weight or space, or what consumption of fuel, would be required for the internal-combustion engines of great power that might, perhaps, ultimately be made to fulfil the onerous requirements of marine work. Engineers and metallurgists may, by working together, succeed some day in overcoming the difficulties of producing large cylinders which will stand the high impulses and great and rapid variations of temperature that occur with internal combustion; but till that is accomplished no great step ahead can be taken. There are no two opinions, however, as to the advantages that might be gained by doing away with the present boilers and their appurtenances, and abolishing with them that very arduous and disagreeable class of labour known as 'marine stoking.'

"The subject of oil-fuel for marine boilers is an interesting one, but I have no time to say more about it than that great practical advance has been made in its use during the last decade, and a consumption as low as 0.9 lb. per indicated horse-power per hour has been regularly realised in mercantile vessels which employ the system of spraying the liquid for combustion by means of hot air. American steamships have used oil-fuel largely during the last three years, under a combined system of high and low-pressure air respectively for desiccating or pulverising the oil before combustion, and for assisting the combustion afterwards. This system has proved highly successful and economical. Vessels of 14,000 tons displacement belonging to the Shell Transport Company have made voyages regularly and successfully from Singapore to this country by the long route of the Cape of Good Hope, and still larger vessels have made equally successful voyages from New York to San Francisco *via* Cape Horn, the use of oil fuel in both cases giving a high economy."

SPEED TRIALS OF THE LUSITANIA.

THE RESULTS OBTAINED ON THE OFFICIAL TRIALS OF THE NEW TURBINE STEAMER OF THE CUNARD LINE.

Engineering.

THE Lusitania represents the greatest step that has ever been taken either in size or power in a merchant ship, apart altogether from the adoption on a great scale of the turbine system of steam propulsion. The remarkably successful outcome of the official trials is, therefore, a matter of great interest and importance, no less to the marine-engineering profession than to the Cunard Company whose courage in making such a tremendous experiment merited the fullest measure of success. The following record of the official speed trials is taken from recent numbers of *Engineering.*

"The Cunard liner Lusitania has, in her official trials, carried out under the direction of the technical staff of the Cunard Company and of the Admiralty representatives, met the most sanguine anticipations of all concerned. At a draught of 30 ft., she has steamed over 26 knots on the measured mile; on a 48 hours' sea run on long measured distances she has maintained a mean speed of 25.4 knots. The contract anticipated a speed of 24½ knots on the round voyage on the Atlantic, and this will be easily achieved.

"Justification for this view is found in the fact that the long-distance trial represented exactly the conditions of the Atlantic voyage. The unprecedented length of the trial precluded 'jockeying.' The course of about 300 miles was traversed four times in alternate directions, so as to eliminate the influence of tide and weather. And thus any speed maintained on such a trial may be continued indefinitely, so long as coal and other supplies are available. It is unnecessary to say that the machinery worked satisfactorily. The general result stated carries conviction from this point of view. Before entering upon this, the most crucial test, at midnight on Monday, July 28, the Lusitania had made several preliminary trials on the Clyde measured mile, not only to tune up the turbine machinery, but to standardise the relation between revolutions, power, and speed, so that a series of trials could be made to determine the coal consumption at various speeds. These economy tests began on Saturday, when, with a large company of guests of the Cunard Company and of Messrs. John Brown and Co., Limited, the owners and builders respectively, the Lusitania left the Clyde for a cruise around Ireland. The vessel was loaded to a draught of 32 ft. 9 in., equal to a displacement of 37,000 tons, and on the cruise the water and coal consumption were taken while the vessel ran for six hours at speeds of 15, 18 and 21 knots respectively. The results were thoroughly satisfactory, but the data obtained were in connection with service requirements rather than scientific purposes; the results on the Atlantic will be more valuable.

"The guests were transferred to the tender at the Mersey Bar on Monday, and thereafter the more exacting tests were entered upon, water and coal consumption data at 23 and 25 knots being taken. On the run to the Firth of Clyde, the starting-point of the full sea-speed trials, the course measured out on the chart was between the Corsewall Light on the Wigtownshire Coast to the Longship Lighthouse at Lands End, and this had to be traversed four times, alternately north and south. The compass bearings gave the distance, which aggregated about 1,200 miles; the trial began at midnight on Monday, and ended about 1 o'clock on Thursday morning. The weather was favourable, with cloudless days and starlight nights; but on both nights north-west winds freshened to forces of six and eight, and although this occurred when the vessel was steaming north, and somewhat increased resistance and slightly reduced speed, it brought consolation in the fact that it prevented fog. The feature of the trial was the uniformity of the speed on both runs south and on the two runs north, the latter being against the wind and tide. The course, as a glance at the chart would show, was divided into three approximately equal parts by the Fodling and Tuskar lights. Compass-bearings taken at these intermediate points proved the uniform rate of steaming. The time taken on the runs south, on Tuesday and on Wednesday, differed by only two minutes; further proof is unnecessary of the great regularity of steam-supply or of turbine efficiency. The speed on four runs was: South, from Corsewall, 26.4 knots; north, from Longship, 24.3 knots; south, from Corsewall, 26.3 knots, north, from Longship, 24.6 knots; mean speed, 25.4 knots. We omit second place decimals, but, in any case, the percentage of error in observation is, with such distance, negligible.

"This is a great performance: it exceeds by two nautical miles per hour any similarly long run made. The truest significance lies in the uninterrupted mechanical precision with which every unit of the machinery worked. The air pressure in the ashpits of the boilers did not at any time reach the maximum of ¾ inch prescribed in the specification by the Cunard

Company. The boiler pressure averaged 186 pounds per square inch, while the pressure at the receiver of the high-pressure turbines varied little from 150 pounds; at the low-pressure receiver it was 3½ pounds. The mean vacuum was 28.2 inches, with an average barometric reading of 29.8 inches. The mean revolution of the four shafts were 188 per minute, and the power, according to the torsionmeter, was 64,600 horse-power. To those not versed in the details of steam-turbine performances the fact is illuminative. The circumferential or tip velocity of the rotors of the low-pressure turbines was 150 feet per second, equal to over 9,000 feet per minute. The general procedure in the machinery department accorded with Atlantic practice, and Tuesday's and Wednesday's performance might to all intents and purposes have been two days running, each equal to over 600 miles, on a voyage to New York. This will certainly be the condition a month or six weeks hence. On returning to the Clyde on Thursday, the vessel proceeded on shorter distance tests between Corsewall and Chicken Rock, the southern extremity of the Isle of Man, and between the Holy Isle and Ailsa Craig, in the Firth, while following this were progressive runs on the measured mile at Skelmorlie, also on the Clyde; these ranged up to 26 knots, as already indicated in our introductory sentence.

"The steering qualities of the vessel have also been tested. When steaming at 15 knots the rudder was put from amidships to hard over, both to port and starboard, in 15 seconds, and the full circle was completed in 8 minutes. Immediately before commencing to turn, the engines were running at the rate of revolution which gave 15 knots. A careful record of revolutions was made on a time basis during the evolution, and it was found at the completion of the circle that the rate of revolution was then 70 per cent. of the rate at 15 knots. The final speed was thus assumed as 10.5 knots, the average speed 13 knots, and the diameter of the circle about 1,100 yards. This for a ship of this great length is a most satisfactory performance; the ship, at 22 knots, made the complete circle in 7½ minutes, with 15 degrees of helm. In ordinary steering the ship answered her helm very rapidly, according to the testimony of the pilot, and her swing was easily checked. Although the weather was very fine, alike on the 36 hours' run around Ireland and on the 48 hours' trial on the deep-sea course, there was sufficient swell on the Atlantic in the first-named trip, and between the Tuskar and the Longship Light on the subsequent runs, to cause pitching and rolling motions to be perceptible and to afford opportunity for repeated records. In respect to these points a large number of observations gave the period of a single roll from side to side as almost exactly 10 seconds; a single pitch occupied 4 seconds. This latter result calls for no remark and the slow rolling indicates that, while the vessel has a satisfactory measure of stability, the long period of roll assures us that there will be the minimum of discomfort to passengers through this cause. . . . . .

"On Thursday the vessel, at slightly less draught, proceeded on a double run between Corsewall Light and the Chicken Rock, at the extreme south of the Isle of Man. The distance between these points is 59 odd miles, and the vessel on the southern run averaged 26.7 knots, and on the northern run 26.2 knots, giving a mean speed of 26.45 knots. This, it must be admitted, is a particularly fine performance, surpassing even the best record made on the measured-mile trials. Following upon this trial further speed tests were made. There were six runs, alternately north and south, between the Holy Isle, on the east coast of Arran, and Ailsa Craig Light; and here again most satisfactory results were achieved. It is not, however, necessary to enter into details regarding these, because they simply bear out the results previously obtained. Several runs were made to determine the speed for given revolutions or power at various draughts; but unless most complete data were published, general results might be misleading. The main point is that the contract speed has been far exceeded, and that when the vessel enters on the Atlantic service on September 7, she is certain to meet every expectation formed by the owners and builders, and, one might also say, by the general public."

ELECTRIC TRACTION IN ITALY.

A DISCUSSION OF THE ECONOMIC IMPORTANCE OF ELECTRIC TRACTION IN ITALY AND THE RESULTS ALREADY ATTAINED.

Philippe Tajani—Revue Economique Internationale.

ELECTRIC traction for trunk line operation has been developed to a greater extent in Italy than in any other country. In America, for instance, where coal is still abundant, electric traction on railways has been applied, so far, only to short suburban lines where the conditions of traffic require frequent and fast train service. In Italy, however, the lack of coal and the abundance of water powers have made the substitution of electric for steam power on railways a matter of national economic importance and electric traction has been developed with a view to the eventual operation of all the railroads of the country by electric power. A recent article by M. Philippe Tajani in the *Revue Economique Internationale* gives an interesting account of the economic considerations which have determined the lines of development of electric traction and the encouraging results which have been attained.

As soon as the possibilities of electric traction were recognized in Italy, the government took steps to conserve all the more important water powers of the country for national uses. Italy is absolutely dependent on other countries for her coal supplies and besides the large drain upon the resources of the country caused by the large purchases of coal for railway purposes, the condition was most unsatisfactory from a military standpoint and also seriously interfered with the internal traffic movements of the country. It was recognized, therefore, that the introduction of electric traction on all the railroads was a matter of the greatest importance economically, in that it would develop the national resources and remove to a large extent the necessity for importing large amounts of coal. The problems of electrification were approached, therefore, with a view to the ultimate operation of all lines by electric power and it was kept in mind that while a reduction of operating expenses was to be desired, it was not the primary consideration. The national aspect of the change from steam to electricity was never lost sight of in the preliminary investigations and in such developments as have taken place, and this fact is to a large extent responsible for the leading position Italy now holds in electric trunk-line operation.

The preliminary investigations of the problem occupied a long period of time and the most important result was the adoption of the three-phase, high-tension alternating current for the transmission of power. They further showed that no radical changes could be made in the actual methods of exploitation of the railroads with the change from steam to electric power. The Italian engineers had begun their work on the assumption that the change would mean the substitution of light, fast trains for the heavier and slower ones used with steam traction, but they were forced to abandon this idea, both on account of the requirements of traffic and a consideration of operating costs. It was seen also that separate locomotives would give better results than would be obtained by placing the motors in the cars.

The only line on which a complete substitution has been made is on the so-called Valentina line, 105 kilometres long, which

has been widely described in the technical press. M. Tajani gives a description of the locomotives, etc., which we omit as being already well known. His discussion of the economic effects of the change, however, brings out some interesting facts, which apparently show that electric traction cannot be expected to make such large reductions in fixed charges and operating expenses as some of its advocates claim. M. Tajani points out, however, that a slight saving has been made and that the results obtained on this experimental line are a satisfactory evidence that electric traction can be installed, in the course of time, on all important Italian railways without adding to the costs of operation.

Concluding, M. Tajani summarizes the principles on which further extensions of the system will be based. "Our programme does not anticipate any astonishing results: we are satisfied if we can operate our railroads by means of our own natural resources at no greater expense than with steam traction. We will not be able to do this on all roads, for in some cases the traffic is too small to make the change from steam to electric power, and in others water powers are not conveniently located for the generation of electricity; but we have already attained one important result in that we can affirm that we have established electric traction on trunk lines under the same conditions as steam power, without lessening the size of trains, without restrictions in operating methods, and at no greater expense."

AUTOGENOUS WELDING WITH ACETYLENE.

THE APPLICATIONS OF THE OXY-ACETYLENE BLOWPIPE AND THE ADVANTAGES OF THE OXY-ACETYLENE OVER THE OXY-HYDROGEN FLAME.

Ernest Schneider—Scientific American Supplement.

A S mentioned by Mr. Kershaw on another page of this issue of THE ENGINEERING MAGAZINE, the calcium-carbide industry has been given a new impetus by the introduction of the oxy-acetylene blowpipe for the autogenous welding of metals. This process has been largely developed on the Continent and is used to a considerable extent in England but as yet it has been adopted in the United States by only two or three firms. Its ad-

vance is likely to be rapid, however, for it offers numerous advantages over the oxy-hydrogen system of autogenous welding. These advantages, and the applications of the oxy-acetylene flame, are interestingly enumerated in an abstract, in a recent number of the *Scientific American Supplement,* of a paper presented before the technical department of the Workmen's Union, Chemnitz, Germany, from which the following extracts are taken.

"The Fouché process with acetylene and oxygen is the latest in the line of autogenous welding. About ten or eleven years ago, as acetylene became known, the industrial world was certain that it would find important applications in the useful arts when employed in combination with oxygen; but the technical difficulties were great. In employing the Daniell burners there was formed at the mouth of the jet, by reason of the high percentage of carbon in the acetylene (92.3 parts by weight), deposits of graphite. When, however, the gas was mixed in the interior of the burner, the unexampled swiftness of kindling of the acetylene-oxygen mixture prevented the employment of such burners. There was always striking back of the flame, as the speed of kindling was 1,000 meters, or 3,280 feet, per second. After this, acetylene was employed under high pressure, as with hydrogen and oxygen, but with little success. At last, however, the Fouché burner enabled the employment of acetylene under the ordinary pressure at which it is usually produced.

"In this welding process the acetylene is, with the aid of a very fine injector in the interior of the burner, drawn from the gasometer by the oxygen under a pressure of one to two atmospheres, so that the mixture enters the jet at the required velocity. This is possible only because the oxygen is in greater quantity than the acetylene, the purpose of which is to prevent striking back of the flame.

"One cubic meter = 35.32 cubic feet of acetylene produces in burning 14,340 calories. At the beginning of combustion the acetylene is decomposed into its elements, carbon and hydrogen, and in this dissociation alone it produces 2,600 calories. The rest of the heat is set free by the combustion of the carbon to carbonic acid, and of the hydrogen to water vapor.

"In theory, one volume of acetylene requires for its perfect combustion 2.5 volumes of oxygen. The flame must, however, have a reducing action; there must be for every volume of acetylene 1.7 of oxygen. There thus results a flame composed of an ordinary short blow-pipe flame and a large secondary one. The first one is the place of dissociation of the acetylene into its components. It has a temperature of about 3,000 deg. C., equal to 5,432 deg. F., and consists only of carbonic oxide and pure hydrogen; both of these being reducing gases. With this short flame the welding of the metal is effected without oxidation.

"Welding by acetylene has important advantages over that with hydrogen, as one cubic meter = 35.32 cubic feet of hydrogen produces in burning only 3,091 calories. Therefore we have for acetylene and oxygen: $1 + 1.7 = 2.7$ cubic meters, $= 14,340$ calories in all $= 5,238$ calories per cubic meter of the gas mixture; for hydrogen and oxygen: $4 + 1 = 5$ cubic meters ($4 \times 3,091$) $= 12,364$ calories in all $= 2,473$ calories per cubic meter of gas mixture; *i. e.*, only about half the results in the first case.

"A further great advantage of the acetylene-oxygen flame is, that the heat is concentrated in a small primary flame, while in the hydrogen-oxygen flame it is more diffused. The heat of the acetylene-oxygen flame is therefore very rapidly given out to the surface to be welded, while the hydrogen and oxygen flame, by reason of its greater size, loses considerably by conduction and radiation.

"The consumption is four volumes of hydrogen, against only 0.59 of acetylene; that is, there is used seven times as much hydrogen as acetylene. But the losses by the oxygen-hydrogen flame are still greater; so that about 1.5 times as much oxygen is necessary as with the acetylene flame. This again increases the hydrogen consumption, so that for the same amount of work about 1.5 times as much oxygen and 10 times as many volumes of hydrogen are required, as where acetylene is used. As the compressed hydrogen costs for freight and other expenses about 1.25 mark per cubic meter, the acetylene, inclusive of depreciation, only about 1 mark, the saving by the employment of acetylene is very evident. In practice, hydrogen costs from 2 to 10 times as much.

"A complete plant for welding consists of: (1) An acetylene generator; (2) a cylinder of compressed oxygen, with reducing valve; (3) a water seal; and (4) a number of welding burners of different sizes. The acetylene generator must produce clean gas, free from air. Especially to be avoided as impurities are phosphureted and sulphureted hydrogen, as these make the iron 'short.' The oxygen cylinder with its reducing-valve is exactly the same as for the oxygen-hydrogen process. The water seal is to prevent oxygen getting into the acetylene apparatus in case of any stoppage, as this would form an explosive mixture. The water seal permits the passage of gas in only one direction. The welding burners are made like an injector; that is, the oxygen coming under pressure exhausts the acetylene under a pressure of about 100 millimeters, equaling say 4 inches of water column. The burners are of various sizes, according to what is required of them.

"As regards the cost of the process, it has been shown that with acetylene it is much cheaper than with hydrogen. It is also possible to weld plates of an inch thick with acetylene, which is not possible with hydrogen. The cost per welded seam is about as follows:

Thickness of sheet.	Oxygen.	Acetylene.	Wages.	Total.	
mm.	pf.	pf.	pf.	pf.	cts.
0.5	11	2.5	2.5	5	1.19
1	14	4	3	7	1.67
2	40	10	6	16	3.81
3	82	22	8	30	7.14
4	120	35	12	47	11.19
5	140	37	15	52	12.38
6	210	62	18	80	19.00
8	400	115	25	140	33.80
10	475	140	30	170	40.50
12	800	250	40	290	69.00

"The cost per seam depends naturally somewhat on the ability of the workmen. The acetylene process of welding is, however, easily learned, and the workman soon is enabled to make seams that are as strong as the solid metal. In any case, such a seam is better than a riveted or brazed one.

"As regards the heat of the flame, there is hardly any material that can withstand it. Brick, pumice-stone, the so-called fireproof brick, even carbide, which is made at a temperature of 3,000 deg. C. (5,432 deg. F.) is melted in a moment; graphite is the only thing that will stand up under it.

"In France there were, at the time this paper was prepared, over 800 firms employing this process. Among the various applications of the process may be mentioned:

(a) Welding iron or copper up to an inch thick.

(1) Replacing riveting for thin sheets.

(2) Replacing soldering and brazing.

(3) Making boilers and reservoirs, and repairing such, up to an inch plate-thickness.

(4) Welding together angle iron and profile iron.

(5) Manufacture of tubes in all dimensions, making both lengthwise and end joints.

(b) Welding flanges, pipe, nozzles, etc., whether the material be iron, copper, brass or what not.

(c) Making steel castings, malleable iron, bronze or brass castings.

(d) Repairing fractures in cast iron retorts, faults in general, and in automobile and bicycle manufacture; doing away with all rivets, bolts, screws, and making the frames in one piece.

(e) General industry—all sorts of repairs, making endless pipe-lines,. etc.

(f) In mechanical and electric construction works, where several pieces are to be united, even where they are of different metals; as for instance in dynamo armatures, and in making joints in conductors.

(g) In ornamental forging, the manufacture of burglar-proof safes, and the manufacture of sheet-iron ware, either enameled or not.

(h) Welding band-saw blades.

(i) Manufacture of articles for the army.

(j) Making cannon, side-arms, etc.

(k) Mending shovels or spades, hoes, forks, etc.

(l) Repairing gear-wheels of steel, brass, or cast iron.

(m) In locomotive building and narrow-gage railway building and in shipyards.

(n) Construction of iron ships and motor boats, and repairing boilers on board.

(o) Manufacture of iron vessels.

(p) Manufacture of steel heaters.

(q) Manufacture of iron window-frames.

Also, in any machine shop or establishment where machinery is used, broken parts may be repaired in place. . . . .

"The oxygen-acetylene process is cheaper to work than any other method of welding, and the cost for plant is comparatively low. It is advantageous even for very small transportable plants, as there may be used an acetylene apparatus not larger than a hydrogen cylinder. For large and very large establishments it is possible, and also advantageous, to make the oxygen in the establishment itself, which brings the price down to one-tenth of that where it must be bought."

MOTOR BUSES IN NEW YORK CITY.

DETAILS OF THE DE DION-BOUTON GASOLENE CARS IN SERVICE ON FIFTH AVENUE, NEW YORK CITY.

Street Railway Journal.

FIFTH Avenue, New York City, is closed to street-car traffic but has had, for a number of years, an omnibus service. Until recently the buses were drawn by horses, but about a year ago the company operating the service began to experiment with motor buses. An order was recently placed for fifteen De Dion-Bouton gasolene cars which are now in service and giving excellent satisfaction. The following details of these cars is taken from the *Street Railway Journal.*

"The body is of the typical 'London Road Car' type, having longitudinal seats inside with a capacity for sixteen passengers, and transverse seats on the top, seating eighteen passengers. The end windows are removable; the side windows are permanently closed, ventilation being obtained by louvers and small ventilating windows in the sides over the large windows. The winding stairway is of very light sheet steel and wood construction, protected by a high railing to prevent accident in case the bus should start suddenly. The body is easily removable, having no connection whatever with any part of mechanism, brake rods, etc., and being fastened to members of the frame with eight carriage bolts.

"The engine is rated at 24 horsepower capacity of the four-cylinder vertical type, with 105 x 130-millimetre cylinders, automatic inlet valves, high-tension ignition, with high-tension distributor and single non-trembling coil, working either from the magneto or dry cells by a two-way switch. The water circulation is by a centrifugal pump driven by a flexible shaft. The carburetor is automatic and heated by shunt pipe from the circulating water. The radiator is of the honeycomb type, having a large capacity and good efficiency. The lubrication is entirely automatic, the oil being forced by a positive gear pump through the crankshaft, which has an oil passage drilled through it, the oil being fed through small holes to the main bearings, crank pins, magneto and pump spindles. After passing through the bearings

the oil drops to the crank case and through a filter to the main oil reservoir which supplies the pump. The consumption of oil is very slight, as the bearings are so constructed as to prevent any oil escaping; the only loss is due to carbonization in the cylinders. The latter are protected from excessive supply of oil by baffle-plates, so that the exhaust is never smoky, nor do the cylinder walls, valves and spark plugs become coated with carbonized oil.

"The gear box is of the sliding-gear type, having four gear ratios and a progressive engagement of gears. The gears are very light and of comparatively small diameter, but this is admissible owing to the large reduction of gear ratio between the driving pinions and the internal toothed driving rings on rear wheels; that is, the gears run at high speed and are subjected to comparatively small torsional stresses and toothed contact pressures. The gears are made of chrome nickel steel, and after four months' operation show no signs of wear. The differential gear and countershaft are incorporated in the gear box, thus conducing to compactness and efficiency, owing to the elimination of extra bearings. The lubrication of the gears is by a gear oil pump fitted to the bottom of the gear box, similar to the one used on the engine, this pump delivering oil to all the bearings and also to a pipe parallel to the gear-shaft and located directly above it. This pipe is provided with holes directly above each of the gears. A constant stream of oil is thus playing upon the latter at all times when the vehicle is in motion, and practically no oil is lost, as special means are provided to prevent its escape.

"The machine is extremely easy to control: two pedals are provided, the right one being the clutch pedal, and the left one actuating the differential hand brake and also closing the throttle. A change-speed lever and an emergency brake are at the right of the driver. Either the differential brake or hand brake will lock the rear wheels on asphalt. The clutch, which is of the single disc metal-to-metal type, is extremely smooth and easy in engagement, and the bus can be started on the top speed owing to the gradual way in which the clutch picks up the load. It runs in oil and is subject to very little wear, although automatic compensation is provided for it. When driving in traffic it is not necessary for the operator to remove either of his hands from the steering wheel, as the clutch, throttle and brake can be operated with his feet. The bus complete with water and fuel weighs 7,900 pounds, increased when loaded to 13,000 pounds. The speed is 12 miles per hour on top speed at 1,000 revolutions per minute. This engine speed can be safely increased to 1,250 revolutions per minute, equivalent to 15½ miles per hour."

WEALTH AND WAGE EARNERS.

A DISCUSSION OF THE RELATIVE EARNINGS OF CAPITAL AND LABOR.

Engineering News.

THE following abstract of an editorial in a recent number of *Engineering News* contains a notable refutation of the Socialist doctrine that vast accumulation and concentration of wealth constitute a public danger. The Socialist aims at a redistribution of wealth, holding that the earnings of capital so far exceed those of labor as to impose an undue burden upon production, a burden the removal of which would result in a vast improvement of the welfare of the people as a whole. But the Socialist bases his opinion, on the superficial evidences of wealth, on the rapid increase in the number of multi-millionaires and of families who lavish large sums of money on ostentatious display. This illuminating presentation of the concrete facts of the case shows clearly the fallacy of the Socialistic premis, that the earnings of capital are largely in excess of those of labor, and that an arbitrary redistribution of wealth involving the overthrow of all economic laws, would improve the condition of labor.

"To a certain extent the question raised by the Socialist is a fair one and deserves fair investigation and an honest answer. Is the rapid increase of wealth a threat against the welfare of the people as a whole? If we concede, for the sake of argument, that payments for the use of capital are a burden—a fixed charge—upon production, it is reasonable to inquire how great has this burden become?

"The United States Census Bureau has recently issued a volume containing information which, it seems to us, may be fairly made use of to test the claims of social-ism. The volume is entitled "Wealth, Debt and Taxation." It gives what is doubtless the most accurate and careful estimate ever made of the total wealth of the United States. The estimate is made for the year 1900 and the year 1904. In the former year the total value of all property in the United States is set at $88,500,000,000. In the latter year the total was $107,000,000,000.

"These figures by themselves mean nothing. They are too huge for comprehension. But if it were possible to draw from them an approximate estimate of the total income which capital, or accumulated wealth, receives, and if we could then compare this income with the total earnings of labor, the comparison might contain some useful lessons.

"Let us take first the task of estimating the total income which capital receives. The census report before us classes the total wealth of the United States as follows:

Real property and improvements...	$62,341,000,000
Live stock	4,074,000,000
Farm implements and machinery...	845,000,000
Manufacturing machinery, tools and implements	3,300,000,000
Gold and silver coin and bullion...	2,000,000,000
Railroads and their equipment.....	11,245,000,000
Street railways, shipping, waterworks, electric light and power systems, telegraph and telephone systems, and canals...........	4,841,000,000
All other property — products of agriculture, manufactures and mines, merchandise, clothing, furniture, carriages, and miscellaneous personal property..........	18,462,000,000

In the first place it must be recognized that not all the wealth in the above table yields any income. The personal property included in the last item is uninvested wealth and yields no income. Likewise the list includes all property belonging to the Government, states, cities and counties. This property, including churches, hospitals, schools, etc., is held for the public benefit and where it produces any income, that income belongs to the public and returns directly to them.

"The total amount of this public property, according to the census schedule, is $7,830,000,000. Adding this to the $18,462,-000,000 worth of personal effects, we have $26,292,000,000 to be deducted from the total of $107,104,000,000, leaving $80,812,-000,000 as the amount of possible income-producing wealth in the United States. Undoubtedly much of this property also produces no income, and much more produces very small returns; on the other hand this is offset by other property which yields large returns. In estimating this average rate of income yielded by property, we may well be guided by the returns on railway property since in that field the accurate statistics of the Interstate Commerce Commission are available. In 1904 the railways of the United States paid to their owners, the railway stock and bond holders, in interest and dividends the sum of $480,000,000 in round numbers, which was an average of 4¼ per cent. on the value of the railways given above.

"It is doubtful whether real estate, which is seen above to constitute the bulk of the property of the United States, would show anything like as large an average return as the railways; but as we are aiming to find the maximum possible drain that capital may be making on the country, we will assume that railway capital represents a fair average of the rate of income produced by all wealth. On this basis $80,-812,000,000 of invested capital at 4¼ per cent. would yield an annual return of $3,534,000,000.

"And now let us turn and study the side of labor. The population of the United States on June 1, 1904, is placed by the census at 81,256,000. It was 76,000,000 in round numbers in 1900. The number of wage earners, or 'persons engaged in gainful occupations,' was given by the census of 1900 as 29,286,000. At the present time, therefore, this number must be swelled to some 31,000,000. We believe, however, that this is in excess of the true number of

wage workers, as it includes those who are past the age of active work and those who are incapacitated for other reasons. A more accurate method, we believe, is to estimate one worker to each 3½ persons, which would give a total of 23,200,000 wage earners in round numbers.

"What are the average annual earnings of these workers? For a guide in answering this question we turn again to the Interstate Commerce Commission railway statistics, and find there that in 1905 there were 1,382,000 men employed on the railways of the United States who received wages and salaries amounting to 840 million dollars, or an average of $608 per employee. Again, the Census report on Manufactures just issued shows that in 1905 there were in round numbers 6,000,000 persons engaged in manufacturing industries who received wages and salaries amounting to $3,186,000,000, or an average of $531 each. Similar statistics are not available for other occupations; but it will probably be generally agreed that the 1½ million persons engaged in professional occupations receive a higher average than this, while the 10½ millions engaged in agricultural and the 5½ in domestic and personal service receive considerably less. All things considered, $450 per annum seems as low a figure as can reasonably be estimated as the earnings of the average worker.

"Multiplying now the total number of wage earners found by these average earnings, we have the total compensation paid to labor in a year at $10,340,000,000.

"If these figures be accepted, then it appears that capital is now receiving about one dollar where labor is receiving three; or, to put it another way, if the annual product of the country is divided into four equal heaps, capital is getting one of these heaps and labor the other three.

"This conclusion can be substantiated, however, in other ways. Everyone familiar with business affairs knows that in almost every industry the annual pay roll is several times as large as the amount which the owners of the business can draw out in profits. In the railway industry exact statistics are available and were quoted above. viz.. annual payments to security holders of $480,000,000 compared with annual payments of 840 million dollars. Here the proportion is slightly less than 1 to 2; but in railway transportation there is a larger investment of fixed capital required per employee than in almost any other industry. The figures above given were: Capital in railways, $11,245,000,000; number of employees, 1,382,000; average investment per employee, $8,140. In contrast with this are the conditions in agriculture where two or three men are required to do the work on a farm worth perhaps $5,000. Remembering that some ten millions of the wage earners of the country are engaged in agriculture, it is an inevitable conclusion that our estimate of 1 to 3 as the proportional distribution to capital and to labor errs if anything by giving a larger return to capital than it actually receives."

In making comparisons between the earnings of capital and the earnings of labor, a common error is to confound capital with wealth and labor with poverty. A large part of the total wealth of the United States is the property of the wage earners themselves. The huge fortunes of the multi-millionaires are in bulk much less than the accumulated savings of people of moderate means. The $3,482,000,000 deposited in the savings banks, the $2,700,-000,000 assets of the life insurance companies, and the $16,615,000,000 worth of farm property all represent widely-distributed wealth belonging to the latter class in which the Socialist can surely see no menace to public welfare. To give the Socialist, however, every chance to prove his claims, let it be supposed that all existing wealth is confiscated and distributed among all wage earners, that under those circumstances it would be as productive of income as at present, and that the workers in the revolutionized social order could reap the same return as now in the shape of wages. The result would be that the average wage earner who now has an income of $450 per annum would receive about $150 additional, an increase which could neither make everybody wealthy, nor abolish poverty, but could only to a very slight degree lighten the task of the worker or increase his comforts.

ABSTRACT SCIENCE AND ENGINEERING.

THE INFLUENCE OF THE INDUSTRIAL DEVELOPMENT OF SCIENTIFIC PRINCIPLES ON THE PROGRESS OF PURE SCIENCE.

Prof. Silvanus P. Thompson—British Association for the Advancement of Science.

PROF. S. P. Thompson's recent presidential address before the Engineering Section of the British Association for the Advancement of Science discussed in a very interesting manner the relation between abstract science and engineering. Touching first upon the vast development of the material resources of civilization, the rapidity of its progress, and the absolute dependence of the whole upon science, directed mainly by the engineer, Prof. Thompson developed his theme along novel lines. The influence of abstract science upon engineering and industrial progress has often been discussed, but very little attention has been paid to the influence of the industrial applications of scientific principles upon the development of abstract science itself. It is to the latter that Prof. Thompson devotes the greater part of his address and the following extracts give an outline of his argument.

"In engineering, above all other branches of human effort, we are able to trace the close interaction between abstract science and its practical applications. Often as the connection between pure science and its applications has been emphasised in addresses upon engineering, the emphasis has almost always been laid upon the influence of the abstract upon the concrete. We are all familiar with the doctrine that the progress of science ought to be an end in itself, that scientific research ought to be pursued without regard to its immediate applications, that the importance of a discovery must not be measured by its apparent utility at the moment. We are assured that research in pure science is bound to work itself out in due time into technical applications of utility, and that the pioneer ought not to pause in his quest to work out potential industrial developments. We are invited to consider the example of the immortal Faraday, who deliberately abstained from busying himself with marketable inventions arising out of his discoveries, excusing himself on the ground that he had no time to spare for money-making. It is equally true, and equally to the point, that Faraday, when he had established a new fact or a new physical relation, ceased from busying himself with it, and pronounced that it was now ready to be handed over to the mathematicians. But, admitting all these commonplaces as to the value of abstract science in itself and for its own sake, admitting also, the proposition that sooner or later the practical applications are bound to follow upon the discovery, it yet remains true that in this thing the temperament of the discoverer counts for something. There are scientific investigators who cannot pursue their work if troubled by the question of ulterior applications; there are others no less truly scientific who simply cannot work without the definiteness of aim that is given by a practical problem awaiting solution. There are Willanses as well as Regnaults; there are Whitworths as well as Poissons. The world needs both types of investigator; and it needs, too, yet another type of pioneer—namely, the man who, making no claim to original discovery, by patient application and intelligent skill turns to industrial fruitfulness the results already attained in abstract discovery.

"There is, however, another aspect of the relation between pure and applied science, the significance of which has not been hitherto so much emphasised, but yet is none the less real—the reaction upon science and upon scientific discovery of the industrial applications. For while pure science breeds useful inventions, it is none the less true that the industrial development of useful inventions fosters the progress of pure science. No one who is conversant with the history, for example, of optics can doubt that the invention of the telescope and the desire to perfect it were the principal factors in the outburst of optical science which we associate with the names of Newton, Huygens, and Euler. The practical application, which we know was in the minds of each of these men, must surely have been the impelling motive that caused them to concentrate on abstract optics their great and exceptional powers of thought. It was

in the quest—the hopeless quest—of philosopher's stone and the elixir of life that the foundations of the science of chemistry were laid. The invention of the art of photography has given immense assistance to sciences as widely apart as meteorology, ethnology, astronomy, zoology, and spectroscopy. Of the laws of heat men were profoundly ignorant until the invention of the steam-engine compelled scientific investigation; and the new science of thermodynamics was born. Had there been no industrial development of the steam-engine, is it at all likely that the world would ever have been enriched with the scientific researches of Rankine, Joule, Regnault, Hirn, or James Thomson? The magnet had been known for centuries, yet the study of it was utterly neglected until the application of it in the mariners' compass gave the incentive for research.

"The history of electric telegraphy furnishes a very striking example of this reflex influence of industrial applications. The discovery of the electric current by Volta and the investigation of its properties appear to have been stimulated by the medical properties attributed in the preceding 50 years to electric discharges. But, once the current had been discovered, a new incentive arose in the dim possibility it suggested of transmitting signals to a distance. This was certainly a possibility, even when only the chemical effects of the current had yet been found out. Not, however, until the magnetic effects of the current had been discovered and investigated did telegraphy assume commercial shape at the hands of Cooke and Wheatstone in England and of Morse and Vail in America. Let us admit freely that these men were inventors rather than discoverers: exploiters of research rather than pioneers. They built upon the foundations laid by Volta, Oersted, Sturgeon, Henry, and a host of less famous workers. But no sooner had the telegraph become of industrial importance, with telegraph lines erected on land and submarine cables laid in the sea, than fresh investigations were found necessary. New and delicate instruments must be devised, means of accurate measurement heretofore undreamed of must be found, standards for the comparison of electrical quantities must be created, and the laws governing the operations of electrical systems and apparatus must be investigated and formulated in appropriate mathematical expressions. And so, perforce, as the inevitable consequence of the growth of the telegraph industry, and mainly at the hands of those interested in submarine telegraphy, there came about the system of electrical and electromagnetic units, based on the early magnetic work of Gauss and Weber, developed further by Lord Kelvin, by Bright and Clark, and last, but not least, by Clerk Maxwell. Had there been no telegraph industry to force electrical measurement and electrical theory to the front, where would Clerk Maxwell's work have been? He would probably have given his unique powers to the study of optics or geometry; his electromagnetic theory of light would never have leapt into his brain; he would never have propounded the existence of electric waves in the ether. And then we should never have had the far-reaching investigations of Heinrich Hertz; nor would the British Association at Oxford in 1894 have witnessed the demonstration of wireless telegraphy by Sir Oliver Lodge. A remark of Lord Rayleigh's may here be recalled, that the invention of the telephone had probably done more than anything else to make electricians understand the principle of self-induction.

"In considering this reflex influence of the industrial applications upon the progress of pure science it is of some signifi-cance to note that, for the most part, this influence is entirely helpful. There may be sporadic cases where industrial conditions tend temporarily to check progress by imposing persistence of a particular type of machine or appliance; but the general trend is always to help to new developments. The reaction aids the action; the law that is true enough in inorganic conservative systems, that reaction opposes the action, ceases here to be applicable, as indeed it ceases to be applicable in a vast number of organic phenomena. It is the very instability thereby introduced which is the essential of progress. The growing organism acts on its environment, and the change in the environment reacts on the organism—not in such a way as to oppose the growth, but so as to promote it. So it is with the development of pure science and its practical applications.

"In further illustration of this principle one might refer to the immense effect which the engineering use of steel has had upon the study of the chemistry of the alloys. And the study of the alloys has in turn led to the recent development of metallography. It would even seem that through the study of the intimate structure of metals, prompted by the needs of engineers, we are within measurable distance of arriving at a knowledge of the secret of crystallo-genesis. Everything points to a very great and rapid advance in that fascinating branch of pure science at no distant date."

THE PUPIN TELEPHONE CABLE OF THE LAKE OF CONSTANCE.

A DESCRIPTION OF THE FIRST SUBMARINE TELEPHONE CABLE ON THE PUPIN SYSTEM, RECENTLY LAID IN THE BODENSEE.

Engineering—Elektrotechnische Zeitschrift.

A RECENT number of *Engineering* contains an interesting discussion of the problem of submarine telephone communication over long distances, and a description of the first submarine cable on the Pupin system, which was recently laid in the Lake of Constance. The latter part of the article is largely drawn from a paper by Dr. Ebeling, published recently in *Elektrotechnische Zeitschrift.*

"Considering that we have had submarine telegraph cables for half a century, the small number of submarine telephone ca-bles may be a matter of surprise to the non-technical public. There is, however, a special difficulty as to the submarine telephone cable. Owing to its construction the telephone cable has greater capacity and smaller inductance than a land line of corresponding resistance and length. But we know, from the researches of Oliver Heaviside, that in order to obtain clear speech, we should have small capacity and high self-induction. Where we are not confronted with insulation troubles, the two conditions may easily be combined. For submarine cables gutta-percha has generally been considered the best insulating material, in spite of its high price and its high inductive capacity. Vulcanised rubber may be used, and has, indeed, answered in a few of the submarine cables of the early days. But rubber is not generally trusted, and air space and paper insulation, so useful and convenient in many instances, and of small inductive capacity, are difficult of application in submarine cables, because we have to provide a suitable impervious sheath, which will also bear a great water pressure.

"If it were not for these mechanical difficulties, the experimental verification by Pupin of Heaviside's deductions should long ago have helped us to long-distance submarine telephone cables. Pupin showed that by adding self-inductance to the line at intervals we can greatly increase the range of clear speech. Long lines of high, or of moderate, capacity exert a damping influence upon the high-frequency alternating-current waves passing through a telephone line. The damping depends upon the capacity, inductance, resistance, length, and leakage of the line—a certain leakage does not do any harm—and the frequency of the currents, and that term of the function, which is more particularly known as the Pupin constant β represents a numerical factor which teaches us over what length of line speech may be possible, or what values must be given to the chief factors—resistance, capacity, and inductance—in order to render speech feasible. When the inductance is very high, waves of all periods will practically be damped to the same amount, and the sounds will be reproduced weakened, but with their natural tones. When the inductance is very small, on the other hand, the overtones and the hissing sounds will be damped, and we obtain loud, but indistinct sounds. The inductance can be increased by enlarging the copper core, when the capacity will likewise rise, however. This arrangement was tried on the English-Belgian cable, laid before these problems were clearly understood. The core was made thick to decrease the ohmic resistance, and the capacity was decreased by increasing the thickness of the gutta-percha. In the English-French cable the capacity was reduced by providing the four cores with a common gutta-percha sheath. All these early cables contain four symmetrically grouped cores to prevent mutual interference between the two pairs.

"When, in 1902, an improved telephone service was wanted between North Germany and Denmark, and also between Cuxhaven and Heligoland, telephone-cable manufacturers were not prepared to supply submarine Pupin cables, with self-inductance inserted at intervals, although land cables of this type had been laid. In 1899, Breisig, of Berlin, showed that the inductance of a cable could also be increased by coiling iron tape in a spiral round the core, over its full length, and later on the Danish engineer Krarup proved that fine iron wire wound directly on the copper core would serve the same purpose. The iron wire need not be insulated from the copper, which may be varnished. Krarup's suggestion was adopted by Messrs. Felten and Guilleaume in the cables of 1903, which were laid in depths of about 20 metres.

"In the four-core Fehmarn-Laalland cable, each of the seven copper strands, of 1.35 millimetres, is coiled with bare iron wire of 0.3 millimetre; over this are placed the impregnated paper insulation, jute, two lead sheaths, compound, and an iron wire armouring. The Cuxhaven-Heligoland cable, 75 kilometres (47) miles in length, is similarly constructed; the Refsnaes-Soelvig cable, laid about the same time, is a gutta-percha cable.

"The comparative weights (in kilogrammes per metre) and total costs (marks or shillings per metre) of some of these cables have been estimated by Breisig as follows:

	Marks.
English-Belgian cable—7.2 kilog.	6.78
Greetsiel-Borkum cable—6.8 kilog.	4.78
Heligoland cable—8.5 kilog.	6.38

"The Greetsiel-Borkum cable (30 kilometres, or nearly 20 miles in length) which is also one of the Felten and Guilleaume cables of 1903, has only half the damping constant of the English-Belgian cable—an important point for cables joined to long land lines—and was considerably cheaper, it will be seen.

"In none of the cases so far mentioned, however, did the depth of the water ex-

ceed 40 metres. The case of the Bodensee cable, laid by Messrs. Siemens and
Halske in 1906, was different. The depth
of the open lake portion between Fried-
richshafen (in Württemberg) and Romans-
horn (in Switzerland) is 250 metres (820
ft.), and considerable difficulties have been
encountered, although the length of the
cable is only 12 kilometres (7.5 miles). Dr.
Ebeling's explanation of the difficulties and
temporary failures, which he gave in his
account of the manufacture and laying of
this cable delivered before a recent meeting
of the Elektrotechnische Verein, was particularly interesting. It is the first submarine telephone cable on the Pupin system, the self-inductance of coiled iron wire
being inserted at intervals of 1 kilometre;
calculations as to the number and distribution of the induction coils required for
particular purposes were given by Breisig
in 1904. The cable comprises seven complete circuits.

"As the water pressure is 25 atmospheres
at a depth of 250 metres, experiments were
made as to the strength of the cables, and
the pressure was in these experiments
raised to 150 atmospheres in order to make
sure that an apparently harmless pressure
might not prove injurious in the course of
time. The cables, provided with a double
lead sheath of 5 millimetres (0.2 in.) thickness, were all crushed; a pressure of 50
atmospheres would, for instance, flatten
out in two hours cables which had borne
that pressure quite well for half an hour.
The ordinary armouring could not prevent
the crushing; but a remedy was found by
coiling a steel wire of 2 millimetres thickness in a spiral round the paper insulation; the spirals being 5 millimetres or
more apart. It was desired to divide the
insulation into water-tight sections by inserting impervious portions in the middle,
between the inductance coils. Originally
the cable was made as lead cable in lengths
of 500 metres, and either an inductance
coil or a section insulator were alternately
interposed between the different lengths.
These parts were thicker than the cable itself. The iron wire coil was common to
the seven circuits, and formed a cylindrical
portion of larger diameter; a section insulator formed a doubly conical piece. The
armouring was continuous all over the

cable, and over those thicker portions.
When the laying of the cable was commenced in the autumn of 1905, however,
one of the splices failed almost at the beginning, and as the season was advanced,
the laying was put off till the following
summer, and certain alterations were made
in the cable. The alteration concerned the
coils and section insulators, which are now
combined. The inductance now consists,
as before, and as in the cables we have
mentioned, of iron wire; but the wire is
coiled in alternating sections round three
or four of the seven pairs of lines. The
coil itself is cylindrical; the larger diameter of this cylinder is then brought down
to the smaller diameter of the cores by
two conical portions; and these conical
portions are the waterproof seals between
different cable sections. The lead sheath
is, in the conical portion, supported by a
conical wire spiral, and the paper insulation is replaced by gutta-percha. These
repairs had to be executed on the lake
shore, and the cable armouring was applied
by hand while the cable was kept under a
tension of 2 tons. During these repairs
the method of splicing was also perfected;
in the case just referred to the lead sheathing had burst over a splice. . . . . .

"As regards the cable, the size of a copper strand corresponds to a diameter of 1.5
millimetre. The resistance of 1 kilometre
of double line was limited by contract to
20 ohms, the total resistance to 40 ohms,
and the actual resistance is 33.5 ohms. The
capacity is 0.039 microfarad per kilometre,
against 0.05 of the contract; the inductance, 0.21 henry, against 0.20; and the
Pupin damping constant 0.0072, against
0.01. The Pupin constant is almost as
small as in the case of the Heligoland
cable, where it is 0.0065; but the copper
diameter of this latter cable is 4 millimetres, and its resistance 3.8 ohms, so that
almost as good a damping has been attained in the Pupin cable with only one-
seventh of the cross section of copper.
This is a very important gain. No doubt
the telephone cable lying in the Lake of
Constance has been costly, but we know
now that submarine Pupin cables can be
made, and how they are to be made, and
the ordinary scale of expenditure must not
be applied to pioneer work."

THE ENGINEERING INDEX

THE KEYSTONE
IN·THE·ARCH·OF·APPLIED·SCIENCE·

The following pages form a descriptive index to the important articles of permanent value published currently in about two hundred of the leading engineering journals of the world—in English, French, German, Dutch, Italian, and Spanish, together with the published transactions of important engineering societies in the principal countries. It will be observed that each index note gives the following essential information about every publication:

(1) The title of each article,   (4) Its length in words,
(3) A descriptive abstract,   (5) Where published,
(2) The name of its author,   (6) When published,

(7) *We supply the articles themselves, if desired.*

The Index is conveniently classified into the larger divisions of engineering science, to the end that the busy engineer, superintendent or works manager may quickly turn to what concerns himself and his special branches of work. By this means it is possible within a few minutes' time each month to learn promptly of every important article, published anywhere in the world, upon the subjects claiming one's special interest.

The full text of every article referred to in the Index, together with all illustrations, can usually be supplied by us. See the "Explanatory Note" at the end, where also the full title of the principal journals indexed are given.

DIVISIONS OF THE ENGINEERING INDEX.

CIVIL ENGINEERING

BRIDGES.

Bascule.
New Bascule Bridge at Duisbourg. Illustrated detailed description of a new bridge costing approximately £24,465. 1200 w. Engr, Lond—Aug. 9, 1907. No. 86474 A.

Blackwell's Island.
The Erection of the Blackwell's Island Bridge. Illustrations, with brief account of the erection of the new cantilever bridge over the East River, at New York.

1600 w. Sci Am—Aug. 10, 1907. No. 86303.

Centering.
Moving the Centering of the Walnut Lane Arch at Philadelphia. Illustrates and describes the method of moving the centering from the finished ring to its position under the twin ring. 600 w. Eng News—Aug. 15, 1907. No. 86394.

Failure.
Failure of Masonry Arch Carrying Erie Canal Over Onondaga Creek, Syra-

We supply copies of these articles. See page 239.

cuse, N. Y. An illustrated account of the collapse on July 30, which has closed the canal to through traffic in one of its busiest seasons. 1700 w. Eng News—Aug. 8, 1907. No. 86313.

Highway Bridge.
Economical Methods of Highway Bridge Construction. Describes wooden falsework adjustable to different heights and adapted to be used over and over again for successive bridge. Ills. 2000 w. Eng Rec—Aug. 24, 1907. No. 86563.

Plate Girder.
A Plate Girder Highway Arch Bridge. Illustrated description of a bridge in Princeton, N. J., consisting of four three-hinge plate girder arch spans. 900 w. Eng Rec—Aug. 17, 1907. No. 86421.

Removals.
Removing a Stone Arch Bridge Over the Northern Railway of France. Illustrates and describes methods used under difficult conditions. 800 w. Eng News—Aug. 1, 1907. No. 86078.

Steel.
The New Railway Bridge over the Neckar at Heidelberg (Die Neue Eisenbahnbrücke über den Neckar bei Heidelberg). G. Lucas, D. R. Müller and G. Trauer. The first part of the serial describes the foundation work for this four-span steel structure. Ills. 1000 w. Serial. 1st part. Deutsche Bau—July 6, 1907. No. 86253 D.

Steel Designing.
The Proportioning of Steel Railway Bridge Members. Henry S. Prichard. Discusses the strength of structural steel, safe working stresses, rules for designing compression members, and provisions for live loads. 9500 w. Pro Engrs' Soc of W Penn—July, 1907. No. 86398 D.

Swing Bridge.
The Double-Deck Swing Bridge at the Upper Harbor at Hamburg (Pont Tournant à deux Etages sur le Port Supérieur de Hambourg). A. Bidault des Chaumes. Illustrated description of this large steel bridge, the upper deck of which carries four lines of rails and the lower is arranged for vehicular and foot traffic. 3000 w. Génie Civil—July 6, 1907. No. 86209 D.

Trestles.
A Pile Trestle Erected with a Pivotal Pile-Driver. R. Balfour. Illustrates and describes a pivotal pile-driver and its use. 1000 w. Eng News—Aug. 15, 1907. No. 86389.

Viaducts.
Armored Concrete Viaduct at Deurne-Merxem. Illustrated detailed description of a structure near Antwerp, constructed on the Hennebique system. 2500 w. Engr, Lond—Aug. 16, 1907. No. 86603 A.

Curved Girder Approach Viaduct of

the Austerlitz Bridge Over the Seine. R. Bonnin. A statement of the essential difference between the curved girder and the straight girder, pointing out the additional forces to which the former is subjected. An appendix gives a general theory of curved girders. Ills. 2000 w. Eng News—Aug. 15, 1907. No. 86386.

Wooden Truss.
Early Wooden Bridges. Illustrates and describes early forms of wooden bridges. The Burr bridge, the Howe truss bridge, etc. 800 w. Ry & Loc Eng—Aug., 1907. No. 86043 C.

CONSTRUCTION.

Arches.
The Application of the Elastic Theory to Certain Forms of Arch. A letter from R. C. Barnett on the application of the elastic theory to arch analysis. 1200 w. Eng News—Aug. 1, 1907. No. 86080.

Breakwaters.
Stone Breakwater Construction at Huron, Ohio. Wilson T. Howe. Describes an example of the modification that has taken place in stone breakwater construction. Ills. 1000 w. Eng News—Aug. 22, 1907. No. 86508.

Buildings.
The Design and Construction of Industrial Buildings. D. C. Newman Collins. An illustrated article discussing materials and methods representing the most recent practice. 7000 w. Engineering Magazine—Sept., 1907. No. 86610 B.

The Calvary Cemetery Mortuary Chapel. A structure to cost about $200,000, mainly of concrete construction, and having architectural features of interest is illustrated and described. 1500 w. Eng Rec—Aug. 3, 1907. No. 86090.

The Hudson Companies' Building, New York. Plans and description of construction details of this large terminal station, occupying two city blocks in New York. 2500 w. Eng Rec—Aug. 3, 1907. No. 86088.

Building Details.
The Construction of the New York Central Office Building, New York. Describes a steel-cage, fire-proof, 20-story building of special design. Ills. 2500 w. Eng Rec—Aug. 24, 1907. No. 86557.

Concrete.
The Chateau des Beaux-Arts on Huntington Bay. Illustrated description of extensive use of concrete in the construction of a summer resort on the south shore of Long Island. 3000 w. Eng Rec—Aug. 17, 1907. No. 86423.

Contracts.
Engineers' Contracts and Specifications from a Contractor's Point of View. James W. Rollins, Jr. Especially discusses the contract as a mutual understanding, and the troubles arising on account of badly-

We supply copies of these articles. See page 239.

drawn specifications. General discussion. 25000 w. Jour Assn of Engng Socs—July, 1907. No. 86395 C.

Dams.

A Buttressed Concrete Dam. A stone dam at Matteawan, N. Y., raised 7 ft. by an addition of buttressed concrete is illustrated and described. 2000 w. Eng Rec—Aug. 24, 1907. No. 86562.

Movable Crest Dams at the Water Power Development of the Chicago Drainage Canal. Illustrated detailed description of movable crests of structural steel and steel plates, designed to furnish a passage for all drift which cannot safely pass through the water turbines. 3500 w. Eng Rec—Aug. 24, 1907. No. 86555.

Reconstructing a Masonry Dam for Increased Depth of Storage. An unusual piece of dam-building, near Lennep, Germany, in connection with the water supply system, is illustrated and described. 600 w. Eng News—Aug. 29, 1907. No. 86629.

Specifications for the Main Dams, Ashokan Reservoir, New York City Water Supply. Gives sections from the specifications, with illustrations and explanatory notes, for a high masonry dam with earth wings. 2000 w. Eng News—Aug. 1, 1907. No. 86074.

The Construction of the Dam of the Nevada-California Power Co. Describes a dam to be built of loose rock puddled on the upper side, across a narrow valley at a high altitude in the Sierra Nevada Mts. Ills. 1200 w. Eng Rec—Aug. 3, 1907. No. 86092.

Excavation.

Retaining the Sides of a large Excavation. Illustrated description of sheeting and bracing used for the Swift & Co. building in New York City, where the excavation was through a deep stratum of fine sand and quicksand, with abundance of ground water. 1500 w. Eng Rec—Aug. 24, 1907. No. 86556.

Failure.

Collapse of Reinforced Concrete Building in Philadelphia. Gives the opinions of engineers, and the views of the coroner's jury as to the cause of failure, with editorial. Ills. 4500 w. Cement Age—Aug., 1907. No. 86482.

Floors.

Economics of Reinforced Concrete Floor Slabs. A. E. Budell. A comparison between the actual cost of the slab floor and the combination tile and beam system. Ills. 800 w. Engng-Con—Aug. 14, 1907. No. 86405.

Foundations.

Notes on Foundations. Discusses the requisites of a good foundation, and the processes used in excavating. 2500 w. Can Archt—Aug., 1907. No. 86572 C.

Ice Problem.

The Ice Problem in Engineering Work in Canada. Howard T. Barnes. Read before the British Assn. An explanation of conditions in the Canadian streams during winter, and some of the means of overcoming difficulties. 2000 w. Engng—Aug. 9, 1907. No. 86471 A.

Pier Failure.

The Locust Point Pier Collapse in Baltimore. Photographs of the wrecked pier with an explanation of the probable cause of the collapse. 600 w. Eng News—Aug. 8, 1907. No. 86307.

Reinforced Concrete.

Design of Reinforced Concrete Structures—Flat Top Culverts. Walter W. Colpitts. The present article discusses the design of flat top culverts on the Kansas City Outer Belt & Electric R. R. 2800 w. Ry Age—Aug. 16, 1907. Serial. 1st part. No. 86438.

Design of Reinforced Concrete Structures—Abutments. Walter W. Colpitts. Describes in detail the design of reinforced concrete abutments adopted by the Kansas City, Mexico, and Orient Ry. Ills. 2500 w. Ry Age—Aug. 23, 1907. No. 86567.

Notes on Reinforced Concrete Designing. Gives appendices to the report of the Joint Committee on Reinforced Concrete, explaining the reasons for certain requirements suggested. 2000 w. Eng Rec—Aug. 3, 1907. No. 86091.

Reinforced-Concrete Roofs for Manufactories (Die Eindeckung der Fabrikdächer in Eisenbeton). L. Geusen. A discussion of the value of reinforced concrete as a roofing material with points on roof design. Ills. 3500 w. Serial. 1st part. Beton u Eisen—July, 1907. No. 86270 D.

Reinforced Concrete Structures on the Kansas City Outer Belt & Electric Railroad. Walter W. Colpitts. Illustrates and describes culverts, abutments, bridges, etc. Map. 800 w. Ry Age—Aug. 2, 1907. No. 86121.

A Reinforced Concrete Water Tower at Anaheim. A tower in Southern California is described in detail. Ills. 1500 w. Eng Rec—Aug. 24, 1907. No. 86559.

See also same title under Civil Engineering, Materials of Construction.

Three Light Reinforced-Concrete Arch Bridges at Venice, California. Paul H. Ehlers. Illustrated description of structures across a lagoon near Los Angeles, Cal., stating the conditions and restrictions. 800 w. Eng News—Aug. 29, 1907 No. 86636.

Rock Excavation.

Method of Excavating Rock in Large Masses. George C. McFarlane. Notes on methods used on the Grand Trunk Pacific Railroad for rocks of extreme hardness. Deep holes and heavy blasts. 2500 w. Eng & Min Jour—Aug. 3, 1907. No. 86103.

Roofs.

A Long-Span Truss Roof: Armory for Squadron C, New York National Guard, New York City. Drawings and description of the general characteristics of the steel riding-hall roof of an armory in Brooklyn. 2500 w. Eng News—Aug. 29, 1907. No. 86631.

Simplon.

The Simplon Tunnel. Francis Fox. A very interesting account of the construction of this great engineering work and matters relating to it. Also discussion and correspondence. Ills. 20500 w. Inst of Civ Engrs —No. 3651. No. 86369 N.

Siphon.

A Reinforced-Concrete Siphon on an Irrigation Canal in Spain. Illustrates and describes the construction of a siphon carrying the main canal across the valleys of the Sosa and the Ribabona, having a total length of about 3,200 ft. 1000 w. Eng News—Aug. 1, 1907. No. 86076.

Timber Trusses.

The Design of Timber Howe Trusses. R. Balfour. Gives a typical design for a ten-panel, 150-ft. span, Howe truss, single-track railway bridge, 25-ft. between lower and upper chords and 16-ft. clear width, explaining the variations from usual practice. 1000 w. Eng News—Aug. 29, 1907. No. 86632.

Tunnels.

The Construction of the Gattico Tunnel of the Santhià-Borgomanero-Arona Railway (Bauausführung des Gattico-Tunnels im Zuge der Santhià-Borgomanero - Arona - Bahn). Gaetano Crugnola. The first part of the serial gives plan and profile of this Italian line and describes the power, shaft, and pumping plants. Ills. 2500 w. Serial. 1st part. Schweiz Bau—July 6, 1907. No. 86254 D.

Warehouse.

The New Warehouse of the Newark Warehouse Company. Illustrates and describes a large 6-story fireproof building in Newark, N. J., combining the facilities of a freight delivery and receiving terminal and a storage warehouse. 4500 w. Eng Rec—Aug. 10, 1907. No. 86315.

MATERIALS OF CONSTRUCTION.

Angles.

Tension Tests of Steel Angles with Various Types of End-Connection. Frank P. McKibben. Read before the Am. Soc. for Test. Mat. Reports tests of 33 steel angles, such as are used for tension members in riveted framed structures. 2500 w. Eng News—Aug. 22, 1907. No. 86510.

Cement.

The Relations between Highly-Basic Blast Furnace Slags and Cement from the Standpoint of Physical Chemistry (Ueber Chemisch-Physicalische Verhältnisse der Hochbasischen Hochofenschlacken und Zemente). Karl Zulkowski. 3300 w. Serial. 1st part. Stahl u Eisen—July 17, 1907. No. 86228 D.

Iron Corrosion.

The Electrolytic Theory of the Corrosion of Iron. Dr. Allerton S. Cushman. Abstract of a paper read at meeting of Am. Soc. for Test. Mat. Reports experimental investigations. Ills. 5500 w. Ir Age—Aug. 8, 1907. No. 86117.

Materials.

On the Study of Material. Editorial on the present day requirements of an intimate knowledge of materials used in engineering work. 2500 w. Eng News —Aug. 29, 1907. No. 86634.

Metal Protection.

Priming Coats for Metal Surfaces: Linseed Oil vs. Paint. F. P. Cheesman. Slightly condensed from a paper read before the Am. Soc. for Test. Mat. Shows that linseed oil by itself is a bad thing. 1500 w. Eng News—Aug. 8, 1907. No. 86308.

Reinforced Concrete.

Investigation of the Thermal Conductivity of Concrete and Embedded Steel and the Effect of Heat Upon Their Strength and Elastic Properties. Ira H. Woolson. Read before the Am. Soc. for Test. Mat. A record of experimental investigations. 2500 w. Eng News—Aug. 15, 1907. No. 86393.

The Slipping Resistance of Steel and Brass in Concrete. H. Burchartz. A report of adhesion tests made with steel and brass rods in concrete. 1200 w. Eng Rec—Aug. 3, 1907. No. 86087.

See also same title under Civil Engineering, Construction.

Tiles.

Ceramic Tiles and Their Use. C. Howard Walker. Gives information regarding the uses of tiles in construction, and designs of methods of obtaining variety in tile work for floors and walls. 2200 w. Br Build—July, 1907. No. 86144 D.

Timber Preservation.

Causes of Decay in Timber. C. W. Berry. Discusses the causes and the use of live steam for treatment in timber preservation. 1500 w. Munic Engng—Aug., 1907. No. 86040 C.

The Preservation of Timber. Gives results, with regard to the penetration of heat into timber during the process of treatment, obtained during investigations carried out for the State Railways of France. 2000 w. Sci Am Sup—Aug. 3, 1907. No. 86068.

We supply copies of these articles. See page 230.

Wood Preservation.

Seasoning of Telephone and Telegraph Poles. Henry Grinnell. From a circular of the U. S. Dept. of Agri. Describes experiments to prolong the service of timbers, particularly those which pertain to the rate of seasoning. 2000 w. Elec Rev, N Y—Aug. 17, 1907. No. 86425.

MEASUREMENT.

Beams.

Test of a Cast Steel Beam. J. V. McAdam. Describes the method of testing devised by the writer. 1500 w. Eng News—Aug. 1, 1907. No. 86077.

Plate Girders.

Distribution of Stresses in Plate-Girders. Cyril Walter Lloyd-Jones. Mathematical study of the relation between stress and strain in a flange-web connection as affecting the distribution of stress in a plate-girder. Appendices. 9000 w. Inst of Civ Engrs—No. 3637. No. 86372 N.

MUNICIPAL.

Gravel Plants.

Gravel Screening and Washing Plants. Brief illustrated descriptions of plants in different localities. 2000 w. Eng News —Aug. 1, 1907. No. 86071.

Pavements.

Cement Filler for Brick Pavements. Requirements and explanations taken from the specifications for laying vitrified brick pavements, approved by the National Brick Manufacturers' Assn. 2500 w. Munic Engng—Aug., 1907. No. 86039 C.

Notes on Tar Macadam. C. F. Wike. Abstracts of remarks offered for discussion at Liverpool, Eng. Does not consider it desirable for steep gradients. Gives specifications, cost, etc. 1000 w. Eng News—Aug. 8, 1907. No. 86311.

Paving Specifications and Inspection. Francis P. Smith. Remarks on the faults of old methods, and the new methods made possible by additional knowledge. 2200 w. Munic Jour & Engr—Aug. 7, 1907. No. 86129 C.

The Asphalt Pavement on the Thames Embankment. Illustrates and describes interesting sheet asphalt work laid on top of the old macadam pavement. 700 w. Eng Rec—Aug. 24, 1907. No. 86558.

River Purification.

River Flushing Plants at Milwaukee, Wis. Describes a system of low-level intercepting sewers, and also flushing plants for purifying fouled rivers. Ills. 2000 w. Eng News—Aug. 1, 1907. No. 86072.

Roads.

The Application of Tar to Macadamized Roads. Thomas Aitken. Extract from a paper read in Scotland. Describes an apparatus for spraying tar, either in a hot or cold state, and its operation, with information in regard to the work. 1500 w. Eng News—Aug. 22, 1907. No. 86513.

Sanitary Code.

The Revised Sanitary Code of the Town of Montclair, N. J. A reprint of this code for a suburb of New York City having a population of about 17,000. Also editorial. 14500 w. Eng News—Aug. 22, 1907. No. 86509.

Sewage.

Chicago's Thirty-ninth Street Sewage Pumping Station. Illustrated detailed description of a station with capacity for handling 1,500,000 gallons a minute. 4500 w. Engr, U S A—Aug. 1, 1907. No. 86130 C.

Sewage Experiments at Matunga, Bombay. Gilbert J. Fowler. Describes work of the sewage experiment station at the Leper Colony, a septic tank installation. The marked septic activity and small sludge residue, under a uniformly high temperature, and the large volume of gas produced are of interest. Ills. 1800 w. Eng News—Aug. 8, 1907. No. 86312.

Sewage Problem of Western Pennsylvania. I. Introductory. Morris Knowles. II. Administration of Pennsylvania Laws Respecting Stream Pollution. F. Herbert Snow. III. Present Condition of Municipal Sewers of Pittsburgh. I. Charles Palmer. Three papers discussed together. 16000 w. Pro Engrs' Soc of W Penn—July, 1907. No. 86396 D.

The Use and the Abuse of Sewage Purification Plants. A. Elliott Kimberly. Condensed from paper read before the Ohio Engng. Soc. Discusses particularly the neglect of such plants in the State of Ohio. 4500 w. Eng News—Aug. 29, 1907. No. 86637.

Work at the Madeleine Sewage Experiment Station, Pasteur Institute of Lille, France. Earle B. Phelps. Information based on two French reports, describing the plant and its operation. 1300 w. Eng News—Aug. 15, 1907. No. 86391.

Sewage Pumping.

New Sewage Pumping Station at Chicago. Athburn W. Meltrose. Illustrated description of the installation of the Thirty-ninth Street pumping station, on the South Side, which will soon be in operation. 3500 w. Power—Aug., 1907. No. 86032 C.

The Efficiency of the Shone System of Pumping. Charles Leslie Cox. Gives data concerning the actual working of this system, with results of tests, carried out on the sewage-pumping plant of the Gosport and Alverstoke Urban District Council. 2200 w. Inst of Civ Engrs—No. 3659. No. 86375 N.

Trade Wastes.

Report on the Disposal of Trade

We supply copies of these articles. See page 239.

Wastes of Reading, Pa. A report of investigations made, giving the conclusions and recommendations. 1200 w. Eng News—Aug. 22, 1907. No. 86512.

WATER SUPPLY.

Boston.
The Metropolitan Water Supply System. Caleb Mills Saville. Illustrated description of reservoirs, aqueducts, and pumping stations belonging to the system that supplies Boston and vicinity. 2000 w. Munic Engng—Aug., 1907. No. 86038 C.

Filtration.
Experiments with a Jewell Filter at the Posen Water Works. E. A. Gieseler. Explains the sources of supply and the remarkable results obtained by mixing in certain proportions the "Herzog" and "Eichwald" water, describing tests made of a Jewell filter. 2500 w. Eng Rec—Aug. 10, 1907. No. 86317.

Irrigation.
A New Egyptian Irrigation Canal. J. B. Van Brussel. Illustrated description of a canal built of riveted steel, and a reinforced concrete service reservoir, forming a part of the scheme for irrigating 125,000 acres in Egypt. 1500 w. Sci Am—Aug. 17, 1907. No. 86418.

Pumping Engines.
See Mechanical Engineering, Hydraulics.

Purification.
New Water Purification Plant at Exeter, N. H. Robert Spurr Weston. Illustrated detailed description of a new plant consisting of a concrete coagulating basin, covered with an arched roof, adjoining two filters and a pipe gallery. Its operation is described. 2300 w. Eng News—Aug. 8, 1907. No. 86309.

Reservoirs.
Stripping Reservoir Sites. A brief summary of report discussing the advisability of stripping the Ashokan reservoir site of New York City's additional supply. Aeration and filtration recommended instead. 2200 w. Munic Jour & Engr—Aug. 7, 1907. No. 86128 C.

Stream Pollution.
The Massachusetts Position Regarding Pollution of Streams by Mill Wastes. Gives the text of the court's statement regarding stream pollution. 2000 w. Eng Rec—Aug. 10, 1907. No. 86316.

Water-Works.
Some Notes on Oriental Water-Works. George A. Johnson. Read before the Am. W.-Wks. Assn. Describes features of Japanese water-works, and of the works at Shanghai, China, Singapore, Straits Settlements, and Bethmangala, and Calcutta, India. 5800 w. Eng Rec—Aug. 24, 1907. No. 86560.

Wood-Stave Pipe.
The Wood-Stave Pipe Line of the Madison River Power Company. W E. Belcher. Describes this pipe line in western Montana. 800 w. Eng & Min Jour—Aug. 24, 1907. No. 86556.

WATERWAYS AND HARBORS.

Bruges.
The Harbor Works at Bruges and the Port of Call at Zeebrugge, Belgium (Les Installations Maritimes de Bruges et le Port d'Escale de Zeebrugge, Belgique). A. Dumas. An illustrated description of the extensive works, breakwater, docks, canal, etc., recently completed. Plate. 4500 w. Génie Civil—July 20, 1907. No. 86212 D.

Zeebrugge and Bruges. An illustrated account of the construction of new docks in Bruges and a ship canal to Zeebrugge, in the effort to restore the lost trade of this once important city. 3300 w. Engr, Lond—July 26, 1907. No. 86187 A.

Channel Improvement.
New Entrance Channel at Saint Nazaire. Map and description of an important engineering work at the mouth of the Loire River, in France. 2000 w. Engr, Lond—Aug. 9, 1907. No. 86473 A.

The West Neebish Channel of the St. Mary's River. An illustrated account of nearly completed work on this channel which will relieve the present situation and give practically a double course from Lake Superior to Lake Huron. 3500 w. Eng Rec—Aug. 3, 1907. Serial. 1st part. No. 86085.

Coast Protection.
Sea Defence Works at Hornsea. Illustrated description of a concrete sea-wall recently constructed, as part of a scheme of protective works planned. 1200 w. Engr, Lond—Aug. 9, 1907. No. 86475 A.

The de Muralt System of Reinforced Concrete Sea Defenses. An illustrated article describing a system evolved by Mr. de Muralt, applicable to all phases of land protective work. 2500 w. Sci Am Sup—Aug. 3, 1907. No. 86066.

The Form and Energy of Sea-Waves. Gerald Otley Case and Frank J. Gray, in *The Contract Jour.* A discussion of ocean mechanics, and coast erosion. 1500 w. Sci Am Sup—Aug. 3, 1907. No. 86070.

Débris Barrier.
Destruction of Débris Barrier No. 1, Yuba River, California. E. C. Murphy. Illustrates and describes this barrier, constructed to hold back the mining débris, giving an account of its destruction and the probable cause. 1500 w. Eng News—Aug. 8, 1907. No. 86306.

Floods.
Floods and Means of Their Prevention in Our Western Rivers. T. P. Roberts.

We supply copies of these articles. See page 239.

Considers the reservoir and the embankment systems and their limitations. Especially discusses conditions at Pittsburgh. 6000 w. Pro Engrs' Soc of W Penn—July, 1907. No. 86397 D.

Harbor Works.
The Contractor's Plant and Methods on the Harbor Work at Gary, Ind. An illustrated article describing the general arrangement of the harbor, its construction, etc. 3500 w. Eng Rec—Aug. 17, 1907. No. 86420.

Rotterdam.
The Quays at Rotterdam (De Kaaimuurbouw te Rotterdam). H. A. Van Ysselsteyn. Illustrates and describes the various works for the protection of Rotterdam harbor within the last twenty years, with particular reference to some important works now under construction. Ills. 13000 w. De Ingenieur—July 20, 1907. No. 86280 D.

Wave Motion.
Progressive and Stationary Waves in Rivers. Vaughan Cornish. The present number considers progressive waves in rivers caused by added water, such as flood waves, tidal bores, etc. 6800 w. Engng—July 26, 1907. Serial. 1st part. No. 86182 A.

ELECTRICAL ENGINEERING

COMMUNICATION.

Electric Waves.
The Propagation of Electric Waves. A critical statement of the difficulties a student encounters in endeavoring to study the phenomena of wireless telegraphy, especially referring to the detachment of the electric lines of force from the oscillator and their propagation into space. Gives the writer's way of explaining the subject. 4800 w. Elec Rev, Lond—Aug. 2, 1907. No. 86359 A.

Oscillations.
Note on an Oscillographic Study of Low-Frequency Oscillating Arcs. J. T. Morris. Read before the British Assn. Gives observations made on direct and alternating current arcs, chiefly with the object of studying the effect of a change in the medium in which the arc is burning, and also to examine the effect on the arc of a transverse magnetic field. Diagrams. 1500 w. Elec Rev, Lond—Aug. 9, 1907. No. 86459 A.

On the Discharge of Condensers Charged by Transformers, and on the Regulation of Resonance Transformers. A. Blondel. Abstract from *L'Eclairage Electrique.* Considers the working of transformers used for the production of oscillations, and investigates the conditions determining the nature of the secondary discharges. 900 w. Elect'n, Lond—Aug. 2, 1907. No. 86363 A.

The Production of High-Frequency Oscillations from the Electric Arc. L. W. Austin. Abstract from *Bul. of Bureau of Standards.* Gives a short account of the work accomplished in this field, describing the apparatus used by the author in his experimental investigations. 4000 w. Elect'n, Lond—Aug. 2, 1907. No. 86362 A.

Perforator.
Kotyra Keyboard Wheatstone Perforator. Illustrated description of a Wheatstone perforator actuated by a typewriter keyboard which can be applied to existing apparatus. 700 w. Elec Rev, Lond—Aug. 16, 1907. No. 86590 A.

Radiotelegraphy.
Hot-Wire Relay for Selective Signalling. Richard Heilbrun. Describes a retardation relay, the essential part of which is a hot wire. 1000 w. Elec Rev, Lond—Aug. 9, 1907. No. 86460 A.

Note on Tuning in Wireless Telegraphy. Sir Oliver Lodge. Read before the British Assn. Suggestions for sending and receiving at tuned stations, describing a system which eliminates the influence of the earth. 1200 w. Elec Engr, Lond—Aug. 9, 1907. No. 86457 A.

The Arc and Spark in Radiotelegraphy. W. Duddell. Discourse before the British Assn. An explanation and discussion of methods of producing Hertzian waves for wireless telegraphy, and their application and development. 5500 w. Elec Engr, Lond—Aug. 9, 1907. No. 86458 A.

Submarine Cables.
The Pupin Telephone Cable of the Lake of Constance. Editorial on some of the difficulties encountered in submarine telephony, and means of overcoming them, giving an account of the work of laying the Pupin cables in the Lake of Constance. 2000 w. Engng—July 19, 1907. No. 85996 A.

Telegraphy.
Wire-Testing. L. M. Jones. Read before the Assn. of Ry. Tel. Supts. On the work at wire-testing offices, the methods used in locating and clearing of trouble, etc. 2000 w. Elec Rev, N Y—Aug. 24, 1907. No. 86546.

Discussion on "The Rowland Telegraphic System," at New York, April 15, 1907. 6000 w. Pro Am Inst of Elec. Engrs—July, 1907. No. 86539 D.

We supply copies of these articles. See page 239.

Telephony.

On the Pupin Mode of Working Trunk Telephone Lines. Sir William Preece. Read before the British Assn. Gives an historical account of the efforts made to overcome the inductive effect, discussing the Pupin method, and giving theoretical reasons for its success. 4800 w. Elect'n, Lond—Aug. 9, 1907. No. 86462 A.

Telephone Troubles (Telephonische Störungen). H. Schreiber and S. Strauss. Discusses several of the accidents peculiar to telephone service with means of prevention. 3200 w. Elektrotechniker—July 10, 1907. No. 86264 D.

The Influence of High-Tension Conductors on Telephone Lines (Ueber den Einflusz der Hochspannungs-Leitungen auf die Betriebs-Fernsprech-Leitungen). F. Schrottke. A mathematical paper giving numerous examples of the effects of both alternating and direct current systems. Ills. 4500 w. Serial. 1st part. Elektrotech Zeitschr—July 18, 1907. No. 86279 D.

The Laying of the Telephone Cable with Self-Induction Coils on the Pupin System in the Bodensee (Ueber das im Bodensee Verlegte Fernsprechkabel mit Selbstinduktionsspulen nach dem Pupinschen System). Dr. Ebeling. Illustrates and describes the cable, its laying and the results obtained. 4000 w. Elektrotech Zeitschr—July 4, 1907. No. 86278 D.

Telephotography.

Telephotography (Fernphotographie). Br. Glatzel. Describes the sending and receiving apparatus of the Korn system, illustrating several examples of transmitted photographs. 2000 w. Verkehrstech Woche u Eisenbahntech Zeitschr—July 20, 1907. No. 86244 D.

Telephotography (La Photographie à Distance). L. Baradel. A review of the development of telegraphic transmission of pictures, illustrating apparatus employed and results obtained. 2500 w. Electricien—July 6, 1907. No. 86202 D.

DISTRIBUTION.

Alternating Current.

Some Notes on Alternating-Current Distribution. A. J. Cridge. Read before the Incor. Munic. Elec. Assn. Shows in the form of curves and concise tables a few of the results obtained recently on the Sheffield (Eng.) system. 3800 w. Mech Engr—July 27, 1907. No. 86158 A.

Current Rectifier.

A Home-Made Alternating-Current Rectifier. Wilmar F. Lent. Explains the essential features involved in the successful action of the electrolytic rectifier, and gives plans for the construction. 2000 w. Sci Am—Aug. 24, 1907. No. 86516.

Mains Records.

Records and Drawings of Electric Mains. W. Mayall Milnes. Describes a scheme drawn up by J. E. Starkie, showing its advantages. Ills. 1500 w. Elec Rev, Lond—Aug. 16, 1907. No. 86591 A.

National Code.

The Influence of the Underwriters' Rules on Electrical Development. C. J. H. Woodbury. Read before the N. Y. Elec. Soc. An account of the work of preparing the National Electrical Code, and showing that it represents a consensus of intelligent opinion. 2800 w. Cent Sta—Aug., 1907. No. 86328.

Rotary Converters.

Hunting in Rotary Converters. Norman G. Meade. Explains the cause and effect of hunting. 1500 w. Elec Wld—Aug. 3, 1907. No. 86151.

Voltage Loss.

Voltage Lost in Direct-Current Wiring. N. A. Carle. Gives charts designed to read directly the volts lost for different strengths of current transmitted over various sizes of wire of different lengths, illustrating their use by examples. 600 w. Power—Aug., 1907. No. 86020 C.

Wiring.

The Fallacies of Free Wiring. G. Basil Barham. A criticism of the system claiming that it does not encourage long hour consumers, and leads to dissatisfaction. 2200 w. Elect'n, Lond—July 26, 1907. No. 86166 A.

DYNAMOS AND MOTORS.

Direct Current.

The Starting, Regulating and Stopping of Continuous-Current Motors. John T. Mould. A discussion of problems and methods giving curves. Ills. 5000 w. Elec Engr, Lond—July 26, 1907. Serial. 1st part. No. 86161 A.

Eddy Currents.

Eddy Current Losses in Alternating-Current Machines with Elliptical Rotating Field (Wirbelstromverluste in Wechselstrommaschinen mit Elliptischem Drehfelde). Reinhold Rüdenberg. A mathematical and theoretical discussion. Ills. 2500 w. Elektrotech u Maschinenbau—July 7, 1907. No. 86250 D.

Electrodes.

Tantalum Electrodes. Günther Schulze. Abstracted from the *Ann. der Physik*. Gives an account of experiments on tantalum electrodes as electrolytic rectifiers. 2600 w. Elect'n, Lond—Aug. 16, 1907. No. 86594 A.

Guarantees.

The Definition and Determination of Guarantees of Efficiency and Drop in Potential in Electrical Machines and Transformers (Definition und Bestimmung der Garantien von Wirkungsgrad und Spannungsabfall bei Maschinen und Transformatoren). Ills. 3500 w. Elektrotech Rundschau—July 17, 1907. No. 86263 D.

We supply copies of these articles. See page 239.

Heating.

Investigations of Heating in Electrical Machines (Untersuchungen zur Frage der Erwärmung Elektrischer Maschinen). Ludwig Ott. Considers heating in the armature and in the field coils, giving formulæ and describing tests to determine the value of coefficients. Ills. 2500 w. Zeitschr d Ver Deutscher Ing—July 20, 1907. No. 86276 D.

Magnet Coils.

Aluminium Coils. Felix Singer. Information concerning a German invention, utilizing the property of aluminium to become covered with a layer of oxide, for winding magnet coils, etc., with bare aluminium wire. 1400 w. Elec Rev, Lond —July 26, 1907. No. 86163 A.

Repulsion Motor.

A Combined Single-Phase and Continuous-Current Series and Compensated Repulsion Motor. E. Danielson. Abstract from *Elektrotechnische Zeitschrift.* Illustrated description. 900 w. Elect'n, Lond —July 26, 1907. No. 86169 A.

Single-Phase.

The Cramp Single-Phase Motor. W. Cramp. A report of tests of this machine. 2000 w. Elec Rev, Lond—Aug. 9, 1907. No. 86461 A.

Synchronous Motors.

Circular Current Loci of the Synchronous Motor. A. S. McAlister. Describes certain simple circular current loci of the synchronous motor which allow its characteristics to be determined equally as readily as does the circular current locus of the induction motor. 3500 w. Elec Wld—Aug. 24, 1907. No. 80544.

The Use of the Synchronous Motor as Phase Compensator. R. E. Hellmund. A study of the relation of capacity of synchronous motors to percentage fuel-load power-factor, discussing a series of cases covering the conditions which occur most frequently in practice. 2000 w. Elec Rev, N Y—Aug. 24, 1907. No. 86545.

ELECTRO-CHEMISTRY.

Copper Corrosion.

Corrosion of Copper and Copper-Alloys. John G. A. Rhodin. Remarks on the close connection between chemical and electrical phenomena, introducing a theory of dissolution of metals, considering the problem more as an electrical one than a chemical. 4000 w. Engr, Lond— July 19, 1907. Serial. 1st part. No. 85998 A.

Electric Smelting.

See Mining and Metallurgy, Iron and Steel.

Hydrates.

Hydrates in Solution. Abstracts of four papers contributed to the general discussion of this subject at a meeting of the Faraday Society. I. Thermochemis-try of Electrolytes. W. R. Bousfield and T. M. Lorry. II. Hydrates in Solution. Dr. J. C. Philip. III. Methods for Determining the Degree of Hydration. Dr. G. Senter. IV. Stability of Hydrates. Dr. A. Findlay. 8500 w. Elect'n, Lond —July 26, 1907. No. 86168 A.

ELECTRO-PHYSICS.

Ether.

On the Motions of Ether Produced by Collisions of Atoms or Molecules Containing or Not Containing Electrodes. Lord Kelvin. Read before the British Assn. A study of the motions of ether. 3000 w. Elect'n, Lond—Aug. 16, 1907. No. 86592 A.

Ionization.

The Nature of Ionization—Ionomania. H. E. Armstrong. Read before the British Assn. The author considers the word "ionization" is being improperly used, "dissociation" being really all that is usually implied. 2500 w. Elect'n, Lond —Aug. 9, 1907. No. 86464 A.

Resistance.

The Variation of Manganin Resistances with Atmospheric Humidity. W. Jaeger and St. Lindeck. A criticism of statements made by Messrs. Rosa and Babcock in an earlier paper. 1200 w. Elect'n, Lond—Aug. 2, 1907. No. 86360 A.

Thermo-Couples.

The Thermoelectromotive Forces of Potassium and Sodium with Platinum and Mercury. Harold C. Barker. Deals with the measurement of the thermoelectromotive forces of potassium-platinum and sodium-platinum couples for varying temperatures. 2000 w. Am Jour of Sci —Aug., 1907. No. 86023 D.

GENERATING STATIONS.

Accumulators.

The Present Status of the Iron-Nickel Accumulator (Etat Actuel de l'Accumulateur Fer-Nickel). M. Jumau. Gives the theory and construction of the Edison cell, a report of elaborate tests and a discussion of the results. Also remarks by M. Janet. Ills. 15000 w. Bull d l Soc Inter d Elect'ns—July, 1907. No. 86223 F.

Central Station.

A Progressive Suburban Central Station at Revere, Mass. Illustrates and describes a power plant developed under unusual load requirements, in a suburban territory. 2200 w. Elec Wld—Aug. 3, 1907. No. 86146.

Electric Power Plant at Lansford, Pennsylvania. W. E. Joyce. Illustrated description of a new plant costing $1,000,-000, built to supply 16 coal mines with power, and for other uses. 2000 w. Eng & Min Jour—Aug. 24, 1907. No. 86554.

Economics.

The Financial Side of the Central Station. A. D. Williams, Jr. Outlines some of the elements affecting the fixed charges and operating expenses, computing approximately what the fixed charges would be upon an investment. 2500 w. Elec Wld—Aug. 3, 1907. No. 86148.

Geneva, Switzerland.
The Reserve and Emergency Station of the City of Geneva (Usine de Réserve et de Secours de la Ville de Genève). M. Elmer. Illustrates and describes a secondary steam installation at the Chèvres hydro-electric station, made necessary by the large fluctuations in the River Rhone. 2000 w. Bull Tech d l Suisse Rom—July 10, 1907. No. 86204 D.

Hydro-Electric.
Power Plant Inside of a Dam on the Patapsco River. Illustrated description of a recently completed plant near Ilchester, Md., which is entirely submerged, the same structure serving as dam and power house. 1500 w. Elec Wld—Aug. 3, 1907. No. 86145.

The Caffaro Electric Power Station. Illustrates and describes the utilization of this Alpine stream to develop power for Brescia and neighboring districts. 1200 w. Engng—Aug. 9, 1907. No. 86468 A.

The Hydro-Electric Plant at Trezzo on the Adda (L'Impianto Idro-Elettrico di Trezzo sull' Adda). An illustrated detailed description of this plant, the first part discussing the dam and other hydraulic works. 2700 w. Serial. 1st part. Il Monit Tec—July 20, 1907. No. 86218 D.

The Kern River No. 1 Power Plant of the Edison Electric Co., Los Angeles. C. W. Whitney. An illustrated detailed description of interesting features of this high-head long-distance transmission plant. 6500 w. Eng Rec—Aug. 10, 1907. Serial. 1st part. No. 86314.

Lagging Currents.
Synchronous Motor Compensation for Lagging Currents. Clarence P. Fowler. Discusses the effects of lagging wattless currents on power station equipment and on the supply circuits. 4000 w. Elec Wld—Aug. 10, 1907. No. 86338.

Power-Factor.
Synchronous Motors for Improving Power-Factor. William Nesbit. Discusses the most efficient point to which the power-factor should be raised, and the advantage derived by installing a synchronous motor in a specific case. Short editorial. 3000 w. Elec Jour—Aug., 1907. No. 86399.

Rates.
Electric Power Tariffs. C. S. Vesey-Brown. Considers the charges made by hydro-electric and steam generating plants, analyzing the different tariffs in operation. 2000 w. Cassier's Mag—Aug., 1907. No. 86007 B.

Storage Batteries.
Improvements in Edison's Secondary Battery. Illustrated description of two inventions by Thomas A. Edison, upon which patents have been recently granted, to improve the operation and increase the efficiency of storage batteries. 1500 w. Sci Am Sup—Aug. 24, 1907. No. 86520.

Turbine Station.
The New Power Station of the South Metropolitan Electric Light and Power Co. Briefly considers power supply in London, and gives an illustrated description of the modern turbine station recently erected at East Greenwich. 2200 w. Elec Rev, Lond—Aug. 2, 1907. No. 86358 A.

The Central Station of the Société du Gaz et de l'Electricité, of Nice, France. C. L. Durand. Illustrated description of a steam turbine plant of considerable size. 2500 w. Elec Rev, N Y—Aug. 10, 1907. No. 86327.

LIGHTING.

Filaments.
Experiments on Osram, Wolfram, Zircon, and Other Lamps. J. T. Morris, F. Stroude, and R. Milward Ellis. Gives results of researches on the physical properties of such lamps, discussing the effect of voltage variation when the lamp is working on direct-current. 1700 w. Elect'n, Lond—July 26, 1907. Serial. 1st part. No. 86165 A.

Incandescent Lamps.
Government Incandescent Lamp Specifications. Gives the main portions of specifications as contained in a recent call for proposals for lamps to be supplied for U. S. Government service. 2000 w. Elec Wld—Aug. 3, 1907. No. 86147.

Developments in Electric Incandescent Lamps. Leon Gaster. A short review of some of the most important improvements in metallic filaments and vacuum tubes. 4500 w. Elec Engrs, Lond—Aug. 16, 1907. No. 86589 A.

Recent Progress in Electric Incandescent Lamps. Lionel Calisch. Briefly reviews recent improvements made to increase the efficiency, concluding that the light of the future will probably be either a luminescent gas or vapor. 1500 w. Cassier's Mag—Aug., 1907. No. 86011 B.

New Illuminants.
New Lights and New Illuminants from the Central Station's Point of View. R. S. Hale. Read before the Ill. Engng. Soc. Suggestions for the handling of the new lamps and points from the writer's experience. 3500 w. Elec Rev, N Y—Aug. 17, 1907. No. 86424.

Resistance.
The Beck "Regler." Illustrates and describes this automatic resistance for maintaining constant current for arc

lamps, electric furnaces, etc. 350 w. Elec Engr, Lond—Aug. 2, 1907. No. 86357 A.

Terminology.

The Concepts and Terminology of Illuminating Engineering. Dr. Clayton H. Sharp. Presidential address delivered at the convention of the Ill. Engng. Soc., Boston. Aims to show the utility of certain ideas and names, outlining the line of development recommended. 4000 w. Elec Rev, N Y—Aug. 10, 1907. No. 86326.

Vacuum Tubes.

The Occlusion of Residual Gas by the Glass Walls of Vacuum Tubes. A. A. Campbell Swinton. Abstract of a paper read before the Royal Soc. describing investigations on this subject. 1400 w. Elect'n, Lond—July 26, 1907. No. 86170 A.

Voltage.

The Case for Low Pressure. A. C. Hanson. Discusses the relative suitability of high and low voltage for different types of lamps and for motors. 2500 w. Elec Rev, Lond—July 26, 1907. No. 86164 A.

MEASUREMENT.

Instrument Transformers.

Instrument Transformers. Charles C. Garrard. Considers the measurement of high potentials, insulation of instrument transformers, terminals of current, testing polarity, etc. Ills. 2500 w. Elec Engr, Lond—July 26, 1907. Serial. 1st part. No. 86160 A.

Meters.

Shunted Type of Graphic Recording Meters. Paul MacGahan and H. W. Young. An illustrated description of the construction, stating its advantages. 1800 w. St Ry Jour—Aug. 24, 1907. No. 86543.

Protometers.

A New Comparison Photometer. Dr. Charles H. Williams. Read before the Ill. Engng. Soc. Describes a portable instrument for measuring the light reflected from surfaces to determine whether one method of lighting is better than another. 1000 w. Am Gas Lgt Jour—Aug. 19, 1907. No. 86435.

Illumination Photometers and Their Use. Preston S. Millar. Read before the Ill. Engng. Soc. General discussion of the aim of illumination photometers, giving illustrated descriptions of eleven types. 7000 w. Am Gas Lgt Jour—Aug. 19, 1907. No. 86436.

Standards.

Report of the British Association Committee on Practical Standards for Electrical Measurements. 4500 w. Elect'n, Lond—Aug. 16, 1907. No. 86593 A.

Synchroscope.

The Synchroscope. S. R. Dodds. Diagram and description of its operation.

1000 w. Elec Wld—Aug. 17, 1907. No. 86432.

TRANSMISSION.

Direct-Current.

Critical Consideration of the High-Tension Direct-Current System of Electric Power Transmission. Abstract of an article in *Elektrotechnic und Maschinenbau*, discussing critically Mr. Highfield's recent paper. 1200 w. Elect'n, Lond—July 26, 1907. No. 86167 A.

Insulators.

Three New Electrical Devices. Illustrates and describes two new insulating devices, and a horizontal support for conductors, recently patented by Louis Steinberger. 1500 w. Sci Am Sup—Aug. 10, 1907. No. 86305.

Lightning Arrester.

The Electrolytic Lightning Arrester. R. P. Jackson. Illustrates and describes the electrolytic or aluminum-cell arrester. 700 w. Elec Jour—Aug., 1907. No. 86401.

Phasing.

Problem in Phasing. J. P. Jollyman. An illustration of a simple method of solving phasing problems. 400 w. Elec Wld—Aug. 3, 1907. No. 86150.

Poles.

Experiments with Concrete Telegraph Poles. G. A. Cellar. Read before the Assn. of Ry. Tel. Supts. Also tests by Robert A. Cummings. Ills. 5500 w. Cement Age—Aug., 1907. No. 86483.

Switch-Gear.

Mechanical Considerations in the Design of High-Tension Switch-Gear. Henry William Edward Le Fanu. Illustrates and describes examples of high-tension switch-gear, which have successfully borne the test of practice. Discussion and correspondence. 20500 w. Inst of Civ Engrs —No. 3589. No. 86368 N.

Wire Outlets.

A Suggestion for a High-Tension Wire Entrance. H. C. Reagan. Illustrates and describes the outlets adopted for the Pittsburg & Butler Street Ry. Co. which give air insulation. 700 w. Elec Ry Rev —Aug. 17, 1907. No. 86441.

MISCELLANY.

Accidents.

Accidents with Electrical Machinery. Information from the last annual report of G. S. Ram, Inspector of Factories in England. 4000 w. Mech Engr—July 27, 1907. No. 86159 A.

Cells.

Selenium Cells. R. A. L. Snyder. Read before Pittsburg Sec. of Inst. of Elec. Engrs. Information concerning selenium and the uses to which it has been applied. Short discussion. Ills. 4000 w. Pro Am Inst of Elec Engrs—July, 1907. No. 86530 D.

We supply copies of these articles. See page 239.

Standard Cells. K. E. Guthe and C. L. von Ende. From the *Physical Rev.* Gives results of experiments on both Clark and Cadmium cells, having regard to the materials used in construction. 2000 w. Elect'n, Lond—Aug. 16, 1907. No. 86595 A.

Dumb-Waiters.

Electric Dumb-Waiter Machines and Systems. E. L. Dunn. Describes briefly the principal features of modern dumb-waiters, referring to the system installed in the new Plaza Hotel, New York City. Ills. 1600 w. Elec Wld—Aug. 3, 1907. No. 86149.

Education.

On the Concentric Method of Teaching Electrical Engineering. V. Karapetoff. Outlines the concentric method of edu-

cation, discussing also the present method. 5000 w. Pro Am Inst of Elec Engrs —July, 1907. No. 86536 D.

The Attitude of the Technical School Toward the Profession of Electrical Engineering. Henry H. Norris. Examines the methods of instruction in technical schools in order to ascertain how the requirements are being met, and to note the progress made. 3500 w. Pro Am Inst of. Elec Engrs—July, 1907. No. 86534 D.

Standardization.

Standardization Rules of the American Institute of Electrical Engineers. Gives rules approved and accepted June, 1907. 12000 w. Pro Am Inst of Elec Engrs— July, 1907. No. 86535 D.

INDUSTRIAL ECONOMY

Agreements.

Working Agreements Between Manufacturers. J. F. Gairns. Reviews briefly some of the more important aspects of the subject of working agreements between firms actually independent but co-operating for mutual benefit. 2000 w. Cassier's Mag—Aug., 1907. No. 86010 B.

American Trade.

American Trade Opportunities and Handicaps in South America. Lewis R. Freeman. Discusses the obstacles that work to the disadvantage of American trade. 4500 w. Engineering Magazine— Sept., 1907. No. 86606 B.

Apprenticeship.

The Training of Apprentices in Foundry Work. Brief account of a successful course of education introduced by the Ingersoll-Rand Co., at their Phillipsburg, N. J., shops. 2000 w. Foundry—Aug., 1907. No. 86014.

Cost Keeping.

An Economical and Practical Cost System. B. A. Franklin. Describes a simple and effective shop cost system for machine builders, the forms used, and its installation in a factory. 3800 w. Am Mach—Vol. 30, No. 32. No. 86197.

Costs.

What Constitutes Cost of Production. F. E. Webner. The eleventh of a series of articles on cost-keeping. Also supplement. 2000 w. Ir Trd Rev—Aug. 1, 1907. No. 86034.

Depreciation.

Depreciation. P. D. Leake. A plea for the study and use of better methods of measuring and providing for depreciation of industrial plants. 4500 w. Mech Engr—July 27, 1907. Serial. 1st part. No. 86157 A.

Dock Business.

Dock Capital, Expenditure, Receipts, and Management. Lee Galloway. Detailed discussion of the business of a dock system. 5400 w. Jour of Acc—Aug., 1907. No. 86521 C.

Education.

Address to the Engineering Section of the British Association at Leicester. Silvanus P. Thompson. Deals with the interaction of science and engineering, education and training of engineers, etc. 9000 w. Engng—Aug. 2, 1907. No. 86365 A.

Shop Work at an Engineering School. W. S. Graffam. Discusses the aim of shop practice in such a school, the needs, etc. 2000 w. Mach, N Y—Aug., 1907. No. 86136 C.

The Foundry Plant of the Siegen School for Iron and Steel Technology (Die Giesserei-Anlagen der Königlichen Fachschule für die Eisen- und Stahlindustrie des Siegener Landes zu Siegen). An illustrated description of the method and equipment for teaching foundry work at this German trade school. 2000 w. Stahl u Eisen—July 3, 1907. No. 86226 D.

The Training in Foundry Work at Pratt Institute. William C. Stimson. An illustrated description of the course of instruction given. 7000 w. Foundry— Aug., 1907. No. 86022.

The Training of Economic Geologists and the Teaching of Economic Geology. John A. Reid. Gives the writer's views of the preparation required and the methods of teaching. 3500 w. Ec Geol—June, 1907. No. 86488 D.

Eight-Hour Day.

The Reduction of the Working Day to Eight Hours (La Réduction de la Journée de Travail à Huit Heures). Maurice

We supply copies of these articles. See page 239.

Alfassa. An exhaustive consideration of its economic effects in French government works, the first number of the serial dealing with the postal and naval departments. 7000 w. Serial. 1st part. Bull d l Soc d'Encour—July, 1907. No. 86205 F.

Government Ownership.
See Railway Engineering, Miscellany.

Industrial Betterment.
Industry and Social Politics (Industrie und Sozialpolitik). R. Krause. A discussion of schemes for industrial betterment now occupying the attention of the German Reichstag. 2500 w. Stahl u Eisen—July 3, 1907. No. 86227 D.

Labor Agreement.
Labor Agreement in the Engineering Trades. Editorial reviewing the terms of agreement settled on March 22, 1907, which are soon to be ratified or condemned by the votes of the whole body of members of the unions. 1900 w. Engng —July 26, 1907. No. 86180 A.

Library System.
The Library System of Stone & Webster. G. W. Lee. A paper describing the library system of a firm having general control of some thirty public service corporations in various parts of the country. The library is at the service of them all, though particularly for the headquarters in Boston. 4500 w. Eng Rec—Aug. 24, 1907. No. 86561.

Lowest Tender.
Ethics of the Lowest Tender. Ian Gorach. Discusses some of the ethical considerations relating to tenders. 1600 w. Elec Rev, Lond—July 26, 1907. No. 86162 A.

Municipal Ownership.
General Conclusions of the Municipal Ownership Committee of the National Civic Federation. A summary of the conclusions reached by the National Civic Federation, with editorial comment. 4000 w. Eng News—Aug. 15, 1907. No. 86390.

Municipal Plants.
Municipal Electric Lighting. Ernest S. Bradford. A review of conditions and statistics in the United States. 3000 w. Munic Jour & Engr—Aug. 21, 1907. No. 86498 C.

Power Plants.
The Central Station and the Isolated Plant. H. S. Knowlton. Discusses the considerations governing the selection of each class of service for industrial purposes. 2800 w. Cassier's Mag — Aug., 1907. No. 86013 B.

Premium System.
Fixing Premium Rates Discussed. Forrest R. Cardullo. Condemns old methods and suggests a new one based on minimum total cost and no limit to possible earnings. 2500 w. Am Mach—Vol. 30, No. 31. No. 86055.

Railway Valuation.
See Improvements, under Railway Engineering, Permanent Way.

Shop Management.
See Management, under Mechanical Engineering, Machine Works and Foundries.

Socialism.
Concerning Wealth and Wage Earners. Editorial making a comparison between the earnings of capital and the earnings of labor, and giving much information of interest. 3000 w. Eng News—Aug. 1, 1907. No. 86079.

Wages.
The Wages Problem (Le Problème des Salaires). Dr. Franz Oppenheimer. A discussion of the economic problems connected with the fixing of wages. 6000 w. Rev ·Econ Inter—July, 1907. No. 86281 H.

MARINE AND NAVAL ENGINEERING

Battleships.
Ventilation and Refrigeration of Ammunition-Holds. Adrien Bochet. Abstract trans. of a paper read at the Bordeaux Int. Cong. Discusses methods of insulating holds, and ventilation by cooled air. 2200 w. Engng—July 26, 1907. No. 86185 A.

Cruiser.
· The New United States Scout Cruiser "Salem." Illustrates and describes what is said to be the ·fastest ship in the U. S. Navy. 900 w. Sci Am — Aug. 17, 1907. No. 86416.

Dreadnought.
The "Dreadnought" ("Dreadnought"). Ernst Müller. A detailed description.
from both technical and naval points of view. 2500 w. Serial. 1st part. Schiffbau —July 10, 1907. No. 86246 D.

Dry-Dock.
Opening of the Large Dry-Dock at League Island. Illustration, with brief description of this dry-dock near Philadelphia. 500 w. Sci Am—Aug. 24, 1907. No. 86515.

Floating Dock for Trinidad. Gives 2-page plate and other illustrations, with description of this self-docking floating dock. 700 w. Engng—July 26, 1907. No. 86178 A.

Present and Prospective Docking Facilities of the Pacific Coast. H. A. Crafts. Brief account of present facilities, and of the large drydock soon to be constructed at Hunter's Point. 1200 w. Sci Am—Aug. 3, 1907. No. 86060.

Electricity.

Electricity on Board Ship. Sydney F. Walker. First of a series of articles describing apparatus used, explaining the principles of the construction, and arrangement of cables, different forms of wireless telegraphic apparatus, etc. 4500 w. Marine Rev—Aug. 1, 1907. Serial. 1st part. No. 86123.

Exhibition.

The International Naval Exhibition at Bordeaux, 1907 (Die Internationale Schiffahrtausstellung in Bordeaux, 1907). W. Kaemmerer. The first part describes the grounds and buildings and commences a description of the exhibits. Ills. 2000 w. Serial. 1st part. Zeitschr d Ver Deutscher Ing—July 13, 1907. No. 86272 D.

Feed Water.

Sea Water and the Use of the Salinometer. Jas. Shirra. Abstract of a paper read before the Inst. of Marine Engrs. Explains the difference between "density" and "salinity," and the objections to the use of sea water for feeding marine boilers. 2000 w. Mech Engr—Aug. 17, 1907. No. 86588 A.

Fulton.

Fulton in France. Henry Harrison Suplee. An interesting illustrated account of Fulton's life in France, from 1797 to 1806, and his experiments with submarines and steamboats. 5500 w. Cassier's Mag—Sept., 1907. No. 86614 B.

Hydroplanes.

A Practical Gliding Craft with Submerged Hydroplanes. Illustrations, with description of a new gliding craft invented by Peter Cooper Hewitt. 2500 w. Sci Am—Aug. 3, 1907. No. 86061.

Indicator Cards.

Combined Indicator Cards. Charles S. Linch. Gives cards from every-day running of the City of Chester, and the Mohican, with particulars of the vessels. 1600 w. Int Marine Engng—Sept., 1907. Serial. 1st part. No. 86574 C.

Internal-Combustion Engines.

Internal-Combustion Engines for Marine Purposes. James Tayler Milton. States the conditions to be satisfied by a successful marine engine, discussing the application of internal-combustion engines, especially dealing with variable speed and balancing. Discussion and correspondence. 20800 w. Inst of Civ Engrs, No. 3678. No. 86370 N.

Kronprinzessin Cecilie.

Kronprinzessin Cecilie. Brief illustrated description of this twin-screw express steamship of the North German Lloyd Steamship Co. 2500 w. Marine Rev—Aug. 22, 1907. No. 86540.

Light-Draft.

Light-Draft Single Screw Passenger Steamer. Illustrates and describes the steel steamer Kalika, intended for service in and around Bombay, India. 4000 w. Int Marine Engng—Sept., 1907. No. 86-576 C.

Light Vessels.

Single Screw Steam Light Vessels. Drawings and description of vessels for the United States Government, built according to the standard specification adopted. Engng—Sept., 1907. Serial, 1st part. No. 86579 C.

Liner.

The New Indian Liner City of London. Illustrated detailed description of the vessel and its equipment. 1200 w. Int Marine Engng—Sept., 1907. No. 86573 C.

"Lusitania."

The Cunard Turbine-Driven Quadruple-Screw Atlantic Liner "Lusitania." A brief review of the progress of marine construction as applied to Atlantic steamships, with full illustrated detailed description of this vessel, and much information relating to its construction and equipment. Also editorial. 35000 w. 9 plates. Engng—Aug. 2, 1907. No. 86364 A.

The Saloons of the "Lusitania." Gives an 8-page supplement of engravings, with description of the artistic furnishings and decorations. 3000 w. Engng—July 19, 1907. No. 85992 A.

The 25-Knot Turbine Liner "Lusitania." Illustrations with brief description of this new liner. 1200 w. Sci Am—Aug. 10, 1907. No. 86301.

Merchant Ships.

Structural Development in British Merchant Ships. J. Foster-King. Read before the Inst. of Nav. Archts. A review of the development, referring especially to the Great Eastern, and to the Mauretania and Lusitania, and other types, and discussing features of design. Ills. 4800 w. Engng—July 19, 1907. No. 85997 A.

Motor Boats.

The "Typhonoid"—A New Type of Motor Boat. M. J. Peltier. Illustrates and describes a new racing boat recently launched at a French shipyard. The invention of M. André Gambin. 500 w. Sci Am—Aug. 24, 1907. No. 86517.

Pallograph.

The Electric Pallograph. Illustrated description of an instrument, designed by Otto Schlick, for measuring and registering the vibrations of steamships. 1300 w. Prac Engr—Aug. 16, 1907. No. 86584 A.

Propulsion.

The Propulsion of Ships by Means of Non-reversible Engines. Drawings and description, from *Rivista Marittima*, of vessels employing non-reversible engines, using four direct propulsion electric mo-

We supply copies of these articles. See page 239.

tors actuated by current from generators operated by the engines. 2300 w. Int Marine Engng—Sept., 1907. No. 86578 C.

Revenue Cutter.
Revenue Cutter Pamlico. Illustrates a vessel for the North Carolina sounds, recently put in service. Also gives a brief account of the U. S. revenue cutter service. 3500 w. Naut Gaz—Aug. 22, 1907. No. 86564.

Schooner.
The Auxiliary Coasting Schooner Northland. Describes the vessel, said to be the largest motor schooner in the world, and the gasoline engines. 1500 w. Int Marine Engng—Sept., 1907. No. 86575 C.

Sectional Construction.
Sectional Work in Ship Construction. J. L. Twaddell. Remarks on the changed conditions in industrial engineering which have caused the subdivision of work and the greater use of machinery. General discussion. 3500 w. Trans N-E Coast Inst of Engrs & Shipbldrs—July, 1907. No. 86491 N.

Shipyards.
The Werf Gusto of A. F. Smulders in Holland. Frank C. Perkins. Illustrated description of this shipbuilding yard. 1500 w. Int Marine Engng—Sept., 1907. No. 86577 C.

Steam Navigation.
Robert Fulton and the Centenary of Steam Navigation. A brief account of the events that led to the development and circumstances connected with it. Ills. 2200 w. Engr, Lond—Aug. 16, 1907. No. 86602 A.

Submarines.
The Present and Future of Submarine Navigation. A. M. Laubeuf. Abstract translation of a paper read at Bordeaux Int. Cong. of Naval Archt. Discusses the field of activity for submarines, the differences between the submarine and submersible types, and related subjects. 3500 w. Engng—July 26, 1907. No. 86183 A.

Trim Curves.
Trim Curves. A. E. Long. Shows how curves of sectional areas at different trims can be obtained, and how their properties can be utilized for trim calculations. Also discussion. 2 plates. 6000 w. Trans N-E Coast Inst of Engrs & Shipbldrs—July, 1907. No. 86490 N.

Turbiners.
New American Turbine Steamers Yale and Harvard. Illustrated description of these boats, with sketch of the company which owns them. 6500 w. Naut Gaz—Aug. 8, 1907. No. 86318.

Resumed Discussion on the Hon. C. Á. Parsons and Mr. R. J. Walker's Paper on "Marine Steam Turbine Development." 8700 w. Trans N-E Coast Inst of Engrs & Shipbldrs—July, 1907. No. 86489 N.

Vibration.
See also Pallograph.

Wrecking Steamer.
A New American Wrecking Steamer Relief. Illustrated description of a vessel built for service in the West Indies. 1000 w. Naut Gaz—June 27, 1907. No. 85338.

Yacht.
The Steam-Yacht Medusa. Illustrations, with brief description. 500 w. Engng—July 26, 1907. No. 86179 A.

MECHANICAL ENGINEERING

AUTOMOBILES.

Commercial Vehicles.
The Fawcett-Fowler Steam-Car. Illustrated description of an interesting heavy steam-motor vehicle. 1500 w. Engng—July 19, 1907. No. 85995 A.

Competition.
The Competition for the Kaiser's Prize (Das Kaiserpreisrennen). M. Ettlinger. Gives tables showing the characteristics of the competing cars and their performances during the test. Ills. 8000 w. Zeitschr d Mit Motorwagen Ver—July 31, 1907. No. 86259 D.

Results of the American Automobile Association's Fourth Annual Tour for the Glidden and Hower Trophies. Brief account of results with illustrations of successful cars. 1500 w. Sci Am—Aug. 3, 1907. No. 86062.

The *Grand Prix* of the Automobile Club of France in 1907 (Le Grand Prix de l'Automobile-Club de France en 1907). Ch. Dantin. Gives details of the cars taking part in this race on the "Dieppe circuit" and illustrates and describes the winners. 3500 w. Génie Civil—July 13, 1907. No. 86211 D.

Construction.
Reliability in Car Construction. F. Strickland. Gives some interesting statistics and deductions from results of the Scottish Reliability Trials. 2000 w. Autocar—Aug. 17, 1907. No. 86582 A.

Daimler-Mercedes.
The Daimler - Mercedes Petrol Cars. Illustrated detailed description of leading features of these chain-driven cars. 1500 w. Auto Jour — Aug. 10, 1907. No. 86449 A.

We supply copies of these articles. See page 239.

Dinin.

The Dinin Electric Automobiles (Les Automobiles Electriques "Dinin"). J. A. Montpellier. Illustrated description of the car, motor, storage batteries, etc. 2500 w. Electricien—July 13, 1907. No. 86203 D.

Dust.

The R. A. C. Dust Trials. Illustrations of some of the special devices entered, and of cars being driven over the prepared surfaces. 400 w. Auto Jour—Aug. 3, 1907. No. 86346 A.

Exposition.

The Display of Motor Cycles and Automobiles at Milan, 1907 (La Mostra del Ciclo e dell' Automobile, Milano, 1907). Begins an illustrated description of the various exhibits at this show. 1800 w. Serial. 1st part. Industria—July 14, 1907. No. 86221 D.

Franklin.

More Horse-Power for Franklins in 1908. An illustrated article describing the improvements introduced in the new models of this car. 2000 w. Automobile—Aug. 22, 1907. No. 86523.

Fuels.

Possible Fuels for Motor Cars. Condensed report of the Motor Union Fuels Committee. 3000 w. Auto Jour—July 27, 1907. No. 86155 A.

Knocking.

Knocking in Motors: Its Cause and Effect. An explanation of the causes and effects. 2000 w. Automobile — Aug. 22, 1907. No. 86522.

Mors.

The Mors Petrol Cars. The present article illustrates and describes the frame and suspension, and the engine. 1500 w. Auto Jour—Aug. 17, 1907. Serial. 1st part. No. 86581 A.

Omnibuses.

Gasoline Buses for New York. Brief illustrated description of vehicles recently placed in service on Fifth Avenue. 1200 w. St Ry Jour—Aug. 24, 1907. No. 86542.

The Motor 'Bus in London. Extracts from the proceedings of meetings held in London to discuss the use of heavy motor vehicles on London streets. 3500 w. Auto Jour—Aug. 3, 1907. No. 86348 A.

Reliability Trials.

Scottish Reliability Trials—Report and Results. An illustrated report of the Scottish Auto-Club, with tabulated results. 5000 w. Auto Jour—Aug. 3, 1907. No. 86347 A.

Skidding.

How Far Skidding is Due to Road Surfaces. Douglas MacKenzie. Read before the Roy. Auto. Club of Gt. B. & Ire. Discusses briefly the materials and construction of non-skidding roads. 1200 w. Automobile—Aug. 15, 1907. No. 86440.

Starting.

Motor Starting Devices for Gasoline Automobiles. Harold H. Brown. Illustrates and describes some ingenious mechanical contrivances that do away with the starting crank. 3000 w. Sci Am Sup—Aug. 3, 1907. No. 86067.

Wheels.

The Rudge-Whitworth Detachable Wire Wheel. Illustrations of the wheel, with remarks on its advantages. 1200 w. Auto Jour—Aug. 17, 1907. No. 86580 A.

COMBUSTION MOTORS.

Design.

Producer Gas Engine. F. W. Burger. Calls attention to points to be considered in designing this type of engine. 2000 w. Engr, U S A—Aug. 1, 1907. No. 86134 C.

Diesel Motors.

The Design of a Two-Cycle, Reversible Diesel Motor (Metodo di Calcolo di un Motore "Diesel" a due Tempi, Reversibile). Alessandro Guidoni. A concrete example, fully worked out, of the method of designing Diesel motors for marine service. Ills. 6000 w. Rivista Marittima—July-Aug., 1907. No. 86215 E + F.

Exhaust Gases.

On the Gases Exhausted from a Petrol Motor. Prof. B. Hopkinson and L. G. E. Morse. Read before the British Assn. Describes investigations dealing with the conditions under which carbon monoxide is formed in a high-speed internal-combustion motor, and the relation between the composition of the exhaust gases, the strength of the mixture, the power developed, and the thermal efficiency. 2500 w. Engng—Aug. 9, 1907. No. 86472 A.

Gas Engines.

See also Marine and Naval Engineering.

The Present Position of Gas and Petrol Engines. Dugald Clerk. Read before British Assn. Considers the difficulties in constructing large gas engines, describes experiments with the object of compounding gas engines and concludes with a few general remarks on petrol engines. 4000 w. Elect'n, Lond—Aug. 9, 1907. No. 86463 A.

Gas Producers.

A New Type of Pressure Producer. Illustrated description of type F Westinghouse producer, with no gas holder, the gas supply being regulated by throttling the steam blast. 1200 w. Engr, U S A—Aug. 15, 1907. No. 86413 C.

Gas Testing.

A Commercial Method of Testing Producer Gas for Sulphur. Randolph Bolling. Outlines briefly Bunte's method, and Drehschmidt's method, and gives a description of the writer's method. 900 w. Eng News—Aug. 1, 1907. No. 86073.

We supply copies of these articles. See page 239.

Ignition.

Bradley's Ignition Device. Illustrates and describes an ignition device for internal-combustion engines, aiming to secure efficient work at high speeds, and capable of giving but a single spark. 1000 w. Mech Engr—Aug. 19, 1907. No. 86454 A.

Eisemann H. T. Magneto Ignition Systems. An illustrated description of the wiring system of this device. 1800 w. Autocar—Aug. 10, 1907. No. 86450 A.

Oil Engines.

Oil Engine for Marine Purposes. F. M. Timpson. Read before the Inst. of Marine Engrs. Describes oil engines, and gives information in regard to applications, especially for marine purposes. 3000 w. Marine Rev—Aug. 1, 1907. No. 86124.

Oil Engines. S. M. Howell. General discussion with illustrated descriptions of types. 2000 w. Mach, N Y—Aug., 1907. No. 86143 C.

Progress in Oil Engines. E. F. Lake. Discusses wherein the oil engine excels the gasoline, its cost of operation and progress. Ills. 2500 w. Engr, U S A—Aug. 15, 1907. No. 86412 C.

Producer Plants.

Producer Gas Plant in Newark, N. J. Edward J. Kunze. Illustrates and describes a plant in which the same gas generated for feeding gas engines is made also to serve the burners of the annealing furnaces of a jewelry factory. 2000 w. Engr, U S A—Aug. 15, 1907. No. 86411 C.

Producer Plant for Lighting and Electric Drive. Illustrated description of a power plant of three vertical engines fed from one producer. 1500 w. Engr, U S A—Aug. 1, 1907. No. 86133 C.

Suction Gas Producer Plant at Rocky Ford, Colo. Illustrated description of an installation for operating a water and light plant. 700 w. Power—Aug., 1907. No. 86030 C.

Rating.

Horse-Power Rating of Gas Engines. Henry C. Hart. Explains a method of calculating the horse-power of any given engine. 1800 w. Gas Engrs' Mag—Aug. 15, 1907. No. 86620 A.

Testing.

The Testing of Large Gas Engines (Ueber Grossgasmaschinen und ihre Untersuchungen). Discusses the indicator diagram of the gas engine, and methods and appliances for testing, gas calorimeters and gas-analysis apparatus, with full calculations of actual tests. Ills. 5000 w. Glückauf—July 13, 1907. No. 86235 D.

HEATING AND COOLING.

Domestic.

Indirect System of Residence Heating. Plans and description of a low pressure gravity circulating system of steam heating, using the indirect system for the principal rooms, and the direct for rooms at a distance from the boiler. 1500 w. Met Work—Aug. 3, 1907. No. 86059.

Graphical Data.

Graphical Tables for Hot-Water Heating, Ventilating, and Chimney Design (Graphische Tafeln zur Berechnung von Warmwasserheizungen, Lüftungen und Schornsteinen). Herr Gramberg. Gives a number of curves showing graphically loss of pressure in pipes, etc., and with examples of their use. Ills. 6000 w. Gesundheits-Ing—July 6, 1907. No. 86265 D.

Hot-Water.

Rapid-Circulation Hot-Water Heating Systems (Warmwasser-Schnellumlaufheizungen). Ed. Meter. Read at the Heating and Ventilating Congress in Vienna, 1907. Illustrates and describes various systems. 6000 w. Gesundheits-Ing—July 20, 1907. No. 86267 D.

The Arrangement and Design of Piping in Hot-Water Heating with Special Reference to the Loss of Heat in Pipes (Anordnung und Bemessung der Rohrleitung bei Warmwasserheizungen mit Besonderer Berücksichtigung der Wärmeverluste im Rohrnetz). Ernst Lucas. A mathematical discussion with elaborate tables. Ills. 6500 w. Gesundheits-Ing—July 13, 1907. No. 86266 D.

Hygiene.

The Hygienic and Economic Importance of Heating and Ventilating Engineering (Die Gesundheitliche und Wirtschaftliche Bedeutung der Heizungs- und Lüftungstechnik). Karl Suwald. Read at the Congress of Heating and Ventilating Engineers at Vienna, 1907. 3500 w. Gesundheits-Ing—July 27, 1907. No. 86268 D.

Mechanical Equipment.

Mechanical Equipment of the New City Hall at Newark, N. J. Plan and description of the mechanical plant for a 4-story and basement building, having an unusual heating and ventilating system. 3500 w. Eng Rec—Aug. 17, 1907. Serial. 1st part. No. 86422.

Plenum System.

The Plenum System of Warm Air Heating for a School or Office Building. J. D. Hoffman. Read before the Am. Soc. of Heat. & Vent. Engrs. Aims to show how such a line of work could be accomplished. 2500 w. Dom Engng — Aug. 17, 1907. No. 86433.

Refrigeration.

Refrigeration as an Auxiliary to the Power Plant. Joseph H. Hart. Discusses the utilization of exhaust steam for the evaporation of ammonia in connection with the absorption process for

We supply copies of these articles. See page 239.

refrigeration. 1500 w. Cassier's Mag — Aug., 1907. No. 86006 B.

HYDRAULICS.

Annular Openings.
The Flow of Water Through Annular Spaces (Strömungsvorgänge in Ringförmigen Spalten). Ernst Becker. A mathematical paper, describing a series of tests to determine the values of coefficients used in the theoretical formulæ. Ills. 5000 w. Zeitschr d Ver Deutscher Ing—July 20, 1907. No. 86275 D.

Centrifugal Pumps.
Centrifugal Pumps. William O. Webber. Several types of pumps and their construction are illustrated and described. 1200 w. Engr, U S A—Aug. 1, 1907. No. 86131 C.

High-Pressure Machinery.
On the Design of Machinery for Very High Pressures. J. E. Petavel. An illustrated article discussing the special difficulties encountered at very high pressures. 1200 w. Engng—July 26, 1907. No. 86173 A.

Pumping Machinery for Dalny Water-Works. Illustrates and describes two sets of pumping engines of the vertical triple-expansion surface condensing type, which work three vertical plunger pumps. 700 w. Engr, Lond—Aug. 2, 1907. No. 86366 A.

Pumping Engines.
Two-Stage Operation of a Large Pumping Engine. A. O. Doane. Describes changes made in one of the engines of the Metropolitan Water-Works of Massachusetts as to enable it to be used in pumping to high service as well as low. Ills. 1000 w. Eng News—Aug. 29, 1907. No. 86630.

Pumping Plants.
Electric Pumping System. Illustrates and describes the system of electrically-driven pumping-plants for water supply and mine drainage recently installed at Walhalla, V. 2500 w. Aust Min Stand—July 12, 1907. No. 86154 B.
See also Civil Engineering, Municipal.

MACHINE WORKS AND FOUNDRIES.

Alloy Steels.
Heat Treatment of Alloyed Steels. E. F. Lake. Discusses methods of annealing, temperatures which are safe, ways of heating and quenching, and the effect of various alloy materials on steel. 4500 w. Am Mach—Vol. 30, No. 31. No. 86053.

Ball Bearings.
Designing a Three-point Ball Bearing. Describes the way to design the bearing and test the balls. 700 w. Am Mach—Vol. 30, No. 31. No. 86057.

Bell Yoke.
Tools for Manufacturing a Locomotive Bell Yoke. L. E. Salmon. Line engravings of a bell yoke and the fixture and small tools used in machining it, with description of work. 1000 w. Am Mach—Vol. 30, No. 33. No. 86385.

Boring Mill.
Work of the Floor-plate Boring Mill. Gives photographs and description of original shop methods at the Crocker-Wheeler Elec. Co.'s works. 700 w. Am Mach—Vol. 30, No. 33. No. 86378.

Brass.
Science Applied to the Brass Industry. Andrew M. Fairlie. Read before the Am. Found. Assn. Gives examples in proof of the benefit of scientific principles and methods applied in the brass foundry. 2000 w. Foundry—Aug., 1907. No. 86021.

Brass Oxidation.
Oxidizing Yellow Brass with the Production of a Black or Brown Color. J. W. Manson. Describes methods used by the writer and found satisfactory. 2500 w. Brass Wld—Aug., 1907. No. 86480.

Brass Works.
The Hygienic Aspect of Brass Melting and Casting. Erwin S. Sperry. On the dangers of metal poisoning and the diseases to which brass smelters are subject, giving rules proposed in England for maintaining brass shops in a sanitary condition. Ills. 3500 w. Brass Wld—Aug., 1907. No. 86479.

British Methods.
Advanced British Manufacturing Methods. Illustrations of interesting examples of repetition work in machine tool manufacture, with notes. 500 w. Mach, N Y—Aug., 1907. No. 86138 C.

Cams.
Notes on Cam Design and Cam Cutting. James L. Dinnany. Calls attention to points to be considered in designing and producing satisfactory cams. Ills. 2000 w. Mach, N Y—Aug., 1907. No. 86142 C.

Simplifying a Difficult Cam Job. States the problem and the solution. Ills. 800 w. Am Mach—Vol. 30, No. 31. No. 86054.

Case-Hardening.
Case-hardening the Alloy Steels. E. F. Lake. Shows that selecting the best alloys, the proper carbonizing materials, and giving the correct heat treatment secures satisfactory results. 3000 w. Am Mach—Vol. 30, No. 32. No. 86199.

Case Hardening (Einsetzen oder Oberflächenhärtung). Discusses the various problems of case hardening, packing, heating, etc., illustrating a case-hardening oven. 4000 w. Zeitschr f Werkzeugmasch u Werkzeuge—July 25, 1907. No. 86262 D.

We supply copies of these articles. See page 239.

Castings.

Aluminum Alloy Founding Practice. Hugh Dolnar. An illustrated article describing mixtures used and precautions taken to secure sound castings. 2500 w. Am Mach—Vol. 30, No. 34. No. 86506.

Casting Under Difficulties. Jabez Nall. Illustrates and describes the pouring a 50-ton forging machine frame with limited melting capacity and insufficient crane power. 1000 w. Foundry—Aug., 1907. No. 86016.

Chilled Castings in Iron. Walter J. May. Suggestions for the making of chilled castings. Ills. 900 w. Prac Engr —Aug. 2, 1907. No. 86349 A.

Iron Castings—Some Causes of Failure in Service. Robert Job. Read before the Am. Soc. for Test. Materials. Gives method pursued in ascertaining defects in locomotive castings and the remedies by which they were overcome. 1200 w. Foundry—Aug., 1907. No. 86017.

Malleable Castings. E. L. Rhead. The present article considers the effects of different ingredients and treatments. 5500 w. Mech Engr—Aug. 3, 1907. Serial. 1st part. No. 86350 A.

Sandless Castings. John H. Shaw. Read before the Am. Found. Assn. Describes the casting of brakeshoes in permanent molds that automatically provide for expression. Ills. 1000 w. Foundry —Aug., 1907. No. 86015.

Crank-Shaft.

Making a Six-throw Crank-shaft Out of a Solid Bar. John L. Bogert. Drawing and description, with cost of manufacture. 900 w. Am Mach—Vol. 30, No. 33. No. 86383.

Cutting Metals.

A Burner for Cutting Metals. M. U. Schoop. Explains the principle and describes the method, giving a summary of precautions which must be taken. 600 w. Elec-Chem & Met Ind—Aug., 1907. No. 86340 C.

Designing.

Graphic and Other Aids in Designing. Luther D. Burlingame. On the use of drawings and samples of various machine parts to aid draftsmen in determining sizes, etc., of component parts of machines. Ills. 1200 w. Am Mach—Vol. 30, No. 35. No. 86622.

Dies.

Dies for Making Square Pans. Julius F. A. Vogt. Sketches and description of the dies and their Use. 1200 w. Am Mach—Vol. 30, No. 34. No. 86507.

Making a Sub-press Die. H. M. Hyatt. Illustrated description of method. 1000 w. Am Mach—Vol. 30, No. 34. No. 86504.

Drying Stoves.

New Drying Stoves for Iron and Steel Foundries (Neuerungen an Trockenkammern für Eisen- und Stahlgiessereien). E. Freytag. Illustrated description of several types, particularly gas-fired stoves. 2000 w. Stahl u Eisen—July 24, 1907. No. 86230 D.

European.

Some Manufactories of Electrical and Other Machinery in Central Europe (Quelques Ateliers de Constructions Électriques et Mécaniques de l'Europe Centrale). Alfred Lambotte. Describes a large number of the more important works in France, Germany, Italy, Switzerland, etc., with notes on the electrical machines, pumps, turbines, engines, etc., produced. Ills. 35000 w. Soc Belge d'Electriciens—July, 1907. No. 86200 E.

Feeds.

Feeds and Feed Mechanisms. John Edgar. Considers the problem of feed variation for the lathe, for the milling machine, and for dependent and independent drives. 3500 w. Mach, N Y—Aug., 1907. No. 86135 C.

Files.

Files and Their Testing (Ueber Feilen und Ihre Prüfung). Illustrates and describes the file-testing machine built by Edward G. Herbert, Manchester, and methods of making tests. 2500 w. Zeitschr f Werkzeugmasch u Werkzeuge— July 5, 1907. No. 86261 D.

Flanging.

A Large Hydraulic Flanging Press. E. A. Dixie. Illustrates and describes how the flanged work for the P. R. R. system is produced by the use of a hydraulic press and cast-iron dies. 1500 w. Am Mach—Vol. 30, No. 33. No. 86381.

Fly-Wheels.

A Wire-wound Wood-rim Fly-wheel. E. E. Clock. Describes the building of these wheels. Ills. 1000 w. Power— Aug., 1907. No. 86031 C.

Foundries.

A Steel Foundry for Lumber Machinery Castings. Illustrated detailed description of a plant in Menominee, Mich., and its equipment. 2000 w. Ir Trd Rev —Aug. 1, 1907. No. 86033.

Messrs. R. Stephenson and Co.'s Foundry, Hebburn. Plan showing the position and arrangement, with illustrations and short description. 800 w. Engng— Aug. 16, 1907. No. 86600 A.

Gauges.

Adjusting Gauges by Peening. W. H. Booth. Suggestions for keeping gauges properly adjusted, stating objections to peening. 800 w. Am Mach—Vol. 30. No. 31. No. 86052.

Gears.

Calculations Respecting Epicyclic Wheel Trains. W. Owen. Gives a rule applicable to both spur- and bevel-geared

We supply copies of these articles. See page 239.

epicyclics, illustrating its application by examples. 1000 w. Am Mach—Vol. 30, No. 32. No. 86198.

Grading Steels.

The Spark Method for Grading Steels. Albert F. Shore. Describes a method depending on the action of oxygen in the air on the more combustible elements contained in different steels, which tend to act explosively when the proper temperature is reached. Ills. 1000 w. Am Mach —Vol. 30, No. 33. No. 86380.

Grinding.

Speeds on the Grinding Machine. H. F. Noyes. A brief study of speed conditions necessary to obtain the greatest output from a grinding machine, giving results of tests. 1500 w. Am Mach— Vol. 30, No. 34. No. 86505.

Vertical Spindle Surface Grinder. Illustrated detailed description of a machine of novel design and construction, utilizing a large cup wheel for rapidly and accurately grinding flat surfaces. 2500 w. Am Mach—Vol. 30, No. 31. No. 86051.

Hammers.

Compound Pneumatic Hammers. Illustrates and describes a recent design of English manufacture, giving interesting details of tests. 900 w. Engr, Lond —July 19, 1907. No. 86001 A.

Hammer Driving.

Compressed Air for Steam Hammers. Frank Richards. Considers the economy and expediency of employing compressed air for driving steam hammers, indicating the conditions under which it is advisable. 2500 w. Compressed Air—Aug., 1907. No. 86402.

Labor.

The Promotion of Employees. J. F. Gairns. Considers the relative importance of the various qualifications for promotion, aiming to give the main considerations which should govern an employer's policy. 4000 w. Cassier's Mag —Sept., 1907. No. 86615 B.

Management.

Profit Making in Shop and Factory Management. C. U. Carpenter. This eighth article of the series considers costkeeping, stock-keeping and the wage system. 4000 w. Engineering Magazine— Sept., 1907. No. 86611 B.

Manganese.

Manganese in Cast Iron. Herbert E. Field. Read before the Am. Found. Assn. Discusses the effects of the use of manganese in increasing carbon and decreasing and neutralizing sulphur. 2000 w. Foundry—Aug., 1907. No. 86020.

Metal Ruptures.

The Formation of Cracks due to Expansion (Wärmespannungen und Riss-

bildungen). Carl Sulzer. Considers ruptures of engine cylinders due to unequal heating, the cracks developed in riveting boiler plates, etc. Ills. 3500 w. Zeitschr d Ver Deutscher Ing—July 27, 1907. No. 86255 D.

Milling Machine.

A Handy Milling Machine. Illustrated detailed description of the Farwell Quick Change Milling Machine. 1200 w. Am Mach—Vol. 30, No. 31. No. 86056.

Modern Practice.

Notes on Modern Engineering Works. R. D. Summerfield. Discusses modern practice under the headings of buildings, equipment and management. 4500 w. Mech Engr—Aug. 17, 1907. No. 86585 A.

Molding.

Entirely a Molding Machine Product. Henry M. Lane. Illustrates and describes the methods followed in making Becker-Brainard milling machine castings. 1200 w. Ir Age—Aug. 1, 1907. No. 86035.

Machine Molding of a Chaff-cutter Mouthpiece Casting. George Buchanan. Illustrated description of the casting and method of molding. 1400 w. Am Mach —Vol. 30, No. 35. No. 86627.

Pattern Making.

Shop Kinks or Short Cuts. and Dont's. Presented to the American Foundrymen's Association by the Cleveland Club of Associated Patternmaker Foremen. 3000 w. Foundry—Aug., 1907. No. 86019.

Patterns.

The Patternmaking of a Locomotive Cylinder. Jno. W. Wolfenden. Illustrated detailed description of method. 900 w. Am Mach—Vol. 30, No. 33. No. 86384.

White Metal Pattern Mixtures. S. T. McMurten. Gives the composition of various mixtures and some useful information. 800 w. Brass Wld—Aug., 1907. No. 86481.

Pattern Shop.

Arrangement of a Pattern Shop in a Confined Space. Oscar E. Perrigo. States the conditions and describes the arrangement. Plan. 1500 w. Foundry—Aug., 1907. No. 86018.

Pneumatic Tools.

Portable Pneumatic Tools. Herbert Bing. Illustrates and describes the improvements effected in the early tools and the most important new tools. 11000 w. Inst of Mech Engrs — July, 1907. No. 86353 N.

The Economic Importance of Compressed-Air Tools (Die Wirtschaftliche Bedeutung der Pressluftwerkzeuge). Alexander Lang. An elaborate analysis of operating costs showing under what circumstances they are economical. 3000 w. Zeitschr d Ver Deutscher Ing—July 20, 1907. No. 86277 D.

Production Records.
A Graphical Wall Record for the Production Department. H. L. Whittemore. Explains a system giving instant oversight of all work in progress, showing its advantages. Diagrams. 3000 w. Engineering Magazine — Sept., 1907. No. 86609 B.

Reamers.
Reamers. Erik Oberg. Deals principally with the purpose of reamers and the manner in which they are used, describing types. 2000 w. Mach, N Y— Aug., 1907. Serial. 1st part. No. 86140 C.

Riveting.
Hydraulic Riveting. E. W. de Rusett. Read at Bordeaux Int. Cong. On the use of hydraulic riveting in the construction of the "Mauritania." Ills. 1200 w. Engng—July 26, 1907. No. 86184 A.

Rods.
The Design of Eccentric Pull and Push Rods. H. M. Prevost Murphy. Gives an analysis of the two most common types. Diagrams. 1200 w. Am Mach—Vol. 30, No. 35. No. 86624.

Safety Appliances.
Safety Appliances on Speed-Frames in Cotton-Mills. Explains the dangers in cotton mills, especially to women and girls, and gives illustrated descriptions of recent safety appliances. 2500 w. Engng—Aug. 9, 1907. No. 86467 A.

Stoneywood.
The Stoneywood Paper Works. Illustrates and describes the interesting engineering features of these works, near Aberdeen, Scotland. 3000 w. Engr, Lond —Aug. 16, 1907. No. 86604 A.

Storage.
Storage in a Large Concrete Machine Shop. L. P. Alford. Illustrates and describes the fire-resisting racks, boxes and other storage devices in the plant of the United Shoe Machinery Co., at Beverly, Mass. 5000 w. Am Mach—Vol. 30, No. 32. No. 86196.

Thread Milling.
The Philadelphia Thread Milling Attachments. Illustrated description of a device which renders engine lathes capable of performing thread milling with facility. 1000 w. Ir Age—Aug. 8, 1907. No. 86115.

Tubes.
Rolling Seamless Tubes. Gives an account of the development of the Mannesmann tubes, and the phenomenon of the central rupture in cross rolling, describing present practice. Ills. 1800 w. Ir Age— Aug. 29, 1907. No. 86621.

Turning.
Turning a 12-inch Gun Roller Path. Gompei Kuwada. Illustrates and describes the method adopted. 1000 w. Am Mach— Vol. 30, No. 31. No. 86058.

Turret Attachment.
A Multiple-Spindle Tailstock for Turret Work on an Engine Lathe. Oscar E. Perrigo. Illustrates and describes a very useful and economical attachment which can be made in the shop. 1500 w. Am Mach—Vol. 30, No. 35. No. 86626.

Ventilation.
The Ventilation of Workshops. The second report of a Departmental Committee to the Home Secretary in regard to investigations made as to the ventilation of English factories, dealing especially with the application of fans for the removal of dust, fumes, steam and other impurities. Also the question of respirators. 3500 w. Mech Engr—Aug. 17, 1907. Serial. 1st part. No. 86587 A.

Welding.
An Improved Electric Welder. A. Frederick Collins. An illustrated description of the Prescott welder, especially designed for welding wire and rods of small cross section. 1200 w. Sci Am— Aug. 3, 1907. No. 86065.

Autogenous Welding. F. C. Cutler. Illustrated description of the oxy-acetylene blow-pipe and its application. 1800 w. Cassier's Mag—Sept., 1907. No. 86618 B.

Wire-Drawing.
Modern Practice in Wire-Drawing Machines. An illustrated review of the trend in the design of machines used for all classes of wire making. 2000 w. Engng— Aug. 9, 1907. Serial. 1st part. No. 86466 A.

Wooden Rolls.
Making Wooden Rolls for Textile Machinery. G. A. Dexter. Brief illustrated description of methods used by the writer. 700 w. Am Mach—Vol. 30, No. 32. No. 86300.

MATERIALS OF CONSTRUCTION.

Alloy Steels.
The Forging of Alloy Steels. E. F. Lake. Showing the effect of temperature, methods of forging, welding and annealing on the production of high-grade forgings. 3000 w. Am Mach—Vol. 30, No. 35. No. 86623.

Aluminum.
Cast Aluminum for Automobile Work. Thomas J. Fay. A discussion of this material, reporting tests, and giving information of interest. 2000 w. Automobile— Aug. 15, 1907. No. 86439.

Annealing.
Annealing High-Speed Steel. Ethan Viall. Gives replies to an inquiry sent to prominent makers of high speed steel as to the best way to anneal their particular brand. 1200 w. Am Mach—Vol. 30, No. 35. No. 86625.

Automobiles.
Gray Cast Iron in Auto Construction.

We supply copies of these articles. See page 239.

Thomas J. Fay. Discusses the uses for which it is suitable, explaining some of its characteristics. 3800 w. Automobile—Aug. 1, 1907. No. 86037.

Castings.

Experiments with Cylinder Irons. Reports experiments made by F. J. Cook with mottled irons for casting cylinders, also with hematites and phosphoric irons for ordinary castings. 1600 w. Ir & Coal Trds Rev—Aug. 9, 1907. No. 86478 A.

Locomotives.

The Use of Steel in Locomotive Construction. F. A. Lart. Discusses the use of the newer and much stronger steel alloys for the construction of locomotives, with a reduction in the proportions and weights of their individual parts. 3500 w. Cassier's Mag—Sept., 1907. No. 86617 B.

MEASUREMENT.

Drop Test.

The Actual Status of the Drop Test on Nicked Bars (Ueber den Gegenwärtigen Stand der Schlagbiegeprobe mit Eingekerbten Stäben). A discussion of this method of testing, outlining the researches which have been made as to its accuracy and utility. 2500 w. Serial. 1st part. Stahl u Eisen—July, 31, 1907. No. 86232 D.

Indicator Cards.

See also Marine Engineering.

Measuring Machines.

Making a Dial Measuring Machine. Oscar E. Perrigo. Illustrated description of the machine and its use. 2500 w. Am Mach—Vol. 30, No. 33. No. 86382.

Pressure Gauges.

History and Construction of Pressure Gauges. C. E. Stromeyer. From the annual memorandum issued by the Chief Engineer to the Manchester Steam Users' Assn. Gives the history of the development of pressure gauges, and describes various instruments. 4000 w. Mech Engr—Aug. 17, 1907. Serial. 1st part. No. 86586 A.

Pyrometry.

The Electrical Measurement of Temperature (Misura Elettrica delle Temperature). A discussion of the theory of electrical pyrometry with brief details of construction and operation of various systems. 2400 w. Elettricita—July 12, 1907. No. 86216 D.

The Thwing Electric Pyrometers. Illustrated detailed description of types of this radiation pyrometer. 1800 w. Ir Age—Aug. 1, 1907. No. 86036.

Testing Machines.

An Electrically Controlled Single-Lever Testing-Machine and Some Torsion Tests. Charles E. Larard. A full illustrated description of the essential parts of the latest of J. Hartley Wicksteed's testing machines. 6000 w. Inst of Mech Engrs—July, 1907. No. 86355 N.

New Machines and Methods for the Testing of Metals (Nouveaux Mécanismes et Nouvelles Méthodes pour l'Essai des Metaux). Pierre Breuil. The first instalment illustrates and describes various types of extensometer. 8500 w. Serial. 1st part. Rev d Mécan—June 30, 1907. No. 86206 E + F.

Water Gages.

Water Level Gages. Robert E. Horton. Illustrates and describes styles of river gages designed by the writer. 1300 w. Eng News—Aug. 22, 1907. No. 86514.

POWER AND TRANSMISSION.

Belts.

Power Transmission by Belts and Ropes (Cinghie e Corde per Trasmissioni). Gino Scanferia. The first part discusses the stretching of leather and the theory of belt design. 3000 w. Serial. 1st part. Il Monit Tec—July 20, 1907. No. 86219 D.

Cableways.

Cableways Used on Shipbuilding Berths. John M. Henderson. An illustrated description of the equipment of the Jarrow Yard, for handling heavy material. 5 plates. 2500 w. Inst of Mech Engrs—July, 1907. No. 86352 N.

Coal Storage.

Some Recent Mechanical Coal Storage Plants. Wilbur G. Hudson. Discusses storage of anthracite and bituminous coal, the systems installed, etc. Ills. 4000 w. Eng News—Aug. 29, 1907. No. 86628.

Costs.

Comparative Cost of Steam and Hydro-Electric Power. William O. Webber. A commentary and supplement to Mr. H. von Schon's discussion, giving cost data from actual experience with working installations. 1000 w. Engineering Magazine—Sept., 1907. No. 86608 A.

Comparative Costs of Gasoline, Gas, Steam, and Electricity for Small Powers. William O. Webber. Gives tables of itemized cost of 1 H. P. per hour on 2, 6, 10 and 20 H. P. plants respectively for gasoline, gas, steam, and electricity. 1200 w. Eng News—Aug. 15, 1907. No. 86388.

Relative Economy of Steam and Gas Power Where Exhaust Steam is Used for Heating. F. W. Ballard. Read before the Ohio Soc. of Mech., Elec., & Steam Engrs. A comparison showing that when the exhaust steam is utilized, steam is more economical. 3000 w. Eng News—Aug. 15, 1907. No. 86387.

Cranes.

A Comparison Between Hydraulic and Electric Cranes (Comparaison entre les Grues Hydrauliques et les Grues Electriques). R. Gasquet. Describes the mechanism of both types and compares their operation, with especial reference to their utility on docks. Ills. 4000 w. Serial. 1st part. Génie Civil—July 27, 1907. No. 86213 D.

Electric Driving.
Plant of the Manz Engraving Company. Illustrated description of a modern engraving and printing plant equipped with individual electric drive. 1500 w. Engr, U S A—Aug. 15, 1907. No. 86409 C.

Rope Driving.
Investigations of Belt and Rope Driving (Versuche mit Riemen und Seiltrieben). Herr Kammerer. Describes the specially constructed machine on which the tests were made, the tests themselves, and the results. Ills. 7000 w. Zeitschr d Ver Deutscher Ing—July 13, 1907. No. 86271 D.

Transport Devices.
Hoisting and Transport Devices in Steelworks and Rolling Mills (Hebe- und Transportmittel in Stahl- und Walzwerksbetrieben). G. Stauber. Gives elaborate illustrations and detail drawings of various types of molten-metal conveyors, cranes for handling ingots, etc. 25000 w. Stahl u Eisen—July 10, 1907. No. 86224 D.

Worm Gearing.
Calculating the Dimensions of Worm Gearing. Ralph E. Flanders. A compilation of rules. Ills. 2500 w. Mach, N Y—Aug., 1907. No. 86137 C.

STEAM ENGINEERING.

Boiler Explosion.
Keir Explosion at Maypole. A report of the investigation of a boiler explosion on Feb. 18, 1907, at the Agricultural Implement Works of A. Jack & Sons, Ltd. 2000 w. Engng—July 26, 1907. No. 86186 A.

Boiler Plates.
The Rupture of Boiler Plates (Zur Frage der Rissbildung in Kesselblechen). Suggests improvements in materials, processes of manufacture, and methods of testing, illustrating their necessity and utility from actual examples of failure. Ills. 3500 w. Stahl u Eisen—July 3, 1907. No. 86225 D.

Boiler-Room Design.
Boiler-Room Design and Equipment. William H. Bryan. An illustrated article discussing in detail the present trend of practice and the controlling ideas which should govern the design. 2500 w. Cassier's Mag—Sept., 1907. No. 86619 B.

Boilers.
The Piping and Fittings for a Tubular Boiler. F. C. Douglas Wilkes. Methods of determining the size of the steam opening, the safety valve, etc., are considered in the present article. Ills. 1800 w. Boiler Maker—Aug., 1907. Serial, 1st part. No. 86042.

Economics.
Notes on Steam Plant Operation. Howard Williams. Considers both land and

marine practice and the strict attention to every possible source of loss or gain needed to secure satisfactory results. 1800 w. Elec Wld—Aug. 3, 1907. No. 86152.

Engine Adjustment.
Adjusting the Sturtevant Compound Engine. Carl S. Dow. An illustrated article giving directions for setting valves and adjusting the governor and drainage systems of the vertical cross-compound type. 2000 w. Power—Aug., 1907. No. 86024 C.

Engine Breakdown.
An Engine Wreck. P. F. Walker. An illustrated account of an accident to a vertical cylinder. 1000 w. Engr, U S A—Aug. 15, 1907. No. 86410 C.

Engines.
Estimation of the Unbalanced Forces in Multi-Cylinder One-Crank Engines. Archibald Sharp. An investigation of the unbalanced forces in an engine of this type. 4000 w. Inst of Civ Engrs—No. 3658. No. 86374 N.

Mary. An illustrated description of a new compound piston drop valve engine recently installed in the Huddersfield Corporation Electricity Works. 2500 w. Engr, Lond—July 19, 1907. No. 85999 A.

Exhaust Steam.
The Utilization of Exhaust Steam by Accumulators and Low-Pressure Turbines (Utilisation des Vapeurs d'Echappement par Accumulateurs et Turbines à Basse Pression). E. Cornez. Describes the system devised by M. A. Rateau, illustrating installations and giving details of working. 4500 w. L'All Indus—July, 1907. No. 86201 D.

Utilization of Exhaust Steam for Power Generation. T. Rich. Concerning exhaust steam turbines and the work for which they are adapted. 1500 w. Ir & Coal Trds Rev—Aug. 9, 1907. No. 86477 A.

Feed Water.
See also Marine Engineering.

Feed-Water Heaters.
Heat Transmission in the Heating Surfaces of Feed-Water Heaters (Der Wärmeübergang an Vorwärmerheizflächen). Paul Fuchs. Gives results of elaborate tests. 3000 w. Zeitschr d Ver Deutscher Ing—July 13, 1907. No. 86273 D.

Flash Steam.
The Flash Steam Generator. H. W. Bolsover. Discusses this method, especially in connection with the design of automobile vehicles, showing it to be applicable for such purposes. 1600 w. Cassier's Mag—Sept., 1907. No. 86616 B.

Flue Gases.
Apparatus for Analyzing Flue Gases. Illustrates and describes Bement's improved Orsat apparatus, and Wilson's portable gas-analysis apparatus. 1200 w. Power—Aug., 1907. No. 86029 C.

We supply copies of these articles. See page 239.

Fuel Losses.
Fuel Losses in Steam Power Plants. George H. Barrus. Discusses the more important losses as indicated by actual tests and examinations. 4500 w. Cassier's Mag—Aug., 1907. No. 86008 B.

Gas Firing.
Producer Gas for Firing Steam Boilers. A. M. Gow. Gives the writer's opinion as to why producer gas is not a suitable fuel for steam boilers. Ills. 2000 w. Power—Aug., 1907. No. 86025 C.

Injectors.
The Operation and Care of Injectors. W. H. Wakeman. A practical discussion of the leading types, explaining their action and the most approved methods of connecting. Ills. 2500 w. Power—Aug., 1907. No. 86027 C.

Isolating Valves.
Tests of Isolating Valves for Steam-Pipes. A brief account of tests made in Germany, taken from Dr. Koehler's illustrated monograph. 1600 w. Engng—July 19, 1907. No. 85993 A.

Liquid Fuel.
Kermode's Liquid-Fuel System Applied to Babcock and Wilcox Boilers. Illustrates and describes an installation of this system which has been successfully at work at the Toula Brass Works in St. Petersburg. 1500 w. Engng—July 19, 1907. No. 85994 A.

Rotary Engines.
Rotary Engines (Rotationskraftmaschinen). Leo Russmann. Illustrates and describes numerous types. 3500 w. Elektrotech u Maschinenbau—July 14, 1907. No. 86251 D.

Safety Appliances.
Safety Appliances in the Engine Room. William Wallace Christie. Calls attention to dangers due to increasing use of power machinery, illustrating results of disasters, and giving illustrated descriptions of special safety appliances for use in the engine room. 3000 w. Cassier's Mag—Aug., 1907. No. 86009 B.

Smoke Abatement.
Boiler Room Economy. Paul M. Chamberlain. Abstract of lecture before the International Assn. for the Prevention of Smoke. Discusses means of securing perfect combustion, losses, and the stoker as a smoke abater. 2500 w. Engr, U S A—Aug. 1, 1907. No. 86132 C.

Stacks.
The Layout and Construction of Steel Stacks. An illustrated article discussing methods of anchoring of self-supporting stacks, details of riveting, and other features of the construction. 2500 w. Boiler Maker—Aug., 1907. No. 86041.

Steam Flow.
The Dynamics of the Flow of Steam in Reciprocating Engines (Dynamique de l'Ecoulement de la Vapeur dans les Machines à Pistons). W. Schüle. A translation from *Zeitschrift des Vereines Deutscher Ingenieure*. Ills. 10,000 w. Serial. 1st part. Rev d Mécan—June 30, 1907. No. 86207 E + F.

Steam Properties.
A Useful Diagram of Steam Properties. H. F. Schmidt and W. C. Way. Gives a diagram from which may be found the physical properties of moist and superheated steam at different pressures and temperatures. 3500 w. Power — Aug., 1907. No. 86028 C.

Turbines.
A Technical Review of the Development of the Steam Turbine and Turbo-dynamo (Dampfturbinen und Turbodynamos in Betriebstechnischer Hinsicht). F. Niethammer. The first part of the serial reviews the progress of the turbine. Ills. 4000 w. Serial. 1st part. Elektrotech u Maschinenbau—July 21, 1907. No. 86-252 D.

Multi-Stage Steam Turbines with Impulse Effect (Mehrstufige Dampfturbinen mit Aktionswirkung). H. Wagner. An illustrated discussion of the action of Rateau and Zoelly turbines. 2000 w. Zeitschr f d Gesamte Turbinwesen—July 10, 1907. No. 86260 D.

The Modern Steam Turbine. Arnold Frean Harrison. A general description of the best known types of this class of prime mover, with some of the reasons for the construction peculiar to each. Ills. 3500 w. Inst of Civ Engrs (Student's Paper, No. 572). No. 86376 N.

See also Marine and Naval Engineering.

MISCELLANY.

Aeronautics.
A German Dirigible Air Ship: the "Zeppelin" (Un Dirigeable Allemand: le "Zeppelin"). G. Espitallier. Describes the ship and a series of elaborate tests to which it was recently subjected. Ills. 2000 w. Génie Civil—July 6, 1907. No. 86210 D.

The International Kite Ascensions. S. P. Fergusson. An illustrated account of the work in meteorological aeronautics carried out by the International Commission. 2000 w. Sci Am—Aug. 10, 1907. No. 86302.

The Use of Hydrolith for the Inflation of Balloons. Dr. G. F. Jaubert. Information in regard to this chemical compound and its manufacture. 1000 w. Sci Am—Aug. 3, 1907. No. 86063.

An Aeronautical Observatory. Dr. Alfred Gradenwitz. Illustrated description of an institution recently erected in the Prussian province of Brandenburg. 1500 w. Sci Am Sup—July 13, 1907. No. 85520.

Santos Dumont's Combined Aeroplane

and Airship. An illustrated account of experimental work near Paris. Refers briefly to recent work of other experimenters. 1400 w. Sci Am—July 6, 1907. No. 85411.

The Use of the Gyroscope in the Balancing and Steering of Aeroplanes. Robert H. Goddard. Describes an apparatus and its application to aeroplanes, explaining the controlling action. Ills. 700 w. Sci Am Sup—June 29, 1907. No. 85318.

Captive Balloons in the German Army and Navy. Dr. Alfred Gradenwitz. Gives an illustrated description of the improved type of balloon now being used, and an account of experiments recently made in locating submarine boats. 1500 w. Sci Am—July 13, 1907. No. 85518.

To the North Pole in a Dirigible Balloon (Au Pole Nord en Ballon Dirigeable). Lieut. Graham. An illustrated description of the balloon in which Wellman will make his second attempt to reach the Pole during the present year. 800 w. Génie Civil—June 8, 1907. No. 85615 D.

Centrifugal Fans.

High-Pressure Centrifugal Fans. A. Rateau. Reports experiments made to determine how far centrifugal fans could be driven under the best conditions as regards mechanical efficiency. 4500 w. Engng—Aug. 16, 1907. Serial. 1st part. No. 86601 A.

Cylinders.

Design of Thick Cylinders. T. A. Marsh. Considers the design of thick cylinders for pressures above 1000-lbs. per square inch, with special reference to hydraulic press cylinders. 1000 w. Mach, N Y—July, 1907. No. 85483 C.

Elevators.

The Growth and Development of the Elevator Industry. Charles H. Kloman. Reviews the rapid development of this industry, showing the changes through which it has passed, and the types to be installed in buildings of 40 stories or more. Ills. 2000 w. Cassier's Mag—Sept., 1907. No. 86613 B.

Energy.

The Energy Problem of the Universe. A. H. Gibson. Considers the forms which the prime mover has passed through and those to which evolution may ultimately tend, the present known sources of energy, etc. 6000 w. Cassier's Mag—Sept., 1907. No. 86612 B.

Gyroscope.

The Gyroscope. C. M. Broomall. From the Pro. of the Delaware Co. Inst. of Science. An explanation of the action of the gyroscope without mathematics. 5000 w. Sci Am Sup—Aug. 10, 1907. No. 86304.

Jute Machinery.

Observations on Present-Day Practice in Jute Preparing and Spinning. D. J. Macdonald. Describes a new method of driving the spindles with the view to saving power, changes in pressing rollers, transmission gear, etc. Ills. 2500 w. Inst of Mech Engrs—July, 1907. No. 86356 N.

Kinematics.

The Newman Kinematic Apparatus. Illustrates and describes a contrivance which simplifies the study of mechanisms. 1000 w. Sci Am Sup—Aug. 24, 1907. No. 86518.

Laboratory.

The New Engineering Laboratory at the City and Guilds of London Institute, Finsbury. Prof. E. G. Coker. Illustrated description. 1500 w. Engng—Aug. 16, 1907. No. 86599 A.

Lubricants.

Deflocculated Graphite. Edward G. Acheson. Gives an account of experimental investigations, and results. 1200 w. Pro Am Inst of Elec Engrs—July, 1907. No. 86537 D.

Moments of Inertia.

A Diagram for Calculating Moments of Inertia. O. A. Thelin. Gives a diagram designed to simplify calculations relative to moments of inertia of sections, explaining its use. 600 w. Am Mach—Vol. 30. No. 33. No. 86379.

Needle Machine.

Automatic Needle-Making Machinery. Photographs and description of an interesting machine for finishing sewing machine needles after they have been shaped. 1200 w. Mach, N Y—Aug., 1907. No. 86141 C.

MINING AND METALLURGY

COAL AND COKE.

Anthracite.

New Supplies of Anthracite Coal. W. E. Joyce. A report of some recent discoveries which upset established geological theories. 1500 w. Eng & Min Jour—Aug. 3, 1907. No. 86106.

Briquetting.

Progress in the Briquetting of Coal (Fortschritte im Kohlenstampfverfahren). A. Thau. Describes and illustrates recent machines and processes for this purpose. 4500 w. Glückauf—July 27, 1907. No. 86240 D.

We supply copies of these articles. See page 239.

Coke.

The Manufacture of Coke from Western Coal. R. S. Moss. Suggestions for coking these coals, showing the necessity of using crushed coal and raising the temperature of the oven very rapidly. Ills. 1500 w. Min Wld—Aug. 17, 1907. No. 86446.

The Carbonization of Durham Coking Coal. Andrew Short. Abstract of paper read before the Newcastle Sec. of the Soc. of Chem. Ind. Describes investigations carried out on a battery of patent coke ovens of the Otto-Hilgenstock type. 1800 w. Ir & Coal Trds Rev—July 19, 1907. No. 86004 A.

Colliery.

The Baggeridge Colliery. James Tonge. An account of the development of a new area of the thick coal of South Staffordshire, England. 2500 w. Mines & Min—Aug., 1907. No. 86100 C.

Collinsville, Ill.

Mine No. 17, Collinsville, Ill. Illustrates and describes the plans and arrangement of a mine having a capacity of 3,000 tons of coal per day, and the washery for preparing it. 1000 w. Mines & Min—Aug., 1907. No. 86097 C.

Crushing Strength.

The Ultimate Crushing Strength of Coal. Joseph Daniels and L. D. Moore. Describes tests made to determine the resisting power of anthracite and bituminous coals, in connection with a study of the size of coal-mine pillars. Ills. 2500 w. Eng & Min Jour—Aug. 10, 1907. No. 86335.

Electrical Equipment.

Allis-Chalmers Machinery Used in Coal and Coke Operations. W. B. Spellmire. Brief illustrated description of the York Run plant of the H. C. Frick Coke Co. 1000 w. Elec Rev, N Y—Aug. 24, 1907. No. 86547.

Explosions.

Colliery Warnings. A detailed examination of the value of colliery warnings, based upon barometric pressure, with an account of prevalent opinions. 4500 w. Col Guard—Aug. 9, 1907. No. 86465 A.

Geology.

Palaeontological Methods and Results in Coal Measure Geology. J. T. Stobbs. Abstract of an address before the So. Staffordshire Min. School Old Boys' Soc. 3500 w. Ir & Coal Trds Rev—July 19, 1907. No. 86003 A.

Peat.

The Technical Utilization of Bogs (Ein Beitrag zur Technischen Ausnutzung der Moore). N. Caro. Discusses the practicability of using peat as a fuel. 1800 w. Glückauf—July 13, 1907. No. 86237 D.

Prospecting.

Variability of Coal Seams. Arthur

Lakes. Explains the cause, and the necessity of many bore holes at varying distances to prove the value of an area. 800 w. Min Wld—Aug. 3, 1907. No. 86114.

Rescue Appliances.

The "Westfalia" Rescue Apparatus (Der Rettungsapparat "Westfalia"). Herr Grahn. Illustrated detailed description of this recent German device, with results of tests. 1500 w. Glückauf—July 6, 1907. No. 86233 D.

Testing.

Recent Testing of Coal Used by the Federal Government in Its Public Buildings at Washington, D. C. J. A. Holmes. Explains methods of sampling at mines and on cars, and gives analyses of coals tested. 1500 w. Mines & Min—Aug., 1907. No. 86098 C.

Washing.

Bituminous Coal Washing. G. R. Delamater. Shows the wonderful growth of the bituminous coal industry, discussing methods of removing impurities, conditions of mining, testing, etc. 6000 w. Mines & Min—Aug., 1907. Serial. 1st part. No. 86095 C.

COPPER.

British Columbia.

Mining in the Boundary Copper Field. Ralph Stokes. Map and illustrated account of the mines and methods. 3500 w. Min Wld—Aug. 3, 1907. No. 86112.

The Britannia Copper Mine, British Columbia. Ralph Stokes. Notes based on a brief inspection of the plant, and data furnished. Ills. 1300 w. Min Wld—Aug. 17, 1907. No. 86445.

The Industrial Outlook of Boundary Copper Field. Ralph Stokes. A review of the development and progress, considering the outlook encouraging. Ills. 1800 w. Min Wld—Aug. 10, 1907. No. 86324.

California.

The Ore-Deposits of Copperopolis, Calaveras Co., California. John A. Reid. The location, history, topography and general geology are reviewed, and the deposits described. 13700 w. Ec Geol—June, 1907. No. 86487 D.

Electrolytic Refining.

Ryan Process for Making Electrolytic Copper. John H. Ryan. Describes the process, giving approximate cost of a plant. 600 w. Min Rept—Aug. 8, 1907. No. 86331.

Lake Superior.

Early History of the Lake Copper Region. A brief interesting account of the earliest information in regard to these deposits, and the development from 1830 to 1870. 1400 w. Min Wld—Aug. 10, 1907. No. 86325.

Nevada.

The Ely Copper Deposits and Their

We supply copies of these articles. See page 239.

Rapid Development. William Starr Bullock. An illustrated account of these extensive deposits and their development. 1500 w. Min Wld—Aug. 10, 1907. No. 86323.

Nicaragua.
Copper Ores in Nicaragua and Their Treatment. Information concerning this metal and its occurrence on the Escondido River and its tributaries. 3000 w. Min Jour—July 27, 1907. No. 86172 A.

Rolling.
The Rolling of Sheet Copper. Describes the style of rolling-mills used, and the general methods used in the manufacture. 3000 w. Ir Age—Aug. 22, 1907. No. 86502.

Smelting.
Briton Ferry Works of the Cape Copper Co. Edward Walker. Illustrates and describes the methods of treatment at these copper smelting works in South Wales. 1000 w. Eng & Min Jour—Aug. 17, 1907. No. 86428.

Negative Results in Pyritic Smelting. G. F. Beardsley. Describes an attempt made to treat, by pyritic smelting, the copper-nickel ores of the Sudbury district, Ontario, Canada. An account of the furnaces used and their construction; the tests made and the results obtained. 1500 w. Eng & Min Jour—Aug. 24, 1907. No. 86549.

Smelting Plants.
The Largest Copper Smelting Plant in the World From a pamphlet issued by the Anaconda Copper Mining Co. describing the Washoe smelter. Ills. 3000 w. Eng News—Aug. 1, 1907. No. 86075.

The Mining and Smelting Equipment of the Canadian Copper Company. David H. Browne. Illustrated description of the plants in operation and under construction for the treatment of copper-nickel ores. Electric power is used, also compressed air and steam. 5000 w. Can Min Jour—Aug. 1, 1907. No. 86127.

Wyoming.
Copper Deposits of Hartville Uplift, Wyoming. Sydney H. Ball. The present number describes the ore deposits. 1700 w. Min Rept—Aug. 8, 1907. Serial. 1st part. No. 86330.

GOLD AND SILVER.

Assaying.
The Litharge Method. B. M. Snyder. Gives the writer's experience with the litharge method of fire-assay. 1600 w. Min Rept—Aug. 1, 1907. No. 86111.

Black Sands.
Black Sands of the Pacific Coast. Dr. David T. Day. An interesting account of the investigations made, with information of methods of analysis and concentration and the valuable by-products found. 4500 w. Jour Fr Inst—Aug., 1907. No. 86343 D.

The Black Sand Problem. F. Powell. A suggestion from personal experience for an effective method of collecting black sand. 1200 w. Eng & Min Jour—Aug. 10, 1907. No. 86333.

Burma.
The Auriferous Deposits of Burma. Malcolm Maclaren. Sketch map showing localities, with brief review of the history of gold exploration in Burma. 2500 w. Min Jour—July 27, 1907. No. 86171 A.

Cobalt.
Vein Formation at Cobalt, Ontario. J. B. Tyrrell. Discusses these ore deposits, giving opinions of writers of note, and the writer's conclusions. Ills. 2000 w. Can Min Jour—Aug. 1' 1907. No. 86126.

The Cobalt Silver Field as an Industry. Ralph Stokes. The first of a series of articles aiming to give facts in regard to present conditions. Ills. Also editorial. 3500 w. Min Wld—Aug. 24, 1907. Serial. 1st part. No. 86571.

Colorado.
Lodes in the Tertiary Eruptions of Colorado. T. A. Rickard. Describes the geology and considers the deposits of gold and silver. Ills. 2000 w. Min & Sci Pr—Aug. 10, 1907. No. 86407.

Cyanide Plant.
A Cheap Form of Cyanide Plant. Charles Hunter. Gives a copy of the contract and specifications for one of the portable plants used in Southern Rhodesia. 1000 w. Inst of Min & Met, Bul 34—July 18, 1907. No. 86496 N.

Egypt.
Gold Mining in Egypt. C. S. Herzig. An illustrated article reviewing the history of the gold-mining industry in Egypt. 2000 w. Min & Sci Pr—Aug. 17, 1907. No. 86524.

Filter Press.
The Kelly Filter Press. D. J. Kelly. Describes a development of pressure filtration for the treatment of slimes and the operation of a plant on the Kelly System. 2000 w. Min Rept—Aug. 22, 1907. No. 86569.

Klondike Placers.
Concentration of Gold in the Klondike. J. B. Tyrrell. Describes the conditions that prevail in the Klondike district, studying the rich placers, their origin and concentration. Ills. 2000 w. Ec Geol—June, 1907. No. 86484 D.

Milling.
Stamp-Mill Reduction-Plant of the New Kleinfontein Company, Limited, Witwatersrand, Transvaal. Edward John Way. An illustrated detailed description of a new 200-stamp mill installation designed to use mechanical means wherever hand-labor could be dispensed with. Appendices. 23500 w. Inst of Civ Engrs—No. 3631. No. 86373 N.

We supply copies of these articles. See page 239.

Mining.

Present Status of the Gold Mining Industry. J. H. Curle. A review of the present conditions in all parts of the world where gold is produced. 3300 w. Min & Sci Pr—Aug. 3, 1907. No. 86189.

Nevada.

Pioche, Nevada. James W. Abbott. Gives the history of the discovery and development of this silver region. Ills. 2500 w. Min & Sci Pr—Aug. 10, 1907. No. 86406.

The Goldfield District, Nevada. Abstract of paper by J. E. Spurr, giving information of the geology of this district. 2000 w. Jour Fr Inst—Aug., 1907. No. 86344 D.

Placers.

The Essential Data of Placer Investigations. J. P. Hutchins. Calls attention to points to be ascertained and precautions to be taken in the examination and valuation of placer ground before exploitation. Ills. 2500 w. Eng & Min Jour—Aug. 24, 1907. Serial. 1st part. No. 86548.

The Nomenclature of Modern Placer Mining. J. P. Hutchins. Discusses the classification of placers, their origin, and methods of exploitation. Ills. 2000 w. Eng & Min Jour—Aug. 17, 1907. No. 86426.

Separator.

Combined Ore Crusher, Pulverizer, Retort and Bullion Separator. A combined arrangement for a continuous operation on any kind of ore is illustrated and described. 800 w. Min Rept—Aug. 22, 1907. No. 86570.

Slimes.

The Utilization of Waste Heat in Slimes Settlement. A. Salkinson. Describes trials made with encouraging results. 2500 w. Jour Chem, Met & Min Soc of S Africa—June, 1907. No. 86448 E.

IRON AND STEEL.

Blast Furnaces.

Charging a Modern Iron Blast Furnace. Bradley Stoughton. Considers the variations required in the charge of the furnace to produce the required results in pig iron. 1200 w. Eng & Min Jour —Aug. 24, 1907. No. 86551.

The Chemistry of the Iron Blast-Furnace. Bradley Stoughton. Traces the successive chemical reactions which occur as iron ore is reduced to pig iron. Ills. 1500 w. Eng & Min Jour—Aug. 3, 1907. No. 86104.

The Operation of the Iron Blast Furnace. Bradley Stoughton. Outlines the working of the furnace when in blast, giving an account of some difficulties encountered. 2000 w. Eng & Min Jour— Aug. 17, 1907. No. 86429.

Blowing Engines.

Parson's Turbine Blower as a Blast-Furnace Blowing Engine (Das Turbinengebläse von C. A. Parsons als Hochofengebläsemaschine). Julius Fürstenau. Enumerates and gives details of the plants at which this blower is used, with results of tests and remarks on accessory devices, etc. Ills. 5500 w. Zeitschr d Ver Deutscher Ing—July 20, 1907. No. 86274 D.

The Parsons Turbo-Blower for Blast Furnaces. Illustrated description of an installation at the Trzynietz furnace plant, as given by Julius Fuerstenau, in the *Zeit. des Ver. Deut. Ing.* 1500 w. Ir Age—Aug. 22, 1907. No. 80499.

Cuba.

The Mayari Iron Ore District of Cuba. Information concerning this remarkable deposit, estimated to contain over 500,000,000 tons, and some of the problems in the handling and treatment of the ore. Maps and Ills. 4500 w. Ir Age—Aug. 15, 1907. No. 86377.

Electric Smelting.

Electrical Smelting of Iron Ore. Brief account of the most successful experimental run of the Héroult electrical smelting plant in Shasta Co., Cal. 600 w. Min Rept—Aug. 8, 1907. No. 86329.

Electric Smelting of Iron Ore in California. Gives a summary of an article describing the application of the Héroult Process to the smelting of iron from the large magnetite deposits. Also an editorial "Electric Heat versus Heat from Fuel.' 2000 w. Elec Chem & Met Ind —Aug., 1907. No. 86339 C.

Impurities.

The Nonmetallic Impurities in Steel. Reviews the investigations by E. F. Law, of London. 1600 w. Ir Age—Aug. 22, 1907. No. 86501.

Lake Superior.

Mining Methods on the Gogebic Iron Range. Reginald Meeks. Describes the exploratory methods, the steel shafts and headframes which are replacing wooden construction, etc. Ills. 1400 w. Eng & Min Jour—Aug. 10, 1907. No. 86332.

The Iron-Ore Mines of the Mesabi Range. Reginald Meeks. An illustrated account of mines opened in 1892, which now produce 62 per cent. of the total ore from the northern ranges. 2200 w. Eng & Min Jour—Aug. 3, 1907. No. 86101.

Metallography.

The Present Condition of Our Knowledge of the Hardening and Cooling Phenomena in Iron-Carbon Alloys (Ueber den Augenblicklichen Stand unserer Kenntnisse der Erstarrungs- und Erkaltungsvorgänge bei Eisenkohlenstofflegierungen). P. Goerens. Illustrated by micro-photographs. 2500 w. Stahl u Eisen—July 24, 1907. No. 86229 D.

We supply copies of these articles. See page 239.

Phosphorus.

The Estimation of Phosphorus in Steel and Iron. J. J. Morgan. Describes a method by which a determination in steel may be made in 30 minutes without the use of any special apparatus. 800 w. Prac Engr—Aug. 16, 1907. No. 86583 A.

Piping.

The Avoidance of Piping (Zur Frage der Vermeidung von Lunkerbildung). Adalbert Obholzer. A description of the results obtained at the Diosgyör steelworks, Hungary, from the use of thermit. Ills. 2300 w. Serial. 1st part. Stahl u Eisen—July 31, 1907. No. 86231 D.

Segregation.

Segregation in Steel. Abstract of a paper by John E. Stead, read at the Engng. Con. of the Inst. of Civ. Engrs. Results of an investigation. 1200 w. Ir Age—Aug. 8, 1907. No. 86116.

Smelters.

The Plants of the Luxemburg Mining, and Saarbrück Smelting Company (Die Anlagen der Luxemburger Bergwerks- und Saarbrücker Eisenhütten-Aktiengesellschaft). F. Schroeder. The first part of the serial reviews the history of this German company from its formation in 1856. Ills. 3000 w. Serial. 1st part. Oest Zeitschr f Berg u Hüttenwesen—July 27, 1907. No. 86249 D.

MINING.

Air Compressors.

A New Development in Air Compressors. Illustrated description of the Nordberg constant speed variable delivery air compressor. 1200 w. Am Mach—Vol. 30, No. 34. No. 86503.

Bore-holes.

The Deviation of Rand Bore-holes from the Vertical. Joseph Kitchin. Gives the results of the surveys of 22 bore-holes, with facts related to the subject. 6000 w. Inst of Min & Met, Bul 34—July 18, 1907. No. 86495 N.

Boring Devices.

The Valve Mechanisms of Hydraulic Boring Devices (Die Steuerungen der Hydraulischen Tiefbohrrichtungen). Frd. Freise. The first part of the serial deals with devices with spindle valves. Ills. 1600 w. Serial. 1st part. Oest Zeitschr f Berg u Hüttenwesen—July 20, 1907. No. 86248 D.

Cable Ways.

The Cable Ways for Culm of the Courl and Scharnhorst Mines of the Harpener Mining Company (Die Drahtseilbahnen für Versatzmaterial der Zechen Courl und Scharnhorst der Harpener Bergbau-Aktien-Gesellschaft). F. Schulte. Describes a system for transporting culm from these two mines to a central dump. Plan. 2300 w. Glückauf—July 13, 1907. No. 86236 D.

Drills.

The Gordon Drill on the Rand. Information from recent Consular Reports in regard to the operation and trials of this drill. 2000 w. Compressed Air—Aug., 1907. No. 86404.

Haulage.

Rack-rail Haulage in Coal Mines. George E. Lynch. An illustrated article showing how the rack-rail and cog-wheel are used to supplement ordinary traction where steep grades occur in mines. 3000 w. Eng & Min Jour—Aug. 3, 1907. No. 86105.

Hydraulic Mining.

Notes on Hydraulic Mining. Discusses the subject of hydraulic gold mining, with special reference to the Cariboo District, British Columbia, and Yukon Territory. Ills. 4000 w. Mines & Min—Aug., 1907. No. 86093 C.

Joplin District.

Ground Breaking in the Joplin District. Doss Brittain. Illustrated article describing method. Holes are drilled with air drills and squibbed before charging. 2500 w. Eng & Min Jour—Aug. 10, 1907. No. 86334.

Magnetic Surveying.

Magnetic Observations in Geological and Economic Work. Henry Lloyd Smyth. Explains the construction of the dial-compass and the magnetometer, and the fields to which each is adapted. Ills. 4000 w. Ec Geol—June, 1907. Serial. 1st part. No. 86486 D.

Magnetometric Prospecting.

The Locating of Ore by Electricity (Das Aufsuchen von Erzen mittels Elektrizität). Herr Petersson. An explanation of the method of locating and determining the extent of ore bodies used by the Electrical Ore Finding Company, London. Ills. 2500 w. Glückauf—July 20, 1907. No. 86239 D.

Mine Dam.

A Mine Dam to Recover Flooded Workings. John H. Haertter. Plan and description of how difficulties were overcome in constructing a dam to permit unwatering flooded coal-mine workings. 1800 w. Eng & Min Jour—Aug. 17, 1907. No. 86431.

Mine Waters.

Mine Water and Mine Fire in the Neu-Diepenbrock III Mine at Selbeck (Grubenwasser und Grubenbrand in dem Erzbergwerk Neu-Diepenbrock III zu Selbeck). D. Hilt. An exhaustive discussion of the causes of the troubles experienced at this German mine. Ills. 5500 w. Glückauf—July 20, 1907. No. 86238 D.

Ore Handling.

A Cripple Creek Ore-handling Plant. S. A. Worcester. Illustrates and describes a single, counterbalanced skip

We supply copies of these articles. See page 239.

hoist, discharging into self-dumping cars, claiming that it adds 33 per cent. to the capacity of a shaft. 1300 w. Eng & Min Jour—Aug. 24, 1907. No. 86552.

Prospecting.
Practical Points for Prospectors. Matt. W. Alderson. Information from actual experience. 1500 w. Min Wld—Aug. 3, 1907. Serial. 1st part. No. 86113.

Pumping.
Electric Pumping Plant at the Gilly Collieries, Belgium. Dr. Alfred Gradenwitz. Short illustrated description. 1000 w. Prac Engr—Aug. 9, 1907. No. 86452 A.

Quarrying.
Granite Quarrying in Aberdeenshire. William Simpson. An illustrated description of these quarries and the methods of working. 6500 w. Inst of Mech Engrs —July, 1907. No. 86354 N.

Sand Pumping.
Pumping Sand by Compressed Air. Lucius I. Wightman. Illustrateu description of a plant at Uttawa, Ill. 1000 w. Compressed Air—Aug., 1907. No. 86403.

Shaft Signals.
Electric Shaft Signals (Die Elektrischen Schachtsignalanlagen). Herr Schmiede. Illustrates and describes various types of bell, methods of installation, etc., giving wiring diagrams. 2000 w. Glückauf—July 6, 1907. No. 86234 D.

Shaft Sinking.
Sinking Through Bad Ground. F. W. Adgate. Describes the sinking of a shaft on the Mesabi range, near Biwabik, Minn. Of interest because of unusual difficulties, and the novel application of the pneumatic process. Ills. 2500 w. Min & Sci Pr—Aug. 10, 1907. No. 86408.

Ventilation of Shafts During Sinking Operations Gives diagrammatic views and description of an arrangement of "Sirocco" plant for ventilating shafts. 1200 w. Ir & Coal Trds Rev—July 19, 1907. No. 86002 A.

Subsidence.
Subsidence in Underground Mines. Alexander Richardson. A summary of investigations, with suggestions as to means of prevention. 4000 w. Eng & Min Jour —Aug. 3, 1907. No. 86102.

Tin.
Tin Mining in Ulu Selangor, Federated Malay States. E. Nightingale. Describes the geological features and methods of mining. 2000 w. Inst of Min & Met, Bul 34—July 18, 1907. No. 86497 N.

Ventilation.
On the Practical Measurement of Air (Ueber die Praktische Ausführung von Wettermessungen). Kurt Seidl. Illustrates and describes a method of air measurement by anemometer readings at a number of points in a specially-made opening, giving formulæ. 1800 w. Glückauf—July 20, 1907. No. 86241 D.

MISCELLANY.

Bismuth.
Electrolytic Refining of Bismuth. Dr. Arnold Mohn. Describes the process used with Mexican lead-bismuth bullion. 1200 w. Elec-Chem & Met Ind—Aug., 1907. No. 86342 C.

Corundum.
Corundum at Craigmont. H. E. T. Haultain. Describes the deposits of Craig Mountain, the discovery, mining operations; and gives information relating to the applications of corundum. Ills. 3500 w. Can Min Jour—Aug. 1, 1907. No. 86125.

Jamaica.
Geology of Jamaica, as Related to Its History. Rossiter W. Raymond. Outlines the general natural features, indicating briefly their effect upon the history of the island. 2000 w. Min & Sci Pr—Aug. 3, 1907. No. 86188.

Lead.
The Daly-Judge Mine and Mill, Park City, Utah. Paul A. Gow, Andrew M. Howat, George S. Kruger, and F. H. Parsons. A description of the veins, methods of working and timbering, and report of production. Ills. 4000 w. Mines & Min—Aug., 1907. Serial. 1st part. No. 86099 C.

Lead Refining.
The Betts Process at Trail, B. C. A. G. Wolf. Describes the electrolytic refining of lead bullion, treatment of gold and silver slimes, and copper sulphate recovery. Ills. 5000 w. Mines & Min— Aug., 1907. No. 86096 C.

Manganese.
Prospects of Indian Manganese Industry. A. Ghose. Shows that India has attained a leading position in the manganese market, and gives information of interest. 3500 w. Jour Soc of Arts—. Aug. 2, 1907. No. 86345 A.

Mercury.
The Use and Care of Mercury. Explains some of the causes of the flouring and sickening of quicksilver, and various methods of remedying the evil. 2500 w. Min & Sci Pr—Aug. 17, 1907. No. 86525.

New Mexico.
Burro Mountains Copper District. William Rogers Wade. Gives briefly the history of the camp and an account of present operations. Ills. 1200 w. Eng & Min Jour—Aug. 24, 1907. No. 86553.

New Zealand.
The Mineral Resources of New Zealand. T. Hilton. Suggestions for the development of the deposits of gold, silver, lead, copper, iron ores, manganese, petroleum, and especially the coal fields.

We supply copies of these articles. See page 239.

6000 w. N Z Mines Rec—May 16, 1907. No. 86153 B.

Nickel Ores.

On the Microstructure of Nickeliferous Pyrrhotites. William Campbell and C. W. Knight. An interesting metallographic study of nickel ores from widely distant localities, with conclusions. Ills. 5000 w. Ec Geol—June, 1907. No. 86485 D.

Ore Deposits.

The Ore Deposits of the Schneeberg near Sterzing in the Tyrol (Ein Beitrag zur Kenntnis der Tektonik der Erzlagerstätten am Schneeberg bei Sterzing in Tyrol). B. Granigg. A contribution to the theory of ore deposits. The first part describes the complex vein system of this Austrian district. Ills. Map. 4000 w. Serial. 1st part. Oest Zeitschr f Berg u Hüttenwesen—July 6, 1907. No. 86247 D.

Tin.

Alluvial Tinstone Deposits of Northern Nigeria. Describes the deposits and the washing and smelting methods used by the natives. 1500 w. Min Jour—Aug. 17, 1907. No. 86597 A.

Electrolytic Refining of Tin. Otto Steiner. Explains the theory and the conditions of importance, describing experiments and methods of working. 4000 w. Elec-Chem & Met Ind—Aug., 1907. No. 86341 C.

RAILWAY ENGINEERING

CONDUCTING TRANSPORTATION.

British Practice.

Note on the Operating of Trains and Locomotives in England (Note sur le Service des Trains et des Machines en Angleterre). MM. Demoulin and Bezier. Illustrates and describes cars and locomotives on British railways and discusses points of difference in British and French operating practice. 12000 w. Rev Gen d Chemins d Fer—July, 1907. No. 86208 G.

Freight Cars.

Freight Car Department. D. T. Taylor. Briefly considers car interchange, inspection, repairs, etc. 3000 w. Pro St Louis Ry Club—July 12, 1907. No. 86005.

Wrecks.

A Disastrous Blast. Map and illustrated account of a blast on May 16, near Lookout Mountain, which wrecked a bridge 600 ft. distant, struck a pile driver 900 ft. distant, killing two men, wrecked a train of 11 freight cars, and did other damage. 500 w. Ry & Engng Rev—Aug. 3, 1907. No. 86119.

Curve Mechanics and the Woodlawn Wreck. T. H. Brigg. A letter from an English writer offering an explanation of the accident in Feb., 1907. 1200 w. R R Gaz—Aug. 23, 1907. No. 86527.

MOTIVE POWER AND EQUIPMENT.

Boiler Tubes.

Causes of Leaks in Locomotive Boiler Tubes. M. E. Wells. Abstract of a paper presented at convention of the Am. Ry. Mas. Mechs.' Assn. Gives a résumé of reported causes and suggested remedies. 1500 w. Eng News—Aug. 29, 1907. No. 86633.

Brakes.

The Braking of Trains at High Speed (Le Freinage des Trains à Grande Vitesse). A. Boyer-Guillon. Outlines the theory of braking, the defects of the common type of air brake and describes the mechanism and advantages of the Maximus brake. Ills. Plate. 4000 w. Génie Civil—July 27, 1907. No. 86214 D.

The Maximus Brake. Illustrated detailed description of an English invention developed in connection with both the vacuum brake and the Westinghouse, which gives a uniform retarding effect. 1600 w. R R Gaz—Aug. 2, 1907. No. 86108.

Brake-Shoes.

Review of the Master Car Builders Brake-Shoe Tests. A report of interesting tests made during the past twelve years at the Purdue University testing plant. 6000 w. St Ry Jour—Aug. 3, 1907. No. 86084.

Car Heating.

Heating and Ventilating of Cars. E. R. Swan. Read before the Am. Soc. of Heat. & Vent. Engrs. Discusses systems that have been tried and gives an illustrated description of the system designed by the writer for the automatic control of the air supply. 6000 w. Heat & Vent Mag—Aug., 1907. No. 86494.

Driving Shoes.

Driving Shoes and Wedges. Editorial statement of methods of repair. 1200 w. Ry & Loc Engng—Aug., 1907. No. 86047 C.

Driving Springs.

Tests of the Live Load on Driving Springs. Charles A. Howard. Illustrates and describes apparatus designed for the purpose of furnishing data to enable the live load to be determined. 1500 w. Ry Age—Aug. 2, 1907. No. 86122.

Dynamometer Car.

100,000 Lb. Capacity Dynamometer Car.

We supply copies of these articles. See page 239.

Illustrated detailed description of a recently completed car for the Pennsylvania Railroad. 4000 w. Am Engr & R R Jour—Aug, 1907. No. 86048 C.

Fire-Boxes.

On Repairing the Tube Plates of Locomotive Fire-Boxes. S. Ragno. On the advantage of such repairs, and the methods used. Ills. 6700 w. Bul Int Ry Cong —July, 1907. No. 86493 E.

Freight Cars.

Forty-Ton Well Wagon. Illustrated description of a special car for carrying heavy loads of awkward size, in use on the Great Northern Railway of England. 600 w. Ry & Loc Engng—Aug., 1907. No. 86045 C.   .

Special Service Wagons; Great Central Railway Company. Illustrates and describes types of cars used on this English railway for carrying special forms of goods which are too bulky or awkward in shape, or too heavy for the usual cars. 1200 w. Plate. Engng—Aug. 9, 1907. No. 86469 A.

Steel, Triple Hopper-Bottom, Self-Clearing Coke Car. Gives illustration, plan, elevation, and sections of new cars for the Pittsburg & Lake Erie R. R. 800 w. Am Engr & R R Jour—Aug., 1907. No. 86050 C.

Locomotives.

A Balanced Compound Locomotive for the Italian State Railroads. Illustrated description of an express locomotive having novel features of exceptional interest 2500 w. R R Gaz—Aug. 2, 1907. No. 86110.

A New Mountain Locomotive. Trans. from *La Nature.* An improved system with increased adherence, used on the Puy de Dome road in France, is illustrated and described. 1800 w. Sci Am Sup —Aug. 3, 1907. No. 86069.

An Interesting Locomotive Conversion. Illustrates and describes the rebuilding of some locomotives on the Great Western Ry., changing 6-wheel engines with tenders into tank locomotives with the 2-6-2 wheel arrangement. 1000 w. Mech Engr —Aug. 3, 1907. No. 86351 A.

Atlantic Locomotives of the Austrian State Railways and the Southern Railway (Atlantik-Lokomotive der Oesterreichischen Staatsbahnen und der Südbahn). R. Sanzin. Illustrated detailed description with dimensions and results of tests. 3000 w. Serial. 2 parts. Verkehrstech Woche' u Eisenbahntech Zeitschr—July 6 and 13, 1907. No. 86242 each D.

Bothwell Convertible Locomotive. Illustrates and describes a reconstructed locomotive aiming at maximum hauling capacity with the least weight. 700 w. Ry & Engng Rev—Aug. 3, 1907. No. 86118.

British-Built Locomotives for India. Illustrations, with descriptions of a new 4-6-0 type express engine for the Bombay, Baroda, and Central India Ry., and a duplex, or Fairlie type locomotive for the Burma Ry. 1500 w. Mech Engr—Aug. 10, 1907. No. 86453 A.

British Tank Locomotives. J. R. Thompson. Illustrated description of a tank locomotive for long-distance service on the Midland Railway of England. 700 w. Cassier's Mag—Aug., 1907. No. 86012 B.

Decapod Locomotive for the Buffalo, Rochester & Pittsburg. Illustrated description of one of the six decapod (2-10-0) locomotives recently built for pusher service. 1200 w. R R Gaz—Aug. 9, 1907. No. 86193.

Locomotives on the Lancashire ˜and Yorkshire Railway. A brief account of the development of the Horwich Works, with illustrated descriptions of engines and motor cars built. 1300 w. Engng—Aug. 9, 1907. No. 86470 A.

Mallet Compound Locomotive for the Erie Railroad. Compares the Erie engine with the Baltimore & Ohio engines, and gives illustrated description of interesting features, and the principal dimensions. 2200 w. R R Gaz—Aug. 16, 1907. No. 86414.

Pacific Locomotive for the Pennsylvania Lines. The heaviest engine of this type ever constructed, having a total weight of 269,200 pounds, is illustrated and described. 500 w. Ry Age—Aug. 23, 1907. No. 86566.

Prairie for the Soo Lines. Illustration, with brief description of engines intended for mixed service. 700 w. Ry & Loc Engng—Aug., 1907. No. 86044 C.

Prairie Locomotive for the Minneapolis, St. Paul Sault Ste. Marie. Illustration, dimensions and particulars of engines for both freight and passenger service. 700 w. R R Gaz—Aug. 2, 1907. No. 86107.

Recent British Locomotives for Abroad. Illustrated description of a 4-cylinder balanced compound locomotive for the Buenos Ayres Gt. Southern Ry., and of a 6-wheeled locomotive crane engine for India. 700 w. Mech Engr—July 27, 1907. No. 86156 A.

Siamese and Westphalians. Illustrates two standard gauge engines built in Germany. One is a freight engine for burning wood; the other of unusually strong construction and special design to meet requirements of the Westphalia Ry. 400 w. Ry & Loc Engng—Aug., 1907. No. 86046 C.

Simple Consolidation Locomotive. Illustrated description of a very interesting locomotive, fitted with Baldwin superheater, being shown at Jamestown Exhibition. 900 w. Am Engr & R R Jour—Aug., 1907. No. 86049 C.

Six-wheel Coupled Locomotive (Metre-Gauge) for the Federated Malay States Railways. Illustration, with particulars. 500 w. Engng—July 26, 1907. No. 86176 A.

Ten Wheeled Coupled Freight Locomotives for the Austrian State Railways (5-5 Gekuppelte Güterzugslokomotive der Oesterr. Staatsbahnen). R. Sanzin. Illustrated detailed description. 3000 w. Oest Wochenschr f d Oeffent Baudienst—July 20, 1907. No. 86269 D.

Ten-Wheel Locomotive for the Canadian Pacific. Engines equipped with the Horsey-Vaughan type of superheater, and representing the use of large cylinders and low boiler pressure in conjunction with superheated steam are illustrated and described. 500 w. Ry Age—Aug. 23, 1907. No. 86568.

The Work of the "Experiments" on the London and Northwestern Railway. On the performance of the first-six-coupled express passenger engines tried on this line, giving records of the work on the southern section, between Euston and Crewe. 2500 w. Engng—July 26, 1907. No. 86174 A.

Motor Cars.

New Steam Motor Cars Designed by F. X. Komeret, Vienna (Neuere Dampfwagen von F. X. Komaret in Wien). C. Guillery. Illustrated detailed description of boiler, engine and car, and the result of competitive tests. 4000 w. Glasers Ann—July 15, 1907. No. 86258 D.

Narrow-Gauge.

The Longest Narrow-Gauge Light Railway in the World. An illustrated detailed description of the Otavi Railway, in German South-West Africa, and its rolling-stock. 2500 w. Engng—July 19, 1907. No. 85991 A.

Overbalancing.

The Overbalancing of Locomotives with Regard to the Dampening Effect of Springs (Das Wogen und Nicken der Lokomotive unter Berücksichtigung der Dämpfenden Wirkung der Federn). W. Lindemann. A mathematical paper. Ills. 3500 w. Glasers Ann—July 1, 1907. No. 86257 D.

Resistance.

The Resistance of Railway Trains. Editorial review of investigations made to determine train resistance. 2000 w. Engng—July 26, 1907. No. 86181 A.

Staybolts.

Manganese Bronze Staybolts. M. Rodrigue. Trans. from *Revue Generale des Chemins de Fer*. Report of tests showing that manganese bronze is superior to copper for staybolts used in high-pressure locomotive boilers. 500 w. R R Gaz—Aug. 9, 1907. No. 86194.

Steam Domes.

Wrong Location of Steam Domes. B. H. Jeffries. States the writer's objections to the present location of the dome, and reasons for desiring a change. 800 w. Ry & Engng Rev—Aug. 3, 1907. No. 86120.

"Tiregraph."

A Locomotive Driving-Wheel Recording Device. Illustrates and describes this recording instrument, called a "tiregraph," by means of which the contour of tires is recorded, and reports tests made and conclusions reached. 1200 w. Sci Am—Aug. 17, 1907. No. 86417.

Train Lighting.

Train Lighting. H. Henderson. Abstract of a paper read before the Newcastle Local Sec. of the Inst of Elec. Engrs. Deals with the lighting of railway carriages by electricity, giving particulars of cost of equipment and maintenance and describing the various systems. 2500 w. Elect'n, Lond—Aug. 16, 1907. Serial. 1st part. No. 86596 A.

NEW PROJECTS.

India.

Indian Railway Property. Information concerning progress on the different lines and recent important undertakings. 1600 w. Engng—July 26, 1907. No. 86177 A.

N. Y., N. H. & H.

New York, New Haven and Hartford Improvement at New Haven. Explains the general features of the improvements, giving illustrated descriptions of types of bridges and other work of interest. 2000 w. Ry Age—Aug. 9, 1907. No. 86336.

Philippines.

The Railroad Development of the Philippines. P. H. Ashmead. An illustrated account of the progress made in construction, and the industrial effects manifest. 3000 w. Engineering Magazine—Sept., 1907. No. 86607 B.

South America.

The Trans-Andine Railroads. Lewis R. Freeman. An interesting illustrated account of difficult railway building, and of the discovery of a new pass through the Andes so open that a broad gauge line can be run through without building a single tunnel. 2500 w. R R Gaz—Aug. 2, 1907. No. 86109.

PERMANENT WAY AND BUILDINGS.

Improvements.

The Value of Railroad Improvements. Morrell W. Gaines. An examination of money earnings from increased train loads, giving a short résumé of the course of traffic and operating conditions since the early eighties. 4000 w. Engineering Magazine—Sept., 1907. No. 86605 B.

Maintenance of Way.

A Technical and Economic Consideration of Materials for Maintenance of Way

We supply copies of these articles. See page 239.

(Sullo Studio Tecnico ed Economico dei Materiali di Manutenzione dell Massicciate Stradali). 1500 w. Serial. 1st part. Rivista Gen delle Ferrovie—July 28, 1907. No. 86217 D.

Rail Corrugation.
The Origin and Production of Corrugation of Tramway Rails. W. Worby Beaumont. Read before the British Assn. Offers an explanation of the origin and production of these corrugations. 2000 w. Elec Engr, Lond—Aug. 9, 1907. No. 86-456 A.

Rail Laying.
A Machine for Laying Rails in Streets. Illustrated description of the Romapac machine for laying and fixing the rails of street railways. 2500 w. Sci Am Sup—Aug. 24, 1907. No. 86519.

Rails.
Steel Rails of Better Quality. Robert Job. Discusses the need of better rails to meet changed conditions, and considers the latest specifications proposed by the American Society for Testing Materials. 2000 w. Ir Age—Aug. 22, 1907. No. 86500.

Signaling.
Standard Convertible Semaphore Signals of the Belgian State Railways. L. Weissenbruch. Illustrated description of the new semaphores and their use. 7200 w. Bul Int Ry Cong—July, 1907. No. 86492 E.

Simplon.
The Italian Approaches to the Simplon Tunnel. Brief review of the negotiations which made this work possible, illustrating and describing the approaches on the Italian side. 3800 w. Engng—July 26, 1907. Serial. 1st part. No. 86175 A.

Spikes.
Holding Power of Railroad Spikes. Roy I. Webber. From Bul. No. 6 of the Univ. of Ills. Engng. Exp. Station. Report of a series of experiments made to determine the resistance to withdrawal of the same type of spike in different timbers, and of different spikes in the same timber, and whether the preservative has any influence. Ills. 7500 w. R R Gaz—Aug. 9, 1907. No. 86191.

Stations.
The Alterations to Ludgate-Hill Station. Illustrates and describes changes to be made to relieve congested traffic. 700 w. Engr, Lond—Aug. 2, 1907. No. 86-367 A.

The Design of Wayside Stations for Single Lines of Railway. Frederick George Royal-Dawson. Considers types of stations designed for India, especially stations designed for the safety of fast trains, and also the development of local traffic. 8300 w. Inst of Civ Engrs, No. 3671. No. 86371 N.

Terminals.
The Land and Water Terminal of the Seaboard Air Line Railway at Savannah, Ga. Description abstracted from an article by W. Faucette. Plan. 1700 w. Eng News—Aug. 22, 1907. No. 86511.

The Pennsylvania New Terminal, New York. The first of a series of illustrated articles describing in detail the great engineering work in progress to furnish a station in New York City to serve the Pennsylvania Lines. 2500 w. Engr, Lond—July 19, 1907. Serial. 1st part. No. 86000 A.

Track Elevation.
Track Elevation on the Milwaukee Division, Chicago & Northwestern Railway. Brief illustrated description of the work. 1000 w. Ry Age—Aug. 16, 1907. No. 86437.

Tracks.
Notes on English Railway Track. Gives opinions based on a recent examination of railways in England and America. 1800 w. Eng News—Aug. 8, 1907. No. 86310.

Turntables.
Standard Turntable Pit: Seaboard Air Line Ry. Philip Aylett. Drawings and description of a pit consisting essentially of a concrete circular wall, in the foundation of which is embedded and bolted the circular rail. 700 w. Eng News—Aug. 15, 1907. No. 86392.

TRAFFIC.

Freight.
Loss and Damage to Freight. An address by Robert L. Calkins before the N. Y. Traffic Club. Discusses some of the causes of losses and delays and the payment of claims, urging the closer relationship and co-operation between different interests. 1200 w. R R Gaz—Aug. 9, 1907. No. 86192.

Preference Freight. An interesting account of the manner of handling this class of freight. 2500 w. Ry & Engng Rev—Aug. 10, 1907. No. 86337.

MISCELLANY.

Abandoned Railway.
A Derelict Somersetshire Railway. An illustrated account of a mineral railway abandoned because of the decay of the industry it was built to serve. 1800 w. Engr, Lond—Aug. 9, 1907. No. 86476 A.

Construction Organization.
See Industrial Economy.

Government Control.
Secretary Taft's Columbus Speech. Extract concerning railroads and their regulation, with editorial comment. 4200 w. R. R. Gaz—Aug. 23, 1907. No. 86528.

Government Ownership.
The Costly Mistake of State Railroads. Information given before the Viceregal Commission on Irish Railways is commented upon editorially, with review of

We supply copies of these articles. See page 239.

E. A. Pratt's book on State Railways, especially in Belgium. 2000 w. R R Gaz —Aug. 9, 1907. No. 86190.

Japan.

The Development of Japanese Transportation Facilities (Die Entwicklung des Japanischen Verkehrswesens). Dr. Wolff. Describes the progress in extension of railway and steamship lines and gives tables showing the development of commerce and traffic during the past twenty years. 4000 w. Verkehrstech Woche u Eisenbahntech Zeitschr—July 27, 1907. No. 86245 D.

Legislation.

Clashes Between Federal and State Authority. Editorial on the recent legislation in the Southern States in regard to railroad rates. 2000 w. R R Gaz—Aug. 23, 1907. No. 86526.

Railroad Legislation in Connecticut. Clarence Deming. A review of the steam and trolley legislation of the session just ended. 1600 w. R R Gaz—Aug. 9, 1907. No. 86195.

Mexico.

Mexican Government Railroad High Finance. A review of railroad development and the approaching consolidation into one great system under the control of the national government. 2000 w. R R Gaz—July 19, 1907. No. 85779.

The Railroads of Mexico. Erdis G. Robinson. Reviews their development. The present number describes the resources and topography of the country. 1800 w. R R Gaz—July 12, 1907. Serial. 1st part. No. 85515.

New Haven Merger.

The New Haven-Boston & Maine System. Gives two maps showing the steam railroad mileage owned by the N. Y., N. H. & H. R. R. and the B. & M. R. R., with remarks on the consolidation. Also editorial. 2000 w. R R Gaz—Aug. 23, 1907. No. 86529.

Reading.

Early Years of the Philadelphia & Reading. C. H. Caruthers. A review of the early history of this line, illustrating and describing early engines used, and other peculiarities of equipment. 7000 w R R Gaz—July 26, 1907. No. 85904.

U. S. Railroads.

Agitation Against the Railroads. Col. H. G. Prout. An address delivered before the Traffic Club of Pittsburg, on the present attitude of the public toward the railroads. 4000 w. Ry Age—July 12, 1907. No. 85555.

STREET AND ELECTRIC RAILWAYS

Atlantic Shore Line.

The Atlantic Shore Line Railway. Illustrates and describes a recently completed new link between York Beach and Kennebunk, connecting the east and west branches of this system, which has a length of about 100 miles. 3500 w. Elec Ry Rev—Aug. 24, 1907. No. 86565.

Cars.

A Radical Design of Semi-Steel Car. Illustrated description of cars for service on interurban systems near Milwaukee. 1200 w. St Ry Jour—Aug. 10, 1907. No. 86320.

Electrification.

Electrification of the New York, New Haven & Hartford. E. H. McHenry. A fully illustrated article describing important features of the system. Also short editorial. 9500 w. R R Gaz—Aug. 16, 1907. No. 86415.

Great Western Railway.—Electric Power and Lighting and the Electrification of the Hammersmith & City Railway. Illustrated detailed description of the new electric supply system that has been installed in London. 4500 w. Elect'n, Lond —Aug. 2, 1907. Serial. 1st part. No. 86361 A.

Great Western Railway. Illustrated detailed description of the new electric power and lighting system, and the electrification of the Hammersmith and City Railway. 15000 w. Tram & Ry Wld— Aug. 1, 1907. No. 86455 B.

Heavy Electric Traction on the New York, New Haven & Hartford Railroad. E. H. McHenry. Explains the reasons for the adoption of the 25-cycle, 11,000-volt system, and the expected results, giving an illustrated detailed description of the system. 7000 w. St Ry Jour—Aug. 17, 1907. Serial. 1st part. No. 86443.

Some Facts and Problems Bearing on Electric Trunk-Line Operation. Frank J. Sprague. Analyzes some phases of problems discussed by engineers, giving comparative facts, and developments in electric locomotive construction, and features of the equipments now commanding attention. Ills. 23000 w. Pro Am Inst of Elec Engrs—July, 1907. No. 86538 D.

The Electrification of the Hammersmith & City Railway Branch of the Great Western Railway. Illustrated detailed description of this recently completed electric traction system in London. 6500 w. St Ry Jour—Aug. 3, 1907. No. 86083.

We supply copies of these articles. See page 239.

The Inauguration of the New Haven Railroad Electric Service. An illustrated description of the first application in the United States of single phase traction to an important trunk railroad. The Cos Cob power station, the overhead trolley system, and the electric locomotives are described. 1500 w. Sci Am—Aug. 3, 1907. No. 86064.

Feeders.

Determining the Size of Feeders. Henry Docker Jackson. Briefly explains methods of calculating used. 1200 w. Elec Ry Rev—Aug. 17, 1907. No. 86442.

Frequency.

The Choice of Frequency for Single-Phase Alternating-Current Railway Motors. A. H. Armstrong. Aims to open a discussion on the relative merits of 25-cycles and a lower frequency, briefly considering the advantages and disadvantages of the present, and of any proposed standard. 1700 w. Pro Am Inst of Elec Engrs—July, 1907. No. 86532 D.

Twenty-five versus Fifteen Cycles for Heavy Railways. N. W. Storer. Presents arguments in favor of each, discussing them in detail. 3500 w. Pro Am Inst of Elec Engrs—July, 1907. No. 86533 D.

Grade Reduction.

Lowering the Grade of a Tunnel at Kansas City, Mo. Abstract of a paper by W. M. Archibald on the reconstruction of the 8th Street tunnel of the Metropolitan St. Ry., at Kansas City, Mo. Ills. 1500 w. Eng News—Aug. 1, 1907. No. 86081.

History.

The Historical Development of Electric Railways From Their Introduction to the Present (Die Geschichtliche Entwicklung der Elektrischen Bahnen vom Ursprung bis zur Neuzeit). Herr Peter. Traces the development of electric traction from W. von Siemen's line established in Berlin in 1879, giving a chronological table of leading events. 5000 w. Glasers Ann—July 1, 1907. No. 86256 D.

Interurban.

Interurban Railway Development Near Milwaukee. Map and illustrated description of the large interconnected system planned, and some of the construction work. 3500 w. St Ry Jour—Aug. 3, 1907. No. 86082.

Italy.

Italian Tramways (Die Straszenbahnen Italiens). D. Liebmann. Discusses the extent and distribution of electric railways in Italy, with general notes on Italian practice. 2000 w. Verkehrstech Woche u Eisenbahntech Zeitschr—July 6, 1907. No. 86243 D.

The Single-Phase Electric Railroad of the Val Brembana (Ferrovia Elettrica Monofase della Val Brembana). Illustrat-

ed description of the electric locomotive built by the Westinghouse Company. 2200 w. Industria—July 14, 1907. No. 86220 D.

Locomotive.

Single-Phase 20,000-Volt Locomotive for Swedish State Railways. Dr. Alfred Gradenwitz. Illustration with description of the normal-gauge locomotive under trial. 700 w. Prac Engr—Aug. 9, 1907. No. 86451 A.

Paris.

The Construction of the New Lines of the Paris City Railways (Ueber die Bauten der Neuen Linien der Pariser Stadtbahn). E. A. Ziffer. Describes recent extensions to the Paris tramway system, tunnels, stations, etc., giving costs. 5000 w. Mitt d Ver f d Förd d Lokal u Strazzenbahnwesens—July, 1907. No. 86222 F.

Power Station.

The Reconstruction of the East St. Louis & Suburban Ry. Power Station. Illustrates and describes additions and re-arrangement. 2500 w. Eng Rec—Aug. 3, 1907. No. 86089.

Shops.

New Shops of the Bangor Railway. Illustrated description of shops built of concrete blocks, and of their arrangement and methods. 1800 w. Elec Ry Rev—Aug. 10, 1907. No. 86322.

Single-Phase.

Single-Phase vs. Direct-Current Railway Operation. Malcolm MacLaren. Discusses features of the single-phase system which the writer feels have been inadequately presented, giving comparisons. 2500 w. Elec Jour—Aug., 1907. No. 86400.

Single-Phase versus Three-Phase Generation for Single-Phase Railways. A. H. Armstrong. Various methods of distribution are given with some of the advantages and disadvantages of each. 1300 w. Pro Am Inst of Elec Engrs—July, 1907. No. 86531 D.

The Pittsburg & Butler Street Railway Co. M. N. Blakemore. Illustrated description of an interurban single-phase electric railway and its equipment. 3300 w. St Ry Jour—Aug. 10, 1907. Serial. 1st part. No. 86319.

Standardization.

Meeting of the Standardization Committee. An account of the meeting of the Committee of Standards of the Am. St. & Int. Ry. Engng Assn., giving a summary of the discussions. 3000 w. St Ry Jour—Aug. 10, 1907. No. 86321.

Tracks.

Wear and Maintenance of Street Railway Track. C. F. Wike. Abstract of a paper read before the Munic. & Co. Engrs. at Liverpool, Eng. Concerning the cost of repairs and renewals at Sheffield, Eng. 1700 w. Eng News—Aug. 29, 1907. No. 86635.

EXPLANATORY NOTE— THE ENGINEERING INDEX.

We hold ourselves ready to supply—usually by return of post—the full text of every article indexed in the preceding pages, *in the original language*, together with all accompanying illustrations; and our charge in each case is regulated by the cost of a single copy of the journal in which the article is published. The price of each article is indicated by the letter following the number. When no letter appears, the price of the article is 20 cts. The letter A, B, or C denotes a price of 40 cts.; D, of 60 cts.; E, of 80 cts.; F, of $1.00; G, $1.20; H, of $1.60. When the letter N is used it indicates that copies are not readily attainable, and that particulars as to price will be supplied on application. Certain journals, however, make large extra charges for back numbers. In such cases we may have to increase proportionately the normal charge given in the Index. In ordering, care should be taken to *give the number* of the article desired, not the title alone.

Serial publications are indexed on the appearance of the first installment.

SPECIAL NOTICE.—To avoid the inconvenience of letter-writing and small remittances, especially from foreign countries, and to cheapen the cost of articles to those who order frequently, we sell coupons at the following prices:—20 cts. each or twelve for $2.00, thirty-three for $5, and one hundred for $15.

Each coupon will be received by us in payment for any 20-cent article catalogued in the Index. For articles of a higher price, one of these coupons will be received for each 20 cents; thus, a 40-cent article will require two coupons; a 60-cent article, three coupons; and so on. The use of these coupons is strongly commended to our readers. They not only reduce the cost of articles 25 per cent. (from 20c. to 15c.), but they need only a trial to demonstrate their very great convenience—especially to engineers in foreign countries, or away from libraries and technical club facilities.

Write for a sample coupon—free to any part of the world.

CARD INDEX.—These pages are issued separately from the Magazine, printed on one side of the paper only, and in this form they meet the exact requirements of those who desire to clip the items for card-index purposes. Thus printed they are supplied to regular subscribers of THE ENGINEERING MAGAZINE at 10 cents per month, or $1.00 a year; to non-subscribers, 25 cts. per month, or $3.00 a year.

THE PUBLICATIONS REGULARLY REVIEWED AND INDEXED.

The titles and addresses of the journals regularly reviewed are given here in full, but only abbreviated titles are used in the Index. In the list below, *w* indicates a weekly publication, *b-w*, a bi-weekly, *s-w*, a semi-weekly, *m*, a monthly, *b-m*, a bi-monthly, *t-m*, a tri-monthly, *qr*, a quarterly, *s-q*, semi-quarterly, etc. Other abbreviations used in the index are: Ill—Illustrated: W—Words; Anon—Anonymous.

Alliance Industrielle. *m.* Brussels.
American Architect. *w.* New York.
Am. Engineer and R. R. Journal. *m.* New York.
American Jl. of Science. *m.* New Haven, U. S. A.
American Machinist. *w.* New York.
Annales des Ponts et Chaussées. *m.* Paris.
Ann. d Soc. Ing. e d Arch. Ital. *w.* Rome.
Architect. *w.* London.
Architectural Record. *m.* New York.
Architectural Review. *s-q.* Boston.
Architect's and Builder's Magazine. *m.* New York.
Australian Mining Standard. *w.* Melbourne.
Autocar. *w.* Coventry, England.
Automobile. *w.* New York.
Automotor Journal. *w.* London.
Beton und Eisen. *qr.* Vienna.
Boiler Maker. *m.* New York.
Brass World. *m.* Bridgeport, Conn.
Brit. Columbia Mining Rec. *m.* Victoria, B. C.
Builder. *w.* London.
Bull. Am. Iron and Steel Asso. *w.* Phila., U. S. A.

Bulletin de la Société d'Encouragement. *m.* Paris.
Bulletin of Dept. of Labor. *b-m.* Washington.
Bull. Soc. Int. d'Electriciens. *m.* Paris.
Bulletin of the Univ. of Wis., Madison, U. S. A.
Bulletin Univ. of Kansas. *b-m.* Lawrence.
Bull. Int. Railway Congress. *m.* Brussels.
California Jour. of Tech. *m.* Berkeley, Cal.
Canadian Architect. *m.* Toronto.
Canadian Electrical News. *m.* Toronto.
Canadian Engineer. *m.* Toronto and Montreal.
Canadian Mining Journal. *b-w.* Toronto.
Cassier's Magazine. *m.* New York and London.
Cement. *b-m.* New York.
Cement Age. *m.* New York.
Central Station. *m.* New York.
Chem. Met. Soc. of S. Africa. *m.* Johannesburg.
Colliery Guardian. *w.* London.
Compressed Air. *m.* New York.
Comptes Rendus de l'Acad. des Sciences. *w.* Paris.
Consular Reports. *m.* Washington.
Deutsche Bauzeitung. *b-w.* Berlin.

239

Domestic Engineering. w. Chicago.
Economic Geology. m. So. Bethlehem, Pa.
Electrical Age. m. New York.
Electrical Engineer. w. London.
Electrical Review. m. London.
Electrical Review. w. New York.
Electric Journal. m. Pittsburg, Pa.
Electric Railway Review. w. Chicago.
Electrical World. w. New York.
Electrician. w. London.
Electricien. w. Paris.
Electrochemical and Met. Industry. m. N. Y.
Elektrochemische Zeitschrift. m. Berlin.
Elektrotechnik u Maschinenbau. w. Vienna.
Elektrotechnische Rundschau. w. Potsdam.
Elektrotechnische Zeitschrift. w. Berlin.
Elettricità. w. Milan.
Engineer. w. London.
Engineer. s-m. Chicago.
Engineering. w. London.
Engineering-Contracting. w. New York.
Engineering Magazine. m. New York and London.
Engineering and Mining Journal. w. New York.
Engineering News. w. New York.
Engineering Record. w. New York.
Eng. Soc. of Western Penna. m. Pittsburg, U. S. A.
Far Eastern Review. m. Manila, P. I.
Fire and Water. w. New York.
Foundry. m. Cleveland, U. S. A.
Génie Civil. w. Paris.
Gesundheits-Ingenieur. s-m. München.
Giorn. dei Lav. Pubb. e d Str. Ferr. w. Rome.
Glaser's Ann. f Gewerbe & Bauwesen. s-m. Berlin.
Heating and Ventilating Mag. m. New York.
Ice and Refrigeration. m. New York.
Industrial World. w. Pittsburg.
Ingenieria. b-m. Buenos Ayres.
Ingenieur. w. Hague.
Insurance Engineering. m. New York.
Int. Marine Engineering. m. New York.
Iron Age. w. New York.
Iron and Coal Trades Review. w. London.
Iron and Steel Trades Journal w. London.
Iron Trade Review. w. Cleveland, U. S. A.
Jour. of Accountancy. m. N. Y.
Journal Asso. Eng. Societies. m. Philadelphia.
Journal of Electricity. m. San Francisco.
Journal Franklin Institute. m. Philadelphia.
Journal Royal Inst. of Brit. Arch. s-qr. London.
Jour. Roy. United Service Inst. m. London.
Journal of Sanitary Institute. qr. London.
Jour. of South African Assn. of Engineers. m.
  Johannesburg, S. A.
Journal of the Society of Arts. w. London.
Jour. Transvaal Inst. of Mech. Engrs., Johannes-
  burg, S. A.
Jour. of U. S. Artillery. b-m. Fort Monroe, U. S. A.
Jour. W. of Scot. Iron & Steel Inst. m. Glasgow.
Journal Western Soc. of Eng. b-m. Chicago.
Journal of Worcester Poly. Inst., Worcester,
  U. S. A.
Locomotive. m. Hartford, U. S. A.
Machinery. m. New York.
Madrid Cientifico. s-m. Madrid.
Manufacturer's Record. w. Baltimore.
Marine Review. w. Cleveland, U. S. A.
Men. de la Soc. des Ing. Civils de France. m. Paris.
Métallurgie. w. Paris.
Minero Mexicano. w. City of Mexico.

Mines and Minerals. m. Scranton, U. S. A.
Mining and Sci. Press. w. San Francisco.
Mining Journal. w. London.
Mining Reporter. w. Denver, U. S. A.
Mittheilungen des Vereines für die Förderung des
  Local und Strassenbahnwesens. m. Vienna.
Motor Wagon. w. Cleveland, U. S. A.
Municipal Engineering. m. Indianapolis, U. S. A.
Municipal Journal and Engineer. w. New York.
Nature. w. London.
Nautical Gazette. w. New York.
New Zealand Mines Record. m. Wellington.
Oest. Wochenschr. f. d. Oeff. Baudienst. w. Vienna.
Oest. Zeitschr. Berg & Hüttenwesen. w. Vienna.
Plumber and Decorator. m. London.
Popular Science Monthly. m. New York.
Power. m. New York.
Practical Engineer. w. London.
Pro. Am. Soc. Civil Engineers. m. New York.
Pro. Canadian Soc. Civ. Engrs. m. Montreal.
Proceedings Engineers' Club. qr. Philadelphia.
Pro. St. Louis R'way Club. m. St. Louis, U. S. A.
Pro. U. S. Naval Inst. qr. Annapolis, Md.
Quarry m. London.
Queensland Gov. Mining Jour. m. Brisbane,
  Australia.
Railroad Gazette. w. New York.
Railway Age. w. Chicago.
Railway & Engineering Review. w. Chicago.
Railway and Loc. Engng. m. New York.
Railway Master Mechanic. m. Chicago.
Revista d Obras. Pub. w. Madrid.
Revista Tech. Ind. m. Barcelona.
Revue de Mécanique. m. Paris.
Revue Gén. des Chemins de Fer. m.- Paris.
Revue Gén. des Sciences. w. Paris.
Revue Industrielle. w. Paris.
Revue Technique. b-m. Paris.
Rivista Gen. d Ferrovie. w. Florence.
Rivista Marittima. m. Rome.
Schiffbau. s-m. Berlin.
Schweizerische Bauzeitung. w. Zürich.
Scientific American. w. New York.
Scientific Am. Supplement. w. New York.
Sibley Jour. of Mech. Eng. m. Ithaca, N. Y.
Stahl und Eisen. s-m. Düsseldorf.
Stevens Institute Indicator. qr. Hoboken, U. S. A.
Street Railway Journal. w. New York.
Technograph. w. Urbana, Ill.
Technology Quarterly. qr. Boston, U. S. A.
Tijds v h Kljk. Inst. v Ing. qr. Hague.
Tramway & Railway World. m. London.
Trans. Am. Ins. Electrical Eng. m. New York.
Trans. Am. Ins. of Mining Eng. New York.
Trans. Am. Soc. Mech. Engineers. New York.
Trans. Inst. of Engrs. & Shipbuilders in Scot-
  land, Glasgow.
Transport. m. London.
Verkehrstechnische Woche und Eisenbahntech-
  nische Zeitschrift. w. Berlin.
Wood Craft. m. Cleveland, U. S. A.
Yacht. w. Paris.
Zeitschr. f. d. Gesamte Turbinenwesen. w. Munich.
Zeitschr. d. Mitteleurop. Motorwagon Ver. s-m.
  Berlin.
Zeitschr. d. Oest. Inq. u. Arch. Ver. w. Vienna.
Zeitschr. d. Ver. Deutscher Ing. w. Berlin.
Zeitschrift für Elektrochemie. w. Halle a S.
Zeitschr. f. Werkzeugmaschinen. b-w. Berlin.

VOL. XXXIV.   NOVEMBER, 1907.   No. 2

COMBINATION AND COMPETITION IN THE STEEL TRADE.

By T. Good.

While present economic conditions in the United States do not in the least presage depression, they do indicate some change in the rate and in the distribution of industrial activity. Any such period in the affairs of a great manufacturing nation means some shifting of interest from strictly domestic markets to export opportunities. The "American invasion" of 1897 is in no likelihood of being repeated, but American manufacturers may look with some interest upon the conditions of the metal trades of Great Britain and the Continent. Mr. Good's well informed review is therefore very timely, for it deals with the most important commercial conditions now prevailing in the steel trade, which in turn underlies all modern engineering industries.—THE EDITORS.

THE movement towards industrial combination, especially in the iron and steel trades, is very pronounced just now. This movement deserves something more than the thoughtless condemnation it too frequently receives. The history of the iron industry is marked throughout by violent fluctuations, and the need for some method by which such fluctuations might be moderated has long been felt. Then, again, Great Britain must recognize that so far as future foreign competition in iron and steel is concerned, one of her prime dangers lies in the circumstance that such competition is likely to be conducted by "trusts" and syndicates more powerful than in the past, while the opposition she may offer will be that of individual firms— that is, assuming that she unduly harasses, either by legislation or the pressure of an intolerant public opinion, the reasonable development of the combination movement among her manufacturers. Moreover, we must not forget that the British manufacturer must meet, not merely combined competitors in the world's markets, but combined workmen at home, and that both capital abroad and labour at home tend to become more strongly organised. For these reasons we would do well to consider the efforts of our iron and steel makers to combine their efforts without prejudice.

Some time ago makers of shipbuilding and structural steel in Scotland and the North of England arrived at an understanding in reference to the regulation of prices and the allocation of orders, and on several occasions there have been more or less loose agreements in various branches of the British iron trade, although, owing to a variety of reasons which need not be specified, trusts, combines, and syndicates have not hitherto dominated British industries to anything like the extent that has obtained in Germany and America. But British manufacturers, despite their inherent conservatism, are being driven by sheer force of circumstances to recognise the need of some measure of organisation. In September last the steel-tube makers entered into a close arrangement for the control of their trade. About the same time there was formed a South Wales Siemens Steel Association. Early this year an Ironfounders' Combine was announced, and now we are informed that several of our very largest steel, engineering, and shipbuilding firms are joining forces in order to combat foreign competition more successfully. In these circumstances it is opportune to review briefly the past history of our iron and steel trades, as affected by competition, on the one hand, and by combination on the other. And we may venture a forecast of the possibilities of the future.

While we have had no very important or permanent trusts of the monopoly kind in Great Britain, it remains a fact that there have been occasions when, and markets where, our iron and steel makers have been practically at the mercy of American trusts and German syndicates. Whether we have been combined or not, we have had to compete with and submit to the trading methods and prices of combined rivals. Syndication, organisation, or whatever we may call it, is upon us, and no matter how strongly we may be opposed, either by conviction or prejudice, to some of the principles or effects of industrial combination, we have not only got to deal with combined labour in our workshops, but we have to meet, in the markets of the world, competitors enjoying all the benefits of closely organised production, sale, and transit. And we should frankly recognise that, whatever may be the merits or demerits of industrial combination—whether it destroys that healthy individualism which has been held to make for true progress or not—the iron and steel industries of Germany and America, with their large syndicates and trusts, have in recent years made greater headway, and are now in a more progressive condition, broadly and generally speaking, than those of this country in which the sacred liberty of the individual has been so fondly cherished.

One of the most valuable object lessons available in industrial combination is that afforded by consideration of the inception, operation, and suspension of co-operative export bounties on German steel. Although a nominal bounty has been granted on certain classes of steel until the middle of this year, for all practical purposes the bounty system has been inoperative since June 30, 1906. But a brief review of the system—a system which may possibly be re-introduced in the near future—will be of interest. Introduced ten years ago by the syndicates controlling the coal, iron, and steel trades of Germany, this system of granting private bounties on steel and steel goods for export has exercised a considerable influence over the German steel trade, and has been largely instrumental in stimulating German competition with Great Britain. While the bounties were high "dumping" was indulged in to our undoubted detriment; as the bounties declined "dumping" diminished and our trade increased; since they have been withdrawn, or have been merely nominal, we have enjoyed a comparative "boom." Of course, there have been other factors influencing the situation, but the bounty system has been an important phase of a clearly organised movement, and as such demands attention.

In order to understand how those bounties were manipulated it is necessary to note that industrial syndication is more complete, and more effective, in the coal, iron, and steel trades of Germany than in any group of industries in any other country—not even excepting America, the home of huge trusts and prodigious things in general. The production, distribution, and price of practically every important article and material of manufacture in Germany, from coal and iron ore to sewing machines and wire, are regulated by some kind of syndicate. Ten years ago there were about ninety syndicates in the German coal and iron trades. Owing to absorption consequent upon improved organisation there are now about half that number. There are separate syndicates governing coal, coke, briquettes, lignite, and iron ore; pig iron, iron and steel castings, and foundry products; tubes, rails, girders, and rods; plates, light sheets, and heavy sheets; there are unions of electrical works, of sewing-machine makers, and of nail manufacturers. Besides syndicates for various products there are in some cases distinct syndicates where local conditions are of a special character. But these local, special, or sectional syndicates are usually united with kindred trade syndicates, or work in harmony with one another. For example, while there are four distinct pig-iron syndicates, enjoying local autonomy, there is a Consolidated Pig-iron Syndicate controlling the local organisation. Then, in all but supreme authority, there is the famous *Stahlwerksverband,* or Steel Syndicate, which has recently been renewed for a term of five years.

These German syndicates are highly organised (but not on the American model) and, generally speaking, efficiently conducted. As evidence of the line of policy they usually adopt it may be mentioned that at a recent meeting of the Stahlwerksverband, when a member proposed a substantial increase in price in consideration of great pressure of orders, a counter-resolution was carried by a large majority to the following effect:—

"That moderate prices are the best guarantee of a continuance of purchasers, and that the mission of a trade syndicate is just as much to prevent an exaggeration of prices during times of activity as to support the courage of manufacturers during times of inactivity." Indeed, the directors of the Steel Syndicate have declared their policy to be "the preservation of the home market from violent fluctuations by providing constant work at moderate profits."

The usual methods of procedure adopted by the German syndicates are for a number of experts to visit each of the works associated, to ascertain its capacity of production and its aptitude for any special class of work. Orders, being received by the central offices, are then allocated and remitted to the several establishments, regard being paid to the situation and circumstances of the various firms so as to avoid waste, overlapping, and unnecessary delay. And not only are prices fixed and orders allocated by syndicated agreement, but payments usually pass through the central offices. Moreover, the syndicate frequently attends to times, methods, and costs of transit, and many other matters, with the greatest possible advantage to its constituents, for its expert officials are better informed than the private manufacturer immersed in the technicalities of his own particular works can hope to be. While different syndicates have different rules—some allowing their members more liberty of action than others—these may be taken as the general outlines of the working of German iron and steel syndicates.

It was in 1897 that the German syndicates inaugurated the export-bounty policy. The coal, the coke, the pig-iron, and other syndicates, and the Half-finished Goods Union, agreed to supply their associated customers—the rolling mills and steel manufacturers—with raw and partly manufactured materials at lower prices when those materials were required for manufacturers for export than when needed for goods for the home market. From time to time the bounty rates, or differences between home and export prices, were increased or decreased in response to the fall or rise of the home demand—the primary object being to effect continuous running of the mills. When the home demand has been brisk the bounties have been low; when it has slackened the bounties have been put up in order to promote "dumping" abroad rather than resort to damping furnaces at home— with results well-known in this country.

Those export bounties sometimes represented a bonus equal to a reduction of one-third of the standard home price for such materials as ingots, blooms, and billets of steel. By these means German iron and steel goods have been sold in Great Britain below actual cost of profitable production. German plates, rails, girders, etc., have been sold in England at from 20s. to 30s. per ton cheaper than in Germany, and at less than our own cost price. Having no effective means of repelling this invasion of our markets we have witnessed the laying of imported tram rails in our streets, while many of our own furnaces, rolling mills, and workers have been idle. In one half-year the German Wire Rod Syndicate, while selling 22,000 tons at home at a profit of £58,000, exported 19,000 tons at a loss of £42,000. With the wisdom or unwisdom of this policy—with the economic effects of this policy either upon the "dumpers" or those who were "dumped" upon —I have no intention to deal, but merely wish to point out the powers and possibilities of organisation under certain fiscal conditions. As recently as 1905 the bounties granted on materials used in the manufacture of goods for export were as much as 1s. 6d. per ton of coal, 4s. 10d. per ton of pig-iron, 15s. per ton of semi-finished steel, and 20s. per ton of shapes—the bounty on steel including that allowed on coal and iron.

During the nine years, 1897-1905, in which the export bounties were in active operation, German exports of iron and steel goods increased from, approximately, 1,000,000 tons a year to 3,000,000 tons; imports decreased from 500,000 tons a year to 400,000 tons; pig-iron production increased from 7,000,000 tons a year to 11,000,000 tons; and steel production from 4,000,000 tons a year to 8,000,000 tons. With these increases there came, of course, increases in employment and wages. In the mining and manufacture of minerals in Germany the increase in the number of workpeople represented 50 per cent., and the increase in wages, per capita, 28 per cent. I do not contend that the whole of this progress has been due to syndication and export bounties, but I do insist that without organisation the progress would not have been so substantial.

Recently, however, with a vigorous home demand taxing the uttermost resources of Germany's furnaces and mills, the need for export bounties and "dumping" has ceased—for a time. But we must not assume that German rivalry has ceased forever, or that it will ever cease for any great length of time. More and more capital is being invested in Germany's iron and steel trades; plants are being extended and equipment improved; large firms are amalgamating for the promotion of economy in production; the Steel Syndicate and its subsidiary organisations have taken a new lease of life; on every hand consolidation of interests is taking place, economies are being effected, and productive capacity is being increased. Should the home demands of our contemporaries decline—as, according to some reports, they are declining—or should their production again largely exceed their consumption, it is possible that the export-bounty system will be re-introduced, and that we may again be involved in a severe struggle. I do not think there is any cause for alarm on this score, but I urge that there is need imperative to strengthen our position in every way possible. And in nothing is there greater strength than in intelligent organisation.

Not less important than German competition is the question of American iron and steel trade prospects. Here we have to consider factors and circumstances much more uncertain than in the case of Germany, for industrial combination in America is unstable in character and inconsistent in policy—at any rate, it has been so hitherto, for the whole history of the American iron industry is a succession of "booms" and panics. Possibly, with improved organisation, the fluctuations will not be nearly so pronounced in the future as in the past—indeed, I believe that alike in Britain, in Germany, and in America, the future development of industrialism in general will be more steady.

The methods of industrial organisation usually adopted in America differ materially from those of Germany. While the Americans aim at one huge corporation, in the form of a limited liability company, the Germans are content to form what may be termed a manufacturers' union, or association. The United States Steel Corporation, formed in the Spring of 1901, has had a somewhat chequered career. It tried to absorb all the steel companies of any importance in America, and to regulate every process of production and method of distribution from mining the iron ore to fixing the price and effecting the sale of the finished product. For some of the iron and steel producing concerns it was obliged to pay big prices, and thus, from the first, the Trust was over-capitalized. And it failed in its primary object—it failed to obtain full control of the American iron trade, and just about one-half of the iron and steel production of the country remained independent.

From the commencement a keen struggle was waged between the Trust and the independent producers. The struggle is not yet ended. When the Trust was inaugurated prices were between 60 and 80 per cent higher than those prevailing in 1897, a year of marked depression, and although the tendency was distinctly towards still higher prices the Trust accepted orders at the then existing figures. The immediate result was that the Corporation mills became booked for many months, and those who needed delivery within any reasonable time were obliged to deal with the independent producers and pay bigger prices. Then followed a period of reaction. Demand relaxed, and delivery not being urgent, the independent producers had, perforce, to drop their prices. At this stage the Trust, having refused to put up prices to the extreme limit on a rising market, declined to follow the course of the falling market. Of course, custom went to the cheapest makers, the Trust got few orders, and was driven into a tight corner. Production was curtailed, extensions arrested, depression prevailed, and there was much dissatisfaction among shareholders who had bought at high premiums.

It has been argued that the cry of over-capitalization in connection with the Steel Trust was exaggerated; that it was fallacious to estimate the capital of any American concern by adding preference and common stock together; and that the common stock merely represented so much goodwill entitling the holders to dividends in good years. But the fact remains that this common stock was, and is now, the subject of much speculation. However, despite many difficulties, and despite many predictions of failure, the directors have managed to place the Trust on what appears to be a solid foundation. Plants are working up to the full limits of their capacity, positively enormous extensions are being made, and most of the known available deposits of high-grade ore in the country have been acquired. The American Trust, like the German Syndicate, seems therefore, to be destined to exercise in the future a huge influence, not only in its own country, but in the iron markets of the world. Hitherto American and German competition in iron and steel—syndicated and bounty-fed competition—has been keen enough; in the future it may possibly be even more keen and relentless, because better organised and more skilfully conducted. In these circumstances it is well, therefore, that British manufacturers are displaying a stronger tendency than usual in the direction of combination for the protection of their rights and the development of their industry. Far better meet combined effort by combined effort, far better organise our methods, improve our policy, cheapen and specialise production, and adopt the co-operative principle in sale and transit, than go on the old-fashioned lines, clinging to ancient customs and antiquated ideas until we are elbowed out of the race.

Although in the past we have been easily relegated from the first to the third position as an iron- and steel-producing country, there is no justification for despair concerning the future. There is no need to discourage extensions and the investment of further capital in British steel industries because those of Germany and America have made greater headway in recent years than our own. Our rivals' progress is not due to the enjoyment of some great monopoly in Nature's bounties or in human skill, but is due to the fact that, having adopted and improved upon our best methods, they have wisely discarded some of our worst ones. We have not been beaten through lack of materials, or men, or money, but through lack of organisation. The idea that this country is played out, or about to be played out, must be dispelled. The growth of international competition, the progress of other nations, should have no terrors for us, but should simply spur us on to improvement, and to a wise organisation and development of our resources.

It is possible that at no very distant date Britain's rivals may have some surplus products which they may seek to "dump" upon her markets, but it is hardly likely that they will ever "dump" below cost price for any great length of time. The last "dumping" of cheap steel was, to a certain extent, a confession of weakness on the part of the "dumpers." Our rivals are now in a stronger position, financially, than they were then. We, too, are in a better position now than then, and, whatever may be the strength of evidence of an approaching collapse of the world's iron "boom," there is just as much evidence that, before current contracts are fulfilled, there will be a renewal of the pressure; and there is no reason why we should not hold our own, whatever may take place. One fact stands out clearly: there is no sign of permanent abatement in the world's demand for iron goods. Steel is being put to an ever-increasing number of new purposes. There is no reason why we should not obtain an adequate share of this growing trade in iron and steel, for despite all that is said about the wonderful natural resources, and about the great enterprise and superlative skill, of our rivals, we in Great Britain can today profitably produce pig iron—the foundation of our steel and engineering trades—just as we can build ships, at a cost below that of other countries. That is the cardinal point, the supreme fact, in the issues involved, and should serve to inspire confidence in Britain's industrial future.

We have coal in abundance, and of good quality; we are not without vast and valuable deposits of iron ore close by our coalfields; we have ample foreign supplies of the richest ore within the easiest possible distance by sea; we have coal, ore, furnaces, and ships practically side by side, while our rivals' raw materials and shipping ports are widely separated; alike in natural resources and in geographical situation we possess some great—indeed, we possess some absolutely unique—advantages; and we have a wealth of capital and a plethora of skilful, efficient, and willing labour; but, to develop and utilise to the best and fullest extent our resources we need intelligent organisation.

It is not a valid argument to say that because trusts, syndicates, and combines in the past, or in other countries have made mistakes and abused their powers, therefore, organisation should not be effected, and cannot be perfected, among our manufacturers. In industrial unity there is commercial strength; and in commercial strength, national prestige and economic welfare. Industrial combination need not clash with public well-being, for, guided by the experiences of our contemporaries, we can avoid their errors, and guard against encroachment upon popular liberties.

PROFIT MAKING IN SHOP AND FACTORY MANAGEMENT.

By Charles U. Carpenter.

IX. THE UPBUILDING OF A SELLING ORGANIZATION.

Mr. Carpenter's series began in January, and the eight articles heretofore presented discussed, first, the general methods of examining into the condition of any manufacturing business so as to discover the existence of waste and loss and to determine the "sticking points"; second, the nature and the working of the "committee system" of factory administration; third, the reorganization of the designing and drafting departments; fourth, the great importance and value of the tool-room as a source and spring of profitable methods; fifth, the general conditions necessary for manufacturing at minimum cost; sixth, methods for fixing standard times for manufacturing operations; seventh, minimizing the time of assembling; eighth, stimulating production by wage, stock, and cost systems.—THE EDITORS.

I N an article as brief as this must be, it is, of course, out of the question to discuss at all fully the broad general question of methods of selling. Again, each business has its own peculiarities which make it necessary to apply special methods. At the same time, much can be said on that question of tremendous importance—the upbuilding of a highly trained, efficient selling force—that will be applicable to a business of any character. Different methods of making different manufactured goods make necessary some change in the plans, but in almost every case the fundamental elements are the same.

Methods of selling manufactured goods may broadly be divided into four divisions:—

First, where the goods are sold direct to the consumer by selling representatives of the manufacturing concern itself. This may be either through the company branch houses, or commission or salaried men, all of whom are employed and paid by the company. In this case, while considerable working capital is tied up and the stock of goods must be heavy, at the same time the selling division is directly under the control of the management and the territories themselves secure a company representation possible in no other way.

Second, selling to exclusive agencies, who themselves employ salesmen to deal direct with the consumer. This plan has much merit, although it is often difficult to devise plans whereby the agencies themselves may be forced to cover their territories and to develop an efficient selling department. It is also important to notice here that in such cases the manufacturing company very often does not come into close contact with the individual members of the selling division of such agencies, and thus their methods are not properly impressed upon the men. A connecting bond between such agency salesmen and the manufacturing concern is of great importance. This becomes especially so in cases of sudden terminations of contract, etc.

Third, selling to jobbers. This plan is susceptible of such variety that it hardly needs consideration. Close connection between the jobber and the manufacturing company is very necessary. Constant and skilful attention must be paid him in order to secure adequate and permanent representation.

Fourth, direct advertising, selling by catalogues, etc. A discussion of this method of marketing product need not be considered here.

Whatever the business, whatever the method of selling, the importance of a highly trained efficient selling division stands out paramount. In developing a selling force, we must consider:

1. Training of salesmen.
2. Training of sales managers.
3. Developing of a proper system, whereby both salesman and sales manager can be properly checked up—the former, to ascertain if he is properly covering his territory, and getting his full quota of business, securing proper prices and terms, and keeping his old and new customers satisfied; the latter, to see if he has the proper grade of salesmen employed, proper control over them, proper methods of training them; also to note if he is securing an adequate volume of business at such prices and with such economy in operating expenses as to guarantee a sufficient profit.

TRAINING OF SALESMEN.

Singularly enough, the majority of concerns today spend tens of thousands of dollars in advertising and in forcing the attention of the public upon their goods—in "creating the demand"—and yet they fail to train their sales employees—the men and women upon whom they must absolutely and finally depend as "closers"—so that they may know the "talking points" of their goods, the best methods of presenting their arguments, and the surest methods of finally "clinching the order." The unknowing sales manager often dismisses the arguments in favor of training of the salesmen with the trite remark "salesmen are born, not made." A mischievous belief! Granted that some men are by nature better fitted for selling than others, those men are only too few. Training would surely improve their efficiency. Unfortunately, however, we manufacturers who market our own goods can find but very few of these "born salesmen," and are obliged to rely upon the "average salesman" for the most of our business getting. Such being the fact, it must be apparent that the average selling division needs badly a scientific and systematic method of training, in order that the large proportion of "average salesmen" may be brought to as high a degree of efficiency as possible. Nor does the advantage to be reaped stop simply with the training of the employees in selling. The meetings which this system calls for, if managed aright, are sure to prove of immense benefit in arousing a healthy and stimulating enthusiasm in the employees affected. An honest desire and intent on the part of these salesmen to "put in the best licks for the House" gives an impulse to their activities that nothing else can supply. This feeling can be instilled into them by a skillful, tactful sales manager. Do not forget that a sullen, listless, or disloyal member of your selling force affects your profits immediately. If you have many such, your department is costly and inefficient. The effect of your thousands spent on advertising is nullified by such conditions. On the other hand, a highly trained, loyal, interested and active selling force is one of the most valuable assets a firm can possess. That such a selling organization can be developed, even from one of a peculiar degree of inefficiency, has been proven by a long personal experience in several lines of business and very close observation of the results obtained in other modern business concerns.

John H. Patterson, the brilliant president of the National Cash Register Company of Dayton, Ohio, was the first business man to grasp the possibilities that lie in the training of the salesman. Through his genius, his company has developed probably the greatest and most efficient selling organization in the business world today. His example has been followed in the manufacturing business by such concerns as the Burroughs Adding Machine Company and the Herring-Hall-Marvin Safe Company, with decided success. Work of this character has been successfully applied to "Hapgoods," the firm doing a large employment business. Indeed, its essential principles can be adapted to the conditions of a business of any character in which the marketing of goods plays a large part.

DEVELOPMENT OF A SELLING SYSTEM.

While each business requires special study and special methods, the general plans of the "Science of a Selling System" can be clearly pointed out so that they may be adapted. The two essential features of this system are "Salesmen's Demonstration Meetings" and "Salesmen's Training Department." While the "Salesmen's Training Department" is in fact the more important feature, I place the "Salesmen's Demonstration Meetings" first, because the training department is, nine times out of ten, the outgrowth of the demonstration meetings. The logical beginning of this system is with these meetings, because the points to be used in the training department are invariably secured from the discussions arising in these meetings. Again, these meetings serve as a gradual introduction for the later training department, and so accustom the selling force to the methods themselves that there arises but little opposition to the training department when it is first proposed. Each salesman, too, feels that he has had some part in the development of the training department, inasmuch as his own arguments are often used, and so thorough support can be elicited in place of the violent opposition that may be expected if any arbitrary methods are used. Woe betide the sales manager and the sales system if he starts at the "other end of the line," and attempts to force these methods upon his selling department before the members are ready for it.

SALESMEN'S WEEKLY DEMONSTRATION MEETINGS.

The points that will be brought out in regard to these meetings need no elaboration, as their merit is self-evident.

First, a time for these meetings must be set and constant attendance insisted upon. The sales manager must invariably be present and take part in discussions of all matters of importance. If possible, some higher official should be present once a month, in order to inject a new interest and new quality of enthusiasm into the salesmen. It will stir up both salesmen and sales manager tremendously if they are compelled to exhibit before someone high in authority.

Second, it must be kept in mind that these meetings are for the training of the salesmen (and incidentally the manager) and helping them over their difficulties, for arousing interest and enthusiasm, for giving the salesmen a chance to "blow off steam" on any trouble they may have that is affecting their efficiency, and for securing from them suggestions for the improvement of the business. A sample weekly programme is given on the next page.

Keeping in mind the main objects of this sales system—the creation of interest and enthusiasm and the training of salesmen—an examination of the succeeding suggested programme leaves an explanation almost superfluous.

PROGRAMME OF SALESMEN'S DEMONSTRATION MEETING.

1. Announcements. By Sales Manager.
2. Description of new products and fields they are designed to fill. By Sales Manager. (Suggestions and criticisms from salesmen requested.)
3. Demonstration of salesmanship. By Salesman—J. H. Smith.
   By Customer—G. R. Brown (Salesman).
   Censors—R. Fowler, H. White.

   (a) Selling the product to the customer whose business is carefully selected and who desires a good article.
   or (b) Selling customer asking for low priced article, a higher priced and more profitable product.
   or (c) Selling second hand product to customer.
   or (d) Selling customer asking for second hand product a new product.
   or (e) Selling customer new product, taking old product in exchange, at profitable allowance figure.
   or (f) Selling customer against strong competition, another salesman entering the demonstration as competitor's salesman.

   NOTE: These demonstrations may be varied by having the same points illustrated as "Company-office Sales," where all stock and other paraphernalia are present, or as sales at the customer's office, where the salesman must depend upon illustrations, samples and catalogs.

4. Discussion of demonstration, first, by the appointed censors, Fowler and White, and second, by each salesman personally.
5. Discussion of week's business; why individual salesmen have not made their quota of sales and difficulties met by salesmen, from blackboard individual-sales record and from individual-sales reports.
6. A talk by the sales manager or some high-grade salesman on general important points of salesmanship, such as:

   Investigation of prospective customer's business and his methods.
   The proper "lining up" of selling arguments so that the "selling climax" may come at the right time.
   Methods of introduction, or "the approach."
   Methods of getting the prospective customer's attention and making demonstrating arguments.
   Methods of using closing arguments and "getting the signature to the order."
   Ways to meet certain arguments and objections of prospective customer.
   Ways to meet competitor's claims and arguments.
   Methods of cultivating a territory.
   Importance of "satisfied user."
   Importance of "knowledge of the business."
   Advantages to be gained by paying close attention to such seemingly small points as tact, dress, industry, perseverance, talking too fast or too much, answering customer's questions quickly, and a multitude of similar matters, perfection in which is so important.

7. Discussion of different competitors' products, their talking points and how to controvert them, their defects and how to prove them.
8. Suggestions and complaints.
9. General subjects such as advertising, etc.

In making announcements and describing new products, much can be done to create a lively interest in the company's affairs on the part of the salesman. A frank and full discussion of new products or proposed new designs will often prevent serious mistakes and will almost invariably result in suggestions that will make the product more marketable.

The suggested variations of "Demonstrations of Salesmanship" are also self-explanatory. Note carefully the appointment of censors. It is also especially desirable to hear comments upon demonstrations from each salesman. If the sales manager handles matters right so that the salesmen enter into this programme in a proper spirit, there need be no fear of salesmen becoming angry over fair criticisms. The "customer" can be selected from the sales office. He must be given to understand that he is not to aid the salesman in any manner, direct or indirect, and that he should bring up all the arguments and objections against buying that he himself has learned from his own customers. In a business where the product covers a broad field, embracing a number of widely different kinds of business, where a discussion of the business system enters into the sales, the "customer" and the character of the business can be selected so that in the course of a short time the demonstrations will cover the entire field and a full line of argument be brought out for each line of business. The character of the business and the conditions surrounding it should be thoroughly understood before beginning. Often a salesman who has had a particularly hard nut to crack will suggest the conditions and himself act the part of customer against a good salesman, in the hope of either "stumping him" or getting some good pointers. Especial attention may well be given to methods of convincing a customer that he should purchase a higher priced and more profitable product. Inasmuch as the selling expense remains the same, or very nearly so, an effective method of accomplishing this will result in a much larger proportion of profit. The most modern concerns pay a great deal of attention to this point and have developed a highly scientific and effective method for accomplishing it. This applies also to a business in which exchanges for old products enters into a large proportion of the sales. This is often a puzzling feature, and unless thoroughly understood results in large hidden losses.

Great interest can be aroused by the introduction of a salesman representing a strong competitor, who is supposed to do his best. Such exhibitions are not only highly instructive, but also inspire salesmen with a confidence in their own goods.

It is important that two demonstrations be often given, one conducted by an old and skilful salesman and one by a newer member of the selling force. This not only aids greatly in the education of the newer salesman, but often acts as a great spur on the older man not to be outdone by the newer ones. In conducting these demonstrations care must be used to see that no slipshod methods be allowed to creep in. The "sales" must be conducted with all the dignity and formality of a real transaction from the beginning, in order that the best form of "approach" or introduction of the subject may be observed.

A better method of instilling selling confidence into a man is hard to devise. After he has had to appear several times before a body of his selling companions and his superior officers he gains confidence rapidly and his attacks of "nerves" are things of the past. Such methods quickly develop the "quitter," for his improvement or elimination from the organization. They certainly show up the "dead wood" quickly.

The talks by sales manager or high-grade salesman upon general selling points, as noted under programme item 6, will prove of great benefit to new salesmen. The points brought forth—taken by a stenographer—are of great value later when organizing a method of training. The items given embrace only a very few of the important topics that may be discussed, but are given to illustrate clearly the nature of the talk suggested.

Programme item 7. While it is not often advisable to instruct salesmen to talk against competitors' goods, I regard it as absolutely necessary that salesmen be thoroughly posted on the character of competitors' products. Very often true statements of defects in such articles become necessary.

Suggestions and Complaints. This section deserves more consideration than can be given it at this point. Nothing is more vital to the progress of a company than the proper and conservative meeting of the actual market demands and the improvement of product so as, if possible, to keep ahead of the demand. No one knows the needs of the market or the advancement of competition as does the salesmen. A systematic plan to secure these suggestions from the body of salesmen will prove of very great value. The same may be said regarding complaints. Legitimate complaints should be "aired," and when the causes are ascertained prompt steps taken to rectify the troubles. Many a firm today prefers to shut its "business ears and eyes" and refuses to hear of troubles or to see perfectly obvious defects which are continually having a disintegrating though hidden effect upon the business and organization.

The details of these meetings should be invariably taken down in shorthand. I have found it of great value to have sufficient copies of each meeting made to allow of their being distributed to the sales managers in all parts of the country.

It is important that the general sales manager consider it his particular duty to read carefully all minutes of these meetings and then to write to each local sales manager, commenting upon them, (in each case mentioning names of salesmen). The effect upon both manager and men is very beneficial.

The quality of the demonstration affords a very good proof of the calibre of the salesmen in each district and thus provides the clearest kind of an index to the quality of salesmen throughout all the points of the organization. Again, the salesmen are impelled to do their level best, knowing that the general sales manager will himself note the character of their work, even though he may be thousands of miles away. By such a simple means the influence and power of the general sales manager will be felt throughout the entire selling organization.

SALESMEN'S TRAINING DEPARTMENT.

While much good can be derived from such weekly meetings, the progress of the men toward high-grade selling is necessarily slow. The influence toward rational methods is not constant enough. Again, constant individual attention should be given the new men at the beginning so that they may have the full benefit of such methods early in their selling career. Experience has shown that the only rational plan for developing salesmen rapidly and upbuilding a strong homogeneous selling department is to develop a strong training department for salesmen. This department should be independent of the influence of any local sales manager, but should be under the direct supervision of the general sales manager. It should be his "selling right hand." All local managers should be thoroughly trained in this department's methods so as to supplement its activity in their own local districts. All salesmen should be trained therein. Particular attention should be paid to the export trade agencies. Some firms establish training departments in the several foreign countries. Personally, I prefer to have even representatives of export agencies taught at the home office, so that they can not only get the best possible course of instruction but also may come under the direct strong influence of the home-office executives.

Scope of Training Department Work.

The first step (and often the most difficult) is to find the proper instructor. No greater mistake can be made than to attempt to use a cheap man. This work requires a man with the widest selling experience, coupled with great tact, patience, and teaching ability; a man whom the salesmen will respect for his ability. The outline of his work will demonstrate the necessity for having a man of sterling ability.

His first work will be to prepare a "Manual" for salesmen. This must contain

    (a) Strong points on general salesmanship.

    (b) A thorough and careful explanation of each product and its adaptation to all different lines of business.

    (c) An exceedingly thorough explanation of the "talking points" or "selling arguments" of each product.

    (d) An analysis of competitors' products and a comparison with the manufacturing company's product.

    (e) A careful and scientific analysis of the best methods of introduction to a prospective customer so as to gain his attention and interest, this forming the "approach."

    (f) Statements of the best methods of marshalling the talking points together so that a demonstration of the product's merits may be made to the customer—this forming the "demonstration and argument."

    (g) A thorough and complete analysis of the best "closing arguments" and discussion of various ways to "get the order signed."

    (h) A full list of the most common objections to making a purchase and ways of meeting these objections. After this plan is worked out, it will astonish many to note how simply the objections to purchasing on the part of prospective customers can be classified. It will also astonish the average sales manager to note how many different and excellent answers can be made to these objections by taking the answers to these by many managers and salesmen in different parts of the country. Whenever a salesman meets with some new form of rebuff, arguments to meet a new condition can quickly be secured by referring the question to the different managers for settlement in their weekly demonstration meetings.

    (i) Much space may well be given to a thorough and logical explanation of best methods of raising a customer desiring a low-priced product to one of higher price, "Raising him up the line," as it is called. Really scientific work can be done along this line. Methods of handling second-hand sales and exchange sales should also be treated fully.

It will be noted that almost all of these invaluable data can be secured from the salesmen demonstration meeting reports.

After the selection of the instructor and preparation of the manual the balance of the work is largely routine.

First, each man—new or old—must be made to learn the manual "backward and forward." No halfway learning can be tolerated. He should then be thoroughly drilled in the "approach," "demonstration" and "closing" arguments under differing conditions along the lines noted in the Programme of Salesmen's Demonstration Meetings. He should be compelled to go through these in the regular demonstration meetings before the entire body of salesmen. This process will require from two to six weeks depending upon the man and the character of the business. He is then started out in a territory and carefully watched. It is well also at times to have him attempt to make a sale at the office so that his methods may be noted. After about a week of this experience the instructor should accompany him on his regular rounds so as to note his methods. Failure to attend to important points may thus be observed. The instructor should then illustrate the proper methods by taking the selling end himself with several prospective customers and closing the sales. After the salesman has been in his territory for a full month the instructor should spend another period with him.

After these men are distributed to different territories the local manager should give them the same attention.

The instructor should visit the several territories from time to time and note closely the salesmen's method of demonstration, both in the weekly salesmen's meeting and before their customers. A constant and close study of weekly sales reports, supplemented by the reports of the weekly demonstration meeting, will clearly indicate the weak spots needing attention.

The local managers should from time to time be called into the home office for conventions. These can be made exceedingly helpful to both company and managers. They are the backbone of the selling division and they cannot be watched, trained, inspired, and worked with too much. Not only must they be driven on the question of sales, but they must also be held responsible for economy in management.

The devising of proper sales systems is a subject in itself that cannot be fully treated here.

It is essential though that mention be made of the two forms from which spring many branches of the system, namely, the Salesman's Daily Reports. One covers sales made and gives the important details as to the customer and his business; the other covers cases of failure and gives the reasons for non-success. The two forms are shown below:

SALESMAN'S DAILY REPORT OF SALES.

Name Customer ....Rating..........Date...........

Character of business and system used....................................

What Sold.......................Date Delivery.........................

Higher Priced Machine needed.................(Date)...................

Duplicate Machine needed......................(Date)...................

Send advertising matter as follows: ....

SALESMAN'S DAILY REPORT OF FAILURE.

Name ....Rating..........Date...........

Character of business and system used....................................

What needed?.......................Why not sold?......................

....Date to return......................

Send advertising matter as follows: ....

Couple these two reports with a comprehensive list of prospective customers, and you have the best foundation for a comprehensive and valuable sales system. Many other forms will naturally be added to these and many different methods used for properly tracing up "prospective customers" noted from these reports.

Through your prospective customer list, coupled with the salesman's daily reports, you can determine whether or not each man is properly covering his territory. If he is not, you can make him do so. The daily reports data enable you to classify these prospective customers so that you may be sure that they receive the proper attention at the proper time and the proper kind of advertising matter pending the next visit of the salesman. These reports are capable of indefinite amplification along lines which will be of immense benefit to the business, especially when used in connection with a sales system along the lines described. Such systems will admirably supplement the factory methods advocated in preceding papers, and will serve to unify the entire plan of organization, business, and method along such logical lines that there can be but one result—Progress!

A selling department built up along such lines is the best guarantee of high prices and good profits—a bulwark of strength against competition, and the strongest possible business foundation, especially in times of industrial depression.

APPLIED ELECTRO-METALLURGY UP TO THE END OF 1906.

By John B. C. Kershaw.

Mr. Kershaw's review began in the October number of THE ENGINEERING MAGAZINE. The first instalment of his article discussed the electro-metallurgical production of aluminum, refined bullion, calcium carbide, calcium, carborundum, copper, diamantine, and graphite. His concluding section deals with the electro-metallurgy of the ferro-alloys, iron and steel, lead, nickel, siloxicon, silicon, sodium, tin, and zinc.—THE EDITORS.

FERRO-ALLOYS.—The application of the electric furnace for producing alloys of iron with silicon, chromium, manganese. tungsten, and vanadium has developed into a large and important metallurgical industry. Since Moisson's early research work, the value of these alloys for the manufacture of special steels has been recognised by expert steel-makers in all countries. The manufacture of ferro-alloys is carried on at present chiefly in France and Switzerland, a cheaply developed water-power being essential for the commercial production of these compounds. In France, MM. Keller, Leleux & Cie. are producing ferro-silicon and ferro-chrome in large amounts at Livet and Kerrousse, while the Société Electro-metallurgique Française devote a portion of their power to the same manufactures, at La Praz and St. Michel. The largest works are, however, to be found in the Haute Savoie, on the borders of Switzerland, where the Société Electro-metallurgique Girod are utilizing 18,000 horse power for the production of a ferro-silicon, ferro-chromium, ferro-tungsten. and ferro-molybdenum, the aggregate output of the three works owned by this company being given by Dr. Hutton as 9,000 tons per annum, and the value as £360,000.

In Germany MM. Goldschmidt & Cie and MM. Krupp are using the aluminium reduction process, in place of the electric furnace, at Essen for producing ferro-alloys free from carbon.

In America the ferro-alloys industry is less developed, the Willson Company, with works at Kanawha Falls and at Holcombs Rock, Va., being the only producers of ferro-chromium; about 3,000 tons are produced in the two works. Rossi is, however, experimenting at Niagara Falls with electric-furnace methods of producing ferrotitanium and a new works has been erected during 1906 at Newmire,

Colorado, by the Vanadium Alloys Co. of New York, for the manufacture of ferro-vanadium. Recent trials of ternary and quaternary
steels, made with the addition of vanadium, have proved that these
steels are specially suited to the demands made by motor-car work,
and it is expected that in time the manufacture of vanadium steel
may become a branch industry of considerable importance.

With regard to the use of ferro-alloys generally, ferro-silicon is
employed as a deoxydizing agent, while the other alloys are employed for introducing the rarer metals into the steel, it having been
found that a more homogeneous product is obtained when the metal
is introduced into the molten steel in the form of an alloy than when
it is introduced in the pure state. All the chrome-steel used for
armour-plate manufacture is now made with the aid of ferro-chrome.

IRON AND STEEL.—The methods of producing iron and steel in
the electric furnace have been developed chiefly by French electro-
metallurgists, a large number of works in France having been rendered idle by the collapse of the boom in the calcium-carbide industry
in 1899-1900, and new applications being required for the water-
power and electric-furnace plant thus made available. The earliest
trials of the electric
furnace for iron and
steel production date
from 1899, and since
that year experi-
mental work has
been carried on con-
tinuously. During
the last three years
the new methods
have attracted the
attention of steel
makers, and it is
now generally re-
çognised that cer-
tain of the methods
and processes have
attained a perma-
nent footing in the
iron and steel indus-
try. The Heroult
and Kjellin methods

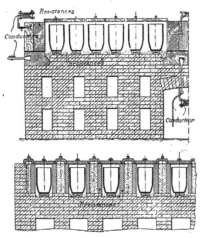

VERTICAL SECTION OF A GIROUD ELECTRIC FURNACE
FOR CRUCIBLES INDIVIDUALLY HEATED.

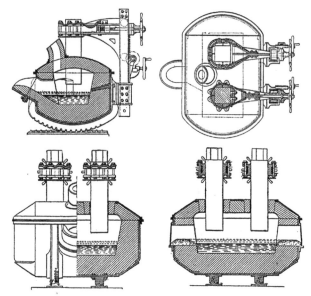

LONGITUDINAL AND TRANSVERSE SECTIONS OF HÉROULT CRUCIBLE FURNACE.

of steel refining by aid of electric heat have shewn the most striking development, and a large number of works in Europe and America are using these methods with satisfactory results.

The Heroult steel-refining furnace is of the crucible type, and the heating is initially effected by means of the electric arc, which forms between the surface of the slagging materials covering the metal and the two massive carbon electrodes which are suspended above it. The slag when molten is used for resistance heating, the carbons being lowered until they touch it. The impurities of the iron are removed by renewing the slag from time to time. The refining operation thus becomes a "washing out" of the impurities of the iron, by treatment with suitable slags. When purified, "carburite" in requisite amount is added, and the crucible is tipped.

Heroult claims that with this furnace iron or steel of any degree of impurity can be refined, and that from the purified metal, a steel

KJELLIN 1,000 HORSE-POWER ELECTRIC FURNACE. TOP AND BOTTOM VIEWS.
The lower showing the motors and gearing by which the furnace is tipped for pouring.

of any desired composition can be produced by the addition of the necessary amount of "carburite" and other ferro-alloys.

The Heroult furnace is now in operation at La Praz and Froges in France, at Kortfors in Norway, at Remscheid in Germany, and at Syracuse, New York. The Remscheid plant has been in operation since February 1906 and is on a smaller scale than the Syracuse plant.

A NEAR VIEW OF THE TOP OF A KJELLIN ELECTRIC FURNACE OF 1,000 HORSE-POWER.
The latest type, of which details have so far been kept secret.

The Kjellin furnace has been developed at Gysinge in Sweden, and differs materially from the Heroult furnace. In place of the use of direct current for combined arc and resistance heating, the Kjellin process utilizes induced currents, and the heating effect is obtained by the rapid changes in the magnetic state of the iron or steel which forms the secondary coil of the circuit. The Kjellin furnace is in reality a large transformer, in which an alternating current of low amperage, but high voltage, is transformed into an alternating current of large intensity but low pressure in the secondary coil of the apparatus. The metal is contained within an annular ring built up of refractory blocks round the primary coil of the furnace, and by varying the current in the primary, the heat developed in the secondary can be regulated as desired. The advantage of the Kjellin furnace

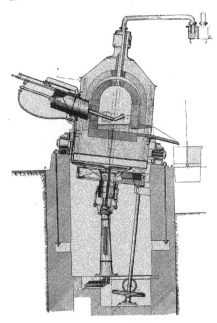

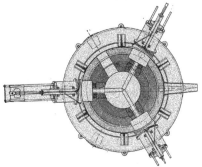

PLAN AND ELEVATION OF STASSANO REVOLVING FURNACE.

are the develop-
ment of the heat
just where it is
wanted, *i. e.,* en-
tirely within the
metal, and conse-
quent small wear
and tear upon the
structure a n d
walls of the fur-
nace, and second-
ly, the absence of
i m p u r i t i e s
picked up from
t h e electrodes
used in all other
methods of ap-
p l y i n g electric
heat. The Kjel-
lin process a n d
furnace are being
worked success-
fully at Gysinge
in Sweden, a t
G u r t m e l l a n
i n Switzerland,
at Krupp's steel
w o r k s in Ger-
many, at Vicers
s t e e l works in
England, and also
at the Araya steel
works in Spain,
while in America
a f u r n a c e
patented by Col-
b y b u t worked
upon the s a m e
principle has been
operated w i t h
successful results
at Philadelphia.

Three electric-furnace methods for the production of iron and steel direct from the ore have been tried upon a small industrial scale in Italy, France and Canada.

The first of these—the Stassano—has not achieved success, although large sums of money were expended upon the trials at Rome and Darfo in Northern Italy. An arc furnace of the rotary type was employed, and the ore was ground and briquetted with the lime and coke, before charging into the furnace. The costs of grinding and briquetting all the raw materials, and the difficulty of maintaining a durable lining to the furnace, were the principal causes of failure.

STASSANO 1,000 HORSE-POWER ELECTRIC FURNACE, FIXED TYPE.

The Keller furnace and process for the production of grey, mottled, and white pig iron from the ore, has been operated at Livet in France for several years, with moderate success. The furnaces are of 1000 horse-power and 308 horse-power capacity, of the two shaft type, with large carbon-block electrodes slung in chains in the centre of each shaft. The heat is obtained by combined arc and resistance heating. A canal connects the hearths of the two shafts, and when filled with molten iron this canal serves as the electrical connecting link between the two portions of the furnace. The ore is crushed roughly, to a size of 2 inches, and is charged with the lime and coke into each shaft of the furnace. The electric power required per short ton of pig iron produced at Livet averages 2,300 kilowatt hours, and

POURING A 1,000 HORSE-POWER STASSANO REVOLVING FURNACE.

it is estimated that with power at $10.00 per electrical horse-power year, a ton of pig iron could be produced by the Keller furnace and process for $11.60. A furnace designed to produce 20 tons of grey iron castings per 24 hours has been erected at Livet, but I am not aware whether it is yet in work.

The third process and furnace are that of Heroult, and the most important trials have been conducted at Sault Sainte Marie, Canada, under the auspices of the Canadian Government. The furnace is a single-shaft furnace of most simple type. The smelting of the ore is carried out by combined arc and resistance heating, the raw materials being charged without grinding into the shaft of the furnace, in which hangs the heavy carbon-block electrode, while the sole plate primes the other electrode. The experiments with this furnace at

Sault Sainte Marie proved that magnetite and titaniferous iron sand could be smelted without difficulty and that charcoal could be substituted for coke, without briquetting. The electric power required per ton of iron was 1,541 kilowatt hours, or less than at Livet, but later trials of the same furnace at Sault Sainte Marie have shown that the larger power consumption is the more correct. The furnace has now been taken over by the Lake Superior Company financing the development of this new industrial centre, and 54 tons of nickel pig have been produced in it from the roasted pyrrhotite ore of the district. This attempt to found a new iron and steel centre in Canada may have most important results upon the development of countries which have iron-ore deposits, but no coal with which to smelt the native ores.

LEAD.—Several attempts to introduce electrolytic or electro-thermal methods for the refining of lead have been made in America, and one such process was worked for some time upon a large scale at Niagara Falls, but the company financing this venture ultimately ended in liquidation. At the present time the Betts refining process, in which lead bullion or raw lead is used as anode material, in a bath of lead fluo-silicate, is in operation at Trail, British Columbia, and at

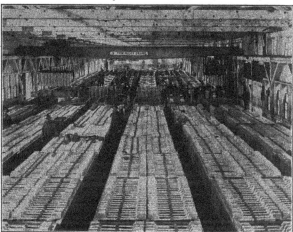

TANK HOUSE AND MELTING ROOM, CONSOLIDATED MINING & SMELTING CO., TRAIL,
B. C.
Capacity 80 tons of lead daily, by the Betts process.

Newcastle, England. The plant at Trail was enlarged in 1906, and consists of 240 vats, each 7 feet in length by 30 inches wide. When charged, each vat contains 20 anodes and 21 cathodes, and the capacity of the plant is stated to be 90 tons of refined lead per day. The separation of the lead from the copper, bismuth, and cadmium contained as impurities in the raw lead, is reported to be almost perfect. Betts has recently proposed to introduce electro-thermal methods for smelting the lead ores, but these proposals do not appear to have yet been submitted to practical trial.

NICKEL.—Nickel is produced by electrolytic or wet methods, by three companies, and at Sault Sainte Marie, Canada, experimental trials have recently been carried out which show that ferro-nickel can be successfully extracted from the ores of the district by the Heroult electric smelting furnace. A permanent installation of the Heroult furnace at this place is therefore possible. As regards the electrolytic methods of extraction, the Hoepfner process is in use by the Allegemeine Elektrometallurgische Gesellschaft of Papenburg.

ELECTROLYTIC LEAD REFINERY OF THE CONSOLIDATED MINING & SMELTING CO., TRAIL, B. C.

Germany. The process depends upon the elctrolysis of mixed solutions of copper, calcium, and nickel chlorides, these being obtained by leaching the roasted nickel ore with a solution of calcium and cupric chlorides.

In America, the Orford Copper Co. have recently commenced to produce electrolytic nickel, u s i n g a s anode material for the vats slabs of n i c k e l sulphide. These are obtained by operation of the "tops and bottoms" process for separat- ing nickel and cop- per sulphides. The electrolyte is nickel- c h l o r i d e solu- tion, w h i l e t h i n sheets of pure nickel are used as cath- odes. The electro- deposited n i c k e l tests 99.5 per cent.

A third electroly- tic process in use at Sault Sainte Marie is stated to be the Hybinette process (United States pa- tent No. 805,969 of

COLLECTORS AT THE BASE OF A STASSANO 1,000 HORSE-POWER ELECTRIC FURNACE.

1905). The electrolyte in this process is a dilute solution of nickel sulphate, to which a small quantity of boric or phosphoric acid has been added. The anodes are made from a ferro-nickel-copper alloy. The cathodes are thin sheets of copper enclosd in porous bags, and held in wooden frames to prevent buckling. The flow of fresh electrolyte is directed into the bags which contain the cathodes, and by maintaining a higher level of liquid in these, the drift of copper ions, dissolved at the anode, towards the cathode compartment is stopped, and only pure nickel is deposited at the cathode. The elec- trolyte becomes continuously richer in copper and iron, and is regen- erated by passing over slabs of nickel, or of a nickel-copper alloy. The copper is deposited and the nickel takes its place, while the iron is removed at a later stage, by oxidation. The solution then contains only nickel sulphate, and is ready for use again in the vats.

SILOXICON.—This is the name given to an electric-furnace product made by Acheson at Niagara Falls, by heating carbon and silicon in a fine state of sub-division and well mixed, to a temperature slightly below that required to produce carborundum. The product is a highly refractory material, and a company has been floated in the United States for the manufacture of siloxicon crucibles, muffles, bricks, etc. The chief difficulty in the manufacture of siloxicon is the regulation of the temperature, since if this be raised too high (above 1,700 degrees C.) the oxygen escapes and carborundum is produced.

SILICON.—F. J. Tone has produced this metal in large amount at Niagara Falls, by heating sand with carbon in an electric furnace of the resistance type. It is essential that the raw materials be finely ground and well mixed, and that the temperature be carefully regulated to prevent formation of carbides. The metallic silicon must be drawn off as formed, the process being continuous, and the metal obtained bright and crystalline. Tone states that the metal may be used as a deoxidizer in the iron and steel industry, and as a substitute

STASSANO 1,000 HORSE-POWER REVOLVING FURNACE.

for aluminium in the "thermit" mixture; but the demand for silicon for these and other purposes does not appear to have developed, and the difficulty at present is to find a market for the product.

SODIUM.—The production of this metal by the electrolysis of the fused hydrate has grown in recent years into an important industry, and the older chemical method of manufacture has now been quite supplanted by the electrolytic method. The Castner cell and process are generally employed. Installations of this cell are now working in

HIGH-TENSION GENERATÖRS, KELLER, LELEUX & CIE., LIVET.
Three three-phase 2,500-kilowatt alternators and two continuous-current units.

England, America, France, and Germany. The manufacture of metallic sodium in England is in the hands of the Castner-Kellner Alkali Co. The plant has recently been transferred from Weston Point to Wallsend-on-Tyne, where a new works has been erected, the power required being purchased from the Newcastle and District Electric Supply Co. at a very low rate.

Ashcroft has patented a cell and process recently by which sodium chloride can be substituted for the hydrate in this manufacture. This process is about to be tried upon an industrial scale in Norway Should the attempt succeed, the cost of metallic sodium which has already been reduced from 4s. to 1s. 8d. per kilogramme by the improvements in the Castner electrolytic process, will be still further reduced.

The sodium produced by the electrolytic method is largely employed for the manufacture of sodium cyanide, and of sodium peroxide. "Oxone" is the trade name given to fused sodium peroxide,

DOUBLE ELECTRIC FURNACE, WORKS OF KELLER, LELEUX & CIE., LIVET.
The metal is melted in an upper furnace and poured into a lower one, in which the operation is finished.

and this product is being advertised and sold in America for the generation of pure oxygen.

TIN.—Electrolytic or electro-thermal methods have not been applied with any success to the extraction or refining of tin, but in a branch industry—namely, "tin stripping"—they have become of considerable value and importance.

In the manufacture of cans, boxes and vessels of all kinds from tin plate, an immense amount of waste occurs with the cuttings, and the recovery of the tin from these has been carried out for some years by electrolysis. The process usually employed was first applied industrially by Goldschmidt at Essen in Germany, and consists in the use of the scrap and cuttings as anode material in a bath of sodium hydrate. Stannic-chloride solution has also been used as electrolyte in the Bergsoe process at Copenhagen. In the former case, only the tin is dissolved at the anode; in the latter case the iron is also attacked, and care is therefore required to prevent the solution of tin-chloride from becoming supersaturated with the iron salt. The chief development of the electrolytic tin-stripping industry has occurred in Germany, but similar factories have also been erected and carried on in Denmark, Austria, England, and America. The chief difficulty in working the process has been to maintain an adequate supply of tin-scrap and cuttings, and some of the works have had to close down from this cause. Purely chemical methods of stripping by means of chlorine gas are also now coming into favour.

This will still further accentuate the difficulty of supplies, since the electrolytic alkali and bleach works will enter the market as purchasers of the tin scrap and cuttings. By this method of stripping, stannic chloride is produced, and not metallic tin. The manufacture of "tin-salts" has already been taken up by some of the electrolytic alkali works in Europe and America.

ZINC.—The attempts to apply electrolytic and electro-thermal methods in the zinc industry have met with only partial success, and the greater proportion of the zinc found in commerce is still produced by the old metallurgical method of distillation.

The coating of iron articles with a protective deposit of zinc is, however, carried on in a large number of works by the electrolytic or wet method, a solution of zinc sulphate being generally employed as electrolyte, with lead anodes. "Electro-galvanizing" as it is called, is then an important branch industry.

As regards the extraction of zinc from its ores, the Hoepfner process is in operation at Winnington in Cheshire and at Hruschau in

HÉROULT ELECTRIC TIPPING FURNACE. THE LOWER VIEW SHOWS THE POURING OF
THE CHARGE.

Austria. By this process, zinc chloride is obtained from the waste liquors of the ammonia-soda process, and is electrolysed in order to obtain metallic zinc and chlorine gas.

A zinc-ore chlorination process, patented by Swinburne and Ashcroft, is operated at Weston Point, England, by the Castner Kellner Alkali Co. Zinc-sulphide ores are treated with hot chlorine gas, and the corresponding chlorides are obtained, but the zinc chloride is sold as such, and is not subjected to electrolysis as described in the patents covering this process.

Electrothermal methods of treating raw zinc and zinc ores are being experimented with by de Lavel and by Ferraris, in Sweden and Italy. The de Lavel furnace has already produced some hundreds of tons of pure zinc from spelter, but I understand that it has not yet been applied with success to the reduction of the ore. At Monte-Poni in Italy, Ferraris is carrying out similar trials with an electric furnace, and has estimated the cost of the process at 40 lire per ton of calamine. In a recent letter he stated, however, that the method has not yet reached the industrial stage of its development.

PRACTICAL EXPERIENCE WITH EXHAUST-STEAM TURBINES.

By Dr. Alfred Gradenwitz.

In these days of discord and strife over the applicability of the steam turbine to marine propulsion, it is refreshing to turn to a consideration of one of the minor fields of applica- tion of this type of prime mover, in which there can be no doubt of its utility and Value. Dr. Gradenwitz's article, with its numerous illustrations taken from the operating records of actual plants, shows clearly the economies which may be effected by the installation of low-pressure turbines for the utilization of exhaust steam, and the advantages which this system possesses, particularly in mining and steel-works plants, in a great Variety of applications.—THE EDITORS.

THE greatest drawbacks encountered in using the exhaust steam from factories and mines for the production of energy have been the unsatisfactory utilisation of low-pressure steam as effected in reciprocating engines, and the intermittent operation of most steam-consuming machines in iron works and mines, which called for the arrangement of some steam accumulator. The secondary engine, furthermore, had to be made entirely independent of the primary engine.

M. A. Rateau of Paris, in connection with his work in turbine construction, has evolved a type of steam turbine which is especially adapted to work with high efficiencies, and which accordingly may be operated by the use of exhaust steam.

That engines designed on the turbine principle are alone suitable for utilising exhaust steam is shown in a lecture recently delivered by Rateau at Brussels, the proof being found in theoretical considerations based on the entropy diagram. A reciprocating engine worked with an initial pressure approaching that of atmosphere, in order conveniently to utilise the whole of the drop in pressure produced by a condensing plant, would in fact require dimensions so enormous as to make its space requirements, weight and cost of installation in proportion to output quite inadmissible. If, on the other hand, a turbine be used, low-pressure steam could very well be employed, the vacuum given by even the most perfect condensers being utilised to the best advantage. Owing to the considerable issuing speed of the steam and the enormous resulting capacity, it would in fact be quite feasible to design a turbine utilising a low-pressure

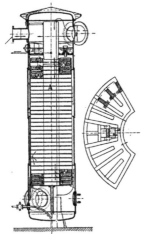

FIG. I.   IRON ACCUMULATOR.

steam flow of several thousands and even tens of thousands of kilogrammes per hour, while remaining within rather moderate dimensions as regards the size of the engine.

As the expansion of steam at low pressure is utilised so satisfactorily in turbines, special interest attaches to the provision of improved condensing plants, ensuring an excellent vacuum, which in the case of a reciprocating engine would be of no avail. The efficiency of turbines in fact is the more satisfactory as the steam pressure is lower, because of the reduction of losses due to the friction of the rotating discs in the steam, and to leakage between the rotating and stationary parts, which two losses are practically proportional to the specific weight of the fluid medium and accordingly are the smaller as the specific volume of the latter is greater.

By the aid of the same diagram of entropy, Rateau then shows that a combination consisting of a reciprocating steam engine (for the high-pressure stage) and a steam turbine (for the low-pressure stage) would be the best means of ensuring an expansion of the steam as perfect as possible. A similar combination, in conjunction with a suitable steam accumulator, would in fact allow of as high efficiency being obtained under average conditions as in the case of the most highly improved gas or petrol engines. This would accordingly afford the most desirable solution of the problem at issue.

As regards in the first place the steam accumulator which is indispensable for regulating the flow of steam issuing from the primary engine, Rateau first used an apparatus made up of cast-iron plates. This, according to Figure I, comprises a vertical sheet-metal cylinder, containing in its interior a pile of annular cast-iron cups filled with water between which the steam is allowed to circulate. Intermittent steam currents coming from the primary engine through the upper pipe are distributed to the cups by the central channel A, while the regularised steam current issues through the lower pipe towards the secondary low-pressure steam engine. The mass of the cast-iron cups and the liquid contained by the latter constitute a heat buffer by the action of which the steam is condensed and accumulated when it arrives abundantly, in order to be formed anew during the time of slackening or discontinuance in the flow of steam issuing from the primary engine. The variations in temperature required for condensing and regenerating the steam correspond to small fluctuations of pressure in the accumulator, the pressure rising when the apparatus is being replenished and falling when it is discharged according to the requirements of the turbine. The maximum fluctuations in temperature controlling the working of this accumulator are as an average 2 to 4 degrees C. and those in pressure 0.10 to 0.15 kilogramme per square centimetre (1.4 to 2.0 pounds per square inch). They may moreover be reduced at will by designing the apparatus of sufficient size. When the pressure in the interior of the accumulator rises above the figure allowed for the back pressure of the primary engine (0.15 to 0.25 kilogrammes per square centimetre), (2.1 to 3.4 pounds per square inch), the steam escapes through a valve controlled at will.

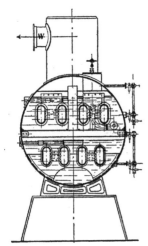

Accumulators of the above type have been constructed for the Bruay mining plant, and the Poensgen and Donetz steel works. Owing, however, to their considerable cost of installation, Rateau subsequently designed another type of accumulator, consisting of the shell of old cylindrical boilers, inside of which some old rails were arranged. These accumulators, the cost of which was evidently reduced to a minimum, have been actually installed at several Spanish and English mines.

However, the best system of accumulator obviously would be a water accumulator, as

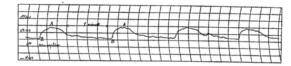

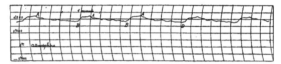

FIG. 3. MANOMETER RECORDS OF ACCUMULATOR PRESSURES.

water possesses a maximum heat capacity, while costing little
or nothing. A difficulty encountered in this connection is the
bad thermic conductivity of this liquid. In order therefore
to impart a large amount of heat to a given mass of water, the
latter should be given a very active circulation. The apparatus constructed by Rateau comprises a horizontal cylinder boiler containing
water, which in the case represented in Figure 2 is divided into two
compartments placed above one another. In the interior of the liquid
are located big horizontal pipes of oval cross section, into which the
steam is thrown, the walls of these pipes being perforated by a number
of holes, so as to lead the steam without any shock to the intermediary
spaces between the pipes. While the steam is arriving the water containing steam bubbles will rise in the intermediary spaces between
the pipes and descend sideways. Circulation is so lively that the water
particles, about every two seconds, come into contact with the steam,
so that practically the whole mass of water takes an efficient part in
the absorption and recuperation of heat. The accumulator is provided
with discharge pipes ensuring an automatic regulation of the water
level.

Graphical records—obtained by means of a recording manometer
—of the results ensured by this type of accumulator (which is nearly
exclusively employed at present) are given in Figure 3. The horizontal line O represents atmospheric pressure, and the interval between
each two vertical curves corresponds to a duration of 15 seconds. The
upper diagram of Figure 3 was obtained on the accumulator of the
Béthune mines when this was fed only by the winding engine. The
abscissa BA corresponds to a whole winding period, lasting about

10 seconds. During 15 seconds, from B to A, the winding engine throwing steam towards the accumulator, the pressure in the latter will rise to 0.11 kilogramme per square centimetre. From A the accumulator receives no further steam, while continually feeding the turbine, so that its pressure decreases until the winding engine in B starts for its next run, thus supplying it with further steam flows.

The lower figure shows the diagram recorded while the accumulator was fed simultaneously from the winding engine and other continually working machines. In this case variations of pressure in the accumulator are seen to be much smaller, not exceeding 0.05 kilogrammes per square centimetre instead of 0.10 kilogramme as before (0.7 pound per square inch instead of 1.4 pounds).

The mutual independence of the primary and secondary engines is ensured by fitting the accumulator with a self-acting valve, which discharges the steam from the primary engine while the turbine does not absorb any. The surplus of steam can be thrown into the atmosphere, but is preferably conveyed to the condenser, thus allowing the accumulator if desired to be worked at a pressure inferior to atmospheric pressure, so that the primary engine utilises a certain vacuum. In the case of a prolonged standstill in the primary engine, such as

FIG. 4. TURBINE AT THE KLEIN RÖSSELN MINES.

FIG. 5. TURBINE OF THE ZOLLVEREIN MINES.

during the hours of meals of the workmen, the live steam from the generators can be drawn on so as to make up for any deficit in the output of the accumulator. For this purpose a self-acting steam expander is provided, which is opened as soon as the pressure in the accumulator falls below a given limit, adjusted for at will. In the place of one expander and a low-pressure steam turbine, a mixed set comprising a high-pressure turbine and low-pressure turbine will finally be found more advantageous, ensuring a far better utilisation of the steam in case the turbine has to be fed directly from the boiler.

The efficiency of low pressure turbines of the Rateau system is about 75 per cent of the energy theoretically available between the working pressures, in the case of units of about 1,000 horsepower. By means of certain improvements this figure may be raised to about 80 per cent.

The first plant fitted up on this system was installed in 1902 in
pit 5 of the Bruay mines. This plant comprises a generating set consisting of a continuous-current turbo-dynamo of 250 to 300 kilowatts,
and an accumulator consisting of four vertical sheet-iron reservoirs
containing 40 tons of cast iron and about 4 tons of water. The
energy produced by this set, outside of the lighting plant, serves to
supply the several electric machines of the mine. Since the inauguration of the accumulator plant, the reciprocating engines which formerly were used in producing electric currents have been shut down,
it being found more advantageous to allow the turbine to work even
by night.

FIG. 6.   TURBINE AT THE GELSENKIRCHEN MINES, ALMA SHAFT.

The first application of the Rateau system in England was made
at the Hucknall-Torkard Collieries near Nottingham. This plant
comprises a steam accumulator consisting of 50 tons of old rails, and a
low-pressure turbine of the Rateau system. The latter is a 200 horse-
power unit rotating at 3,000 revolutions, and is coupled to a rotary-
current generator yielding current at 500 volts, which is utilised to
a great extent for actuating a winding capstan, the turbine having
been substituted for a reciprocating steam engine, the coal consumption of which was rather high. According to reports published by
Mr. W. Morris, engineer-in-chief of the mine, no other means would
have allowed as economical results to be obtained as have been
secured by using a low-pressure turbine.

Figure 4 next represents a plant installed at the mines of Klein-Rosseln, Germany, comprising two generator sets which utilise the exhaust steam from two winding engines of a steam consumption of about 13,000 kilogrammes per hour. This accumulator is of the water-accumulator system. The generator set (constructed by Brown, Boveri & Co.) has an output of about 370 kilowatts, the speed of rotation being 3,000 revolutions. The consumption of the turbine is 17 kilogrammes per kilowatt-hour (that is 11.3 kilogrammes per horse-power-hour). The initial pressure is 1.3 kilogrammes per square

FIG. 7. ACCUMULATOR OF THE POENSGEN STEEL WORKS AT DÜSSELDORF.

FIG. 8.   ACCUMULATOR, STEEL COMPANY OF SCOTLAND.

centimetre and the final pressure .10 kilogramme per square centimetre, the latter being obtained by means of a Balcke condenser giving
a vacuum of 90 per cent; the current produced is three-phase current.

Another application of the system has been made at the Zollverein
Mines, Figure 5. This is the most extensive low-pressure turbine
plant ever installed at a mine, comprising three primary engines, two
winding engines of the double twin type, and a ventilating engine,
the aggregate consumption being 16,000 kilogrammes of steam per
hour. A steam accumulator of 50 tons weight in connection with a
Balcke condenser was installed at this plant. The current-generator
set of an output of 1,000 kilowatts consists of a low-pressure turbine
and a 3-phase alternator constructed by Brown, Boveri & Co. This
engine, with a vacuum of 92.32 per cent and a feeding pressure of

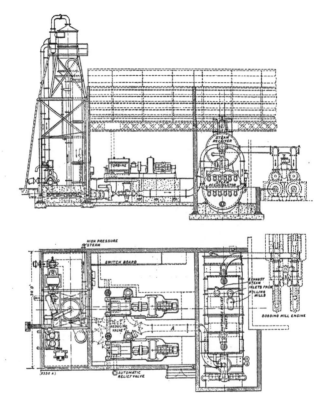

FIG. 8. PLAN AND LONGITUDINAL SECTION OF ACCUMULATOR AND EXHAUST-STEAM TURBINE INSTALLATION, STEEL COMPANY OF SCOTLAND.

1.094 kilogramme per square centimetre, shows a consumption of 14.77 kilogrammes per kilowatt hour when working at full load, which consumption is reduced to 14.34 kilogrammes per kilowatt-hour (or 9.6 kilogrammes per horse-power-hour) in the case of overloads of 70 per cent, the feeding pressure being 1.155 and the vacuum of the condenser 8.29 per cent.

After installing this plant, a live-steam engine of 600 kilowatts, which formerly ensured the electric service of the pit, could be shut down, while providing for the operation of an electric compressor of 400 kilowatts. In spite of this increase in energy, a saving of 10,000 kilogrammes of steam per hour could be ensured from boilers being shut down. In view of these excellent results, the managers of the mine have decided on installing an identical plant in another pit of the Gutehoffnungshütte concern.

Of other low-pressure turbine plants installed in German mines (the constructors being Messrs. Balcke & Co., the German representatives of Prof. Rateau) the plant recently inaugurated at the Alma pit of the Gelsenkirchen mines (Figure 6) should be mentioned. In this plant a number of tests have been made on a low-pressure turbine of an output of 400 kilogrammes, the results bearing out completely the previsions of the constructors. An efficiency of 66 per cent was recorded.

The plant installed at the Poensgen steel works of Düsseldorf (Figure 7) in connection with an existing central condensing plant comprises an accumulator of the cast-steel type which contains about 120 tons of cast steel; the several machines of this steel works of which the largest are a reversible mill and hammers, are arranged in connection with this accumulator, the aggregate amount of steam consumed being 11,500 kilogrammes per hour.

The current-generating set is a low-pressure turbine turning at a speed of 1,400 revolutions and driving two continuous-current dynamos direct-coupled to the turbine. The vacuum of the condenser is about 66 centimetres mercury and the total amount of energy recovered by this set is 650 kilowatts. This energy is used for lighting purposes and for operating the roll trains and auxiliary machines.

The increase in the consumption of these machines, as compared with their working in connection with the central condensing plant has been found to be only 1,700 kilogrammes of steam per hour, and the saving ensured by condensing was 15 per cent; it now exceeds 40 per cent after the inauguration of the low-pressure turbines.

The Rateau system of low-pressure turbines worked by exhaust steam has further been applied at the Rombach steel works, which already comprised a powerful gas-engine central station, fed by blast-furnace gases. Still, the Rateau plant has been found so advantageous in operation, that after eight months' working the managers of the steel works decided on erecting another plant comprising a 550-kilowatt electricity-generating set.

FIG. 10. TURBO-DYNAMO, MINES OF ROCHE-LA-MOLIERE, FIRMINY.

The first installation comprises four water accumulators of two stories each, each 2.6 metres in diameter and 7.5 metres in length, of a total capacity of 100 tons of water. These accumulators are fed by two three-cylinder reversible rolling-mill engines, which formerly worked on central condensing plants, with a total consumption of 25,000 kilogrammes per hour. As the primary engines are worked by

FIG. 11. ACCUMULATOR OF THE MINES AT DONCASTER.

superheated steam, the amount of water discharged from the accumulators is practically nil. Each of the two electricity-generating sets has an output of 550 kilowatts and supplies continuous current at 230 volts, which is utilised in the first place for operating the electric roll trains in conjunction with the energy generated at the gas-motor central power plant.

The first plant erected in an English steel works is that of the Steel Company of Scotland near Glasgow. In Figure 8 is shown the accumulator constructed by P. J. Mitchell, which comprises a double-story shell 3.5 metres in diameter and 10 metres in length, surmounted by a tank 2.75 metres in diameter and 8 metres in length. The six steam-feeding pipes are distinctly visible in the figure. This accumulator contains 55 tons of water and receives 20,000 kilogrammes of steam per hour, which is supplied by several primary engines, viz. a two-cylinder twin engine of 1.1 metre diameter and 1.35 metre stroke driving a reversible roughing-down roll train, a similar engine driving a finishing roll train, and four steam hammers. The generating set comprises two units of 450 kilowatts each, rotating at 1,500 revolutions and yielding continuous current at 230 volts.

This current is utilised for lighting purposes, as well as for operating several plants, of which the continuous roll trains are the most important and use the greatest amount of energy.

In order to give an idea of the results ensured with this low-pressure turbine plant, it may be said that all the steam engines of the steel works, apart from those driving the roughing-down roll trains and hammers could be shut down. Some very careful tests were made during the first six months of operation of the first 450-kilowatt unit, when an aggregate saving of 4 per cent on the whole of the coal consumption of the works was secured. The managers, by utilising this unit, further avoided the necessity of purchasing six boilers, while saving the payment of the stokers required to operate these. In view of the satisfactory results, the second unit was installed.

In the Schoeller & Co. steel works at Ternitz the exhaust steam from 14 hammers of 200 to 10,000 kilogrammes, of a pass, and two small feeding pumps of an aggregate consumption of 5,000 kilogrammes per hour, is utilised for the operation of low-pressure turbines. The accumulator is made up of a double-story shell of 2.75 metres diameter and 8.5 metres length, surmounted by a buffer tank 1.5 metres in diameter and 4 metres in length which contains 33 tons of water. The current-generator set of an output of 300 horse-power yields continuous current at 220 volts.

FIG. 12. ACCUMULATOR OF THE BÉTHUNE MINES.

The plant utilised at the Roche-la-Molière mines (Figure 10) is especially interesting in so far as a very interesting problem, viz. that of effecting the pumping of a pit with the steam from the hauling engine, that is to say without any expense as regards fuel, has been solved there under the most satisfactory conditions. This plant comprises a water accumulator of 25 tons consisting of an old boiler shell receiving the exhaust steam from a hauling engine and continually working auxiliary motors. The total amount of steam used up per hour by these various machines is about 4,500 kilogrammes, of which 3,500 as an average corresponds to the hauling engine. The current-generating set comprises a low-pressure turbine and a three-phase alternator of an output of 250 kilowatts at 1,800 revolutions. This

FIG. 13. TURBO-COMPRESSORS OF THE BÉTHUNE MINES.

set will yield 1 kilowatt with a consumption of steam of 13 kilogrammes per hour, the feeding pressure being 1 kilogramme in absolute value per square centimetre and the final pressure 0.1 kilogramme per square centimetre. The latter is obtained by means of a barometric condenser of the Balcke system which with refrigerated water gives a vacuum of 90 per cent. The energy produced by the turbine serves to actuate the condensing plant and two centrifugal mining pumps operated by electro-motors. These pumps, which are constructed on the Rateau system, will lift 100 cubic metres and 60 cubic metres per hour respectively, to 125 and 220 metres respectively.

After installing this plant it was found possible during the day operation, while working the pumps and without any additional consumption of steam, to pump out the water from two pits, which operation had to be carried out formerly by hand, the resulting saving being 13,000 francs during 9 months. Several tests have shown that though the accumulator was not impervious to heat, the losses due to conduction and radiation did not exceed 2 per cent of the weight of the steam.

The plant installed at the Doncaster mines (Figure 11) is interesting in so far as the energy produced by the aid of exhaust steam there serves for actuating a set of centrifugal sinking pumps.

The instance of the Béthune mines (Figures 12, 13) derives additional interest from the fact that it affords the first examples of an installation of multicellular high-pressure turbo-compressors on the Rateau system. The accumulator contains 36 tons of water and is fed by an intermittently acting hauling engine and ventilating motor, and a continually working pump motor, the steam supplied by these three engines amounting to 6,000 kilogrammes per hour. The low-pressure turbines, as above mentioned, serve to operate a centrifugal turbo-compressor drawing in 60 cubic metres of air and compressing it to a final pressure of 7 kilogrammes in absolute value per square centimetre.

These centrifugal compressors are of quite recent invention and were discussed at the engineering conference of the Institution of Civil Engineers held on June 21. These pumps would seem to be destined to prove extremely useful, the more so as they are susceptible of considerable improvement, while preferable in many cases to the ordinary reciprocating or piston compressors. They are especially suitable in mining plants, in which compressed air must be used for operating the hauling engines, and where a direct coupling with the low-pressure turbine is quite easy.

The system of low-pressure turbines for utilising the exhaust steam in connection with a steam accumulator not only affords a means of considerably improving any existing plants, but of fitting up new plants on quite modern lines for maximum efficiencies. While a concentration of the different operations in a single electrically driven plant will be found suitable in many cases, the considerable drawbacks to such an arrangement cannot be left out of account. Any serious breakdown in the central station is, in fact, bound to result in a discontinuance of the whole service. It will therefore in many cases be found preferable, instead of electrifying a given plant, to improve its output by the installation of the system above outlined.

STEAM PRODUCTION FROM THE CHEAPER GRADES OF ANTHRACITE.

By William D. Ennis.

The first part of Mr. Ennis' discussion, published last month, dealt with the physical and chemical characteristics of the small sizes of anthracite and their commercial importance. This concluding section discusses the mechanical problems of air supply, grate and heating surfaces, stoking, draft, etc., which affect their economical use in the boiler furnace.—THE EDITORS.

LOW-GRADE coals offer relatively high mechanical resistance to the passage and diffusion of the air supplied. They naturally burn more slowly, and consequently, for the burning of a stated amount of such coal, either excessive draft or excessive area of grate may be required. As a rule, both are required. With soft coal, a rate of combustion of 30 pounds per square foot of grate per hour is not considered excessive; with a good quality of anthracite, 15 pounds to 18 pounds per square foot of grate per hour would be a corresponding figure. Fine sizes of anthracite would be limited to a rate of 6 to 10 pounds, unless the draft were greatly increased. By increasing the draft to 1 inch of water and upward, rates of combustion can also be increased, until in the locomotive, with draft intensities as high as 2, 3 and 4 inches, rates of combustion of 60, 80, and 100 pounds are reached. The locomotive boiler is, however, always worked several times beyond the rating which is customary for stationary boilers, viz., 3 to 4 pounds of water evaporated per square foot of heating surface per hour. At such a rate of evaporation, the stationary boiler, burning soft coal at a 30-pound rate of combustion, should have a heating surface equal to at least 65 times its grate surface. Burning ordinary anthracite at a rate of 20 pounds, the heating surface would be over 40 times the grate surface. Burning fine sizes at 15-pounds rate, with draft somewhat forced, the grate surface would be rather more than 1/40 the heating surface. As the quality of fuel is decreased, the relative amount of grate surface must always be increased. This leads to wide fireboxes on locomotives, and to narrower ratios of grate and heating surfaces in stationary practice; the reason being that decreased rates of combustion produce a decreased amount of heat to be taken up by the boiler heating surface. Whether it is desirable to reduce greatly the amount of heating surface is doubtful. The tendency of business is always toward intensified operation, increased production. Year after year, we run our mills harder, load our engines more heavily, and demand higher rates of evaporation and combustion at the boilers. The average plant burns 50 per cent more coal per square foot of grate per hour than it did twenty years ago, and the demand is for still more. This demand is a corollary of industrial prosperity. As facilities for burning low-grade fuels improve, so that higher rates of combustion become common, the opportunity of saving in cost by skimping the boiler heating surface will look less tempting. At the moment, it would seem that a ratio of heating surface to grate surface of 40 to 1 was ample for any hard coal of grade below pea; ten, or even five, years hence, ratios of 50 to 1 or even 60 to 1 may prove to be the thing, and power plants unfortunately equipped on a 40 to 1 basis will be obliged to get along at relatively low rates of combustion, or to force their boilers beyond the economical obsorbing capacity of the heating surface. In other words, they will have to ride on the caboose while their competitors travel on the Limited, or else will be obliged to see their dollars pour out of the top of the chimney.

The following figures are given* as representing the draft desirable for the specified rates of combustion, per square foot of grate per hour, using No. 1 buckwheat coal:

Rate of Combustion.	Draft in Inches of Water.
10	0.30
15	0.45
20	0.70
25	1.00

With soft coal, at high rates of combustion, the usual difficulty is (1) an excess of air supply and (2) either an insufficient amount of heating surface or the short-circuiting of the heating surface by the hot gases, either of which conditions results in an unduly high temperature of flue gases. With low-grade hard coals, the flue-gas temperature is not so apt to be excessive; besides the loss due to excess of air, the principal source of waste is from incomplete combustion due to insufficient intensity of draft. . The two conditions of insufficient draft intensity and surplus air supply are by no means incompatible.

Hard coals are generally hand-fired. The size of any particular grade is fairly uniform; there are no large lumps to break up; ordinary furnaces and grates are adapted to all but the finest sizes; there is no need of special devices for consuming the hydrocarbon, since the total volatile ranges under 7.7 per cent of the total combustible.

* The *Electrical World*, September, 1906, p. 633.

It requires a little more skill, with hard coal,* to fire so as to "keep up steam," but not so much skill to avoid wasteful and inefficient combustion, as in the case of soft coal.   The fire must be kept level, well covered, free from air holes, and the whole surface in as nearly as possible the same condition as to stage of combustion.   Cleaning and "poking up" should be as infrequent as possible, although large clinkers must not be allowed to stop the movement of air.   The thickness of the fire and the amount of draft must be mutually regulated in accordance with each other so as to approach as closely as possible to the theoretically desirable amount of air supply.   If the fire is too thick, the air supply will be deficient, combustion incomplete, and shooting blue flames will appear over the surface of the fuel.   The furnace temperature will be low, and the whole furnace will have a "dead" look.   If the fire is so thin or broken as to permit of an excessive supply of air, the low percentage of CO_2 in the flue gases is the best indication of the low efficiency.

The difficulty of keeping the fire free from clinker without almost incessant raking and poking has been the cause of the introduction of various shaking, dumping, and rocking grates, by means of which, without opening the fire doors, the bed of coal may be agitated and finely broken up, in sections, one section at a time being operated so as to avoid chilling the furnace.   There is of course some cracking of the fire even with these grates, resulting in a large excess of air for a short time at least.   The loss of fuel to the ash is somewhat increased, as a rule, over that experienced with careful hand firing, the increased loss seldom ranging below 2 or 3 per cent of the total heat in the fuel.   This is a small matter as compared with the loss due to the interference, by clinkers, with the desirable amount of air supply, or even with the loss due to cleaning a non-clinkering coal through the fire door.

The necessary size of air space in the grate bars depends entirely upon the amount of tendency to clinker.   It is usually considered that 30 to 50 per cent of air opening is desirable, and that the air spaces should never exceed ⅜ inch in width; but even this width has been found insufficient to avoid clogging by clinker, and Professor

---

* A coal is known as "hard," or anthracite, when the "fixed" or non-gaseous carbon is from 92.3 per cent upward of the total weight of combustible in the fuel. Thus, if coal on analysis shows 80 per cent fixed carbon, 5 per cent hydro-carbon, 7 per cent ash, 8 per cent moisture, then the combustible part of the coal is 85 per cent of the total, and the fixed carbon being 80 ÷ .85 = 94 per cent of the combustible, the coal is anthracite. Coal in which the fixed carbon constitutes from 87.5 to 92.3 per cent of the total combustible is semi-anthracite. Semi-bituminous coal has a fixed-carbon ratio of from 75 to 87.5 per cent; true bituminous coals have a ratio below 75 per cent. It is worthy of note, therefore, that soft coal differs from hard in containing less fixed carbon, and more hydrocarbon or volatile matter. It is this relatively high percentage of volatile hydrocarbon that causes most of the difficulties in burning soft coal.

Kent reports one instance in which a 1/2-inch space clogged fast from fusing ash even with 1/2 inch of draft. On the other hand, tapering holes of 1/2-inch diameter in flat grates, giving only 25 per cent of air space, have been favorably reported, and in one case, with a non-clinkering coal, 1/8-inch air spaces aggregating 15 per cent of the total grate area gave ample air supply. The only general rule that can be laid down is that the air space should be made as narrow as is possible, consistent with the tendency to clinker of the coal that is to be used. The loss of coal to the ash pit will be always in simple proportion to the size of the air openings.

The usual liberal tables for chimney dimensions to suit given capacities of steam boilers are based on a fuel consumption of 5 pounds per horse power per hour. With anthracite coal, the usual allowance of 1/3 square foot of grate surface per horse power would permit of a rate of combustion of 15 pounds per square foot per hour. Ordinary chimneys permit of somewhat higher rates of combustion than this; but nevertheless there is no question but that under any reasonable limitations of chimney cost or height, the maximum possible rate of combustion is soon reached. Even with soft coal, it cannot greatly exceed 40 pounds of fuel per square foot of grate per hour, with chimneys as usually designed. If we take the normal consumption as 20 pounds, the maximum corresponds, therefore, to a 100 per cent overload. The chimney, therefore, besides being a remarkably inefficient machine, is remarkable also for its sharply limited overload capacity. The fan is more efficient, has a much wider overload capacity, and is absolutely independent of atmospheric conditions.

Why, then, should not the fan be applied universally where high draft intensities are required? Probably the principal, if not the only, reason is that the chimney has no parts to break down or wear out, while the fan has many such parts. Fan installations are therefore frequently made duplicate; and where not in duplicate, it is good practice to keep on hand a very large and complete assortment of spare parts. Any machine with moving parts must wear out and break down, eventually. Fans for draft, while cheaper, more efficient, and more controllable to conditions, involve almost a certainty of interruption to the operation of the plant, sometime. The fan, moreover, consumes some fuel and some labor is required to attend to it. The chimney *appears* to consume neither fuel nor labor; but the chimney will not burn fine sizes of coal at reasonable rates of combustion, and will not respond to heavy overload requirements. Each device—fan and ·chimney—possesses merits and demerits which the other lacks. Why not, therefore, combine the two, obtaining the merits of both and eliminating the demerits of either? The chimney can be used for normal conditions, to ensure production and abolish operating expense; the fan can take care of overloads, variations in the weather and the fuel, and may assist the chimney in providing draft for cheaper fuels. The chimney, then, is the apparatus, and the fan becomes a refinement, or economizing device attached to it.

This combination of chimney and fan could be made by placing the fan between furnace and chimney, producing induced draft, and providing a by-pass so that the gases would not pass through the fan when the latter was not being used. For reasons which will be explained, there are advantages in using the fan as a *forced* blast fan rather than for induced draft, discharging from the fan into closed ash pits. By properly and mutually adjusting the speed of the fan and the position of the back damper, or in other words, by controlling the gas-outlet area in proportion to the amount of air supplied for combustion, *balanced draft* can be produced in the furnace—a condition in which, while the air is moving at its usual or any desired velocity through the fuel, and consequently burning it, the exit of the gases from the furnace is so retarded that within the latter there may exist a state of pressure—very slight, it is true, but still a pressure instead of a suction.

When the gases in the furnace of a steam boiler are under suction, they naturally follow the shortest possible course to the outlet, into the smoke breeching. The result is that many portions of the furnace and boiler heating surface are not impinged upon by these gases. The rapid, direct movement through the furnace "short circuits" some of the ends and corners. With "balanced draft," the gases are puffed back, filling every portion of furnace and setting, coming in contact with every square inch of the heating surface, and passing the heating surface at a relatively less velocity than when under suction. It has been known for some time that there were certain important economical advantages in "balanced draft," and various devices, said to be more or less crude, have been employed to effect it. It has been made the subject of commercial promotion within a comparatively short time. This feature of the system we shall not discuss, proposing rather to investigate the principles underlying "balanced draft" than the devices by which it is proposed to produce it.

There are many theories advanced to account for the increase in efficiency that is usually secured by the installation of balanced draft. Some of these have been disputed and cannot be said to be definitely established. Others, which we will consider first, offer obvious reasons for the good results obtained.

1.—Avoidance of short circuiting. It is a well-known fact that good results always follow the application of any device which causes the gases to more fully and thoroughly impinge against the heating surface. Tubular boilers are more efficient than plain cylinder boilers; locomotive boilers are surprisingly efficient, when the enormous overloads are considered, largely because of the small tubes used. The application of retarders to the tubes of horizontal boilers frequently effects a large saving. In one series of tests conducted by me, the saving in fuel due to the installation of retarders was 9.6 per cent. (Trans. A. S. M. E., XXVIII, 4, 875.) All of these high efficiencies are due to the thorough impingement of the heated gas against the heating surface in all of its parts.

2.—A large amount of air is drawn through leaky fronts, doors, tops, and through cracked settings, under ordinary draft conditions. Besides this, the bricks composing the settings are themselves porous and permit air to pass through them. This can be remedied to some extent by painting the setting with paint made from specially "bodied" linseed oil, which provides an absolutely impervious skin, sometimes remaining intact even when the brickwork cracks underneath it. The infiltration of air due to these causes is serious; so serious, in fact, that in analyzing flue gases it is necessary to take the samples of gas as close to the furnace as possible, in order to avoid getting a highly diluted sample on account of air leakage through the setting. Under balanced draft, there can be no infiltration of air into the furnace, through cracks or interstices, the tendency being rather for leakage to be in the other direction. Further, there is no dash of cold air when the fire doors are open, the draft through the fire door being outward rather than inward. The entrance of cold air into the furnace, from any of these causes, always checks combustion and causes waste of heat. With balanced draft, this waste is avoided. Furthermore, influx of cold air against highly heated boiler surfaces results in rapid deterioration of the boiler. The high temperature due to good combustion is far less injurious than the irregular heat and frequent blasts of cold air on heated surfaces that are inseparable from natural draft.

3.—The factor of overwhelming importance in the economical operation of steam boilers is the regulation of air supply. With hand firing and hand control of damper, the position of the last ought, theoretically, to be changing constantly in accordance with the condition of the fire and the demand for air. In practice, under the best operation, it is probably changed at the moment of firing and once more between intervals of firing. Under the usual mode of operation in the vast majority of plants, the damper is touched only when the steam goes "down" or the pop valves begin to sing. With damper regulators, control is automatic in accordance with the steam pressure, rather than in accordance with the requirements of the fire. With balanced draft and a stoker the speed of which is controlled by variations in the steam pressure, the condition of the fire, the pressure of steam, the position of the damper and the quantity of air supplied are all under simultaneous, concurrent control, resulting in a variation of air supply precisely in accordance with the variation in rate of combustion, without any annoying and wasteful modifications due to the infiltration of cold air through the boiler setting. Balanced draft automatically controls the air supply in accordance with the demands imposed by varying rates of combustion, in a manner superior to that practised by the best fireman and not to be at all compared with that practised by the average fireman. Even if natural draft could give equally efficient combustion, balanced draft more perfectly maintains the conditions of efficient combustion during variations in rate of combustion.

4.—Assuming these merits conceded, the principal claim made for balanced draft has now to be considered. This is, that with "balance" in the furnace, the quantity of air supplied, accidentally or otherwise, is less than with natural draft.

If this be true, certain advantages must follow:

a.—The percentage of CO_2 in the flue gases will be higher and the efficiency vastly increased, due to the reduction of excess air.

b.—The quantity of air being decreased, its velocity in passing through and out of the boiler setting will also be decreased, giving more time for the absorption of heat by the boiler and resulting consequently in more complete absorption, lower temperature of flue gases, and higher efficiency.

c.—The necessary intensity of draft for the mechanical texture and condition of the coal can be produced without reference to the actual quantity of air supplied, so that sufficient *force* of draft can be provided for burning cheap fuels without at the same time providing any increase in the *quantity* of air supplied.

These advantages are so important, that the question whether balanced draft decreases the quantity of air supply is worthy of serious consideration. In the first place, under first-class operation, very fine regulation of air supply has been obtained with natural draft. Some tests quote results of flue-gas analysis showing CO_2 upward of 14 per cent, corresponding to an excess air of 30 per cent and a heat loss due to air of 13 per cent. Now the maximum possible percentage of CO_2, without any excess air whatever, is only 21 per cent, corresponding to a heat loss of about 10 per cent, or a 3 per-cent saving over the test quoted. It would be ridiculous to claim, therefore, that balanced draft could show any result greatly superior to those of such tests. It is not to be doubted that with natural draft, equally good results, or results nearly equally good, may be obtained under proper auspices. The question before us, however, is whether under ordinary operating conditions, balanced draft is more likely to produce results more nearly correct.

In order to produce a definite intensity of draft, there must be a definite velocity of air through the fuel. With any kind of draft this velocity is produced by a difference in pressure in the ash pit and furnace. With forced draft, chimney draft, induced draft, or balanced draft, the (positive) pressure in the furnace must be less than that in the ash pit, in order that air may flow from the latter to the former through the fuel. For the same fuel, the same difference of pressure must be maintained, whatever system of draft is used. Further, the egress of the hot gases from the furnace depends upon a fixed difference in pressure in the furnace and at the outlet of the smokestack; and this difference is also fixed without regard to the system of draft employed. In any of the systems, therefore, there is a definite resistance to the passage of air through the fire, and another definite resistance to the passage of the gases out of the furnace. The sum of these resistances must be overcome by the total draft. The relation of these various pressures to the external pressure of the atmosphere is of no consequence, but only the mutual relations of the pressures themselves. It is difficult to see, therefore, how the percentage of excess air is necessarily diminished by the use of balanced draft, although in practice we know that it almost invariably *is* diminished.

The point of maximum efficiency in a steam-boiler furnace is reached when all the carbon in the coal is burned to CO_2, resulting in a gas which contains, by volume, about 21 per cent of CO_2, no excess air being present. In practice, there is always some excess air, and the percentage of CO_2 is always less than 21; even under the very best conditions about 30 per cent of air in excess of requirements is present. This excess is necessary and is always found, because of the imperfect mixing of the fuel and air in the furnace. With balanced draft, all the conditions are present for perfect mixing; sustained high temperature, slow movement, thorough diffusion, and that suffusion which results from the condition of pressure in the furnace and leads to rapid and intimate mingling of the molecules. From the superiority of mixing alone, balanced draft results in better combustion, as evidenced by high CO_2, and *permits* of cutting down the amount of air supplied *because* that air is entirely mingled with the gas from the fuel and not subjected to such stratification as prevents the mingling. Decreased air supply, with balanced draft, is due not to the relations of pressure and suction in ashpit and furnace, but to the fact that the condition of pressure in the furnace results in complete mixture of air and gas. The result is the same; the amount of air supplied may be decreased, without risking the formation of CO and the savings claimed to follow decreased air supply must be conceded to be theoretically as well as practically demonstrable.

Evolution is on the side of balanced draft. With intensified industrial activity we are continually forced to meet heavier overloads and more severe service conditions. We may add new boilers to our plants, or burn a better grade of fuel; but the existing chimney sharply limits the extent of progress. Let the time-honored chimney be "boosted" by the more modern fan, the two forming a tandem team, and we have vastly augmented the capacity of our entire plant at the lowest possible cost. We can burn cheaper coal and can make more steam. The combination of a fan with an existing chimney costs far less than a complete installation for either chimney, forced, or induced draft. Of course, men can and should fire more skilfully; air supply can be controlled by damper regulation; cheaper coal can be burned by other devices; high CO_2 and economical combustion can be obtained in other ways; but in no other way so surely, so quickly, so easily, and so cheaply. The plant that just manages to get along with run-of-mine coal and natural draft can easily use the 30 per-cent cheaper rice or barley coals with greater efficiency. The fine-sized coals are cheaper per B. T. U. to begin with, but cannot be burned with chimney draft. Help out the chimney with the fan, and not only the benefit of the cheap fuel, but also the benefit of more efficient combustion, is readily obtainable. With New York prices as they are at the present day, no power producer with a reasonably good load factor can afford to burn other fuel than the finer sizes of buckwheat coal, for steam generation.

A NEW MINERAL INDUSTRY—THE MANUFACTURE OF RADIUM.

By Jacques Boyer.

THE manufactory lately installed at Nogent-sur-Marne by M. Armet de Lisle bears but little resemblance to ordinary establishments; whole carloads of divers minerals are there treated for an ultimate product consisting of a few minute particles of radium salts. The photographs presented herewith show these modern alchemists at their tasks, upon which they labor as hard as their predecessors of the Middle Ages. Is not pure radium bromide, for instance, a new and most remarkable philosopher's stone, seeing that a single kilogramme of this precious substance would be worth 400,-000,000 francs at the present time?

It is well-known that, as a matter of fact, radium has not yet been isolated in a metallic state, but exists only in the form of salts (chlorides, bromides or sulphates) possessed of a greater or less degree of activity, secured by stopping the operations of manufacture at certain determined points. The radio-active salts may be obtained throughout a scale of activity adjusted to the object for which they are desired. Thus, after the series of manipulations known under the name of the "gross treatments" the activity (taking uranium as unity) reaches 50 to 60; the first fractionings beyond this raise it to 1,000 and the final ones carry it to 2,000,000.

At the Nogent works the following minerals are treated:

Pitchblende (oxide of uranium associated with a large number of foreign substances) which is found principally at Joachimsthal and at Przibram in Bohemia, at Rezbanya in Hungary, in parts of Sweden, Canada, and in Colorado.

Autunite (a double phosphate of uranium and calcium) and pyromorphite (phosphate of lead) from the environs of Autun in France.

Chalcolite (double phosphate of uranium and copper) from Bohemia.

Carnotite (vanadate of uranium) from Portugal and Utah.

Thorianite (oxide of uranium and thorium) of which deposits exist in Ceylon.

Residues from the preparation of uranium, like those of Joachimsthal in which M. and Mme. Curie made the first discovery of radium.

On their arrival at the works all these minerals (except the pitchblende residues, which have already been pulverized for their preceding chemical treatment) are subjected to grinding. For this, the first of the mechanical operations, a variety of machinery is used—among other apparatus a jaw-crusher, a ball-mill and a stamp-mill. Following this, the manipulations differ for each mineral and a very complicated series of processes is to be met in their methodical progress through the works; for the Nogent establishment extracts all the radio-active substances (radium, polonium, actinium, uranium, thorium) which are found in the minerals it treats. Nor are the by-products neglected. The brief description following outlines the method employed in the case of pitchblende, or rather the pitchblende residues from which uranium has been removed. The "gross treatment" is carried on in wooden vats and in steel vats with agitators. One ton of residues requires five tons of chemicals and fifty tons of wash-water. The scheme of the consecutive reactions is as follows:

The residues from the uranium minerals contain sulphates of almost all the metals, and as the sulphate of radium is the least soluble of all, this property is used for its separation, through the medium of a progressive series of washings and of attacks, sometimes with acids, sometimes with alkaline solutions, and sometimes with water. These manipulations carry away each time the more soluble metals. The sulphate of radium always remains at the bottom of the tank and is recovered at the end of the series of manipulations which are thus conducted:

1.—The residues are mixed with concentrated hydrochloric acid, by which the larger part of the sulphates is dissolved. The solution is decanted and the solid residuum, which is then washed with water, retains the radium.

2.—To this deposit carrying the radium is added a boiling concentrated solution of carbonate of soda in order to transform into carbonates any remaining sulphates which resisted the preceding attack. Workmen stir the mixture with great wooden paddles and finally decant it and wash the residuum with large quantities of water.

3.—The residuum is treated with dilute hydrochloric acid. The solution now carries the radium. It is filtered and precipitated with sulphuric acid and thus is obtained crude sulphate of barium, radium-bearing and exhibiting an activity varying from 30 to 60. From 8 to 15 kilogrammes of these sulphates are recovered for each ton of original residues treated.

4.—To purify these crude sulphates (in which are found small quantities of lime, lead, iron, and actinium) they are boiled with a concentrated solution of carbonate of soda.

5.—The radium-bearing carbonate of barium treated with hydrochloric acid, furnishes radium-bearing chloride of barium.

6. — After washing out, this radium-bearing chloride of barium is precipitated by carbonate of soda and the carbonate of barium thus obtained is transformed into bromide by hydrobromic acid.

GRINDING RADIUM-BEARING MINERALS AT NOGENT-SUR-MARNE.

This impure bromide of radium, showing an activity of 50 to 60, is obtained in the proportion of one to two kilogrammes for each ton of uranium-mineral residues treated. The whole course of these delicate operations requires about two months and a half.

Next in order come the fractionings, which have for their purpose the preparation of radium-bearing bromides of barium of constantly increasing richness in radium. To secure this, the mixture of bromides is submitted to a series of crystallizations, first from pure water, afterwards from water to which hydrobromic acid has been added. Advantage is thus taken of the difference in solubility between the two bromides, the radium salt being less soluble than the barium. The mixed bromides are therefore dissolved in hot distilled water, and the solution is then brought to the saturation point at boiling temperature. It is then allowed to crystallize by cooling in a capsule, at the bottom of which the fine crystals are collected and ultimately are separated from the supernatant liquid by decantation. These crystals exhibit an activity five times as great as that of the bromide in solution. By repetition of similar operations, both upon the liquor and the separated crystals, products extremely rich in radium are at last secured. The following diagram shows schematically the progress and relation of the successive manipulations:

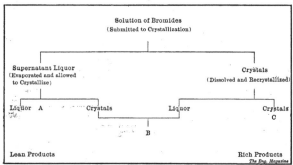

SCHEME OF FRACTIONAL CRYSTALLIZATIONS FOR CONCENTRATION AND PURIFICATION OF RADIUM SALTS.

We have thus to consider four products:

(1) A liquor, A, and (2) crystals, B, derived from the supernatant liquor of the first crystallization, and (3) a liquor, B, and (4) crystals, C, obtained by recrystallization from this first-obtained series of crystals themselves. These last crystals (which we might designate as the "C" series) are naturally the most active. The products obtained from the supernatant liquor of this process of crystallization are possessed of an activity about equal to that of the crystals of what we might call the "B" series obtained from the liquor of the first crystallization. These two lots are therefore combined. Each grade of the radio-active crystals is then submitted anew to a similar treatment, and the fractioning is carried as far as desired by like methods. However, when the "tailings" become too lean, and no longer show anything but an insignificant degree of activity, they are thrown away. And, on the other hand, the concentrates, very rich in radium, are set aside as finished.

The first set of fractionings constitute the last series of manipulations which reach the scale of industrial operations, and proportionately as the process advances the volume of material operated upon

IN THE FRACTIONING LABORATORY.

VATS AND TIPPING KETTLES FOR THE PRELIMINARY TREATMENT.

307

CONDENSING THE RADIUM EMANATION ON VARIOUS SUBSTANCES BY MEANS OF LIQUID AIR.

LABORATORY FOR THE MEASUREMENT OF RADIO-ACTIVITY.

decreases; so much so that by the time we reach a scale of activity 1,000 times that of uranium, each ton of residue originally treated is represented by only 30 grammes of bromide of radium. The operations are then transferred to the laboratory and treatment by the same general scheme is continued until after a large number of fractionings, which involve serious technical difficulties, a final product is secured consisting of one or two milligrammes of radium bromide of an activity 2,000,000 times that of metallic uranium.

Beyond this, the laboratories of the Nogent works are occupied with the analysis of minerals, of the various products of manufacture, and of the residues. Microscopic examinations are also conducted, and, as some of the photographs show, accurate measurements are made of the radio-activity and the emanations of the products; sundry subjects also are radio-energized.

In order to determine precisely the radio-activity of solid substances (radio-active minerals or salts

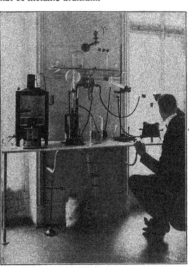

EXAMINATION OF MINERAL WATERS FOR RADIO-ACTIVE PROPERTIES.

of radium of varying concentration, etc.) use is made of the process indicated by P. Curie, which requires as essentials three pieces of apparatus—a plate condenser, a piezo-electric quartz sphere and an electrometer.

The lower plate of the condenser is raised to a high potential by means of a small accumulator battery, and the upper plate is connected with the electrometer and the quartz. Normally, as is known, no current passes between the two plates of the condenser, but in the neighborhood of a radium salt the air separating the two becomes a good conductor and allows the electricity to flow. Advantage is taken

FILTER PRESS FOR URANIUM SALTS.

of this current to deflect the needle of the electrometer, and this deflection is counterbalanced by an equal and contrary current obtained from the piezo-electric quartz. This compensating current, directly measurable, furnishes the measure of the radio-activity of the substance under examination, the instrument having first been standardized on a salt of known activity.

If it is required to determine the radio-active properties of a mineral water (an operation frequently demanded since it has been shown that many springs employed medicinally exhibit radio-activity) a considerable quantity of the water is boiled, and the gas thus obtained (after drying by the ordinary chemical methods) is introduced into a closed condenser which is then substituted for the plate condenser.

Finally, in these same works various substances are radio-energized, particularly medicines in which this treatment seems to excite curative properties. In this procedure the material to be treated is placed in a bulb and with the aid of a partial vacuum the emanation disengaged from a solution of bromide of radium is condensed upon it. This having been done the bulb is plunged into liquid air and sealed with a lamp flame when by electroscope tests it appears that the condensation is complete. Substances thus endowed with radio-activity retain the property for a long time.

PUBLIC LIGHTING SERVICE, CORPORATE AND MUNICIPAL.

A STUDY OF THE COST OF ELECTRIC STATIONS

By *Judson H. Boughton.*

Mr. Boughton's article is of special value because of the practical answers it gives to questions concerning the cost and the working of small and medium plants. These are just the installations for which it is often impossible to make an exhaustive expert study of conditions and design, such as would be undertaken by a great central station, and for which indeed preliminary figures may have to be made without the employment of a specialist. There is no attempt here to make "every man his own electrical engineer." No work of any importance should be entered upon without competent professional advice and supervision. But the clear presentation of average figures, compiled from a large number of plants working under a great variety of conditions, will suffice in many cases to give an approximate idea of the outlay and the returns to be expected, and will serve perhaps to indicate many promising opportunities for new undertakings. And in view of the persistence of the agitation for municipal ownership, the comments upon privately and municipally operated lighting stations, presented at the end of the article, may be studied with profit.—THE EDITORS.

THE design, construction and operation of electric-lighting plants to serve efficiently American cities of from 10,000 to 200,000 inhabitants, presents a diversity of problems with the differing local conditions encountered. There is, however, a set of general requirements and facilities for meeting them, including the generation from steam power, using coal as the fuel, of current to supply from 50 to 2,000 lamps, which is represented by the great majority of installations in this service. The experience gained from these plants is of much value, and deductions made from their performance are very generally applicable to the design and operation problems of the majority of plants of this class.

Although perhaps not of greatest importance, the question of plant location is the one which logically invites first attention, after the exact station capacity and distribution requirements and the fuel-transportation facilities offered have been investigated. Other things being equal, the site which will permit of the most economical distribution of current over the entire area to be served should be selected, but often this location is not accessible to railroads or docks, and if it be selected the coal-hauling account may become a much larger item than the transmission losses throughout the entire system. This is quite as true in small installations as in larger plants. Accessibility to fuel supply should, then, be the first consideration in the average case, and following this, efficient transmission, with due reference to probable future expansion of lines and service. The ideal is realized when the aggregate of the interest on the site investment, the fuel-haulage account, and the transmission losses, is minimum.

Of buildings, it may be said that they are never too large and seldom too well built, although there are of course plants in which all elements of the ultimate objective—commercial efficiency—are disregarded by the designers. Substantial foundations should be built not only for each piece of machinery, but for the building, although the latter is to be regarded as a cover and not as a support for the former. Fire-proof construction has been found to be cheaper than insurance protection, which even though commanding high rates never fully covers the real loss sustained. Natural light is very desirable, as is good ventilation, and in modern construction windows are cheaper than walls. The general lay-out of the building should of course be such as to provide for the economical handling of fuel and ashes, and for the distribution of steam and transmission of power with least possible loss. Provision for expansion, and for handling and duplication of apparatus, should not be overlooked.

The machinery of the lighting plant may be considered in three distinct groups; the boilers and accessories, the engines, steam pipes and condensers, and the dynamos and switchboard. Concerning boilers for this service, it may be said that no one type is better than every other type for every set of conditions. High efficiency in the production of steam can be obtained only by the careful selection, setting, and firing of boilers which are adapted to the fuel available and to the other local conditions of service.

For smaller plants any well designed and constructed and properly set horizontal tubular boiler should give satisfactory results. In larger plants the water-tubular type is generally preferred on account of its greater steaming capacity, decreased floor space, and somewhat greater efficiency. In the economical operation of any type of boiler the fireman is the most important factor, and the exercise of training and intelligence in firing will save many times the amount of apparently high wages. A defective setting may likewise dissipate a considerable amount of heat which should be utilized.

By way of performance-comparison of the two general types of boilers, Isherwood submits the following figures, which are here reduced to a horse-power basis:

TABLE A. AVERAGE COSTS HORIZONTAL TUBULAR BOILERS, ACCESSORIES, CONNECTIONS AND SETTING.

Tubular Boilers, H. P.	30	50	60	70	80	90	100	125	150
Aver. Size of Boiler	44"x12'	54"x14'	54"x16'	60"x14'	60"x16'	66"x15'	66"x16'	72"x16'	78"x18'
Aver. Cost, Boiler and Fixtures, Full Front	$405	$595	$670	$725	$810	$940	$975	$1,010	$1,370
Aver. Iron Stack and Raising	60	95	98	110	120	125	130	150	240
Aver. Boiler Setting, Masonry	185	240	260	297	300	305	310	385	450
Aver. Hauling	15	15	15	20	30	30	30	30	40
Aver. Cost of Boiler Set	$665	$945	$1,043	$1,162	$1,270	$1,500	$1,440	$1,585	$1,900
Aver. Feed-Water Heater	55	90	90	95	20	40	9	20	20
Aver. Boiler-Feed Pump	55	70	70	70	90	10	10	45	45
Aver. Injector	14	14	14	18	18	20	20	24	28
Aver. Piping, including Engine Connections	75	105	115	130	140	185	195	245	310
Aver. Pump and Heater Foundations	14	18	18	18	21	21	21	21	28
Aver. Total Cost, Boiler, Stack, Heater, Pump, Set and Connected	$878	$1,242	$1,350	$1,593	$1,659	$1,876	$1,926	$2,210	$2,701

Water-tube—11 square feet heating surface, 3.3 pounds of coal; relative economy 100, relative rate of steaming 100.

Horizontal Tubular—16 square feet of heating surface, 4.0 pounds of coal; relative economy 91, relative rate of steaming 50.

The average cost of different sizes of boilers of the horizontal tubular type (which type has been more generally used), especially in smaller plants, is shown in Table A, which includes also data representing the cost of accessories and connections, and of setting. Five to seven per cent is a fair reduction to make from these figures for each boiler set in battery. Although always greater, the cost of the different types and makes of water-tubular boilers varies so much that a similar tabulation is of little value.

A supply of pure feed water, holding no lime or other matter in suspension or solution, and proper draft, are elements of importance in efficient boiler operation which must be provided for in the location and construction of the plant. Table B shows the sizes of stacks in varying heights which are usually built for various boiler capacities, and C and D represent the general dimensions and average costs of brick chimneys and iron stacks respectively Some progress has been made in the adaptation of automatic stokers and forced draft to lighting-plant service, but the greatest success has been met in installations of larger capacity than those here considered.

Table B. Sizes of Chimneys for Steam-Boilers.

Square Chimney, Sides of Square	Round Chimney, Diameter in Inches	Area, Square Feet	Effective Area, Square Feet	Commercial Horse-Power of Boilers									
				50 ft.	60 ft.	70 ft.	80 ft.	100 ft.	110 ft.	125 ft.	150 ft.	175 ft.	200 ft.
16x16	18	1.77	.97	20	25	30	....	....	....	....	....	....	....
19x19	21	2.41	1.47	35	40	40	....	....	....	....	....	....	....
22x22	24	3.14	2.08	50	55	60	60	....	....	....	....	....	....
24x24	27	3.98	2.78	65	70	80	85	....	....	....	....	....	....
27x27	30	4.91	3.58	85	90	100	110	....	....	....	....	....	....
30x30	33	5.94	4.48	....	115	125	135	....	....	....	....	....	....
32x32	36	7.07	5.47	....	140	150	165	180	....	....	....	....	....
35x35	39	8.30	6.57	....	....	180	200	220	....	....	....	....	....
38x38	42	9.62	7.76	....	....	220	230	260	270	....	....	....	....
43x43	48	12.57	10.44	....	....	....	310	350	365	390	....	....	....
48x48	54	15.99	13.51	....	....	....	....	450	470	500	550	....	....
54x54	60	19.64	16.98	....	....	....	....	565	590	630	690	750	....
59x59	66	23.76	20.83	....	....	....	....	700	730	780	850	920	980
64x64	72	28.27	25.08	....	....	....	....	835	875	935	1,020	1,100	1,180
70x70	78	33.18	29.73	....	....	....	....	....	1,035	1,100	1,215	1,300	1,400
75x75	84	38.48	34.76	....	....	....	....	....	1,200	1,300	1,420	1,530	1,630
80x80	90	44.18	40.19	....	....	....	....	....	....	1,500	1,640	1,770	1,000·
86x86	96	50.27	46.01	....	....	....	....	....	....	....	1,880	2,000	2,200

314

TABLE C. BRICK CHIMNEYS.

Approx. Horse-Power.	Height, Feet.	Diameter Flue, Inside.	Outside Dimensions, Base Square.	Outside Wall.		Cost, Fire Brick Lining, ½ Height.	Cost, Concrete Foundation.	Total Cost Chimney.
				No. Brick.	Cost @ $14 per M.			
85	80	25 in.	7 ft. 5 in.	32,000	$448	$60	$90	$598
135	90	30 "	8 " 3 "	40,000	560	82	144	786
200	100	35 "	9 " 10 "	65,000	910	118	198	1,226
300	110	43 "	10 " 2 "	75,000	1,050	190	252	1,492
450	120	51 "	11 " 2 "	87,000	1,218	261	306	1,785
750	130	61 "	12 " 6 "	131,000	1,834	334	360	2,528
1,000	140	74 "	13 " 11 "	151,000	2,114	432	414	3,060
1,650	150	88 "	15 " 1 "	200,000	2,800	482	468	3,750
2,500	160	110 "	17 " 10 "	275,000	3,850	720	525	5,095

The cost of labor and fuel, the two principal items of expense in boiler operation, are indicated by tables E, F, and G, which represent an average of the conditions found in the United States. Table E shows the cost of coal per hour to generate 100 horse power of steam under ordinary conditions, with coal varying in price from $0.75 to $4.00 per ton, and with an evaporation of from 5 to 11 pounds of water per pound of coal. Thirty-five pounds of water at 212 degrees are treated as one boiler horse power, allowance being made for loss in delivering steam to the engine. The average evaporation is between 7 and 8 pounds of water per pound of coal. The corresponding

TABLE D. IRON STACKS.

Horse-Power.	Height, Feet.	Diameter, Inches.	No. of Iron.	Price, Stack Complete.
25	40	16	12 and 14	$60
...	40	18	12 " 14	70
...	50.	18	12 " 14	85
75	50	20	12 " 14	90
...	50	26	12 " 14	105
...	60	22	12 " 14	110
100	60	24	12 " 14	125
...	60	26	12 " 14	135
...	60	28	12 " 14	150
125	60	28	10 " 12	190
...	60	32	10 " 12	205
150	60	34	12 " 14	165
200	60	36	10 " 12	215
225	60	38	10 " 12	230
250	60	42	10 " 12	260
300	60	46	10 " 12	290
400	60	52	10 " 12	340

cost of labor per hour to generate 100 horse power of steam is shown in Table F, figures representing the operation of from one to five boilers being given. It will be observed that the maximum capacity of one fireman is reached in a three-boiler battery, both in the eight-hour and twelve-hour runs. The cost of labor per day to fire from one to five boilers of from 75 to 300 horse power is shown for both single and double shifts in Table G, from which the most efficient arrangement, from the standpoint of labor alone, will be apparent.

TABLE E. COST OF COAL PER HOUR TO GENERATE 100 HORSE-POWER OF STEAM UNDER ORDINARY CONDITIONS.

Cost per ton	$0.75	$1.00	$1.25	$1.50	$1.75	$2.00	$2.50	$3.00	$3.50	$4.00
5	.26	.35	.44	.53	.61	.70	.88	1.05	1.23	1.40
6	.22	.29	.36	.44	.51	.58	.73	.87	1.02	1.16
7	.19	.25	.31	.36	.44	.50	.63	.75	.88	1.00
8	.17	.22	.28	.33	.39	.44	.55	.66	.77	.88
9	.15	.20	.25	.30	.35	.40	.50	.60	.70	.80
10	.13	.18	.23	.27	.32	.36	.45	.54	.63	.72
11	.12	.16	.20	.24	.28	.32	.40	.48	.56	.64

(Left row labels: Pounds of Water Evaporated per Pound of Coal)

TABLE F. COST OF LABOR PER HOUR TO GENERATE 100 HORSE-POWER OF STEAM UNDER ORDINARY CONDITIONS.

	Eight-Hour Run.				Twelve-Hour Run.			
Size of Boiler, Horse-Power	75	100	200	300	75	100	200	300
No. of Boilers, 1	$0.28	$0.21	$0.11	$0.08	$0.19	$0.14	$0.07	$0.05
No. of Boilers, 2	0.15	0.11	0.06	0.07	0.10	0.07	0.04	0.05
No. of Boilers, 3	0.09	0.14	0.07	0.06	0.06	0.09	0.05	0.04
No. of Boilers, 4	0.15	0.11	0.09	0.07	0.10	0.07	0.06	0.05
No. of Boilers, 5	0.16	0.12	0.07	0.09	0.11	0.08	0.05	0.06

TABLE G. COST OF LABOR PER DAY TO GENERATE STEAM UNDER ORDINARY CONDITIONS, USING COAL.

	Single Shift.				Double Shift.			
Size of Boiler, H.P.	75	100 to 150	200 to 250	300	75	100 to 150	200 to 250	300
No. of Boilers, 1	$1.67	$1.67	$1.67	$1.67	$3.34	$3.34	$3.34	$3.34
No. of Boilers, 2	$1.67	$1.67	$3.34	$3.34	$3.34	$3.34	$5.00	$5.00
No. of Boilers, 3	$1.67	$3.34	$3.34	$3.34	$3.34	$5.00	$5.00	$5.00
No. of Boilers, 4	$3.34	$3.34	$3.34	$5.00	$5.00	$5.00	$6.68	$6.68
No. of Boilers, 5	$3.34	$3.34	$3.34	$5.00	$5.00	$5.00	$6.68	$8.34

More of a problem is presented in the selection of an engine properly adapted to local conditions than in the choice of the boiler, or in fact, of any other piece of station machinery. Questions of first and of maintenance cost and of operating efficiency are encountered, and with the many types and makes of engines on the market, the solutions offered often seem perplexingly diverse. The tendency is toward the use of direct-connected apparatus, which as now designed is regarded as standard. This arrangement, however, means in small units somewhat higher first cost, higher maintenance, and lower operating efficiency; in large units where operating economy demands the use of Corliss valve gears, and consequent low speed, the first cost of both engine and dynamo is considerably greater. In any case the choice is to be made with proper consideration for first and maintenance cost and operating efficiency of engine and dynamo together, if the most satisfactory results are to be obtained.

In general it may be said that the best engine of any type is not too good. Although the design and proportions of each type have become standardized, it does not follow that the machine in question has been properly built and that the various parts fit and are in alignment. Engines which will operate efficiently and satisfactorily for continued periods can not be produced without adequate shop facilities. Ordinarily, the investment represented by the Corliss engine with the necessary shafting, pulleys, and belting may be taken as double that of the high-speed machine. Then come the questions of efficiency, flexibility, and durability.

TABLE H. AVERAGE PRICE AND SIZES OF CORLISS SINGLE-CYLINDER ENGINES.

Size.	R. P. M.	Horse Power at 80 Pounds. Cut-off.		Horse Power at 100 Pounds. Cut-off.		Price F. O. B. Cars.
		1-5.	1-4.	1-5.	1-4.	
14x36	85	85	100	107	130	$1,700
14x42	82	95	112	120	145	1,800
16x36	82	105	125	135	160	1,950
16x42	78	115	140	150	180	2,000
18x42	78	150	175	190	235	2,350
18x48	75	160	200	210	250	2,600
20x42	75	175	210	225	270	2,600
20x48	72	190	230	250	300	2,850
22x42	75	210	250	270	325	3,000
22x48	72	230	280	300	360	3,300
24x48	70	270	320	345	415	4,000
26x48	70	315	380	400	485	4,650
28x48	68	360	425	460	550	5,150
28x54	68	400	480	525	620	5,300
30x48	68	410	490	525	630	5,800
30x60	62	465	560	600	700	7,000

In the first connection it may be assumed for single-cylinder engines that a total of 3 pounds of coal per horse-power hour will be required by the Corliss, and from 4 to 5 pounds by the high-speed type. Except, however, in the case of large units where direct connection is possible, a considerable part of the gain in fuel economy will be absorbed by power transmission between engine and dynamo. The steam pressure used is an element of considerable importance in fuel economy, as is shown by Curve A representing the results of a series of tests made by me of a 500 horse-power cross-compound. The difference is more marked, naturally, in multiple-cylinder engines where more expansion is possible, but in any case the efficiency increases materially with increasing pressures, within certain limits.

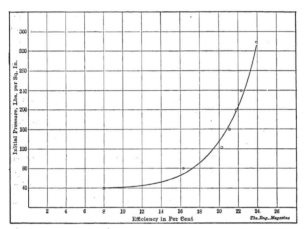

CURVE A. EFFICIENCY OF COMPOUND CORLISS ENGINE OF 500 HORSE POWER.

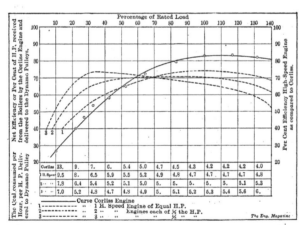

CURVE B. EFFICIENCY OF HIGH-SPEED ENGINES BELTED DIRECT, AS COMPARED WITH THE CORLISS ENGINE DRIVING SHAFTING AND BELTS.

TABLE I. CORLISS COMPOUND CONDENSING ENGINES.

Capacity Horse Power	Speed, Rev. per Minute	Steam Pressure, Pounds	High Pressure Cylinder	Low Pressure Cylinder	Stroke	Cost	Steam Pressure, Pounds	High Pressure Cylinder	Low Pressure Cylinder	Stroke	Cost	Foundation and Erecting	Weight Complete, pounds
200	80	80	14	28	42	$4,565	120	13	26	42	$4,465	$1,050	60,000
300	75	80	18	34	42	5,700	120	16	32	42	5,500	1,025	85,000
400	75	80	20	38	48	7,300	120	18	36	48	7,100	1,250	110,000
500	75	80	22	42	48	8,480	120	20	40	48	8,280	1,400	140,000
600	75	80	24	46	48	10,100	120	22	44	48	9,900	1,675	170,000

The largest factor, however, in determining the engine to be used is the load to be carried. The load curve, which may be anticipated very approximately, should be studied most carefully. Curve B represents the relative performance of high-speed and Corliss engines operating under similar conditions, platted from data secured by Mr. B. J. Arnold. It will be observed that the efficiency of the high-speed is greater up to about 60 per cent of the rated load, beyond which point the advantage gained by the superior steam distribution of the Corliss is increasingly marked. From these curves, which are characteristic, it will be apparent that, other things being equal, the Corliss is better adapted to loads exceeding the rating, whereas the use of the high-speed is advisable for loads varying within the rated capacity. Aside from the matter of type, the question of number and size and units must be determined from the general, regular fluctuations of the load curve during the twenty-four hour or other daily run.

The commercial sizes of Corliss engines, the usual speed, horse power at 20 and 25 per cent cut-off, at both 80 and 100 pounds pressure, together with the average price F. O. B. cars, are indicated in Table H. The prices will, naturally, vary with different makes and from time to time with the general fluctuations of the market, a comment which applies also to Table J, page 320, representing in addition the average cost of the foundations, piping, and erecting and of the heater, feed pump and injector with connections. Table I shows the horse power, speed, cylinder dimensions, and cost of compound Corliss condensing engines (including condens-

TABLE J. PRICE, CORLISS-SINGLE-CYLINDER ENGINE, SET AND CONNECTED.

Size of Cylinder.	Horse Power 80 Lb. ¼ Stroke.	Cost of Engine.	Cost of Foundation.	Cost of Erecting.	Cost of Piping.	Total Cost.	Approx. H.P. of Engine.	Heater. Price, Brass Tubes.	Feed Pump. Price, Pump Alone.	Feed Pump. Price, with Connections.	Injector. Price, Injector Alone.	Injector. Price, with Connections.
14x36	100	$1,700	$275	$175	$165	$2,315	20	$47	$70	$84	$20	$77
14x42	110	1,800	300	90	175	2,475	35	74	80	93	20	82
16x36	125	1,950	325	210	180	2,665	60	107	80	140	20	109
16x42	140	2,000	350	225	190	2,765	70	134	127	190	28	110
18x36	155	2,150	375	240	90	2,905	80	134	127	196	28	111
18x42	175	390	400	250	210	3,210	100	161	127	196	30	126
18x48	200	2,600	425	260	20	3,595	110	161	127	196	30	135
20x42	210	2,600	90	90	90	3,600	125	90	143	253	30	135
20x48	90	2,850	525	275	250	3,900	140	200	143	253	38	166
22x42	250	3,000	550	90	310	4,166	150	90	143	253	38	166
22x48	280	90	60	325	340	4,565	160	332	143	253	38	166
24x48	90	4,000	90	375	90	5,465	170	332	143	253	38	191
26x48	380	4,650	800	440	90	6,450	90	332	143	268	45	191
28x48	425	5,150	90	500	800	7,650	225	332	143	268	45	191
28x54	480	5,300	1,050	575	950	7,875	250	……	143	……	45	……
30x48	490	5,800	90	60	1,070	8,670	300	……	143	……	55	……
30x60	560	7,000	490	700	1,140	10,240	400	……	187	……	65	……

320

ers) and their weight and average cost of setting and erecting. The prices given apply to both tandem and cross-compounded engines, the cost of the former being less than 10 per cent lower in smaller sizes and sometimes greater in large sizes. The average erection and total cost of condensers is separately indicated in Table K.

TABLE K. AVERAGE PRICE, CONDENSERS.

H.P. Single Cylinder.	H.P. Compound.	Cost.	Erection.	Total.
40	50	$260	$90	$350
60	100	380	125	505
100	120	490	160	650
120	150	540	175	715
150	200	610	200	810
200	250	670	230	900
250	325	775	250	1,025
320	425	975	325	1,300

The cylinder dimensions, horse power, speed and average cost of high-speed single-cylinder engines of standard make are shown in Table L, which includes also the average cost of the sub-base, foundation, and erecting. Table M incorporates similar figures for high-speed compound engines. This type completes the list of those available for the service in question, to which triple or quadruple-expansion engines are not adapted except in an occasional station of large capacity. Although much progress has been made recently in the development of the steam turbine, its application to this class of service has not as yet become general enough to produce serviceable

TABLE L. HIGH-SPEED SINGLE-CYLINDER ENGINES.

High-Speed Engine—H.P...	50	75	100	125	150	200	250
Size of Cylinder, Inches....	9x10	10x12	12x12	13x14	15x14 / 17x16	18x16	19x18
Steam Pressure	100	100	100	100	100	100	100
Revolutions per Minute....	300	300	290	275	245	225	200
Cost delivered, F. O.B......	$695	$890	$1,085	$1,260	$1,595	$2,010	$2,800
Sub-Base	45	50	50	70	80	90	250
Engine Foundation	65	75	80	95	110	140	200
Superintendence—Labor ...	70	70	70	70	75	85	100
Handling	10	15	15	17	20	25	35
Total Cost of Engine Set up on Foundation......	$885	$1,100	$1,300	$1,512	$1,880	$2,350	$3,385

figures for comparison with the records of the older reciprocating types. The first cost of turbines with necessary auxiliaries, including the generator, up to 1,500 kilowatt capacity, ordinarily exceeds that of the corresponding reciprocating apparatus, but for larger units there is a slight difference in favor of the turbine equipment. From the performance records already established, the total operating and

TABLE M. HORSE-POWER AND PRICES, HIGH-SPEED ... ENGINES.

Horse-Power at 80 Pounds Steam Pressure.

	60	80	100	125	150	180	250	300	350	400
High Pressure Cylinder	8 in.	9 in.	10 in.	11 in.	12 in.	13 in.	15 in.	16 in.	17 in.	20 in.
Low Pressure Cylinder	13 "	16 "	18 "	19 "	20 "	22 "	25 "	28 "	30 "	36 "
Stroke	12 "	12 "	12 "	14 "	16 "	16 "	16 "	18 "	18 "	18 "
Price	$1,070	$1,300	$1,400	$1,600	$2,145	$2,470	$2,730	$3,380	$3,900	$4,290
Sub-Base	120	120	120	140	140	150	160	200	250	300
Total	$1,190	$1,420	$1,520	$1,830	$2,285	$2,620	$2,890	$3,580	$4,150	$4,590

maintenance costs of turbine installations appear to be about equal to those of the reciprocating plants of corresponding outputs, except in the case of very large units, where there is a slight difference in favor of the turbine. The efficiency of the turbine is more nearly constant with varying loads, making it especially adaptable to electric and lighting work, and its greatly reduced floor space and head room requirements make it adaptable where reciprocating machinery could not be installed.

As in the case of the engine, the load curve of the station is the most important factor in determining the general type and even the make of dynamo to be used. The characteristics of each machine under all conditions of operation should be carefully studied and the dynamo which is best adapted to the load to be carried and to the general arrangement of apparatus should be chosen. Although varying somewhat among different makes of the same type, the first cost is of minor relative importance. The accuracy of rating, the general construction, and wearing qualities, should by no means be disregarded. The relation between mechanical and electrical quantities and the equivalent of both in lamp candle power are shown in convenient form in Table N.

The switchboard should be designed to handle the full capacity of the station quickly and safely, and care should be exercised in its insulation and erection. Proper efficiency is as important in the distribution of energy throughout the mains and feeders as in its generation from the coal, which should be selected on the basis of the heat units it contains rather than the pounds avoirdupois it represents.

The accompanying tabulation of rates prevailing in thirty odd representative American cities, taken at random, indicates the trend of progress in municipal lighting. The marked

TABLE N. CONVERSION TABLE OF ELECTRICAL UNITS. (1 AMPERE IN 50 VOLT TRANSFORMER = 1 16 CANDLE-POWER LAMP. 1 AMPERE IN 100 VOLT TRANSFORMER = 2 16 CANDLE-POWER LAMPS.)

Volts.	1,000	2,000	3,000	4,000	5,000	Pressure.
1 Horse Power =	3/5	3/10	1/5	3/20	1/8	Amperes
1 Ampere =	1 2/3	3 1/3	5	6 2/3	8 1/3	Horse-Power
1 Horse Power =	12	11 1/2	11	10 1/2	10	16 Candle-Power Lamps.
1 Ampere =	20	40	60	80	100	16 Candle-Power Lamps.

reduction in rates during the past ten years has resulted both from increased competition and from decreased cost of apparatus and greater economy of operation. In each community special problems are presented, sometimes involving conditions entirely foreign to the actual subject of lighting. Other things being equal, power and light should be produced with greatest economy by private enterprise rather than by changing municipal governments whose available financial resources and general business aggressiveness are more limited.

LIGHTING BY PRIVATE CONTRACT.

City.	Population.	No. of Lamps.	Schedule.	Cost per Lamp per Year. 1896.	1906.
Auburn, N. Y.	31,000	417	all-night	$87.50	$68
Albany, N. Y.	100,000	727	"	150	98.55
Akron, O.	55,000	396	moonlight	77	80
Atlanta, Ga.	125,000	1,002	all-night	120	75
Binghamton, N. Y.	51,000	411	"	109.50	84
Buffalo, N. Y.	400,000	2,892	"	127.75	75
Council Bluffs (a)	25,000	....	moonlight	164.25	70
Clinton, N. Y.	2,000	27	"	....	40
Colo. Springs	12,000	250	"	160	66
Canton, O.	45,000	289	"	70	43
		110	all-night	....	62
Camden, N. J.	84,000	672	"	146	109.50
Cincinnati, O.	326,000	946	"	84	72
		4,504	(overhead)...		60
Dubuque, Ia.	42,000	389	moonlight	65	64.75
Denver, Colo.	150,000	1,248	"	150	60
Fort Wayne, Ind.	50,000	383	"	125	70
Harrisburg, Pa.	75,000	555	all-night	85.50	47
Indianapolis, Ind.	225,000	1,600	moonlight	85	74
Leavenworth, Kan.	25,000	107	all-night	96	84
Manchester, N. H.	70,000	600	"	136	90
Mobile, Ala.	50,000	287	moonlight	76.50	75
Quincy, Ill.	40,000	400	"	75	54.50
Rochester, N. Y. (b)	200,000	3,200	all-night	104	78 (single)
			"	....	74 (pairs)
Troy, N. Y.	74,000	460	"	....	119.50
Terre Haute, Ind.	60,000	500	"	70	70
Washington, D. C.	327,000	990	moonlight	182.50	85
Average				110.70	89.00

(a) Plant located in Omaha, which city is also supplied.
(b) Water power. New contract dating from July, 1907, at $57.95.

AVERAGE COSTS OF LIGHTING BY ARC AND INCANDESCENT LAMPS UNDER VARIOUS DAILY SCHEDULES.

Length of Time Burning.	Cost 16 Candle-Power Lamp per Hour.	Cost 2,000 Candle-Power Arc per Hour.	Cost 1,200 Candle-Power Arc per Hour.
½ Hour each day	$2.00	$0.16	$0.14
1 " " "	1.12	.08½	.07¼
2 Hours " "	.62	.05	.04¼
3 " " "	.46	.04	.03½
4 " . " "	.37	.03½	.03
5 " " "	.32	.03	.02½
6 " " "	.28	.02¾	.02¼
7 " " "	.26	.02½	.02⅛
8 " " "	.25	.02¼	.02
9 " " "	.24	.02⅛	.01⅞
10 " " "	.22	.02	.01¾

The experiment of municipal ownership of lighting properties which has now been quite extensively tried has not proven actually successful. Some few municipally operated plants have produced current continuously at comparatively low cost, but in most cases the properties have ultimately fallen, through political changes, into the hands of incompetents—with disastrous results. In some instances honest management has been maintained, but such vital elements as depreciation have been wholly unprovided for during the period of apparently successful operation, resulting in impaired service and in greatly increased operating costs, and in some cases in the abandonment of plants.

Statistics show that the electric-lighting plant, if properly designed, installed, and managed, offers opportunities to the private owner for large returns on investment, and as a municipal institution, a means of reducing materially one of the largest items of city expense. On the other hand it affords unlimited chances in operation for numerous small leaks, which, taken in the aggregate, may easily counteract the elements of efficiency in design. Reports covering a large number of plants throughout the United States and Canada show that 12.6 per cent are losing money, that with 10.8 per cent the receipts barely cover operating and fixed charges, and that dividends of from 2 to 50 per cent are being earned by the remaining 76.6 per cent.

The elements to be considered are many, but it is to be remembered throughout the design, construction, and operation of the lighting plant that the aggregate efficiency to which each element must be made to contribute in maximum degree, is to be measured ultimately in commercial units, whether the stockholders appear as individuals or as organized members of an urban community.

A REVIEW OF THE WORLD'S TIN-MINING INDUSTRIES.

By A. Selwyn-Brown.

THE steady rise in the value of tin and the comparatively slow expansion of the world's output of tin ore have given rise to apprehension. Substitutes for tin for many industrial purposes have been sought in vain, and the highest chemical skill has been engaged in discovering the most efficient means of diminishing the amount of the metal utilized in tinning processes, as well as in recovering tin from scraps. These fields of research have yielded well and tin is now used with but little waste, while several patented metallurgical processes are successfully employed in recovering large quantities of tin annually from waste tin scrap.

It is well known that tin is the most sparsely distributed of all the useful common metals. Deposits of iron, lead, copper, zinc, and aluminium occur in almost every country, but there are only a few restricted areas in which tin occurs in sufficient quantities to permit it to be profitably mined. Although much capital has been expended in prospecting for tin-bearing ores in the United States and Mexico, little more than traces of the metal have been found. There is not a single important tin-producing mine in operation in any part of North America. Prospecting is still carried on in Mexico, North and South Carolina, and Alaska, with the view of developing tin-mining industries, but the prospects at present are not very promising.

Not only over the vast mineral areas of North America is there an absence of valuable tin fields, but no tin-mining operations of importance are being carried on in Africa, nor, excepting in Cornwall and in a small district in Saxony and Spain, in Europe. Furthermore, some of the districts which have hitherto supplied the world's tin wants are showing unmistakable signs of exhaustion at a not distant date. Notwithstanding these conditions and the complete failure of the attempts to find a substitute for tin, the demand is increasing and must continue to do so in proportion to the growth of the world's commerce and wealth. What, then, is the outlook of the tin-mining industries? It is proposed in this article to review the conditions on the principal tin-mining fields and to indicate briefly how the world's increasing demands will continue to be met.

THE WORLD'S TIN PRODUCTION.

The most important tin-mining countries are the Malay Peninsula, Bolivia, Australia, and England. They rank in importance in the order named. Statistics generally place England above Australia, but it should be borne in mind that the world's greatest tin market is in London and much foreign tin ore is sent to English smelters for reduction and some of it is credited to English production.

The world's tin output in 1906 was estimated at 79,000 long tons and its New York value was 35 cents per pound. During the past three years tin production was vigorously stimulated and the output was as given in the following Table I :

TABLE I. THE WORLD'S TIN PRODUCTION.

(In Long Tons.)

	1904.	1905.	1906.
England	4,132	4,468	4,920
Straits	60,827	58,547	58,438
Australia	4,846	5,028	6,888
Banka	11,363	9,960	9,300
Billiton	3,215	2,715	1,950
Bolivia	11,867	12,500	14,700
Totals	96,250	93,218	96,196

The present value of tin in New York is between 35 and 36 cents per pound, with a good demand and tendency to go higher. The production this year promises to fall a little short of last year's.

In the above table the prominent position occupied by the Straits Settlements as tin-producing centres is plainly emphasized.

THE STRAITS SETTLEMENTS.—The Straits Trading Company, an English corporation the shares of which are largely in Chinese hands, has a large tin smelter at Singapore which reduces the largest part of the tin mined in the Orient, and much of the tin credited to the Straits is really mined over a larger area than the Malay Peninsula.

The tin fields in the Federated Malay States have been mined regularly during the past three centuries. At the present time they produce between 40,000 and 50,000 tons of tin per annum. The returns last year amounted to 48,616 long tons as compared with 50,991 tons in 1905. The principal mines are in Perak, Selangor, Negri Sembilan and Pahang. There are discrepancies between the Tin Mining Association's figures and those furnished by the Government. The latter estimate of the production of these States in the past four years is given in Table II.

TABLE II. MALAY STATES TIN PRODUCTION.

	1903.	1904.	1905.	1906.
Perak	29,014	29,527	29,711	28,988
Selangor	18,926	19,978	19,272	17,864
Negri Sembilan ..	5,684	5,643	5,662	5,172
Pahang	1,684	1,877	2,320	2,294
Totals...........	55,308	57,025	56,965	54,318

The Malay tin-bearing formations are chiefly shallow placers. Tin veins occur in granitic rocks in some of the mountain ranges and a few small veins are being worked. They are, however, low-grade and not very profitable. More than 90 per cent of the ore is won by Chinese and natives, by primitive methods of sluicing and hand panning the surface alluvial soil. The figures in Table II exhibit a tendency of the output to decline. This is due to two main causes, the exhaustion of the easily worked surface deposits and the scarcity of Chinese coolie labor.

In the height of the industry's prosperity the average value of the tin deposits in the Federated Malay States, and in the islands of Billiton and Banka, was about 6 per cent of black tin ore per ton of earth mined. Today the average value of the material mined is not more than 1 per cent.

The tin-mining leases are chiefly owned by Chinese merchants residing in Singapore, Penang, and other important trading centres in the peninsula, or by Chinese companies domiciled in China. Chinese coolies, expressly recruited in China for the work, do most of the mining and washing of the ore. Although China possesses a large population, experience has demonstrated very clearly that only a comparatively small proportion of the strong and virile single men competent to undertake steady laboring work are prepared to leave their homes and brave the hazards of distant emigration, in response to the call for their services in foreign lands. Migration has always been distasteful to the Chinese, who are essentially a home-loving people and possess a rich and beautiful country. Industrial conditions in their homeland, too, have been such as to detain them in China. During the past few years there has been a large measure of prosperity in Oriental countries. China has enjoyed a good share of this and greatly extended the national railroads, canals, and harbor and river improvements which absorbed a large amount of labor. Owing to the steadiness and reliability of the Chinese coolie, there was strong competition in securing his services for Oriental countries adjacent to China. In addition to the Oriental demand, the South-African gold fields absorbed several hundred thousand Chinese laborers, and large numbers were introduced by the western railroad companies into the United States either directly or by way of Hawaii, Mexico, and Canada. In consequence of the shortage of labor in the Malay States which was caused by these conditions the tin-mining industries have been severely crippled. Barely half the coolies required by the mines are obtainable. The miners take advantage of the mine owners' embarrassments and frequently go on strike. Such troubles have lately led to a great extension of the leasing system, and mine owners are letting out large portions of their properties to coolies on annual leasing agreements. This system, experience has shown, is one of the most effective yet devised for keeping the mines in operation during the present dearth of labor.

SIAM.—In addition to the tin fields of the Federated Malay States, valuable tin deposits are worked in adjacent countries including Siam, Annam, Cambodia, Burma and China. The mines in Siam and China are the largest of these outside producers. Tin is being mined in the valley of the Nam Sak river in northern Siam, at Bangtaphan, Lakon, Ratburi, Kelantan, Tringannu and other localities on the east coast, and at Puket, Kedah and Renong on the western coast. The present annual output is about 5,000 tons. The tin-mining industries of Siam, according to the recent reports of geological investigation, are capable of undergoing a large expansion.

CHINA.—Large quantities of tin ore have been annually mined in China for many centuries, but as no statistics are kept it is difficult to determine the present position of that country as a tin producer. There is no doubt, however, that the annual production of tin is still large. The principal tin-mining centers are in the provinces of Yunnan and Kwangsi. The annual yield of the Yunnan mines is estimated by the local authorities at from 5,000 to 6,000 tons.

AFRICA.—The occurrence of tin ore has been reported from several districts in Africa and it is possible that as that country becomes developed, tin mining will become one of its industries. Specimens of tin ore have been discovered in the Congo Free State in the vicinity of the Lualaba river. Alluvial, or placer, tin deposits were found by an exploring expedition in the Uwet district in Southern Nigeria in 1905, and prospecting operations carried on by an English company at Bautchi, in Northern Nigeria, resulted in the discovery of several low-grade tin-bearing veins in granite country. Reports of a favorable nature regarding the prospects of these deposits have been made by experts and there are prospects of European capitalists being induced to develop them.

EUROPE.—Notwithstanding the high prices ruling for tin during several years past, the tin production of Cornwall fails to exhibit signs of any important expansion. But fresh capital is being attracted to Cornish tin-mining investments; a number of old mines are being reopened and are resuming mining operations, and mines like the well-known Dolcoath, that have produced regularly for many years, are now being worked to their utmost capacity and vigorously developed. The mines of Cornwall are reported to have produced 7,201 tons of tin ore last year, which on being smelted yielded a little under 5,000 tons of metallic tin. This is about the same amount as they produced in 1905.

Several English and French companies have been formed to re-work some of the old Spanish tin mines. The most important of these is the Sultana mine in the province of Orense. In this property there is a large vein carrying tin in association with arsenical pyrites. The assay value of the ore is low and the tin is difficult to reduce owing to the presence of the sulphides.

SOUTH AMERICA.—The most promising tin deposits in South America are in the Andean districts in Bolivia. In the past six years a large amount of English, American, and Chilean capital has been expended in developing the Bolivian tin mines, and as a result tin exports to Europe are rapidly increasing. The amount exported last year was 14,700 tons. This ranked Bolivia next to the Malay States in tin production.

Tin mining is pursued in Bolivia under many disadvantages. The mines are mostly high up in the Andes in almost inaccessible positions, the climate is exceptionally bad owing to the high altitudes, mining titles are insecure through the mal-administration of the law, water for mining purposes is difficult to conserve, and mining labor is inefficient and very difficult to obtain.

The most important deposits are in the Quinsa Cruz district, and are situated at an altitude of 15,500 feet above sea level. At such an elevation only native Indians can do laboring work, and they are few in numberes and are disinclined to work steadily, even under the stimulation of high rates of pay. It is only with difficulty that men brought up in low-lying districts can walk about on the mines in this district. The ore deposits occur chiefly in porphyry veins traversing a series of clay schists, bearing a somewhat strong resemblance to the famous Mount Bischoff deposits in Tasmania. The value of the ore varies between 3 and 10 per cent of metallic tin per ton. An extensive and expensive metallurgical plant is required.

The Chorolque mine is at present the largest producer in the Quinsa Cruz district. It is yielding about 200 tons of tin per month. Large mining and metallurgical plants are being installed at the Llallagua, Potosi, and Monte Blanca mines, and these are expected to become important producers. The Oploca and Chocaya mines are also increasing their production, and the Bolivian tin-ore production consequently promises to reach record figures during the next few years.

TASMANIA.—The greatest individual tin producer in the world is the Mount Bischoff mine at Waratah, Tasmania. This famous mine has been in continuous operation during the past thirty-four years, and during that period the shareholders in the company owning it have received $10,395,000 in dividends, or $866 per each original $5 share. Hitherto the ore won from this mine came from a purely detrital deposit formed as the result of the weathering and decay of a series of stanniferous quartz- and topaz-porphyry dikes traversing a series of greatly metamorphosed schists which form the summit of Mount Bischoff, a prominent hill lying between the Waratah and Arthur rivers. This detrital, or placer, deposit has nearly been worked out. Lately the monthly yield of tinstone has been reduced from 100 tons to 70 tons. At a recent meeting of the directors of the company in Launceston, Tasmania, the mine manager stated that he will be able to keep the mill supplied with sufficient amount of placer material to yield an average of 70 tons of tinstone per month for the next seven years. He will be able to do this by cleaning up the old workings, mining away material left standing to support old workings, and re-picking material from the dumps. At the end of seven years a different class of material will have to be dealt with.

At the same meeting it was determined to re-form the company. The present company's charter was drawn for a period of thirty-five years. This term will elapse early in 1908. It is proposed to draw up a new charter for a period of one hundred years and enlarge its scope so that the company may purchase or operate other tin-mining properties on lease.

In addition to the detrital deposits, the Mount Bischoff mine contains inexhaustible low-grade sulphide tin deposits in numerous quartz veins and in quartz- and topaz-porphyry dikes which run throughout the company's leases and traverse the West Bischoff and other adjoining mines. These tin-sulphide veins also carry arsenical pyrites in large quantities, making mining dangerous through the frequent firing of the workings, and greatly complicating the metallurgical treatment of the ore. Large sums of money have already been expended in experimenting with metallurgical processes for treating low-grade tin-sulphide ore, but an efficient and cheap reduction process is still being sought.

Adjacent to the Mount Bischoff there are several smaller tin mines in successful operation, and alluvial tin is being won from the Waratah and Arthur rivers in the vicinity of the mines. Tin mining is also being vigorously pursued in many other districts in Tasmania. The metal occurs practically all over the island and many of the mines are well established and profitable. The Blue Tier mines, situated on the west coast of the island on the range of mountains bearing the same name, are the most noteworthy after the Mount Bischoff mine. The ore is found in a series of quartz veinlets varying in width from a mere thread up to several inches, which traverse granite deposits and are known to miners as "stockwork" tin formations. The ore is of low-grade, but it has been amply shown that when the mines are operated on a large scale with suitable machinery the tin can be profitably won. The Anchor and Australasian mines are the chief producers. The average value of the ore worked is 5 pounds of black tin per ton of granite. The total cost of mining, concentrating, and dressing the ore is 75 cents per ton. The Anchor mine is worked as a quarry, employs 120 men, and yields about 20 tons of tin per month. The Mount Lyell Company has recently acquired a large number of tin-mining leases in the Blue Tier field and is making preparations for mining tin on a large scale. It is very probable, also, that the Mount Bischoff company will soon undertake the development of a number of the smaller mines on the field.

The Heemskirk tin field, situated between Mount Lyell and Mount Bischoff, is at present attracting much attention. It is an old field, but has hitherto been worked on a small scale only. Tin occurs in the mines in the form of veins, many of which are very rich. The Cumberland, Orient, and St. Clair mines are now in course of development. The ore in these mines runs from 3 to 10 per cent of tin per ton.

There are many placer tin mines in operation in Tasmania, including the Briseis and New Brothers Home, and two successful tin dredges.

The Briseis and New Brothers Home placer mines are owned and operated by an English company. Their combined output last year was 1,408 tons of black tin, containing 1,031 tons of metallic

tin. To this amount, the Briseis contributed 912 tons valued at
$840,615. It is estimated by the company's engineers that the tin-
bearing drifts at present developed in the Briseis mine exceed
2,000,000 cubic yards of ore.

AUSTRALIA.—The tin production accredited to Australia last year
amounted to 9,000 tons; but it was doubtless much larger, as the
estimates are based chiefly on export statistics and take no account of
the tin used in local industries. One mine alone, the Vulcan, produces between 6,000 and 7,000 tons yearly. Tin ore is dispersed
over a wider area in Australia than in any other country, and there
is little doubt that when capital is invested in the tin mines Australia
will rapidly take rank as the leading tin-producing country. The
principal mines now in operation are in the eastern states of New
South Wales and Queensland and in West Australia.

The most valuable tin mine in New South Wales, and one of the
greatest on the continent, is the Gundle mine, the property of the
Gundle Tin Mining Company, Sydney. The mine is situated on
Mount Gundle, a prominent peak on the North Coast range, about
30 miles west of Port Macquarie, and a little to the south of the
Hastings river. The vein is 30 feet in width and averages over 10
per cent tin. The Conrad mine, near Inverell, in the New England
district, is another important mine. It is also a vein-mining property
averaging about 5 per cent tin. The ore, however, is complex and
carries good values in silver, copper, lead, and zinc, in addition to
tin. At present some trouble is being experienced in metallurgically
dealing with the ore.

A feature of the New South Wales tin-mining industry is the
great success of the tin dredges. At present more than thirty dredges
are successfully engaged in tin mining in New South Wales, and
about a dozen are operating, or are in course of construction, in
Queensland. Both centrifugal and bucket dredges are engaged in
tin mining, one kind being more efficient in one class of work and
the other in another. As a rule the tin-bearing formations are
composed of pure sand which has been derived from the decay of
stanniferous granites. The centrifugal dredges are found to give the
best results in such a material; but where the ground is hard and
clayey, bucket dredges are the best. The average value of the
material treated by the dredges is 2.5 pounds of tin per cubic yard,
valued at 75 cents. Mining amounts to 10 cents per cubic yard; so
the resulting profit is about 65 cents per yard.

The tin-mining industries in New South Wales are undergoing a rapid expansion and there are indications that European capitalists are beginning to recognize their value.

Several successful tin-mining dredges are operating on the Stanthorpe tin field in Southern Queensland. The greatest activity in that State, however, is on the Herberton tin field in North Queensland. The famous Vulcan mine, at Herberton, is yielding between 6,000 and 7,000 tons of tin per annum and paying its fortunate owners large dividends. The Smith's Creek and Stannary Hills mines are also developing into large regular tin producers. During the past few months a large and very rich tin vein has been cut in the Tornado mine, at Irvinebank and this property promises to become as important a producer as the Vulcan. Considerable progress is being made on the tin fields in the far north of Queensland and in the Northern Territory.

The principal tin-mining centres in West Australia are in the Greenbushes district, on the north coast. The tin-mining industries there are giving employment to about 1,000 men at present, but English capital is being directed to the field and within the next few years the industry promises to expand largely and afford employment to a much greater number of miners.

CONCLUSION.—The principal tin-producing centres have now been cursorily surveyed. It will be noticed that the active tin-mining fields are few in number, and generally speaking, are not in a very healthy condition, notwithstanding the high value of tin and the large existing demand for it. The alluvial deposits in the Malay States are approaching exhaustion. The deposits in the Dutch East Indies, on the islands of Banka and Billiton, are in a similar position. Scarcity of coolie labor is severely handicapping the tin-mining industries in those centres. Bolivia is advancing its production, and for some years to come will maintain a large annual output. Finally, however, consumers of tin will have to depend upon Australia and Tasmania for the principal part of their tin supplies. Africa may develop into a tin-mining country of importance. It is possible, also, that some of the mountain districts of South Eastern Asia may become centres of profitable tin-mining enterprises, and capital may open up tin-bearing veins in some of the old alluvial tin fields in the Malay States.

The proved tin fields of Australia, however, are ample to supply the world's wants, and their slow development is due solely to the scarcity of capital. As the demand for tin becomes still keener this will, doubtless, be satisfactorily remedied by the introduction of European capital into tin-mining enterprises.

EDITORIAL COMMENT

THE discussion of the Quebec Bridge disaster abstracted from *Engineering News*, which appears in our Review pages opposite, probably strikes very close to the true explanation of the failure. At the moment of the wreck, indeed, A9R rather than A9L seems to us to have led in the collapse; but if so, it would appear to be because the partial failure of the left-hand truss some days previously had already thrown an undue proportion of the load upon the right-hand member—a proportion which it bore until the "last straw" was added, and then, by sudden buckling, threw back again upon the already crippled A9L.

Examination of the wreck suggests fear that the compression members generally were under-proportioned; but whether or not that impression is well founded, even the most casual inspection excites amazement—almost consternation—at the comparative lightness of the latticing and the insufficiency of the riveting. The whole system by which the huge built-up members ought to have been connected, bound, and stiffened so as to act as units, was —flimsy; that is the only adequate word. As *Engineering News* shows, by reducing the scale of the latticing protionately down to the case of a strut of ordinary dimensions, the provision made in these "huge columns" did not at all reach the limits of practice common in the case of trusses of very moderate size. Under such circumstances it does not seem necessary to appeal to any mystery of knowledge as to the action of materials under stress, lying in that range beyond experience but within experiment in which this structure was projected. The great vision by which vast engineering work has heretofore been carried safely into the region where "accepted formula" must be carried forward and enlarged, seems to have been lacking. The most deplorable thing is that it was not swift enough to perceive error and to act in time, even when deadly danger was clearly revealed in the incipient collapse of the work.

* * *

THE New York street-railway scandal follows the insurance scandal, the Mercantile Bank shame jostles both, and one wonders indeed how many more of our once-great and honored captains of industry belong to the thieves' company. Europe is not to be blamed for a searching distrust of American "finance," nor America for dallying with almost any political nostrum which promises honesty in high places. But property has always tempted the plunderer, and he will always steal when and what he can, whether the opportunity is given him by the trusting vote of a spellbound citizen electorate, or of a plausibly solicited majority of stock certificates. The strongest deterrent to the thief is light. Publicity makes it much harder for him to get away with the goods—or if he has got away with them, it at least (providing there is still a vigorous, healthy public opinion on the subject of keeping stolen property) spoils most of his pleasure in the fruits of his crime. State ownership at best would but substitute one set of frailties for another. Public examination and inspection—not retrospective, but current and continuous—could far more easily be kept in incorruptible hands. The very conditions make corruption far less probable. This way lies salvation.

THE STRENGTH OF LARGE COLUMNS.

A DISCUSSION OF THE NECESSITY FOR MORE ACCURATE DATA ON WHICH TO BASE THEIR
DESIGN, AS EVIDENCED BY THE FAILURE OF THE QUEBEC BRIDGE.

Engineering News.

LAST month we reviewed in these columns an elaborate analysis, taken from a number of *Engineering News*, of the available evidence as to the point of initial failure of the Quebec bridge. The deduction made from the facts available at the time the article was written was that the failure of some compression member in the anchor arm of the cantilever was the initial cause of the wreck and that all the indications pointed most strongly to chord A9L. Subsequent investigations have served only to offer additional evidence as to the correctness of this theory, which seems now to be generally accepted as the true explanation of the cause of the disaster. *Engineering News* has since extended its analysis in an endeavor to determine why this member should have failed and reaches the conclusion, suggested in the former article, that our deficient knowledge of the strength of large columns was primarily responsible for the disaster.

It is first pointed out that there are only two possible explanations of the failure of chord A9L. The first is that this member had some inherent weakness due to defects of manufacture or to accident in handling and transportation; the other is that it was typical of all members of its class, and that while it was the first to fail, the other members were in no less serious danger. All the evidence points to the latter explanation as the true one. It is admitted that chord A9L had suffered accidents of one sort or another in handling and trans-

portation. It is inconceivable, however, that it could have been put into the structure in a seriously damaged condition without attracting notice, or that the bend discovered in it just before the wreck could have been of long standing.

"And now let us proceed a step farther. Study of the wreck at the bridge site points strongly to the conclusion that when lower-chord member 9A in the left truss was at the point of failure, *some compression member in the opposite truss was in almost the same condition.*

"To show this, let us for a moment assume the contrary case and suppose that 9A was the only weak member in the bridge and that all others had the margin of safety for which they were designed. If this had been true, then the left-hand truss would have failed appreciably before the right-hand. Destruction of the cross-bracing would of course quickly wreck the left truss also; but almost certainly it would have held together several seconds longer than the left-hand truss and would then have gone down on top of the other. At least, the wreck would bear evidence of such a sequence in failure of the two trusses.

"What the wreck actually indicates, however, is that the failure of the two trusses was probably nearly simultaneous. If anything, the wreckage of the anchor-arm lies to the right of the center line of the bridge; its center of gravity in falling appears to have gone eastward, perhaps 8 to 10 feet.

The bottom chord of the left-hand truss is plainly visible near the whole length of the anchor-arm. The bottom chord of the right-hand truss is buried under wrecked web members nearly all the way.

"If chord member 9A on the left failed first, then almost simultaneously a member in the opposite truss, perhaps the chord member directly opposite, went, too. It went because it was also overloaded, and the stresses produced by the falling truss opposite almost instantly buckled it like its mate on the other side.

"Here we have at least two compression members designed for a compression stress of 24,000 pounds per square inch and failing at a load of some 16,000 pounds, not more than half the elastic limit of the metal in tension. Why did these members fail? It was not weakness at the ends. The construction of these members at the joints was worked out with greatest care to properly distribute the stress and the workmanship was of the highest excellence. It was not faulty material. The steel of this bridge has been subjected to a more terrible ordeal than any testing laboratory ever devised. It has been torn, twisted, bent and crushed in every possible way; but its fractures indicate it to be throughout just the ductile, homogeneous metal which the designers aimed to secure. Further, even brittle material would not explain a failure under a compression by static load. We may admit perhaps a possibility that incomplete field riveting may have contributed to weaken the members; but from any evidence thus far uncovered, it does not seem to have been in any degree the chief cause.

"All other causes of failure being found untenable, one is forced to the conclusion that these members failed from lack of stiffness. The more one studies this huge scrap-heap on the south shore of the St. Lawrence, the more one is impressed with the comparative weakness of the means adopted by the designers to stiffen and bind together the massive plates of which these columns were built up into a single member.

"Let us look for a moment at the construction of the compression chord members of this bridge. They consist substantially of four massive plate webs or girders, each made up of four rolled plates, stitch-riveted together to form one built-up plate nearly 4 inches thick. Each of these is finished with an angle at top and bottom, and these four long thin girders are supposed to be stiffened and bound together so as to act as one member by latticing on the top and bottom made of 4 × 3 inch 8½ pound angles.

"The point to which we would direct attention is the small size of the angles which are supposed to lattice these four massive webs into one member compared with the size of the webs themselves and, of course, of the load the whole member has to carry. Near the ends of the member, it is true, a long, heavy cover plate replaces the latticing angles; and it is very noticeable in the wreck how vastly more efficient a plate covering is than angle latticing to hold members in place against heavy punishment.

"Of course it may be urged in defense of this design that comparatively little material is needed to hold members rigidly in line when subject to direct compression. In theory this is true; and it would be true in practice if all the parts of a compression member were absolutely true, free from all initial strain, and if the load upon them were applied with absolute exactness. Such conditions do not and cannot obtain with even the best workmanship. All our plates and angles have to be straightened after coming from the rolls. Shearing, punching and riveting all set up initial stresses. Every joint in our structural work is far from mathematical perfection. There is no doubt, then, that the stiffening lattice or web that binds the parts of a strut together ought to be in proportion to the parts which it holds.

"But the evidence that carries most conviction is the evidence of actual trial, and it is this evidence that indicates the latticing of these chord members to be inadequate. Lower-chord member 9A at Quebec was found bowed out of line on Aug. 27 *because the latticing and the rivets that held it were not strong enough and stiff enough to hold the four webs in place and make them act together.* And if the latticing was not strong enough for its duty here—even granting previous injury—was it strong enough for its duty elsewhere?

"The Quebec bridge, like any other bridge, was no stronger than its weakest main member. All its main compression members—posts and chords alike—were made up of thin webs which independently of each other were too flexible to have borne any end compression of consequence. All of them, therefore, depended absolutely for their rigidity and stiffness to resist compression without flexure and failure upon the lattice bars by which they were connected, and these in turn upon the rivets that held them. The same thing is true, undoubtedly, of all our struts and columns in all our bridges and buildings; but it is well to have the fact brought clearly and forcibly home to the comprehension of every engineer who has anything to do with structural work.

"Was this latticing on these huge columns strong enough for the work it had to do? What does the profession know about the stresses in column latticing? Where in engineering literature has its proportioning been carefully and soundly discussed? If practical experience shows, as may be claimed, that ordinary practice in latticing is good and safe, how shall we safely bridge the gap between the strut in an ordinary bridge designed to carry a load of say a million pounds and these huge columns carrying more than twenty times as much?

"Should the latticing be proportioned to the cross-section of the members which it holds in line? Then these 8½-pound angles on a member with 781 square inches cross-section would correspond to single lacing bars only 1 inch $\times$ ⅛ inch on a strut one-fortieth as large, say of 20 square inches cross-section, or a pair of 15-inch 33-pound channels.

"Again, how should the proportions of lattice-bars be varied with the variation in the length of the member? Are the lattice-bars to be considered like the web-members of a truss, since their office is to hold the several ribs they connect from bending? If so, the long column like the long span bridge would seem to require latticing heavy in proportion to the column's length.

"Still again, what relation does the great depth of these webs bear to the security of the latticing? We know that a thin flat plate has little stiffness to resist end compression. In broad plates secured by latticing only at their edges and subjected to end compression, is there possibly some deforming action near the center of the plate which does not occur in the comparatively narrow plates used in ordinary columns?

"What, if any, were the computations by which the designers of this bridge assured themselves of the sufficiency or supposed sufficiency of this latticing? On what assumptions were their computations made and can these assumptions bear the criticism of the profession?

"Such are some of the questions which surely arise wherever engineers are studying the failure of this huge structure; and an answer to which cannot be given without venturing on debatable ground. And to answer with entire certainty these questions which are troubling the profession we must have something more and better than mere computations of the stresses in column latticing, important though these are at the present juncture. It will not be possible, we believe, to reassure either the public or the engineering profession as to the actual strength of such huge columns until actual tests are made upon them. . . .

"How many thousands of engineers, as they realize at times the heavy responsibilities that they carry for the safety of life and property, have longed for records of tests of full-size columns such as they are obliged to use in their works. The designers of the Quebec Bridge had to build up a column of steel plates and shapes to carry a load at the ninth panel of the lower chord of 22,000,000 pounds. They had for a guide the accepted theories of column stresses based on tests made a generation ago on columns of trifling dimensions.

"We do not mean to indict the soundness and general reliability of the accepted column formulas, for members of ordinary size where past experience serves as a check and safeguard. We do believe, however, that for the columns of massive dimensions such as are being more and more called for by present-day demands in bridge and building construction, theory needs to be supplemented by actual tests."

It is suggested that money should be forthcoming immediately to build a testing machine large enough to investigate these matters.

FOREST PLANTING FOR RAILROAD PURPOSES.

DETAILS OF THE WORK UNDERTAKEN BY THE PENNSYLVANIA RAILROAD TO SECURE A
FUTURE TIMBER SUPPLY.

E. A. Sterling—Engineering News.

IN a review in these columns in THE
ENGINEERING MAGAZINE for May,
1907, some details were given of the
enormous consumption of timber, particularly for ties, by the railroads of the United
States, and it was pointed out that the
rapid exhaustion of the forest reserves of
the country imposes upon all consumers of
timber the necessity for economy in its use.
The increasing scarcity of all grades of
timber and the steady advance of prices
are phenomena of long standing, but until
recently they have had little or no effect in
influencing individual consumers to use
less wasteful methods, and practically the
whole burden of the task of conserving and
replenishing the timber supplies of the
United States has fallen upon the Forestry
Departments of the State and National
governments. It is encouraging, however,
to note that in the last few years some of
the larger corporations of the country have
begun the planting of forests on waste and
unproductive land in order to ensure a future supply of timber for their own use,
and the immediate future will no doubt see
a large extension of private enterprise in
this direction. The following abstract
gives some interesting details of the forest
planting by the Pennsylvania Railroad as
described by their forester, Mr. E. A. Sterling, in a recent number of *Engineering
News.*

"The Pennsylvania Railroad has probably
done more forest planting than any other
eastern corporation, and some of the cost
figures on this work may be interesting,
and possibly give an incentive for similar
work by other land-owning companies.

"The work was started in 1902 with the
object of determining the feasibility of
utilizing, for the production of timber suitable for cross-ties and fence posts, land
acquired and held for railroad purposes.
In straightening and widening the main
line, and to avoid expensive overhead or
undergrade crossings and damages by reason of stream or drainage interference on
new low-grade lines, the company in many

instances purchased entire farms. These
farms, ranging from 20 to 200 acres in extent, are the lands on which planting has
been done.

"Between 1902 and 1907 black locust was
the species exclusively used, and during
this time 1,679,320 trees were planted. The
average cost of planting was a little less
than $20 per 1,000, but as the work was ex-
perimental and entirely new this cost cannot be considered representative. In the
spring of 1907 it was decided to try other
species, since it was found that the 'locust
borer' was working serious injury in the
plantations established, and that the locust
was not adapted to all the situations available for planting.

"The following table shows the various
species and the total number of trees
planted in the spring of 1907:

Species.	No. of seedlings planted.
Red oak	252,154
Black locust	175,716
European larch	6,970
Pin oak	4,570
Scotch pine	3,500
Tamarack	3,000
Chestnut	2,316
Total	448,226

"The 448,226 trees listed above were
planted during April and May at three
points on the main line between Harrisburg and Altoona, on ten typical farms
which were acquired for railroad purposes.
The cost items represent, with a fair
amount of accuracy, the expense of private planting on such scattered areas and
under existing labor conditions. At Altoona, where Italian laborers were employed and the trees planted 6 by 6 feet
apart in small holes prepared with a mattock, the cost was $5.12 per 1,000. This
cost increased to a maximum of $7.85 at
Mt. Union, where American labor was
used and the trees were set 4 to 6 feet
apart in furrows plowed 5 feet apart; while
at Newton Hamilton, where the same crew
planted in furrows, the cost dropped to an
average of $4.82 per 1,000, the expense of

furrowing being included in both cases. The difference in cost was due largely to extra expense at Mt. Union in clearing the ground from brush, and in resetting fences. Marked variations in the cost of planting may always be expected, but the average of $6.04 per 1,000 may be considered a fair approximation of the expense for reforesting old fields where there is not much brush to hinder operations.

"The seedlings planted ranged from one to three years old, and were purchased from commercial nurseries at an average cost of $5.25 per 1,000, making the total cost $11.29 per 1,000. With the usual spacing of 6 by 6 feet, or 1,210 trees to the acre, the cost of such work per acre would be about $15. The Pennsylvania Railroad has established a large forest nursery at Hollidaysburg, near Altoona, in order to grow its own plant material. With home-grown seedlings, the total cost of establishing forest plantations should not exceed $10 or $12 per acre, if the work is done on a large scale.

"In the choice of species, much depends on the character of the ground and on the use which is to be made of the timber when mature. A railroad company is naturally most interested in cross-ties and fence posts, and the fastest growing timber suitable for these purposes would be selected. The most desirable trees, however, are often of discouragingly slow growth, which necessitates a compromise between growth and quality; for example, red oak is being planted instead of the more durable white oak, because of its faster growth, and on the assumption that treatment with preservatives will be in general use by the time the trees mature. Scotch pine is being used with the same end in view. In the light of European experience, both red oak and Scotch pine woods, if creosoted, and protected from mechanical abrasion under the rail, will have double the life of untreated white oak. Such species as pin oak and tamarack are used in wet situations, where the more valuable trees would not flourish."

HYDRAULIC POWER IN FRANCE.

DETAILS OF THE EXISTING HYDRO-ELECTRIC DEVELOPMENTS AND OF THE WATER POWERS WHICH STILL REMAIN UNUTILIZED.

Engineering..

A RECENT number of *Engineering* gives an interesting review of the importance of hydraulic power in France. It is estimated that the nominal power developed by steam engines reaches a total of about 8,900,000 horse power, which would correspond to only about 3,860,000 horse power, actually effective throughout the entire year. The following details, therefore, of the possibilities of hydro-electric power development throw an interesting light on the effect the utilization of her immense water powers would have on French industry.

"In France, the power available at different periods of the year in the various water-courses has not been gauged with the degree of accuracy which has obtained in other countries, and it is not possible, therefore, to arrive at an exact figure for the hydraulic power that can be turned to account. The country, however, is favoured in the matter of water-courses and

waterfalls, and differences of level in the case of several rivers ensure for these a rapid flow. The rainfall is generally regular, and a constant supply for the hydraulic stations may be relied on. The highest waterfalls are found in the High Savoy, Savoy, Isère, and High Alps districts. In the Lower Alps and the Pyrenees there are also waterfalls that might be utilised, but they are of less importance than those of the former districts. In many other parts of the country there are mountains and hills, among which the valleys act as reservoirs for a number of streams, having an appreciable head of water and a regular flow; such are the central plateau of France, the Jura, the Vosges Mountains, and Normandy.

"Previous to the invention of the steam-engine, the natural sources of power were utilised to some extent, and windmills and water-wheels were frequently met with, distributed over the whole country. The former have now almost completely disappeared, but water-wheels, on the other hand, are still utilised, the principal reasons for this being that water-power is more reliable than wind-power, and in the second place, the cost of installation of a large turbine plant involves a heavy initial outlay, which requires time to write off. Nevertheless, a number 'of water-wheel plants which were used mainly for corn-grinding are gradually ceasing running, and are being superseded by modern installations for the production of current for power and light distribution.

"On the non-navigable rivers throughout France there are at the present time 46,000 hydraulic installations of all kinds, the total power available of which is 500,000 horse-power; on the navigable rivers there are 1,500 such installations, with a total of 80,000 horse-power. The very small average power of the installations is due to the existence of a number of water-wheels still running. These are generally in streams in which the head of water is low. Streams capable of giving power at low head of water are generally distinguished by the term *houille verte* (green coal), in contrast to the *houille blanche* (white coal), representing mountain torrents and waterfalls from which power can be derived. The mountainous districts are those in which are found the highest figures in hydraulic power utilised. Thus in the Isère Department alone there are employed about 40,000 horse-power; over 30,000 in Savoy; 22,000 to 25,000 in the Lower Pyrenees; over 20,000 in the High Savoy, and almost the same figure in the Higher Pyrenees. The Departments of the Vosges and Jura show also high figures. The Rhône Department, although it is not in a mountainous district, compares favourably with the preceding, owing to a station at Jonage, which utilises through a canal a difference of level in the flow of the Rhône.

"The first step in the utilisation of waterfalls in France on scientific and commercial lines dates back to 1871, when an installation was put down at the junction of the Valserine with the Rhône. The company was originally a French one, but the undertaking subsequently passed into English hands. At that time the power could not be distributed as is now possible, and the situation of the company was not satisfactory; good results were only obtained in 1898, when a new French company took over the business and generated current for power transmission. When once the movement for power transmission was fairly started, developments were rapid, and particularly in the Alps district, in the valleys of which—such as that of the Romanche—there are successions of generating-stations, one taking the water from the tail-race of a station at a higher level. In the region between the Alps and Lyons, the power utilised in the manufacture of aluminium alone has been increased during the year 1906 by 5,000 horse-power, and that for carbide of calcium production by a like amount. At the present time there are in this same region 35,000 horse-power in use for aluminium; 25,000 horse-power for carbide of calcium; 15,000 horse-power for potassium or sodium chloride; and 22,000 horse-power for metallurgical purposes, such as the manufacture of steel. The town of Lyons receives current generated at Moutiers, at a distance of 180 kilometres (112 miles). The higher heads of water available were first utilised: Mr. Bergès started with the Chapareillan station, where the head is 600 metres (1,968 feet). In the same district the Epierre station utilises a head of 550 metres (1,805 feet). More recently the lower heads, averaging from 50 to 150 metres (165 feet to 490 feet), have been much sought after from watercourses in which the flow is comparatively heavy, mainly, perhaps, because the higher heads are combined with less regularity in the flow and with difficulty in arranging for regulating reservoirs.

"Among the individual stations, we may briefly mention that of Livet, on the Romanche, in the Isère, of a maximum capacity of 15,000 horse-power; those of Avignonnet and Champs, on the Drac, of 7,000 to 8,000 horse-power each. The Drac could alone generate 100,000 horse-power in the same department—that of Isère. In the above-mentioned stations the head is only from 18 to 37 metres (00 feet to 120 feet). In the Higher Savoy there is the Pont du Risse station, on the Giffre, where the head is about 71 metres (233 feet), with a capacity of 9,000 horse-power; and several others where the height of the fall is considerable. In the centre of France—at Clermont-Ferrand—there is the Sioule station, of 6,000 horse-power, with a head of 25 metres (82 feet)only; in the same district a station on the Vézère supplies the town of Limoges. The *houille blanche* available in the Mediterranean region is rapidly being developed. The Jonage station, near Lyons, above referred to, belongs to the *houille verte* class, and the 16,000 horse-power it produces is now all utilised.

"Much hydraulic power, however, remains to be regulated in the vicinity and to the south of St. Etienne, and in all the districts between Lyons and the Pyrenees. The same may be said of the Higher Alps Department, where a total of 300,000 horse-power could be obtained in the dry season, and 500,000 horse-power with an average flow. The corresponding figures for Savoy would be 320,000 and 650,000 horse-power, and those for the Maritime Alps, north of Nice, 70,000 and 140,000 horse-power. Mr. Tavernier estimates that the whole district comprised between the Alps, the Rhône, and the Mediterranean could yield 3 million horse-power in the dry season, and 5 million horse-power with an average flow.

"It is estimated that in the south-eastern districts of France, and in the departments which adjoin the Pyrenees, there are available 1,300,000 horse-power in the dry season. In the Jura, Vosges, and in the centre of France there would be a further total of 900,000 horse-power; while for various other departments of the country the total is estimated at 1,400,000 horse-power.

"The total hydraulic power for the whole of France may therefore be estimated to reach from 9 to 10 million horse-power, available on the shaft of the turbines, or about the amount now produced by heat-engines. A large portion of this total has still to be turned to account by hydro-electric development."

A PROJECT FOR A TRANSALPINE WATERWAY.

AN ESTIMATE OF THE POSSIBILITY AND COST OF ESTABLISHING A CANAL OVER THE ALPS FROM VIENNA TO TRIEST.

Jos. Riedel—Oesterreichische Wochenschrift für den Oeffentlichen Baudienst.

A PROJECT which has long been the dream of the people of central Europe is the establishment of a waterway over the Alps to serve as a means of communication with the Mediterranean or the Adriatic. The project has received most attention in Austria, having been brought into prominence first about the close of the last century when the subject of canals was in the forefront of political discussion and there began an era of great activity in canal-building in France and Germany, to give the people of mid-Europe communication with the North Sea. Recently the project was considered by the International Navigation Congress and a special report has been presented to that body. The following details of the estimated cost are taken from an abstract of this report by Herr Jos. Riedel in a recent number of *Oesterreichische Wochenschrift für den Oeffentlichen Baudienst.*

In the examination of the economic feasibility of the scheme the question of the probable traffic is first taken up. It is pointed out that the establishment of a canal from Vienna to Triest, connecting the Danube and the Adriatic, is a problem essentially distinct economically from that of a waterway from Venice or Genoa west to the Lake of Geneva or north to the Lake of Constance and the Rhine. Vienna already has communication by water with the Mediterranean through the Danube and the Black Sea, and a canal from Vienna to Triest would have to compete with this route. On the other hand a waterway from Venice through the Po to the Lake of Geneva and thence to Basel and the Rhine would lead to Strassburg and Mannheim and consequently be much more favorably situated for the securing of traffic than the Vienna-Triest line.

It is estimated that the average freight-movement per head of population in the interior of Europe amounts to half a ton yearly. Of this fully one-half consists of raw material or package freight, for the distribution of which water transportation would be preferred to rail. By the opening up of branches a good waterway through the Alps would soon have an influence over traffic for a distance of 60 kilometres on either side. It would eventually serve an area of 100,000 square kilometres and a population of 20,000,000, drawing therefrom a yearly traffic of 6,000,000 tons of package freight. An average haul of 300 kilometres and a profit of 2 centimes per ton-kilometre would result in a net annual profit of about 38,000,000 francs. On this basis, and taking into account the indirect social and political benefits such a waterway would confer, the economic feasibility of the project would be established beyond a doubt, even assuming an initial expenditure of 1,500,000,000 francs.

The technical difficulties in the way are not so easy of solution. The report enumerates as the problems connected with the technical possibility of the project the condition of the already navigable rivers and canals, the building of the open sections of the waterway, the establishment of numerous tunnel sections, the motive power to be adopted, the overcoming of the great differences of level, and the procuring of a sufficient supply of water at high altitudes. So far as the open sections of the canal are concerned, one of the great difficulties would be to secure a line free from curves of small radii. On this account it would be much more satisfactory to cut through an elevated table-land than to follow the windings of valleys, and from this point of view the line from Triest of Fiume over the Karst to the Danube would be preferable to one which would have to cross the Central Alps. In neither case, however, could a canal be built without resorting to considerable tunneling.

A considerable section of the report is given up to a consideration of the tunnels which would be necessary. On account of the probable size of the traffic turn-outs could not be used; these have proved unsatisfactory even in large open canals and a system of double tunnels would have to be resorted to. A boat 67 metres long, 8.2 metres wide, 1.8 metres draught, and of 600 tons capacity, such as is proposed for the Danube-Oder canal, would require a tunnel 12 metres wide and at least 80 square metres in section or for a double tunnel 160 square metres. The type of boat in use on French canals, 38.5 metres long, 5 metres wide, and 1.8 metres draught would require a section of about 70 square metres or for the double tunnel 140 square metres. It might be found more economical to use boats only 4 metres wide and of 200 tons capacity, such as are said to give good results on the Volga where the boats are usually towed in pairs.

Several examples of the cost of driving large size tunnels are given. A tunnel on the Marne-Saone canal, 4,820 metres long and 66 square metres in section, was driven through marl and chalk in six years and cost 2,500 francs per metre. A canal at Condes, 103 square metres in section and 307 metres long was driven in three years at a cost of 3,140 francs per metre. Judging from the cost of these and other tunnels, a tunnel through the Alpine strata of 60 square metres section would cost 3,107 francs per metre. It is estimated that on the line from Triest to Vienna there would be 22 tunnels of a total length of 9,320 metres. The total length of the line would be 512 kilometres and the total estimated cost 1,143,055 francs per kilometre.

THE DEVELOPMENT OF RADIO-TELEGRAPHY.

A REVIEW OF THE INVENTION AND DEVELOPMENT OF RADIO-TELEGRAPHY FROM A SCIENTIFIC STUDY TO AN APPLIED ART.

Report of the Select Committee on the Radio-Telegraphic Convention.

THE Select Committee, appointed by the British Government to consider the advisability of adhesion to the international Radio - Telegraphic Convention signed at Berlin in November, 1906, have completed their labors and submitted their report. A large part of the latter deals with matters more or less political, but apart from this the report contains a historical record of the development of wireless telegraphy, both scientific and political, which is unsurpassed by any previous publication. The following extracts taken from the section dealing with the "invention and development of radio-teleg-- raphy from a scientific discovery to an ap- plied art" will be of general interest.

"Wireless telegraphy or radio-telegraphy means the transmission of signals by elec- trical energy between two points which are not connected by a wire or other metallic conductor. The term is specially used— and exclusively used in the convention—in connection with instruments employing the Hertzian waves, so called from their dis- coverer, Heinrich Hertz. Before the dis- covery of the Hertzian wave other meth- ods were used experimentally for sending messages across space without conjunction other than the earth: (1) by earth induc- tion, using two parallel or nearly parallel wires erected, for example, on either side of a stream, each end being fitted with tele- graphic apparatus; (2) by electric mag- netic induction; and (3) by a combination of the two former. This means was taken up by Sir (then Mr.) W. Preece in 1885 (then electrician to the Post Office), and a system was set up of telegraphic commu- nication without direct connecting wires between places not more than six or seven miles apart. A wireless telephone service has been established by Mr. Gavey on this basis between Holyhead and the Skerries, and recently telephonic speech has been ex- changed between the two stations with the greatest ease.

"The induction systems are not, how- ever, suitable for communication over a long distance, nor for communication with ships, and the methods now used for that purpose are all based on the employment of the Hertzian waves. The means em- ployed in transmitting messages across space without physical connection, consists broadly of two stations, a transmitting sta- tion and a receiving station. At each of these stations a wire or wires must be erected in the air by means of a mast, tow- er, balloon, kite, etc. This structure of wires is known as the antenna or aerial. Speaking generally, the greater the range intended to be covered the higher or more extensive must be the antenna or aerial. The energy is generated by a battery or generator worked by a steam, gas, oil, or other power engine, and is communicated to the aerial by the transmitting apparatus proper.

"The receiving apparatus is also connect- ed with the aerial. There are many forms of receiving apparatus in use, and it may be said that one of the principal differences between the several systems of wireless telegraphy consists in the various forms of receiving apparatus. In some the appa- ratus automatically records the dots and dashes of the Morse alphabet, but in most forms a telephone is attached, and the op- erator hears the dots and dashes, and this permits of more rapid working than any re- cording instrument yet devised.

"With the most powerful stations very great distances have been covered, and sig- nals have actually been transmitted across the Atlantic, but in the present condition of radio-telegraphy communication has not been effectively maintained with regularity over so great a distance, especially by day. The maximum effective range for regular communication by day or night may be put, possibly, at 1,000 or 1,500 miles, but this state of affairs is changing almost month by month. Wireless messages are frequent- ly sent from Poldhu to Gibraltar and Mal- ta. As regards the ordinary ship and shore stations, the effective range may be said to vary between 50 and 300 miles, possibly rather more, according to the power of the station.

"The signals are usually sent out with equal energy in all directions. A message sent out, for instance, from the high-power station at Poldhu, Cornwall, and intended for ships in mid-Atlantic, has been equally received at Nauen, near Berlin, and at the Dutch station at Scheveningen, or at Gib- raltar. Attempts have been made to devise methods of directing the waves, or, at least, of sending them with the greatest energy in one particular direction, but so far only with moderate success. The Admiralty are unable to determine the direction in which a wireless message is coming under certain conditions.

"The main defect which was first found in the working of wireless telegraphy, re- sulted in part from this inability to control direction. As every receiving station with- in range of a transmitting centre could read all the messages sent from that cen- tre, it was not possible for more than two stations in a given area to interchange sig- nals at a given time without mutual interference. It was impossible, of course, for the ether within a wide radius thus to be monopolised by two persons if the invention was to be of real practical value. It was also obvious that if two sets of apparatus could be so accurately adjusted the one to the other as to ensure reciprocal radiation, whilst other apparatus not so adjusted would not be affected, the practical problem would be solved. It is in the direction of 'syntony' or the 'tuning' of apparatus so as to induce 'selectivity' between them that recent scientific research and experiment have been chiefly concentrated; and, as will appear later, the more or less complete manner in which this has been achieved between different systems constitutes one of the difficulties placed before the committee.

"The committee do not wish to express any opinion on controversial questions of priority and patent right, but it appears to be generally admitted that the Hertzian waves were first experimentally applied for practical telegraphic purposes in 1895-96 by Mr. Marconi. Many inventions and improvements by Mr. Marconi and others have appeared since at a rapid pace, and there are now several 'systems' or 'methods' in existence, whilst the number of patents is very great.

"The principal systems in use are: Marconi—The first practical system in the field. The first Marconi British patent was applied for on June 2, 1896, and accepted on July 2, 1897. De Forest—American in origin; acquired in Great Britain, and it is said for all parts of the world except North America, by a British company (the Amalgamated Radio-Telegraph Company). Fessenden—American. Lodge-Muirhead—British, both as regards invention and owner-

ship; first patent applied for May 10, 1897, accepted Aug. 10, 1898. Telefunken ('far-sparking')—German, an amalgamation of the systems of several different inventors (Slaby, Arco, Braun, Siemens). Rochfort —French. Branly-Popp—French. Poulsen —Danish, of recent invention, the rights in which for all parts of the world have also been acquired by the Amalgamated Radio-Telegraph Company. Shoemaker—American. Massie—American.

"The use of radio-telegraphy is rapidly increasing. A large number of instruments are already in operation in various parts of the world. Conflicting evidence was given as to the exact number in commercial use, but this is partly explicable by the fact that conditions are rapidly changing, and that there are in existence a number of experimental stations and other stations not open for commercial work. A relatively small proportion of the existing stations are on British territory. The greatest number are in the United States. There are also a considerable number in Italy and Germany, a few in Holland, Belgium, Spain, and France, and several for non-commercial purposes in Russia. According to the most recent returns, it would appear that there are 186 commercial stations throughout the world, of which some 63 are Marconi and 123 are other than Marconi. There is a preponderating number of Marconi stations in Great Britain, Italy, and Canada, but in all other parts of the world, with few exceptions, the stations are on other systems. The most important stations are those situated on the south coast of the United Kingdom, which is, owing to its geographical situation and the predominant part it plays in the maritime business of the world, the chief focus of ocean commerce."

ELECTRIC LOCOMOTIVES FOR MOUNTAIN GRADES.

A DISCUSSION OF THE ADVANTAGES OF ELECTRIC TRACTION OVER STEAM ON THE MOUNTAIN
DIVISIONS OF RAILWAYS.

Engineering Record.

THAT the advantages of the electric over the steam locomotive for service on heavy grades, which have been so conclusively demonstrated on Italian railways, are being recognized by American engineers is evidenced by the

fact that the Great Northern Railway has adopted electric traction for its Cascade Tunnel section and that larger projects are under serious consideration by other roads. The difficulties of steam traction on the mountain sections of American railroads have long been accepted as irremediable since not even the Mallet type of locomotive with its large draw-bar pull and great horse-power capacity succeeded in lowering materially the high operating costs on heavy grades or in removing the dangers, delays and limitations of railway operation in mountain sections. A recent editorial in *The Engineering Record*, however, looks for a rapid and immediate development of electric traction on mountain grades on account of the general recognition of the advantages outlined in the following extracts.

"Electric operation on mountain grades appears to offer many advantages of greater importance than a generally admitted small saving in the expense of fuel, though this in itself may prove a most attractive feature under certain conditions and in certain localities. Admitting that electric power produced in hydro-electric stations costs less per ton-mile hauled than coal inefficiently burned under the abnormal overload conditions obtaining in steam locomotive operation on mountain grades, this item in itself will seldom show a sufficient return on the capital required for electrification to make it attractive to the steam railroad management.

"It is a well known fact that the figures given in annual reports for locomotive maintenance running from six to ten cents per locomotive-mile do not apply to those locomotives operating under heavy grade conditions, and a maintenance charge ranging from ten to eighteen cents per locomotive-mile is not uncommon in this class of work. It is contended that the electric locomotive can perform the same service with a much less maintenance charge than is required in steam locomotive operation; while the 50,000-mile endurance run of the New York Central locomotive extended but a year, it showed a maintenance charge of less than one and one-half cents per locomotive-mile, and it is probable, according to electrical specialists, that an electric locomotive properly proportioned for service can be maintained for much less than a steam locomotive, performing the same service. Just what the gain will be in this direction is not apparent from operating figures, and it will be interesting to see results from the New York Central electric zone after electric locomotives have been in service operation for a year or more. The saving in the maintenance charge of locomotives is reflected more generally than merely in labor and material for necessary repairs, as a low maintenance account means great reliability, and freedom from breakdowns and delays, so that if all expenses due to failure of locomotives in service were charged directly to maintenance of locomotives this item would assume formidable proportions. Hence, a saving in this item is of far-reaching importance and should prove a valuable asset to the electric locomotive on heavy-grade work, where the conditions of operation are very exacting as regards performance of the motive power.

"The great advantages offered by the introduction of the electric locomotive seem to lie more in the direction of increasing the tonnage capacity of the single track, commonly met with on heavy grade divisions, and the general reduction in cost per ton-mile resulting from the handling of heavier trains at higher speeds with a certainty of operation not shared to an equal extent by the steam locomotive. Where the capacity of the electric motor is properly proportioned to the work which it has to do, there is no doubt of its being a highly efficient and extremely reliable piece of apparatus, requiring little or no attention; only periodic inspection, and with complete independence of the round houses, coaling towers, water tanks, and ash pits, required for the frequent repairs and grooming of the steam locomotive. This freedom of action and reliability must be reflected in a greatly increased daily mileage, a lower crew expense, and a considerable curtailment of the false mileage common to steam locomotive operation.

"The electric locomotive is merely a transforming or converting piece of apparatus and can draw an unlimited supply from its distant stationary power house; hence its output is limited only by mechanical considerations, and the result is a type of locomotive capable of delivering an enormous power if designed to do so. Its motive power can be sub-divided into several motor units without loss in efficiency or increase in dead weight, thus permitting a construction which can utilize the entire weight of the locomotive upon the drivers and still keep within the recognized limits of weight upon each driving axle. Not only can the motive power in one locomotive structure be divided to suit the convenience of construction, but two or more locomotives can be coupled together and operated by a single motorman in the cab of the leading locomotive. While the steam locomotive, and especially the larger types as represented by the Mallet compound, is seriously handicapped by the difficulty and expense of stoking, the electric locomotive can be operated in groups of two or more by a single operator who has under his complete control the 2,000 or 3,000 horse power concentrated in each of the several locomotives of the group. The possibilities opened up in this direction are enormous.

"A steam locomotive is considered large when its boiler is capable of sustaining an output of from 1,500 to 2,000 indicated horse power. The electric locomotive, on the other hand, as represented by the 3,400 type now operating on the New York Central, can give a sustained output of 2,200 horse power rated with ample margin of overload in excess of this. It should be well understood also, that this 2,200 horse power output of the motors is obtained with a total weight of something less than 100 tons, of which 68 tons is upon the drivers, and this locomotive was designed for a specified duty and does not in any way represent the maximum possibilities of electric locomotive construction.

"It seems entirely practicable to construct an electric locomotive or a group of electric locomotives capable of delivering any drawbar pull permitted by the strength of the drawheads, and, moreover, this drawbar pull can be delivered at any speed desired, in this respect far exceeding the possibilities of steam locomotive construction. This means that instead of operating trains downgrade at a maximum speed of from 25 to 30 miles an hour and up grade at a maximum speed of from 6 to 12 miles per hour, it is possible with electric locomotives to operate at any speed up grade that is consistent with the alignment of the track. In other words, the same schedule speed can be maintained up grade and down grade, thus greatly facilitating the movement of trains on heavy grade sections and vastly increasing the tonnage of the tracks.

"Air brake equipments on all cars have contributed towards increasing the size of trains and safety in operation on mountain grade sections, but such operation is still handicapped by the limitations and dangers attending the holding of a heavy train on a downgrade of long extent. As the electric locomotive is a convertible piece of apparatus, changing electricity into mechanical power and vice versa with equal efficiency, it offers a means of relieving the air brake entirely or in part, and performing the functions of braking by returning electricity to the line. Elimination of airbrake shoe troubles and overheated tires, will be appreciated by steam railroad operators as soon as this valuable feature of the electric locomotive becomes well understood, and aside from the economic value of electric braking, it offers much towards the safety of carrying passenger trains over mountain grade sections."

THE PROGRESS OF THE SUBMARINE BOAT.

A REVIEW OF THE RESULTS OF THE TESTS RECENTLY CARRIED OUT BY THE UNITED STATES NAVY DEPARTMENT.

Engineering.

THE recent announcement that an order had been placed by the United States Navy Department for seven submarine boats of the "Octopus" type recalls the elaborate and searching tests made in the early summer of the present year to determine the best available type of submarine boat for the proposed extension of this arm of the Navy. Three designs were submitted to the Board: the "Octopus" of the well-known Holland type, a boat of 273 tons displacement, launched in 1906 by the Electric Boat Company of New York and embodying the latest conceptions of this design of boat, the "Lake," and a small model of the sub-surface type. The report submitted to the Navy Department shows the unanimous opinion of the Board to have been in favor of the adoption of the "Octopus" type. The results of the tests as given in the report are summarized in an editorial review in a recent number of *Engineering,* which also contains some interesting details of the progress of the submarine boat in other navies.

"The Octopus, like the great majority of submarine craft, is driven on the surface by internal-combustion engines, and the consumption is such as to insure that, with a fuel storage supply of 4,000 gallons, the radius of action will be 700 miles. Electric motors are used for propulsion when submerged. The surprisingly good speed of 10.03 knots was realised as a mean of three measured-mile runs with the conning tower of the vessel 10 feet under the surface. This is a splendid performance. On the surface the maximum speed was 11.57 knots, and the mean 11.02 knots. As regards diving, the vessel went down at an angle of 8 degrees to a depth of 26 feet within 40 seconds; she immediately returned to the surface, remained there under observation for five seconds, and dived once more. The complete evolution was carried out in about a minute and a half. This facility of disappearance is of the greatest importance from the point of view of fighting efficiency, and the result is, therefore, most interesting. Again, the time taken to disconnect the gasoline engines and to couple the electric motors was only 12 seconds, five seconds for the former, and seven seconds for the latter operation. As to manœuvring, the vessel, when awash, made a complete circle in 3 minutes 40 seconds, the diameter of the circle being about 200 yards. Running on the surface, with only one screw, she made a half circle to starboard in 1 minute 35 seconds, and a half circle to port in 2 minutes 40 seconds; in the latter instance the screw propeller was working against the rudder. Going full speed ahead when awash, the vessel was able to reverse her direction of propulsion in 52 seconds. As to endurance, the boat was required to remain 24 hours submerged at a depth of 200 feet, and it was computed that only 1-45th of the total air supply was exhausted, which suggests a long radius of action under water.

"The Lake boat also did well in remaining under water for this lengthy period, but in the other tests the Octopus proved superior. The Holland boats are fitted with an automatic device for blowing out the tanks when submerged, in order that the vessel may rise to the surface from any predetermined depth, for which the apparatus is set to come into action. The mechanism was set to be effective at 40 feet, and when this depth was reached 30 tons of water were blown out of the ship in 18 seconds, the total time taken for the test, including the immersion of the boat, being 48 seconds. Another important trial was made in connection with submarine bell signals from the 'mother' ship, and by this means it was possible for the commander of the fleet to communicate to the various vessels when submerged. It was also found that wireless telegraphy could be used on the Octopus when on the surface and awash. The masts were 30 feet high, and the antennæ, 50 feet long, consisted of four strands of wire. Under these conditions it is anticipated that the range of communication will be 40 miles.

"These facts, which were evolved by the Government tests, again prove the practicability of submarine navigation, which has frequently been demonstrated, although actual data have been withheld. The data further establish the efficiency of the Holland type of submarine boat, and 'that she is equal to the best boat now owned by the United States or under contract.' The results suggest that a larger boat than the vessel referred to, which is of 273 tons displacement, would be a superior weapon. As to the Lake type of boat, the Commission report:—'1. That the type of submarine boat as represented by the Lake is, in the opinion of the Board, inferior to the type as represented by the Octopus. 2. The closed superstructure of the Lake, with the large flat deck which is fitted to carry water ballast, and to contain fuel tanks and air flasks, which is an essential feature of the Lake boat presented to us for trial, is inferior to the arrangement on board the Octopus for the same purposes, and also, in the opinion of the Board, is detrimental to the proper control of the boat. 3. The hydroplanes, also an essential feature of the Lake boat presented to us for trial, were incapable of submerging the boat on an even keel. They are, therefore, regarded as an objectionable incumbrance.'

"As regards the design known as the sub-surface type of boat, the Commission very properly reported that it could not be compared with submarine boats, being of an entirely different type. In this class of boat the machinery, magazines and habitable quarters are enclosed in a submerged hull, from which there is communication to a surface hull through conning-towers or armoured-tubes, the two hulls being joined, pretty much like the booms of a girder, by web plating or cellular structure. The surface hull is used only for the accommodation of the guns and the gear for controlling propulsion and navigation. In other words what would be considered the upper deck of an ordinary ship is separated from the sub-structure with the exception of tubes for communication, so that in action damage to the upper part would not affect or endanger the lower hull with the machinery and magazines. The system is ingenious, but is, as the Navy Board point out, analogous to the torpedo-boat or torpedo-boat destroyer. The sub-surface boat does not afford that advantage of invisibility which is the great desideratum met by submarine craft, and therefore its potentiality for damage is not so great. There was only submitted to the Board a quarter-size model, and consequently it was impossible to make a satisfactory comparison, even with the performance of torpedo-boats and torpedo-boat destroyers. The Commission, however, point out that so far as their observation went there was no reason to doubt that the guarantee made as to speed, etc., would not be carried out. The sub-surface boat is less vulnerable than the torpedo-boat, requires fewer men, and has a larger steaming radius, but she has less speed and greater draught. The President of the Commission, Captain Adolph Marix, took exception to the general pronouncement that the tests of the sub-surface model 'did not develop that boats of this type, built of a size suitable to render their qualities available, are equal to the best torpedo-boats now owned by the Government.' In the opinion of the Captain, the smallest size of sub-surface boat fitted with a regular torpedo-tube, and built to

give a speed of 15 knots, would be a weapon of great value, additional to any now owned by the Government, and that this value could be enhanced by the rapidity with which they could be constructed, and the ease with which they could be transported.

"The Board were not called upon to pronounce as to the strategical or tactical advantages of the submarine boat. This was scarcely necessary in view of the general consensus of opinion in all Admiralties in favor of the type, and the large number now being built by the various Powers. In the recent Dilke return it was shown that there are already in existence 117 vessels of a submarine type, and that there are building 86; while the programmes of many Powers, in addition to the United States, anticipate very considerable additions. We, in Britain, have 37 completed and 11 on order, and the size has steadily advanced from 122 tons displacement to 400 tons, the power of the machinery having increased in the same period from 160 to 800 horse-power on the surface, while the power of electric motors for propulsion under the surface has increased from 70 brake horse-power to well over 200 brake horse-power. These vessels, as we have already indicated, have been evolved from the Holland type, and long experimental research has been carried out, with the result that no dubiety exists as to the efficiency of the type, while at the same time there is achieved that homogenity which is so important in the training of the *personnel* to secure a sufficient supply of men during action, and to maintain the highest efficiency, which is dependent on thorough experience in these vessels more than in any other. France has forty boats completed, and fifty-nine are in course of construction. Here there is great diversity of opinion as to the best type of vessel, and the boats vary in size from 21 to 560 tons displacement. The tendency, however, is all in favour of vessels of large size, most of the boats under construction being of about 400 tons displacement. Russia has twenty vessels completed, and has on order eight. Germany has moved more slowly in this matter, but she has completed her first vessel and has two building; the estimates for the next three years provide about

£350,000 for submarine construction. Italy has four vessels building, and two in course of construction. Japan has seven built, while the United States have in commission eight vessels, and four in course of completion, all of the Holland type." As noted above, seven additional boats of this type have been ordered for the United States Navy since this review of the trials was written.

THE RELIABILITY OF MOTOR CARS.

A DISCUSSION OF THE TROUBLES DEVELOPED IN THE MOTOR CARS COMPETING IN THE RECENT RELIABILITY TRIALS OF THE SCOTTISH AUTOMOBILE CLUB.

Engineering..

A FACTOR which has been largely instrumental in the rapid development of the technology of automobile construction is the attention which has been paid to the gathering of accurate data on the performance of cars during the numerous international competitions held by several of the larger automobile clubs. Of these contests, those held annually by the Scottish Automobile Club have probably been of more value to the motor-car industry than any others, in that they are not purely speed competitions but are designed to test all the qualities which are essential to the perfect touring car. The trials extend over five days, and are held over a course of 750 miles of highland and mountain road. The worst of the many bad hills encountered are utilised for special tests of starting, stopping and climbing power. Since every detail of the behavior of every car is carefully noted by official observers, it is obvious that the conclusions which may be drawn from the final report are of the utmost importance as indicating the details of automobile construction which still require attention from an engineering point of view.

In the Scottish reliability trials of the present year there were 96 starters. Of these, Great Britain contributed 57, France 24, the United States seven, Italy two, Germany, Belgium, Austria and Holland one each, and there were in addition two hybrid cars, partly French and partly English. The following discussion of the troubles which developed in the competing cars during the trials is taken from a review in *Engineering* of the recently published official report.

"It is beneficial to study the performances of the cars from an engineering point of view. The most valuable lessons are to be learnt from the various troubles that were experienced on the road, all of which are noted faithfully and impartially in the official report. We will first take the steam-cars, of which there were but two, both of a well-known American make. Many engineers have a preference, or at least a sentimental liking, for the steam-car, and its silence is certainly attractive to other purchasers; but we fear steam vehicles will have to justify their existence more conclusively than on the trials in question before they ever become serious rivals to the petrol type. It may not be fair to judge their merits on the performances of two only, but on official trials every car must stand or fall by its official performance. The first of the steamers, rated at 20 horse-power, withdrew on the third day, having gone through a sea of troubles, mostly connected with burner, valves, or steam-pipes. It had to fill up with water five times, procure more petrol, and—curiously for a steamer—it failed once on a hill. The second steamer, rated at 30 horse-power, completed the trials, but lost more marks than any one of the petrol-cars, excepting two. In this case it seems to have been the vaporiser and feed-pump which gave most trouble, although on one day there were apparently five engine stops pure and simple. To make matters fair for steam-cars competing, every vehicle was allowed two minutes in which to restart after a stop. On the morning starts, after the first day, marks were deducted for more than this delay in starting, and the steamers were always penalised, their average starting time being over six minutes.

"Turning to the performances of the petrol-cars, we find thirteen of these, or about 14 per cent., had to withdraw from the contest. This may seem at first a large proportion of the total, but the uselessness of proceeding with anything of the nature of a contest when hopelessly out of the running, and the exceedingly severe nature of the trial, must be taken into account. The conditions were far more arduous than a private tourist need encounter, and in spite of the fact that nearly 20 per cent. of the starters made non-stop runs, one must not be too hard upon those which did not. Every one of the ten small cars in Class I., costing, for chassis and tyres complete, not more than 200*l.*, finished the course, which is highly creditable. Of the vehicles, which were withdrawn, three cars gave up owing to stripped gear, and another to vague 'gear troubles.' Kindred causes, accounting for two more withdrawals, were broken ball-race and differential and broken driving-pinion. A seized gear-shaft and a broken brake-drum shaft put another two out of action. Of the remainder, one got broken spokes and a bent axle when taking a sharp turn, one withdrew on account of a broken spring, one owing to a broken gear-striking fork, one because the clutch-leather burned out, and the last because 'pins dropped out of a driving-shaft.' It will be seen that gear trouble of one sort or another caused the bulk of the serious mishaps in the contest; and makers might well take this lesson to heart, for driving skill of the highest order was presumably employed; and if gears are stripped by good drivers, what can be expected when novices are at the wheel? The burnt-out clutch does not necessarily prove anything, except that the clutch slipped; and although, of course, a slipping clutch is one of the minor points which may cause so much annoyance to car-drivers, and should be more unusual than it is, one is bound to sympathise with the makers on this occasion. The failure of the wheel and axle at a bend in the road is vastly more serious than any of the other troubles which led to withdrawal, and, again presuming reasonably good driving, is inexcusable. The car to which this accident happened was of Continental manufacture.

"A careful analysis of the minor troubles which were met with, with a view to their elimination, should be made by every firm engaged in the manufacture of cars. There are in the official report notes of every 're-

grettable incident,' no matter how trivial, which occurred in running something like 67,000 car-miles. Most information would be gained by comparing the frequency and gravity of any particular trouble with the type of mechanism involved; but to do this requires more details of the cars than are given in the report. However, in the majority of cases it should not be beyond the power of any manufacturer reasonably well acquainted with the work of his competitors. Neglecting the steam-cars, the tribulations of which have already been referred to, we find broadly three great fields in which attention to details seems urgently wanted if the lesser troubles of the motorist are to be eliminated. The carburettor and petrol supply together are responsible for the bulk of the involuntary road-stops; ignition troubles of all sorts run a fairly close second; and the water-circulation system comes third. These three sources of annoyance overshadow all others. The petrol-engine, as a piece of mechanism, seems almost to have reached perfection; for although the engine-stops are fairly numerous, they are mostly due apparently to changing gear on hills, and mechanical defects in this important part of a car seem to have been practically non-existent.

"Defects of, or troubles in connection with, the carburettor and petrol supply were responsible for something like fifty road-stops, wasting an aggregate of seven hours, although more than half of the failures took less than five minutes apiece to rectify. Nearly twenty were due solely to choked carburettors or petrol-pipes. This surely is a state of affairs that is capable of remedy. A useful comparison might be made of the chokes occurring with pressure-fed as against gravity-fed carburettors, and the relative freedom of multiple-jet as against single-jet devices. Probably the most prolific source of chokes is dust, either originally in the petrol or finding admission when the tank is refilled. Until carburettors and pipes are designed so that they will not choke, the motorist can do little more than exercise care in the exclusion of dust; but as we must presume this to have been done on the trials, the number of chokes demands the attention of manufacturers.

"Ignition troubles caused about thirty

road-stops, aggregating about five hours; though here, again, the length of the average stop only amounted to a few minutes. In eleven cases the sparking plugs alone were apparently to blame. The reliability of these simple little articles is yet far from perfect, although the price charged by good makers is more than enough to cover the best of workmanship and material. Sparking-plugs are much better than they formerly were, especially as regards sooting-up, but mechanically they are often very bad. The temperature at which they work, and the voltage they have to stand, both are adverse to reliability; but as a comparatively cheap plug will show often an almost indefinite life, the conditions are not insurmountable.

"Troubles with the cooling-water system, leakages, etc., caused more than a dozen stops, aggregating nearly four hours, without counting the numerous cases where extra water had to be obtained for no apparent cause. Leaks seem to have been a prolific source of trouble, and the rigidity of radiator fastenings, pipe connections, etc., no doubt accounts for the occurrence of many leakages. The radiator in every car ought to be cushioned from the frame in such a way that it is protected, as far as possible, from vibrations and shocks, and the pipes from it ought also to be flexibly attached. From the report it is impossible to gather which type of radiator or circulating system it was that occasioned most trouble.

"It is impossible to say from the particulars given of the equipment of the contesting cars which is the most popular form of ignition. Most of the cars were fitted with two alternative forms: the ordinary battery and coil, and some type of magneto. The cheaper vehicles had accumulator and coil only, and many others had only either high-tension magneto, or low-tension magneto. When alternative ignitions were provided, the high-tension magneto seemed the most popular, but the records are frequently indefinite as to the type used. One might deduce conclusions by comparing the frequency of ignition troubles with the type of ignition employed, but whether they would be accepted by the interested parties is another matter.

"The tyre troubles we will not enter into, as so much of pure chance is involved in the question. It is sufficient to say that, on the whole, the tyres behaved as well in their way as the cars, and no car was penalised for tyre delays, as in no case did any competitor lose more than an hour on this account during the whole of the trial."

THE PRESENT POSITION OF THE LARGE GAS ENGINE.

A DISCUSSION OF THE CONSTRUCTIVE DIFFICULTIES WHICH HAVE PREVENTED ITS RAPID
ADVANCE IN ENGLAND.

Dugald Clerk—British Association for the Advancement of Science.

WHILE the gas-engine industry in Great Britain is in a condition of sound commercial prosperity, the production is limited, for the most part, to motors of small powers. With regard to the large gas engine, British engineers have assumed a conservative attitude, and while considerable experience has been accumulated and British practice in large gas-engine construction has developed steadily, there still remains a number of constructive difficulties which must be overcome before the gas engine of high power can compete with the reciprocating steam engine. The following abstract outlines these difficulties as described by Mr. Dugald Clerk in a paper read before the recent meeting of the British Association for the Advancement of Science.

"Engines of small and moderate powers are built in large quantities; their difficulties have been thoroughly overcome and they have attained to an almost fixed type. The larger part of the British gas-engine industry is occupied with such motors, generally under 100 horse power per cylinder. The turnover in Britain at present of such engines is at the rate of some 300 engines per week. It is generally recognized that these engines are as reliable as the best steam-engines of similar dimensions and much more economical in fuel consumption. The smaller engines mostly use coal gas, and the larger producer gas evolved by means of modern suction producers using anthracite for fuel.

"Experience in the construction and design of the large gas-engine is accumulating. They are better understood in Britain than they were even three years ago. It is a remarkable fact, however, that engines which attained a reputation for success upon the Continent were not at first successful here. This is shown by the fact that the Koerting, Oechelhauser, and Cockerill engines had all to be modified in their construction by the British engineers who undertook their manufacture here. This is also true of the Diesel oil-engine. Alterations have been made in England to fit it for the conditions of practice here. All these engines have been much improved in the last few years, and they are now, no doubt, better able to compete with the steam-engine with regard to reliability and freedom from breakdown.

"Large gas-engines of English design have also been built in greater numbers, principally by the Premier Company, Messrs. Crossley Bros., Limited, and the National Gas-Engine Company. English designers have always felt the desirability of keeping down cylinder dimensions as much as possible, and in this Continental designers have recently shown a strong tendency to follow them. This trend is due to a more general recognition of two facts: practical difficulties with large-diameter cylinders due to unequal expansions, resulting in cracking, and a better appreciation of the fact that increase in cylinder and other dimensions requires an increased expenditure in metal and workmanship in greater proportion than increase of power obtained. The large gas-engine really presents two distinct problems. The first is to build engines of large power which will continue to run effectively and economically for long periods without breaking down, and the second is to build such engines at costs sufficiently moderate to enable the engines to effectively compete with the large steam-engines in the matter of first cost. British engineers recognised for some time that the first part of the problem has been solved to some extent on the Continent; but many of them have felt that this solution has involved weights of material and costs of construction which are almost prohib-

itive, considering the moderate powers obtained. In fact, English engineers consider the large gas-engine as it at present exists both too heavy and too costly for its power. Personally, I do not believe that sound and continued commercial success can be looked for with really large gas-engines until some better solution be found for their present constructive difficulties. Apart from the questions of the engines themselves there are other difficulties which prevent the equal competition of gas-engines with steam-engines for powers, say, greater than 400 horse power or 500 horse power. Coal gas is too expensive a fuel for large engines. Producer gas, evolved by the suction producer from anthracite, air, and steam, effectively meets the wants of medium-sized engines up to 200 horse power, but the cost of anthracite handicaps engines of larger size, and equal competition will not be possible until better bituminous fuel producers are designed than those which at present exist. The work on the Continent has not aided the solution of the bituminous fuel producer problem. Practically all the large Continental gas-engines are operated with blast-furnace gas. Some success has been attained in Britain as the result of strenuous and praiseworthy efforts by Dr. Mond, Messrs. Crossley, and others, but it cannot yet be said that an entirely satisfactory bituminous producer has appeared. In my view no bituminous fuel producer can be considered really satisfactory until it attains simplicity, lightness, and the fewness of parts of the anthracite suction producer which now forms so large a British industry. Returning, however, to the engine difficulties, the large gas-engine may be considered as combining the difficulties of hydraulic engineering work at considerable pressures with those proper to a boiler furnace or flue. The possible pressures to be resisted by such engines cannot be taken as less than 400 pounds to 500 pounds per square inch, and a heat flow through the cylinder and combustion chamber walls has to be provided for greater than that of most boiler furnaces. It is obvious that here we have contradictory conditions involved, which become rapidly onerous with increase of dimensions. Thick castings are required to stand the high pressures, but to allow free heat flow from the flame within the cylinder to the water in the water-jacket calls for thin castings. Engines of small dimensions do not involve any serious conflict, but where metal is required of about 3 inches thickness to resist internal pressures the temperature difference between the flame and water side of the metal becomes serious, and great stresses are set up which ultimately lead to the cracking of the castings. Great attention has been paid to this phenomenon of cracking, and in existing large gas-engines the difficulty has been partly met by skilful design and special quality of metal used. Although much ingenuity and skill has been spent in this direction, yet it is found that a dimension limit is very soon reached. Cylinders, for example, of 51 inches diameter have been found to be too large. Nothing but the highest skill in designing and the greatest care in the choice of material and workmanship enables such cylinders to withstand for any length of time the severe treatment to which they are exposed."

A NEW METHOD OF CUTTING METALS.

A DESCRIPTION OF A SYSTEM FOR CUTTING HEATED METALS BY MEANS OF A JET OF OXYGEN, RECENTLY DEVELOPED IN FRANCE.

Léon Guillet—Le Génie Civil.

ON account of its portability and convenience of handling, oxygen has recommended itself for several industrial applications connected with the metal trades. Welding is cheaply and easily performed by means of the oxy-hydrogen or the oxy-acetylene blowpipe, and in many other processes the use of oxygen has come to be a recognized practice, particularly in cases where an easily transported apparatus is of advantage. One of the most important of these applications is in the cutting of metals, for which oxygen is successfully used in several works in France. The following interesting description of the process is taken from a paper by Dr. L. Guillet in a recent number of *Le Génie Civil.*

The system as at present developed is comparatively recent. Since 1901 a jet of oxygen has been employed at the Say Refineries, not for actual cutting but for breaking down old pipes and I-beams. The metal was first heated to the temperature of fusion by means of a blowpipe and then, discontinuing the application of the heating gas, a stream of pure oxygen was later applied to the heated metal. The operation was in two parts, first heating of the metal and later burning it away. It was possible in this way to break up the metal into different parts, but the operation produced a very large and heavy cut with many blisters and it was found impossible to secure a clean separation by this method. It has been successfully applied, however, to breaking down metallic concretions in blast furnaces.

The method as at present developed is a great improvement on the old. The apparatus consists of two blowpipes which operate simultaneously; the first, an ordinary oxy-hydrogen or oxy-acetylene blowpipe, heats the metal to the desired temperature; the other, attached to the first and following it at a distance of a few millimetres, projects upon the heated metal a jet of oxygen under pressure. By this means the metal which is heated by the first blowpipe, burns under the influence of the oxygen; there is formed an oxide more fusible than the metal itself, which is driven forward by the strength of the oxygen jet. The resulting cut is extremely clean, with no appearance of blistering.

In view of the numerous applications which may be made of the method a number of different devices have been developed for various kinds of work. Cutting along a straight line, as for steel sheets, armor plate, etc., is accomplished by the use of a traveling carriage, carrying the blowpipe, which is driven by an endless screw. In this apparatus also the blowpipe is mounted on a swinging arm which permits it to trace out arcs of circles. Several of these devices are used for cutting armor plate at the French arsenals and navy yards.

A rolling device, consisting of a blowpipe mounted on a crosshead carrying an adjustable roller at each end, may be used for making either straight or beveled cuts. This apparatus is also provided with a swinging arm for the cutting of circles or spirals. The device is largely used by the Chemin de fer du Nord for cutting T-rails.

For cutting pipes, the blowpipe is mounted on an annular ring which is clamped to the pipe and about which the blowpipe revolves. By means of this device pipes up to 300 millimetres in diameter may be cut in place. A universal portable apparatus is so arranged as to admit of cutting out any desired section, the blowpipe being guided by cams of special shapes. Other devices may be applied to cutting circular openings of 30 to 350 millimetres diameter in plates of any thickness, to cutting shafts up to 130 millimetres diameter, and lastly for cutting openings of any radius in steel pipes for the purpose of making elbows, the bevel being cut in the end of the other pipe by the same machine. For the preparation of two pipes of 150 millimetres diameter in this way the time taken was four minutes, a great improvement on the 30 to 40 minutes required by other methods.

M. Guillet gives some interesting examples of the rapidity of this method of cutting metals, in which lies its chief value. An armor plate, 160 millimetres thick, was cut through for a length of one metre in ten minutes. A similar length in a plate 20 to 30 millimetres thick can be cut in less than five minutes at a cost of not more than one and one-half francs. To cut a man-hole 300 by 400 millimetres in a plate 20 to 30 millimetres thick takes from four to five minutes. A pipe fitting, 150 by 150 millimetres, in a pipe 5 millimetres thick takes three to four minutes and costs 12 to 15 centimes, while the same operation could not be performed by hand in less than 35 to 40 minutes.

In one case a steel stairway was cut through in a quarter of an hour. A workman cut through 130 T-irons, 200 millimetres high, imbedded in cement, in three hours. The system has been largely used in the French navy yards for the breaking up of ships and has been found very satisfactory in many applications. It is particularly useful in deriveting plates: the head of a rivet can be burned off in less than twelve seconds without injuring the plate itself.

The maximum thickness which has been cut by this method up to the present is 210 millimetres in armor plates, but a thickness of 300 millimetres has been attained in round shafts. M. Guillet has made a number of tests to determine whether the cutting by this method has any effect on the physical properties of the metals. Under the microscope the steel appeared quite unchanged and other tests by shock, etc., showed its strength to be absolutely unimpaired. On this account and considering the cheapness and rapidity of the process, he looks for a wide and growing use of the system in the near future.

A CO-OPERATIVE SYSTEM OF TECHNICAL EDUCATION.

DETAILS OF A CO-OPERATIVE COURSE OF ENGINEERING INSTRUCTION AT THE UNIVERSITY OF CINCINNATI.

Herman Schneider—Society for the Promotion of Engineering Education.

THE following extracts from a paper read before the annual meeting of the Society for the Promotion of Engineering Education describe an interesting experiment in technical education which is just entering upon its second year at the University of Cincinnati. Some six years ago the writer, Mr. Herman Schneider, began an investigation to discover, if possible, what requirements the graduates of an engineering school should fulfil and the best means of training students to obtain the desired results. The co-operative course described below is the result of his researches. The outcome of the experiment, which seems to offer a solution of many of the difficulties of both students and the employers of graduate engineers, will be watched with great interest alike by the governing bodies of engineering schools and by industrial managers.

"This course is so planned that the students taking it work alternate weeks in the engineering college of the university and at the manufacturing shops of the city. Each class is divided into two sections alternating with each other, so that when one class is at the university the other is at the shops. In this way the shops are always full-manned, and thus the manufacturers suffer no loss and practically no inconvenience by the system. There are two facts on which it is desired to place especial emphasis so that there may be no misunderstanding about this work. First, the entrance requirements for this course are precisely the same as for the regular four-year course. Secondly, the university instruction under the co-operative plan is just as complete, thorough, broad, and cultural, as the four-year course. As a matter of fact, it is broader and more cultural. Let there be no misunderstanding about this. The course is not a short-cut to a salary.

"The length of the course is six years. During this time the students work alternate weeks in the shops of the city during the scholastic year, and in the summer full time. They are given one week vacation at Christmas and two or three weeks during the summer. The practical work at the shops is as carefully planned as the theoretical work at the university, and in all cases the students follow, as nearly as possible, the path of the machines manufactured, from the raw material to the finished product sold. For instance, at the Bullock Electric Company the students spend the first year in the foundry, the next year and a half in the machine shop, the next two years in the graduate-apprentice course, covering the commutator, controller, winding, erecting, and testing departments, and the subsequent time in the drafting-room and sales office. A contract is signed in triplicate by the student, the university, and the firm. This contract has a blank space to be filled out with the shop work the student is to receive during his six-year course. In all cases the dean of the engineering college and the professor of electrical, chemical, or mechanical engineering, as the case may be, confer with the manufacturers in planning this course of shop work, so that the young men get a logically and carefully planned shop and business training.

"The students are paid for their services on a scale of wages beginning at ten cents an hour and increasing at the rate of one cent an hour about every six months. The rate at which the first class started was lower than this, but on account of the quality of work which the young men did in the shops, the manufacturers made a voluntary increase which is equivalent to about four hundred dollars for the whole course. The student's total earnings in the six years will amount to about two thousand dollars.

"Young men desiring to enter this course are required to go to the shops in June or July preceding their entrance to the university. We believe that certain men are mentally, physically, and temperamentally adapted to engineering, and that the process of elimination which comes through this summer work weeds out the weaker ones and leaves us a residue which can be depended upon for results. . . . . . .

"A comparison of the work of the four-year freshmen with that of the six-year freshmen during the past year is worth a moment's consideration. The six-year co-operative students, although working but one-half the time of the regular students, have done three-quarters of the work of these regular students, including all the mathematics and sciences of the freshman year; their average grades are 25 per cent. higher than those of the four-year freshmen. As a matter of fact, they have taken all the university work excepting three hours of English and three periods of shop work, and, of course, they have received much more shop work at the city plants than they would have covered at the university.

"This course applies to the departments of electrical, chemical, and mechanical engineering. It has not been found feasible to establish a co-operative course in civil engineering because the local conditions will not permit. The aim of the course is not to make a so-called pure engineer; it is frankly intended to make an engineer for commercial production. For the investigation spoken of in the forepart of this paper disclosed the fact that a great majority of the engineering graduates are employed in commercial production, whereas the present college courses do not contemplate anything other than that the college graduate will become a so-called pure engineer.

"I regret that the time will not permit a

further exposition of this course, for the experiences which we have had in one year's operation and in gathering our second class have taught us a great many things which I believe would be of interest to you. I should like, for instance, to point out in detail some of the following features: The advantage to the student of the extra two years of time for the assimilation of his theory; the effect of his shop work on his theoretical work; the solution of the problem of proper exercise for the student; the knowledge he obtains of the labor problem, and of time as being the very essence of commercial production; the fact that this course resolves itself down to a training in commercial production with a university training in the underlying science—for you will note that of the six years, four are spent in practical shop work, and that in the two years' actual time spent at the university the coöperative student gets more than the regular student does in four years."

"It is believed this system of education will furnish to the manufacturer a man skilled both in theory and in practice, and free from the defects concerning which so much complaint is made. It is not held, 'of course, that this method of training will supply full fledged engineers, aged 23 years, or thereabouts; but it is believed that it will provide a better preparation, a stronger foundation, for the successful practice of engineering. The qualifications which the engineering graduate should possess will be more nearly attained. He will be just as thoroughly grounded in the fundamental

principles of science as he is under present conditions, but he will have greater facility in applying them to practical problems. He will be much more highly specialized, but not at the sacrifice of fundamentals. A knowledge of the achievements in other fields of engineering will result from his constant association with the best practice in electrical, mechanical, structural, and chemical engineering as exemplified in the construction of the co-operating works, in their methods of power generation and transmission, and in their processes of manufacture—his attention having been called to these details in the class room, and his observation of them having been checked by searching questions thereon. He will become familiar with business methods by constant contact with business conditions, supplemented by class room instruction and practical talks on business law. He will obtain a knowledge of men by working intimately with all sorts and conditions of men in his gradual rise through the various departments of the co-operating plants. The cultural part of his education will be planned to make him a man of good address and broad sympathies.

"And, finally, a combination of these conditions will teach him to do his best naturally and as a matter of course. It will start him on his life work with a symmetrical and uniform fundamental development which will continue evenly and make him a skilled engineer, a safe business man, and a broadly tolerant and intelligent citizen."

THE SITUATION ON THE RAND.

A DISCUSSION OF PRESENT ECONOMIC CONDITIONS ON THE RAND AND SUGGESTIONS FOR THEIR IMPROVEMENT.

John Yates—Chemical, Metallurgical and Mining Society of South Africa.

THE gold-mining industry on the Rand is at present in a more unsettled and depressed condition than at any other period in its history. Reports of the imminent exhaustion of the higher-grade ore have created an acute financial depression and the consequent shrinkage in the value of the shares of the mining companies is causing great public alarm in both South Africa and England.

The situation is further complicated by the ever-present labor problem, in which the repatriation of the Chinese and an extensive strike are the most prominent factors, and by the unknown character of the new government which has recently come into power. All the presidential addresses of the South African technical societies which are just coming to hand emphasize the necessity for prompt and thorough revision of the technical and economic conditions of the gold-mining industry before confidence and prosperity can be restored. The following extracts from the presidential address of Mr. John Yates before the Chemical, Metallurgical and Mining Society of South Africa give an interesting review of the present situation and outline the means by which it can be improved.

"And now you may be inclined to put to me that all absorbing question of the moment, how can the present acute depression, a depression which is wrecking the fortunes of so many, be removed and the future welfare of the country best assured? Remedies have been put forward from time to time, and they are again obtaining publicity before the Industrial Commissions at present sitting, and as it is largely reiteration, permit me to put them briefly: Reduce rents and bring down the cost of living; pull down freight charges, for cheap railway charges contribute largely to the development and prosperity of a country; lay out our new mines on an adequate scale and not on the kindergarten lines of the past; have large joint mills of even 1,000 stamps with centralised workshops, power station, and staffs to match; put down the minimum number of shafts; look upon labour saving devices with suspicion, for their promises are often better than their performances; afford our miners the opportunity of learning both the principles and practice of their profession; engage .the Kafir on two years' contracts, and pay him a less wage for the work he does; give the miner and the Kafir each his proper sphere of work, the former using his intellectual gifts as a superintendent controlling the physical power of the coloured worker:—I submit that to have white men expending their energy on physical work which the much cheaper coloured labour is quite capable of performing is, on the face of it, absurd from a business point of view. Pay our miners a reasonable wage and let the scheme of pay be such as to encourage the men to develop and exhibit skill. See that underground supervision is made more effective, that small stopes are run by hand labour and large ones by ordinary sized machines; see that the healthiness of our working places receives due attention and insist on the miner using water jets, and plenty of water generally, to keep down dust. Workmen permanently resident in the country are desirable and, as suitable material is available, all that is necessary is its education and training. The married man has always been our best worker, so let us take care that our wage rate will suffice for his living here in comfort with his family and so induce him, if not a son of the country, to at least make it the land of his adoption. On the subject of our unskilled labour supply, the cause of such wordy strife here and such perturbation in Great Britain, I can only say that in view of the scarcity of natives practically ever since these goldfields were opened, the apparent impossibility of securing a sufficiency under the conditions hitherto obtaining, the finding of the many Commissions that have investigated this subject, the inability of the industry to forge ahead without an adequate complement, in view of these and other reasons I consider the Chinese an unfortunate necessity, but a necessity all the same, their importation has been an expensive experiment for the mines, and it is sincerely to be hoped that no unreasoning bias against this alien labour will be allowed to cripple the mines and the revenue of the country. Cool reasoning surely counsels that we will do well to retain the Chinese until the expected substitute has materialised, and if it so happens that this great importation experiment results in converting unbelievers to the actuality of our labour scarcity and gives rise to measures which will assure our mines an adequate supply of natives on suitable terms then I think it will be admitted that the silk-clad, luxury-loving, bicycle-riding, heathen Chinese has served a good purpose, and our mining houses will probably look upon the money as not having been spent in vain.

"I have here a classification of the mines of the Witwatersrand for September (1906), based upon their yield per ton milled. Out of the 65 mines then milling, 24 yielded not more than 30s., whilst the remaining 41 gave over 30s. Out of the total 65 mines 21 yielded between 25s. and 30s. and 20 gave between 30s. and 35s. Now these figures call for most earnest consideration. We are, as sensible business men,

at the present day working the rich, selected areas which offer us a profit, such areas
as do not promise a due return for our capital and enterprise we naturally endeavour
to avoid, though our efforts in this direct-
tion are not always successful. It may be
assumed in a general way that most of the
ground which offers an adequate profit under present economic conditions is being
worked, and as this comprises only a small
number of mines it follows that we have
an immense tonnage of low grade banket
on or near the outcrops as well as our
great and almost intact deeper horizons of
the reef, and all this reef, owing to the
high cost of working, the average low
grade, or the large capital which has to be
risked, as yet remains largely unproved and
unworked. Here I would say that I do not
think that we have anything like conclu-
sive proof of the falling away of value in
depth so far as this field is concerned; in
judging this matter allowance must be
made for the fact that selection of stopes
has been taking place, and still continues,
in our existing properties. Are these low
grade areas and deep zones to remain un-
exploited, is the country content, for instance, that only the present mines should
work and working gradually fall out exhausted leaving the country to face serious
financial straits only a few years hence?
Surely not, and yet if this is not to be the
case something of the nature of an economic revolution will have to be engineered
to effect a lowering of costs, the augmenting of profits and the restoring of the confidence of the investing public. I have
shown you how nearly 40 per cent. of our
present mines, our selected rich areas, are
not yielding more than 30s. per ton milled,
and it is of interest to note that it has been
recently estimated that 30s. per ton milled
is probably the average value of all the unworked reef on the Rand. It is in essentially low grade rock that this country is
rich, and it is the many golden miles of
this rock that the country has to look forward to for sustenance in the years of the
future. Our engineers have recently expressed the opinion that the best interests
of the gold industry and the country call
for a reduction of costs to about 16s. per
ton milled. Now let us make every allowance for ulterior motives on their part, for

even our mining engineers and financiers
are not without them, let us examine their
statement as critically as possible, and
there will, I think, nevertheless remain a
large majority of us who will agree with
them that herein lies our present and future salvation, the return of prosperity, the
solution of our white labour troubles, and
the solid foundation upon which it will be
possible to build the future of the country.
Only by expansion is it possible, only by economic reform is expansion attainable, then,
I say, let us have such reforms. And be it
remembered that the working of very low
grade ore will confer much greater benefit
upon the country than upon the shareholders, the reward of the latter will be small
indeed, but so long as profit is forthcoming
will our gold beds be worked. Do not let
our attention be distracted from the broader issues by such a comparatively small
consideration, for instance, as labour ratios,
this after all affects but a few of the many
people in the land; do not let us, for the
sake of the minority, sacrifice the majority
by refusing to avail ourselves to the full
of the country's rich asset of native labour,
but let us earnestly and honestly assist in
erecting an economic framework in the
country which will stand as a fitting support for all future time. Our farming
should be developed with all rapidity, our
base metal and other industries should be
encouraged, our present population must
be provided with the wherewithal to live
and settle in the country, and to do all this
calls for the reforms mentioned. The country is called upon to select one of two paths
—down the one there is a distant vista of
a numerous and prosperous people with
farming their chief vocation, substantially
supported by the mining of gold, coal and
base metals: down the other is seen a
country undeveloped and forsaken of all
but a few farmers, farmers without a market and content merely to live, a country
of deserted towns and bankrupt in money
and prestige, a country from which its very
sons have fled to where progress holds
sway. These are no idle visions, but they
are, I believe, faithful outline pictures of
the logical outcome of the adoption of the
two policies open to the country; is it possible that we will ever see the latter one?
Let us hope not."

The following pages form a descriptive index to the important articles of permanent value published currently in about two hundred of the leading engineering journals of the world—in English, French, German, Dutch, Italian, and Spanish, together with the published transactions of important engineering societies in the principal countries. It will be observed that each index note gives the following essential information about every publication:

(1) The title of each article,   (4) Its length in words,
(3) A descriptive abstract,   (5) Where published,
(2) The name of its author,   (6) When published,

(7) *We supply the articles themselves, if desired.*

The Index is conveniently classified into the larger divisions of engineering science, to the end that the busy engineer, superintendent or works manager may quickly turn to what concerns himself and his special branches of work. By this means it is possible within a few minutes' time each month to learn promptly of every important article, published anywhere in the world, upon the subjects claiming one's special interest.

The full text of every article referred to in the Index, together with all illustrations, can usually be supplied by us. See the "Explanatory Note" at the end, where also the full title of the principal journals indexed are given.

DIVISIONS OF THE ENGINEERING INDEX.

CIVIL ENGINEERING

BRIDGES.

Concrete Arch.
The Walnut Lane Bridge, Fairmount Park, Philadelphia. An illustrated description of the methods employed in constructing a concrete arch of 233 feet span. 4000 w. Eng Rec—Aug. 31, 1907. No. 86687.

Germany.
Some Notable German Bridges. F. C. Kuntz. Illustrated descriptions of beautiful bridges at Harburg, and at Mainz

and other points on the Rhine. 1700 w. Sci Am—Aug. 31, 1907. No. 86677.

Highway.
Methods and Cost of Constructing Concrete Highway Bridges by Day Labor in Green County, Iowa. 1700 w. Engng-Con—Sept. 4, 1907. No. 86830.

Quebec.
The Fall of the Quebec Cantilever Bridge. An illustrated article giving the facts about the structure and its collapse, so far as at present known, with inter-

We supply copies of these articles. See page 397.

esting editorial. 10000 w. Eng News—Sept. 5, 1907. No. 86761.

The Quebec Bridge Collapse. An illustrated article, with report by H. M. Mackay on the collapse of the steel superstructure. 1500 w. Ir Age—Sept. 5, 1907. No. 86746.

The Quebec Bridge Disaster. An illustrated account of the collapse of this great cantilever bridge during construction, with short discussion of the possible cause. 3000 w. Engng—Sept. 6, 1907. No. 87076 A.

The Quebec Bridge. A brief illustrated description of this great cantilever structure, with editorial discussion of the collapse during construction. 4500 w. Engr, Lond—Sept. 6, 1907. No. 87082 A.

The Collapse of the Quebec Bridge. A statement of the facts thus far brought to light concerning this accident and the local conditions at the time. 5000 w. Eng Rec—Sept. 7, 1907. No. 86792.

The Fall of the Quebec Bridge. Gives drawings and essential data, illustration and editorial in regard to this failure. 6500 w. Eng News—Sept. 12, 1907. No. 87003.

The Quebec Bridge. Frank W. Skinner. An illustrated account of this important engineering work. Written previous to the disaster. 5700 w. Engng—Sept. 13, 1907. Serial. 1st part. No. 87210 A.

The Fall of the Quebec Bridge. Illustrations with brief notes. 500 w. Engng—Sept. 13, 1907. No. 87214 A.

The Quebec Bridge Disaster. An illustrated account of the fall of the cantilever bridge across the St. Lawrence, with editorial. 3500 w. Sci Am—Sept. 14, 1907. No. 86895.

The Fall of the Quebec Bridge. Editorial on two facts brought out recently which add to the information concerning the cause of the collapse. Drawings. 1500 w. Eng News—Sept. 19, 1907. No. 87117.

The Suspended Falsework of the Cantilever Arms of the Quebec Bridge. Illustrated description of the suspended falsework used in the erection. 2000 w. Eng Rec—Sept. 21, 1907. No. 87154.

Reinforced Concrete.

The New Stanford Bridge, Worcestershire. Illustrated description of the longest-span ferro-concrete arch hitherto built in Great Britain. 2000 w. Engng—Aug. 23, 1907. No. 86727 A.

Skew.

Skew Bridge of Reinforced Concrete—Wabash Railroad. A bridge over the Sangamon River, near Decatur, Ill., is illustrated and described. 700 w. Ry Age—Sept. 6, 1907. No. 86800.

Steel.

The Kafue River Bridge on the Cape to Cairo Railroad. Brief illustrated account of the longest bridge in Africa, a steel structure of the open-lattice girder type. 1200 w. Sci Am Sup—Sept. 28, 1907. No. 87254.

Substructure.

The Substructure of the Dearborn Street Bridge, Chicago. Brief description of this bridge and of the character of the river bottom, which presents peculiarly difficult conditions, with illustrated description of cofferdams and construction work. 2500 w. Eng Rec—Sept. 14, 1907. No. 87024.

Suspension.

An Eye-Bar Suspension Bridge over the Aare from the Knechteninseli in Bern (Kettensteg über die Aare beim Knechteninseli in Bern). F. Ackermann. An illustrated description of the design of a steel foot-bridge of 55 metres span. 1200 w. Schweiz Bau—Aug. 31, 1907. No. 86960 D.

Viaducts.

Test of the Reinforced Concrete Viaduct of the Richmond & Chesapeake Bay Ry. An illustrated account of the tests made of a viaduct designed for a train of cars each weighing 150,000 lbs., with a factor of safety of 4. 1000 w. Eng Rec—Sept. 21, 1907. No. 87156.

The Erection of Las Vacas Viaduct. An illustrated article describing methods used in erecting the narrow gauge single track viaduct on the Guatemala Ry. 2000 w. Eng Rec—Sept. 14, 1907. No. 87029.

The Genesee River Viaduct, Erie R. R. Illustrated detailed description of a viaduct with a steel plate girder superstructure, 3,121 ft. long and about 120 ft. in height from the water level. 1200 w. Eng Rec—Aug. 31, 1907. No. 86692.

The New Steel Viaduct Between Kansas City, Mo., and Kansas City, Kan. Explains the conditions and gives an illustrated detailed description of this recently completed work. 4500 w. Eng News—Sept. 26, 1907. No. 87243.

Reinforced Concrete Viaduct—Richmond & Chesapeake Bay Railway. Illustrated detailed description of a viaduct of unusual type, being built on an electric line in Virginia. The Kahn system of reinforcing is used. 1400 w. Ry Age—Sept. 27, 1907. No. 87287.

CONSTRUCTION.

Arches.

Symmetrical Masonry Arches—Coefficients for Reactions and Moments at the Supports. Malverd A. Howe. Gives an analysis of stresses, and methods of finding the moments at the supports and coefficients. 1200 w. R R Gaz—Aug. 30, 1907. No. 86662.

We supply copies of these articles. See page 307.

Beam Deflection.

The Calculation of the Deflection of Beams by Graphical Integration (Le Calcul des Flèches des Poutres par l'Intégration Graphique). E. Aragon. A theoretical and mathematical description of the method. Ills. 7000 w. Serial. 2 parts. Génie Civil—Aug. 17 and 24. No. 86927 each D.

Chimney Collapse.

Fall of a 150-Ft. Reinforced Tile-and-Concrete Chimney at La Crosse, Wis. An illustrated account by W. E. Miller, with information from several other sources. 2000 w. Eng News—Sept. 5, 1907. No. 86760.

Concrete.

How to Prevent Failure in Concrete Construction. Dr. W. Michaelis, Jr. Gives the writer's views on the merits and limitations of cement and concrete and on the causes of failure in concrete construction, with suggestions for the prevention of such failures. 9000 w. Jour W Soc of Engrs—Aug., 1907. No. 87138 D.

Concrete Pipe.

Methods and Cost of Constructing a Reinforced Concrete Pipe for Carrying Water Under Pressure. Chester Wason Smith. Describes the construction of about 6,000 ft. of reinforced concrete pipe for carrying water under pressure, giving results attained and cost. Ills. 3000 w. Engng-Con—Sept. 11, 1907. Serial. 1st part. No. 86894.

Reinforced Concrete Pipe for Carrying Water Under Pressure. Chester Wason Smith. Describes the construction of about 6,000 ft. of pipe intended to carry water under pressure; also gives results and figures of cost. Ills. 4500 w. Pro Am Soc of Civ Engrs—Aug., 1907. No. 86639 E.

Concrete Towers.

Reinforced Concrete Towers. D. W. Krellwitz. Illustrates and describes the reinforced-concrete towers, 142 feet high, built to carry a transmission line across the Welland Canal, near St. Catherines, Ontario. 1500 w. Pro Am Soc of Civ Engrs—Aug., 1907. No. 86638 E.

Dams.

The Stresses on Masonry Dams. Editorial review of a recent paper by Prof. Karl Pearson. 2000 w. Engng—Sept. 13, 1907. No. 87212 A.

Completion of the Cross River Dam, Croton Water-Works System. An illustrated account of this masonry dam which is to impound about 9,000,000,000 gallons of water for the increased supply of New York City. 3000 w. Eng Rec—Sept. 14, 1907. No. 87025.

Earth Pressures.

The Bracing of Trenches and Tunnels,

with Practical Formulas for Earth Pressures. J. C. Meem. Treats of the sheathing and bracing of trenches and tunnels in dry and water-bearing strata. Ills. 6000 w. Pro Am Soc of Civ Engrs—Aug., 1907. No. 86640 E.

Excavating.

Hints on Handling Wheeled Scrapers. Suggestions for operating scrapers so as to increase the amount of earth excavated. 1500 w. Engng-Con—Aug. 28, 1907. No. 86651.

Foundations.

Machinery and Methods for the Excavation of Sky-Scraper Foundations. T. Kennard Thomson. An illustrated description of methods used in construction of these foundations, and modern machinery for carrying on the work. 4500 w. Engineering Magazine—Oct., 1907. No. 87280 B.

Method of Constructing the Foundations for the Trust Company of America Building, New York City. Maurice Deutsch. An illustrated description of foundation work for a 25-story building in Wall St. 2800 w. Sch of Mines Qr—July, 1907. No. 86648 D.

Foundations. An informal discussion of the best system of construction on an alluvial deposit, materials, setting of concrete, etc. 3800 w. Pro Am Soc of Civ Engrs—Aug., 1907. No. 86644 E.

Foundation Work and Ground-Water Protection for the Enlargement of the Merchants' and Manufacturers' Bank at Berlin (Fundierung und Grundwasser-Abdichtung für den Erweiterungsbau der Bank für Handel und Industrie zu Berlin). Th. Gesztessy. An illustrated mathematical description of the design. 2500 w. Beton u Eisen—Aug., 1907. No. 86980 F.

Lighthouse.

The Fastnet Rock Lighthouse. An illustrated description of this granite structure on the S. W. coast of Ireland, and its erection, as described by C. W. Scott. 3500 w. Engr, Lond—Sept. 13, 1907. No. 87219 A.

Material Handling.

Some Ancient Methods of Lifting Stones and Timber. Clement E. Stretton. Describes mechanical contrivances used by the ancients. Diagrams. 2000 w. Sci Am Sup—Sept. 14, 1907. No. 86898.

Reinforced Concrete.

The McNulty Building, New York. A ten-story and basement, small-column, reinforced-concrete building on 52d St., New York City, is illustrated and described. 1200 w. Eng Rec—Sept. 14, 1907. No. 87026.

The Mushroom System of Reinforced

We supply copies of these articles. See page 397.

Concrete Building Construction. A. S. Malcolmson. Illustrated description of the system invented by C. A. P. Turner. 2300 w. Engng-Con—Sept. 4, 1907. No. 86829.

Roads.

Road Intersections Along the Grand Boulevard and Concourse, New York City. Describes the transverse roads under construction at Tremont Ave. and Kingsbridge Road as typical of all that are to be built. Ills. 1600 w. Eng Rec —Aug. 31, 1907. No. 86693.

Rules.

Revision of French Rules. Gives final version as applied to reinforced-concrete construction, with list of corrections made. 2500 w. Cement Age—Sept., 1907. No. 87136.

Sewers.

Constructing a Sewer Under the Brooklyn Subway. Describes the construction of a curved by-pass 290 ft. long, explaining the conditions and requirements. Ills. 2000 w. Eng Rec—Aug. 31, 1907. No. 86689.

Stresses.

Higher Structures. C. Derleth, Jr. Gives a discussion of the literature and a survey of American and English treatment for the analysis of complex framed structures, with remarks on the latest edition of the work of Professors Merriman and Jacoby. 3500 w. Cal Jour of Tech—Aug., 1907. No. 87097.

Tanks.

Strength of Air Pressure Tanks. Ira J. Owen. Formulæ and specifications for the construction of steel sprinkler pressure tanks, discussing causes of explosion. Ills. 2500 w. Ins Engng—Sept., 1907. No. 87223 C.

Tunnels.

Cost and Methods of Construction of Tunnel for Water Pipes Under Mystic River, Boston, Mass. Caleb Mills Saville. An illustrated description of the work with tables of itemized costs. 2500 w. Eng News—Sept. 26, 1907. No. 87245.

Subaqueous Tunnel at Detroit. Illustrated description of the method of construction, which consists of dredging a channel across the river, laying sections of tube in the ditch, filling around and over them with concrete, and then pumping out the water. 1500 w. Ry & Loc Engng—Sept., 1907. No. 86705 C.

The Present Condition of the Battery Tunnel Under the East River. From a report by George S. Rice to the Public Utilities Commission, detailing the present status of the work. 1500 w. Eng News—Sept. 19, 1907. No. 87116.

· Underground Workings in New York City. H. T. Hildage. Considers especially work in soft ground and the methods used, illustrated by the large undertakings in progress in New York City. 6000 w. Engineering Magazine—Oct., 1907. No. 87283 B.

Underpinning.

A Combined Underpinning and Sheeting Job. Describes complicated work in the construction of a six-story steel-cage bank building at the S.-W. corner of Murray St. and Broadway, New York City. Ills. 2000 w. Eng Rec—Sept. 7, 1907. No. 86786.

Warehouse.

Newark Warehouse of the Central Railroad of New Jersey. Photograph, drawings, and description. 700 w. R R Gaz—Aug. 30, 1907. No. 86659.

Wind Pressure.

The Chance of a Hurricane. A. P. Trotter. A discussion of the maximum wind pressure to be adopted in investigating the effect of wind on buildings and engineering structures. 3000 w. Engng —Aug. 23, 1907. No. 86726 A.

MATERIALS OF CONSTRUCTION.

Brick Walls.

Strength and Fire Protection of Brick Walls. Alfred Stone. A discussion of the carrying capacity of brick walls, their ability to stand upright and resist wind pressure and vibration, and the fire resistance of brick walls of certain thicknesses. 2700 w. Ins Engng—Sept., 1907. No. 87222 C.

Cement.

Notes on the Le Chatelier Test. W. Lawrence Gadd. Gives facts in connection with the behavior of rotary-made cement, based on hundreds of tests by the Le Chatelier method, and showing that under present conditions it is possible for a false conclusion to be drawn. 2800 w. Engr, Lond—Sept. 13, 1907. No. 87216 A.

Simple Tests of Portland Cement. W. Purves Taylor. Gives simple tests, that can be made by small consumers, for soundness, time of setting, purity, and strength. 1500 w. Cement—Aug., 1907. No. 86833 C.

Concrete.

Effect of Steam Curing on the Crushing Strength of Concrete. A report of investigations made to determine the efficiency of this method, with conclusions. 1500 w. Eng News—Sept. 5, 1907. No. 86757.

The Permeability of Concrete and Methods of Waterproofing. Richard H. Gaines. Gives the result of an investigation being carried on by the N. Y. Board of Water Supply in their effort to perfect the materials and methods of construction for the new aqueduct for the city. Also editorial. 5000 w. Eng News —Sept. 26, 1907. No. 87250.

The Equipment and Work of the U. S. Government Concrete Testing Laboratory at St. Louis, Mo. Richard L. Humphrey. Information in regard to the tests being carried out. Ills. 1800 w. Engng-Con—Sept. 25, 1907. No. 87260.

Concrete Columns.

Concrete Column Tests. Arthur N. Talbot. Presented at meeting of the Am. Soc. for Test. Mat. Gives a summary of the results of tests made at the Univ. of Illinois. Ills. 3000 w. Cement Age—Sept., 1907. No. 87135.

Strength Tests of Hooped Concrete Columns. Abstract of results reported by Prof. A. N. Talbot to the American Society for Testing Materials. Ills. 2500 w. Eng News—Sept. 26, 1907. No. 87249.

Corrosion.

See Electrical Engineering, Electro-Chemistry.

Fireproofing.

Terra Cotta Fireproofing. A. S. Atkinson. On its wide application in building construction, giving statistics of fireproof construction in New York City. Ills. 2500 w. Ins Engng—Sept., 1907. No. 87224 C.

Forest Planting.

Some Cost Figures on Forest Planting for Railroad Purposes. E. A. Sterling. Gives some cost figures in regard to work done by the Pennsylvania Railroad. 900 w. Eng News—Aug. 29, 1907. No. 87157.

Mortar.

Mortar and Concrete Mixtures. William Challoner. Discusses lime and cement mortars and the conditions under which the materals are available, the methods of improving the binding qualities, the advantages conferred by the use of trass, cost, etc. 2500 w. Engr, Lond —Sept. 6, 1907. No. 87085 A.

Reinforced Concrete.

An Experimental Study of Reinforced Concrete in Compression. Translation from *Beton und Eisen*, of an article by Messrs. Schinke and Loeser, describing experiments made on small prisms and columns with the object of studying the effect of lateral reinforcing on the compressive resistance of concrete. 2000 w. Cement—Aug., 1907. No. 86832 C.

Ferro-Concrete and Examples of Construction. J. S. E. De Vesian. Read before the British Assn. Refers to the ancient use of concrete and the present condition of the old material, and discusses the theory of ferro-concrete. 1700 w. Ir & Coal Trds Rev—Aug. 23, 1907. No. 86742 A.

Steel Columns.

Stresses in Lattice Bars of Built Columns. Thomas H. Johnson. A study offered as a tentative suggestion, stating the assumptions. 1000 w. Eng News—Sept. 26, 1907. No. 87247.

Timber.

Some Problems in Wood Utilization in the United States. R. S. Kellogg. Considers the reduction of waste, economy in use, etc. Discussion. 3500 w. Jour W Soc of Engrs—Aug., 1907. No. 87137 D.

The Results of Tests on the Strength of Timber Made at the Imperial Forestry Institution at Mariabrunn (Ueber die an der k. k. forstlichen Versuchsanstalt Mariabrunn gewonnenen Resultate der Holzfestigkeitsprüfungen). Gabriel Janka. The first part of the serial describes the results establishing a relation between specific gravity and compressive strength. Ill. 4000 w. Serial. 1st part. Zeitschr d Oest Ing u Arch Ver—Aug. 9, 1907. No. 86971 D.

Timber Preservation.

Specifications for Creosoting Oregon Fir Piling and Bridge Timber, A. T. & S. F. Ry. A copy of the specifications of this company. 3000 w. Ry & Engng Rev—Sept. 21, 1907. No. 87173.

MEASUREMENT.

See under Civil Engineering, Construction, and Materials of Construction.

MUNICIPAL.

Manila.

The New Water and Sewerage Systems of Manila. Illustrates and describes interesting work under construction in the Philippines, including a rubble concrete masonry dam creating a reservoir with a capacity of about 2,000,000,000 gals., and a combination gravity and pumping sewerage system. 2500 w. Eng Rec—Sept. 14, 1907. No. 87027.

Pavements.

Brick Pavements in Paris, Ill. Statistics from specifications, with comments and remarks on the construction. 4000 w. Munic Engng—Sept., 1907. No. 87133 C.

Pavements. An informal discussion of whether the paving materials of the present will be used in the construction of the pavements of the future. 4000 w. Pro Am Soc of Civ Engrs—Aug., 1907. No. 86645 E.

On Better Systems of Pavements for Streets and Roads from the Point of View of Efficiency and Economical Maintenance (Sui Migliori Sistemi di Pavimentazione delle Strade Comunali e Provinciali dal Punto di Vista della loro Efficacia e della piu Economica Manutenzione). Nicolo Valente. 4000 w. Rivista Gen d Ferrovie—Aug. 11, 1907. No. 86904 D.

Roads.

See Civil Engineering, Construction.

Sanitation.

The Work of the Sanitary District of Chicago; That Already Accomplished and Yet Contemplated Below the Controlling Works at Lockport. Isham Randolph. Ills. Discussion. 5000 w. Jour W Soc of Engrs—Aug., 1907. No. 87141 D.

Sewage.

Comparative Résumé of the Sewage Purification Tests at Columbus, Ohio. George W. Fuller. With discussion. 24700 w. Jour Assn of Engng Socs—Aug., 1907. No. 87225 C.

The Use and Abuse of Sewage Purification Plants. A. Elliott Kimberly. Read before the Ohio Engng. Soc. Discusses the question of protection afforded by such plants to rivers and bodies of water, and the causes of failure when this protection is not accomplished. 5400 w. Eng Rec—Aug. 31, 1907. No. 86691.

Sewers.

See Civil Engineering, Construction.

WATER SUPPLY.

Artesian Wells.

Artesian Wells in Australia. Editorial on the great artesian basin of Australia; its geology, characteristics, its great value, etc., and the legislation controlling it. 2000 w. Engng—Sept. 6, 1907. No. 87080 A.

Canal Sections.

The Effect of Changes in Canal Cross-Sections Upon Rate of Flow. F. W. Hanna. Develops formulæ for solving problems often occurring in irrigation work. 1500 w. Eng News—Sept. 26, 1907. No. 87246.

Carbonic Oxide.

The Importance of Carbon Dioxide in Water-Supply Projects (Die Bedeutung der Freien Kohlensäure im Wasserversorgungswesen). Hartwig Klut. A discussion of its hygienic and technical effects. 7700 w. Gesundheits Ing—Aug. 10, 1907. No. 86976 D.

Concrete Pipe.

See Civil Engineering, Construction.

Dams.

See Civil Engineering, Construction.

Filtration.

The Circular Tanks at the Lancaster Filtration Plant. Illustrates and describes the coagulating and softening tanks of unusual design and operation. 1200 w. Eng Rec—Sept. 14, 1907. No. 87030.

The Water Filtration Plant of Harrisburg, Pa. Illustrated description of the new plant for filtering Susquehanna river water and its operation. 3000 w. Munic Engng—Sept., 1907. No. 87132 C.

Hydraulic Formulæ.

Hydraulic Formulæ: Development and

Discussion. Develops a General Equation of discharge, and, from it, derives the various formulæ for flow through orifices and over weirs of the shapes usually considered. 6500 w. Sch of Mines Qr—July, 1907. No. 86647 D.

Intake.

An Infiltration Water-Works Intake Under the Ohio River. Describes an intake built of well point strainers buried in the sand bottom of the Ohio River which has been in successful operation nearly a year. Ills. 1600 w. Eng Rec—Aug. 31, 1907. No. 86688.

Irrigation.

Evaporation Losses in Irrigation. Samuel Fortier. Reviews studies recently made to determine the losses by evaporation in irrigation in the state of California. Ills. Also editorial. 4800 w. Eng News—Sept. 19, 1907. No. 87113.

Irrigation and Drainage Problem in Stanislaus Co., Cal. C. S. Abbott. Describes the irrigation system of the Modesto-Turlock district, the harm caused by too much water, and methods used to remedy the trouble. Ills. 1500 w. Cal Jour of Tech—Aug., 1907. No. 87095.

Irrigation in Western Canada. A report of the first Canadian Irrigation Convention, recently held at Calgary, Alberta. 3000 w. Engng—Aug. 30, 1907. No. 86872 A.

Pipe Line.

A 7-Ft. Steel Pipe Line at St. Louis. Describes the construction of this line, 19,634 ft. in length, in connection with the improvements to the water supply system of St. Louis, Mo. 1500 w. Eng Rec—Sept. 7, 1907. No. 86787.

Pipe Sizes.

A Solution of the Problem of Determining the Economic Size of Pipe for High-Pressure Water-Power Installation. Discussion of paper by Arthur L. Adams on this subject. 4000 w. Pro Am Soc of Civ Engrs—Aug., 1907. No. 86641 E.

Pipe Tunnel.

See under Tunnels, Civil Engineering, Construction.

Pumping Tests.

See Mechanical Engineering, Hydraulics.

Purification.

Direct and Indirect Methods of Electrical Purification of Water. Henry Leffmann. Considers some of the processes that have come under the writer's notice, giving an illustrated description of a plant in operation on the Schuylkill River, Philadelphia. 1800 w. Jour Fr Inst—Sept., 1907. No. 87089 D.

The Water Purification and Softening Works at New Orleans, La. Illustrates and describes two plants for filtering and softening the Mississippi River water.

1500 w. Eng Rec—Aug. 31, 1907. No. 86690.

Rates.
Rates for Water Service. Dabney H. Maury. Presidential address to the Am. Water-Works Assn. Discusses the basis for fair rates. 2000 w. Munic Engng—Sept., 1907. No. 87134 C.

Reservoirs.
A Colorado Mountain Reservoir. R. M. Hosea. Deals with the Sugar Loaf reservoir near Leadville, giving its history, describing the construction, and giving cost. Ills. Discussion. 7000 w. Jour W Soc of Engrs—Aug., 1907. No. 87140 D.

Stream Pollution.
Court Decision Against the Pollution of a Brook in Dracut, Mass., by the American Woolen Co. An explanation of the case with the decision. 3500 w. Eng News—Sept. 5, 1907. No. 86756.

Supply.
Water Supply. An informal discussion of whether it is better policy to control water-sheds, or to rely upon filtration. 8800 w. Pro Am Soc of Civ Engrs—Aug., 1907. No. 86643 E.

Tanks.
See Civil Engineering, Construction.

Water Softening.
McKeesport's Water Softening Plant. Illustrated detailed description of a 10,000,000-gallon municipal plant under construction, to use the caustic lime and soda ash process. 2500 w. Munic Jour & Engr—Sept. 4, 1907. No. 86827 C.

The Beissel Water-Softener. Illustrates and describes an automatic water-softening apparatus. 1700 w. Engng—Aug. 30, 1907. No. 86870 A.

WATERWAYS AND HARBORS.

Appalachian Streams.
Southern Appalachian Streams. Charles E. Waddell. Considers their characteristics and prominent features, their utilization as power producers, etc. Ills. 3500 w. Jour Fr Inst—Sept., 1907. No. 87087 D.

Docks.
The First Lake Dock of Steel. Dwight E. Woodbridge. Gives elevation and sections, with description of details of the structure to be built at Two Harbors, Minn. 2000 w. Ir Age—Oct. 3, 1907. No. 87377.

Drainage Canal.
The Chicago-St. Louis Waterway. W. Frank M'Clure. An illustrated article giving information of recent developments with reference to this project. 1500 w. Sci Am—Sept. 21, 1907. No. 87125.

Dry Regions.
The Procuring of Water in Dry Regions (Die Gewinnung von Wasser in Trockenen Gegenden). Hermann Haedicke. An illustrated description of methods of procuring water where wells or surface water are not available. 3300 w. Gesundheits Ing—Aug. 3, 1907. No. 86975 D.

French Canals.
Canal Transportation in France. From the report of Consul-General Robert P. Skinner covering the co-operation of State and communities in the construction of French canals, the cost, and the comparative freight charges by canals, rivers and railways. 1200 w. Eng Rec—Sept. 14, 1907. No. 87031.

Georgian Bay Canal.
Completion of the Surveys for the Montreal, Ottawa, and Georgian Bay Ship Canal. J. A. McDonald. Information concerning this project based on the recently completed survey. 1300 w. Eng News—Oct. 3, 1907. No. 87426.

Salton Sea.
The Story of Salton Sea. An interesting account of the recent outbreak in the Colorado River on the southern border of California. 2500 w. Builder—Aug. 24, 1907. Serial. 1st part. No. 86712 A.

Transalpine Waterway.
A Transalpine Waterway (Eine Transalpine Wasserstrasse)? Jos. Riedel. Reviews the history of the project for a waterway through the Alps and examines its feasibility. 4000 w. Oest Wochenschr f d Oeffent Baudienst—Aug. 10, 1907. No. 86968 D.

MISCELLANY.

Channel Tunnel.
The Project for a Railway-Tunnel Connection between England and France under the English Channel (Das Projekt der Eisenbahn-Tunnelverbindung zwischen England und Frankreich unter dem Canal la Manche). E. A. Ziffer. Outlines the various projects which have been put forward and describes the tunnel recently suggested. Ills. 5000 w. Mitt d Ver f d Förd d Lokal u Strassenbahnwesens—Aug., 1907. No. 86930 F.

Earthquake.
The Effects of the San Francisco Earthquake of April 18. 1906, on Engineering Constructions. Continued discussion of reports of committees of the San Francisco Association of members of the Am. Soc. of Civ. Engrs. 15400 w. Pro Am Soc of Civ Engrs—Aug., 1907. No. 86642 E.

Wagon.
Strength Tests of a Contractor's Dumping Wagon. Summary abstract of a report of tests made by Prof. W. K. Hatt, at Purdue University, of a special type of dumping wagon. Ills. 1000 w. Eng News—Sept. 19, 1907. No. 87115.

We supply copies of these articles. See page 397.

ELECTRICAL ENGINEERING

COMMUNICATION.

Radio-Telegraphy.

Atmospheric Absorption of Wireless Signals. R. A. Fessenden. Reports results of some experiments recently made. 600 w. Elec Rev, Lond—Sept. 6, 1907. No. 87058 A.

Radio - Telegraphic Convention. Extracts from the report of the Select Committee of the House of Commons on the progress thus far. 2000 w. Elec Engr, Lond—Sept. 6, 1907. Serial. 1st part. No. 87055 A.

The Radio-Telegraphic Station at Nauen, Germany. Illustrated detailed description of an interesting station on the Telefunken system. 2000 w. Elec Rev, N Y—Sept. 7, 1907. No. 86784.

Wireless Telegraphy (La Télégraphie sans Fil). P. Janet. A discussion of the scientific principles on which wireless telegraphy is based. Ills. 6000 w. Mem Soc Ing Civ de France—June, 1907. No. 86909 G.

Wireless Telegraphy and its Most Recent Advances (Die Drahtlose Telegraphie und ihre Jüngsten Fortschritte). Otto Nairz. A brief review of the development of wireless telegraphy, particularly describing the recent advances made in Poulsen's researches. Ill. 4000 w. Serial. 2 parts. Verkehrstech Woche u Eisenbahntech Zeitschr—Aug. 3 and 10, 1907. No. 86949 each D.

Radio-Telephony.

Wireless Telephony for the United States Navy. Herbert T. Wade. Explains the construction and operation of the De Forrest system, illustrating the instruments. 1600 w. Sci Am—Sept. 28, 1907. No. 87252.

Telautograph.

The Cerebotani Telautograph (Der Telautograph von Cerebotani). Hans Dominik. An illustrated detailed description of the mechanism, diagrams of connections, etc. 4000 w. Verkehrstech Woche u Eisenbahntech Zeitschr—Aug. 31, 1907. No. 86952 D.

Telegraphy.

A "New" Sounder. Illustrated description of Vyle's direct-reading polarized sounder. 1000 w. Elec Rev, Lond—Sept. 6, 1907. No. 87061 A.

Telephony.

Distortion in Telephonic Transmission. Louis Cohen. Maintains that the resistance varies with the frequencies. 600 w. Elec Wld—Sept. 21, 1907. No. 87169.

Capacity in Telephone Lines (Della Capacità in Telefonia). R. Salvadori. A theoretical discussion of its causes, effects, methods of measurement, etc. Ills. 3500 w. Elettricità—Aug. 9, 1907. No. 86900 D.

DISTRIBUTION.

Fuses.

Fuses. Dr. Georg J. Meyer. Abstract translation from *Elek. Zeit.* Gives results of investigations made of the heating and arcing of fuses. 2200 w. Elect'n, Lond—Aug. 30, 1907. No. 86863 A.

Notation.

Notation for Polyphase Circuits. Charles H. Porter. Gives a notation based essentially on lettering every junction and terminal on the diagram of connections, and on the use of two subscripts with every symbol of current or electromotive force or vector representing them. 2000 w. Elec Jour—Sept., 1907. No. 87130.

Regulations.

Electricity in Factories. A draft of the regulations proposed by the Home Office (England) for the generation, transformation, distribution, and use of electrical energy in premises. 2200 w. Elec Engr, Lond—Aug. 30, 1907. No. 86860 A.

The Generation, Transformation, Distribution, and Use of Electric Energy. A discussion of draft regulations sent out for examination by suppliers and users of electrical energy, with the object of securing a complete code under the Factory and Workshop Act in regard to works where electric energy is used. 1800 w. Elec Engr, Lond—Sept. 6, 1907. No. 87054 A.

Shop Equipment.

The 220-Volt Direct Current System of Distribution. Explains the advantages of this system for shop equipment. 1300 w. Elec Age—Aug., 1907. No. 86701.

Wiring Risks.

Electric Wiring Risks. Frank Broadbent. Notes on American and English practice. 2800 w. Elec Rev, Lond—Sept. 6, 1907. No. 87059 A.

DYNAMOS AND MOTORS.

Alternating-Current.

General Theory of Alternate-Current Motors. H. Görges. Abstract from *Elek. Zeit.* Explains the action of single and polyphase alternate-current motors, giving a classification based on the nature of the rotor currents. 2000 w. Elect'n, Lond—Sept. 13, 1907. No. 87209 A.

We supply copies of these articles. See page 397.

Armature Design.
The Influence of the Teeth and Grooves on the Armatures on Dynamo Operation (Der Einfluss der Zähne und Nuten auf die Wirkungsweise der Dynamoanker). Reinhold Rüdenberg. A theoretical and mathematical paper. Ills. 5000 w. Serial. 1st part. Elektrotech u Maschinenbau—Aug. 4, 1907. No. 86966 D.

D. C. Generators.
Three-Wire Direct-Current Generators. B. T. McCormick. Read before the Can. Elec. Assn. at Montreal. Gives briefly the principles of operation of three-wire generators, calling attention to some of their advantages. 1000 w. Elec Rev, N Y—Sept. 28, 1907. No. 87303.

Dynamo Dimensions.
A Method of Determining the Leading Dimensions of Large and High-Speed Continuous Current Dynamos. H. M. Hobart and A. G. Ellis. Describes a method based on a general relation between the leading dimensions, armature strength, number of poles, rated output and speed, and the commutating quality. 1200 w. Elec Rev, Lond—Sept. 6, 1907. Serial. 1st part. No. 87060 A.

Induction Motors.
Fractional Pitch Windings for Induction Motors. C. A. Adams, W. K. Cabot, and G. Æ. Irving, Jr. Reports an investigation made to develop a method by means of which the effects of such windings may be calculated. 4000 w. Pro Am Inst of Elec Engrs—Aug., 1907. No. 87143 D.

Losses.
The Experimental Determination of the Losses in Motors. C. F. Smith. Abstract of a paper read before the Manchester Loc. Sec. of the Inst. of Elec. Engrs. Describes the methods in use for obtaining a separate determination of the iron and friction losses, in both continuous and alternating-current motors, and discusses the advantages and disadvantages of the various methods. 2500 w. Elect'n, Lond—Aug. 30, 1907. Serial. 1st part. No. 86865 A.

Railway Motors.
See under Motors, Street and Electric Railways.

ELECTRO-CHEMISTRY.

Cells.
Mercurous Sulphate, Cadmium Sulphate, and the Cadmium Cell. G. A. Hulett. From the *Phys. Rev.* Describes a series of experiments made on cadmium cells, and gives the results of some experiments with regard to the ageing of standard cells. 4000 w. Elect'n, Lond—Sept. 6, 1907. No. 87066 A.

Conductivity.
On the Conductivity of Electrolytes in Pyridine and Other Solvents. Kenneth

Somerville Caldwell. A report of experimental work carried out in Leipzig under the direction of Prof. A. Hantzsch. 800 w. Elec Engr, Lond—Aug. 30, 1907. No. 86859 A.

Corrosion.
The Electrolytic Theory of the Corrosion of Iron. Dr. Allerton S. Cushman. A study of the real cause of iron rust. 3000 w. Sci Am Sup—Aug. 31, 1907. Serial. 1st part. No. 86681.

Corrosion of Iron. Abstracts of papers by Dr. Allerton S. Cushman, and by Dr. William H. Walker, Anna M. Cederholm, and Leavitt N. Bent, with editorial comment. Deals with the causes of corrosion and means of overcoming the troubles due to it. 5500 w. Elec-Chem & Met Ind—Sept., 1907. No. 86839 C.

The Corrosion of Iron: Rusting. An outline of some of the experiments and applications of the electrolytic theory, as given in papers by Allerton S. Cushman, and Prof. W. H. Walker, with interesting comparison of the various theories. 3300 w. Eng News—Sept. 26, 1907. No. 87244.

Electrolysis.
Influence of Iron in Copper Electrolysis. E. L. Larson. Gives results showing the retardation of varying amounts of iron, and the effect of other influences. 1200 w. Eng & Min Jour—Sept. 7, 1907. No. 86807.

Electrolytic Assaying.
The Results of Researches on Electrolytic Analysis (Elektroanalytische Forschungsergebnisse). Franz Peters. A review of the progress of knowledge and methods of the electrolytic determination of metals, the first part of the serial giving a general, historical introduction to the subject. 7000 w. Serial. 1st part. Glückauf—Aug. 17, 1907. No. 86940 D.

Electro-Metallurgy.
Applied Electro-Metallurgy Up to the End of 1906. John B. C. Kershaw. A summary of the various electro-metallurgical industries, describing briefly the processes and methods in use. Ills. 3000 w. Engineering Magazine—Oct., 1907. Serial. 1st part. No. 87281 B.
See also Mining Engineering, Iron and Steel.

Gold Deposition.
Heavy Gold Deposits and a Good Solution for Producing Them. 2000 w. Brass Wld—Sept., 1907. No. 87187.

Gold Recovery.
The Electrolytic Recovery of the Gold from Rolled Gold-Plate Scrap. Describes the method of electrolytic refining used because of its economy. Ills. 2500 w. Brass Wld—Sept., 1907. No. 87185.

Nitrogen.
The Fixation of Nitrogen. Norman Whitehouse. An account of a research

We supply copies of these articles. See page 397.

carried out to obtain a workable commercial process by which nitrogen of the atmosphere could be obtained combined as ammonia or nitrate. 2500 w. Elec-Chem & Met Ind—Sept., 1907. No. 86838 C.

Electrothermic Combustion of Atmospheric Nitrogen. Abstract of a paper by F. Howles, giving an interesting résumé of the chief processes employing electric discharges through air. 3000 w. Elec-Chem & Met Ind—Sept., 1907. No. 86836 C.

The Formation of Compounds of Nitrogen and Oxygen from the Air by Electricity (Die Darstellung von Stickstoff-Sauerstoff-Verbindungen aus Atmosphärischer Luft auf Electrischem Wege). Dr. B. Springfeldt. A description of various devices and processes for the fixation of atmospheric nitrogen. Ills. 1500 w. Serial. 1st part. Elektrochem Zeitschr—Aug., 1907. No. 86948 G.

Oxygen.

The Commercial Preparation of Oxygen and Hydrogen (Per la Preparazione Industriale dell' Ossigeno e dell' Idrogeno). An illustrated description of a new device for the production of hydrogen and oxygen by the electrolysis of water. 1700 w. Serial. 1st part. Il Monit Tech—Aug. 10, 1907. No. 86906 D.

ELECTRO-PHYSICS.

Conductors.

Forces in the Interior of an Electric Conductor. Edwin F. Northrup. Abstract from *Phys. Rev.* Describes experiments for showing the forces existing in the interior of conductors when traversed by a current, and considers their magnitude theoretically. 4500 w. Elect'n, Lond—Sept. 13, 1907. No. 87206 A.

Mechanical Action of Currents on Conductors. P. Bary. Abbreviated translation from *L'Eclair Elec.* Describes experiments which show that when a current passes through a conductor certain mechanical forces are set up causing a stricture effect. Ills. 3500 w. Elect'n, Lond—Sept. 6, 1907. No. 87064 A.

Magnets.

The Design of Plunger Magnets. C. P. Nachod. Shows the relation between the pull and the diameter of the plunger, and other properties of the magnets. 800 w. Elec Wld—Sept. 21, 1907. No. 87168.

GENERATING STATIONS.

See also under Street and Electric Railways.

Azores.

Electricity in the Azores. Illustrated description of the three hydro-electric stations which furnish electricity for lighting and power. 1200 w. Engr, Lond—Aug. 30, 1907. No. 86875 A.

Charges.

Some Fundamental Principles Underlying the Sale of Electrical Energy. Clarence P. Fowler. Discusses the main features upon which rates should be based, illustrating by an example. 3000 w. Elec Wld—Sept. 7, 1907. No. 86845.

Excitation.

Some Notes on the Excitation of Power Units. Discusses direct-current generators, and alternators, and methods of excitation. 1500 w. Elec Rev, Lond—Sept. 13, 1907. No. 87205 A.

Hydro-Electric.

The Hydro-Electric Development of the Great Northern Power Co. The first of a series of articles giving an illustrated detailed description of this development on the St. Louis River, near Duluth, Minn. 5500 w. Eng Rec—Sept. 7, 1907. Serial. 1st part. No. 86785.

The Hydro-Electric Plant at Kakabeka Falls, Ont. Illustrated detailed description. 2500 w. Can Elec News—Sept., 1907. No. 86820.

The Hydro-Electric Plant of the Canadian Niagara Power Company. A brief illustrated description of the plant and its operation. 3000 w. Can Elec News—Sept., 1907. No. 86822.

The Hydro-Electric Plant of the McCall Ferry Power Co. Illustrated detailed description of the design and methods of construction for a plant with a nominal capacity of 100,000 h. p. 6500 w. Eng Rec—Sept. 21, 1907. No. 87153.

The McCall Ferry Hydro-Electric Power Plant on the Susquehanna River. Illustrated detailed description of the construction of this large plant, and the methods of carrying out the work. The whole plant is essentially a concrete structure. 7500 w. Eng News—Sept. 12, 1907. No. 87000.

The Brusio Hydro-Electric Plant and the Transmission of Power in Lombardy (Impianto Idroelettrico di Brusio e Trasporto di Forza in Lombardia). An illustrated description of an interesting plant in Italy, the construction of which was much complicated by its inaccessibility. 2000 w. Serial. 1st part. Elettricità—Aug. 16, 1907. No. 86901 D.

Hydro-Electric Plants on the Mediterranean Coast (Installations Hydro-Electriques de L'Energie Electrique du Littoral Méditerranéen). E. de Marchena. Illustrates and describes power plants and transmission lines in the southern part of France. 12000 w. Mem Soc Ing Civ de France—July, 1907. No. 86911 G.

Hydro-Electric Power Developments in France (Les Usines Hydro-Electric en France). G. de Lamarcodie. An article descriptive of the immense development of electrical energy from water powers,

We supply copies of these articles. See page 397.

outlining the uses to which it is put in the several parts of the country. Ills. 8000 w. Rev Gen d Sci—Aug. 30, 1907. No. 86920 D.

Loads.

A Suggestion for the Improvement of the Small Power Load. H. S. Hatfield. Considers the sub-meter system, illustrating and describing a type of meter. 1000 w. Elect'n, Lond—Sept. 13, 1907. No. 87207 A.

Oil-Engine Driven.

Oil-Engine Driven Power Plant of the Pittsfield Electric Company. Illustrated description of a plant driven by Diesel engines. 1500 w. Elec Wld—Sept. 7, 1907. No. 86842.

Rio de Janeiro.

Electrical Developments in Rio de Janeiro. Plans and description of the Rio das Lages power plant, and information of the possible developments of electrical power and lighting. 2000 w. Elec Rev, Lond—Sept. 6, 1907. No. 87057 A.

Storage Batteries.

The Application of the Storage Battery to Lighting, Power, and Railway Service. J. M. S. Waring. Considers means of controlling the charge and discharge of batteries; and the results accomplished by the batteries in their various applications. Discussion. Ills. 7000 w. Jour W Soc of Engrs—Aug., 1907. No. 87139 D.

The Trend of Storage Battery Development. L. H. Flanders. Considers only storage batteries designed for stationary use, describing an installation. Ills. 2200 w. Elec Jour—Sept., 1907. No. 87131.

Substations.

A Note on Sub-Station Cable Arrangement. W. Pleasance. Notes on methods of securing a regular arrangement of cables inside the sub-station to avoid confusion, waste, and dangerous troubles. 600 w. Elec Engr, Lond—Sept. 6, 1907. No. 87056 A.

Sub-Station Troubles. A. Wohlgemuth. Considers troubles caused by a breakdown at the generator plant, and causes and effects of local troubles. 2400 w. Power —Oct., 1907. No. 87265 C.

Three-Phase Power Supply of the Manchester Corporation. Gives a typical sub-station plan for alternating-current supply only, describing the method. Ills. 1800 w. Elect'n, Lond—Aug. 23, 1907. Serial. 1st part. No. 86722 A.

Synchronizing.

Synchronizing. Paul MacGahan and H. W. Young. Illustrates and describes types of synchronizing devices. 3000 w. Elec Jour—Sept., 1907. No. 87129.

LIGHTING.

Arc Lamps.

Recent Developments in Metallic Flame Arc Lamps. G. Brewer Griffin. Describes direct-current lamps for operation in series, designed for street-lighting service. Ills. 2500 w. Elec Rev, N Y—Sept. 14, 1907. No. 87022.

The Flaming Arc Lamp. J. H. Hallberg. Reviews the development and present efficiency of arc lamps. 2000 w. Elec Rev, N Y—Sept. 14, 1907. No. 87013.

New Developments in Arc Lamps and High-Efficiency Electrodes. George M. Little. A comparison between metallic and carbon arcs. 2500 w. Sci Am Sup— Aug. 31, 1907. No. 86682.

New Methods of Suspending Arc Lamps (Neuerungen auf dem Gebiete der Bogenlampen—Aufhangungen). R. Osterburg. Illustrates and describes a large number of devices. 3000 w. Serial. 1st part. Elektrotech Zeitschr—Aug. 15, 1907. No. 86990 D.

Diffused Light.

The Elements of Inefficiency in Diffused Lighting Systems. Preston S. Millar. Read before the Ill. Engng. Soc. Considers the elements of inefficiency in lighting systems of this character, basing most of the conclusions upon measurements of illumination intensity. 3000 w. Pro Age—Sept. 2, 1907. No. 86665.

Factories.

Factory Lighting. Abstract of a paper by A. P. Biggs, read before the Ohio Elec. Lgt. Assn. Considers the sources of light now available and their merits, cost, etc. 4000 w. Eng Rec—Sept. 21, 1907. No. 87151.

Germany.

Electric Lighting in Germany. Dr. Phillip G. Klingenberg. Gives a review of the general aspects of electric lighting in Germany. 4500 w. Elec Rev, N Y— Sept. 21, 1907. No. 87159.

Helion Lamps.

The Helion Light. Walter G. Clark. An account of the Helion filament by one of the discoverers, stating the advantages claimed. 1800 w. Elec Rev, N. Y.—Sept. 14, 1907. No. 87015.

The Helion Lamp. Walter G. Clark and Herschel C. Parker. Describes the development of the Helion filament and the method of making it, explaining its characteristics. 1800 w. Sch of Mines Qr —July, 1907. No. 86649 D.

Illumination.

Recent Advances in Artificial Lighting. An illustrated article considering recent developments in lamps for out-door use, or for very large interiors. Arc and vacuum-tube electric lamps. 5500 w. Eng News—Sept. 12, 1907. No. 87002.

Some Artistic Requirements of Artificial Illumination. Bassett Jones, Jr. Discusses the considerations that must gov-

We supply copies of these articles. See page 397.

ern the design of an architectural lighting scheme. 2500 w. Elec Rev, N Y—Sept. 14, 1907. No. 87018.

The Engineering of Show-Window Illumination. J. R. Cravath and V. R. Lansingh. Discusses the lighting of windows for the purpose of displaying goods at the best advantage. 3300 w. Elec Wld —Sept. 7, 1907. No. 86843.

The Status of Illuminating Engineering. Dr. Louis Bell. Discusses the requirements of illuminating engineering as a profession. 1500 w. Elec Rev, N. Y— Sept. 14, 1907. No. 87009.

Incandescent Lamps.

Incandescent Lamp Development During the Last Twenty Years. S. E. Doane. A résumé of interesting features in the development. 3500 w. Elec Rev, N Y— Sept. 14, 1907. No. 87010.

The Recent Incandescent Lamp Developments and Their Significance. Francis W. Willcox. A review of recent improvements in metallic filaments, showing that many important changes are in progress. Ills. 4500 w. Elec Rev, N Y—Sept. 14, 1907. No. 87011.

The Effect of Higher-Efficiency Lamps on Station Output and Income. A symposium from central station men in the United States and Great Britain. 2500 w. Elec Rev, N Y—Sept. 14, 1907. No. 87014.

The "Turn-Down" Electric Lamp. A. Frederick Collins. Reviews the various types of "turn-down" lamps invented, explaining their construction. Ills. 1500 w. Sci Am—Sept. 21, 1907. No. 87126.

The Sirius Colloid Lamp. Paul McJunkin. Describes this lamp and its behavior. 800 w. Elec Rev, N Y—Sept. 14, 1907. No. 87016.

Mercury Vapor.

The Mercury Vapor Lamp as a Factor in Electricity Supply Development. Discusses the merits of the mercury vapor lamp and its practical application, illustrating some forms recently brought out. 2000 w. Elec Rev, Lond—Aug. 23, 1907. No. 86721 A.

The Mercury Vapor Lamp. Percy H. Thomas. Gives an outline of the form and characteristics of this illuminant, stating its advantages and limitations. 3500 w. Elec Rev, N Y—Sept. 14, 1907. No. 87021.

The Development of the Mercury-Vapor Lamp (Die Entwicklung der Quecksilberlampe). Dr. Br. Glatzel. Outlines the development of the mercury-vapor lamp, illustrates and describes various existing types, and discusses their utility. 3000 w. Verkehrstech Woche u Eisenbahntech Zeitschr—Aug. 24, 1907. No. 86950 D.

Moore Light.

The Moore Light and Illuminating En-

gineering. D. McFarlan Moore. An illustrated article describing the Moore tube-lighting system. 2000 w. Elec Rev, N Y —Sept. 14, 1907. No. 87023.

Nernst Lamps.

The Nernst Glower and the Present Status of the Nernst Lamp. Otto Foell. Describes characteristics of the Nernst lamp and details of new applications. Ills. 3500 w. Elec Rev, N Y—Sept. 14, 1907. No. 87020.

Rates.

The Sale of Electricity for Lighting Purposes Commercially Considered. L. E. Buckell. Suggestions for electric light salesmen, based on conditions in England. 1500 w. Elec Engr, Lond—Sept. 13, 1907. No. 87180 A.

Residences.

Economical House Lighting. Van Rensselaer Lansingh. Considers the lighting of houses where good, economical illumination and artistic effects are desired. Ills. 2000 w. Elec Rev, N Y—Sept. 14, 1907. Serial. 1st part. No. 87012.

Standards.

Primary, Secondary, and Working Standards of Light. Dr. Edward P. Hyde. Read before the Ill. Engng. Soc. Considers practical questions in connection with the relationship of primary, secondary, and working standards of light. 1700 w. Pro Age—Sept. 2, 1907. No. 86664.

Tungsten Lamps.

The Economy of the Tungsten Lamp. Alfred A. Wohlauer. A critical discussion of the relations between initial cost of the lamp, its life and its efficiency, price of energy, etc. 2000 w. Elec Wld—Sept. 7, 1907. No. 86844.

MEASUREMENT.

Alternator Tests.

Report on Test of 1,000 K. W. B. T. H. Turbo-Alternator Set Made at Contractors' Works, Rugby. Illustration, with results of an official test on a Curtis steam-turbine, driving a two-phase alternator. 600 w. Elec Engr, Lond—Sept. 13, 1907. No. 87179 A.

Frequency Meters.

Induction Type Frequency Meters. Paul MacGahan and H. W. Young. Brief illustrated description of the Graphic Recording Frequency Meter and its operation. 2000 w. Elec Age—Aug., 1907. No. 86702.

A New Frequency Meter for Wireless Telegraphy (Ein Nèuer Frequenzmesser der Strahlentelegraphie). Eugen Nesper. Illustrated detailed description. 2000 w. Verkehrstech Woche u Eisenbahntech Zeitschr—Aug. 24, 1907. No. 86951 D.

Galvanometer.

Observation of Alternate Currents with

We supply copies of these articles. See page 397.

the String Galvanometer. Gyözö Zemplén. Trans. from *Phys. Zeit.* Describes the writer's work with a small string galvanometer in measuring alternating currents. 900 w. Elect'n, Lond—Sept. 6, 1907. No. 87065 A.

High Frequency.

A Universal Instrument for High-Frequency Measurements (Ein Universal-Messinstrument der Hochfrequenz-Technik). Eugen Nesper. Illustrated detailed description of instrument and method of use. 2800 w. Serial. 1st part. Elektrotech Zeitschr—Aug. 29, 1907. No. 86991 D.

Hysteresis.

On a Method of Plotting the Hysteresis Loop for Iron with an Application to a Transformer. Dr. Gisbert Kapp. An explanation of method. 800 w. Elec Age—Sept., 1907. No. 87262.

Instruments.

Instruments for Testing Direct-Current Machinery. E. S. Lincoln. Gives an outline of important tests and instructions for carrying them out. 2000 w. Power—Sept., 1907. No. 86675 C.

Electrical Measuring Instruments. Explains properties which may be utilized in designing ammeters, voltmeters, and wattmeters. 1500 w. Ry Loc Engng—Sept., 1907. No. 86706 C.

Photometry.

Photometric Units. Preston S. Millar. Discusses the nomenclature used in photometry in the United States, and the system of units adopted at the Geneva Congress. 1700 w. Elec Rev, N Y—Sept. 14, 1907. No. 87019.

Photometry at the Bureau of Standards. Dr. Edward P. Hyde. Illustrated description of the photometric equipment and methods of testing and investigating. 3500 w. Elec Rev, N Y—Sept. 14, 1907. No. 87017.

The Application of the Spherical Photometer (Zur Anwendung des Kugelphotometers). R. Ulbricht. An illustrated mathematical paper on the use of the Ulbricht photometer. 4500 w. Elektrotech Zeitschr—Aug. 8, 1907. No. 86988 D.

Switchboard.

A Testing Switchboard. Describes the general arrangement of a board designed for testing direct-current motors and dynamos. Ills. 1000 w. Elec Rev, Lond—Aug. 30, 1907. No. 86861 A.

TRANSMISSION.

Concrete Towers.

See Civil Engineering, Construction.

European Lines.

Some Examples of Power Transmission in Central Europe (Quelques Transports d'Energie de l'Europe Centrale). Alfred Lambotte. Illustrates and describes hydro-electric, steam and gas power stations and transmission lines in Central Europe, the information being collected during an extensive tour. 40000 w. Soc Belge d'Elec'ns—Aug., 1907. No. 86912 E.

High Voltage.

A Year's Operation of the Highest Working Voltage in the World. F. E. Greenman. Read at the Mich. Elec. Assn. Con. Brief account of the operating of the Grand Rapids-Muskegon Power Company's transmission line at 72,000 volts. Ills. 1200 w. Elec Wld—Sept. 14, 1907. No. 87036.

Insulators.

High-Tension Insulators, from an Engineering and Commercial Standpoint. C. E. Delafield. Read at Montreal meeting of the Can. Elec. Assn. Discusses the requirements of insulators for high voltages. 3000 w. Elec Rev, N. Y—Sept. 28, 1907. No. 87301.

Lightning Protection.

Discussion on "Lightning Phenomena in Electric Circuits," "Protection Against Lightning, and the Multigap Lightning Arrester," and "New Principles in the Design of Lightning-Arresters" at New York, March 29, 1907. Papers by Charles P. Steinmetz, D. B. Rushmore, D. Dubois, and others, are discussed. 8000 w. Pro Am Inst of Elec Engrs—Aug., 1907. No. 87147 D.

Line Calculations.

The Calculation of Transmission Lines. Harold Pender. Gives formulæ and tables devised to furnish a simple and accurate method for making the necessary calculations. 1500 w. Elec Age—Sept., 1907. No. 87261.

Lines.

The 50,000-Volt Line of the Taylor's Falls-Minneapolis Power Transmission. Illustrated description of this pole line 40.6 miles long, designed to carry the total present capacity of the plant. 1300 w. Elec Wld—Sept. 7, 1907. No. 86841.

Wood Pole Line Construction. A. B. Lambe. Discusses in detail the main and standard points in wood pole line construction. 6500 w. Can Elec News—Sept., 1907. No. 86821.

Single-Phase Catenary Line Construction. G. D. Nicoll. Read before the Cent. Elec. Ry. Assn. Deals with a catenary construction for the operation of high-voltage alternating-current electric railways. 1200 w. Elec Ry Rev—Sept. 28, 1907. No. 87299.

Transformers.

Abnormal Primary Current and Second Voltage on Placing a Transformer in Circuit. Trygve Jensen. Gives results of investigations made of the values of the primary current and the secondary e. m. f. when the primary circuit of a transformer is closed at different pre-determined time-phase positions of the supply e. m. f. 1600 w. Elec Wld—Sept. 14, 1907. No. 87033.

Outline of the Characteristics of Constant Potential Transformers. George A. Burnham. Outlines briefly the important factors and considers the action of the simple transformer. Diagrams. 1800 w. Elec Wld—Sept. 7, 1907. No. 86846.

The Design of Transformers as Influenced by the Employment of Alloy-Sheets. Dr. Robert Pohl. Abstracted from *Elek. Zeit.* Proves that the use of alloy-sheets for smaller sizes of transformers leads to increased efficiency, reduced cost and weight of active material. 600 w. Elect'n, Lond—Sept. 13, 1907. No. 87208 A.

The de Faria Electrolytic Transformer. O. de Faria. An illustrated description of design, construction, and operation, with results of tests. 3200 w. Bull Soc d'Encour—July, 1907. No. 86918 G.

MISCELLANY.

Argentina.
Electrical Development in Argentina. Lewis R. Freeman. Facts elicited from prominent residents of Mendoza during a recent visit in regard to power development possibilities. Ills. 1800 w. Elec Wld —Sept. 14, 1907. No. 87034.

Matter.
The Kinetic Theory of Matter (Die Kinetische Theorie der Materie). Dr. G. Jäger. An exposition of the kinetic theory of matter as held by the physicists in the present controversy between physicists and chemists. 4500 w. Zeitschr d Oest Ing u Arch Ver—Aug. 16, 1907. No. 86973 D.

Peru.
Electricity in Peru. Emilio Guarini. An illustrated article giving an account of the electrical development and the favorable conditions for the application of electricity. 2000 w. Elec Wld—Sept. 14, 1907. No. 87035.

Plant Growth.
Effect of Electricity Upon Plants. J. H. Priestly. Abstract of a paper read before the British Naturalists' Soc. Gives results of experiments with the overhead discharge system of electrification; also with earth currents. 3000 w. Elect'n, Lond— Sept. 6, 1907. No. 87063 A.

INDUSTRIAL ECONOMY

Apprenticeship.
A Model Apprenticeship System. Magnus W. Alexander. On the General Electric Company's system as established at Lynn, Mass. 2200 w. Ir Age—Sept. 19, 1907. No. 87103.

Arbitration.
Arbitration: Compulsory and Voluntary. Discusses the failure of compulsory arbitration in Australia. 2500 w. Min & Sci Pr—Sept. 7, 1907. No. 86892.

Canadian Branches.
United States Manufacturers in Canada. Edward Porritt. Condensed from Boston *Transcript.* On the causes that have led to the establishment of branches of American industrial concerns in Canada. 2800 w. Ir Age—Sept. 26, 1907. No. 87229.

Coal Lands Tax.
The Proposed Tax on Anthracite Coal Lands. Editorial discussion of this tax, showing it to be unjust. 1500 w. Mines & Min—Sept., 1907. No. 86769 C.

Commerce of U. S.
Commerce of the United States by Principal Ports and Sections, 1907. Some recent statistics and comparisons showing the trend of the import and export trade. 1800 w. Sci Am Sup—Aug. 31, 1907. No. 86680.

Copper Industry.
The Copper Situation in the United States. James Douglas. Gives a review of the sources of supply and demand and a study of the present position and economic outlook. 6000 w. Engineering Magazine—Oct., 1907. No. 87275 B.

Cost Keeping.
A Perfect Cost System. F. E. Webster. The twelfth and final article of a series on cost keeping. 2000 w. Ir Trd Rev—Sept. 5, 1907. No. 86774.

Education.
Commercial Training Workshops. Charles Prescott Fuller. Gives an outline of a course of training which the writer believes would best fit the youths for the exacting requirements of the future. 2500 w. Am Mach—Vol. 30, No. 37. No. 86887.

Education for Industrial Workers. Arthur D. Dean. A plea for a machine trade school giving a suggested outline. 3500 w. Mach, N Y—Sept., 1907. No. 86749 C.

Co-operative Courses in Engineering at the University of Cincinnati. Herman Schneider. Read before the Soc. for the Pro. of Engng. Ed. An account of a course in which students work alternate weeks in the engineering college of the university and at the manufacturing shops

We supply copies of these articles. See page 397.

' of the city. 2000 w. Am Mach—Vol. 30, No. 37. No. 86883.

Education of Mining Engineers. T. A. Rickard. Address delivered at the opening of the new Mining building of the University of California. 2500 w. Min & Sci Pr—Aug. 31, 1907. No. 86795.

Opening of the New Mining Building of the University of California. A report of the opening of the Hearst Memorial Building, with illustration. 2700 w. Min & Sci Pr—Aug. 31, 1907. No. 86793.

The National Engineering School of Mexico. Mansfield Merriman. Illustrates and describes the building and gives the history of this engineering college, undoubtedly the oldest institution for engineering instruction on the western continent. 3300 w. Eng News—Sept. 19, 1907. No. 87112.

Eight-Hour Day.

Probable Effect of an Eight-Hour Shift on Colliery Work. Gives tables showing the probable distribution of labor, estimated working costs, etc., at the Northumberland and Durham Collieries should the Eight Hours Bill be placed on the Statute Book. 1500 w. Ir & Coal Trds Rev—Aug. 23, 1907. No. 86744 A.

Exposition.

The Jamestown Exposition. C. A. Graves. An illustrated description of the exhibits relating particularly to mineral resources and mining. 4200 w. Mines & Min—Sept., 1907. No. 86770 C.

Gold Production.

Prosperity; Its Relation to the Increasing Production of Gold. Alex. Del Mar. An examination of the circumstances which denote prosperity or adversity and the effect of money circulation. 5000 w. Engineering Magazine — Oct., 1907. No. 87276 B.

Strikes.

Peace in Labor Affairs. Editorial discussion of the strike in the British shipbuilding trade. 2300 w. Engng—Aug. 23, 1907. No. 86731 A.

The Strike Situation in San Francisco. Henry K. Brent. Describes the conditions under which strike-breakers worked, discusses causes of the strike, etc. 5500 w. St Ry Jour—Sept. 21, 1907. No. 87171.

The Common Law on Strikes. A summary of court decisions on labor troubles, taken from the last report of the Commissioner of Labor. 2500 w. Ir Age—Sept. 26, 1907. No. 87230.

MARINE AND NAVAL ENGINEERING

Armor Plating.

The Development of Armored War Vessels. J. H. Morrison. The present article reviews the use of armor plating in the United States. Ills. 2800 w. Sci Am Sup—Aug. 31, 1907. Serial. 1st part. No. 86679.

Astronomy.

Navigation by Celestial Observation. Stephen P. M. Tasker. The present number shows how to locate one's self by day or night. 1500 w. Int Marine Engng—Oct., 1907. Serial. 1st part. No. 87198 C.

Boiler Cleaning.

Cleaning Marine Boilers. Lieut. H. C. Dinger. A general description of latest methods, especially those in use in naval service. 3500 w. Int Marine Engng—Oct., 1907. No. 87196 C.

Buoys.

Mooring-Buoy for the "Lusitania" and "Mauritania." Illustrates and describes the largest mooring-buoy that has ever been made. 500 w. Engng—Aug. 23, 1907. No. 86728 A.

Regarding Spar Buoys. C. H. Claudy. An illustrated article on the service and care of spar buoys. 1000 w. Sci Am—Sept. 21, 1907. No. 87123.

Coaling Vessel.

Self-Propelling Coaling Vessel. Illustrates, and briefly describes, a vessel for coaling steamships, with automatic self-registering weighing machine. 700 w. Sci Am Sup—Sept. 21, 1907. No. 87127.

Cruiser.

Scout Cruiser Salem. Illustrated description of the first cruiser to be launched in the United States equipped with Curtis turbines. 1400 w. Marine Rev—Sept. 12, 1907. No. 86899.

The American Scout Cruiser Salem. Illustrated detailed description of the vessel and its equipment. 1200 w. Int Marine Engng—Oct., 1907. No. 87197 C.

Electricity.

The Present Status of Electricity on Shipboard (Ueber den Heutigen Stand der Schiffselektrotechnik). C. Schulthes. Describes the various applications of electricity on shipboard, illustrating plants, distribution systems, etc. 2000 w. Serial. 1st part. Elektrotech Zeitschr—Aug. 1, 1907. No. 86989 D.

Fire Protection.

The Clayton Apparatus (Der Clayton Apparat). F. Heintzenberg. Illustrates and describes this system of fire-protec-

We supply copies of these articles. See page 397.

tion on shipboard. The SO_2 generator and the method of installation and use. 3800 w. Serial. 1st part. Schiffbau—Aug. 28, 1907. No. 86956 D.

Historical.
A Century of Steamship Construction (Hundert Jahre Dampfschiffahrt). Conrad Matschoss. A review of steam navigation from Robert Fulton to the present day. Ills. 6000 w. Zeitschr d Ver Deutscher Ing—Aug. 17, 1907. No. 86983 D.

Lake Freighter.
Bulk Freighter William M. Mills. Illustrated description of one of the largest vessels on the lakes. 1000 w. Marine Rev—Sept. 5, 1907. No. 86825.

Lusitania.
The New Turbine Liner "Lusitania." Illustrations, with information of interest in regard to this and other fast vessels. 2000 w. Sci Am—Sept. 14, 1907. No. 86896.

Motor Boats.
The Kiel Motor Boat Exhibition (Die Kieler Motorboot-Ausstellung). A brief general description of the exhibits. 2000 w. Schiffbau—Aug. 14, 1907. No. 86-954 D.

Racing Boat.
New Racing Launch. Illustration, with description, of a racing launch of the torpedo boat type designed by André Gambin. 900 w. Engr, Lond—Sept. 6, 1907. No. 87086 A.

Salving Submarines.
A Study of Some Processes and Methods of Salving and Refloating Submarines (Etude de quelques Procédés et Méthodes de Sauvetage et de Renflouage des Navires Sous-Marines). M. Dibos. Considers the difficulties and dangers of submarine navigation, and the salving methods employed. 12000 w. Mem Soc Ing Civ de France—July, 1907. No. 86910 G.

Submarines.
Submarine Boat. Brief illustrated description of certain improvements in means of governing invented by John M. Gage. 600 w. Marine Rev—Sept. 5, 1907. No. 86824.

Survey Vessel.
Indian Survey Vessel "Palinurus." Illustrates and describes a steel single-screw steamer recently built for the Government of India, intended for marine survey work. 1000 w. Engng—Sept. 13, 1907. No. 87211 A.

Torpedo Boats.
German Torpedo Boat Construction. Information concerning these vessels, with illustration and brief description of the latest destroyer, fitted with turbines. 900 w. Engr, Lond—Sept. 13, 1907. No. 87218 A.

Train-Ferry.
Train-Ferry Steamer "Lucia Carbo" for the Entre Rios Railway. Illustrated description of a steamer designed to convey trains about 50 miles on the River Plate, South America. 500 w. Engng—Sept. 6, 1907. No. 87078 A.

MECHANICAL ENGINEERING

AUTOMOBILES.

Cadillac.
The 20 H. P. Cadillac Car. Illustrated description of a well-designed four-cylinder, vertical engined car. 1500 w. Autocar—Aug. 24, 1907. No. 86715 A.

Carbureters.
The Dentan Carbureter. Sections, with brief description of a novel carbureter. 1200 w. Auto Jour—Sept. 14, 1907. No. 87178 A.

Some Mysteries of the Carbureter. Charles B. Hayward. Explains causes of some of the troubles and means of correcting them. 3000 w. Automobile—Sept. 12, 1907. No. 86889.

Clutches.
The Friction Clutch in Automobile Practice. Forrest R. Jones. Discusses some of the undesirable properties in the types of clutches used. 2000 w. Automobile—Sept. 26, 1907. No. 87289.

Commercial Vehicles.
The Stoltz Steam Truck (Dampflastwagen Bauart Stoltz). Herr Pflug. Illustrates and describes the truck, engine, boiler, etc., and gives details of performance under various conditions. 2000 w. Verkehstech Woche u Eisenbahntech Zeitschr—Aug. 31, 1907. No. 86953 D.

Design.
Fundamental Principles in the Construction of Automobiles (Konstruktionsgrundlagen für den Bau von Kraftwagen). Ernst Valentin. Discusses materials of construction, transmission gears, etc., and illustrates various types of cars. 5000 w. Zeitschr d Ver Deutscher Ing—Aug. 24, 1907. No. 86986 D.

Ignition.
A Chapter on Magneto Ignition. Elmer G. Willyoung. Extract from Lecture No. 4, Correspondence School of Motor Car Practice. Explanatory. Ills. 1700 w. Automobile—Aug. 29, 1907. No. 86654.

We supply copies of these articles. See page 397.

The Leitner Ignition and Lighting System for Motor Vehicles. Illustrates and describes a modification of the Leitner train-lighting system suitable for road motors of all kinds. 1000 w. Engng—Aug. 23, 1907. No. 86729 A.

Industrial Locomotives.

Gasoline Locomotives and Motor Cars. Illustrated descriptions of a gasoline locomotive, and a gasoline motor car recently built. 600 w. Eng News—Sept. 5, 1907. No. 86759.

Magneto.

A New Gearless Magneto. Illustrated description of the Van Raden gearless high tension magneto. 900 w. Autocar—Sept. 14, 1907. No. 87200 A.

Maja.

The Maja Car. Illustrates and describes a car produced by the German Daimler Company. 1000 w. Autocar—Sept. 14, 1907. No. 87202 A.

Mass.

The 18 H. P. and 24 H. P. Mass Cars. Illustrated description of cars gaining gold medals in the recent Scottish Reliability Trials. 1000 w. Autocar—Aug. 31, 1907. No. 86855 A.

Motobloc.

The Motobloc Petrol Cars. An illustrated description of these French-built cars. 2000 w. Auto Jour—Sept. 14, 1907. Serial. 1st part. No. 87177 A.

Peerless.

Peerless 1908 Presentation. Illustrated discussion of details of the new models. 1400 w. Automobile—Sept. 26, 1907. No. 87290.

Porthos.

The Six-Cylinder Porthos Car. Illustrated detailed description. 1200 w. Autocar—Sept. 14, 1907. No. 87201 A.

Reliability.

The Reliability of Motor Cars. Editorial on the value of the Scottish trials, and the performance of the cars. 2000 w. Engng—Sept. 26, 1907. No. 87079 A.

Speed.

The Effect of Weight Upon Speed. A report of tests made on the Brooklands track, showing the small difference made in the speed of a racing car by considerable variation in weight. 9000 w. Auto Jour—Sept. 7, 1907. No. 87050 A.

Springs.

Construction of Motor Vehicle Springs. J. G. Rumney. Read before the Soc. of Auto. Engrs. Calls attention to the methods which should be followed in the calculation of springs so as to guard against excessive deflection. 2500 w. Automobile—Aug. 29, 1907. No. 86652.

Stability.

A Formula Relative to a Condition of Stability of Automobiles and Especially of the Motor Omnibus (Formule Relative à une Condition de Stabilité des Automobiles et Spécialement des Autobus). Georges Marié. A theoretical and mathematical discussion of the conditions which have to be fulfilled to prevent skidding, overturning, etc. Ills. 12000 w. Mem Soc Ing Civ de France—June, 1907. No. 86908 G.

Steering.

A Criticism of Steering Gears in General Use. Louis T. Weiss. A criticism of the action of the connecting rod with suggestion of a way to prevent the dropping of the rod if the springs should break. 800 w. Automobile—Sept. 26, 1907. No. 87291.

Westinghouse.

The Westinghouse Petrol Cars. Illustrates and describes the leading features of these cars, and also some of the minor details. 2500 w. Auto Jour—Sept. 7, 1907. No. 87049 A.

Wind Resistance.

Wind Resistance. Douglas Leechman. Discusses some effects of wind pressure, the importance of form, and related subjects, as applied to motor vehicles. 3000 w. Auto Jour—Aug. 24, 1907. No. 86713 A.

Wind Screens.

The Effect of Wind-Screens on Speed. An illustrated account of experiments on the Brooklands race track. 2000 w. Auto Jour—Aug. 24, 1907. No. 86714 A.

COMBUSTION MOTORS.

Alcohol.

Alcohol for Operating Engines. Gives a synopsis of an interesting report prepared by Charles E. Lucke and S. M. Woodward, on official tests of internal combustion engines on alcohol fuel. 4500 w. Ir Age—Sept. 26, 1907. No. 87231.

Tests of Internal-Combustion Engines on Alcohol Fuel. Charles Edward Lucke and S. M. Woodward. A report of researches made to determine whether the gasoline and kerosene engines at present on the American market can run on alcohol as fuel; and what improvements might be desirable in the design of engines manufactured especially for alcohol. Ills. 31200 w. U S Dept of Agri, Bul. 191—Sept. 14, 1907. No. 87285 N.

Does Alcohol Attack Metallic Surfaces? G. Lessard. Trans. from La Vie Automobile. An examination of the effect of alcohol, and the effect of impurities contained in it. 1200 w. Automobile—Aug. 29, 1907. No. 86653.

Gas Analysis.

Rapid and Accurate Gas Analysis. Edwin Barnhart. Illustrates and describes an apparatus giving a high degree of accuracy, and a minimum of time. A complete analysis of coal or producer gas re-

We supply copies of these articles. See page 397.

quires about twenty minutes. 1200 w.
Elec-Chem & Met Ind—Sept., 1907. No.
86835 C.

Gas Cleaning.
Gas Engines. Informal discussion of
what is the best apparatus and most economical system for cleaning producer or
furnace gas to be used in gas engines;
and to what extent ordinary producer gas
is used in gas engines, and the results
obtained by removing tar or soot. 3500
w. Pro Am Soc of Civ Engrs—Aug.,
1907. No. 86646 E.

Gas Engines.
Gas Engine Breakdowns. Michael
Longridge. From the annual report to the
British Engine and Boiler Insurance
Company. 5000 w. Mech Engr—Aug. 31,
1907. No. 86858 A.
Pressures and Work in a Gas Engine.
Cecil P. Poole. Gives formulæ for pres-
sures developed by compression, combus-
tion, and expansion; work, horse-power,
and mean effective pressure. 2000 w.
Power—Sept., 1907. No. 86676 C.
Novelties in Large Gas Engines
(Neuerungen an Grossgasmaschinen).
Herr v. Hansdorff. Relates principally to
the two-cycle Körting engines produced
by Pokorny and Wittekind of Frankfurt.
Ills. 4000 w. Zeitschr d Ver Deutscher
Ing—Aug. 17, 1907. No. 86985 D.

Gas Power Plants.
Mains and Fittings for Gas-power
Plants. L. L. Brewer. Considers the
proper design of mains, special fittings,
and related subjects. 2000 w. Power—
Sept., 1907. No. 86670 C.
The Modern Gas Engine Power Plant
(Die Moderne Gasmaschinenzentrale).
M. Langer. Discusses the fundamental
considerations in the design of a gas-
engine power plant, with particular reference to plants for the utilization of the
waste gases in ironworks. 5000 w. Stahl
u Eisen—Aug. 14, 1907. No. 86933 D.

Gas Producers.
Gas Generation and the Development
of Producer Plants (Ueber Gaserzeugung
und die Entwicklung der Generatorgas-
anlagen). J. Schmidt. A history of the
generation of gas for lighting and power
purposes from 1867 to the present time.
Ills. 2500 w. Serial. 1st part. Elektrotech u Polytech Rundschau—Aug. 14,
1907. No. 86964 D.

Indicator Diagrams.
Indicator Diagrams from a Canadian
Crude-oil Engine. H. Addison Johnston.
Briefly describes the Johnston crude-oil
engine and examines indicator diagrams
taken. 900 w. Power—Sept., 1907. No.
86669 C.

Oil Engines.
Novel High Compression Oil Engine.
Alfred Gradenwitz. An illustrated de-
tailed description of the Trinkler motor.

1000 w. Mach, N Y—Sept., 1907. No.
86748 C.
The De La Vergne Oil Engine. Illustrated description of a new engine claiming absolute safety, reliability, economy,
high speed, small space and light weight.
1500 w. Ir Age—Sept. 26, 1907. No.
87228.
See also Electrical Engineering, Generating Stations.

Water Gas.
Water Gas (Ueber Wassergas). H.
Dicke. Describes the Dellwik-Fleischer
system of water-gas generation, the first
part of the serial describing the generator
and giving details of operation, efficiency,
losses, and costs. Curves. Ills. 3000 w.
Serial. 1st part. Stahl u Eisen—Aug. 14,
1907. No. 86932 D.

HEATING AND COOLING.

Auditorium.
Cooling an Auditorium by the Use of
Ice. John J. Harris. Describes means
used to cool the auditorium of the Scranton High School during commencement
exercises. 500 w. Heat & Vent Mag—
Sept., 1907. No. 87189.

Factories.
The Ventilation of Factories. Editorial
review of the Departmental Committee
report to the Home Secretary, on investigations made to remove injurious dust
in the trades. 2500 w. Engng—Aug. 23,
1905. No. 86732 A.

Furnace.
Fan Furnace System of Heating a Hospital. Drawings and description of an
example of the use of a fan with a warm
air furnace. 1200 w. Met Work—Sept. 7,
1907. No. 86826.

Heat Losses.
Methods of Estimating Heat Losses
from Buildings. Charles L. Hubbard.
Presents, in tabular form, the information contained in Reginald Pelham Bolton's heat loss chart, with figures of several additional forms of construction.
1200 w. Heat & Vent Mag—Sept., 1907.
No. 87188.

History.
Heating from Roman Times to 1870.
Hermann Vetter. Translation of a paper
appearing in *Gesundheits-Ingenieur*. 2800
w. Met Work—Sept. 28, 1907. Serial. 1st
part. No. 87274.

Hot Water.
The Reck System of Heating by Hot-
Water Circulation (Die Warmwasser-
heizung "System Reck" mit Wassermisch-
ung). An illustrated description of the
system and of recent installations. 4500
w. Gesundheits Ing—Aug. 24, 1907. No.
86977 D.

Houses.
The Heating of Houses in France.

We supply copies of these articles. See page 397.

Charles F. Hauss. Describes the hot-water system as applied to an apartment house, and to a dwelling house. Plans. 2500 w. Met Work—Sept. 14, 1907. No. 86893.

Refrigeration.

Refrigerating Plant of Washington Brewing Company. Illustrates and describes a modern plant in Columbus, Ohio, designed for economical operation. 2500 w. Engr, U S A—Sept. 16, 1907. No. 87046 C.

The Cooling Plant for Walter Baker & Co., Ltd., Dorchester, Mass. Describes a plant embodying advanced ideas in refrigerating engineering. Ills. 2000 w. Eng Rec—Sept. 21, 1907. No. 87155.

Shops.

The Heating and Ventilation of Machine Shops. Charles L. Hubbard. Discusses methods of air distribution, showing the application of hot-blast heating to machine shops and other buildings of similar construction. 2500 w. Mach, N Y—Sept., 1907. No. 86747 C.

Vacuum System.

The Vacuum Heating System in the Godfrey Block, Grand Rapids. Brief description, with plans, of a departure from the usual arrangement of connections and radiation for vacuum heating. 1600 w. Eng Rec—Sept. 14, 1907. No. 87032.

HYDRAULICS.

See also Civil Engineering, Water Supply.

Centrifugal Pumps.

Even Diffusers on Centrifugal Pumps (Glatter Diffuser bei Zentrifugalpumpen). J. Novak. A theoretical and mathematical discussion of their design. Ills. 2500 w. Zeitschr f d Gesamte Turbinenwesen—Aug. 30, 1907. No. 86979 D.

Contraction.

On the Free Discharge of Liquids from Orifices with Defective Contraction (Ueber den Freien Ausfluss von Flüssigkeiten an Mündungen bei Unvollkommener Kontraktion). A. Jarolimek. Considers the three cases of a circular opening, a long narrow slit, and an annular orifice. Ills. 3000 w. Zeitschr d Oest Ing u Arch Ver—Aug. 9, 1907. No. 86972 D.

Friction.

The Friction of Moving Liquids (Sugli Attriti dei Liquidi in Movimento). Pietro Alibrandi. A mathematical and historical treatment of the subject, reviewing various investigations. 10000 w. Serial. 1st part. Ann della Società d Ing e d Arch Ital—Aug. 1, 1907. No. 86903 F.

Pipe Velocity.

The Velocity of Water in Conduits. From *Zeit. des Ver. Deut. Ing.* Describes the installation of J. M. Voith, of Heidenheim for conducting turbine tests, and the methods. 1200 w. Engng—Aug. 23, 1907. No. 86733 A.

Pumping Plants.

Mechanical Tests of Pumping Plants Used for Irrigation. R. P. Teele. A comparison of gasoline and steam outfits and of centrifugal, reciprocating, and air lift pumps. 1500 w. Engr, U S A—Sept. 2, 1907. No. 86709 C.

Pumping Plant for the New South Dock, Cardiff. An illustrated detailed account of the machinery of a plant constructed to deal with an unusually large quantity of water. 1300 w. Engng—Sept. 6, 1907. No. 87077 A.

Turbine Governing.

Notes on the Governing of Hydraulic Turbines. Robert S. Ball. Read before the British Assn. A discussion of details of governing mechanisms for hydraulic turbines. Ills. 4000 w. Engng—Aug. 23, 1907. No. 86737 A.

Mechanical Details of the Governors of the Water-Wheel Equipment of the Kern River Plant No. 1. Arnold Pfau. Describes the generators and the exciter units, explaining the method of regulation. Ills. 2200 w. Am Mach—Vol. 30, No. 38. No. 87105.

MACHINE WORKS AND FOUNDRIES.

Automobile Works.

The New Manufacturing Plant of the George N. Pierce Co., Buffalo, N. Y. Howard S. Knowlton. An illustrated description of this new automobile manufacturing establishment, built on the site of the Pan-American Exposition grounds. 4800o w. Eng Rec—Sept. 7, 1907. No. 86789.

Bevel Gears.

A French Machine for Cutting Bevel Gears. G. Eude. Trans. from *Revue de Mécanique.* Illustrated description of a machine for cutting bevel gears at a single cut. 2000 w. Am Mach—Vol. 30, No. 36. No. 86810.

Brass Waste.

Brass Waste in Machine Shop and Foundry. Walter J. May. Briefly considers mechanical wastes which affect the costs of shops and foundries. 1200 w. Prac Engr—Sept. 13, 1907. No. 87203 A.

Cupolas.

A Continuous Cupola Tuyere System. Illustrated description of a cupola equipped with Knoeppel tuyeres, which are designed to introduce a large volume of air into the furnace at a low pressure, insuring a low melting point, and uniform operating conditions. 1000 w. Ir Trd Rev—Sept. 5, 1907. No. 86776.

Cutlery Works.

Manufacture of Cutlery at Solingen. Report by C. F. Johnston on the industrial methods adopted at Solingen, Germany, giving details on sanitation and conditions of labor. 3000 w. Mech Engr —Aug. 24, 1907. No. 86719 A.

Cutting Metals.

Cutting Metals by Oxygen (Le Coupage des Métaux par l'Oxygène). L. Guillet. A method of cutting large masses of metal is described, which consists of heating along the cutting line to a welding heat with a blowpipe and following the latter with a jet of pure oxygen which forms a very fusible oxide. Ills. 2000 w. Génie Civil—Aug. 10, 1907. No. 86925 D.

The Cutting of Iron and Steel by Means of Oxygen (Das Zerschneiden von Eisen- und Stahlmassen mittels Sauerstoff). Arthur Dohmen. Describes the system of cutting iron and steel by a blowpipe and a jet of oxygen and illustrates many examples of the use which may be made of the system. 2200 w. Schiffbau—Aug. 28, 1907. No. 86957 D.

Dust Prevention.

Improved Methods of Dust Prevention in the Grinding Trades. Samuel R. Bennett. Describes apparatus, explaining the principles for a wet method of "racing" grindstones when first hung, and an improved hood for catching dust, etc., in the processes of dry grinding. Ills. 2200 w. Jour Soc of Arts—Aug. 30, 1907. No. 86849 A.

Engine Works.

The Minneapolis Steel and Machinery Co. Illustrated description of a factory for the manufacture of Corliss and gas engines. 2800 w. Ir Trd Rev—Sept. 5, 1907. No. 86773.

Electric Ovens.

Electrically Heated Re-heating and Tempering Furnaces (Der Glüh- und Härteofen mit Elektrisch Geheiztem Schmelzbad). L. M. Cohn. A paper dealing principally with small furnaces, read before the Elektrotechnische Verein in Vienna. Ills. 2800 w. Serial. 1st part. Elektrotech u Maschinenbau—Aug. 18, 1907. No. 86967 D.

File Works.

Making Swiss Files in America. An illustrated description of the plant and methods of manufacture of the American Swiss File & Tool Co., Elizabethport, N. J. 3000 w. Mach, N Y—Sept., 1907. Serial. 1st part. No. 86750 C.

Fire Protection.

The Prevention of Fires in Industrial Establishments (La Prevenzione degli Incendi negli Stabilimenti Industriali). A description of the fire-protection arrangements and apparatus of a Milan firm, which are a model of completeness and efficiency. 3500 w. Elettricità—Aug. 30, 1907. No. 86902 D.

Flanged Fittings.

Patterns for Flanged Pipe Fittings. F.

W. Barrows. Explains how patterns and core boxes are designed, made, combined and recorded for standard and extra heavy flanged pipe fittings. Ills. 6500 w. Am Mach—Vol. 30, No. 38. No. 87107.

Forging.

Forging a Lathe Boring Tool. J. F. Sallows. Directions, with diagrams. 500 w. Mach, N Y—Sept., 1907. No. 86752 C.

Foundries.

Some New Steel Foundries. Illustrates and describes plant and methods used at different foundries in England, United States and Germany. 4500 w. Ir & Coal Trds Rev—Aug. 23, 1907. No. 86743 A.

Iron Foundry of the Hill Clutch Co. Illustrated description of a gray iron plant in Cleveland, O., designed for the production of power transmission machinery castings. 1500 w. Foundry—Sept., 1907. No. 86818.

The Foundry Using Moulding Machines of the Aplerbecker Ironworks, Brügmann, Weyland and Co. in Aplerbeck (Die Giesserei für Formmaschinenbetrieb der Aplerbecker Hütte, Brügmann, Weyland and Co. in Aplerbeck). An illustrated description of this foundry in which all moulding is done by machines. 2500 w. Stahl u Eisen—Aug. 7, 1907. No. 86931 D.

Malleable Cast Iron Foundries (Einiges über Tempergiessereien). W. Müller. Discusses the production of malleable cast iron in crucible, cupola, converter and reverberatory furnaces. Ills. 3200 w. Stahl u Eisen—Aug. 28, 1907. No. 86934 D.

Furnaces.

Side-Fired Furnaces for Annealing, etc. Walter J. May. Illustrates and describes various types of side-fired furnaces and annealing ovens. 1000 w. Prac Engr— Aug. 30, 1907. No. 86856 A.

Gauges.

A New Swedish Combination Gaging System. E. A. Dixie. An illustrated account of a new system of gauging and of originating shop sizes. 1800 w. Am Mach—Vol. 30, No. 38. No. 87104.

Screw Gauges. Extracts from the Engineering Standards Committee's report on British Standard Systems for limit gauges for screw threads, with editorial. 1800 w. Engr, Lond—Sept. 13, 1907. No. 87220 A.

Gear Changing.

A Four Speed Change-Gear Device. B. F. Landis. Illustrates and describes an arrangement, limited to four speeds, which has proved to be efficient. 1200 w. Am Mach—Vol. 30, No. 39. No. 87241.

Gears.

Machine-molded Teeth on Large Cast Gears. An illustrated article showing how the teeth of gears are machine-

We supply copies of these articles. See page 397.

molded from a simple tooth block, and some large herring-bone gears. 2500 w. Am Mach—Vol. 30, No. 37. No. 86884.

Grinding.

Grinding Cutters for Rotary Planers. John Riddell. A wood grinding machine which automatically grinds every tooth alike and gives them all the correct clearance is illustrated and described. 700 w. Am Mach—Vol. 30, No. 39. No. 87242.

Jigs.

A Milling Jig for Connecting-rods and Straps. Grover Roy. Describes the construction and use of a convenient jig. Ills. 900 w. Am Mach—Vol. 30, No. 38. No. 87111.

Lapping.

Lapping. E. A. Johnson. Describes methods used in lapping metals. Ills. 1700 w. Am Mach—Vol. 30, No. 37. No. 86886.

Lathe.

The Lo-swing Lathe and Some of Its Work. Illustrated description of a lathe and its work as used in a Newark, N. J., shop. 1000 w. Am Mach—Vol. 30, No. 36. No. 86811.

Metallography.

Metallography in Practice (Metallographie en Practijk). P. D. C. Kley. An illustrated discussion of several of the industrial applications of metallography. 6000 w. De Ingenieur—Aug. 24, 1907. No. 86993 D.

Milling.

A New Milling Machine Dividing Head. Illustrated description of the Becker-Brainard universal index and spiral head. 1200 w. Ir Age—Sept. 5, 1907. No. 86745.

Milling Attachments for the Lathe. Oscar E. Perrigo. Illustrates and describes attachments by means of which many jobs usually performed on a milling machine may be done on the lathe. 1500 w. Am Mach—Vol. 30, No. 38. No. 87110.

Modern Machinery.

Modern Machinery and Its Future Development. H. L. Brackenbury. Read before the British Assn. An account of the forces which have affected the development. Remarks are limited to small machinery, covering the same ground as the engine lathe. 4000 w. Engng—Aug. 30, 1907. No. 86873 A.

Molding.

An Air Cooled Gas Engine Cylinder. E. F. Lake. Methods of molding and pattern-making for light weight automobile castings. Ills. 3500 w. Foundry—Sept., 1907. No. 86816.

Oerlikon.

The Oerlikon Works. A brief illustrated description of one of the most important factories in Switzerland. 1000 w. Engr, Lond—Aug. 23, 1907. No. 86741 A.

Pattern Shops.

Electric Power in a Pattern Shop. A. B. Homo. Indicates some of the points of superiority as experienced by the writer. 1100 w. Am Mach—Vol. 30, No. 38. No. 87106.

Pipe Machinery.

Some New Stoever Pipe Machinery. Illustrates and describes a pipe threading and a pipe bending machine recently brought out. 2000 w. Ir Trd Rev—Sept. 5, 1907. No. 86775.

Press Tools.

Tools for Making Can Tops. W. H. Sirius. Illustrates and describes an interesting combination of press tools. 2000 w. Am Mach—Vol. 30, No. 38. No. 87108.

Propellers.

Laying Out a Propeller. J. S. Watts. Describes a method of making a working drawing of a propeller. Diagram. 800 w. Mach, N Y—Sept., 1907. No. 86751 C.

Records.

A System for Caring for Changes in Machine Parts. E. R. Vator. Describes a process of recording drawings and changes and method of locating a piece used. 1200 w. Am Mach—Vol. 30, No. 37. No. 86885.

Riveting.

How to Drive Steel Rivets by Hand. Henry Mellon. Considers driving up the furnaces of a two-furnace marine boiler the plates of which are ⅜ inch thick. 800 w. Boiler Maker—Sept., 1907. No. 86667.

Safety Appliances.

Safety Appliances on Drawing-Frames in Cotton-Mills. Describes the details of the drawing process, showing the points where risk is entailed, and the methods of guarding against these risks. Ills. 1700 w. Engng—Aug. 30, 1907. No. 86869 A.

Screw Machines.

The Camming of Automatic Screw Machines. C. L. Goodrich and F. A. Stanley. Illustrates and describes the method of determining cam angles of Pratt & Whitney automatics by calculation or graphically to give desired feeds per turn of spindle. 3800 w. Am Mach —Vol. 30, No. 37. Serial. 1st part. No. 86882.

Solders.

Tests of Solders (Ueber Versuche mit Lötmitteln). Adolf Lippmann. Discusses a large number of types of solders and describes the results of tests made upon them. Ills. 3500 w. Serial. 1st part. Elektrotech Zeitschr—Aug. 29, 1907. No. 86992 D.

We supply copies of these articles. See page 397.

Speeds.

Speeds and Feeds in Geometric Progression. John Parker. Explains how speeds and feeds for machine tools are calculated in geometric progression by logarithms and shows typical examples. 2000 w. Am Mach—Vol. 30, . No. 39. No. 87237.

Threads.

Repairing Defective Threads. D. S. Cole. Describes briefly a method of repairing by adding the necessary new stock in the form of a wire soldered in a groove. 800 w. Am Mach—Vol. 30, No. 39. No. 87238.

A Compensating Thread-cutting Device. Ethan Viall. Illustrated description of the device and its use. 700 w Am Mach—Vol. 30, No. 39. No. 87240

Tool Works.

The Canada Tool Works. H. R. C leigh. An illustrated detailed descript of the enlarged plant of the John tram & Sons Company at Dundas, C rio. 3800 w. Ir Age—Sept. 12, No. 86879.

Twist Drills.

The Making and Testing of Drills. Fred H. Colvin. Illustra describes some of the tools used, of testing and results obtained. Am Mach—Vol. 30, No. 39. Nc

Welding.

The Oxy-Acetylene Blowpip Lightfoot. From a paper pr the Int. Acetylene Assn. Dea types of blowpipe in use and tical application. 1600 w. Ir 19, 1907. No. 87102.

Worm Gearing.

Hobs for Worm Gears. Suggestions on the design tions of the hob. 1500 w. Sept., 1907. No. 86754

MATERIALS OF CO

Alloy Investigations.

The Thermic Metho Professor Tammann a the Constitution of of Göttingen hermique du Rech s te

ngr,
ational
cription
a group
Eng Rec

Blowers
Haute Pres-
rates and de-
bo-compressor,

machines, and
10000 w.
113 E

ric
t.
del
07.

ve
e
ext
ter-t
19, No.

ooling Towers. Re-
O. Schmitt, read be-
n Assn. of Engrs. on
for the Circulation Wa-
ndensing Plants." Deals
of towers, discussing their
2000 w. Engr, Lond—
No. 87221 A.

downs.

reakdowns and Their Lessons.
ongridge. Information of in-
the annual report for 1906, in
id the safe working of power
ts. Is. 7000 w. Mech Engr—Aug.
1907. No. 86717 A.

ine Foundations.

Steam engine Foundations. R. T. Strohm. Considers the requirements, loads suited to various soils, kind of material to be used, and things to guard against. 2000 w. Power—Oct., 1907. No. 8727 C.

Fuels.

Burnin Cheap Grades of Fuel. Kingsley Willins. Points on the firing with anthracite dust and soft coal. 700 w. Elec Wld-Sept. 7, 1907. No. 86847.

The Choice of a Bituminous Coal. R. H. Kuss. Discusses the problem of selecting coal, considering the availability, cost, adatability, smoke, refuse, and treatment 2500 w. Eng Rec—Aug. 31, 1907. Nc 86694.

Steam Production from the Cheaper

ly copies of these articles. See page 397.

Grades of Anthracite. William D. Ennis.
An illustrated discuss n of methods of
burning the small and heap grades, giving much information n regard to low-
grade fuels. 2700 w. ngineering Magazine—Oct., 1907. Serl. 1st part. No.
87282 B.

Fuel Specifications.

The Purchase of Cal Under Specifications. J. E. Woodell. Presented at
meeting of Am. Soc. fr Test. Mat. Considers methods of sanling and analysis,
variations in heating lues, etc. 6000 w.
Mines & Min—Sept., 107. No. 86771 C.

Fuel Testing.

The Fuel-Testing Pnt of the Geological Survey. C. 1. Wlkinson. Illustrations and particulars c the plant in Virginia. 800 w. Eng Min Jour—Sept.
21, 1907. No. 87167.

The Swiss Goverment Fuel-Testing
Establishment in Zürh (Die Eidgenössische Prüfungsanstaltüür Brennstoffe in
Zürich). Heinrich Tschsler. Describes
in detail the various arts of the establishment and the appaitus installed. Ills.
4000 w. Schweiz B1—Aug. 24, 1907.
No. 86959 D.

Furnace Lining.

The Lining of Stea: Boiler Furnaces.
William Kavanagh. uggestions of importance in securing eonomical operation
of the plant. 1000 w Elec Wld—Sept.
7, 1907. No. 86848.

Gas Analysis.

Flue-gas Analysis: s Value. *Joseph
W. Hays. Discussesits importance in
determining furnace ficiencv. 2000 w.
Power—Oct., 1907. N. 87267 C.

Governor.

Some Peculiarities f the Rites Governor. S. H. Bunne. Discusses some
peculiarities in the ation of this governor, the causes and evention. 1500 w.
Engr, U S A—Sept. 21907. No. 86708 C.

Indicators.

Indicator Practice. rthur Curtis Scott.
Illustrates and descres types of indicators and reducing ears, methods of
taking diagrams, andof combining diagrams. 3500 w. Engi U S A—Sept. 16,
1907. No. 87048 C.

Indicators (Neuerugen an Indikatoren). A. Wagener. llustrates and describes several newl developed devices
and gives a technical liscussion of their
use. 7500 w. Zeitscl d Ver Deutscher
Ing—Aug. 31, 1907. o' 86987 D.

Lubrication.

The Design of Rir Lubricating Devices (Die Bearbeitun der Ringschmierlager). C. Volk. Illstrated description
of various types. 200w. Zeitschr d Ver
Deutscher Ing—Aug 10, 1907. No.
86982 D.

Oil Fuel.

Technical Aspects of Oil as Fuel. Franz
Erich Junge. Discusses liquid fuels, their
treatment and the formation of combustible mixtures, with regard to their application in power plants. 3000 w.
Power—Oct., 1907. Serial. 1st part. No.
87266 C.

Pumps.

Pump Troubles and Their Remedies.
H. Jahnke. Discusses some probable
causes of steam pump troubles and their
correction. 1800 w. Power—Oct., 1907.
No. 87269 C.

Pump Valves.

How to Set the Valves of a Duplex
Pump. F. F. Nickel. Gives simple directions, with explanatory uiagrams, including also the effect and treatment of the
cross-exhaust. 1200 w. Power—Oct.,
1907. No. 87264 C.

Safety Appliances.

Modern Safety Devices for Prime Movers. C. C. Major. The important features of various engine-stop and speed-
limit systems, and their application, are
illustrated and described. 3500 w. Power
—Oct., 1907. No. 87270 C.

Saturated Steam.

Physical Properties of Saturated Steam.
Fred R. Low. Explains the steam tables;
why temperature varies with the pressure,
and what is meant by latent heat, saturated steam, etc. 4500 w. Power—Sept.,
1907. No. 86668 C.

Steam Consumption.

A Graphical Table for the Determination of Steam Consumption in Large
Plants by Condenser Measurements
(Ueber eine Graphische Tabelle zur Bestimmung des Dampfverbrauches grösserer Aggregate aus dem Kondensat).
Hans Neubauer. Gives curves and tables
and describes a method of measuring
steam consumption in large plants with a
central condenser plant. Ills. 3000 w.
Oest Zeitschr f Berg u Hüttenwesen—
Aug. 17, 1907. No. 86944 D.

Superheating.

The Latest Research on the Specific
Heat of Superheated Steam. Robert H.
Smith. An account of recent investigations carried out at Munich, giving curves
showing the new results. 2000 w. Engr,
Lond—Aug. 23, 1907. No. 86739 A.

Superheated Steam for an Underground Pumping Engine at the St. Pankraz Mine at Nürschan (Heissdampf
beim Betrieb einer Unterirdischen Wasserhebmaschine der St. Pankraz-Zeche
in Nürschan). Julius Divis. Discusses
the problem of using superheated steam
underground and illustrates and describes
the Nürschan plant, giving details of operation. 4000 w. Oest Zeitschr f Berg

We supply copies of these articles. See page 397.

Speeds.

Speeds and Feeds in Geometric Progression. John Parker. Explains how speeds and feeds for machine tools are calculated in geometric progression by logarithms and shows typical examples. 2000 w. Am Mach—Vol. 30, . No. 39. No. 87237.

Threads.

Repairing Defective Threads. D. S. Cole. Describes briefly a method of repairing by adding the necessary new stock in the form of a wire soldered in a groove. 800 w. Am Mach—Vol. 30, No. 39. No. 87238.

A Compensating Thread-cutting Device. Ethan Viall. Illustrated description of the device and its use. 700 w. Am Mach—Vol. 30, No. 39. No. 87240.

Tool Works.

The Canada Tool Works. H. R. Cobleigh. An illustrated detailed description of the enlarged plant of the John Bertram & Sons Company at Dundas, Ontario. 3800 w. Ir Age—Sept. 12, 1907. No. 86879.

Twist Drills.

The Making and Testing of Twist Drills. Fred H. Colvin. Illustrates and describes some of the tools used, methods of testing and results obtained. 5000 w. Am Mach—Vol. 30, No. 39. No. 87239.

Welding.

The Oxy-Acetylene Blowpipe. Cecil Lightfoot. From a paper presented to the Int. Acetylene Assn. Deals with the types of blowpipe in use and their practical application. 1600 w. Ir Age—Sept. 19, 1907. No. 87102.

Worm Gearing.

Hobs for Worm Gears. John Edgar. Suggestions on the design and proportions of the hob. 1500 w. Mach, N Y—Sept., 1907. No. 86754 C.

MATERIALS OF CONSTRUCTION.

Alloy Investigations.

The Thermic Method of Analysis of Professor Tammann and the Researches on the Constitution of Alloys at the University of Göttingen (La Méthode d'Analyse Thermique du Professeur Tammann et les Recherches sur la Constitution des Alliages de l'Université de Goettingen). A. Portevin. Ills. 6000 w. Rev d Métal —Aug., 1907. No. 86916 E + F.

Alloy Steels.

Nickel Steel. E. F. Lake. Gives much information in regard to the effect of nickel, the treatment, and applications. 2500 w. Mach, N Y—Sept., 1907. No. 86753 C.

Boron Steels (Les Aciers au Bore). L. Guillet. Describes the influence of boron on the metallographic and physical properties of normal, tempered, an-

nealed and case-hardened steels. Ills. 5000 w. Rev d Métal—Aug., 1907. No. 86915 E + F.

Recent Researches on Ternary and Quarternary Vanadium Steels (Nouvelles Recherches sur les Aciers au Vanadium Ternaires et Quarternaires). L. Guillet. Describes the results of tests to determine the effects of different quantities of vanadium on the properties of the alloys. 4000 w. Rev d Métal—Aug., 1907. No. 86914 E + F.

Bearing Metals.

Bearing Metal for Automobile Use. Thomas J. Fay. Extract from a book on Materials for Automobile Construction. 4500 w. Automobile—Sept. 19, 1907. No. 87118.

Bronze.

Structural Bronze for Automobile Parts. Thomas J. Fay. Extracts from "Materials for Automobile Construction." Gives tests of this material showing its strength, with information concerning its production. 3000 w. Automobile—Sept. 5, 1907. No. 86823.

Ferro-Alloys.

Ferro-Alloys for Foundry Use. E. Houghton. Discusses methods of treatment to secure definite chemical and physical properties in iron, giving analyses of various alloys. 4000 w. Foundry —Sept., 1907. No. 86819.

Magnetic Alloys.

On the Magnetic Properties of Heusler's Alloys. J. C. McLennan. Abstracted from the *Phys. Rev.* A report of investigations made of these alloys. 1000 w. Elect'n, Lond—Sept. 6, 1907. No. 87067 A.

Rods.

The Effect of Antimony on High-Brass Rods. Erwin S. Sperry. Shows that over .02 per cent. of antimony in high brass will cause it to crack in rolling, and demonstrates that there is still a lack of uniformity in cathode copper. 900 w. Brass Wld—Sept., 1907. No. 87186.

Test Bars.

Tests of Some Brass and Bronze Bars. Harry B. de Pont. Gives results of a series of experiments to ascertain their chemical and physical properties. 1700 w. Foundry—Sept., 1907. No. 86817.

Thermit Steel.

Thermit Steel. Gustav Reiniger. A report of tests made with the object of deciding on the possibility of producing thermit steel of the same strength as the hard steels at present produced. Ills. 4000 w. Jour U. S. Art—July, 1907. No. 87094 D.

Tungsten.

Tungsten and Its Use as a Hardener of Steel. J. Forrest Lewis. Gives figures

We supply copies of these articles. See page 397.

regarding the increasing production of ores of this metal and its use in the manufacture of "high-speed" steel. 2000 w. Ores & Metals—Sept. 5, 1907. No. 86814.

MEASUREMENT.

Accuracy.

On Accuracy in Mensuration and Calculation. Defines accuracy as distinct from sensitiveness, and discusses the advantage of expressing approximate results in measuring in few figures. 4500 w. Engr, Lond—Sept. 6, 1907. No. 87084 A.

Belts. .

The Determination of Length of Belts and Diameters of Cone Pulleys. Mathematical. 400 w. Prac Engr—Sept. 6, 1907. No. 87051 A.

Cylinders.

Test of Tube-Connected Cylinders. S. F. Jeter. Gives a cylinder head designed by the writer and results of a series of comparative tests made between that form and the regular bumped head. Ills. 700 w. Power—Sept., 1907. No. 86674 C.

Graphite Analysis.

A Convenient Means of Determining the Ash in Graphite. S. S. Sadtler. Describes the method used. 700 w. Jour Fr Inst—Sept., 1907. No. 87088 D.

Indicator.

Toolmakers' Universal Test Indicator. J. D. Stryker. Illustrated description. 600 w. Am Mach—Vol. 30, No. 38. No. 87109.

Pressure-Gauges.

The Accuracy of Pressure-Gauges. An account of tests made at the National Physical Laboratory in England. 1500 w. Engng—Aug. 23, 1907. No. 86734 A.

Pyrometry.

Optical Pyrometry. Dr. L. Holborn. Read before the British Assn. Discusses the method of measuring high temperatures based on the distribution of energy in the spectrum, and the increase of energy between the wave-lengths as the temperature rises. 4000 w. Engng—Sept. 6, 1907. No. 87081 A.

High Temperature Measurements. Thomas C. McKay. Indicates the industrial applications of electrical and radiation thermometers, describing types. 1500 w. Cal Jour of Tech—Aug., 1907. No. 87096.

Regulations.

Weights and Measures Regulations. Information concerning the new regulations of the British Board of Trade, which come into force Oct. 1. 1907. 2500 w. Ir & Coal Trds Rev—Sept. 13, 1907. No. 87183 A.

Testing Methods.

Modern Methods of Mechanical Testing of Metallurgical Products (Les Mé-

thodes Modernes d'Essais Mécaniques des Produits Métallurgiques). L. Guillet. A description of the evolution of testing methods, the most modern developments, and the conclusions which may be drawn from them, beginning with tension tests. Ills. 8000 w. Serial. 1st part. Rev Gen d Sci—Aug. 15, 1907. No. 86919 D.

Weighing Machine.

Machine for Weighing the Forces on a Cutting Tool. John F. Brooks. Read before the British Assn. Illustrated description of an apparatus for measuring the three component forces which result in the formation and separation of a chip of metal in a lathe, explaining the principle of its operation. 900 w. Engng—Aug. 23, 1907. No. 86730 A.

POWER AND TRANSMISSION.

Air Compressors.

A Month's Output of Air Compressors. Frank Richards. Gives information showing the rapid growth of air-power and the many applications. 1700 w. Compressed Air—Sept., 1907. No. 86710.

Reciprocating Air - Compressors. W. Reavell. Deals with the features which are necessary to obtain high efficiency. 1200 w. Compressed Air—Sept., 1907. No. 86711.

Barge Crane.

A 60-Ton Barge Crane of Variable Range (Ponton-Bigue de 60 Tonnes, à Portée Variable). Ch. Dantin. Illustrated description of a crane in use on the Suez Canal. 2000 w. Génie Civil—Aug. 31, 1907. No. 86928 D.

Belts.

Pull of Belts. V. C. Wynne. Discusses tension of belts, sag, weight, and methods of maintaining tension. 1500 w. Engr, U S A—Sept. 16, 1907. No. 87047 C.

Compressed Air.

Compressed Air and the Kinetic Theory of Gases. J. H. Hart. Shows how this theory makes clear obscure points. 2000 w. Power—Oct., 1907. No. 87268 C.

Cranes.

Electric Cranes. H. H. Boughton. Reviews the rapid development of electric cranes. stating their advantages, giving particulars of comparative tests of electric and hydraulic cranes. 3800 w. Elect'n, Lond—Sept. 6, 1907. Serial. 1st part. No. 87062 A.

The Design of Under-Carriages for Jib Cranes. E. G. Fiegehen. The present article considers types of under-carriages, axles, and gearing. Ills. 2000 w. Prac Engr—Aug. 23, 1907. Serial. 1st part. No. 86716 A.

Dock Cranes.

Modern Dock and Harbor Cranes (Moderne Werft- und Hafenkrane). Bruno Müller. An illustrated description of the

Speeds.

Speeds and Feeds in Geom
gression. John Parker. Ex
speeds and feeds for machin
calculated in geometric pr
logarithms and shows typic
2000 w. Am Mach—Vol.
N   237.

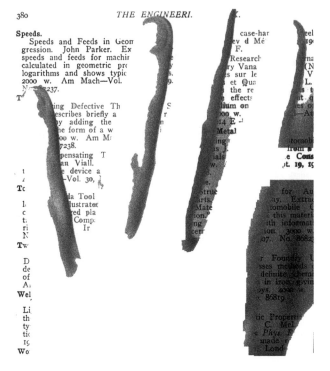

case-har   eel
ev d Mé   19

F.

Research   na
ry Vana   (N
s sur le   V
et Qua   L.

the re   s t
effects   at
lium on   es o
00 w.   At
14 E

Metal

ing   tomob
s J   iron a
ials   e Cons
w.   t. 19, 19

ing Defective Th
escribes briefly a
y adding the
e form of a w
00 w. Am M
238.

pensating T
an Viall.
e device a
Vol. 30,

la Tool
lustrate
red pla
Comp
Ir

stru
arts
Mate
ion
ng
em

for A
y. Extra
tomobile C
this materi
ith informat
ion 3000 w
7. No. 8682

r Foun
sses me
definite
in iron
ys. 4000 w
86819

tic Properti
C. Mel
Phys.
made
Lond

T

T

l
c
t
ri
N

Tw

D
de
of
A
Wel

Li
th
ty
ti
19
Wo

increasing of tho lernes d'Essais Mécaniques des
al and its 1an- P étallurgiques). L. Guillet. A
igh-speed" w. de of the evolution of testing
-Sept. 5, 1 814. m most modern developments,
a lusions which may be drawn
SUREMEN fr eginning with tension tests.
I Serial. 1st part. Rev Gen
n Mensur l- d 15, 1907. No. 86919 D.
accurac t
and dis We hine.
sing ap or Weighing the Forces on
in few a ol. John F. Brooks. Read
Sept. b British Assn. Illustrated de-
s an apparatus for measuring
nponent forces which result
of L on and separation of a chip
Pu lathe, explaining the prin-
B eration. 900 w. Engng—
No. 86730 A.

d C AND TRANSMISSION.
h s.
o Output of Air Compressors.
t ds. Gives information show-
d growth of air-power and
plications. 1700 w. Com-
-Sept., 1907. No. 86710.

ing Air - Compressors. W.
als with the features which
y to obtain high efficiency.
ompressed Air—Sept., 1

Barge Crane
n-Bigue de
le). Ch.
on of a cra ie
000 w. G ng.
86928 D.

V. uss-
s neth-
o w.
7047 C.

ic The-
ws how
e points.
. 87268 C.

Boughton.
it of elec-
ntages, giv-
tive tests of
anes. 3800 w.
907. Serial. 1st

r-Carriages for Jib
ehen. The present
es of under-carriages,
Ills. 2000 w. Prac
07. Serial. 1st part.

Harbor Cranes (Mo-
Hafenkrane). Bruno
ted description of the

Speeds.

Speeds and Feeds in Geometric Progression. John Parker. Explains how speeds and feeds for machine tools are calculated in geometric progression by logarithms and shows typical examples. 2000 w. Am Mach—Vol. 30, No. 39. No. 87237.

Threads.

Repairing Defective Threads. D. S. Cole. Describes briefly a method of repairing by adding the necessary new stock in the form of a wire soldered in a groove. 800 w. Am Mach—Vol. 30, No. 39. No. 87238.

A Compensating Thread-cutting Device. Ethan Viall. Illustrated description of the device and its use. 700 w. Am Mach—Vol. 30, No. 39. No. 87240.

Tool Works.

The Canada Tool Works. H. R. Cobleigh. An illustrated detailed description of the enlarged plant of the John Bertram & Sons Company at Dundas, Ontario. 3800 w. Ir Age—Sept. 12, 1907. No. 86879.

Twist Drills.

The Making and Testing of Twist Drills. Fred H. Colvin. Illustrates and describes some of the tools used, methods of testing and results obtained. 5000 w. Am Mach—Vol. 30, No. 39. No. 87239.

Welding.

The Oxy-Acetylene Blowpipe. Cecil Lightfoot. From a paper presented to the Int. Acetylene Assn. Deals with the types of blowpipe in use and their practical application. 1600 w. Ir Age—Sept. 19, 1907. No. 87102.

Worm Gearing.

Hobs for Worm Gears. John Edgar. Suggestions on the design and proportions of the hob. 1500 w. Mach, N Y—Sept., 1907. No. 86754 C.

MATERIALS OF CONSTRUCTION.

Alloy Investigations.

The Thermic Method of Analysis of Professor Tammann and the Researches on the Constitution of Alloys at the University of Göttingen (La Méthode d'Analyse Thermique du Professeur Tammann et les Recherches sur la Constitution des Alliages de l'Université de Goettingen). A. Portevin. Ills. 6000 w. Rev d Métal —Aug., 1907. No. 86916 E + F.

Alloy Steels.

Nickel Steel. E. F. Lake. Gives much information in regard to the effect of nickel, the treatment, and applications. 2500 w. Mach, N Y—Sept., 1907. No. 86753 C.

Boron Steels (Les Aciers au Bore). L. Guillet. Describes the influence of boron on the metallographic and physical properties of normal, tempered, an-

nealed and case-hardened steels. Ills. 5000 w. Rev d Métal—Aug., 1907. No. 86915 E + F.

Recent Researches on Ternary and Quarternary Vanadium Steels (Nouvelles Recherches sur les Aciers au Vanadium Ternaires et Quarternaires). L. Guillet. Describes the results of tests to determine the effects of different quantities of vanadium on the properties of the alloys. 4000 w. Rev d Métal—Aug., 1907. No. 86914 E + F.

Bearing Metals.

Bearing Metal for Automobile Use. Thomas J. Fay. Extract from a book on Materials for Automobile Construction. 4500 w. Automobile—Sept. 19, 1907. No. 87118.

Bronze.

Structural Bronze for Automobile Parts. Thomas J. Fay. Extracts from "Materials for Automobile Construction." Gives tests of this material showing its strength, with information concerning its production. 3000 w. Automobile—Sept. 5, 1907. No. 86823.

Ferro-Alloys.

Ferro-Alloys for Foundry Use. E. Houghton. Discusses methods of treatment to secure definite chemical and physical properties in iron, giving analyses of various alloys. 4000 w. Foundry —Sept., 1907. No. 86819.

Magnetic Alloys.

On the Magnetic Properties of Heusler's Alloys. J. C. McLennan. Abstracted from the *Phys. Rev.* A report of investigations made of these alloys. 1000 w. Elect'n, Lond—Sept. 6, 1907. No. 87067 A.

Rods.

The Effect of Antimony on High-Brass Rods. Erwin S. Sperry. Shows that over .02 per cent. of antimony in high brass will cause it to crack in rolling, and demonstrates that there is still a lack of uniformity in cathode copper. 900 w. Brass Wld—Sept., 1907. No. 87186.

Test Bars.

Tests of Some Brass and Bronze Bars. Harry B. de Pont. Gives results of a series of experiments to ascertain their chemical and physical properties. 1700 w. Foundry—Sept., 1907. No. 86817.

Thermit Steel.

Thermit Steel. Gustav Reiniger. A report of tests made with the object of deciding on the possibility of producing thermit steel of the same strength as the hard steels at present produced. Ills. 4000 w. Jour U. S. Art—July, 1907. No. 87094 D.

Tungsten.

Tungsten and Its Use as a Hardener of Steel. J. Forrest Lewis. Gives figures

We supply copies of these articles. See page 397.

regarding the increasing production of ores of this metal and its use in the manufacture of "high-speed" steel. 2000 w. Ores & Metals—Sept. 5, 1907. No. 86814.

MEASUREMENT.

Accuracy.

On Accuracy in Mensuration and Calculation. Defines accuracy as distinct from sensitiveness, and discusses the advantage of expressing approximate results in measuring in few figures. 4500 w. Engr, Lond—Sept. 6, 1907. No. 87084 A.

Belts.

The Determination of Length of Belts and Diameters of Cone Pulleys. Mathematical. 400 w. Prac Engr—Sept. 6, 1907. No. 87051 A.

Cylinders.

Test of Tube-Connected Cylinders. S. F. Jeter. Gives a cylinder head designed by the writer and results of a series of comparative tests made between that form and the regular bumped head. Ills. 700 w. Power—Sept., 1907. No. 86674 C.

Graphite Analysis.

A Convenient Means of Determining the Ash in Graphite. S. S. Sadtler. Describes the method used. 700 w. Jour Fr Inst—Sept., 1907. No. 87088 D.

Indicator.

Toolmakers' Universal Test Indicator. J. D. Stryker. Illustrated description. 600 w. Am Mach—Vol. 30, No. 38. No. 87109.

Pressure-Gauges.

The Accuracy of Pressure-Gauges. An account of tests made at the National Physical Laboratory in England. 1500 w. Engng—Aug. 23, 1907. No. 86734 A.

Pyrometry.

Optical Pyrometry. Dr. L. Holborn. Read before the British Assn. Discusses the method of measuring high temperatures based on the distribution of energy in the spectrum, and the increase of energy between the wave-lengths as the temperature rises. 4000 w. Engng—Sept. 6, 1907. No. 87081 A.

High Temperature Measurements. Thomas C. McKay. Indicates the industrial applications of electrical and radiation thermometers, describing types. 1500 w. Cal Jour of Tech—Aug., 1907. No. 87096.

Regulations.

Weights and Measures Regulations. Information concerning the new regulations of the British Board of Trade, which come into force Oct. 1. 1907. 2500 w. Ir & Coal Trds Rev—Sept. 13, 1907. No. 87183 A.

Testing Methods.

Modern Methods of Mechanical Testing of Metallurgical Products (Les Mé-

thodes Modernes d'Essais Mécaniques des Produits Métallurgiques). L. Guillet. A description of the evolution of testing methods, the most modern developments, and the conclusions which may be drawn from them, beginning with tension tests. Ills. 8000 w. Serial. 1st part. Rev Gen d Sci—Aug. 15, 1907. No. 86919 D.

Weighing Machine.

Machine for Weighing the Forces on a Cutting Tool. John F. Brooks. Read before the British Assn. Illustrated desciption of an apparatus for measuring the three component forces which result in the formation and separation of a chip of metal in a lathe, explaining the principle of its operation. 900 w. Engng—Aug. 23, 1907. No. 86730 A.

POWER AND TRANSMISSION.

Air Compressors.

A Month's Output of Air Compressors. Frank Richards. Gives information showing the rapid growth of air-power and the many applications. 1700 w. Compressed Air—Sept., 1907. No. 86710.

Reciprocating Air - Compressors. W. Reavell. Deals with the features which are necessary to obtain high efficiency. 1200 w. Compressed Air—Sept., 1907. No. 86711.

Barge Crane.

A 60-Ton Barge Crane of Variable Range (Ponton-Bigue de 60 Tonnes, à Portée Variable). Ch. Dantin. Illustrated description of a crane in use on the Suez Canal. 2000 w. Génie Civil—Aug. 31, 1907. No. 86928 D.

Belts.

Pull of Belts. V. C. Wynne. Discusses tension of belts, sag, weight, and methods of maintaining tension. 1500 w. Engr, U S A—Sept. 16, 1907. No. 87047 C.

Compressed Air.

Compressed Air and the Kinetic Theory of Gases. J. H. Hart. Shows how this theory makes clear obscure points. 2000 w. Power—Oct., 1907. No. 87268 C.

Cranes.

Electric Cranes. H. H. Boughton. Reviews the rapid development of electric cranes, stating their advantages, giving particulars of comparative tests of electric and hydraulic cranes. 3800 w. Elect'n, Lond—Sept. 6, 1907. Serial. 1st part. No. 87062 A.

The Design of Under-Carriages for Jib Cranes. E. G. Fiegehen. The present article considers types of under-carriages, axles, and gearing. Ills. 2000 w. Prac Engr—Aug. 23, 1907. Serial. 1st part. No. 86716 A.

Dock Cranes.

Modern Dock and Harbor Cranes (Moderne Werft- und Hafenkrane). Bruno Müller. An illustrated description of the

We supply copies of these articles. See page 397.

dock cranes exhibited at the Milan Exposition by the firm of Bechem and Keetman, Duisburg. 2000 w. Schiffbau—Aug. 14, 1907. No. 86955 D.

Electric Driving.
The Electric Driving of Ring Spinning Machines. Enrico Bignami. An account of tests made by Messrs. Brown, Boveri & Co. and the results. Ills. 1500 w. Elec Rev, Lond—Aug. 23, 1907. No. 86720 A.

Electricity for Cement Plants. J. B. Porter. Read before the Assn. of Am. Portland Cement Mfrs. Points out some of the special advantages of the electric drive in the manufacture of cement, both in economy and flexibility. 2500 w. Eng Rec—Sept. 21, 1907. No. 87152.

Elevators.
A Positively Controlled, High-Speed Electric Elevator. Illustrates and describes a mechanism developed by Charles Richardson Pratt. 1200 w. Eng News—Sept. 26, 1907. No. 87251.

Notes on Elevators. E. R. Carichoff. Discusses both hydraulic and electric elevators, their safety, economy, efficiency, etc. 2500 w. Elec Rev, N Y—Sept. 21, 1907. No. 87158.

The Electric Elevator Equipment for a Tall Office Building. Illustrates and describes an installation to meet unusual conditions in an 18-story building at No. 1 Wall St., New York City. 1500 w. Eng Rec—Sept. 21, 1907. No. 87150.

Failures.
Failures of Power-Generating Plant. Editorial review of the report of Michael Longridge, for 1906, on the casualties of steam engines, gas and oil engines, and electrical machinery. 2000 w. Engng—Aug. 30, 1907. No. 86871 A.

Lifting Magnets.
A New Line of Lifting Magnets for Use with Cranes. Illustrated description of new designs to meet requirements. 900 w. Eng News—Sept. 26, 1907. No. 87248.

Power Plants.
Test of Power Plant of Home Gas Company, Redlands, California. W. F. Durand. An illustrated description of the plant, with a discussion of the acceptance test made by the writer. 3500 w. Engr, U S A—Sept. 2, 1907. No. 86707 C.

The Power Plant of the Elgin National Watch Works. An illustrated description of a new power plant to supply a group of factory buildings. 3000 w. Eng Rec—Sept. 14, 1907. No. 87028.

Turbo-Compressors.
High-Pressure Centrifugal Blowers (Ventilateurs Centrifuges à Haute Pression). A. Rateau. Illustrates and describes various types of turbo-compressor,

comparing them with piston machines, and discussing their advantages. 10000 w. Rev d Métal—Aug., 1907. No. 86913 E + F.

High-Pressure Centrifugal Blowers (Kreiselgebläse für Hohen Druck). A. Rateau. An illustrated description of their development and of various types and installations. 7000 w. Zeitschr d Ver Deutscher Ing—Aug. 17, 1907. No. 86984 D.

STEAM ENGINEERING.

Boiler Cleaning.
See Marine and Naval Engineering.

Boiler Explosions.
Three Notable Boiler Explosions. Illustrated account of the explosion on the steamer "City of Trenton"; the explosion that destroyed a power plant at Marion, Ohio; and the explosion at the St. Joseph's Orphan Asylum, Philadelphia. 1000 w. Locomotive—July, 1907. No. 87093.

Boilers.
An English Boiler Improvement. Illustrates and describes the Temperley-Cockburn system of steam extraction and progressive heating in water-tube boilers. 2300 w. Ir Age—Sept. 19, 1907. No. 87101.

Cooling Towers.
The Capacity of Cooling Towers. Reviews a paper by C. O. Schmitt, read before the S. African Assn. of Engrs. on "Cooling Towers for the Circulation Water of Surface Condensing Plants." Deals with four types of towers, discussing their efficiency. Ills. 2000 w. Engr, Lond—Sept. 13, 1907. No. 87221 A.

Engine Breakdowns.
Engine Breakdowns and Their Lessons. Michael Longridge. Information of interest from the annual report for 1906, in regard to the safe working of power plants. Ills. 7000 w. Mech Engr—Aug. 24, 1907. No. 86717 A.

Engine Foundations.
Steam - engine Foundations. R. T. Strohm. Considers the requirements, loads suited to various soils, kind of material to be used, and things to guard against. 2000 w. Power—Oct., 1907. No. 87271 C.

Fuels.
Burning Cheap Grades of Fuel. Kingsley Williams. Points on the firing with anthracite dust and soft coal. 700 w. Elec Wld—Sept. 7, 1907. No. 86847.

The Choice of a Bituminous Coal. R. H. Kuss. Discusses the problem of selecting the availability, cost, adaptability, smoke, refuse, and treatment. 2500 w. Eng Rec—Aug. 31, 1907. No. 86694.

Steam Production from the Cheaper

Grades of Anthracite. William D. Ennis. An illustrated discussion of methods of burning the small and cheap grades, giving much information in regard to low-grade fuels. 2700 w. Engineering Magazine—Oct., 1907. Serial. 1st part. No. 87282 B.

Fuel Specifications.

The Purchase of Coal Under Specifications. J. E. Woodwell. Presented at meeting of Am. Soc. for Test. Mat. Considers methods of sampling and analysis, variations in heating values, etc. 6000 w. Mines & Min—Sept., 1907. No. 86771 C.

Fuel Testing.

The Fuel-Testing Plant of the Geological Survey. C. T. Wilkinson. Illustrations and particulars of the plant in Virginia. 800 w. Eng & Min Jour—Sept. 21, 1907. No. 87167.

The Swiss Government Fuel-Testing Establishment in Zürich (Die Eidgenössische Prüfungsanstalt für Brennstoffe in Zürich). Heinrich Trachsler. Describes in detail the various parts of the establishment and the apparatus installed. Ills. 4000 w. Schweiz Bau—Aug. 24, 1907. No. 86959 D.

Furnace Lining.

The Lining of Steam Boiler Furnaces. William Kavanagh. Suggestions of importance in securing economical operation of the plant. 1000 w. Elec Wld—Sept. 7, 1907. No. 86848.

Gas Analysis.

Flue-gas Analysis: Its Value. Joseph W. Hays. Discusses its importance in determining furnace efficiency. 2000 w. Power—Oct., 1907. No. 87267 C.

Governor.

Some Peculiarities of the Rites Governor. S. H. Bunnell. Discusses some peculiarities in the action of this governor, the causes and prevention. 1500 w. Engr, U S A—Sept. 2, 1907. No. 86708 C.

Indicators.

Indicator Practice. Arthur Curtis Scott. Illustrates and describes types of indicators and reducing gears, methods of taking diagrams, and of combining diagrams. 3500 w. Engr, U S A—Sept. 16, 1907. No. 87048 C.

Indicators (Neuerungen an Indikatoren). A. Wagener. Illustrates and describes several newly-developed devices and gives a technical discussion of their use. 7500 w. Zeitschr d Ver Deutscher Ing—Aug. 31, 1907. No. 86987 D.

Lubrication.

The Design of Ring Lubricating Devices (Die Bearbeitung der Ringschmierlager). C. Volk. Illustrated description of various types. 2000 w. Zeitschr d Ver Deutscher Ing—Aug. 10, 1907. No. 86982 D.

Oil Fuel.

Technical Aspects of Oil as Fuel. Franz Erich Junge. Discusses liquid fuels, their treatment and the formation of combustible mixtures, with regard to their application in power plants. 3000 w. Power—Oct., 1907. Serial. 1st part. No. 87266 C.

Pumps.

Pump Troubles and Their Remedies. H. Jahnke. Discusses some probable causes of steam pump troubles and their correction. 1800 w. Power—Oct., 1907. No. 87269 C.

Pump Valves.

How to Set the Valves of a Duplex Pump. F. F. Nickel. Gives simple directions, with explanatory diagrams, including also the effect and treatment of the cross-exhaust. 1200 w. Power—Oct., 1907. No. 87264 C.

Safety Appliances.

Modern Safety Devices for Prime Movers. C. C. Major. The important features of various engine-stop and speed-limit systems, and their application, are illustrated and described. 3500 w. Power —Oct., 1907. No. 87270 C.

Saturated Steam.

Physical Properties of Saturated Steam. Fred R. Low. Explains the steam tables; why temperature varies with the pressure, and what is meant by latent heat, saturated steam, etc. 4500 w. Power—Sept., 1907. No. 86668 C.

Steam Consumption.

A Graphical Table for the Determination of Steam Consumption in Large Plants by Condenser Measurements (Ueber eine Graphische Tabelle zur Bestimmung des Dampfverbrauches grösserer Aggregate aus dem Kondensat). Hans Neubauer. Gives curves and tables and describes a method of measuring steam consumption in large plants with a central condenser plant. Ills. 3000 w. Oest Zeitschr f Berg u Hüttenwesen— Aug. 17, 1907. No. 86944 D.

Superheating.

The Latest Research on the Specific Heat of Superheated Steam. Robert H. Smith. An account of recent investigations carried out at Munich, giving curves showing the new results. 2000 w. Engr, Lond—Aug. 23, 1907. No. 86739 A.

Superheated Steam for an Underground Pumping Engine at the St. Pankraz Mine at Nürschan (Heissdampf beim Betrieb einer Unterirdischen Wasserhebmaschine der St. Pankraz-Zeche in Nürschan). Julius Divis. Discusses the problem of using superheated steam underground and illustrates and describes the Nürschan plant, giving details of operation. 4000 w. Oest Zeitschr f Berg

We supply copies of these articles. See page 397.

u Hüttenwesen—Aug. 24, 1907. No. 86945 D.

See also Railway Engineering, Motive Power and Equipment.

Thermodynamics.

The Action of Steam During Expansion. Fred R. Low. Explains why steam exerts pressure; why pressure varies with volume and temperature; and the isothermal and adiabatic conditions. 3000 w. Power—Oct., 1907. No. 87272 C.

Turbine Lubrication.

Best Method of Supplying Oil to Turbines. Thomas Franklin. Presents the advantages of the isolated oil-filtering and cooling plant, and points to be looked to in its design and operation. 2500 w. Power—Sept., 1907. No. 86673 C.

Turbines.

Compounding Piston Engines with Turbines. Prof. A. Rateau. From a paper upon the "Development of Exhaust Steam Turbines" presented to the Soc. of Belgian Engrs., describing the advantages and answering the objections to the system. 5000 w. Power—Oct., 1907. No. 87273 C.

Combined Steam Turbines (Zur Frage der Kombinierten Dampfturbinen). W. Jasinsky. A mathematical discussion of turbines constructed partly on the impulse and partly on the reaction principle. Ills. 1600 w. Serial. 1st part. Zeitschr f d Gesamte Turbinenwesen—Aug. 30, 1907. No. 86978 D.

Turbine Tests.

See Electrical Engineering, Measurement.

MISCELLANY.

Aeronautics.

The Aeroplane Experiments of M. Louis Bleriot. Capt. Ferber. An illustrated résumé of what has been accomplished to date in France with aeroplanes of various types, including that invented

by Prof. Langley. 2200 w Sci Am Sup—Sept. 14, 1907. No. 86897.

The First British Military Airship. Illustrations of the "Nulli Secundus" and report of its trial. 1500 w. Auto Jour—Sept. 14, 1907. No. 78176 A.

The Problem of Flight and Its Solution by the Aeroplane (Le Problème de l'Aviation et sa Solution par l'Aéroplane). J. Amengaud. Discusses the principles of air-ship design and reviews the work of Santos-Dumont, Ponton, Renard, and many other inventors and experimenters. 9500 w. Bull Soc d'Encour—July, 1907. No. 86917 G.

The Problem of Flight and the Resistance of the Air (Le Problème de l'Aviation et la Résistance de l'Air). E. Maleire. The first part of the serial begins a theoretical discussion of the mechanics of aerial navigation. Ills. 4800 w. Serial. 1st part. Génie Civil—Aug. 10, 1907. No. 86926 D.

Flame.

Flame Problems and Chemical Science. Prof. A. Smithells. Abstract of president's address to the Chemical Section of the British Assn. Gives a summary of the present state of knowledge with discussion of the subject in general. 6000 w. Engng—Sept. 13, 1907. No. 87215 A.

Guns.

The Dynamics of Long Recoil. A. G. Greenhill. Mathematical discussion of an ideal case, showing the nature of the calculation required in the production of a new design of long recoil. 3000 w. Engr, Lond—Aug. 23, 1907. No. 86740 A.

Rotating Discs.

The Stresses in Swiftly Rotating Discs (Die Nebenspannungen in Rasch Umlaufenden Scheibenrädern). A. Stodola. A theoretical and mathematical paper. Ills. 5000 w. Zeitschr d Ver Deutscher Ing—Aug. 10, 1907. No. 86981 D.

MINING AND METALLURGY

COAL AND COKE.

Anthracite.

The So-Called New Supplies of Anthracite. Harry W. Althouse. Showing that recent discoveries do not indicate the existence of coal beds in unexpected places. 3000 w. Eng & Min Jour—Sept. 14, 1907. No. 87008.

Briquettes.

A Brief Record of Progress in Fuel Briquette Manufacturing at Detroit. W. S. Blauvelt. 2000 w. Am Gas Lgt Jour —Sept. 23, 1907. No. 87184.

British Coalfields.

Geological Research in the Coalfields. Information from the Geological Survey of Great Britain in regard to the work in 1906. 3000 w. Col Guard—Aug. 30, 1907. Serial. 1st part. No. 87072 A.

The Geology of the Coatbridge and Motherwell and Glasgow Districts. Describes these coalfields of Great Britain. 5500 w. Ir & Coal Trds Rev—Aug. 30, 1907. No. 86876 A.

British Collieries.

A Mining Engineer's Notes of a Visit

We supply copies of these articles. See page 397.

-to England (Bergmännische Reisebriefe aus England). Martin Baldauf. The first number of the serial begins a description of some collieries near Cardiff, comparing the methods used with those in use in Germany. Ills. 1600 w. Serial. 1st part. Oest Zeitschr f Berg u Hüttenwesen—Aug. 31, 1907. No. 86946 D.

Coke Handling.
The Quenching and Handling of Coke in Gas Works (Extinction et Manutention du Coke dans les Usines à Gaz). A. Thibeault. Illustrates a method of quenching and handling coke by a current of water. 4000 w. Génie Civil—Aug. 3, 1907. No. 86924 D.

Coking.
Advantages of the By-product Coking Process. Ernest Bury. Abstract of paper read before the British Inst. of Gas Engrs. Discusses it mainly as a gas-making apparatus. 3500 w. Min Wld—Sept. 7, 1907. No. 86782.

Deposits.
The Coal Deposits of the World. A survey of the existing more or less developed coal-deposits of the world, omitting China. 1200 w. Ir & Coal Trds Rev —Aug. 30, 1907. No. 86877 A.

Electric Powers.
Electric Power in the Coal and Coke Industry. W. B. Spellmire. Brief account of applications made of electric power and the operations at the Yorkrun plant. 1000 w. Ir Age—Sept. 12, 1907. No. 86881.

India.
Coal Mining in India. Information from the annual report issued by W. H. Pickering, chief inspector of mines in India, for the year ending Dec. 31, 1906. 3000 w. Col Guard—Aug. 30, 1907. No. 86868 A.

Kentucky.
Coal Mining in Kentucky. Describes the deposits of both the eastern and western fields, giving statistics of the production, labor employed, etc. 2000 w. Min Wld—Sept. 7, 1907. No. 86783.

Mine Roofs.
Roof-Weights in Mines. H. T. Foster. Read before the Inst. of Min. Engrs. Views from observations in longwall workings, dealing with the movement of the roof, from its first disturbance to the final settling. 2000 w. Col Guard—Sept. 6, 1907. No. 87074 A.

Mining.
The Breaking Down of Undercut Ledges of Coal by Hydraulic Pressure in the Saar District (Hereingewinnung Unterschrämter Kohlenstösse mittels Hydraulischer Sprengarbeit auf Gruben des Saarbezirks). W. Mentzel. Illustrates and describes the method and the hand pump with which the hydraulic pressure

is applied, giving costs. 3500 w. Glückauf—Aug. 3, 1907. No. 86937 D.

Montana.
The Montana Coal Situation. R. P. Tarr. Gives the past output, the extent of the deposits, their character and composition. 1000 w. Eng & Min Jour— Sept. 21, 1907. No. 87166.

Peat.
The Manufacture of Peat-fuel in Michigan. Francis J. Bulask. Illustrated description of the only "wet process" in use in America, with a brief outline of the industry. 1400 w. Power—Sept., 1907. No. 86671 C.

Pennsylvania.
Workable Coal Seams of Western Pennsylvania. William Seddon. Concerning the Pittsburg coal seam and various seams found below it; the thickness, extent, and comparative value. 1600 w. Eng & Min Jour—Sept. 21, 1907. No. 87165.

Westphalia.
Westphalian Collieries. An illustrated article giving impressions of a two days' visit. 3500 w. Can Min Jour—Sept. 1, 1907. No. 86762.

COPPER.

Bingham, Utah.
Mining the Porphyry Ore of Bingham. Walter Renton Ingalls. An illustrated description of these deposits in Utah, and the methods of mining. Two companies expect to produce about 80,000,000 pounds of copper yearly. Plate. 6000 w. Eng & Min Jour—Sept. 7, 1907. No. 86806.

Milling the Porphyry Ore of Bingham. Walter Renton Ingalls. Illustrated description of the mills and methods at Garfield, Utah, comparing the methods of the Utah and the Boston mills. 5000 w. Eng & Min Jour—Sept. 14, 1907. No. 87004.

Mining at Bingham, Utah. Robert B. Brinsmade. Gives the history and geology of the region, and methods of stripping and mining copper ores with steam shovels. Ills. 3500 w. Mines & Min— Sept., 1907. Serial. 1st part. No. 86772 C.

Converters.
Modern Hydraulically - Operated Copper Converters. G. B. Shipley. Describes the converter equipment for the Mammoth Copper Company at Kennet, Cal., which consists of two hydraulically-operated stands and eight shells, 96 inches in diameter by 150 inches long. Ill. 2000 w. Min Rept—Sept. 12, 1907. No. 87044.

Greenwater, Cal.
The Greenwater Mining District, Cal. O. M. Boyle, Jr. Information concerning the copper resources of this district. Ills. 1200 w. Cal Jour of Tech—Aug., 1907. No. 87098.

We supply copies of these articles. See page 397.

Idaho.

The Seven Devils and Snake River Districts. George D. Reid. Brief illustrated description of this copper-bearing section. 1200 w. Eng & Min Jour—Aug. 31, 1907. No. 86685.

Industry.

See Industrial Economy.

Nevada.

The Genesis of the Copper Deposits of Yerington, Nevada. E. P. Jennings. Describes the interesting geological features. 1000 w. Can Min Jour—Sept. 1, 1907. No. 86763.

Ore Deposits.

Some New Points in the Geology of Copper Ores. James F. Kemp. Read before the Can. Min. Inst. Discusses secondary enrichment, and recently acquired information in regard to the geology of copper ores, and the minerals with which they are associated. 2200 w. Min Jour —Aug. 31, 1907. No. 86866 A.

Roasting Furnace.

The McDougall Roasting Furnace. L. S. Austin. Illustrates and describes a McDougall roasting plant and its operation. 1500 w. Min & Sci Pr—Aug. 31, 1907. No. 86797.

Santo Domingo.

Metallic Sulphides in the Tuffs of Santo Domingo. F. Lynwood Garrison. Outlines the geology of Santo Domingo, and describes the copper minerals found in the tuffs, discussing their characteristics and origin. Ills. 5000 w. Min & Sci Pr— Sept. 7, 1907. No. 86891.

Smelting.

Lead and Copper Smelting at Salt Lake. Walter Renton Ingalls. An illustrated article describing the smelting plants and methods, the character of the ore supply, etc. 3500 w. Eng & Min Jour—Sept. 21, 1907. Serial. 1st part. No. 87160.

Smelting in the Boundary Copper Field, B. C. Ralph Stokes. Illustrates and describes the principal smelters, stating the characteristics of the Boundary field. 3000 w. Min Wld—Sept. 21, 1907. No. 87192.

Sudbury, Ont.

The Sudbury Nickel-Copper Field, Ontario. Ralph Stokes. Remarks on last year's production with a review of the characteristics of the region and its development. Ills. 3000 w. Min Wld— Sept. 28, 1907. Serial. 1st part. No. 87295.

GOLD AND SILVER.

Alaska.

The Great Treadwell Mines on Douglas Island. Arthur Coe Spencer. Information in regard to the deposits, development, production, geology, etc., of these gold mines. Ills. 1200 w. Min Wld— Sept. 21, 1907. Serial. 1st part. No. 87194.

Alluvials.

The Occurrence of Alluvials in Hungary and Transylvania. Louis Horváth. Trans. from *Mon. Zeit.* On the prospect of working these gold-bearing alluvials on a large scale by means of dredges. 3500 w. Min Jour—Sept. 7, 1907. Serial. 1st part. No. 87071 A.

Deep Leads of Victoria, or the Cainozoic Buried Auriferous River Deposits. H. L. Wilkinson. Deals with the buried auriferous gravel deposits, the principles determining their value from the economic standpoint. 1600 w. Inst of Min & Met. Bul 35 — Aug. 15, 1907. No. 87236 N.

Deep Alluvial Leads in the Mount Ida District, Otago. Moses Brown. From *Mount Ida Chronicle.* A study of the deposits of this district. 3300 w. N Z Mines Rec—July 16, 1907. Serial. 1st part. No. 87234 B.

Bullion Refining.

The Clean-Up, Melting, and Refining of Gold Bullion. Gerard W. Williams. Describes methods used on the Rand. 2500 w. Min & Sci Pr—Aug. 31, 1907. No. 86796.

Colorado.

Some Gold and Tungsten Deposits of Boulder County, Colorado. Waldemar Lindgren. Describes deposits in the southwestern part of the county. 3500 w. Ec-Geol—July, 1907. No. 87090 D.

Cyaniding.

Electrolytic Recovery of Gold in Cyanide Solutions. Berrih Neumann. Abstract translation from *Zeit. für Elektrochemie.* Describes experiments made to determine the current efficiency in the electrolytic recovery of gold from cyanide solutions. 900 w. Min Rept—Sept. 12, 1907. No. 87043.

Gold Mining.

Gold Mining and Gold Production. Prof. John Walter Gregory. (Cantor Lecture.) This first lecture considers alluvial gold mines, the source of gold, washing and sluicing, dredging, methods, etc. Ills. 6500 w. Jour Soc of Arts— Sept. 13, 1907. Serial. 1st part. No. 87199 A.

Hydraulic Mining.

Débris from Hydraulic Mining in California. William W. Harts. Some account of the work done by the California Débris Commission. 1600 w. Min Rept— Sept. 19, 1907. No. 87191.

Hydraulic Mining in Cariboo. Douglas Waterman. An illustrated description of this auriferous deposit and the method used in mining. 1800 w. Min & Sci Pr— Sept. 7, 1907. No. 86890.

We supply copies of these articles. See page 397.

The Hydraulic Equipment of the Old Channel Mines. John M. Nicol. Describes mines that have been worked for 30 years, especially the present plant and methods of working. 3000 w. Min & Sci Pr—Sept. 14, 1907. No. 87121.

Launders.

Flow of Water Carrying Sand in Suspension. F. K. Blue. An experimental study of factors which influence the relation between the velocity required and the proportion of sand in suspension. 2500 w. Eng & Min Jour—Sept. 21, 1907. No. 87162.

Mexico.

Old Methods in Mexico. Extract from a report by E. Tilmann, describing conditions at Guanajuato as they appeared forty years ago. 2500 w. Min & Sci Pr —Sept. 21, 1907. No. 87259.

The Mines of El Doctor. T. D. Murphy. An illustrated article giving the location of this district, with a review of its history and geology, and the methods of mining and treating the silver bearing ores. 4500 w. Min & Sci Pr—Aug. 24, 1907. No. 86655.

New Mexico.

The Lordsburg Mining Region, New Mexico. Fayette A. Jones. Gives the history of the region, its geology, and report by E. A. Jones, silver, copper, lead and miscellaneous minerals. 1600 w. Eng & Min Jour—Sept. 7, 1907. No. 86808.

New Zealand.

Notes on Waihi Ore Treatment. Ralph Stokes. Describes the characteristics of the gold ores, and their requirements, calling attention to the honey-comb tube mill liners, the independent placing of the amalgamating house, and the vacuum slimes plant. Ills. 2800 w. Jour Chem, Met & Min Soc of S Africa—July, 1907. No. 87175 E.

Ontario.

The Montreal River Silver District. Reginald Meeks. An illustrated report of a new district, 60 miles from Cobalt, and covering an area of about 80 square miles, where native silver has been found. 2500 w. Eng & Min Jour—Sept. 21, 1907. No. 87164.

Queensland.

The Premier Goldfield of Queensland. John Plummer. An account of the discovery and development of "Charter's Towers." Ills. 1200 w. Min Wld—Sept. 14, 1907. No. 87042.

Rand.

The Past, Present, and Future of the Rand. H. C. Behr. Abstract of the president's valedictory address to the Transvaal Inst. of Mech. Engrs. Briefly considers the utilization of the power of Victoria Falls and the applications to mining. Also other improvements possible.

1500 w. Min Jour—Sept. 7, 1907. No. 87068 A.

Mining Conditions in South Africa. J. B. Pitchford. A lecture delivered to the mining students of the Univ. of Cal. Deals with the value of ore, labor conditions, mining plants and methods, and costs of mining and milling. Ills. 6500 w. Mines & Min—Sept., 1907. No. 86765 C.

Labor Conditions on the Mines. H. H. Johnson. Abstract of the president's valedictory address to the Transvaal Inst. of Mech. Engrs. Discusses methods of reducing labor costs, and other labor questions. 2500 w. Min Jour—Sept. 7, 1907. No. 87069 A.

The Commercial Aspect of Rand "Profits." George A. Denny. An examination of the situation, the working costs, and future outlook. 3500 w. Min Jour— Aug. 24, 1907. Serial. 1st part. No. 86723 A.

Inaugural Address. Prof John Yates. Discusses mining practice of the present and the future outlook, education, health of miners, depression in the mining industries, etc. 8500 w. Jour Chem, Met & Min Soc of S Africa—July, 1907. No. 86999 E.

The Origin of Gold in the Rand Banket. J. W. Gregory. States the theories advanced of the genesis of the Rand gold, describes the rocks, and investigates the evidence of the different theories, deciding in favor of the marine placer theory, giving reasons. 18000 w. Inst of Min & Met, Bul No 35—Aug. 15, 1907. No. 87235 N.

Santo Domingo.

Gold Mining in Santo Domingo. F. Lynwood Garrison. An illustrated article giving information in regard to the placer gold deposits, and the probable presence of platinum. 2500 w. Eng & Min Jour—Sept. 14, 1907. No. 87006.

Victoria.

Walhalla Goldfield, Victoria. H. Herman. An illustrated account of changes and progress during the last six years. 2000 w. Aust Min Stand—Aug. 14, 1907. Serial. 1st part. No. 87232 B.

IRON AND STEEL.

America.

The Manufacture of Steel and Wrought Iron in America. Bradley Stoughton. An illustrated review of the various processes and their characteristics, the expense and quality, the uses to which the products are applied, etc. 5000 w. Engineering Magazine—Oct., 1907. No. 87278 B.

Analysis.

The Determination of Chromium in Steel in the Presence of Tungsten (Ueber

We supply copies of these articles. See page 397.

die Chrombestimmung im Stahl, inshesondere bei Anwesenheit von Wolfram). G. v. Knorre. A technical discussion of various methods, giving results. 3200 w. Stahl u Eisen—Aug. 28, 1907. No. 86935 D.

Blast Furnaces.

The Irondale Furnace Near Seattle, Wash. H. Cole Estep. Gives the history and illustrated description of this blast furnace on the Pacific Coast. 2500 w. Ir Trd Rev—Sept. 19, 1907. No. 87119.

Cold Blast Charcoal Furnace in Bald Eagle Valley, Pa. Illustrated description of an old blast furnace of the Curtin Iron Co., still in operation. 1600 w. Ir Trd Rev—Sept. 12, 1907. No. 86888.

Blast Furnaces at Vajda-Hunyad. Illustrated description of works in the south-east of Hungary, controlled by the Government. 1200 w. Engr, Lond—Sept. 6, 1907. No. 87083 A.

Electric Furnaces.

Electric Smelting Furnaces (Elektrische Schmelzöfen). Hermann Wilda. A descriptive and historical review of their development. Ills. 5000 w. Elektrotech u Polytech Rundschau—Aug. 7, 1907. No. 86963 D.

Electric Smelting.

The Electrothermic Reduction of Iron Ores. Albert E. Greene and Frank S. MacGregor. Results of an experimental investigation carried out at the laboratory of the Massachusetts Inst. of Technology on the electrothermic reduction of iron ores containing titanium. Ills. 3300 w. Elec-Chem & Met Ind—Sept., 1907. No. 86840 C.

The Production of Pig Iron in the Electric Furnace (Die Erzeugung von Roheisen im Elektrischen Ofen). B. Neumann. Illustrates and describes various types of furnace and reviews recent progress and investigation in electric smelting. 3300 w. Stahl u Eisen—Aug. 28, 1907. No. 86936 D.

Gas Fuel.

The Use of Producer Gas in the Chemical Industries. Oskar Nagel. States the general advantages of producer firing, and discusses its application to reverberatory furnaces, and to lime-kiln firing. Ills. 900 w. Elec-Chem & Met Ind—Sept., 1907. No. 86837 C.

Gogebic.

Progress on the Gogebic Range. Report of conditions at the various iron mines, the output, development, improvements, etc. 2500 w. Ir Trd Rev—Sept. 19, 1907. No. 87120.

Industry.

The Iron Industry in 1906 (Das Eisenhüttenwesen im Jahre 1906). B. Neumann. A statistical review of iron pro-

duction in the leading countries of the world during 1906, with particular reference to Germany's position in the iron trade. 6300 w. Glückauf—Aug. 31, 1907. No. 86942 D.

Manganese Ore.

The Occurrence of Manganese Iron Ore at the Elisenhöhe at Bingerbrück (Das Manganeisenerzvorkommen der Grube Elisenhöhe bei Bingerbrück). Herr Jüngst. A detailed description of the geology of this deposit in Germany, illustrated with plans and numerous sections. 2500 w. Glückauf—Aug. 10, 1907. No. 86939 D.

Metallography.

Metallography (La Metallografia). Giulio Revere. A complete review of metallographic researches and knowledge, beginning with a consideration of iron-carbon alloys. Ills. 3300 w. Serial. 1st part. Il Monit Tech—Aug. 20, 1907. No. 86907 D.

Mexico.

The Iron Deposits of Cerro de Mercado, Durango, Mexico. Edward Halse. Historical and descriptive sketch. 1800 w. Min Jour—Sept. 17, 1907. No. 87-070 A.

Osmondite.

A New Iron-Carbon Phase, Osmondite. Henry M. Howe. An account of the discovery of this new and important iron-carbon phase, showing the results, and studying the properties. 3000 w. Elec-Chem & Met Ind—Sept., 1907. No. 86-834 C.

Production.

American Pig Iron Production. H. V. Luty. A review of the production since the beginning of 1905, showing the capacity of the United States, and the future increases and prospects. 2500 w. Ir & Coal Trds Rev—Sept. 13, 1907. No. 87-182 A.

Rolling Mills.

A Bar Iron and Light Rail Mill. Plan and description of interesting features of a mill at Roanoke, Va. 2200 w. Ir Age—Sept. 12, 1907. No. 86880.

A Rolling Mill of the Pacific Coast. H. Cole Estep. Gives the history and a brief illustrated description of the plant of the Seattle Steel Co. 2000 w. Ir Trd Rev—Sept. 26, 1907. No. 87255.

The Bethlehem Steel Company's New Plant. Illustrated description of the 28-in. rail mill and the 28-in. structural mill at South Bethlehem, Pa. 3500 w. Ir Age—Sept. 26, 1907. No. 87227.

Slag.

Iron Slag. Dr. Theodor Koller. On the utilization of this waste product. 2700 w. Sci Am Sup—Sept. 21, 1907. No. 87128.

We supply copies of these articles. See page 397.

Vanadium Steel.
The Effect of Vanadium in Steel. E. T. Clarage. Explains the effect on nitrogen, oxygen, and carbon and other important effects. 2000 w. Ir Trd Rev—Sept. 26, 1907. No. 87256.

MINING.

Analysis.
Technical Methods of Analysis. W. A. Seamon. Abstracted from *W. Chem. & Met.* A criticism of modern methods, with suggestions. 2500 w. Min & Sci Pr—Aug. 24, 1907. No. 86657.

Boring.
Deep Boring at Barlow, Near Selby. H. St. John Durnford. Read before the Inst. of Min. Engrs. Some notes on a deep boring recently put down. Short discussion. 2500 w. Col Guard—Sept. 6, 1907. No. 87075 A.

Boring Devices.
A New Boring Chisel (Ein Neuer Erweiterungsmeissel). Anton Pois. The first part of the serial discusses the problem of chisel boring generally, and illustrates several types of bits. 1700 w. Serial. 1st part. Oest Zeitschr f Berg u Hüttenwesen—Aug. 31, 1907. No. 86-947 D.

Compressed Air.
Use of Compressed Air in Mining. Jos. H. Hart. Explains the value of hydraulic compression for mining uses. 1700 w. Min Wld—Sept. 14, 1907. No. 87041.

Economy.
Economy in Mining Operations. Thomas E. Lambert. Suggestions for the operating of mining properties. Ills. 2500 w. Min & Sci Pr—Sept. 14, 1907. No. 87122.

Electric Hoisting.
Electricity Applied to Hoisting (L'Electricité Appliquée à l'Extraction). L. Creplet. A discussion of the general problems of electric hoisting and the applications of various types of motor, giving the advantages of the use of electric power. Ills. 8000 w. All Indus—Aug., 1907. No. 86923 D.

The New Electric Hoisting Plant on the Ilgner System at the Salomon Mine at Ostrau, Moravia (Die Neue Elektrische Förderanlage Patent Ilgner am Salomon-Schachte in Mähr.-Ostrau). J. Havlicek. The first part of the serial discusses the design of Ilgner hoists and illustrates the arrangement of the machines at Ostrau. 1600 w. Serial. 1st part. Oest Zeitschr f Berg u Hüttenwesen—Aug. 3, 1907. No. 86943 D.

Electric Plant.
A Novel German Mining Electric Accumulator Plant. Frank C. Perkins. Brief description, with illustration, of the storage battery plant at the mining plant of the Gewerkschaft "Carlsfund" at Gross-Rhuden near Seesen, Germany. 600 w. Min Rept—Sept. 26, 1907. No. 87292.

Excavator.
A Stripping Excavator with a Conveyor to Form the Waste Bank. Illustrated description of an interesting combination of excavating and conveying plant designed to meet special conditions in the stripping of coal lands. 1200 w. Eng News—Sept. 12, 1907. No. 87001.

Explosives.
Modern Explosives (Les Explosifs Modernes). H. Schmerber. A review of the preparation, manufacture, and use of modern high-power explosives, in view of recent explosions on shipboard, the first part of the serial outlining the general principles of the theory of explosives. Ills. 3000 w. Serial. 1st part. Génie Civil—Aug. 31, 1907. No. 86929 D.

Finance.
Mining Stocks and Mining Engineers. Francis C. Nicholas. Gives reasons why mining engineers should have more intimate relations with bankers and brokers who have mining stock for sale. 2000 w. Min Wld—Sept. 7, 1907. No. 86781.

Granby, Mo.
Mining and Smelting at Granby, Missouri. Edwin T. Perkins. Describes the geology and deposits of this zinc-lead district; how the ores are mined, treated and smelted and the mines developed. 2500 w. Eng & Min Jour—Aug. 31, 1907. No. 86683.

Locating Claims.
State Provisions for Location of Lode Claims. R. S. Morrison. Explains the procedure necessary in Nevada to secure a claim. 2500 w. Ores & Metals—Sept. 5, 1907. No. 86813.

Locomotives.
Unique European Electric Mining Locomotives. Frank C. Perkins. Illustrates and describes types used in Austria, Germany, and Belgium. 2500 w. Min Wld—Sept. 7, 1907. No. 86780.

Machinery.
The Mechanical Engineering of the Mine. Charles C. Christensen. An illustrated summary of machinery that has become an important factor in the mining, treatment, and dressing of ore. 5500 w. Engineering Magazine—Oct., 1907. No. 87277 B.

Milling.
The Humphreys' Quartz Mill. C. E. Humphreys. Brief illustrated description of a mill embodying the pounding principle of the stamp-mill and the grinding-amalgamating principles of the arrastre. 900 w. Min Rept—Sept. 26, 1907. No. 87293.

Mules.
Care of Mine Mules. Dr. I. C. New-

We supply copies of these articles. See page 377.

hard. An illustrated article giving information in regard to feed and feeding, harnessing, shoeing and care of hoofs, stables, etc. 3000 w. Mines & Min—Sept., 1907. No. 86766 C.

Plant.
A New Mine Plant in Germany. B. F. Hirschauer. Illustrated description of an interesting installation at the Thiederhall mines, near Brunswick. 1500 w. Elec Rev, N Y—Sept. 28, 1907. No. 87302.

Reinforced Concrete.
Reinforced Concrete in Mining Operations. Joseph H. Hart. Calls attention to the chief applications thus far. 1500 w. Min Wld—Sept. 28, 1907. No. 87296.

Shaft Plumbing.
· A New Method of Shaft Connection. Henry Briggs. Discusses the causes of deflection of the wires in shafts, and proposes a new method, stating the advantages. 1800 w. Eng & Min Jour—Sept. 14, 1907. No. 87005.

Shaft Sinking.
Sinking of Bentley Colliery. J. W. Fryar and Robert Clive. Read before the Inst. of Min. Engrs. An account of difficult sinking through quicksand. 3000 w. Col Guard—Sept. 6, 1907. No. 87073 A.

Sinking Pumps.
Improvements in the Vertical-plunger Sinking Pump. A. H. Hale. Discusses defects in the design of sinking pumps suggesting improvements. Ills. 1200 w. Power—Sept., 1907. No. 86672 C.

Stores Keeping.
Keeping Account of Supplies. Matt. W. Alderson. The present article is introductory to a discussion of methods of accounting for mining supplies. 1200 w. Min & Sci Pr—Aug. 31, 1907. Serial. 1st part. No. 86794.

Tailings Disposal.
Dumping Residue at Kalgoorlie. M. W. von Bernewitz. An illustrated article explaining conditions and showing the necessity of using conveyors. 900 w. Min & Sci Pr—Sept. 21, 1907. No. 87258.

Timbering.
· Setting Timber in Anthracite Coal Mines. John H. Haertter. Gives reasons why it would be impracticable to have a special timber corps instead of having the props placed by the miners. Ills. 3300 w. Eng & Min Jour—Aug. 31, 1907. No. 86686.

Timbering Mines in the Joplin District, Missouri. Doss Brittain. Illustrates and describes methods of timbering during sinking and driving through soft and loose ground in use in the lead and zinc mines of this region. 2000 w. Ores & Metals—Sept. 5, 1907. No. 86812.

Square-Set Mining and a Modification of It. Claude T. Rice. Outlines a method proposed to better square-set practice.

3500 w. Min & Sci Pr—Sept. 21, 1907. No. 87257.

Unwatering.
The Unwatering of Mines in the Anthracite Region. R. V. Norris. An illustrated description of pumping methods used, their cost, etc. 4000 w. Engineering Magazine—Oct., 1907. No. 87284 B.

Vancouver Island.
Metalliferous Mining on West Coast of Vancouver Island. Notes, with illustrations, of various mining properties on the island. 7500 w. B C Min Rec—July, 1907. No. 86815 B.

Ventilation.
Concrete Overcasts in Coal Mines. John H. Haertter. Illustrates and describes methods of construction, and advantages of concrete and steel over wood in ventilating coal mines. 2500 w. Eng & Min Jour—Sept. 7, 1907. No. 86809.

Test of a Mine Fan at the Zinc-Lead Mine, Neu-Diepenbrock III (Untersuchung eines Grubenventilators auf der Zink- und Bleierzgrube Neu-Diepenbrock III). Describes the method of testing and gives the results in tables and curves. 2000 w. Glückauf—Aug. 3, 1907. No. 86938 D.

Wire Ropes.
A Few Notes on Results of Tests of Worn Ropes. W. Martin Epton. Read before the Trans. Inst. of Mech. Engrs. Gives results of ninety complete tests of worn wire ropes. Also short discussion. 2000 w. Min Jour—Aug. 31, 1907. No. 86867 A.

Württemberg.
The History of Mining in Württemberg (Württembergs Erzbergbau in der Vergangenheit). Axel Schmidt. A historical review of silver, copper, cobalt, nickel, and iron mining industry in this district of ' Germany. 5000 w. Glückauf—Aug. 17, 1907. No. 86941 D.

MISCELLANY.

Antimony.
Notes on Antimony Smelting. G. Pautrat. From *Revue di Chimie Ind.* Describes the plant and processes employed at a smelter in Mayenne, France, for the production of antimony metal and oxides from stibnite ore. 1500 w. Eng & Min Jour—Sept. 14, 1907. No. 87007.

Asbestos.
Asbestos; Its Mining, Preparation, Markets and Uses. E. Schaaf-Regelman. An illustrated review of the evolution of this industry and the production and uses. 3500 w. Engineering Magazine—Oct., 1907. No. 87279 B.

Asbestos Mining in the United States. J. S. Diller. Brief accounts of producing districts. 1500 w. Min Wld—Sept. 21; 1907. No. 87193.

We supply copies of these articles. See page 397.

Borax.

The Borax Deposit of Salinas, Near Arequipa, Peru. A. Jochamowitz. Describes this deposit, considering its origin and constitution. 1500 w. Min Jour—Aug. 24, 1907. No. 86725 A.

Cornwall.

The Quantity of Tin, Copper, and Other Minerals Produced in Cornwall. Statistics concerning the output as determined by published records. 700 w. Ir & Coal Trds Rev—Aug. 30, 1907. No. 86878 A.

Diamonds.

The Origin and Occurrences of the Diamond. Prof. T. W. E. David. A review of the occurrence of diamonds in matrix, with a discussion of the genesis of Oakley Creek diamonds in New South Wales. 3300 w. Min Jour—Aug. 24, 1907. No. 86724 A.

The Origin of Diamonds. Dr. F. W. Voit. Read before the Geol. Soc. of S. Africa. Discusses the origin of the diamond in pipes, basing the theory on the manner in which the diamond can be artificially produced. 2500 w. Min Jour—Sept. 14, 1907. No. 87181 A.

Exposition.

See Industrial Economy.

Fluorspar.

Fluorspar. Ernest F. Burchard. Considers its character, occurrence, preparation, and uses, market conditions, and production in 1906. 1500 w. Min Rept—Sept. 26, 1907. No. 87294.

Gypsum.

Montana Gypsum Deposits. J. P. Rowe. Describes the deposits in different localities, developments, mills, geological formation, etc. Ills. 2500 w. Mines & Min —Sept., 1907. No. 86767 C.

Laboratory.

Laboratory of the Coffeyville Zinc Works. E. W. Buskett. Illustrated description of laboratory equipment. 1000 w. Eng & Min Jour—Sept. 21, 1907. No. 87163.

Lead Assay.

Some Practical Notes on the Lead Assay. C. A. Cooper. Discusses the treatment of a medium grade concentrate and a zinc concentrate low in lead. 2000 w. Min Wld—Sept. 21, 1907. No. 87195.

Lime Burning.

A New Regime in Lime-Burning Furnaces (Sopra un Nuovo Regimé nei Forni per la Cottura della Calce). Frederico Carini. A paper presented before the Milan Chemical Society, discussing hearth, regenerative, and other furnaces. 2800 w. Serial. 1st part. L'Industria — Aug. 25, 1907. No. 86905 D.

Mexico.

The West Coast of Mexico. Dwight E. Woodbridge. Notes on the resources of

the country, the transportation, business methods, and labor conditions. 2700 w. Eng & Min Jour—Aug. 31, 1907. No. 86684.

New Caledonia.

New Caledonia and Its Minerals. G. M. Colvocoresses. Information in regard to the geography, geology, and mineral resources of an interesting island in the South Pacific, its discovery, exploration and settlement. Ills. Nickel, chrome and cobalt ores are the chief minerals exported. 2500 w. Eng & Min Jour—Sept. 21, 1907. No. 87161.

New Zealand.

The Mining Industry of New Zealand. P. E. Cheal. Some suggestions for its development. 6800 w. N Z Mines Rec—June 17, 1907. No. 86854 B.

Oil.

Oil in the State of Vera Cruz. Ezequiel Ordonez. Describes the conditions prevailing in this part of Mexico, and shows that they extend over quite a zone of the Gulf-coast lands. Ills. 2000 w. Min & Sci Pr—Aug. 24, 1907. No. 86656.

Pyrite.

The Oxidation of Pyrite. Alexander N. Winchell. An account of experimental investigations with pure water, charged only with air, to determine the rate of oxidation under favorable conditions. 1300 w. Mines & Min—Sept., 1907. No. 86768 C.

Rutile.

Occurrence of Rutile in Virginia. Thomas Leonard Watson. Gives results of field and laboratory study of the occurrence of rutile in the Virginia area. Ills. 2800 w. Ec Geol—July, 1907. No. 87092 D.

Tin.

Tingha Tin Dredging Industry (N. S. W.). P. J. Thibault. Introduction to articles dealing with the tin dredging plants in operation on this field. 1200 w. 'Aust Min Stand—Aug. 7, 1907. Serial. 1st part. No. 87099 B.

Turquoise.

Turquoise in the Burro Mountains, New Mexico. Edward R. Zalinski. The geology of the turquoise deposits, the composition, genesis, etc., are considered. Ills. 8000 w. Ec Geol—July, 1907. No. 87091 D.

Zinc.

Open-pit Zinc Mine at Webb City, Missouri. F. Lynwood Garrison. Illustrations, with brief description of features of interest. 600 w. Eng & Min Jour—Aug. 17, 1907. No. 86427.

The Treatment of Zinc Ores in Colorado. F. W. Traphagen. Briefly discusses the several methods of treatment. 2500 w. Mines & Min—Aug., 1907. No. 86094 C.

We supply copies of these articles. See page 397.

The Zinc Smelting Works of Swansea, Wales. Edward Walker. An illustrated description of six smelters which produce about 450 tons a week, mostly from custom ores and partly from ores imported from Italy and Africa. 1200 w. Eng & Min Jour—July 27, 1907. No. 85954.

Zinc-Lead.

The Kelly Mine, New Mexico, and Treatment of Its Ores. Woolsey McA. Johnson. An illustrated article describing the geology of the deposits of zinc-lead, their mining, concentration, etc. 2000 w. Min Wld—Aug. 17, 1907. No. 86444.

The Lead and Zinc Fields of Southwestern Wisconsin. George E. Edwards. Describes this district lying in Iowa, Illinois, and Wisconsin, and the deposits of lead and zinc, methods of mining, etc. 2500 w. Min Wld—Aug. 17, 1907. No. 86447.

Zinc-Lead Pigments.

Zinc-Lead Pigment Plant of United States Smelting Company, Canon City, Colorado. Pierce Barker. Brief description of a plant for the manufacture of zinc-lead pigment using a modification of the Lewis and Bartlett process. 700 w. Min & Sci Pr—Sept. 5, 1907. No. 86798.

Zinc Milling.

Differentiation by Leaching in the Wisconsin Zinc Region. H. A. Wheeler. Describes changes in this district by leaching action. 1500 w. Eng & Min Jour—Aug. 17, 1907. No. 86430.

Zinc Ores.

Zinc Ore and Manufactures Therefrom. W. G. Scott. Information concerning the ores of zinc and the uses made of it. 2500 w. Min Wld—Sept. 14, 1907. No. 87040.

RAILWAY ENGINEERING

CONDUCTING TRANSPORTATION.

Track Reversal.

Reversal of Track Running on the New York Central. An account of recent changes in running of trains and in signaling on the electric lines of the N. Y. C. & H. R. 1200 w. R R Gaz—Aug. 30, 1907. No. 86660.

Waste of Energy.

The Waste of Energy in Railroad Operation. D. C. Buel. Abstract of a paper presented before Trav. Engrs.' Assn. Calls attention to the channels through which energy is wasted. 2500 w. Ry & Engng Rev—Sept. 7, 1907. No. 86801.

MOTIVE POWER AND EQUIPMENT.

Air Brakes.

Variable Pressure Mechanism for Air Brakes. Illustrates and describes a mechanism operated and controlled by the friction of the shoe with the wheel, varying the shoe pressure without wasting the air pressure. The new brake is called the "Maximus." 1000 w. Sci Am—Aug. 31, 1907. No. 86678.

Locomotive Design.

Curiosities of Locomotive Design. From Development of the Locomotive Engine, by Angus Sinclair. Illustrates and describes some odd inventions in this field. 1800 w. Ry & Loc Engng—Sept., 1907. No. 86704 C.

Locomotives.

A New Continental Locomotive Type. Illustrated description of the Pacific (4-6-2) type compound engine for the Paris-Orleans Ry. 800 w. Mech Engr—Aug. 24, 1907. No. 86718 A.

Goods Locomotive for the Lancashire and Yorkshire Railway. Illustrates and describes interesting particulars of the 1000th locomotive built by this railway company. Plate. 800 w. Engng—Aug. 23, 1907. No. 86736 A.

Recent Locomotives and Carriages on the L. & Y. Railway. Brief illustrated account of the Lancashire and Yorkshire Railway Locomotive Works and of engines and cars built there. 1700 w. Engr, Lond—Aug. 30, 1907. No. 86874 A.

British 4-Cylinder Locomotives. Charles S. Lake. General discussion of the advantages and disadvantages of these engines, with illustrations and brief descriptions of recent types. 1600 w. Ry Age—Sept. 27, 1907. No. 87286.

Recent British Locomotives for Abroad. Illustrated description of locomotives lately shipped to Ceylon, South America, and Brazil. 1000 w. Mech Engr—Aug. 31, 1907. No. 86857 A.

Pacific Locomotive for the Pennsylvania Lines West. Illustrated description of a 4-6-2 locomotive, which is the heaviest passenger engine that has been built up to the present time for any road. 1500 w. R R Gaz—Aug. 30, 1907. No. 86663.

The Locomotives of the Atchison, Topeka and Santa Fe Railway. Map showing the extent of this system, with illustrated descriptions of types of locomotives used and account of the service. 3000 w. Engr, Lond—Sept. 13, 1907. No. 87217 A.

Mallet Articulated Compound Locomotive, 0-8-8-0 Type. Full illustrated detailed description of the first of an order of three pushing locomotives for the Erie

We supply copies of these articles. See page 397.

R. R., which break all records of weight, size and power. 2000 w. Am Engr & R R Jour—Sept., 1907. No. 86803 C.

A Note on Compound Locomotives. M. Maurice Demoulin. An examination of data relating to compound locomotives, with personal opinions. 3500 w. Engr, Lond—Aug. 23, 1907. Serial. 1st part. No. 86738 A.

The Development of the Bavarian Locomotive and its Representatives at the Nürnberg.Exhibition, 1906 (Die Entwicklung des Bayerischen Locomotivbaues und dessen Erzeugnisse auf der Jubiläums- und Landesausstellung in Nürnberg,1906). Julius Weil. A historical review and a description of recent productions. 1800 w. Serial. 1st part. Elektrotech u Polytech Rundschau—Aug. 28, 1907. No. 86965 D.

Locomotive Testing.

Some Important Results from the Pennsylvania Railroad Testing Plant at St. Louis. A convenient table containing results obtained from locomotives tested at the St. Louis Fair. 800 w. Am Eng & R R Jour—Sept., 1907. No. 86804 C.

Milan Exposition.

Railway Rolling Stock at the Milan Exposition (Le Matériel Roulant des Chemins de Fer à l'Exposition de Milan). L. Georges. The first number of the serial gives an elaborate, detailed description of the various types of locomotives shown, illustrated by many plates and drawings. 20000 w. Serial. 1st part. Rev Gen d Chemins d Fer—Aug., 1907. No. 86922 G.

Repairs.

Co-operation Between the Operating and Mechanical Departments. A. W. Wheatley. Suggestions for decreasing the cost of locomotive repairs. 1000 w. Am Engr & R R Jour—Sept., 1907. No. 86805 C.

Smoke.

Eliminating Smoke with Locomotives Burning Soft Coal. Abstract of a report presented at meeting of the Traveling Engrs.' Assn. Gives the recommendations of the committee. 2000 w. Eng News—Sept. 19, 1907. No. 87114.

Smoke Box.

Locomotive Smoke-Box Arrangements. E. W. Rogers. Gives examples of typical types of smoke-box arrangements which are in use to-day, and the methods of laying out this part of the boiler. Ills. 1600 w. Boiler Maker—Sept., 1907. No. 86666.

Superheating.

Superheated Steam Locomotives in Germany. An illustrated review of the use of superheated locomotives on the Prussian State Railways, the improvements made, and a brief description of the latest locomotives. 2700 w. Mech Engr—Sept. 14, 1907. No. 87204 A.

Superheated Steam Passenger Locomotive, Series B¾, of the Swiss State Railways (Heissdampf-Personenzugslokomotive Serie B¾ der Schweizerischen Bundesbahnen). M. Weiss. An illustrated detailed description of two locomotives recently put in service. 2300 w. Schweiz Bau—Aug. 3, 1907. No. 86958 D.

Washing Boilers.

Modern Method of Washing Locomotive Boilers. E. J. Harris. Describes the Miller system and the improvements introduced by D. W. Cunningham. 1600 w. Pro Iowa Ry Club—June 14, 1907. No. 87226 C.

NEW PROJECTS.

Current Construction.

Current Railway Construction. Notes on the new works in progress in Great Britain and Ireland. 1800 w. Engng—Aug. 23, 1907. No. 86735 A.

Cut-Off.

A New Transcontinental Cut-Off for the Southern Pacific and the Santa Fe. Map showing present location of lines and indicating the probable cut-off, with explanatory notes. 2200 w. R R Gaz—Aug. 30, 1907. No. 86661.

Florida.

Florida East Coast Railway Extension. Howard Egleston. Map, with brief illustrated review of the work done, and an account of the work necessary to complete the line. 13000 w. Engng-Con—Sept. 4, 1907. Serial. 1st part. No. 86831.

Splügen.

The Splügen Railway. From *Schweizerische Bauzeitung.* Information in regard to this proposed line. 3900 w. Bul Int Ry Cong—Aug., 1907. No. 86852 E.

PERMANENT WAY AND BUILDINGS.

Buildings.

The Cost of Four Frame Depots. The first of a series of articles on the cost of railway buildings. Itemized cost of labor and materials will be given, also standard plans and bills of materials. 2000 w. Engng-Con—Aug. 28, 1907. Serial. 1st part. No. 86650.

Coal Yard.

The Coal Yard at Cologne Station (Eifeltor). C. Guillery. Trans. from *Zeit. des Ver. deut. Ing.* Illustrated description of a plant for the supply of coal to locomotives. 2000 w. Bul Int Ry Cong —Aug., 1907. No. 86851 E.

Culverts.

A Method of Tunneling High Railroad Embankments for Culverts. Describes reinforced concrete culverts built in tunnels driven through the embankments. Ills. 1500 w. Eng Rec—Sept. 7, 1907. No. 86791.

Easement Curves.

A Unit Table for Talbot's Spiral. Gives

We supply copies of these articles. See page 397.

a table, based on Talbot's Spiral, compiled by G. A. Kyle, adapted to any length up to 400 ft. varying by whole feet. 3000 w. Eng Rec—Sept. 7, 1907. No. 86788.

Grade Crossings.

Elimination of Grade Crossings in New York Central Electric Zone. Illustrates and describes the High Bridge and Morris Heights stations where overhead bridges are used. 1500 w. Eng Rec—Sept. 7, 1907. No. 86790.

Rail Corrugation.

The Lack of Perfect Regularity of the Rolling Surface of Certain Hard-Steel Rails and the Inconveniences which Result from it (Note sur le Défaut de Nivellement Parfait de la Surface de Roulement de Certains Rails en Acier Dur et les Inconvénients qui en Résultent). E. Perroud. Observations on rail corrugation on the Northern Railway of France. Ills. 6600 w. Rev Gen d Chemins d Fer—Aug., 1907. No. 86921 G.

Shops.

Car Repair Shops at East Decatur, Ill. Illustrated description of shops for the Wabash Railroad. 2500 w. Ry Mas Mech —Sept., 1907. No. 87100.

The New Springfield (Mo.) Shops of the Frisco System. Plan and description of a large locomotive repair plant under construction for the St. Louis and San Francisco System. 2500 w. Ry Age— Sept. 6, 1907. No. 86799.

Stations.

The Construction and Arrangements of the New German-Russian Frontier Station at Skalmierzyce with Particular Reference to Its Mechanical Equipment (Bauund Betrieb des Neuen Preussisch-Russischen Grenzbahnhofs Skalmierzyce in Maschinentechnischer Beziehung). Hans A. Martens. Ills. 6000 w. Glaser's Ann— Aug. 15, 1907. No. 86969 D.

Terminals.

Proposed Improvement of Passenger and Freight Terminals at Buffalo. Report, with illustrations, of a proposed union passenger station and freight terminals, explaining the complex railroad situation at Buffalo. 3500 w. Ry & Engng Rev—Aug. 31, 1907. No. 86699.

Ties.

Kimball Concrete-Steel Ties. Illustrates and describes a type of reinforced-concrete tie now being tested on the Chicago & Alton and the Pere Marquette. 700 w. Ry Age—Sept. 20, 1907. No. 87190.

Transfer Bridge.

Electrically Operated Transfer Bridge. Illustrated description of the transfer bridge of the N. Y. C. & H. R. R. Co., installed at points where freight-cars are transferred from water to land, or land to water. 1000 w. Elec Age—Sept., 1907. No. 87263.

Turntables.

Modern Turntables. Remarks on changes required by the increase in length and weight of locomotives, with illustrated descriptions of recent types for different lines. 1200 w. Ry Age—Sept. 13, 1907. Serial. 1st part. No. 87045.

TRAFFIC.

Freight.

The Railroad Situation. Address of W. W. Finley before the American Association of Freight Traffic Officers, at Chicago. 3500 w. Ry & Engng Rev—Sept. 21, 1907. No. 87174.

On the Reform of Freight Traffic on the Prussian State Railways (Ueber die Reform des Güterverkehrs auf den Preussischen Staatseisenbahnen). Herr Schwabe. Describes the improvements in rolling stock and in methods of handling freight and conducting transportation which have taken place in recent years. Ills. 6000 w. Zeitschr d Oest Ing u Arch Ver—Aug. 30, 1907. No. 86974 D.

Rates.

Export Railway Rates in Austria-Hungary, Italy, etc. Particulars concerning the special tariff system of Austria-Hungary, Italy, and one or two other countries. 2000 w. Engng—Sept. 13, 1907. No. 87213 A.

MISCELLANY.

Legislation.

Railroad Legislation In Pennsylvania. A summary of some of the less important acts recently passed, with some details of the Railroad Commission Law. 1000 w. RR Gaz—Aug. 30, 1907. No. 86658.

Regulation.

Railroad Regulation. Walker D. Hines. Address before the Traffic Club of New York. Considers the laws already passed sufficient for the correction of every real evil. 2000 w. Ry Age—Sept. 27, 1907. No. 87288.

Review.

The Railways in 1905-1906. The results of working the railways during 1905 in France, England and Germany. 6800 w. Bul Int Ry Cong—Aug., 1907. No. 86853 E.

Southern Railway.

Relation of Railways to the Public. W. W. Finley. From an address at Atlanta, Ga. A brief review of transportation conditions in the Southern States, and the policy of the Southern Railway. 5500 w. Ry & Engng Rev—Sept. 7, 1907. No. 86802.

U. S. Railways.

Statistics of the Railways of the United States for 1906. An interesting summary from Poor's Manual, showing the remarkable railway activities. 2500 w. Eng News —Sept. 5, 1907. No. 86758.

We supply copies of these articles. See page 397.

STREET AND ELECTRIC RAILWAYS

Austria.

The Open Lines· of the New Alpine Railways (Die Offenen Strecken der Neuen Alpenbahnen). J. Zuffer. Describes particularly the permanent way, the tunnels, bridges, etc., giving plan and profile and tables of details of construction. Ills. 4000 w. Zeitschr d Oest Ing u Arch Ver—Aug. 2, 1907. No. 86970 D.

Chicago.

Pay-As-You-Enter Cars and Dispatching System in Chicago. An illustrated account of changes to be made in Chicago as given in a report from T. E. Mitten. 2500 w. Elec Ry Rev—Sept. 21, 1907. No. 87172.

Electrification.

Middletown-Berlin Electrification of the New York, New Haven & Hartford Railroad Company. An illustrated description of the electrification of a branch line, 9.53 miles in length. 1200 w. St Ry Jour—Sept. 21, 1907. No. 87170.

The Installation of Electric Traction on the New York Terminal Section of the New Haven Railroad. An introductory explanation with illustrated description of the installation between Woodlawn Junction and Stamford. 11300 w. Eng News —Sept. 5, 1907. No. 86755.

Reports of the Swiss Commission for Studying the Electrification of the Railways. W. Wyssling. (Abstract.) On the quantity of power required to operate the Swiss railways by means of electric traction. 7600 w. Bul Int Ry Cong—Aug., 1907. No. 86850 E.

Indiana.

The Evansville and Eastern Traction System. Features of interest on this new line in Indiana are illustrated and described. 1800 w. St Ry Jour—Sept. 28, 1907. No. 87297.

Locomotives.

Metropolitan Railway Electric Locomotives. Illustrated description of the electric locomotives for passenger and freight service on the Metropolitan Railway of London. 1800 w. St Ry Jour—Sept. 7, 1907. No. 86778.

Electric Locomotives. H. L. Kirker. Address before the Grand Trunk Ry. Literary & Scientific Inst. of Port Huron, Mich. 2500 w. Elec Age—Aug., 1907. No. 86703.

Electric Locomotives on Mountain Grades. Editorial discussion of the advantages of electric traction over steam on mountain grades of railways. 1500 w. Eng Rec—Aug. 17, 1907. No. 87233.

Los Angeles, Cal.

Track and Roadway of the Pacific Electric and Los Angeles Interurban Railways. Map showing the territory served by these companies, with an illustrated description of recent track work. 1200 w. Elec Ry Rev—Aug. 31, 1907. No. 86695.

Manchester, Eng.

Manchester Corporation Tramways. Indicates the scope of this undertaking, illustrating special track work, rolling stock and other features of interest, and describing methods of operation. 2500 w. Tram & Ry Wld—Sept. 5, 1907. No. 87052 B.

Motors.

Commutating-Pole Direct-Current Railway Motors. E. H. Anderson. Briefly reviews the important developments in railway motors, discussing ·in detail the motor named and stating its advantages as compared with the non-commutating pole type. 3000 w. Pro Am Inst of Elec Engrs—Aug., 1907. No. 87144 D.

Regeneration of Power with Single-Phase Electric Railway Motors. William Cooper. Describes a system of regenerating power which has been thoroughly tested. 2700 w. Pro Am Inst of Elec Engrs —Aug., 1907. No. 87142 D.

1200-Volt. and Commutating-Pole Direct-Current Railway Motors. E. H. Anderson. Read before the Cent. Elec. Ry. Assn. Reviews briefly features in the development of railway motors, explaining the advantages of commutating-pole motors. 3000 w. Elec Ry Rev—Sept. 28, 1907. No. 87300.

The Characteristic Curves of Rotating-Current Motors with Change-Speed Gear for the Requirements of Electric Traction (Ueber die Charakteristischen Kurven von Drehstrommotoren mit Stufenregelung der Umdrehungszahl für die Bedürfnisse der Elektrischen Traktion). Dr. W. Kummer. A theoretical and mathematical paper. Ills. 3000 w. Serial. 1st part. Schweiz Bau—Aug. 31, 1907. No. 86961 D.

Power Station.

Cos Cob Power Station, New York, New Haven & Hartford Railroad Company. Illustrates and describes this station, of interest as a part of the first installation of single-phase electric equipment for the operation of a trunk line railway. 3500 w. St Ry Jour—Aug. 31, 1907. No. 86697.

Rails.

Standardization of Rail Sections of the German Street & Interurban Railway Association. Gives a report recently pre-

We supply copies of these articles. See page 397.

sented at Mannheim, Germany, in which fifteen sections are proposed for standardization. 1500 w. St Ry Jour—Sept. 14, 1907. No. 87039.

Reconstruction.

Reconstruction of the South Side Elevated Railroad, Chicago. Illustrates and describes the work of changing grade and alignment, and other improvements. 2500 w. Elec Ry Rev—Aug. 31, 1907. No. 86696.

Rules.

Report of Committee on Interurban Rules. A report of the rules adopted at the Kingston meeting of the Street Railway Association of the State of New York, Sept. 21, 1907. 12000 w. St Ry Jour—Sept. 28, 1907. No. 87298.

Service Tests.

Service Tests on Columbus City Cars Operated Singly and in Two-Car Trains with Multiple Unit Control. Gives a summary of tests made by the Ohio State University's Department of Electrical Engineering. 2000 w. St Ry Jour—Aug. 31, 1907. No. 86698.

Shops.

Concrete Shops and Car Houses at Nashville. Illustrated description of shops representing the most advanced ideas in fireproof construction and in arrangement for facilitating car repairs. 1200 w. St Ry Jour—Sept. 14, 1907. No. 87038.

Signaling.

Track-Circuit Signaling on Electrified Roads. L. Frederic Howard. Traces the development of track-circuit apparatus during the last ten years, showing the relations existing between the track-circuit system and the propulsion system on electrified roads. Ills. 4000 w. Pro Am Inst of Elec Engrs—Aug., 1907. No. 87145 D.

The Electric Signals on the Paris Metropolitan Railway. A Soulier. Translated from *L'Ind. Elec.* Describes the two types of the Hall automatic block system installed; the track circuit system, and the pedal system. Ills. 2000 w. Elect'n, Lond —Aug. 30, 1907. No. 86864 A.

Standardization.

Report of the German Street and Interurban Railway Association's Committee on Standardization. Abstract translation. 3300 w. St Ry Jour—Sept. 7, 1907. No. 86779.

Station.

The Double-Deck Surface and Tunnel Station of the Hudson Companies in Jersey City. Outlines the extensive engineering works in progress to connect Jersey City and Manhattan Island, and gives an illustrated description of the underground station, below the present terminal of the Pennsylvania Railroad in Jersey City. 1800 w. Sci Am—Sept. 21, 1907. No. 87124.

Stubai.

The Stubai Valley Railway (Die Stubaitalbahn). An illustrated description of this Alpine railway which was the first to use high-frequency alternating current. 1800 w. Serial. 1st part. Elektrotechniker —Aug. 25, 1907. No. 86962 D.

Substations.

Substations of the Los Angeles Railway. Brief illustrated description. 1200 w. Elec Ry Rev—Sept. 14, 1907. No. 87037.

Tacoma, Wash.

The Pacific Coast Traction Company, Tacoma, Washington. An illustrated account of a new line to American Lake from Tacoma, built mostly on private right of way. 1000 w. St Ry Jour—Sept. 7, 1907. No. 86777.

Third Rail.

Third Rail Design. E. Goolding. States the necessary requirements and gives illustrations of types in use with critical notes. 1300 w. Tram & Ry Wld—Sept. 5, 1907. No. 87053 B.

Train Testing.

Improved Interurban Train Testing Apparatus. Sidney W. Ashe. An illustrated article describing improvements which greatly simplify the method of making up results. 1200 w. St Ry Jour— Aug. 24, 1907. No. 86541.

Transmission Systems.

High-Voltage Direct-Current and Alternating-Current Systems for Interurban Railways. W. J. Davis, Jr. Also discussion at Chicago. 4400 w. Pro Am Inst of Elec Engrs—Aug., 1907. No. 87148 D.

Trunk Lines.

Electric Trunk Line Operation. Briefly considers the three electric systems proposed for the operation of trunk lines and the advantage of each. 2500 w. Elec Age —Aug., 1907. No. 86700.

Discussion on "Some Facts and Problems Bearing on Electric Trunk-Line Operation," at New York, May 21, 1907. Discussion of paper by Mr. Frank J. Sprague. 16000 w. Pro Am Inst of Elec Engrs—Aug., 1907. No. 87146 D.

Tunnels.

The Selby Hill Street Railway Tunnel, St. Paul, Minn. Illustrates and describes recently completed work which greatly reduces a steep grade on an important street railway line. 3500 w. Eng Rec—Sept. 21, 1907. No. 87149.

Vienna-Baden.

The Rolling Stock of the Single-Phase Vienna-Baden Inter-Urban Line. Illustrates and describes the car equipment, giving full details of the controlling and braking systems. Adapted to use on both single-phase and direct-current circuits. 3000 w. Elect'n, Lond—Aug. 30, 1907. No. 86862 A.

We supply copies of these articles. See page 397.

EXPLANATORY NOTE—THE ENGINEERING INDEX.

We hold ourselves ready to supply—usually by return of post—the full text of every article indexed in the preceding pages, *in the original language*, together with all accompanying illustrations; and our charge in each case is regulated by the cost of a single copy of the journal in which the article is published. The price of each article is indicated by the letter following the number. When no letter appears, the price of the article is 20 cts. The letter A, B, or C denotes a price of 40 cts.; D, of 60 cts.; E, of 80 cts.; F, of $1.00; G, of $1.20; H, of $1.60. When the letter N is used it indicates that copies are not readily obtainable and that particulars as to price will be supplied on application. Certain journals, however, make large extra charges for back numbers. In such cases we may have to increase proportionately the normal charge given in the Index. In ordering, care should be taken to *give the number* of the article desired, not the title alone.

Serial publications are indexed on the appearance of the first installment.

SPECIAL NOTICE.—To avoid the inconvenience of letter-writing and small remittances, especially from foreign countries, and to cheapen the cost of articles to those who order frequently, we sell coupons at the following prices:—20 cts. each or twelve for $2.00, thirty-three for $5, and one hundred for $15.

Each coupon will be received by us in payment for any 20-cent article catalogued in the Index. For articles of a higher price, one of these coupons will be received for each 20 cents; thus, a 40-cent article will require two coupons; a 60-cent article, three coupons; and so on. The use of these coupons is strongly commended to our readers. They not only reduce the cost of articles 25 per cent. (from 20c. to 15c.), but they need only a trial to demonstrate their very great convenience—especially to engineers in foreign countries, or away from libraries and technical club facilities.

Write for a sample coupon—free to any part of the world.

CARD INDEX.—These pages are issued separately from the Magazine, printed on one side of the paper only, and in this form they meet the exact requirements of those who desire to clip the items for card-index purposes. Thus printed they are supplied to regular subscribers of THE ENGINEERING MAGAZINE at 10 cents per month, or $1.00 a year; to non-subscribers, 25 cts. per month, or $3.00 a year.

THE PUBLICATIONS REGULARLY REVIEWED AND INDEXED.

The titles and addresses of the journals regularly reviewed are given here in full, but only abbreviated titles are used in the Index. In the list below, *w* indicates a weekly publication, *b-w*, a bi-weekly, *s-w*, a semi-weekly, *m*, a monthly, *b-m*, a bi-monthly, *t-m*, a tri-monthly, *qr*, a quarterly, *s-q*, semi-quarterly, etc. Other abbreviations used in the index are: Ill—Illustrated: W—Words; Anon—Anonymous.

Alliance Industrielle. *m.* Brussels.
American Architect. *w.* New York.
Am. Engineer and R. R. Journal. *m.* New York.
American Jl. of Science. *m.* New Haven, U. S. A.
American Machinist. *w.* New York.
Annales des Ponts et Chaussées. *m.* Paris.
Ann. d Soc. Ing. e d Arch. Ital. *w.* Rome.
Architect. *w.* London.
Architectural Record. *m.* New York.
Architectural Review. *s-q.* Boston.
Architect's and Builder's Magazine. *m.* New York.
Australian Mining Standard. *w.* Melbourne.
Autocar. *w.* Coventry, England.
Automobile. *w.* New York.
Automotor Journal. *w.* London.
Beton und Eisen. *qr.* Vienna.
Boiler Maker. *m.* New York.
Brass World. *m.* Bridgeport, Conn.
Brit. Columbia Mining Rec. *m.* Victoria, B. C.
Builder. *w.* London.
Bull. Am. Iron and Steel Asso. *w.* Phila., U. S. A.

Bulletin de la Société d'Encouragement. *m.* Paris.
Bulletin of Dept. of Labor. *b-m.* Washington.
Bull. Soc. Int. d'Electriciens. *m.* Paris.
Bulletin of the Univ. of Wis., Madison, U. S. A.
Bulletin Univ. of Kansas. *b-m.* Lawrence.
Bull. Int. Railway Congress. *m.* Brussels.
California Jour. of Tech. *m.* Berkeley, Cal.
Canadian Architect. *m.* Toronto.
Canadian Electrical News. *m.* Toronto.
Canadian Engineer. *m.* Toronto and Montreal.
Canadian Mining Journal. *b-w.* Toronto.
Cassier's Magazine. *m.* New York and London.
Cement. *b-m.* New York.
Cement Age. *m.* New York.
Central Station. *m.* New York.
Chem. Met. Soc. of S. Africa. *m.* Johannesburg.
Colliery Guardian. *w.* London.
Compressed Air. *m.* New York.
Comptes Rendus de l'Acad. des Sciences. *w.* Paris.
Consular Reports. *m.* Washington.
Deutsche Bauzeitung. *b-w.* Berlin.

397

Domestic Engineering. *w.* Chicago.
Economic Geology. *m.* New Haven, Mass.
Electrical Age. *m.* New York.
Electrical Engineer. *w.* London.
Electrical Review. *w.* London.
Electrical Review. *w.* New York.
Electric Journal. *m.* Pittsburg, Pa.
Electric Railway Review. *w.* Chicago.
Electrical World. *w.* New York.
Electrician. *w.* London.
Electricien. *w.* Paris.
Electrochemical and Met. Industry. *m.* N. Y.
Elektrochemische Zeitschrift. *m.* Berlin.
Elektrotechnik u Maschinenbau. *w.* Vienna.
Elektrotechnische Rundschau. *w.* Potsdam.
Elektrotechnische Zeitschrift. *w.* Berlin.
Elettricità. *w.* Milan.
Engineer. *w.* London.
Engineer. *s-m.* Chicago.
Engineering. *w.* London.
Engineering-Contracting. *w.* New York.
Engineering Magazine. *m.* New York and London.
Engineering and Mining Journal. *w.* New York.
Engineering News. *w.* New York.
Engineering Record. *w.* New York.
Eng. Soc. of Western Penna. *m.* Pittsburg, U. S. A.
Far Eastern Review. *m.* Manila, P. I.
Fire and Water. *w.* New York.
Foundry. *m.* Cleveland, U. S. A.
Génie Civil. *w.* Paris.
Gesundheits-Ingenieur. *s-m.* München.
Giorn. dei Lav. Pubb. e d Str. Ferr. *w.* Rome.
Glaser's Ann. f Gewerbe & Bauwesen. *s-m.* Berlin.
Heating and Ventilating Mag. *m.* New York.
Ice and Refrigeration. *m.* New York.
Industrial World. *w.* Pittsburg.
Ingenieria. *b-m.* Buenos Ayres.
Ingenieur. *w.* Hague.
Insurance Engineering. *m.* New York.
Int. Marine Engineering. *m.* New York.
Iron Age. *w.* New York.
Iron and Coal Trades Review. *w.* London.
Iron and Steel Trades Journal *w.* London.
Iron Trade Review. *w.* Cleveland, U. S. A.
Jour. of Accountancy. *m.* N. Y.
Journal Asso. Eng. Societies. *m.* Philadelphia.
Journal of Electricity. *m.* San Francisco.
Journal Franklin Institute. *m.* Philadelphia.
Journal Royal Inst. of Brit. Arch. *s-qr.* London.
Jour. Roy. United Service Inst. *m.* London.
Journal of Sanitary Institute. *qr.* London.
Jour. of South African Assn. of Engineers. *m.* Johannesburg, S. A.
Jour. of U. S. Artillery. *b-m.* Fort Monroe, U. S. A.
Jour. W. of Scot. Iron & Steel Inst. *m.* Glasgow.
Journal Western Soc. of Eng. *b-m.* Chicago.
Journal of Worcester Poly. Inst., Worcester, U. S. A.
Locomotive. *m.* Hartford, U. S. A.
Machinery. *m.* New York.
Madrid Científico. *t-m.* Madrid.
Manufacturer's Record. *w.* Baltimore.
Marine Review. *w.* Cleveland, U. S. A.
Men. de la Soc. des Ing. Civils de France. *m.* Paris.
Métallurgie. *w.* Paris.
Minero Mexicano. *w.* City of Mexico.

Mines and Minerals. *m.* Scranton, U. S. A.
Mining and Sci. Press. *w.* San Francisco.
Mining Journal. *w.* London.
Mining Reporter. *w.* Denver, U. S. A.
Mittheilungen des Vereines für die Förderung des Local und Strassenbahnwesens. *m.* Vienna.
Motor Wagon. *w.* Cleveland, U. S. A.
Municipal Engineering. *m.* Indianapolis, U. S. A.
Municipal Journal and Engineer. *w.* New York.
Nature. *w.* London.
Nautical Gazette. *w.* New York.
New Zealand Mines Record. *m.* Wellington.
Oest. Wochenschr. f. d. Oeff. Baudienst. *w.* Vienna.
Oest. Zeitschr. Berg & Hüttenwesen. *w.* Vienna.
Plumber and Decorator. *m.* London.
Popular Science Monthly. *m.* New York.
Power. *m.* New York.
Practical Engineer. *w.* London.
Pro. Am. Soc. Civil Engineers. *m.* New York.
Pro. Canadian Soc. Civ. Engrs. *m.* Montreal.
Proceedings Engineers' Club. *qr.* Philadelphia.
Pro. St. Louis R'way Club. *m.* St. Louis, U. S. A.
Pro. U. S. Naval Inst. *qr.* Annapolis, Md.
Quarry *w.* London.
Queensland Gov. Mining Jour. *m.* Brisbane, Australia.
Railroad Gazette. *w.* New York.
Railway Age. *w.* Chicago.
Railway & Engineering Review. *w.* Chicago.
Railway and Loc. Engng. *m.* New York.
Railway Master Mechanic. *m.* Chicago.
Revista d Obras. Pub. *w.* Madrid.
Revista Tech. Ind. *m.* Barcelona.
Revue de Mécanique. *m.* Paris.
Revue Gén. des Chemins de Fer. *m.* Paris.
Revue Gén. des Sciences. *w.* Paris.
Revue Industrielle. *w.* Paris.
Revue Technique. *b-m.* Paris.
Rivista Gen. d Ferrovie. *w.* Florence.
Rivista Marittima. *m.* Rome.
Schiffbau. *s-m.* Berlin.
Schweizerische Bauzeitung. *w.* Zürich.
Scientific American. *w.* New York.
Scientific Am. Supplement. *w.* New York.
Sibley Jour. of Mech. Eng. *m.* Ithaca, N. Y.
Stahl und Eisen. *s-m.* Düsseldorf.
Stevens Institute Indicator. *qr.* Hoboken, U. S. A.
Street Railway Journal. *w.* New York.
Technograph. *yr.* Urbana, Ill.
Technology Quarterly. *qr.* Boston, U. S. A.
Tijds V h Kljk. Inst. V Ing. *qr.* Hague.
Tramway & Railway World. *m.* London.
Trans. Am. Ins. Electrical Eng. *m.* New York.
Trans. Am. Ins. of Mining Eng. New York.
Trans. Am. Soc. Mech. Engineers. New York.
Trans. Inst. of Engrs. & Shipbuilders in Scotland, Glasgow.
Transport. *w.* London.
Verkehrstechnische Woche und Eisenbahntechnische Zeitschrift. *w.* Berlin.
Wood Craft. *m.* Cleveland, U. S. A.
Yacht. *w.* Paris.
Zeitschr. f. d. Gesamte Turbinenwesen. *w.* Munich.
Zeitschr. d. Mitteleurop. Motorwagon Ver. *s-m.* Berlin.
Zeitschr. d. Oest. Ing. u. Arch. Ver. *w.* Vienna.
Zeitschr. d. Ver. Deutscher Ing. *w.* Berlin.
Zeitschrift für Elektrochemie. *w.* Halle a S.
Zeitschr. f. Werkzeugmaschinen. *b-w.* Berlin.

Note—Our readers may order through us any book here mentioned, remitting the publisher's price as given in each notice. Checks, Drafts, and Post Office Orders, home and foreign, should be made payable to THE ENGINEERING MAGAZINE.

Bins.

The Design of Walls, Bins and Grain Elevators. By Milo S. Ketchum. Size, 9 by 6½ in.; pp., xiv, 394. Ills., 260. Tables, 40. Price, $4. New York: The Engineering News Publishing Company; London: Archibald Constable and Company.

Those familiar with Prof. Ketchum's book on Steel Mill Buildings will welcome this pioneer treatise on bin design, which is characterized by the same thoroughness, clearness, and logical and systematic arrangement displayed in the former volume. Since a thorough understanding of the theory of the retaining wall is essential to a correct understanding of the theory of the pressure of granular materials in bins, the author devotes considerable space to a discussion of the various retaining-wall theories and their application to actual design. The second part of the book is devoted to the design of bins for the storage of coal, ore, etc., and the concluding section takes up in detail the design of grain bins and elevators. The extensive use of reinforced concrete in this class of work is recognized and considerable space is given to a discussion of the theory of reinforced concrete construction and to the development of the formulæ necessary for use in the design of walls, bins and grain elevators. Appendices give a glossary of masonry terms and sets of standard specifications for stone-masonry, concrete and steel construction. A valuable feature of the book is to be found in the tables of costs of actual structures which are included wherever possible and analyzed so thoroughly as to be of the greatest assistance and value. For practical data and scientific and theoretical accuracy, Prof. Ketchum's book can be recommended to the student and practicing engineer alike.

Civil Service Examinations.

Manual of Examinations for Engineering Positions in the Service of the City of New York. By Myron H. Lewis and Milton Kempner. Size, 9 by 6 in.; pp., about 600. Ills. Price, $5. New York: The Engineering News Publishing Company.

The main body of this book is divided into eight parts covering respectively, examinations for the positions of axeman; chainman and rodman; leveler; transitman and computer; assistant engineer—Rapid Transit Commission; assistant engineer—general, aqueduct, docks, sewers, and highways; draftsman and draftsman's helper; inspector—buildings, masonry and carpentry, steel, regulating, grading, and paving, and sewers; and two appendices giving useful formulæ for surveyors and civil engineers. Under each section are given sets of previous examination questions for the various positions and a number of typical questions with full answers. Two general appendices to the volume give information and previous examination papers of the civil service of the United States, State of New York, Cities of Buffalo, Boston, New Orleans, etc. The book is likely to be of value to examiners in setting papers, either for the civil service or in technical schools, and to students preparing for examinations for the various positions covered.

Steel Castings.

Open Hearth Steel Castings. By W. M. Carr. Size, 8 by 5½ in.; pp., 118. Ills. Cleveland, O.: The Penton Publishing Company.

A series of practical papers on the methods involved in the manufacture of open-hearth steel castings by the basic and acid processes, compiled from articles written by the author for *The Iron Trade Review* and *The Foundry*.

BOOKS RECEIVED.

Die Ausnutzung der Wasserkräfte. By E. Mattern. Size, 9½ by 6½; pp., 260. Ills. Leipzig: Wilhelm Engelmann.

Das Praktische Jahr des Maschinenbau-Volontärs. By F. zur Nedden. Size, 8½ by 6½ in.; pp., 234. Berlin: Julius Springer.

Der Thermodynamik der Dampfmaschinen. By Fritz Krauss. Size, 8 by 5½ in.; pp., 144. Ills. Price, M. 3. Berlin: Julius Springer.

The Sanitary Evolution of London. By Henry Jephson. Size, 9 by 6 in.; pp., 440. Price, $1.80. Brooklyn, N. Y.: A. Wessels Company.

Details of Mill Construction. By Hawley Winchester Morton. Size, 12½ by 9½ in.; Plates, 25. Boston, Mass.: Bates and Guild Company.

Computation and Mensuration. By P. A. Lambert. Size, 7½ by 5 in.; pp., 92. Ills. Price, 80 cents. New York: The Macmillan Company.

Concrete Country Residences. Second Edition, 1907. Size, 12¾ by 10 in.; pp., 168. Ills. Price, $1. New York: The Atlas Portland Cement Company.

Quick Rules for Heating and Ventilating. By Wm. J. Baldwin, Jr. Size, 6 by 4½ in.; pp., 23. Price, $1. New York, N. Y.: Wm. J. Baldwin, Jr.

Annual Report of the Department of Public Works of the City of Buffalo, N. Y. Size, 9 by 6 in.; pp., 1094. Buffalo, N. Y.: Department of Public Works.

Fifth Annual Report of the State Board of Public Roads, Rhode Island, 1907. Size, 9 by 6 in.; pp., 39. Ills., 58. Providence, R. I.: Department of Public Roads.

Report of the State Engineer and Surveyor, State of New York, 1905. Size, 9 by 6 in.; pp., 775. Ills. Maps. Albany, N. Y.: State Engineer and Surveyor.

The Six-Chord Spiral. By J. R. Stephens. Size, 7½ by 5 in.; pp., 68. Ills. Cloth, $1.25; boards, $1. New York: The Engineering News Publishing Company.

Specifications for Street Roadway Pavements. By S. Whinery. Size, 9 by 6 in.; pp., 56. Price, 50 cents. New York: The Engineering News Publishing Company.

Power Stations and Power Transmission. By George C. Shaad. Size, 9½ by 6½ in.; pp., 150, v. Ills. Price, $1. Chicago: American School of Correspondence.

Annual Reports of the Department of the Interior, 1905; Commissioner of Education, Vol. 2. Size, 9 by 6 in.; pp., 657 to 1400. Washington, D. C.: Department of the Interior.

Report of the Fire Department of the City of New York for the Quarter and Year Ending December 31, 1906. Size, 10 by 7 in.; pp., 319. New York: Fire Department.

The Finances of Gas and Electric Light and Power Enterprises. By William D. Marks. Fourth Edition. Size, 7 by 5½ in.; pp., 540. Price, $4. New York: William D. Marks.

The Elements of Mechanics. By W. S. Franklin and Barry Macnutt. Size, 8½ by 6 in.; pp., xi, 283. Ills. Price, $1.50. New York: The Macmillan Company; London: Macmillan and Company, Limited.

The Problem of Flight, a Text-Book of Aerial Engineering. By Herbert Chatley. Size, 9 by 6 in.; pp., 119. Ills. London: Charles Griffin and Company, Limited; Philadelphia, Pa.: J. B. Lippincott Company.

The Thermo-Dynamic Principles of Engine Design. By L. M. Hobbs. Size, 8 by 5½ in.; pp., 143. Ills. Price, $1.75. London: Charles Griffin and Company, Limited; Philadelphia, Pa.: J. B. Lippincott Company.

Sixth Annual Report of the Metropolitan Water and Sewerage Board, being Public Document No. 57, Massachusetts. Size, 9 by 6 in.; pp., vii, 253. Ills. Boston, Mass.: Metropolitan Water and Sewerage Board.

The Mineral Industry, its Statistics, Technology and Trade during 1906. Volume XV. Edited by Walter Renton Ingalls. Size, 9 by 6 in.; pp., xxiv, 954. Ills. Price, $5. New York and London: Hill Publishing Company.

Report on British Standard Systems for Limit Gauges for Screw Threads. Prepared by the Engineering Standards Committee and edited by the Secretary, Leslie S. Robertson. Size, 13 by 8 in.; pp., 19. Price, 1/-. London: Crosby Lockwood and Son.

Papers and Reports relating to Minerals and Mining comprising Statement by the Minister of Mines; Report on the Goldfields; Report on Coal-Mines; State Coal-Mines. Size, 13 by 8½ in.; pp., 200. Ills. Maps. Wellington, New Zealand: Department of Mines.

Substitution of the Electric Locomotive for the Steam Locomotive. By Lewis B. Stillwell and H. St. Clair Putnam. Reprint from Volume XXVI of the Transactions of the American Institute of Electrical Engineers. Size, 9 by 6 in.; pp., 301. Ills. Price, $1. New York: American Institute of Electrical Engineers.

Report of the Fuels Committee of the Motor Union of Great Britain and Ireland, Together with the Evidence of Witnesses, and Memoranda on Petroleum, Petroleum Spirit, Paraffin, Shale, Benzol, Alcohol, etc., and a Bibliography. Size, 11 by 8½ in.; pp., 27. Price, 1/-. London: The Motor Union of Great Britain and Ireland.

Lubrication and Lubricants. A Treatise on the Theory and Practice of Lubrication. By Leonard Archbutt and R. Mountford Deeley. Second Edition, thoroughly revised and enlarged. Size, 8½ by 5½ in.; pp., xxx, 538. Ills. Price, $6. London: Charles Griffin and Company, Limited; Philadelphia, Pa.: J. B. Lippincott Company.

Supplement to the Report of the State Engineer and Surveyor, State of New York, 1905. History of the Canal System of the State of New York, together with Brief Histories of the Canals of the United States and Canada. By Noble E. Whitford. Two Volumes. Size, 9 by 6 in.; pp., 1547. Ills. Albany, N. Y.: State Engineer and Surveyor.

VOL. XXXIV   DECEMBER, 1907.   NO. 3

A RISING INDUSTRIAL PROBLEM; THE NEW APPRENTICESHIP.

By George Frederic Stratton.

We have repeatedly expressed the belief that the provision of "the mechanic of the future" was one of the rising problems of modern industry, too long postponed and soon imperatively to be met. Further reference to this view is made in our editorial pages in this issue. We wish here to call attention especially to the light thrown by Mr. Stratton on the present attitude and policy of the trade unions toward the efforts to recruit the army of trained workers. His interviews with representative New England trade-union leaders, which have been confirmed as printed and are authoritative, show that the old antagonism to the trade school still prevails in the United States, though some change of policy has induced a change of expression so far as public-school manual training is concerned. On the other hand, a most cheering condition is revealed in the present disposition of the unions to encourage shop training for the young mechanic. The manufacturer who is at last becoming interested in the question of training the workmen of the future is thus assured of an open opportunity, free (at least for a while) of union opposition.—THE EDITORS.

FROM the commencement of factory systems, in the early part of the nineteenth century, until the last decade, the boys of the two greatest industrial countries—America and England—have been sadly neglected. The estimation of the value of the apprentice which obtained in the seventeenth and eighteenth centuries almost entirely disappeared upon the introduction of machinery and organized factories. The boy was used, it is true, but only as a factory "cub" and because he could perform some mechanical operation as well as a man, and at greatly reduced cost. He was hired at no better rate of pay than apprentices had obtained, and he was taught nothing but some single operation. In fact, as machines were improved and became more and more automatic, younger boys than ever before—and even girls—were put at factory work and the boy found himself in competition with child labor.

In some shops a few apprentices were still taken on, but their condition was no better than that of the machine boys. The regular mechanics did not look favorably upon them and, as it was nobody's business to teach them, they learned but little. They were the butt of every workman and were considered an unmitigated nuisance by busy foremen; and the employers, hearing of no good results from them, gradually lost all interest in the system (such as it was) so that during the middle of the nineteenth century it was very unusual, indeed, to find a single apprentice in many of the large shops.

The increase in the size of factories, followed by the consolidation of small industries into great central organizations, created a demand for skilled mechanics, foremen, and expert workmen on outside installations and repairs, which could not be met; and the old factory proprietor who had been puzzled by undeveloped and unsettled conditions and who had devoted himself to meeting those conditions as they arose—without much far-reaching preparation for the future—gave way to the industrial manager with great capital, scientifically adjusted organization, and broad, comprehensive views of the future as well as of the present.

Thus it is that the example set ten, eight, or six years ago, by a few great companies, in devising new apprentice systems worthy of the name, is being followed today by a large number of industrial managers, and the interest excited by these systems *per se,* is largely increased by the absolute necessity for obtaining a much greater supply of good, all-round mechanics in every trade.

Some of the manufacturers' associations, notably the National Metal Trades Association and the Machine Tool-Builders' Association, have formulated conditions for apprenticeship contracts upon broad and comprehensive lines for the use of their members.* The present tendency of large employers, however, is to devise special systems which, while embracing practically all of the ideas in the association schedules, contain some additions and modifications, and a review of those systems will show that in almost all cases the conditions and requirements have been studied and met, with the same common-sense progressiveness which so eminently distinguishes American captains of industry.

The system of the New York Central Railroad Company is exceedingly comprehensive and interesting. It starts a boy as an ordinary shop apprentice, on a four-years term; but as he develops his

---

* The conditions officially approved by the National Metal Trades Association and recommended to its members for adoption were quoted in full in connection with Mr. O. M. Becker's article "A Modern Adaptation of the Apprenticeship System" in THE ENGINEERING MAGAZINE for November, 1906.

abilities and characteristics, he obtains the opportunity to switch off onto other departments, providing he shows indications of becoming more valuable in those departments than in the shops. Thus the shop apprentice, in two or three years, may be transferred to the drafting-room, to the engineering department, or to some division superintendent's department.

A certain amount of night study in various branches is insisted upon; literature and the best of illustrated lectures are continually furnished, and the young men are also transferred, when willing, from one of the company's shops to another, thus affording them varied and valuable experiences.

Complete card records covering the entire course are kept and filed in the Superintendent of Apprentices' office, and every year the following searching list of questions regarding each boy is answered by the foreman or department superintendent, and sent to the apprentice superintendent :—

1. Does he work overtime on drawings or problems?
2. Is he the type of boy we wish to have in our employ?
3. Is his attitude toward his employer good?
4. Does he spend his time well outside of shop hours?
5. Have you (or has the shop instructor) succeeded in gaining his confidence—that is, would he come to you first 'in trouble of any kind?
6. Can you recommend him, at present, to start in the company's drafting-room, or will he qualify during the present year?—(Give probable date.)
7. What is his strongest point, or for what type of work is he best fitted?
8. What is his weakest point, or for what type of work is he least fitted?
9. Does he live at home, or board?
10. What is his address?

Each year two or three of the best of the apprentices are selected for a two-years technical course, in the line for which they show the greatest promise, and as the expenses are paid by the company, the incentive to excel and to obtain the scholarship with its opening into fields of higher possibilities is very great.

While this plan and the plans of many other employers are meeting with very gratifying success, there are others which are meeting with failure, and which are, for that reason, as valuable for study as those which are more fortunate or of better design.

The apprentice contract of one great industrial company contains the following clause :—

"The compensation paid to employees by the ——— Company for services, covers inventions, and the undersigned hereby agrees, in consideration of such compensation, that every invention pertaining to the business of the Company conceived or developed by him during such employment, is the property of the ——— Company, and he agrees to make assignment of such inventions to the said Company, on request; the Company to pay the expense of securing the patents."

It is perhaps needless to say that, although this company has splendid equipment and its arrangements in all other respects are excellent for apprentices, it is getting but very few young men to sign that contract. Whatever may be the true view as to the fairness of that clause, it is certain that parents generally are by no means willing that a mortgage shall be given upon their boys' possible ability or chance of inventing something of great value, in return for a recompense averaging 8¼ cents per hour for the four-years course.

The causes of another partial failure are more involved, and perhaps the result is even more unfortunate for the company which suffers by it. The conditions under which its apprentices are received are exceptionally good and liberal. The course is four years. In addition to a rate of 6, 8, 10, and 12 cents per hour for each year, respectively, a bonus of $25 is given at the end of the second year, $50 at the end of the third year, and $100 at the expiration of the term. The instruction covers a very large field, as the works are extensive and contain every description of machine tool, besides a great number of special machines. The boys, also, receive such schooling as is adapted to their needs—arithmetic, elementary mathematics, and drawing sufficient to enable them to read other drawings with facility. The time occupied by this schooling is paid for by the company.

But when the young man has completed his course the failure comes; and in order to understand it clearly it will be necessary to glance at the company's policy regarding all of its employees. The shop is an open one. When any new man is taken on his hourly rate is fixed according to a schedule established by the manager, for all classes of work. The rate, at the commencement, is usually low, but the start is always accompanied by a promise of a raise in a few weeks, if the man's work is satisfactory—a promise which is kept. After that, increases of pay of 2 cents per hour are granted yearly, if the man is worth it, until the limit of good pay is reached.

Now, the way this system operates with the apprentice is this: when his time is out he is receiving 12 cents per hour. If he stays on with the company he is usually given, at once, an advance to 16 or 18 cents, and thereafter the usual raise of 2 cents per hour each succeeding year until he is earning the pay of a first-class mechanic.

This is not satisfactory to the young man. No matter how good a mechanic he is, he has got to serve another seven years before he can command the normal good wage of 30 cents per hour. He is a journeyman as soon as he is out of his time, and he feels that he should, at once, receive a higher rate than 18 cents per hour, which is much less than hundreds of men are getting all about him on machine jobs—men who do not pretend to be machinists and who have never worked, as he has, for four years at a rate averaging 9 cents per hour. Accordingly, he looks for another job (and gets it, with little or no difficulty) and the company loses the man it has spent four years in educating. Of all the apprentices who have finished their course with the company, but 22 per cent have remained with it; a result due to having placed the young journeyman on the same footing as any entirely new employee, and subject to the same inelastic system.

The difficulty, of course, arises from the difference in the points of view between the manager and the man. The manager considers that the young man has received a liberal education and training, and that the company has had, so far, but little profit from his work in return. The quick raise from 12 to 18 cents per hour seems to him to be adequate and even liberal. Moreover, he is naturally unwilling that his established system of periodical and fixed increases of wages up to a set limit for all employees—a system which has worked satisfactorily with thousands of hands for some years—should be broken by a different rate schedule for the apprentice graduate. He feels (and perhaps with good reason) that two methods of advancing pay rates would induce much discontent.

On the other hand, the young journeyman argues that his services for the previous four years have been well worth the wages he has received—that the account, so far, is squarely balanced—and that he should now be considered a mechanic and paid according to his skill and general usefulness. He is by no means willing to make his new start at 18 cents per hour, and under a system which will permit only an annual raise of 2 cents per hour.

"And so——" as Mr. Dooley says, "there yez ar-re!"

In addition to these various and generally very excellent plans for training apprentices in the shops, there has been, in the past decade, a remarkable movement among the authorities of technical schools and the public schools for the establishment of manual-training classes as a part of the regular education. Among industrial managers, however, there appears to be considerable diversity of opinion as to the practical utility of these classes. One who has devoted much

thought and study to the development of an efficient apprentice system said to me:—

> "The technical schools are excellent so long as they confine the studies
> to the principles of the trades they are teaching. But they, and the
> grammar schools, can do little or no good to a boy by dabbling in man-
> ual training. Neither I nor any other manufacturer would lessen an
> apprentice course by three months on account of a boy's having been
> through manual training in his school. It is far better for him to put in
> that time in rounding out his education; we'll teach him all about the
> tools."

Another said:—

> "The schools should teach mechanical drawing to every boy who
> expects to go into an industrial occupation—not in order to make him a
> draughtsman, but so that he can easily and correctly read regular draw-
> ings—a thing which three-fourths of our machinists cannot do, excepting
> in the case of simple details. These drawing studies would save a lot
> of valuable time when the boy begins to learn his trade. The manual
> training will not, I think, amount to much when he gets up against the
> real conditions."

Some recent communications to the National Society for the Promotion of Industrial Education* give the views of other prominent
manufacturers:

R. K. LeBlond, president, R. K. LeBlond Machine Tool Co., Cincinnati, Ohio:—

> "I think it of just as great importance that hands and brain should
> be educated technically as any other. Any man forfeits a great deal
> of pleasure and usefulness in this life when his education lacks a con-
> structive course. I do not favor evening schools of any description,
> while I realize they are an absolute necessity and splendid results are
> obtained from them. I have always held that it is just as necessary
> to good health and good results that a certain amount of time be given
> to recreation as to work, study and sleep."

Henry Hess, Hess-Bright Manufacturing Co., Philadelphia, Pa.:—

> "I favor preparatory trade school work under public-school auspices,
> but at present in this country these schools usually run to fads. What
> I have seen of the work turned out and 'pointed to with pride' by the
> amateurs who run them, is of a most abominably slouchy character which
> not only fails of its first object but also has a decidedly deleterious influ-
> ence on the scholars, in giving them the idea that careful and conscien-
> tious work is not material. I do not favor trade schools under the
> auspices of manufacturing concerns as a substitute for the other kind."

---

* The quotations are extracted from Bulletin No. 3 of the Society—"A Symposium on
Industrial Education," containing the contributions of thirty-two employers and labor
leaders.

Richard Moldenke, secretary, American Foundrymen's Association, Watchung, N. J.:—

> "I believe that the only hope we have to keep this nation in the front, industrially, is to push industrial education with might and main—and not wait very long before beginning. All trades should be taught. The trade school should teach its students the principles of the respective trades in question, together with enough practical manipulation to make the student self-supporting from the start, after leaving the school. It should also give him a general education, so as not to get the student into grooves."

Frank C. Caldwell, vice president, H. W. Caldwell and Son Company, engineers, founders and machinists, Chicago:—

> "Owing to union restrictions and a number of other reasons, there is very little opportunity for boys to acquire the special training necessary to fit them as practical workmen in some specific trade. We certainly think there is a promising field for a number of such schools. * * * * We are quite certain that the graduates of such schools would be much better equipped as all-round workmen than the average apprentice at the end of his term, or the average workman who has picked up his trade in actual practice. We are inclined to favor the general school as distinguished from the special school."

Any discussion of this vital problem of Industrial Education of course involves the consideration of the attitude of organized labor towards it. And here we are at once confronted with a situation which appears to be exceedingly complicated and difficult; and which is by no means made clearer by the recent change of heart which appears to have affected some of the labor leaders. In 1904 Mr. John Mitchell expressed the following opinion for publication:—

> "It is a regrettable fact that a large number of the graduates from the industrial schools are imbued with a hostile spirit toward trade unionism. In many cases the instruction is of such a nature as to fail to promote sympathy on the part of the boys for the doctrines and customs of labor organizations. This defect should be remedied. No line of cleavage should separate the shop-taught man and the school-taught man. The boys at an industrial school should learn, not a trade alone, but methods for bettering their condition within the trade. I do not believe that graduates of industrial schools will permanently remain outside the trade-union movement; but much needless friction and bitter feeling might be avoided if their instruction were of such a nature as to create in them a sympathetic feeling towards the great trade-union movement."

Mr. Samuel Gompers, in 1905, forcibly expressed the following opinion:—

"It is not only ridiculous, but positively wrong, for trade schools to continue in their turning out of botch workmen, who are ready and willing, upon their 'graduation,' to take the places of workmen at far below the wages prevailing in the trade. With practically half the toiling masses of our country unemployed the continuance of the practice is tantamount to a crime."

Two years ago the Mason Builders' Union of Boston passed resolutions forbidding any of their members to teach in a trade school.

The late Col. Auchmuty, head of the New York Trade Schools, made this statement shortly before his death:—

"The American boy has no rights which organized labor is bound to respect. If he is taught his trade in a trade school he is refused admission to nearly all trade unions."

Again, in the annual report of the New York Bureau of Labor, for 1905, the secretary of one of the unions is quoted as saying:—

"I believe in all journeymen and apprentices being connected with the unions. If a boy becomes a full-fledged mechanic in a technical school he doesn't know anything about unions; nor would he have any sympathy with their rules and regulations."

A large number of other quotations might be cited showing the antagonistic feelings of many union leaders, two, three, and four years ago to trade schools. Now, however, a recent request of the National Society for the Promotion of Industrial Education for the latest opinions on the subject brings forth the following:—

John F. Tobin, general president, Boot and Shoe Workers' Union, Boston:—

"I am in favor of public education conducted at the public expense, wherein an opportunity is given in the practical workings of any given trade. * * * * * It is frequently said that trade unionists are opposed to industrial education, but this is not true. The opposition which appears amongst trade unionists is because they have in mind the particular private enterprises that have been conducted solely for profit."

John Golden, general president, United Textile Workers of America, Fall River, Mass.:—

"I can safely say that organized labor is not, and never will be opposed to industrial education properly controlled and scientifically administered.* * * * Rest assured that the labor unions will, at all times, oppose trade schools of the character I have mentioned" (corporation schools—"scab hatcheries," Mr. Golden calls them) "but will further a movement for more opportunities for industrial and technical education."

John Fitzpatrick, president Chicago Federation of Labor says:—

"I am in favor generally of industrial education. The form I favor is that of the preparatory and practical. I believe that all trades can be taught. * * * * * I favor preparatory trade school work, under public auspices, but do not favor trade schools conducted by manufacturing concerns."

Henry Abrahams, Secretary Central Labor Union, Boston, Mass.:—

"Personally, I am of the opinion the trade school is coming, hence we must recognize the inevitable. If we were to take the children of twelve and for two years teach them the use of tools they would find themselves better fitted for the battle of life. There should be evening schools for men in engineering, electricity, plumbing. I am opposed to trade schools run by private corporations."

I. B. Armstrong, master mechanic, L. A. No. 3662, Knights of Labor, Lynn, Mass.:—

"It would seem that, in order to obtain the greatest results or benefits from trade schools, experience in the trade should be made a condition or prerequisite for admission to trade schools, with the possible exception of the graduates of the public schools in the locality. The limiting of applicants to the school to those already engaged in the business would be necessary in order to prevent the over-crowding of the industry. I do not favor trade schools conducted by manufacturing concerns.'

Emma Stehagen, secretary Women's Trade Union League, Chicago, Ill.:—

"I am in favor of industrial education if carried on in the proper manner, by which I mean, under the auspices of the public schools and giving practical teaching. The trades union movement stands for the uplifting of the worker, and I believe an industrial education is one of the aids of trade unionists. If schools are conducted by manufacturing concerns * * * * they are to be deprecated."

Although these opinions are decided, and in some instances forcible, they are by no means clear excepting as regards the unanimous opposition to instruction courses founded and conducted by the manufacturer. And it was for the special purpose of obtaining an insight into the underlying influences which produce this antagonism to such schools, that I secured interviews with and explanations from several prominent union leaders as to their attitude. The first was with Mr. Abrahams of the Central Labor Union, who expressed himself as being deeply interested in the educational problem and desired to put his views in writing. He says:—

> "The trade school is here, and here to stay! Such being the fact, we wage-earners, whose children are to be the workers of the future, are compelled, from the stand-point of self-preservation, to sense this question in all its lights.
>
> "We find that there are many of the large corporations establishing what are termed 'trade schools.' In these establishments the learner is taught but a part of the craft—the school being established for profit to the concern and not for the benefit of the community or the uplifting of the race—and this should be of sufficient weight to prevent the approval of their aims and objects by organized labor.
>
> "Again, there is a suspicion in the minds of some that their object is to build up a wage class.
>
> "We want no class lines in America!
>
> "Regarding the reason raised by some of our business firms, that they cannot secure sufficient skilled help, it is due, in many cases, to the conditions existing in their plants, viz:—long hours and small wages.
>
> "As we are to have these schools, I for one am of the opinion that such institutions run by the State, with no view of profits, but whose sole object shall be the welfare of the student, do not contain the objections that can be urged against private institutions of this nature.
>
> "In most, if not all, schools run in the interests of a corporation a boy has to work so long for nothing. As fast as these boys become proficient at any branch the skilled workmen are discharged. Again, those pupils who are not skilled, in the event of strikes for better conditions, can be used as strike-breakers—as was recently done in one case here in Massachusetts."

"What would be the standing, Mr. Abrahams," I asked, "of a boy graduated from a public industrial training school?" The answer was not instantly forthcoming and I added:—"Would he at once take his place as a journeyman?"

"If our committee were satisfied," he answered, "that he was a competent workman he could obtain admission to the union."

"And if he cannot satisfy your committee as to his ability, can he obtain employment, at low wages, so as to acquire experience?"

"Not in a union shop. Of course he can go into an open shop and make any arrangement he pleases."

I went from him to Mr. Tobin, president of the Boot and Shoe Workers' Union, and one of the deepest thinkers on labor problems in New England. He displayed much bitterness (as indeed did every labor man I interviewed) against the trade schools which are run by private parties and for a set tuition fee for the course.

"Such schools," he said, "are for the sole purpose of private gain. A fixed charge, to be paid by the pupil, is required for the opportunity to apply himself to whichever machine, used in the shoe trade, he desires to learn. The pupil applies himself to this machine, more or less regularly, for a few weeks and the tendency is for him to conclude that he has mastered the machine at an early date and is now qualified to work for wages. In this conclusion he is encouraged by those in charge of the school, as the tendency is to make a place for another pupil who is willing to pay from $25 to $100 for the privilege of learning. The quantities of operators turned out, rather than the quality, is the aim."

"Just what objection is made by labor leaders to the new apprenticeship system which is being established in some great factories?"

"As regards apprenticeship in the shoe trade, the idea of any apprentice system, covering a period of years, is ridiculous. An apprentice who applies himself to the task of learning all the various branches of the shoe trade, instead of being better equipped by the wide range, is positively handicapped—for the reason that he would not be able to acquire the necessary skill and speed to compete successfully with specialists on one branch. In other words, in most mechanical trades the subdivision is so great as to require speed as the principal factor, coupled with skill; which latter is comparatively easy to acquire providing the pupil has a natural aptitude for the line of work in which he has engaged."

"What is your opinion as to the usefulness of manual training in the public schools?"

"I say that an education of that kind tends to train and fix the mind, and better fits the pupil for any calling. The time devoted to manual training better equips the pupil for mental training."

"What chance does a boy so trained have of getting employment in a union shop?"

Mr. Tobin's fine eyes twinkled good-humoredly. "Well," he replied, with some little hesitation, "the locals settle that matter. In Brockton, for instance, they insist upon six-months experience in some other shop before admitting to the shops there."

"And where can he get that six-months experience?"

A shrug of the shoulders and a slight smile preceded the answer: "That is difficult to tell!"

The secretary of the Machinists' Union, at Providence, was extremely bitter against what he called the "catch-penny trade schools." "Look at that!" he exclaimed, as he savagely turned the pages of a magazine and placed his finger upon an advertisement. I read as follows:—"You can get $5.00 a day after a two to three months' course in the plumbing, bricklaying and plastering or electrical trades."

"I'm expecting," said the secretary, "to see an ad. like that about
machinists, at any moment! And the half-baked, so-called 'mechanics'
they turn out are crowding into the trades and throwing out good
men who have served apprenticeship and had large experience."

"What about the new apprenticeship systems which are in vogue
in some of the great machine shops—the Pratt and Whitney shops,
for instance, or the big locomotive shops?"

"Look here!" he exclaimed, "every one has a motive, and the
motive of those big corporations is to obtain, first, a supply of young
men, under contract, for four years at rascally low pay, and secondly, to build up a large class of workers so as to have a choice—
and also reduce wages."

"How do you feel about manual training in the public schools?"

"Oh, well—that, of course, is all right. I think it is a good thing
for any boy to have some acquaintance with tools. It makes him a
handy man about the house, you know! Apart from that it is one of
the 'fads'—harmless, and about as useless. The unions haven't any
quarrel with public-school training. No one supposes that two or
three hours a week—monkeying with a few nice tools—is going to
make any difference to a boy's start at learning a trade."

Interviews with six other union men, in the machinist, foundry,
pattern-making, tool-making and other trades, elicited opinions practically identical with those already quoted. The problem of industrial
education resolves itself, in the minds of the union men, into three
distinct methods :—the public-school manual training, which they view
with a more or less friendly and tolerant feeling; the commercial
trade school against which they are extremely bitter; and the four-
years apprentice systems of the great corporations, to which they
seem to object on general principles. It appears, however, that the
objection to the latter is based upon past experience with apprentices
who were really cheap producers, rather than upon any clear understanding and appreciation of the present fine systems. Curiously
enough, the view of the unions is most graphically set forth by a
manufacturer, Mr. Robert H. Jeffrey, vice president of the Jeffrey
Manufacturing Co., Columbus, Ohio. He says:—

> "The desire to take advantage of the apprentices, in order to secure
> the most production from them, has caused the manufacturers to fail in
> giving them proper instructions to secure mechanical knowledge and
> efficiency."

The whole situation is further complicated by the charge that the
unions hamper shop-apprentice systems by severe restrictions placed
upon the number of apprentices to be allowed in any one shop. This

is denied by the union leaders, who emphatically assert that in almost every trade the number of apprentices does not nearly reach the number permitted by the union rules. And their assertion seems to be borne out by some very surprising and quite authentic figures.

The United States census of 1900 shows that the percentage of apprentices to mechanics was but 2.45, while the strictest union allows 7 per cent; many allow 12 per cent. Again, the report of the Massachusetts Commissioner of Labor for 1906 shows that the percentage of apprentices in open shops in that State was no greater than in the union shops.

A thorough investigation made at the instance of the American Social Science Association showed that, out of forty-eight trade unions, having a membership of half a million, twenty-eight with 220,000 members had no restrictions upon apprenticeship. In ten unions, with membership of 197,000, the apprentice limits were fixed at from 7 per cent to 15 per cent. In the other ten the question of apprenticeship was left to the locals.

In the last Trade Census of Massachusetts it is shown that in none of the building trades were there one-half, and in many cases not over one-quarter, as many apprentices as union rule allowed. In 1903 one of the delegates to the Annual Convention of the Pennsylvania Association of Master House Painters and Decorators stated that, after a personal investigation among at least six-hundred master painters, he found that not an average of one in fifteen had a single apprentice in his business, and that the larger the workshop or business, the greater was the dislike of the masters to teaching boys. Also, in corroboration of all this, Professor Thomas Sewall Adams, of the University of Wisconsin, states that only about one strike in three hundred is caused by disputes relating to the apprentice.

This I know; that, whatever their outward expressions of opinion may be, there is among union men a deep, underlying feeling that it is better for the embryo mechanic to be trained in the shops and, in a measure, under the influence of the unions, than in outside schools of any kind. Whenever they become convinced of the absolute good faith of the employers who are introducing the new apprentice systems —when they see that the boys are being trained into the very best mechanics they are capable of becoming, instead of being used, at low wages, as producers—then, I believe, the unions will look with full favor upon the shop apprentice, although it may be expected, of course, that they will always attempt to place and enforce restrictions upon the number employed.

ORGANIZATION PROBLEMS IN STEAM-RAILROAD ELECTRIFICATION.

By Howard S. Knowlton.

Conditions demanding the electrification of steam railways, or dictating the construction of new "heavy" electric lines, will press with constantly increasing force upon the established trunk lines and compel a continuous rise in the rate at which mileage is converted from steam to electric operation. This process will play a large part in the economic and the manufacturing development of the immediate future as well as in the modification of railway affairs. Some of its most interesting and most clearly settled elements—mechanical, as well as functional—will be discussed in the following papers by specialists who are active in the work of reconstruction.—THE EDITORS.

THE electrification of any considerable or important portion of a modern steam railroad is a task of far greater magnitude than is generally appreciated by those outside the organization responsible for the success of the work. To the layman it appears merely a matter of dollars intelligently expended to change the motive power of a railroad from steam to electricity; but to the experienced railroad official the problem bears an entirely different aspect. The selection of suitable equipment in face of the progress being made in all branches of applied electricity is a most difficult question in itself, demanding the highest ability in the consulting engineers retained. The maintenance of an unimpaired service during the transition period is a monumental task; the adjustment of the working personnel to the new conditions; the modification of records and creation of new systems of reports to meet the altered situation in the operating department; the proper installation and adjustment of the power plant, transmission, distribution, and rolling-stock equipment; the protection of the traveling public, safeguarding of the train service, and provision for special repairs and an entirely new kind of inspection, throw a burden upon the railroad company and its advisory experts which can be carried only by a carefully developed organization of trained men, employing the most advanced methods known to the transportation and engineering professions.

At the present stage of electrification development in the steam-railroad field, it is not feasible to present any detailed organization diagram applicable to the general problem, for experience has yet to show the strength and the weakness of the methods thus far applied. It was my privilege, however, in the past summer, to make a careful study of the operating organizations of three of the most interesting electrifications in the United States, two of them being in connection with heavy terminal and suburban service, and the third an interurban section of a trunk line. In each one of these cases the organization was so different from that in force in the others that the importance of defining the conditions of the general problem became clear, and this article has been prepared in the hope that it will be suggestive as to the relations and possibilities existing and latent in the working system which must be maintained in the successful electrification of a piece of steam-railroad property.

In dealing with problems of organization it is important to realize how profoundly a change in motive power modifies present conceptions of standard railroad practice. It has been well said that the roots of the steam locomotive penetrate every nook and corner of the modern railroad structure, and it is emphatically the case that the displacement of the steam machine by the electric locomotive and the motor car means, in the long run, a revolution in transportation methods, both from the standpoint of the railroad employee and from that of the public. Many of the methods and facilities suitable to steam-railroad operation remain equally acceptable in the *régime* of the electric motor, but in matters of power supply, maintenance, inspection, train handling, and construction, many new conditions must be faced with plans of action depending primarily upon the fundamental differences between steam and electric motive power, and secondarily upon the peculiarities of the local railroad situation.

The essential differences between electrified and steam service have been emphasized so often that they need not be elaborated here. In brief, they are greater capacity to handle traffic, increased comfort, and safety—all being in favor of electric motive power. It remains to be seen whether the actual cost of electrical operation on a converted steam road can be made materially less than that of steam for the same service. If a larger traffic can be handled more safely and comfortably for a given expense, it matters little if the total cost of operation increases, provided the rates are maintained at a level giving a reasonable profit on the investment. This, however, is aside from the divergence between steam and electric practice which stands out so prominently when one views the subject from the railroad standpoint.

The separation of construction work from operating problems is not difficult to carry out in the early stages of electrification. At the time when the change from steam to electricity is begun, the questions at issue are almost entirely constructional in nature. Operation proceeds on the former steam-locomotive basis for many months. If a railroad company from the beginning adopts the policy of properly separating its construction expenses from its operating costs, and adheres to this policy as its electrification proceeds, endless confusion will be avoided. As the use of electric motors begins and extends, the importance of the construction work gradually tends to decrease, until at last, when a complete electrified section or division is in service, the problem is purely one of operation, including maintenance, so far as that particular equipment and trackage are concerned. This general range of conditions must be faced in carrying through the work, and the organization to handle it on a large scale must be both constructional and operating in its nature if the railroad company desires the fullest control of the situation at every hour of the time required to make the change in motive power. In other words, a large amount of engineering work must be handled, part of it being in the operating field and part in the domain of design and installation. No matter how well qualified any contractor or group of contractors may be to carry out the physical changes in the property, these changes are so far-reaching in electrification and so insistent in demanding the fullest resources of the existing steam organization for successful completion with minimum disturbance of traffic, that the railroad company ought never to yield to any other authority the right of direct supervision within its territory.

The importance of electrification, even of a small section of road, is so great in its relation to future development, that the work as a whole should be in responsible charge of one or more vice-presidents of the railroad company. Where the work involves costly terminal changes in the heart of large cities, or far-reaching modifications in the trunk-line equipment and operating methods, it may be necessary for one officer of this rank to give practically his whole business time to it. In certain cases of the electrification of branch tracks or parallel routes supplementary to the existing trunk-line service, the immediate personal attention of the vice president may not be as essential at every stage of the development. Wherever the work be undertaken, though, many details ought to be deflected from the shoulders of the highest executive officer charged with the responsibility of the task, and allowed to fall upon the organization, which may well be headed by an officer granted supreme authority in the execution of the vice-president's policy and decisions. The title of this officer may be electrical superintendent, assistant general manager in charge of electrification, chief engineer of electric traction, or something else determined by the existing organization and operating conditions.

Fundamentally there is need of an organization, newly created, whose business it is to solve all the problems connected with the actual installation and assembly of the electrical equipment which cannot be handled by the regular organization of the steam road, and to settle those questions in operation and maintenance which are new to the employees of the road on the basis of steam-locomotive operation. Certain very important problems in the design of details fall within the scope of such an organization no less than other important questions in the actual handling of the equipment on the road. They could not be handled by the old staff of the road, for they are almost entirely peculiar to electric service.

It is clearly the part of wisdom to disturb the existing machinery of railroad business as little as possible in carrying out a policy of electrification. So far as the original rules governing train-service employees can be retained, well and good. The probabilities are that in a well-rounded scheme an additional instruction book or two will be necessary to supplement the knowledge of train-service employees. The character of such instructions will obviously depend much upon the character of equipment selected for the service of the road, but the indications are clear that, sooner or later, the train-service men must be given, in comprehensive form, a working handbook covering in detail the methods of handling the electrical rolling stock in regular and emergency operation, outlining the methods of inspection on the road peculiarly electrical in nature, and possibly specifying the essential emergency handling of the distribution system on the right of way. Just how far each class of employees in the general service should be expected to understand the workings of the auxiliary electrical apparatus—auxiliary from the standpoint of train movement—is an unsettled question. Experience may show that it is enough if the electric-locomotive runner understands the general operating rules of the road and the details of his machine; it may show that some knowledge of the rolling stock is desirable for the men in the distribution end, as well as perfect familiarity with the handling of switches and lines, trolley or third-rail appurtenances, with sub-stations and their machinery, if the equipment includes them; or it may show that the best plan is the publication in a single rule book of all the vital operating features of the electrical equipment, from the power house to the motors. There is scarcely any doubt that as electrified operation extends, the importance of giving each man these essentials in black and white will become imperative.

At almost every turn in dealing with the altered conditions in routine operation which result from the greater use of electric motive power, the question arises, "To what extent shall the engineering organization have jurisdiction over the actual operation of the road?" Upon the answer to this question largely hinges the form of the organization itself. Viewing the matter in the broadest way, it is hard to escape the conclusion that sooner or later the engineering staff must relinquish the direct supervision of rolling-stock operation as such. It may never have been granted supervisory power in this direction, but if it has been, the responsibility for actual train movement is almost certain to be removed from its shoulders. We have here involved the broad question of how far the duties of an engineering staff in any large industrial organization shall be held responsible for pure operation. In the last analysis, the work of the engineer is cast in the field of providing the requisite facilities at the lowest cost consistent with economical service, and of analyzing the actual working of those facilities with a view toward their proper maintenance and maximum economical improvement. Advisory supervision rather than actual handling of equipment and its operators is the logical duty of the engineering organization. Jurisdiction over train movements, then, would seem to belong to the regular operating department headed by the general superintendent.

The character of the equipment is certain to influence the form of the organization to a very large extent. In a system whose power supply is drawn from one or more alternating-current generating plants, with high-tension transmission, and direct-current third-rail distribution from rotary-converter or motor-generator sub-stations, it is out of the question to expect as simple an organization as would naturally be the case on a purely alternating-current system in which no sub-stations are installed. On any electrified steam road the external supply of power calls for an entirely new branch of the operating organization. Power-house employees to the number of several hundred may easily be added to the company's pay roll; line, sub-station and distribution forces add more men who were never before needed in heavy railroad service, and in some cases maintenance of way in the electrical zone calls for a specially developed staff of foremen and subordinates. In some cases the importance of the electrification demands the creation of a new office filled by one or more men who devote their whole time to the broad supervision of the power supply, from the high-tension bus bars in the generating plant to the collecting devices on the electric locomotives and motor cars. Such officials may be designated supervisors of power, load dispatchers, or by other suitable titles. Their work in a system which needs them can be made of extraordinary importance as regards the maintenance of continuous service, for in them will presumably be lodged the absolute operating control of every ampere of current supplied for transportation service.

Another new branch of the operating organization must be the group of electrical inspectors of rolling stock and the shop forces required to handle repairs on the electrical equipment. One of the most important questions in electrification is whether the electric locomotives (and motor cars, if any) shall be inspected and repaired in the existing shops of the company, or whether new shops and inspection facilities shall be required. There is so little in common between an electric and a steam locomotive that in the opinion of those best qualified to know, the establishment of new and separate facilities is almost a necessity, at least in the early stages of electrification. As the work extends, however, and as fewer steam locomotives are maintained in a given zone, it is sure to become a question how far the old shops can be utilized in the repairs of the new rolling stock. Such a problem cannot be settled off-hand without knowledge of the conditions of operation on a given road. Much depends upon the design of the existing shops, their crane and hoisting facilities, arrangement of inspection pits, machine tools, and store-room facilities. It is certain that many changes in the shop arrangement will sooner or later be needed to handle the repairs of electrical rolling stock with speed and economy, and can these be effected without marring the usefulness of the shops as hospitals for the remaining steam locomotives? Possibly so, for repairs on trucks, frames, and bearings: but the probabilities are that sooner or later complete new shops will be necessary at some central point, designed specifically for the intensified repair work demanded by the new rolling stock, worked as it is ultimately sure to be to the fullest extent of its capacity. However easily a railroad may start off in the use of its electric locomotives, the time will come when the system is forced to the limit of its transportation output, and every hour's delay in making repairs will be regarded unfavorably. Spare parts will be substituted and the rolling stock kept on the road in every possible case, in place of the old policy of holding a whole car or locomotive in the shop pending the repair of a single element of its detail equipment. The new shop facilities may start in a very limited way, or they may be built larger than are at first necessary, and in a permanent location, according as circumstances dictate. The ultimate electrification must always be kept in mind, however, in fixing such a location.

Let us now gather together these considerations and see where they may lead in the organization. We can fairly assume that the train-service organization as a whole will be substantially as before

electrification took place. In place of the engine runner and the fireman, the electric locomotive takes the motorman and his helper; the
conductors, baggagemen, brakemen, signal men, station agents, and
many other employees have practically the same duties as always.
Power supply, maintenance, and inspection are new (or at least different) elements from former steam-locomotive days. Where shall
the organization take account of these?

Experience seems to show so far that the train service must remain
under the jurisdiction of the general superintendent and his subordinates, the division superintendents. It is not so clear what the final
place of the electric power and maintenance forces shall be. The
structure of the original steam organization must affect this. Given
a vice president in charge of all maintenance, and possibly construction, it would seem that sooner or later the power supply and the electrical inspection and repair work must settle under his jurisdiction.
These questions of the final authority of different officials and departments cannot be settled except by experience, but it is clear that
the electrical organization must for a time, at least, be held responsible
for the installation and proper working of the new equipment. In
certain mechanical matters it may be possible to secure the aid of the
regular mechanical organization; in other cases, the electrical organization will broadly include both electrical and mechanical supervision.

In a general way, the head of the engineering department, certainly
during the construction period, will presumably be in charge of the
operation of the power-supplying system and the maintenance of electrical rolling stock. Construction matters will also probably be under
his jurisdiction. Depending upon the character of equipment, the organization will be divided into sub-engineering departments. All the
power supply may be under the jurisdiction of a department head responsible to the chief engineer; in another case, maintenance of rolling-stock equipment, the operation of power equipment, its maintenance and inspection, may be each separately divided. The magnitude of the work is a measure of the scope of subdivision required.
The purpose of the new engineering organization must ever be to
solve those problems which could not be solved by the existing steam-
railroad organization and to co-operate with the regular operating department in securing, first, as good service as was given before electricity came into the field—and better if possible; second, to point out
and secure economies in the routine operation of the new motive
power; and third, to analyze the daily conditions and service experiences with a view toward improving apparatus and methods for the
benefit of future extensions of electrified service.

REINFORCED-CONCRETE BUILDING LAWS; THEIR DIFFERENCES AND DEFICIENCIES.

By H. C. Hutchins.

BUILDING laws controlling reinforced-concrete construction have been prepared by a number of cities in the United States, and it is the purpose of this article to compare them, and to point out the need of a standard building code which should be adopted by all cities.

A great many differences are to be noticed in the codes. Some are complete and definite in their specifications, while others are indefinite and refer everything of importance to the building inspector's decision. This is not a good feature; the laws should specify the requirements clearly, and not leave matters to the decision of men whose tenure of office depends upon political conditions.

The table shown on pages 422 and 423 is compiled from the building laws of New York (where the borough of Manhattan and the borough of Brooklyn have laws differing in many respects), Buffalo, N. Y., Boston, Mass., Chicago, Ill., Denver, Colo., Louisville, Ky., Toledo, O., San Francisco, Cal., Rochester, N. Y., and also from the regulations of the National Fire Protection Association and the Royal Institute of British Architects. Laws are in preparation by Philadelphia, Washington, Pittsburg, Detroit, Cincinnati and New Orleans.

MIXTURE AND TESTING.

It will be observed that a majority of cities call for a concrete mixture of one part of cement, two parts of sand, and four parts of broken stone, gravel, broken brick, slag, or cinders. In some cities cinder concrete is prohibited, except by special permit from the building inspector. Limestone should not be allowed, for it has a tendency to disintegrate under the action of heat, and render the concrete of little value in case of a severe fire in the building.

Cinder concrete is liable to contain sulphur, which would have an injurious effect upon the reinforcing steel. It should not be used in building work unless the percentage of sulphur is very low, and then only when the cost of rock concrete is so high as to prohibit its use. Rock concrete is so much better that it should always be speci-

TABLE SHOWING REQUIREMENTS OF BUILDING CODES.

	New York, Borough of Manhattan.	Borough of Brooklyn.	Buffalo, N. Y.	Boston, Mass.	Chicago, Ill.	Denver, Col.	Louisville, Ky.	Toledo, Ohio.	San Francisco, Cal.	Royal Institute of British Archit'ts.	National Fire Protection Association.	Rochester, N. Y.
Kind of cement.........	Portland	Portland	Portland	Portland	Portland	Portland	Natural or Portland	Portland	Portland	Portland	Portland	Portland
Test ten- { 1 day in air. (per sq.in.) sion, lb. { Neat briquette	300	175			200	200		300			300	
Test ten- { 1 day in air. sion, lb. { 6 days in water	500	500	500		500	500		500			500	
Test ten- { 1 day in air. sion, lb. { 27 days in water	600	600	600		600	600		600			600	
Tests { Fineness, con- { stancy of volume etc.	A. S. C. E.	Special rules	A. S. C. E.		A. S. C. E.	A. S. C. E.	A. S. T. M. A. S. C. E.	A. S. C. E.	A. S. C. E.		A. S. C. E.	
Concrete mixture.........	1-2-4	1-2-4	1-2-5	1-5	1-3-5	1-2-3	1-2-4	1-2-4	1-3-5 and 1-2-4		1-3-5	1-2-5
Method of mixing.........	Wet	Wet 2 in.	Wet	None	Wet	Wet		Wet	Wet		Wet	2
Size of { Foundations, mixing, { columns, stone, in. { slabs.	¾	¾	¾		¾	2 2 ¾	2 1½ 1	¾	2 1 1	¾	¾	
Concrete { Compression test { after 28 days { per sq. in., lb.	2000	2000	2000	2000				2000	2000	2400	2000	
Extreme fibre stress per sq. in. for concrete in compression, lb......	500	500	500	500 to 600	500	500	700	500	500	600	500	
Concrete in tension per sq. in., lb......	None	None	None	None	None	None	None	None	None	None	None	None
Concrete in direct compression, lb.........	350	400	350	350	350	450	700	350	450	500	350	230
Shear in concrete per sq. in., lb...	50	75	50	60	75	50	30	50	75	60	50	
Adhesion of concrete to steel, lb...	50	75	50	60 Beams & slabs 1/12 col. 1/8	75	75	30	50	75	100	50	
Ratio of modulus of elasticity...	1/12	1/12	1/12		1/12	1/12		1/12	1/12	1/15	1/12	
Column stress, { Horiz. lb. { hooping	350	600 to 750							700		6000S 350C	

Table of comparative reinforced-concrete specifications (page printed sideways).

	Special formula	1000C 8000S								
Column stress, Spiral hooping, lb.	Special formula	1000C 8000S	……	……	……	……	700	……	……	……
Columns, ratio of height to least diameter.	r	r	120 times least rad. of gyr.	r	r	r	r	r	……	r
Insulation—Columns, in.	2	2	1½	……	Diam. of bars.	2	1½	2	……	4 times diam. of bar to 4″
" —Beams, in.	1½ times diam. of bars to 1½ in. max.	1½ times diam. of bars to 1½ in. max.	1½	1¼	1¼	1½	1½	2	……	4 times diam. of bar to 4″
" —Slabs, in.	1½ times diam. of bars to 1¼ in. max.	……	¾	¾	1¼	1¼	1¼ times diam.	1½	……	……
" —Walls, in.	Diam. of bars to 1 in. max.	……	¾	¾	½	1	1	1	……	……
Tensile stress per sq. in. for steel, lb.	16000	16000	16000	½ of elastic limit 10000	½ of elastic limit 10000	12000 to 18000	½ of elastic limit 10000	15000 to 17000	……	¾ of ultimate strength 10000
Shear stress, for steel, lb.	10000	10000	10000	Least width of col.	Least width of col.	10000	Least width of col.	17000	……	Least diam. of col.
Spacing of hoops in columns.	Least width of col.	1½ times least width of col.	……	……	……	Least width of col.	……	……	……	……
Minimum thickness of concrete slabs, in.	3½	3½	½ span of girder	5 times depth of slab	5 times width of girder	10 times width of girder	3½	……	……	……
Width of slab allowed for figuring tee beams.	10 times width of beam	10 times width of beam	10 times width of beam	Twice load 1/160 def.	Twice load 1/160 def.	3× load no sign of failure	5 times thickness slab	……	……	……
Deflection allowed on load test.	Twice load 3× load no sign of failure	Twice load 3× load no sign of failure	Twice load no sign of failure	……	……	……	Twice load 1/160 def.	……	……	……
Bending moments—Beams.	W L / 8	W L / 8	Contin. W L / 10	W L / 8	W L / 8	W L / 8	W L / 8	……	……	W L / 8
Bending moments—Continuous slabs.	W L / 10	W L / 10	W L / 10	W L / 12	W L / 12	W L / 10	W L / 12	……	……	W L / 10
B. M. for slabs supported on all four sides.	W L / 20	W L / 20	W L / 10	W L / 20	W L / 20	W L / 20	W L / 20	……	……	W L / 20
Maximum percentage steel in beams.	20	……	……	20	20	20	20	20	……	20
Minimum percentage of steel in columns.	……	1/100	……	……	……	……	……	……	……	……
Time of removal of centering for columns.	……	……	……	4 days	……	……	8 to 28 days	……	……	1 to 60 days Depending on weather conditions
Time of removal of centering for beams.	……	……	……	12 days	……	10 to 15 days	……	……	……	7 days
Time of removal of centering for slabs.	……	……	……	6 days	14 to 21 days	10 to 15 days	……	……	……	……

423

fied for columns, beams, and slabs, using cinders only for filling between the wooden sleepers in floors, for grading under basement
floors, or for grading roofs. Cinder concrete has been used for floor
slabs between steel beams, when the spans are short, but it does not
possess the strength of rock concrete and its use should be limited
by the codes.

THE POST AND OFFICE BUILDING, GRAND CENTRAL TERMINAL, NEW YORK CITY.
SHOWING CONCRETE FIREPROOFING OVER STEEL CONSTRUCTION.
Reed & Stem and Warren & Wetmore, associated architects; the Roebling Construction Co.,
contractors.

The quality of the rock is important, and should be carefully
looked into on all work. Rock containing much mica should be prohibited, for the particles of mica offer cleavage planes for the concrete
to shear on. A good quality of crushed granite or other hard and
sound stone not containing lime, mica, sulphur, phosphorus, or other
elements in quantities liable to injure the steel or the strength of the
concrete should be used. Gravel may be used when it is clean and
free from clay or loam, which tend to form inert spots in the concrete.

All stone for foundations of concrete should be broken into sizes
that will pass through a 2-inch ring in any direction. For floor slabs,
beams, girders, columns, or walls, the size should be between ¼ inch
and ¾ inch, so that when the concrete is poured wet the particles of

stone can pass between the reinforcing metal without clogging the flow, and the concrete can reach every part of the forms.

TESTING.

Most cities refer the standards of testing the cement to the specifications of the American Society of Civil Engineers, or those of the American Society for Testing Materials. These standards have been carefully worked up, and are excellent. Samples of the cement should be tested at frequent intervals during the construction of any concrete work, so that any defects in quality may be discovered, and the cement rejected if not up to the required standards. Too little attention is paid to the testing of the materials of construction in concrete at the present time. Concrete is used by thousands of contractors who never test their cement at all, simply throwing it in as it comes from the mill and sacrificing all tests and analyses to the desire for "speed of construction." Some contracting firms make a special point of thi‧ advertise their construction records widely.

INTERIOR OF BUILDING FOR AMERICAN OAK LEATHER CO., CINCINNATI, OHIO, SHOWING 50-FOOT SPAN IN REINFORCED CONCRETE.
The Ferro-Concrete Construction Co., Cincinnati.

Failures of concrete construction have been due solely to extreme speed of construction and a lack of care in making tests of the cement and concrete. There is no reason why structures of this character should fail, provided the work is in the hands of intelligent engineers who test the cement and concrete and see that the work is done according to the specifications. Important concrete construction requires the services of trained experts, and should not be left

MAIN FLOOR CORRIDOR, COOK COUNTY COURT HOUSE, CHICAGO. SHOWING CONCRETE
CEILING CONSTRUCTION.

Holabird & Roche, architects; the Roebling Construction Co., contractors.

solely to the contractor, as his main interest is to erect the building
in the shortest possible time so as to increase his profits. This craze
for "speed of construction" has produced failures which have hurt the
concrete business, by producing a feeling of distrust in the minds of
prospective builders, and the steel construction advocates have used
these failures to produce business for themselves by advertising them
widely.

INSULATION.

Insulation for the reinforcing steel calls out many differences. Some codes specify that there shall be a thickness of concrete equal to the diameter of the bars used. This rule is dangerous, for if very small rods are used, down to $\frac{1}{4}$ inch or less, the insulation becomes very thin. According to that ruling, if expanded metal were used for reinforcing the thickness underneath the metal would be not much thicker than a mere coat of plaster, which would afford very slight protection against a hot fire heating and damaging the steel reinforcement. A rule requiring a certain definite minimum thickness of one inch, as called for by the National Fire Protection Association, The Royal Institute of British Architects, and others, affords a proper protection against damage by fire.

In the insulation of the reinforcing bars of columns and beams, the rules vary from $1\frac{1}{2}$ to 4 inches, which latter seems excessive. Two inches should be specified, outside of all reinforcing steel, and no account should be taken of this covering in figuring the stresses, as in case of fire it would be liable to spall off and become useless for carrying loads.

STRESSES IN CONCRETE.

A majority of codes specify that 500 pounds per square inch is high enough to figure the extreme fibre stress for compression of concrete in beams and slabs. The British Architects call for 600 pounds. Concrete in tension is rejected by all the above codes, and should be, for the steel in tension in beams does not begin to take its stress properly until the concrete has developed fine cracks, and therefore has failed by tension before the steel begins to be stressed.

Concrete in direct compression takes values of 350 to 700 pounds per square inch. The former is too low for 1—2—4 mixture testing from 2,000 to 2,400 pounds per square inch. A factor of safety of four would allow 500 to 600 pounds, the former of which is not excessive for columns properly hooped.

COLUMNS.

The greatest differences in the codes is in the question of allowable stresses for columns. The requirements range from 350 to 1,000 pounds per square inch on the concrete inside the hooping. Many codes do not specify any values for allowable stress, leaving it to the judgment of the building inspectors. A more uniform range of stresses should be specified. Columns limited to 350 pounds per square inch soon become excessively large in high buildings. For example let us take the case of a warehouse, with 200 pounds per square foot live load on floors and 100 pounds dead load, twelve stories high, with floor panels of 16 by 20 feet (or 320 square feet) bearing on each column, comparing the rule of 350 pounds with 500 pounds and 750 pounds per square inch on the concrete inside the hooping, figured for the reduction of live load allowed by the New York Code as follows:

> "For the roof and top floor the full live loads shall be used; for each succeeding lower floor, it shall be permissible to reduce the live load by five per cent (5%), until fifty per cent (50%) of the live loads fixed by this section shall be used for all remaining floors."

No column diameter is taken less than 1/12 of the height, making the roof columns 12 by 12 inches. For the purposes of this comparison the weight of the column itself is neglected, as it forms only a small percentage of the total weight. The column sizes are taken to

THE GRETSCH BUILDING, OF REINFORCED CONCRETE, BROOKLYN, N. Y.
Benj. Finkensieper, architect; P. R. Moses, consulting engineer; Turner Construction Co., constructors.

FACTORY BUILDINGS IN REINFORCED CONCRETE, FOR THE NATIONAL CASH REGISTER
CO., DAYTON, O.
Built by the Ferro-Concrete Construction Co.

the nearest inch, fractional dimensions being disregarded. With columns figured at 350 pounds for a load of 928,000 pounds, the size in
the basement becomes 51 by 52 inches, or 18.4 square feet., which is
5¾ per cent of the total area of the floor panel. At 500 pounds the
size is 43 by 43 inches, or 12.8 square feet., which is 4 per cent of the
total area of the floor panel. At 750 pounds, the size is 35 by 36
inches, equal to 8.75 square feet, or 2½ per cent of the total area of
the floor panel.

The tests made at the Watertown Arsenal show that on columns
reinforced with vertical rods and made of 1—2—4 concrete, the ultimate strength ranged from 1,990 pounds to 3,160 pounds per square
inch, with an average of 2,350 pounds. Taking the lowest value, and
using a factor of safety of four, we get 1,990 ÷ 4 = 497, or practically 500 pounds for the allowable stress per square inch on this
form of column.

Concrete in columns gives notice of approaching failure by the
spalling off of pieces of its covering over the reinforcing steel. Total
failure produces a breaking out of the concrete through the spaces
between the hooping. This indicates the desirability of decreasing
the size of the spaces between the hooping, which is done by winding
the column outside the vertical rods with a spiral hooping composed

SAND BINS AND DRYER HOUSE OF REINFORCED CONCRETE FOR J. B. KINGS & CO.,
HEMPSTEAD HARBOR, L. I.

Turner Construction Co.

of wire or flat bands. Flat bands offer a broader surface to the concrete with less thickness than round-wire hooping, leaving less space
for the concrete to break out between the coils of the spiral.

M. Considere's experiments upon the strength of spirally hooped
columns show that spirally hooped columns with vertical rods will
stand pressures of 4,550 pounds to 5,400 pounds per square inch
before total failure. With a factor of safety of four, loads of 1,100
to 1,200 pounds per square inch on the concrete inside the hooping can
be used. One building department code allows 1,000 pounds per
square inch on the concrete and 8,000 pounds per square inch on the
steel. This rule, however, does not appear in the code, but was
learned by consulting the department engineer. Using this rule for
the case of the twelve-story building, we find that the largest column
required, using for vertical reinforcing eight (8) 1¼-inch rods, as
follows:

Total load	928,000	lb.
Load carried on steel rods, 8 by 1.22 by 8,000 lb.....	78,080	"
Load to be carried by concrete.....................	849,920	"
Area concrete required at 1,000 lb. per sq. in........	849.92	sq. in.
Diameter of concrete column required.............	32.9	inches.

This form of column will give the smallest size, and the size can be decreased further by using larger vertical reinforcing rods, as the steel is allowed 8,000 pounds per square inch, while the concrete has only 1,000 pounds per square inch.

The hooping of concrete columns must be made in such a way that every vertical rod is prevented from moving outward. This may be done best by making the columns circular or octagonal in shape, as in Figures 1 and 2.

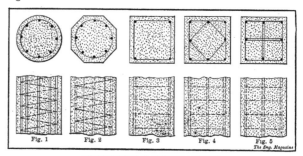

REINFORCEMENT OF CONCRETE COLUMNS.

Spirally hooped columns should be made either round or octagonal. Horizontally hooped columns may be made with any of the arrangements of hooping shown above, and should have the vertical rods attached by a light wire to each rod, so that the steel will remain in place in the forms during the process of pouring the concrete. When the column becomes large enough to require eight vertical rods, as in Figures 4 and 5, an extra hoop binding the intermediate rods together should be used, for otherwise these rods are not confined sufficiently to prevent a displacement of an inch or more before the hooping acts upon them; and then it would be useless, for the concrete enclosed within the hooping would be distorted to a dangerous extent.

The spacing of the hooping in columns is specified in a majority of cities as not to exceed the least width of the column. Many contracting firms use a standard spacing of 12 inches, varying the size of wire where necessary. The codes themselves do not specify clearly the exact stresses and arrangement allowed on columns, and as a result columns have been passed by building departments where the stresses on the concrete have been as high as 1,200 pounds per square inch. The question of the proper stresses to place upon both the steel and concrete should be governed by some definite rule regarding the minimum amount of steel allowed in the column. One code only has a rule governing this, and that is :—

> "The total cross-sectional area of the reinforcing bars shall in no case be less than 1/260 of the total cross-sectional area of the column."

When the hooping is used in columns, it should be of sufficient size and strength to prevent it becoming bent, for if it is not stretched tightly around the rods, it cannot act properly in binding them.

Many codes require a pipe sleeve to be placed to cover the joints of the vertical rods, ⅛ inch larger than the rods, so that it can be grouted with cement mortar. A better method would be to specify sleeve nuts, which would insure positive action when the joint is called upon to resist tension caused by wind pressure. This is the most satisfactory method, but is more expensive than grouted pipe sleeves, or the method of overlapping the rods and clamping them, or winding the overlapping ends with wire.

The ratio of height of column to its least diameter varies from 1/12 to 1/15. An additional rule should be added for building work, specifying that no concrete column shall be less than 12 inches in diameter, so that the pouring and ramming may be done in a satisfactory manner.

Great care should be observed in cleaning out from the column forms all loose pieces of wood, tile, or other rubbish, as these weaken the column and have contributed to cause failure.

There is a diversity of opinion regarding the stresses allowed in the vertical steel rods in concrete columns. It should be governed by the ratio of the modulus of elasticity of concrete to the modulus of elasticity of the steel. If the concrete is allowed 500 pounds per square inch and the ratio allowed is 1/12, then 6,000 pounds per square inch can be allowed safely upon the steel. The modulus of elasticity, or the stress, divided by the strain for concrete varies for different mixtures. In comparison, it varies from 1,800,000 to 2,500,-000. For mild steel the modulus is 30,000,000. The differences in the ratios specified between steel and concrete in different codes arise in this way, as the ratio may vary from 1/12 to 1/17.

STRESSES ALLOWED IN STEEL.

The stresses allowed in the reinforcing steel are much the same as for structural steel work, allowing 16,000 pounds per square inch for tension, and 10,000 pounds per square inch for shear. In some codes the stress is specified as 1/3 of the elastic limit, or as 1/4 of the ultimate strength. Under these last two rules high-carbon steel, with

REINFORCED-CONCRETE BUILDING OF THE EMPIRE AMERICAN BANK, SEATTLE.
The lower portion is occupied by tenants while the upper stories are still under construction.
Ferro-Concrete Construction Co.

high values for the elastic limit and ultimate strength, can be used to advantage. Twisting the steel has been shown by experimental tests to increase its strength from 30 per cent to 40 per cent, and twisted steel is used in some localities under stresses of 22,000 pounds per square inch. The building codes should go more into detail on these points, and fix definitely the stresses allowable upon mild steel, high carbon, deformed and twisted bars.

CONCRETE SLABS.

The minimum thickness of slabs and the percentage of steel allowed should be specified by all codes. This has been done in very few cases, as the table shows.

REINFORCED-CONCRETE WAREHOUSE FOR EASTERN STATES REFRIGERATING CO., JERSEY
CITY.

Kirkhan & Paulette, architects. Built by the Turner Construction Co.

Slabs should not be less than 3 1/2 inches thick, for below this thickness the resistance to shock becomes so little as to cause danger of holes being broken out of the slab by any considerable weight falling upon it. The depth of the slab should be governed by the span between the beams, and should not be thinner than 1/30 of the span. Beams and girders should have a depth of not less than 1/15 of the clear span.

The percentage of reinforcing steel in the slabs or beams should be limited. As many codes specify that the beam shall be equally balanced, so that the strength of the steel in tension shall determine the strength of the beam, the maximum percentage of steel allowed in the beam should cause the beam to fail by pulling the steel reinforcement apart before the crushing limit of the concrete in the top part of the beam is reached. This condition will be realized if the percentage of steel in tension is kept at or below 1 per cent for 1—2—4 rock concrete.

BENDING MOMENTS.

The majority of codes specify that beams and girders shall be figured as being simply supported at their. ends, no allowance being made for continuous action by top bars at the supports. This gives the bending moment for a uniform load equal to WL ÷ 8, where W = total weight supported, including the weight of the beam or girder, L = length of span in inches, the result being in inch-pounds.

For slabs, continuous action over the supports is allowed, limited to WL ÷ 10, and in some cases WL ÷ 12. When the slab is reinforced in both directions, so that its load is distributed on beams on all four sides of the floor panel, a bending moment equal to WL ÷ 20 is allowed.

MOMENT OF RESISTANCE.

The codes specify the following assumption on which the moment of resistance shall be based:

(a) The bond between the concrete and steel is sufficient to make the two materials act together as a homogeneous solid.

(b) The strain in any fibre is directly proportionate to the distance of that fibre from the neutral axis.

(c) The modulus of elasticity of the concrete remains constant within the limits of the working stresses.

(d) The dimensions of such a beam or girder and its reinforcement shall be determined and fixed in such a way that the strength of the metal in tension shall measure the strength of the beam or girder.

(e) If the concrete in compression, including the allowable concrete in adjoining floor construction, does not afford sufficient strength for the purpose, the compression side of the beam or girder in question shall be reinforced with metal.

(f) The tensile strength of the concrete shall not be considered.

From these assumptions, the formulæ for determining the stresses in reinforced-concrete slabs, beams, girders, arches, and other parts are to be derived.

In order that the bond between the concrete and steel be sufficient to make the two materials act together, the steel should not be painted or covered with oil or any substance tending to keep the concrete from adhering to the steel. The bond is improved by using deformed and twisted sections, and rigidly attached members reaching up into the concrete. All rust or mill scales must be removed from the steel before placing, as these will prevent proper adhesion of the concrete to the steel. The deformations of the bars must be of such a nature as not to impede the flow of concrete into them, and should have rounded edges, so that no sharp angles are formed in the concrete. This furnishes an additional reason for limiting the size of the broken stone used, as large pieces would allow hollow spaces to occur over

INTERIOR OF TOP FLOOR OF REINFORCED-CONCRETE BUILDING FOR THE MURPHY VARNISH CO.
Showing roof and skylight construction. Turner Construction Co.

THE MURPHY VARNISH COMPANY'S BUILDING, OF REINFORCED CONCRETE, NEWARK, N. J.

Howard Chapman, architect; built by the Turner Construction Co.

the cups and deformations of the steel, and form voids in the body of the concrete, thereby decreasing its strength.

The second assumption (b) is similar to the assumption made for computing steel and wooden beams. While it is not exactly true, owing to shearing stresses set up in the concrete and the action caused by the reinforcing metal, it is close enough for all practical purposes and simplifies the formulæ greatly. Within the stresses allowed with a factor of safety of four, it may be used with perfect safety.

The third assumption (c) is not quite true, when the beam is stressed up to its breaking load, as the curve of the stress divided by the strain is more nearly parabolic than straight; yet within the limits of stress specified by the codes it is on the safe side, and reduces the complication of formulæ.

The fourth assumption (d) is an excellent one, as the stresses in the steel do not vary to any great degree, and is a better measure of the strength of a beam than to figure the beam so that it breaks by compression of the concrete firt. The value of the compression may vary, on account of the mixture not being quite uniform in character, and this rule obviates any excessive loading upon the concrete in compression.

COLD-STORAGE BUILDING OF REINFORCED CONCRETE, ROTH PACKING CO., CINCINNATI.
Separate walls of reinforced concrete and of brick, with cork insulation. Plaster finish on
exterior. Ferro-Concrete Construction Co.

The fifth assumption (e) is self-evident, and must be carefully determined when figuring beams. It is not economical to reinforce the concrete in compression, and it should not be done unless the limits fixed as to size of beam allowed require it.

The last assumption (f) has been widely discussed, and a variety of opinions regarding the computation of the tensile resistance of concrete in structures have been expressed. Experiments have shown that fine cracks, barely visible to the eye, occur on the lower side of beams even before the load is applied. These cracks prevent the concrete from acting properly in tension, and for this reason the concrete should not be figured for tension stresses. If these cracks do not occur the factor of safety is increased, but the possibility of their occurring is sufficient to prohibit the use of tension stresses. These cracks do not interfere with the action of the steel, and they may be caused partly by the concrete shrinking slightly during the process of setting, the reinforcing steel holding the concrete so that the cracks due to contraction are small and occur at frequent intervals.

REINFORCED-CONCRETE BUILDING FOR THE HODGMAN RUBBER CO., TUCKAHOE, N. Y. W. L. Stoddart, architect. Built by the Concrete Steel & Tile Construction Co., on the Kahn system.

All the above assumptions, while not exactly accurate, err on the side of safety, and reduce the formulæ to simple terms.

CENTERING.

Very few codes specify the time of removal of centering further than to specify that the forms shall remain in place until the concrete has become strong enough to sustain itself. The rule should be more definite, specifying, the minimum number of days for the different parts of the work. It is considered best to remove the column forms first, so that the columns may dry out better, but no load should be allowed to come upon them from the beams and girders for at least three weeks. The sides of the beam forms should be designed so that they can be removed, allowing the air to dry the beams. The careless removal of centering causes most of the accidents in concrete construction. A good system is to have all the centering posts placed on wedges, which can be taken out slowly, and the concrete work allowed to support itself for a day or so with the centering dropped a couple of inches below it, so that in case of settlement there will be no floor panels dropping upon those below and causing a general wreck of the building.

Concreting in freezing weather is prohibited by several codes, and should be by all, unless the work is protected properly from the action of frost. This can be done by covering the newly poured concrete with boards in such a manner as to leave at least 3 inches air space, then laying tar paper over these, and then a layer of sawdust, hay, straw, or manure. The under side of the concrete should at the same time be heated by the use of salamanders, and the sides of the building should be enclosed with canvas or other suitable material to keep the heat in. This must be continued until the concrete has hardened thoroughly, and all danger of freezing is past.

Frozen concrete is unsafe, and all portions which become frozen must be chopped out and replaced. The use of salt to lower the freezing point of the water may be adopted, but the salt must not be used in larger quantities than $1\frac{1}{4}$ pounds per bag of cement. If a good surface finish is desired, however, salt must not be used, as it produces surface efflorescence.

When for any reason the concreting has to be stopped, it must be done in places where the shear in the concrete is the least, as at the center of beams and slabs. Columns can be poured up to the bottom ·of the beams and then stopped until the beams and slabs are poured, which must be done at one operation.

WALLS.

There is a great diversity of rules concerning concrete walls. Some cities are very conservative, and do not allow any concrete wall to be built which is thinner than the brick-wall specifications. This is unfair to concrete construction, for a concrete wall has a greater bearing value per square inch than a brick wall, and is also one solid piece if poured at one operation, whereas brick walls are not. The concrete

THE INGALLS BUILDING, CINCINNATI; A TALL BUILDING IN REINFORCED CONCRETE.
Elzner & Anderson, architects. The Ferro-Concrete Construction Co.

wall, when reinforced, has a capacity of resistance against outside pressure caused by wind, earth, water, or other substances pressing against it, which a brick wall never can have.

The codes should give concrete walls a greater value, allowing reduced thickness from the brick-wall specifications. This should apply, especially to walls for concrete skeleton construction. Some codes allow walls of reinforced concrete from 6 to 9 inches, depending upon the floor they enclose, while others allow nothing less than 12 inches to be used in skeleton construction.

THE NEW TRAYMORE HOTEL, ATLANTIC CITY, N. J.

Built of reinforced concrete on the Kahn system. Price & McLanahan, architects.

There seems to be a need of a national code of laws gotten up by experts, the rules for which shall be the result of careful study and exhaustive tests on all forms of concrete construction. The American Society of Civil Engineers has a committee which is now at work upon this question, and when its report is made, it will be of interest to all cities and those engaged in concrete construction. At present the codes do not cover the subject with direct printed rulings on all points in question, although the points not covered can be settled by recourse to the building inspector. A number of cities have no codes on reinforced concrete at all, and others report that they are preparing them.

A BAVARIAN MUSEUM OF INDUSTRIAL SAFETY DEVICES.

By Dr. Alfred Gradenwitz.

The United States and Great Britain are still far behind the Continental nations in safeguarding the lives and health of industrial workers but there are many encouraging signs of awakening interest. In the United States, the exposition of safety devices held last winter under the auspices of the American Institute of Social Science with its resultant plans for the foundation of a permanent "Museum of Security," in Great Britain, the work of the Royal Commission on Safety in Mines, and in both countries proposals for the enactment of laws for the protection of workmen evidence a public recognition of the need for better conditions, second only in importance to the vast amount of private work being done by large employers of labor. Dr. Gradenwitz's article describes one of the most important museums devoted to industrial hygiene in Germany where the cause of industrial betterment has been largely furthered by such institutions. It is noteworthy that the exhibits described fully bear out the contention in our editorial pages of a few months ago, that the educational value of such museums lies not only in "teaching the workman to value his own life," but also in impressing upon employers that the utmost precaution for safety and sanitary conditions in factories is the truest economy.—THE EDITORS.

TO protect workmen against the many dangers attending almost all branches of industrial occupation is a necessity, the importance of which is now fully appreciated, as evidenced by the good work done by modern social legislation. The tendency has appeared most markedly in Continental practice, and offers these precedents which may be of value for study and comparison under the more active interest now awakening in Britain and America. It therefore seems desirable to illustrate the results of some of these endeavours made to increase the safety of operation in factories and other industrial works by means of permanent exhibitions, foremost among which stands the Royal Bavarian Workmen's Museum at Munich.

This museum, which in its present shape was inaugurated at the end of last year, is intended to further any efforts made in the field of workmen's protection, while affording a comprehensive view of present achievements in the prevention of accidents, industrial hygiene, sanitary habitation and alimentation, and any other institution apt to raise the social level of workmen. The museum further demonstrates the use of safeguards for operatives using all sorts of machinery, and undertakes to examine the possibilities of such protective devices. Lectures on workmen's protection, industrial hy-

giene, and related topics further serve to popularise the subject, a
thorough knowledge of which is made accessible by a special library
connected with the museum.

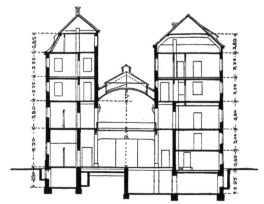

GENERAL SECTIONAL ELEVATION OF THE MUSEUM.

The museum building was recently erected in the Pfarrstrasse,
and occupies all the rooms of the ground and first floors in addition
to part of the basement. The general arrangement and dimensions
will be explained by the schematic views. From the vestibule a flight
of stairs leads to an exhibition hall 27 metres in length and 6 metres
in width, and to the engine room on the ground floor and to show-
room in the basement. The first floor contains a similar exhibition
hall of the same dimensions, and a gallery 2 metres in width surrounding the machinery hall, whence access can be had to the administration rooms, the officers' room, and the library and reading rooms.

The building is constructed of rammed concrete to the top of the
basement walls, and brick masonry above that. All the ceilings,
beams, columns, girders and roof framing are made of concrete-steel.
The exhibition halls, galleries and administration rooms are lined with
linoleum, while the greater part of the machinery hall is provided with
a plank lining 7 centimetres in thickness secured to a set of anchored
beams. The belting is arranged below the machinery hall, one-third of
which is covered with double T-girders with simple T-rails screwed
to the latter at average distances of 70 centimetres, removable planks
9 centimetres in thickness being fitted in the intervening spaces. These

planks are replaced at will by all sorts of machines secured to supports of equal height, which are thus independent of any foundation.

In the asphalted courtyard there is a track leading from the street through a sliding gate to the centre of the machinery hall. Beside the sliding gate and close to this track there is a hand-operated elevator for the transport of boxes, etc., to the basement. The machinery hall is covered over with a Morier vault below the concrete-steel frame, the runway of the crane being secured to the abutments of the latter by means of concrete-steel brackets. The travelling crane of 3-tons capacity covers the hall between galleries and can be run the entire length of the hall. Ample illumination is provided by side windows of modern construction, in addition to three double sky-lights 7 metres in length and 4 metres in width, which are fitted with ventilation valves operated from the gallery. Snow-melting devices connected to the central heating plant are provided between the two glass surfaces 1 metre distant from each of these sky-lights. All the rooms of the building are connected to a central low-pressure heating plant fed from two boilers; the building is lighted by electricity throughout.

The many machines installed in the museum are driven from three lines of shafting, while a number of recent additions are operated on the individual electric system. All belts are protected by wooden boxes or perforated sheet-metal cases, while the electromotors, as well as their starters and cut-outs, are enclosed in ventilated glass cases, through which they can be inspected by visitors.

The department given up to the operation of steam boilers is especially interesting. This contains some strikingly deformed steam-boiler parts, calling to mind certain grave steam-boiler disasters and illustrating the factors liable to prove fatal to boilers, such as the use of defective material, deficient construction, want of water in the boiler, slow corrosion due to unsuitable feed water, boiler-scale deposits and the resulting superheating and bulging of the boiler sheets, action of the sulphurous acid contained in the combustion gases, corrosive action of the carbonic-acid bubbles evolved on heating the feed water, etc. Cracked water gauges illustrate the necessity of designing a safeguard against the water and steam blowing out and the glass débris flying about; the destructive effects of water shocks are illustrated by the cross-head of a destroyed steam engine. Water gauges with divers protective mantles are exhibited in a special room, which also contains signalling apparatus, giving an alarm to the engineer in charge of the boiler whenever the water level through

neglect in superintendence or irregularity in operation has fallen
below a minimum. Valves which in the case of steam pipe fractures
are closed automatically, thus preventing the issue of steam, as well

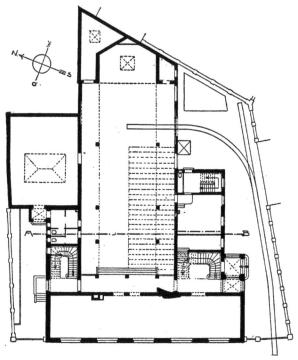

GROUND FLOOR OF THE MUSEUM.

The exhibition rooms occupy the T-shaped center. At the left is the auditorium. A small
workshop is between the stairways on the right, and a drafting room in the acute
angle at the top of the plan. The dotted lines across part of the exhibition
gallery indicate the power transmission system below.

as a number of fixtures ensuring the safety of boiler operation, such
as manometers, etc., are other exhibits of this interesting section.

The electrical department contains in the first place high-tension
fuses, especially in connection with telephone plants. Those ex-

hibited by a Munich firm will protect such plants against lightning and high-tension currents, which are conveyed to the ground conductor through toothed plates. Protection against loose high-tension

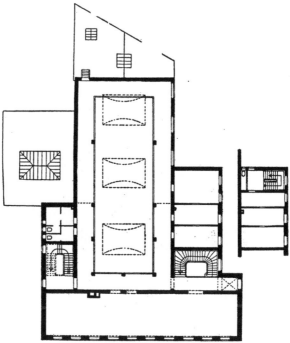

UPPER FLOOR OF THE MUSEUM.

The exhibition space occupies the cross arm of the T and the galleries around the vertical stem. The three small rooms at the right of the main plan are the library, the officers' room and the business office. The detached section on the right is a plan of the mezzanine, containing a kitchen and two chambers.

conductors hanging down from the poles is afforded by two short-circuit and ground-circuit devices which can be substituted for the more expensive and less reliable catching nets. In one of the constructions exhibited, the short and ground circuit is effected by a three-armed angular lever, which on the rupture of a wire will catch with a clamp connected with the ground conductor, while in another case, short-circuits are produced by a ring of copper wire. This apparatus is used in connection with the self-acting high-tension switch installed in the central station, a relay being disengaged on the rupture of a high-tension conductor, thus disconnecting the latter from the machine and cutting out any current from the latter.

EXHIBITION OF DEVICES FOR SAFEGUARD AGAINST ACCIDENTS.

The same firm exhibits diagrams showing some recent designs of transformer cabins, in which the high-tension and low-tension parts are each entirely self-contained, and are not accessible simultaneously, thus excluding any risk of mistake. Another interesting apparatus is an automatic lightning arrester comprising a rotary carbon rod and a solenoid actuated by the magnetising current produced by lightning, which, by turning the carbon rod round, will break the electric arc; after which (the magnetising current having disappeared) the whole apparatus returns to its initial position, thus being ready for dealing with another discharge. Other exhibits are rubber boots and gloves for persons engaged in high-tension work, insulating tools, etc. An interesting type of automatic fire alarm dependent on the expansion of metal springs under the influence of heat will likewise be found here. At a given temperature, which is adjusted at will, this expansion increases to such an extent as to cause an electric contact to be closed, working an alarm bell.

A special department is reserved for safeguards to be used in connection with acetylene, with a view to eliminating any risk of explosion in acetylene-gas plants.

An extensive group of exhibits includes safeguards for motors and shafting, intended either to allow the engine to be started without danger, or to prevent any access to moving machine parts, or finally to stop the machinery immediately at a critical moment, or to eliminate any risk of an unexpected starting. The first purpose is served by switching levers doing away with the necessity of pulling or barring the fly-wheel of the steam engine past the dead point of the crank. Many varieties of starting devices for explosion motors, intended to prevent any danger arising from a premature starting or igniting, are also shown. In the case of motors of up to 14 horse-power these devices take the shape of convenient hand-operated cranks which are thrown out of gear automatically as soon as the fly-wheel is in advance of the motion of the crank, or as soon as there is a recoil. Protection against projecting parts of shafting, which might result in danger to the men engaged in sewing belts or lubricating bearings, is afforded by smooth, round wood planking. Other devices do away with the necessity of putting belts by hand on revolving pulleys or throwing them off the latter. These devices enable the heaviest belts to be handled by mechanical means, the pulley carrying the belt being brought up to the one to which it is to be shifted, and the belt changed when both are in motion.

An apparatus exhibited by Mr. Cejka of Munich comprises a trip fitted to a steam-engine cylinder and connected to the steam valve. This is disengaged by electrical means, closing the admission valves, opening the exhaust valves and actuating a fly-wheel brake that stops the engine within about 4 seconds. This apparatus is combined with a device for preventing water shocks in steam cylinders, which is used with good results in nearly all of the large Munich breweries.

A Munich architect illustrates by means of a set of diagrams his mode of construction for isolating the foundations of heavy machines and engines, otherwise liable to transmit heavy vibrations to the ground and accordingly to cause nuisance to the neighbours.

A special department is set apart for apparatus, designed to obviate dangers of fire or explosion. This comprises all kinds of fireproofing (asbestos, etc.), several types of safety lamps, and the explosion-proof vessels of the Salzkottem factory, in which, all openings being covered with a fine wire mesh, the flame is prevented from flashing back to the interior of the vessel. In addition to the electric

EXHIBITS OF APPLIANCES FOR INDUSTRIAL HYGIENE.

fire alarm above described, an apparatus called autopyrophone, which is based on a similar principle, also serves the purpose of signalling the outbreak of a fire. Special features are automatic fire extinguishers, in which the bolt of a sprinkler (consisting of a readily melting bismuth alloy) is caused to melt by the increase of temperature, thus allowing the water to escape. The same section comprises all kinds of rescue devices, as well as a collection of apparatus intended for use as first aid in case of accidents. An apparatus designed by Prof. Stockmeier is intended for demonstrating the explosiveness of the dust of aluminium bronze under certain mechanical conditions. Artificial-respiration apparatus deserves especial interest because of the grave mining accidents that have occurred in recent years.

. Another section contains safeguards for the mining and quarrying industry, both in extraction and in transportation and handling. The extraction of raw material is illustrated by several photographs of quarries and by models that show the working in stages of the basalt quarry of Messrs. Staud & Co. Other models illustrate the methods of electric blasting used in limestone and sandstone quarries, collieries, and slate quarries. An interesting model of a magazine for explosives is remarkable for its light roof framing and wooden walls, which can be partially hinged up so as to prevent any excessive resistance to the high gas pressure arising in the case of explosions. Apparatus for thawing frozen dynamite cartridges, viz. double-walled sheet metal reservoirs containing tepid water between the walls, are likewise found in this section. The most recent progress in the manipulation of explosives is illustrated by a collection of "safety explosives," (e. g., carbonite, etc.), which do not ignite fire-damp. Another exhibit of interest is a safety belt and rope grip for use in climbing in the mountains and in working on high buildings, roofs, towers, etc., which, while in no way interfering with the movements of the person, protects the latter against any danger of falling. Safeguards for the transport of materials are illustrated by interesting models of self-acting inclined planes, one of which comprises an automatic barrier, intended for preventing the transporting vehicle from going beyond the top of the incline. The ascending truck automatically opens the barrier, which after its passage returns of its own accord to its position of rest, thus locking it. Another model is a truck automatically locking the passage by electrical means, as soon as the lower station (receiving station) has advised the upper station (transmitting station) that the empty trucks are ready for starting.

Safety devices for the handling of materials are represented by a

SAFETY METHODS AND APPLIANCES FOR BUILDING OPERATIONS.

model of a protective frame to be fitted over the mouth of clay-crushing machinery, as well as by diagrams illustrating the protection of mills, crushers, etc.

Of special interest among safety devices for metal-working machines are those pertaining to emery wheels; owing to their high speed of revolution and resultant risk of bursting, these are liable to cause the gravest accidents to operators. They are therefore fitted to the shafts by means of pressure discs and nuts with the intermediary of elastic plates, in addition to protective sleeves of various constructions. In the grinding of tools there is the risk of the tool being carried along between the grind stone and the support, thus jamming the hands. This risk is avoided by a movable grindstone support, which in such cases automatically recoils. Presses and punching machines likewise offer much danger in operation and especial interest attaches to the safeguards of such machines that are shown at the Munich Museum. In an eccentric press exhibited by Mr. Schuler, some protection against the descending ram is afforded by the very construction of the press, while an S-shaped protective grating prevents any involuntary approach of the hands within the dangerous range of the ram. In another eccentric press this protection is afforded by not allowing the machine to be thrown into gear before a protective grating has been placed in front of the piston.

In an adjoining room will be found stopping devices for punching levers and sheet-metal shears, protective linings for the projecting parts of chucks, safeguards for use in connection with lathe gearing, and an apparatus allowing machines fitted with claw clutches to be thrown out of gear nearly instantaneously at full load. Safeguards for the textile and related industries will be found in neighbouring rooms. Owing to the considerable dimensions of such machines, many of them had to be exhibited in the shape of models or diagrams. One of these models represents the lock of a beater, which can be opened only while the machine is stopped, thus precluding any injury arising from the beater. The Baudoin belt applier, viz. a vane loosely fitted on the transmission shaft for applying heavy belts to the rotating belt pulley, is likewise on view. The operation of weaving mills is represented only by shuttle guards of which many varieties will be found.

The leather and clothing industries are scarcely represented, one of the most remarkable exhibits coming under this heading being a punching machine constructed by Mr. Moenus of Frankfort a/M., which is provided with an automatically rising and dropping protective grating to guard against the approach of the matrix, and a

TWO VIEWS OF THE MACHINERY HALL.

453

centrifugal washing machine fitted with a safety lock and brake. A callender model most instructively shows how the hands, in introducing the cloth to be callendered into the rolls, can be guarded by an axle moving on two idlers near the mouth of the rollers and carrying a loose socket, which, though not applied directly to the cloth, owing to the motion of the idlers, partakes in the rotation but is stopped on any contact.

Another department is given up to safeguards for lifts and elevators, protecting the latter in the case of breakage of the carrying cables, or locking the lift pit, or finally, protecting the gearing. Safeguards of the first kind belong mainly to the catching device type, being based on the principle that on the parting of the carrying cable or belt a spring connecting with the lift is disengaged and drives the catching claws or wedges immediately into braking position. Another arrangement provides a suspension on both sides by means of a horizontal equilateral double lever, which in case of rupture or excessive tension to even one cable, is caused to rotate, throwing a catching device into gear. This lift also exhibits a double-leaf shaft-gate connected with the driving gear. Models of shaft-gates in the form of sliding doors which will open or lock the shaft opening gradually as the lift approaches or departs from the loading floor, are exhibited by several firms. In the case of express lifts, automatic reversals are liable to occur owing to the momentum of the cage; Prof. Ernst of Stuttgart therefore prefers a gear controlled by an operator, which would do away with any automatic adjustment at the various storeys. Safety cranks for capstans are likewise on show.

Another group of exhibits relates to wood-working machinery; these protective devices however are demonstrated mainly in the shape of photographs. Of special interest is a combined saw-milling and drilling machine with adjustable table, the stationary frame of which contains the cleaving wedge guide, in addition to which the machine is fitted with a protective hood for the circular saw and a protective cap for the milling of curved pieces. A number of protective hoods for circular saws are exhibited by leading German firms, and safeguards for milling machines are represented quite as fully.

Safeguards for use in the food-stuff industry are exhibited in a room adjoining the lift department. Mixing and kneading machines and meat-mixing machines are shown, in which all toothed wheel gearings are enclosed and the covers are bolted. Similar mixing machines, by the way, are used in place of crushing mills, in the manufacture of pasteboard. Dairy machines with all kinds of protective devices are further features of this department, as well as photographic illustrations of dough rollers, candy machines fitted with protective devices, and exhausters for snuff mills. An interesting exhibit is a safeguard against the explosion of mineral-water bottles, and a barrel tightener exhibited by a well-known Munich brewery.

Safeguards for agricultural machinery, such as steam threshers and fodder cutters, are likewise very fully represented, as are all kinds of apparatus for disinfecting the localities in which workmen are occupied.

The exhaustion of tar and acid vapours is illustrated by means of models, while several diagrams and photographs illustrate the great strides recently made in the illuminating-gas industry, from a hygienic point of view, especially retort-charging machines and coke extinguishers, which remove any fumes without nuisance to the neighbours and workmen. A number of articles forming part of the personal equipment of workmen, such as respirators, protective spectacles, smoke masks, asbestos clothing to guard against combustion by incandescent metals or hot gases, are further exhibits of the Museum.

A few rooms are set apart for sanitary habitation, illustrating by means of models, photographs, and diagrams, the housing of workmen. Much space has been devoted to sanitary buildings provided by the employer, while communal solutions of the housing problem are likewise fully represented. The ventilation, heating and lighting of workshops is illustrated by models and diagrams, exhausting plants being amongst the most interesting of this class of exhibits. Washing and bathing facilities designed for use by great numbers at a time are likewise very fully represented by diagrams and photographs. A related group of exhibits comprises all kinds of apparatus connected with the alimentation problem, such as meal heaters for factories, kitchen stoves, cooking boxes (fireless stoves). A number of institutions for the welfare of workmen, such as apprentice schools, are likewise illustrated in this department of the Museum.

Another department finally contains an extensive collection of models relating to the protection of workers in the building trades, such as safeguards for scaffolds, etc. Safeguards for use in connection with canal digging are represented by two instructive models in which the excavated soil is prevented from falling back into the pit, while passers-by are protected by a railing. Another model which will be found interesting is the Schaaf safety belt enabling the wearer to maintain a vertical suspended position while in no way interfering with his movements.

THE PROTECTION OF STEAM PIPES FROM ACCIDENT.

By Arthur Herschmann.

In the preface of Dr. Gradenwitz's article preceding we referred at length to the growing conviction that every possible safeguard should be thrown about the mechanical employments. Mr. Herschmann reviews briefly and comprehensively the results accomplished in protecting one particular engineering operation from the greatest risk to which it has been exposed. The short study is interesting for the ingenuity it discovers, and important because (as the author points out) the enormous extension in the use of steam power brings almost everyone within reach of the danger in question, and hence within the number of those immediately interested in the fuller introduction of the protective devices described.— THE EDITORS.

NORMAL activity tends to enlargement and improvement. Since Savery invented the atmospheric engine, to be followed by Watt's low pressure engine in 1768, the field of steam has expanded to so great an extent as to affect almost everybody in his daily work, directly or indirectly. The path of development of its use has been marked by improvements in the generation of steam and conversion into other forms of useful energy.

In the year 1801 Evans invented the high-pressure engine. The compound engine followed about 1850, the triple-expansion engine in 1880, and about 1890 the steam turbine became popularly known. Simultaneously, steam pressures have been constantly increased so as to obtain increased economy of coal consumption, and we find that today pressures of 150 pounds per square inch are in common use and that locomotive and marine boilers are operated at 250 to 300 pounds. The superheating of steam has also been known for a great many years as a means of obtaining increased economy.

Superheated steam of very high pressure became possible in sustained operation only with improved materials for the construction of boilers and engines and improved design of the latter. Manufacturing and testing have become standardized, and moreover systems of careful supervision have been evolved by highly organized boiler-inspection companies which contribute very largely to the high degree of comparative safety of which we may be proud.

The big boiler with its fire, the engine with moving parts and flywheel, are always in the public mind; less so the piping. Yet the latter, with its component parts such as fittings, presents a source of danger which is equally great in the aggregate.

The danger of a failing steam pipe, or of other damage in the steam line permitting the sudden escape of steam, naturally is greatest with the highest pressure, as every pound of pressure represents increased capacity for devastation at the time of sudden release. The heat energy liberated to work damage is represented not only by the steam itself, but even more particularly by the heated water contained in the boiler which passes into sudden evaporation as soon as the pressure is suddenly taken off its surface. This water necessarily is at a temperature above 212 degrees, so that the surplus heat will be freed to generate more steam very quickly. The work of devastation will be helped by concussion, owing to water which is mechanically trapped by the steam when an explosion occurs and which enters the scene of accident very much like a projectile. The attendant shock is of course felt by the boiler, and if the seams of the latter are strong enough to re-

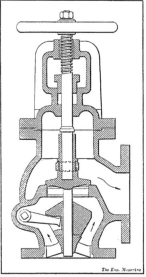

The Eng. Magazine

SECTIONAL VIEW OF THE H & M VALVE. Has no auxiliary pipes or pilot valves, and is not actuated by a piston. The valve is controlled by the flow of steam around the cone-shaped body, the upper valve closing when the steam pressure of its boiler falls lower than that of the other boilers with which it is connected. The upper valve revolves on its stem and cushions itself in seating and at the upper limit of travel. The position is indicated on the outside. H M Valve Co., New York.

sist it, chances are that it will be quickly bared of water and "burned out."

But more important than the material damage which is likely to follow a pipe failure, is the danger to life.

Boiler rooms may still be found where the men must necessarily crawl through tight places to work a certain valve, or where a steep ladder must be climbed to reach certain fittings which might as well be worked from the firing floor. But in the best designed boiler room little opportunity exists to reach the exits safely should a large pipe

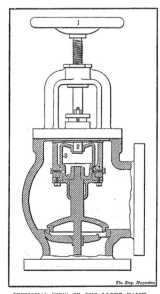

The Eng. Magazine

SECTIONAL VIEW OF THE LOCKE VALVE.

If a flue break or other accident occur, the Valve
closes, cutting the boiler out. Nor can it be cut
in again until its pressure equals that in the
main. The valve is held slightly above its seat as
shown, the steam being shut off by the closely
fitting extension; thus chattering is avoided
and the seats preserved for hand closing.
Locke Regulator Co., Salem, Mass.

burst and permit the boilers
to empty their contents.

Let us assume that we are
in a boiler room 60 feet long,
20 feet high, and 40 feet deep.
It is a 1,000 horse-power
plant, four units, and we find
that, after the boiler volumes
have been deducted, a free air-
space remains of 29,000 cubic
feet. Let us suppose that
only a supply line from the
header bursts at a pressure of
150 pounds. The velocity of
outflow expanded is 1,478
feet. (For comparison it
may be noted that the muzzle
velocity of a rifle ball is about
2,000 feet per second.) The
discharge of steam for the 5-
inch area is about 65,000 cubic feet per minute, so that in
less than 30 seconds the room
will be filled with suffocating
steam. Think of the proportions of your boiler room, of
those on board ship or in
mines where pipe galleries exist, and make a comparison!

In view of this terrible danger many provisions are made
to render pipes safe. The design of the piping is studied with the
greatest care, so as to avoid initial strain. Special care is taken to provide for expansion, loops being frequently resorted to, and pitch is
given so as to facilitate the flow of condensation. To minimize the
possibility of rupture, piping of this class has even been strengthened
by steel and copper wire winding. All these devices tend to reduce the
chances of a pipe failure; but as experience proves that the failure can
not be rendered an impossibility, means must be provided to check the
rush of steam and isolate the defective line in case of the break of
a pipe or fitting.

For the last fifty years valves have been on the market which could be operated mechanically from a distance, using levers, gears and chains, and more recently these devices have been designed to be actuated electrically. It is, however, plain that such a device will be a partial protection, at best. Serious accidents happen quickly and without the least warning; they leave no time to think and act, and must be averted automatically.

In the pipes of the average power plant steam travels at the rate of a mile per minute, or about that—say 90 feet per second for maximal average speed. Should a pipe break the velocity of outflow will be 900 feet per second, in round figures. Whereas pressure differences are but slowly transmitted, the above shows clearly that the great difference in velocity is effected at the moment of the explosion, so that the quickly traveling steam may be made to engage a valve disk in its path and "shut the door behind it." This sounds very simple, but in actual practice is most difficult of accomplishment.

A close study of the conditions of plant operation shows that such a valve must answer very conflicting requirements. Modern engines have high piston speeds; the point of cut-off may change very suddenly, and consequently if the safeguard valve were responsive to such a sudden rush of steam, it would close prematurely. This would seriously endanger a plant, and on board ship for instance would render a steamer unmanageable just at the criti-

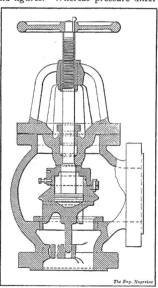

SECTIONAL VIEW OF THE COLDWELL-WILCOX VALVE.

Any accident such as the blowing out of a tube or header, or anything causing a counter current through the valve, leads to its instant and automatic closing, preventing any passage of steam between the main header and the defective boiler or part. Coldwell-Wilcox Co., Newburgh, N. Y.

cal moment of manœuvering. On the other hand, however, it is necessary that even a comparatively slight pipe break should close the protective valve. To meet this conflicting condition such a valve must not be too heavy in its movement and I believe should be kept "off" the closing point by artificial means.

My experience seems to support the theory that gravity and the assisting ejector action of the flowing steam are reliable principles for a safe law of action which such devices shall obey and be made entirely dependable in practical plant operation.

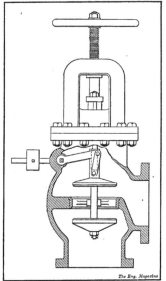

SECTIONAL VIEW OF THE LAGONDA VALVE.
Works instantly either way, depending for its action not on differences of pressure but on the flow of the steam. No pistons or pilot valve and but few moving parts. The double-beat valve, closing upon either side of a double seat, is very nearly counterbalanced by the lever and weight; with a slight flow of steam it rises to mid position as shown, and so remains. A sudden rush caused by a break beyond will carry the lower valve up against its seat; a reverse rush caused by accident to the boiler will close down the upper valve. Either shuts off the boiler from the main. The weights of valve and counterpoise, the distance apart of the valve seats, and the strength of the springs are so adjusted as to prevent the valve from closing under ordinary fluctuations of normal working. The Lagonda Manufacturing Co., Springfield, Ohio.

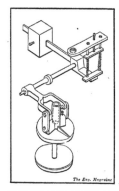

COUNTERBALANCE OF THE LAGONDA VALVE.

It should be possible also to adjust such valves to suit the nature of the plant, and these adjustments should be rendered "fool-proof." The internal moving parts should be few, and so arranged that they cannot easily corrode, rust together, or tighten up. The safe action of the device must be independent of the degree of opening of main valve. If the main valve is combined with the automatic valve, then the closing spindle of the former should be made to "stir" the latter every time it is worked up and down. Once the valve has closed automatically, it must stay closed, otherwise it might re-open on a group of people who are likely to rush to the scene of accident. It may advantageously show its position from the outside. It should be double-acting—that is, protect both the boiler and the line. It must possess special strength to resist the shocks to which it may become exposed. These are but few of the most important requirements which such devices should meet to be desirable.

The testing of such devices in the shop affords a means for their reliable adjustment but does not adequately represent their true value in case of emergency when the human mind might be so affected as to upset clear reasoning and quick manual operation.

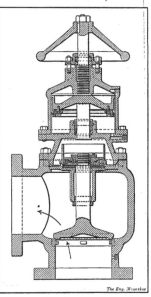

SECTIONAL VIEW OF THE FOSTER VALVE.

Worked by hand, either directly or from a distance, but will shut off automatically in case of any accident breaking pipes or connections. The main valve shown is operated by a pilot valve. Any sudden drop of pressure below a predetermined limit relieves the pressure in the chamber of this pilot, a spring immediately closing one valve and opening another, thus permitting the boiler steam to pass into the piston chamber of the combination valve. The piston, being of larger area than the large valve, will instantly close this latter against its seat and prevent passage of steam in either direction.

Foster Engineering Co., Newark, N. J.

In the United States this subject has in the last few years been given considerable attention and there are probably not many experienced engineers who do not recommend pipe protection. In France automatic valves have been compulsory since 1886 when the explosions of Eurville and Marnaval liberated steam

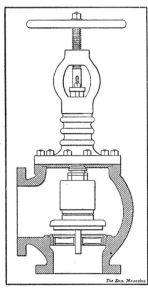

The Eng. Magazine

SECTIONAL VIEW OF THE NEW BEDFORD
VALVE.

Raising the hand wheel lifts the cylinder attached to the stem, the disc remaining closed
until the pressure in the boiler equals that in
the main and closing again at once if the
boiler pressure falls below that of the steam
line. The action is noiseless and positive; if
the pressure falls, from accident or other
cause, the valve closes. New Bedford
Boiler & Machine Co., New
Bedford, Mass.

which killed 30 people and seriously injured 60 others. Marine boilers are not governed by
this law. In other European
countries such laws are now being considered, and in Norway
a regulation exists governing
the use of special shut-down
valves on marine boilers and
their piping.

Whether the installation of
such safeguard devices proves
demonstrably profitable or not,
the investment represented by
them should be made out of the
profit and loss account and it
should be considered in the
same way as advertising, and
insurance of other kind. Actually the insurance which they
provide against interruptions
of various kind may be expressed in dollars and cents.

That pipe protection saves
life is no longer questioned by
those in a position to know of
the occurring pipe failures although these may not be reported in the daily press. To
such persons it is a matter of
surprise to notice that occasionally the question is still raised
if it is not a safe gamble just
"to depend on the piping." The tendency of the times is to be more
and more exacting that every known precaution shall be taken to
protect life and limb. Legislation may soon put these requirements
into mandatory form; but purely economic (to say nothing of humanitarian) reasons impel the far-seeing manager or owner to be in advance of legislation in such matters as this.

PULVERIZED COAL AND ITS INDUSTRIAL APPLICATIONS.

By *William D. Ennis.*

I. THE CHARACTERISTICS AND PREPARATION OF THE FUEL.

In a preceding series of articles in this magazine, Professor Ennis discussed the advantages to be secured by burning in the boiler furnace the small sizes of anthracite—that is, those grades which include the very fine coal naturally produced at the breakers, but requiring special grate and draft arrangements to make their firing practically successful. In this article and a succeeding one he takes up an extension of this idea—that of artificially reducing the fuel (though in this case it is bituminous coal) to a still finer state and blowing it, mixed with air, into the furnace. The practice has already found advantageous application in the cement and other industries, and offers large hope for relief from the smoke problem through its introduction into power-plant and similar work. The present installment of the review covers the preparation of the fuel, and a following one will take up the actual firing and the devices for that purpose.—THE EDITORS.

FINELY powdered coal has been used as a fuel in various industrial operations for many years. It is familiar to the foundryman as an essential constituent of many high-priced facings. As fuel, it has been quite widely employed in annealing and heating furnaces, and more generally in connection with rotary cement kilns. In the last named, the requirements for a hot fire and long flame, with ease of accessibility to the kiln for repairs, early led to the adoption of either oil or natural gas as the only fuels commercially possible. Where natural gas is generally obtainable nothing can compete with it; but under the ruling prices in the eastern and middle States, the use of fuel oil involves a cost of operation about double that obtainable with pulverized coal. The ash from this latter fuel affords no complication, entering into the cement "clinker" and thus eventually commanding a price of something like $10.00 per ton. It is stated that pulverized coal was used in rotary puddling furnaces in the arsenal at Woolwich as early as 1873. The bibliography of the subject leads back as far as 1831, to Henschel, and suggests the names of Wegener (1891), Unger, Friedeberg, De Camp, Ruhl, R. Schwartz-kopff, M'Cauley (1881), Whelpley, Storer (1876), Westlake, Crampton (1873), Stephenson, and Bryan Donkin. The use of powdered coal in forges and furnaces of various kinds is now becoming general and rapidly increasing. For these applications, a high percentage of ash is not generally found to be detrimental, and the grinding is seldom finer than from 24 to 30 mesh. Wherever powdered coal thus displaces oil, the saving in fuel cost is about one half. In replacing coke or coal with powdered coal, savings must be looked for, if at all, in the reduced labor cost, higher efficiency and increased output. That these considerations, together with that of convenience, have led to a widespread use of fuel oil would seem to indicate that they are of considerable moment.

The application of pulverized coal to steam-boiler operation, although offering well realized advantages, has progressed until recently very slowly. The common view of the matter, that pulverizing is destined to solve the problem of utilization of the waste products of the mine, does not seem likely to be realized. That coal in a finely triturated condition permits of such perfect mixture with air as to furnish the conditions for ideal combustion is a far more vital phase of the subject. The most successful steps taken toward the utilization of mine dust and refuse have been in connection with briquetting processes, which, however, present no advantages over the combustion of ordinary coal with relation to control of air supply. Briquetting has been found to be most successful with bituminous or semi-bituminous coal. It appears to offer the only method yet suggested for the disposal of waste slack or of the refuse from the operations incidental to the marketing of anthracite. The coal is ground, mixed with heated pitch, and moulded to shape under a pressure of from ½ ton to 6 tons per square foot. The grinding need not be as fine as that necessary for pulverized coal. The briquettes may be made in a shape permitting of compact storage, but firing is always by hand. Over two hundred patents relating to briquetting processes have been issued in the United States.

Pulverized coal is physically interesting. A lighted candle can be plunged into it without other result than the extinguishing of the candle; but if the coal is sifted over a flame from above, in a thin stream, it flashes like gunpowder. A recent British report on bituminous dust describes microscopic examination as indicating that the individual particle resembles, not a solid, so much as a bubble of explosive gas. Each grain consists of hollow spores having gas within and about it. This dust contains from 25 to 67 per cent of ash, and from 18.75 to 30.54 per cent of volatile matter.

Under all circumstances, the storage and handling of soft coal is attended with some degree of fire hazard. This is particularly the case with low-grade coals, running high in volatile matter. When pulverized, these coals present a still more greatly increased hazard.

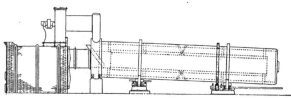

THE RUGGLES-COLES DRYER

Consists of two concentric shells of heavy steel plate, solidly connected in the centre by six heavy cast-iron arms, with swinging arms between the centre and the end to allow for expansion. The inner cylinder at the feed end extends through the stationary head and connects with the hot-air flue from the furnace. At the discharge end a revolving head fitted with lifting buckets raises the material and discharges it through the central casting. The cylinders are fitted with two heavy rolled steel tires ground true and revolving on chilled iron bearing wheels. The Ruggles-Coles Engineering Co., N. Y.

The processes of pulverizing and drying should be kept in a separate building, and the usual precautions taken against the possibility of dust explosions. Storage should be in non-combustible covered bins

CROSS SECTION OF RUGGLES-COLES DRYER.

of small capacity, safely located and well ventilated. The danger at the burners is considered by the insurance authorities to be rather less than is the case with oil fuel. The pulverizing rooms must of course be well ventilated. No fires or lights must ever be permitted. In all cases of proposed installation, the plans should be fully worked out in advance and submitted to the criticism of the companies insuring the works.

The best coal for pulverizing is one containing upward of 30 per cent of volatile matter. Pulverizing processes, as exemplified in the United States at the present time, are not adapted to anthracite, or even to semi-bituminous coals. In this respect, pulverizing is strongly contrasted with the present·state of producer-gas development, the one requiring highly volatile coal, the other a coal comparatively low in volatile. Anthracite dust might be burned if mixed with dust from gas coal; but it is not itself sufficiently gassy in its composition to maintain a flame in the furnace. Mr. S. H. Hall* reports experiments showing that coal containing less than 20 per cent of volatile matter would not ignite. A sample analyzing 35 per cent volatile ignited readily, and a somewhat less volatile coal ignited without difficulty when the combustion chamber was heated in advance. Among several works using pulverized coal for fuel, one expressed a preference

* Thesis, Sibley College, Cornell University, 1903.

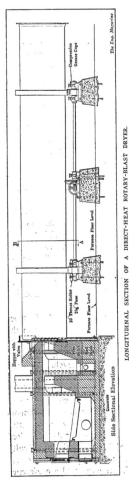

Side Sectional Elevation

Furnace Floor Level

Hopper with Valve

10 Thrust Boiler 1½ Fans

Furnace Floor Level

Compression Grease Cups

The Eng. Magazine

LONGITUDINAL SECTION OF A DIRECT-HEAT ROTARY-BLAST DRYER.

Direct-heat, radt-contact type. The wet matter a and the furnace gases enter the shel at the higher end; as the ryed shel revo ves, the matt r a is repeated y caught by the inter or shelves, rai ed a mbst to the highest oint of rotation, and show re through the hot gases, unt it is e ther dry, at the ower end. For coal rying a spot a furnace is use with an air cham er betwen com ust on cham er and shell. The forc - rat latt t is batche to both ass t and air cham er, and by regi at ng the air su de to these two ot ts; the temp crature of the gases enter ng the shel can be c ode y contr e . Amer can Process Co., New York.

for West Virginia slack, but was actually using a coal containing from 30 to 33 per cent of volatile, 5 to 58 per cent of fixed carbon, and 9 to 13 per cent of ash. Sulphur was permitted to run as high as 6 per cent (this was in cement kilns) and the heating value of the coal was from 13,200 to 14,000 B. T. U. This coal was considered too low in volatile for the best results. Another works, which has been using pulverized coal for steam production for nearly five years, reports an average composition of .82 per cent moisture (as fired), 35.13 per cent volatile, 54.70 per cent fixed carbon, 9.35 per cent ash, and .99 per cent sulphur. This coal yields 14,080 B. T. U. per pound. A series of tests made at Detroit, in steam-boiler operation, specifies the following percentages in the coal used (Westmoreland, Youghiogheny);—fixed c a r b o n, 57.40; volatile, 34.47; ash, 7.50; moisture, .63. The sulphur was 1.74 per cent of the total. This coal tested 13,315 B, T. U.

Tests made in 1904 at the works of the Erie Malleable Iron Co., Erie, Penna., where pulverized coal has been successfully used for

both heating furnaces
a n d steam boilers
during the past five
years, gave the fol-
lowing composition
for the fuel after pul-
verizing; fixed car-
bon, 53.7; volatile,
35.5; moisture, 2.4;
ash, 8.4. The heating
value per pound was
13,152 B. T. U. At
Lebanon, Pennsyl-
vania, the American
Iron & Steel Manu-
facturing Company
u s e s (for furnaces
only) pulverized coal
showing fixed carbon
58.49, volatile 31.97,
moisture .52, a s h

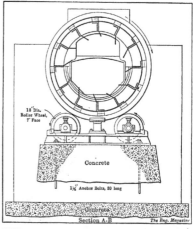

CROSS SECTION OF THE AMERICAN PROCESS COMPANY'S DRYER.

9.02, sulphur 1.34, heating value 13,721 B. T. U. Still another works shows 46.7 per cent of fixed carbon, 32.3 of volatile. 3.5
of moisture (too high for good results) and 17.5 of ash. The detrimental result of the high percentage of moisture is shown by the ash
analysis, which was, 2.74 per cent volatile, 1.78 fixed carbon, 95.48
incombustible; a very poor result for pulverized coal. The main point
to be observed from all these analyses is the evident effort toward a
relatively high percentage of volatile matter. A long-flaming gas
coal is what is needed.

Practice has also become somewhat standardized with regard to
the grade of coal. In all of the examples quoted, the slack coal is
used. This probably costs less to pulverize than nut or run of mine.
at any equal stage of dryness. Unfortunately, the slack always contains more moisture than the other grades, and therefore always involves extra expense, either for drying, or for the greatly increased
amount of power necessary to pulverize coal not thoroughly dried.
It is therefore unsafe to accept too confidently the claim frequently
made, that pulverizing offers a cheap and certain method for the
economical utilization of mine waste. One of the works to which
reference is made buys the 1¼ coal for furnace work, and uses the

screenings for grinding, In winter, on account of the excessive moisture in slack, and insufficient dryer capacity, nut is purchased to augment the supply from its furnace screenings; yet this works does not aim to reduce the percentage of moisture below 2½. Mr. W. H. Bryan, in discussing pulverized coal,* states that the slack may be used, at a cost for grinding of 35 cents per ton, which cost will be offset by the saving in labor at the fire room. This would make powdered coal the cheapest of all fuels in St. Louis, at the ruling market prices. Mr. Bryan's premise that pulverized coal may be produced from slack or mine waste is correct; that it is universally made from slack at the present time, is scarcely to be claimed. That it can be produced at an expense that will be entirely offset by the labor saved does not seem yet to have been demonstrated.

AN INSTALLATION OF THE RUGGLES-COLES DIRECT-HEAT DRYER.

The process of pulverizing consists of (1) a preliminary grinding of the coal, if necessary, to bring it down to about nut size, (2) drying, (3) pulverizing proper. From the pulverizer, the coal is delivered to storage bins, and from thence fed to the furnaces by some suitable device which regulates the rate of feeding and in some cases the supply of air. The preliminary crushing referred to is unnecessary with some coals and with some forms of pulverizer. When necessary, the operation is performed by means of the ordinary type of roll crusher or cracker. The reduction of the large lumps thus effected facilitates the drying and permits of a proper regulation of the feeding of the coal to the pulverizer.

* American Machinist, July 12, 1906, p. 52.

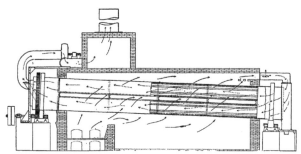

LONGITUDINAL SECTION OF FOUR-COMPARTMENT DRYER.
The C. O. Bartlett & Snow Co., Cleveland, Ohio.

The object of drying the coal is threefold: it prevents the material from forming pasty snowball-like masses, thus interfering with its conveying from stage to stage of the process; it permits of more perfect feed-regulation at the furnace; and it keeps the small particles separated, thus permitting of perfect air mixture and perfect combustion within the furnace. Moreover, the finely divided coal is much safer with respect to spontaneous combustion if thoroughly dry. The operation of drying is of course cheapened if the coal is first stored under cover sufficiently long to drive off part of the moisture, but this is rarely done. The cost of fuel for drying, with coal at $3.00 per ton, should never exceed 6 cents per ton, and coal is rarely re-handled at as low a cost as this.

Direct heat is practically the only drying method now practiced, steam dryers having long since been abandoned. With a material like coal, steam dryers would have certain advantages with regard to safety; but these are overweighed by other considerations, mainly those of capacity and economy. As ordinarily designed, coal dryers are externally fired cylinders. sometimes enclosed in brickwork, special precautions being taken to prevent contact of the coal with the flame. In certain cases, the

CROSS SECTION OF C. O. BARTLETT & SNOW DRYER.

THE AMERICAN PROCESS COMPANY'S DIRECT-HEAT ROTARY DRYER INSTALLED.

hot gases, diluted with air, are permitted to reach the coal at fixed stages of the operation. The figures illustrate commercial forms of dryer largely used. In the one shown on page 469, after the heated air has been used for heating the outside of the cylinder, it is drawn through an opening in the arch over the cylinder at the rear end, through a steel flue into the rear-end breeching, and through the cylinder, coming in direct contact with the material, passing out through the exhaust fan and carrying the moisture with it. In drying material which will not permit of passing the furnace gases directly through, the air ducts in the brickwork are brought together at the rear end and connected to the rear breeching, and then drawn through the cylinder in contact with the material. The walls of the dryers are always built with air ducts. The four-compartment direct-heat rotary dryer of the type illustrated on page 469 has an air or gas space between each two compartments. This increases the heating surface, and the compartment arrangement of the cylinder divides the load so as to decrease the strain on the structure of the dryer cylinder. The Bartlett-Snow dryers are made in various sizes, the Number 6 having a capacity of 6 tons per hour, and the Number 10 of 10 tons; both drying from 10 or 12 per cent down, say, to 1 per cent. The former has a cylinder 48 inches by 27 feet, the latter, 60 inches by 28 feet. The horse power required to revolve the dryer and operate the fan is stated to be 5½ for the Number 6 machine and 8 for the Number 10.

The Cummer dryer, shown on page 471, avoids the compartment arrangement, passing the gases directly into the cylinder through hooded openings as shown. A sufficient number of openings is put into each cylinder to allow about three-fourths of the entire volume of gases to pass into it. The balance of the gases enters the cylinder through the rear end. The result is that there is comparatively little circulation at the rear end of the cylinder where the material is practically dry and often dusty, and consequently there is little or no dust blown out at the stack.

The Schwartzkopff dryer, not illustrated, has four compartments, as in the one shown on page 469. All of these dryers feed the wet material in, continuously, at the front, or cooler end, of the cylinder, the hottest gases thus coming in contact with the dryest material. The American Process dryer, (pages 466-467) follows the reverse order.

SELF-CONTAINED DRYER WITH EXHAUST-FAN.
F. D. Cummer & Son Co., Cleveland, Ohio.

The fuel consumption for drying has been found by various users of pulverized coal to range from 1 to 1½ per cent of the weight of coal dried. Taking 10 per cent as the initial, and 1 per cent as the final, amount of moisture in the coal, each ton of 2,000 pounds would contain 200 — 20 = 180 pounds of water, to evaporate which requires 180 × 966 = 173,880 heat units. With coal containing 12,000 B. T. U., a fuel consumption of 1 per cent or 20 pounds, would indicate a dryer efficiency of 72 per cent: or a fuel consumption of 1½ per cent would evidence an efficiency of 48 per cent. These are reasonable figures. Occasionally the coal leaves the dryer in a more or less dry condition than is represented by 1 per cent of moisture. At one cement works a sample taken by me showed only .33 per cent. Another works, using pulverized coal under steam boilers, ran at about 2½ per cent of moisture. This was admittedly bad practice, and was due to insufficient dryer capacity. The fuel used in drying may be coal, pulverized coal, or coke. The last named is frequently used in order to decrease the labor necessary for firing the dryer, to avoid the possible accumulation of soot, and to avoid any danger from long flames.

With insufficient dryer capacity, the use of low-grade slack coal is often impracticable, especially in winter when the coal may be received dripping wet and is usually required for immediate use. One works, as previously stated, was found burning nut coal, mainly on account of a lack of dryer capacity. A successful installation involves, necessarily, long dryers of ample size.

The exhaust gases from the dryers carry away more or less dust, which, from considerations of safety, economy, and hygiene, must be collected and saved. This is usually done by providing a dust chamber in connection with the exhaust fan. Such a chamber may be arranged for either on top or at one side of the dryer.

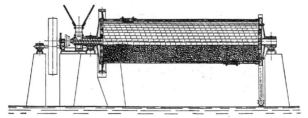

LONGITUDINAL SECTION OF DAVIDSEN TUBEMILL.
A pulverizer, not a coarse grinder. Made in three sizes—27¼ by 9½, 33½ by 12½, and 35 by 13 ft., requiring 27, 80 and 120 horse power respectively.   F. L. Smidth & Co., N. Y.

An important point in connection with drying concerns the volume of gases. If the gas moves in small volume and consequently at low velocity, it may at its initial high temperatures absorb and retain large proportions of moisture without exceeding a humidity limit of 70 to 80 per cent; but as it cools it may actually reach the dew point and deposit moisture on the material at the later stages of the operation. Large volumes of gas (excess air from combustion) and short lengths for transit are the remedies for this difficulty.

The operation of pulverizing is the present stumbling block in the way of standardized practice with regard to the application of powdered coal. Pulverizers give more or less trouble, the cost of operation is high, especially for power and repairs, and the pulverizers themselves expensive. Originally, the grinding of materials was done by means of burrstones or stamp mills; the former being used for soft substances, the latter for hard. At present, these forms of machinery are not employed for coal, which is extremely hard and slippery,

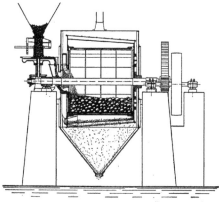

LONGITUDINAL SECTION OF LINDHARD "KOMINUTER."
A low-speed coarse grinder, combining good qualities of ball mill and tube mill. F. L. Smidth & Co., N. Y.

and requires very fine grinding. The power consumed, and the capacity of the grinder, both depend upon the material, its nature, size. and hardness, the mesh desired, the uniformity of grinding necessary, and the percentage of moisture in the material. Pulverizers or grinding mills may be divided into several well-defined classes. One of the best known is the tube mill. a strong cylinder revolving on an horizontal axis, containing a quantity of flint pebbles, from 1¼ inch to 3 inches in size. The coal is fed in at one end, hammered by the pebbles, and taken out fine at the other end. This form of mill requires very little repairs. and probably grinds finer than any other. I have taken samples of the pulverized product from which only 1.96 per cent

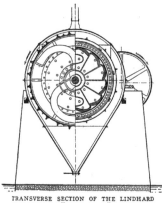

TRANSVERSE SECTION OF THE LINDHARD "KOMINUTER."

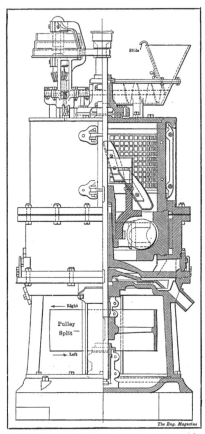

ELEVATION AND HALF SECTION OF THE FULLER-LEHIGH
PULVERIZER MILL.
Lehigh Car, Wheel and Axle Works, Catasauqua, Pa.

was left on a 100-mesh screen, a n d only 3.22 per cent on a 200-mesh screen. The great drawback of t h e tube mill is the power required. O n e mill, 60 inches by 17 feet, running at 23½ revolutions per minute, used nearly 60 horse power. Another, 60 inches by 20 feet u s e d 6 5 h o r s e power on coal, and r a t h e r m o r e on cement slurry, running up to 74 horse power on clinker. A 54 inch by 16 foot tube mill consumed 42 to 47 horse power. The power consumption can be greatly reduced by using special heads to permit of filling the cylinder up nearly to the top with coal and pebbles and thus balancing the load. In one case, this a r r a n g e m e n t was claimed to h a v e doubled the output while saving 40 per cent of the power.

Centrifugal roller and ring mills, with screen separation, are made by the Raymond Brothers Impact Pulverizer Co., the Schwartzkopff

Coal-Dust Firing Syndicate, t h e Kent Mill Co., and the Bradley Pulverizer C o . Typical mills of this type a r e shown. One, the Griffin, has a sin- g l e roll only. Others have two, three and f o u r rolls. The capac- ity is given at 2 tons per h o u r with a power con- sumption of 20 to 25 horse power. A three-roller mill of the same make is stated to turn out 4 to 6 tons per h o u r w i t h 35 horse power. A test run made at one

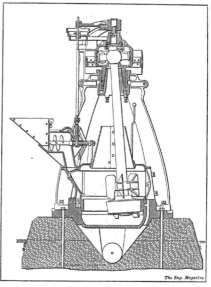

VERTICAL SECTION OF THE GRIFFIN MILL.
The Bradley Pulverizer Co., Boston, Mass.

works showed these mills, with the auxiliary dryer, but without con- veying or feeding apparatus, to consume 84 horse power for a capacity of 5 tons per hour. This is equivalent to 4.2 per cent of the output in equivalent horse power. The repairs in this particular instance were about $20 per month on each mill. The Clark pulverizer, shown on page 476, is a three-roller mill adopted by the Schwartzkopff syn- dicate. The Raymond mill has four rollers, and is claimed to give 4 to 5 tons per hour from slack coal, with 60 to 65 horse power, when grinding 95 per cent to 100-mesh. It can be adjusted for almost any degree of fineness. The three-roll Kent mill (page 477) has powerful springs pushing the rolls against the ring. The ring and three rolls are the only wearing parts inside the housing, and a life of 3,000 hours is claimed for these. This mill has a claimed capacity of 7 tons per hour to 40 mesh at 20 to 25 horse power. One user reported capacities ranging from 1 ton at 80 mesh to 7 tons at 24 mesh, but the fineness determinations do not seem to have been reliable, as a sample claimed to be Number 30 was found upon examination to be 50 per cent Number 100 and 32 per cent Number 200. One set of rolls was stated to have run 5 years, 10 hours each working day. All these mills fracture the coal rather than cutting or rubbing it. All require external screen separation, with return of the consequent coarse particles for regrinding. The statements as to power consumption are apt to mislead, especially in the absence of data as to the uniformity of fineness.

Another type of mill, largely used, has hardened balls revolving in cast metal races at high speed. The Fuller-Lehigh (page 474) is an example. It has a claimed capacity of 3½ to 4 tons per hour, 95 per cent 100-mesh, at 32 to 35 horse power. The repairs in one case averaged less than $12 per month. An actual test gave an output of 9,810 pounds, 87½ per cent 100-mesh, per hour. Actual operation at another works gave an output of 5,000 pounds per hour at 30 to 40 horse power. This mill is largely used in cement works, and grinds fine. The Stroud mill is radically different from the others, depending for its effect on a cutting or shredding action. This mill will take coal in sizes up to 2½ inch, and will handle coal containing as high as 10 or 12 per cent of moisture, although at greatly reduced capacity and efficiency. It is stated to produce from 7½ to 12½ tons per hour. Like the Fuller-Lehigh and the roller and ring mills already described, it requires external screen separation, but the same manufacturers also build an air-separation mill, which is the type recommended by them where fineness is necessary, as in the case of coal. The air separation mill, as shown on

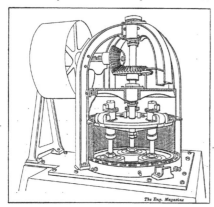

SECTIONAL VIEW OF CLARK PULVERIZER.
Schwartzkopff Coal-Dust Firing Syndicate.

page 478, h a s
a settling cham-
ber from which
the finest particles
only are drawn
off. It is stated
to produce 4 to 5
tons of dust per
hour at an ex-
penditure of 40
to 45 horse pow-
er. The degree
of fineness can of
course be perfect-
ly regulated by
the speed of the
exhaust fan. A
fineness of .90 to
95 per cent 100-
mesh is claimed.

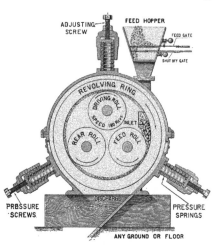

SECTIONAL VIEW OF THE KENT MILL.

Kent Mill Co., N. Y.

The economi-
cal limit of fine-
ness is reached when the cost of finer grinding equals the saving by
increased efficiency in the furnace. The finer the better, disregarding
this cost. Some consumers state 40 mesh to be sufficiently fine, but 60
to 80 mesh is more commonly used in steam-boiler practice. Much of
the uncertainty as to power consumption and capacity of the various
mills would be avoided if accurate data were obtained as to fineness
and uniformity of fineness. The cost of grinding increases greatly
as the fineness, or the uniformity, is increased. An outside figure for
power consumption, sufficiently safe for preliminary calculations, is
1 horse power for 100 pounds of coal per hour. This may be decreased 50 per cent or even more, in practice. Coal is hard to grind.

The handling of the powdered coal is usually done by screw conveyors and ordinary belt and bucket or chain and bucket elevators.
Leaks in elevator housings and in conveyor troughs are dangerous.
It is not safe to store the coal in large quantities or for long times, so
that small isolated storage bins should be used, and these emptied or
the contents rehandled at regular intervals. The loss of weight in
storage is not as high as with ordinary coal, but the loss of heating
value is greater. Bins must have steeply sloping bottoms, the angle

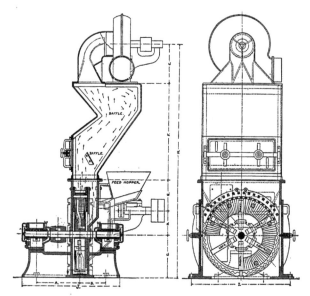

SIDE AND END SECTIONAL VIEWS OF STROUD AIR-SEPARATION MILL.

which the surface of the coal makes with the vertical being only from 34 to 38 degrees, when dry. Wetter coal lies at a wider angle. The specific gravity of the coal is from 1.3 to 1.35. The conveyors to the feeding bins over the boilers should have end overflows, discharging into return conveyors, in order to avoid packing and breakdowns in case of the shutting off of the downtakes to the hoppers. On account of the problems involved in safe storage, it is preferable to have drying and pulverizing apparatus in duplicate units so that the coal may be ground as nearly as possible just as it is wanted for consumption. The same condition prohibits any widespread marketing of powdered coal by shipment from central points of preparation, an otherwise desirable arrangement.

THE FILING OF CORRESPONDENCE IN A MANU-
FACTURING BUSINESS.

By Sterling H. Bunnell.

THE ENGINEERING MAGAZINE was foremost in the recognition of the new profession of "Industrial Engineering"—that is, of the peculiar field for technical specialists afforded by the development and refinement of systematic, organized manufacturing. While this lies largely within the domain of mechanical practice, it includes in addition much that is drawn from economics and much that is taken from commercial principles. But in turn, the genius of the engineer has remodelled economic and commercial methods by scientific study and adaptation, and given them back to the business world better than he received them. Mr. Bunnell's article is concerned with one of these collateral elements in the conduct of modern scientific manufacturing—one which at first sight appears to belong to the office rather than the shop, but which, as he shows, is so intimately bound up with the successful prosecution of the business as a whole that it deserves the serious attention of the works manager. —THE EDITORS.

THE old system of filing letters alphabetically according to their writers, and making copies of all replies in consecutive order by dates in tissue copying books, has long since been found inadequate to the needs of modern offices. The single advantage of that form of filing is the certainty of preserving a copy of every letter written, provided only that the invariable rule of copying before mailing has been observed; for even faulty indexing cannot do more than delay the search for a certain letter. Against the loss or misplacement of letters received, the system affords no protection except by care in filing and consulting files. This being so, it is a question whether there is any real advantage in effecting greater security against loss of letters written than against loss of letters received; while the massing of correspondence from several firms which happen to have names commencing with the same combination of letters is extremely objectionable.

The modern system of filing letters vertically has practically taken the place of former methods. The vertical file has the advantage of convenience in use, first, because its divisions are merely paper folders which can be inserted in or removed from the file with all their contents, when desired, and second, because it is much more convenient to handle and inspect letters filed on edge in such a way that the sheets can be separated like the pages of a book and easily examined without removal. The vertical file is universally used to receive all correspondence both into and out of the office. Carbon or tissue copies of each letter written are filed consecutively according to dates, with the letters received, one folder being given exclusively to the letters from and to each correspondent. By making the various forms used in the conduct of the business, shipping lists, specifications, contracts, and the like, on sheets the size of the usual letter sheet, it is possible to have but a single index for the entire correspondence.

The indexing of the vertical file is carried on in one or both of two ways. The alphabetical system is continued from the old form of filing, and for many purposes is most convenient. For very extensive correspondence, the numerical system is preferred, the name of each correspondent and the number assigned to his file being indexed on cards for ready reference. All letters received are opened by a clerk regularly designated, and each is given the proper number of the correspondence, nearly all of which are readily supplied from memory by the clerk accustomed to the work, while a few require looking up or entering as new names appear. In dictating replies, the writer gives first the correspondent's number, and no further attention is necessary, as the file clerk has only to follow the number in putting each sheet in its proper place.

In applying a filing system to a large manufacturing business modifications in details become desirable. Most of the correspondence, particularly the most important from a business point of view, originates by letters or other notifications giving knowledge of possible sales for the product of the concern. A correspondence is thereupon opened which may or may not amount to anything. Such correspondence is best placed in a separate file belonging to the sales department, and in general is most easily filed alphabetically, a new folder with tab pasted on carrying the name for indexing purposes. Letters and replies may accumulate to a considerable extent, but eventually the inquiry either comes to nothing, in which case the papers may be transferred to a case with other lost causes, or the correspondence develops into a sale, often followed by the signature of a contract. The entire file of letters is now transferred to the proper persons for investigation and decision upon the credit of the customer and the general acceptability of the order which the salesman has secured. The correspondence next goes to the construction department, and being given a number, is placed in the general file. It should be noted that all files in the office belong to the general scheme, except that for convenience certain sections dealing with inquiries are given to the sales department, others to the purchasing agent, and so on as found most convenient. In planning the arrangement, it is necessary only to take care that letters from one correspondent on two entirely different topics shall not be so separated and filed that the entire information wanted cannot be secured by searching in one place. This is easily accomplished by a system of cross-reference, a sheet being placed in the general file whenever letters on different topics from one correspondent are for convenience placed in different sections of the file.

The construction department's general file becomes very bulky as time goes on, particularly so if the business involves work running over a term of months, or years. Having first copied the signed contract and other valuable papers and returned them to the fire-proof vault, the engineering department proceeds to make drawings, and corresponds with the purchasers from time to time with regard to the various features of the work and the questions that arise. When finally the articles are shipped, packing lists and other files are made out, and copies placed in the file with the papers belonging to the transaction. Erecting men's reports pertaining to the particular work are filed in the same way, and the system remains undisturbed until the final acceptance is received and the construction department turns the matter over to the treasurer for collecting the balance due and closing the affair. The papers pertaining to the sale and delivery of machinery may amount to only a few sheets, while with a building operation or large contracting work the correspondence may run into thousands of pages. In such cases, it is convenient to make a little sub-file for each special order, supplying separate folders for correspondence with the purchaser himself, with consulting engineers, with sub-contractors and the like, so that one or more of these folders may be taken out as desired without involving the necessity of going through all the papers to find letters pertaining to any one party.

The later filing of letters from purchasers requires special study. It happens frequently in some lines of business that several different names are connected with the work as principals between the first inspection and the final conclusion. Thus the sale of engines for an office building comes up first as an inquiry from a sub-contractor for work on the building. A little later the names of the architect, consulting engineers, general contractor, and real-estate company owning or handling the property appear, and correspondence from all these belongs to the same work and should be filed together. After the building is finished and equipped and turned over to the purchaser, a new name may appear for the building and another corporation take hold, and all previous names be quickly forgotten. It is therefore necessary for the engineering department, which has furnished machinery to such a combination, to keep a record of the present name under which the job is known, and this may be done by a simple cross-reference in the letter file between the old and new names, with an appropriate entry in the serial record of shop orders. This being carefully kept up, reports or information can be readily obtained from the manufacturer, and the good will of the customer retained to great advantage.

There are several important divisions of correspondence aside from those of the sales and construction departments. Purchasing supplies brings in many quotations and accompanying letters, which are best filed by classes of material offered. Letters from persons seeking employment are conveniently filed by themselves in folders grouped by the class of work the applicant desires. In this way, reference can quickly be had to the names of persons seeking employment in offices or shops to fill vacancies in the several trades, which would not be the case if the letters were filed by the names of the individuals writing them.

Whatever system of filing is used, there is no safeguard against the loss of letters carelessly placed in the wrong files, other than attention on the part of the filing clerk. It is therefore absolutely necessary that the files be consulted by but one responsible clerk, or in the case of sections assigned to the departments, by one person in charge of each section, and no one else. With proper care given to the handling of the correspondence files in their daily use, the vertical system of filing, and classification according to subjects rather than according to names, it is possible for one man to handle an amount of business which would be very much beyond the capacity of the most able man working with the old system of alphabetical files.

The universal use of the typewriter and of rapid copying methods has much increased the amount of correspondence by facilitating the writing of letters, in many cases to the disadvantage of the business world by multiplying useless verbiage. At the same time, the free use of correspondence methods has made it possible to supply manufactured products to purchasers at vastly increased distance, and to study their needs in a way which could not have been done with any less easy method of obtaining and transferring information.

MECHANICAL REFRIGERATION ADAPTED TO RAILWAY TRANSPORTATION.

By Dr. Joseph H. Hart.

Dr. Hart's suggestions have very great interest and possible importance for both the railway and the mechanical manufacturing world. Their adoption would mean, on the one hand, the betterment of economy, of convenience, of time, and of capacity in cold-storage transportation; on the other, it would mean a large and profitable expansion in the installation and the operation of refrigerating plant. They deserve, therefore, the careful consideration of traffic managers and of machinery builders.—THE EDITORS.

THE quantity of perishable freight requiring transportation, and the utilization of refrigeration for its preservation in the United States, are daily increasing by leaps and bounds, and the limit is not yet in sight. This tremendous increase in freight tonnage is doubtless responsible for a considerable amount of the imperfection existing in the refrigerating methods in operation today; but there is no doubt that, as a whole, the application of mechanical refrigeration to transportation is in a most chaotic condition. With mechanical refrigeration perfected as an efficient process, and with capabilities in the production of refrigeration (and even of ice, at a much less cost than is required even in handling and storing the same from natural sources), still the great bulk of all perishable products is preserved in transit by means of natural ice and ordinary refrigerating methods.

The difficulties inherent in the application of mechanical refrigeration to this process are more imaginary than real, and the cause of the arrested development along this line is to be found more in the unusual and unnecessary processes which have been attempted than in any inherent difficulty in the matter. Further, mechanical refrigeration-machine manufacturers have found an expanding market with many new applications always at hand and generally a large excess of orders; and hence, the introduction along lines that have presented apparent difficulties have been neglected in favor of movement along lines of least resistance. It can be said that there exists no inherent difficulty in the thorough and efficient application of mechanical refrigeration to railroad practice and its necessities, although the lines ordinarily pursued appear to discover many. A general consideration of the entire problem from the viewpoint of the railroad engineer, and from that of the refrigerating expert as well, is necessary for a full understanding of the difficulties involved; but when these are once considered and thoroughly understood, the solution of the problem is comparatively easy.

Thus the problem from a mechanical-refrigeration standpoint is not difficult. Refrigerating units of capacity much smaller than that involved in a refrigerator car are in practical and efficient operation today. Unusual temperatures are not required, and the quantity of heat involved is not prohibitively small per car. Further, while under some circumstances water and space and power enter as important problems from the viewpoint of the individual or automatic refrigerating machine, from the broad standpoint of railroad practice in general these factors become of minor importance. Thus from the railroad point of view the conditions which are of prime importance are primarily reliability, space, first cost, and care. Under this latter head, of course, come cost of maintenance and the depreciation factor; but these are the only necessary considerations from the railroad point of view. From the shipper's point of view—and, of course, incidentally, that of the railroad as well—the question of ventilation and moisture enter as important factors.

Now to understand fully the difficulties involved and their relative importance in this particular problem of railroad refrigeration, the best results can be obtained by a complete review of the present methods in operation, their faults and inadequacy, and the various attempted improvements now under way. Thus today on railroads we have the following refrigerating systems:—

The great majority of cars are cooled by the insertion of ice in suitable compartments at various stations en route. This ice, in the great majority of cases, is natural ice collected in ice houses from suitable localities along the railroad and in some cases is shipped by them from suitable collecting localities to convenient distributing points or car-filling sheds. In some cases, comparatively few, this ice is produced by mechanical refrigeration, from the operation of large units with considerable efficiency; and there is no doubt that this practice is bound to increase as it presents the simplest and most obvious way out of the difficulty. Ice can certainly be manufactured and used much more cheaply than it can be harvested and stored by the ordinary methods.

The next step in the process is the attempt to develop automatic refrigerating machines for individual car operation. At the present date it can be said, in a general way, that almost without exception such installations are more or less complete failures. The problem is attacked in an unsatisfactory manner, and the inherent conditions are such that failure becomes almost inevitable. To begin with, the automatic refrigerating machine itself is not today in a condition which would warrant any such application. A large number of automatic refrigerating-machine plants have been organized but almost without exception these have proved inefficient and the companies have gone to the wall. The difficulties inseparable from the problem are such that almost every automatic refrigerating machine in existence today requires constant care in its maintenance or at least a large amount of attention. Under these conditions to complicate the matter further by constant vibration and jolting, such as are incident in railway practice; to operate under such conditions that the supplies of water and power are often inaccessible or subject to accidental stoppage; where space is limited and first cost is an important factor, and where, further, the very worst of care and attention becomes almost inevitable—this is a combination with which few engineers would care to struggle. Thus the majority of attempts to develop refrigerator cars involving their own individual refrigerating unit, have been limited to persons not wholly familiar with the entire situation or fitted to judge of the conditions. The possibility under the present conditions of an automatic machine being developed to fill this field satisfactorily and become general in its application is extremely remote.

Now, in order to see the lines along which engineering progress should continue with best hopes of success, it is necessary to consider further one or two of the attempts which have already been made. As one illustration we have the remarkably efficient process known as air cooling in use by the United Fruit Company in the shipment of bananas. Large refrigerating plants are constructed, at convenient distributing stations, which cool air to a suitable temperature. This air is injected into the cars and allowed to fill the interstices of the cargo until it is all thoroughly chilled. The cars are then partially closed, and radiation prevented as much as possible, and under these conditions the cargo is carried on to the next distributing point, where a new supply of cold air is injected and suitable cooling produced. This, of course, represents one of the minor and easiest applications of mechanical refrigeration in this field, but it has been accomplished along scientific lines. The temperature required is a comparatively mild one, the quantity of refrigeration held in the car units is reasonably large, and this method was adopted because it supplied the ventilation rendered absolutely necessary by the nature of the refrigerated product itself. Bananas require for their preservation a temperature of about 65 degrees F., with free ventilation. This has been secured by keeping the cars open to a certain extent through suitable outlets, so that the air can be changed as desired. Today bananas can be shipped considerable distances without fresh refrigeration, and the process is extremely efficient.

The application of this process elsewhere from a refrigerating point of view is limited on account of temperatures required and quantity of refrigeration necessary with other products. Thus meat requires a much lower temperature and the refrigerating units are in many cases much smaller. This means that radiation is much greater and the storage capacity of the car for cold more limited as well. The problem, however, becomes simply one of heat units and insulating materials. And there is no doubt that many other products will soon be shipped by the same process now used in the banana trade.

Automatic machines are divided into two classes, the single-car-unit machine and the train-unit machine. Both of these are inefficient in comparison to stationary units, and unnecessarily bring in a number of additional features which tend to diminish efficiency. Thus all the difficulties inherent in electric lighting of trains by single-car units or by total-train units enter again here, and in ten-fold degree. The refrigerating machine is not so simple that it can be carried around and operated under its present condition with the minimum only of care and attention that it will almost certainly receive here. The air-cooling device is limited in its applicability by the very nature of the process, from a refrigerating point of view and from the nature of the duty required. In order to see what methods are applicable to this service we must review the situation briefly from the standpoint of the refrigerating engineer.

There is no doubt (and it becomes almost unnecessary to make the statement) that the ammonia machine is in all probability the only machine capable of adaptation to this work under present conditions. Ammonia occupies a position in refrigeration analogous to that of water in power production; it is generally recognized as the best substance for the conveyance of heat. Now, ammonia machines operate in two ways, utilizing either direct expansion or indirect (or brine) cooling. Further, the direct expansion may be used to cool a substance directly, or indirectly by means of air. Further, the indirect brine system, in turn, may be used to cool directly or indirectly by means of air. As a further development come the two types, (namely, the absorption machine and the compression machine) in the production of mechanical refrigeration—the compression machine with its great efficiency for mild refrigeration, and the absorption machine with its capability of utilizing exhaust steam and its increased efficiency in the production of extreme cold. These special characteristics, however, are not important to our present consideration; the installation in any given case must be adapted to the conditions at hand, choosing that which will show the best efficiency. The probable outcome is the development of a correlated system composed of units of both types, utilizing the exhaust steam from the compressor to operate the absorption plant. Thus whatever system is used, ammonia is the operating substance, and refrigeration from it can be applied in a large number of ways. Its mode of utilization, of course, depends to a certain extent upon its production, cost, and efficiency. Thus, there is no doubt that the large-unit absorption plant, or possibly a combination plant, operates much more efficiently in the production of refrigeration than does a small one, and that in a general way the larger the unit involved the more efficient and cheaper becomes the process of development. This is a consideration which cannot be neglected in railroad work any more than in any other, and the attempt to divide refrigerating apparatus into car units or train units is a step in the wrong direction, with diminished efficiency as the result. The tendency is irrevocably towards larger and larger units, and undertakings which clearly recognize this prime condition in economy and embody it in their operations have an advantage at the very start over their competitors, which in general can rarely be overcome. In all probability the single-car refrigerating unit and the train-system unit will never come into general use, and their isolated application will diminish more and more with lapse of time.

Now, with the large refrigerating unit and the production of refrigeration in enormous quantities at fixed points accepted as a prime condition, the problem simplifies itself somewhat and becomes one merely of method of using the refrigeration when produced. This in reality is the situation today. Large ice houses exist, and furnish at stated intervals refrigeration which continues, but gradually becomes weaker, until the next refrigerating station is reached. The use of ice and especially natural ice, outside of the question of cost, is objectionable in itself. Ice is rather inconvenient to handle in its natural state and after storage is dirty; it occupies considerable room in the refrigerating car and is clumsy in its adaptation, and, further, it requires considerable manual attention in its installation. The amount of refrigeration involved is practically greater than that involved in any other space of equal dimensions, since the latent heat of ice is large. The large quantities of ice employed, however, in refrigerating cars are not absolutely necessary and the quantity of refrigeration carried can be cut down considerably, whereas, further, the advantage gained from heat concentration in the ice is often lost by the injudicious packing and the absence of fore-cooling. Thus it is universally recognized that ice is extremely objectionable in the natural form, its cost is high, and it cannot compare with mechanical refrigeration. In its application, moreover, there is a loss all along the line which seems absolutely unnecessary. There is a loss in its production, there is a loss in its handling, and there is a loss in its refrigerating effect. Seldom if ever are cars packed with ice and allowed to cool off to the ice limit and then repacked before shipment. They come in, are loaded up and then carried on—whether the refrigerators are full or not makes little difference. The space occupied is extremely large and inconveniently placed. The cost of handling is enormous and the moisture always present is a very objectionable feature.

The utilization of air as shown in the United Fruit Company's practice, although manifestly desirable in many features, is practically impossible in many other applications, on account of the quantity of heat involved. The only answer to the problem is the utilization of brine in a method similar to that of air as exemplified and utilized in the best installations today. Refrigerating units could be increased enormously. The brine could be produced and stored at minimum cost with maximum ease of handling. Refrigerator cars could be so constructed as to utilize all available space—often an impossibility under the present system—and to produce much superior refrigeration. Thus often a large portion of the car is set off for an ice chamber, in present-day work. With brine, this chamber could be thinned out to include all walls, floor, and ceiling, except the door, and even this might be rendered available. Good insulation could be installed. The carrying capacity of the car both for merchandise and for refrigerating material or brine could be much increased, so that the refrigerating capacities actually available would be larger than they are now. Fore-cooling could be secured readily by rapid installation, and the brine changed just before starting. If necessary, additional brine reservoirs could be carried on trains as a safeguard in an emergency, and if desirable, their contents could be pumped through the system by the engine.

No attempt at modern insulation for refrigeration is made today in ordinary refrigerator cars. With suitable insulation a layer of cold brine, if cooled off sufficiently and made to occupy a small layer completely around the car (and, if necessary, in small partitions in the interior) would carry fully as much refrigerating power as the present ice system and would allow much more space. Extremely cold brine could be readily produced by the absorption machine and it could be handled with a great saving over present-day methods. The cars could empty almost automatically into reservoirs immediately under the track. Pipes could supply the fresh refrigerant from a reservoir in the plant by almost the same process, or possibly an automatic one from above, and the train could be stored at once with cold brine, wait a short time to cool off, could be partially or completely discharged, and then refilled, in an interval of time depending simply upon the size of the discharge pipes and the capacity of the plant. The cost of installation of the entire system would be much less, even including machinery, than it is now, and the saving in almost every department would be large. The utilization of exhaust steam in an absorption plant and in large units, as large almost as any now in existence, would present new features for development in the utilization of the live steam for power and other purposes, and the limit of increase in efficiency by this process can be hardly foretold.

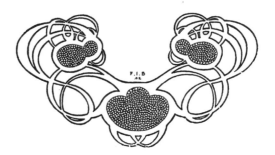

PROFIT MAKING IN SHOP AND FACTORY MANAGEMENT.

By C. U. Carpenter.

X. EFFECTIVE ORGANIZATION IN THE EXECUTIVE MANAGEMENT.

With this issue Mr. Carpenter closes his series which began in January last—a study of effective works-management methods, marked throughout by the clear sight, the fair mind, the direct dealing and the strong vitality of the author. He has been compelled to the writing, amid incessant and insistent claims upon his time by his active work as president and manager of a great manufacturing company, by his keen interest in the promotion of better ideals in industrial organization. It is largely a labor of love, freely devoted to the advance of his profession. And it bears throughout the stamp of tried, practical success.

The articles previously published in Mr. Carpenter's series discussed, first, the general methods of examining into the condition of any manufacturing business so as to discover the existence of waste and loss and to determine the "sticking points"; second, the nature and the working of the "committee system" of factory administration; third, the reorganization of the designing and drafting departments; fourth, the great importance and value of the tool-room as a source and spring of profitable methods; fifth, the general conditions necessary for manufacturing at minimum cost; sixth, methods for fixing standard times for manufacturing operations; seventh, minimizing the time of assembling; eighth, stimulating production by wage, stock, and cost systems; ninth, the upbuilding of an efficient selling organization.—THE EDITORS.

THE preceding articles appearing in the series which began in this magazine last January have dealt somewhat fully with the defects existing in the average manufacturing concern as regards its organization and its selling and factory methods and systems, and have indicated some methods of overcoming these troubles that have been found effective in everyday practice in the upbuilding of run-down concerns.

This discussion would not be complete unless the executive division, whether it consists of one man or twenty, were given some treatment. For indeed the troubles—the defects, both in organization and methods, that may be met with in selling force or factory— exist in particular strength in many an executive division, and cause infinite trouble. The small jealousy that impels one man or one group of men to underrate the value of work done by others, the lack of unity of purpose which often leads one set of men to block surreptitiously the good work of another group, often works incalculable loss.

The larger and more complex the executive end of a business, the more important does it become that great care be used in organizing it in such a manner that each member of this group shall be allowed to do his own work without interference from others, and yet, at the same time, that each member shall bring to the business and to the most serious problems in the business his best judgment on the business as a whole, and pour into a common reservoir his reason for objection, his particular plan, his best thought, his enthusiasm, his best self, for "the good of the business."

As long as the executive force work in unison, the balance of the organization will generally do likewise. Let there be discord and lack of harmony amongst the heads of the business, and there will be an instant lining up of forces in opposition one to the other from one end of the working body to the other. This lack of harmony in the executive division arises for reasons similar to those that disintegrate the balance of the organization and cause it to lose so much of its latent power. The only cure, in my opinion, is the old one of "getting together." The general ideas already expressed in earlier numbers of this series relative to the formation of committees are particularly applicable here.

It is my intention to illustrate briefly the work of a group of executives, showing how the work of each one is related to and interlaced with the work of all. Before doing so, however, it is advisable to consider a few suggestive forms of reports from different sections of the organization which are very necessary in order that a full grasp upon the business may be held by those managing it. Of course no attempt to outline fully any system of reports generally applicable could be successful. The general forms of the ones suggested, however, are adaptable to many differing conditions.

Executive Reports from Selling Division.—As a fundamental form of report invariably necessary I submit one (page 492) which shows how much business *must be done* in each line of product and in each territory. This must show the volume of sales required in detail. As against this there must be set the allowable factory costs, together with allowable costs of extras of all character. In addition to this, there must be carefully calculated out the allowable selling expense, including all items, such as salesmen's and managers' salaries, commissions, traveling expenses, advertising, etc. There must further be shown the allowable general expense, such as rent, insurance, taxes, telephone, telegraph, office salaries, etc. In all cases allowable percentages should be carefully worked out.

The "allowable amounts" must be calculated from close knowledge, first, of how much profit the concern should make; second, of how much profit should come from each territory considering the possibilities of the business and expense of conducting it. When these computations are made for each territory, covering selling prices, factory costs, selling and general expenses, together with percentage calculations, showing the proper relationship of all of these items, you have a solid foundation upon which to work and from which to drive for business. Of course this should be carried much further within the selling division. Each salesman should have his record to strive for. His showing should be based upon the same idea.

	%	N. Y. Branch.	%	Phila. Branch.	Etc.	
Required Sales. Amounts (Classified).				·		
Factory Costs. Amounts (classified) .. All other Cost Items Classified						
Gross Profits						
Selling Expenses (Classified). (a) Salesmen, salaries.. (b) " commissions (c) " expenses .. (d) Mngrs., salaries ... (e) " expenses .. (f) Advertising						
Total Selling Expense						
General Expenses (Classified). (a) Rent, Insurance, Taxes .... (b) Office Salaries ..., (c) Telegraph and telephone .... (d) Miscellaneous ..... (e) Legal, etc.						
Total General Expense						
Total All Expense...						
Net Profit Required..						

STANDARD SELLING RECORD.

This Record is used, first, for showing Required Sales and Allowable Expenses with percentages; second, for showing Actual Sales Made and Actual Expenses Incurred and percentages. The use of the same form facilitates comparisons.

Having these data, the next step naturally is to supply the exact information as to sales record and expenses. The sales record, as far as the selling division is concerned, should, of course, be made up daily, the expense items being roughly calculated upon a percentage basis. The monthly sales record should, however, be complete. It should follow exactly the lines of the standard selling record illustrated above, the sales data being taken from the sales records, the expense data directly from the accounting department. This leaves in the mind of each manager of a selling division no iota of doubt as to what *must* be done, and by comparison he sees how much he has gained or how far he has fallen behind. Such reports provide the

FACTORY EFFICIENCY AND OUTPUT REPORT.

Comparisons are weekly or monthly results with averages of a revolving period of six months.

Week Ending	No. employees	Output		Hours Worked		PAY ROLL					Material Purchased		Inventory	Miscel. Charges	Factory Balances	Costs					Increases or Decreases	
		No. of Machines	Values	Total	Over Time	Direct Labor	Piece Work	Day Work	Indirect	% Jobs reaching Standard Time	Direct	Indirect	(Subject to proper classification)	(classified)	(classified)	Material	Labor (classified)	General	Wages	Total	Where?	Why?

executives with accurate and usually much needed indices of sales conditions. They are especially valuable where the business done is in the nature of long-time contracts under which deliveries are made for long periods after the actual sales are recorded.

Other selling-division reports, such as those showing the distribution of sales, the covering of territories, the development of the selling efficiency of the organization, etc., and the condition of competition, would naturally be made and need no discussion here.

Factory Reports. — In preceding numbers of this magazine, both in this series and that appearing in 1902, the question of factory reports has been quite fully dealt with. From the data secured through the suggested methods many interesting and important reports are derived. These are:

1. Factory Output and Efficiency Report.

This report may be adapted to meet the requirements of departmental efficiency reports also. It should compare the results for various periods with the results of like previous periods, either as a total or an average. By noting the number of employees, pay rolls, and material purchases, and comparing the figures with values of output and inventories, a very accurate idea may be secured as to the relative efficiency of the entire shop or any department.

2. Progress-of-Work Report.

To every executive managing any shop, and particularly those working

Name and Location of Customer.	Description of Order (Briefly).	Factory Order No.	Date of Order	Must Erect by	Must Ship by	What is Shipped	Drawings % Done.	Material % Received	Condition of Orders (Classified)	
									In these columns place the different classifications of each contract, showing % finished	General.

DATE................

FACTORY PROGRESS REPORT (WEEKLY).

Signed—Committee (by Sec'y).

upon long-time delivery contracts a weekly progress-of-work report will prove invaluable. If this report were to be made up from the cost records, the task would be a great one. It can be handled very easily and simply, however, by the committees. Each foreman should have a keen knowledge of the progress of his portion of any job. These men, together with a bright secretary, in one of their regular meetings can easily and quickly formulate such a progress-of-work report as suggested. The advantages of such reports are many.

First, they provide an alert executive with a most important index as to progress of work and enable him to "pound the shop" for any work that he can note is falling behind.

Second, they force upon the attention of the foreman individually the progress of each job in his own department. They *compel* him to accumulate a very useful knowledge of his own work—oftentimes lacking.

Third, they place the factory in possession of data making it possible to give the selling department delivery dates which can be met.

Fourth, they are an invaluable aid to the treasurer, enabling him to predict with some degree of certainty when his collections will come due. Especially valuable are they in cases of large contracts calling for partial payments as work progresses.

Such reports, covering a very wide variety of work, have for a long time been made out by our factories on Saturday morning, arriving on my desk the following Monday. Meetings with interested heads of financial and sales divisions immediately follow, and as a result there exists the closest possible touch between the several divisions upon the one important point—Production.

Executive Report.—The report of paramount importance is the one illustrated on this page. It should be produced monthly from the books of the concern. The several divisions are self-explanatory.

DATES............ from............ to...... .....

	New York Branch.	%	Philadelphia Branch.	%	And all other Selling Divi'ons.
Deliveries (Classified).					
Total Deliveries					
Factory Costs (Classified).					
Total Factory Costs.......					
All other Cost Charges (Classifieu), *e. g.,* Delivery—Finishing.					
Total All Other Costs....					
Total Delivered Costs.....					
Gross Profit					
Deduct Selling Expense (Classified).					
Salesmen, salaries and commissions					
(Depmts.) (a)					
(b)					
(c)					
Salesmen, expenses					
(Depmts.) (a)					
(b)					
(c)					
Mngrs., salaries					
" expenses					
Advertising					
Total Selling Expense.....					
General Expenses (Classified).					
Office Salaries					
Rent, Insurance and Taxes..					
Telegrams, Postage, etc......					
Stationery and Printing.....					
Light, Heat and Power......					
Maintenance					
Adjustments and Losses					
Legal					
Miscellaneous					
Total General Expense....					
Total Selling and General Expense					
Profit (black). Loss (red)					

EXECUTIVE PROFIT AND LOSS REPORT.

Comparing this with the selling-division reports, there is provided a full index of business conditions. Couple it with factory reports already mentioned, and a monthly balance sheet follows.

Having now described in a general manner the detail, I can perhaps best illustrate the value of the methods advocated by quoting a programme of action taken "from actual life" by a committee of executives of a very large concern. The committee is formed of the president, the general manager, the treasurer, the sales manager, and the factory manager. The understanding is general that heads of departments shall frequently be called into their conferences. A typical program is as follows:

> President; acts as chairman.
>
> General Manager.—Critical discussion of business conditions, with especial reference to those sections in which the records show a decided falling off in business. Discussion with sales manager as to the reasons for this condition. Discussion develops that a new form of competition has arisen in that particular region. Steps are taken immediately to meet this before it grows to greater strength.
>
> General discussion of executive report. An analysis shows where the greatest profits in the business lay and where the greatest losses occurred. Discussion of expense items shown on executive report.
>
> Sales Manager.—(a) Reports on sales conditions in various territories and orders coming therefrom.
>
> (b) Competition; what must be done in way of new methods or new product to meet it.
>
> (c) Progress of training department.
>
> (d) Expense.
>
> Factory Manager.—(a) Discussion of factory output and efficiency.
>
> (b) The progress-of-work report.
>
> (c) Report on progress of new factory methods and inventions.
>
> Treasurer.—A discussion of financial outlook with especial reference to the future deliveries as outlined on progress-of-work report.

In this concern, formerly handicapped severely by a lack of knowledge of such important points, a revolution was worked through the adoption of the simple and direct system.

And so I maintain that the adoption of these simple and natural methods will change an organization burdened with jealousies, handicapped by misunderstanding, held back by lack of knowledge, lack of progress and general inefficiency, to one comparatively free from personal enmities, led by a sense of certitude as to actual condition in field and factory; constantly impelled, even in the smallest details, toward a greater degree of efficiency and surcharged with the feeling of true enthusiasm—the will to work "for the good of the company."

EDITORIAL COMMENT

IT is doubtful if there has been heretofore any spasm of business depression which was so much the result of a state of mind, with so little really the matter, as that which now hangs half-poised on the balance of the times. Every fundamental of prosperity, and stable prosperity, is sound. We have not been carried away by any inordinate temper of inflation—over-trading, over-building, over-expansion. Indeed, we have scarcely provided amply enough for the care and handling of the enormous stores of natural wealth piled up by our miners, farmers, planters, producers in every field. Nor is there any slightest menace from any unsound currency system. Every event and every phenomenon confirms the conclusions formulated by Mr. Del Mar in our October issue :—"The foundations of our welfare are solid. Aside from a remediable constriction in the monetary field, there is nothing in the situation to warrant any apprehension."

It is a "psychological panic." The blame for the abnormal state of the public mind, it has been wisely said, may best be fixed later.* At present the pressing need is to cure the disorder. The fear is individual. The individual depositor fears that others may precipitate a disastrous run on his bank—and heralds his fear by trying to get first in line. The individual bank fears losses through disturbed conditions—and by calling loans and denying accommodations so as to build up its reserve, precipitates the very condition it dreads. The individual manufacturer fears cancellation of orders by his customers, and by cancelling his own purchases he plunges the whole community into the depression all are trying to avoid. A wholesome and wholly practicable remedy lies in active, instant, frank conference between financial, commercial and industrial leaders. The sense of sympathy, support and security which co-operation brings would put the whole fund of courage at the back of each timid individual, and the fear which is working the damage would vanish as swiftly as it came.

This is no time for hoarding of funds, for retrenchment in manufacturing, for piling up of bank reserves. Not for years has any such opportunity been afforded for money actively employed to make more money for its owners. Whether loaned to well assured business or invested in sound securities, the certain reward is almost unprecedented for those who have the wisdom and the courage to seize the opportunity now. Later, when everyone sees it, the opportunity will be gone.

No great courage is necessary ; only common-sense. If—at first perhaps locally and then more widely—those who have the money and those who have business will get together and determine that accommodation shall if necessary be stretched, rather than restricted, and that operations shall be continued at the highest possible point, rather than curtailed—the trouble will disappear like mist before the sun.

To manufacturers especially it should be plain that retrenchment begets retrenchment, and the wave swings back and forth with rising intensity at each return. The maintenance of business at the highest schedule which can possibly be managed, at almost any temporary cost, is better than contraction with its attendant loss of the "intangible assets" of organization, personnel and goodwill—intangible, but the most valuable possession of any enterprise. Better to borrow $100,000 at 10 per cent —or 20 per cent—for three months

* Cordial acknowledgement is made to Mr. H. L. Dougherty for many suggestions herein.

than to let a plant investment of $1,000,000 lie idle for a year and more, to be with difficulty and cost brought back to efficiency at the end of that time.

Such a movement of co-operation between financiers and business men should be started simultaneously throughout the country, and probably would be most immediately effective in the smaller cities and the distinctively manufacturing centres. The point is to make a common fund of the underlying confidence everyone feels in the essential soundness of conditions throughout the country—to pile up our faith in the future and in one another until it rolls in an uplifting wave of realized prosperity, which will sweep back the floods of doubt and pessimism beyond any power for evil. · The admonition can be put very briefly : Get ˙together. Do it now !

* * *

THE three great factors in the profitable employment of capital in manufacturing are materials, superintendence, and labor. Of these, the first —natural resources and products of the soil—have generally been a matter of national solicitude and often the cherished object of national economic policy. The second—scientific works-management—has enlisted the earnest study of some of the clearest-sighted and most ingenious engineers and has practically developed into a new profession and created a new literature. But the cultivation of the third—of the volume and grade of labor needed to support and extend the great manufacturing industries of our day and of the day just ahead—this (in England and America, at least) has been left almost wholly to chance.

Some few of the largest of our industrial establishments foresaw, years ago, the necessity for training a new generation of workers; now the need is becoming widely manifest and is beginning to take a prominent place in the discussions of employers. Interest in the subject has grown remarkably since, just a year ago, we published Mr. O. M. Becker's study of existing and possible apprenticeship systems under modern conditions. And as employers are turning favorably to plans for industrial education, the attitude of organized labor toward the movement, so well set forth in Mr. Stratton's article leading this number, becomes of active importance.

It is of much interest to compare British conditions, as proved in the recent Engineering Trades Agreement of which a digest appears in our Review of the Engineering Press this month (pages 504-507). It is noticeable that no fixation by the unions of the proportion of apprentices to journeymen is conceded by the employers, and that industrial education is more than "tolerated" by British labor organizations—it is solicited, and employers are asked to promise to encourage the novice in obtaining it.

The political economy of trade unionism has rarely been broad-minded or far-sighted ; it is therefore the more interesting and hopeful to see it in this instance laying so safe and wise a course. The pressure of competition in the engineering trades is no longer wholly domestic, but to an increasing extent becomes international. Germany is a rival to be reckoned with ; and the part which systematic, specialized industrial education is playing in her expanding commercial programme was impressively shown in the thoughtful studies by the late Louis J. Magee published in these pages January-March, 1906.

The American mechanic, of all concerned, is least in position to allow any fraction of American mechanical supremacy to pass to outsiders through decadence of the rank-and-file personnel that have made American machinery and American industries the admiration of the world.

THE QUEBEC BRIDGE DISASTER.

OPINIONS OF THE LEADING BRITISH TECHNICAL JOURNALS AS TO THE CORRECTNESS OF
THE DESIGN OF THE COMPRESSION MEMBERS.

Engineering—The Engineer.

THE two preceding issues of THE ENGINEERING MAGAZINE have contained reviews of articles published in *Engineering News,* which were representative of American sentiment on the failure of the Quebec Bridge. It is interesting to compare with these the conclusions reached by the leading representatives of the technical press of Great Britain as shown by the following extracts from *Engineering* and *The Engineer.*

The editorial comments of *Engineering* are based on a report by Mr. Frank W. Skinner, who made a personal investigation at the scene of the wreck. Stress is laid upon the fact that the chord member which was responsible for the initial failure was found bowed 1½ to 2 inches out of line three days before the wreck, while carrying little more than one-half the load for which it was designed. A heavy responsibility is placed upon the local engineering staff for the appalling loss of life. The clear duty of the engineer in charge of erection, in view of the unmistakable signs of crippling in the strut, was to close down the work immediately pending a thorough investigation of the cause.

"Mr. Skinner's article fully establishes the fact that bad design, and not bad material, was responsible for the catastrophe. The conditions at the time of the collapse were, he states, everywhere normal or favourable, and the material and workmanship were both excellent of their kind; whilst the substructure, even after the ac-

cident, was found in practically perfect condition. Obviously, therefore, the fault must have lain in the general or detailed design of the superstructure. There are no doubt a number of points in which this general design is open to criticism, but the really fatal error appears to lie in the system of bracing adopted for the lower chord members.

"The bracing or latticing of these compression members was absurdly light according to English ideas. Taking the lighter strut, No. 8, first into consideration, few engineers on this side of the Atlantic would care to try and hold such thin and deep plates in line by merely latticing them across the top and bottom. The addition of a solid diaphragm-plate running from end to end at mid-depth would undoubtedly have made a better job. Presumably the reason why some such arrangement was not adopted lay in the supposed difficulty of transferring across to the pins the load carried by this plate; and hence the adoption of the design criticised. Although unmechanical, we are not, however, prepared to say that it was actually dangerous in this instance; but there are good grounds for believing that chord member No. 8, and, *a fortiori,* Nos. 9 and 10, were, on the other hand, in a somewhat parlous state. The bracing here was, in short, incapable of developing the full bending strength of the web-plates. If those of member No. 8 were laid flat side by side, over supports 55 feet apart, they could sustain a uniformly-distributed load of about 53 tons, with a stress in the metal of some 10 tons per square inch. Treating the complete strut in the same way, a load of 40 tons would cause a stress of about 10¾ tons per square inch in the bracing members. Each 4-inch by 3-inch angle would then, it is true, take a load of only about 7½ tons on an area of 2.48 square inches; but as this is applied eccentrically through one web only, the actual stress in the metal would reach the figure stated. As a matter of fact, the rivets might, indeed, go first, since these are of ⅞ inch diameter only. The bending of these bracing bars under their eccentrically applied load would, moreover, allow the webs to get out of line, thus accentuating the effect of their light scantlings.

"Unless the load on a strut is applied absolutely centrally, a bending moment is developed, causing an increase in the stress on the metal. In the case of the chord members, this additional stress on the plating is not in itself serious, since the length of the member is only about ten times its width. The same cannot, however, be said as to the stresses due to shear, which are simultaneously developed on the bracing by an eccentrically-applied load."

A mathematical discussion is given of the effect an eccentrically-applied load would have on this member and it is shown that the bracing used was quite incapable of developing the full compressive strength of the strut. "This will be evident when it is remembered that the stress on the bracing is directly proportional to the eccentricity, whilst the direct stresses on the plate-webs increase much less rapidly when the eccentricity is increased. Thus with zero eccentricity of the thrust the direct stress in the case of chord member No. 8 is about 13 tons under the combined action of dead load, live load, and wind load. With an eccentricity of 1.5 inch the direct stress would only be increased by some 25 per cent., making a total of, say, 16½ tons per square inch.

"The bracing, on the other hand, with the same eccentricity, would be subject to a stress of 24 to 25 tons per square inch— that is to say, it would be strained up to practically its ultimate strength, if constructed out of 26-ton steel, as in this case.

"Chord member No. 9 is longer than No. 8, and the load on it is greater, so that a lesser eccentricity would probably be fatal to it, were it ever called upon to take its intended maximum load. At the time of the disaster the stress here is said to have been only about 8 tons per square inch, and a greater eccentricity than above calculated might then be needed to endanger the bracing. At the best, however, the margin of safety appears very small, and the construction adopted is certainly unmechanical. In the river-arm the struts are still longer, and the bracing has here, accordingly, a still smaller margin of safety. It is to be presumed that the Commission which has been appointed to report on the failure will test some large models of these struts. It is to be hoped that such tests will not only include experiments on the models centrally loaded, but that the investigation will be extended to cover their behaviour under eccentrically applied thrusts. We are convinced in that case that the bracing will prove wholly inadequate to develop the full strength of the specimens. This view receives additional confirmation in Mr. Skinner's article, where he notes the circumstance that in the case of the struts wrecked by the fall the bracing gave way before the struts crippled.

"A different hypothesis as to the immediate origin of the catastrophe is also suggested by Mr. Skinner. He notes here that the lower chord members near the pier did not merely abut on the main pin, but were continuous across the latter, thus introducing the bugbear of so many American bridge engineers—viz., indeterminate stresses. To English ears the suggestion that a mechanically designed detail such as this was responsible for the accident sounds somewhat strange, since it is perfectly well known that riveted structures make by far the best bridges, in spite of a theoretic possibility of uncalculated secondary stresses. It is impossible, for instance, to calculate with accuracy the stresses arising in a common plate-girder, yet every railway engineer knows that such girders give less trouble under traffic than any other type of small-span metal bridge.

"This insistence that all structures shall be statically determinate seems to be the special fad of one particular school of

American bridge engineers. The idea rests on a theoretical basis only, and has no support from the teachings of experience. We know of no case in which a structure has failed merely through being statically indeterminate, whilst the rigidity gained has great practical advantages; and where such structures have failed through defective material or overloading, the wreck has not been practically instantaneous, as at Quebec, but has given time for the majority of those endangered to get clear.

"Further, it should be noted that in actual practice American pin bridges are no more free from secondary stresses than properly-designed riveted trusses. Bearing in mind the frictional resistances involved, it is inconceivable that the pins actually rotate in the eyes of the links at each change of load on the structure. In fact, if they did, serious wear might be expected, and in the absence of such rotation the bars coupled up are to all intents and purposes as rigidly connected as if riveted, and suffer from the same secondary stresses. As to the objection taken to rigidity of the joint between the chords across the main pin at Quebec, it would seem that it might be equally well raised against every other joint of the lower chord. Indeed, in the old days, pin-jointed compression chords were somewhat in favour, until experience proved decisively the superior advantages of constructing such chords in continuous lengths. Again, it seems well established that the strut which failed buckled sideways, and not vertically. If secondary stresses due to the rigidity of the joint past the main pin had anything to do with this, the member should have yielded in a vertical plane, and the contrary fact would *per se* suffice to confute the suggestion, were it not already sufficiently discounted by the teachings of experience."

The comments of *The Engineer* are in a different strain though touching, in the main, the same points.

"The bridge fell down because it was not strong enough to stand up. It is useless to ask why it fell down. The question is why it was too weak? There are four answers. The design was defective, or the material was not good, or the workmanship was bad; or, lastly, any two or all of the three may have been combined. Now, the whole American technical Press is agreed that the materials and the workmanship were excellent. It follows, then, that the bridge was not strong enough because the design was not good. There is no such thing as an absolutely good design for a bridge. Whether designs are good, better, or best, depends on the way in which they fit in with the ruling conditions; and it must be always kept in mind that a bridge which is on paper, all that can be desired, is not by any means certain to be the best when it is called into existence to span a ravine or a wide river. We may be told that all this is quite well known to engineers. Perhaps so; yet the Quebec Bridge has fallen down. Our contemporaries cannot for the moment get away from the belief that the design of the bridge is responsible for its downfall. Yet they admit that it was designed by the most competent men in the United States, and, to avoid the necessary implication, the hint is now thrown out that the behaviour of long and large struts is different from that of shorter or smaller struts in a way which has not hitherto been recognized.

"Our knowledge of the stiffness of struts is mainly of an empirical nature. The formulæ applied to them are complicated, and not entirely trustworthy. We are not in a position to assert of struts, as we are of beams, that the general formulæ for small sizes apply equally to those of great dimensions. Hence the argument that if the design was faulty it was faulty because precise knowledge of certain phenomena is lacking, has something to support it—at all events, for the time being. We may remark here that the design of the Forth Bridge presented also its quota of uncertainties. In that case the engineer, Sir Benjamin Baker, took no half measures. He threw science to the winds and built to cover his ignorance. What the bridge lost in elegance it gained in the certainty that no ordinary forces could overthrow it. There are times when it is the highest science to make the 'coefficient of uncertainty' enormously large.

"It is now confidently stated that the first failure of the Quebec Bridge occurred in the bottom boom of the shore span. This was built up of long plates secured to each other in groups, and then carried in the bridge on enormous turned pins fitting carefully bored holes. The lateral stiffness of such a structure was comparatively small, and was augmented to the amount deemed requisite by cross bracing. The boom, however, buckled horizontally, and the bridge collapsed. This is the theory as far as it goes at present. We do not dispute its accuracy because we have no contradictory information; but it is worth notice that in this case one side of the bridge obviously fell first. Yet there is no slewing of the structure such as might have been expected, to be gathered from the photographs which have reached us. Let us ask ourselves whether a built-up strut óf this kind is or is not.likely to be the best? The assumption made is that if links and pins are suitable for tension they must be just as good for compression, always taking for granted that they are stiff enough laterally. This is, however, begging the question. The onus is shifted from the links to their stiffeners. When a bar is in tension all defects such as bends or kinks, or such like deformities, tend to elimination; the pull straightens them out.

But with members in compression the effect of out of truth of line always tends to aggravation; the kink or bend will get worse, not better. In practice the trained engineer—in this country, at all events—meets the difficulty by providing a far larger margin of safety in the members of his bridge than he thinks enough for the tension members. The engineers of the Forth Bridge took no chances. The compression booms are mighty tubes. To calculate the principal stresses in any bridge, large or small, is a very simple affair. Graphic statics and a knowledge of the parallelogram of forces will suffice. But in all bridges there are secondary stresses. For these, in most cases, it is only possible to estimate. The secondary stresses are held by American bridge engineers to be less prominent and more easily dealt with in pin than in riveted structures. This, no doubt, and questions connected with labour, have had much to do in giving a popularity to the pin bridge in the United States which it has never enjoyed in any other country."

ELECTRIC POWER FROM BLAST FURNACES.

AN EXPOSITION OF THE ADVANTAGES OF A POOLING SYSTEM FOR ITS ECONOMIC DISTRIBUTION.

B. H. Thwaite—Iron and Steel Institute.

THE successful application of electric power to the driving of heavy rolling mills in several continental steel works, described in previous issues of THE ENGINEERING MAGAZINE, is a fact of great economic significance. The heat and power efficiency of rolling mill plants has always been very low. The steam boiler plant necessary to satisfy maximum requirements of rolling mills is usually large, and usually also intermittent in its working. The replacement of coal by blast-furnace gas as a boiler fuel was a great step in advance, but even under present circumstances condensation in steam pipes and frequent closing down of rolling mills for week ends or longer periods, during which blast-furnace gas is going to waste, still entail great losses. Steel works can be run at the highest possible efficiency only when all the blast-furnace gas is utilized for the generation of electric power for use within the works and, if

above the requirements of the plant, for sale to surrounding industries. Technical difficulties in the way of the general gas-power electrification of rolling mills have been for the most part overcome, but the heavy expenditure necessary makes the reform impossible for firms of small or moderate financial resources. At the recent meeting of the Iron and Steel Institute, however, Mr. B. H. Thwaite outlined a system of pooling blast-furnace gas in iron-making districts which would materially lessen the burden and risk of the electrification of rolling mills and prevent any waste of power. The scheme has been tried and has proved successful in some districts in Germany and in at least one district in England.

"The basis of this programme is to pool the waste furnace gases from all the furnaces of an iron-making district independently of the ownership of such furnaces. The energy (electrically transformed) of the different furnaces would be transmitted to a central distributing and transforming station, in which the current would be transformed to the voltage to satisfy different customers. Of course, there may be cases where the furnaces are so concentrated as to justify the delivery of the *furnace gases* to a pooling or central station for conversion into electrical energy, but usually it would be found to be more practical to transform the gas into electrical energy on the site of the ironworks, for transmission to a distributing central station.

"The first call on this power would, of course, be the satisfaction of the internal demands of the iron and steel works, the balance of power remaining being available for external distribution over a wide area, the extent of which, owing to the familiarity with and knowledge of the resources and safety of high-pressure transmission systems, is being yearly increased. When the author put forward the first long-distance electrical transmission project of over 100 miles in 1892, the line pressure then suggested was considered dangerous in the extreme. To-day, power is being easily and profitably transmitted over varying distances at very much higher pressures than those suggested in 1892. In 1892 the pressure specified was 20,000 volts; to-day, transmissions of 250 miles with 88,000 volts pressure are in daily service.

"The pooling programme proposed would be provided by a separate and distinctive joint-stock electric power organisation, on the lines of the county electric power companies, which sprang from the author's 1892 proposition, with the exception that a proportion of the output is preferentially allocated to the satisfaction of the power demands of the iron and steel works from which the gases are drawn. The balance could be transmitted to any group of customers whose power demands justified the cost of distributing mains and associated transforming stations, but a more appropriate field of demand for an all-the-year supply would be the satisfaction of the electric energy requirements of the many profitable electro-chemical and electro-fusion processes.

"Included amongst these processes, in which the demand for power would correlate with its production at the blast furnace, or from January 1st to December 31st, there are the production of high-class steel from selected scrap, special alloys, such as ferro-silicon and ferro-titanium, ferro-chromium, besides the production of carbides (silicon and calcium). Some of these processes could take the current just as supplied, in large or small proportions, and at any moment, so that the leveling up of the load or the demand could be readily effected—obviously an important factor in the economic production and sale of electricity.

"In the selection of the customers it would be advisable to secure the class of demand for operations such as mine pumping, electrolytic, and like processes, on which the supply could be turned on or off whenever available, so that the distribution of the current could be arranged to fill up any gaps in the demand.

"In large districts the power-driving requirements of the factories would provide a demand running into thousands of horse power, although the direct driving of important machinery will secure a great saving of (and consequent reduced demand for) power compared with belt driving and its associated power losses. The satisfaction of electric transport power requirements is becoming an increasing factor both for water and rail transport services. By this pooling system the tall stacks of the furnaces would become centres from which would radiate the benefits of the cheap power, and the advantages would be reciprocal. Besides removing the many smoke-stack polluters of the atmosphere, the additional profit would increase the stability of the iron industry of this country.

"The flexible demand for power which it would be the aim of the power organisation to secure would be appropriate for the intermittent periods of mill idleness, whether temporary or prolonged; for instance, the sudden release of power for external services at the week-end periods of mill idleness (or during longer periods of compulsory idleness due to the falling off or cessation of orders) would be available for one or other of these flexible power demands.

"Another and important argument in favour of the pooling policy is the fact that some 4,000 to 5,000 kilowatts of energy is the minimum demand of many of the most desirable of the electro-chemical and other operations, so that in a single works, without a drastic reorganisation of the plant, it would be impossible to secure this output of electric energy; whereas, by the pooling of the available power from several works, this power would be obtained with little change from existing conditions. Of course, the power available entirely depends upon the extent the gas engine displaces the steam engine to satisfy the internal power demands of the works.

"One of the advantages of this policy of pooling is obvious: it would not only permit such an association of power production as to justify the cost of mains to central stations, but it would provide an irresistible incentive to the thorough electrification of all the iron and steel power machinery, and justify the expenditure in securing the maximum possible efficiency of the blast-furnace gas. . . . . . .

"The advantages of the pooling policy will, of course, depend upon the individual circumstances of different works. The mills are probably standing some 23 per cent. of the total hours of a week, and the power waste associated with this idle week-end period will be available. Besides, the displacement of the steam rolling and other machinery by gas and electric motors will provide power all the week, and the auxiliary gas generators can be drawn upon to balance the power required for the mills. . . . . . .

"Re-capitulating the advantages of pooling, it may be said that besides facilitating the electrification of the rolling mills by the reduction in capital cost of plant, it will secure, in addition to the advantages of electrical driving, a substantial increase in the proportion of the furnace power gas available, by the suppression of the mill steam boilers. The pooling proposition will constitute an ideal combine, securing an all-round benefit to the ironmaster and the power consumer, and tend to promote the establishment of new electrical industries and thus enlarge the field of demand for power; it will financially justify expenditure in apparatus and on technical changes to secure the fullest possible economy in the use of furnace gas in the works. Of course, any available gas from associated coke oven plant, subject to certain mixing operations as will reduce the proportion of hydrocarbon in the combined gases, will be available for the pooling station. The economy resulting from the direct production of power, with a rationally devised plant, has been abundantly proved. A well-designed and built gas engine, using pure and waterless furnace gas and associated with a high-class electrical generating machine, is claimed to be able to develop the hours of a kilowatt year for 66s. 8d., assuming, of course, that no charge is made for furnace gas. This economic figure of cost to an ironmaster is unapproachable by any other system of power generation, barring water power, which is often located in such awkward if not almost inaccessible positions as to destroy the economic advantages of such power, *per se.*"

THE BRITISH ENGINEERING TRADES' AGREEMENT.

AN OUTLINE OF THE NOTABLE AGREEMENT RECENTLY ENTERED INTO BY THE EMPLOYERS AND WORKMEN IN THE ENGINEERING TRADES OF GREAT BRITAIN.

Engineering.

THE following extracts from an editorial in *Engineering* on the recently signed agreement between employers and employees in the engineering trades of Great Britain give an outline of what is probably the most fair and impartial labor agreement ever entered into by any considerable organization either of employers or of workmen. The conferences which preceded the signing of the agreement were marked by an earnest desire for co-operation and a better mutual understanding and the result promises well for industrial peace in the trades directly concerned. It is anticipated that before long the example will be followed in many other trades and that it will have a permanently beneficial effect on British industries.

"Perhaps the most significant feature of the existing relations between masters and men in the engineering trades is the fact that settlements are now national, in place of local, as they were up to a few years ago. The change has, of course, resulted from the development of the Engineering Employers' Federation, which was the natural outcome of the tactics followed by the different trade unions during the decade which ended in 1898. The unions first saw the importance of a national, rather than a local, organisation, which, so long as the employers were not themselves similarly associated, conferred on them the enormous advantage of localising any dispute which might arise, the strain on their funds by a local strike being minimised by contributions from the districts where members were still at work. As a consequence, the conditions regulating the relations between employers and their workmen varied enormously from one district to another. Shop customs which were accepted by the men without demur in one locality were utterly repudiated in others, and it followed, therefore, that work of a particular character might cost substantially more in one district than in another. When the great fight on the eight-hour day question arose in 1897 the Employers' Federation determined that this state of affairs must be abolished and all districts stand on an equal footing as regards conditions of employment. This resolution was embodied in the terms of settlement agreed to in 1898, which since that date have governed the relations between the two parties in the engineering trades. That agreement, as was natural, showed some trace of the bitter warfare of the preceding months; but, in spite of that, it undoubtedly led to an improvement in the relations between the parties to it. . . . . . .

"A noticeable feature of the new agreement is the omission of all direct or indirect reference to the strife of 1897. The preamble states that the representatives of the Employers' Federation on the one hand, and of the engineering trade unions on the other, being convinced that their mutual interests will be best served and their rights best maintained by a mutual agreement, have decided to adopt measures to avoid friction and stoppage of work. The first clause declares that the employers shall not interfere with the proper functions of trade unions, and that the latter shall, in their turn, not interfere with the employers in the management of their business. The terms here used are, no doubt, somewhat vague, but matters are, to a certain extent, elucidated in the succeeding clauses.

"With respect to the employment of workmen, the unions have gained a concession in terms rather than in actual fact. The employers, whilst reserving their right to employ non-unionists, now positively recommend their members not to object to employ trade unionists, whereas in the 1898 agreement they merely did 'not advise their members to object to union workmen.' They further agree that no workman shall, as a condition of employment, be called upon to state whether or not he is a society man. On the other hand, the unions reaffirm their previous recommendation to their members not to object to work with non-unionists. As now worded, this agreement as to the employment of workmen puts both sides on an identically equal footing.

"In the matter of piece-work and overtime the men have obtained more important concessions. The Federation agrees definitely that the day rate shall always be guaranteed irrespective of a man's piece-work earnings, and overtime and night-shift allowances shall be paid, in addition to the piece-work prices, at the same rates as prevail for time wages. Further, it is provided that all balances and wages shall be paid through the office. With the exception, perhaps, of the payment of balances through the office, the above conditions have long constituted the established practice in important districts, but they now become universal. Systematic overtime is deprecated, and it is agreed that no union workman shall be required to put in more than 32 hours' overtime in four weeks, exception being made in the case of break-down work, repairs, trial trips, and other urgent or emergency cases. In the old agreement the time-limit was 40 hours in place of 32, so that the men have gained a concession of 8 hours. On the other hand, the union officials have shown themselves reasonable in recognising that overtime is often unavoidable, and, indeed, in a statement to their members, they declare that a closer examination of the subject has led all parties to realise that it is difficult to avoid special overtime in their particular trade. Here we probably have an instance of the benefit of the closer intercourse now subsisting between the unions and the employers. Unreasonable demands are most generally the outcome of ignorance. The average man is a fairly decent individual, not given to demanding the unattainable; but he, not unnaturally, desires some other evidence of the impracticability of his desires than the mere *ipse dixit* of interested parties.

"In the old agreement it was declared that there should be no limitation to the number of apprentices. In the new clause dealing with this matter the employers refuse to fix any definite proportion between journeymen and apprentices, but agree that the question of the proportion in the whole federated area shall be open for discussion. The precise meaning of this is not as clear as is desirable, but presumably gives the unions the right of calling attention to cases in which the proportion is far in excess of the general average. The employers also agree that the apprentice shall be given facilities for acquiring a practical knowledge of his trade, and shall be encouraged to obtain a theoretical knowledge thereof. Here the employers are probably taking an enlightened view of their own interests. Cases have not been uncommon in which apprentices have been far from fairly treated, and kept at one class of work for unduly prolonged periods. In the end such procedure, however immediately profitable, has proved far from satisfactory, and it was, moreover, hardly honest or honourable. An apprentice, in view of the fact that he is to be taught a trade, is paid less than he could obtain for his services in the unskilled labour market, and a fair training in his selected craft is distinctly a part and, indeed, the more important part, of the remuneration to which he is justly entitled. The recommendation that he shall, as far as possible, be encouraged to obtain theoretical knowledge is merely an official recognition of the existing practice of numbers of the best firms. Many of these definitely require their apprentices to attend evening classes,

and still more offer rewards for successful work at these. It is sometimes maintained that after a hard day's work in the shops a youth is not in a fit condition to do further work in the evening. This, however, is generally the plea of outsiders who have no personal experience of the system. Physical tiredness, unless approaching absolute exhaustion, is not a bar to effective head-work, and the youth, if not at a class, is probably employing his leisure in some form of distraction at least equally exhausting, and possibly actually pernicious. The classes, moreover, generally cease during the summer months—that is to say, during the period of the year in which it may fairly be claimed that a lad shut up in the shops all day, will best spend his evenings in the open air.

"The final and, no doubt, the most important clauses of the agreement deal with the subject of avoiding future serious disputes. In the old terms of settlement it was laid down that in the first instance grievances should be brought before an employer by a deputation of his workmen, and only in case of disagreement were the union officials to be called in. The employers stated, and we believe truly, so far as the mass of them were concerned, that no workman should be victimised for forming part of such a deputation; but the fear of action of this character is undoubtedly strong amongst workmen generally, and there are, no doubt, sufficient exceptions amongst the masters to render this apprehension not entirely without substantial foundation. In the new agreement all grounds for distrust are removed, as the workmen can at choice either deal directly with their employer by deputation, as hitherto, or through their trade-union official, without this preliminary. It is further provided that local conferences on matters in debate can be demanded by either party, such conferences to meet within twelve days of the receipt of the application by the other party. Failing settlement by this, a central conference shall be arranged as soon as possible between the executive of the Federation and the central authorities of the unions affected. No employer who refuses to employ union workmen is eligible to sit as a member of these conferences. In no case, it is provided, shall there be stoppage of work, either partial or general, until the completion of the above procedure; but work is to proceed under current conditions. This latter provision was, of course, included in the old terms of settlement.

"It will be noted that the new arrangement, like the old, tends to increase very much the powers of the central executives of the trade unions, as compared with what they were ten years ago. In the old days strikes were undoubtedly sometimes started by ambitious and fluent, but often inexperienced, logal organisers, sometimes much against the wishes of their official superiors. It is still, no doubt, possible, for the same thing to happen, as no penalty is directly prescribed for a breach of the above conditions; but presumably such action would lead to a general lock-out, making the risks too great to be lightheartedly undertaken. In some trades, we believe, where somewhat similar agreements exist, large sums, as caution-money for the due observance of the terms, have been deposited by both parties with trustees."

ELECTRIC TRACTION ON BRITISH RAILWAYS.

A COMPARISON OF THE COST OF STEAM AND ELECTRIC OPERATION ON SOME OF THE MORE IMPORTANT ELECTRIC RAILWAYS OF GREAT BRITAIN.

Philip Dawson—Street Railway Journal.

A RECENT number of the *Street Railway Journal* contains an interesting review of the electrical equipment of main line railways in Europe by Mr. Philip Dawson, whose contributions to the literature of electric traction are well known on both sides of the Atlantic. We extract from it the following authoritative comparison of operating expenses and other results of the electrification of several British railways.

"It is practically impossible to compare directly the cost of electricity and steam per train mile and obtain anything like satisfactory results, for the reason that the introduction of electrical haulage on a suburban line will at once alter the size and type of train adopted, and in all probability lead to shorter train units being used and run at more frequent intervals than was the case with steam.

"Furthermore, whereas for suburban service there is very little difference, as far as steam locomotive costs are concerned, in hauling a train having a capacity of, say, 250 passengers, as compared with the cost of hauling a train seating, say, 700 passengers, with electricity on the multiple unit system, a unit train, say, of three cars, once adopted, would according to requirements, be made up in multiples of three and the cost per train mile for locomotive power would be practically strictly proportional to the number of units of which a train is composed. Still, it may be interesting to compare some published results on the train mile basis for steam and electricity on the same road.

"The Mersey Railway is the only one of what may be called main line systems which has been completely electrified in this connection and working for a sufficient number of years so that figures are available to warrant fairly accurate conclusions being drawn from the results obtained. It is also a good example to quote, since electricity superseded steam on a given day when the whole system was converted from steam to electric haulage. There is little use in considering the increase of traffic resulting from electrification in this case, because the line when operated by steam gave a fairly rapid and frequent service. It is a short distance traffic line, and there existed no particular reason why electricity should largely increase the traffic.

"But the financial results as regards cost of operating and the cost of locomotive power per train mile operated by steam and electricity, as well as the cost of maintenance of permanent way, fully bear out the expectation of those engineers who have made electric haulage on railways a serious study. The figures show that with the increased speed which electrification renders possible, with slightly smaller and therefore more convenient train units, the train mileage can be enormously increased without increasing the total operating costs, and that notwithstanding the very much larger train mileage, the repairs of permanent way do not, as might be expected, increase, but are rather lower in their total annual cost than when steam is the propelling power.

"The following table affords a highly interesting comparison of steam and electric operation on the Mersey Railway for the years 1901 and 1905, respectively:

Items.	1901 with Steam.	1905 with Electricity.
Locomotive cost per train mile in pence	13.653	6.29
Train lighting and cleaning per train mile in pence..........	1.665	0.580
Repairs and renewals of carriages and wagons per train mile in pence	1.719	1.075
Train miles run 311,360		829,898
Total expenditure £64,662		£69,036
Maintenance of permanent way. £6,055		£3,793

"The total gross receipts for the half year ending June 30, 1901, were £38,327, while those for the half year ending June 30, 1906, were £47,129, or an increase of 23 per cent.

"Another comparison of Mersey figures is between the year 1901, operated by steam, and the year 1906, operated by electricity, which entirely confirms the figures given in the previous table; thus, in 1901 the total expenditure, including everything, was £64,662, the train mileage being 311,360 train miles; in 1906 the total expenditure, including everything, was £70,930 and the train mileage 829,188 train miles. It will thus be seen that an increase in train mileage of 167 per cent. increased the total expenditure by only 10 per cent.

"In comparing the cost per train mile of steam and electricity it must be borne in mind that steam trains were, on the average, composed of seven cars, having a capacity of 350 passengers. Now, when operated by electricity, according to the time of day, the number of cars per train varies from two to five, and the average seating capacity of an electric can be taken as 55 passengers, as compared to the average seating capacity of steam cars of just under 50 passengers.

"These figures clearly demonstrate the great benefit which railways will get by electrification of their suburban systems. It shows that they will, owing to the high rate of acceleration, be able to greatly increase the average speed, and therefore the number of trains, and thus to utilize better the existing facilities, both as regards

tracks and termini. Consequently they will largely augment their earning capacity without increasing their present total working expenses. And as Mr. Aspinall (the general manager of the Lancashire & Yorkshire Railway) has always maintained, the object of electrification is not so much to decrease the expenditure of a railway company as to increase its earning power. Nothing but a greatly accelerated and more frequent service can enable them to hold their own, and it is only electric traction that will provide this acceleration. That frequent and rapid service does, in most cases, provide an efficient remedy is clearly shown by the satisfactory results obtained both in the case of the Lancashire & Yorkshire Railway and of the North Eastern Railway. The latter is a particularly good example, because the electric lines between Newcastle and Tynemouth suffer from the most acute tramway competition. Notwithstanding this fact, the profits are considerably greater now on this railway than they were in the days before there existed any electric tramway competition. This is clearly shown in the following table:

Items.	Half Year Ending 1903. Steam Operation.	Half Year Ending 1905. Electric Operation.	Percentage of Increase.
Gross earnings......	£129,000	£151,000	17.1
Running costs........	£42,761	£47,779	11.7
Ratio of costs to receipts	33.2%	31.8%	....
Loco. costs per train mile	14.5d.	6.75d.	....
Passengers carried ..	2,844,000	3,548,000	24.8

"In reading the above table it must be borne in mind that the figures given as running costs for steam operation, are, if anything, too low. An exact figure is unobtainable owing to the difficulty of differentiating between the costs for local service and the total locomotive costs for the whole service. But even on the supposition that the comparison is a fair one, it affords a most satisfactory proof of the advantages to be obtained from electrification, particularly in view of the fact that electric tramway competition has very largely increased of late, which shows that the electrification of the railway has not only enabled it to hold its own but also to increase its traffic.

"Having thus considered the actual results as regards working expenses which are obtainable and authenticated, it may be interesting to study some figures specially compiled for London conditions and in which an attempt has been made to compare on a similar basis the cost of operating a service under similar conditions.

Items.	——STEAM—— Cost per train mile, 7-car train, with seating capacity of 520 passengers.	——ELECTRIC—— Cost per train mile, 3-car train with capacity of 210 passen- gers in London.
Fuel or current........	6.96d.	5.00d.
Water, oil, waste and stores	0.90d.	0.90d.
Maintenance and repairs	3.41d.	0.85d.
Wages	3.25d.	1.58d.
Lighting	0.85d.	included in current

Ratio of locomotive costs of electricity to steam 48.5 per cent (including current and capital charges on conduction and distributing system).

"From the above table it will be seen that on the train mile basis the locomotive costs for electric traction, maintenance and capital charges in electrical distribution included, are less than half the cost per train mile if operated by steam. Against this, however, would have to be placed the fact that the capacity of the steam train is more than double that of the electric train under consideration. But there is the additional important consideration that the costs given for electric service are for an average acceleration more than double that obtained with steam, and consequently that the average speed of traveling with the electric train is nearly double the maximum that the local steam trains, with present stops, can attain.

"It must be borne in mind that it would be impossible with steam haulage to increase effectually the speed of these local trains. To make a fair comparison, one must take as the average speed for electric trains the same speed as that of the present steam trains. On that basis the same number and power of motors that are required to drive the three-car train as given in the previous table, would be amply sufficient to operate a seven-car train, and if the trailer and rolling stock were the same as that at present in use with steam haulage, the carrying capacity of the train should easily equal that of the present steam trains. The table previously given would then very closely represent the comparative cost of operating by steam and electricity under the same conditions.

"From this it will be seen that, when compared on the same basis, the cost of hauling suburban trains by means of electricity is less than half that of operating by steam, with the additional advantage already cited thrown in, of the saving of time effected at termini, owing to the use of motor cars and multiple unit control.

"There remains another and most important factor to take into consideration, and that is the better use of the seating capacity per hour if run by smaller and more frequent trains instead of fewer and more capacious trains."

MUNICIPAL OWNERSHIP OF PUBLIC UTILITIES.

A COMPARISON OF THE COST OF STEAM POWER IN MUNICIPAL AND PRIVATELY-OPERATED PLANTS.

John W. Hill—Central States Water-Works Association.

ONE of Mr. Judson H. Boughton's tables in the November number of THE ENGINEERING MAGAZINE showed the average cost of street lighting to be much less under private contract than under municipal ownership of electric-power plants. Mr. Boughton remarked that while municipal plants have in some few instances showed satisfactory results for a time, they have eventually fallen, through political changes, under the disastrous management of incompetents. That this is bound to happen in the management by a municipality of almost any public utility is held by Mr. John W. Hill in a recent address before the Central States

Water-Works Association. Taking the operation of steam-power plants in municipal water-works and in private enterprises as a particular example, he gives the following interesting comparison of costs.

"During the past few months I have had occasion to collect statistics on the cost of steam power, and in support of the claim of better management by private corporations, advanced in this paper, it is found that even in the few cities having the highest type of triple-expansion pumping engines, and accessories to match, and in which the contract test and annual duties of the machinery are the best ever attained, the cost of power per year is greater than in the large well-managed steam power plants, owned and operated by the private electric and manufacturing corporations, and a comparison of statistics divided as to cost of fuel, labor, repairs and stores, show the excess cost to be principally in the item of labor.

"In the following table, the figures represent the annual costs per indicated horsepower of prime movers: Coal figured at $2.50 per ton, excepting Hamilton, $2.20 per ton. City Water-Works, and Philadelphia Power, 8,760 hours; Pittsburg Power, 8,666 hours; Hamilton, 8,000 hours.

Nature of power.	Fuel.	La-bor.	Re-pairs.	Sup-plies.	Total.
City pumping works ... (1)	$15.70	$24.62	$5.96	$2.74	$49.02
" (2)	17.44	25.19	5.25	3.58	51.48
" (3)	16.70	24.01	1.28	2.05	44.04
" (4)	16.75	21.90	1.85	1.88	42.38
" (5)	16.82	26.19	1.17	1.59	45.77
" (6)	21.39	15.64	9.62	1.24	47.90
" (7)	18.89	....	4.14	...	50.03
" (8)	27.64	26.44	9.55	5.44	69.08
New York.. (9)	25.55	13.13	5.90	1.42	46.00
Philadelphia (10)	20.51	6.61	1.52	1.32	29.96
Pittsburg ...(11)	19.51	4.42	3.55	2.50	29.97
Hamilton ..(12)	16.90	5.18	0.88	0.64	23.62

"It is interesting to note that the total cost of power by a private corporation, No. 12 on the list, is less than the cost of labor alone in the city water-works power, Nos. 1, 2, 3, 5, and 8. The figures are from the last annual reports. The costs of labor, repairs, and supplies are not known in detail for No. 7, but can roughly be stated as $25 for labor, $4.14 for repairs and $2 for supplies.

"In each city pumping station, the engines considered are from the best known and highest class builders, and the waterworks profession is accustomed to point to the cities considered as examples of excellent water-works management.

"The average cost of labor in the eight city power stations is $23.62 per indicated horse-power per year for 8,760 hours, while the average cost of labor in the private corporation stations is $7.33, or about one-third the cost in the municipally-owned and operated stations, or to state the matter in different form, the city uses three men to do one man's work.

"Engineers generally recognize the modern high duty triple-expansion pumping engine as the highest type of steam power, and the service of pumping at constant speed, against a steady head, to reservoirs, the 'optimum' condition for high running

duty. Moreover, the long runs of pumping engines, without interruption for Sundays are calculated to favor the annual economy, when compared with steam engines working under a constantly varying load, and, in all but traction and electric lighting stations, stopped altogether for Sundays and holidays.

"The duties obtained on the trials of Engines 1 to 6, inclusive, are the highest in the history of pumping machinery, a fact well attested by the annual charge for fuel; indeed the fuel costs, on the average, are as good for the city-managed works, as for the private-managed works. But when you come to the labor charge, there is where the politician comes to the front in great shape. There could be no political advantage in being wasteful of fuel, but there is a decided advantage in future elections in being wasteful of labor. Two men can cast two ballots, and three can do better, while the coal burned in the furnace cannot vote, and there is therefore no advantage in being prodigal with it.

"After allowing for the favorable conditions of modern water-works steam pumping service, even then the cost of power in the best private works is less than in the best water-works under municipal management.

"Instances can be multiplied of power costs under municipal and private control which will verify the low cost for labor for private control and the high cost for municipal control. And if compared on the volume of business transacted, it can be demonstrated in almost any if not all cities that the cost per unit of work performed or business conducted is much greater for public than for private business.

"The least objectionable of public works which can be conducted by the public are the building and maintenance of sewers and sewage disposal works, the cleaning of streets and the collection and disposal of garbage, although in some large cities this work is now the subject of annual contract. Water-works, electric and gas works, trolley lines, and steam and hot water heating systems, works employing large numbers of people, and requiring skilled assistance, should be the subject of private construction and control under contract or franchise which will properly safeguard the interests of the public and prevent extortion or poor service.

"Cities have been and can be robbed by public service corporations, but only with the aid and connivance of public officials. Standing alone, the public service corporations will be compelled to meet the obligations imposed on them by their franchises, and give the public the required service at fair rates. The legislature can pass laws and the city councils can pass ordinances giving away public rights, and the public service corporations may be the beneficiaries thereby, but the officials chosen by the people to represent them are the culpable parties in such transactions."

RAILROAD REGULATION AND POLITICS.

A DISCUSSION OF THE POLITICAL ASPECT OF THE AGITATION FOR FURTHER GOVERNMENT CONTROL OF THE RAILROADS OF THE UNITED STATES.

Walker D. Hines—Traffic Club of New York.

TO the impartial student of national questions it must be apparent that the present anti-railroad agitation has long since passed from an honest endeavor to safeguard the public interest and has developed into a general attack, promoted and fostered for political purposes. An impressive exposition of this fact is contained in an address recently delivered by Mr. Walker D. Hines, general counsel for the Santa Fe, before the Traffic Club of New York, from which the following extracts are taken.

"The interstate commerce act as now amended regulates in a thorough-going way the relations between railroad companies and shippers, prohibits unjust discriminations and preferences and all unreasonable rates and in all respects enjoins the utmost publicity. The commission is given power to enforce all these regulations and is given practically unlimited power to inquire into the affairs of the railroad companies. Most of the states have similar laws and commissions with similar powers. In addition there are numerous laws, both federal and state, regulating the equipment of railroad companies and their relations with their employes. In matters of train schedules, speed, etc., the railroad companies are subject to numerous local regulations. In matters of taxation railroad companies are subject to exceptional burdens. In addition, railroad companies are subject to especially strict liabilities which are being constantly and rigorously enforced by the courts.

"Under such conditions there is no serious wrong which a railroad company can do that is not susceptible of substantial correction, and generally such correction can be accomplished by public officers without expense to the individual. Therefore, it would seem that all the railroad situation requires is for the public authorities to enforce the numerous laws already in effect. Strange to say, however, the principal talk is about further railroad regulation. Strangers to the fact would assume that railroads are free from any substantial public control and are in an exceptionally lawless attitude, and that the paramount necessity is for the inauguration of an original policy of public control. What is the explanation of this anomalous condition? Why is it that the most strictly regulated enterprise in the land is singled out for so much additional legislative restriction?

"We find the explanation in the fact that railroad regulation is the political field most easily cultivated and most fruitful in results to the politician. It is human nature to be distrustful and critical with respect to great wealth, and this distrustful and critical sentiment has generally settled especially upon the railroad, for that is the thing which the people see and with which they deal to a much greater extent than with any other form of capital. In every community the railroad is the most tangible and obvious form of accumulated wealth. The champion of the people who wishes to pose as St. George and slay the dragon for the pretended protection of the people finds that the railroad furnishes the most visible and striking material out of which that dragon may be constructed. Coupled with this availability of the railroads to political exploitation is the further fact that the advantages of such political exploits are merely temporary. Such political advantage is a commodity which is consumed in the using. The politician who scores by securing railroad regulation feels that to continue his success he must score again by securing still further railroad regulation. The other politicians not identified with the railroad regulation obtained feel that they must find other ways to score in the same game so as to restore their prestige. Consequently it is more strictly true of railroad regulation than of any other form of legislation that generally speaking every act of legislation stimulates its author to additional activity and inspires his rivals and imitators with the highest competitive zeal. Competition is said to be the life of trade, but competition among the politicians is proving almost the death of the railroad trade.

"One of the arguments for the Hepburn bill was that that bill merely conferred upon the interstate commerce commission the same powers that were already exercised by state commissions. Yet the mere fact that congress thus adopted the policy of the states inspired those states to go much further and to seek to devise all sorts of additional anti-railroad policies. The local statesmen, seeing the political advantages which accrued to national leaders by reason of the Hepburn act, straightway sought to secure corresponding advantages to themselves by devising new forms of local regulation.

"But the agitation has not been confined to the states. The clamor for additional national legislation is greater than it was before the Hepburn act was passed. The passage of that act was urged because it was said rates were extortionate. Now it is claimed that it is impossible to tell whether rates are extortionate or not until further legislation can be had which shall regulate the capitalization and the valuation of the railroads.

"Another striking illustration of the political competition in the matter of railroad regulation is shown by the unwillingness of governors and legislators to leave to the state commissions the administration of the subjects within the jurisdiction of the commissions. Apparently there is fear in some quarters that the commissioners might get political credit

which others are anxious to appropriate. Thus we find in states where the commissions have ample power to reduce rates and have the machinery for making thorough investigation to ascertain whether rate reductions are proper that the legislators with the approval of the governors rush in to make reductions on their own account and without investigation. Governor Hughes' veto of the 2-cent rate bill in New York was a striking exception which proved the rule.

"Another prominent indication that this agitation has now reached a peculiarly political stage is shown in the character of legislation regulation and notably in the case of reductions in passenger rates. Passenger travel is a luxury to a much greater extent than is the transportation of freight. It is absolutely necessary for people to have fuel and food and clothing, but it is rarely absolutely necessary for people to travel. Therefore, from the public standpoint, passenger rate reduction is even more unimportant than are freight rate reductions. Passenger business is far less profitable to the railroads than is the freight business. Therefore, from the railroad standpoint, passenger rate reduction is much more unjust than freight rate reduction. Yet, although passenger rate reduction is, from any point of view, the least justifiable form of rate reduction, it is the form which has proved most popular with legislatures. The only explanation is that the legislators appreciate that a reduction in passenger rates is more obvious to the voter and therefore has more political advantage. Another form of legislation which has been exceedingly popular is legislation at the instance of or for the benefit of organized labor, where again the appeal could be made in the most tangible form to the voter."

A state of political competition for anti-railroad legislation has been reached in which every possible means is being taken to create worse feeling and to prevent co-operation between the railroads and the public. For several years the railroads have been struggling to handle a traffic which has increased much more rapidly than their facilities could be extended, and, considering the relative difficulties, have attained a degree of efficiency higher than has been realized in any other industrial enterprise. But in the present anti-railroad agitation no credit is given the railroads for what they have accomplished; on the contrary, every minor deficiency is seized upon and utilized to excite hostile feeling. The public has been talked into an extremely hysterical frame of mind by the large numbers of active and intelligent politicians who are devoting a great part of their time to furnishing the suggestion necessary to fire the public imagination and to intensify and magnify any germs of railroad evils which may exist.

"I do not, of course, mean to say that all railroad laws have attained a condition where they are not susceptible of improvement from the public standpoint. I do contend, however, that at the present time railroad laws are so far reaching as adequately to cover all substantial evils, and that the public has been brought to such a state of mind, and the politicians have developed such an enthusiastic rivalry that any further legislation, even though theoretically wise, when considered in itself, will cause a vast amount of harm by stimulating further imitative and competitive activity on the part of politicians in general, and at the same time would add little if anything to practical safeguards which the public already enjoys.

"In a word, I say to each of you as good citizens, and in the public interest fully as much as in the railroad interest, do your best to discourage the professional maker of anti-railroad laws, and do your best to encourage the administrator of the laws already made, to study the facts thoroughly and then enforce the laws accordingly. What the public needs is less anti-railroad talk and less anti-railroad tinkering and a fair investigation of the facts and intelligent enforcement of existing laws."

A TIDAL-POWER AIR-COMPRESSING PLANT.

DETAILS OF THE DESIGN OF A PLANT FOR COMPRESSING AIR BY TIDAL POWER TO BE ERECTED AT SOUTH THOMASTON, MAINE.

William O. Webber—The Engineer.

SOME months ago a review in these columns gave the general features of the problem of compressing air by tidal power with reference to a project for the erection of a large plant at South Thomaston on the Maine coast. Recently more detailed information of the design adopted has been communicated to *The Engineer* by Mr. Wm. O. Webber, the author of the former paper, from which the following extracts are taken.

Diagrams are shown, giving the details of the lock and inlet and outlet gates, a general plan and sections of the complete construction at the same point, and a profile of the location at South Thomaston. "At this point the maximum tide is 10.6, mean tide 9.4 and minimum tide 7.9 feet, giving, respectively, 5,000, 4,000 and 3,000 horse-power.

"In the dam, where the main channel is navigable, will be a lock for vessels, 40 feet wide, 200 feet long, and 28 feet deep. On either side of this lock will be one or more sets of shafts, each making a unit, or compressor, of 1,000 or more horsepower.

These shafts will be sunk into the rock to a depth of 203.5 feet below mean low water, the down-flow shaft being 15.75 feet in diameter, and the up-flow shaft 35.65 feet in diameter. The in-flow gates will be five in number and 10 feet wide, the out-flow gates six in number and 10 feet 8 inches wide.

"The water on entering the inflow gates will swing them open, pass down the down-flow shaft at a velocity of 16 feet per second, drawing in air through about 1,500 half-inch inlet tubes. Arriving at the bottom of the shaft, the combined air and water will flow in both directions horizontally, the air separating from the water until all of the air is accumulated in the separating chambers. The water will then flow up the up-take shaft at a velocity of three feet per second, and out through the out-flow gates.

"The air entrapped in the air chamber is then under a head of water 195.5 feet high, varying with the height of the tides. This compressed air is then led up the up-flow shaft in a 14-inch pipe. At the top of the gates these 14-inch pipes are united into a

30-inch pipe, which conveys the air ashore. "The air thus compressed will contain about one-sixth the moisture that is in the atmosphere from which the air is drawn, owing to the absorption of the moisture in the compressed air by the surrounding water. The dryness of this air makes it particularly adaptable for transmission to considerable distance, in pipes, without undue friction. For instance, the whole 5,000 horsepower could be transmitted a distance of 1 mile, in 30-inch pipe, with a loss of only 1.5 pounds pressure, or 10 miles, in a 48-inch pipe, with only a loss of 2.5 pounds pressure.

"This air can be used cold, without danger of freezing in expanding, in steam engines or rock drills. A test was made on an 80-horsepower Corliss engine, in which the entering air was 5.3 degrees F., and the exhaust minus 40 degrees, and continued for 10 hours without the slightest sign of frost in the exhaust passages and pipes of the engines. A marked economy, however, is obtained by preheating this air immediately before using it in motors, as raising the air to 370 degrees will practically double the volume of the air, and, instead of requiring 3 to 4 pounds of coal per horsepower per hour, as air receives heat about six times as easily as water, these results

can be obtained at an expenditure of from ½ to ⅝ pound of coal per horsepower per hour.

"As there are no working parts in the compressor, there is no depreciation, or operating expenses, to be taken into account, excepting watchman to prevent depredations on the plant, keep ice and floating timbers from permanently obstructing the inlet racks, and operate the boat lock. Therefore the cost per horsepower is practically represented by the interest on the original investment and the wages of these watchmen. The cost of original construction will amount to about $100 per horsepower.

"There are numerous places, all practically situated between the 40th and 50th parallels of latitude, in both the northern and southern hemispheres, where the tides are of sufficient magnitude to make this plan commercially feasible, the necessary requirement being a tidal basin, of considerable size, connected with the ocean by a comparatively narrow outlet. Each acre of such basin, under a 9-foot tide, is capable of producing 5 horsepower. It is not commercially feasible to develop such a plant with a basin containing much less than 200 acres, or requiring a length of dam exceeding 3 feet per acre of pondage."

HIGH PRESSURE TURBO-COMPRESSORS.

A DISCUSSION OF THE ADVANTAGES OF THE RATEAU CENTRIFUGAL COMPRESSOR WITH A DESCRIPTION OF TYPICAL INSTALLATIONS.

A. Rateau—Revue de Métallurgie.

IN his article on exhaust steam turbines in the November number of THE ENGINEERING MAGAZINE, Dr. Gradenwitz mentioned in passing the high-pressure turbo-compressors developed by M. A. Rateau of Paris, whose work on low-pressure steam turbines has met with such a degree of success. The following details of the advantages of turbo-compressor operation are taken from a paper by the inventor in a recent number of the *Revue de Métallurgie*.

The turbo-compressors are exactly analogous to the centrifugal pumps devised by M. Rateau, with the addition of a device for cooling the air. In the first turbo-compressors cooling was accomplished by a

device similar to a surface condensing plant placed between each pair of successive compressor units. Recently, however, a more satisfactory system has been devised in which fresh water is introduced into the partitions and the diffuser vanes. The effect of this large increase in the cooling surface has been to reduce greatly the heating of the air and the compression is accomplished even closer to the isothermic cycle than is the case with reciprocating compressors. Efficiency tests conducted by M. Rateau on several recently-constructed machines indicate that very high pressures can be attained with an efficiency of 60 to 63 per cent., which is very near the result obtained in the best reciprocating compressor practice and it is probable that this will be improved upon.

Besides having an efficiency about equivalent to that of reciprocating compressors, the turbo-compressor possesses several peculiar advantages. One of the most important of these lies in the fact that for units of equal power the advantage in space is largely on the side of the turbocompressor. M. Rateau instances the case of the Commentry steel works where a blowing engine occupying 164 square metres of floor space was replaced by a turbocompressor of the same power taking up an area of 7 square metres. Besides the saving in foundations and space, the almost entire absence of vibration was a notable result of the installation of the centrifugal compressor.

On account of their smaller size and comparative simplicity of design turbocompressors are relatively cheap when compared with reciprocating compressors. Further, their working expenses are much lower; there are no complicated parts to get out of order and the consumption of lubricating oil is very small, since the bearings of the shaft are the only parts subject to friction.

The air current yielded by turbo-compressors is perfectly regular and there are none of the pulsations which necessitate the installation of compensating reservoirs when reciprocating machines are used. The output may be regulated readily within wide limits by slightly varying the speed or by opening or closing a gate in the suction or compression conduits, but in spite of this elasticity, a constant output at a constant pressure can be maintained.

The possibility of coupling turbo-compressors directly to electric motors and steam turbines without any intermediary gearing is of great advantage on account of the lowering of friction losses in this method of driving. As mentioned by Dr. Gradenwitz last month, the exhaust steam from engines, operating low-pressure Rateau turbines can thus be utilized for the production of compressed air. This feature is of especial value in mining plants and metallurgical works, where compressed air can be produced by means of the exhaust steam from the winding engines and rolling-mill engines respectively, with the aid

of a steam accumulator. In this special field, the utilization of exhaust steam for compressing air, the reciprocating compressor obviously cannot compete with the centrifugal type.

M. Rateau gives several examples of turbo-compressor plants now in operation in Europe. Many plants for the compression of air by the utilization of exhaust steam are in course of construction, modeled after the pioneer plant at the Béthune mines. In this installation the exhaust steam from the winding engine is used to compress air up to six or even seven atmospheres. The plant which has been in regular operation since May, 1906, consists of four multicellular ventilator units traversed by the air in series. These units are arranged in groups of two each on two parallel shafts, each of which is driven by a low-pressure turbine. A high-pressure turbine fitted to one of the shafts is idle under normal conditions and is used only as a reserve in the case of interruption of the exhaust steam supply. When such an interruption occurs an automatic apparatus opens the entrance conduit of the boiler steam to the high-pressure turbine, the exhaust of which is conveyed to the low-pressure turbines. Another automatic device distributes the load uniformly between the two shafts. Between each two consecutive compressor units are cooling devices to which cold water is supplied by a small centrifugal pump mounted on the same shaft. Running at 5,000 revolutions per minute, this plant yields 900 litres of air per second at a pressure of 7.13 kilograms per square centimetre above atmosphere, the highest pressure yet reached by turbo-compressors. The efficiency of the first three units is as high as 70 per cent.

A turbo-compressor at the Chasse blast furnaces yields 7.2 cubic metres of air per second (atmospheric pressure) at a pressure of 30 centimetres of mercury. The plant is so designed that in the event of any irregularity in the working of the blast furnaces it is capable of yielding four cubic metres per second at a pressure of 60 centimetres of mercury. This feature is attained by dividing the compressor into two identical units which may be coupled up either in parallel or in series. The speed of rotation is 3,000 revolutions per minute.

THE UTILIZATION OF PEAT.

A REVIEW OF RECENT ADVANCES IN THE USE OF PEAT IN THE GAS PRODUCER FOR POWER GENERATION WITH RECOVERY OF BY-PRODUCTS.

Electrochemical and Metallurgical Industry.

THE development of the gas producer and gas engine has to a great extent changed the problem of the utilization of peat. In the old problem the object was to convert the peat into a satisfactory fuel for domestic purposes and for use in the boiler furnace, and to produce the latter at such a price that it could be shipped to industrial centres and compete with coal. A later problem aimed at the utilization of peat in by-product coke ovens, but within recent years the most satisfactory method of setting free the vast amounts of energy stored up in the peat bogs of the world has been shown to be the use of peat in the gas producer for power generation, with recovery of by-products. An interesting description of the results which have been attained, based on lectures by the German investigators, Drs. Frank and Caro, who have been largely responsible for the development of process, is given in a recent number of *Electrochemical and Metallurgical Industry.*

The composition of peat varies greatly in respect to its content of water, ash, nitrogen, etc., but the purely organic substance has the fairly uniform composition of carbon, 60 per cent., hydrogen, 5 per cent., and oxygen, 35 per cent. The water and ash are, of course, deleterious elements, the nitrogen may be used for the production of useful nitrogen compounds, but the carbon and hydrogen content represents a large amount of stored-up energy, since both of these elements when oxidized set free considerable quantities of heat.

The difficulties in recovering this energy are due to both the composition and structure of the peat. Normally peat contains, when first cut, 80 to 90 per cent. of water, a large part of which must be removed before the peat is of any value as fuel. Natural drying takes a long time and produces a fuel with little heat value per unit of volume, while attempts to press the water out mechanically have proved unsuccessful owing to the peculiar structure of the substance. Better results have been obtained by crushing, kneading, and mixing the peat and then compressing it into quick drying blocks of "machine peat," but it must be recognized that peat is an inferior fuel and cannot compete with coal except under very exceptional circumstances.

Machine peat when quite pure and low in ash can be used with some success in by-product coke ovens for the production of coke, with recovery of acetic acid and ammoniacal liquor, but such undertakings could be financially successful in only a very limited number of places and under only the most favorable circumstances.

"The processes which we have considered so far relate to the application of peat as fuel, either in the form of peat fuel or peat-coke fuel. In any case, it is important first to dry and compress the peat for these purposes down to somewhat like 20 per cent. of water. Further, if one wants to produce good peat coke, one is restricted to the use of peat low in ash as raw material. Such processes may be profitable at certain places under local conditions, but they do not represent the solution of the general peat problem.

"Prof. Frank attempts this solution in another way. Instead of trying to subject peat to a treatment that would make it available for shipment to industrial centers, he says we must endeavor to bring industry and commerce into the peat deposit districts. Peat areas should be considered just like waterfalls, as centers of stored-up energy. In his first paper (1897) Frank considered the erection of large generating stations in the peat districts with reciprocating engines, the peat being used as fuel under boilers. But even in those early days Frank considered that it might be possible to gasify the peat in gas producers and use the gas directly in gas engines. This idea was taken up more strongly in his second paper (1903), since the large gas engine operated by producer gas, blast-furnace gases and coke-oven gases had made its appearance in the meantime in the industry. In this paper he estimates that in a gas-engine plant it is possible to produce about twice the amount of power which can be produced from the same amount of peat in a steam-engine plant.

"In this paper (1903) it is also mentioned that the generation of a suitable producer gas from lignite and peat has been proven practical, that lignite gas engines are operating at the Mansfeld Gewerkschaft, near Eisleben, and on the Sophien mine, near Meuselwitz, and that peat gas engines have been built by the Deutz Gas Engine Co., by Körting Bros., by Julius Pintsch, by the Oberursel Machine Co., and others. The Deutz Gas Engine Co. reported that when using air-dried peat, containing 16.57 per cent. water, 1 horse-power-hour requires 1.27 kilogram of peat, while according to a later statement the same company has succeeded in reducing the quantity of peat required for 1 horse-power-hour to 0.85 or 0.9 kilogram.

"But what is still more important, it is by no means necessary in such processes to dry the peat to such an extent that it contains only somewhat like 20 per cent. of water. It is quite possible to gasify peat containing 50 to 55 per cent. of water. Moreover, in this process, it is possible to recover almost the total nitrogen contained in the peat in form of ammonia.

"In the dry distillation of peat only a small amount of the nitrogen is changed into ammonia, as is proven by the fact that peat coke thus produced contains considerable amounts of nitrogen; as an example the following analysis of peat-coke is given by Caro: 86.41 per cent. C, 1.57 H, 1.26 N, 7.09 O, 3.67 ash.

"On the basis of the well-known Mond process, Dr. Caro has worked out a new method for gasifying peat in a mixture of air and overheated steam in excess. This process has been tried with Irish peat on the Mond works in Stockton, and it has been found that almost the total amount of nitrogen in the peat is changed into ammonium sulphate, which can be easily sold as fertilizer.

"At the Stockton works the output from 100-kilogram peat, calculated as free from water and containing somewhat more than 1 per cent. nitrogen, was 2.8 kilograms ammonium sulphate and 250 cubic meters of producer gas with a calorific value of 1,300 calories.

"Dr. Caro gives the following results obtained in tests in Winnington, England, where there is a large Mond gas producer plant. The Mond gas producers which were available there were partly used for gasifying peat. The peat gas was supplied to the gas engines which were otherwise operated with Mond gas, and ammonium sulphate was recovered in the same works.

"The engineer in charge of the gas engine plant did not know whether he received Mond gas or peat gas, because all gas came through the same supply mains. He did not even find the difference in the operation of the gas engines.

"Six hundred and fifty tons of peat were gasified in the whole. The composition of the peat substance, assumed free from water, was ash 15.2 per cent., volatile substances 43.8 per cent., nitrogen 1.62, total carbon 56.3, fixed carbon 34.2, with a calorific value of 5,620 large calories. The peat was used in different conditions, mostly with an average content of 40 per cent. water, and 1,780 cubic meters of gas with a calorific value of 1,360 large calories were obtained per ton of water-free peat substance. Besides this there were obtained 118 British pounds, 55 kilograms, ammonium sulphate per ton of water-free peat. I state again that this was not ammonia gas, but real salt which was weighed.

"The gas was partly used for generating the steam required for the gas producer process, partly for heating the ammonium sulphate solution, and besides this an excess of gas was obtained, namely, for each ton of water-free peat, gas was obtained giving 480 horse-power-hours in gas engines.

"In this plant the cost of the treatment of 100 tons of peat (the weight being calculated on the basis of water-free peat) was $50, including wages ($1 to $1.25 per man per day), repairs, etc. Further, for the production of the ammonium sulphate, sulphuric acid, costing $41.25 (at $7.50 per ton), was used. Finally, if the amortization is taken as $33.75 (at 10 per cent.), the total cost is $125. On the other hand, from these 100 tons of water-free peat ammonium sulphate in the amount of about $325 was obtained. This shows a good profit, especially if it is considered that the gas is supplied to the gas engines in absolutely pure condition.

"It has been found that the segregation of dust particles in a gas which has been freed from ammonia, takes place with much greater speed and intensity than in peat gas containing tar. One cubic meter of gas contained only 0.16 gram of tar, and the gas engines operated with this gas very well and without trouble. The content of hydrogen in the gas does not vary more than ½ of 1 per cent. in the maximum.

"The cost of the 480 horse-power-hours produced was, of course, very low. If we do not take into account the profit from the ammonium sulphate the cost of the electric horse-power-hour was less than 0.125 cent.

"Finally, peat gas produced in this way is not only suitable for generation of power, but may be used for all kinds of heating purposes if proper furnaces are employed. Its use in steel works recommends itself, since it is absolutely free from sulphur."

THE TESTING OF WINDING ROPES.

RECOMMENDATIONS FROM THE REPORT OF THE TRANSVAAL COMMISSION ON THE USE
OF WINDING ROPES IN MINE SHAFTS.

Iron and Coal Trades Review.

THE Commission appointed by the Lieutenant-Governor of the Transvaal in 1905 to inquire into the use of winding ropes, etc., in mine shafts has recently issued its report. The following extracts, taken from the *Iron and Coal Trades Review,* give the recommendations of the Commission with regard to the testing of winding ropes both before installation and during service.

"In dealing with the important matter of how best to ascertain the true strength of a new rope, the weight of evidence is in favour of a tensile test to destruction of a sample of the whole rope. It is true that some manufacturers state that the true breaking load can be calculated from the aggregate strength of the wires, but there still exists a considerable diversity of opinion amongst these experts as to the percentage reduction that should be made from this total to take account of the fact that the wires, when laid up in the form of rope, are not in a position to give out their full efficiency. One manufacturer advises a deduction of 8½ per cent. whilst others state 10 per cent. to 12 per cent. There is also a lack of agreement as to whether or not these deductions should vary according to the construction of the rope. As a practical acknowledgment of the inaccuracy of these assessments for use generally, it is found that the majority of the manufacturers, who export ropes to the Transvaal prefer to give the strength of the rope as the aggregate of the strength of its wires, and allow the purchaser to make any deduction he may think proper.

"Mr. Epton, after a large number of tests of new ropes of various manufacture and construction, found that a difference of from 3 to 16 per cent. existed between the aggregate strength of the wires and the breaking load of the rope. Thinking that this loss of efficiency might disappear after the rope had been well tightened in the course of a few weeks' use, some further tests were made under these conditions. The results point to the fact that this efficiency is not regained. Seeing that it is highly desirable in the interests of safety that the real breaking load of a new rope should be known in preference to the theoretical ultimate strength, the Commission is led to the conclusion that the only correct method is to test a whole sample of each new rope before it is put into use; that is to say, each new rope should be delivered to the purchaser accompanied by a test certificate, which, if not obtained in the country of origin, should be procured from the Government Testing House in the Transvaal, duplicates being allowed for separate portions of the same coil.

"As to the length of rope required for such a test, opinions vary, some witnesses saying that a length of 2 feet is sufficient, while others consider that it is necessary to use at least 10 feet, but from experiments made in the Transvaal Mines Department Mechanical Laboratory, very little difference appears to exist even in these limiting cases. The results of comparative tests between long and short samples of two ordinary winding ropes are made public by Mr. Epton. In one case there was practically no difference, and in the other case, in which a length of 4 feet 4 inches was compared with a length of 14 feet 10 inches, a difference of 900 pounds in 50 tons was found, this being in favour of the short test length. Several different methods of holding the whole rope specimen during test appear to be in vogue, but the best and most frequently used in public test houses is to open out, clean and tin the wires, and then run on a soft white-metal conical seal. In the Mines Department Laboratory it is the practice to use an ordinary commercial white-metal which has a melting point of about 375 degrees Fahrenheit, and has been found not to damage the wires in the least degree. It is considered a proper practice to apply to the individual wires of a rope other standard tests, such as tension, torsion, and bending tests. These are very useful to mark down a very inferior wire, or to show up large departures from uniformity. To determine the strength of a winding rope during its working life, the evidence taken appears to show that the best method is to re-cap the rope at intervals of from three to six months, and to have a portion of the rope cut off for testing purposes. It is also strongly advised to thoroughly examine the wires at the same time for corrosion or any other deteriorating influence by means of a magnifying glass or microscope. It would be more correct to make the interval between successive 're-cappings' dependent on the 'duty' of the rope, but this course is obviously impracticable. The Mining Regulations of the Transvaal provide for regular re-capping, not more than six months being allowed to elapse between successive re-cappings in the case of a round rope, and three months in the case of a flat rope. A portion of the rope cut off in length not less than half the circumference of the pit-head sheave, must be sent to the Government Mechanical Laboratory for testing. The practice followed

in the Laboratory is to furnish a certificate of the actual breaking load of the specimen, and at the same time to call the attention of the mine manager to any defects, such as brittleness, corrosion, want of lubrication, etc. This practice is in every way commendable, but we would recommend that the six months' period should apply also to flat ropes, which are no more difficult to examine than round ropes.

"With reference to the utility of an elongation test of the whole rope, from time to time, during its working life, opinions seem to be about equally divided on this point. Some witnesses think it would be an advantage at stated intervals to ascertain the stretching power of the rope when hanging in the shaft, by applying its usual working load or something in excess of this. Others think that no useful purpose would be served, and that certainly loading in excess of the normal conditions might do damage. Professor Joseph Hrabak, of Pribram, carried out some interesting experiments to determine the elongation of wire ropes under known loads, and thus to find out the modulus of elasticity. He refers fully to these in his book 'Die Drahtseile,' and from the description of his tests it appears that the data were arrived at by loading the cage with men in many deep-level vertical shafts, and noting the elongation thus produced in the rope. In a 3,000-foot shaft the elongation amounted to about 1 inch per man, and it is, therefore, capable of being measured reliably and accurately. The modulus of elasticity of a winding rope will vary during its working life, being small at first, afterwards increasing to something nearer the value of the modulus of the wire itself, and then, if the rope is kept in use long enough, it will probably decrease. This last effect, if observed, is to be taken as a danger signal, pointing to the deterioration of the rope.

"On the whole, the Commission considers that the practice of making elongation tests of the whole rope is to be recommended, but that the ordinary working load should not be exceeded to more than the extent of doubling the authorised load of the skip or cage. It is also strongly recommended that when a winding rope is taken off and its life considered ended for the particular purpose for which it was installed, a sample from the most worn place should be tested. The results of these tests should prove of the greatest service to the person who has had charge of the rope, as they would enable him to see how far his judgment was sound concerning its condition. It is not recommended that this practice should be made compulsory, for the reason that it would probably involve in most cases the wrecking of any utility that the rope might still possess for purposes other than that of transporting persons.

"A very important question arises in this section of the Commission's investigations, as to what loss of strength a rope has sustained owing to the removal by friction, etc., of a certain definite proportion of the cross-sectional area of the wires. To come to any sound conclusion regarding this matter, it is necessary to know what support is afforded to any individual wire by the gripping tendency of the adjacent wires. Mr. Epton, in his evidence, submitted the results of tests conducted by him of two rope samples, 26 inches in length between the grips, each specially prepared by having ten of the outside wires cut through. These tests showed that the supporting effect of the sound wires on the broken wires amounted practically to nothing. These samples were new or only slightly worn.

"Of ninety-one complete tests of worn ropes carried out in the Mines Department Mechanical Laboratory, 33 show that the laid-up wires have a greater strength than when tested individually at the most worn part, and 49 show the opposite, whilst eight give almost the same results, whether tested as separate wires, or in the form of a rope. It is interesting to note that the great majority of the ropes which tested higher than the aggregate of their wires, were of the simple construction, viz., six strands of seven wires, 6.6.1. All these results go to show that in a short length of rope—2 or 3 feet—the strength of the rope cannot fairly be assumed to amount to more than the aggregate strength of the wires, estimated in each case at the position of greatest wear, although these positions may not all occur at exactly the same point in the rope. Mr. Epton stated that in only five cases out of over 400 had he found ropes to break at a much lower load than was estimated from external appearances, the estimation having been made by assuming that there was no lateral support between adjacent wires. In these cases there were excessive internal corrosion and wear, and the first-named defect would probably have been quite as pronounced in a sample cut off from the end of the rope."

THE VENTILATION OF THE BATTERY TUNNEL.

A DESCRIPTION OF THE MACHINERY AND METHODS FOR THE VENTILATION OF THE BROOKLYN EXTENSION OF THE NEW YORK SUBWAY.

The Engineering Record.

THE problems connected with the ventilation of the Battery tunnels under the East River which will be opened for traffic within a few months were, owing to the length of the tubes and the character of the service to be handled, among the most difficult ever encountered in tunnel operation. They have therefore been carefully studied and the design finally selected is an example of the most advanced practice in tunnel ventilation. The following details of the ventilating plant and methods is taken from a recent number of *The Engineering Record.*

"The subaqueous connection has been built as two single-track parallel tubes, running from 26 to 28 feet apart on centers and about 6,750 feet in length. The tubes are cylindrical cast-iron shells, lined with concrete, having an inside diameter of $15\frac{1}{2}$ feet and they were projected toward the center of the river with gradients of 3 1/10 per cent, which are united at the lowest point near the center of the river by a vertical curve, with the tops of the tubes about 80 feet below mean high water level in the river."

Under ordinary conditions it is expected that the piston action of trains running at high speed and on close headway will suffice for all ordinary requirements for ventilation but it was thought advisable to install a complete system of mechanical ventilation for continuous use if necessary, and for ready use in case of emergency, such as the stalling of a train, etc. The system decided upon is that of blowing air into the tubes through jets or nozzles in the direction of motion of the trains, a system which experience has shown to be most effective for use in tunnels both ends of which must remain open. One of the great difficulties in the installation was the smallness of the nozzles which could be used in the restricted clearance space. Another difficulty was the procuring of suitable locations for the blowing stations. These were finally located, the one for the New York end in Battery Park, and the one for the Brooklyn side in Joralemon St., some three blocks back from the water front.

"After a careful study of the problem, the engineers in charge of the design decided that a volume of from 45,000 to 50,000 cubic feet per minute should be introduced into either tube for the results desired with provisions for a slight increase over this amount for emergency conditions. With the space available for nozzles, however, the areas of the orifices were limited somewhat, which factor, of course, largely influenced the capacities of blowers selected. The nozzles, although designed as large as possible for the spaces available are but 15⅜ inches wide by 4 feet total height, the upper corners being cut off by the upper portion of the tube shells, and have each a total cross sectional area of orifice of 5 square feet. Each tube thus has a total capacity of nozzle orifice of 10 square feet, each of the nozzles being supplied with air from the blowers through a converging connection from a delivery duct of 18 square feet cross section. The blowers selected as best adapted to this duty with these particular sizes of connections were centrifugal downward-discharge fans with 7-foot wheels, 4 feet in width. Operated against a static pressure of 0.33 ounce blast, each fan has a capacity for delivering 46,500 cubic feet per minute. The blowers are all ¾-housed steel plate centrifugal fans, and 2 units are installed in duplicate at each blower station, it being thought that with this provision of reserve capacity, no combination of conditions will ever arise in which at least one unit cannot be kept in

running order in each station. Each of the blowers is direct-connected to a 75-h.p. interpole motor which is designed for operation on the 600-volt current from the third rail of the Subway propulsion system, and is fitted with control apparatus for a speed variation of from 300 to 413 revolutions per minute. The source of current supply has, also, been made doubly reliable by the extension of an independent feeder for this purpose from the substation of the Subway lines nearest to the blower stations, and a direct connection to the third rails in the adjoining tubes, although the latter will serve as an emergency connection only and will not be used under ordinary conditions. They are also fitted with automatic starting switches arranged for distant control, so that the blowers may be placed in operation upon an instant's notice from the office of one of the despatchers of the Subway lines. . . . . .

"The duct connections from the blowers to the nozzles are simply arranged, consisting in general of a vertical rectangular duct of No. 18 galvanized iron, which extends from the delivery outlets of each group of two fans directly downward to the tube through which air will normally be blown, with cross branches to the other sides of the shafts to connect with the reserve nozzles for reversal of the flow in the tubes, if this may be necessary in emergency. The down delivery ducts are of 18 square feet cross section which is ample to carry the maximum volume that can be delivered through the nozzles, and they divide over either tube, through a Y-connection, into two branches of 9 square feet section which supply the two nozzles, the branches and connection to the nozzles being similarly proportioned and shaped on either side in order to equalize the flow from either nozzle. The blowers deliver through connections of 16 square feet cross section which join at the top of the shaft into the down delivery ducts, special flap dampers being used at the joining to permit either blower desired to deliver independently, while in case it is desired to operate both blowers together, the dampers will assume a central position and thus not interfere with the delivery of either unit. The choice of these sizes of ducts is due to the fact that the capacity of the system is limited by the areas of the nozzle orifices, upon which the design of the blowers was based; the effect of the nozzles is such that even with the second blower in operation, the volume delivered through them would be increased but to a slight extent, probably not over 20 per cent. The control of the delivery to the nozzles in either tube is had through sets of louvre dampers in the ducts at points immediately beyond the joining of the cross branch to the vertical duct, which dampers are interconnected and so actuated that the blowers cannot deliver into the nozzles of both tubes at the same time; these damper mechanisms in the two stations are furthermore fitted for distant electric control, by means of which the interlocking arrangement is secured, and which also permits of instantly changing the direction of flow in both tubes from a distant point. . . . .

"The distant control mechanism is of the electro-pneumatic type, the electrically-controlled pneumatic cylinders built by the Union Switch & Signal Co. for actuating signal mechanisms being used. As these cylinders are of the single-acting type, two are installed at each blower station, one for throwing the dampers in either position. They are mounted side-by-side on the side of the duct near the dampers and have their pistons connected to a rocker arm, pivoted at the middle, which carries a bell-crank so connected to the louvre dampers as to always maintain one closed while the other is open. The angles which the louvres assume when closed were so chosen in relation to the directions of flow that but about 60 degrees of movement of the vanes is required, and in order that the cylinders or levers may not stick in a partly opened position, positive throw spring attachments have been applied to the levers of either louvre which will always ensure completion of the motion of opening or closing, provided the cylinders throw the mechanism more than half way over. The electric air valves that control the pneumatic cylinders are actuated by the regular 16-volt direct-current supply for the signal system, a double-throw switch at the Bowling Green interlocking tower controlling both sets of cylinders; when thrown in one

direction, it actuates simultaneously the cylinders in both blower stations for the direct flow, while in the other direction, the cylinders for the reverse flow. In this way the adjustments of the sets of dampers are positively interlocked and it is thus rendered impossible through mistake to direct the blast into both ends of either tube at one time. No provision has been made for directing the blast in the same direction through both tubes at the same time.

"The provisions for ventilation thus made enable a great variety of conditions to be met. Under ordinary conditions, it is, as above stated, expected that the air in the tubes will be maintained sufficiently clear for the comfort of the trainmen and passengers by the piston-action of the trains passing through, while, if in hot summer weather, more ventilation should prove desirable for cooling purposes, the operation of one of the blower units for either tube will add considerably to the effect of the trains. It is, however, in the case of a serious emergency that the full value of the ventilating equipments would be realized. While there is no danger from fire in the tunnels, as cars of metallic construction only are to be operated on the Brooklyn extension, the smoke that would result from a burnt-out motor or other derangement of the electrical apparatus of the propulsion system could incite a serious panic amongst passengers; such smoke or obnoxious gases in case of a stalled train can, with the blowers in operation, be rapidly dissipated, and, by virtue of the provisions for reversal of flow, in such a manner as to be least offensive to the people at the scene of the accident. By means of the system of telephones in the tubes, the trainmen may very quickly advise the despatcher if the reversal of flow is necessary, which is accomplished in the office of the latter by the mere throwing of a single-pole switch. Furthermore, if the blowers should happen not to be in operation, they could be instantly started by the mere throwing of another switch at the same point, one for either blower station, the blower motors being, as above stated, fitted with automatic starting equipments with distant control."

The following pages form a descriptive index to the important articles of permanent value published currently in about two hundred of the leading engineering journals of the world—in English, French, German, Dutch, Italian, and Spanish, together with the published transactions of important engineering societies in the principal countries. It will be observed that each index note gives the following essential information about every publication:

(1) The title of each article,   (4) Its length in words,
(3) A descriptive abstract,   (5) Where published,
(2) The name of its author,   · (6) When published,

(7) We supply the articles themselves, if desired.

The Index is conveniently classified into the larger divisions of engineering science, to the end that the busy engineer, superintendent or works manager may quickly turn to what concerns himself and his special branches of work. By this means it is possible within a few minutes' time each month to learn promptly of every important article, published anywhere in the world, upon the subjects claiming one's special interest.

The full text of every article referred to in the Index, together with all illustrations, can usually be supplied by us. See the "Explanatory Note" at the end, where also the full title of the principal journals indexed are given.

DIVISIONS OF THE ENGINEERING INDEX.

CIVIL ENGINEERING

BRIDGES.

Abutments.
Discussion, Design and Specifications for a Reinforced Concrete Bridge Abutment. T. M. Fyshe. Drawings with detailed description of the design, with related information. 8000 w. Can Soc of Civ Engrs—Oct. 17, 1907. No. 87854 N.

Anchorages.
Concrete Anchorage in the Brooklyn Bridge. Charles M. Ripley. On the Manhattan anchorage for Brooklyn Bridge, No. 3, and the contractor's plant and methods used. Ills. 2200 w. Cement Age—Oct., 1907. No. 87859.

Arch.
The Construction of the 175th Street Arch, New York City. Describes methods used in building an elliptical brick arch, of 66-ft. span and 20-ft. rise, built on a 9 deg. skew. Ills. 1500 w. Eng Age—Oct., 1907. No. 87859.

Failure.
The Ponts-de-Cé Disaster (La Catas-

We supply copies of these articles. See page 559.

523

trophe des Ponts-de-Cé). Jean Phizey. A description of the failure of this railroad bridge on August 4, by which a passenger train was precipitated into the River Loire. Ills. 4800 w. Génie Civ—Sept. 14, 1907. No. 87611 D.

Girders.

Continuous Girders. Gives calculations for determining the design of a type of railway overbridge. Ills. 2500 w. Engr, Lond—Sept. 27, 1907. No. 87525 A.

Lattice Girder.

Strengthening an Old Lattice Girder Bridge. George Jacob Davis, Jr. Notes on the method used in computing the stresses in the Leonard St. bridge, Grand Rapids, Mich., and its failure and strengthening. Ills. 1200 w. Eng Rec—Sept. 28, 1907. No. 87307.

Manhattan.

The Manhattan Bridge Pedestal. Outline description of this bridge across the East River, at New York, with illustrated detailed description of the design and construction of the pedestals. 2500 w. Eng Rec—Oct. 19, 1907. No. 87784.

Quebec.

The Erection of the South Cantilever Arm of the Quebec Bridge. Illustrates and describes details of erection of the south cantilever arm, not previously published. 3500 w. Eng Rec—Sept. 28, 1907. No. 87308.

The Collapse of the Quebec Bridge. Frank W. Skinner. Gives results of personal observations at the site of the Quebec Bridge disaster, attributing the accident to bad design. Also long editorial. Ills. 6800 w. Engng—Sept. 20, 1907. No. 87342 A.

The Compression Members of the Quebec Bridge. Editorial discussion, with letters bearing on the Quebec disaster from engineers of prominence. 9000 w. Eng News—Oct. 3, 1907. No. 87424.

The Quebec Bridge Wreck. Folding plate, with notes. 1800 w. Eng News—Oct. 10, 1907. No. 87552.

Theodore Cooper on the Quebec Bridge and Its Failure. The official statement of the consulting engineer for the Quebec bridge, to the Canadian Investigating Commission. 11500 w. Eng News—Oct. 31, 1907. No. 87992.

Rebuilding.

Calf Killer Bridge, Nashville, Chattanooga & St. Louis Ry: Detailed description of methods used in rebuilding this bridge with reasons for the re-erection. 2000 w. Eng Rec—Sept. 28, 1907. No. 87305.

Reinforced Concrete.

See under Viaduct.

Suspension.

See under Manhattan.

Viaducts.

The Nelson Street Viaduct at Atlanta. Illustrated description of a reinforced concrete structure of ten flat arch spans. 1000 w. Eng Rec—Oct. 12, 1907. No. 87570.

The Erection of the Moodna Creek Viaduct. Illustrates and describes the overhead traveler used which is of rather peculiar construction. 2000 w. Eng Rec—Oct. 26, 1907. No. 87936.

Ferro-Concrete Viaduct for the Cala Mineral Railway, Seville. Mateo Clark. Explains the conditions, and gives an illustrated description of the construction. 2000 w. Engng—Oct. 18, 1907. No. 87957 A.

CONSTRUCTION.

Beams.

Strain in Beams of Rectangular Section with Curved Axes (Die Anstrengung Stabförmiger Träger mit Gekrümmter Mittellinie). Carl Pfleiderer. A mathematical paper in extension of a previous paper in the same publication. Ills. 3600 w. Zeitschr d Ver Deutscher Ing—Sept. 21, 1907. No. 87652 D.

See also under Reinforced Concrete.

Buildings.

See under Reinforced Concrete.

Cement.

The Regulation and Control of Cement Construction. E. S. Larned. A discussion of the use of cement blocks with suggestions for specifications and uniform instructions. 3500 w. Cement—Sept., 1907. No. 87398 C.

Column Stresses.

Imperfect Butt-Joints in Columns, and Stresses in Lattice-Bars. Henry S. Prichard. Gives an analysis of the bending stresses so produced, illustrating by problems. 2000 w. Eng News—Oct. 3, 1907. No. 87425.

Concrete.

Concrete Construction. E. S. Larned. Read before the Am. Portland Cement Mfrs. Discusses its regulation and control. 3500 w. Ir Age—Oct. 10, 1907. No. 87537.

Sound Concrete Construction. J. T. Noble Anderson. Information based on experience. 2800 w. Engr, Lond—Oct. 4, 1907. Serial. 1st part. No. 87595 A.

Contractor's Plant.

An Unusual Plant for Constructing a Submarine Tunnel at Chicago. Illustrated description of a plant used in connection with the construction of the lake section of a new tunnel for the water-works system. An aerial tramway is a special feature. 5500 w. Eng Rec—Oct. 26, 1907. No. 87935.

Dams.

Masonry Dams. Thomas G. Bocking.

Gives examples illustrating the constructional practice of the five continents. Diagrams. 1500 w. Engr, Lond—Sept. 27, 1907. No. 87528 A.

Hydraulic Excavation and Dam Building at the Croton and Lyons Dams in Michigan. William G. Fargo. Illustrates and describes the plant and the work. 2500 w. Eng News—Oct. 24, 1907. No. 87872.

Crocker's Reef Dam Across the Hudson at Ft. Edward. Herbert Spencer. Illustrates and describes this rock-fill dam built in connection with the improvement of the Champlain Canal. 1500 w. Eng Rec—Oct. 5, 1907. No. 87447.

A Collapsible Steel Dam Crest, Bear River, Near Garland, Utah. J. C. Wheelon. Illustrated detailed description of a collapsible steel dam crest invented by the writer. It can be attached to either masonry or timber dams. 1000 w. Eng News—Oct. 3, 1907. No. 87421.

Filling.

Fills Made from Cars Running on a Suspended Cableway. Illustrated description of a suspension bridge system of making earth fills. 1500 w. Eng News—Oct. 10, 1907. No. 87550.

Fireproofing.

Fireproof Construction. M. M. Sloan. Discusses the requirements and special features of such structures. 3000 w. Archs & Bldrs' Mag—Oct., 1907. Serial. 1st part. No. 87974 C.

Foundations.

Foundations. Continued discussion of the best system for heavy structures on alluvial ground; the use of iron or steel when in contact with water; and the strength of concrete under stated conditions. 6500 w. Pro Am Soc of Civ Engrs—Sept., 1907. No. 87715 E.

Groynes.

Ferro-Concrete Groynes Near Brighton. Illustrated description of work on the foreshore to the east of Brighton, where the high chalk cliffs have suffered from erosion by the sea. 1200 w. Engr, Lond—Sept. 27, 1907. No. 87527 A.

Lookout Tower.

The Outlook Tower of Beinn Bhreagh, the First Iron Structure Built of Tetrahedral Cells. T. W. Baldwin. Illustrated description of an interesting experimental structure recently built by Dr. A. G. Bell at his summer home in Cape Breton. 1000 w. Sci Am—Oct. 5, 1907. No. 87474.

Reinforced Concrete.

The Reinforced Concrete Work of the McGraw Building. William H. Burr. Illustrated detailed description of a true reinforced-concrete strucutre, designed to afford the greatest possible resistance to

the vibration of heavy machinery. 5800 w. Pro Am Soc of Civ Engrs—Oct., 1907. No. 87881 E.

Method and Cost of Constructing Reinforced Concrete Kiln House of Separately Molded Members, Edison Portland Cement Co. Gives data relating to a new kiln house at New Village, N. J., with illustrations. 2500 w. Engng-Con—Oct. 2, 1907. No. 87406.

Cast Concrete-Steel Members. Illustrated description of kiln houses of the Edison Portland Cement Co., where the members were cast and assembled like fabricated steel. 800 w. Ins Engng—Oct., 1907. No. 87856 C.

New Buildings of the King Plaster Mills, Staten Island. Five-story reinforced-concrete structures, occupying irregular areas, designed to support floor loads from 200 to 600 lbs. per square foot, are illustrated and described. 2000 w. Eng Rec—Sept. 28, 1907. No. 87313.

The New Plaster Mill of the American Gypsum Co. Describes a fireproof reinforced concrete building near Port Clinton, O. Ills. 1800 w. Eng Rec—Oct. 12, 1907. No. 87568.

Method and Cost of Building a Reinforced Concrete Car Barn. Illustrated detailed description, with costs, of a barn at Easton, Pa. 1100 w. Engng-Con—Oct. 23, 1907. No. 87888.

Report of the Special Committee on Reinforced Concrete of the Engineers' Club of St. Louis Embodied in the Building Ordinances of the City of St. Louis. 5500 w. Jour Assn of Engng Socs—Sept., 1907. No. 87863 C.

An Analytic and Graphic Method of Determining the Section of Reinforced Concrete Beams (Ein Weg zur Analytischen und Graphischen Behandlung des mit Eisen Armierten Betonquerschnittes). Rudolf Pokorny. A mathematical paper on the design of reinforced-concrete members. Ills. 3200 w. Zeitschr d Oest Ing u Arch Ver—Sept. 6, 1907. No. 87645 D.

The Design of Rigid Frames (Beitrag zur Berechnung Steifer Rahmenkonstruktionen). Otto Leuprecht. A mathematical paper with worked-out examples. Ills. 2200 w. Serial. 1st part. Beton u Eisen —Sept., 1907. No. 87655 F.

See also under Anchorages and Abutments and under same title, under Bridges.

Reservoir Leak.

Repairing a Remarkable Leak in a Reservoir Embankment at Providence, R. I. Extract from report of Otis F. Clapp, describing the work. 600 w. Eng News —Oct. 17, 1907. No. 87740.

Retaining Wall.

The Melwood Avenue Retaining Wall,

We supply copies of these articles. See page 559.

Pittsburg. Describes the construction of a reinforced-concrete retaining wall about 220 ft. long which has a maximum height at the center of 64 ft., exclusive of the 4-ft. parapet wall. Ills. 1500 w. Eng Rec—Sept. 28, 1907. No. 87309.

Roads.
See under Municipal.

Scraper Work.
Methods and Cost of Moving Earth in the Fresno Scrapers in Arizona, and the Cost of Trimming Slopes. An account of work in grading a railroad bed. 1500 w. Engng-Con—Oct. 2, 1907. No. 87407.

Sewers.
See under Municipal.

Shingles.
Asbestos Shingles. Charles H. Stringer. Explains the advantages of the use of composition of fibre and cement. 1200 w. Ins Engng—Oct., 1907. No. 87855 C.

Tunnel Lining.
Lining a Tunnel with Concrete. Illustrates and describes work on the Southern Ry. in Indiana. 2000 w. Eng Rec—Oct. 12, 1907. No. 87565.

Tunnels.
The Reconstruction of the Eighth Street Tunnel, Kansas City, Mo. Describes work for the Metropolitan Street Railway Co., which required changing the grade in the tunnel. 1700 w. Eng Rec—Oct. 5, 1907. No. 87451.

The Detroit River Tunnel. An illustrated description of the methods used in sinking steel shells in a dredged trench. 3000 w. Eng News—Oct. 31, 1907. No. 87984.

See also under Contractor's Plant.

Underpinning.
Underpinning Adjacent to the Silversmiths' Building, New York. Explains conditions adjacent to the site for this 20-story, steel-cage building, and illustrates and describes methods of supporting the adjoining buildings. 3000 w. Eng Rec—Sept. 28, 1907. No. 87310.

MATERIALS OF CONSTRUCTION.

Cement.
Qualities and Methods of Using Cement. S. B. Newberry. Abstract from a paper before the Ohio Bldrs. Sup. Assn. Considers qualities of various cements and methods of use. 2000 w. Munic Engng—Oct., 1907. No. 87393 C.
See also under Slag Cement.

Concrete.
Results of a New Series of Tests on the Shearing Strength of Plain and Reinforced Concrete (Ergebnisse einiger neuen Versuche über den Scherwiderstand Reiner und Armierter Betonprismen). J.,E. Brik. Tests made on small square-section blocks with longitudinal reinforcement. Ills. 2000 w. Oest Woch-

enschr f d Oeffent Baudienst—Sept. 7, 1907. No. 87641 D.

Paints.
A New Blue-Black Iron Paint as a Protective Covering. F. J. R. Carulla. Read before the Ir. & St. Inst., at Vienna. Describes an English patented process, utilizing "waste pickle" liquors. 800 w. Mech Engr, Lond—Sept. 28, 1907. No. 87510 A.

Slag Cement.
Tests of the Strength of Slag Cement (Versuche über die Festigkeit von Schlackenbeton). G. Kaufmann. Gives results of compression and shear tests, with applications to formulæ. Ills. 3000 w. Beton u Eisen—Sept., 1907. No. 87654 F.

Timber.
Maximum Length of Service for Poles, Ties and Timbers. Edward J. Weihe. Presents facts in regard to the structure, seasoning, and preservative treatment. 2800 w. St Ry Jour—Oct. 12, 1907. No. 87775.

Timber Preservation.
The Steaming of Timber. Octave Chanute. Read before the Nat. Assn. of Wood Preservers. Considers that condition of the wood determines whether it is advisable to steam or not. 1800 w. Munic Engng—Oct., 1907. No. 87394 C.

Results of Recent Work of the Timber Tests by the Forest Service, United States Department of Agriculture. W. Kendrick Hatt. 3000 w. Am Ry Engng & Main of Way Assn, Bul 85—March, 1907. No. 87887 N.

A New Injection Process of Timber Treatment (Nouveau Procédé d'Injection des Bois). M. F. Lantier. Describes methods and apparatus for treating timber by the Rüping creosoting process. Ills. 5000 w. Rev Gen des Chemins de Fer—Sept., 1907. No. 87602 G.

MEASUREMENT.

Beams.
A Note on the Effect of the Time Element in Loading Reinforced Concrete Beams. W. K. Hatt. Read before the Am. Soc. for Test. Mat. Records general results of tests made. Ills. 1200 w. Eng News—Oct. 24, 1907. No. 87874.

Strength of Materials.
The Work of Deformation as a Measure of the Strength of Materials (Ueber die Deformationsarbeit als Mass der Beanspruchung). Rudolf Girtler. A mathematical demonstration with concrete examples. Ills. 3500 w. Zeitschr d Oest Ing u Arch Ver—Sept. 13, 1907. No. 87646 D.

Surveying.
Triangulation and Traverse Work in the Bronx, New York City. Notes from

a paper by Edward H. Holden, read before the Munic. Engrs. of the City of New York, explaining some of the methods followed in the work. 2500 w. Eng Rec —Oct. 5, 1907. No. 87450.

Surveying Tapes.

Invar (Nickel-Steel) Tapes on the Measurement of Six Primary Base Lines. Owen B. French. Reports investigations made of the properties of invar tapes, with results of work and statement of advantages. 7700 w. Pro Am Soc of Civ Engrs—Oct., 1907. No. 87882 E.

Telemeter.

Commandant Gérard's Telemeter. Illustrated description of an instrument for measuring distances from the observer to an object. 1500 w. Sci Am—Oct. 26, 1907. No. 87879.

Testing Materials.

Investigations of Structural Materials by the United States Geological Survey. Richard L. Humphrey. An illustrated account of the work at the testing laboratories at St. Louis. 2500 w. Cement— Sept., 1907. No. 87397 C.

Government Work in the Testing of Materials. Editorial discussion of how far the Federal Government may advantageously carry this work. 2000 w. Eng News—Oct. 17, 1907. No. 87742.

Test of a Reinforced Concrete Beam. Guy O. Fraser. A detailed account of the test representing actual working conditions. 2800 w. Cal Jour of Tech—Oct., 1907. No. 87981.

Compression Tests of the French Government Commission. Leon S. Moisseiff. Describes compression tests on longitudinally reinforced columns. Ills. 2500 w. Cement—Sept., 1907. No. 87396 C.

MUNICIPAL

Drainage Canal.

A Favorable Report on the Calumet Drainage Canal. Gives the projects under discussion in connection with the sewage disposal of this district of Chicago, and the opinion of Mr. Rudolph Hering. 2500 w. Eng Rec—Oct. 26, 1907. No. 87937.

Municipal Planning.

City Planning and Replanning. Editorial discussion of points deserving attention. 2000 w. Eng News—Oct. 31, 1907. No. 87991.

Pavements.

Pavements. Continued discussion of whether paving materials of the present will be used in the future. 7000 w. Pro Am Soc of Civ Engrs—Sept., 1907. No. 87716 E.

Roads.

The Construction of Macadam Roads. Austin B. Fletcher. A brief description and discussion of the processes and es-

sential features. Also data as to costs. Ills. 18000 w. U S Dept of Agri—Bul. 29. No. 87867 N.

Methods and Cost of Building Macadam Roads in the State of Washington. From a paper by A. L. Valentine, read recently before the Pacific N.-W. Soc. of Civ. Engrs. 2000 w. Engng-Con—Oct. 2, 1907. No. 87408.

Sewage Disposal.

A New Zealand Method of Sewage Treatment for Isolated Buildings. John Mitchell. Extracted from "Some Thoughts Upon Rural Sanitation and Suggestions for a Betterment." 1800 w. Eng News —Oct. 31, 1907. No. 87990.

Sanitation in the Suburbs of Paris (Assainissement de la Banlieue de Paris). Describes principally the Suresnes sewerage siphon under the Seine and outlines other projects for the purification of the waters of the river. Ills. 2400 w. Génie Civ—Sept. 7, 1907. No. 87610 D.

Sewage Purification.

The Sewage Purification Plant at Reading, Pennsylvania. Illustrated detailed description of a new installation consisting of a sewage screening device, a septic tank, a sprinkling filter, and a final settling basin. 5000 w. Eng Rec—Oct. 5, 1907. No. 87441.

Sewers.

A Large Sewer Project in St. Paul, Minn. Plan and description of the construction of a deep drop shaft and tunnel as part of a system to serve a manufacturing district. 1200 w. Eng Rec—Oct. 5, 1907. No. 87443.

Trenches.

Back-Filling Trenches. George C. Warren. Read at Detroit, before the Am. Soc. of Munic. Imp. Discusses some difficulties in connection with back-filling of trenches in city streets both as regards efficiency and economy. 3000 w. Eng Rec—Oct. 5, 1907. No. 87448.

WATER SUPPLY

Concrete Pipes.

The Testing of Clay and Concrete Pipes. Abstracts of parts of a report of Burchartz and Stock of the royal building material testing department in Gross-Lichterfelde upon tests of clay and cement pipes. 1600 w. Munic Engng—Oct., 1907. No. 87395 C.

Small Cement Pipe for Irrigation. Albert Eugene Wright. Data on the construction, cost and strength. 2000 w. Cal Jour of Tech—Oct., 1907. No. 87980.

Percolating Beds.

The Time of Passage of Liquid Through Percolating Beds. William Clifford. A report of investigations, giving tabulated results. 800 w. Eng News —Oct. 17, 1907. No. 87739.

We supply copies of these articles. See page 559.

Pipe Friction.

Experiments with Submerged Tubes 4 Ft. Square. C. B. Stewart. Describes experiments made at the University of Wisconsin on the effects of changing the length of the tube and of modifying the entrance conditions. Also outlines a plan for future experiments. Ills. 2000 w. Eng Rec—Sept. 28, 1907. No. 87312.

Purification.

Electrolytic Treatment of Water for Technical Uses. Dr. Herman Fleck. Reports of electro-filtration tests made at Denver, Colo. 1000 w. Min Rept—Oct. 17, 1907. No. 87799.

Water Purification at St. Louis, Mo. Edward E. Wall. An illustrated description of the plant and process for clarifying and purifying the water supply, and the conditions previous to its use. 7000 w. Pro Am Soc of Civ Engrs—Sept., 1907. No. 87712 E.

Water Supply. Informal discussion on whether it is better to control wa₁er-sheds or rely on filtration to purify the water. 2500 w. Pro Am Soc of Civ Engrs—Sept., 1907. No. 87714 E.

Examination of Water Purification Plant at Owensboro, Ky. Philip Burgess. Gives report of an examination made of a recently completed plant, with special reference to results obtained. 2000 w. Eng Rec—Sept. 28, 1907. No. 87306.

The Water Purification Plant of Harrisburg, Pa. Describes the operation of the filters and the apparatus used. Ills. 3500 w. Munic Engng—Oct., 1907. No. 87392 C.

Water Softening and Purification Plant of the Pennsylvania Railroad at Hartsdale, Ind. Illustrated description of a recently installed plant. 1400 w. Ir Trd Rev—Oct. 10, 1907. No. 87539.

Sterilization.

New Apparatus for the Sterilization of Drinking Water by Heat. Illustrates and describes the system of operation of the Forbes water-sterilizing system. 4000 w. Eng News—Oct. 31, 1907. No. 87987.

Typhoid Fever.

Engineering Studies of a Typhoid Outbreak at the State Hospital, Trenton, N. J. From the report of the State Sewerage Commission showing it to be due to the pollution of a portion of the water supply. Also a report by Dr. Henry Mitchell attributing the outbreak to personal contact. 4500 w. Eng News—Oct. 3, 1907. No. 87422.

Waste.

Water-Waste Investigations in Philadelphia. Reports investigations undertaken to measure the amount of water delivered by each pump to determine whether the pumps were operated economically; to ascertain whether the reservoirs were leaking; and to determine just what becomes of the water after it leaves the reservoirs. 2700 w. Eng Rec —Oct. 5, 1907. No. 87442.

WATERWAYS AND HARBORS.

Danube.

The Sulina Mouth of the Danube. An account of the radical changes made, including the elimination of curves, shortening this branch eleven miles, deepening the channel and removing obstructions, etc. 2500 w. Engr, Lond—Sept. 20, 1907. No. 87345 A.

Docks.

New Graving Docks and Shipyard on the River Tees. Illustrates and describes this dock scheme and proposed improvements. 1800 w. Engr, Lond—Oct. 18, 1907. No. 87964 A.

Dredges.

A New Flexible Connection for Suction Pipes of Dredges. Illustrates and describes a device designed to replace rubber hose and obviate its disadvantages. 1000 w. Eng Rec—Oct. 19, 1907. No. 87789.

Dredging.

A Sea-Going Bucket Dredge. Dr. Alfred Gradenwitz. Illustrated description of the Fedor Solodoff, a suction-pipe type with floating conduits, for service on the River Don and on the Sea of Azoff. 1600 w. Int Marine Engng—Nov., 1907. No. 87810 C.

The Excavation of the West Neebish Channel in the St. Mary's River, Michigan. Illustrated description of important work to facilitate navigation. 3000 w. Eng News—Oct. 10, 1907. No. 87546.

Flood Prevention.

Floods and Means of Their Prevention in Our Western Rivers. T. P. Roberts. Discussion of this paper. 17800 w. Pro Engrs' Soc of W Penn—Oct., 1907. No. 87860 D.

Flood Protection.

Flood Protection in Hungary (Der Hochwasserschutz in Ungarn). E. von Kvassay. Discusses the importance of regulating the rivers of Hungary and the works which have been undertaken for that purpose. Ills. 4400 w. Serial. 1st part. Oest Wochenschr f d Oeffent Baudienst—Sept. 7, 1907. No. 87642 D.

Groynes.

See under same title under Construction.

Havre.

The Development of the Port of Havre. Map, plans and description of the works carried out under the 1895 scheme, and the proposals for future extensions. 3000 w. Engr, Lond—Oct. 11, 1907. No. 87839 A.

Levee.

New Methods for Closing a Crevasse in a Mississippi River Levee: the Live Oak Crevasse, Louisiana. Mary W. Mount. The manner of effecting the closing is illustrated and described. 2200 w. Eng News—Oct. 24, 1907. No. 87873.

Mississippi Valley.

The Deep-Water Route from Chicago to the Gulf. Map and illustrations, with information and discussion of proposed improvements of the waterways of the Mississippi valley. 4000 w. Nat Geog Mag—Oct., 1907. No. 87710 C.

Nile.

The English on the Nile (Die Engländer am Nil). Franz Ritter. A description of the problem of regulating and utilizing the waters of the Nile and the works accomplished and in progress. 3600 w. Serial. 1st part. Zeitschr d Oest Ing u Arch Ver—Sept. 20, 1907. No. 87647 D.

Tennessee River.

Water-Power: By-Product of Navigation Improvement. Charles H. Baker. An account of the improvement of the Tennessee River and the project for the establishment of navigation over shoals, and the development of water-power in connection. Ills. 3000 w. Mfrs' Rec—Oct. 10, 1907. No. 87533.

U. S. Waterways.

President Roosevelt on Improvement of National Waterways. Extracts from address at Memphis, with editorial. 4500 w. Eng News—Oct. 10, 1907. No. 87553.

MISCELLANY.

Computing.

On Computing. Russell Tracy Crawford. Presents practical suggestions, some of which have not previously appeared. 3500 w. Cal Jour of Tech—Oct., 1907. No. 87982.

ELECTRICAL ENGINEERING

COMMUNICATION.

Oscillators.

On the Elementary Theory of Electric Oscillators. J. A. Fleming. An explanation of the theory of the oscillator for a few of the simple and most practically important forms of open and closed radiating circuit. 3500 w. Elect'n, Lond—Sept. 27, 1907. Serial. 1st part. No. 87514 A.

Radio-Telephony.

Long Distance Wireless Telephony. R. A. Fessenden. An account of the system in operation between Brant Rock and New York—nearly 200 miles—discussing the possibilities. Ills. 2200 w. Elect'n, Lond—Oct. 4, 1907. No. 87590 A.

Relays.

Alternating-Current Relays on the Ferraris Principle (Zur Frage der Wechselstrom-Relais nach dem Ferraris-Prinzip). R. David and Dr. K. Simons. A mathematical paper discussing their efficiency. Ills. 2000 w. Elektrotech Zeitschr—Sept. 26, 1907. No. 87661 D.

Telegraphone.

The Latest Forms of the Telegraphone (Die Neuesten Formen des Telegraphons). E. Hytten. Illustrated description of the latest applications of Poulsen's device. 1800 w. Elektrotech Zeitschr—Sept. 5, 1907. No. 87656 D.

Telephone Siwtchboards.

Multiple Telephone Switchboards with Central Battery and Luminous Signals (Les Commutateurs Téléphoniques Multiples à Batterie Centrale et à Signaux Lumineux). A. Le Vergnier. Principally devoted to a description of installations on the Siemens & Halske system. Ills. 3800 w. Génie Civ—Sept. 28, 1907. No. 87612 D.

Telephotography.

Recent Developments in Picture Telegraphy. Dr. Alfred Gradenwitz. An illustrated report of tests made of the Korn apparatus, and description of the invention of H. Carbonelle. 1000 w. Sci Am—Oct. 26, 1907. No. 87878.

Wire Testing.

Wire Testing. L. M. Jones. Briefly considers the methods which have given satisfactory results. 2000 w. R R Gaz—Sept. 6, 1907. No. 87901.

DISTRIBUTION.

Insulators.

The Design and Testing of Electrical Porcelain. Dean Harvey. Describes methods and processes in preparing porcelain insulators. Ills. 3000 w. Elec Jour—Oct., 1907. No. 87561.

DYNAMOS AND MOTORS.

Auxiliary Poles.

The Influence of Auxiliary Poles on the Operation of Dynamos and Motors (Ueber den Einfluss der Wendepole auf die Wirkungsweise von Generatoren und Motoren). Hermann Zipp. A mathematical paper. Ills. 3500 w. Elektrotech u Maschinenbau—Sept. 8, 1907. No. 87632 D.

D. C. Motors.

Some Notes on the Care and Working of Continuous-Current Motors. Deals

We supply copies of these articles. See page 559.

with the brushes, commutators, and points requiring attention in connection with these motors. 1200 w. Prac Engr—Sept. 20, 1907. No. 87331 A.

Induction Motors.
The Heyland Method of Starting and Regulating, and of Compensation for Phase Shifting of Induction Motors (Das Heylandsche Verfahren zum Anlassen und Regulieren und zur Kompensation der Phasenverschiebung von Induktionsmotoren). F. G. Wellner. Ills. 2500 w. Zeitschr d Ver Deutscher Ing—Sept. 7, 1907. No. 87649 D.

Starters.
A New Automatic Method of Starting Motors. Dr. Martin Kallmann. Abstract from *Elek. Zeit.* An account of the application of iron resistances to the starting of motors, describing various forms. 1200 w. Elect'n, Lond—Oct. 11, 1907. No. 87835 A.

Synchronous Motors.
Excitation Characteristics of the Synchronous Motor. A. S. Langsdorf. A mathematical study stating the assumptions. 800 w. Elec Wld—Oct. 19, 1907. No. 87804.

Turbo-Generators.
Continuous-Current Turbo-Generators (Gleichstrom-Turbogeneratoren). H. I. C. Beyer. A discussion of the points of design in which they differ from the normal type of dynamo. Ills. 4000 w. Serial. 1st part. Elektrotech u Maschinenbau—Sept. 29, 1907. No. 87635 D.

ELECTRO-CHEMISTRY.

Electro-Metallurgy.
Applied Electro-Metallurgy Up to the End of 1906. John B. C. Kershaw. This second article gives a review of the ferroalloys, iron and steel, lead, nickel, siloxicon, silicon, sodium, tin, and zinc processes. Ills. 3000 w. Engineering Magazine —Nov., 1907. No. 87968 B.

Phosphorus.
Production of Phosphorus in the Electric Furnace. From an article by George W. Stose, in a recent publication of the U. S. Geol. Surv. Reviews old methods of making phosphorus, and gives information in regard to recent processes. 2200 w. Elec-Chem & Met Ind—Oct., 1907. No. 87503 C.

ELECTRO-PHYSICS.

Hysteresis.
Magnetic Hysteresis Phenomena. M. O. Bolser. Describes a device for showing the cyclic change in reluctance when a portion of a transformer core is revolved synchronously with the alternation of the flux. 1100 w. Elec Wld— Sept. 28, 1907. No. 87320.

Induction Coils.
The Secondary Current of the Induc-

tion Coil. H. Clyde Snook. A discussion of the currents flowing in the secondary of an induction coil, as observed by means of a Duddell High Frequency Oscillograph. Ills. 2200 w. Jour Fr Inst—Oct., 1907. No. 87759 D.

Oscillations.
The Poulsen Arc as a Means of Obtaining Continuous Electrical Oscillations. J. A. Fleming. Abstract of a paper read before the Phys. Soc. Describes experiments on the Poulsen arc, and discusses results. 3500 w. Elect'n, Lond—Sept. 20, 1907. No. 87339 A.

GENERATING STATIONS.

Blast Furnace Gas.
The Economic Distribution of Electric Power from Blast Furnaces. B. H. Thwaite. Read before the Ir. & St. Inst. at Vienna. Describes an application of the pooling system proposed by the writer. Ills. 3500 w. Mech Engr, Lond— Sept. 28, 1907. No. 87509 A.

Central Stations.
Recent Developments in Steam Turbine Power Station Work, with Special Reference to the Fort Wayne & Wabash Valley Traction Company Spy Run Station. J. R. Bibbins. Describes a particular design of station for steam turbine power, the construction and operation. Ills. 6800 w. St Ry Jour—Oct. 19, 1907. No. 87818.

The Park Royal Generating Station of the Great Western Railway, London. Illustrated description. 3200 w. Elec Wld —Oct. 19, 1907. No. 87803.

Spring Street Station of the Columbus Railway and Light Company. Illustrates and describes a 6,500 kw. station for railway service. 1500 w. Engr, U S A— Oct. 15, 1907. No. 87731 C.

Operating Features of the Fifty-ninth Street Station of the Interborough Rapid Transit Company. Describes the operation of a station built on the unit system. 3500 w. St Ry Jour—Oct. 12, 1907. No. 87781.

Gas-Power Central Station of the Duquesne Light Co., Pittsburg, Pa. Norman C. MacPherson. Read before the Pittsburg Branch of the A. I. E. E. Detailed description of the equipment and operation. 4000 w. Eng Rec—Oct. 26, 1907. No. 87938.

The New Power Plant of the Lowell (Mass.) Electric Light Corporation. Illustrated description of the new turbine power station under construction. 2200 w. Elec Wld—Oct. 5, 1907. No. 87485.

A Central Power Station in the Whitewood, S. D., Mining District. Illustrates and describes a plant at Deadwood, built to supply electric current for power in the mines and mills in its vicinity, and for lighting in Deadwood, Lead and other camps. 1800 w. Eng Rec—Sept. 28, 1907. No. 87311.

Power Plant of the Gulfport and Mississippi Coast Traction Co. Earl F. Scott. Illustrated description of a plant designed to furnish light and power for several towns and for interurban electric traction. 2200 w. Eng Rec—Oct. 19, 1907. No. 87787.

A Modern Central Station (Ein Modernes Elektrizitätswerk). Dr. Müllendorff. Illustrates and describes the steam-turbine station of the city of Eberswalde, Germany. 3500 w. Glasers Ann—Sept. 1, 1907. No. 87637 D.

See also Street and Electric Railways, under Long Island R. R., New York Central R. R. and West Jersey and Seashore R. R.

Construction.

Modern Power Station and Electrical Construction. W. A. Haller. Deals especially with the constructional features pertaining to the installation of turbo-generators and their supplementary apparatus. Ills. General discussion. 6000 w. Jour Assn of Engng Socs—Sept., 1907. No. 87862 C.

Hydro-Electric.

Three Low-Head Hydro-Electric Developments in Michigan. Illustrates and describes hydro-electric developments of the Grand Rapids-Muskegon Power Co. in the southern and western part of Michigan. 7500 w. Eng Rec—Oct. 19, 1907. Serial. 1st part. No. 87783.

Loch Leven Water Power Works. Illustrated description of a scheme to furnish electric power for the manufacture of aluminium. 1000 w. Engr, Lond—Oct. 4, 1907. No. 87597 A.

Recent Water-Power Plant in Switzerland (Neue Wasserkraftanlagen der Schweiz). S. Herzog. A series of articles illustrating and describing a number of hydro-electric developments, the first being a plant on the Rabiusa at Chur. 2000 w. Serial. 1st part. Zeitschr f d Gesamte Turbinenwesen—Sept. 10, 1907. No. 87636 D.

The Caffaro Hydro-Electric Plant (Ancora dell' Impianto Idroelettrico del Caffaro). S. Herzog. Illustrated detailed description of this Italian plant. 2100 w. Serial. 1st part. L'Industria—Sept. 15, 1907. No. 87613 D.

Isolated Plant.

The Electric Light and Power Plant in the "Warehouse of the West" in Berlin (Die Elektrische Licht- und Kraftanlage im "Kaufhaus des Westens" zu Berlin). R. Zaudy. Illustrates and describes a plant of about 650 horse-power. 2500 w. Elektrotech Zeitschr—Sept. 26, 1907. No. 87660 D.

London.

London's Electricity Supply. Explains conditions in London and briefly discusses the schemes proposed at different times and rejected. 2800 w. Engr, Lond—Sept. 27, 1907. No. 87526 A.

Parallel Operation.

The Influence of Damping on the Parallel Operation of Alternating-Current Machines (Der Einfluss der Dämpfung auf die Schwingungsvorgänge in Parallel Geschalteten Wechselstrommaschinen). Fritz Emde. A mathematical discussion. Ills. 4800 w. Elektrotech u Maschinenbau—Sept. 22, 1907. No. 87634 D.

Sub-Stations.

The Connection of Cables to Switchboards in Underground Sub-Stations. W. Pleasance. Brief notes on some of the small but important details of underground sub-stations. Ills. 800 w. Elec Engr, Lond—Sept. 20, 1907. No. 87333 A.

Switchboards.

Switchboards for Small Stations. E. T. Mug. Suggestions for the design and arrangement. Diagrams. 1400 w. Elec Rev, N Y—Oct. 5, 1907. No. 87440.

The New Switchboard of the International Electric Company in Vienna (Die Neue Schaltanlage der Internationalen Elektrizitäts-Gesellschaft in Wien). Oskar Spitzer. Illustrated description. 1600 w. Elektrotech u Maschinenbau—Sept. 1, 1907. No. 87631 D.

Turbo-Generators.

Curtis Turbines in Railway Service. August H. Kruesi. Read before the Am. St. & Int. Ry. Engng. Assn. Reports results of three recent tests and discusses the advantages of turbine generators. 2500 w. St Ry Jour—Oct. 19, 1907. No. 87817.

LIGHTING.

Arc Lamps.

Invention of the Enclosed Arc Lamp. L. B. Marks. Gives the history of the progress in solving this problem. Ills. 1200 w. Sibley Jour of Engng—Oct., 1907. No. 87865 C.

Cities.

Public Lighting Service, Corporate and Municipal. Judson H. Boughton. A study of electric stations, giving unit costs of plant and operation from many installations. 3500 w. Engineering Magazine—Nov., 1907. No. 87972 B.

Illumination.

Calculation of Illumination from Inclined Light Sources. J. S. Codman. Mathematical. 800 w. Elec Rev, N Y—Oct. 5, 1907. No. 87439.

Incandescent Lamps.

Developments in Electric Incandescent Lamps. Leon Gaster. Reviews some of

We supply copies of these articles. See page 559.

the more important improvements. 3800 w. Cent Sta—Oct., 1907. No. 87563.

See also under Nernst Lamp and Tantalum Lamp.

Mechanical Equivalent.

The Mechanical Equivalent of Light. Reviews the measurements of Dr. C. V. Drysdale and A. C. Jolley, their methods, and the results. 1400 w. Engng—Oct. 18, 1907. No. 87959 A.

Nernst Lamps.

The Value of the Nernst Lamp to the Central Station. A. E. Fleming. Explains some of the advantages of the Nernst system. 2200 w. Cent Sta—Oct., 1907. No. 87562.

Singer Building.

The Electric Illumination and Wiring of the Singer Building. Illustrated description of the plant and the installation. 3500 w. Elec Age—Oct., 1907. No. 87868.

Tantalum Lamp.

The Deformation of the Tantalum Filament when Used with Alternating Current. F. Stroude Stud. Brief note on the observed deformation and its cause, with suggestion for improving the tantalum lamp. 700 w. Elec Engr, Lond—Sept. 20, 1907. No. 87332 A.

MEASUREMENT.

Alternating Current.

The Application of the Alternating-Current Arc in Electrical Measurements (Die Verwertung des Lichtbogen-Wechselstromes in der Messtechnik). C. Heinke. A mathematical paper discussing the application of the high-frequency arc to the determination of alternating-current coefficients. Ills. 4000 w. Elektrotech Zeitschr—Sept. 19, 1907. No. 87659 D.

Condensers.

Simultaneous Measurement of the Capacity and Power Factor of Condensers. Frederick W. Grover. Deals with the determination of the power factor of condensers to be used for precision measurements of inductance and capacity. 13800 w. Bul Bureau of Stand—Aug., 1907. No. 87701 N.

Insulation Resistance.

A Method of Testing D. C. Networks for Insulation Resistance During Working. Daniel Shirt. Describes test, illustrating by example. 600 w. Elec Rev, Lond—Oct. 11, 1907. No. 87831 A.

Meter Bridge.

A Home-Made Slide-Wire Electrical Bridge. Henry C. Ter Meer. Illustrates and describes a method of constructing a modified meter bridge. 1000 w. Sci Am Sup—Oct. 19, 1907. No. 87743.

Motor Testing.

The Commercial Testing of Series Motors. Gives schemes of connections for

testing, with particular reference to those designed for diverter control. 700 w. Elec Rev, Lond—Sept. 20, 1907. No. 87338 A.

Permeability.

The Permeability of Alloyed Irons at High Flux Densities. E. A. Watson. Describes a method of testing small samples at high flux densities, and gives results obtained on alloyed irons at densities such as commonly occur in armature teeth. 1500 w. Elect'n, Lond—Oct. 18, 1907. No. 87955 A.

Photometry.

On the Determination of the Mean Horizontal Intensity of Incandescent Lamps. Edward P. Hyde and F. E. Cady. Gives method and results of recent experiments to study the accuracy of the rotating lamp method. 4000 w. Bul Bureau of Stand—Aug., 1907. No. 87700 N.

Resistance Coils.

Resistance Coils and Comparisons. C. V. Drysdale. Read before the British Assn. at Leicester. An illustrated account of the principal types of coils and methods of measurement is given, with description of some new forms. 1000 w. Elect'n, Lond—Sept. 27, 1907. Serial. 1st part. No. 87516 A.

Transformer Testing.

On Transformer Indicator Diagrams. T. R. Lyle. Abbreviated paper, read before the Physical Soc. Describes the use of the author's wave tracer and analyzer. 2500 w. Elect'n, Lond—Oct. 11, 1907. No. 87834 A.

Units.

A New Determination of the Ratio of the Electromagnetic to the Electrostatic Unit of Electricity. E. B. Rosa and N. E. Dorsey. Gives work undertaken to make a new determination of the value of V, using the method of capacities. Ills. 21000 w. Bul Bureau of Stand—Aug., 1907. No. 87702 N.

Voltage Drop.

A Graphical Method of Determining the Voltage Drop in Power Distribution Systems. T. L. Kolkin. Explains method, dealing with cases in common practice. 1500 w. Elec Rev, Lond—Oct. 11, 1907. No. 87832 A.

TRANSMISSION.

Cable Breakdown.

The Accident to Sydney Tramway Feeders. An illustrated account of a breakdown caused by the dragging of a feeder cable by a ship's anchor. 1200 w. Elec Rev, Lond—Oct. 4, 1907. No. 87589 A.

Cables.

Practical Notes on Plumbed Joints. W. Pleasance. Instructions in regard to

We supply copies of these articles. See page 559.

lead-covered cables. 1800 w. Elec Engr, Lond—Oct. 4, 1907. No. 87588 A.

Conduits.

A Method of Laying Vitrified Conduits for Electrical Cables, and a Collapsible Core for Forming Conduits in Concrete Without the Use of Vitrified Ducts. F. Lavis. Describes work in the Pennsylvania Railroad tunnels and methods used. Ills. 1200 w. Eng News—Oct. 3, 1907. No. 87423.

Direct Current.

The Fleury Direct-Current Transmission System. D. Kos. A discussion of the principal features and its possibilities. Also editorial. 8500 w. Elec Wld—Oct. 26, 1907. No. 87949.

Grounded Neutral.

Earthing the Neutral Point. E. V. Shaw. Explaining the advantages of the automatic earth switch invented by the writer. 1200 w. Elec Rev, Lond—Oct. 18, 1907. No. 87954 A.

The Grounded Neutral, With and Without Series Resistance, in High-Tension Systems. Paul M. Lincoln. Discusses, from the viewpoint of the operating engineer, the advantages and disadvantages of a grounded neutral. 4000 w. Pro Am Inst of Elec Engrs—Sept.; 1907. No. 87-719 D.

Experience with a Grounded Neutral on the High-Tension System of the Interborough Rapid Transit Company. George I. Rhodes. Explains conditions which led to the grounding of the neutral, and reports successful results. 1500 w. Pro Am Inst of Elec Engrs—Sept., 1907. No. 87720 D.

The Grounded Neutral. F. G. Clark. Gives reasons for grounding in the design of a high-tension installation, the reasons for and against the introduction of resistance, and facts relating to the operation. 2000 w. Pro Am Inst of Elec Engrs—Sept., 1907. No. 87721 D.

High Tension.

Recent Italian High-Tension Experiments. Frank C. Perkins. Brief illustrated account of interesting and instructive experiments with pressures ranging from 160,000 volts to 320,000 volts. 1000 w. Prac Engr—Sept. 20, 1907. No. 87-330 A.

Line Constants and Abnormal Voltages and Currents in High-Potential Transmissions. Ernst J. Berg. Gives equations for determining these constants, and methods of calculating phenomena by their help. 3500 w. Pro Am Inst of Elec Engrs—Sept., 1907. No. 87718 D.

Losses.

The Losses in Heavy Alternating Current Cables. Abstract translation of an article by E. Stirnimann in the *Elektro-*

technische Zeit, describing experiments on the increased resistance offered to alternating currents by cables of large section. 1000 w. Elec Rev, Lond—Oct. 11, 1907. No. 87830 A.

Minneapolis.

System at Minneapolis for Distributing the Energy Transmitted from Taylor's Falls. Illustrates and describes the provisions made in Minneapolis for receiving and distributing the electrical energy. 1800 w. Elec Wld—Oct. 5, 1907. No. 87486.

Niagara.

The Transmission Plant of the Niagara, Lockport and Ontario Power Company. Ralph D. Mershon. Illustrated detailed description of line construction and stations. 4500 w. Pro Am Inst of Elec Engrs—Sept., 1907. No. 87717 D.

Phase Transformation.

Unbalanced Loads in Two-Phase to Three-Phase Transformation. Bernh. F. Jakobsen. Records a general solution of a problem in phase transformation. 1000 w. Elec Wld—Oct. 12, 1907. No. 87709.

Resistance.

No-Load and Short-Circuit Resistance of Alternating-Current Cables (Leerlauf- und Kurzschlusswiderstand von Wechselstromkabeln). Carl Breitfeld. A mathematical and theoretical discussion. Ills. 4500 w. Elektrotech u Maschinenbau—Sept. 15, 1907. No. 87633 D.

Sweden.

The Tofwehult Westerwik Transmission System, Sweden. Arvid Westerberg. An illustrated article outlining interesting details of an installation for transmitting energy from the waterfall at Tofwehult, where small power is utilized with regard to low charges. 1500 w. Elec Wld—Sept. 28, 1907. No. 87319.

Thury System.

See under Direct Current.

Transformers.

Wiring and Connections for Constant Potential Transformers. George A. Burnham. Diagrams and description of the core-type and shell-type transformers and the electrical connections. 2800 w. Elec Wld—Oct. 5, 1907. No. 87488.

Wire Spans.

Tension and Sag in Wire Spans. Harold Pender. Gives charts devised to obtain a graphical solution of equations deduced by the author in an article published Jan. 12, 1907. 1300 w. Elec Wld—Sept. 28, 1907. No. 87321.

The Suspension of Wires (Ueber den Durchhang von Freileitungen). G. Nicolaus. Discusses particularly telegraph lines, showing the tension which should be placed on the wires to prevent too free swing, with formula. Ills. Tables and

We supply copies of these articles. See page 559.

curves. 3200 w. Serial. 1st part. Elektrotech Zeitschr—Sept. 12, 1907. No. 87658 D.

MISCELLANY.

Electrical Phenomena.
A Presentation of Two Theories of Electrical Phenomena (Deux Théories des Phénomènes Electriques en Présence). P. Harckman. An explanation of the conflicting electronic and ionic theories of electrical phenomena. Ills. 6000 w. Soc Belge d'Elec'ns—Sept., 1907. No. 87600 E.

Insulating Varnishes.
Concerning Insulating Varnishes. A. R. Warnes. Gives results of investigations made to find a reliable insulating varnish. 2500 w. Elec Rev, Lond—Sept. 20, 1907. No. 87337 A.

Possibilities.
The Possibilities of Electrical Development. R. Borlase Matthews. Read before the Birmingham & Dist. Elec. Club. A discussion of the methods that should be employed in England. 6000 w. Elec Engr, Lond—Oct. 11, 1907. No. 87828 A.

INDUSTRIAL ECONOMY

Apprenticeship.
Industrial Education. W. B. Russell. Gives details of the apprentice system of the New York Central lines. 4500 w. Pro Am Soc of Mech Engrs—Oct., 1907. No. 87772.

Cost Determination.
Cost Determining. W. B. Koller. Describes a system of interest to the large as well as the small manufacturer. 1200 w. Ir Age—Oct. 10, 1907. No. 87535.

Higher Costings. Arthur Winder. A criticism of systems of determining costs and an explanation of a better method. 4000 w. Cassier's Mag—Oct., 1907. No. 87389 B.

Education.
College and Apprentice Training. John Price Jackson. Discusses the relation of the student engineering courses in the industries to the college technical courses. 6000 w. Pro Am Soc of Mech Engrs—Oct., 1907. No. 87773 C.

The Russell Sage Laboratory. Rensselaer Polytechnic Institute. Describes the mechanical and electrical equipment. 1500 w. Eng News—Oct. 31, 1907. No. 87986.

The Training of Engineers on the "Sandwich" System. L. D. Coueslant. Explains the working of the apprenticestudentship scheme of the Sunderland Technical College. 2500 w. Engng—Oct. 11, 1907. No. 87837 A.

The Electrotechnic Institute of the Technical University in Carlsruhe (Baden). Stanley P. Smith. Gives a general idea of the curriculum and of the examinations. Ills. 2500 w. Elect'n, Lond—Oct. 11, 1907. No. 87833 A.

The Education of Engineers for Operation and Management in the Technical High Schools of Germany (Die Wirtschaftliche Ausbildung der Maschinen-Ingenieure für Betrieb und Verwaltung an den Technischen Hochschulen Deutschlands). Walter Conrad. 3400 w. Serial. 1st part. Zeitschr d Oest Ing u Arch Ver—Sept. 6, 1907. No. 87644 D.

Electric Railways.
The Electric Railway Situation of Today. Henry J. Pierce. Address before the Street Railway Assn. of the State of N. Y., and the Empire State Gas and Electric Assn. 3500 w. Elec Ry Rev—Oct. 5, 1907. No. 87490.

Mexican Labor.
Present Labor Conditions in Mexico. A. H. Tays. An illustrated article describing the characteristics of the Mexican peon, his mode of life and efficiency as a workman, and the bad results following the raising of wages in Sonoro. 2200 w. Eng & Min Jour—Oct. 5, 1907. No. 87461.

Municipal Ownership.
Municipal Ownership of Public Utilities. John W. Hill. From a paper before the Cent. States W.-Wks. Assn. Arguments opposed to municipal ownership and operation. 3000 w. Eng News—Oct. 10, 1907. No. 87549.

Public Utilities.
Control by State Commissions. Alexander C. Humphreys. From an address at Madison Sq. Garden, N. Y., Oct. 1. Discusses the attempts to reform public service corporations. 2500 w. R R Gaz —Oct. 4, 1907. No. 87922.

See also under Regulation, Railway Engineering, Miscellany.

Shop Management.
See under Management, Mechanical Engineering, Machine Works and Foundries.

Steamship Lines.
The Consolidated Steamship Lines. Information concerning the recent consolidation by Charles W. Morse, of the lines on the Atlantic and Gulf coast. Ills. 1200 w. R R Gaz—Oct. 4, 1907. No. 87925.

We supply copies of these articles. See page 550.

Steel Trade.

Combination and Competition in the Steel Trade. T. Good. A review of the market and of manufacturing conditions in England, Germany, and America. 3500 w. Engineering Magazine—Nov., 1907. No. 87966 B.

Trade Cycles.

Trade Cycles. Herbert Foster. Considers the periodic rise and fall in prices, wages, etc., and gives charts tending to show the operation of a general law. 1500 w. Ir Age—Oct. 24, 1907. No. 87876.

MARINE AND NAVAL ENGINEERING

American Navy.

The American Fleet from an English Point of View. Archibald S. Hurd. Discusses the strength and weaknesses of the American navy. Considers it the second greatest naval force in the world, but lacking in officers and men. Ills. 2500 w. Cassier's Mag—Oct., 1907. Serial. 1st part. No. 87384 B.

Armor.

Modern Armor and Armor-Piercing Projectiles. H. J. Jones. Discusses some of the problems relating to this subject, giving much information. 11000 w. Jour Am Soc of Nav Engrs—Aug., 1907. No. 87767 H.

Battleships.

The Results of the Russo-Japanese War in their Effect on the Further Development of War Ships (Die Ergebnisse der Russisch-Japanischen Seekrieges in ihrer Wirkung auf die Weiterentwickelung im Kriegsschiffbau). G. Neudeck. Ills. 6000 w. Serial. 2 parts. Schiffbau—Sept. 11 and 25, 1907. No. 87630, each D.

Condenser Corrosion.

The Experiments Made by Mr. Uthemann to Discover a Process for Preventing the Corrosion of Copper and Brass by Sea-Water Under the Conditions Found in the Surface-Condensers of Marine Steam-Engines. B. F. Isherwood. Trans. from *Le Génie Civil.* 7000 w. Jour Am Soc of Nav Engrs—Aug., 1907. No. 87761 H.

Electrical Equipment.

See under Mauretania.

Engine Bearings.

Difficulties Experienced with the Main Engine Bearings on Some of Our Latest Vessels. A. F. H. Yates. Gives data collected in the case of five vessels. 9000 w. Jour Am Soc of Nav Engrs—Aug., 1907. No. 87764 H.

Gas Engines.

The Present Status of Marine Gas Engineering. Peter Eyermann. Deals with the internal-combustion engine as at present used in all kinds of services on salt water as well as on fresh-water vessels. Ills. 6500 w. Jour Am Soc of Nav Engrs —Aug., 1907. No. 87762 H.

Gun Mounts.

Comparative Trials of Elevating Gears for Rapid-Fire Gun Mounts. John F. Meigs. An illustrated account of trials. 1500 w. Jour Am Soc of Nav Engrs—Aug., 1907. No. 87763 H.

Hydroplane.

The Crocco and Ricaldoni Hydroplane Boat. Illustrations, with brief description. 500 w. Engng—Oct. 4, 1907. No. 87593 A.

Mauretania.

The Electrical Equipment of the SS. "Mauretania." In this issue the system of power distribution, electrical generating plant, fans, motors for dismantling the turbines, cranes, lifts, etc., are described. 3000 w. Elect'n, Lond—Sept. 27, 1907. Serial. 1st part. No. 87515 A.

Model Basin.

A Brief Description of the Functions and Operation of an Experimental Model Basin. R. H. M. Robinson. Ills. 5000 w. Pro U S Naval Inst—Sept., 1907. No. 87757 H.

Revenue Cutter.

U. S. Revenue Cutter Itasca. Charles S. Root. Describes the U. S. S. Bancroft and its rebuilding and refitting for training cadets of the line and the engineer corps. Ills. 2800 w. Jour Am Soc of Nav Engrs—Aug., 1907. No. 87765 H.

Ship Building.

The Moran Company. H. Cole Estep. Illustrated description and history of this Pacific Coast ship building plant. 2000 w. Marine Rev—Oct. 3, 1907. No. 87436.

Steamboats.

Steamboat Architecture on the Western Rivers. Melville H. Kiel. Illustrates and describes the characteristic features of the boats of the Mississippi and its tributaries. 4000 w. Marine Rev—Oct. 3, 1907. No. 87437.

The Steamer Maryland. George Jenkins and A. E. Woodruff. Describes the hull and propelling machinery of a new twin screw steel transfer boat for service between Cape Charles and Norfolk. Ills. 1600 w. Int Marine Engng—Nov., 1907. No. 87806 C.

The Vessel of the Future. Arthur R. Siddell. Considers certain structural re-

We supply copies of these articles. See page 559.

forms for the improvement of large passenger steamers. 2800 w. Int Marine Engng—Nov., 1907. No. 87808 C.

New Swiss Lake Steamers. Trans. from *Zeit. des Ver. deut. Ing.* Illustrates and describes the Blümlisalp for Lake of Thun and the Rhein for the Lake of Constance. 1000 w. Int Marine Engng —Nov., 1907. No. 87807 C.

Steamships.
The Steamships Delaware and Pawnee. Charles S. Linch. Illustrates and describes these latest additions to the fleet of the Clyde Steamship Co. 2000 w. Int Marine Engng—Nov., 1907. No. 87811 C.
· See also under Turbine Steamers.

Submarines.
The Progress of the Submarine Boat. Editorial review of the report issued by the Special Board appointed by the United States Navy Department to make tests as to the mechanical efficiency of different types of submarine boats. 2000 w. Engng —Aug. 16, 1907. No. 87363 A.

The Submarine Fleet of France. Brief illustrated descriptions of types of submarines and submersibles belonging to France. 1200 w. Naut Gaz—Oct. 3, 1907. Serial. 1st part. No. 87480.

Suevic.
Repairing the Suevic. Illustrated description of the bow for this White Star liner which was wrecked on a reef off the Lizard. 800 w. Engr, Lond—Sept. 27, 1907. No. 87529 A.

Symington.
William Symington and the Beginnings of Steam Navigation. Robert Cochrane. An illustrated account of his work in the early development of steam navigation. 5500 w. Cassier's Mag—Oct., 1907. No. 87390 B.

Torpedo Boat.
Brazilian Torpedo Boat. Plate, with brief description. 300 w. Engr, Lond— Oct. 4, 1907. No. 87598 A.

Turbine Steamers.
The Turbine Steamship Camden. A. F. H. Yates. Illustrated description of the vessel with report of trials. 1200 w. Jour Am Soc of Nav Engrs—Aug., 1907. No. 87766 H.

Builders' Trials of Curtis Turbine Steamer Creole. Charles B. Edwards. Illustrated description of vessel and equipment, with report of trials. 2500 w. Jour Am Soc of Nav Engrs—Aug., 1907. No. 87768 H.

MECHANICAL ENGINEERING

AUTOMOBILES.

Bearings.
Bearing Castings for Automobiles. E. F. Lake. Remarks on alloys introduced to meet the requirements of the automobile industry, discussing the qualities of the ingredients used in the making of castings for bearings. 3300 w. Foundry —Oct., 1907. No. 87355.

Carbureters.
The Carbureter and Its Functions. Charles E. Duryea. Read before the Soc. of Auto. Engrs. A discussion of the requirements and of the faults of past and present designs, describing the Duryea carbureter. 5000 w. Automobile—Oct. 3, 1907. No. 87413.

Change Gears.
Automobile Change Gears and Their Journals. Henry Hess. Read before the Soc. of Auto. Engrs. An illustrated discussion of how such gears and journals should be designed, and an analysis of the construction of many of the leading types. 3800 w. Am Mach—Vol. 30, No. 44. No. 87976.

Clement.
The 25-35 H.P. Clement Car. The first part describes the engine and the clutch. Ills. 800 w. Autocar—Oct. 5, 1907. Serial. 1st part. No. 87583 A.

Commercial Vehicles.
Motor Cars for Municipal Work. Harry W. Perry. Illustrates various vehicles used for city work, such as gasoline, electric and steam motors for fire apparatus, street sprinklers and sweepers, refuse wagons, police patrol and ambulance, etc. 2300 w. Munic Jour & Engr—Oct. 2, 1907. No. 87383 C.

Swiss Motor Trucks (Schweizerische Motorlastwagen). A. Vogt. The first instalment of the serial illustrates and describes the "Soller" gasoline truck for heavy service. 2000 w. Serial. 1st part. Schweiz Bau—Sept. 28, 1907. No. 87629 B.

Construction.
Future Motor Car Construction. M. C. Krarup. Analyzes the present features and gives a forecast of probable developments. 4500 w. Ir Age—Oct. 10, 1907. No. 87536.

Electromagnetic Control.
Electromagnetic Control of Motor Vehicles. Describes the invention of Arnoldo Paolo Zani, and its application. Diagrams. 1500 w. Elec Engr, Lond—Oct. 11, 1907. No. 87829 A.

Hillman-Coatalen.
The 25 H.P. Hillman-Coatalen Car. Illustrated description. 1200 w. Autocar —Oct. 12, 1907. No. 87821 A.

We supply copies of these articles. See page 559.

Lubrication.

The Nature of Lubrication. Considers the perfectly lubricated bearing, the effect of varying load, and related subjects. 1600 w. Autocar—Oct. 19, 1907. No. 87951 A.

Napier.

The New 30-H.P. Napier Car. Illustrates and describes an interesting model. 1000 w. Auto Jour—Oct. 12, 1907. No. 87820 A.

Pennsylvania.

Type C Pennsylvania "50." Illustrates and describes the distinctive features of the 50 h.p. Pennsylvania car for 1908. 1600 w. Automobile—Oct. 3, 1907. No. 87414.

Pumps.

Idiosyncrasies of Friction - Driven Pumps—Their Prevention and Cure. Considers the troubles that may occur when this means of avoiding overheating is used on motor cars. Ills. 1700 w. Auto Jour—Sept. 28, 1907. No. 87505 A.

The Pittler Rotary Pump and Motor at the Olympia Exhibition. Illustrated detailed description of this rotary engine and its application to an omnibus chassis. 1800 w. Engng—Sept. 27, 1907. No. 87521 A.

Rover.

The Rover Cars. Illustrates and describes the leading characteristics of these cars. 2000 w. Auto Jour—Oct. 5, 1907. Serial. 1st part. No. 87582 A.

Siddeley.

The 30 H.P. Siddeley Car. Illustrated detailed description of this British-built car. 1000 w. Autocar—Sept. 28, 1907. Serial. 1st part. No. 87507 A.

Steering Gears.

The Design of an Automobile Steering Gear. E. W. Graham. Discusses some of the problems to be solved and the details of the design. Ills. 3500 w. Am Mach—Vol. 30, No. 40. No. 87380.

Test.

Test of a White Motor Car at the University of California. Arthur B. Domonoski. Brief description of the car with report of test. 2000 w. Cal Jour of Tech—Oct., 1907. No. 87983.

Tires.

Elastes — Latest Developments. Illustrated description of improvements made in this unpuncturable core for pneumatic tires. 900 w. Auto Jour—Sept. 28, 1907. No. 87506 A.

Valveless.

The 20-H.P. Valveless Car. Illustrated detailed description of an unusual car. 1700 w. Auto Jour—Oct. 5, 1907. Serial. 1st part. No. 87581 A.

Weigel.

The 25-H.P. Weigel Car. Illustrated description of a new model of medium power. 1600 w. Auto Jour—Sept. 21, 1907. No. 87329 A.

COMBUSTION MOTORS.

Exhaust Gases.

On the Gases Exhausted from a Petrol Motor. Prof. B. Hopkinson and L. G. E. Morse. Read before the British Assn., at Leicester. Describes investigations made of the conditions under which carbon monoxide is formed in internal-combustion motors, and related matters. 2500 w. Elec Engr, Lond—Oct. 4, 1907. No. 87587 A.

Gas Cleaning.

Gas Engines. Informal discussion on the best system for cleaning gas to be used in gas engines. Ills. 4000 w. Pro Am Soc of Civ Engrs—Sept., 1907. No. 87713 E.

Gas Engines.

Scope for the Use of Large Gas-Engines for Generating Electric Power in England. Leonard Andrews. Brief discussion giving capital and running cost. 1500 w. Elec Engr, Lond—Oct. 18, 1907. No. 87953 A.

See also under same title under Marine and Naval Engineering; under Central Stations, Electrical Engineering, Generating Stations; under Pumping Plants, Mechanical Engineering, Hydraulics.

Gas Engine Tests.

Researches on the Influence of Mixture Ratio on Gas-Engine Operation (Versuche an der Gasmaschine über den Einfluss des Mischungsverhältnisses). A. Nägel. Illustrates and describes tests, made on an 8 horse-power Körting engine and gives results. 13000 w. Serial. 2 parts. Zeitschr d Ver Deutscher Ing—Sept. 7 and 14, 1907. No. 87648, each D.

Gas Producers.

See under Fuels, Steam Engineering.

Gas Turbines.

New Gas Turbine and Centrifugal Air Compressor. Illustrations and information in regard to a new machine on trial in Paris. 1400 w. Sci Am Sup—Oct. 12, 1907. No. 87557.

Governing.

Speed Regulation of Internal Combustion Engines. Explains the method of governing in use on the Diesel crude oil engine. 1500 w. Engr, U S A—Oct. 1, 1907. No. 87327 C.

Ignition.

Contact Method of Gas Engine Ignition. E. J. Edwards. Mathematical discussion of the design of a sparking equipment. 1500 w. Elec Wld—Oct. 19, 1907. No. 87805.

Liquid Fuel.

Liquid Fuel for Internal-Combustion Engines. R. W. A. Brewer. Read before the Soc. of Engrs. Considers the use of heavy and light oils, vaporization, supply, price, etc. 5500 w. Mech Engr—Oct. 12, 1907. No. 87827 A.

Losses.
Thermal and Power Losses in Internal-Combustion Engines. A. H. Burnand. Describes experimental investigations made. 3000 w. Engng—Oct. 4, 1907. Serial. 1st part. No. 87592 A.

Oil-Engines.
Pumping with Oil-Engines in India. Editorial on the experiments carried out by the Government of Madras to ascertain the actual cost of raising water under different conditions. 1800 w. Engng—Sept. 20, 1907. No. 87344 A.

HEATING AND COOLING.

Blower Systems.
More Data Concerning Fan Heaters. E. T. Child. Reports investigations made of the steam requirements and temperatures obtained in pipe coil heaters. 2000 w. Met Work—Oct. 5, 1907. No. 87434. See also under Schoolhouses.

Buildings.
Heating and Ventilating the Commercial National Bank Building, Chicago. Plan and detailed description of plant for an 18-story office building. 3500 w. Eng Rec—Oct. 26, 1907. Serial. 1st part. No. 87939.

The Mechanical Ventilation and Warming of St. George's Hall, Liverpool. Charles R. Honiball. Describes an installation for warming and ventilating that has been in successful working since 1851. Ills. 4500 w. Heat & Vent Mag —Oct., 1907. No. 87747.

The Heating and Ventilating Plant of the Post-Office Savings Bank Bureau at Vienna (Die Heizungs- und Lüftungsanlage im k. k. Postsparkassenamtsgebäude in Wien). A. Puppel. Ills. 2700 w. Gesundheits-Ing—Sept. 21, 1907. No. 87-639 D.

Charts.
Charts Showing the Performance of Hot-Blast Coils. Burt S. Harrison. With explanatory notes. 700 w. Heat & Vent Mag—Oct., 1907. No. 87748.

Electrical Plant.
The Electrical Heating Plant of the Biltmore Estate. Charles E. Waddell. States the conditions which led to the installation of an electrical plant, giving an illustrated description, and an account of its performance in daily service. 1200 w. Elec Wld—Oct. 5, 1907. No. 87487.

Hot Air.
Improved Hot-Air Heating (Verbesserte Luftheizung). Herr Boehmer. Discusses hot-air heating as applied to small buildings, describing heaters and systems. Ills. 4500 w. Gesundheits-Ing—Sept. 21, 1907. No. 87640 D.

Hotel Ventilation.
Cooling Public Rooms in a Chicago Hotel. Illustrated detailed description of air cooling and ventilating plant for rooms in the Auditorium Annex at Chicago. 3000 w. Ice & Refrig—Oct., 1907. No. 87432 C.

Lecture Halls.
Heating and Ventilating the United States Naval Academy. Showing method of ventilating lecture halls having stepped floors. Ills. 1200 w. Heat & Vent Mag —Oct., 1907. No. 87746.

Schoolhouses.
Blower System of Schoolhouse Heating. Plans, with description. 1500 w. Dom Engng—Oct. 12, 1907. No. 87577.

Steam Heating.
Air Valves for Steam Heating Systems. W. H. Wakeman. The first of a series of illustrated articles giving details of why and where air valves are used. 1800 w. Dom Engng—Oct. 26, 1907. Serial. 1st part. No. 87918.

Heating System of the St. Francis Home, Detroit, Mich. Describes a scheme of heating combining hot blast for public rooms, and direct radiation for smaller rooms and for auxiliary heating. 4500 w. Eng Rec—Oct. 19, 1907. No. 87785.

Temperature Regulator.
Device for Automatically Controlling the Heating of a House. Illustrated description of the "sylphon regitherm," a device invented by Weston M. Fulton, explaining its operation. 1000 w. Sci Am Sup—Oct. 5, 1907. No. 87476.

Tunnel Ventilation.
Ventilation of the Battery Tunnels of the New York Subway Extension to Brooklyn. Plan and description of a system of mechanical ventilation under difficult conditions. The arrangements of the blower stations and the apparatus and duct systems are shown. 3000 w. Eng Rec—Oct. 5, 1907. No. 87444.

HYDRAULICS.

Centrifugal Pumps.
Centrifugal Pumps. E. F. Doty. Explains the necessary elements of the centrifugal pump, briefly reviewing its history and discussing the theory and its relation to present day practice. 2500 w. Engr, U S A—Oct. 1, 1907. No. 87326 C.

High Pressure Centrifugal Pumps (Les Pompes Centrifuges à Haute Pression). André Hallet. Gives a mathematical demonstration of why simple centrifugal pumps are not suitable for high lifts and describes and discusses the designs of several compound systems. Ills. Serial. 1st part. 4500 w. All Indus—Sept., 1907. No. 87607 D.

Dense-Air Pumping.
Operating Reciprocating Steam Pumps

We supply copies of these articles. See page 559.

by the Dense-Air System. Snowden B. Redfield. Outlines the advantages and disadvantages of the system, showing under what conditions it will prove economical. 4500 w. Compressed Air—Oct., 1907. No. 87558.

Friction.
See under Civil Engineering, Water Supply.

Pumping.
See under Oil Engines, Mechanical Engineering, Combustion Motors; and under Unwatering, Mining and Metallurgy, Mining.

Pumping Engines.
Prevention of the Short-Stroking of Direct-Acting Duplex Pumping Engines. A. P. Blackstead. An explanation of the different ways this may be accomplished. 600 w. Eng News—Oct. 31, 1907. No. 87989.

Pumping Plants.
A Weber Gas-Power Pumping Plant. Illustrated description of a novel equipment for operating the water and light plant of Rockyford, Colo. 1500 w. Ir Age—Oct. 24, 1907. No. 87877.

Pumps.
Pumps and Pumping Machinery. Joseph H. Hart. Suggestions on the choice of a type for a particular purpose, and the method of installation. 2300 w. Elec Rev, N Y—Oct. 19, 1907. No. 87790.

Sluice Flow.
The Flow of Water Through Sluices. T. E. Thain. Brief discussion of cases and features requiring consideration. 500 w. Prac Engr—Oct. 11, 1907. No. 87826 A.

MACHINE WORKS AND FOUNDRIES.

Ball Bearings.
Ball and Roller Bearings. J. F. Springer. Discusses the use of separators to reduce friction. 1200 w. Ir Age—Oct. 17, 1907. Serial. 1st part. No. 87737.

Bell Metal.
The Effect of Iron on Bell Metal and a Flux for Introducing It. Charles H. Poland. Describes the writer's experience. 1000 w. Brass Wld—Oct., 1907. No. 87857.

Bevel Gears.
Cutting Bevel Gears with a Rotary Cutter. H. P. Fairfield. Illustrated description of the process of forming the teeth of a bevel gear by milling them with a rotary cutter. 3000 w. Mach, N Y—Oct., 1907. No. 87469 C.

Boiler Making.
Pneumatic Tools for Boiler Shops. Charles Dougherty. Illustrated descriptions with discussion on the design, operation and care. 1500 w. Boiler Maker—Nov., 1907. Serial. 1st part. No. 88000.

Layout of a Large Y Connection. Emmet S. Hegerty. Illustrated description. 1500 w. Boiler Maker—Nov., 1907. No. 87998.

Boring Bars.
A Boring, Reaming, and Facing Bar—How It Was Made. E. John. An illustrated description of how a large bar was made. 1800 w. Am Mach—Vol. 30, No. 44. No. 87979.

Brass Casting.
Methods of Casting Yellow Brass. C. Vickers. Considers the treatment necessary to make yellow brass castings. 1800 w. Foundry—Oct., 1907. No. 87356.

Case-Hardening.
The Case-Hardening of Mild Steel. C. O. Bannister and W. J. Lambert. Read at Vienna meeting of the Ir. & St. Inst. Reports results of investigations made of the microstructure of cemented bars, of the depth of hardness, and a few cases of the carbon content. Micrographs. 1600 w. Ir & Coal Trds Rev—Sept. 27, 1907. No. 87706 A.

Case-Hardening. G. Shaw Scott. Abstract of paper read before the Ir. & St. Inst., at Vienna. An account of research work. 3000 w. Mech Engr, Lond—Sept. 28, 1907. No. 87511 A.

Castings.
The Foundry Department and the Department of Engineering Design. William A. Bole. Notes on the design and manufacture of large and complicated castings. 3800 w. Pro Am Soc of Mech Engrs—Sept., 1907. No. 87429.

Faults of Iron Castings. Forrest E. Cardullo. Considers the causes of blowholes, green sand and dry sand molding, the support of cores, etc. 1800 w. Mach, N Y—Oct., 1907. Serial. 1st part. No. 87468 C.

Cement Plants.
See under Electric Driving, Power and Transmission.

Drop-forging.
Drop-forging Methods and Interesting Work. Describes methods showing difficult forgings and their dies. Ills. 1800 w. Am Mach—Vol. 30, No. 41. No. 87545.

Dust Removal.
Dust Removal in a Brass Foundry. Walter B. Snow. Read at Philadelphia meeting of the Am. Brass Found. Assn. Illustrates and describes mechanical means of removing harmful dust. 2000 w. Heat & Vent Mag—Oct., 1907. No. 87749.

Exhibition.
The Engineering and Machinery Exhibition, Olympia. Joseph Horner. Observations on machines exhibited, with illustrations. 5500 w. Engng—Sept. 27, 1907. Serial. 1st part. No. 87520 A.

We supply copies of these articles. See page 559.

Miscellaneous Exhibits at Olympia. Illustrates and describes types of piston-rings for packing; winding engines for colliery work; steam and gas-engine indicators, etc. 1800 w. Engng—Oct. 18, 1907. No. 87956 A.

Files.

Examining and Testing Files. Oscar E. Perrigo. Considers the properties of files, especially the characteristics of good files, and the methods of testing. 2500 w. Mach, N Y—Oct., 1907. No. 87470 C.

Fly Cutters.

The Fly Cutter as a Reference Gage. An illustrated article showing additional novel uses of the fly cutter. 1000 w. Am Mach—Vol. 30, No. 43. No. 87870.

Foundries.

A Model Pacific Coast Iron Foundry. H. Cole Estep. An illustrated account of the rapid reconstruction of the plant of the Olympic Foundry Co., which was wrecked by a landslide. 1800 w. Foundry—Oct., 1907. No. 87352.

A Foundry for Bench Work. W. J. Keep and Emmet Dwyer. Brief illustrated description of the new foundry of the Michigan Stove Co., Detroit. 1000 w. Pro Am Soc of Mech Engrs—Oct., 1907. No. 87884.

The Iron Foundry of the Firm of H. Bovermann's Successors (Die Eisengiesserei der Firma H. Bovermann Nachf.). Illustrated description of this malleable and gray-iron foundry. 3500 w. Stahl u Eisen—Sept. 4, 1907. No. 87614 D.

Foundry Blower Practice. Walter B. Snow. Considers types of blowers and foundry practice. 6000 w. Pro Am Soc of Mech Engrs—Oct., 1907. No. 87886 C.

Driving a Foundry Fan. Walter J. May. Calls attention to points to bear in mind. 800 w. Prac Engr—Oct. 4, 1907. No. 87584 A.

Foundry Furnaces.

Melting Iron for Foundry Purposes. E. L. Rhead. Deals with appliances for melting. Ills. 2500 w. Mech Engr—Oct. 19, 1907. Serial. 1st part. No. 87952 A.

Foundry Practice.

Melting Iron for Castings. Bradley Stoughton. Discussion of cupola practice, methods of charging and calculating mixtures, tuyere area, fuel, etc. 5300 w. Foundry—Oct., 1907. No. 87351.

Early History of Iron Founding. Robert Buchanan. From a presidential address to the S. Staffordshire Ir. & St. Inst. 4800 w. Ir & Coal Trds Rev—Oct. 4, 1907. No. 87708 A.

Gear Design.

Design of Helical and Herringbone Gears. Charles H. Logue. Explains how helical gears are designed by using simple formulæ and their application to the herringbone gear. 2200 w. Am Mach—Vol. 30, No. 43. No. 87869.

Gears.

Some Memories of English Gears. W. H. Booth. Remarks on early methods of design and the forms and proportions used. 2500 w. Am Mach—Vol. 30, No. 40. No. 87382.

Interference in Involute Gears. C. C. Stutz. Gives a graphical demonstration, formulæ and diagrams. 800 w. Am Mach —Vol. 30, No. 41. No. 87544.

Gear Shaper.

A Novel Gear Shaper. Frank C. Hudson. Illustrates the machine, explaining its working. 700 . . Am Mach—Vol. 30. No. 40. No. 87351.

Grinding.

The Grinding of Woodworking Tools. C. C. Bosworth. Considers the grindstone and other abrasives, their peculiarities and their efficiencies. 3300 w. Woodcraft—Oct., 1907. No. 87402.

Economies and Advantages of Grinding. H. Darbyshire. Considers the selecting of proper wheels, keeping them in good condition, and using them at the right pressure and speed. 4500 w. Am Mach—Vol. 30. No. 42. Serial. 1st part. No. 87736.

Grinding Machines.

A New Universal Grinding Machine. Illustrated description of a machine being shown at the Olympia exhibition. 2200 w. Engr, Lond—Sept. 27, 1907. No. 87531 A.

Universal Grinding Machines (Universal-Schleifmaschine). Illustrated detailed description of machine and methods of use. 4500 w. Serial. 3 parts. Zeitschr f Werkzsugmaschinen u Werkzeuge—Sept. 5, 15 and 25, 1907. No. 87643 each D.

Lathes.

Brummond's Screw-Cutting Lathes. Illustrated description. 2000 w. Engng— Sept. 20, 1907. No. 87343 A.

Reciprocating Attachment for Engine Lathes. E. Beck. Illustrated detailed description. 1400 w. Am Mach—Vol. 30. No. 44. No. 87978.

Five Years' Progress in Engine Lathe Design. Illustrates and describes work of the Lodge & Shipley Machine Tool Co., Cincinnati, O. 1600 w. Ir Age—Oct. 31, 1907. No. 87995.

The Bullard Vertical Turret Lathe. Illustrated description of a machine designed to perform rapidly all kinds of turret lathe chucking work. 4000 w. Ir Age —Oct. 3, 1907. No. 87376.

Lead Scrap.

Tea Lead. Describes the method of the Chinese in the manufacture of tea-lead, and discusses its value as scrap lead. 1100 w. Brass Wld—Oct., 1907. No. 87858.

We supply copies of these articles. See page 559.

Management.

Profit Making in Shop and Factory Management. Charles U. Carpenter. This ninth article of a series discusses the up-building of a selling organization. 4500 w. Engineering Magazine—Nov., 1907. No. 87967 B.

The Drafting Room, Its Location and Work. Oscar E. Perrigo. The first of a series on shop management and cost keeping. 3300 w. Ir Trd Rev—Oct. 3, 1907. Serial. 1st part. No. 87405.

Shop Efficiency. H. W. Jacobs. Gives the efficiency records of various workmen where the bonus system has been adopted, giving the results of the system. 1600 w. Am Engr & R R Jour—Oct., 1907. No. 87415 C.

Material Handling.

Overhead Tramrails for the Foundry. A. W. Moyer. Illustrated description of uses made of this labor-saving device in foundry work. 2200 w. Foundry—Oct., 1907. No. 87354.

Milling Machine.

A New Brown & Sharpe Machine Tool. Illustrated description of the No. 3 vertical spindle milling machine with constant speed drive. 1800 w. Ir Age—Oct. 31, 1907. No. 87997.

Molding.

Molding a Cast Steel Runner. H. J. McCaslin. Illustrates and describes the method used. 1800 w. Foundry—Oct., 1907. No. 87353.

Molding a Screw Propeller in Loam. Joseph F. Hart. Illustrates and describes methods used. 1400 w. Am Mach—Vol. 30. No. 43. No. 87871.

Molding Curved Pipe in Dry Sand. Illustrates and describes a method of making curved pipe, vertically, in dry sand. 900 w. Am Mach—Vol. 30. No. 44. No. 87977.

Molding Sand.

Molding Sand. Alexander E. Outerbridge, Jr. Considers the improvement of molding sand by mechanical treatment, and reduction in cost. Ills. 2500 w. Pro Am Soc of Mech Engrs—Oct., 1907. No. 87769.

Paper Making.

See under Electric Driving, Power and Transmission.

Patterns.

Patterns for Repetition Work. E. H. Berry. Discusses requirements and conditions, and details of pattern work. Ills. 7500 w. Pro Am Soc of Mech Engrs—Oct., 1907. No. 87885 D.

Pneumatic Tools.

See under Boiler Making.

Power Values.

Power Values for Machine Tools in Groups. L. P. Alford. Giving the h. p.

required for 150 machine tools, showing how these values were obtained. 2000 w. Am Mach—Vol. 30. No. 44. No. 87975.

Roller Bearings.

See under Ball Bearings.

Saws.

Some Circular Saw Kinks and Short Cuts. James F. Hobart. Discusses the adaptability of the common saw table and gives an example of its efficient use for angular work. 1800 w. Wood Craft—Oct., 1907. No. 87403.

Screw Machines.

Speeds and Feeds for Screw Machine Work. C. L. Goodrich and F. A. Stanley. A record of speeds and feeds employed in the automatic screw machine. 1000 w. Am Mach—Vol. 30. No. 41. No. 87542.

Shops.

The Arthur Koppel Company's New Works. Illustrated description of the new plant for the manufacture of portable railways and industrial cars, and its equipment. 2500 w. Ir Age—Oct. 10, 1907. No. 87534.

Soldering.

Experiments on Soldering. Adolf Lippmann. From *Elek. Zeit.* Investigations on the degree of acidity of fluxes and the resulting effect on metals, and the testing of soldering materials. 3300 w. Elec Rev, Lond—Sept. 20, 1907. Serial. 1st part. No. 87336 A.

Speed Changing.

A Variable-Speed and Feed Mechanism. T. M. Lowthian. Illustrated detailed description of a new device for obtaining a multiplicity of changes for either speed or feed gears. 700 w. Am Mach—Vol. 30. No. 42. No. 87735.

Speed-Changing Devices on Machine Tools (Die Umlaufzahlenreihen bei Werkzeugmaschinen). Franz Adler. Illustrated description of various types with a discussion of their design. 8000 w. Serial. 1st part. Zeitschr d Ver Deutscher Ing—Sept. 21, 1907. No. 87651 D.

Tempering.

The Hardening of Steel. L. Demozay. Read Before the Ir. & St. Inst. at Vienna. Gives an analysis of the part played by the different factors under which the changes in the metal take place. 4500 w. Ir & Coal Trds Rev—Sept. 27, 1907. No. 87704 A.

Vaults.

The Largest Armor Plate Vault. Illustrated description of methods of manufacture at the works of the Bethlehem Steel Co. 1200 w. Ir Age—Oct. 31, 1907. No. 87996.

Worm Gear.

Collier's Ball Worm Gear. W. H. Booth. Illustrated description of a curious form of worm gear devised for the purpose of a direct drive in automobiles, as well as for a speed-change gear. 1000 w. Am Mach—Vol. 30. No. 42. No. 87734.

MATERIALS OF CONSTRUCTION.

Alloy Steels.

Vanadium Steel. E. F. Lake. Information in regard to the characteristics of a new steel for machine construction. Ills. 1800 w. Mach, N Y—Oct., 1907. No. 87467 C.

Vanadium Steel. Trans. from *Revue Industrielle.* Information concerning vanadium and the characteristics of vanadium steel. 1800 w. R R Gaz—Sept. 20, 1907. No. 87912.

Vanadium Steel. J. Kent Smith. Describes this metal giving its history and its beneficial effects on steel. General discussion. 8800 w. Pro Engrs' Soc of W Penn—Oct., 1907. No. 87861 D.

Boron Steels. Dr. Leon Guillet. Reports micrographic investigations and mechanical tests of normal and quenched case-hardened and annealed steels, giving theoretical and commercial conclusions. Plates. 3000 w. Jour Ir & St Inst—No. II, 1907. No. 87361 N.

Copper Steels. Pierre Breuil. A study of the alloys of iron and copper, the preparation of the ingots, their chemical composition, segregation, treatment; mechanical tests including tensile tests, shock tests on notched and unnotched bars, hardness and torsional tests, with investigation of steels containing 1 per cent of carbon. Ills. 17000 w. Jour Ir & St Inst—No. II, 1907. No. 87357 N.

Cast Iron.

Cast Iron as Cast and Heat Treated. W. H. Hatfield. A report of research work, discussing the decomposition of carbides, the nature of annealing carbon, and the influence of casting temperatures. Plates. 3800 w. Jour Ir & St. Inst—No. II, 1907. No. 87358 N.

Ferro-Alloys.

Special Ferro-Alloys for the Foundry. E. Houghton. Read before the British Found. Assn. Considers ferro-manganese and spiegel, silicon-spiegel, ferro-silicon, ferro-chrome and ferro-phosphorus, giving analyses of blast furnace, and of electrical furnace alloys. 3500 w. Ir & Coal Trds Rev—Sept. 20, 1907. No. 87350 A.

Hardened Steels.

Hardened Steels. Percy Longmuir. Read before the Ir. & St. Inst at Vienna. Indicates briefly some variables in the case of carbon steels. Ills. 1800 w. Mech Engr—Oct. 5, 1907. No. 87585 A.

The Testing of Hardened Steel with Consideration of Spherical Bodies (Prüfverfahren für gehärteten Stahl unter Berücksichtigung der Kugelform). R. Stri-

beck. Discusses the testing of ball bearings and hardened steels in general, giving results. Ills. 5000 w. Serial. 1st part. Zeitschr d Ver Deutscher Ing—Sept. 14, 1907. No. 87650 D.

Initial Stresses.

The Stresses Remaining in Metals after Cooling (Ueber Bleibende Spannungen in Werkstücken infolge Abkühlung). E. Heyn. A mathematical discussion of the initial stresses set up in metals during the contraction due to cooling. Ills. Serial. 2 parts. 8000 w. Stahl u Eisen—Sept. 11 and 18, 1907. No. 87615 each D.

Metallography.

Some Micro-Structural Considerations. John Magee Ellsworth and Thomas J. Fay. Extract from paper read before Soc of Auto. Engrs. On the defects revealed by micro-photographs, considering normal and abnormal steels, etc. 3500 w. Automobile—Oct. 17, 1907. No. 87754.

The Researches of G. Cartaud on the Passage of Metals from the Liquid to the Solid State (Les Recherches de G. Cartaud sur le Passage de l'Etat Liquide à l'Etat Solide). F. Osmond. Reviews his published work and the researches left uncompleted at his death. Illustrated by 72 microphotographs. 6700 w. Rev de Métal—Sept., 1907. No. 87603 E + F.

Mild Steel.

Further Experiments in the Ageing of Mild Steel. C. E. Stromeyer. Read at Vienna meeting of the Ir. & St. Inst. Gives tests of steel plates and results, with micrographs, and deductions drawn. 4800 w. Ir & Coal Trds Rev—Sept. 27, 1907. No. 87705 A.

Steel Impurities.

The Non-Metallic Impurities in Steel. E. F. Law. Gives results of observations, considering sulphide of iron, sulphide of manganese, silicate of iron, silicate of manganese and oxide of iron, discussing their effects. Plates. 3700 w. Jour Ir & St. Inst—No. II,. 1907. No. 87359 N.

Tool Steels.

A Study of Hardness in Tool Steels (Etude de la Dureté dans les Aciers à Outils de Tour). M. Demozay. A discussion of the effects of heat treatment on alloy steels of various compositions in their relation to hardness and durability. Ills. 5500 w. Rev de Métal—Sept., 1907. No. 87605 E + F.

MEASUREMENT.

Anemometer.

A Recording Anemometer. J. Rogers Preston. Prize paper read before the Inst. of Heat. & Vent. Engrs., London. Illustrated description. 2000 w. Met Work—Oct. 12, 1907. No. 87572.

Pyrometry.

Optical Pyrometry. Dr. Charles Féry.

We supply copies of these articles. See page 559.

Read before the British Assn. Illustrates and describes the Le Chatelier radio-pyrometer as modified by the writer. 1500 w. Engng—Oct. 18, 1907. No. 87961 A.

Steam Flow.

Cylinder Port Velocities. Jacob H. Wallace. Presents a method for determining the velocity of flow through the ports of a steam engine. 1500 w. Pro Am Soc of Mech Engrs—Oct., 1907. No. 87770.

Test Indicator.

Inspecting Tools with the Test Indicator. J. H. Boulet. Illustrates and describes a number of methods of testing and measuring various classes of tools. 1800 w. Am Mach—Vol. 30. No. 42. No. 87733.

Testing Materials.

See under Hardened Steels, Materials of Construction.

POWER AND TRANSMISSION.

Air Compressor.

1,200 Horse-Power Air Compressor. Illustrated description of a plant recently installed at the Seaham Colliery. 1500 w. Engr, Lond—Oct. 11, 1907. No. 87841 A.

Cranes.

Electrical Rolling Mill Transporters. Illustrates and describes electrically driven special transporting devices recently built in Austria. 1800 w. Engr, Lond— Oct. 18, 1907. No. 87965 A.

Notes on Boat and Anchor Cranes. Drawings showing details of electrically operated cranes for handling boats and anchors on warships, representing practice of the past ten years. 1000 w. Int Marine Engng—Nov., 1907. Serial. 1st part. No. 87809 C.

Crane Trolleys.

Formulas for Force Required to Move Crane Trolleys. John S. Myers. Briefly considers the conditions that should be taken into account and the values of factors in the calculations, developing formulæ. 1000 w. Mach, N Y—Oct., 1907. No. 87471 C.

Electrical Equipment in Cement Plants. J. B. Porter. Points out the special advantages of the electric drive in the manufacture of cement both in economy and flexibility. 2500 w. Cement—Sept., 1907. No. 87401 C.

Power Required to Drive Individual Machine Tools. H. B. Emerson. Gives data gathered by the writer. 800 w. Am Mach—Vol. 30. No. 41. No. 87543.

The Electric Driving of Paper Machines (Commande Electrique des Machines à Papier). M. Orban. Discusses the special difficulties in the way of applying electric power to the driving of paper machines, and the manner in which they

have been overcome. Ills. 4000 w. Soc Belge d'Elec'ns—Sept., 1907. No. 87601 E.

Elevators.

Electric Elevators. William Baxter, Jr. Illustrates and describes the Cutler-Hammer elevator controller. 1200 w. Engr, U S A—Oct. 1, 1907. No. 87328 C.

Elevators for Industrial Plants. E. R. Carichoff. Brief discussion of the electric, hydraulic and other systems of operation and means of meeting the requirements. 800 w. Elec Rev, N Y—Oct. 26, 1907. No. 87891.

High Speed Elevators. Charles R. Pratt. Describes the type of elevator selected for the new Singer building and the tower of the Metropolitan Life Insurance building in New York, both of which exceed 40 stories. Also short discussion by Orman B. Humphrey. 3000 w. Pro Am Soc of Mech Engrs—Oct., 1907. No. 87771.

Temporary Elevators for the Erection of the Singer Building Tower. Illustrated description of an electrically operated drum elevator with provisions for raising the headworks as needed. 1800 w. Eng Rec—Oct. 12, 1907. No. 87566.

Friction Driving.

Power Transmission by Friction Driving. W. F. M. Goss. An extension of an earlier study of the application of friction wheels to shaft driving. Ills. 6000 w. Pro Am Soc of Mech Engrs—Sept., 1907. No. 87430 C.

Overhead Tramways.

See under Material Handling, Machine Works and Foundries.

STEAM ENGINEERING.

Boiler Furnaces.

Cylindrical Boiler Furnaces. Illustrated review of the development of the modern suspension furnace and the methods of manufacture. 2500 w. Boiler Maker— Oct., 1907. No. 87322.

Boiler Inspection.

Rules for Boiler Inspection. Formulated by the Massachusetts Board of Boiler Rules. 3000 w. Boiler Maker—Nov., 1907. No. 87999.

See also under Railway Engineering, Motive Power and Equipment.

Boiler Management.

Boiler Room Economies. James Smith. This subject is discussed in connection with central-station operation, considering briefly fuel analysis, fuel gas analysis, auxiliary apparatus, feed water, firing, superheating, load factor and radiation. 1200 w. Elec Wld—Oct. 5, 1907. No. 87489.

Boilers.

Novelties in Steam Boilers (Neuerungen auf dem Gebiete des Dampfkessel-

We supply copies of these articles. See page 559.

wesens). Illustrates and describes various recent types of boilers, superheating devices, mechanical stokers, etc. 7000 w. Glückauf—Sept. 7, 1907. No. 87624 D.

Condenser Corrosion.
See under Marine and Naval Engineering.

Energy Diagram.
A New Energy Diagram for Steam. Henry F. Schmidt. Presents the advantages of this diagram and the methods used in its construction, giving examples showing its value. Inset. 3500 w. Elec Ry Rev—Oct. 19, 1907. No. 87812.

Engine Design.
Design of Engines for the Use of Highly Superheated Steam. Max E. R. Toltz. An illustrated discussion of the design of steam engines for very high temperatures. 1500 w. Pro Am Soc of Mech Engrs—Sept., 1907. No. 87431.

Engine Foundations.
Engine Foundations. Discusses the preparation of the ground, how to obtain the main center lines of the engine, etc., in the present number. Ills. 1800 w. Prac Engr—Sept. 27, 1907. Serial. 1st part. No. 87508 A.

Engine Speed.
Inertia Pressure at a Rational Measure of Engine Speed. Sanford A. Moss. Aims to show that the magnitude of initial engine pressure is the principal function of the engine speed which needs to be taken into consideration. 2200 w. Sibley Jour of Engng—Oct., 1907. Serial. 1st part. No. 87866 C.

Exhaust Steam.
See under Turbines.

Feed Water.
The Purification of Boiler Feed Water. Louis Waefelaer. Considers methods of treatment and apparatus used, testing, care, etc. Ills. 7000 w. Jour S African Assn of Engrs—Sept., 1907. No. 87950 F.

Fuels.
Steam Production from the Cheaper Grades of Anthracite. William D. Ennis. This second part of the writer's discussion deals with the mechanical problems of air supply, grate and heating surfaces, etc., which affect the economy. 4000 w. Engineering Magazine—Nov., 1907. No. 87970 B.

The Rational Utilization of Low Grade Fuels. F. E. Junge. Discusses how to use inferior coals most efficiently, with special consideration of the application of gas producers. 9800 w. Pro Am Soc of Mech Engrs—Oct., 1907. No. 87883 C.

See also under Locomotive Coals, Railway Engineering, Motive Power .and Equipment.

Fuel Testing.
The Fuel-Testing Plant of the United States Geological Survey at Norfolk, Va. Herbert M. Wilson. A report of the work carried out, and description of plant. 3000 w. Eng News—Oct. 10, 1907. No. 87547.

Governing.
The Speed Regulation of Steam Engines. Sterling H. Bunnell. Illustrates and describes various types of governors. 2500 w. Cassier's Mag—Oct., 1907. No. 87388 B.

Governor.
A Balanced Automatic Engine Governor. George P. Hutchins. Illustrated description of the improved design of the American ball governor. 1700 w. Engr, U S A—Oct. 15, 1907. No. 87732 C.

Injectors.
The Injector and Some Requirements for Its Successful Operation. Paul E. Capraro. Explains the principle on which its operation is based, illustrates and describes types, discussing injector troubles and their remedies. 2000 w. Engr, U S A—Oct. 1, 1907. No. 87325 C.

Pumping Engines.
See under same title under Mechanical Engineering, Hydraulics.

Smoke.
The Smoke Problem. David Townsend. Considers the difficulties and the attempted solutions of the smoke problem. 3000 w. Cassier's Mag—Oct., 1907. No. 87391 B.

Practical Smoke Abatement. William Nicholson. Abridged paper, read before the Sanitary Inspectors' Assn. Refers to conditions in London, reviewing action taken to stop unnecessary smoke, and discusses coal smoke, the effects, the cost, practical abatement, smoke from boilers, furnaces, fireplaces, smoke law and its amendment and administration. 3500 w. Elec Engr, Lond—Sept. 29, 1907. Serial. 1st part. No. 87513 A.

Superheating.
See Engine Design, Steam Engineering; and under Superheating, Railway Engineering, Motive Power and Equipment.

Turbines.
A Steam Turbine Economy Test. Reports a test made on a 7,500-kw. Westinghouse-Parsons steam turbine at the Waterside Station, No. 2, of the New York Edison Company. 1200 w. Ir Age—Oct. 10, 1907. No. 87538.

The Sturtevant Steam Turbine. Illustrated detailed description. 900 w. Eng News—Oct. 31, 1907. No. 87985.

Parsons Type Steam Turbines. Charles H. Naylor. A mathematical study of the flow of steam through a turbine of the reaction type. 2500 w. Engr, Lond—Oct. 4, 1907. No. 87599 A.

Some Practical Points in Steam Turbine Construction; with Particular Reference

We supply copies of these articles. See page 559.

to the Parsons Type. St. John Chilton. Read before the Am. St. & Int. Ry. Engng Assn. Deals with constructional details. Ills. 3300 w. St Ry Jour—Oct. 19, 1907. No. 87815.

Practical Experience with Exhaust-Steam Turbines. Dr. Alfred Gradenwitz. Illustrates and describes successful applications of low-pressure turbines, using exhaust steam, showing the advantages of the system. 3500 w. Engineering Magazine—Nov., 1907. No. 87969 B.

See also under Central Stations and Turbo-Generators, Electrical Engineering, Generating Stations; and under Turbine Steamers, Marine and Naval Engineering.

Valve-Gear.

Recke-Ruston Positive Valve-Gear. Brief description, with 2-page plate. 600 w. Engng—Sept. 27, 1907. No. 87522 A.

MISCELLANY.

Aeronautics.

The First British Military Airship. Harold J. Shepstone. Illustrations, with description and general information. 1500 w. Sci Am—Oct. 5, 1907. No. 87472.

The Gammeter Orthopter—A Beating-Wing Flying Machine. H. C. Gammeter. Illustrated description. 1000 w. Sci Am—Oct. 12, 1907. No. 87555.

New European Flying Machines. Illustrates and describes a number of machines recently brought out. 1700 w. Sci Am—Oct. 26, 1907. No. 87880.

Aerial Navigation in Marine Service (L'Aeronautica al Servizio della R. Marino). Ettore Cianetti. A discussion of the possibility of using balloons, etc., in naval warfare. 3300 w. Rivista Marittima —Sept., 1907. No. 87662 E + F.

Liquid Air.

The Commercial Liquefaction of Air and the Extraction of Oxygen from the Atmosphere (La Liquéfaction Industrielle de l'Air et l'Extraction de l'Oxygène de l'Atmosphère). E. Mathias. A description of various processes and machines. Ills. 11700 w. Rev Gen des Sciences— Sept. 15, 1907. No. 87606 D.

Minting Machinery.

Minting Machinery. Illustrates and describes machinery made for the Royal Mint, at Ottawa. 8000 w. Engr, Lond— Sept. 27, 1907. No. 87532 A.

Rubber Manufacture.

Manufacture of Mechanical Rubber Goods. An illustrated article describing the reclaiming process, and also the manufacturing methods employed at the Mercer Rubber Co.'s plant in New Jersey. 2500 w. Sci Am—Oct. 5, 1907. No. 87473.

Shock Phenomena.

Character of the Vibrations due to Shock Deduced from an Examination of Fractures (Caractères des Vibrations Accampagnant le Choc Déduits de l'Examen des Cassures). Ch. de Freminville. Ills. 15000 w. Rev de Métal—Sept., 1907. No. 87604 E + F.

MINING AND METALLURGY

COAL AND COKE.

Briquetting.

Coal Briquetting in the United States. Edward W. Parker. A report of the present condition of this industry in the different states, illustrating the machines used. 11400 w. Bul Am Inst of Min Engrs—Sept., 1907. No. 87729 D.

British Coalfields.

The Dover Coalfield in England. Edward Walker. An illustrated account of the district and the measures taken to work it. 1800 w. Eng & Min Jour—Oct. 12, 1907. No. 87576.

British Columbia.

A New Coal Field in British Columbia. Arthur Lakes Sketches and description of a valuable field of non-coking lignitic-bituminous coal, near the junction of the Similkameen and Tulameen rivers. 700 w. Min Wld—Oct. 5, 1907. No. 87483.

Coking Qualities.

Why Do Some Coals Coke and Others Not? F. C. Keighley. Read before the Coal Min. Inst. of Am. Considers the influence of geological conditions and physical characteristics of the coal upon coking qualities. 6200 w. Mines & Min—Oct., 1907. No. 87455 C.

Explosions.

Coal-Dust Explosions in Collieries. James Ashworth. Discusses the influence of water and steam on coal dust, as illustrated in the Wingate Grange disaster. Ills. 2800 w. Cassier's Mag—Oct., 1907. No. 87385 B.

The Fire-Damp Explosions at the Reden and Klein-Rosseln Mines in Saarreviere (Die Schlagwetterexplosionen auf den Gruben Reden und Klein-Rosseln im Saarreviere). Friedrich Okorn. Discusses their causes and effects. Ills. 3000 w. Serial. 2 parts. Oest Zeitschr f Berg u Hüttenwesen—Sept. 21 and 28, 1907. No. 87618 each D.

Germany.

The Mining Operations on the Lignite Deposits between Kölzig, Weisswasser, Muskau and Templitz in the Niederlausitz with Special Reference to Their In-

We supply copies of these articles. See page 559.

fluence on the Prevention of Spontaneous Combustion of the Coal (Der Bergwerksbetrieb auf dem Braunkohlenvorkommen zwischen Kölzig, Weisswasser, Muskau und Teuplitz in der Niederlausitz unter besonderer Berücksichtigung seines Einflusses auf die Verhütung der Selbstentzündung der Kohle. Herr Riegel. Ills. 8400 w. Glückauf—Sept. 7, 1907. No. 87621 D.

Handling.
Arrangements for Handling Coal Output. Floyd W. Parsons. Illustrates and describes simple mechanical methods for efficiently handling coal at mines under various conditions. 2200 w. Eng & Min Jour—Oct. 19, 1907. No. 87795.

Longwall Mining.
Mechanical Conveyors as Applied to Longwall Mining. J. I. Thomas. Describes methods and apparatus employed at Vintondale, Pa. 3500 w. Mines & Min —Nov., 1907. No. 87853 C.

Mine Gases.
Mine Gases and Methods of Preventing Explosions. H. E. Gray. Discusses gases found in coal mines, their identification and means for preventing accidents. 3000 w. Eng & Min Jour—Oct. 26, 1907. No. 87897.

Mining.
Coal Mining in Michigan. Lee Fraser. An account of the development and the difficulties, method of mining. etc. Maps. 1600 w. Eng & Min Jour—Sept. 28, 1907. No. 87315.
See also under Longwall Mining.

Mining Plant.
A Modern Illinois Coal Plant. An illustrated description of the surface equipment of Mine B of the Saline County Coal Co., near Harrisburg, Illinois. 2000 w. Mines & Min—Oct., 1907. No. '87453 C.

Peat.
The Utilization of Peat for Power Purposes with the Recuperation of By-products. An article based on five papers by Dr. A. Frank, with letter and editorial. 5000 w. Elec-Chem & Met Ind—Oct., 1907. No. 87502 C.
Organization of the American Peat Society. A report of an organization effected at a meeting at the Jamestown exposition, with abstracts of papers presented. 3000 w. Eng News—Oct. 31, 1907. No. 87993.

Pennsylvania.
Pennsylvania Mines in 1906. An abstract of the report of the Chief of the Department of Mines of Pennsylvania for the year 1906. 4800 w. Mines & Min—Oct., 1907. No. 87460 C.

Rescue Stations.
Underground Rescue. or Refuge Stations (Unterirdische Rettungs- bezw.

Fluchtstationen). J. Mayer. Discusses the advantages, location, and design of underground stations to serve as places of refuge in case of accident or as working bases in rescue work. 3600 w. Serial. 2 parts. Oest Zeitschr f Berg u Hüttenwesen—Sept. 14 and 21, 1907. No. 87617 each D.

Testing.
See under Fuel Testing, Mechanical Engineering, Steam Engineering.

COPPER.

British Columbia.
The Emma Mine. Frederic Keffer. An account of this low-grade mine of the Boundary District. Ills. 1600 w. Can Min Jour—Oct. 15, 1907. No. 87756.

Converter Hood.
The Laist & Tanner Movable Converter Hood. L. S. Austin. Brief illustrated description of a new invention. 400 w. Min & Sci Pr—Sept. 28, 1907. No. 87410.

Flotation Process.
Concentration Upside Down. Walter Renton Ingalls. Illustrated description of the process of A. P. S. Macquisten which causes sulphide minerals to float while quartz and other gangue minerals sink. 4400 w. Eng & Min Jour—Oct. 26, 1907. No. 87892.

Germany.
Operations and Tendencies of Modern Mansfield. P. A. Wagner and J. S. G. Primrose. An illustrated account of new processes and treatment of ores at these German copper mines. 2200 w. Eng & Min Jour—Oct. 12, 1907. No. 87573.

Mexico.
The Greene Mining Interests in Mexico. Reynolds Coleman. An illustrated account of the valuable copper, gold, and silver properties, and other interests. 18000 w. Min Wld—Oct. 26, 1907. No. 87947.

Montana Deposits.
Copper Deposits of the Belt Formation in Montana. M. Collen. Describes the geology of this region and gives the writer's views in regard to the genesis of the copper ores. 1200 w. Ec-Geol—Sept., 1907. No. 87899 D.

Moss Copper.
Moss Copper on Matte. E. L. Larison. Describes the Ducktown, Tenn., mattes and the formation of metallic copper, giving information in regard to the mosses. 800 w. Min Wld—Oct. 5, 1907. No. 87484.

Nevada.
The Copper Mines of Ely, Nevada. Walter Renton Ingalls. Gives the history, geology and general mining conditions of the district. Ills. 6500 w. Eng & Min Jour—Oct. 12, 1907. No. 87574.
The Productive and Earning Capacity of Ely. Walter Renton Ingalls. An illus-

We supply copies of these articles. See page 559.

trated account of this property which has an assured production for many years. Much of the copper will be produced for 7 cents per pound. 3000 w. Eng & Min Jour—Oct. 19, 1907. No. 87791.

Ore Deposits.
Origin of Copper Impregnations in Sedimentary Rock. H. E. Blake. Refers to an article by Prof. Arthur Lakes on this subject, and gives the writer's explanation. 1000 w. Min Rept—Oct. 17, 1907. No. 87800.

Precipitation.
Precipitation of Copper from Chloride Solutions by Means of Ferrous Chloride. Gustave Fernekes. Gives analyses of mine waters, and an experimental study of copper precipitation therefrom. 1500 w. Ec-Geol—Sept., 1907. No. 87900 D.

Sheet Copper.
The Rolling of Sheet Copper. Describes the characteristics of sheet copper mills, and the general methods. 3800 w. Met Work—Oct. 5, 1907. No. 87435.

Smelters vs. Producers.
I. The Mutual Relations and Grievances of the Smelting Trust and the Ore Producer. E. M. de la Vergne. Discusses the wrongs and remedies. II. A Response to the Address of Senator E. M. de la Vergne. Franklin Guiterman. Also general discussion. 12000 w. Am Min Cong —1906. No. 87368 N.

Smelting.
Copper Smelting in Utah. Robert B. Brinsmade. Illustrates and describes the practice at the Utah Consolidated, the Bingham Consolidated, and the Yampa smelters. 5500 w. Mines & Min—Nov., 1907. No. 87849 C.

Washoe Reduction Works at Anaconda, Montana. Describes the McDougal roasters, the large reverberatory smelters, and the utilization of waste heat. 1600 w. Mines & Min—Oct., 1907. No. 87459 C.

Smelting Works of the Consolidated Mining and Smelting Company of Canada, Limited, at Trail, B. C. J. M. Turnbull. Illustrated description of the plant and its equipment. 3000 w. Can Min Jour —Oct. 1, 1907. No. 87438.

Washington.
Copper Deposits of Washington. Albert W. McIntyre. Gives information which indicates the presence of copper deposits of value. 6000 w. Am Min Cong—1906. No. 87373 N.

GOLD AND SILVER.

Assaying.
The Effect of High Litharge in the Crucible-Assay for Silver. Richard W. Lodge. Shows that the use of a large excess of litharge in the assay of some ores will give results for silver that are uneven as well as low. 1200 w. Bul Am

Inst of Min Engrs—Sept., 1907. No. 87724 C.

Chlorination.
Chlorination of Gold-Ores; Laboratory Tests. A. L. Sweetser. Detailed description of laboratory tests in connection with the chlorination process. 2500 w. Bul Am Inst of Min Engrs—Sept., 1907. No. 87726 C.

Comstock Lode.
The Great Comstock Lode. G. McM. Ross. Outlines the history of the discovery and development of this mining district and the output of gold and silver. 3000 w. Min & Sci Pr—Oct. 12, 1907. No. 87753.

Cyaniding.
Recent Improvements in the Cyanide Process. F. L. Bosque. Reviews the development of the process and recent advances. 3000 w. Am Min Cong—1906. No. 87366 N.

Recent Advance in Cyanidation in Mexico. J. Leslie Mennell. Gives an account of the increasing use of this process for the extraction of gold and silver. 2500 w. Min Wld—Oct. 26, 1907. No. 87943.

The Cyanide Process at the Portland (Colo.) Mill. Regis Chauvenet. General remarks on the history of this process, describing the present practice at this mill. 2200 w. Min Rept—Oct. 24, 1907. Serial. 1st part. No. 87920.

Distribution.
The Geological Distribution of Gold. T. A. Rickard. Considers the relation of geology to the gold industry, some mistakes made, recent advances, etc. 4000 w. Am Min Cong—1906. No. 87371 N.

Goldfield District.
The Combination Mine. Edgar A. Collins. The present number gives an account of the early development and the geologic structure of this Nevada mine. Ills. 2500 w. Min & Sci Pr—Sept. 28, 1907. Serial. 1st part. No. 87409.

Mexico.
Mines of the Taviche District, Oaxaca, Mexico. A. E. Place and H. L. Elton. Describes these mines, and the veins bearing the silver sulphide ores, which are remarkably free from faulting and usually appear at the surface. 1500 w. Eng & Min Jour—Oct. 5, 1907. No. 87462.

Milling.
The Milling Practice at El Oro Mine, Mexico. E. Burt. Describes the treatment and the tube mills. Ills. 2500 w. Min Wld—Oct. 26, 1907. No. 87945.

The Desert Mill. A. R. Parsons. Detailed description of 100-stamp mill and power-plant at Millers, Nevada, for milling ores from the Tonopah mines. Ills. 4500 w. Min & Sci Pr—Oct. 19, 1907. No. 87889.

We supply copies of these articles. See page 559.

Nevada.

The Osceola Gold District of Nevada. Gives the history and a description of this district. Ills. 1600 w. Min Rept—Oct. 3, 1907. No. 87501.

Nile.

The Nile as a Mining River. Alexander Del Mar. Reviews the ancient history of mining in this region. 3800 w. Min & Sci Pr—Oct. 12, 1907. No. 87752.

Nova Scotia.

The Development of an Ore Shoot in Nova Scotia. E. Percy Brown. Gives an outline of the history of the Brookfield mine. 1600 w. Can Min Jour—Oct. 15, 1907. No. 87755.

Gold Measures of Tangier, Nova Scotia. George A. Packard. Describes characteristics of a district where the mines have been worked by small owners, or lessees. Ills. 2000 w. Min & Sci Pr—Oct. 5, 1907. No. 87540.

Placers.

Cause of Fine Gold in San Juan River, Utah. A. Lakes. Explains the probable origin of the gold and the cause of its extreme fineness. 900 w. Min Rept—Oct. 3, 1907. No. 87500.

Rand.

Working Costs of the Mines of the Witwatersrand. Discussion by R. N. Kotzé, of the paper by Ross E. Browne. 5500 w. Jour S African Assn of Engrs—Aug., 1907. No. 87504 F.

Servia.

Gold in Eastern Servia. Douchan Ivanovitch. Abstract translation. Brief account of this region, the geology, mineralogy and the mining industry. 1600 w. Min Jour—Oct. 5, 1907. No. 87591 A.

Slimes Treatment.

Recent Improvement in Slimes Treatment. D. J. Kelly. Read before the W. Assn, of Tech. Chem. & Met. Describes methods of filtration used for separating the gold solutions from slimes. 4500 w. Mines & Min—Oct., 1907. No. 87454 C.

IRON AND STEEL.

Austria.

The Austrian Iron Industry. Wilhelm Kestranek. Read before the Ir. & St. Inst., at Vienna. Reviews the history of the industry during the last twenty-five years. 4000 w. Engng—Sept. 27, 1907. No. 87523 A.

Austrian Deposits.

The Erzberg of Eisenerz. H. Bauerman. Read before the Ir. & St. Inst., at Vienna. Describes this ore mountain of the Eastern Alps, and its deposits, especially the iron ore workings. 2800 w. Engng—Sept. 27, 1907. No. 87524 A.

Blast Furnaces.

Address on the Effect of Air and Moisture on Blast-Furnaces. Joseph

Dawson. Reprint of an address at meeting of ironmasters on June 11, 1800. 7500 w. Jour Ir & St Inst—No. II, 1907. No. 87362 N.

Blow-Holes.

Blow Holes in Steel Ingots. E. von Maltitz. Gives the writer's views and experience in regard to the formation of blow-holes, their prevention, and related matters. 14600 w. Bul Am Inst of Min Engrs—Sept., 1907. No. 87722 D.

Charcoal-Iron.

A By-Product Charcoal-Iron Plant at Marquette, Mich. Plan and description. 2500 w. Eng Rec—Oct. 12, 1907. No. 87571.

Laboratory.

The Creation of a Laboratory for a Modern Ironworks (Wie muss das Hauptlaboratorium eines Neuzeitlichen Eisenhüttenwerks Beschaffen. sein). August Kaysser. Discusses the various departments necessary, the work to be done in each, apparatus, etc. Ills. 5300 w. Serial. 2 parts. Stahl u Eisen—Sept. 11 and 18, 1907. No. 87616 each D.

Lake Superior.

Developments in the Menominee Range. An illustrated account of the mines of this Lake Superior district and their equipment. 2200 w. Ir Trd Rev—Oct. 17, 1907. No. 87750.

Lapland Ores.

The Geology and Origin of the Lapland Iron Ores. Dr. O. Stutzer. Describes the geology of Lapland, and considers the more important occurrences of iron ore, discussing their origin. Plates and bibliography. 32000 w. Jour Ir & St Inst—No. II, 1907. No. 87360 N.

Metallurgical Chemistry.

Application of the Laws of Physical Chemistry in the Metallurgy of Iron. Baron H. von Juptner. Read before the Ir. & St. Inst. at Vienna. Deals with the doctrine of chemical equilibrium as applied to metallurgical chemical processes. 5800 w. Ir & Coal Trds Rev—Sept. 27, 1907. No. 87703 A.

Meteoric Iron.

Steel and Meteoric Iron. Prof. Frederick Berwerth. Read before the Ir. & St. Inst. at Vienna. Aims to show in a general way that meteoric iron and steel works steels are results of essentially similar chemical and physical causes. 3500 w. Engng—Oct. 4, 1907. No. 87594 A.

Minnesota.

The Cuyuna Iron Range. Newton H. Winchell. Describes the geological structure. 2500 w. Ec-Geol—Sept., 1907. No. 87898 D.

Iron Ore in Crow Wing County, Minn. Dwight E. Woodbridge. An account of recent discoveries of merchantable ore.

We supply copies of these articles. See page 559.

1500 w. Eng & Min Jour—Oct. 26, 1907. No. 87894.

Smelting.

Zinc Oxide in Iron-Ores, and the Effect of Zinc in the Iron Blast-Furnace. John J. Porter. Explains the difficulties that arise from the presence of zinc. 2500 w. Bul Am Inst of Min Engrs—Sept., 1907. No. 87725 C.

Steel Works.

The Forges and Mines of the Hungarian State. Plan and illustrated description of the Diosgyor steel works, their equipment, management, etc. 6000 w. Engr, Lond—Sept. 20, 1907. No. 87347 A.

The Krupp Works in Germany. J. B. Van Brussel. An illustrated description of a group of plants comprising blast furnaces, steel mills, gun and armor plate shops. 4000 w. Ir Trd Rev—Oct. 3, 1907. No. 87404.

Trade.

See under Steel Trade, Industrial Economy.

MINING.

Accidents.

The Prevention of Mine Accidents. Edward W. Parker. Remarks on the dangers, especially in coal mining, and the need of some action in the United States to correct the abuses and causes. 2000 w. Am Min Cong—1906. No. 87372 N.

Prevention of Accidents in Mines. P. J. Slevin. Read before the Connellsville Min. Inst. Discusses some causes of accidents and means for guarding against them. 2200 w. Mines & Min—Oct., 1907. No. 87456 C.

Australia.

Mining in Australia. W. J. Loring. Deals with the states in order, giving a description of present mining conditions. 2000 w. Min & Sci Pr—Oct. 19, 1907. No. 87890.

Boring.

The Theory of the Movement of the Flushing Stream in Bore Holes (Die Theorie der Bewegung des Spülstromes in Bohrlöchern). Richard Sorge. A theoretical and mathematical discussion. Ills. 7000 w. Glückauf—Sept. 28, 1907. No. 87626 D.

Cage Guides.

The Arrangement of Cage Guides for Modern Collieries. J. S. Barnes. Drawings with description of working details of both rope and steel rail guides for winding coal at high speeds and heavy loads. 2500 w. Ir & Coal Trds Rev—Oct. 4, 1907. No. 87707 A.

Canada.

Canadian Mining Intelligence. H. Mortimer Lamb. Discusses briefly the mineral production of British Columbia, the gold output of Nova Scotia, the Canadian mint, and gold dredging on the Saskatchewan. 2500 w. Mines & Min—Oct., 1907. No. 87458 C.

Concentration.

Milling Practice at the Granadena Mill. S. F. Shaw. Describes the method of treatment of the complex ore from this Mexican mine. Ills. 1200 w. Eng & Min Jour—Oct. 5, 1907. No. 87465.

Tails and Sludge Milling at the Old Judge Mine, Oronogo, Mo. W. D. Glenn. Describes a new method of treatment, equipment, and pulp flow. 1000 w. Mines & Min—Nov., 1907. No. 87845 C.

Mill Building Progress in the Joplin District. Claude L. Watson. Illustrated detailed description of the Oakwood mill, Webb City, Mo. 1500 w. Mines & Min—Nov., 1907. No. 87844 C.

Damage Litigation.

Recent Litigation Involving Questions of Alleged Damage from Tailings, Tailings Waters, and Smelter Fumes. Prof. F. W. Traphagen. Reviews cases before the courts of Montana and the investigations made by experts. 1500 w. Am Min Cong—1906. No. 87374 N.

Diamond Drilling.

The Diamond Core-Drill in Prospecting. Lewis T. Wright. Presents some of the advantages and disadvantages. 2000 w. Min & Sci Pr—Oct. 12, 1907. No. 87751.

Drainage.

Draining the Panther Creek Basin. H. H. Stoek. Illustrated description of tunnel 13 miles long for draining the mines of the Lehigh Coal and Navigation Co. 2200 w. Mines & Min—Nov., 1907. No. 87846 C.

Drainage Districts.

Mine Drainage Districts. D. W. Brunton. Introductory to a discussion of the question of creating mine drainage districts, followed by discussion. 7000 w. Am Min Cong—1906. No. 87367 N.

Electric Hoisting.

The Electric Hoist at the Hermann Mine at Eisleben (Die Elektrische Fördermaschine auf "Hermannschacht" bei Eisleben). L. Hoffmann. Illustrates and describes an installation on the Ilgner system. 1800 w. Glückauf—Sept. 7, 1907. No. 87625 D.

Some New Flywheel Storage Systems. A. P. Wood. Read before the Manchester Sec. of the Inst. of Elec. Engrs. Describes some recent patents in connection with electric winding plants, discussing the three-phase system, the cascade flywheel storage system, and others. Ills. 3800 w. Elec Engr, Lond—Sept. 20, 1907. No. 87334 A.

We supply copies of these articles. See page 559.

Electrical Machinery.

Fire and Explosion-Proof Electrical Mining Machinery. Briefly considers switchgear, motors, motor controlling resistances, transformer and switch oil, etc. 2500 w. Elec Rev, Lond—Sept. 20, 1907. No. 87335 A.

Explosives.

Explosives in Coal Mines. E. J. Deason. A review of the British regulations of the past ten years and of the Woolwich tests. Ills. 6500 w. Ir & Coal Trds Rev—Sept. 20, 1907. No. 87349 A.

The Testing of Safety Explosives (Ueber die Prüfung von Sicherheitssprengstoffen). Herr Beyling. Discusses the composition of the various classes of safety explosives, the conditions they should fulfil and methods of testing. Tables. 6000 w. Glückauf—Sept. 7, 1907. No. 87620 D.

Haulage.

Construction of Tracks in Coal Mines. M. S. Hachita. Urges that more attention be given to building haulage roads, the laying out of frogs, switches, etc. 3500 w. Eng & Min Jour—Oct. 5, 1907. No. 87466.

Self-Acting Endless-Rope Haulage at the Glückhilf Mine at Hettstedt (Bremsbergförderung mit Seil ohne Ende auf "Glückhilfschächte" bei Hettstedt). L. Hoffmann. Illustrates and describes the haulage installation, signals, etc. 1600 w. Glückauf—Sept. 7, 1907. No. 87622 D.

Hoisting.

Transvaal Commission Report on the Use of Winding Ropes and Safety Appliances in Pit Shafts. Abstract. 11000 w. Ir & Coal Trds Rev—Oct. 11, 1907. No. 87842 A.

Report of the Transvaal Commission on the Use of Winding Ropes, Safety Catches and Appliances in Mine Shafts. An important statement in regard to safe practice in deep mine hoisting. The full report will be given. 6000 w. Eng News —Oct. 31, 1907. Serial. 1st part. No. 87988.

The Use of Winding Ropes, Safety Catches, and Appliances in Mine Shafts. Abstract of the report of commission appointed by the Governor of the Transvaal to investigate this subject. 4800 w. Col Guard—Sept. 27, 1907. Serial. 1st part. No. 87519 A. .

See also under Electric Hoisting.

Joplin District.

The Joplin Zinc District. R. L. Herrick. An illustrated general description of the mining and milling methods in use in the region. Plate. 12000 w. Mines & Min—Nov., 1907. No. 87843 C.

The Yellow Dog Mine and Mill. R. L. Herrick. Illustrated general description of methods of mining sheet-ground, hoist-

ing by endless conveyor, and methods of milling at Webb City, Mo. 2000 w. Mines & Min—Nov., 1907. No. 87847 C.

Handling "Sheet-Ground" in the Joplin District. J. H. Polhemus. Illustrates and describes methods of mining and milling as practiced by the American Zinc, Lead, and Smelting Co. 2000 w. Mines & Min—Nov., 1907. No. 87848 C.

Mining in the Joplin District. Ch. Guengerich. Brief account of conditions in this region which furnishes 50 per cent of the zinc ore smelted in the United States. 1800 w. Am Min Cong—1906. No. 87369 N.

See also under Concentration.

Magnetic Separation.

The Separation of Tin-Oxide from Wolfram. Amos Treboar and Gurth Johnson. Brief account of experimental work . 600 w. ' Inst of Min & Met, Bul. 36—Sept. 19, 1907. No. 87560 N.

Metal Mining.

The Development of the Metal Mining Industry in the Western States. Waldemar Lindgren. An outline of the gold and silver mining, considering placer mining, gold-quartz mining, silver mining, reduction works, etc. 5000 w. Am Min Cong —1906. No. 87364 N.

Mine Signals.

Electric Signal System for Mines. Carl L. C. Fichtel. Illustrates and describes a new system installed in some of the copper mines of the Lake Superior region. 900 w. Eng & Min Jour—Oct. 26, 1907. No. 87893.

Open-Cut.

Nickel Mining in New Caledonia. G. M. Colvocoresses. Describes the methods of mining nickel ores from large and small open workings or quarries on the sides of the mountains, and the method of handling the ore. Ills. 4000 w. Eng & Min Jour—Sept. 28, 1907. No. 87314.

Ore Feeders.

Notes on Feeders, with a Description of a New Driving Device. D. J. Pepler. Describes Pepler's Operative Device to Challenge Feeder. 1500 w. Jour Chem, Met, and Min Soc of S Africa—Aug., 1907. No. 87823 E.

Pumping.

See under Unwatering.

Shaft Sinking. .

Shaft Sinking by the Freezing Process. Sydney F. Walker. Describes the vertical pipe method used in England, and a new ring process employed in Germany. 4500 w. Eng & Min Jour—Oct. 12, 1907. No. 87575.

Care of the Plant in Sinking by Refrigeration. Sydney F. Walker. Causes of interruptions in the operation of this system are discussed, and methods of

We supply copies of these articles. See page 559.

correcting them. 2500 w. Eng & Min Jour—Oct. 26, 1907. No. 87895.

The Hydraulic Shaft Borer (Der Hydraulische Schachtbohrer). Herr Schlüter. Describes a device for boring full-size shafts, with details of power requirements, labor, costs, etc. Ills. 5300 w. Glückauf—Sept. 7, 1907. No. 87619 D.

Timbering.

Re-timbering the Kearsarge Shaft. Lee Fraser. Describes work made necessary by a fire. Sections. 700 w. Min & Sci Pr —Oct. 5, 1907. No. 87541.

Unwatering.

Unwatering by Means of an Inclined Skip. Douglas Muir. States the conditions and describes the work. Ills. 900 w. Eng & Min Jour—Oct. 19, 1907. No. 87793.

Unwatering Plant for the Lindal Moor Mines. Illustrated description of the pumping plant for unwatering these iron mines. 1500 w. Engng—Oct. 11, 1907. No. 87838 A.

See also under Drainage.

Ventilation.

Deficiencies in Mine Ventilation. Thomas A. Mather. Read at Min. Inst. at Latrobe, Pa. The causes are discussed and the relative degrees of responsibility of the different people contributing to ventilation. 2800 w. Mines & Min—Oct., 1907. No. 87457 C.

MISCELLANY.

Alaska.

Alaska and Its Possibilities. J. T. Cornforth. Brief account of placer gold districts, other minerals found, and matters relating to the development of the country. 1500 w. Am Min Cong—1906. No. 87375 N.

Arsenic.

The Arsenic Industry of Cornwall. Illustrated description of the process of manufacture. 2000 w. Sci Am Sup—Oct. 5, 1907. No. 87475.

Asbestos.

The Asbestos Industry of Quebec. Ralph Stokes. Illustrates and describes the mines and quarries in the present article. 2200 w. Min Wld—Oct. 19, 1907. Serial. 1st part. No. 87801.

Asia.

A Journey to Central Asia. A. Adiassewick. General description with special reference to the great mineral wealth. 11000 w. Inst of Min & Met, Bul. 36— Sept. 19, 1907. No. 87559 N.

Blow-outs.

The Blow-out. F. Lynwood Garrison. Illustrates and describes interesting geological formations of this character in Arizona and Nevada. 800 w. Min & Sci Pr—Sept. 28, 1907. No. 87412.

Borax.

Death Valley Borax. O. M. Boyle, Jr. Brief account of the deposits and the condition of the industry. Ills. 1200 w. Cal Jour of Tech—Sept., 1907. No. 87428.

Cement.

Advances in Cement Technology, 1906. Edwin C. Eckel. Considers the growth of the cement industry, kilns and kiln practice, and matters related. 3500 w. Cement —Sept., 1907. No. 87399 C.

Statistics of Cement Industry in 1906. L. L. Kimball. Considers the general conditions, the production, with report by states. 7000 w. Cement—Sept., 1907. No. 87400 C.

Chloride Process.

A New Application of Chlorine in Metallurgy. Charles E. Baker. Read at N. Y. meeting of the Am. Elec.-Chem. Soc. Deals with the manner of treating sulphide ores. 1800 w. Min Rept—Oct. 24, 1907. No. 87919.

Diamonds.

Diamonds in the United States. Douglas B. Sterret, in Mineral Resources of the U. S., 1906. Describes a discovery by M. J. Cooney, on the Feather River, California, and other finds on record. 1200 w. Min Rept—Oct. 10, 1907. No. 87578.

Kimberlite Dykes and Pipes. Dr. F. W. Voit. Read before the Geol. Soc. of S. Africa. Describes peculiarities of these diamond mines, studying the origin of the dykes and pipes, and the character of the rock. 2500 w. Min Jour—Sept. 21, 1907. Serial. 1st part. No. 87341 A.

Exhibition.

Minerals at the Jamestown Exposition. Joseph Struthers. An illustrated brief description of the exhibits of the different states. 4000 w. Eng & Min Jour—Oct. 19, 1907. No. 87794.

Leadville.

Genesis of the Ores of Leadville. S. F. Emmons. Abstracted from Bul. of the U. S. Geol. Surv. A study of geological formation and the ore deposits of this region. 5500 w. Min & Sci Pr—Sept. 28, 1907. No. 87411.

Lime Analysis.

Notes on the Estimation of Caustic Lime. Edw. H. Croghan. Facts obtained during a series of tests and analyses on certain burnt limes. 4000 w. Jour Chem, Met, & Min Soc of S Africa—Aug., 1907. No. 87822 E.

Mexico.

The History of Mining Developments in Mexico. T. C. Graham. Brief review. 2500 w. Min Wld—Oct. 26, 1907. No. 87946.

Mexico: Its Geology and Natural Resources. Robert Thomas Hill. An interesting illustrated study of the geology,

We supply copies of these articles. See page 559.

mineralization, etc. 5000 w. Min Wld—Oct. 26, 1907. No. 87942.

The Economic Progress of Mexico Since Early Times. Charles C. Schnatterbeck. Illustrated historical review. 8000 w. Min Wld—Oct. 26, 1907. No. 87940.

Some Notes on a Journey Through Mexico. Dwight E. Woodbridge. Illustrated description of the country, its transportation facilities, etc. 4000 w. Min Wld—Oct. 26, 1907. No. 87944.

Traveling on the West Coast of Mexico. Dwight E. Woodbridge. Illustrated description of the region and of the business and mining conditions. 3500 w. Eng & Min Jour—Oct. 5, 1907. No. 87463.

Mineral Resources on Kansas City, Mexico & Orient Railway. James W. Malcolmson. Extract from a recent report of this railway, describing the mines in Mexico. 2500 w. Min Wld—Oct. 26, 1907. No. 87948.

Characteristics of Some Mexican Mining Regions. Robert T. Hill. Discussion of relationships between geographic features and economic resources in Chihuahua, Sonora, and Western Sierra Madre. Ills. 3500 w. Eng & Min Jour—Oct. 5, 1907. No. 87464.

See also under Tin; and under Mexican Labor, Industrial Economy.

Monazite.

The Mining of Monazite. Douglas B. Sterrett. Describes methods of working the deposits of this thorium mineral. 2000 w. Sci Am Sup—Oct. 19, 1907. No. 87744.

Nickel.

The Occurrence of Nickel in Virginia. Thomas Leonard Watson. Describes in some detail these nickeliferous pyrrhotite deposits at the Lick Fork openings. Maps. 4000 w. Bul Am Inst of Min Engrs—Sept., 1907. No. 87730 C.

Nitrogen.

A New System for the Fixation of Atmospheric Nitrogen. F. Savorgnan di Brazza. Illustrates and describes a process based upon the property of calcium carbide to fix nitrogen at high temperatures. 1500 w. Sci Am—Oct. 12, 1907. No. 87554.

Oil Fields.

Oil Field on Buffalo Creek, West Virginia. Frank W. Brady. Information in regard to the location of wells, methods and production. Ills. 2700 w. Mines & Min—Nov., 1907. No. 87851 C.

Economic and Technical Notes on the Wietze Petroleum District (Wirtschaftliche und Technische Mitteilungen über den Wietzer Erdölbezirk). Herr. Dobblestein. Describes the oil-wells, methods, etc., and gives statistics of production of this most important oil district in Germany. Ills. 3500 w. Glückauf—Sept. 7, 1907. No. 87623 D.

Ore Deposition.

The Interaction Between Minerals and Water Solutions, with Special Reference to Geologic Phenomena. Eugene C. Sullivan. Describes work undertaken in the chemical laboratory of the U. S. Geol. Survey, in an attempt to apply chemical methods to the investigations of geologic processes, especially of the secondary deposition of ores. 4000 w. Min Jour—Sept. 21, 1907. Serial. 1st part. No. 87340 A.

Peru.

The Mineral Wealth of the Province of Santiago de Chuco, Peru. F. Málaga Santolalla. Describes this region of the Andes, and the deposits of silver, copper, lead, and other minerals. 3000 w. Min Jour—Sept. 28, 1907. Serial. 1st part. No. 87518 A.

Petroleum.

The International Petroleum Congress. David T. Day. An account of the gathering of petroleum experts and chemists, and the progress of the industry, especially in Roumania. Ills. 2200 w. Eng & Min Jour—Oct. 26, 1907. No. 87896.

Radium.

A New Mineral Industry—The Manufacture of Radium. Jacques Loyer. An illustrated outline of the methods used at Nogent-sur-Marne. 1800 w. Engineering Magazine—Nov., 1907. No. 87971 B.

Rocky Mountains.

Certain Features of the Rocky Mountain Region. Horace F. Evans. Describes the general features of the Rocky Mountain systems, especially north of the 49th parallel. 1800 w. Min Wld—Oct. 19, 1907. Serial. 1st part. No. 87802.

Sakhalin.

The Mineral Resources of the Island of Sakhalin. Reviews the history and the characteristics of the political situation, geographical and geological features and present conditions, discussing the useful mineral deposits and their importance. Ills. 3500 w. Min Jour—Sept. 28, 1907. No. 87517 A.

Smelting Regulations.

German Smelting Regulations. Gives regulations pertaining especially to zinc smelters. 1200 w. Min Rept—Oct. 17, 1907. No. 87798.

Tantalum.

Tantalum. Abstract translation of an article by Paul Nicolardat in *Revue Scientifique*, on the properties of this metal. 2000 w. Sci Am Sup—Oct. 5, 1907. No. 87478.

Tin.

The Tin Deposits of Mexico. H. G.

We supply copies of these articles. See page 559.

Bretherton. Describes the Paso de Sotos district. 1000 w. Min Wld—Oct. 26, 1907. No. 87941.

Tin Mining in Siam. K. Van Dort. An illustrated account of the methods of mining and working. 3000 w. Eng & Min Jour—Oct. 19, 1907. No. 87792.

A Review of the World's Tin-Mining Industries. A. Selwyn-Brown. Information relating to the production, sources, mining conditions and prospects. 4000 w. Engineering Magazine—Nov., 1907. No. 87973 B.

Geology and Mining of the Tin-Deposits of Cape Prince of Wales, Alaska. Albert Hill Fay. An illustrated description of geographic and climatic conditions and of the geology and mining prospects. 4000 w. Bul Am Inst of Min Engrs—Sept., 1907. No. 87728 C.

Utah.

The Mining and Mineral Resources of Utah. John Dern. Considers a few of the leading mining districts. The minerals of importance are those of gold, silver, copper, and lead, but other deposits of value are found. 4000 w. Am Min Cong—1906. No. 87365 N.

Vanadium.

The Present Source and Uses of Vanadium. J. Kent Smith. Shows its great industrial importance, its effect on steel, and considers the present sources of supply. 1200 w. Bul Am Inst of Min Engrs—Sept., 1907. No. 87723 C.

Wisconsin.

Mining and Mineral Resources of Wisconsin. W. O. Hotchkiss. Describes the geological formations of the state, the ore deposits and their development. Lead and zinc ores are principally considered, but iron and other metallic minerals are of value. 1200 w. Am Min Cong—1906. No. 87370 N.

RAILWAY ENGINEERING

CONDUCTING TRANSPORTATION.

Accident.

Shrewsbury Railway Accident. An illustrated account of a serious accident in England. 900 w. Engng—Oct. 18, 1907. No. 87960 A.

Non-Stop Run.

Longest Non-Stop Run on Great Western Railway. Charles Rous-Marten. An account of a run from Paddington to Fishguard over the new route of the Anglo-Irish mail service. 1500 w. Engr, Lond—Sept. 27, 1907. No. 87530 A.

MOTIVE POWER AND EQUIPMENT.

Air Brakes.

The K Triple in Modern Brake Service. John P. Kelly. Remarks on present requirements of air brakes, with illustrated description of this improved design and its working. Also discussion. 11800 w. Pro Cent Ry Club—Sept. 13, 1907. No. 87864 C.

The Westinghouse Air Brake Company. —Quick-Service Triple Valve. An illustrated description of the features of Type K triple valve, outlining what it is designed to accomplish. 7000 w. Pro Ry Club of Pittsburgh—May, 1907. No. 87-711 C.

Air Pipes.

Broken Air Pipes. G. W. Kiehm. Suggestions for repairing pipes in the air-brake equipment. 1500 w. Ry & Loc Engng—Oct., 1907. No. 87318 C.

Boiler Inspection.

New York Laws for Locomotive Boiler Inspection. Laws requiring thorough in-spection to be made of all boilers and fittings of steam locomotives used on railroads. 2500 w. Boiler Maker—Oct., 1907. No. 87323.

Locomotive Boiler Inspections in New York State. Gives the amended requirements that went into effect Sept. 1st, with the regulations for inspecting, testing, etc. 3000 w. R R Gaz—Sept. 13, 1907. No. 87907.

Dynamometer Cars.

New Dynamometer Car for the Pennsylvania Railroad. Illustrated description of the car and its equipment and operation. 2500 w. Eng News—Oct. 17, 1907. No. 87741.

North-Eastern Railway Dynamometer Car. Illustrated description of a car for this English road. 1200 w. Engr, Lond—Oct. 4, 1907. No. 87596 A.

Fireboxes.

Methods of Obtaining Flexibility in Locomotive Fireboxes. Shows the advantage of flexibility and considers methods of securing it. Ills. 2500 w. Boiler Maker —Oct., 1907. No. 87324.

Grade Assistance.

Cable-Assisted Trains Upon a Scottish Railroad. Brief illustrated description of a novel means of assisting trains up a severe gradient in use at the Cowlairs Incline in Scotland. 600 w. Sci Am Sup— Oct. 5, 1907. No. 87479.

Locomotive Coals.

Locomotive Coals. A Jacobsen. An account of results of tests made by the Danish State railways. Ills. 4000 w. Engr, Lond—Oct. 18, 1907. No. 87962 A.

We supply copies of these articles. See page 559.

Locomotive Performance.

Recent Locomotive Work on the French Northern Railway. Charles Rous-Marten. An account of runs showing the ordinary work on this railway. 3000 w. Engr, Lond—Sept. 20, 1907. No. 87346 A.

Locomotives.

Curiosities of Locomotive Design. Angus Sinclair. From "Development of the Locomotive Engine." Illustrates and describes peculiar designs. 1800 w. Ry & Loc Engng—Oct., 1907. No. 87316 C.

Seth Wilmarth's Locomotives. C. H. Caruthers. An account of works once existing in Boston, Mass., and the engines built. Ills. 18000 w. R R Gaz—Sept. 27, 1907. No. 87916.

The Development of the American Locomotive. Report of the Committee on Science and Arts on the contribution of the Baldwin Locomotive Works. Ills. 5500 w. Jour Fr Inst—Oct., 1907. No. 87758 D.

Locomotive Exhibit at the Milan Exposition. Abstract translation of an article by M. L. Georges, in the *Revue Generale*. Ills. 1600 w. Ry & Engng Rev—Oct. 5, 1907. No. 87497.

Locomotives at the Jamestown Exposition. Illustrated description of the exhibits of the Baldwin Locomotive Works and the American Locomotive Company. 1500 w. Ry Age—Oct. 4, 1907. No. 87494.

Details of Mallet Articulated Compound Locomotive. Detailed description of the interesting special features. Ills. 4000 w. R R Gaz—Oct. 4, 1907. No. 87923.

Pusher Locomotives of the Mallet Duplex Type for the Erie Ry. Illustration with description. 1000 w. Eng News—Oct. 3, 1907. No. 87420.

Pacific Locomotive for the Lake Shore and Michigan Southern. Compares them with the prairie engines which have been used up to the present, and gives an illustrated description. 2500 w. R R Gaz—Sept. 6, 1907. No. 87902.

Pacific Type Locomotives. Illustrated description of new engines for freight and passenger service on the Richmond, Fredericksburg & Potomac R. R. 600 w. Am Engr & R R Jour—Oct., 1907. No. 87419 C.

Four Cylinder Simple Express Locomotives; Great Western Railway. The latest type of locomotive for hauling express passenger traffic on this line is illustrated and described. 500 w. R R Gaz—Oct. 4, 1907. No. 87924.

Compound Ten-Wheel Locomotives for the Buenos Ayres Western Railway. Illustrated description of compound passenger and freight locomotives, having special features of interest. 600 w. R R Gaz —Sept. 6, 1907. No. 87904.

Royal Bavarian 4-4-2. Illustration with brief description of a type in use on the continent of Europe. 500 w. Ry & Loc Engng—Oct., 1907. No. 87317 C.

An Interesting Locomotive. Illustrates and describes an example of the latest construction of light passenger engines for the Austrian State Railways, to serve main line feeders. 600 w. Ry & Engng Rev—Oct. 5, 1907. No. 87498.

Express Locomotive for the Prussian State Railways. Illustrated description of the latest type of six-coupled express locomotive, fitted with the Schmidt type of smoke-tube superheater. 1500 w. Plate. Engng—Oct. 11, 1907. No. 87836 A.

Locomotive Tests.

A Method of Plotting Locomotive Characteristics. Lawford H. Fry. Gives a diagram showing a system of plotting the results of locomotive tests to give a series of characteristic curves which cover the whole range of operation. 2000 w. Am Engr & R R Jour—Oct., 1907. No. 87416 C.

The Results of Tests on Tank Locomotives Using Superheated Steam (Versuchs- und Betriebsergebnisse mit Heissdampf-Tenderlokomotiven). Herr Müller. Gives tables and curves showing the performance of the 10-wheel locomotives of the Prussian state railways. Ills. 3500 w. Verkehrstech Woche u Eisenbahntech Zeitschr—Sept. 7, 1907. No. 87627 D.

Motor Cars.

Steam Motor Cars. Illustrates and describes motor cars recently put in service on the Intercolonial Railway of Canada. 1000 w. Am Engr & R. R. Jour—Oct., 1907. No. 87417 C.

Mountain Railway.

The Puy-de-Dome Mountain Railway. Illustrated description of this railway in France, in which the locomotive is assisted by the Fell friction-grip, centre-rail system. 1000 w. Eng News—Oct. 10, 1907. No. 87551.

Refrigerator Cars.

Ventilating Refrigerator Cars. Illustrates and describes the Garland Ventilator and the principle on which it is based. 1500 w. Ice & Refrig—Oct., 1907. No. 87433 C.

Shops.

New Railway Shops (Neuere Eisenbahnwerkstätten). Herr Müller. A series of articles descriptive of the shops of the Prussian state railways, the first dealing with the Oppum shops at Krefeld. Ills. 7000 w. Serial. 1st part. Verkehrstech Woche u Eisenbahntech Zeitschr—Sept. 14, 1907. No. 87628 D.

The Shops of the Swiss State Railways and Their Extension (Les Ateliers des Chemins de Fer Fédéraux, à Yverdon, et

We supply copies of these articles. See page 559.

leur Agrandissement). M. G. Guillemin. Illustrated description of buildings and equipment. 4950 w. Bull Tech d l Suisse Rom—Sept. 25, 1907. No. 87608 D.

Steam v. Electricity.

Steam Locomotive versus Electric Locomotive. Max Toltz. Aims to show that the steam locomotive properly improved is .by far more economical than the electric locomotive. Ills. Discussion. 18300 w. Pro N Y R R Club—Sept. 20, 1907. No. 87760.

Steel Cars.

All-Steel Passenger Cars. Illustrates and describes the cars to be operated in the tunnels and subways of the Hudson Companies in New York city and vicinity. 1500 w. Am Engr & R R Jour—Oct., 1907. No. 87418 C.

Superheating.

Are We Superheating Along Correct Lines? Editorial discussion of this subject. 1100 w. Ry & Engng Rev—Oct. 5, 1907. No. 87499.

Tires.

Causes of Defects and Failures of Steel Tires. George L. Norris. Read before the W. Ry. Club. An illustrated discussion of this subject. 2500 w. R R Gaz—Oct. 25, 1907. Serial. 1st part. No. 87931.

Wheel Loads.

Greater Loads on Rails. H. V. Wille. Gives accurate data showing the great increase in weight in rolling stock since 1885. 1700 w. Ir Age—Oct. 3, 1907. No. 87379.

Wheel Stresses.

Pressure of Locomotive Wheels Against the Rails. George L. Fowler. A report of recent investigations of lateral stresses. 500 w. R R Gaz—Sept. 20, 1907. No. 87910.

Valve Gear.

Express Passenger Engine, Midland Railway—The Valve Gear. Illustrates and describes the valve gear of the latest passenger engine for this line. 500 w. Engr, Lond—Sept. 20, 1907. No. 87348 A.

NEW PROJECTS.

Brazil.

The Survey of the Madeira and Mamore R. R. in Brazil. Ernest H. Liebel. Gives a general description of the territory through which the proposed line will pass, and reports the experiences of the engineers on the preliminary exploration survey. 4000 w. Eng News—Oct. 24, 1907. No. 87875.

Great Western Ry.

New Route Between Bristol and Birmingham. An illustrated description of a new direct route scheme of the Great Western Ry. 2800 w. Engr, Lond—Oct. 11, 1907. No. 87840 A. .

Northern Pacific.

The New Tacoma-Tenino Line of the Northern Pacific. H. Cole Estep. An illustrated description of this section of the Northern Pacific. 1000 w. R R Gaz—Oct. 11, 1907. No. 87927.

Philippines.

Philippine Railroad Building with Filipino Builders. Outlines the work and gives an illustrated description of construction work. 3500 w. R R Gaz—Sept. 13, 1907. No. 87909.

Portland and Seattle. .

The Portland & Seattle Railway. H. Cole Estep. Illustrates and describes a road being built to get a direct low-grade freight between eastern Washington and the coast. 2000 w. R R Gaz—Sept. 27, 1907. No. 87917.

South Africa.

The Amabele - Butterworth Railway, South Africa. Brief description of this . new line and its construction. Its length is 78 miles. 1500 w. Eng Rec—Oct. 5, 1907. No. 87445.

Southern Railway.

The Southern's New Line from Jasper, Ind., to French Lick. Illustrated description of the construction of a new line, 24.8 miles in length. 1400 w. R R Gaz—Sept. 6, 1907. No. 87903.

Western Pacific.

Engineering Features of the Western Pacific Railway. George P. Low. Map, profile, and illustrations, with description. 4500 w. Eng Rec—Sept. 28, 1907. No. 87304.

PERMANENT WAY AND BUILDINGS.

Car Dumping.

New Car-Dumping Devices (Neuere Wagenkipper). Georg v. Hanffstengel. Illustrated detailed description of three installations for dumping coal and ore cars bodily, recently developed in Germany. 4400 w. Zeitschr d Ver Deutscher Ing—Sept. 28, 1907. No. 87653 D.

Concrete Ties.

Experiments with Reinforced Concrete Ties in America. W. M. Camp. Describes some of the experiments made with these ties. Ills. 4000 w. Bul Int Ry Cong—Sept., 1907. No. 87427 E.

Earth Slides.

Earth Slides. From a bulletin contributed by H. Rohwer, to the Am. Ry Engng. & Main. of Way Assn. Describes slides coming to the notice of the writer and the remedies. 1700 w. Eng Rec—Oct. 5, 1907. No. 87446.

The Otavi Railway in South Africa. The Longest 24 in. Gage Railway in the World. Illustrates and describes a recently completed road of light construction. 2000 w. Eng News—Oct. 10, 1907. No. 87548.

We supply copies of these articles. See page 559.

Rails.

Wheel-Loads and Rail-Weights. George Sherwood Hodgins. Shows the interrelation of rail weight and wheel load, giving a comparison of the rise and fall of rail sections with the development of the locomotive. Ills. 2500 w. Cassier's Mag—Oct., 1907. No. 87386 B.

Scales.

The Standard Track Scale. Illustration and brief description of a 600-ton suspension steel frame scale. 700 w. Ir Age—Oct. 3, 1907. No. 87378.

Shops.

See under Motive Power and Equipment.

Terminals.

A Few Points on Railroad Terminals, with Special Reference to Roundhouses. R. E. W. Hagarty. Deals with the general layout of terminal yards and discusses some of the structures. 2500 w. Trans Eng Soc, Univ of Toronto—No. 20. No. 88132 N.

Timber.

See under Civil Engineering, Materials of Construction.

Track Reversal.

Track Work Involved in the New York Central's Reversal of the Direction of Traffic. Describes the work involved. 1800 w. R R Gaz—Sept. 13, 1907. No. 87908.

Tunnels.

See under Civil Engineering, Construction.

Widening.

Widening Operations on the Great Central Railway. An account of extensive improvements on this British railway. 1000 w. Engr, Lond—Oct. 18, 1907. No. 87963 A.

Yards.

The Pitcairn Yard of the Pennsylvania Railroad. Plan and description. 4000 w. Eng Rec—Oct. 12, 1907. No. 87569.

TRAFFIC.

Bills of Lading.

The Uniform Bill of Lading. R. L. Ardrey. Analysis of the hearing before the Interstate Commerce Commission. 2000 w. Ir Age—Nov. 7, 1907. No. 88138.

Car Efficiency.

Car Efficiency. Addresses by E. B. Boyd, and Arthur Hale, before the Traffic Club of Chicago. 5000 w. R R Gaz—Nov. 1, 1907. No. 88028.

Demurrage.

Reciprocal Demurrage. W. Heyward Drayton. Address before the Traffic Bureau of the Illinois Mfrs. Assn. Considers the questions of non-furnishing of cars, and the failure to move a car a specified number of miles per day. 1600 w. Ry Age—Oct. 4, 1907. No. 87495.

Freight Rates.

Do Reductions in Freight Rates Benefit the Masses? From a pamphlet concerning the effects of reductions of freight and passenger rates issued by Joseph M. Brown. Shows that reduction in freight rates brings no benefit to the masses of the people. 4000 w. Ry Age—Oct. 4, 1907. No. 87496.

Steam vs. Electric.

Electric Railway Competition. Ray Morris. Gives a summary of information obtained, discussing the results shown and the development of interurban lines. 2500 w. R R Gaz—Oct. 11, 1907. No. 87929.

MISCELLANY.

Algerian Railways.

African Railroads in Algeria and Tunis. From *Moniteur Industriel*. Reviews the history of these lines. 2000 w. R R Gaz—Oct. 4, 1907. No. 87926.

Australian Railways.

Australian Railways (Les Chemins de Fer Australiens). Paul Privat-Deschanel. Gives a history of their development and a general description of permanent way. stations, rolling stock, traffic, new projects, etc. Ills. Serial. 1st part. 5500 w. Génie Civ—Sept. 7, 1907. No. 87609 D.

British Railways.

British Railway Economics. Editorial discussion of the returns issued by the Board of Trade. 3000 w. Engng—Oct. 18, 1907. No. 87958 A.

Italian Railways.

Travel Notes from Italy, Referring Especially to the Milan Exposition, 1906 (Reisebeobachtungen aus Italien und insbesondere von der Mailänder Ausstellung 1906). Herr Cauer. The first part of the serial discusses the transportation facilities of the exposition and describes some of the exhibits. Ills. 4500 w. Serial. 1st part. Glaser's Ann—Sept. 15, 1907. No. 87638 D.

Regulation. .

The Railroad Problem. Robert Mather. From an address before the Chicago Assn. of Commerce. Discusses the subject of state regulation. 5800 w. R R Gaz—Oct. 18, 1907. No. 87930.

Union Pacific.

The Growth of the Union Pacific and Its Financial Operations. Thomas Warner Mitchell. An account of its organization and reconstruction, its growth and activities. 14000 w. R R Gaz—Sept. 13, 1907. No. 87906.

Water vs. Rail.

The Complementary Nature of Water Transportation as Allied to Transportation by Rail. J. T. Harahan. An address before the deep waterway convention, at Memphis, Tenn. 1700 w. Ry Age—Oct. 11, 1907. No. 87580.

We supply copies of these articles. See page 559.

STREET AND ELECTRIC RAILWAYS

Car Barn.
See under Reinforced Concrete, Civil Engineering, Construction.

Cars.
Pressed Steel Pay-as-You-Enter Cars for Montreal. Illustrated detailed description of new cars. 800 w. St Ry Jour—Oct. 5, 1907. No. 87492.

Rolling Stock for the Washington, Baltimore & Annapolis Electric Railway. E. P. Roberts. Illustrated description. 2500 w. Elec Ry Rev—Oct. 12, 1907. No. 87579.

Catenary Lines.
Recent Improvements in Catenary Line Construction and Methods of Installation. Gives details of the latest work of the General Electric Co. in line construction, describing designs and methods of installation. 4000 w. St Ry Jour—Oct. 26, 1907. No. 87932.

Control Apparatus.
Recent Improvements in Control Apparatus for Railway Equipments. F. E. Case. Describes new features of multiple-unit and hand control. 1500 w. St Ry Jour—Oct. 19, 1907. No. 87816.

Croydon.
Croydon Corporation Tramways. History and illustrated description of these lines and their equipment. 3800 w. Tram & Ry Wld—Oct. 3, 1907. No. 87825 B.

Depreciation.
Where Maintenance Ends and Depreciation Begins. J. H. Neal. Read before the Am. St. & Int. Ry. Acc. Assn. Discusses the relations between maintenance and depreciation. 1200 w. St Ry Jour—Oct. 19, 1907. No. 87814.

Equipment Depreciation and Renewal. William Mahl. Gives statements bearing on equipment renewals and replacements, discussing the difficulties of keeping equipment accounts. 2500 w. R R Gaz—Oct. 11, 1907. No. 87928.

Electrification.
The Field for Electricity on Steam Railways. Frederick Darlington. Reviews the progress of electricity as a motive power on railroads, showing its superiority and reliability for certain kinds of work, and also its economy. 3500 w. Ry Age—Oct. 18, 1907. No. 87796.

Equipment.
Electrical Equipment of Main Line Railways in Europe. Philip Dawson. Discusses progress abroad in heavy electric traction, particularly in the field of single-phase motors. Ills. 6800 w. St Ry Jour—Oct. 12, 1907. No. 87782.

Erie R. R.
Single-Phase Electric Motive Power on the Rochester Division of the Erie Railroad. W. N. Smith. Illustrated detailed description. 9000 w. Eng Rec — Oct. 12, 1907. No. 87567.

Single-Phase Electric Traction on the Rochester Division of the Erie R. R. W. N. Smith. Illustrated detailed description of a noteworthy installation on the single-phase system. 9000 w. Eng News—Oct. 17, 1907. No. 87738.

European.
See under Equipment.

Freight Traffic.
Light Freight Handling by Electric Lines. P. P. Crafts. Read before the Am. St. & Int. Ry. Assn. Considers methods of making the freight traffic successful 2500 w. St Ry Jour—Oct. 26, 1907. No. 87933.

Freight Service on Electric Railways. H. H. Polk. Read before the Am. St. & Int. Ry. Assn. Discusses methods, rates, equipment and related matters. 3000 w. St. Ry Jour—Oct. 26, 1907. No. 87934.

Locomotives.
Electric Locomotives of the Pennsylvania Railroad. Illustrated description of the two direct-current locomotives now undergoing tests. 1200 w. R R Gaz—Sept. 20, 1907. No. 87913.

See under Equipment, N. Y., N. H. & H. R. R., New York Central R. R., Long Island R. R., and West Jersey and Seashore R. R., and West Shore R. R.

Long Island R. R.
Operating Features of the Long Island City Power Station of the Pennsylvania Tunnel and Terminal Railroad Co. Several articles on the equipment and electrical operation of the Lond Island R. R. Ills. 12000 w. St Ry Jour—Oct. 12, 1907. No. 87778.

New York Central R. R.
General Features of the New York Central Electrification. A collection of articles dealing especially with the operating practice of stations, sub-stations, etc., and of methods of conducting transportation. Fully illustrated. 30600 w. St Ry Jour—Oct. 12, 1907. No. 87776.

New York City.
Rapid Transit Conditions in New York. Diagrams and interesting information from the report for 1906 of the Board of Rapid Transit Railroad Commissioners for the City of New York. 3500 w. R R Gaz—Sept. 20, 1907. No. 87914.

We supply copies of these articles. See page 559.

N. Y., N. H., & H. R. R.

Operation of Electric Locomotives by the New Haven Railroad. On the operation of the combined single-phase, direct-current locomotives. 8500 w. St Ry Jour —Oct. 12, 1907. No. 87779.

Power Plants.

See under Central Stations, Electrical Engineering, Generating Stations.

Rails.

The Use of the T-Rail in Cities. C. Gordon Reel. Read before the Am. St. & Int. Ry. Assn. Gives an account of satisfactory use at Kingston, N. Y. Diagrams. 1000 w. St. Ry Jour—Oct. 26, 1907. No. 87994.

Service Plant.

Line and Track Service Plant, Brooklyn Rapid Transit Company. Plans and description of interesting details of an elaborate plant under construction. 1500 w. Elec Ry Rev—Oct. 5, 1907. No. 87482.

Shops.

New Shops and Car Houses at Knoxville—Shop Practice. Illustrated article giving details regarding these shops and the shop practice. 1500 w. St Ry Jour—Oct. 5, 1907. No. 87491.

Signals.

Solenoid Signals on the Manhattan Elevated. Illustrates and describes the semaphore block signals recently put in service on curves. 800 w. R R Gaz—Sept. 20, 1907. No. 87911.

Single Phase.

General Notes on the Development and Future Prospects of the Application of Single-Phase Current to Electric Railroads (Allgemeine Gesichtspunkte über die Entwicklung und die Aussichten des Einphasenstrom-Bahnbetriebes). A. Heyland. Considers the problem both theoretically and practically. 3200 w. Serial. 1st part. Elektrotech Zeitschr—Sept. 12, 1907. No. 87657 D.

Steel Ties.

Steel Tie and Concrete Construction on Utica and Mohawk Valley Railway System, Utica, N. Y. M. J. French. Illustrates and describes recent work, giving cost. 2500 w. St Ry Jour—Oct. 12, 1907. No. 87774.

Steel Ties in Germany. Dr. Ing. N. C. A. Haarmann. Reports the increasing use of steel ties, illustrating and describing the writer's "hook plate." 1200 w. R R Gaz—Sept. 29, 1907. No. 87915.

Structures.

The Influence of the Design of Railway Structures on Economy of Operation. H. T. Campion and William McClellan. Notes on materials and design, particularly in connection with the smaller systems. Ills. 3500 w. St Ry Jour—Oct. 19, 1907. No. 87819.

Subways.

Paris Subway Stations. George B. Ford. Illustrates and describes some of the entrances to subway stations in Paris, comparing them with work in London and New York. 1600 w. Am Archt—Oct. 15, 1907. No. 87452.

The Market Street Subway, Philadelphia. Illustrated detailed description. 2500 w. Eng Rec—Oct. 12, 1907. No. 87564.

Systems.

Which Is the Best Electric Railroad System? C. L. de Muralt. Editorial letter aiming to explain the differences between the three rival electric railroad systems. 3500 w. R R Gaz—Sept. 13, 1907. No. 87905.

Timber.

See under Civil Engineering, Materials of Construction.

Track Maintenance.

Care of Electric Railway Tracks. George L. Wilson. Read before the Am. St. & Int. Ry. Assn. Suggestions for the care of interurban and city tracks. 3000 w. Eng Rec—Oct. 19, 1907. No. 87788.

Track Reconstruction.

Track Reconstruction by the Chicago Electric Traction Lines. Illustrates and describes the three types of construction in the work already done, and the general characteristics of the work. 1200 w. Elec Ry Rev—Oct. 5, 1907. No. 87481.

The Rehabilitation of the Tracks of the Chicago Street Railway. An illustrated description of the general features of the work required by the city council. 3500 w. Eng Rec—Oct. 19, 1907. No. 87786.

Trucks.

Long Wheel-Base Trucks. R. L. Acland. Read before the Munic. Tram. Assn., at Manchester, Eng. Presents the merits of these trucks for street cars. Ills. 3000 w. Elec Engr, Lond—Sept. 27, 1907. No. 87512 A.

Practical Views on Trucks for Electric Motor Service. Franklyn M. Nicholl. Discusses the practicability of the principles of the M. C. B. truck in their application to electric service requirements. 800 w. Elec Ry Rev—Oct. 19, 1907. No. 87813.

West Jersey & Seashore R. R.

The Power Station Practice of the West Jersey & Seashore Railroad. With related articles describing the electrical operation of this line. Ills. 9000 w. St Ry Jour—Oct. 12, 1907. No. 87780.

West Shore R. R.

Electrical Operation on the West Shore Railroad. Illustrated account of the operation between Utica and Syracuse, N. Y. 5000 w. St Ry Jour—Oct. 12, 1907. No. 87777.

We supply copies of these articles. See page 559.

EXPLANATORY NOTE—THE ENGINEERING INDEX.

We hold ourselves ready to supply—usually by return of post—the full text of every article indexed in the preceding pages, *in the original language*, together with all accompanying illustrations; and our charge in each case is regulated by the cost of a single copy of the journal in which the article is published. The price of each article is indicated by the letter following the number. When no letter appears, the price of the article is 20 cts. The letter A, B, or C denotes a price of 40 cts.; D, of 60 cts.; E, of 80 cts.; F, of $1.00; G, of $1.20; H, of $1.60. When the letter N is used it indicates that copies are not readily obtainable and that particulars as to price will be supplied on application. Certain journals, however, make large extra charges for back numbers. In such cases we may have to increase proportionately the normal charge given in the Index. In ordering, care should be taken to *give the number* of the article desired, not the title alone.

Serial publications are indexed on the appearance of the first installment.

SPECIAL NOTICE.—To avoid the inconvenience of letter-writing and small remittances, especially from foreign countries, and to cheapen the cost of articles to those who order frequently, we sell coupons at the following prices:—20 cts. each or twelve for $2.00, thirty-three for $5, and one hundred for $15.

Each coupon will be received by us in payment for any 20-cent article catalogued in the Index. For articles of a higher price, one of these coupons will be received for each 20 cents; thus, a 40-cent article will require two coupons; a 60-cent article, three coupons; and so on. The use of these coupons is strongly commended to our readers. They not only reduce the cost of articles 25 per cent. (from 20c. to 15c.), but they need only a trial to demonstrate their very great convenience—especially to engineers in foreign countries, or away from libraries and technical club facilities.

Write for a sample coupon—free to any part of the world.

CARD INDEX.—These pages are issued separately from the Magazine, printed on one side of the paper only, and in this form they meet the exact requirements of those who desire to clip the items for card-index purposes. Thus printed they are supplied to regular subscribers of THE ENGINEERING MAGAZINE at 10 cents per month, or $1.00 a year; to non-subscribers, 25 cts. per month, or $3.00 a year.

THE PUBLICATIONS REGULARLY REVIEWED AND INDEXED.

The titles and addresses of the journals regularly reviewed are given here in full, but only abbreviated titles are used in the Index. In the list below, *w* indicates a weekly publication, *b-w,* a bi-weekly, *s-w,* a semi-weekly, *m,* a monthly, *b-m,* a bi-monthly, *t-m,* a tri-monthly, *qr,* a quarterly, *s-q,* semi-quarterly, etc. Other abbreviations used in the index are: Ill—Illustrated: W—Words; Anon—Anonymous.

Alliance Industrielle. *m.* Brussels.
American Architect. *w.* New York.
Am. Engineer and R. R. Journal. *m.* New York.
American Jl. of Science. *m.* New Haven, U. S. A.
American Machinist. *w.* New York.
Annales des Ponts et Chaussées. *m.* Paris.
Ann. d Soc. Ing. e d Arch. Ital. *w.* Rome.
Architect. *w.* London.
Architectural Record. *m.* New York.
Architectural Review. *s-q.* Boston.
Architect's and Builder's Magazine. *m.* New York.
Australian Mining Standard. *w.* Melbourne.
Autocar. *w.* Coventry, England.
Automobile. *w.* New York.
Automotor Journal. *w.* London.
Beton und Eisen. *qr.* Vienna.
Boiler Maker. *m.* New York.
Brass World. *m.* Bridgeport, Conn.
Brit. Columbia Mining Rec. *m.* Victoria, B. C.
Builder. *w.* London.
Bull. Bur. of Standards. *qr.* Washington.

Bulletin de la Société d'Encouragement. *m.* Paris.
Bulletin of Dept. of Labor. *b-m.* Washington.
Bull. Soc. Int. d'Electriciens. *m.* Paris.
Bulletin of the Univ. of Wis., Madison, U. S. A.
Bulletin Univ. of Kansas. *b-m.* Lawrence.
Bull. Int. Railway Congress. *m.* Brussels.
Bull. Tech. de la Suisse Romande. *s-m.* Lausanne.
California Jour. of Tech. *m.* Berkeley, Cal.
Canadian Architect. *m.* Toronto.
Canadian Electrical News. *m.* Toronto.
Canadian Engineer. *m.* Toronto and Montreal.
Canadian Mining Journal. *b-w.* Toronto.
Cassier's Magazine. *m.* New York and London.
Cement. *m.* New York.
Cement Age. *m.* New York.
Central Station. *m.* New York.
Chem. Met. Soc. of S. Africa. *m.* Johannesburg.
Colliery Guardian. *w.* London.
Compressed Air. *m.* New York.
Comptes Rendus de l'Acad. des Sciences. *w.* Paris.
Consular Reports. *m.* Washington.

VOL. XXXIV.   JANUARY, 1908.   No. 4.

THE MECHANICAL MANAGEMENT OF THE WORLD'S STOCK OF GOLD.

By Alex. Del Mar.

The larger undertakings in which the engineer is concerned are intimately connected with problems of finance. But above all he is interested in the application of practical economy and common sense in the physical operations of business. Recent events have called attention to a striking opportunity for the exercise of these qualities to prevent the constant, wasteful, and inconvenient transport of actual gold coin or bullion to and fro across the Atlantic, sometimes making voyage after voyage in the same unopened packages, when a far more scientific method of balancing the gold stock should be adequate for all purposes of international commerce and exchange. Mr. Del Mar's interesting proposal has thus the elements of timeliness and of practicability to recommend it to the attention of an engineering and industrial audience.—THE EDITORS.

THE precious metal, whose supply to the world now amounts in value to something over a million dollars per diem, is nevertheless so mechanically managed that it fails to subserve fully the tremendous interests which depend upon it. It is to this purely mechanical problem, and not to any other relation of gold except so far as such relation may flow from the mechanism, that the present article will be devoted.

Gold is used not only for the purposes of money, but also in the arts; and as the very first feature of the problem concerns the quantity to be managed, it becomes necessary at the outset to estimate the relative quantities reserved for money, and used up in the arts; next to consider whereabouts the Stock lies; and finally to offer such a plan for mobilising it, or some material portion of it, as will render it most subservient to the interests of the commercial world.

PROPORTION USED IN THE ARTS.

This subject has engaged the attention of numerous enquirers; the usual method being to address circular letters to the jewelry, dentistry, gilding, and other trades using up gold, and to make conjectural allowances for such as fail to answer the enquiries fully. This method is so defective that it has never been even approximately successful. To cover the ground completely, it should embrace the entire commercial world; and the replies should be comprehensive and truthful. But no official has authority to compel replies even in his own country, much less in foreign ones; and no manufacturer or consumer of gold in the arts cares to disclose to others the extent of his business operations. Enquiries of this character have therefore always ended in failure. The writer of this article, once an officer of the United States Treasury Department, has had considerable practical experience in the matter; and he is satisfied that no return of the sort ever made, either before or since, is worth the paper it is written upon. Such returns seldom amount to more than 25 per cent of the production; and even when eked out by conjecture, rarely exceed 35 or 40 per cent.

The only reliable method of obtaining such information is, from the entire quantity of the metal reported produced, to deduct the quantity of the discernible Stock on hand. The quotient must show the amount consumed in the arts, plus the quantity lost by attrition or accident, hidden in forgotten hoards, or exported to the uncommercial world. The quantity of gold hoarded is too small to deserve consideration. The results obtained by this method prove that the usual estimates do not embrace more than about one-half of the quantity really consumed in the arts, or otherwise lost to commerce. In other words, the supplies of gold to the visible Stock, after deducting recoinages, seldom exceed 25 per cent of the reported production; and at one period, 1828 to 1838, did not amount to more than 18 per cent. The details, which embrace the entire period, by short intervals, from 1675 to the present time, will be found in the writer's "History of the Precious Metals." These proportions include both gold and silver, because until recent years the Stocks on hand were always reported or estimated in both metals combined. While it is not believed that the proportion of gold consumed in the arts, etc., to the whole amount produced, is quite so great as that of both metals combined, it is nevertheless certain that it is not much below it; a belief founded upon the coinages of new gold and upon the reported Stocks of that metal on hand.

GOLD IN SIGHT.

Reverting now to this subject, there is to be discerned today in the various treasuries, banks and other depositaries of the commercial world, a Stock of gold coin and bullion equal in value to about three thousand three hundred million dollars. Without discussing that fur-

ther sum which is estimated to be in circulation outside of the treasuries and banks—usually a matter of conjecture, and always one of
dispute—it becomes necessary to set forth where this Stock is deposited, as preliminary to the discussion of how it may best be mechanically managed.

GOLD STOCK OF THE COMMERCIAL WORLD AT LATEST DATES.

	Millions of dollars.
United States Treasury, November 1, 1907	916
Bank of France, November 14	541
State Bank of Russia, November 5	508
Bank of Austro-Hungary, November 7	229
Bank of Italy, October 20	167
Bank of England, November 14	159
Imperial Bank of Germany, November 9	146
Bank of Spain, November 10	78
Bank of the Netherlands, November 9	38
Bank of Naples, October 10	35
Bank of Scotland, September 21	26
National Bank of Denmark, October 31	24
National Bank of Belgium, November 7	21
Royal Bank of Sweden, September 30	21
National Bank of Roumania, November 2	20
Bank of Ireland, September 26	16
Switzerland, banks of issue	13
National Bank of Switzerland, November 7	12
Germany, local banks, October 31	11
Bank of Sicily, October 10	9½
Bank of Norway, September 30	9
National Bank of Bulgaria, October 13	6
Bank of Portugal, October 16	5½
Bank of Finland, October 15	5½
National Bank of Servia, September 30	3
National Bank of Greece, September 30	0½
	3,020

Add for omitted banks and depositaries in the United States
and for depositaries in Canada, Mexico, Central and South
America, the European and American Colonies, Turkey,
Egypt, India, Japan, and the commercial ports of China, say, 280

Grand total of Gold in treasuries and banks.............. 3,300

Of the 916 millions of Gold Stock reported by the United States
Treasury, 749 millions were "Trust Funds"; in other words, they
were held against an equal sum of gold certificates at issue. These
certificates and the Clearing House certificates largely figure in the
bank accounts as "gold." On the one hand, they are as good as gold,
because gold can be obtained for them on demand. On the other hand,
to add them to the gold in the Treasury would be to count the same
metal twice over. From this Table it appears that the United States
Treasury holds nearly 30 per cent of all the gold in treasuries and
banks throughout the commercial world. The United States Treasury
is therefore in an excellent position to undertake any reform in the

"Mechanical Management of the Gold Stock" that may promise to render it more effective. Taken in connection with the enormous crops and mineral and manufacturing production of the country, it also affords a guarantee of financial strength that can scarcely fail to attract universal attention.

It is a popular delusion that the present method of managing this Stock leaves it substantially free to flow whither it may be demanded in exchange; but in fact it is attached, by means of numerous and often hidden ligatures and restrictions, to the country which has succeeded in securing its possession. For example, in Great Britain, the coinage of gold, or the conversion of gold bullion, which is *not* legal tender, into gold coins which *are* legal tender, is nominally free, but really restricted. The Mint will not receive for coinage less than £10,000 worth (about $50,000) in any one deposit. This rule practically throws the entire Mint purchases of gold into the Bank of England, which, although obliged to receive it at a fixed price, is nevertheless at liberty to send it first to its own refiners and assayers, with the result that the depositor is subjected both to expense and delay; conditions that, apart from a legally unrestricted Bank rate of interest, seriously hamper the free movement of the metal, either for coinage or export. In France, Germany, Austria, Russia, etc., there are other but somewhat similar restrictions. In all these countries, as well as in the United States, the laws require a certain amount of gold to be held as reserves against Government or bank issues, or for other purposes. Some of the countries named retain large sums of gold as a War Chest. Without going into details, which a lack of space forbids, the general result is that so far from being free to respond to the demands of trade, the world's available Stock of Gold is so tied up, that a demand from one country to another, even in exchange for securities or commodities offered at a depreciation of 25 to 30 per cent, is responded to with the greatest reluctance, difficulty, and delay. Recent events have shown what great efforts and sacrifices were required to draw from Europe to the United States an amount of this metal equivalent to less than a single year's production of the mines of the latter country alone. It is quite warrantable to estimate that no conceivable circumstances would suffice to invite across the ocean so much as a single year's production of all the mines of the world—the fact being that gold enters into the arts and the coinages of the commercial world almost as soon as it is produced; and that once thus engaged, it is, as the case stands, almost impracticable to withdraw it in any large amounts from the countries of consumption or employment.

There are some minor, though not unimportant, obstacles to the mobility of the Gold Stock. The expenses of shipment, including freight, insurance, interest and recoinage, are by no means negligible obstacles. The recent depreciation in cotton and wheat bills reflects only a portion of these obstacles; they are also reflected in the fall of shares and other securities.

REMEDY PROPOSED.

It is believed that there is no insuperable difficulty in the way of remedying this immobility of the Gold Stock and of rendering it amenable to the demands of legitimate commerce. That there *are* difficulties must be freely admitted. As the case stands, each great commercial State requires, and must have, command of the means to liquidate its paper issues in gold; but there is no essential reason why the whole Stock of gold should not be rendered more readily amenable to such sacrifices as commerce and credit are able or willing to concede in the interest of restoring their own equilibrium. The honor and resources of each State are pledged to the payment of its obligations, and they will therefore have to be paid. But there is no necessity for each State selfishly to accumulate gold, and to render it difficult for its neighbors to obtain it when required by the exigencies of legitimate commerce—and to obtain it, too, without being obliged to sell securities and products at bankruptcy prices.

The remedy proposed is purely mechanical. It is to mobilize the entire Stock of gold held by the contracting States, by means of issuing, against such Stock, certificates of deposit, which shall be made legal-tenders in all of the contracting States, except at the Treasury of the State of issue. Each State shall substitute such certificates in place of the gold, for all purposes for which the gold itself is now employed; and shall undertake to pay them on demand. The security afforded by such certificates would be just as good as—nay, even better than—that of the gold itself. The expense entailed and time lost in conveying the metal to and fro across the ocean and of recoining it, would be avoided; and in case of urgent demand from either side, or as between the first-class powers (the only ones which need at yet to be considered in this connection) the certificates would respond to the demands of commerce and of exchange, with a celerity and certainty that cannot be imparted to the metal itself.

Sovereign States have frequently admitted into their monetary circulation, with full legal-tender power, the coins of other States. Therefore there can be no objection in principle to their admitting the coin certificates of deposit herein proposed. The chief coins of Spain were formerly full legal tenders in the United States; those of Portugal were legal tenders in England, and those of England are today full legal tenders in Portugal and several other States. A Convention, commonly known as the Latin Monetary Union, dated December 23, 1865, taking effect August 1, 1866, was entered into between France, Belgium, Switzerland and Italy, which provided for a uniform system of coinage based upon the franc and for the admission to full legal-tender privileges in all of the contracting States of all the gold and certain of the silver coins of each State. In 1867 this Convention was joined by the States of the Church and by Greece and Roumania. On September 20, 1872, a similar Convention, commonly known as the Scandinavian Union, was made between Denmark, Sweden, and Norway, providing for a uniform system of coinage based upon the kroner and for the admission to full legal-tender privileges in all of the contracting States of all the gold and certain of the silver coins of each State. These provisions were adopted in the Danish, Swedish, and Norwegian laws of May 23, 1873, May 30, 1873, and June 4, 1873, and by the Treaties of May 27, 1873, and October 16, 1875, which went into effect April 1, 1876, and were rendered obligatory from January 1, 1877. Many other instances of this kind are mentioned in the writer's "History of Monetary Systems," and "The Science of Money." Already we have an International Postal Union and Money Order system which deals in credits amounting to hundreds of millions per annum. During the present year, alone, the United States Post Office Department reports having transmitted over 72 million dollars (equivalent to gold) to other countries, in the form of drafts or bills of exchange drawn upon foreign postal departments. This is a sum substantially equal to the whole amount of gold imported during the late financial flurry. Why not, then, a system of International legal-tender certificates, backed by deposits of gold coin to the full amount of their issue, but, unlike the coin, full legal-tenders in each and all of the contracting States? When wanted at home, as a basis for other issues, they could be locked up in the Treasury; when wanted abroad they could be used at once without expense of carriage or of recoinage.

Whatever objections to this plan may arise from considerations connected with the guaranteed domestic issues of each State are at once disposed of by the reply that the same objections are equally valid in respect to the coins or bullion now held against such issues. There are, in fact, only three difficulties in the way. One is, where shall the Stock be deposited? Virtually a mechanical consideration; for it is not to be supposed that the United States, Great Britain, France, Germany, Austro-Hungary, Italy, or Russia could find reason to distrust one another for the sake of a sum of gold that would not equal a tithe of the current contracts which would be voided by a declaration of war.

The second objection is the political one of observing such a contract in case of hostilities. But as the payment of national obligations, the rights of neutrals, and many other equally important matters, are now by convention preserved from the consequences of war, there can be no great difficulty in arranging this one. Besides, as the full legal-tender quality of the proposed certificates would depend not merely upon the disposition of any two States which might go to war, but upon the Treaty obligations and rights of all the contracting States, there need be no fear that the Convention would not be scrupulously observed by all of the signatory powers. The third objection is that of denominations. This has been already solved by a scale which enables large sums in pounds-sterling, francs, marks, florins, roubles and dollars, to be expressed in national integers of equivalent gold weights.

The only single unanswered difficulty is: Where shall the Stock be deposited? The solution of this problem lies on the surface. Let each State keep its own Stock; let it be agreed by Treaty that each State shall only be at liberty to issue International Certificates of Deposit equal to its own reserved Stock of gold coins held in the public Treasury; and let the Certificates be valid only when countersigned and registered by the Commissioners of all the contracting States.

There need be no difficulty in providing an adequate penalty for the infraction of so solemn and important a convention. The plan herein outlined would virtually provide a Bank of the World; and its promise of influence in securing the peace of that world should be great enough to sweep away any objections to its adoption that may be raised by either class interest, or diplomatic intrigue.

DETAILS OF MANAGEMENT.

Assuming that the public advantages of this plan may commend it to the favorable consideration of Government, certain details of management may be connected with it that would still further enlarge its scope, augment its usefulness, and procure for it the support of those numerous interests, without whose co-operation it is often difficult to secure attention for mechanisms which, like the present one, are proposed for the public welfare.

Selecting for illustration only one of the details alluded to, it is to be remarked that the present system of remitting small sums of money to places within the country are cumbrous, expensive, and unsafe. It is a choice, on the one hand, of a postal order, which in thousands of towns and villages cannot be obtained at all, and which, where attainable, costs far too much, both in time lost and in traveling expenses and fees; or, on the other hand, of a remittance by letter of coins or notes and postage stamps; always an unsafe procedure.

The aggregate of money orders required in the near future within the United States is estimated by the Post Office officials at about 400 or 500 million dollars per annum. The amount likely to be remitted in money and stamps is probably as great. It is believed that with additional facilities the money orders would readily increase to 1,000 millions, with a corresponding increase in trade, which trade is now wholly neglected for the lack of remitting conveniences. The remittances of money and stamps would of course decrease, but not nearly to the extent that the money orders would increase. A somewhat similar remark is applicable to the small remittances between this and foreign countries. It has been estimated that were foreigners in South America and the Orient provided with a convenient method or mechanism for remitting money to the United States, especially from interior towns, it would result in doubling the export trade of manufactured goods and machinery to those countries.

To provide this mechanism, and without any other expense than paper and printing, all that is necessary is to print, say, 50 million dollars of the proposed Certificates of Deposit in small sums, leaving a blank space on each certificate, after the "Promise to pay (so many) dollars," in which space the name and address of a payee may be written by the remitter. This would at once convert the Certificate into a Bill of Exchange, *e. g.,* on the United States Treasury. When made thus available in distant countries such Certificates would be eagerly sought after and command an extensive circulation abroad, with a corresponding accession of gold at home. After payment at the Treasury, the Certificates could be destroyed and re-issued in blank (or to Bearer,) as before. They should be engraved in several languages, side by side.

When it is recalled that nearly 300,000 letters containing money or postage stamps are annually handled in the Dead Letter Office, for lack of proper addresses or for other reasons, it will be perceived how much benefit to the public from employed remittances, and profit to the Government from unclaimed remittances, would follow such a reform in the Management of the Gold Stock.

THE MANAGEMENT OF PRODUCTION IN A GREAT FACTORY.

By George F. Stratton.

DIVISION OF RESPONSIBILITY AND AUTHORITY IN THE GENERAL ELECTRIC COMPANY'S SHOPS.

Mr. Stratton's article is the first of a small but significant group to appear in THE ENGINEERING MAGAZINE, giving a graphic idea of the production methods of a few of the leading manufacturing plants in the United States. The establishments selected will all be of world-wide reputation and interest on account of technical and commercial success. The descriptions will be by men intimately familiar with the internal life and movement of the shops. They will not attempt to go so far into detail that the large perspective is lost, but they will concentrate the organic essentials into a clear, easily grasped picture of the whole. They will show how, in these vast works, an army of men is kept in orderly and productive employment on a multitude of operations, all moving harmoniously to the completion of the desired product. The next paper expected in the series will describe the manufacturing system of the Westinghouse Electric & Manufacturing Company.—THE EDITORS.

IN the great manufacturing plant at Lynn, Mass., where more than ten-thousand hands are employed, visitors who satisfy the manager that they are something more than curious sight-seers are readily admitted and furnished with guides. These visitors may be officials of other great plants—engineers, managers, or superintendents—or they may be proprietors or employees of very modest shops, working perhaps on specialties. But whoever they are, the questions which they put to the guides with the greatest frequency and interest relate to the management. What impresses a man first, and with apparently the greatest force, is the control of so large a force of workers, and the movement of great quantities of material and parts in such rhythmical order and succession as to keep every man at work and working smoothly.

One of the guides, a young and exceedingly bright engineer, has a set formula for reply to questions on this subject. "The manager," he will say, "is at the top. He takes all the kicks and all the credit. The engineering department designs all the work—tells the men what to make; the superintendent of production tells them how much, or how many to make; and the mechanical superintendent tells them how to make it, and sees that they do it."

This is concise. Its conciseness needs no qualifying adjective. Visitors who are themselves engaged in large affairs recognize this and often, with twinkling eyes, appreciate the futility of seeking detailed information during the necessarily hurried tour through the thirty or forty shops comprising the plant. Smaller men are awed by it. They recognize its sweeping comprehensiveness and are undoubtedly hungry for details, but, fortunately for the guide, searching questions are not easy to formulate and the trip usually ends unsatisfactorily to them, as far as concerns insight into the management.

"I would far rather," said one such man to me, as we were watching the dense crowd of workers pouring into the streets at the noon hour, "I would far rather know how these men are handled without confusion, how work is furnished for each one of them without waste of time, and how the stuff is pushed through the various shops, than listen to the most exhaustive description of the plant and the product."

As a matter of fact very many of the employees, even above the grade of workman, do not understand the plan of management. They know to whom to look for orders, or rules on discipline, in their own particular line of work, but beyond that they have little opportunity to acquaint themselves. Moot questions are referred to other departments, and again to others, in apparently a hopelessly confusing manner. Yet there is really no confusion, and seldom anything but a ready and prompt determination of the responsibility—or non-responsibility—(the lines of which are sharply drawn)—of each department.

It is hardly within the scope of this article to do more than briefly mention the administrative department; the board of directors (elected directly by the stockholders and immediately responsible to them for the entire conduct of the business) and the president (appointed by the directors), who usually attends to the finances of the company and frequently dominates its financial policy.

The story of the interesting details of factory management really begins with the vice-president, and although systems vary in different plants, that of the company already alluded to may be taken as a type.

There are two vice-presidents. The first is the head of the commercial system, which comprises the distribution of the manufactured products, the establishment and conduct of sales agencies, and all consequent operations. He is also in control of the engineering department—auxiliary to which and subordinate to it is the draughting force. The engineers design the machines and apparatus to be manufactured. They are responsible for the efficiency of such products and, when required under contract, for the proper installation and

DIAGRAM OF MANUFACTURING ORGANIZATION, GENERAL ELECTRIC COMPANY'S LYNN SHOPS.

testing of special apparatus. Theirs is purely the scientific end of the business. They have nothing to do with the manufacture except to watch closely, as they do, to see that their drawings and instructions are rigidly followed, and that no materials other than those they specify are used. They have no concern in the sales, or the profit and loss. Their concern is entirely with the reputation of the goods; upon which, of course, rest their own personal reputations.

The second vice-president is at the head of the production department—a general charge involving the plant, its equipment, and the entire line of manufacture. Briefly, he makes the goods;—the first vice-president sells them. The second is also in control of the experimental department which covers all mechanical tests and investigations, and the securing of inventions and patents.

The real executive head of the factory proper is the general manager. Although, in many manufacturing plants this official is also in charge of the sales department, in the system we are describing the manager has nothing whatever to do with the commercial end. His responsibility commences with the receipt of orders from the commercial department and ends when those orders are shipped out. That sounds simple, but it involves much. The buildings and equipment must be designed and arranged with the closest study of economy in handling men and materials; the supplies must be purchased with the keenest of foresight and acumen; the working force must be handled with discriminating firmness; the wage schedules must be scrutinized continually and minutely, and every item of shop expense viewed with suspicion and religiously compared with preceding percentages. The manager is responsible in a great measure for manufacturing at prices which will ensure successful competition with other shops. Upon him largely depend the figures on the profit and loss account.

Right at this point it might be well to dwell for a moment upon the irrepressible friction which seems to occur in many large engineering establishments between the engineers and the manager or superintendent. I use the phrase "seems to occur" advisedly. There is rarely any real friction. What seems to be such is simply the arguing by the two departments, from different points of view. As previously stated, the engineers are concerned solely with the efficiency and reputation of the machines built; the manager is concerned chiefly with the cost, and the ever-looming spectre of loss instead of profit. The one department is a healthy check upon the other; a check upon what might develop, on the one hand, into an inordinate extravagance of detail and expensive finish; on the other hand into a disastrous cheese-paring economy, with its concomitants of poor workmanship and inferior material.

Immediately under the general manager are the mechanical superintendent, the purchasing agent, the superintendent of production, and the factory accountant.

The mechanical superintendent has charge of the shops and machinery, and the discipline of the men. He has four assistants: A tool expert, who designs special machines and tools needed in the equipment and who continually watches new machine tools which come on the market, reporting upon their capacities and efficiencies; a superintendent of construction, who builds and repairs the shops. This man has a standing force of nearly one hundred carpenters, masons and plumbers—and at times when a new shop is being erected his force is largely increased; the third assistant is the engineer of power, heat, and lighting. His duties comprise the care of the steam and electric power generators; the over-seeing of the steam engineers, the firemen, and the electricians; and the proper arrangement of steam pipes and lighting wires, etc.; the last assistant sees to the discipline of the hands; the proper working of the time-recording systems; the inquiry into absentees; the engagement of new hands; the orderliness and cleanliness of the shops, and the general sanitary arrangements.

The mechanical superintendent has nothing to do with fixing any wages; they are always determined by the various foremen, after consultation with the general manager. His duties and responsibilities are comprised in the purchase or building of proper equipment, the hiring of proper men, and the direction of those men in carrying out the orders of customers received through the superintendent of production.

The purchasing agent purchases everything required in factory or office—from a lead pencil to ten-thousand tons of coal or pig iron. He has a force of assistants and a system of records by which comparisons with past prices are readily made. He is under the immediate control of the general manager.

The superintendent of production* is also in the general manager's department. He receives shipping or manufacturing orders from the commercial department and distributes shop orders to the various foremen, and to the head shipper. A very important duty of his department is that of keeping track of the stock of finished products ready for shipment, and of such products in course of construction, so as to enable the manager to determine readily the condition of affairs all through the factories, and the possibility of immediate or early shipments when required. It is to this department that the shop foremen come for orders to keep their men going; and upon its intel-

---

* The term "superintendent of production" suggests a range of authority which hardly exists. Although the entire productive energy of the plant is engaged upon orders issued by him alone, and although his proper distribution of these orders is depended upon to keep every department working smoothly and harmoniously, he has no authority to enforce them. That rests with the mechanical superintendent. In effect, he works in co-operation with this officer, both being subject to the manager but not either one to the other.

ligent distribution of orders largely depends the steady and friction-
less flow of work through the numerous shops. No department chief
could so quickly and effectually bring disorder and confusion upon
the entire plant as the superintendent of production.

The shop-accounting department, also under the general manager,
is in three sections. One keeps account of the purchases of supplies,
of shop expenses, direct and indirect, of costs of construction and
equipment, and of shipments. Another, called the cost division, keeps
an elaborately systematized record of the cost of manufacturing. The
third section is the paymaster's—where day and piece work is figured
and the pay-rolls are made up.

Then, a little lower in the scale, come the foremen; several of
whom are in charge of five or six hundred hands each, with perhaps
ten or a dozen assistant foremen under them.

Such, briefly, are the personnel and duties of the various departments of this plant. If the passage of an order through the various
shops is followed, it will clearly show the inter-relation of one department with another.

Assuming that a contract has been obtained for, say, a special,
high-powered, turbo-generator, and that the engineers have prepared drawings of all details; an order comes from the commercial
department to the superintendent of production. It is numbered, and
directs him to build a machine according to specifications carrying the
same number. From that moment, with every detail of that intricate
machine, with every cent's worth of material used in it, with every
stroke of labor expended upon it—that same number is used. When
that engine is finished, the shop records, properly tabulated, show
to a fraction the cost of designing and draughting, of labor, and of
material put into it.

The superintendent of production at once issues shop orders (all
similarly numbered). Some go to the pattern shop for such parts
as will be required, others to the steel, iron, and brass foundries,
others again to the machine shops. The work may be commenced
in a dozen different shops simultaneously, and in all probability the
mechanics engaged on such work will never see the engine assembled
—or any part of it, excepting that which they themselves have finished up. In the meantime, the mechanical superintendent has been
in consultation with his assistant—the tool expert. It may be that,
for this special engine, some new tool or tools will be required. That
matter is settled. It is also possible that additional shop room may be
required—a setting-up floor, or some change in one of the buildings.

The superintendent of construction is called upon, his mill draughtsmen set to work and the necessary addition, or change, at once put under way. With the shop orders sent out to the foremen are blueprints of the detail drawings of all the parts. As soon as they are received the foremen are required to notify the superintendent, at once, of any materials or small tools which will be needed. If satisfied that their requirements are correct, he in turn puts in a requisition to the purchasing agent, who takes proper steps to supply the wants.

When the castings commence to come from the foundries, and are distributed among the various machine and finishing shops, the engineers are constantly on guard. They are as solicitous as any parent can be in the training of his child. Many are the little disputes with foremen, which have to be referred to the superintendent, and perhaps to the manager. There are blueprints everywhere, frequently thumbed and worn so much that a half-dozen successive copies may be called for by some foremen.

In due time the parts come—it may be from eight or ten different shops—into the assembling room, and the engine is set up. Steam pipes are connected and, with the engineers in charge, exhaustive tests are made. When all is satisfactory it is taken down and shipped, and a small force of men under an expert foreman follow it to its destination, where they install it permanently. A final test is then made by the company's engineers and the representatives of the customer.

In the meantime, the cost department has been collecting the details of the cost. That of the labor is obtained from the paymaster; that of the material, from the purchasing agent and from the various stock rooms of the shops where the work has been done. The system used is complete and reliable, the result showing the actual factory cost of the engine, but not including the factory expense, which is figured under a logically determined percentage ratio by the accounting department.

In addition to these regularly constituted authorities there are various committees for various purposes. In this company's manufacturing operations there are two standing committees, to whom matters too momentous—or of too general consequence to the entire plant to be decided by any one department chief—are referred.

One is the manufacturing committee, before which questions of the desirability of taking up new lines of manufacture or discontinuing old ones, are decided. It is formed by strong representatives from the sales department and from the engineering department, with the mechanical superintendent and the superintendent of production. Frequently shop foremen are called into its meetings for detail information.

The other committee is known as the shop plant committee. It decides upon questions of radical changes in the arrangement of existing shops, and the location and equipment of new ones. Its members are the general manager, the mechanical superintendent and his assistants, and the principal shop foremen.

In addition to the irregular and perhaps infrequent meetings of these standing committees, special meetings are frequently called by the general manager to discuss questions of shop policy. The members may be of the office force, or, more frequently, a few of the foremen directly interested, and occasionally of the workmen— selected, perhaps, for their skill on special lines or, which is more likely, for their judgment of and influence with their fellow workmen.

As has been already intimated, there is but scant knowledge of the completeness of this systematic organization, even among many of the office employees; still less among the workmen. This is unfortunate, and is the direct cause for the inveterate growling about "red tape" so prevalent among workmen in highly organized plants. If these men could understand that the trifling entries they are compelled to make upon time cards and piece-work slips are of the first importance—if they could see that, when a man changes from one job to another, there is a vital reason for a record of that change— if they could understand the lines upon which each chief of department is working, and the motives which actuate when apparent conflicts occur—their interest in the whole plant, and in their individual work, would hardly fail to increase largely.

If they knew that the trifling data recorded by themselves each day, at an expenditure of perhaps five minutes of time and often a woeful expenditure of profanity, were the means, and often the only means, of enabling the officials to determine accurately costs of production, and hence to secure the confident ability to figure on future close contracts—those men would see that the prosperity of the company, and the continuance of steady work for themselves, must depend upon "red tape"; or, distinctly stated, upon a system of high organization which can have its root and its fruit only in a strict adherence to clearly outlined divisions of responsibility and authority, and in accurate recording of the minutest details.

PULVERIZED COAL AND ITS INDUSTRIAL APPLICATIONS.

By William D. Ennis.

II. MODES OF FIRING AND CONSTRUCTION OF THE FURNACE.

In this installment Professor Ennis concludes his review, begun last month, of the industrial applications of pulverized coal. The preceding article discussed the characteristics and preparation of the coal, illustrating and describing the leading types of drying, grinding and conveying devices applicable to its economic use as a boiler fuel. In the following pages Professor Ennis describes methods and apparatus for firing, and gives details of the cost of installation and operation. In conclusion he points out that the commercial value of pulverized coal is not limited to its use in the boiler furnace but that it is destined to have a widespread application in the metallurgical industries, where its cheapness will enable it to replace oil and gas for the firing of certain classes of industrial furnaces with no loss of convenience or rapidity of operation.—THE EDITORS.

THE device for feeding pulverized coal into the furnace is a matter of importance. To some extent, this importance has been over-emphasized, in discussions of storage-bin stirrers, swing sifters, air preheating, etc. The essential features of a good feeder are: positive action, ensuring an uninterrupted supply of fuel to the furnace; provision for a sufficient length of flame before striking cold surfaces; the holding of the dust in suspension until each particle has been consumed; intimate mixture of the dust with air, and definite provision against the flushing down of large quantities of dust at a time. The last requirement prohibits all forms of purely gravity feeding. The necessity for suspension of the particles, producing what is practically gaseous combustion, leads to the usual practice of spraying the fuel into the furnace, either mechanically or by means of air pressure. Mechanical feeders are used in connection with natural draft, usually without any attempt at automatic regulation of the air supply. Air-blast feeders give forced draft, and lend themselves readily to air regulation. With an air blast against a refractory bridge wall or arch (sometimes of firebrick faced with carborundum), high efficiencies may be obtained, but the blow-pipe action of the flame is highly destructive.

An early suggestion for the use of pulverized coal was to provide a small independent pulverizer directly in front of each boiler, grinding as needed and discharging directly to the furnace. This eliminated storage and conveying problems, but did not meet the requirements as to drying, concentration of plant, and saving in labor. A later suggestion was to retain the central pulverizing plant, using an air injector for feeding. The principal objections to this method of feeding arose from defective regulation, the high air pressure required, and the necessity for augmentation of the blast air by natural draft. Probably the simplest satisfactory feeding arrangement introduced was that of providing short screw conveyors, which received the coal through down spouts from the overhead bins and delivered it to blast pipes leading into the furnace. These prevented "flushing down,"

APPARATUS FOR BURNING POWDERED COAL INSTALLED IN THE POWER PLANT OF THE EDISON ILLUMINATING CO., DETROIT, MICH.
A 365 horse-power Babcock & Wilcox boiler equipped with the system of the International Combustion Corporation, Buffalo, N. Y.

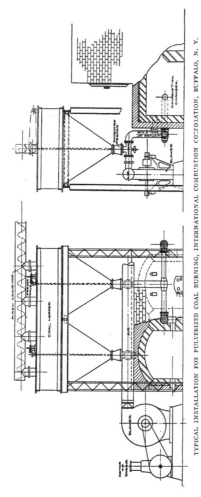

TYPICAL INSTALLATION FOR PULVERIZED COAL BURNING, INTERNATIONAL COMBUSTION CORPORATION, BUFFALO, N. Y.

and in cement kilns gave a well-distributed flame. The International Combustion Company's feeder takes the fuel from similar down-spouts and delivers it in controllable quantities to an air-blast pipe. The admission of fuel and air is simultaneously controlled by variations in the steam pressure, by means of an apparatus similar to a damper regulator. An air pressure of from 3 to 5 ounces is used, and while the air thus supplied through a 3-inch pipe to a 125 horse-power boiler is insufficient for complete combustion, the possibility of regulation to the extent of the capacity of a pipe of this size, with supplementary unregulated air supply through the ash pit, permits of fairly economical results. One man can attend to an entire battery of boilers, together with the fan for the air blast.

A modified form of the screw-conveyor method of feeding is covered by the

SCHWARTZKOPFF COAL-DUST BURNER OR STOKER. FRONT AND BACK VIEWS.
Schwartzkopff Coal-Dust Firing Syndicate.

U. S. Letters Patent No. 784,307, issued to J. V. Culliney, involving a self-contained feeding box having one forward conveyor carrying the coal to the air-blast inlet, and one return conveyor which carries back any dust which may have escaped the action of the blast. This feeder is used by the American Iron & Steel Manufacturing Co. at Lebanon, Penna.

For large applications, the individual-pulverizer method is sometimes still retained, as in the case of the Aero Pulverizer Company's apparatus, consisting of a three-stage pulverizer and fan, direct-connected to a motor, receiving the coal in nut size and blowing the powder direct into the furnace or kiln.

Among mechanical feeders, the Schwartzkopff is probably best known. This, illustrated above, consists of a brush which practically sweeps the dust into the furnace. The brush is composed of a steel spindle double-encased with cast-iron sleeves carrying triple flat bristles. It runs in phosphor-bronze bushings, and is rotated by a belt drive at about 850 revolutions per minute. From the brush spindle a reduction drive is taken on each side of the apparatus. On the driven or pulley side, a leather link V belt gives rotary motion to a five-pronged agitator spindle in the upper portion of the feeder, and upon the opposite side a pair of gear wheels actuates an eccentric pulley which imparts reciprocating motion through an adjustable rod to a rocking lever. By means of a regulating handle and gear of the worm and sector type, a block is caused to slide up and down this rocking lever in such a manner that it may either, by remaining central with the fulcrum, receive no reciprocating motion, or by moving in a downward direction, may receive the maximum motion permitted by the gear relations. The sliding block is attached by a link and pins to the lever on the spindle of a so-called vibratory plate, which forms a movable shutter within the cast-iron sections of the feeder. This mechanically operated plate, upon which the dust fuel rests, closes against the front plate of the feeder when at rest, or when the end of travel of the rocking lever is reached. By the revolution of the brush and the geared drive, this vibratory plate may open and close about 550 times per minute, the amount of such opening being controlled by setting the sliding block in the required position on the rocking lever. The rate of feeding of the coal is of course proportional to the amount of opening.

The Welles feeder, shown on the next page, consists of a hopper having a semi-cylindrical bottom in which is cut a longitudinal slot. The pulverized coal is continually stirred or agitated to prevent pack-

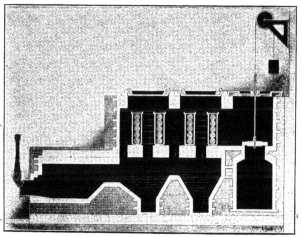

ANNEALING FURNACE BURNING POWDERED FUEL.
Schwartzkopff Coal-Dust Firing Syndicate.

ing, and drops as required into the pockets of a revolving shaft below the slot. The quantity delivered to these pockets is determined by the position of an adjustable sleeve. As the pocket shaft revolves, the coal is discharged into the feeding chamber and thence into the furnace. Air for feeding is admitted through the valve into the feeding chamber, and low-pressure air for combustion enters through the end of the Y fitting on the discharge pipe below the feeding chamber.

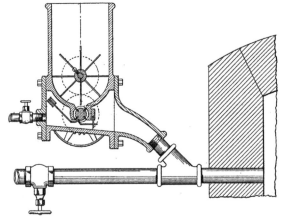

FEEDING MECHANISM OF THE WELLES FUEL-SAVING SYSTEM.

Actual practice with powdered coal under steam boilers has been limited. It has proven a great labor saver. The labor required for the operations of crushing and drying is not great. At one plant, two men take care of 5 tons per hour. At another small plant, running on a 10-hour basis, the change to powdered coal saved 21 cents per ton in fire-room labor. The cost of fire-room labor in the power houses of manufacturing plants runs from 25 to 75 cents per ton, with an average of not much under 50 cents excepting in large plants running 24 hours daily. Powdered coal should certainly save the larger part of this labor expense.

Coal-dust firing leads to efficient combustion and consequently to high furnace temperatures. The back walls of the settings will be white-hot continuously, and in horizontal tubular boilers, trouble may be experienced with the back tube ends. In one works it was found necessary to protect these with malleable iron thimbles, as well as to discard fusible plugs, which melted out hollow, leaving only a thin wall immediately in contact with the water, too weak to withstand the pressure from inside. The entire furnace must be lined with fire-brick. The destructive action of the flame on the brickwork has been much lessened, however, by the device introduced by the International Combustion Company, of two opposing flames entering from the side walls of the furnace and thus not impinging on the brickwork.

A COMPLETE PLANT FOR THE PREPARATION OF POWDERED FUEL.
Installed by F. L. Smidth & Co. for the Glens Falls Portland Cement Co. One kominuter and one tubemill. The coal-grinding department is very free from dust.

The ash gives some trouble. It is apt to fuse on the tube ends, or to form a tough slag, not easily handled. With a non-fusing ash however, most of the incombustible matter will be deposited in the back end, from which it may easily be cleaned out, in the form of a light-colored uniform powder, bone dry, which has found an application in the arts as a parting sand for foundrymen. One sample of ash had a specific gravity of 2.50, a sliding angle with the vertical of 41 degrees, and analyzed as shown on the next page:—

Si	....	50.40
Fe₂O₃	....	16.07
Al₂O₃	....	28.25
CaO	....	3.30
MgO	....	.76
Sulphates	....	1.13

Pulverized coal is smokeless after the brickwork is once heated. It can be burned with just sufficient draft to keep it in suspension. There is no fuel-bed resistance to overcome. While the equipment involved is expensive and elaborate and costly in maintenance, so also are many forms of mechanical stoker. There are no grates to renew. In case of a bad breakdown in the pulverizing plant, boilers can be quickly hand-fired with ordinary coal. Pulverized coal will bring a boiler up to pressure as quickly as gas or oil. The loss by fuel in the ash is very small, one sample of ash showing only ¼ of 1 per cent of combustible matter.

It is obvious that a finely divided dust, floated into the furnace,

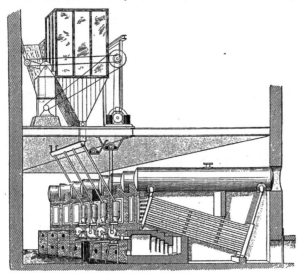

SCHEMATIC DIAGRAM OF "CYCLONE" POWDERED COAL-BURNING SYSTEM.

E. H. Stroud & Co., Chicago.

FURNACE FOR HEATING PLOW BEAMS, BURNING POWDERED COAL.
Welles Fuel-Saving System, Chicago. Average weight of metal heated per pound of coal was 13.05 pounds. Average cost of fuel per 100 beams was 21 cents.

permits of more perfect mixture of air and fuel, and consequently of a decreased quantity of excess air and consequent higher furnace efficiency, than any form of solid fuel. That this has not been thoroughly demonstrated in practice on a wide scale is due to the hitherto limited application of pulverized coal. Tests have been made showing fairly high percentages of carbon dioxide; in one, this ran from 10.8 to 15 per cent, with gases at 450 degrees at the uptake, the boiler working at just about rated capacity. The average CO_2 throughout the test was 14.50, oxygen being reported at 4.25 per cent. The rate of combustion was equivalent to 21 pounds per square foot of grate per hour. Another test gave 14.34 to 14.47 per cent of CO_2, the exit gas being at 463 degrees. These are good results, even if not at all remarkable. The point is that with pulverized coal it should be easy to get good results practically all the time, while with ordinary coal it is only under perfect conditions and for short intervals that we can expect to get anything like 14 per cent of carbon dioxide. The fire from powdered coal, as viewed through a peep-hole, resembles an oil fire, being, if possible, rather better distributed. As with oil, the air supply should be cut down almost to the point of smoking, and to effect this, tight ash pits with good doors are necessary. To give room for the flame from the highly volatile coal used, at least eight feet of length must be provided for combustion, and if the ordinary boiler setting does not furnish this, a "dutch oven" should be built out. The bridge wall has no other object, with this fuel, than to contract the flames up against the heating surface.

The first cost of a pulverizing plant depends entirely upon the degree of elaboration and duplication of equipment desired. For a single-unit system comprising dryer, pulverizer, conveyors, overhead bins and feeders, without outside storage, having a capacity of 4 tons per hour, the cost was $7,000. This is equivalent to $4.38 per horse power. A tube-mill installation having a capacity of 75 tons per day cost $70,000, or $23.33 per horse power. This involved duplicate equipment in pulverizers and driving engines, and a very extensive system of distribution over a wide yard area. A British estimate* for a plant having a capacity of 500 tons per week (say 1,400 horse power) was £2,880, or practically $10 per horse power. On account of the large size of some of the units obtainable, a plant for a small daily production would be relatively much more expensive. It is probably safe to say that a pulverizer plant should not in any case be installed at a works using less than 20 tons of coal per day, unless the conditions are unusual.

The cost of operation has been variously estimated. One writer† puts it at 35 cents per ton, and thinks that this amount is entirely offset by the saving in fire-room labor. A series of tests made at one plant, covering power and labor only, showed a cost of $0.2476 per ton for pulverizing, drying, conveying and firing. A cement works reported a cost from car to kiln of 35 cents, exclusive of fixed charges. My own approximate estimate, based on an actual output of 6 tons per hour, would be as follows:—

	COSTS PER HOUR.			
	Power.	Labor.	Repairs.	Total.
Dryer, 9 horse power @ $.02	$0.18			
Dryer Fan, ½ horse power @ $.02	.01	$0.60	$0.01	
Dryer Fuel, 1½ per cent. @ $3.00	.27			
Pulverizer and Screen or Fan, 90 horse power @ $.02	1.80		.06	
Conveyors, say 12 horse power @ $.02	.24			
Feeders and Blast Fan, say	.20	.10	.01	
	$2.70	$0.70	$0.08	$3.48

* Address of W. R. Harrison before Leeds University Engineering Society, December 10, 1906, on Dust Fuel Stokers.
† W. H. Bryan, *American Machinist*, July 12, 1906, p. 52.

This is equivalent to $0.58 per ton, to which must be added the fixed charges. The British writer previously referred to estimates the cost at 15.37 d. per ton, including fixed charges, but he uses a cost for power of £4 per year, against our cost of $60, which I think more nearly approaches the average in manufacturing plants in the United States. Using our figure for cost of power, Mr. Harrison's estimate would be increased to about 46 cents per ton. A large iron works, using tube mills, found that the total expense, including fixed charges, was not over 60 cents per ton. My approximate estimate would therefore appear to be on the safe side. It is not safe, as a general rule, to assume that a cheaper grade of coal may be used by resort to powdering. This has frequently been found to be a delusion. On the other hand, pulverized coal should not be rejected on account of the high cost of installation alone. This cost is pretty closely comparable with that of mechanical stokers, when the overhead storage bins and expensive building construction necessary for stoker installations are considered.

A SMALL FORGING FURNACE, AND A DOUBLE FURNACE WITH PRELIMINARY HEATING CHAMBER ON ONE SIDE AND FORGING FIRE ON THE OTHER.
Both burning powdered fuel on the Welles system.

To sum up the situation with regard to steam boiler practice, it would seem that the expense of pulverizing is not entirely offset by the saving in fireroom labor, under usual conditions. Assuming, however, a loss of 20 cents per ton due to the cost of pulverizing, this is a small percentage of the cost of the coal as compared with the percentage to be saved by the more efficient combustion which powdered fuel induces. This more efficient combustion now so widely sought in boiler-furnace operation seems to be more readily obtainable by the use of pulverized coal than by any other method at present offered.

A POWDERED-FUEL BURNING BOLT FURNACE ON THE WELLES FUEL-SAVING SYSTEM.
Designed especially for making short heats on the ends of rods for bolt heads or similar forging work; 22 inches wide by 27 inches deep; length of heats, 3 inches.
Welles Fuel Saving System, Chicago.

In industrial applications aside from steam production, powdered coal presents great opportunities for saving.

One interesting metallurgical application is represented on page 585. This shows a plow-beam furnace on the Welles system in which 425 beams weighing 13,158 pounds were heated in 10 hours by 1,008 pounds of coal, costing when pulverized $1.80 per ton. This furnace replaced one which formerly used coke. With coke the production was 350 2¾-inch beams in 10 hours, the fuel consumption being 1,528 pounds, at a cost of $5.10 per ton.

A large forging furnace also equipped with the Welles powdered-coal system is stated to have been in operation for nearly a year and to have effected a saving of 67 per cent in weight of fuel consumed. It was also claimed that the output was increased 50 per cent owing to the better regulation of heat. The furnace is used for heating ingots running in sizes as large as 15 inches square by 8 feet long.

In general, the slight additional cost for pulverizing is more than offset by the relatively higher price in most localities of either fuel oil or natural gas. In such special locations, for applications in which ordinary coal is impracticable, no fuel is as cheap as pulverized coal. Considerations of speed in production have led in metallurgical and other establishments to the widespread use of oil or gas even at a cost per heat-unit twice as great as that of coal. Powdered coal would be equally convenient and quick, and far more economical, for the operation of varnish kettles, annealing furnaces, drop forge furnaces, brass foundry crucibles, enameling furnaces, cable-coating machines, flanging furnaces, steam-hammer furnaces, rivet and bolt heaters, etc. The only possible competitor of powdered coal in these applications would be industrial gas. For the steam automobile, the dispensing of dust fuel from established depots would effect a saving of about 90 per cent in the cost of fuel, would permit of perfect regulation from outside, continuous and automatic feeding of fuel, high efficiency, smokelessness, and cleanliness.

EXAMPLES OF FAN-DRAFT TOWERS.

Above, a steel-encased tower of 50,000 gallons capacity; below, a wooden tower of 45,000 gallons capacity; both with chimney connections. George J. Stocker, St. Louis, Mo.

COOLING-TOWER PRACTICE.

By Dr. Jos. H. Hart.

COOLING TOWERS, as an engineering development, have passed the experimental stage. Their application is rapidly increasing and their standardization of form is steadily going on. Owing to the novelty of the subject to the majority of engineers, and the growing importance of the device with its wider and wider application, a complete review of the entire field is not inappropriate here.

The term "cooling tower" is descriptive of the nature and purpose of the device. It consists essentially of a tower-like structure as used for the cooling (at least theoretically) of any material. Its application, to this date, has been confined exclusively to the cooling of hot water. Cold water is a necessary condition for operation in many industries of the present day. In power plants, or in factories where the production of power is quite large, cold water is essential for the operation of the steam condenser. In refrigerating plants cold water is also used in the operating of the condenser, if of the compression type, and in the operation of both the condenser and absorber, if of the absorption type. In gas-engine operation, water is required for circulation to keep the cylinder walls comparatively cool so that the benefits of lubrication may be secured. The quantity approximates 8 or 10 gallons per horse power per hour, and with large engines dependent upon city service for supply this would be an appreciable addition to the cost of operation. Well waters are further often ill suited to this purpose on account of contained salts which deposit in the water spaces and impair the efficiency of the cooling. In either case, the use of a cooling tower eliminates 90 to 95 per cent of the difficulty. While these applications represent the chief demands today for cold water as an absolute necessity, without which the plant cannot operate without great loss in efficiency, and in some cases not at all, there are many other applications in which similar conditions are steadily approaching.

Wherever cold water in large quantities is needed, and whenever this water becomes unserviceable when heated, a cooling tower becomes an absolute necessity for the best economic operation even when

OPEN TYPE COOLING TOWER FOR 50 GALLONS A MINUTE.
For cooling water for a gas-engine installation at the plant of Chas. Mundt & Sons, Jersey City, N. J. Edwin Burhorn, N. Y.

cold water can be obtained for much less than the ordinary rates holding in the large cities.

In general, there are two kinds of cooling towers which have been developed as the result of experiments and use. These are called, respectively, the closed-tower type and the open or atmospheric type. Of each of these two types there are many variations on the market, but the differences appear chiefly in the details of construction and cost of erection. These two types are essentially the same in many respects; they both consist of square or rectangular towers rising into the air from 20 to 60 feet, and both have water pipes extending to the top of them, with various sprinkling devices at these points. The water in all cooling towers is cooled primarily by evaporation into the surrounding air. The closed type differs from the open one, in theory at least, only in the fact that the sides are closed, and the contact of the water with the air is accomplished by means of a power fan situated at the base. The atmospheric type, on the other hand, has the walls of the tower entirely open and admits air from all sides into immediate contact with the water. All cooling towers invariably have pans or obstructions interposed in the path of the water, with the object of

A BURHORN COOLING TOWER AT THE PLANT OF THE JEFFERSON ICE CO.

detaining it in its passage downward and dividing it into smaller units, with the object of promoting thorough and continuous contact with the air at all points. The modifications of design and construction distinguishing various towers in every case consists essentially in the variations in the sprinkling device at the top, in the details of the obstructions in the path of the water used for detaining and subdividing it, in the collecting pan at the bottom, and in the situation of the pan, if one is used, for producing greatest efficiency. Open-type cooling towers are sometimes partially covered with wire netting or provided with special appliances to prevent high winds from carrying off a large portion of the water in its downward fall.

CLOSED-TYPE COOLING TOWER, LEHIGH PORTLAND CEMENT MILL, MITCHELL, IND.

Henry R. Worthington, Inc., N. Y.

The principle back of the cooling tower, as has been said, is based upon the absorbing power of air for water vapor and the consequent cooling of the water from which this vapor is obtained. Cooling by this process is much greater than any other; the latent heat of water enters as a factor here with remarkable results. A pound of water changed to steam absorbs 966 B. t. u. It does this whether heated or not, provided only the change to vapor is accomplished; and if the heat is not supplied otherwise, the change to vapor which demands it is accomplished by absorption of heat from surrounding bodies, thus producing in them a cooling effect. Thus water subjected to a strong wind undergoes rapid evaporation, and its consequent cooling follows as an immediate result. It has been shown recently that the great bulk of natural ice is produced by this process, and very little by conduction of heat.

Cooling towers, then, depend for their operation absolutely upon the extent of the vaporization produced in the water during its passage through the tower. Thus, about 5 per cent of water must be lost by evaporation in a single passage through the tower, to cool the residue from 120 degrees to 80 degrees F. Thus cooling towers do not absolutely eliminate the demand for water by plants in which they are operated; they merely diminish the amount required.

Their efficiency, therefore, depends primarily upon the cost, availability, and temperature of the natural water supply, and further, in a general way, upon the cost of fuel and labor for the maintenance of the pump, and (if the tower be of a closed type) of the power fan. Interest on first cost and cost of maintenance must also be considered. Where plants demand a large amount of cold water and are situated in large cities, without a supply immediately available at a low

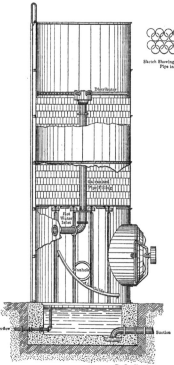

Sketch Showing Interlocking Pipe in Tower

The Eng. Magazine.

SECTIONAL VIEW OF WORTHINGTON COOLING
TOWER, FAN TYPE.

WOODEN FAN-DRAFT TOWER OF 100,000 GALLONS CAPACITY.
George J. Stocker, St. Louis, Mo.

cost, the towers are undoubtedly very economical, absolutely cutting the consumption of water down to only 5 or 10 per cent of its former amount; and when very large quantities are considered at ordinary supply rates, the cost of operation and installation of a tower compares very favorably with the reduction in the water bill. A saving of 50 per cent in this item is not uncommon in many large plants where such appliances have been installed, and this figure may rise much higher when the installation is under the charge of a competent engineer.

It may be assumed in a general way that a tower 20 by 20 by 40 feet high, has usually a cooling capacity of 300 gallons per minute, from 120 degrees to 80 degrees F. This is the approximate performance of the average cooling tower, but it is dependent absolutely upon hygrometric conditions in the atmosphere. Conditions vary so widely that no general statement can have much value. If the air contains a large amount of moisture, or rather has a relatively high humidity, the cooling-tower loses much of its efficiency. By humidity is meant the ratio of the amount of water vapor in the air to that which it could contain without precipitation at that temperature. This, of course, represents the power of the air to absorb water vapor and produce cooling effects. It is only fair to say that on hot moist days, when the humidity is high, the cooling tower diminishes very greatly in efficiency; a machine of this type will have the water scarcely below.

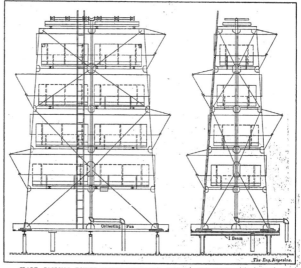

HART COOLING TOWER WITH SECTIONAL TRAYS AND SPRAY PREVENTERS.
B. Franklin Hart, Jr., & Co., N. Y.

100 degrees F. in these instances, thus undergoing a loss in efficiency of about 50 per cent. It must be understood, in fact, that cooling towers depend absolutely for their effect upon the absorptive power of air for water vapor to produce the cooling, and not at all upon the radiation of heat or conduction into the atmosphere. Thus the operation of such a tower is much more efficient on a dry, hot day than it is upon a cold wet one.

In comparing the relative efficiency of the two designs, it was formerly supposed that the closed type was superior. The cooling effect, however,

OPEN TYPE COOLING TOWER EQUIPPED WITH
HART SPRAY PREVENTERS.

depends merely upon the quantity of water changed to vapor in the atmosphere. This later depends upon the relative efficiency in this process of a unit of air, and upon the number of

TOP OF AN ALBERGER COOLING TOWER, SHOWING THE DISTRIBUTING SYSTEM.

CLOSED-TYPE COOLING TOWERS AT THE POWER HOUSE OF THE EAST ST. LOUIS &
SUBURBAN RAILWAY.

Alberger Condenser Co., N. Y. The towers are to take care of 50,000 lb. of steam per hour
and give 26 in. vacuum with air at 80 degrees F. and 70 per cent. relative humidity.

units of air operated upon per unit of time. The closed type, with a
powerful fan, was supposed to convey much more air over the surface
of the water than was the case in the open or atmospheric type, where
this transfer of air was dependent upon natural causes. However, it
has been found that air in contact with water rapidly attains a saturated state, and thus the air from the fan in its passage over a great
amount of the water is absolutely ineffective in producing vaporization, since it becomes saturated almost immediately upon contact with
the water. Hence it is recognized today that the atmospheric type is
superior in efficiency, from an operative point of view, in the cooling

mechanism. The fact that the operation of the fan requires power, and that this latter must be considered as a cost factor in the operation of the plant, has enabled the open cooling tower to compete successfully with its rivals.

The closed type of tower may triple, or even quadruple, the first cost of a tower of the atmospheric or open type; and hence, its efficiency being less, it would soon cease active competition if it

FAN-COOLED TOWER, ROCKLAND LIGHT & POWER CO., ORANGEBURG, N. Y.

Alberger Condenser Co., N. Y.

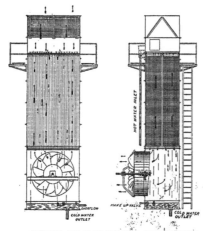

SECTIONS AND PLAN, WHEELER COOLING TOWER.

were not for other conditions. T h e s e conditions center about cost of maintenance, reliability, and various considerations of a more or less urgent character, which in general hold against the open or atmospheric type.

In point of maintenance, there is no doubt that the closed type is much superior in constructive details and in its general development. It costs more to erect, and this cost is not altogether on account of design. It must, from the nature of the design, be made up of much heavier material and is, in general, a much stronger and more permanent tower. With its closed sides it is not nearly as subject to weather conditions and the deteriorating effects which result from these.

So far as concerns reliability in performance, a cooling tower of the open or atmospheric type has a certain definite capacity for cooling water under any given set of weather conditions, which natural capacity cannot be exceeded, and its performance is thus dependent to a greater extent upon the hygrometric conditions existing in the atmosphere at different times. Thus the quantity of air coming into contact with the water is dependent upon the wind, and its efficiency varies with its humidity. Under certain conditions, the closed tower may become more efficient because it is more adaptable; it is not dependent upon the wind for a supply of fresh air, and the ill-effects due to a bad hygrometric state of the atmosphere can be offset by measures which reduce them to a minimum in this type. Under such conditions, a speeding up of the fans, with the resultant increase in air supply, will often compensate for a high hygrometric state, thus rendering the operation of the tower more nearly uniform.

OPEN TYPE COOLING TOWER FOR 600 GALLONS PER MINUTE.
For a plant using about 5,000 lb. of steam per hour. Wheeler Condenser & Engineering
Co., N. Y.

Now we come to a further objection to the open type of tower
which often becomes of paramount importance. This is due to conditions existing in large cities and is certainly one for serious consideration. An open tower of ordinary type loses a large amount of water
as waste when a high wind is blowing, because it escapes from the
tower through the sides. This is sufficient to wet thoroughly much
surrounding property and in some cases this has been the cause of
damage suits; in any case, it invariably hastens the deteriorating
effects of water and moisture on the surrounding property. In some
forms of construction this difficulty is met and removed by providing
the tower with systems of spray guards.

Thus we see that each type represents a different field with its own advantages and disadvantages. The closed type is efficient and reliable—not as efficient at its best as the open type, but more reliable. It costs much more to install, but complicated conditions are not apt to arise in its maintenance, other than those of a mechanical nature. The open-type tower, on the other hand, requires no machinery in its operation, and is the most efficient under ordinary circumstances. The result has been, as would be supposed, that large and efficient plants, with considerable capital and knowledge back of them, have almost invariably installed the closed type. Smaller units, in which cost is an important consideration, have installed the open type. Thus large power stations, operating trolley and lighting systems in the large cities very generally have cooling towers of the closed type. Small

COOLING TOWERS FOR 320,000 POUNDS OF STEAM PER HOUR.
Two sections of five fan-cooled towers, each unit having two 120-inch fans requiring 23 brake horse power. Installed at the Liverpool Corporation Electricity Supply Works by the Wheeler Condenser & Engineering Co.

refrigerating plants almost invariably adopt the open one. Plants midway in this category are variable in their choice. Sufficient to say here, merely, that both types are in active and efficient operation today; that their use and application is daily becoming more and more extended; that their principles are becoming more and more understood, and their design is tending toward a general standardization. The most successful forms on the market now, are comparatively few, and a process of elimination is continually going on.

THE THIRD-RAIL PROBLEM.

By A. D. Williams, Jr.

Mr. Williams' paper takes up certain physical problems of steam-railway electrification in the same spirit that characterized Mr. Knowlton's discussion of organization problems in our preceding issue. This is the spirit of frank and full recognition of the difficulties, and their clear definition so that they may more speedily be recognized and solved. The great movement (in which these and any other difficulties are but questions of detail to which an answer will be found) will progress all the more surely and speedily thereby.—THE EDITORS.

ONE of the questions to be answered after a decision has been reached to adopt electric traction, is the method to be used for delivering the electric current to the train or locomotive. Two methods are at present available for this purpose, each having some advantages and disadvantages; these are the third rail and the overhead trolley wire. To those who are unfamiliar with the actual operating conditions of main-line railroad operation the problem is simplicity itself; the railroad executive must, however, take many items into consideration before reaching a decision on this simple little question. Many of the conditions brought forward by the long-headed railroad man may seem to have but a trivial bearing upon the subject at issue, and while they are of great importance on the main highways of traffic, these items are not considered by outsiders. The cost of a mistake, however, will be enormous and a serious handicap to the road.

In the past, electric traction has been limited to a very narrow field; tramways in cities and towns, urban and interurban service using multiple-unit and individual motor cars, and some electric locomotives, are in use. In all of these lines single or double-track operation exists using a special size of dwarf car, and the problem is of a comparatively simple nature when compared with the difficulties to be surmounted prior to the successful application of electric traction to the full-size requirements of a trunk line or the problems of a heavy terminal traffic. In this latter field it is impracticable to adopt the multiple-unit cars, except for the suburban zone passenger traffic, and the main difficulties lie in the satisfactory solution of the problems arising from the through passenger service and freight traffic. Of necessity this traffic must be handled by a locomotive of some type. The change from steam to electric operation must be effected with the minimum amount of delay to traffic, and the installation must be made at points where the existing facilities are already taxed to the breaking point. For these reasons the system adopted must be one about which there is no tinge of experiment, and the price of a failure cannot be computed in dollars and cents.

In dealing with heavy traffic it is of the utmost importance that it be kept moving, and the only excuse which the railroad man will consider valid for the absolute interruption of the flowing stream is the total loss of the roadbed and track. When traffic is blocked the first duty is to open a way past the obstruction and to get things moving. This is more important than anything else. After this has been done other tracks can be cleaned up. In doing this, temporary work of the most hasty character is required, and the system of electrical distribution must be such as will lend itself readily to this purpose. This system of distribution must be devised to permit of all kinds of irregular operation and total reversal of the direction of traffic; it must not interfere with the operation of signals or obscure their visibility, particularly on lines where high schedule speeds are required. The conductors supplying currents to the trains and locomotives must be so located that all turnouts and cross-overs, no matter how infrequent the likelihood of their use, can be negotiated at all times with the train under complete control. No conductor that has a gap which must be jumped complies with the conditions, and the conductor which cannot be readily restored to working conditions after a wreck does not meet the requirements. The conductor must be susceptible of emergency repair with materials which may be readily carried by the wrecking equipment, without adding unduly to the outfit. In this connection it must be borne in mind that the scene of a wreck cannot be selected in advance, and the amount of temporary work required cannot be limited by written orders and instruction books. The trouble must be handled right on the ground; and there will be no opportunity to request bids on the material required.

Another point to be considered is the obstruction offered by the electric conductors to wrecking operations, and the facility with which portions of it may be torn out and re-established. Large gangs of men and heavy cranes must be used in wrecking operations, and the danger of the work must not be increased by mysterious electrical displays, otherwise it will be impossible to keep the labor on the job.

In terminals and yards, where dense traffic must be handled, it is absolutely essential that the electric conductor should add as little as possible to the ordinary complications and dangers inseparable from present methods of operations. Cumbersome and makeshift constructions should not be countenanced. In such places trains must be made up and handled, frequently with urgent haste, at night as well as by day, and a conductor so located that it interferes with or increases the hazard of operations is matter out of place.

The improper location of the conductor also produces serious troubles with regard to maintenance-of-way operations. Track must be continually lined in and levelled up. This work must be carried on under a traffic which cannot be slowed up or stopped for a gang of men to get their tools out of the way, and is never absolutely suspended. On a one- or two-track road, with infrequent trains, this is one sort of a proposition; on a four-track line, with seven or eight hundred train movements, the problem is not a simple one at the best.

The operation of a line dealing with all classes of traffic and handling hundreds of trains per day is a vastly different problem from that met in the handling of a dense traffic of a single class, as on the elevated lines and subways of large cities; and these differences are much greater than those between the little single-track interurban road and the latter. It is in these differences that the difficulties lie. In the latter two cases there is the added advantage in handling the problem that a dwarf variety of equipment is to be dealt with, and in addition the roads are usually designed with the idea of electric operation. In the application of electric traction to an existing railroad, designed for steam operation, there are many obstacles to overcome. The electric equipment itself may not cost as much as the alterations necessary to permit the use of electricity. Most of these lines were laid out years ago, and the gradual growth in the size of cars and locomotives has encroached on the originally ample clearance lines, until today there are many roads that are unable to admit heavy freight cars and the large private and Pullman cars to certain portions of their lines; the clearance between tracks has been shaved to the limit and equipment is frequently damaged at tight places. Necessarily, the only successful method of eradicating the trouble is the expensive one of revising the line, and while this is not an impossible task, it would be an extremely costly proceeding. The principal difficulty in the way of such revision, naturally, lies in its cost; in addition there is a certain lack of initiative on the part of those in authority and a fear of departing too far from the antiquated standards of the past.

A peculiar fact in regard to many railroad standards (?) is the number of different standards, the same road having several standards for certain constructions, each in use on separate portions of the line and all out of harmony with one another. This is due to corporate consolidations which have not as yet coalesced throughout. In many cases these out-of-date standards are blindly followed in absolutely new construction, where unbiased consideration and the ability to view matters with a broad perspective would show the futility of clinging to the past and the advantages to be gained by throwing deadwood overboard. Unfortunately, there is sometimes a certain lack of co-operation between the various departments of railroads, each department and sub-department striving for its own individual record without regard for others or for the ultimate result. When such departmental jealousies are rife the interests of the road suffer.

In the location of the electric conductor several things must be considered, particularly clearances; two bodies cannot occupy the same space at the same time. The clearance question is one of the most difficult problems the railroad is called upon to solve. Clearances which were ample for the equipment of forty years ago are inadequate for the equipment of today, and on some of the older roads the maximum equipment line is so close to existing structures that further encroachment, by additions to either structures or equipment, is impossible.

The maximum equipment line is supposed to represent the composite cross-section of the largest locomotive, the largest freight car, the largest private car or Pullman, the largest wrecking crane, the largest snow plow, in running condition with markers in place; this cross section being increased by a sufficient amount to cover the side sway permitted by the truck bolsters, lopsidedness due to a broken spring, eccentric loading bringing the car down on its side bearings, journal play and wear. The effect of curvature of the line must also be considered. Necessarily, any conductor conveying current to a locomotive or car must be outside of this line by an amount sufficient to provide a working clearance. The current collectors must be capable of reaching this conductor and of being housed within the maximum equipment line.

The table on page 658 gives the location of the third rail on a number of installations; of the American examples given the first two and the last two are the only ones which deal with full-sized railroad equipment. As will be noted, the Baltimore & Ohio has made a change in the location of the third rail, moving it out 6 inches. This was done in order to clear equipment that was kept off of this portion of the road by the previous location of the third rail. When the under-running third rail was first proposed it was suggested that the location suggested, 2 feet 4¼ inches from the gauge line, be adopted as a standard, as it was supposed that this was outside of the maximum equipment line; but in applying electric traction to another division it was found advisable to move the third rail 3¾ inches further out in order to clear equipment. This change would prevent an uninterrupted interchange of traffic between sections of road equipped with these rails. The move outward was forced by the fact that the original location cuts into the equipment line, and this has caused rigid inspection to be made of all trains coming into the electric zone.

TABLE GIVING THE LOCATION OF THIRD RAIL ON VARIOUS ROADS.

Name of Railroad.	Top of Third Rail above Top of Running Rail. Inches.	Center line of Third Rail outside of Gauge Line. Inches.
Baltimore & Ohio (Old)	1¾	24
Baltimore and Ohio (New)	3½	30
New York, New Haven & Hartford*	1½	Center line track
Paris-Orleans Railway	7⅞	25⅝
Milan-Gallararate Railway	7½	26 9/16
Mersey Railway (Liverpool)	4½	22
North Eastern Railway	...	19¼
Paris-Versailles Railway	7⅞	25⅝
Fayet-Chamonix Railway	9	23
Wannseebahn (Berlin)	12⅝	33½
Albany & Hudson	6	27
Aurora, Elgin & Chicago	6 5/16	20⅛
Lackawanna & Wyoming Valley	6	20⅜
Grand Rapids, Grand Haven & Muskegon	5¾	20⅜
Seattle & Tacoma	7½	20
Columbus & Newark	6	27
Columbus, London & Springfield	6	27
Metropolitan Elevated (Chicago)	6¼	20⅛
Lake Street Elevated	6½	20⅛
South Side Elevated	6¾	20⅛
Southwestern Elevated	6½	20⅛
Brooklyn Elevated	6	22¼
Kings County Elevated	5¼	19½
Boston Elevated	6	20⅜
Manhattan Elevated	7½	20¾
Central London	1½	Center of track
New York Central Under-running	2¾†	28¼
West Shore Div. New York Central	2¾†	32

While insulation troubles lie entirely within the province of the electrical engineer, the breakage of insulators is another question. If

* Abandoned by order of the Superior Court of the State of Connecticut, dated June 13, 1906.

† The last two are under-running rails and the dimension marked is from top of running rail to contact surface of third rail.

it were possible to prevent equipment from striking and breaking insulators there would be very little third-rail leakage. This leakage may, however, reach a serious figure when a large number of the insulators are cracked.

In connection with maintenance of way, the third rail adds a little to the trouble in keeping the track in line, as it obstructs one side. In addition it reduces the efficiency of the track gangs and adds a little to the expense of keeping them supplied with tools, and the incidental fireworks it furnishes demoralize the labor gangs. It would seem that with the third rail secured to the same ties that carry the running rails, their relative locations, with regard to each other, would be easy to maintain. The third rail is, however, not secured to every tie, but only to every fifth or sixth tie; and while the lateral variations are but slight there is the opportunity for considerable vertical play. When a train moves along a track each pair of wheels causes a wave in the track, the vertical amplitude of which depends upon the elasticity and softness of the road-bed. Deflections of as much as 2 inches are not uncommon on lines laid with 100-pound rails and broken-stone ballast. A third rail is affected by such waves, and has a certain amount of motion with regard to the track. With a top-contact third rail this motion does not affect the shoe contact to any extent owing to the large amount of play allowed for. With an under-running third rail a different condition occurs, as the motion of the shoe is limited, owing to the fact that it was considered necessary to use a shoe which would permit equipment designed for use on the under-running third rail to pass on the lines equipped with a top-contact protected third rail. To fulfil this condition, the normal position of the shoe is horizontal, and the top-contact rail forces it up, the under-running rail pressing it down. This limitation of the shoe motion must result in great variations in shoe pressure, which are made apparent by the sparking that takes place.

At stations the public must be looked out for, owing to the fact that many people indulge the habit of poking at things with umbrellas, canes, etc. An umbrella with a steel rod possesses interesting attributes as a pointer around a live third rail, and its convenience for the purpose will be felt in the claim department. As it is practically impossible to protect the shoe and its holder by a "fool-proof" guard, it becomes necessary to construct special station platforms at the level of the car platform. Where low platforms are in existence it is necessary to provide a dummy third rail, which is only a partial protection to the public.

From an operating standpoint the third rail is a menace and a constant source of danger to the trainmen, particularly at night. Circumstances often require that a trainman should go back with a red light to protect the rear of his train. Without the third rail there was a certain amount of risk; with it there is a positive danger. A broken leg with the opportunity to get run over by a following train is one of the probable rewards of obedience to the rules. This danger is a vivid reality to all experienced trainmen who have had the opportunity to observe working conditions on a third-rail line. A few accidents to the rear trainman will cause such men to become chary of risking their lives by going back at night. The number of railroad employees killed in the discharge of their duty is large, and many laws have been enacted forcing the railroads to adopt safety appliances. In view of this it would certainly be unfortunate for a road to adopt a method of electric traction which will ultimately be prohibited by law; in fact, the third rail has already been placed under the ban of the courts in Connecticut, owing to the large number of accidents due to its use. The third rail is bad enough along the line, but when installed in a yard it becomes a constant source of danger to all who are employed in such yards, obstructing all of the fairways and reducing the speed with which trains can be sorted and made up. Supposedly dead sections come to life the instant any gap is bridged, and the results of such mishaps have been spectacular in some cases and startling in others. In one case a laborer was electrocuted eight miles from the nearest live third rail, owing to a train over-running a gap.

In sorting yards the men will quickly learn the dangers of the third rail, and instead of riding their cars to the last minute and holding them will be forced for their own protection to drop off at the first sign of danger. The results of their caution are bound to be felt in a large increase in the number of minor smashes which occur. Individually each item is small but the aggregate will be large. The railroad pays for the cars that get smashed, but the employee has to pay for his own repairs. The natural results will be that the man will look out and get clear without trying to prevent impending smash ups.

In connection with wrecking operations the third rail presents considerable difficulty. The first result of a wreck is a short circuit cutting off all current in that block until the third rail has been cleared or re-established. This is a good point, but it makes it absolutely necessary to depend upon steam for handling the wrecking equipment. In addition another extra must be got out to handle the third rail. The third rail weighs from 40 to 70 pounds per yard and cannot be put in until the track has been placed. Brackets and insulators are required about every 10 feet, and it takes considerable time to line the third rail in and get it covered. In the meantime traffic will have to be towed past the gap by steam locomotives, which will mean keeping a number of such machines on hand for emergency use. It may be necessary to handle all passing traffic on one track; this would be troublesome on a four-track road and worse on a double-track line. Any attempt to run a track around the wreck would be handicapped by the necessary third-rail construction; it would not be impossible, but it would require so much time that it would not be feasible. During the entire time the wreck was being cleaned up current would have to be kept off of that section of the road, as it would be impracticable to run the risk of stampeding the large labor gangs required by any spectacular electrical displays.

One of the worst features of the third rail arises from the fact that it is necessary to have turnouts and cross-overs and ladder tracks. These gaps are often as long as 150 feet. A double ladder track with double slip switches is one of the necessities of any yard or terminal, and in some cases triple ladders are used. These tracks cannot be operated by any third rail yet devised. The terminals for heavy traffic are of necessity of the stub-end or through type. The loop terminal, that happy solution of the little, dwarf multiple-unit road's trouble, is impracticable for full sized equipment. Furthermore the loop system is not adapted to holding trains in the terminal, often for several hours, as is necessary at night on all roads.

Another trouble liable to occur on a road is for a number of trains to get bunched, owing to an open drawbridge, for instance. When fifteen or twenty trains get closed up tight on each other it is very liable to lead to trouble at the rotary converters.

In dealing with the gaps at switches it has been necessary to use an overhead conductor. When this is of the low-tension variety and the supports have to span several tracks, supporting a conductor over each of them, it is extremely difficult to devise such a support as will not obstruct all views of signals. These overhead conductors must overlap the third rail at each end of the gap in order to guard against all possible chances of a train becoming stalled at such a point, for though the ordinary operating speeds at such switches and cross-overs may be ample to carry a train over the gaps, there is always the possibility of an emergency arising when trains must be under absolute control. In railroad parlance this means the ability to stop immediately, and to start from such a stop, as well as the ability to increase the speed. The necessity of the overhead conductor has been recognized in the largest installation of third-rail handling standard railroad equipment.

A feature not heretofore mentioned is the fact that at certain points it is necessary to locate derails and dwarf signals; the latter are rendered nearly invisible by the third rail. The derail cannot be done away with as it is an absolute necessity at many points, and should it be over run it will naturally result in a tearing out of the third rail.

In connection with bridge work the third rail will cut many of the floor-beam bracket plates, particularly on through plate-girder bridges, and the further it is located from the gauge line of the track the further into these brackets it will cut. This causes considerable trouble in the design of new structures and necessitates extensive re-modelling in existing structures.

Where cross-overs occur on curves considerable care must be used in order to prevent the contact shoes coming in contact with a running rail in passing from one track to another. The mechanism boxes for switches, derails, torpedo machines, etc., must also be very carefully placed in order to avoid all danger of their being touched by these shoes. These troubles are aggravated by the fact that the shoe is held in position by a spring, and the failure of this spring is liable to let the shoe drop sufficiently to ground on any metal object close to the running rail.

A consideration of the foregoing disadvantages will certainly indicate that the third rail, either over- or under-contact, leaves much to be desired in regard to its assistance in the introduction of electric traction, except on those lines handling special dwarf equipment.

THE FUNCTIONS OF THE ENGINEER AS A SALESMAN

By Sterling H. Bunnell.

The entry of the engineer into the work formerly belonging wholly to the "business man" is the main spring of the movement which within comparatively few years has completely transformed manufacturing. His is the chief influence making it now an applied science—or a systematized application of many sciences. But there is yet in many minds a strong (and often wholesome) disposition to fix a great gulf between professional and commercial work—to shudder at any attempt of the engineer to pass from the mechanical to the business department. Mr. Bunnell shows that these two fields, so far at least as the manufacture of machinery is concerned, must overlap to a wide extent; and in this joint territory the conscientious technical specialist may properly and profitably bring into the sales department many of the functions of the consulting engineer.—THE EDITORS.

ONE of the most successful salesmen of large steam-power machinery, with whom a single contract of a hundred thousand dollars was no uncommon affair, used to say, half in jest, that ignorance was an important part of a salesman's equipment. The statement could hardly be said to be proven by any lack of knowledge on the part of its author, but contained a germ of truth nevertheless. The mistake of many educated and experienced professional men, when attempting to make sales, has been their inability to see an important personal side of the transaction, entirely aside from considerations of good fellowship and social attention. The intending purchaser has his own ideas of the general principles and details of the apparatus which he desires to obtain; and he is not informed as to other details—perhaps misinformed in general. It is the salesman's province to ascertain exactly the requirements of the work to be done, to appreciate and consider carefully the desires of the purchaser, and to supplement his lack of knowledge without slighting and disregarding the value of the information which he has acquired in his business and which has produced the opportunity for the sale.

With the increase in the number of technically educated young men, the former universal doorway by which the higher grades of the profession were entered, via the drafting room, no longer serves to accommodate the crowd. The attractiveness of the selling side of engineering is considerable. For one thing, the young salesman immediately commences to increase his acquaintance with men of attainments in all lines, and his general knowledge of business methods.

The probability of his own financial success is thereby greatly increased. His knowledge of human nature is enlarged, and if he be tactful, agreeable, and (above everything) quick and alert to take advantage of·all the bits of information that crowd upon him in his daily intercourse with the business world, he cannot fail to succeed. The most common mistake of the recent technical graduate is a desire to teach the men whom he addresses, instead of a retentive state of mind, prepared to learn of every one. There is so little exact knowledge of practical conditions that one must be wary of attempting to reduce to figures the statements made by men of loose habits of thought, and equally careful neither to doubt openly, nor to accept without qualification, statements of conditions and results which appear to be entirely contrary to scientific principles.

The engineering salesman is frequently called upon to act independently of the consulting engineer. In steam power-plant work for instance, the great majority of the sales consists of plants comprising one, two, or three units each of engines and boilers of less than two-hundred horse power, the cost of such a plant being small, and the conditions of its successful operation being so well-known that the special ability of the high-priced consulting engineer is not appreciated. In such a steam plant, however, there are many accessories which go far toward making the plant a success or failure as they are well or badly designed. The steam piping, if turned over to a working steam fitter to design and construct, is likely to be a serious source of danger; the drainage of condensed water may be poorly provided for, and a considerable quantity of hot water will probably be thrown away instead of going back to the boilers to reduce the coal consumption. The engines may be inconveniently placed; the foundations may be either too weak, or unnecessarily strong and expensive; and in a dozen other ways the best of engines and boilers may form part of a poor outfit. The purchaser of a plant of this character feels able to discriminate between the various propositions made to him, to examine the catalogues, cuts, and specifications submitted, and with the precaution to deal only with builders of reputation, to satisfy himself of obtaining well constructed and satisfactory units. If, now, among the salesmen who address him upon these several propositions, he meets one whose knowledge evidently extends considerably beyond the details of the apparatus he offers, and who is able to make up and submit special sketches and plans showing the plant complete, to make slight improvements in the arrangement to suit conditions which the purchaser has overlooked, to determine the size of accessories and to include them if desired in his estimate of the cost, and to advise as to convenient sources of supply for apparatus which he is not justified in furnishing as part of the contract—this man is nearly sure to carry off the job for his company, even at a considerably higher price than that submitted by others.

Such a salesman has the same advantages over the man without technical ability, when dealing with the most expert consulting engineers. While in this case he is not expected, and in general must not attempt, to make suggestions in regard to changes in details which have been carefully designed with a full knowledge of conditions which the newcomer cannot hope to grasp complete in a cursory view of the situation, his readiness to grasp the desired features of the apparatus, his appreciation of reasons for the specifications of the consulting engineer, and the confidence which his explanations of the details of the construction he offers will inspire, in the mind of the purchaser's engineer, of the undoubted substantial and satisfactory character of the work that will be delivered, will be quite as important as the matter of price in determining the choice among several bidders.

In many other classes of work the chance for the engineering salesman is excellent. In the utilization of waste products, in novel methods of manufacture of well-known articles, in producing new and desirable forms of common materials, and often in arranging details of machinery for already standard processes, no consulting engineer skilled in the particular work in hand can be had, for the reason that the processes or conditions are entirely new. The inventor or promoter of the process is most likely anything but a man of engineering ability; usually he is a practical business man, or a visionary enthusiast, incapable of producing a complete working plant on the lines he has vaguely in mind, and often aware of his inability to do so. The engineering salesman, meeting such a man for the purpose of selling him his power plant or his transmission machinery, or of obtaining the order for constructing his special apparatus, is very likely, if he is awake to the opening presented, to carry off in the end the contract for the complete plant. From his acquaintance with the possibilities of local shops, or of men or corporations of ability in special details of the proposed installation, he is able to collect the necessary data, to make plans and to assemble details, to modify sizes and arrangements, and to supplement the general knowledge of the inventor or promoter with his own special technical ability, so that he can offer the prospective purchaser a well-defined equipment for his total requirements, covered by a single contract and for a single stipulated price. The purchaser is glad to have but one concern to deal with, and that one which can offer him the services of so skilful an engineer as the man who has assisted him in producing the complete design; while the commission which the salesman secures for his office on the apparatus which he has collected from other parties, and which involves no expense to his office except the collection and transmission of the purchase price, will be a very desirable item of income.

Even after the salesman has obtained his order, his usefulness to his home office is by no means ended. Being familiar with local conditions, he is able to give useful advice as to methods of shipment, points of delivery, local facilities for handling the shipment, and labor skilled and unskilled which may be had on the spot for the purpose of completing the erection of the machinery. In some cases, the salesman, though by no means a workman, can demonstrate the use of machinery and equipment which he has placed, to very great advantage, as he already has the respect and confidence of the purchaser. Being skilled in business affairs, he is able to place local contracts for handling, construction, supplies necessary for erection, and the like, and peculiarly in a position to promote a feeling of goodwill on the part of the purchaser's organization and managing officials. When it comes to a working test demonstrating the capacity or efficiency of the construction, the salesman can undertake the nominal direction of the work, his presence and friendship smoothing many an otherwise, rough place in the path of the testing expert, who is generally regarded by the purchaser as more or less of a sharper rather than a scientific investigator. Finally when delivery and test is completed, the engineering salesman, by his ability and his undivided efforts toward the satisfaction of the purchaser with the finished work, has nine out of ten points gained in closing the affair with not only the full payment of the balance due, but a feeling of goodwill which will go very far toward securing the next order for an increase in the capacity of the successful working plant.

THE RAIL HEAD AT RHODESIA BROKEN HILL, CAPE-TO-CAIRO RAILWAY.

RAILWAY PROGRESS IN THE DARK CONTINENT.

By J. Hartley Knight.

The following article carries forward the current summary of African railway construction which Mr. Hartley Knight has presented from time to time in these pages. The significant suggestion for an engineering audience is that where the railway goes, a vigorous demand soon arises for electrical, power, and mining machinery and the whole range of mechanical equipment and supplies.—THE EDITORS.

TWO years have passed since I was permitted in these pages to review the status of railway construction in Africa in an article dealing more particularly with the line between Cape Town and Cairo. During that period railway engineering has made great strides in all parts of the Dark Continent and a vast amount of work in the same direction remains to be attempted in the near future.

At the time at which I wrote, the Cape-to-Cairo railway—which continues to loom large in the popular fancy—was well on the way to its present terminus at Rhodesia Broken Hill, about 2,000 miles north of Cape Town on the coast. This point was successfully reached in the middle of 1906, the first through train arriving at rail head on

LOOKING THROUGH THE KAFUE RIVER BRIDGE, CAPE-TO-CAIRO RAILWAY.

BRIDGE OVER THE KAFUE RIVER, CAPE-TO-CAIRO RAILWAY.

June 24 of that year, and the line being open for public traffic on September 1 following. Between the Victoria Falls and Rhodesia Broken Hill, a bridge more than a quarter of a mile in length had to be constructed to carry the line across the Kafue River. The bridge in question is the longest (as the bridge at the Victoria Falls is the highest) in Africa, and was put together by means of pontoons in record time, the construction of the line itself—north and south of the Kafue—being proceeded with in the meantime. The gauge of the line, it is worth recalling, is 3 feet 6 inches, and although the track is single there are numerous sidings en route. For the time being the work of construction is at a standstill, the state of the money market not permitting the promoters of the line to proceed; but it is not anticipated that the rail head will remain much longer at Rhodesia Broken Hill. It will be remembered that the late Mr. Alfred Beit, the African magnate, left over five million dollars for the development of the scheme "known as the Cape-to-Cairo railway," and as soon as the gentlemen appointed to act as "railway trustees" under his will come to a decision in regard to the allocation of the money, construction will be immediately proceeded with. Already the next section of the line has been surveyed—to a point on the Congo border known as Bwana M'Kubwa, a little over a hundred miles north of Rhodesia Broken Hill—and all concerned are ready to proceed as soon as they get the word to go, the idea being to carry the line northwards (as foreshadowed in my article of two years ago) until it reaches the Katanga copper fields and thus connects with the line known as the Benguella railway, now being built from Lobito Bay, in Portuguese West Africa, to the aforesaid district of Katanga, some 900 miles in a slightly north-easterly direction. It remains to be seen what is going to be done in regard to carrying the other "fork" of the line up to and alongside Lake Tanganyika in German territory, but I am inclined to think that that route will be abandoned—for some years to come at all events.

Meanwhile, it is gratifying to report that excellent progress is being made in regard to the Benguella railway which, as I need scarcely point out, is likely to have an important bearing on the great Cape-to-Cairo trunk line. The object of the Benguella railway is the development of the reputedly enormously rich mineral area known as Katanga, in the Congo Free State, and, incidentally, the exploitation of other mineral deposits en route. The scheme is that of Mr. Robert Williams, who holds an important concession from the King of the Belgians to work the copper mines; but the leading spirits, so far as railway construction is concerned, are Mr. J. Norton Griffiths, of the London firm of Griffiths & Co., and Mr. Edward Robins, chief resident engineer at Lobito Bay and local representative of the well-known consulting engineers Sir Dogulas Fox and Partners. Notwithstanding the fact that the railway company is in Portugal, the line is regarded as a British enterprise. At the moment of writing the most difficult portion of the line—that between Lobito Bay and the Katingue plateau, 1,400 metres above the sea—is nearing completion; and when once this is accomplished all will be plain sailing, as the remainder of the track, for upwards of 700 miles, offers absolutely

CATUMBELLA BRIDGE, BENGUELLA RAILWAY.

no difficulty from the engineering point of view. Indeed, Mr. Norton Griffiths quite recently assured me that he fully anticipated that when once the summit of the plateau is gained plate-laying will be proceeded with at the rate of a mile a day. Bridges will not be necessary as there are no rivers to be negotiated; and "the nature of the country"—to quote Mr. Griffiths—"is indeed, from now onwards so kindly that really hardly any earthworks are required for the construction of a serviceable line; it is merely a question of laying rails on the flat." The gauge of the line is 3 feet 6 inches; the rails weigh 60 pounds per yard and the steel sleepers 70 pounds each. As has been indicated, the initial part of the undertaking has been anything but a picnic to those chiefly concerned. The spanning of the various ravines which cross the ascent to the Lengue Gorge, for example, entailed some fine engineering work, t h e railway hereabouts having a gradient of 1 in 16. No. 1 viaduct consists of five equal spans of 18 metres on a gradient of 1 in 20.75; No. 2 viaduct is made in three spans and

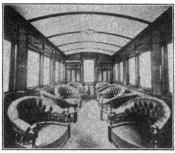

TYPE OF SALOON CARRIAGE, BENGUELLA RAILWAY.

is built on a gradient of 1 in 16.5; whilst No. 3 viaduct has four 18-metre spans, built on curve of 150-metres radius. Owing to the rapid rise in the levels of the country the contractors decided to make use of the Riggenbach rack system, as extensively used in Switzerland,

the length of which is about two kilometres. The gradient is about 1 in 16 and the sharpest curve 100-metres radius. A t time of writing some 8,000 men (including 1,500 Indian coolies imported from Natal) are employed on the line; and as showing the progress made it may be mentioned that during the twelve months ended last June, 300,000 cubic yards of material were excavated, largely consisting of decomposed granite.

Before leaving West Africa one may be permitted to glance at railway progress, present and prospective, in the British

ST. PEDRO BRIDGE AT THE TOP OF LENGUE GORGE, BENGUELLA RAILWAY.

colony of Nigeria. In August last it was announced in the House of
Commons that it had been decided to authorize the immediate construction of a pioneer railway, of 3 feet 6 inches gauge and 400 miles long,
from Baro, the highest convenient point on the perennially navigable
reaches of the Niger, to Zungeru *via* Bida, and thence in a north-easterly direction to Zaria and Kano.   Ultimately the line will connect at
Zungeru with that commencing at the coast port of Lagos, and as it
will tap the huge cotton district its importance to commerce can
hardly be over-estimated.   It is anticipated that the line will take
four years to build, the supervision being in the hands of the well-
known Canadian engineer Sir Percy Girouard, who has had considerable experience in railway engineering in Egypt and the Soudan.
On latest advices work on the new railway is now in active progress,
the rails for the first 20 miles of line having already been landed from
England, and over 2,000 workmen being engaged on the earthworks.
The section now under construction is the first portion of the new
system and is 120 miles in length, running from the Niger northwards towards Zaria, the point of the important junction to the
Banchi highlands and Lake Chad.   It is expected that Zaria will be
reached within 18 months; and during next summer another section
of 150 miles of rail will be delivered on the Niger.   Sometime during
1909 railway communication will be established between the coast

LENGUE VIADUCT, BENGUELLA RAILWAY.

and Zaria from Lagos—a distance of 700 miles—Kano itself being
reached in 1910. The railway from Lagos is now being carried to
Ilorin which place it is expected in June 1908, a total distance of
250 miles.

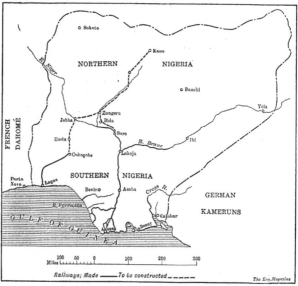

SKETCH MAP OF THE NIGERIA RAILWAY.

Coming further south, but still on the western side of Africa,
must be chronicled the completion of the light railway running from
the coast port of Swakopmund, in German West Africa, to the copper
district of Tsumeb, known as the Otavi railway. The work was
planned and carried out by Messrs. Arthur Koppel, of London and
Berlin. The railway, which runs for 350 miles, is the longest 2-foot
gauge line in the world. Construction was commenced in November,
1903, but was greatly retarded owing to the outbreak of the Herero
rebellion. Seven months later the contractors engaged a number of
Italian workmen and things went well for a time, but the big strikes
then prevailing in Italy affected the workers on the Otavi line and
construction was again seriously interrupted. Scarcity of water,

TRACK LAYERS AT WORK, AND A GROUP OF NATIVE WORKMEN WAITING FOR RATIONS,
OTAVI RAILWAY.

which had to be carried on ox waggons a distance of 30 or 40 miles, was also a drawback and, boring experiments failing, special water trains had to be run up from Swakopmund. Then the natural features of the country traversed by the line were unfavourable to rapid construction, a height of 3,600 feet having to be climbed in a distance of only 68 miles, at the same time keeping within the maximum gradient of 15 in 1,000. I understand that the rails used on the Otavi line weigh 30 pounds per yard and are fastened to steel sleepers throughout; the trucks have a capacity of 10 tons of 1,000 kilos, and

are of the double-bogie type. The locomotives are of 120 and 150 horse-power capacity, t h e speed varying between 9 miles and 15½ miles per hour. In its course the line passes over one hundred s t e e l bridges of 5½ to 22 y a r d s span.

"COVERED GOODS WAGON," OTAVI RAILWAY.

Another light railway, the partial completion of which is already proving a factor of considerable importance in the development of the Dark Continent, is that known as the Western Oases railway, which will link up the once prosperous, but more latterly, much neglected, rich green patches in the Libyan desert. Take a map of Egypt, and some 300-odd miles south of Cairo and on the Nile you will see a place called Farshut. That is the starting point of the Western Oases Railway. It may interest the curious to know that the concession for the line is for a period of 70 years; that the Egyptian Government has given a provisional guarantee of interest and the concessionaires, the Corporation of Western Egypt, possess for 20 years a monopoly of railway construction between any of the four western oases and the valley of the Nile, and "within or between" any of these four oases. From Farshut the course runs in a south-westerly direction to a place called Meteng, then turns and drops, almost due south, to Kharga. A detailed survey of the first section of the course was made early in 1906, by Messrs. Kincaird, Waller, Manville & Dawson, consulting engineers, whose report, a very favourable one, showed that the only engineering difficulty to be encountered en route was that presented by the steepness of the descent into Kharga. The engineers were convinced, however, that the line could be laid here, using no worse a gradient than 1 in 25—a gradient which, although steep, is not, I believe, without precedent in narrow-gauge railway construction. As a matter of fact the worst gradient has been improved from 1 in 25 to 1 in 40, as a result of which the freight capacity of a train will be doubled. The question of gauge was, for a time, something of a stumbling block, but eventually, on the recommendation of the consulting engineers, it was decided to adopt the standard narrow gauge used on the Delta Railway and other Egyptian lines, namely, .75 metres (equal to 2 feet 5½ inches). This determined, the work of construction was rapidly proceeded with, the Corporation acting as their own contractors. The line, which connects with the Egyptian State Railway at a point near Farshut on

ON THE WESTERN OASES RAILWAY. A TRAIN OF CAMELS HAULING RAILWAY MATERIAL, AND A BIT OF SUB-GRADE FORMATION ON THE LIBYAN DESERT.

THE WELL AT BERES, KHARGA OASIS, WESTERN OASES RAILWAY.

the Nile, has been carried across a tract known as the Cultivation and up the Wady Sahmoud onto the plateau (the Libyan Desert) and at time of writing the first section of the line is rapidly approaching completion. All the formation work has been finished on the scarp (a long descent into the oasis) to the terminus at Kharga town. The full length of the track is 194 kilometres. The weight of the steel rails used is 36 pounds per lineal yard, these rails being laid on creosoted timber sleepers in the Nile valley and in the oasis, and on steel sleepers across the desert. The railway is being laid as a single line, sidings to enable trains to pass being provided for the present at intervals of about ten miles. Base buildings have already been erected and headquarters established on the edge of the desert; telephone equipment being completely laid to railhead. The headquarters of the railway, I am advised, will shortly be fixed at Kharga instead of as at present at Qara, at the Nile valley end. The transport of material and machinery by pack camel from Farshut to the Kharga oasis was an interesting business. A depot was established at El Der, some 110 miles from Farshut and 20 miles to the north-east of Kharga village, where are the remains of a Roman fortress. While the greater portion of the material was transported by pack camel, certain heavy details such as boilers, drill frames, etc., had necessarily to be transported on wheels by means of draft camels. In order to expedite matters 30 special camels were purchased and trained, while the rougher portions of the desert track were improved as far as possible and a good zig-zag road cut on the cliff descent from the desert plateau into the oasis. The operations in question will, it is anticipated, do not a little in the repeopling of the large area comprised in the concessions, and render the communities dwelling therein not as at present merely self-sustaining, but capable of a large production of commodities suitable for exchange with those of western Europe.

CLEANING A LOCOMOTIVE ON THE WESTERN OASES RAILWAY.

In the space at my command it is possible to refer only briefly to the progress which has lately been made in railway engineering in other parts of Africa than those already indicated. Thus slow but steady progress has to be reported in connection with the railways of the Congo Free State, and trains now run regularly between Stanleyville and Ponthierville in less than 5½ hours, including stoppages. In Abyssinia, the line has reached Harar, but all hope of its extension must apparently be abandoned—for the time being at any rate—owing to the existing political deadlock. The locomotives on this line, by-the-bye, are built to carry 13 tons of water, which enables them to run one hundred kilometres without the necessity of stopping for rewatering. In British Central Africa railhead has now reached Mile 62½, and it is confidently expected that Blantyre will be gained at the end of the present year. The section of the line between Chiromo and Port Herald has been in constant use and has greatly

expedited and facilitated trade generally. In French Africa a good
deal of activity prevails and the iron horse is advancing in the Ivory
Coast, in Dahomey and elsewhere; and preparations for railway construction in Algeria on a somewhat extensive scale are in contempla-
tion. Progress has at last to be chronicled in regard to the Mrogoro
railway in German East Africa, Colonial Secretary Dernburg, during
his recent official tour in those parts, being enabled to travel over the
line from Dar-es-Salam to Mrogoro, the inland terminus. The contractors for the line are Messrs. Phillippe Holzmann & Co. In another part of German Africa, on the western side of the continent, a
section of the new Lüderitz Bay railway, about 30 kilometres in
length, was lately opened for military transport.

A VIADUCT ON THE ABYSSINIAN RAILWAY.

Coming, in conclusion, to the sub-continent, quite a number of
branch and other lines there have been or are being constructed.
Inter-communication with the several British colonies there is rapidly
becoming more general, and the network of lines running in all directions south of the Zambesi is slowly but surely increasing. Not the
least important of these is the line which is being built by the Portuguese from Lourenço Marques to the border of Swaziland, which
is British territory, through which country it will eventually be extended to the Transvaal—presumably to the existing rail-head at

ON THE SWAZILAND RAILWAY.   WORK ON THE UMBELOZI RIVER, AND A CUT AT
MATOLLA.

Ermelo. The line when completed will form an alternative route
from Delagoa Bay to the great gold fields of the Rand, and will
mean a saving of eight or nine hours on the journey between the
points mentioned. The line, which has been described by a competent
authority as "a model of engineering skill" will run through Mbabaan.

the new capital of Swaziland, and past Lake Christie, thus effecting a saving of 60 miles on the route. It is a matter for regret that the Transvaal authorities, for some occult reason, do not show any particular enthusiasm for the Swaziland railway, as it is called, nor display any great desire to push on with the "linking up." They appear to be more in favour of spending a large sum of money in reducing the curves and gradients on the existing line between Pretoria and Lourenço Marques. Swaziland—as the "party" most affected—avers that this if carried out would be the falsest of false economy. And I am not so sure that Swaziland is very far out in her reckoning. The connection is surely bound to come, sooner or later, and there is hardly any doubt that the scheme, when completed, will be of immense benefit to Lourenço Marques as well as to Swaziland—the well-named "Cinderella" of the many States in the great sub-continent.

DJIBOUTI STATION ON THE ABYSSINIAN RAILWAY.

CAN RAILROAD COLLISIONS BE REDUCED TO A THEORETICAL MINIMUM?

By Harold Vinton Coes.

"The only way by which the disregard of signals and rules can be prevented is by making it physically impossible. The engineer must have no discretion as to whether he will stop or run past a block signal; he must be stopped by a mechanical device the release of which is beyond his control." That this statement in the editorial columns of THE ENGI-NEERING MAGAZINE for December, 1904, can be repeated after a lapse of three years with added emphasis is amply proved by the evidence given by Mr. Coes of the increasing number of railway collisions and corresponding sacrifice of human lives. Mr. Coes maintains that if railway officials still refuse to equip their locomotives with such mechanical safety devices, the value of which has been proved on the express tracks of the New York Subway, it can only be because they lack the most ordinary humanitarian sentiments and not from motives of econommy.—THE EDITORS.

IN this article, one phase of railroad accidents only is dealt with, viz., collisions, as it is into this class especially that the author has been examining. From the engineer's standpoint this type of accident is particularly interesting as it affords him many complex problems to study, the satisfactory solution of which would be of great benefit to mankind.

When one analyzes the reports of the Interstate Commerce Commission and the Statistics of Railroads, he is astonished by the appalling showing of figures and tables—statements of facts which cannot be denied. If he is of a questioning turn of mind or interested in the subject from a humane standpoint, he naturally asks himself how much of this is true, how much due to the rapid growth of the great railroad systems, and how can this seemingly unnecessary slaughter and maiming be reduced to a theoretical minimum?

Mr. Carroll W. Doten, of the Massachusetts Institute of Technology, in his admirable paper entitled "Recent Railway Accidents in the United States," read before the American Statistical Association, shows in Table I, next page, that the increase in casualties from collisions for the whole country between 1895 and 1903 was 270 per cent. He says, "By comparing this table with the following one in which density of traffic is measured in train-miles per mile of line, it will be seen that the amount of traffic does not afford an explanation of this increased liability to collisions. An increase of 305 per cent between 1898 and 1903 is observed in Group II where traffic is densest and 300 per cent in Group X where traffic is least dense.

TABLE I. PASSENGERS AND EMPLOYEES.

Casualties per Million Train-Miles—Collisions.

Group Year.	I.	II.	III.	IV.	V.	VI.	VII.	VIII.	IX.	X.	U.S.
1891...	2.5	3.4	3.0	4.3	6.8	2.3	3.0	2.8	3.3	5.0	3.4
1892...	3.0	4.0	2.6	5.0	5.9	2.5	1.9	2.4	2.4	2.3	3.3
1893...	4.1	4.3	2.5	1.6	4.1	2.0	1.5	1.9	2.4	2.7	3.1
1894...	1.6	3.0	3.6	2.6	3.1	2.1	0.6	1.2	2.2	2.1	3.2
1895...	1.5	2.6	1.7	2.0	2.6	1.2	1.2	1.6	4.0	3.4	2.0
1896...	1.3	2.2	1.6	4.1	6.3	1.1	1.1	2.1	3.6	2.3	2.2
1897...	1.4	2.0	1.8	2.7	3.7	2.0	2.0	2.2	3.2	2.8	2.2
1898...	2.6	2.0	2.5	2.9	3.0	1.4	2.2	3.5	2.3	2.7	2.3
1890...	4.0	3.3	2.6	3.6	3.4	1.4	3.7	2.0	3.3	4.5	3.1
1900...	3.3	3.6	3.8	7.0	6.9	2.3	5.0	2.4	3.5	3.5	3.6
1901...	1.4	6.6	3.7	6.4	6.6	2.3	4.7	3.1	5.0	4.5	4.4
1902...	3.7	8.1	6.5	8.4	6.7	4.5	6.0	4.3	7.6	8.0	6.1
1903...	3.4	8.1	9.2	15.0	7.9	5.1	8.4	9.7	6.5	9.8	7.4

TABLE II. DENSITY OF TRAFFIC IN THOUSANDS OF TRAIN-MILES PER MILE OF LINE.

Group	I.	II.	III.	IV.	V.	VI.	VII.	VIII.	IX.	X.	U.S.	U.K.	Can.
1891..	7.3	9.9	6.1	3.3	3.6	3.9	2.8	3.1	2.0	3.0	3.6	16.0	3.1
1892..	7.7	10.7	6.3	3.2	3.8	4.3	2.8	3.1	2.9	2.9	3.7	16.1	3.0
1893..	8.1	10.6	6.5	3.3	3.6	4.3	2.9	3.5	2.9	2.9	3.8	15.6	2.9
1894..	7.5	9.5	5.8	3.1	3.2	3.8	2.5	3.0	2.8	2.6	3.4	15.9	2.8
1895..	7.0	9.5	5.7	3.2	3.2	3.4	2.3	2.8	2.8	2.5	3.3	16.0	2.5
1896..	7.3	9.8	5.8	3.4	3.2	3.4	2.2	2.9	2.8	2.6	3.4	16.6	2.7
1897..	7.1	9.6	5.4	3.4	3.3	3.7	2.2	2.9	3.0	2.4	3.3	17.2	2.7
1898..	7.2	10.0	5.6	3.3	3.5	4.0	2.4	3.3	3.0	2.5	3.4	17.5	3.0
1899..	7.2	9.8	5.8	3.3	3.6	4.0	2.9	3.2	3.0	2.7	3.4	18.2	2.9
1900..	7.3	9.4	5.9	3.3	3.4	3.9	2.4	3.1	2.7	2.4	3.4	18.4	3.0
1901..	6.9	9.3	6.0	3.2	3.6	3.9	2.6	3.2	2.8	2.6	3.4	18.1	2.9
1902..	7.1	9.3	6.1	3.3	3.7	4.0	2.6	3.3	2.9	2.8	3.4	18.0	3.0
1903..	7.4	9.3	6.2	3.4	3.8	4.1	2.9	3.2	2.8	2.9	3.5	18.6	3.2

NOTE.—The groups here referred to are those ten geographical divisions of the United States, which the Interstate Commerce Commission has adopted for statistical purposes. This divides the railroads according to the territory through which they run, and takes into account density of population, topography of the country, and the character of the industries served by the railways.

Official figures for the year ending June 30, 1905, show that 196 passengers were killed and 3,336 were injured in railway collisions. The same authority gives the total number of passengers carried as 738,834,667. Hence one was killed in every 3,769,819 and one injured in every 221,443, by collision accidents alone. The total of passengers killed during the same year by railway accidents of every class was 537; the number killed in collision (196) was therefore 36.5 per cent of the whole. In the year ending June 30, 1906, 418 passengers were killed in railway accidents of all classes, 120 of these being by collisions; the loss of life in collisions during this year was thus 28.7 per cent of the whole. The table at the top of page 635 covers these same periods, but includes all persons killed or injured—employees, trespassers, etc., as well as passengers.

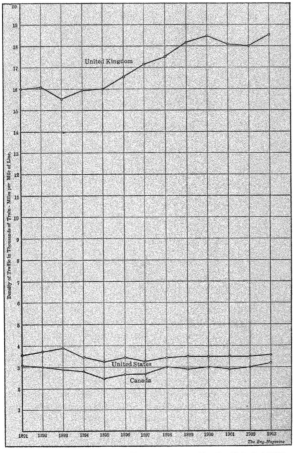

CHART I. SHOWING DENSITY OF TRAFFIC PER MILE OF LINE IN THE UNITED
KINGDOM, UNITED STATES AND CANADA, 1891-1903.

TABLE III. COLLISIONS, DAMAGES TO CARS, ENGINES AND ROADWAY.
Authority of Interstate Commerce Commission.

	Number.	Loss.	Persons Killed.	Persons Injured.	Number.	Loss.	Persons Killed.	Persons Injured.
	Year Ending 1906—June 30.				Year Ending 1905—June 30.			
Collisions, rear	1,722	$1,720,365	169	2,427	1,493	1,463,012	152	2,085
" butting ...	866	1,599,568	251	2,733	707	1,451,906	304	2,453
" trains sep- arated..	901	359,156	9	375	972	440,495	11	369
Collisions, miscellane- ous	3,705	1,640,669	175	2,379	3,052	1,493,641	141	2,204
Total.........	7,194	$5,319,758	604	7,914	6,224	4,849,054	608	7,111

Recently two accidents due to collisions were reported on railways in the eastern United States. The first was on October 1, at Providence, R. I., on the N. Y., N. H. & H. R. R. The Knickerbocker Limited was run into by the Shore Line Express and three persons were seriously injured. The former train was one of the crack extra-fare trains on the road, and yet it was not exempt. The second wreck was on the Worcester, Nashua and Portland division of the Boston & Maine R. R. Eight people were injured in this accident. The collision was due, it was reported, to a thick cloud of smoke obscuring the signal. This is only an instance of what takes place every few days, and even kills railroad presidents, as in the case of Samuel Spencer on the Southern R. R. What can we do to stop this? How can we safeguard human life, and at the same time get our trains over the road on time? There is no question but that the railroad companies have all they can do to meet the increased demands on their traffic facilities. How can we reduce the collision rate? Who are to blame? The engineers, or the signalmen, or the system? These are questions that are on every railroad official's tongue and are interesting very many of the thinking public at the present time.

Mr. Doten in his comments on the table given above as Table II says: "This table has a certain significance on its own account. It shows how much denser railway traffic is in the United Kingdom than in the United States and Canada. 18,600 train-miles per mile of line means that on the average 51 trains pass over each mile of line in the United Kingdom every twenty-four hours, Sundays included. It also shows that the density of traffic in the United States measured in this way has not been increasing, but has rather fallen off since the early nineties. (See Chart I.) This signifies that the growth of traffic measured in train-miles and passenger-miles, has been distributed over longer lines and bunched in heavier trains, which would seem to indicate that the train-despatcher's duties are no greater than before. But whether the fault is with the brain that issues the orders, or in the one that receives and executes them, does not matter so long as the results are the same."

Let us now investigate the block-signal system and see if that throws any light on our previous figures.

Block signals may be classified in three ways :—*

1. Depending upon the manner in which their day indications are shown.

2. Depending upon the manner in which they are controlled and operated.

3. Depending upon what they control.

In the first classification there are :—

I. Banner signals, whose indications are displayed by a revolving banner.

II. Disc signals, whose indications are displayed by a movable disc in front of a fixed background, and

III. Semaphore signals, whose indications are displayed by the positions of an arm moving in a plane at right angles to the track. In all types under one class the night indications are displayed by colored lights.

In the second classification there are :—

I. Manual, the signal being controlled and operated by man power.

II. Controlled manual, the signal being operated manually and constructed so as to require the co-operation of the signal-man at both ends of the block.

III. Automatic, the signal being operated by power which is controlled entirely by the presence or absence of a train in the block, or by the condition of the track.

An absolute block system is one which never allows more than one train in the same block at the same time.

A permissive block system is one which may allow more than one train in the same block at the same time, provided the trains are moving in the same direction and the second train has been warned by signal that another train is in the block.

The signals must not only warn the engine driver that there is a train in the adjacent block, but they must be so spaced as to give him

---

* This classification is due to Mr. W. G. Foster and appeared in the *Electric Journal* for July, 1907.

room enough to bring his train to a stop, before entering the block. It is when this principle is overlooked that a collision occurs.

Blocks vary in length from 4,000 to 12,000 feet. It is difficult to arrange the signals so as to obtain uniform spacing and at the same time place them where the engineer can obtain an unobstructed view of them on approach. The minimum length of block depends upon the distance in which the train at maximum speed can be stopped. Hence if the signal is placed so as to give a good view, but at the same time is spaced back in order to do this (so that the length of the block is equal to or less than the minimum) then the above principle is violated and the block is dangerous.

While riding on a locomotive on one of the eastern American roads using signals of the banner type, I noticed that the signal flashed to danger when the train was about 150 feet away from it. This peculiarity or defect caused a serious wreck on the road a little later, the engineer thinking that his train had set the signal while in reality it had been set by a local freight on the cross-over just beyond.

The non-automatic or manually controlled signal is usually operated from block towers or cabins, each tower controlling a certain section or sections, the signals repeating themselves either on a dummy board or by sode signals. This is well providing the signal man sets the right signals and has not been worked too many hours; or the engineer is not dead, sick, color-blind, or wilfully runs by the signal. In other words, the element of the individual enters into this system and in fact the system depends upon this element for its successful operation.

The automatic block signal is at present a permissive system, for if a signal is out of order it assumes the stop position, as it is constructed to do; so that traffic may not be entirely suspended for several hours. This type of signal depends upon a track circuit for its operation. Certain rail joints at the end of each section are insulated from each other by removing the fish plate and inserting one of the many non-metallic track bonds now in use, the intermediate rails being bonded together, so as to complete the circuit by a No. 8 galvanized-iron wire. A battery is then connected across the rails at one end and an electro-magnet or relay across the other, as shown on the next page.

The pivoted armature falls away from the relay by gravity and must be energized to hold it up. The wheels of the train bridge the track as at "a-b" and thus complete the circuit through the engine and rails cutting out the relay, which drops its armature, thereby operating the

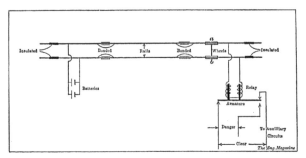

DIAGRAM OF OPERATION, AUTOMATIC BLOCK SIGNAL.

auxiliary circuits which control the signal and thus set it to danger. Hence it is readily seen that a broken rail will ordinarily interrupt the circuit, thereby demagnetizing the relay and consequently setting the signal to danger. The automatic block, therefore, serves three purposes; it warns an engineer that there is a train occupying the block ahead of him, or that there is a defective rail, or that the signal apparatus is out of order. If the signal does not set to clear within a reasonable length of time, the engineer then knows that one or the other of the last two conditions is causing its non-operation, or that the train ahead has broken down. But if the engineer does not obey this warning, or if he is dead, sick, asleep from overwork, or color-blind (this latter ailment applying particularly at night when colored lamps are used) or if the signal is obscured by fog or smoke and he runs past the place where he thinks it ought to be and enters the block ahead— then all the effectiveness of the automatic block-signal system is lost, and we again have to depend upon the human element as to whether the signal does its work or not. The last paragraph may be summed up by the statement that no signal on its own account ever stopped a train.

It is, therefore, plain that all of the present signal systems depend upon fallible human nature for their operation, or at least for their correct interpretation, and hence that collisions can and do occur with appalling frequency on all roads using signal systems of any kind. This may be seen from Chart II. A perusal of the records shows that these catastrophes are not due as a rule to the signal systems themselves, but are directly traceable to any one of the human fallibilities mentioned above.

In other words, we must extend our automatic block-signal system one step farther and make it absolute; we must eliminate the dangerous personal elements; we must take the control of the locomotive out of the hands of the engineer and make it impossible under normal conditions for him to run past a block. As an instance to illustrate this point, recently on the Erie Railroad an engine of the camel-back type, (i. e. using a Wooten firebox and thus separating the engineer from the fireman) lost its engineer by an accidental blow on the head, and if it had not been that the fireman noticed that the train was running past a station, a disastrous collision might have occurred. So great is this danger on this type of locomotive that the legislatures of

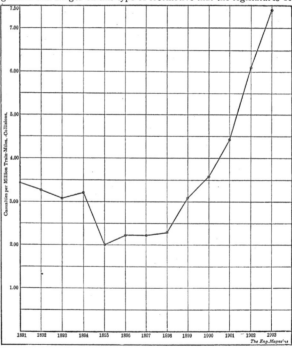

CHART II. SHOWING CASUALTIES PER MILLION TRAIN-MILES DUE TO COLLISIONS IN THE UNITED STATES, 1891-1903.

one or two of the States are advocating carrying an additional man in the cab. However, any operating official knows how little room there is in one of these cabs for an extra man, and the inadvisability of carrying one.

The conservative traffic and operating official will say that this is too radical a step, that "if a block system is made absolute it will be impossible to get trains through on time." This reasoning however is

AN ANALYSIS OF 14 RAILROAD COLLISIONS OCCURRING DURING THE MONTHS OF JANUARY, FEBRUARY AND MARCH, 1907.

Class of Collision.	Kind of Train.	Killed.	Injured.	Damage to Engines Cars and Roadway.	Cause.
Rear End	F & F	2	2	$5,250	Engineman ran past two automatic block signals set against him.*
Butting	P & P	4	107	7,745	Regular east bound passenger train ran past meeting point.
Miscel.	P & P	0	6	9,567	Misplaced switch at station. Target of switch covered with snow.*
Rear End	P & P	1	15	10,100	Approached station (8 P. M.) not under control; faulty flagging.
Rear End	F & F	2	5	11,000	Excessive speed, and failure of standing train to flag.
Rear End	F & P	0	11	11,025	False clear block (manual controlled block, communication from cabin to cabin by bell code, mistake in bells).*
Butting	F & F	0	3	11,100	Eastbound ran past signal at meeting point at 4 m. p. h.*
Butting	P & F	0	11	11,170	Westbound passenger train disregarded wait order.
Rear End	P & F	0	16	12,000	Passenger train unexpectedly stopped (3 A. M.); run into at rear by freight.
Miscel.	P & F	5	3	12,400	Extra freight entered main track in face of fast train disregarding automatic indicator at switch.
Butting	F & F	1	3	15,933	Northbound train disregarded meeting order. The order read "meet engine 567." At the meeting point engine 565 was on the side track and this was taken for 567. N. bound passed full speed.
Miscel.	P & F	0	3	21,900	Passenger train disregarded distant and home automatic block signals; ran into side of freight train. Fireman had called clear signals to engineman without seeing signals.*
Butting	P & F	9	8	48,500	Passenger train ran past block signal, and freight disregarded time limit.*
Butting	P & F	2	12	11,300	Westbound extra freight (2 A. M.) neglected to head in at entrance to side track; engineman asleep.
Total		14	26	205	198,990

Accidents marked with * bear directly on the subject.

false, for the engineer, knowing that, even if he cannot see the signals, his train will be brought to a stop if the signal is set to danger, will be able to keep his train up to speed between blocks, instead of crawling along and feeling for the signal. Again the official will say: "The engineer will get to depending too much upon this device and will, therefore, relax his vigilance." This can readily be taken care of by making the automatic stop self-recording, so that the superintendent will know at the end of the day's run just how many times the train has been stopped by this method.

The most powerful argument that the railroad companies will use against this step is the cost of equipping the roads with such a system. Let us inquire into this. The Twentieth Annual Report of the Interstate Commerce Commission for the year ending June 30, 1906, shows that the cost to the railroad companies of collisions, excluding damages to property and indemnities paid to or on account of persons killed or injured, was $10,659,189, and if there were any method of obtaining the figures for the damages and indemnities, it would swell the figure to many millions more. Is not this worth saving? Why spend hundreds of thousands of dollars in trying to secure further economies in the rolling stock, while there is this big leak at the other end?

The above mentioned report also states that there were in operation in the United States up to June 30, 1905, 48,357 locomotives. These, excepting 947, were classified as: passenger 11,618; freight 27,869, and switching, 7,923. Deducting the 7,923 and the 947 from the sum total we have 39,487 locomotives in effective operation for long-distance runs. A conservative estimate shows that the cost of equipping a locomotive with a device absolute in its operation including the track connections and connections to signal would be $1,000. Hence the total cost of equipping the locomotives of all the roads would be $39,487,000. Assuming that the new system would save, if it were at all efficient, 50 per cent of the $10,000,000 paid out last year for collisions, and taking the cost of the system in round numbers at $40,-000,000, the device would pay for itself (exclusive of the amount saved on the damage and indemnity bill) in eight years, or pay 25 per cent on the original investment, the maintenance being charged off on the amount saved on the damage and indemnity bill.

Let us assume another point of view and examine the income figures of the railroads and see if they can stand the expense, apart from the above figures. The total income of the railroads, covered by the Twentieth Annual Report of the Interstate Commerce Commission, was $920,221,859. Against this amount was charged for interest, rents, betterments, taxes, and operating and miscellaneous expenses, the sum of $590,386,554, and as dividends the sum of $229,406,598, leaving a surplus for the year of $100,428,707. From these figures it looks as though the railroads could well afford to safeguard human life more closely in the matter of collisions.

In general, then, the following conclusions may be safely stated:

That with our present signal systems collisions seem to be on the increase, as may be seen from Chart I, and that apparently the railroads are unable to cut the rate down to its approximate theoretical minimum.

That all systems depend upon the human element for their successful operation or at least interpretation, and that as man is fallible he must be superseded by tireless and infallible automatic machinery.

That some system of automatic stops should be used in connection with our present signal systems, so constructed as to perform the same functions as the engineer, viz, to shut off the power and to apply the brakes, at the same time retaining the air in the auxiliary reservoirs after the brakes have been applied, so that traffic will not be delayed by waiting for the train line to build up to normal pressure; and that the connections from the signal system to the trips on the locomotive be so mechanically constructed as to operate successfully in ice or snow storms, or whenever the signal system is in operation.

That the public should guard its interests in this respect, if the railroads do not see fit to do so, by appealing to the proper source, namely, Congress, and have such laws passed as will attain the desired results, as was done in the case of the air brake, the automatic coupler, the block-signal system, and the car-heating system.

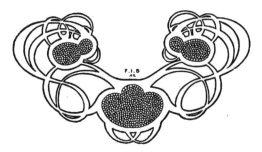

MINING DEVELOPMENTS IN NEVADA.

By A. Selwyn-Brown.

The review following covers the conditions up to the end of 1907 and is guided by the author's intimate personal knowledge of the districts. It is especially interesting in view of the present prominence of the Nevada mining regions in the public eye. Mr. Selwyn-Brown's references to individual properties, we need hardly emphasize, contemplate physical conditions only—not financial conditions. He is concerned with ore values alone, without favor and without prejudice to share values. A similar study of the Nevada copper regions will follow next month.—THE EDITORS.

THE great Tonopah gold mine was discovered on one of the prominent mesas, or barren mountains which are so numerous in the desert areas in Nevada and Arizona, in May, 1900. Nevada at that time was just beginning to awaken from the deep economic lethargy into which she was plunged after the flooding of the Comstock mines in the early eighties, and the closing down of nearly every mine of importance on the other mining fields through the fall in the value of silver. Although gold was discovered and mined in many places in the State in the early days, Nevada was generally looked upon as a purely silver producing State, in consequence of the extraordinary richness of the silver mines on the Comstock, Eureka, Candelaria, Austin, Pioche, and other famous fields. It was, however, a difficult matter for the discoverers of the Tonopah mine to finance it. The investing public was skeptical when told of rich gold discoveries in a remote part of the desert region in the south-eastern portion of Nevada. Had the mine not carried rich shipping ore from the grass roots down, the mining history of Nevada today would not be so interesting as it is. The mine, to a large extent, paid for its own development. Early in 1902 Philadelphia capitalists, led by J. W. Brock, purchased a controlling interest in it at a high figure, and at once set about the building of the Tonopah railroad. A steady rush of prospectors then set in which resulted in the rapid development, during the years 1904-7, of important gold mines in many other localities. During 1905, and until the early part of the present year, the wonderful reports of the rich finds and yields of some of the gold mines in these new camps attracted world-wide attention, and resulted in the investment of a large amount of capital. This influx of outside capital greatly stimulated prosperity in the mining camps, particularly in Goldfield, Bullfrog, and Manhattan, and led to extensive share-dealing operations not only in local centers, but on the San Francisco, Philadelphia, New York, and other stock exchanges, with the consequent rapid soaring of share values. Dollar shares like those of the Tonopah, Tonopah Extension, Goldfield Consolidated, Shoshone Consolidated, and other mining companies advanced in some instances above $18. The general monetary stringency commenced to extend to Nevada, however, in August. In November the local banks suspended payments and a general slump in share values set in, with the result, that business on the stock markets is at present practically at a standstill. This unhealthy financial situation is accompanied by a train of other economic ills, including the closing down of many mines and construction works, the reduction of wages, and restriction of general business affairs. Labor troubles and a serious increase in crimes of violence are beginning to add to the State's embarrassment. With such conditions prevailing, there need be little wonder that some of the less reflective people interested in the State's progress opine that Nevada is again entering a period of economic quiescence like that following the Comstock boom. My own experience and investigations on the principal mining fields do not induce me to view the present disordered conditions in Nevada pessimistically.

TOPOGRAPHY.

A large portion of southern Nevada has an elevation above sea level varying between 2,000 and 6,000 feet. The elevation of the Tonopah and Goldfield districts is between 5,000 and 6,000 feet. These districts are traversed, as is indeed the whole State, by innumerable mountain ranges and peaks with intervening valleys trending, generally, in a south-westerly direction toward the Sierra Nevada. The mountains, as a rule, do not rise much more than 400 or 500 feet above the level of the valleys, but in places some prominent peaks of great elevation are met with. In the Toyabe and Toquima ranges north of Tonopah there are several peaks covered with perpetual snow. The mountains are, generally, gently undulating; but in places, owing to changes in rock formations and other causes, the ranges are formed of numerous sharp peaks and ridges which the prospector finds dangerous, or impossible, to climb and prospect. Most of the ranges and valleys are bare of timber; but in places, the high peaks and the ravines running down their sides are covered with stunted pine trees which are known to miners by the Mexican name of *piñon*. The only vegetation in the mineral districts is the dreary, strong-smelling sagebrush, which is able to thrive in the dryest deserts, and is of inestimable value to miners by serving as food for their trusty burros, and, when wood is not procurable, as fuel.

Water is scarce throughout the mining districts. In places small streams running down the gulches from snow-capped mountain peaks yield supplies of drinking water to many neighboring camps. Some mining centers are dependent upon the scanty supply of some mountain or desert spring. Such springs, however, are numerous, and prospectors rarely go more than 15 or 20 miles before meeting with one. As mining camps develop and capital becomes available, ample water supplies for all purposes are obtainable by tapping the great underground rivers below the desert valleys by means of artesian wells. Almost every valley in the State in the vicinity of the snow-capped mountains carries artesian water streams. At Tonopah, Goldfield, and Bullfrog, artesian water is now being employed in the large milling and cyanide plants.

GEOLOGY.

Nevada is a huge tableland supported by the Sierra Nevada and outliers of the Rockies. As these great mountain chains have risen, the intervening rock formations have been squeezed and faulted, and extraordinary volcanic phenomena have been forced into activity. Evidence of volcanic action is everywhere visible in Nevada, and thermal springs, even today, are playing important parts in bringing about geological changes. Volcanic rocks must be common in such a region. They cover very large areas in the State, and constitute the most important formations in the mining districts. Granitic rocks occur to a less extent, and are not so important to the miner as the volcanic rhyolites, andesites, and porphyries. In some localities, as Reveille, lead and copper lodes occur in limestones and other sedimentary rocks. Coal and mineral oil deposits are also found in some of the sedimentary areas.

MINERALS.

The whole of Nevada is mineralized. There is hardly a settlement in the State which does not possess a mineral deposit of some kind. In addition to gold, silver, copper, and lead, deposits of sulphur, zinc, bismuth, antimony, tungsten, nickel, iron, mercury, arsenic, salt, borax, and other minerals are being developed. In many districts valuable deposits of many kinds of gem-stones also occur, the most important being turquoise and opal. Sapphires, garnets, and diamonds have been found in the gold wash in several places.

MINERAL STATISTICS.

Mining statistics have never been accurately compiled in Nevada and today there is a noticeable variance between even the smelting companies' returns and the Government bullion-tax statistics. The mineral production of the State can only be estimated approximately. The Goldfield mines from January 1, to November 23, this year, shipped 112,081 tons of ore valued at $11,208,100. The miners' strike is likely to interfere with the output for a few weeks, but it is safe to estimate that the production this year will be at least a third as large again as that in 1906. A conservative estimate is 127,000 tons of ore, for a return of $12,700,000. The returns from Tonopah will be somewhat less owing to the mines on that field being worked in a more conservative manner than those at Goldfield, which are to a great extent operated by a large number of lessees working small leases. The Tonopah mines this year will yield 200,000 tons of ore valued at $10,000,000. The smaller fields will return about 100,000 tons of gold ore valued at $4,000,000. The gold and silver yield of Nevada in 1907 will consequently amount to about 427,000 tons of ore valued at $26,700,000. In nearly all Nevada ores the ratio of silver to gold is about two to one. It is necessary to bear this in mind when reading of the returns from the gold mines, which are usually given in dollars per ton, and these figures include both the gold and silver contents of the ore.

The following table, which has been carefully compiled from the most reliable statistics available and rather understates the real production, will give a good idea of the distribution of mineral wealth in Nevada. The production of other metals than gold and silver, is not included.

GOLD AND SILVER PRODUCTION IN NEVADA, 1850-1907.

District.	Mineral.	Returns.
Comstock	Silver, gold	$640,000,000
Eureka	Silver, lead, gold	90,000,000
Pioche	Silver, gold	40,000,000
White Pine	Silver	25,000,000
De Lamar	Gold	15,000,000
Austin	Silver	60,000,000
Belmont	Silver	10,000,000
Tybo	Silver	6,000,000
Tuscarora	Silver, gold	25,000,000
Candelaria	Silver	18,000,000
Aurora	Gold	13,000,000
Pine Grove	Gold	5,000,000
Jefferson	Silver, gold, lead	3,000,000
Cortez	Silver	3,000,000
Silver Peak	Gold, silver	3,000,000
Reveille	Silver	2,500,000
Silver Star	Gold, copper	1,250,000
Tonopah	Silver, gold	29,000,000
Goldfield	Gold	26,700,000
Newer Fields	Gold and silver	5,000,000
Total		$1,033,350,000

To this enormous value of $1,033,350,000, yielded by the gold and silver mines of Nevada since the Comstock rush in 1850 to the end of 1907, must be added the value of the large quantities of copper, lead, iron, zinc, and other metallic ores and of the sulphur, borax, salt and other minerals mined. This would bring the grand total of the mineral production of Nevada to date to about $1,400,000,000. This is a splendid showing when the comparatively small amount of capital that has been invested in the development of the State is kept in view. It bears eloquent testimony to the fact that mining investments in Nevada, on the average, have been unusually successful.

TONOPAH.

The mining area comprised in the Tonopah goldfield is about seven miles square. Mining operations have proved the whole of this area to be mineralized, and it is quite probable that future mining operations will show that the Tonopah vein systems extend to, and form part of, those now being successfully worked on Lone Mountain, which is situated 17 miles west of Tonopah, and at Ray, 12 miles north of Tonopah. The principal development work to date has been done on the first discovered mines, namely, the Tonopah, Montana, Midway, Belmont, North Star, and Tonopah Extension mines.

In parts of the Tonopah field there are no less than six formations overlying the lower ore-bearing formations. As a consequence of this, prospecting work must be carried on under the supervision of geological experts. The correct placing of mining shafts that must be sunk several thousand feet before crosscutting for a vein is undertaken through hard volcanic rock formations, requires the exercise of a high order of scientific acumen. Hitherto, very few mistakes have been made in the planning of mining work, and there are few important fields in any country that can show such successful results as Tonopah. The results here attained are due, probably, as much to the conservative business methods of the directors of the mining companies operating on the field as to the skill of the engineers in charge of the mines. But for the backing received from the directors, the bold work carried on in some of the mines might not have been brought to a successful issue.

The first successful mine developed at Tonopah was the Tonopah mine on which the first discovery of mineral in the district was made by a Californian prospector named Butler, in May, 1900. The vein in this property outcropped for a short distance on the surface, and carried ore sufficiently rich to bear all the expenses of shipping to the smelters in California and Utah. This mine is now developed to a depth of 1,000 feet, and lateral developments underground exceed five miles in length. The value of the ore at present blocked out for stoping is estimated at $20,000,000. A milling plant has recently been erected by the company at a cost of $400,000 at Millers' Siding, a small railroad center situated some 12 miles west of Tonopah. This plant is now treating about 2,000 tons of the company's ore weekly.

The next mine of importance to strike gold was the Montana, which lies at the eastern boundary of the Tonopah leases. The vein was covered by several hundred feet of rhyolite, a volcanic formation of great extent in Nevada. The bearings of the Tonopah vein being accurately taken, a shaft was located and sunk at a point calculated to be on the line of strike of the vein eastward. Results showed the accuracy of the work. This has developed into a property, second only to the great Tonopah mine. It is shipping 1,000 tons of ore to the mills weekly.

Strikes were afterwards made in the Jim Butler, Midway, West End, Belmont, North Star, California, and other properties. The West End mine has been developed to a depth of 500 feet. It has a vein over 60 feet in width which yields ore of both milling and shipping grade. The Jim Butler mine has been developed to the same depth as the West End. It has a strong ore-bearing vein over 6 feet in width. In addition to large reserves of milling ore awaiting treatment, this mine is shipping about 150 tons of high-grade ore to the smelters weekly.

An extension of the Tonopah vein has been met with in the low ground west of the Tonopah mine in the McNamara and Tonopah Extension mines, which are situated about a mile away from the Tonopah mine. Further west, the Ohio, McKane, Red Rock, Great Western, Home, Little Tonopah, and Golden Anchor mines are sinking to catch the extensions of one or more of the veins now being successfully worked in the developed mines. None of these mines has been developed to sufficient depth yet. The McNamara and Extension have shipped a large amount of ore, but the others have not got into the ore-bearing formations.

The western extensions of the veins have been struck in the Boston, North, Star, Belmont, Rescue, and other mines. The Belmont and North Star are already in the shipping class and have large ore reserves opened up.

The Indiana, California, Crown Point, and other properties are operating on the southern portion of the field on cross veins; and there is a tendency for miners to prospect to a large extent on the northern part of the field.

Tonopah, as was intimated above, is not a poor man's field. Owing to the heavy covering of the ore-bearing andesite formations by several hundred feet of rhyolite and other volcanic rocks, which are expensive to sink shafts through, the field can only be developed with the aid of capital. The work already done has proved the value and extent of the field. The veins are large, regularly formed, and productive, and, at depth, are uniformly rich. The district has a healthy climate and there are no hindrances to mining developments. Electric power for the mills is obtainable from generating stations utilizing the hydraulic power of the streams in the Sierra Nevada, and several sources of water for domestic and mining purposes have been developed close to the town. The mining population is about 7,000.

In the vicinity of Tonopah a number of important mining centers are being developed. Hannapah, Bellehelen, Reveille, Golden Arrow, Clifford, and Silver Bow are promising gold centers eastward of Tonopah. The Manhattan field, situated about 20 miles north of Tonopah, is the most important northern camp. About 50 promising mines are under development here. Several mines are shipping high-grade ore. Recently two small milling plants were erected at Manhattan and they are now operating on milling ore from the principal mines. Atwood, Orizaba, Golden, Midas, Wonder and Fairview are other important northern camps. .

Silver Peak, Blair and Candelaria are the important mining centers west of Tonopah. Valuable mines are in course of development in each camp.

GOLDFIELD.

Gold was accidentally discovered at a watering place in the desert about 20 miles south of Tonopah in 1903. Some rich returns were obtained by the early prospectors and the field has been extensively developed with most satisfactory results. In 1905, gold to the value of $3,500,000 was obtained from nine of the principal mines. Last year the production amounted to about $10,000,000, and this year it promises to amount to not less than 127,000 tons of ore valued at $12,700,000. More than eighty important mines are in course of development at Goldfield. Many of the mines have been developed by lessees and the fortunate ones have rapidly acquired large fortunes. Speculation in shares assumed large proportions until temporarily checked by the financial stringency in November. The following table gives the ore yield of the field for each month during 1907, and the total stock sales on the Goldfield stock exchange for the corresponding periods. The figures to November 23 are official, for the balance of the year they are estimated.

Period Ending.	Tons Shipped.	Stock Sales.
Jan. 26	8,440	$3,311,563
Feb. 23	4,989	2,480,070 .
Mar. 30	2,323	3,959,415
Apr. 27	512	4,381,119
May 25	4,283	2,567,821
June 29	11,193	1,586,340
July 27	14,687	1,590,312
Aug. 31	17,724	1,772,549
Sept. 28	14,393	1,180,187
Oct. 26	20,237	728,325
Nov. 23	11,470	104,866
Dec. 31	9,000	1,500,000
Total...................	119,251	$25,162,567

In addition to this Goldfield stocks are actively traded in to a very large extent on the San Francisco, Tonopah, Reno, Chicago, New York, Boston, Philadelphia and other stock exchanges.

BULLFROG.

The Bullfrog goldfield is situated on a direct line about eighty miles south-west of Tonopah. During the past three years several important mines have been developed. The rhyolite which is connected with the gold deposits at Goldfield and Tonopah extends south to Bullfrog, and is the prevailing formation covering the surface over a large area of the district. The topographical characteristics of the district are in most respects similar to those of the more celebrated northern goldfields.

The most valuable mining property at present developed is the Shoshone Consolidated, situated on Montgomery Mountain. This mine has been developed to a depth of more than 600 feet, and opened up with many miles of tunnels and other lateral works. Seven distinct ore-bearing veins have been cut, traced and developed.

Among other important mines on the field the Tramps Consolidated, National Bank, Gibraltar, Mayflower, Gold Bar, Keane Wonder, and Gold Center are prominent. Gold occurs in these mines, as at Tonopah and Goldfield, in regular, well-defined quartz veins, and in irregular bonanzas formed in the rhyolite by mineralized volcanic waters permeating and depositing their contents in the creviced zones in the rocks, which had been shattered by seismic disturbances and the intrusion of volcanic dikes.

Numerous smaller gold-mining centers are being opened up in the vicinity of Bullfrog. The most promising of these are Johnnie, Indian Camp, Monte Cristo, Gold Crater, Kawich, Gold Mountain, Cuprite, Montezuma, Lida, and Palmetto.

WESTERN FIELDS.

Throughout the state a great number of mining camps are being developed more or less successfully. On the main railroad west of Tonopah, gold, silver, and copper deposits of great extent are being opened up in the vicinity of Sodaville. Further north, prospectors are busy developing mining leases recently granted on the old Indian reservation on the western shore of Walker Lake. Some rich gold and copper finds are being made in that locality. At Yerington, further north, in Lyon County, a number of developing mines promise to become steady producers. The Douglas, Wilson, Mohawk, Ludwig and Bluestone are among the leading properties.

There is also a large amount of mining work being carried on in the vicinity of Virginia City. An extensive pumping plant has been erected on the Ward shaft by several of the Comstock companies which have amalgamated with the view of unwatering the deep workings in the old Comstock mines. There are immense low-grade ore bodies in all these mines that can now be profitably mined. A new gold rush recently set in to the Ramsey district which is situated about twenty miles northeast of Virginia City. Gold is being mined in several of the claims and there is every promise of a permanent camp.

NORTHERN FIELDS.

Much less attention has been given by prospectors to the northern districts in Nevada, but development work on a small scale is being carried on at Olinghouse, a gold camp north of Winnemucca. The old silver camps at Austin, Eureka, Bullion, Tenabo, Mineral Hill and Cortez are being reopened by large Utah smelting companies.

EASTERN FIELDS.

During the past three years a number of Colorado and Utah mining men have prospected a large area in White Pine and Lincoln Counties which adjoin the Utah State line. Many successful discoveries have been made. The most important, probably, is the discovery of immense copper deposits in several of the old silver mines at Ely.

From this review, the healthy condition of Nevada mining will be clearly apparent, and it will be easily understood how this State ranking very low as a gold producer a few years ago, has since 1905 forged ahead in gold mining so as to rival Alaska closely in its annual gold yield, and promises within the next two or three years to assume the position Colorado has held for some years as the greatest gold State in the Union. Nevada is not now dependent upon one great mining center and one mineral product as it was in the Comstock days. There is consequently, little likelihood of the present financial stringency seriously arresting its great mining developments.

EDITORIAL COMMENT

Have Courage and Common Sense!

IN our December issue we spoke of the present business disturbance as a "psychological panic"—a trouble due to a state of mind which mental courage and common sense would be certain to dispel. In a sense this is true of every period of depression. But the peculiar difference is that while usually a panic collapse in business results from the realization of conditions long existent and grown cumulative in their effect, the crisis now is caused by the *failure* to realize conditions of soundness that make depression unreasonable and its long continuance impossible.

As Mr. Del Mar points out in his article leading this issue, we have in our hands one-third of the entire gold stock of the world—more than France, England and Germany combined. Instead of the slender stock of gold with which we faced the panic of 1893, we hold secure in our Treasury a thousand millions. Instead of the outflow of sixty-six millions which characterized that period, we have had within two months an influx of nearly a hundred millions, and the golden stream is still pouring into our ports. More gold—new gold—is coming from the mines of the world at the rate of more than a million dollars a day, one-quarter of it from our own States and territories. There is no slightest shadow of menace to the money standard—that great cause of all panics—but in place of it we have currency all "as good as gold," and in ample volume. Our per capita circulation is, in fact, now larger than ever before in the history of the Nation. Our banks are strong beyond peradventure. And our stores of crops and staples—of natural products which the peoples of the other Continents must have—make us creditor to all the world, and insure prosperity for every transportation, manufacturing, and commercial interest, in the profitable work of gathering them and getting them to market.

Such faults of overtrading and overspeculating as we have committed have been liquidated and expiated. Unsound schemes and unsound financiering have been wiped out. The projects and the methods of plungers and promoters of watered stocks have been widely and deeply discredited. They cannot, for years to come, command either credit or capital for such speculation in stocks as precipitated the present unsettled conditions of business. We are on sure foundations, with fundamental wealth in such profusion as the world has never heretofore known, with technical skill and industrial capacity and system which have been the envy of the nations—and nothing lacking but co-operative determination, to carry forward our manufactures and our commerce with unparalleled prosperity. There is coming need in modern civilization for everything we can make and carry and sell. Time lost now—productive capacity unemployed—are lost forever. It is a time for steady speed ahead, for persistently seeking and if need be *commanding* business—not for faint-hearted pause and frightened retrenchment. The years ahead are golden—even more golden than those recently passed. "Speak to the people that they go forward!"

The Ashokan Dam "Scandal."

THE so-called "investigation" of the contract for building the Ashokan Dam has now developed a situation which demands the immediate and watchful scrutiny of the entire press, and all the commercial, civic, and taxpayers' associations of Greater New York.

The building of this great dam is the first and most important step towards the construction of the greatest water-supply system that has ever yet been undertaken anywhere. The aggregate cost of the entire work will be somewhere between $160,000,000 and $250,-000,000, and of course many years of time will be required for its completion. Because of the magnitude of the undertaking, and especially because of the absolute necessity for keeping partisan politics and "political grafters" out of the work, the Legislature of New York publicly pledged Mayor McClellan to the appointment of three Commissioners of the Board of Water Supply to be selected and nominated by the three leading commercial organizations of Greater New York. In accordance with this obligatory public pledge, the Mayor was fortunate enough, for the city, to secure the services of Mr. J. Edward Simmons, nominated by The Chamber of Commerce, of which he is President, Mr. Charles N. Chadwick, nominated by The Manufacturers' Association of New York, and Mr. Charles A. Shaw, nominated by The Board of Fire Underwriters. These are all men of the highest probity of character, of spotless integrity, distinguished by long and successful business careers, and throughout the recent and scandalous newspaper campaign of insinuation, ridicule, and misrepresentation, not even the boldest of the political intriguers back of this so-called "investigation" has dared to give utterance to a suspicion of the straightforward and open honesty of the Commissioners in the matter of letting this contract.

In explanation of their action, on the first day of the so-called "investigation," the Commissioners presented a dignified and exact statement of all the essential facts and circumstances leading up to the letting of the contract. A careful reading of this statement, followed by close study of the official documents and all the evidence that has since been brought out by the Commissioners of Accounts, serves to satisfy us that the Commissioners of the Board of Water Supply not only acted conscientiously, but also with orderly procedure and deliberation, and manifestly with the greatest care to see that the best interests of the city might be served. And in coming to a conclusion, they not only had, in formal written opinions, the high warrant of the very best engineering advice that they could command, but they also had an exhaustive written opinion from the Corporation Counsel, proving that, legally and technically, they would be obeying the manifest purpose of the State Legislature in letting the work to the contractor who, in their best judgment, could best serve the city's interests.

These important official documents are all in available typewritten form for those who may desire to read them; and the signed statement of the three Commissioners, above referred to, has been wisely issued *in printed form*, so that it may be obtained at once, by either personal or written application to the Secretary of the Board, No. 299 Broadway, New York. We strongly commend a reading of this printed document by every one who may be in any way concerned, because it failed of publication in the newspapers, and in this issue we have not available space for repeating the essential facts embodied in the statement.

The point of immediate importance is, that the so-called "investigation" has not developed one essential fact which is contrary to, or in conflict with, the perfectly frank and open statements which the Commissioners of the Board of Water Supply have made, filed, and held open for public inspection, from the very beginning of the transaction. On the other hand, the conduct of the inquiry, from beginning to end, indicates a settled purpose to discredit the Commissioners before the public, and if possible through news-paper notoriety, misrepresentation and ridicule, to drive them out of office, and thus open the way for the Mayor to secure the appointment of commission-ers who will manage the great work as he personally may desire—which is the direct opposite of the known pur-pose of the law creating the Board of Water Supply!

In view of these developments, we are highly gratified to observe that at least three of the most influential of the New York newspapers have already sounded a note of warning. Thus, in an admirable editorial review of the facts of the case, published in its issue of December 7, *The Evening Post* con-cludes as follows:

"So far as the inquiry itself is concerned, it is regrettable that it has been accom-panied by an unusual crop of rumors that politics and a partisan endeavor to control the whole $160,000,000 expenditure are be-hind it all; and that the decision is to be filed at once, if it has not already been writ-ten. So far-reaching a question can only be settled after full deliberation. This we must believe to be the intention of the in-vestigators. Since a partisan inquiry would be little less than a crime, it is regrettable, too, that the Commissioners of Accounts have declined to allow the Water Commis-sioners to cross-examine the former's ex-perts or to submit their statements for ex-amination."

In like manner, *The New York Tribune* in its issue of December 10 says:

"The members of the Board of Water Supply were specifically authorized by law to select not the lowest bid, but that which they considered the best bid; and they did it. They were carefully selected for their places with a view to their exercising their discretion and judgment for the city's wel-fare; and they did it. They were expressly directed to 'select the bid or proposal the acceptance of which will, *in their judgment,* best secure the efficient performance of the work'; and they did it, with the re-inforcement of their own judgment by the deliberate and expert judgment of the best technical authorities at their command.

That the propriety of their action should be called into question, save on the basis of something more than has appeared in the course of the investigation, must be re-garded as strange, disappointing and dis-couraging."

And in its issue of December 14, *The Outlook* concludes a careful review of the facts in this wise:

"It is right that the expenditure of mil-lions of the city's money should be safe-guarded; but that expenditure already has the protection of a Board of Water Sup-ply, chosen upon the advice of civic asso-ciations, and the protection of expert engi-neers. Only the evidence of fraud or gross neglect should be allowed to distract the energies of these public servants from their work. Such evidence has not yet been forthcoming. What most needs to be in-vestigated is the real cause for this investi-gation."

The Engineering Index.

THE attention of the users of THE EN-GINEERING INDEX is called to several changes, beginning with this number, which we hope will add greatly to the value of the Index by making the in-formation concerning the literature of any subject more complete and more easily available. In Mechanical Engin-eering, the items referring to the de-sign of machines and machine parts, as distinguished from operation or man-ufacture, have been placed under the new sub-head "Machine Elements and Design"; while items referring to con-veying and handling machinery are now gathered in a new sub-section headed "Transporting and Conveying." In the Mining and Metallurgy section, new sub-heads are provided for, "Lead and Zinc" and "Minor Minerals," and for "Ore-Dressing and Concentration." Minor improvements are the inauguration of a more elaborate system of cross-refer-ences than has been possible hitherto, and a typographical scheme by which it is hoped the cross-references will be made fully self-explanatory.

HEAVY RAIL SECTIONS IN AMERICA.

AN IMPARTIAL VIEW OF THE CONTROVERSY BETWEEN THE STEEL MAKERS AND THE
RAILROADS REGARDING STEEL RAIL SPECIFICATIONS.

Engineering.

SOME months ago a review in these columns discussed the steel rail situation in the United States and outlined possible improvements in the processes of manufacture. The subject is growing in importance daily and the following impartial comments on the controversy between the buyers and makers of rails, taken from a recent editorial discussion in *Engineering,* will be read with great interest.

A certain number of rail failures is always to be expected, but the figures published by the railway systems and derived from other sources show that the failures of rails in service are out of all proportion to the duty demanded of them. Taking the statistics published by the Railway Commissioners of the State of New York, it is found that whereas the failures in that State during the first three months of 1905 numbered 1,331, during the corresponding period in 1907 they numbered 3,014; in other words, there was one failure for every seven miles of track in 1905, and one for every three miles in 1907. This increase is coincident with the rapid substitution of 100-pound rails for the lighter sections and the statistics show that the heavy section rails are responsible for more than their due share of failures. Another interesting fact is disclosed when the failures are classified according to age, namely, that the figures for the material rolled in the years 1904-7 shows a much higher number of failures than that rolled during earlier years. There seems to be no doubt,

taking the most lenient view of the case, that, whether the question of the section, or the quality of recent product of the rail mills compared with that of former years is considered, the number of failures has lately been inordinately large. The question of who is to blame is the present cause of controversy, the rail makers blaming the section and the railroads blaming the processes of manufacture.

"Commencing with the question of chemical composition, the point which will at once strike anyone familiar with English practice is the high percentage of phosphorus allowed in American rails. It was round this point that the discussion centred on the occasion of the reading of Mr. A. L. Colby's paper on 'American and Foreign Rail Specifications,' at the London joint meeting, in 1906, of the Iron and Steel Institute and the American Institute of Mining Engineers. Mr. Colby advanced the plea that, in order to allow America to compete for European rail orders, the specifications on this side of the Atlantic should be so relaxed that the higher phosphorus rails rolled in America would be accepted over here. Although doubtless such a move would be a most practical way of proving the existence of good feeling between the Old and the New World, none of the authorities over here seemed disposed to accept the invitation. The British standard rail calls for the following composition:—Carbon, 0.35 to 0.5 per cent.; manganese, 0.7 to 1 per cent.; silicon, not to exceed 0.1 per cent.; phosphorus, not to exceed 0.075 per cent.; and sulphur, not to exceed 0.08 per cent. The limits in American practice, as given by Mr. Colby, vary with the weight of section. Phosphorus is not to exceed 0.1 per cent.; silica not to exceed 0.2 per cent.; while carbon varies between the limits of 0.35 and 0.60 per cent.; and manganese, 0.70 to 1.10 per cent., both according to section, the larger quantities being for the heavier rails. These limits are quoted from manufacturers' standard specifications, and, to quote Mr. Colby verbatim: 'To obtain steel rails with a guaranteed phosphorus content lower than 0.10 per cent. is now impracticable from the majority of American rail-makers.' While even then comparatively troublesome, the situation has of late become still more acute, until the railroads have made a definite stand against high-phosphorus rails. It is stated that, owing to the enormous requirements of the last few years, American mills have scarcely been able to supply demands, and, since 1901, have practically refused to be bound by specifications, or to give guarantees of service, as were once given; the most they will concede now being the replacement of defective rails. The rail-manufacturers for a time had the whip-hand, and for the greater part the railways have had to put up with what has been given them. Lately, however, owing to the railroads withholding orders until assured that they would get what they wanted, the manufacturers have been attempting to justify their attitude. They claim that good Bessemer low-phosphorus ores in the American continent are becoming worked out. The natural retort is that the process must be suited to the ores obtainable, instead of the railroads being forced to accept rails of a quality endangering the lives of passengers. This is a view which the steel firms have up to recently refused to consider. Apparently the safety of the traveling public does not concern them. If, the railroads claim, the Bessemer process will not give rails with lower phosphorus than 0.1 per cent. with the ores obtainable, the open-hearth basic process must be employed. At present, however, the United States is largely a Bessemer country. In 1905 nearly 12½ million tons of Bessemer and low-phosphorus pig were produced in the United States, compared with just over

4 million tons of basic pig. The total production of rails in 1905 amounted to 3,375,929 tons, of which all but 183,264 tons were by the Bessemer process. The manufacturers are not at present in a position to supply open-hearth rails if any large demand be made for them, and a certain amount of time must elapse before they can possibly do so. At the present time there are only a few mills which can supply open-hearth rails, but new open-hearth basic plants are being erected, and others are contemplating conversion from Bessemer to the open-hearth process. It is satisfactory to find that the steel-makers are at last moving in this direction, but it must be admitted that they are not accepting the situation very gracefully. This, after all, is only human, for they must expend a good deal of capital in the process of conversion, which in the end will only benefit the railways; whereas if they had been able to continue to foist successfully on the railroads high-phosphorus rails, the work would practically have resulted altogether in returns on present capital alone.

"The next point of consideration is that of piping, the failures in a large number of cases being due to small pipes, which develop under heavy traffic. In this connection the manufacturers claim that they cut off as much crop as is necessary to get down to sound material. It appears that of recent years the amount cropped off has often only been 10 per cent., whereas it is largely held that from 20 to 30 per cent. must be discarded to get down to good steel. The buyers assert that the manufacturers now discard less, with the purely commercial object of increasing the output of their mills. When it is considered that piping and segregation occur to such a marked degree in steel ingots, and that phosphorus is one of the constituents especially liable to segregation, rails from the upper portion of an ingot, of course, run the risk of being of porous material and especially high in phosphorus. The manufacturers up to the present have refused to be bound to any definite percentage of discard.

"In addition to the ordinary defects the ingots of the steel companies have of late years been growing more porous and less homogeneous in the ceaseless endeavours of the manufacturers to run their mills to their utmost capacity. Pouring immediately after filling from the ladle has the effect of preventing any slag carried into the ladle from the converter or furnace from rising to the surface, and tends to make the metal in the ingot mould still less homogeneous. Then comes the question of rolling, the number of passes being very much reduced in the modern processes, not only because of 'improved' methods, but also because of the fact that the ingot remains the same size while rails now are of 100-pound section against 65-pound of a few years ago. The rail has therefore less work put into it, and the wearing part of the rail is finished at a higher temperature than formerly, resulting in a rail of poorer qualities. The manufacturers claim, however, that all improvement in process and correspondingly increased output has resulted from cutting down the waste time between stages, and is not due to, for instance, quick rolling at high temperatures. The increased speed, they claim, is not due to their desire for high dividends, but to the fact that a high finishing temperature is necessary, owing to the particular form of section of American rails. The heavy head, the thin web, and the wide and thin base necessitate high finishing temperatures. Otherwise it would be impossible to roll the base or bottom flange. This, perhaps, is the soundest of all the excuses advanced by the steel companies in their attempts at self-justification, the others being merely shifts to enable them to pass off lightly the present serious state of affairs. So strongly placed are they that even now, when the gravity of the situation has been brought prominently before the world, the buyers do not seem able to enforce their full demands, and the manufacturers still appear to be masters of the situation.

"The ideal rail is one on which all the above-mentioned points have a very material bearing. Phosphorus must be low, with, if anything, a tendency to a lower percentage as the percentage of carbon increases. While a rail containing up to 0.1 per cent. phosphorus may be serviceable with low carbon, if the carbon is increased the rail becomes unduly brittle, unless the phosphorus is decreased. In order to obtain good wearing rails, there is a tendency to use a higher percentage of carbon than formerly, and it therefore follows

that the phosphorus must be present in smaller amounts. The rolling must be so managed that sufficient work is put into that part of the rail which will be subjected to the greatest wear—*i. e.*, the head—which must not only be sound in material, but well worked and of close structure. The holding of rails before the last pass or two, so as to finish at a proper temperature merely produces a rail with a hard skin and of poor internal structure; and when the skin is worn away the soft material inside wears rapidly, even if defects have not caused failure before this stage is reached. The spraying of the heads of rails during rolling is also therefore at best an unsatisfactory expedient. Bearing on this point is the question of section, which should be of such form that the process of rolling and working of the material may be carried on in the manner best calculated to provide a good wearing rail. It should not be necessary to have to resort to annealing, which destroys the beneficial effect of rolling."

The editorial gives a review of the work and influence of the various technical societies in attempting to arrive at a solution of the problem and then concludes as follows:

"In the meantime the manufacturers take credit to themselves for the fact that although the trade is organised so as virtually to result in monopoly, the price of rails has been maintained at a constant figure for the last four or five years. They take care not to mention the economies they have resorted to during this time, with such disastrous results to the buyers. The situation at the present moment can not be considered as satisfactory, the most rational specifications—*i. e.*, those favoured by the American Society of Civil Engineers and the American Railway Engineering and Maintenance-of-Way Association—being rejected by the manufacturers. That buyers have just cause of complaint we have clearly shown above, and that American rails have no good reputation abroad at the present time is emphasised by their shrinkage in exports to Europe of recent years. In the first seven months of 1906, for instance, rails to the value of 22,422*l.* were exported to Europe, as compared with 245*l.* worth in the corresponding period of this year.

"Improvements may be sought, as we have shown, in several directions. The Sandberg process of 'added silicon' tends, as was shown at the Engineering Conference held in London this summer, to the production of sounder metal. Open-hearth steel may ultimately displace Bessemer for rail material, and there is, without doubt, a strong desire to effect the substitution at present only partially possible. Vanadium steel rails are to be tried on some systems, such as the Atchison, Topeka, and Santa Fé. Until, however, the manufacturers make such concessions as will ensure a product of a quality considered by practical engineers as the minimum allowable, the railroads will be put to considerable expense and trouble, while the public, falling between the two stools, travel at uncomfortably high risk."

THE ELECTRIC POWER SUPPLY OF PARIS.

A PROJECT FOR THE DEVELOPMENT OF THE WATER POWERS OF THE RHONE AND THE TRANSMISSION OF THE ELECTRIC POWER TO PARIS.

G. de Lamarcodie—Revue Générale des Sciences.

IN THE ENGINEERING MAGAZINE for November, 1907, these columns contained a description of the immense water powers awaiting development in France, in which the district of the Rhone was mentioned as a possible source of some five million horse-power of electrical energy. A recent number of the *Revue Générale des Sciences* contains a description of a project for the utilization on a large scale of the power of the Rhone itself, and the transmission of electrical energy to Paris, a distance of 450 kilometres. An outline of the scheme as given by M. G. de Lamarcodie after a report by M. Harlé, one of the engineers interested in the project, is contained in the following abstract.

Commenting on the other notable long-distance transmission systems of the world, M. de Lamarcodie points out that, while electrical energy is not at the present time transmitted over so great a distance as 450 kilometres, projects are on foot, notably at Stockholm and the Victoria Falls, for transmissions over much greater distances. There is nothing in the experience of those responsible for the present long-distance transmissions, up to 350 kilometres, to indicate that the project is impracticable.

The demand for power in Paris is next considered. It would be possible to dispose of power to the underground railways and a cheap supply of power would induce the steam railways to use electric traction over their suburban lines. These demands, together with the demands for power for lighting, tramways and manufacturing, would insure a consumption of about 700 million kilowatt-hours per annum. It is

not to be expected that this figure would be reached at once, but that the demand will increase as consumers become assured of regular and efficient service.

The Rhone was chosen as the source of power only after careful consideration of the other possible localities. In these, notable Savoy and Upper Savoy, the streams are too swift for the establishment of large generating stations and besides a great many of the waterfalls are already utilized to supply local demands for power. The Rhone has two distinct advantages: first, its discharge, which never falls below 130 cubic metres per second; and second, the rapidity of its fall which, in the 22½ kilometres between the Swiss frontier and Génissiat bridge, amounts to 65 metres. At least 80,000 horse-power could be generated at low water and about double this during 300 days out of the year, while in time of flood the discharge reaches 1,250 cubic metres per second. Further, the Rhone could be utilized for the supply of power to Paris without prejudice to the requirements of local industries, whose demands are already satisfied from other sources.

It is proposed to erect a very high dam at the lower end of the narrowest part of the Rhone gorge. The most convenient spot for the dam seems to be the neighborhood of Génissiat. A difference of level between this point and the Swiss frontier of about seventy metres can be counted on. The Lac du Bourget will be utilized as a compensating reservoir to do away with variations in the regimen of the Rhone due to variations in the operation of the plant. It was at first intended to transmit the energy by direct current at 120,000 volts on the Thury system, but subsequent changes in the law relating to the transmission of electrical energy have removed the advantages of the direct-current transmission to a great extent and it is now proposed to use three-phase alternating current at the same voltage, transmitting to Paris the 70,000 kilowatts proposed over two lines of slightly different lengths. The current will be generated at 10,000 to 20,000 volts, stepped up to 120,000 for transmission, and stepped down to 10,000 to 15,000 for distribution in Paris. The cables will be carried on steel transmission towers spaced 100 metres apart. On account of the regulations concerning the transmission of alternating current, the transmission line will have to be situated entirely on private property. It is expected that the loss in the line will not exceed 12½ per cent. and that a regular annual transmission of 500 million kilowatt-hours can easily be maintained, taking into account only the mean supply of water in the Rhone.

M. de Lamarcodie considers that the carrying out of the project is not only practicable but necessary from an economic point of view. A supply of cheap power would be of immense advantage to the industries of Paris, and this the project offers. The great consideration urged in opposition to the scheme is the possibility of an interruption to the service due to accident, but the promoters have taken care to guard against this by the provision of a double line. All things considered, M. de Lamarcodie regards the project well worthy of development and one which should receive hearty financial support.

INDUSTRIAL BETTERMENT IN AMERICA.

A SUMMARY OF THE LINES ALONG WHICH THE INDUSTRIAL BETTERMENT MOVEMENT HAS DEVELOPED IN THE UNITED STATES.

H. F. J. Porter—American Machinist.

THE industrial betterment movement in the United States is of comparatively recent origin. It is only within the last ten years that the subject has been given any attention and only within the last five that its merits have been generally acknowledged. During the latter period, however, its growth has been extraordinarily rapid, as employers of labor have come to recognize that the loyalty of employees, to be gained not by policies of philanthropic and patronizing paternalism, but by a common-sense consideration for their health, safety and recreation, is a most important factor in commercial success. The gradual growth of the movement has been shown from time to time, in articles in the pages of THE ENGINEERING MAGAZINE, but the following summary of the lines along which it has developed, taken from a recent paper by Mr. H. F. J. Porter in *American Machinist*, will prove of interest.

"As men do not engage in business from altruistic motives, modern methods of factory management cannot be based on sentiment, but on sound economic principles. The object of any industrial enterprise is to make money for its promoters. In order to do this, expensive machinery is bought and placed in the factory where it will best do its work; it is protected from dust, kept well oiled and in good repair. Just so must the human beings who operate the machines be kept in good condition, physically, morally and mentally, if the employer is to receive the greatest return for the wages he pays. Good light, air, cheerful surroundings, cleanliness and provision for the reasonable degree of comfort which prevents over-fatigue and induces good work—all these things have a commercial value, for, the more favorable the conditions surrounding work, the better will be the class of workmen attracted, the greater their efficiency and the more permanent their service.

"A quarter of a century ago certain Dutch and German manufacturers introduced into their establishments many innovations in factory management tending toward the safety, health, comfort and happiness of their employees, changes which redounded to the benefit of both workers and work. Such instances, rare in Europe, were unknown in the United States. Ten years ago comparatively few men here gave any thought to such matters. But Amerlcans have become as great travelers as their English cousins. Being quick to recognize good things and to turn them to account, American business men have, within the past five years, adopted many foreign industrial betterment ideas, changing and enlarging upon them, in order to adapt them to their economic requirements.

"Massachusetts manufacturers were the first to interest themselves in this direction. They were quickly followed by others in the Middle Western and Western States, and later on, by Southern industrialists. To-day some form of industrial betterment finds a place in our more successful factories, machine shops, foundries, mills, etc., until America is now foremost in such work. Improved factories are greater in number in the United States than elsewhere, and also have greater diversity in their social, educational and recreational features. The wide spread of these ideas is due not to the fact that successful concerns have adopted them, but because it has been noted that concerns which adopt them become successful."

Perhaps the initial indication of interest in industrial betterment on the part of an employer is to be found in the generally manifested desire for order and cleanliness in and about the works. New buildings are usually built with large window space to admit plenty of light and air. Many firms have gone very far in beautifying the surroundings of their work-people and have found the result extremely satisfactory. The most important phase of the movement, however, is the increased interest in safeguarding the operatives from accident in the more or less hazardous trades.

"A great menace to our industrial system is the destruction of life which results from the use of machines operated at a high rate of speed. The number of fatalities would be beyond belief but for statistics, which tell how many are killed and maimed in various industries. Census Bulletin No. 83 shows that the number of deaths by accidents and violence in the United States in 1900 was 57,513. These figures are increasing annually. Thus our peaceful vocations cost more lives every two days than all that we lost in battle during our war with Spain.

"In 1899, the New York Bureau of Labor attempted to gain as complete a record as possible of all accidents for three months in industries employing about one-half of the factory workers in the State. During this period confessedly incomplete returns showed 1,822 accidents. On this basis all the factories in the State would in twelve months show 14,576 accidents. But these figures are undoubtedly far below the facts. Many trades, not particularly dangerous, reported 44 accidents for every thousand employees, while some extra-hazardous trades reported only 16 to a thousand employees. As there is a disposition to conceal accidents, it is safe to assume that no employer reported more than actually occurred, so that the Commissioner of Labor reasonably inferred 44 to a thousand to be more nearly correct than 16 to a thousand.

"On this basis there would be upward of 332,000 factory employees killed or injured annually in this one great division of the industrial army in the United States. Undoubtedly the safeguarding and protection of life and limb is a most important phase of industrial betterment. Factory legislation during the past five or six years has done much to reduce the danger of accidents, but there are still far too many of them. Employers are not eager to place safety devices upon already expensive machinery; good operatives frown upon anything which tends to restrict the output. It sometimes happens that a serious accident is necessary, as an object lesson, before workmen will appreciate or use a device installed for their own protection. There are places, however, where labor accidents for years have been reduced to a minimum by protective improvements, long before their use was required by law.

"As great financial loss may result from the ill health of employees, great care is taken in accepting only those who are physically sound. Those already employed, who are affected by tuberculosis, are especially provided for, while new applicants are not accepted. Alcoholism, the bane of efforts to secure regularity of attendance, is fought by elimination and rejection of its victims. The hygiene of the buildings receives attention; wash rooms with individual basins are often provided; shower baths in foundries are occasional; sanitary closets are common. The dust from grindstones and similar apparatus is carried off to prevent its being inhaled by the operatives. Ventilated clothes lockers are considered necessary. One factory gained the regard of its employees and drew to it the best element from neighboring works by putting a steam pipe under its clothes lockers. Workers arriving on a wet morning changed their wet clothes and shoes for others dry and warm, and at night found their street suits thoroughly dried for their return home.

"Medical service, ranging from simple emergency cases and first-aid classes up to hospitals with the finest equipment, is provided in the modern factory and shop. Almost all improved establishments have some provision for the care of accidents and cases of sudden illness. Usually it takes the form of a room fitted up as a miniature hospital, with a trained nurse in charge. More elaborate systems of medical service are installed in large plants where the work is hazardous in character. The Colorado Fuel and Iron Company has a splendid hospital at Pueblo, in addition to simpler forms of medical relief in the various mining camps. The Illinois Steel Company, at South Chicago, maintains a hospital with constant medical and surgical attendance."

Other features of the movement, not so widespread, are the establishment of suggestion systems and of long-service pension systems, the building of workmen's cottages where transportation facilities are bad, and the establishment of libraries, educational systems and club-houses and gymnasiums by some of the larger corporations.

"Business men, hearing of this much talked about and written about industrial betterment, question if it is really worth the time, thought and money spent upon it. They do not realize that the absence of these features also costs money, through their getting an inferior quality and less quantity of output. Results fully justify the expenditure, which may be either great or small, according to the number of employees and the kind of betterment work which is done. Employers who have had experience with it are enthusiastic, because better work and greater output show the definite, tangible effect it has upon the men. Working people approve because of the advantages it places at their disposal. It is, however, a failure, something worse than useless, if forced upon employees. It must exist only by reason of a need, or an expressed desire, for it upon the part of the men themselves and must be free from the least taint of exploitation or paternalism, which American workmen universally and rightly resent. Where industrial betterment ideas are properly initiated and carried out, the beneficial effect is incalculable, for they reach not only the individual but the outside world. They raise the standard of the man; the quality of his work improves and the consumer gets a better article for his money.

"There is a force in nature which continually makes for progress. Give this force the opportunity to exert itself through the initiative of the working organization. Help the employees to help themselves. Their appreciation of the benefits accruing to them will bring a hundred-fold return to the management.'

THE CALIBER OF SEACOAST GUNS.

A DISCUSSION OF THE RELATIVE MERITS OF INCREASED CALIBERS WITH LOW VELOCITIES
AND SMALLER CALIBERS WITH HIGH VELOCITIES.

Major H. L. Hawthorne—Journal of the United States Artillery.

A QUESTION of considerable importance to the coast defense of the United States is argued by Major H. L. Hawthorne in the current number of the *Journal of the United States Artillery.* It is proposed by the Ordnance Department to replace the present 12-inch high-velocity guns with guns of 14-inch caliber and lower velocity, on account of the effect of ero-

sion on the life of the former type. As the following extracts will show, Major Hawthorne reaches conclusions directly opposite to those accepted by the Ordnance Department.

"In the gradual development of seacoast high-power guns in the United States, increase in power was at first sought in increase of caliber until the 12-inch gun was produced. This type in turn underwent a gradual increase of power through an improvement in the quality of steel, the elongation of the bore and a more rapid device for manipulating the breech mechanism. In the attainment of this increase of power, a higher and higher muzzle velocity was sought, which appeared for a time to be limited by the bursting effect of powder, the maximum being controlled by the rate of burning and the strength of construction of the gun. So long as such factors only entered into the problem of increased velocities, there was no reason why progress in that direction should not have continued; but a serious menace to further advance appeared in the form of erosion produced by the high temperature and volume of the gases from smokeless powders in the bore of the gun, seriously affecting its accuracy, but, what was still more important, so reducing the life of the gun that the movement in the direction of increased initial velocities has been compelled to meet this destructive influence by expedients, or seek for the desired increased power, not by greater velocities but by larger calibers and longer life of guns by lower velocities. Many elements enter into the consideration of this question, apart from the simple increase in power, which would avail little, if all or some of them are sacrificed. In order to present an introductory basis for a discussion of this question, there will be presented at once the claims and counter claims as to a solution of it, by an increase in velocity without change of caliber, or an increase in caliber with lower velocity."

Major Hawthorne lists seventeen of the claims made for the 12-inch gun with 2,250 and 2,500 feet per second muzzle velocity and the replies of the advocates of the 14-inch gun with 2,150 feet per second muzzle velocity. There are dealt with the varied questions of rapidity of fire, ease of handling ammunition, ease of ramming, accuracy and power, accepted practice, mechanical details, cost of gun and charge, life, etc., the arguments covering practically all the controversial points and being presented without bias. Major Hawthorne then proceeds to give his own views as to the advisability of making the change.

"The recent recommendation of the Ordnance Department that high velocities and heavy powder pressures be abandoned, and that we return to the old system of firing heavier projectiles with smaller velocities is a complete acknowledgment of defeat in the struggle between armor and gun. Although the heavier projectile with lower velocities may for a time bring the gun once more to the front, the advantage may be but temporary and an improvement in the resisting power of armor, or its increase in thickness may again force higher velocities on the gun, and this time may find the construction of the gun, with its lighter weight and short barrel, unequal to any advance in that direction. Recourse must again be had to the 12-inch type of gun construction with an increase in velocity, thus finding ourselves where we have now arrived, but with a great expenditure of time and money on guns no longer of value.

"The problem is, therefore, to meet the evil effects of gun erosion and continue on that line until every possible method of prevention has been tried."

The most severe erosion of the bore takes place near the breech in the neighborhood of the seat of the projectile. It decreases towards the centre of the bore and tends to increase towards the muzzle. One theory to account for this is that the metal of the bore is worn away by the rush of the white hot gases through the openings left when the shell is rammed home into place; an obvious remedy would be to devise some more perfect method of obturation of the base of the projectile. Another theory is that the intense heat of the explosive gases, circulating in currents of enormous velocity, raises the inner surface of the gun almost to a melting temperature and renders it extremely liable to erosion by the corrosive action of the gases. This theory would suggest the advisability of reducing the temperature of the gases by the use of some inert volatile solid in connection with the powder. The eroding action of the gases is increased by the use of oxidizable metal in the inner tube of the guns. It might be found possible to use high-speed steel for these tubes. Its cost is very high and it is difficult to work, but the advantages of longer life might be found to outweigh the difficulties.

Major Hawthorne refers to tests made at Sandy Hook, the Springfield Armory, and other places to determine the amount of erosion under various conditions and then gives a summary of the possible means by which it may be reduced or prevented.

"Summarized, the various corrective methods that have been experimented with, more, or less, and the deductions therefrom, and from examination of the erosive action in guns of high power, it is thought that beneficial results will follow from:—

1. Some special form of rotating band, acting to close the bore when the projectile is rammed home and keeping it seated during the passage of the shot down the bore.

2. The elimination of the forcing cone, keeping the lands to their full thickness to the seat of the projectile.

3. Uniform twist rifling.

4. A lower oxidizable metal for the tubes of guns, and one capable of better resistance against great pressures and intense heat.

5. Reduction of powder pressures and greater length of gun, up to 45 calibers.

6. A change in the composition of the propelling charge, by which no loss in velocities is incurred, but which will bring about a reduction of temperatures of the evolved gases.

7. Reduction of the density of loading, by increasing capacity of the powder chamber.

"With the elimination of erosion," he concludes, "the change to lower velocities and large calibers is unnecessary, and the high qualities of the 12-inch, model of 1900, can be retained, with the following resultant advantages over the larger guns:

1. Greater danger spaces.

2. Sufficient striking energy and size of explosive charge.

3. Extreme range.

4. Greater rapidity of fire.

5. Greater accuracy.

6. Simpler methods of breech closure and shot handling.

7. Less cost.

8. Superiority of accuracy over guns afloat."

THE DEVELOPMENT OF MARINE TRANSPORT.

A BROAD VIEW OF THE INFLUENCE OF THE INCREASING SIZE OF SHIPS ON THE PROVISION OF DEEP WATERWAYS AND OF HARBOR AND DOCK ACCOMMODATIONS.

Sir William Matthews—Institution of Civil Engineers—Engineering.

IN his recent presidential address before the Institution of Civil Engineers, Sir William Matthews touched but lightly on the mechanical and structural engineering features of marine transport, but gave a broad and interesting review of the development of harbor and dock accommodations and of deep waterways, made necessary by the increasing size and draught of steamships. The following abstract taken from the editorial columns of *Engineering* will give an idea of the scope and interest of the address.

"The general question of over-sea transport has, in view of recent developments, a special interest at the present day. For some occult reason our daily Press is much more alive to real or imaginary progress made in other countries than it is to similar events at home. We read much about the developments of foreign ports and harbours, whilst little is said as to still greater advances made within the United Kingdom. The mere fact of its transmission by cable, indeed, seems to bestow upon news of this character a fictitious importance out of all proportion to its intrinsic value, and hence data as to the work of the Harbour Commissioners at our principal ports, though appearing, it is true, in the local Press, rarely get transcribed into the metropolitan papers. The new President's address goes far to fill this void, and to show that our various port authorities are rising to their responsibilities, so far as financial considerations allow. These have been most seriously strained by the advent of the giant ship, which, on the very day of the delivery of the address, was signalised by the last of them, the Mauretania, exceeding all previous speed records for Atlantic liners. In preparation for this and her sister-boat and other great liners, Sir William pointed out that the entrance to the Mersey is being dredged so as to admit, at all stages of the tide, ships of 1,000 feet in length and requiring 40 feet depth of water. As matters stand, a depth of 28 feet is even now available at low water, as against 11 feet in 1890. At the same time the dock-floors have been lowered from 20 feet to 30 feet depth, and a new graving dock built 925 feet long by 94 feet wide at the entrance. At Southampton provision is being made for berths 35 feet deep at low water, which will ultimately be increased to 40 feet. At Newport the new lock is 1,100 feet long by 100 feet wide at entrance, with 44 feet 6 inches depth over the sill. London, unfortunately, in spite of its standing as the first port in the kingdom, can as yet show no corresponding efforts to meet the requirements of the giant ships. This is no doubt mainly due to financial considerations, since owners of moderate-sized vessels not unnaturally resent the imposition of dues, the sole effect of which will be to facilitate the competition—already sufficiently serious—of these enormous steamers. Nevertheless, the Conservancy are engaged in constructing a channel giving a depth of 30 feet right up to Gravesend. Sir William points out in his address that Brunel fully realised the advantages, in the matter of economy of working, of the large ship. The mistake made in the case of the Great Eastern was an undue confidence in the forthcoming of cargoes sufficient to load her, since all the advantages of the big boat are lost if she has to wait in harbour to collect freight. As matters stand, the biggest boats are still unsuitable for any trade but that between Europe and the United States, and would probably prove ruinous to their proprietors if engaged in any other traffic.

"Much has been heard of late years of the great development of Antwerp as a port for ocean-going steamers, and undoubtedly the port authorities there have been most enterprising. From Sir William Matthews' address, however, it would seem that there is still much to be done there before the port can fairly vie with Liverpool or Southampton, since in the approaches at low water the limit of draught is 16 feet, though in the port proper there are berths giving 26 feet to 33 feet depth alongside the river quays. Though the Suez Canal at present restricts the draught of ships trading to the East, the port authorities there are, in some instances, making provision for the advent of very large steamers. The great economy with which these can be run may, perhaps, do

something to restore the popularity of the Cape as a route to the East. At present the limit of draught for the Suez Canal is 27 feet, which, however, may be increased to 30 feet by 1909. Even then, however, the heavy dues may make the Cape route preferable for large cargo-boats.

"Referring to the Panama Canal, the President noted an important novelty in the methods adopted by the Americans to facilitate its construction. This has no relation to any of the machinery employed—which, though large, is not essentially new in character—but to the extraordinary and quite unprecedented attention paid to sanitary and hygienic considerations. Here, indeed, would seem to be the dawn of a new era in the conduct of engineering works in tropical climates. The Panama Railway, it used to be said, passed through a cemetery of the Chinese navvies who had been engaged in its construction, and the death-roll of engineers engaged in West Africa has also been exceptionally heavy. It is these extraordinary dangers to health which have hitherto rendered commercially impossible the due development of the enormous natural wealth of many tropical climes. Thanks to experimental biology, however, the origin of these malarious and other similar diseases is being gradually elucidated; and to the Americans belong the credit of being the first to make intelligent and systematic use of the facts so far acquired. From a merely pecuniary standpoint the sanitary campaign at Panama will probably save millions of money—in an indirect fashion, it is true; but from the contempt for expert knowledge which seems the characteristic peculiarity of some of our native politicians, it will, we fear, be a long time before the American example is followed in similar circumstances by our own authorities.

"Some exceedingly interesting data were advanced by Sir William Matthews concerning the extraordinary depths at which material may be moved by wave action and ocean currents. Stones weighing one pound have been, he records, drifted into lobster pots sunk in 20 to 30 fathoms of water. At Peterhead blocks 41 tons in weight have been shifted by the waves, though situated 37 feet below low water of spring tides. In the earlier portion of his address he referred with admiration to the enormous natural resource and great courage of the pioneers in the engineering profession, an appreciation which is, no doubt, in part due to a keen fellow-feeling. Like them, in the absence of data and adequate scientific theories, the harbor engineer must trust to his judgment, and not to mathematics for the proportions which he gives his structures. What will stand well at one place may prove totally inadequate at another. At Dover, for instance, the highest wave recorded in the course of fourteen years' observation has been 18 feet, whilst at Peterhead they reach 40 feet. Again, many mistakes have been made by harbour engineers in endeavouring to improve the approaches to sandy estuaries by training the currents. The situation and probable effect of any wall has had to be fixed by judgment rather than by formula, and, indeed, it is now recognised that what may be called the natural method of maintaining a good depth at low water must be supplemented by mechanical means. The improvement of the Mersey has been largely due to the introduction of the pump-dredger, and the same implement has proved its value at Durban and other ports suffering from sandy bars. The work of

the harbour engineer is therefore empirical to an extent unusual in other branches of the profession, and he can accordingly, with a peculiar sympathy, appreciate the labours of the pioneers of the industry.

"In concluding his address Sir William gave some valuable data as to recent advances in the matter of protecting life at sea. Although the submarine sound-signals were introduced but three or four years ago in a commercial form, they are now, he states, in use on all the principal Atlantic liners. By their means the bearings of an approaching vessel or a suitably equipped lightship can be accurately located in the densest fog. Submarine bells have therefore been applied at the approaches to New York, Liverpool and Cherbourg, and Trinity House is, Sir William says, now taking steps to fit them to a number of important lightships. Lighthouses, he remarks, have had their value increased by the substitution of incandescent oil-lamps for the ring burners previously used; whilst where the electric light is in use the adoption of the 'flame' arc has substantially augmented the penetration of the luminous beam during the prevalence of thick weather."

THE NEW BRITISH PATENT LAW.

A SUMMARY OF SOME OF ITS MORE IMPORTANT AND NOVEL PROVISIONS.

J. A. Law—Journal of the Society of Arts.

THE new British patent law which comes into effect on January 1, 1908, contains many alterations of the previous patent regulations of the utmost importance to manufacturers, inventors and others. It is impossible in a short review to notice more than the most striking innovations, and the following extracts from an article by Mr. J. A. Law in a recent number of the *Journal of the Society of Arts* give a summary only of the more important provisions relating to the granting, working and conveying of patents. The provisions of the Act are retroactive.

The most important provision is that respecting the compulsory working of patents in Great Britain. "If a patented article or process be manufactured, or carried on exclusively, or mainly outside the United Kingdom, then, unless the patentee

prove that the patented article or process is manufactured or carried on to an adequate extent in the United Kingdom, or give satisfactory reasons why it is not so manufactured or carried on, the Comptroller may make an order revoking the patent forthwith, or he may make an order revoking it after a specified interval if the patented article or process be not in the meantime adequately manufactured or carried on within the United Kingdom; but in the latter case, if the patentee give satisfactory reasons for the failure so to manufacture or carry on within the prescribed time, the Comptroller may extend the period by not more than one year. To obtain such an order, application must be made to the Comptroller at least four years from the date of the patent, and one year from the passing of the Act; moreover, any decision of the Comptroller is to be subject to an appeal to the High Court, and no order is to be made that will be at variance with any treaty, convention, arrangement, or engagement with any foreign country or British Possession.

"This provision, which, according to Mr. Lloyd-George, is the pith of the Act, introduces a very important alteration into the law relating to patents. It is true that under the previous law it was possible that a patent might be revoked on a somewhat similar ground, namely, that the reasonable requirements of the public with respect to the patented invention had not been satisfied; but this was only as an alternative to the grant of a compulsory license, and no patent has, at any rate in modern times, been revoked on such a ground. Whether the policy of revoking patents for a lack of working in this country is sound or not, is at any rate very debatable, and the writer has previously advocated the placing in the hands of the Comptroller, in lieu of the power to revoke, the power to grant to suitable applicants licenses to work patents that have not been worked in this country for a given time after their grant."

With a view to cheapening the procedure, petitions to the Board of Trade for the grant of compulsory licenses or the revocation of patents are to be referred by the Board of Trade to the High Court instead of to the Judicial Committee of the Privy Council. "It is also enacted that the reasonable requirements of the public shall not be deemed to have been satisfied if, by reason of the default of the patentee to manufacture to an adequate extent, and supply on reasonable terms the patented article, or any parts thereof necessary for its efficient working, or to carry on the patented process to an adequate extent, or to grant licenses on reasonable terms, any existing trade or industry, or the establishment of any new trade or industry, in the United Kingdom, be unfairly prejudiced, or the demand for the patented article, or the article produced by the patented process be not reasonably met; or if any trade or industry in the United Kingdom be unfairly prejudiced by the conditions attached by the patentee, before or after the passing of the Act, to the purchase, use, or hire of the patented article, or to the using or working of the patented process.

"In any contract made after the passing of the Act in relation to the sale or lease of, or license to use or work, any article or process protected by a patent, any condition will be null and void that will either (a) prohibit or restrict the purchaser, lessee, or licensee from using any article or class of articles, whether patented or not, or any patented process, supplied or owned by any person other than the seller, lessor, or licensor, or his nominee; or (b) require the purchaser, lessee, or licensee to acquire from the seller, lessor, or licensor, or his nominees any article or class of articles not protected by the patent. The provision, however, is not to apply if the seller, lessor, or licensor, prove that at the time when the contract was entered into the purchaser, lessee, or licensee had the option of purchasing the article or obtaining a lease or license on reasonable terms, and without such objectionable conditions; and if, also, the contract entitled the purchaser, lessee, or licensee to relieve himself of his liability to observe the objectionable conditions on giving the other party three months' notice in writing, and on payment of such compensation as may be fixed by an arbitrator appointed by the Board of Trade.

"Any contract relating to the lease of or license to use or work any patented article or patented process, whether made before or after the passing of the Act, may, when the article or process is no longer protected by a patent, be determined by either party on giving three months' notice in writing to the other party; but in the case of any contract made before the passing of the Act, such determination is subject to the payment of compensation.

"It is no secret that the section in question was chiefly aimed at the methods of the American Trusts, one of which, according to Mr. Lloyd-George, has practically compelled the boot industry in this country to employ their machines exclusively, and has obtained contracts with the manufacturers such that if an improvement on the invention to which the contracts relate, were taken up at the end of nineteen years, the trade would be bound to take the improvement, and renew their leases for another twenty years, and so forth, apparently *ad infinitum*. It was consequently necessary in the opinion of the Government to protect this very powerful industry, and to say that this country really could not stand the use of a privilege conferred by the Crown for the purpose of hampering a whole trade.

"In addition to the present ground of opposition, the grant of a patent is to be opposable on the ground that the invention has been claimed in a complete specification that will be dated prior, but not more than fifty years prior, to the patent the grant of which is opposed, or on the ground that the nature of invention, or the manner in which it is to be performed, is not sufficiently or fairly ascertained in the complete specification. The former of these two new grounds is to enable the grant of a patent to be opposed on the basis of an application made in this country, and founded, under the International Convention, on a prior application made abroad.

"Any person who would have been entitled to oppose the grant of a patent, or the successor in interest of such a person, may, within two years from the date of the patent, apply to the Comptroller for its revocation on any ground on which the grant of the patent might have been opposed; but the leave of the Court is requisite when an action for infringement or proceedings for revocation are pending in the Court. The decision of the Comptroller is to be subject to appeal to the Court.

"Any ground on which a patent may be revoked by the Comptroller, or, as an alternative to the grant of a compulsory license on the ground that the reasonable requirements of the public with respect to the patented invention have not been satisfied, is to be available as a defence to an action for infringement, or as a ground for revocation by petition to the Court.

"In an action for infringement a defendant entitled to present a petition to the Court for the revocation of the patent may alternatively counter-claim for revocation.

"Hitherto, to obtain the prolongation of a patent, it has been necessary to present a petition to the King, and the petition has been referred to the Judicial Committee of the Privy Council, with the result that the obtaining of an extension of the term of a patent has been a very costly matter. After December 31st, 1907, a patentee may, at least six months before the time at which the patent would ordinarily expire, present a petition for its extension to the Supreme Court. It is, therefore, to be hoped that the cost of securing such extensions will be materially reduced.

"Prior publication of an invention or any part thereof, if the matter published were derived or obtained from the patentee and published without his knowledge and consent, is no longer to be ground for the voidance of a patent, provided that the patentee, after learning of the publication, applied for and obtained protection for his invention with all reasonable diligence.

"No damages can be recovered for infringement of a patent granted after December 31st, 1907, from any person proving that at the date of the infringement he was unaware, and had no reasonable means of making himself aware, of the existence of the patent. The application to a patented article of a word indicating that a patent has been obtained is not to be deemed to constitute notice of the existence of the patent, unless the year and number of the patent be added. . . . . .

"A patent granted to an inventor in lieu of a patent revoked on the ground of fraud, is to bear, instead of the date of revocation, the date of the revoked patent, but no action is to be brought for any infringement of the patent so granted committed before the date of the actual grant of the substitutional patent. The alteration in the dating of the substitutional patent is important from the point of view of priority.

"Under the present law the grant of a patent to joint applicants gives a joint interest which passes to the survivor or survivors, unless there has been a severance of that interest, and each patentee can retain any profits he may make by working the invention himself, though it is questionable whether if he grant licenses, he will not be accountable to his co-patentees for the royalties obtained. Under the new law, unless otherwise specified in the patent, joint grantees are to be treated for the purpose of the devolution of the legal interest as joint tenants, but, subject to any contract to the contrary, each, whilst being entitled to use the invention for his own profit without accounting to the others, is not to be entitled to grant a license without their consent, and on his death his beneficial interest is to devolve on his personal representative as part of his personal estate."

THE UTILIZATION OF LOW-GRADE FUELS.

A CONSIDERATION OF THE ECONOMIC IMPORTANCE OF CONSERVING FUEL SUPPLIES AND OF THE APPLICATIONS OF THE GAS PRODUCER.

F. E. Junge—American Society of Mechanical Engineers.

THE paper of most general interest presented at the recent annual meeting of the American Society of Mechanical Engineers was that by Mr. F. E. Junge on "The Rational Utilization of Low-Grade Fuels." The paper gives a broad view of the economic and political importance of husbanding national resources and shows that the desired result can best be attained by the general application of the gas producer to the utilization of the immense deposits of low-grade fuels which are at present undeveloped and of other materials which now go to waste. The following brief abstract gives an outline of Mr. Junge's argument.

"The question whether an exhaustion of what we have termed our irreplaceable fuel resources is a danger for the life and prosperity of future generations, can only be discussed on the basis of theoretical prognostications and speculative arguments. The other question, whether for the benefit of present activities it is wise to economize in the methods of utilization of these resources, cannot be answered but in the affirmative.

"It is but a matter of political prudence for a nation to exploit the low-grade fuel materials of the country, such as peat, dust coals and refuse, if they can be used for the generation of heat, light and power, instead of wasting anthracite and coke, and to reserve the latter coals for more profitable and important uses in the metallurgical and other industries. An efficient utilization of coal, generally speaking, tends towards the preservation of national values, making a country self-supporting and independent of the world's markets. It also aids the prevention of hygienic abuses which, if not amended, are apt prematurely to weaken the earning capacity and the industrial activity of a nation.

"From whatever point of view we look at the problem, it remains a matter of the greatest economic importance to find methods and means for utilizing the enormous stretches of lignite and peat lands, especially those located in the neighborhood of large undeveloped bodies of rich ore which abound in remote districts of the United States and elsewhere, and either to transform the raw coals into some form of available energy which can be transmitted over long distances at reasonable cost, or to refine the low-grade fuels into superior products such as briquettes, or coke, or chemicals, that they may serve as a basis for other industries to grow upon and prosper. The question which remains to be settled then is not *whether* we should use the inferior classes of coal, but *how* we can use them most efficiently."

Lignites and peats should not be called low-grade coals on account of their low carbon, high moisture and high ash contents in the face of recent developments. Culm banks and other waste, obtained largely in coal mining, are the only materials which can rightly be called low-grade fuels, since their contents of *both* fixed carbon and volatile hydro-carbons is small. In boiler work, ash reduces the heating power of coal and actually obstructs the process of combustion to such an extent that an ash content of 40 per cent. will absolutely destroy the value of a fuel for raising steam. In producer work these drawbacks are actually turned to advantage. High ash content will promote an even flow of material through the producer when properly treated, if the temperature is not allowed to reach a point where clinker and slag will be formed. Of course, a reduction of the incombustible matter by washing or briquetting is preferable since high quality and quantity of combustibles in a fuel make for more regular and efficient working of the producer plant and a gas of more uniform composition. In cases where it is necessary or desired to use coal in its raw state, however, high ash content cannot be regarded as a limiting condition when producers are employed. In Germany mine culm, containing 65 per cent. of ash and only 25 per cent. of combustible matter is gasified in the Jahns type of ring producer with entire success.

Within certain limits moisture also is an assistance rather than a detriment in producer work. Besides acting as a preventative of excessive temperatures, moisture, when sufficiently heated, will enrich the gas by the addition of hydrogen and oxygen. Moisture is harmful only when large quantities of it are contained in the gas as produced. There is an economic limit for each material, however, which must not be surpassed.

The introduction of by-product coking was an event of great importance, resulting as it did in an increase of from five to ten per cent. in the yield of coke and a return from the sale of by-products varying from 75 cents to $1 per ton of coke made. But the process has not been so generally adopted as its value warrants. Taking the case of England, "if the total quantity of coke made in the United Kingdom for metallurgical purposes is reckoned at 10,-000,000 tons at an average price of $3.30 per ton, the general adoption of by-product coke ovens would result in a saving of from $1,750,000 to $3,500,000 derived from the increased yield of coke, while up to $10,000,000,000 could be derived from the sale of by-products, provided that the intrinsic value of the latter would remain the same in the future as it is now." This has been the case in Germany where the annual gain from the sale of by-products is about $10,000,000, and the same process is now being applied to the utilization of lignite and peat. In Bavaria the latter is subjected to destructive distillation in the Ziegler furnace and yields an excellent metallurgical coke, gas for heating, lighting and power purposes, sulphate of ammonia, tar oil, creosote and paraffin. The utilization of lignite in a similar manner is also increasing rapidly in Germany since it was discovered that lignite tar oils can be used as fuel for Diesel and other oil engines.

The use of lignite briquettes is also increasing rapidly in Germany. "The heat value of brown coal briquettes ranges from 7,700 to 9,600 B. t. u. per pound, compared with an average of 4,900 B. t. u. per pound of raw lignite containing 45 per cent. water. Their heat density is such that up to three tons or 60,000,000 B. t. u. can be stored in a space of 100 cubic feet, hence their commercial distribution range is almost double that of the raw coal. They are an ideal producer fuel on account of the regularity of form and composition. The briquetting tests of the United States Geological Survey show that the Dakota lignites can be treated as successfully as the German brown coal, a fact which will vastly extend the territory which these fuels control."

In discussing the utilization of peat, Mr. Junge refers to the results obtained by Dr. Caro which were fully described in these pages last month. Of mine culm, wash banks, etc., he says:

"The rational utilization of these materials is of great importance for collieries, where they are available in enormous quantities, and where they have formed hitherto a real nuisance to the works management. Owing to excessive ash contents these banks could not be burnt under boilers, nor could they be dumped back into the mines on account of the danger of causing self-ignition of the remaining coal deposits. So they were either stored up in huge piles in the neighborhood of the pit or, where territorial limitations prevented this, they were transported by rail into neighboring dumping grounds, being thus absolutely useless and causing heavy expenditures. There are two possibilities of utilizing these low-grade coals: one is to gasify them in Jahns ring producers where their fuel value is utilized, the 25 to 30 per cent. combustibles yielding a gas free from tar and well suited for heating and power purposes. . . . . . . Another method is that developed by Dr. N. Caro, of Berlin, Germany. It is based on the observation that wash banks and other waste contain more nitrogen than that which corresponds to their coal contents. In Westphalian collieries it was found that wash banks, the coal contents of which show on analysis about 1.2 per cent. of nitrogen, contain up to 1 per cent. of nitrogen, though their total contents of combustible matter is only 25 to 30 per cent. Dr. Caro has succeeded in gasifying this material in producers of the Mond type especially equipped for the purpose, and besides getting a suitable gas he gains about 80 per cent. of its total nitrogen contents in the form of sulphate of ammonia."

Mr. Junge points out that very small sizes of coal can be used in specially designed producers or that they may be briquetted, when they will compete with the very best grades of coal. In conclusion he sums up the case for the gas producer thus:

"One fact stands out clearly: it is that, according to the present state of our knowledge, the rational utilization of coals of high volatile contents requires the adoption of gas producers with by-product recovery and the distribution of heat, light and power from gas-driven central stations to the neighboring districts, a scheme which is feasible only when operating on a large scale and where staple markets for the disposal of goods lie within the commercial distribution radius of the plant. Fuels of high ash contents, on the other hand, such as mine culm and other waste of low heat value, must be used at the spot in producers specially equipped for the purpose. Dust coals and similar fuels can either be gasified in producers particularly designed for their use, or they may be transformed into briquettes, whereupon competition becomes possible with the best grades of coal for all manner of application. In all cases the employment in the electrical central station of large gas engines is a logical supplement to the gasification of coals in producers and is the only means so far available for attaining maximum industrial economy in the operation of plants of some magnitude."

THE EFFICIENCY OF THE GAS ENGINE.

A RECORD OF EXTENSIVE EXPERIMENTAL INVESTIGATIONS ON THE INDICATED POWER AND MECHANICAL EFFICIENCY OF THE GAS ENGINE.

Bertram Hopkinson—Institution of Mechanical Engineers.

THE recent paper by Prof. Bertram Hopkinson, read before the Institution of Mechanical Engineers, is a most important contribution to the literature of the gas engine, clearing up, as it does, a great many points in connection with gas-engine efficiency that have hitherto been obscure and puzzling. The paper represents a vast amount of careful experimental work, and Prof Hopkinson's descriptions of his apparatus and methods are extremely interesting and suggestive. It is impossible here, however, to do more than indicate briefly his more important conclusions.

"In the report of the Committee of the Institution of Civil Engineers on the Efficiency of Internal Combustion Engines, the following remark occurs: 'It would be desirable but for one circumstance to calculate the relative efficiency only from the indicator horse-power. But it appears that in the case of gas engines, and especially gas engines governed by hit-or-miss governors, the indicator diagrams do not give as accurate results as is generally supposed. The diagrams vary much more than those of a steam engine with a steady load, and the mean indicator horse-power, from the diagrams taken in a trial, may, it appears, differ a good deal from the real mean power.'

"This statement is fully borne out by the tests of the Committee, which show that the mechanical efficiency taken as the ratio of brake to indicated power varied from 80 per cent. to 94 per cent. in the three engines tested. These engines were of similar type, but of different sizes, and whereas the smallest of 5 horse-power showed a mechanical efficiency of 90 per cent. the intermediate engine of 20 horse-power showed a lower efficiency of 80 per cent. The Committee remarked that these values were obviously incorrect, and the values adopted by them for the mechanical efficiency were obtained by running the engine light and making an estimate of the indicated horse-power under these conditions. Assuming that the mechanical loss is constant at all loads, the indicated power at full load can be determined by adding the power absorbed at no load to the brake power. The mechanical efficiencies of the three engines found in this way were respectively 0.86, 0.866, and 0.888.

"These results are just what would be expected; the mechanical efficiency showing a slight improvement with the size of the engine.

"The opinion of the Committee quoted above is obviously important and may be expected to have a widespread effect in gas engine testing. It throws doubt upon many of the efficiency tests on gas engines which have hitherto been made and published. Moreover, the method which the Committee themselves adopted for getting the indicated power from the brake power seems to require further investigation before it can be accepted as accurate. It may, no doubt, be assumed on the evidence of steam engine tests that under given conditions of lubrication the friction is practically independent of the pressure in the engine. But where as in the steam engine the whole of the mechanical losses are to be ascribed to friction, that is not the case in the gas engine, in which a considerable amount of power is wasted in pumping and is usually included in the mechanical losses. Moreover, with a given supply of oil, lubrication conditions in the steam engine are practically constant, but in the gas engine that is by no means the case. Great changes can take place in the temperatures of the cylinder walls in a comparatively short time, and this will affect the viscosity of the oil and therefore the work spent in friction. The author therefore determined to undertake an investigation with the object of finding whether the indicator power of the gas engine does, in fact, vary so much, and is so difficult of determination, as the report of the Committee referred to suggests. If it were found that the indicated power could be accurately determined directly, it was further desired to test, by direct comparison of brake and indicated power, the validity of the Committee's method of getting the mechanical efficiency. Briefly, the conclusions reached are:—

"(1) If precautions are taken to keep the pressure of the gas supply constant, the diagrams given by the engine are remarkably regular, and, whether the engine be missing ignitions or not, it is possible, by the use of a sufficiently accurate indicator, to obtain the indicated power from diagrams within 1 or 2 per cent. It seems probable that the difficulty experienced by the Committee was due either to the essential defects, for this purpose, of the ordinary form of indicator, or to casual variations in the gas supply per suction, due perhaps to variations in the gas pressure at the engine.

"(2) The difference between indicated horse-power and brake horse-power is rather less than the horse-power at no load under the same conditions of lubrication, mainly because of the difference in the power absorbed in pumping. In the particular engine tested by the author the error from this cause in obtaining the indicated power would amount to about 5 per cent. The friction is substantially constant from no load to full load, provided that the temperature of the cylinder walls is kept the same, but the influence of temperature is very great."

The engine used in the tests worked on the Otto cycle, with hit-and-miss governing, and was designed for a maximum output of 40 brake horse-power. Its cylinder dimensions were 11½ by 21 inches and the speed 180 revolutions per minute. The compression space was 407 cubic inches, compression ratio, 6.37, and compression pressure, 175 pounds per square inch absolute.

The result of Prof. Hopkinson's investigations show that the confusing results obtained have been due almost entirely to the inaccuracy of the ordinary indicator for gas-engine tests when high compressive charges are used. There is given a discussion of the extent to which the various sources of error in the ordinary indicator, friction, inertia of the moving parts, elasticity of the cord, friction of the pencil, the natural period of vibration of mechanism, and especially backlash, affect the accuracy of measurement in gas engine testing, an extent much greater than in steam engine work, owing to the much lower pressures in the latter case. Prof. Hopkinson's first care was to devise a satisfactory indicator.

"The author determined to get a new design of indicator of the piston and spring type with optical magnifying mechanism. In the form finally adopted, after a considerable amount of experimenting, the spring consists of a straight piece of steel strip held as an encastred beam in a steel frame. A piston slides in a bore communicating with the engine, the axis of this bore being at right angles to the spring and passing through its center. The pressure on the piston deflects the spring and so tilts a small mirror about an axis at right angles to the bore, the pivots of this mirror being carried on a steel frame. To give the other motion to the mirror the whole apparatus—

straight spring and mirror with its pivots—is positively connected to an eccentric on the crank axle by which it is rocked about the axis of the bore, thus giving the piston motion of the diagram without the possibility of any lost motion. This instrument is practically indestructible, and it has been left open to the engine for considerable periods without giving it any attention. The vertical deflection is accurately proportional to the pressure, so that the diagrams can be integrated with a planimeter. Finally, the period of oscillation is only about 1-700th of a second with such strengths of spring as were used in the mechanical efficiency tests. The indicator is very easily calibrated by dead weights. The diagrams used in these measurements were photographed, but for many purposes it has been found sufficient to observe them direct by means of a telescopic arrangement by which they are projected as a bright line of light on to a transparent screen with vertical and horizontal scales. It is easy to plot the diagram on to a piece of squared paper, and its area can thus be obtained within 5 per cent. without the trouble of photography."

Tests with this indicator established the fact that, "given a sufficiently accurate indicator and constant conditions, the indicated horse-power of a gas engine may be determined from diagrams with an accuracy which is probably superior to that attainable in the steam engine." Prof. Hopkinson emphasizes the necessity for keeping conditions constant in making tests and shows the effect of changes in pressure outside the cylinder and of changes in the temperature of the cylinder walls and piston on the accuracy of the measurement of indicated horse-power, outlining experiments undertaken to measure the extent of the effect produced by such changes.

POWER TRANSMISSION BY PRODUCER GAS.

A PROPOSAL FOR THE TRANSMISSION OF POWER GAS IN PIPE LINES FROM COAL MINES TO CENTRES OF INDUSTRIAL ACTIVITY.

Charles E. Lucke—Cassier's Magazine.

A N interesting discussion of the practicability of transmitting power by producer gas is given by Prof. Charles E. Lucke in a recent number of *Cassier's Magazine.* The idea of locating large central producer plants in the vicinity of cities to supply power to local industries is not new, but Prof. Lucke goes a step farther and proposes that coal should be gasified at the mine and the gas transmitted in pipe lines to industrial centres. After reviewing the development of power transmission from such purely mechanical means as shafts, gears, etc., through hydraulic pressure to compressed air, Prof. Lucke says:

"During the period of evolution power transmission methods with their differentiation or adaptation to special situations, culminating in the compressed air system, electrical progress was very rapid, and electrical methods of transmitting power, by reason of their great economy, adaptability and freedom of limiting distances, compared with its forerunners, enabled it to completely outshadow all these competitors.

Electrical power transmission is equally feasible through only a few feet or hundreds of miles, though, of course, with somewhat unequal losses, and may convey small amounts of energy or the enormous outputs of great waterfalls without difficulty. Electrical transmission, by reason of its good features, has brought the stored energy of human nature from waterfalls to industrial communities. It has likewise removed innumerable shafts and belts from our factories and shops, distributed or transmitted power of a central engine or plant to any point fixed or movable, as in cars or cranes, where work was to be done. So successful has been the applications of electrical apparatus to the transmission of power that among engineers it has come to be regarded as pre-eminently the best method of transmitting power; but no matter how good a machine or piece of apparatus may be, the history of engineering progress seems to indicate that some time and somewhere a better one will be found; if not better for all applications, certainly better for some. Engineering practice really consists in the process of differentiation, applying the principle that what is best in one place or time will not necessarily be so in another,' or that 'something new may always find a useful field if essentially good.' "

Dr. Lucke comments on the development of high-pressure, long-distance pipe lines for the transmission of natural gas in various sections of the United States. "What was intended to be a convenience for postponing the inevitable has proved to be a practical method of transmitting power not by the transmission of fluid pressure, but by the transmission of concentrated energy in the form of gas fuel in pipes. Compressed fuel gas may generate power precisely as dry cylinder expansion does with compressed air; but it may also generate several times more, by its combustion in the gas engine, the exhaust heat from which would serve to reheat the compressed gas before its expansion. The comparatively limited supply of natural gas has prevented such a proposition as the above as a commercial proposition becoming a matter of general interest. As, however, the power-gas producer has, under development and improvement, become as reliable and rugged a piece of apparatus as the average boiler, requiring labor no more skillful to handle, the possibility of this method of power transmission by producer gas generated from coal is a matter that at least requires investigation.

"The present method of transmitting power from the coal mine as an origin of energy involves, outside of the mine plant itself, coal-loading machinery, freight car transportation over railroad systems, coal unloaders and reloaders, and the ultimate combustion of the coal in boilers or producers. The gas producer is capable of transforming a fairly large percentage of the energy in the coal to stable fuel energy in producer gas, and the gas-engine driven compressor is capable of pumping this gas at any desired pressure through absolutely unlimited distances by means of relays of the same sort of machine along the line. The pipe line would thus become a substitute for the freight car, and might even be laid on the same right of way, and instead of delivering intermittently carloads of coal, discharge continuously energy in

the form of compressed combustible gas, this applying, however, only so much of the coal transportation as would be required for power or heat. This pipe line could be tapped at any point for fuel gas or for power, which could be sold by meter with equal facility to both large and small consumer, just as it is at present done with electricity, and with one point decidedly in favor of the pipe line. There is practically no chance of a break in a reasonably well laid pipe line situated, for example, on the right of way of a good railroad, and service might safely be guaranteed, whereas at the present time, by reason of atmospheric condition, it is impossible to guarantee continuity of service over high-tension electrical transmission lines.

"If the fuel energy in the gas under transmission be alone considered, and the possibility of obtaining work by its expansion before combustion, the weight of gas to be transmitted may be taken at almost 80 per cent. of the weight of coal per horse-power transmitted, which comparison gives directly data for comparing the cost of transmission of the gas in pipes with coal in cars. At one of the Pennsylvania mines, anthracite buckwheat, selling in New York for $2.40 per ton, is sold for 60 cents, or one-quarter as much, indicating that the cost and profits of transportation represented three times the cost of mining, and that if such gas could be piped for less than $1.80 per ton the proposition would be profitable. In transporting oil the pipe line has shown itself to be a successful competitor of the car, and the analogy suggests the possibility of working the coal mine without railroad connections at all, just as the isolated oil wells have worked. Mines too high up, too low down or too small for railrad spurs might easily install producers and feed the main railroad pipe-line, receiving compensation by meter just as the ultimate consumer pays.

"Not only does the gas producer in connection with the gas-engine driven compressor offer a means of possible economical power transmission from the coal mine to the consumer, but it offers, like electricity, a similarly attractive prospect for the mill or factory having many isolated buildings, where ordinarily electric transmission will be the only thing considered.

A central gas-producer plant, located convenient to the coal supply, may be arranged to furnish gas under its own blast pressure to any part of the mill where power is wanted and where it may be secured by the installation of a gas engine. The gas producer may then economically feed as many gas engines situated as close to the work as may be desired. There are many cases of group driving where this method would furnish power in just as convenient a form as electrical transmission systems, and in such cases this system will prove very much more economical, because there is a loss every time energy is transformed. In the case of electric-transmission systems, the rotary motion is converted into electrical energy at the dynamo with a loss, transmitted over lines with another loss, converted back into rotary motion at the motors with another loss. The gas engine, placed at the point of group drive, may replace the group driving motor with the elimination of all other losses, and may do so in a way infinitely better than could be accomplished by steam, because there is no

pipe-line loss, and because the small gas engine for each individual group drive would have practically the same thermal efficiency as a main engine driving a central dynamo. Furthermore, many manufacturing plants require heat for some of their processes, and there is no more convenient form of heating than producer gas burners. They can be controlled to maintain temperatures varying only a few degrees by means of thermostatic variations, and the gas can be the same as used for the engine.

"It would appear, therefore, to be perfectly feasible to use producer gas as a means of power transmission both for short distances around factory plants or for long distances from coal mines to towns or cities, and the prospects of economical transmission are very good. It may be that practicable as such power transmission certainly is, it will not find economical application; but this possibility seems to be somewhat remote, and that such a method of transmitting power will find a place, situation or set of conditions where it can be usefully applied is almost certain."

ECONOMY OF POWER IN CRUSHING ORE.

A STATEMENT OF THE MORE IMPORTANT POWER LOSSES IN ORE CRUSHING AND THE PRINCIPLES ON WHICH GREATER ECONOMY IS TO BE ATTAINED.

Ernest A. Hersam—Mining and Scientific Press.

THE following abstract of an article on power economy in crushing ore, lately contributed to the *Mining and Scientific Press,* will prove of great interest to managers of mills and smelters. Power waste in crushing, while very large, is probably more difficult to locate and remedy than in any other operation connected with milling and smelting. Mr. Hersam's paper outlines the various ways in which power is consumed without doing useful work and his suggestions for improvement are of great practical interest.

"First among the general losses is the power lost as pure friction. Friction can be either in the machine, in the ore, or even, when regarded in a certain way, within the body of the single pieces of the ore itself, as a kind of plastic deformation. All these true frictional losses are encountered throughout. They extend from the shafting and belting, which convey power from

the prime mover, to the ore itself at the moment of breaking. Thus, friction must be considered as unavoidable as any part of the work done, but it is a part that produces no desired effect, and first indicates a lack of economy, due to improper construction or imperfect conditions.

"The second cause of loss arises from power given out in vibration in the foundation and elsewhere around the machinery, in supports and hangers, and disappearing in many ways. This energy is scattered in every direction. Foundations themselves, in some cases, consume much power; and the constant vibration everywhere, from the ground up to the dust-covered cobwebs on distant rafters, all are taking power to maintain the motion. Vibrations of other kinds than this, through the air as well as through the foundation and supports, take away a small part of the energy constantly. Were other more trifling losses to receive separate attention, there could be included sound waves, heat waves, air pulsations, and even light, electricity, and other forms of energy, resulting from an unstable position of the pieces of ore in the machine, or caused elsewhere, as by the jarring of supports and foundations. Such wasted forms of energy being in part the result of friction, and in part of impact, are found everywhere, where there is motion, from the shafting, bearings, and belting, to the mass of ore itself. For the most part, however, the losses occur within the ore that is undergoing crushing.

"Third among the general losses, is that due to the unavoidable deformation of the ore. This is not precisely a part of the general friction. It is work done in the interior of the unbroken fragments, and is over and above that work which should be necessary to overcome mere cohesion of the molecules in producing an actual rupture of the particles. It is the way in which much power generally goes; and although in many cases it is a loss that can be much reduced by taking advantage of the natural conditions, there is no hope of avoiding it entirely.

"Fourth among the general losses could be named that which results from producing dust or fine below the size sought. The extent to which this becomes serious depends upon the subsequent process of treatment. To a certain degree, the production of dust signifies actual work done, however fine the dust may be. It is often to be regarded as work done improperly, however, if not an entire waste of the power expended upon it.

"The friction of the machinery depends in general upon the construction of the machine and the condition in which it is kept. On the whole it is relatively a small part of the entire loss. Crushing machinery is not different from other kinds in principle; the same loss of power could occur in conveying motion for any purpose. The loss, however, would be increased two or three times by bad management or neglect, or correspondingly reduced by care and careful construction. Matters that should not escape attention, but that are neglected sometimes in regions remote from supplies and inspection, are: Protection of bearings from dust, the use of suit-

able lubricants and of proper belting, the alignment of shafting, the wear of gearing, and the care of bearing parts. Correct construction in the beginning often saves much cost in the end, but no construction can withstand ill use or neglect.

"There have been cases where a change of lubricants has reduced the loss of power 30 per cent., although conditions had been supposed correct before. The change of temperature between winter and summer has its influence, and though the power lost in this way is too variable to state definitely, it is important if lubricants are not suited to the climate and the season. Whenever a bearing is running hot, it represents power wasted. To run bearings even perceptibly warm, and particularly in cold places, is too costly to go unnoticed. When machinery is first started it does not run easily. This is influenced by the nature of the lubricants, the load, the condition of the belting, the dust, the duration of a previous period of idleness, and other factors. With good construction and management the power lost through friction in the machinery may be made less than 10 per cent. of that expended in other ways. It is a visible loss many times, and is one evident to the mechanic."

The loss of power among the ore particles is much greater, the actual loss which usually appears in some form of friction being multiplied by an improper use of a machine. It varies greatly with the speed of the machine and with different methods of feeding and operating. The two forms of pure friction and molecular friction or deformation are to be reckoned with.

Surface friction is always accompanied by strain and deformation below the surface and the consequent molecular rearrangement which manifests itself in heat consumes power. All such friction is lessened by a free discharge of the crushed ore as it becomes reduced to the size wanted, but this condition is difficult to obtain. "To crush ore, the masses must be retained between the approaching surfaces that administer the force; but while too great an amplitude in the motion of the pressure surfaces tends uselessly to compress the fragments, grinding these upon one another, and distributing the force in many directions, too little amplitude, on

the other hand, fails to do more than de-
form some of the particles, and failing to
break these, it wastes whatever energy is
so expended. In a reciprocating jaw-
crusher, at each revolution, whatever the
amplitude, the jaws are certain to come to
rest against the elastic strain of ore. The
power must suffer loss each time the pressure is applied and released. Thus the loss
in working would be most reduced by having exactly the right amplitude for the
quantity of ore retained between the jaws
at any one time, and by having sufficient
voids between the pieces to provide a place
for the crushed particles to occupy as soon
as produced. This can be done by making
the size of the original pieces uniform and
comes from sizing before crushing."

Under stress ore will attempt to adapt
itself without breaking to new positions. It
cannot be worked under pressure like a
plastic material, but there is a tendency for
it to do this and it is this quality which
consumes power. In the amount of power
consumed the time element plays an important part. With plenty of time many
kinds of rock will bend under stress while
a sharp blow will result in fracture. Knowledge of the power consumed by this viscosity is limited. All that can be said practically upon the subject is that "much additional total power is required in running
slowly upon certain kinds of tough rock;
but, on the other hand, brittle or highly
elastic rocks show little compensation for
the mechanical loss there is in high speed."

Another important form of deformation
is elastic deformation in which the ore al-
ters and then again regains its shape. The
power retained in this way is given back
again when the pressure is released, but it
may be returned at such a time and in such
directions as to be lost for all practical purposes. This loss of power through elasticity varies with different machines and
with different speeds of operation. Rapidly
acting machines are usually economical
with regard to plastic deformation, but
may or may not be so far as elastic deformation is concerned. Mr. Hersam gives a
discussion of the relative economy of
crushing devices in this connection and
then passes to a consideration of the real
work accomplished in crushing and the
methods of measuring it, concluding thus:
"The attainment of fullest economy in
crushing will be found in a speed that is
sufficiently high for tough ore, an amplitude that is adequate for elastic ore, and
from a moderation of both speed and amplitude for brittle ore. Whenever the reduction in size is intended to be great, the
fine must be protected from further crushing either by water suspension, repeated
sizing, or abundant opportunity for rearrangement of the compressed ore by the
provision of interstitial space that accom-
panies uniform sizing. It depends upon
producing the effect of crushing rather
than of grinding, particularly upon coarse
sizes; and of employing hard and non-
elastic surfaces in contact with the ore. It
demands solid foundations for machinery,
care, cleanliness, and protection of bearings, tested lubricants, and the avoidance
of all unnecessary irregularity of shape in
high-speed running parts designed for the
construction."

SCIENCE IN THE IRON AND STEEL INDUSTRIES.

A DISCUSSION OF THE COMMERCIAL VALUE OF CHEMISTRY IN IRON AND STEEL MANUFACTURE.

C. H. Risdale—Cleveland Institution of Engineers.

THE following abstract of the presidential address of Mr. C. H. Risdale before the Cleveland Institution of Engineers emphasizes the importance of chemistry in the commercial management of iron and steel works. As Mr.
Risdale points out, the development of
methods of production in the iron and steel
industries along the lines of increased size
of plant, rapidity of process and excellence
of mechanical detail in the handling of
materials has about reached its limit of
economy and future cost reduction must
depend upon less obvious refinements in
the processes of manufacture, the successful accomplishment of which must depend,
to a great extent, on the assistance of the
chemist.

The dominant feature of chemistry is precision. All chemical combinations are in exact proportions—law, order, system, run through all, and must be observed in every little detail. This in laboratory practice alone renders analysis possible, and the nearer it is approached in manufacture the better. Works processes are essentially chemical operations on a large scale, and profit depends on the quantitative as well as the qualitative success of these. For instance, 1 per cent. less waste may mean all the difference between profit and loss. Is the attainment of right conditions made a *sine quâ non*, or are they over-ruled by interests adverse to them? Are the re-agents, viz., the materials, sufficiently pure and uniform for the purpose? Have the operators (those who control the system and conduct the working, the directors, managers, and men) sufficient chemical training, or the right attitude of mind to appreciate its full significance? Is the apparatus (the plant) designed for carrying out the reactions in the best possible way? In laboratory practice a chemist often rigs up his own apparatus, and though the manufacture of the plant is the engineer's special business, none the less the chemist should be called in to see that it fulfils chemical requirements. In certain branches of the engineering and chemical manufacture the degree of accuracy attained is very great, and corresponding advantage has been felt, but this has not yet been sufficiently regarded as a necessity in iron and steel manufacture, and most of the hindrances and difficulties met with here are due to the want of regularity and precision. It is to the little things that closer attention must be given. From how many items in the cost sheet can 2s. 6d. or 3s. a ton now be saved? Many—apart from materials—do not exceed 3s. or 4s. altogether. In recent years engineers have tried by sheer force to increase and thus cheapen production. By the free introduction of up-to-date machinery materials are handled on a large scale rapidly and cheaply, and manual labour has been very much replaced and reduced almost to a minimum. Probably in some directions this has been carried too far. Again, efforts have been made to improve the organisation of large concerns, and reduce management and other fixed charges by amalgamation. In present cir-

cumstances what item can be cut right out or reduced very largely? It seems clear that almost the only remaining prospect of economy and progress in future is a closer approach to the conditions of the laboratory, viz., finer working and greater regularity all through, and works must be re-organised more on these lines. Has not this an advantage, besides the sentimental, as compared with expenditure on fresh plant and machinery? Suppose that by re-modelling a works with a 200,000 tons yearly output, at a cost of £100,000, 2s. 6d. or 3s. can be saved per ton, and orders taken at lower prices. The yearly interest and depreciation on the borrowed money will probably be £15,000, an additional burden of 1s. 6d. per ton (apart from maintenance) on the costs before profit can be made. But if successful other works will quickly do the same, and though the margin of advantage may vanish, the burden will remain.

In improvement from closer adherence to chemical requirements, in system and quality of work, orderly, exact habits, care, and the balance of departments, the utmost is got out of existing plants with little or no capital outlay. Also the qualities of mind developed in the staff and workmen will be still there ready for further advance —a valuable reserve when other firms become educated to the same pitch. Many outside circumstances which at first sight have no connection with chemistry become of great concern. For instance, trade customs, the arrangement of works, labour questions, etc.

In commercial management frequently materials of widely different qualities are bought for the same purpose, for instance, a number of good and poor cokes, and uniformity of the mixture used, depending on them being delivered *pro rata*, may be upset by irregular deliveries. Again, traffic questions, internal and external, trucks of old date being cleared out of the order required by uniformity to avoid demurrage, or being blocked through limited siding accommodation, are all factors. In transit much of the small material—consisting of ironstone, coke, and coal largely of shale, and of ore, limestone, lime, etc., of dry soil and other impurities—is shaken down to the bottom; samples from the tops generally show too good results. Yet how many works have proper facilities for sampling them after they have been tipped, or for holding them back if found to be unsatisfactory. The general dislocating effects of stoppages and holidays are well known; as also the effects of rain through wet coke, ore, etc., on blast furnaces; wind on kilns, etc. These affect precision of working, and upset the smooth run. Manufacturers to be successful have always to be contriving not only "fool-proof," but (to coin a term) "chance-proof" methods, which will go smoothly amid all such occurrences.

In the second part of his address Mr. Risdale dwelt upon the sources and results of irregularities in the main sections of the industries. After dealing at length with the raw material supply he proceeded to show that while the producer of raw materials takes very little trouble to control scientifically and regularise his product, it is not so with the pig-iron maker. He takes a great deal of pains to avoid and counteract irregularities. Probably at all blast furnace plants which serve steel works, and a great many which do not, the aid of chemists is enlisted, and materials are more or less watched and products tested. Mr. Risdale enumerated most of the improvements that have been adopted at blast furnaces to aid regular working.

Taking the founder, Mr. Risdale remarked that he expects the iron to flow well, so that the castings will be clean and sound, and of the right strength, etc. His irregularities in material are due largely to the custom, only slowly being overcome in this country, of buying pig by fracture and not by analysis. Founders do not use the chemist enough, and go too much by rule of thumb.

The steel-maker takes the iron a step further and makes a still more uniform product. He is no more exempt from irregularities than other men. Standing midway in the chain of industries, he is the butt of both ends, and has to bear the brunt of those originating with the producer of raw material and the iron-maker, as well as the exactions of the purchasers. These last expect every single piece—not every train or truck load as with raw material—to be right in quality, etc., and perfectly sound. No average is allowed. For finished sections the narrow limits of variation are exactly specified by engineers or standard associations, and it is in the makers' own hands to meet them. The customer expects the steel to stand whatever treatment he may choose. He simply condemns whatever does not suit him, and lays all faults on the steel-maker. Therefore in no branch of the iron and steel trades are more pains taken to guard against and overcome variation. Huge mixers are used, and chemists and inspectors employed freely, while all sorts of tests and checks are used at every stage.

The steel user starts with material as nearly as possible sound and good, which provided the mechanical and physical laws are not violated, may be worked up into good finished product. He does little or nothing to prevent any ills, yet he disclaims all responsibility for these, and often considers himself ill-used. The users rarely keep their own chemists, or anyone whose duty it is to see that the steel gets the treatment which suits it best. Perhaps the greatest crop of irregularities occur during heat treatment of the steel. Mr. Risdale dealt at length with the defects that show themselves in steel and how they should be treated.

In concluding, Mr. Risdale dealt with the need for standards not only for coal, ironstone, ores, etc., but also for minor things, such as lubricants, boiler composition, alloys, etc. The greatest curse of the industries is irregularity. The absence of standards permits untrue statements about them. If everything were bought and sold to well-defined standards much of the irregularity would be removed, and one could rely in buying on always getting for a given grade exactly the same quality of article. Then many of the fluctuations in price from competition of varying qualities would vanish, and markets be steadier. Standardisation of raw materials will have to come if progress is to be made with the further products. Then the relative responsibility of the different sections of the industries should be clearly decided. Mr. Risdale, in saying a few words about the works of chemists, suggested that there was needed an "institute of practically experienced chemists," to which no examination can admit but only ability proved in practical work over a long period.

THE ENGINEERING INDEX

THE KEYSTONE
IN·THE·ARCH·OF·APPLIED·SCIENCE

The following pages form a descriptive index to the important articles of permanent value published currently in about two hundred of the leading engineering journals of the world—in English, French, German, Dutch, Italian, and Spanish, together with the published transactions of important engineering societies in the principal countries. It will be observed that each index note gives the following essential information about every publication:

(1) The title of each article,	(4) Its length in words,
(3) A descriptive abstract,	(5) Where published,
(2) The name of its author,	(6) When published,

(7) We supply the articles themselves, if desired.

The Index is conveniently classified into the larger divisions of engineering science, to the end that the busy engineer, superintendent or works manager may quickly turn to what concerns himself and his special branches of work. By this means it is possible within a few minutes' time each month to learn promptly of every important article, published anywhere in the world, upon the subjects claiming one's special interest.

The full text of every article referred to in the Index, together with all illustrations, can usually be supplied by us. See the "Explanatory Note" at the end, where also the full title of the principal journals indexed are given.

DIVISIONS OF THE ENGINEERING INDEX.

CIVIL ENGINEERING

BRIDGES.

Arches.
See Concrete, and Reinforced Concrete.

Blackwell's Island.
The Manhattan Approach of the Blackwell's Island Bridge. Illustrated detailed description of the massive steel viaducts connecting the anchor spans with the city streets at grade. 3000 w. Eng Rec—Nov. 23, 1907. No. 88540.

Caissons.
Lowering a Large Pneumatic Caisson with Screw Rods. Illustrates and describes manner of building and lowering the caissons for the piers of the 4-track Schuylkill River Bridge, at Philadelphia. 700 w. Eng Rec—Nov. 23, 1907. No. 88539.

See also same title under Construction.

Cantilever.
The New Homberg Bridge. Illustrated description of the largest structure of its kind in Germany. Joins Homberg and

We supply copies of these articles. See page 715.

Ruhrort. Plates. 1200 w. Engr, Lond—Nov. 1, 1907. No. 88239 A.

See also under Blackwell's Island, and Quebec.

Compression Members.

The Compressive Member. Horace E. Horton. Comments and criticisms bearing on the Quebec bridge disaster. 3000 w. Ry & Engng Rev—Nov. 2, 1907. No. 88074.

The Lacing of Compression Members. C. T. Morris. Brief explanation of the theory and its application. Mathematical. 900 w. Eng News—Nov. 7, 1907. No. 88112.

Concrete.

Design for the Henry Hudson Memorial Bridge: A Concrete Rib Arch of 703 ft. Span. Illustrates and describes the design proposed, with editorial comment. 4000 w. Eng News—Nov. 21, 1907. No. 88505.

The New Bridge Over the Wissahickon at Philadelphia. J. A. Stewart. Illustrated description of the Walnut Lane bridge, and the method of construction. 1500 w. Sci Am—Nov. 30, 1907. No. 88593.

Design.

Bridges and the Art Commission. Montgomery Schuyler. Considers the attitude of the Commission in regard to the Henry Hudson Memorial, and the Connecting Railroad bridge across the East River at New York. Ills. 2500 w. Archt Rec—Dec., 1907. No. 88591 C.

Lift-Bridge.

New Cowing Lift Bridge at Jefferson St., Cleveland. Brief illustrated description. 600 w. Ir Trd Rev—Oct. 31, 1907. No. 88007.

Piers.

The Reconstruction of the Anchor Piers of the Poughkeepsie Bridge. Illustrates and describes work made necessary to endure heavier traffic. 2500 w. Eng Rec—Nov. 9, 1907. No. 88192.

Quebec.

The Phoenixville Testimony in the Quebec Bridge Inquiry. Gives the testimony of the Phoenix Bridge Co.'s engineers before the Canadian Government Commission, with editorial. 21500 w. Eng News—Nov. 28, 1907. No. 88618.

See also Compression Members.

Reinforced Concrete.

The Long Key Viaduct. William Mayo Venable. Illustrated description of a two-mile reinforced-concrete railway viaduct. 3500 w. Eng Rec—Nov. 23, 1907. No. 88537.

Empiricism and Error in Arch Design. Charles W. Comstock. Considers erroneous and unscientific statements relative to the design of reinforced-concrete

arches. 3000 w. Eng News—Nov. 28, 1907. No. 88617.

Shearing Stresses.

Railroad Bridges with One Standard Gauge Track (Ponts de Chemins de Fer à une Travée et à Voie Normale). E. Pentecôté. An algebraic determination of the maximum shearing stress on the bearings and at the centre of the principal girders, and of the zone of alternating shearing stresses for a certain standard rolling load. Ills. 5000 w. Rev Gén des Chemins de Fer—Oct., 1907. No. 88317 G.

Steel Truss.

Bridge of Chicago & Northwestern, Pierre, S. D. Illustrates and describes a new bridge across the Missouri River. 600 w. Ry Age—Nov. 1, 1907. No. 88073.

Suspension.

The Hinged Type of Suspension Bridge with Stiffening Girders (Pont Suspendu à Poutres Raidissantes et à Articulation Médiane). M. Gisclard. An elaborate discussion of the superiority of the hinged over the rigid type of stiffened truss, using as illustration the Brooklyn Bridge. Ill. 18000 w. Ann d Ponts et Chauss—1907—III. No. 88306 E + F.

Swing Bridges.

Two Unique Electrically Operated Australian Swing-Bridges. Illustrations with short description of two bridges at Sydney. 1000 w. Elec Rev, N Y—Nov. 23, 1907. No. 88515.

Trestles.

Wooden Railway Trestles. R. Balfour. A critical discussion of the design and construction. Ills. 2500 w. Can Engr—Nov. 1, 1907. No. 88005.

Viaducts.

See Reinforced Concrete.

CONSTRUCTION.

Beams.

Continuous Beams. A. W. Connor. Gives a few applications of the theory to problems. 900 w. Trans Eng Soc, Univ of Toronto—No. 20. No. 88134 N.

The Vierendeel System for Long-Span Reinforced-Concrete Beams (Eisenbetonträger für grosse Spannweiten, System Vierendeel). A mathematical discussion. Ills. 2500 w. Serial. 1st part. Beton u Eisen—Oct., 1907. No. 88354 F.

Buildings.

On the Influence of the Use of Iron and Steel on Modern Architectural Design. Victor D. Horsburgh. Considers the present use of these materials, their influence, and states the conclusions reached. 9000 w. Jour Roy Inst of Brit Archts—Oct. 19, 1907. No. 88268 B.

See also Cellars, Factories, Pier Shed, Reinforced Concrete, Singer Building, and Steel Buildings, under CONSTRUCTION.

Caissons.

Some Notes on Pneumatic Caissons for Bridges and Buildings. T. Kennard Thomson. An illustrated description of caissons and methods of using them, with related matters. 4000 w. Trans Eng Soc, Univ of Toronto—No. 20. No. 88120 N.

Cellars.

Water-Proof Cellar Construction. Colbert A. MacClure. Considers the "tar and felt" and the "hydrolithic" cement systems. Discussion. 10000 w. Pro Engrs' Soc of W Penn—Nov., 1907. No. 88527 D.

Coal Pocket.

A Modern Retail Coal Pocket. Charles J. Steffens. Illustrated description of the arrangement of conveying machinery and coal pockets at a Hoboken, N. J., plant. 2200 w. Eng Rec—Nov. 2, 1907. No. 88048.

Concrete Blocks.

Cost of Molding and Laying Concrete Blocks for a Factory Building. Information in regard to · factory buildings at Grand Rapids, Mich. 1000 w. Engng-Con—Nov. 6, 1907. No. 88149.

Dams.

See same title under WATER SUPPLY.

Docks.

See same title under WATERWAYS AND HARBORS.

Excavation.

Methods and Costs of Excavating Granite in Open Cuts on the Grand Trunk Pacific R. R. 4500 w. Engng-Con—Nov. 27, 1907. No. 88610.

Factories.

Advances in Factory Construction. L. P. Alford. Showing growth in reinforced concrete construction and other details. Ills. 2500 w. Am Mach—Vol. 30, No. 45. No. 88173.

Reinforced Concrete in Factory Construction. Frederick A. Waldron. Gives direct cost and structural comparisons between mill, steel and reinforced-concrete factory buildings. Ills. 2500 w. Am Mach—Vol. 30, No. 45. No. 88170.

The New Leipzig Cotton Mill in Leipzig-Lindenau (Neue Leip'ziger Baumwollspinnerei in Leipzig-Lindenau). Em. Haimovici. An illustrated description of the design and construction of this reinforced-concrete factory building. Plate. 2200 w. Beton u Eisen—Oct., 1907. No. 88356 F.

See also Concrete Blocks, under CONSTRUCTION.

Foundations.

Heavy Foundations for the New Steel Works at Gary, Ind. Illustrates and describes the plant and methods used in the construction. 4000 w. Eng Rec—Nov. 2, 1907. No. 88046.

A Consideration of the Earth's Surface in Its Relation to Building Construction. Owen B. Maginnis. Considers the ingredients that compose the earth's crust in relation to foundation work in building. 1700 w. Archts & Bldrs' Mag—Nov., 1907. No. 88536 C.

Governor's Island.

The Enlargement and Reconstruction of Governor's Island. Illustration, with brief account of plans to enlarge the island by 101 acres, reclaimed by means of sea walls and dredging, and other changes. 1200 w. Sci Am—Nov. 30, 1907. No. 88594.

Piers.

See same title under WATERWAYS AND HARBORS.

Pier Shed.

Reconstructing a New York Pier Shed. The work on Pier 36, North River, is illustrated and described. 1000 w. Eng Rec—Nov. 2, 1907. No. 88049.

Piles.

See Concrete Piles, under MATERIALS OF CONSTRUCTION.

Reclamation Work.

Land Reclamation in Holland. An account of past and present work. 2500 w. Engr, Lond—Nov. 15, 1907. Serial. 1st part. No. 88587 A.

See also Governor's Island, under CONSTRUCTION.

Regulations.

New Reinforced Concrete Regulations in Philadelphia. Regulations approved by the Bureau of Building Inspection on Oct. 8. 2800 w. Eng Rec—Nov. 2, 1907. No. 88047.

Reinforced-Concrete Building Laws; Their Differences and Deficiencies. H. C. Hutchins. A comparison and criticism of the chief established codes. Ills. 5000 w. Engineering Magazine—Dec., 1907. No. 88624 B.

Reinforced Concrete.

A· Ten-Story Reinforced Concrete Building. Illustrated description of a warehouse in Pittsburg. 2200 w. Eng News—Nov. 7, 1907. No. 88113.

Method and Cost of Building a Reinforced Concrete Garage. Illustrated general description with itemized costs. 1100 w. Engng-Con—Oct. 30, 1907. No. 88034.

A Concrete Building for a Chocolate Factory. Illustrated description of a structure at Stamford, Conn., with reinforced-concrete saw-toothed roof and framework, and cement-block walls. 2000 w. Eng Rec—Nov. 9, 1907. No. 88193.

Test of Reinforced-Concrete Roof. J. Ernest Franck. Illustrations, with account of test of the Borough of Hammersmith Public Baths and Wash-Houses, and the reasons for using reinforced con-

We supply copies of these articles. See page 715.

crete. 2000 w. Jour Roy Inst of Brit Archts—Oct. 19, 1907. No. 88269 B.

Faults of Reinforced Concrete Design and Construction. H. F. Porter. Discusses causes of failure, and gives recommendations for securing successful work. 2500 w. Engng-Con—Nov. 13, 1907. No. 88266.

See also Beams, Factories, and Regulations, under CONSTRUCTION; Sewers, under MUNICIPAL; Reservoirs and Water Towers, under WATER SUPPLY; and Pier, under WATERWAYS AND HARBORS.

Reservoirs.
See Reservoirs and Water Towers, under WATER SUPPLY.

Roofs.
See Reinforced Concrete, under CONSTRUCTION.

Singer Building.
The Structural Features of the Singer Building, New York. Illustrated detailed description. 3500 w. Eng Rec—Nov. 16, 1907. No. 88405.

Stacks.
The World's Greatest Stack. W. C. Capron. Illustrated account of a stack in course of construction at Great Falls, Montana. 1000 w. Sib Jour of Engng—Nov., 1907. No. 88489 D.

Steel Buildings.
Modern Steel Buildings. John M. Ewen. On the construction and also the methods of business organization necessary for rapid and efficient work. Ills. 6000 w. Trans Eng Soc, Univ of Toronto—No. 20. No. 88127 N.

The German-American Insurance Building, New York. Illustrates and describes interesting details of a 20-story steel-cage structure in New York City. 2000 w. Eng Rec—Nov. 9, 1907. No. 88196.

The Largest Single Office Building in the World. Illustration of the 33-story structure of the City Investment Company in New York, with descriptive notes. 1000 w. Sci Am—Nov. 23, 1907. No. 88500.

Erecting a Heavy Steel Building with Jinniwinks. The conditions in connection with the extension work on the Lincoln Wharf power house, Boston, are explained, and the methods adopted. 1400 w. Eng Rec—Nov. 2, 1907. No. 88053.

See also Buildings and Singer Building, under CONSTRUCTION.

Syphon.
See same title under WATER SUPPLY.

Tunnels.
The Tunnels Under the North and East Rivers. William Yale. Short illustrated account of the 12 tunnels under construction, the methods and information of interest. 1800 w. Yale Sci M—Nov., 1907. No. 88413 C.

New Boring Machines for Tunneling. Discusses the claims and defects of several newly invented rock-boring machines. 2000 w. Eng & Min Jour—Nov. 23, 1907. No. 88544.

Construction Methods in the Second Bergen Hill Tunnel of the Lackawanna R. R. Describes work in progress in New Jersey. 2500 w. Eng Rec—Nov. 9, 1907. No. 88198.

The Mechanical Plant for the Construction of the Tauern Tunnel in Austria. Notes from a paper by Karl Brabbée, before the Soc. of Austrian Engs. & Archts. Outlines the work and describes the installation of machinery. 2500 w. Eng Rec—Nov. 2, 1907. No. 88050.

The Construction of the Gattico Tunnel (Les Travaux du Tunnel de Gattico). An illustrated description. 3000 w. Génie Civil—Oct. 5, 1907. No. 88325 D.

See also same title under WATER SUPPLY.

Waterproofing.
See Cellars, under CONSTRUCTION.

MATERIALS OF CONSTRUCTION.

Cast Iron.
Tests of Cast Iron Beams. Elmer H. Fish. Brief account of tests made and results. 1200 w. Jour Worcester Poly Inst—Nov., 1907. No. 88261 C.

Cement.
Classification and Uses of Cement. Samuel S. Sadtler. Considers the phenomena of setting, composition, cost and details of use. 2800 w. Jour Fr Inst—Nov., 1907. No. 88254 D.

The Cement Industry and Consumers (L'Industrie des Cements et des Chaux Hydrauliques devant les Consommateurs). Henry Le Chatelier. A discussion of the attitude of consumers to the cement industry in France, the need for standards of composition, tests, etc. 7000 w. Rev de Métal—Oct., 1907. No. 88314 E + F.

Cement Testing.
Some Problems of a Cement Inspecting Laboratory. R. S. Greenman. Discusses problems of a laboratory which inspects cement proposed for use on work. 1800 w. Cement Age—Nov., 1907. No. 88215.

Concrete.
The Action of Sea Water on Concrete. From report of a committee of the Supts. of Bridges and Buildings Assn., giving opinions from replies to a circular letter. 1500 w. R R Gaz—Nov. 8, 1907. No. 88158.

Concrete Piles.
The Simplex System of Concrete Piling. Thomas McKellar. An illustrated detailed description of this method. 3800 w. Jour Assn of Engng Socs—Oct., 1907. No. 88493 C.

Concrete Piles. Charles R. Gow. An account of methods of construction used, with reasons for their adoption. Ills. 4000 w. Jour Assn of Engng Socs—Oct., 1907. No. 88492 C.

Paints.

Rust Prevention. L. M. Stern. (Abstract.) A treatise on the preservation of iron and steel by paint. 6500 w. Ir Age—Nov. 21, 1907. Serial. 1st part. No. 88509.

Reinforced Concrete.

Compression Tests of the French Government Commission. Reports tests made of 12 reinforced-concrete prisms. Ills. 2800 w. Cement—Oct., 1907. No. 88210 C.

Timber Preservation.

Requirements for Treating Wood Paving Blocks. George W. Tillson. Discusses the proper treatment to prevent natural decay and maintain stability in size. 3300 w. Munic Engng—Nov., 1907. No. 88030 C.

Forestry and the Preservation of Wood. Charles E. Koons, Sr. Quotes from reports of the U. S. Forestry Dept. and briefly considers processes in use for wood preservation. 3500 w. Pro St Louis Ry Club—Nov. 8, 1907. No. 88621.

See also same title, under MINING AND METALLURGY, MINING.

MEASUREMENT.

Surveying.

See Surveying, under MINING AND METALLURGY, MINING; and Surveying, under RAILWAY ENGINEERING, NEW PROJECTS.

MUNICIPAL.

Garbage Disposal.

A Project for a Garbage Burning Plant with Drying and Mud-Separating Devices for Pforzheim City (Projek einer Müllverbrennungsanstalt mit Klärschlammtrocknungsanlage für die Stadt Pforzheim). Herrn. Herzberger and Morave. Ills. 7300 w. Gesundheits-Ing—Oct. 5, 1907. No. 88358 D.

Gary, Ind.

Water Supply and Sewerage of the New Industrial Town of Gary, Ind. Describes the water supply, taken from Lake Michigan, and the combined system of sewerage disposal adopted. 2000 w. Eng News—Nov. 7, 1907. No. 88114.

Pavement Repair.

Municipal Paving Repair Plant. An illustrated account of a plant in operation at New Orleans, La., with details of cost and other data. 3800 w. Munic Jour & Engr—Nov. 6, 1907. No. 88152 C.

Pavements.

See Timber Preservation, under MATERIALS OF CONSTRUCTION.

Roads.

Road Hardening and Dust Prevention about Boston. Reviews papers on this subject read at meeting of the Massachusetts Highway Association. 4000 w. Eng Rec—Nov. 23, 1907. No. 88541.

Rescuing Our Roads. George Ethelbert Walsh. Discusses the road problem, describing the manner in which it has been solved in certain localities. 2500 w. Sci Am—Nov. 9, 1907. No. 88159.

Bridle Roads in the West Indies. H. C. Huggins. Pertaining to the methods of construction and maintenance of natural-soil hillside bridle roads. 4 Plates. 3000 w. Soc of Engrs—Nov. 4, 1907. No. 88487 N.

Sewage Disposal.

Colwyn Bay Sewage Works. Brief illustrated description of the scheme to clear the sea front from contamination. 1500 w. Engr, Lond—Nov. 1, 1907. No. 88237 A.

Sewage Disposal Plant of Reading, Pa. Illustrates and describes the apparatus for screening and pumping, the septic tank, sprinkling filters, sedimentation basin, &c. 3500 w. Munic Jour & Engr—Nov. 6, 1907. No. 88151 C.

Report on Sewage Disposal of the Calumet District Through the Chicago Drainage Canal. Gives report of Rudolph Hering on the relative advantages of disposing of the Calumet District sewage by dilution through the drainage canal and by sprinkling filters. 2200 w. Eng News—Nov. 7, 1907. No. 88116.

See also Gary, Ind., under MUNICIPAL; and Pumping Plants, under MECHANICAL ENGINEERING, HYDRAULICS.

Sewage Filters.

Studies of Sewage Distributors for Trickling Filters. C.-E. A. Winslow, Earle B. Phelps, C. F. Story, and H. C. McRae. The general problem, methods of analysis, calculating efficiency, &c., are considered. Ills. 12000 w. Tech Qr—Sept., 1907. No. 88077 E.

Sewage Sterilization.

The Sterilization of Treated Sewage. Gives an outline of investigations described in a recent bulletin of the U. S. Bureau of Plant Industry. 4500 w. Eng Rec—Nov. 16, 1907. No. 88409.

Sewer Discharges.

A Short Method of Recomputing Sewer Discharges for a Changed Value of n in Kutter's Formula. An explanation of the method evolved. 800 w. Eng News—Nov. 21, 1907. No. 88504.

Sewers.

Reinforced Concrete Sewers in Staten Island. Detailed description, with illustrations. 2800 w. Eng Rec—Nov. 2, 1907. No. 88052.

Trenches.

Back Filling Trenches. George C. War-

We supply copies of these articles. See page 715.

ren. Considers the importance of and suggests practicable methods for overcoming the difficulties. 3300 w. Munic Enging—Nov., 1907. No. 88031 C.

WATER SUPPLY.

Air Valves.

Improvised Air Valves for Water Lines. Frank Martin. Illustrates and describes a float valve and its working. 800 w. Power—Nov., 1907. No. 88014 C.

Artesian Wells.

Camden's Artesian Water Supply Is Not Derived from the Delaware River by Infiltration. Prof. Oscar C. S. Carter. Explains reasons for the author's opinions. Ills. 4000 w. Jour Fr Inst—Nov., 1907. No. 88253 D.

The Artesian Water Supply of Australia. C. O. Burge. Information in regard to the artesian area, the quality and possible uses of the water. 2000 w. Eng Rec—Nov. 16, 1907. No. 88411.

Australia.

See Artesian Wells, under WATER SUPPLY.

Dams.

The Hauser Lake Steel Dam in the Missouri River, near Helena, Mont. Illustrated detailed description of an interesting structure. 1300 w. Eng News—Nov. 14, 1907. No. 88553.

Methods and Cost of Making Two Hydraulic Fill Dams. An illustrated article, giving data and information of the building of the earthwork. 3000 w. Engng-Con—Nov. 13, 1907. No. 88267.

See also Tarare, France, under WATER SUPPLY.

Filtration.

Report on the Filtration of the Croton Water Supply, New York City. With editorial. 4000 w. Eng News—Nov. 21, 1907. No. 88506.

The Filtration of the Croton Water Supply of New York. Plan, sections, and description of the proposed filter plant. 4000 w. Eng Rec—Nov. 23, 1907. No. 88538.

Sand Filtration of Water Supply. Andrew Williamson. The present number gives an account of the construction of modern sand filters. Later articles will review the working practice. 4500 w. Engng—Nov. 1, 1907. Serial. 1st part. No. 88234 A.

See also under Springfield, Mass., under WATER SUPPLY.

Gary, Ind.

See same title, under MUNICIPAL.

Irrigation.

Distribution of Irrigation Water from Canals in the Yakima Valley, Washington. S. O. Jayne. Information from Bul, 188, U. S. Dept. of Agriculture, describ-

ing the work. 3500 w. Eng News—Nov. 7, 1907. No. 88117.

The Irrigation Canal from the Hérault and the Projects for Canals from the Rhone (Le Canal d'Irrigation Dérivé de l'Hérault et les Projets de Canaux Dérivé du Rhone). Francis Marre. An account of irrigation projects in the wine districts of Southern France. Ills. 3000 w. Génie Civil—Oct. 5, 1907. No. 88326 D.

Orifices.

The Discharge of a Circular Orifice (Débit d'un Orifice Circulaire). Casimir Monteil. A mathematical investigation of the degree of approximation obtained by the formula commonly used by engineers. Ills. 6000 w. Ann d Ponts et Chauss—1907, III. No. 88310 E + F.

Pipe Lines.

The Fundamental Principles in the Design of Pipe Lines (Grundlagen für die Berechnung der Wasserleitungen). Eduard Sonne. A mathematical review of hydraulic principles. Ills. 6600 w. Zeitschr d Ver Deutscher Ing—Oct. 12, 1907. No. 88363 D.

See also Corrosion, under ELECTRICAL ENGINEERING, ELECTROCHEMISTRY.

Pollution.

See Stream Pollution, under WATER SUPPLY.

Pumping.

See same title, under MECHANICAL ENGINEERING, HYDRAULICS,

Purification.

Removal of Manganese from Deep Well Water. Trans. from Jour. für Gasbeleuchtung und Wasserversorgung. Describes an interesting experience in the water supply of Arad, in Hungary. 1100 w. Eng Rec—Nov. 9, 1907. No. 88197.

See also Filtration, under WATER SUPPLY.

Rand.

The Rand Water Board Plant and Works. An illustrated series of articles giving information in regard to the public water supply. 2500 w. Engr, Lond—Nov. 1, 1907. Serial. 1st part. No. 88236 A.

Reservoirs.

Method and Cost of Building a Reinforced Concrete Reservoir. General description, with itemized costs. 2500 w. Enging Con—Nov. 6, 1907. No. 88148.

See also Tarare, France, and Water Towers, under WATER SUPPLY.

Springfield, Mass.

The Springfield Water Works. Elbert E. Lochridge. An account of troubles with the Ludlow water and the filtration system; also other features of the new supply. Ills. Discussion. 9000 w. Jour N Eng W Wks Assn—Sept., 1907. No. 88598 F.

We supply copies of these articles. See page 715.

Statistics.

Water Works Statistics for the Year 1906, in Form Adopted by the New England Water Works Association. Compiled by Charles W. Sherman. 2500 w. Jour N. Eng W Wks Assn—Sept., 1907. No. 88596 F.

Stream Pollution.

State Protection of the Purity of Inland Waters. R. Winthrop Pratt. Discusses laws for securing pure water supplies and unpolluted streams in connection with the work of the State health departments. General discussion follows. 11700 w. Jour Assn of Engng Soc—Oct., 1907. No. 88490 C.

The Prevention of Stream Pollution by Strawboard Waste. Earle Bernard Phelps. Describes the manufacture of strawboard, discussing the problem of the disposal of the waste liquor, giving laboratory and field investigations, and treatment recommended. Ills. 9000 w. Tech Qr—Sept., 1907. No. 88076 E.

Water Pollution in the Oswego and Upper Hudson River Drainage Areas. Extracts from recent reports of the N. Y. State Dept. of Health. 2500 w. Eng News—Nov. 7, 1907. No. 88110

Syphon.

The Construction of a Syphon in Upper Egypt. E. Neuhaus. Illustrated detailed description of the construction of a syphon consisting of mild-steel pipes, with heads and wings in masonry. 4000 w. Engr, Lond—Oct. 25, 1907. No. 88108 A.

Tarare, France.

The Water Supply of the City of Tarare (Ville de Tarare—Alimentation en Eau Industrielle). M. Pascalon. A brief description of the dam, reservoir, etc., for the supply of water from the River Turdine. 2500 w. Ann d Ponts et Chauss—1907, III. No. 88307 E + F.

Tunnels.

New Water Supply Tunnels at Chicago. Illustrates and describes recent additions to the tunnel system for the purpose of improving the water supply of the southern portion of the city. 3500 w. Eng News—Nov. 14, 1907. No. 88554.

Water Meters.

A Note on Water Meters (Note sur les Compteurs d'Eau). M. Dariès. Deals mathematically with the general design of the disc type and illustrates and describes all the leading devices. 22000 w. Rev de Mécan—Sept. 30, 1907. No. 88318 E + F.

Water Towers.

Note on the Design of Cylindrical Reservoirs (Beitrag zur Berechnung Zylindrischer Reservoire). Emil Reich. A mathematical discussion. 2000 w. Beton u Eisen—Oct., 1907. No. 88355 F.

The Hard-Fussach Water Tower (Der Wasserturm in Hard-Fussach). Mich. Heimbach. Illustrated description of the design and construction of this reinforced-concrete structure for the storage of drinking water. 1500 w. Beton u Eisen—Oct., 1907. No. 88353 F.

WATERWAYS AND HARBORS.

Barge Canal.

The New York Barge Canal. Lindell Theodore Bates. A review of past history of the Erie Canal and an illustrated description of the character of the work in progress. 2000 w. Yale Sci M—Nov., 1907. No. 88412 C.

Bruges.

Bruges Harbor (Der Seehafen von Brügge). Illustrated description of the new harbor works and ship canal. 1600 w. Serial, 1st part. Deutsche Bau—Oct. 26, 1907. No. 88344 D.

Canal Gradients.

The Effect of Changes in Canal Grades on the Rate of Flow. F. W. Hanna. Derives a surface curve formula based on Bresse's tables that will have general application. 1800 w. Eng News—Nov. 21, 1907. No. 88503.

Canals.

See Barge Canal, Bruges, Great Britain, Panama Canal, and U. S. Waterways, under WATERWAYS AND HARBORS.

Docks.

First Steel Ore Docks on the Great Lakes. Describes the steel dock being built at Two Harbors, Minnesota. 2500 w. Ir Trd Rev—Nov. 28, 1907. No. 88606.

The Proposed Dock System of Newark. Plan and brief description of proposed dock and canal system for improving the shipping facilities. 1200 w. Eng Rec—Nov. 16, 1907. No. 88410.

Dredges.

The Dredger "Affonso Penna." Illustrated description of a dredge built for work at the foot of Rio de Janeiro, Brazil. 1000 w. Engng—Oct. 25, 1907. No. 88100 A.

French Ports.

The Projects for the Improvements of French Ports (Les Projets de Réforme du Régime des Ports Français). P. Maurice. The first instalment commences a discussion of the inferiority of French ports and the possible remedies. Ills. 2400 w. Serial. 1st part. Génie Civil—Oct. 19, 1907. No. 88330 D.

Genes Harbor.

The Visit of the Association of Bridge, Highway and Mining Engineers to Gênes Harbor (Visite de l'Association des Ingénieurs des Ponts et Chaussées et des Mines au Port de Gênes). M. Voisin. Describes projected enlargements and

We supply copies of these articles. See page 715.

improvements. Plates. 6000 w. Ann d Ponts et Chauss—1907, III. No. 88308 E + F.

Great Britain.
Navigable Waterways of Great Britain and Ireland (Voies Navigables de la Grande Bretagne et d'Irelande). Quinette de Rochemont. Gives a review of past and present conditions, with a complete table of financial, technical and traffic details of all the canals in Great Britain and Ireland. 12000 w. Ann d Ponts et Chauss—1907, III. No. 88305 E + F.

Harbor Improvements.
Harbor Works at the Beginning of the Twentieth Century (Le Costruzioni Marittime al Principio del XX° Secolo). Luigi Luiggi. Describes modern types of breakwaters, docks, etc., and discusses the most suitable forms for proposed extensive improvements in Italian harbors. 10500 w. Ann d Soc d Ing e d Arch Ital —XXII., No. 6-7. No. 88301 F.

Italian Harbors.
See Harbor Improvements, under WATERWAYS AND HARBORS.

Newark, N. J.
See Docks, under WATERWAYS AND HARBORS.

Panama Canal.
Annual Report of the Isthmian Canal Commission to the Secretary of War. Condensed Report, with illustrations. 10700 w. Eng News—Nov. 28, 1907. No. 88616.

Pier.
The Great Government Pier at the Jamestown Exposition. E. Dabney Lunis.

Illustrations, with brief notes on its construction. 500 w. Sci Am Sup—Nov. 9, 1907. No. 88162.

Shore Protection.
See Reclamation Work, under CIVIL ENGINEERING, CONSTRUCTION.

South America.
Water Power Possibilities in South America. Lewis R. Freeman. An illustrated article showing the ideal conditions for hydro-electric plants. 2300 w. Power —Nov., 1907. No. 88016.

Stream Flow.
The Equations of the Empirical Laws of Stream Flow (Les Equations des Lois Empiriques de l'Hydraulique Fluviale). M. Fargue. A mathematical discussion of the relation between the windings of rivers and their depths. Ills. 6000 w. Ann d Ponts et Chauss—1907, III. No. 88309 E + F.

U. S. Waterways.
The Revival of Internal Waterways in the United States. George Ethelbert Walsh. Calls attention to the natural system of waterways in the United States, and discusses proposed canal developments and river improvements. 3000 w. Cassier's Mag—Oct., 1907. No. 87387 B.

Water Powers.
I. Computation of the Values of Water Powers, and the Damage Caused by the Diversion of Water Used for Power. Charles T. Main. II. Damage Caused by the Diversion of Water Power. Clemens Herschel. III. Water Rights. Richard A. Hale. Three papers discussed together. 20700 w. Jour N Eng W-Wks Assn— Sept., 1907. No. 88597 F.

ELECTRICAL ENGINEERING

COMMUNICATION.

Radio-Telegraphy.
The Clifden Station of the Marconi Wireless Telegraph System. Illustrated description of this station in Ireland. 600 w. Sci Am—Nov. 23, 1907. No. 88499.

Regular Wireless Service Between America and Europe. Reginald A. Fessenden. A letter correcting misstatements which have been published in the daily press. 4000 w. Sci Am Sup—Nov. 16, 1907. No. 88246.

The Naval Wireless Telegraph Station at Sitka, Alaska. H. C. Gearing. General information relating to coast stations, with illustrated description of station named. 1600 w. Elec Wld—Nov. 16. 1907. No. 88433.

Some Observations on the Poulsen Arc. J. A. Fleming. A paper read before the

Physical Soc., London. A record of observations in connection with a study of the arc method of exciting oscillations. 4200 w. Elec Rev, N Y—Nov. 9, 1907. No. 88199.

Telephone Cables.
The Pupin Telephone Cable, with Self-Induction Coils, Laid by Siemens and Halske Co. in the Bodensee (Ueber das von der Siemens & Halske A.-G. in Bodensee verlegte Fernsprechkabel, mit Selbstinduktionsspulen nach dem Pupinschen System). Illustrated description of manufacture and laying of the cable. 2000 w. Elektrotech u Maschinenbau—Oct. 6, 1907. No. 88321 D.

Telephony.
Power Work as Related to Telephone Communications. Thomas Lambert. Explanation of the requirements of the ser-

We supply copies of these articles. See page 715.

vice and the importance of the power plant. General discussion. 7000 w. Jour W Soc of Engrs—Oct., 1907. No. 88523 D.

Telephotography.
Electrical Transmission of Pictures. Illustrated description of the apparatus devised by Dr. Arthur Korn. 1000 w. Elect'n, Lond—Nov. 8, 1907. No. 88470 A.

DISTRIBUTION.

Current Rectifiers.
Characteristics of Circuits Employing a Mercury Arc Rectifier. O. S. Beyer and A. J. Loppin. A report of recent investigations and results. Diagrams. 800 w. Stevens Ind—Nov., 1907. No. 88218 D.

Insulation.
Some Notes on Insulation and Insulation Testing. S. M. Hills and T. Germann. (Abstract.) Considers the requirements of good insulators, and the importance of testing. 1000 w. Elec Engr, Lond—Nov. 8, 1907. No. 88466 A.
See also same title, under TRANSMISSION.

Switches.
Circuit-Interrupting Devices. F. W. Harris. The first of a series of important articles on the design, application and operation. 1800 w. Elec Jour—Nov., 1907. Serial, 1st part. No. 88080.

Wiring.
Wiring with Iron Conduit. Louis J. Auerbacher. Treats only of the unlined conduit, illustrating and describing methods of installing. 1600 w. Elec Wld—Nov. 2, 1907. No. 88061.

DYNAMOS AND MOTORS.

Alternators.
Distribution and Breadth Coefficients of Alternators. N. Stahl. Mathematical. Gives formulae, with explanation. 1200 w. Elec Wld—Nov. 9, 1907. No. 88190.
See also same title, under GENERATING STATIONS.

Asynchronous.
The Separation of the Losses in Asynchronous Machines (Zur Trennung der Verluste in Asynchron Maschinen). W. Linke. A theoretical and mathematical discussion. Ills. 3200 w. Elektrotech Zeitschr—Oct. 3, 1907. No. 88368 D.

Brushes.
Experiments on the Identification of Carbon Brushes for Dynamos (Essais ayant pour But l'Identification des Balais de Charbon pour Dynamos). The results of an elaborate series of tests to establish means of determining the properties of carbon brushes for ease in identification of a suitable material when replacements are necessary. Ills. 12000 w. Bul Soc Int d Elecns—Aug., Sept., Oct., 1907. No. 88315 F.

Commutation.
The Neutral Line of the Cummutator Cylinder (Zur näheren Erläuterung der neutralen Kommutierungszone). C. L. R. E. Menges. An endeavor to clear up certain obscure points. Ills. 4500 w. Elektrotech Zeitschr — Oct. 31, 1907. No. 88376 D.

Commutator.
What Is a Commutator? George S. Hodgins. An explanation. 1500 w. Ry & Loc Engng—Nov., 1907. No. 88002 C.

D. C. Dynamos.
The First American Gramme Ring Dynamo. G. S. Moler. Illustration, with brief account of a dynamo built in 1875. 1000 w. Sib Jour of Engng—Nov., 1907. No. 88488 D.

D. C. Motors.
Direct-Current Motors, Their Action and Control. F. B. Crocker and M. Arendt. The present article deals with the general principles of these motors. In succeeding articles various types of motors and methods of control will be considered in detail. 3000 w. Elec Wld—Nov. 2, 1907. Serial, 1st part. No. 88059.
The Starting, Regulating and Stopping of Continuous-Current Motors. J. T. Mould. Abstract of a paper read before the Assn. of Engrs.-in-Charge. Briefly considers methods of control. 2500 w. Elect'n, Lond—Nov. 8, 1907. No. 88471 A.

D. C. Turbo-Generators.
Direct-Current Turbo-Generators. H. I. C. Beyer. Outlines the constructive features of such machines, giving solutions of many electrical and mechanical problems involved. Ills. 3500 w. Elec Wld —Nov. 16, 1907. No. 88431.

Electric Power.
See Electric Driving, under MECHANICAL ENGINEERING, POWER AND TRANSMISSION.

Large Generators.
Discussion on "The Practicability of Large Generators Wound for 22,000 Volts," at New York, March 22, 1907. 7500 w. Pro Am Inst of Elec Engrs— Oct., 1907. No. 88259 D.

Railway Motors.
The Tramway Motor Question. Alex. R. McCallum. Read before the Dick-Kerr Engng Soc. Considers the relative merits of series, shunt, and separately excited motors for ordinary tramway work. 2500 w. Elec Engr, Lond—Nov. 15, 1907. No. 88567 A.
Electric Braking of Series Motors for Continuous and Alternating Current in Electric Railways, and Particularly Mountain Lines (Die Verfahren der Elektrischen Bremsung von Seriemotoren für Gleichstrom und Wechselstrom bei Elektrischen Bahnen und besonders bei

We supply copies of these articles. See page 715.

Elektrischen Bergbahnen). W. Kummer. 2500 w. Serial, 1st part. Schweiz Bau—Oct. 26, 1907. No. 88346 B.

Synchronous Motors.
A Graphic Calculator. Charles I. Young. Suggests a graphic chart for detetrmining quickly the power-factor improvement obtainable by the use of synchronous motors. 1700 w. Elec Jour—Nov., 1907. No. 88082.

Turbo-Generators.
See D. C. Turbo-Generators, under DYNAMOS AND MOTORS.

ELECTRO-CHEMISTRY.

Cell Arrangements.
The Best Multiple-Series Arrangement. Hugh S. Brown. Gives a method of determining the best multiple-series arrangement for any given number of cells. 1000 w. Elec Rev, N Y.—Nov. 23, 1907. No. 88514.

Corrosion.
The Relation between Polarization and the Corrosion of Iron Pipes by Stray Currents. Irving Langmuir. An account of experimental investigations and results. 4800 w. Stevens Ind—Oct., 1907. No. 88216 D.

Electro-Metallurgy.
See also same title, under MINING AND METALLURGY, IRON AND STEEL; and Lead Smelting and Zinc Smelting, under MINING AND METALLURGY, LEAD AND ZINC.

Electro-Plating.
The Leuchter Process of Reducing or Enlarging Models; and Its Connection with the Manufacture of Sterling-Silver and Plated Ware. Describes a process of great value in the manufacture of ornamental metal. Ills. 2000 w. Brass Wld—Nov., 1907. No. 88520.

A Beautiful and Durable Black Arsenic Deposit Without Use of Cyanide. J. Horton. Describes a method of producing a steel black deposit resembling gunmetal. 1200 w. Brass Wld—Nov., 1907. No. 88519.

Nitric Acid.
The Extraction of Nitric Acid from the Air by Means of the Electric Arc (Gewinnung von Salpetersäure aus Luft bei deren Behandlung mittels elektrischer Flamme). I. Moscicki. An illustrated description of the author's method and apparatus. 2700 w. Serial, 1st part. Elektrotech Zeitschr—Oct. 17, 1907. No. 88371 D.

Standard Cells.
Preliminary Specifications for Clark and Weston Standard Cells. F. A. Wolff and C. E. Waters. Based upon work done at the Bureau of Standards. 6500 w. Bul Bureau of Stand—Oct., 1907. No. 88251 N.

ELECTRO-PHYSICS.

Condensers.
Losses in Condensers with Fixed Dielectric and Their Damping in High-Frequency Circuits (Verluste in Kondensatoren mit Festem Dielektrikum und ihre Dämpfung in Hochfrequenzkreisen). Walther Hahnemann and Leonid Adelmann. A mathematical paper. Ills. 2300 w. Serial, 1st part. Elektrotech Zeitschr —Oct. 10, 1907. No. 88370 D.

Electric Arc.
Observations on the Electric Arc. W. L. Upson. Describes investigations made on arcs burning between metal electrodes, and between a metal and carbon electrode, in air, in hydrogen and in coal gas. Ills. 3500 w. Elect'n, Lond—Oct. 25, 1907. Serial, 1st part. No. 88093 A.

On the Electric Arc Between Metallic Electrodes. W. G. Cady and H. D. Arnold. Investigations of the change from the glow discharge to the arc between terminals of iron, platinum and copper and other metals. 9500 w. Am Jour of Sci—Nov., 1907. Serial, 1st part. No. 88029 D.

Magnetic Field.
A Standard Magnetic Field. R. Gans. Trans. from the *Phys. Zeit.* Mathematical investigations. 2000 w. Elect'n, Lond —Nov. 15, 1907. No. 88574 A.

Resistance.
The Variation of Manganin Resistances with Atmospheric Humidity. Edward B. Rose. A reply to the expressed views of Drs. Jaeger and Lindeck, that such variations are unimportant. 3800 w. Elect'n, Lond—Nov. 15, 1907. No. 88572 A.

GENERATING STATIONS.

Alternators.
A Phenomenon of Revolving Field Generators. F. Punga and W. Hess. Investigates the conditions under which an e. m. f. may be produced in the shaft of a revolving field machine. 1400 w. Elec Wld—Nov. 16, 1907. No. 88430.

The E. M. F. Generated in the Shafts of Alternators and a Suggested New Type of Low-Voltage Alternator. F. Punga and Dr. W. Hess. Calls attention to troubles experienced with bearings, offering an explanation. 1000 w. Elect'n, Lond—Oct. 25, 1907. No. 88095 A.

See also Compounding Alternators, under GENERATING STATIONS.

Central Stations.
Peoria Station of the Illinois Traction System. A new station equipped with turbines, water tubes and stokers is illustrated and described. 3300 w. Engr, U S A—Nov. 15, 1907. No. 88429 C.

Power Plant of the Pittsburg and Butler Street Railway Company. H. C. Reagan. Illustrated description of a turbine

We supply copies of these articles. See page 715.

plant and a single-phase system of transmission. 3000 w. Engr, U S A—Nov. 1, 1907. No. 88035 C.

A Steam Turbine Power Station. C. L. Vickery. Illustrated general description of the new Williamsburg Power Station of the Brooklyn Rapid Transit Co. 2500 w. Trans Eng Soc, Univ of Toronto —No. 20. No. 88122 N.

Steam-Driven Transmission Stations of the Societa Lombards, of Milan, Italy. Brief illustrated description of reserve stations. 2000 w. Elec Wld—Nov. 9, 1907. No. 88188.

Redondo Power Station of the Pacific Light & Power Company. Illustrated description of a large steam generating station in California. 1500 w. Elec Ry Rev —Nov. 2, 1907. No. 88084.

The Hochelaga Power Station of the Montreal Street Railway Company. Illustrated detailed description of a direct-current reciprocating-engine power plant. 2000 w. St Ry Jour—Nov. 9, 1907. No. 88200.

The Central Station of the International Naval Exposition at Bordeaux (La Station Centrale d'Electricité de l'Exposition Internationale Maritime de Bordeaux). Ch. Dantin. Illustrated description of the steam and electrical installations. 3000 w. Génie Civil—Oct. 19, 1907. No. 88329 D.

The Berlin Electric Installations at the Beginning of 1907 (Die Berliner Elektricitäts-Werke zu Beginn des Jahres 1907). K. Wilkens. The first part of the serial describes the steam plant in the generating station. Ills. 3200 w. Serial, 1st part. Elektrotech Zeitschr—Oct. 3, 1907. No. 88367 D.

The Electric Plants of the Rhine Valley Canal (Die Elektrizitätswerke am Rheintalischen Binnenkanal). Leopold Pasching. Illustrates and describes the central-station and sub-station installations for electric canal traction. 1500 w. Serial. 1st part. Elektrotech Zeitschr— Oct. 17, 1907. No. 88372 D.

See also Improvements, under GENERATING STATIONS.

Chili.

The Utilization of Water Power in Chili. A report by Prof. A. E. Salazar on the proposed electrification of a railway. 5000 w. Elect'n, Lond—Nov. 1, 1907. No. 88230 A.

Compounding Alternators.

A Method of Compounding Alternators. J. Rezelman and J. Perret. Abstract translation from *L'Eclairage Electrique*. Explains the principle employed and gives details of a three-phase turbo-generator. 1000 w. Elect'n, Lond— Nov. 1, 1907. No. 88229 A.

Hydro-Electric.

The Necaxa Power Works. The present article gives an illustrated description of the dams and channels of this plant in Mexico. 3000 w. Engr, Lond—Nov. 8, 1907. Serial, 1st part. No. 88483 A.

The Hydro-Electric Plant of the Black Hills Traction Company. Samuel H. Lea. The utilization of the Redwater River in South Dakota is illustrated and described. 1600 w. Eng Rec—Nov. 16, 1907. No. 88407.

The Works of the Vancouver Power Company (Ltd.). E. B. Hermon. Illustrated descriptive account of the hydro-electric power development. 7700 w. Trans Eng Soc, Univ of Toronto—No. 20. No. 88128 N.

The Hydro-Electric Development in the St. Mary's River at Sault Ste. Marie, Mich. Illustrated detailed description of the low-head hydro-electric generating station built in the rapids at the head of St. Mary's River. 3500 w. Eng Rec— Nov. 2, 1907. No. 88051.

The Rio de Janeiro Power Scheme. Illustrated detailed description of proposed plans for developing certain water power and transmitting the electric current over a distance of 51 miles to the city. 2200 w. Engr, Lond—Oct. 25, 1907. No. 88106 A.

Typical European Low-Head Hydro-Electric Power Plants. Charles H. Mitchell. Describes the main features of three typical hydro-electric plants operating under low heads, located in France, Italy and Switzerland. Ills. 3500 w. Trans Eng Soc, Univ of Toronto—No. 20. No. 88126 N.

The Caffaro-Brescia Transmission Plant (Die Kraftübertragungsanlage Caffaro-Brescia). The first instalment begins a description of the generating station at Caffaro. Ills. 2000 w. Serial, 1st part. Elektrotech Rundschau—Oct. 24, 1907. No. 88348 D.

See also Chili, under GENERATING STATIONS; and South America and Water Powers, under CIVIL ENGINEERING, WATERWAYS AND HARBORS.

Improvements.

Power Plant Improvements at El Paso. Outlines interesting work in carrying out necessary changes under hard conditions. Ills. 2500 w. St Ry Jour—Nov. 23, 1907. No. 88533.

Management.

Economic Considerations on the Management of Plant. W. H. Patchell. Pres. address before the Assn. of Engrs.-in-Charge. 5800 w. Elec Engr, Lond—Nov. 15, 1907. No. 88566 A.

Sub-stations.

The "Krummestrasse" Transformer

We supply copies of these articles. See page 715.

Installation of the Berlin Elevated and Underground Railways (Das Umformerwerk "Krummestrasse" der Berliner elektrischen Hoch- und Untergrundbahn). Herr Idelberger. Illustrated detailed description. 1300 w. Serial. 1st part. Elektrotech Zeitschr—Oct. 31, 1907. No. 88375 D.

See same title, under STREET AND ELECTRIC RAILWAYS.

Switchboards.

Electrically-Operated Switchboards. B. P. Rowe. Considers their advantages, reliability, &c. Ills. 3000 w. Elec Jour—Nov., 1907. Serial, 1st part. No. 88083.

Tariffs.

A Discussion of Various Methods of Charging for Electric Energy. E. Richards. Brief explanation of the various methods, and their effects upon load factor. 2200 w. Trans Eng Soc, Univ of Toronto—No. 20. No. 88123 N.

Turbo-Generator Tests.

Tests of Large Turbo-Generators in the United States. Gives test results from two different types of turbine, each of which is practically the largest unit of its kind. Ills. 1500 w. Engr, Lond—Nov. 1, 1907. Serial, 1st part. No. 88238 A.

Wind Power.

Windmill Electricity Works in Denmark. A brief account of the utilization of the wind for power purposes. There are now more than 30 windmills driving electric generators. 1200 w. Elec Rev, Lond—Nov. 1, 1907. No. 88227 A.

LIGHTING.

Illumination.

Interior Electric Illumination. A. T. Beauregard. Discusses some of the fundamental principles governing the production and use of artificial light and its effects on the eye. 7000 w. Trans Eng Soc, Univ of Toronto—No. 20. No. 88124 N.

The Illumination of the Building of the Edison Electric Illuminating Company of Boston. Louis Bell. L. B. Marks and W. D'A. Ryan. Read before the Ill. Engng Soc. Illustrated detailed description. 8000 w. Elec Rev, N Y—Nov. 2, 1907. No. 88045.

Moore Light.

Discussion on "Light from Gaseous Conductors Within Glass Tubes—the Moore Light," at New York, April 26, 1907. 8000 w. Pro Am Inst of Elec Engrs —Oct., 1907. No. 88257 D.

Photometers.

The Sensitiveness of Photometers. Lancelot W. Wild. Describes experiments made to determine the relative sensitiveness of various types. 1200 w. Elect'n, Lond—Nov. 8, 1907. No. 88468 A.

Stages.

Electricity Applied to Stage Lighting. Jno. H. Kliegl. Illustrates and describes modern installations. 2500 w. Elec Rev, N Y—Nov. 16, 1907. No. 88270.

Theatres.

See Stages, under LIGHTING.

Train Lighting.

See same title, under RAILWAY ENGINEERING, MOTIVE POWER AND EQUIPMENT.

MEASUREMENT.

Central-Station Tests.

Convenient Tests for Central Station Operators. W. M. Hollis. Describes convenient methods of making measurements needed in central station practice with ordinary instruments. 1200 w. Elec Wld —Nov. 2, 1907. No. 88060.

Dielectric Strength.

See Insulation, under TRANSMISSION.

Inductance.

Self-Inductance of a Solenoid of Any Number of Layers. Louis Cohen. Formulæ are given for closer approximation than is possible with the Maxwell formula. 400 w. Elec Wld—Nov. 9, 1907. No. 88189.

On the Measurement of Mutual Inductance by the Aid of a Vibration Galvanometer. A. Campbell. Describes vibration galvanometer methods. 1500 w. Elect'n, Lond—Oct. 25, 1907. No. 88094 A.

Iron Losses.

Measurement of Iron Losses in Alternating-Current Operation (Messung der Eisenverluste im Wechselstrom-Betriebe). Johann Sahulka. A description of a new method used by the author. Ills. 2300 w. Elektrotech u Maschinenbau—Oct. 20, 1907. No. 88352 D.

Testing Generators.

Artificial Loading of Large High Voltage Generators. N. J. Wilson. Discusses the chief points to be taken into consideration in designing the required apparatus. 2000 w. Elec Jour—Nov., 1907. No. 88081.

Units.

A Comparison of the Various Methods of Determining the Ratio of the Electromagnetic to the Electrostatic Unit of Electricity. E. B. Rosa and N. E. Dorsey. 5800 w. Bul Bureau of Stand—Oct., 1907. No. 88250 N.

Wattmeters.

Effects of Short Circuits on the Drag Magnets of Watt-hour Meters. A. A. Radtke. Briefly describes experiments made, stating conclusions. 700 w. Elec Wld—Nov. 16, 1907. No. 88432.

Wattmeters for Overload Charges (Compteurs d'Energie Electrique pour Tarif à Dépassement). M. Aliamet. Dis-

We supply copies of these articles. See page 715.

cusses the design of wattmeters for use in cases where a yearly contract is made for a certain amount of energy, with a special charge for amounts in excess of a stated maximum. Ills. 4000 w. Electricien—Oct. 5, 1907. No. 88319 D.

TRANSMISSION.

Aluminium.

Aluminium as a Substitute for Copper for Electrical Transmission Purposes. John B. Sparks. An investigation to determine the conditions under which the substitution of aluminium for copper tends to more economical results. 2000 w. Elec Rev, Lond—Nov. 15, 1907. Serial, 1st part. No. 88570 A.

Balancers.

Balancers Versus Three-Wire Dynamos. B. Frankenfield. Explains the operation of motor balancers, comparing them with three-wire dynamos, to the advantage of the former. 2000 w. Elect'n, Lond—Nov. 8, 1907. No. 88469 A.

Cable Constants.

Constants of Cables and Magnetic Conductors. Ernst J. Berg. Deductions and a discussion showing when approximate formulæ can properly be used. 2500 w. Pro Am Inst of Elec Engrs—Oct., 1907. No. 88256 D.

Cables.

See Insulation, under TRANSMISSION.

Direct Current.

The Modern Three-Wire Continuous-Current System. W. A. Toppin. On the popularity of this system in Great Britain, and the methods of testing. 1000 w. Elec Rev, Lond—Oct. 25, 1907. No. 88090 A.

Insulation.

The Dielectric Strength of Insulating Materials and the Grading of Cables. Alexander Russell. Read before the Inst. of Elec. Engrs. Discusses the laws of disruptive discharge; the methods of measuring the dielectric strengths; and the electric stresses on the insulating materials of a single-core cable. 3000 w. Elec Engr, Lond—Nov. 15, 1907. Serial. 1st part. No. 88568 A.

Insulators.

High Tension Insulators, from an Engineering and Commercial Standpoint. C. E. Delafield. Discusses the features to be desired in such insulators and the qualities of porcelain insulators. 3000 w. Cent Sta—Nov., 1907. No. 88185.

See also same title, under DISTRIBUTION.

Lightning Protection.

Defective Lightning Conductors. Discusses faults, their causes and the importance of a good earth, and a periodical test. 2800 w. Elec Rev, Lond—Nov. 8, 1907. No. 88467 A.

The Protection of Buildlings from Lightning. Alfred Hands. Lecture delivered at the Sch. of Military Engng. Describes cases of lightning damage, explaining the conclusions reached, and the methods advised for protection. 4200 w. Elec Engr, Lond—Nov. 15, 1907. Serial. 1st part. No. 88569 A.

A Comparison of Various Types of Lightning Arresters with Reference to Their Efficiency (Vergleich verschiedener Starkstrom-Blitzableiter in Bezug auf ihre Wirksamkeit). F. Neesen. A record of extensive tests of their practical efficiency. Ills. 4500 w. Elektrotech Zeitschr—Oct. 3, 1907. No. 88369 D.

Losses.

Units Used in Distribution. W. A. Tappin. Deals with losses in direct-current systems. 2500 w. Elec Engr, Lond—Nov. 1, 1907. No. 88226 A.

An Analysis of the Distribution Losses in a Large Central Station System. L. L. Elden. Considers a typical system, giving much information of value. 2500 w. Pro Am Inst of Elec Engrs—Nov., 1907. No. 88599 D.

Niagara.

Niagara Power Transmission. T. H. Hogg. A brief description of the line between Niagara Falls, Ont., and Syracuse, N. Y., and of some of the difficulties of its construction. Ills. 3500 w. Trans Eng Soc, Univ of Toronto—No. 20. No. 88121 N.

Rhone-Paris.

The Project for the Transmission of Power from the Rhone to Paris (Le Projet d'Adduction à Paris des Forces Motrices du Rhone). G. de Lamarcodie. Reviews other long-distance transmissions and outlines the main features of the 450-kilomètre Rhone-Paris project. Ills. 5000 w. Rev Gén d Sci—Oct. 15, 1907. No. 88316 D.

Rotary Converters.

The Starting of Rotary Converters. George I. Rhodes. Describes several methods stating the relative advantages of each. 3500 w. Elec Age—Nov., 1907. No. 88548.

Synchronous Converters.

Discussion on "Motor-Generators *vs.* Synchronous Converters, with Special Reference to Operation on Long-Distance Transmission Lines," at New York, March 22, 1907. 15500 w. Pro Am Inst of Elec Engrs—Oct., 1907. No. 88258 D.

Transformers.

See Sub-stations, under GENERATING STATIONS.

Voltage Rise.

Conditions Producing a Higher Voltage Than Normal at the Receiving End of a Line. Charles Jenkins Spencer. Describes the conditions. 1200 w. Elec Age—Nov., 1907. No. 88547.

We supply copies of these articles. See page 715.

MISCELLANY.

Agriculture.
Electric Power and Light Plants in Agriculture with Reference to the New Plant of the Libnitz Estate at Rügen (Elektrische Kraft- und Lichtanlagen in der Landwirtschaft mit Berücksichtigung der neuen Anlage auf Rittergut Libnitz auf Rügen). E. W. Lehmann-Richter. Describes the generation of power and its applications. Ills. 3500 w. Elektrotech Zeitschr—Oct. 24, 1907. No. 88373 D.

INDUSTRIAL ECONOMY

Apprenticeship.
The Rising Industrial Problem; the New Apprenticeship. George Frederic Stratton. Gives interesting facts concerning shop conditions and the attitude of trade unionism towards apprenticeship systems and trade schools. 6000 w. Engineering Magazine—Dec., 1907. No. 88622 B.

Correspondence Filing.
The Filing of Correspondence in a Manufacturing Business. Sterling H. Bunnell. Describes a system suited to the needs of the production department. 1700 w. Engineering Magazine—Dec, 1907. No. 88628 B.

Cost Estimation.
The Estimation of Costs in General Engineering Work. F. T. Clapham. Discusses the difficulties and gives an example of a system. 4000 w. Mech Engr—Nov. 9, 1907. No. 88465 A.

Education.
Engineering Laboratory Instruction. W. W. F. Pullen. Read at meeting of the Assn. of Teachers in Tech. Inst. Considers laboratory instruction generally, and also in the technological laboratory. 3500 w. Mech Engr—Nov. 16, 1907. No. 88577 A.
Mining Engineering Education in the United States. Victor C. Alderson. Extract from paper read at Joplin meeting of Am. Min. Cong. Considers causes resulting in a low general standard for mining education and recent improvement in teaching methods. 1500 w. Min Rept—Nov. 14, 1907. Serial, 1st part. No. 88414.

Industrial Betterment.
The Industrial Betterment Movement. H. F. J. Porter. How it has improved conditions of employees, and increased prosperity. 4000 w. Am Mach—Vol. 30. No. 45. No. 88176.

Patent Law.
Patents and Designs.—The New Act. J. A. Law. Treats of a new act, in England, embodying alterations in the law relating to patents and designs, of great importance to inventors, manufacturers and others. 5000 w. Jour Soc of Arts —Nov. 8, 1907. No. 88494 A.

Safety Devices.
A Bavarian Museum of Industrial Safety Devices. Dr. Alfred Gradenwitz. Illustrates and describes exhibits at this German museum for safeguarding dangerous occupations. 3500 w. Engineering Magazine—Dec., 1907. No. 88625 B.

Secrecy.
Secrecy in the Arts. James Douglas. Urges the banishment of secrecy of methods and engineering data and outlines a policy of freedom. 6000 w. Bul Am Inst of Min Engrs—Nov, 1907. No. 88639 C.

Shop Management.
See Management, under MECHANICAL ENGINEERING, MACHINE WORKS AND FOUNDRIES.

Wages.
Wages and the Cost of Living in the United States and Canada. Dudley W. Walton. Gives comparative tables and general information. 2000 w. Elec Rev, Lond—Nov. 15, 1907. No. 88571 A.

MARINE AND NAVAL ENGINEERING

Ammunition Holds.
Ventilation and Refrigeration of Ammunition Holds. Adrien Bachet. Abstract translation of paper read before the Bordeaux Int. Cong. of Nav. Archts. Methods of ventilating with cooled air are considered. 2000 w. Marine Rev—Nov. 7, 1907. No. 88178.

Buoy.
The Bredsdorff Stranding Buoy. Illus-trated description of a novel buoy for facilitating communication between a stranded ship and the shore. 700 w. Sci Am—Nov. 16, 1907. No. 88242.

Davits.
Appliances for Manipulating Lifeboats on Sea-Going Vessels. Axel Welin. States the principal requirements of an ideal system of davits and gives an illustrated description of the Welin Quadrant Davit.

We supply copies of these articles. See page 715.

1500 w. Trans Am Soc of Nav Archts and Marine Engrs, No. 10—Vol. 15, 1907. No. 88446 N.

Destroyers.
The New Ocean-Going Destroyers. Illustrated description of H. M. torpedo-boat destroyer "Corsack," with report of trials. 2000 w. Engng—Nov. 8, 1907. No. 88478 A.

Development.
Address of Sir William Matthews, President of the Institution of Civil Engineers. With editorial. 5000 w. Engng —Nov. 8, 1907. Serial. 1st part. No. 88480 A.

Dredges.
See same title, under CIVIL ENGINEERING, WATERWAYS AND HARBORS.

Fire Boats.
Fire Boat Protection. Edward F. Croker. The design and construction considered from the viewpoint of the fireman and the engineer. Ills. 3300 w. Sci Am Sup—Nov. 23, 1907. No. 88501.

Internal-Combustion Engines.
Ship Propulsion by Internal Combustion Engines. A. Vennell Coster. An illustrated discussion of the advantages, objections, applications and success attained. 6000 w. Cassier's Mag—Nov., 1907. Special No. No. 88286 D.

Magnetism.
The Magnetic Properties of Ships (Schiffsmagnetismus). Dr. H. Meldan. A review of the historical and modern methods of compensation and of the sources and kinds of ship magnetism. Ills. 4000 w. Schiffbau—Oct. 9, 1907. No. 88341 D.

Mauretania.
The Trials of the "Mauretania." Editorial on the official trials. 2000 w. Engng—Nov. 8, 1907. No. 88479 A.

The Cunard Turbine-Driven Quadruple-screw Atlantic Liner, "Mauretania." Fully illustrated detailed description of this vessel, its construction, equipment and crew, with an account of all matters relating to it and of the works where it was built. 48000 w. Engng—Nov. 8, 1907. No. 88477 A.

Mine Laying.
A New Mine Laying Steamer. Illustrated description of the Capt. A. M. Wetherill, a steamer for harbor service in connection with planting submarine mines. 1500 w. Int Marine Engng—Dec, 1907. No. 88517 C.

Monitors.
Some Early History Regarding the Double-Turreted Monitors Miantonomah and Class. William T. Powell. Concerning vessels built 30 years ago and still in service. 1500 w. Plates. Trans Am Soc of Nav Archts and Marine Engrs, No. 14—Vol. 15, 1907. No. 88450 N.

Motor Boats.
Motor Boats for Naval Service. L. S. Adams. Discusses the principal objections to their use and the advantages. Ills. 5000 w. Trans Am Soc of Nav Archts & Marine Engrs, No. 5—Vol. 15, 1907. No. 88441 N.

High Speed Motor Boats for Pleasure Use. Henry R. Sutphen. Illustrations, with brief descriptions, of boats of moderate power, with remarks on high-speed boats. 1200 w. Trans Am Soc of Nav Archts & Marine Engrs, No. 6—Vol. 15, 1907. No. 88442 N.

Some Observations on Motor-Propelled Vessels and Notes on the Bermuda Race. William B. Stearns. Considers details of design and arrangement, and some engine and fuel problems. 4000 w. Trans Am Soc of Nav Archts & Marine Engrs, No. 7—Vol. 15, 1907. No. 88443 N.

Oil Fuel.
Oil Fuel on Shipboard. Graydon Hume. Read before the Inst. of Marine Engrs. Discusses its relative value as a heat producer, and the economy to be expected from its application. 5000 w. Naut Gaz —Nov. 7, 1907. No. 88179.

The Importance of Liquid Fuels in the Construction, Operation and Profitableness of a Transatlantic Liner (Die Bedeutung der flüssigen Feuerung für Konstruktion, Betrieb und Rentabilität eines Transatlantischen Schnelldampfers). Ernst Foerster. Ills. 6500 w. Serial. 1st part. Schiffbau—Oct. 9, 1907. No. 88,-340 D.

Refrigeration.
See same title, under MECHANICAL ENGINEERING, HEATING AND COOLING.

Repairs.
Two Instances of Unusual Repairs to Vessels. S. W. B. Ferguson, Jr. An account of repairs to the bottom of the collier Nero after grounding; and to Lightship No. 68, which was seriously corroded and eaten away by electrolysis. Ills. 2000 w. Trans Am Soc of Nav Archts & Marine Engrs, No. 12—Vol. 15, 1907. No. 88448 N.

Resistance.
An Experimental Investigation of Stream Lines Around Ships' Models. D. W. Taylor. Diagrams and description of methods of investigation at the U. S. Ex. Model Basin. 1500 w. Trans Am Soc of Nav Archts & Marine Engrs, No. 1— Vol. 15, 1907. No. 88437 N.

Some Experiments on the Effect of Longitudinal Distribution of Displacement Upon Resistance. Prof. Herbert C. Sadler. Gives results of a partial investigation made at the University of Michigan. Plates. 1500 w. Trans Am Soc of Nav Archts & Marine Engrs, No. 2— Vol. 15, 1907. No. 88438 N.

We supply copies of these articles. See page 715.

Revenue Cutters.
Two New Revenue Cutters for Special
Purposes. C. A. McAllister. Brief descriptions of Nos. 16 and 17, with drawings and explanation of the special work
for which they were designed. Ills. 4500
w. Trans Am Soc of Nav Archts & Marine Engrs, No. 8—Vol. 15, 1907. No.
88444 N.

Sailing Ships.
Wooden Sailing Ships. B. B. Crown-
inshield. Reviews the development of
this type of vessel, with particular reference to those built before the Civil War.
Ills. 5000 w. Trans Am Soc of Nav
Archts & Marine Engrs, No. 13—Vol. 15,
1907. No. 88449 N.

Shipbuildling.
New Works on the Clyde. Illustrated
detailed description of the new works
of Yarrow & Co. Ltd. and their equipment. 8500 w. Engr, Lond—Nov. 15,
1907. No. 88586 A.
Building a Transatlantic Liner. An illustrated account of the building of the
"Kronprinzessin Cecilie." 2000 w. Sci
Am Sup—Nov. 30, 1907. No. 88595.

Steamboats.
Robert Fulton and the Sidewheel
Steamboat. J. H. Morrison. Reviews
information available relating to paddle-
wheels and their uses, and Fulton's application of them to the "Clermont." Ills.
4000 w. Sci Am Sup—Nov. 2, 1907. No.
88019.

Steam Engines.
The Marine Type of Beam Engines.
R. C. Monteagle. Gives a general idea
of the design of this type of engine, illus-
trating by drawings. 4000 w. Tech Qr
—Sept, 1907. No. 88075 E.

Steamships.
The Hamburg - American Steamer
Kronprinzessin Cecilie. F. C. Guenther.
Illustrated detailed description. 2500 w.
Int Marine Engng—Dec, 1907. No. 88-
516 C.

Steamship Trials.
Test of the S. S. Governor Cobb. Prof.
W. S. Leland and H. A. Everett. A re-
port of a test-run by the Mass. Inst. of
Tech. 900 w. Trans Am Soc of Nav
Archts & Marine Engrs, No. 9—Vol. 15,
1907. No. 88445 N.
See also Mauretania, and Destroyer,
under MARINE AND NAVAL ENGI-
NEERING.

Steam Turbines.
Ahead and Astern Turbines and Superheating in Turbine Steamers (Marschtur-
binen, Rückwärtsturbinen und Ueberhit-
zung bei Turbinenschiffen). Felix Langen. The first part of the serial deals
with the steam consumption of turbine
steamers. 2000 w. Serial. 1st part.
Schiffbau—Oct. 9, 1907. No. 88343 D.

Submarines.
Submarines of Battleship Speed. Mason
S. Chace. Discusses possible features of
designs; the powers needed to obtain
high speeds, and various improvements
desirable to increase the usefulness of
these vessels. Plates. 7000 w. Trans
Am Soc of Nav Archts & Marine Engrs,
No. 4—Vol. 15, 1907. No. 88440 N.

Warships.
Further Tactical Considerations Involved in Warship Design. A. P. Nib-
lack. An outline of the problem of handling a ship in action. 2000 w. Trans
Am Soc of Nav Archts & Marine Engrs,
No. 3—Vol. 15, 1907. No. 88439 N.

MECHANICAL ENGINEERING

AUTOMOBILES.

Acetylene Lighting.
Acetylene Light for Automobiles. Eugene Bournonville. Brief description of
acetylene generators appropriate for automobile use. 1200 w. Automobile—Nov.
7, 1907. No. 88141.

Adams.
The 10-H. P. Adams Car. Illustrates
and describes a 4-seated, side-entrance
model. 900 w. Auto Jour—Oct. 26, 1907.
No. 88085 A.

Albion.
The 24-H. P. Albion Car. Illustrated
detailed description of this British car.
2000 w. Autocar—Oct. 26, 1907. Serial.
1st part. No. 88087 A.

American Automobiles.
Tabular Story of the American Auto-

mobile. Gives tables on the basis of cost,
with introductory remarks. 3500 w. Automobile—Nov. 28, 1907. No. 88608.

Argyll.
The New 40-H. P. Argyll. Illustrated
detailed description. 2500 w. Autocar—
Nov. 9, 1907. No. 88456 A.

Austin.
The 1908 Austin Cars. Illustrated description of a chassis having a 6-cylinder
engine. 600 w. Auto Jour—Nov. 16,
1907. Serial. 1st part. No. 88564 A.

Beeston-Humber.
The 1908 Beeston-Humber Cars. Illustrates and describes the special features.
600 w. Auto Jour. Nov. 9, 1907. Serial.
1st part. No. 88454 A.

B. S. A.
The B. S. A. Petrol Cars. Special fea-

We supply copies of these articles. See page 715.

tures, with illustrations, of these models. 700 w. Auto Jour. Nov. 9, 1907. Serial. 1st part. No. 88457 A.

Cadillac.

The 20-H. P. Cadillac Car. Illustrated description. 600 w. Auto Jour—Nov. 9, 1907. Serial. 1st part. No. 88460 A.

Commercial Vehicles.

Commercial Vehicles of the Garden Show. W. F. Bradley. Illustrations, with descriptive notes. 1000 w. Automobile—Nov. 7, 1907. No. 88140.

See also Motor Cabs, under AUTOMOBILES.

Coventry-Humber.

The 1908 Coventry-Humber Cars. The new model is a 10 h.p. vehicle. The special features are illustrated and described. 500 w. Auto Jour—Nov. 9, 1907. Serial. 1st part. No. 88455 A.

Cylinders.

Battle of the Cylinders: Six v. Four. F. B. Stearns. A comparison of the advantages and disadvantages. 1500 w. Automobile—Oct. 31, 1907. No. 88006.

Daimler.

The 1908 Daimler Cars. Illustrated description of new features. Line-axle models, and the adoption of worm gearing on chain-driven cars. 800 w. Auto Jour—Nov. 9, 1907. Serial. 1st part. No. 88452 A.

Darracq.

The New Darracq Models for 1908. Illustrated description of several changes made in the vehicles of this company. 1800 w. Auto Jour—Nov. 2, 1907. No. 88223 A.

Exhibition.

Motor Car Exhibition at Olympia. Illustrated review of the exhibits. 2800 w. Engr, Lond—Nov. 15, 1907. Serial. 1st part. No. 88588 A.

The Motor-Car Exhibition at Olympia. An illustrated review of the exhibits. 3500 w. Engng—Nov. 15, 1907. Serial. 1st part. No. 88581 A.

Fore-Carriage.

The Pullcar—A Petrol-Driven Fore-Carriage. Illustrated description of a self-propelled fore-carriage to which any type of carriage may be attached. 2000 w. Auto Jour—Nov. 2, 1907. No. 88222 A.

Friction Transmission.

A Motor Cab with Friction-Disc Transmission (Fiacre Automobile à Transmission par Phateaux de Friction). A. Le Vergnier. Illustrated description of a car in use in Berlin. 1400 w. Génie Civil—Oct. 19, 1907. No. 88328 D.

Gasoline.

The Gasoline Automobile. Forrest R. Jones. Illustrates and describes features of modern cars. 5500 w. Cassier's Mag—Nov., 1907. Special No. No. 88287 D.

Hillman-Coatalen.

The Hillman-Coatalen Petrol Cars. Illustrated description of the leading characteristics. 600 w. Auto Jour—Nov. 9, 1907. Serial. 1st part. No. 88459 A.

Ignition.

The New Nelmelior Magneto. Illustrated description. 1500 w. Autocar—Oct. 26, 1907. No. 88086 A.

The Lodge Igniter. Illustrates and describes an invention of Sir Oliver J. Lodge, shown at the Olympia exhibition. 1200 w. Autocar—Nov. 9, 1907. No. 88,462 A.

Concerning the Progress of Ignition. Charles B. Hayward. An illustrated review of advances made in the quality of materials and application of principles involved. 3300 w. Automobile—Nov. 21, 1907. No. 88510.

Motor Cabs.

Taximeter Motor Cabs in America. Harry W. Perry. Illustrated descriptions of the Delahaye and the Darracq vehicles. 1000 w. Sci Am—Nov. 9, 1907. No. 88,161.

Motors.

The Rollason Six-Stroke Petrol Engine. Gives details of construction and tests of a new type of petrol engine. 2200 w. Autocar—Nov. 9, 1907. No. 88463 A.

Engines of the 1908 Models. Charles B. Hayward. Brief illustrated descriptions of types shown at the Madison Sq. Garden exhibition. 3000 w. Automobile—Nov. 7, 1907. No. 88139.

Perfection.

The Perfect Automobile: A Layman's Idea. G. H. Godley. Outline design and description of the writer's idea. 2000 w. Automobile—Nov. 21, 1907. No. 88511.

Sheffield-Simplex.

The Sheffield-Simplex Petrol Cars. A new name and new design for the "Brotherhood" cars. Illustrated description. 600 w. Auto Jour—Nov. 9, 1907. Serial. 1st part. No. 88453 A.

Shock Absorbers.

Types of Shock Absorbers for Automobiles. Howard Greene. Illustrated description of types. 2800 w. Sci Am—Nov. 9, 1907. No. 88160.

Siddeley.

45-Horse-Power Siddeley Motor-Car. Illustrated detailed description of this car for the Queen, which will be exhibited at Olympia. 1600 w. Engng—Nov. 1, 1907. No. 88235 A.

Testing.

Tests of Automobiles (Messungen an Motorwagen). A. Heller. An illustrated description of electrical methods of rating motor cars. 3300 w. Zeitschr d Ver Deutscher Ing—Oct. 5, 1907. No. 88,361 D.

We supply copies of these articles. See page 715.

Thornycroft.
The New 30-H. P. and 45-H. P. Thornycroft Cars. Illustrated descriptions. 700 w. Auto Jour—Nov. 9, 1907. Serial. 1st part. No. 88458 A.

Tires.
Quick-Change Tires to the Fore. W. F. Bradley. Illustrates and describes types of quick-change rims, and of dismountable rims. 2500 w. Automobile—Nov. 28, 1907. No. 88609.

Turner-Miesse.
The 30-H. P. Turner-Miesse Steam Car. Illustrated detailed description. 1000 w. Autocar—Nov. 9, 1907. No. 88461 A.

COMBUSTION MOTORS.

By-Product Producers.
By-Product Recovery Gas Producer Plants. H. A. Humphrey. Treats of the recovery of sulphate of ammonia from the wastes of the gas producer, showing the success attained by Dr. Ludwig Mond. Ills. 3000 w. Cassier's Mag—Nov, 1907. Special No. No. 88282 D.

Fly-wheels.
A Fan Fly-wheel for Gasolene Engines. E. J. Bartlett. The design and construction of a combination of fan and fly-wheel is illustrated and described. 1000 w. Am Mach—Vol. 30. No. 47. No. 88498.

Gas Engines.
The Jacobson Automatic Gas Engine. Illustrated description of an engine designed for close regulation when working on producer gas. 1100 w. Engr, U S A —Nov. 15, 1907. No. 88427 C.

Recent Applications of Gas Power. J. R. Bibbins. Reviews interesting features developed in the application of the internal-combustion engine. 2200 w. Cassier's Mag—Nov, 1907. Special No. No. 88289 D.

The Development of the Large Gas Engine in America. E. T. Adams. An illustrated review. 3000 w. Cassier's Mag—Nov, 1907. Special No. No. 88281 D.

The Gas-Power Situation in Germany. F. E. Junge. A discussion of recent progress in the design and construction of large gas engines. Ills. 4000 w. Cassier's Mag—Nov, 1907. Special No. No. 88285 D.

Large Gas and Steam Engines. W. H. Booth. The present article considers gas engines. Ills. 4500 w. Cassier's Mag—Nov, 1907. Serial. 1st part. Special No. No. 88290 D.

Chief Points of Difference Between the Gas Engine and the Steam Engine. William H. Booth. Shows that they differ fundamentally in the essentials of practice. 2500 w. Power—Nov, 1907. No. 88011 C.

Gas-Engine Peculiarities as Compared with Those of the Steam-Engine. Horace Allen. Shows that faults in the design or management of gas engines, or gas producers, have caused condemnation. 2800 w. Elec Engr, Lond—Nov. 15, 1907. No. 88565 A.

On the Indicated Power and Mechanical Efficiency of the Gas-Engine. Prof. Bertram Hopkinson. Gives an account of tests made to determine whether the indicator power of the gas engine does vary, and is so difficult to obtain as reported. Gives also results of tests for mechanical efficiency. Ills. 8800 w. Inst of Mech Engrs—Oct. 18, 1907. No. 88088 N.

Control of Internal Combustion in Gas Engines. Charles Edward Lucke. Examines the conditions under which constancy of effort may, or may not, be obtained with constancy of setting of the governor and valve gear. 3500 w. Pro Am Soc of Mech Engrs—Mid-Nov, 1907. No. 88601 C.

The Influence of Mixture Ratios on the Utilization of Heat in the Gas Engine (Der Einfluss des Mischungsverhältnisses auf die Wärmeausnutzung in der Gasmaschine). Gustav Mees. Records in numerous curves the results of extensive tests. Ills. 5500 w. Zeitschr d Ver Deutscher Ing—Oct. 5, 1907. No. 88,362 D.

The Dependence of the Utilization of Heat in the Gas Engine on the Mixture Ratio (Die Abhängigkeit der Wärmeausnutzung der Gasmaschine vom Mischungsverhältnis). K. Kutzbach. A discussion and amplification of a paper by Gustav Mees. Ills. 8300 w. Zeitschr d Ver Deutscher Ing—Oct. 19, 1907. No. 88,364 D.

See also Gas Plant Testing, Gas vs. Steam, and Historical Review, under COMBUSTION MOTORS; Internal-Combustion Engines, under MARINE AND NAVAL ENGINEERING; Blast-Furnace Gas and Blowing Engines, under MINING AND METALLURGY, IRON AND STEEL; and Gas Power, under MINING AND METALLURGY, MINING.

Gas Plant Testing.
Duty Test on Gas Power Plant. G. J. Alden and J. R. Bibbins. A report of the duty test of the gas power plant of the Norton Co., Worcester, Mass. 6000 w. Pro Am Soc of Mech Engrs—Mid-Nov, 1907. No. 88603 C.

Gas Producers.
Atkinson Automatic Suction Gas Producer. Illustrated detailed description. 1600 w. Engr, U S A—Nov. 15, 1907. No. 88428 C.

The Utilization of Low-Grade Fuels in the Gas Producer. C. T. Wilkinson.

Gives accurate data of tests upon four such fuels, discussing the results. 1000 w. Cassier's Mag—Nov., 1907. Special No. No. 88284 D.

Power Gas from Bituminous Coal. Elbert A. Harvey. An illustrated review of the progress and features of the bituminous producer. 3000 w. Cassier's Mag—Nov., 1907. Special No. No. 88292 D.

The Suction Gas Producer. F. J. Rowan. An illustrated explanation of the principles and construction, giving results of working. 6000 w. Cassier's Mag—Nov., 1907. Special No. No. 88291 D.

See also By-Product Producers, Gas Plant Testing, Gas vs. Steam, and Producer Gas, under COMBUSTION MOTORS.

Gas vs. Steam.

Test of a Producer Gas Plant. J. H. Alexander. Reports comparative tests of steam and gas plants as to efficiencies, first cost and cost of operation. 1800 w. Engr, U S A—Nov. 1, 1907. No. 88038 C.

Producer Gas for Power. S. L. Fear. Compares the efficiency of the suction gas producer plant and the steam plant, giving a short description of the former. Ills. 2500 w. Trans Eng Soc, Univ of Toronto—No. 20. No. 88130 N.

Heat Losses.

A Study of the Heat-Losses in a Gasoline Engine. Arthur J. Ward. A report of tests made. 2000 w. Stevens Ind--Oct, 1907. No. 88217 D.

Historical Review.

Historical Review of the Development of the Internal-Combustion Engine. Henry Harrison Suplee. Photographs of inventors and illustrations of engines. 7000 w. Cassier's Mag—Nov, 1907. Special No. No. 88279 D.

Oil Engines.

See Internal-Combustion Engines, under MARINE AND NAVAL ENGINEERING.

Producer Gas.

Producer-Gas Composition and Its Influence on the Performance of Suction-Producer Plants. Godfrey M. S. Tait. Considers the most desirable composition for the operation of gas engines. Ills. 2200 w. Cassier's Mag—Nov, 1907. Special No. No. 88288 D.

The Present Status of the Producer-Gas Power Plant in the United States. Robert Heywood Fernald. An illustrated review of development and report of tests made by the U. S. Geol. Surv., giving views of manufacturers, owners, and operators, and much information. General discussion. 22500 w. Jour W Soc of Engrs—Oct, 1907. No. 88522 D.

See also Gas Producers, under COMBUSTION MOTORS; and Producer Gas, under POWER AND TRANSMISSION.

HEATING AND COOLING.

Hot-Air Heating.

Gravity Air Heating in a Library Building. Illustrated description of an interesting furnace installation in a Furman University building, at Greenville, S. C. 2000 w. Met Work—Nov. 9, 1907. No. 88144.

Indirect Heating.

Indirect Heating in a Detroit Factory. Illustrated detailed description of a fan system with arrangements for internal circulation. 2000 w. Met Work—Nov. 2, 1907. No. 88032.

Refrigeration.

The Absorption Ice Plant. Heywood Cochran. Read at meeting at the Jamestown Exposition. Reviews the history of the absorption system. 2500 w. Ice & Refrig—Nov, 1907. No. 88213 C.

Laying Up an Ice Plant. William S. Luckenbach. Suggestions applying principally to the absorption system. Ills. 3000 w. Engr, U S A—Nov. 1, 1907. No. 88039 C.

The Influence of Refrigeration on the World's Work. John E. Starr. Read at meeting at the Jamestown Exposition. Reviews what has been achieved and gives predictions for the future. 3800 w. Ice & Refrig—Nov, 1907. No. 88211 C.

The Transportation of Refrigerated Meat to Panama. Roland Allwork. An illustrated description of steamship refrigeration. 4000 w. Trans Am Soc of Nav Archts & Marine Engrs, No. 11—Vol. 15, 1907. No. 88447 N.

Pipe Line Refrigeration. R. H. Tait. Read at meeting at the Jamestown Exposition. On the possibility of enlarging the revenue producing power of plants. 2000 w. Ice & Refrig—Nov, 1907. No. 88214 C.

Compression Ice Plants Using Ammonia as a Refrigerant. Thomas A. Shipley. Read at meeting at the Jamestown Exposition. Considers general problems of ice plants, the quantity, quality and economy. 4000 w. Ice & Refrig—Nov, 1907. No. 88212 C.

The Mechanical Equipment of the North American Cold-Storage Building, Chicago. The mechanical plant for a modern, fireproof, cold-storage warehouse is illustrated and described in detail. 3000 w. Eng Rec—Nov. 16, 1907. No. 88408.

See also same title, under RAILWAY ENGINEERING, TRAFFIC.

Regulating Valves .

Steam Flow Through Regulating Valves in Low-Pressure Steam Heating (Ueber den Dampfdurchgang durch Regulierventile in Wiederdruckdampfheizungen). Gives in curves and tables the re-

We supply copies of these articles. See page 715.

sults of very extensive tests. Ills. 4000 w. Serial. 1st part. Gesundheits-Ing— Oct. 12, 1907. No. 88359 D.

Steam Heating.
See Regulating Valves, under HEATING AND COOLING.

Window Leakage.
Window Leakage and Its Effect on the Amount of Radiation. Burt S. Harrison. A report of a test made, with charts, and an explanation of their use. 2200 w. Heat & Vent Mag—Nov, 1907. No. 88521.

HYDRAULICS.

Centrifugal Pumps.
Centrifugal Pumps. E. F. Doty. Gives some reasons why modern designs are more efficient. Ills. 2000 w. Engr, U S A—Nov. 1, 1907. No. 88036 C.
Centrifugal Pumps. E. F. Doty. Discusses causes of lost energy and lift. 1200 w. Engr, U S A—Nov. 15, 1907. No. 88426 C.

Orifices.
See same title, under CIVIL ENGINEERING, WATER SUPPLY.

Pumping Machinery.
The Pumping Machinery Employed in Mining. Jas. H. Hart. Reviews briefly the types of pumps and considers the available improvements. 1300 w. Min Wld—Nov. 2, 1907. No. 88070.

Pumping Plants.
The Chicago Avenue Pumping Station. Henry A. Allen. Illustrates and describes a fine station, with capacity of 75,000,000 gallons in 24 hours. 2500 w. Engr, U S A—Nov. 15, 1907. No. 88425 C.
The Use of Small Pumping Plants in Connection with Sewerage Systems. Irving T. Farnham. Illustrated description of the Newton (Mass.), pumping plant. General discussion. 10000 w. Jour Assn of Engng Socs—Oct, 1907. No. 88491 C.

Water Meters.
See same title, under CIVIL ENGINEERING, WATER SUPPLY.

MACHINE ELEMENTS AND DESIGN.

Ball Bearings.
Ball Bearings in a Marble Plant. Charles Prescott Fuller. An illustrated account of how the use of ball-bearings scored a success in marble machinery. 2200 w. Am Mach—Vol. 30. No. 45. No. 88167.
Progress with Ball and Roller Bearings. S. S. Eveland and Henry Hess. Reviews the progress made in substituting rolling for sliding friction. 3500 w. Am Mach—Vol. 30. No. 45. No. 88172.

Jigs.
Locating and Boring Holes in Drill Jigs. C. L. Goodrich. Various methods are illustrated and described. 3500 w.

Am Mach—Vol. 30. No. 47. Serial. 1st part. No. 88495.
See also same title, under MACHINE WORKS AND FOUNDRIES.

Roller Bearings.
See Ball Bearings, under MACHINE ELEMENTS AND DESIGN.

Shears.
The Design of Heavy Shears (Bau Schwerer Scheren). E. Kühne. An illustrated mathematical discussion. 3200 w. Serial. 1st part. Zeitschr f Werkzeug— Oct. 5, 1907. No. 88349 D.

Thrust Bearing.
Largest Anti-Friction Thrust Bearing Ever Made. Describes in detail the design, construction and test of a roller thrust bearing carrying a load of 150,000 pounds. Ills. 2500 w. Am Mach—Vol. 30. No. 45. No. 88163.

MACHINE WORKS AND FOUNDRIES.

Bearings.
Hot Bearings. E. Kistinger. Considers their causes and means of avoiding them. Ills. 2000 w. Mach, N Y—Nov, 1907. No. 88065 C.

Boiler Making.
The Making of the Marine Boiler (Die Herstellung der Schiffskessel). Walter Mentz. An illustrated description of the various machines and processes used. 1200 w. Serial. 1st part. Schiffbau— Oct. 9, 1907. No. 88342 D.

Brazing.
The Brazing and Reinforced Brazing of Metals. Illustrates and describes examples with and without reinforcement. 2500 w. Am Mach—Vol. 30. No. 45. No. 88165.

Buildings.
See Factories and Reinforced Concrete, under CIVIL ENGINEERING, CONSTRUCTION.

Case-Hardening.
A Process and Apparatus for the Cementation of Iron and Steel (Procédé et Appareil pour la Cémentation des Objets en Fer et en Acier). Outlines a recently developed process for case-hardening by means of contact with a hydro-carbon gas. Ills. 3400 w. Serial. 2 parts. La Métallurgie—Oct. 9 and 16, 1907. No. 88323, each D.

Castings.
Sash Weights Cast in Chills. Illustrated description of novel equipment for their rapid production. 1200 w. Foundry—Nov, 1907. No. 88024.
See also Foundry Furnaces, under MACHINE WORKS AND FOUNDRIES.

Cement Plant.
A Plant for Making White Portland Cement and Using Producer Gas for Fuel. Detailed description. 1300 w. Eng News—Nov. 7, 1907. No. 88115.

Cupolas.

Foundry Cupola and Iron Mixtures. W. J. Keep. Describes a special form of cupola and its operation, and discusses the composition of iron mixtures. 7700 w. Pro Am Soc of Mech Engrs—Nov, 1907. No. 88078.

Cutting Tools.

Standard Shapes for Cutting Tools. William H. Taylor. Giving detailed instruction for forging and grinding tools according to F. W. Taylor's standard shapes and sizes. Ills. 4000 w. Am Mach—Vol. 30. No. 45. Serial. 1st part. No. 88164.

See also High-Speed Steel, under MATERIALS OF CONSTRUCTION.

Floor Plates.

Machine Operations Over Iron Floor Plates. John Riddell. An illustrated explanation of how portable machine tools are used on floor plates and description of a new rail floor construction. 1700 w. Am Mach—Vol. 30. No. 48. No. 88613.

Forging.

See Ingot Casting, under MINING AND METALLURGY, IRON AND STEEL.

Foundry Design.

Foundry Design and Equipment. A. R. Bellamy. Abstract paper read before the Manchester (Eng.) Assn. of Engrs. Discusses whether a foundry can be successful as a part of a manufacturing business; what determines the decision to build a foundry; and the points to be considered. Ills. 5500 w. Mech Engr—Nov. 16, 1907. No. 88575 A.

Foundry Furnaces.

Converter vs. Small Open Hearth. W. M. Carr. Discusses the advantages and disadvantages of these two processes for the manufacture of steel castings. 1800 w. Foundry—Nov, 1907. Serial. 1st part. No. 88026.

See also Cupolas, under MACHINE WORKS AND FOUNDRIES.

Foundry Materials.

Specifications for Iron and Fuel, and Method of Testing Foundry Output. R. Moldenke. Discusses the value of specifications and the standards prevailing, commending the set recently adopted by the Am. Found. Assn., and the Am. Soc. for Test. Mat. 2500 w. Pro Am Soc of Mech Engrs—Nov, 1907. No. 88079.

Foundry Practice.

Foundry Notes. Suggestions for the making of successful castings. 1200 w. Prac Engr—Nov. 1, 1907. No. 88225 A.

Gates and Rises or Sink-Heads. P. R. Ramp. Brief explanation of methods in vogue. Ills. 700 w. Foundry—Nov, 1907. No. 88022.

Production of Malleable Fittings. Illustrates and describes methods of manufacture practiced at the Morse Iron Works,

Erie, Pa. 3000 w. Foundry—Nov, 1907. No. 88020.

The Choice and Use of Fluxes for Soft Metals. Information in regard to fluxes for lead, tin, zinc and alloys. 3300 w. Brass Wld—Nov, 1907. No. 88518.

Gages.

See same title, under MEASUREMENT.

Galvanizing.

Old and New Methods of Galvanizing. Alfred Sang. Discusses hot galvanizing, pickling, the cold process, zinc dust, sherardizing, vapor galvanizing, and various applications. 10000 w. Pro Engrs' Soc of W Penn—Nov, 1907. No. 88528 D.

Tinning of Metals. Abstract of a special report on dangerous or injurious processes in the coating of metal with lead, or a mixture of lead and tin. Miss A. M. Anderson and T. M. Legge. 4800 w. Mech Engr—Nov. 16, 1907. Serial. 1st part. No. 88576 A.

Gear Cutters.

The Gould & Eberhardt Spur Gear Generator. Henry R. Cobleigh. Illustrated detailed description of the mechanism and its operation, stating also the advantages of the hobbing system. 2500 w. Ir Age—Nov. 7, 1907. No. 88136.

Grinder.

The Leland Universal Grinder. Illustrated detailed description of a machine for both wet and dry grinding. 2000 w. Ir Age—Nov. 7, 1907. No. 88135.

Grinding.

Some Grinding Problems. H. Darbyshire. Discusses the choice of a wheel for different shapes and texture of material, and related subjects. 5000 w. Engng —Nov. 15, 1907. No. 88579 A.

Jig Boring.

Economical Jig Work on the Milling Machine. E. A. Johnson. Illustrates and describes the tool equipment of the miller for jig boring. 1200 w. Am Mach—Vol. 30. No. 48. No. 88615.

Jigs.

See same title, under MACHINE ELEMENTS AND DESIGN.

Lapping.

Lapping Flat-Work and Gage Jaws. F. E. Shailor. An illustrated article giving suggestions of value. 2000 w. Mach, N Y—Nov, 1907. No. 88064 C.

Lathe.

An Automatic Lathe of Unique Design. A tool with worm-driven spindle and cam-operated carriage is illustrated and described. 1200 w. Am Mach—Vol. 30. No. 45. No. 88168.

Machine Tools.

See same title, under POWER AND TRANSMISSION.

Management.

Location of the Pattern Shop. Oscar

We supply copies of these articles. See page 715.

E. Perrigo. The first of a series of articles relating to operation and management. Ills. 1800 w. Foundry—Nov, 1907. Serial. 1st part. No. 88023.

The Machine Shop; Its Place in the Plant. Oscar E. Perrigo. Second of a series of articles on shop management and cost-keeping. Ills. 2500 w. Ir Trd Rev—Nov. 7, 1907. No. 88143.

The Organization of a Jig and Tool Department. John Younger. Describes briefly the system adopted in a large motor factory in England. 1200 w. Engng Oct. 25, 1907. No. 88104 A.

Profit Making in Shop and Factory Management. C. U. Carpenter. This tenth and last article of the series deals with the effective organization in the executive division. 2000 w. Engineering Magazine—Dec, 1907. No. 88630 B.

Molding.
Molding a Retort Casting. Jabez Nall. An illustrated account of difficulties with a poorly designed pattern and how they were overcome. 2500 w. Foundry—Nov, 1907. No. 88021.

Molding Machines.
Machine Molding for Railroad Castings. E. Ronceray. Illustrates and describes machines used. Discussion follows. 4500 w. Pro N Y R R Club—Oct. 18, 1907. No. 88264.

Molding-Machine Patterns and Work. Shows the adaptability of molding-machines to castings weighing from a few ounces to over a ton. Ills. 2500 w. Am Mach—Vol. 30. No. 45. No. 88166.

Some Limitations of the Molding Machine. E. H. Mumford. Considers limitations of machine functions. 3500 w. Pro Am Soc of Mech Engrs—Mid-Nov, 1907. No. 88604.

Molding Sand.
The Treatment and Handling of Molding Sand in Foundries (Ueber Aufbereitung und Beförderung des Formsandes in Giessereien). J. Kraus. The first part illustrates and describes various grinding and disintegrating devices and methods for separating out particles of iron. 2800 w. Serial. 1st part. Stahl u Eisen—Oct. 16, 1907. No. 88334 D.

Pipe Founding.
See Wrought Pipe, under MACHINE WORKS AND FOUNDRIES.

Planer.
A New Chandler 36-in. Planer. Illustrated description of a medium weight planer of special type. 2000 w. Ir Age—Nov. 28, 1907. No. 88631.

Safety Devices.
See same title, under INDUSTRIAL ECONOMY.

Shaping.
A New Method of Shaping Metals. Il-

lustrated description of a process of roll and hammer swaging. 1200 w. Am Mach—Vol. 30. No. 45. No. 88169.

Shops.
The Green Engineering Co.'s Factory for the Manufacture of Mechanical Stokers. Illustrated detailed description of the plant at East Chicago. 2500 w. Ir Trd Rev—Nov. 28, 1907. No. 88607.

See also Shipbuilding, under MARINE AND NAVAL ENGINEERING; and Steel Works, under MINING AND METALLURGY, IRON AND STEEL.

Tempering.
Apparatus for Heat Treating Metals. Articles by various authors describing new appliances and methods. Ills. 6500 w. Am Mach—Vol. 30. No. 45. No. 88177.

Thermit Process.
See Welding, under MACHINE WORKS AND FOUNDRIES.

Vault Manufacture.
A Record-Breaking Armor Plate Vault. Illustrates and describes the design and construction of a very large vault. 2000 w. Mach, N Y—Nov, 1907. No. 88063 C.

Welding.
The Thermit Welding Process and Its Possibilities for Quick Repairs on Heavy Ordnance. Brief discussion of the application of this process to military requirements. Ills. 2200 w. Jour U S Art—Sept, 1907. No. 88294 D.

See also Boiler Repairs, under STEAM ENGINEERING.

Wrought Pipe.
Manufacture and Properties of Wrought Pipe. F. N. Speller. On the making, testing, characteristics, etc. Ills. 3000 w. Trans Eng Soc, Univ of Toronto—No. 20. No. 88133 N.

MATERIALS OF CONSTRUCTION.

Alloy Steels.
Titanium in Steel and Iron. Charles V. Slocum. Read before the Pittsburgh Found. Assn. Gives results of experimental investigations concerning the improvement to both steel and iron by the use of ferro-titanium. 2200 w. Ir Age—Nov. 14, 1907. No. 88249.

See also Improvements, under MATERIALS OF CONSTRUCTION.

Brass.
Brass Castings for Automobiles. E. F. Lake. Discusses the alloys used for special motor car parts. Ills. 2300 w. Foundry—Nov, 1907. No. 88025.

Cast Iron.
A Volumetric Study of Cast Iron. Henry M. Lane. Ills. On the volumetric composition of cast iron and its effect on properties. 1700 w. Pro Am Soc of Mech Engrs—Mid-Nov, 1907. No. 88602.

High-Speed Steel.

Actual Results with High-Speed Steel. Fred H. Colvin. General review of results obtained. 1500 w. Am Mach—Vol. 30. No. 45. No. 88174.

Improvements.

The Advance in Materials for Machinery. E. F. Lake. General review of improvements made in metals, new metals, and alloys. 3500 w. Am Mach—Vol. 30. No. 45. No. 88175.

Lead Pipe.

Lead as a Material for Service Pipes. C. Powell Karr. Reviews the history of lead pipes, and discusses questions relating to their use. 2000 w. Met Work—Nov. 2, 1907. No. 88033.

Malleable Iron.

Malleable Cast-Iron. Bradley Stoughton. Describes the process of manufacture, the properties, uses, etc. 3800 w. Sch of Mines Qr—Nov., 1907. No. 88-531 D.

MEASUREMENT.

Efficiencies.

See Gas Plant Testing, and Gas vs. Steam, under COMBUSTION MOTORS.

Gages.

Methods of Using Those Swedish Gages. Illustrates and describes appliances for using them in limit-gage work, as inside and outside caliper gages and in jig and fixture work. 1000 w. Am Mach—Vol. 30. No. 47. No. 88497.

Hardness.

An Instrument for Testing Hardness. Albert F. Shore. Illustrated description of an instrument for determining the relative and quantitative hardness of all metals. 4500 w. Am Mach—Vol. 30. No. 46. No. 88271.

Test Bars.

Casting Test Bars in Iron. Walter J. May. Suggestions for the making. 800 w. Prac Engr—Nov. 1, 1907. No. 88224 A.

Testing Castings.

See Foundry Materials, under MACHINE WORKS AND FOUNDRIES.

Testing Machine.

An Electrically Controlled Single-Lever Testing Machine and Some Torsion Tests. C. E. Larard. Describes the electrical method of control adopted in a new testing machine. Ills. 800 w. Elect'n, Lond—Nov. 8, 1907. No. 88472 A.

Water Meters.

See same title, under CIVIL ENGINEERING, WATER SUPPLY.

POWER AND TRANSMISSION.

Air-Compressors.

High-Speed Air-Compressors. Illustrated description of a new and simple type of small air-compressor. 800 w. Engng—Nov. 15, 1907. No. 88585 A.

Air Compressors (Luftcompressoren). Hans Wunderlich. A mathematical and theoretical discussion of the design of reciprocating compressors. Ills. 1700 w. Serial. 1st part. Elektrotech Rundschau —Oct. 16, 1907. No. 88347 D.

See also Electric Driving and Turbo-Compressors, under POWER AND TRANSMISSION.

Belt Tightening.

Automatic Belt-Tightening Devices (Selbstspannende Riemengetriebe). Rudolf Hundhausen. Illustrates and describes a number of arrangements of springs and levers for tightening belts. 2000 w. Zeitschr f Werkzeug—Oct. 5, 1907. No. 88350 D.

Chain Driving.

What Has Been Done in Chain Driving. F. L. Morse. The development of the high-speed chain drive is briefly considered. Ills. 900 w. Am Mach—Vol. 30. No. 45. No. 88171.

Electric Driving.

Electrically Driven Air Compressors. Andrew Floyd Bushnell. Illustrated brief descriptions of plants in Montana. 1000 w. Eng & Min Jour—Nov. 2, 1907. No. 88055.

Electric Driving in English Spinning Mills. T. Sington. Discusses the advantages to be derived by the adoption of electricity, especially referring to England. 3500 w. Elec Wld—Nov. 9, 1907. No. 88191.

Hoisting Machinery.

See Cranes, under TRANSPORTING AND CONVEYING.

Machine Tools.

Driving Mechanism for Machine Tools. T. M. Lowthian. Illustrated description of a unique form of drive used in England. 700 w. Am Mach—Vol. 30. No. 46. No. 88272.

Horse-Power Values for Machine Tools. L. P. Alford. Considers percentages of friction loads, losses by electrical driving, and power curves. 900 w. Am Mach—Vol. 30. No. 46. No. 88273.

Producer Gas.

Power Transmission by Producer Gas. Charles E. Lucke. Considers the use feasible. 1500 w. Cassier's Mag—Nov, 1907. Special No. No. 88293 D.

Tidal Pumps.

Power from Tides. William O. Webber. Illustrates and describes a proposed plant for utilizing the power of tides by means of compressed air. 700 w. Engr, U S A—Nov. 1, 1907. No. 88040 C.

Tunneling Machinery.

See Tunnels, under CIVIL ENGINEERING, CONSTRUCTION.

Turbo-Compressors.

The General Electric Centrifugal Air

Compressor. Sanford A. Moss. Illustrated detailed description. 2500 w. Ir Age—Nov. 14, 1907. No. 88248.

Centrifugal Air Compressors for Low Pressures. Sanford A. Moss. Illustrated description, with some of the uses for which they are adapted. 1200 w. Am Mach—Vol. 30. No. 48. No. 88614.

STEAM ENGINEERING.

Accidents.
See Boiler Failures and Steam Pipes, under STEAM ENGINEERING.

Boiler Efficiency.
The Nature of True Boiler Efficiency. Walter T. Ray and Henry Kreisinger. Presents the more important results of experiments made, and gives a few laws governing the rate of heat absorption by boilers. Discussion. Ills. 14800 w. Jour W Soc of Engrs—Oct, 1907. No. 88525 D.

Boiler Failures.
Bulletin of the Accidents to Steam Apparatus during 1905 (Bulletin des Accidents d'Appareils à Vapeur survenus pendant l'Année 1905). Gives in tabular form a list of accidents, the nature of the installation, the circumstances and consequences of the accident and the probable cause. 4500 w. Ann d Ponts et Chauss —1907—III. No. 88311 E + F.

Boiler Inspection.
Regulations for the Management and Inspection of Steam Boilers (Regolamento per l'Esercizio e per la Sorveglianza delle Caldaie a Vapore). Standard rules adopted by the Italian Electrical Association. 2400 w. Serial. 1st part. Elettricita—Oct. 25, 1907. No. 88303 D.

Boiler Plant.
A Unique Boiler Plant. H. Cole Estep. Illustrated description of the Moran Co.'s plant for burning all the refuse from a Pacific-coast sawmill. 900 w. Power —Nov, 1907. No. 88012 C.

Boiler Plates.
Heat-Stresses and Formation of Cracks. Abstract translation of an article by Carl Sulzer, in the *Zeit. des Ver. Deut. Ing.* reporting investigations of a special case, with remarks on similar experience. Ills. 3800 w. Locomotive— Oct, 1907. No. 88619.

Boiler Repairs.
New Methods of Effecting Boiler Repairs. Harry Ruck-Keene. Abstract of a paper read at the Olympia Exhibition. Describes the oxy-acetylene and electric processes of effecting repairs by welding in place. Ills. 1700 w. Ir & Coal Trds Rev—Nov. 8, 1907. No. 88484 A.

Boilers.
Boiler Blow-Off Connections. R. T. Strohm. Illustrates and describes surface and bottom blow-offs. 1600 w. Elec Wld— Nov. 2, 1907. No. 88062.

The Parker Steam Generator. Report of the Franklin Institute on the invention of John C. Parker. Ills. 3000 w. Jour Fr Inst—Nov, 1907. No. 88252 D.

Boiler Scale.
Apparatus for Preventing the Formation of Coherent Scale in Boilers. Mr. Gölsdorf. Brief illustrated description. 500 w. Bul Int Ry Cong—Oct, 1907. No. 88119 E.

Coal Handling.
See same title, under TRANSPORTING AND CONVEYING.

Engines.
Some Features of Modern Corliss Engines. A. K. Spotton. Brief discussion of changes made necessary in the design of direct-connected engines. Ills. 1700 w. Trans Eng Soc, Univ of Toronto— No. 20. No. 88125 N.

See also Gas Engines, under COMBUSTION MOTORS; Steam Engines, under MARINE AND NAVAL ENGINEERING; and Hoisting Engines, under MINING AND METALLURGY, MINING.

Exhaust Steam.
See Turbines, under STEAM ENGINEERING.

Fuels.
The Combustion of Fuels. Gives tables which save much work in computations. 1200 w. Power—Nov, 1907. No. 88015 C.

Pulverized Coal and Its Industrial Applications. William D. Ennis. The first of two articles, considering the characteristics and preparation of the fuel, grinding, drying, conveying, etc, in the present number. Ills. 4000 w. Engineering Magazine. Dec, 1907. Serial. 1st part. No. 88627 B.

See also Boiler Plant, under STEAM ENGINEERING; and Oil Fuel, under MARINE AND NAVAL ENGINEERING.

Heat Radiation.
Heat Radiation Not Directly Proportional to Temperature Difference. Walter T. Ray. The practical engineering importance of Steffan and Boltzmann's radiation law is shown. 1200 w. Engr, U S A—Nov. 1, 1907. No. 88037 C.

Plant Design.
The Ratio of Heating Surface to Grate Surface as a Factor in Power Plant Design. Walter S. Finlay, Jr. Gives results of an investigation. 2000 w. Pro Am Inst of Elec Engrs—Nov, 1907. No. 88600 D.

Plant Efficiency.
See Gas vs. Steam, under COMBUSTION MOTORS.

Plants.
A Noteworthy, Small Steam Plant. J. E. Kent. Describes a small plant in Michigan which is moderate in cost,

and low in maintenance and operation expenses. 1000 w. Power—Nov., 1907. No. 88013 C.

See also Central Stations, under ELECTRICAL ENGINEERING, GENERATING STATIONS.

Steam Pipes.

The Protection of Steam Pipes from Accident. Arthur Herschmann. An illustrated study of automatic self-closing valves. 1700 w. Engineering Magazine —Dec., 1907. No. 88626 B.

Steam vs. Gas.

See Gas vs. Steam, under COMBUSTION MOTORS.

Superheating.

See Turbines, under MARINE AND NAVAL ENGINEERING; and Superheating, under RAILWAY ENGINEERING, MOTIVE POWER AND EQUIPMENT.

Thermodynamics.

The Transformation of Heat into Work. Prof. Sidney A. Reeve. Describes graphically the conversion of various forms of energy, and gives a simple introduction to the temperature—entropy diagram. 4400 w. Power—Nov, 1907. No. 88010 C.

Early History of the Mechanical Equivalent of Heat. Reviews the development of the modern theory of the mutual convertibility of heat and mechanical energy. 4500 w. Locomotive—Oct, 1907. No. 88620.

Turbines.

Steam Turbines (Turbinas de Vapor). D. Alvaro Llatas. A general theoretical and mathematical discussion of their design and operation with notes on the leading features of the leading types. Ills. 6000 w. Serial. 2 parts. Revis Tech Indus—Aug. and Sept., 1907. No. 88300, each D.

The Development of Exhaust-Steam Turbines (Développement des Turbines à Vapeur d' Echappement). A. Rateau. A full discussion of their theory, design and operation, with descriptions of recent installations of the Rateau Turbine. Ills. 11000 w. Revue de Mécanique—Oct. 31, 1907. No. 88377 E + F.

The Determination of the Efficiency of Steam Turbines Without Measurement of Steam Consumption or Power (Die Bestimmung des Wirkungsgrades von Dampfturbinen ohne Dampfverbrauchs- und Leistungsmessung). Felix Langen. Detailed description of method. Tables. 2800 w. Zeitsch f d Gesamte Turbinenwesen—Oct. 19, 1907. No. 88360 D.

See also Central Stations and Turbo-Generator Tests, under ELECTRICAL ENGINEERING, GENERATING STATIONS; and Steam Turbines, under MARINE AND NAVAL ENGINEERING.

TRANSPORTING AND CONVEYING.

Aerial Tramway.

Aerial Wire Ropeway. Illustrated description of a ropeway for conveying coal and shale across a deep valley. 600 w. Engng—Oct. 25, 1907. No. 88099 A.

Coal Handling.

An Unusual Coal and Ash-Handling Equipment for a Power Station. Illustrated detailed description of the mechanical installation for the works of Armour & Co., of Chicago. 2000 w. Eng Rec— Nov. 9, 1907. No. 88195.

See also Ore Handling, under TRANSPORTING AND CONVEYING; and Coal Pocket, under CIVIL ENGINEERING; CONSTRUCTION.

Conveyors.

See Coal Handling and Ore Handling, under TRANSPORTING AND CONVEYING; and Hoisting, under MINING AND METALLURGY, MINING.

Cranes.

Floating Cranes. Trans. from *Prometheus*. An illustrated review of dock hoisting devices. 1500 w. Sci Am Sup— Nov. 2, 1907. No. 88018.

Power Required for Cranes and Hoists. Ulrich Peters. Gives formulae and methods of calculating the power required. 2500 w. Mach, N Y—Nov, 1907. No. 88066 C.

Electric v. Hydraulic Cranes. Gives examples showing the superiority of electric driving for crane work on the score of cost. 1400 w. Elec Rev, Lond—Oct. 25, 1907. No. 88091 A.

A 150-Ton Crane at the Shipyards of John Brown at Clydebank (Grue de 150-Tonnes des Chantiers Navals John Brown à Clydebank). Illustrated detailed description. Plate. 2000 w. Génie Civil— Oct. 12, 1907. No. 88327 D.

Dock Machinery.

A Modern Ship-Loading Installation (Eine Moderne Schiffs-Verladeeinrichtung). F. Stierlin. Illustrates and describes a recent installation at the French colony of New Caledonia. 2500 w. Serial. 1st part. Schweiz Bau—Oct. 26, 1907. No. 88345 B.

Dredges.

See same title, under CIVIL ENGINEERING, WATERWAYS AND HARBORS.

Hoisting.

See same title, under MINING AND METALLURGY, MINING.

Ore Handling.

Ore Handling Plant at South Bethlehem, Pa. Illustrated description of the car dumper and other features. 2500 w. Ir Age—Nov. 14, 1907. No. 88247.

Note on the Handling and Storage of Materials in Metallurgical Works (Quel-

· *We supply copies of these articles. See page 715.*

ques Notes sur le Mouvement et l'Emmagasinement des Matières Premières dans les Usines Métallurgiques). Illustrates and describes coal and ore storage devices, conveyors, cranes, hoists, cardumpers, etc., in use in Europe. 6000 w. All Indus—Oct, 1907. No. 88322 D.

Steam Shovels.
The Development and Importance of Steam Shovels (Die Entwicklung und Bedeutung der Dampfschaufeln). R. Richter. A discussion of their design and use and a description of various types. Ills. 7500 w. Zeitschr d Ver Deutscher Ing—Oct. 26, 1907. No. 88365 D.

MISCELLANY.

Aeronautics.
The Long Distance Balloon Races from St. Louis. An illustrated account. 1500 w. Sci Am—Nov. 2, 1907. No. 88017.

A New American Aeroplane. John Miller Bonbright. The latest invention of J. W. Roshon is described. Ills. 600 w. Sci Am—Nov. 16, 1907. No. 88243.

The Zeppelin Air-Ship. Editorial discussion of the construction and performances of this balloon. 1200 w. Engng—Oct. 25, 1907. No. 88102 A.

M. Henri Farman's Progress. Brief illustrated account of his successful flights at Issy on his aeroplane. 600 w. Auto Jour—Nov. 2, 1907. No. 88220 A.

The Esnault-Pelterie Flying Machine. Illustrations of flying machine and motor, with brief notes. 600 w. Auto Jour—Nov. 2, 1907. No. 88221 A.

The Latest French Aeroplanes and Their Records. Illustrated descriptions of the Pelterie monoplane, the Bréquet gyroplane, and the flights of M. Henri Farman. 300 w. Sci Am—Nov. 16, 1907. No. 88244.

Mechanics.
The True Principles of Mechanics. Sidney A. Reeve. Discusses the use of formulæ which are erroneous approximations and the necessity of teaching the principles of the science. 2000 w. Engr, Lond—Oct. 25, 1907. No. 88107 A.

MINING AND METALLURGY

COAL AND COKE.

Coke Drawing.
Coke Drawing Machines and Other Machinery for Use at the Ovens in the Manufacture of Coke. Walter W. Macfarren. An illustrated description of the progress of coke manufacture and the machines used. Discussion. 16500 w. Pro Engrs' Soc of W Penn—Nov, 1907. No. 88526 D.

Coke Ovens.
A Battery of Belgian Type Coke Ovens. Illustrated description of a new plant near New Salem, Pa. 2500 w. Ir Trd Rev—Nov. 14, 1907. No. 88298.

Coke Production.
Coke Production in 1906. Reviews the annual report of the United States Geological Survey. 2500 w. Ir Age—Nov. 7, 1907. No. 88137.

Coking By-Products.
Recovery of Benzol from Coke Oven Gases. An illustrated article showing the benefit derived from the recovery of the benzol, etc., direct, instead of by the distillation of the tar. 1200 w. Ir & Coal Trds Rev—Oct. 25, 1907. No. 88109 A.

Italy.
The Fossil Fuels of Italy and the Lignite Workings at Ribolla and Casteani in Grosseto Province (Ueber die fossilen Brennmaterialien Italiens und die Braunkohlenwerke Ribolla und Casteani in der Provinz Grosseto). Karl Stegl. The first part gives a record of Italian production

of coal. Geological map. 2800 w. Serial. 1st part. Oest Zeitschr f Berg- u Huttenwesen—Oct. 19, 1907. No. 88338 D.

Low-Grade.
See Gas Producers, under MECHANICAL ENGINEERING, COMBUSTION MOTORS.

Montana.
Montana's Great Coalfields and Its Collieries. Floyd W. Parsons. Discusses the resources, development, transportation troubles, etc. Ills. 3000 w. Eng & Min Jour—Nov. 23, 1907. No. 88546.

Peat.
The Ziegler System of Peat Utilization. Describes the Ziegler plant at Beuerberg, in Upper Bavaria, which was opened in 1906, and its products. Ills. 4000 w. Engng—Nov. 15, 1907. No. 88580 A.

Supplies.
Our Coal Supplies. Edward W. Parker. Abstract of paper read before Am. Min. Cong. A discussion of how long the supply of coal will meet the increasing demands of commerce. 1500 w. Min Wld —Nov. 23, 1907. No. 88563.

The Fuel Situation in the Northwest. Floyd W. Parsons. Discusses the probability of another coal famine and the causes. Map. 2500 w. Eng & Min Jour —Nov. 2, 1907. No. 88058.

West Virginia.
Coal Mining in Southern West Virginia. Floyd W. Parsons. An illustrated de-

We supply copies of these articles. See page 715.

scription of this field and the methods of mining. Yields steam coal of superior quality. 2500 w. Eng & Min Jour—Nov. 9, 1907. No. 88184.

Wyoming.

The Coal-Mining Situation in Northern Wyoming. Floyd W. Parsons. Information concerning the quantity and quality of the coal, and its development. One seam shows small quantities of gold and silver. Ills. 3500 w. Eng & Min Jour—Nov. 16, 1907. No. 88404.

COPPER.

Australia.

New Copper Field in Australia. John Plummer. Brief account of the Mount Cangai field, in New South Wales. 1200 w. Min Wld—Nov. 16, 1907. No. 88422.

The Blue Bell Copper Mine. Lionel C. Ball. Gives the history, report of development, and description of the ore-bodies. 2800 w. Queens Gov Min Jour—Oct. 15, 1907. No. 88605 B.

Blast-Furnace Charging.

Furnace Charging. G. F. Beardsley. Refers to observations made at a copper plant where it became necessary to provide for the handling of an increased tonnage. 1200 w. Min & Sci Pr—Nov. 9, 1907. No. 88263.

British Columbia.

Some Ore Deposits of Similkameen District, B. C. Arthur Lakes. Illustrated description of this district and its mineralization with iron and copper ores, and gold-bearing pyrite. 1000 w. Min Wld—Nov. 16, 1907. No. 88424.

California.

The Copper Belt of California. Herbert Lang. Describes the interesting copper-bearing formation on the western flank of the Sierra Nevada range. Map. 5000 w. Eng & Min Jour—Nov. 16, 1907. Serial. 1st part. No. 88400.

Cold Rolling.

The Manufacture of Tubes and Shapes in the Cold (Fabrication à Froid des Tubes et Profiles). Pierre Breuil. Illustrates and describes the cold rolling and pressing of copper, aluminium and zinc as practiced by the Société Française de Métallurgie. 4500 w. Serial. 1st part. Génie Civil—Oct. 5, 1907. No. 88324 D.

Converters.

Modern Copper Converters, Hydraulically Operated. G. B. Shipley. Describes an installation being made at Kennet, Cal. 2200 w. Elec-Chem & Met Ind—Nov., 1907. No. 88435 C.

Extraction.

An Improved Method of Separating Copper. John A. Haralson. Diagrams and description of an apparatus for extracting metals from ore by leaching and precipitation, especially adapted for copper. 2000 w. Min Wld—Nov. 9, 1907. No. 88207.

North Carolina.

The Union Copper Mines of North Carolina. Francis C. Nicholas. An account of mines near Salisbury, N. C., where the oil separating process is being installed. 1800 w. Min Wld—Nov. 16, 1907. No. 88423.

Smelter Smoke.

Smelter Smoke, With a Discussion of Methods for Lessening Its Injurious Effects. L. S. Austin. Gives conclusions regarding the injury resulting. Ills. 3000 w. Min & Sci Pr—Nov. 23, 1907. No. 88611.

Smelter Stacks.

See Stacks, under CIVIL ENGINEERING, CONSTRUCTION.

Smelting.

See Blast Furnace Charging, under COPPER; and Copper, under ORE DRESSING AND CONCENTRATION.

Trade.

The Copper Situation. Horace J. Stevens. Reviews the past and present conditions, and thinks while the industry is sound, it will not receive a boom for a year or perhaps much longer. 2000 w. Min Jour—Oct. 26, 1907. No. 88097 A.

GOLD AND SILVER.

Assaying.

Experiments in Fire Assaying at the Redjang Lebong Mine, Sumatra. G. B. Hogenraad. Gives analyses of the ores, and an account of a new flux which is very satisfactory and cheap. General discussion. 4000 w. Jour Chem, Met, & Min Soc of S Africa. Sept, 1907. No. 88451 E.

Cobalt.

The Deloro Mining and Reduction Company. An illustrated article giving information in regard to present and past methods employed, and describing the plants for smelting Cobalt ores. 3000 w. Can Min Jour—Nov. 15, 1907. No. 88486.

Colombia.

Gold Mining in Colombia. Juan de la C. Posada. Information in regard to early and recent methods. Ills. 1200 w. Eng & Min Jour—Nov. 2, 1907. No. 88057.

Costa Rica.

Geology and Development of Aguacate Mines, Costa Rica. R. A. Crespi. Map and account of the gold mines and their development. 2000 w. Min Wld—Nov. 9, 1907. No. 88206.

Cyaniding.

History of Cyanidation. Philip Argall. Read before the Colorado Sci Soc. A review. 3500 w. Min & Sci Pr—Nov. 23, 1907. Serial. 1st part. No. 88612.

We supply copies of these articles. See page 715.

Dredging.

Dredging Beach Gravel Deposits Near Nome. John Power Hutching. Illustrates and describes methods tried and difficulties met. 4000 w. Eng & Min Jour—Nov. 23, 1907. No. 88542.

Nicaragua.

Mining in Nicaragua. A review with description of the mines and their development. 3000 w. Min Jour—Nov. 16, 1907. Serial. 1st part. No. 88578 A.

Peru.

The Gold and Silver-Bearing Veins of Huamachuco, Peru. F. Málaga Santolalla. Describes the country and the deposits. 2000 w. Min Jour—Nov. 9, 1907. No. 88475 A.

Philippines.

See Placers, under GOLD AND SILVER.

Placers.

Gold Placers of Nueva Ecija, in the Philippines. Gives the recent report of Mr. Maurice Goodman. 1500 w. Min Jour—Oct. 26, 1907. No. 88096 A.

Rand.

See same title, under CIVIL ENGINEERING, WATER SUPPLY.

Sea-Bottom.

The Presence of Gold and Silver in Deep-Sea Dredgings. Luther Wagoner. Reports assays made of deep-sea dredgings which indicated that the deep-sea bottom carries more gold than near the shore line. 300 w. Bul Am Inst of Min Engrs —Nov, 1907. No. 88636 C.

IRON AND STEEL.

Assaying.

The Determination of Tungsten in Steel in the Presence of Chromium (Ueber die Bestimmung von Wolfram im Stahl bei Gegenwart von Chrom). F. Willy Hinrichsen. On results of investigations made at the Imperial German Institute for Testing Materials. 1400 w. Stahl u Eisen —Oct. 2, 1907. No. 88332 D.

Austria.

See Steel Works, under IRON AND STEEL.

Blast-Furnace Charging.

The Charging of Blast Furnaces. E. H. Messiter. Describes the ore-bedding system designed by the writer. Ills. 4000 w. Min & Sci Pr—Oct. 26, 1907. No. 88009.

Blast-Furnace Gas.

The Utilization of the Waste Gases of Blast-Furnaces and of Coke Ovens in Metallurgical Works. Leon Greiner. Shows their importance as sources of motive power. Ills. 4000 w. Cassier's Mag—Nov., 1907. Special No. No. 88-283 D.

The Blast Furnace as a Center of Power Production. B. H. Thwaite. Considers the benefits that might be secured, compares water and gas power, and re-

lated matters of interest. Ills. 6000 w. Cassier's Mag—Nov, 1907. Special No. No. 88280 D.

Blast-Furnace Gas as the Sole Source of Power in a Modern Steelworks (Le Gaz de Haut-Fourneau comme Source Unique d'Energie dans une Usine Sidérurgique Moderne). Francois Limbourg. Gives the results obtained in an actual plant which formerly employed some steam power. 4000 w. Rev de Métal— Oct., 1907. No. 88313 E + F.

Blast-Furnace Gas as a Sole Source of Motive Power in Modern Ironworks. F. Limbourg, in the Revue de Metallurgie. Gives particulars of the consumption of gas according as gas engines are used exclusively, or in conjunction with steam engines and turbines. 1800 w. Ir & Coal Trds Rev—Nov. 1, 1907. No. 88241 A.

Blast Furnaces.

The Formation of Accretions and of Accumulations of Graphite on the Blast-Furnace Hearth (Ueber die Entstehung von Bodensauen und Graphitansammlungen in Hochofengestellen). Ills. 2300 w. Serial. 1st part. Stahl u Eisen— Oct. 16, 1907. No. 88335 D.

Blowing Engines.

Blowing-Engines Driven by Blast-Furnace Gas; Details of Cylinders. Illustrates and describes particulars of engines of Continental make. 1200 w. Engng—Oct. 25, 1907. No. 88101 A.

Copper-Clad Steel.

Manufacture and Use of Copper-Clad Steel. Concerning the process perfected by J. Ferreol Monnot and the applications. 2300 w. Eng News. Nov. 14, 1907. No. 88556.

Duplex Process.

The Duplex Steel Making Process. B. C. Lauth. Plan and description of a proposed arrangement of plant, patented by T. S. Blair, Jr. 2000 w. Ir Age—Nov. 21, 1907. No. 88508.

Electro-Metallurgy.

The Electro-Metallurgy of Iron and Steel. S. Dushman. An illustrated article, discussing the extensive use of the electric furnace. 3500 w. Trans Eng Soc, Univ of Toronto—No. 20. No. 88131 N.

Electric Induction Furnaces and Their Application in the Iron and Steel Industry (Elektrische Induktionsöfen und ihre Anwendung in der Eisen- und Stahl-Industrie). V. Engelhardt. The first part of the serial gives a historical review of furnaces built on the induction principle. Ills. 1400 w. Serial. 1st part. Elektrotech Zeitschr—Oct. 31, 1907. No. 88374 D.

See also Lash Process, under IRON AND STEEL.

Germany.

The History of the Iron Industry in

the Harz District (Ueber die Geschichte der Eisenindustrie im Harz). Herr Geyer. 4000 w. Stahl u Eisen—Oct. 2, 1907. No. 88331 D.

See also Steel Works, under IRON AND STEEL.

Hollow Ingots.

Press for the Manufacture of Hollow Ingots. The invention of H. Harmet is illustrated and described. 1000 w. Mech Engr—Nov. 9, 1907. No. 88464 A.

Ingot Casting.

The Production of Steel Ingots for Large Crank-Shafts. A. Wiecke. An illustrated article dealing with the treatment of the steel from the pouring until completely set, giving a brief description of the process of forging a shaft. 2200 w. Ir & Coal Trds Rev—Nov. 8, 1907. Serial. 1st part. No. 88485 A.

Lash Process.

The Lash Steel Process and the Electric Furnace. An examination of this process and its working. 1200 w. Elec-Chem & Met Ind—Nov., 1907. No. 88434 C.

Metallography.

The Metallography of Pig Iron (Zur Metallographie des Roheisens). E. Heyn. and O. Bauer. A record of the results of investigations made at the Imperial German Institute for Testing Materials. Ills. 3300 w. Serial. 1st part. Stahl u Eisen —Oct. 30, 1907. No. 88336 D.

Ore Handling.

See same title, under MECHANICAL ENGINEERING, TRANSPORTING AND CONVEYING.

Pig Iron.

See Metallography, under IRON AND STEEL.

Rolling Mills.

New Saucon Plant of Bethlehem Steel Co. Illustrated description of new works at Bethlehem, Pa., comprising a rail mill, structural mill, and beam mill, the latter built under the Grey patents. 5800 w. Ir Trd Rev—Nov. 7, 1907. No. 88142.

The Illinois Steel Company's New Rail Mill. Illustrated description of a recent addition to the South Chicago Works. Plate. 2500 w. Ir Age—Nov. 28, 1907. No. 88632.

Scandinavia.

The Geological Relations of the Scandinavian Iron Ores. Prof. Hjalmar Sjögren. A geological study of the ores, their distribution, origin, classification, etc. 23000 w. Bul Am Inst of Min Engrs— Nov, 1907. No. 88633 D.

Steel Works.

Austrian Steel and Engineering Works. A review of the important works, with a survey of the available raw materials. 4000 w. Engr, Lond—Nov. 15, 1907. No. 88589 A.

The Frederich-Alfred Steelworks at Rheinhausen (Die Friedrich-Alfred-Hutte zu Rheinhausen). An elaborately illustrated descriptive article on a modern German works. 7000 w. Stahl u Eisen— Oct. 9, 1907. No. 88333 D.

LEAD AND ZINC.

Joplin District.

History of Smelting in the Joplin District. Doss Brittain. An illustrated account of the development of an important industry. 5500 w. Eng & Min Jour—Nov. 9, 1907. No. 88180.

The New Sheet Ground of the Joplin District. Doss Brittain. Explains the term "sheet ore," giving the theory of the formation, and an illustrated description of methods of mining and milling. 4500 w. Min Wld—Nov. 9, 1907. No. 88204.

Lead-Mining Chronology.

Chronology of Lead-Mining in the United States. W. R. Ingalls. 5000 w. Bul Am Inst of Min Engrs—Nov, 1907. No. 88637 C.

Lead Smelting.

The Electrolytic Treatment of Galena. Edward F. Kern and Herbert S. Auerbach. Discusses the metallurgical treatment of lead sulphide ores, and reports investigations made of the electrolytic treatment. 4500 w. Sch of Mines Qr—Nov, 1907. No. 88532 D.

Milling.

See Zinc Milling, under ORE DRESSING AND CONCENTRATION.

Missouri.

Sheet-Ground Mine in Southwest Missouri. D. T. Boardman. Illustrated description of mines at Webb City. 2000 w. Eng & Min Jour—Nov. 9, 1907. No. 88,-183.

Wales.

Lead and Zinc Mines of North Wales. Edward Walker. Describes the mines, their management, production, etc. Ills. 3000 w. Eng & Min Jour—Nov. 9, 1907. No. 88182.

Zinc Assaying.

Assay of Zinc. Evans W. Buskett. Describes methods, discussing sources of error and means of guarding against them. 1500 w. Mines & Min—Nov, 1907. No. 87850 C.

Zinc Oxide.

Zinc Oxide: Its Properties and Uses. W. G. Scott. 2500 w. Min Wld—Nov. 9, 1907. Serial. 1st part. No. 88208.

Zinc Pigments.

Manufacture of Zinc Pigments. Evans W. Buskett. Description of the process used at Coffeyville, Kansas, for making zinc oxide and leaded zinc. 1400 w. Mines & Min—Nov, 1907. No. 87852 C.

We supply copies of these articles. See page 715.

Zinc Smelting.

Physical Factors in the Metallurgical Reduction of Zinc Oxide. Woolsey McA. Johnson. The reduction of zinc oxide and physical conditions affecting the efficiency are discussed. 3000 w. Bul Am Inst of Min Engrs—Sept, 1907. No. 87727 C.

Electric Zinc Smelting. Frederick T. Snyder. Describes the electric stack plant for the extraction of zinc with cost as compared with the retort plant. 1500 w. Min Rept—Oct. 24, 1907. No. 87921.

MINOR MINERALS.

Aluminium.

See same title, under ELECTRICAL ENGINEERING, TRANSMISSION.

Arsenic.

See Cobalt, under GOLD AND SILVER.

Barytes.

Geology of the Virginia Barite Deposits. Scotia. W. Spencer Hutchinson. Illustrated account of the deposits and workings. 900 w. Eng & Min Jour—Nov. 2, 1907. No. 88056.

Geology of the Virginia Barite-Deposits. Thomas Leonard Watson. Gives illustrated detailed description of the occurrence, and the conclusions deduced. 7000 w. Bul Am Inst of Min Engrs—Nov, 1907. No. 88635 C.

Cadmium.

Cadmium. Paul Speier. Its recovery, cadmium alloys and their use, and the production are considered. 3000 w. Min Jour—Nov. 9, 1907. No. 88476 A.

Magnesia Test.

Quantitative Field-Test for Magnesia in Cement Rock and Limestone. Charles Catlett. Gives a method worked out by J. J. Porter. 1700 w. Bul Am Inst of Min Engrs—Nov., 1907. No. 88634 C.

Marble.

See Quarrying, under MINING.

Oil Fields.

Relations of Oil Fields and Seismic Zones. M. L. C. Tassart. Trans. from *Comptes Rendus*. Gives results of comparisons made. 1000 w. Min Wld—Nov. 9, 1907. No. 88205.

Pyrites.

The Origin of Deposits of Pyrites. A. B. Wilmott. Deals with occurrences in Ontario. 3500 w. Can Min Jour—Nov. 1, 1907. No. 88069.

Talc.

The Henderson Talc Mine. Illustrated description of this mine at Madoc, Ontario. 1200 w. Can Min Jour—Nov. 1, 1907. No. 88067.

MINING.

Accidents.

Mining Accidents in 1906. Information from the General Report relating to mines and quarries in the United Kingdom. 4000 w. Col Guard—Oct. 25, 1907. No. 88-098 A.

Aerial Tramway.

See same title, under MECHANICAL ENGINEERING, TRANSPORTING AND CONVEYING.

Air Compressors.

See same title, under MECHANICAL ENGINEERING, POWER AND TRANSMISSION.

Apex Law.

A Broad Apex. The decision of the U. S. Supreme Court in the case arising out of litigation at Bingham, Utah. Short editorial. 5000 w. Min & Sci Pr—Nov. 9, 1907. No. 88262.

A Remedy for the Law of the Apex. James Douglas. Read before the Am. Min. Cong. Recommends a voluntary contract for neighboring mine owners to apply to their surface the common-law rule. 1200 w. Eng & Min Jour—Nov. 23, 1907. No. 88545.

Australasia.

Mining in Australasia. H. L. Wilkinson. Information concerning the present condition of the mining industry in the several states. 2000 w. Min & Sci Pr—Nov. 16, 1907. No. 88512.

Blasting.

Can an Electrically-Ignited Shot Hang Fire? William Maurice. Shows that retarded ignition may occur in the electric detonator. 1400 w. Col. Guard—Nov. 1, 1907. No. 88233 A.

Drills.

The Electric-Air Drill. William L. Saunders. Details of construction and results are given. Ills. 3000 w. Bul Am Inst of Min Engrs—Nov., 1907. No. 88-638 C.

Drilling.

See Excavation, under CIVIL ENGINEERING, CONSTRUCTION.

Electric Equipment.

An Electrically Equipped Lead Mine. Henry Floy. Illustrated detailed description of the installation of the St. Louis Smelting & Refining Co., at St. Francois, Mo. 2500 w. Eng News—Nov. 14, 1907. No. 88555.

Engineering Ethics.

To Young Men About to Become Mining Engineers. Courtenay De Kalb. Informal address before the students in the University of California, on some elements of success. 5000 w. Min & Sci Pr—Nov. 2, 1907. No. 88147.

Gas Power.

Gas Power as a Factor in Mine Economics. Archibald Burnett. Aims to show that, with slight modifications, gas engines could be used profitably in mining for all

We supply copies of these articles. See page 715.

purposes except tramming. 3800 w. Eng & Min Jour—Nov. 16, 1907. No. 88401.

Hoisting.
Hoisting in the Yellow Dog Mine Near Webb City, Mo. Doss Brittain. Illustrated description of the conveyor-hoist, operated by electric motors, which is used. 1000 w. Eng & Min Jour—Nov. 16, 1907. No. 88402.

Hoisting Engines.
Winding Engine with Tandem Drums. Illustrated description of a fine direct-acting Corliss winding engine. Plate. 800 w. Engr, Lond—Nov. 1, 1907. No. 88240 A.

Long Range vs. Short Range Trip Gears for Winding Engines. R. H. Collingham. Gives results of efforts to arrive at a solution. Favorable to the short-range trip gear. 2700 w. Engr, Lond—Nov. 8, 1907. No. 88481 A.

New Safety Devices for Steam Hoisting Engines (Neuere Sicherheitsvorrichtungen für Dampffördermaschinen). J. Iversen. An illustrated detailed description of a new system of braking. 5500 w. Oest Zeitschr f Berg- u Hüttenwesen—Oct. 5, 1907. No. 88337 D.

Ore Handling.
See same title, under MECHANICAL ENGINEERING, TRANSPORTING AND CONVEYING.

Pumping Machinery.
See same title, under MECHANICAL ENGINEERING, HYDRAULICS.

Quarrying.
The Marble Quarries of Carrara. Day Allen Willey. Illustrated description of the deposits and the methods of working. 2000 w. Sci Am—Nov. 16, 1907. No. 88245.

Shaft Boring.
The Hydraulic Shaft-Boring Machine. Translated from *Glückauf*. Illustrated description of the present form of this machine and its operation. 2000 w. Col Guard —Nov. 1, 1907. No. 88232 A.

Surveying.
Mine Surveying. C. E. Morrison. With special reference to shaft surveying. 3500 w. Sch of Mines Qr—Nov, 1907. No. 88530 D.

Timber Preservation.
Prolonging the Life of Mine Timbers. John M. Nelson, Jr. Abstract of Forest Service Circular No. 111, giving results of a series of experiments made to determine the best method of prolonging the life of mine timber. 1600 w. Min Rept—Nov. 7, 1907. No. 88203.

Tunneling.
See Tunnels, under CIVIL ENGINEERING, CONSTRUCTION.

Ventilation.
An Improved System for Ventilating Mines. George Marie Capell. Illustrated description of an improved system of fan ventilation, patented Oct. 8, 1907. 1500 w. Min Wld—Nov. 16, 1907. No. 88421.

ORE DRESSING AND CONCENTRATION.

Copper.
The Steptoe Valley Mill and Smelter. Walter Renton Ingalls. Illustrates and describes a mill of 4000-tons daily capacity, and a 1000-ton smelter. 3200 w. Min & Min Jour—Nov. 2, 1907. No. 88054.

Crushing.
Economy of Power in Crushing Ore. Ernest A. Hersam. Discusses principles that underlie economy in ore-crushing and pulverization. 7500 w. Min & Sci Pr— Nov. 16, 1907. No. 88513.

See also Tube Mills, under ORE DRESSING AND CONCENTRATION.

Elmore Process.
See North Carolina, under COPPER.

Hand-Jigs.
Jigging by Hand. Arthur C. Nahl. Illustrated description of jigs used in Mexico which were found serviceable. 1600 w. Min & Sci Pr—Nov. 2, 1907. No. 88146.

Slimes Treatment.
Slimes Agitation by Compressed Air. George G. Lyle. Illustrates and describes the Solis agitator which has given excellent results. 700 w. Min Rept—Nov. 21, 1907. No. 88561.

Tube Mills.
The Hardinge Conical Mill. H. W. Hardinge. Describes the sizing action of a conical pebble mill. 1500 w. Eng & Min Jour—Nov. 16, 1907. No. 88403.

Zinc Milling.
Improvements in Milling Missouri Zinc Ores. W. E. Ford. Discusses whether or not it would pay to treat further the tailings. 3000 w. Eng & Min Jour—Nov. 9, 1907. No. 88181.

Sludge-Mill Practice in the Missouri-Kansas District. Otto Ruhl. Describes the use of circle-tanks and of concentrating tables and the practice in saving fines. 2000 w. Min Rept—Nov. 7, 1907. No. 88202.

See also Joplin District, under LEAD AND ZINC.

MISCELLANY.

Alloys.
The Phenomena of Solidification and of Transformation in Alloys (Les Phénomènes de Solidification et de Transformation dans les Alliages). A. Portevin. An application of the phase law to a system of two bodies. Ills. 4000 w. Rev de Métal—Oct., 1907. No. 88312 E + F.

Bulgarian Law.
The Mining Law of Bulgaria. Trans-

We supply copies of these articles. See page 715.

lated from the French. Gives law as ratified Jan. 16, 1906. 2500 w. Min Jour—
Nov. 2, 1907. Serial. 1st part. No. 88231 A.

Geological Survey.
Relations of Geological Survey to Mining Industry. George Otis Smith. Read
before the Am. Min. Cong. Briefly considers the present work as a basis for
statements regarding further possibilities.
4000 w. Min Wld—Nov. 23, 1907. No.
88562.

Mine Assessments.
Iron Mine Assessments in Minnesota.

Dwight E. Woodbridge. A brief account
of the methods of the tax commission in
levying assessments, and of the results, so
far as they affect individual mine operators. 1400 w. Eng & Min Jour—Nov.
23, 1907. No. 88543.

Review.
The Metal Trades in 1906 (Das Metall-
hüttenwesen im Jahre 1906). B. Neumann. Reviews production and prices of
lead, copper, tin, zinc, gold, iron, coal, etc.,
for 1906 and of silver for 1905. 7500 w.
Serial. 2 parts. Glückauf—Oct. 12 and
19, 1907. No. 88339 each D.

RAILWAY ENGINEERING

CONDUCTING TRANSPORTATION.

Electrification.
Organization Problems in Steam-Railroad Electrification. Howard S. Knowlton. Considers the changes in methods
made necessary by the change in motive
power. 3000 w. Engineering Magazine—
Dec., 1907. No. 88623 B.

Signalling.
British Practice with Distant Signals.
H. Raynar Wilson. Describes present
practice. 2000 w. Ry. & Engng Rev—Nov.
16, 1907. No. 88417.

Electric Signalling Installations at
Euston and Crewe Stations. Illustrated
detailed description. 2000 w. Elect'n,
Lond—Oct. 25, 1907. Serial. 1st part. No.
88092 A.

Audible Distant Signals on the Great
Western. Illustrated description of the
apparatus and working of the installation
on this English railway. 1500 w. R R
Gaz—Nov. 15, 1907. No. 88274.

Automatic Block Signals on the Philadelphia & Western. A description of the
apparatus installed. 2000 w. R R Gaz—
Nov. 22, 1907. No. 88551.

The Philadelphia & Western Railroad's
Automatic Electric Block Signal System.
Illustrated detailed description. 2200 w.
St Ry Jour—Nov. 2, 1907. No. 88072.

Note on the New Iron Posts Used by
the Belgian State Railway for Wire
Transmissions for Operating Signals. L.
Weissenbruch and L. Kirsch. Illustrated
description. 800 w. Bul Int Ry Cong—
Oct, 1907. No. 88118 E.

Train Lighting.
See same title, under MOTIVE POWER
AND EQUIPMENT.

MOTIVE POWER AND EQUIPMENT.

Brakes.
The Vacuum Brake and Slack Brake
Gear. An account of valuable experiments

on the effect of slackness in brake-rigging.
1800 w. Engng—Oct. 25, 1907. No. 88-
103 A.

Brake Valves.
The H 6 Brake Valve. G. W. Kiehm.
Detailed description. 1200 w. Ry & Loc
Engng—Nov., 1907. No. 88003 C.

Car Design.
Transportation by Rail. Abstract of an
address by J. D. Twinberrow discussing
some of the points in locomotive and car
design now occupying the attention of
railway engineers. 3300 w. Engr, Lond
—Nov. 15, 1907. No. 88590 A.

Construction Cars.
New Outfit Cars on the Atchison, Topeka & Santa Fe. Illustrated detailed description. 1000 w. Ry Age—Nov. 15,
1907. No. 88420.

Locomotive Axles.
The Determination of the Loads on
Locomotive Axles (Ueber die Bestimmung der Achsbelastungen bei Lokomotiven). A. Kutschera. A mathematical
and theoretical discussion of methods of
computing axle loads. Ills. 3200 w. Serial. 1st part. Zeitschr d Oest Ing u
Arch Ver—Oct. 11, 1907. No. 88357 D.

Locomotive Design.
See Car Design, under MOTIVE POWER
AND EQUIPMENT.

Locomotive Driving.
Locomotive Engine-Driving. A detailed
account of how a locomotive is driven.
1200 w. Engr, Lond—Nov. 8, 1907. No.
88482 A.

Locomotive Failures.
Engine Failures. J. F. Whiteford. A
discussion of what constitutes a failure,
methods of reporting and comparing, and
for improving conditions. 2000 w. Am
Engr & R R Jour—Nov., 1907. No.
88042 C.

Locomotive Maintenance.
The Relation Between the Condition of

We supply copies of these articles, See page 715.

Motive Power and Its Repair. Clive Hastings. Endeavors to show the advantage of a systematic method of running and repairing engines. 2200 w. R R Gaz—Nov. 15, 1907. No. 88277.

Locomotives.

Rogers 4-6-2 A. & W. P. Illustrated description of a Pacific type engine. 700 w. Ry & Loc Engng—Nov., 1907. No. 88004 C.

Oldest Canadian Locomotive. Photograph and description of the "Samson," built in 1837. 800 w. Ry & Loc Engng—Nov., 1907. No. 88001 C.

Ten-Wheel Locomotive with Superheater for the Canadian Pacific. Illustrated description of engine with Vaughan superheater. 800 w. R R Gaz—Nov. 22, 1907. No. 88550.

Decapod Locomotive with Combination Chamber. Illustrates and describes engines recently built for the Buffalo, Rochester & Pittsburgh Ry. 700 w. Ry Mas Mech—Nov., 1907. No. 88297.

Twelve-Wheel Freight Locomotive. Illustrated description of 4-8-0 type locomotives for the Norfolk & Western Ry. 900 w. Am Engr & R R Jour—Nov., 1907. No. 88044 C.

Pacific Type Locomotive. Illustrated description of an engine for heavy passenger service on the New York, New Haven, & Hartford. 900 w. Am Engr & R R Jour—Nov., 1907. No. 88043 C.

Pacific Locomotives for the New York, New Haven & Hartford. Illustrates and describes these 4-6-2 engines, the first with trailing wheels to be built for this road. 1200 w. R R Gaz—Nov. 8, 1907. No. 88153.

Balanced Compound Atlantic Locomotive for the Chicago, Milwaukee & St. Paul. Illustrated description. 700 w. R R Gaz—Nov. 15, 1907. No. 88278.

New Locomotives for New York, Chicago, & St. Louis. Illustrates and describes types for passenger, freight, and switching service. 1300 w. Ry Age—Nov. 15, 1907. No. 88419.

Four-Wheel Switching Locomotive. Illustrated description of an engine for use in the yards of a manufacturing company. 500 w. R R Gaz—Nov. 8, 1907. No. 88156.

Eight-Coupled Mineral Locomotives; Hull and Barnsley Railway. Illustrated description. 600 w. Engng—Nov. 15, 1907. No. 88582 A.

Locomotive Stokers.

The Crosby Locomotive Mechanical Stoker. Illustrated detailed description. 2000 w. Ry Mas Mech—Nov., 1907. No. 88296.

The Hayden Locomotive Stoker. Illustrated description of a stoker used on the Erie R. R. 800 w. Ry Age—Nov. 8, 1907. No. 88209.

Locomotive Tests.

Tests of the Resistance of a 3/3 Coupled Locomotive (Untersuchung der Widerstand einer 3/3-gekuppelten Lokomotive). R. Sanzin. Describes the method and results. Ills. 4000 w. Zeitschr d Ver Deutscher Ing—Oct. 26, 1907. No. 88-366 D.

Steam vs. Electricity.

Comparative Performance of Steam and Electric Locomotives. Albert H. Armstrong. A study of steam and electric locomotives with technical comparison of their performance, showing reasons for the electrification of steam lines. 7500 w. Pro Am Inst of Elec Engrs—Oct., 1907. No. 88253 D.

Discussion, by J. E. Muhlfeld, of the paper by Max Toltz, "Steam Locomotives *versus* Electric Locomotives." 3000 w. Pro N Y R R Club—Oct. 18, 1907. No. 88265.

Superheaters.

The First Steam Superheaters. Charles R. King. Illustrated descriptions of early types. 2000 w. R R Gaz—Nov. 1, 1907. No. 88027.

Superheating.

The Application of Highly Superheated Steam to Locomotives. Robert Garbe. This first article of a series considers the advantages of superheated steam, size of superheaters, hauling capacity of superheated steam locomotives, etc. 3500 w. Engr, Lond—Oct. 25, 1907. Serial. 1st part. No. 88105 A.

Tires.

Steel Tires—Causes of Defects and Failures. George L. Norris. Discusses defects in steel, conditions of service, etc. Ills. General discussion. 8800 w. Pro W Ry Club—Oct. 15, 1907. No. 88219 C.

Train Lighting.

Methods of Electric Lighting for Railway Trains. Dugald C. Jackson. Gives a summary of results and data obtained in a series of tests of different apparatus. Discussion. 12500 w. Jour W Soc of Engrs—Oct., 1907. No. 88524 D.

NEW PROJECTS.

Construction.

See Construction Cars, under MOTIVE POWER AND EQUIPMENT; and Excavation, under CIVIL ENGINEERING, CONSTRUCTION.

Grading Contracts.

No-Classification Contracts for Railroad Graduation. F. Lavis. An account of a case where it was necessary to resort to arbitration to settle differences in regard to a contract on the no-classification basis. 3500 w. Engng-Con—Nov. 6, 1907. No. 88150.

We supply copies of these articles. See page 715.

Nevada Northern.
The Nevada Northern Railway. Illustrations, with brief description of a new line of importance in the development of the copper-mining industry of Nevada. 800 w. Ry Age—Nov. 22, 1907. No. 88557.

Surveying.
Railway Surveys. Andrew F. Macallum. Notes and suggestions for the work. 4000 w. Trans Eng Soc, Univ of Toronto —No. 20. No. 88129 N.

PERMANENT WAY AND BUILDINGS.

Car Dumping.
See Ore Handling, under MECHANICAL ENGINEERING, TRANSPORTING AND CONVEYING.

Construction.
See Reconstruction, under PERMANENT WAY; and Construction, under NEW PROJECTS.

Earth Slides.
Earth Slides on Railway Work. H. Rohwer. Abstract of a paper in Bul. No. 90, of the Am. Ry. Engng. & Main. of Way Assn. 1500 w. Eng News—Nov. 21, 1907. No. 88507.

Maintenance of Way.
Why Efficient Track Work is Skilled Labor. W. M. Camp. Read at convention of Roadmasters & Main. of Way Assn. Presents arguments to show the need of skill in the work. 2500 w. Ry & Engng Rev—Nov. 16, 1907. No. 88418.

Oil House.
A Modern Oil House. Illustrated description of a building for the Great Northern Ry. at St. Paul, Minn. 2000 w. Ry Mas Mech—Nov., 1907. No. 88295.

Rails.
Heavy-Rail Sections in America. Editorial discussion of the causes of the large number of rail failures and the efforts being made to improve the rails. 4000 w. Engng—Nov. 15, 1907. No. 88584 A.

Rail Stresses.
Lateral Thrust of Car Wheels Against the Rail. George L. Fowler. From advance sheets of a report made to the Schoen Steel Wheel Co. Illustrated account of tests made. 3500 w. R R Gaz —Nov. 15, 1907. No. 88276.

Reconstruction.
Double-Track Work Through Eagle River Canyon, Denver & Rio Grande R. R. Brief illustrated description. 1200 w. Eng Rec—Nov. 16, 1907. No. 88406.

The Rex-Red Cliff Double Track Construction on the Denver & Rio Grande R. R. Illustrated description of interesting details of construction. 1200 w. Eng News—Nov. 21, 1907. No. 88502.

Six-Tracking and Reconstruction of the Harlem River Branch of the New York, New Haven and Hartford R. R. Illus-

trated description of improvements in progress. 3500 w. Eng Rec—Nov. 9, 1907. No. 88194.

Roundhouses.
An Important Development in Roundhouse Design and Operation. Illustrates and describes improvements in design and operation introduced at Dilworth, Minn., on the Northern Pacific Ry. 1200 w. Am Engr & R R Jour—Nov., 1907. No. 88041 C.

Signalling.
See same title, under CONDUCTING TRANSPORTATION.

Timber Preservation.
See same title, under CIVIL ENGINEERING, MATERIALS OF CONSTRUCTION.

Tunnels.
See same title, under CIVIL ENGINEERING, CONSTRUCTION.

TRAFFIC.

Car Efficiency.
Commissioner Clark on the Freight Car Situation. Hon. E. E. Clark's address at Chicago, Oct. 25. 2000 w. R R Gaz— Nov. 8, 1907. No. 88154.

Explosives.
The Bureau of Explosives. From a report of the committee on transportation of explosives to the American Ry. Assn., Oct. 30, 1907. 8500 w. R R Gaz—Nov. 8, 1907. No. 88157.

Refrigeration.
The Present Status of Mechanical Refrigeration in Railroad Work. Jos. H. Hart. Discusses present conditions, the reasons, and the application of mechanical refrigeration in this field. 2000 w. R R Gaz—Nov. 22, 1907. No. 88549.

Mechanical Refrigeration Adapted to Railway Transportation. Dr. Joseph H. Hart. Reviews refrigerating systems in use, and explains the economy of large plants charging brine-cooled cars. 3000 w. Engineering Magazine—Dec., 1907. No. 88629 B.

MISCELLANY.

American Railways.
Some Phases of the American Railroad Problem. Stuyvesant Fish. Address at Columbia Univ. Reviews the conditions under which early railways were constructed and their history. 5500 w. Sch of Mines Qr—Nov., 1907. No. 88529 D.

European Railways.
Observations on European Railroads. A. T. Perkins. On the methods, appliances, and facilities of foreign railways. Ills. Discussion. 6500 w. Pro St Louis Ry Club—Oct. 11, 1907. No. 88008.

French Railways.
French Railway Economics. Editorial review of conditions in France and the *personnel* of the principal companies. 2200 w. Engng—Nov. 15, 1907. No. 88583 A.

STREET AND ELECTRIC RAILWAYS

Accounting.

Electric Tramway Accounting and Finance. The first of a series of articles considering the general characteristics of such accounts. 2200 w. Elec Rev, Lond —Nov. 1, 1907. Serial. 1st part. No. 88228 A.

Brakes.

The Determination of the Correct Braking Power to Be Applied to Electric Cars and Locomotives. H. M. Prevost Murphy. Develops formulæ and illustrates their use. 1600 w. Elec Ry Rev—Nov. 23, 1907. No. 88560.

Foundation Brake Gear Design for Electric Railway Cars. Fred Heckler. Read before the Cent. Elec. Ry. Assn. Discusses the brake problem, mechanical requirements, etc. 2000 w. Elec Ry Rev —Nov. 23, 1907. No. 88558.

Foundation Brake-Gear Design for Electric Railway Cars. Fred Heckler. Read at meeting of Cent. Elec. Ry. Assn. Discusses the proper proportion between air pressure, cylinder piston area and leverage, the mechanical requirements, etc. Ills. 3500 w. St Ry Jour—Nov. 30, 1907. No. 88670.

Brooklyn.

The Livingston Street Railway Improvement for Relieving Brooklyn Rush-Hour Traffic. An account of arrangements for diverting some of the cars from Fulton St. between the Borough Hall and Flatbush Ave. 1500 w. St Ry Jour—Nov. 16, 1907. No. 88416.

Cars.

High-Speed Cars for the Aurora, Elgin & Chicago Railroad. Illustrated description of well-built passenger cars. 700 w. Elec Ry Rev—Nov. 16, 1907. No. 88436.

Conduit System.

The "Robrow" Conduit System as Adopted for Benares Tramways. A proposed installation in India of a shallow conduit system with surface collection is illustrated and described. 3000 w. Tram & Ry Wld—Nov. 7, 1907. No. 88473 B.

Electric Tramways for Benares. Illustrated description of the Robrow shallow-conduit system adopted. 2500 w. Elec Engr, Lond—Nov. 22, 1907. No. 88701 A.

Controller.

The Electrical Controller. W. B. Kouwenhoven. An explanation of its operation. 2800 w. Ry & Loc Engng—Dec., 1907. No. 88683 C.

Electrification.

The Fort Dodge, Des Moines & Southern Railway of Iowa. Illustrates and describes the electrification of a steam road,

with new interurban construction. 1600 w. St Ry Jour—Nov. 30, 1907. No. 88669. See also same title, under RAILWAY ENGINEERING, CONDUCTING TRANSPORTATION.

Europe.

Alternating-Current Electric Traction on European Railways (La Traction Electrique par Courant Alternatif Simple sur les Chemins de Fer en Europe). M. Henry. The first part of the serial discusses the Siemens-Schuckert single-phase system. Ills. Serial. 1st part. 2800 w. Electricien—Oct. 12, 1907. No. 88320 D.

Fares.

Interurban Fares. Theodore Stebbins. Read before the Am. St. & Int. Ry. Assn. Discusses the conditions that should form a basis for fixing the rates. 3800 w. R R Gaz—Nov. 15, 1907. No. 88275.

Fire Car.

Fire Car for the South Side Elevated Railroad. Illustrated description. 600 w. Elec Ry Rev—Nov. 9, 1907. No. 88187.

Germany.

The Munster-Schlucht Electric Railway. E. Ommeganck. Illustrates and describes a combined rack and adhesion scheme. 1300 w. R R Gaz—Nov. 29, 1907. No. 88647.

Interurban.

Interurban Railways. Hugh J. M'Gowan. Abstract of address before the Nat. Business League of America. Considers the present and future development. 1600 w. Elec Ry Rev—Nov. 30, 1907. No. 88675.

The Indianapolis & Louisville 1,200-Volt Railway. Illustrated description of this high-speed interurban line. 2500 w. Elec Ry Rev—Nov. 30, 1907. No. 88674.

Legislation.

Electric Railway and Public Service Legislation. Henry J. Pierce. From an address at Madison Square Garden, New York City, Oct. 1, 1907. 2000 w. R R Gaz—Nov. 29, 1907. No. 88645.

Locomotives.

Electric Locomotives for the Pennsylvania Railroad. Illustrates and describes the engines being tested on the West Jersey & Seashore division with a view to determining the track stresses and other questions in heavy electric traction. 1000 w. R R Gaz—Nov. 22, 1907. No. 88552. See also Steam vs. Electricity, under RAILWAY ENGINEERING, MOTIVE POWER AND EQUIPMENT.

Maintenance.

Analysis of the Cost and Methods of Electric Railway Maintenance. Albert

We supply copies of these articles. See page 715.

Herrick. Abstract of paper read before the Cent. Elec. Ry. Assn. Considers causes that lead to high maintenance costs. 3000 w. Elec Ry Rev—Nov. 23, 1907. No. 88559.

Motor Cars.
The Single-Phase Thomson-Houston Motor Cars (Automotrici A. E. G. Thomson-Houston per Trazione Monofase). Illustrated detailed description of the motor cars in use on the Blankenese-Ohlsdorf line in Germany. 2000 w. Serial. 1st part. Elettricita—Oct. 11, 1907. No. 88-302 D.

Motors.
See Railway Motors, under ELECTRICAL ENGINEERING, DYNAMOS AND MOTORS.

Motor Suspension.
The Suspension of Motors on Tramcars. Brief discussion of the unsatisfactory methods in use, and the remedy. 900 w. Elec Engr, Lond—Oct. 25, 1907. No. 88089 A.

Power Stations.
See Central Stations and Hydro-Electric, under ELECTRICAL ENGINEERING, GENERATING STATIONS.

Quebec.
The Montmorency Division of the Quebec Railway, Light & Power Company. Illustrated description of a combined steam and electric line. 2500 w. St Ry Jour—Nov. 16, 1907. No. 88415.

Rack-Rail.
See Germany, under STREET AND ELECTRIC RAILWAYS.

Rail Bonding.
Contact Resistance in Connection with Rail Bonding. A study made with special reference to compression bonds. 1500 w. Jour Worcester Poly Inst—Nov., 1907. No. 88260 C.

Rail Corrugation.
Report on Rail Corrugation, Presented at the Eleventh Annual Meeting of the German Street and Interurban Railway Association. Ills. 3500 w. St Ry Jour —Nov. 9, 1907. No. 88201.

Corrugations on the Upper Surface of Rail Heads. Abstract report of a committee presented to the German Tramways & Light Rys. Assn. 2500 w. Elect'n, Lond—Nov. 15, 1907. No. 88573 A.

Track Construction and Corrugation. An illustrated description of experiences in Boston and Philadelphia. 4000 w. Tram & Ry Wld—Oct. 3, 1907. No. 87824 B.

The Hardness of Corrugated Rails. George L. Fowler. A report of investigations made which eliminate variations in hardness as a cause of rail corrugation. 1200 w. St Ry Jour—Oct. 5, 1907. No. 87493.

Rail Corrugation. A. L. C. Fell. Abstract of a report to the London County Council on the possible causes and suggesting remedies. 4200 w. Elec Engr, Lond—Oct. 4, 1907. No. 87586 A.

Storage Yard.
The New Sullivan Square Elevated Storage Yard of the Boston Elevated Railway. Plan, sections, and description. 1200 w. St Ry Jour—Nov. 23, 1907. No. 88535.

Sub-Stations.
Portable Sub-Station for Rochester and Sodus Bay Railway. B. C. Amesbury. Illustrated description of a sub-station for use at any points on the line which may temporarily require special power. 900 w. St Ry Jour—Nov. 2, 1907. No. 88071.

See also same title, under ELECTRICAL ENGINEERING, GENERATING STATIONS.

Switzerland.
Electric Traction in Switzerland During 1907 (La Trazione Elettrica in Isvizzera nell' Anno 1907). S. Herzog. A review of progress and a discussion of new projects. 1700 w. Serial. 1st part. Industria—Oct. 13, 1907. No. 88304 D.

Terre Haute, Ind.
Recent Interurban Improvements at Terre Haute. An illustrated account of improvements in power supply and service and in local lignting. 2000 w. Elec Ry Rev—Nov. 9, 1907. No. 88186.

Third Rail.
Standard Location for Third Rail Conductors. Abstract of committee report presented at the Oct. meeting of the Am. Ry. Assn. Ills. 1200 w. R R Gaz—Nov. 8, 1907. No. 88155.

Track Construction.
Some Recent Street Railway Track Construction. Describes recent methods of construction in Chicago, and in Indianapolis, Kansas City and Toronto. Ills. 2500 w. Eng News—Nov. 7, 1907. No. 88111.

Track Construction in Some Southern Cities. Illustrates and describes recent work in San Antonio, Houston, and Chattanooga. 1500 w. St Ry Jour—Nov. 23, 1907. No. 88534.

Vienna-Baden.
The Vienna-Baden Electric Railroad (Die elektrische Bahn Wien-Baden). Leo Kadrnozka. The first part of the serial describes the location and profile of the line. Ills. 2500 w. Serial. 1st part. Elektrotech u Maschinenbau—Oct. 20, 1907. No. 88351 D.

Wales.
Llandudno and Colwyn Bay Tramways. Illustrates and describes a line in Wales to connect sea-side resorts and some interurban territory. 1400 w. Tram & Ry Wld—Nov. 7, 1907. No. 88474 B.

EXPLANATORY NOTE—THE ENGINEERING INDEX.

We hold ourselves ready to supply—usually by return of post—the full text of every article indexed in the preceding pages, *in the original language*, together with all accompanying illustrations; and our charge in each case is regulated by the cost of a single copy of the journal in which the article is published. The price of each article is indicated by the letter following the number. When no letter appears, the price of the article is 20 cts. The letter A, B, or C denotes a price of 40 cts.; D, of 60 cts.; E, of 80 cts.; F, of $1.00; G, of $1.20; H, of $1.60. When the letter N is used it indicates that copies are not readily obtainable and that particulars as to price will be supplied on application. Certain journals, however, make large extra charges for back numbers. In such cases we may have to increase proportionately the normal charge given in the Index. In ordering, care should be taken to *give the number* of the article desired, not the title alone.

Serial publications are indexed on the appearance of the first installment.

SPECIAL NOTICE.—To avoid the inconvenience of letter-writing and small remittances, especially from foreign countries, and to cheapen the cost of articles to those who order frequently, we sell coupons at the following prices:—20 cts. each or a book of twelve for $2.00; three books for $5.00.

Each coupon will be received by us in payment for any 20-cent article catalogued in the Index. For articles of a higher price, one of these coupons will be received for each 20 cents; thus, a 40-cent article will require two coupons; a 60-cent article, three coupons; and so on. The use of these coupons is strongly commended to our readers. They not only reduce the cost of articles 25 per cent. (from 20c. to 15c.), but they need only a trial to demonstrate their very great convenience—especially to engineers in foreign countries, or away from libraries and technical club facilities.

Write for a sample coupon—free to any part of the world.

CARD INDEX.—These pages are issued separately from the Magazine, printed on one side of the paper only, and in this form they meet the exact requirements of those who desire to clip the items for card-index purposes. Thus printed they are supplied to regular subscribers of THE ENGINEERING MAGAZINE at 10 cents per month, or $1.00 a year; to non-subscribers, 25 cts. per month, or $3.00 a year.

THE PUBLICATIONS REGULARLY REVIEWED AND INDEXED.

The titles and addresses of the journals regularly reviewed are given here in full, but only abbreviated titles are used in the Index. In the list below, *w* indicates a weekly publication, *b-w*, a bi-weekly, *s-w*, a semi-weekly, *m*, a monthly, *b-m*, a bi-monthly, *t-m*, a tri-monthly, *qr*, a quarterly, *s-q*, semi-quarterly, etc. Other abbreviations used in the index are: Ill—Illustrated: W—Words; Anon—Anonymous.

Alliance Industrielle. *m.* Brussels.
American Architect. *w.* New York.
Am. Engineer and R. R. Journal. *m.* New York.
American Jl. of Science. *m.* New Haven, U. S. A.
American Machinist. *w.* New York.
Annales des Ponts et Chaussées. *m.* Paris.
Ann. d Soc. Ing. e d Arch. Ital. *w.* Rome.
Architect. *w.* London.
Architectural Record. *m.* New York.
Architectural Review. *s-q.* Boston.
Architect's and Builder's Magazine. *m.* New York.
Australian Mining Standard. *w.* Melbourne.
Autocar. *w.* Coventry, England.
Automobile. *w.* New York.
Automotor Journal. *w.* London.
Beton und Eisen. *qr.* Vienna.
Boiler Maker. *m.* New York.
Brass World. *m.* Bridgeport, Conn.
Brit. Columbia Mining Rec. *m.* Victoria, B. C.
Builder. *w.* London.
Bull. Bur. of Standards. *qr.* Washington.

Bulletin de la Société d'Encouragement. *m.* Paris.
Bulletin of Dept. of Labor. *b-m.* Washington.
Bull. Soc. Int. d'Electriciens. *m.* Paris.
Bulletin of the Univ. of Wis., Madison, U. S. A.
Bulletin Univ. of Kansas. *b-m.* Lawrence.
Bull. Int. Railway Congress. *m.* Brussels.
Bull. Tech. de la Suisse Romande. *s-m.* Lausanne.
California Jour. of Tech. *m.* Berkeley, Cal.
Canadian Architect. *m.* Toronto.
Canadian Electrical News. *m.* Toronto.
Canadian Engineer. *m.* Toronto and Montreal.
Canadian Mining Journal. *b-w.* Toronto.
Cassier's Magazine. *m.* New York and London.
Cement. *m.* New York.
Cement Age. *m.* New York.
Central Station. *m.* New York.
Chem. Met. Soc. of S. Africa. *m.* Johannesburg.
Colliery Guardian. *w.* London.
Compressed Air. *m.* New York.
Comptes Rendus de l'Acad. des Sciences. *w.* Paris.
Consular Reports. *m.* Washington.

Deutsche Bauzeitung. *b-w.* Berlin.
Domestic Engineering. *w.* Chicago.
Economic Geology. *m.* New Haven, Conn.
Electrical Age. *m.* New York.
Electrical Engineer. *w.* London.
Electrical Review. *w.* London.
Electrical Review. *w.* New York.
Electric Journal. *m.* Pittsburg, Pa.
Electric Railway Review. *w.* Chicago.
Electrical World. *w.* New York.
Electrician. *w.* London.
Electricien. *w.* Paris.
Electrochemical and Met. Industry. *m.* N. Y.
Elektrochemische Zeitschrift. *m.* Berlin.
Elektrotechnik u Maschinenbau. *w.* Vienna.
Elektrotechnische Rundschau. *w.* Potsdam.
Elektrotechnische Zeitschrift. *w.* Berlin.
Elettricità. *w.* Milan.
Engineer. *w.* London.
Engineer. *s-m.* Chicago.
Engineering. *w.* London.
Engineering-Contracting. *w.* New York.
Engineering Magazine. *m,* New York and London.
Engineering and Mining Journal. *w.* New York.
Engineering News. *w.* New York.
Engineering Record. *w.* New York.
Eng. Soc. of Western Penna. *m.* Pittsburg, U. S. A.
Far Eastern Review. *m.* Manila, P. I.
Fire and Water. *w.* New York.
Foundry. *m.* Cleveland, U. S. A.
Génie Civil. *w.* Paris.
Gesundheits-Ingenieur. *s-m.* München.
Giorn. dei Lav. Pubb. e d Str. Ferr. *w.* Rome.
Glaser's Ann. f Gewerbe & Bauwesen. *s-m.* Berlin.
Heating and Ventilating Mag. *m.* New York.
Ice and Refrigeration. *m.* New York.
Industrial World. *w.* Pittsburg.
Ingenieria. *b-m.* Buenos Ayres.
Ingenieur. *w.* Hague.
Insurance Engineering. *m.* New York.
Int. Marine Engineering. *m.* New York.
Iron Age. *w.* New York.
Iron and Coal Trades Review. *w.* London.
Iron and Steel Trades Journal *w.* London.
Iron Trade Review. *w.* Cleveland, U. S. A.
Jour. of Accountancy. *m.* N. Y.
Journal Asso. Eng. Societies. *m.* Philadelphia.
Journal of Electricity. *m.* San Francisco.
Journal Franklin Institute. *m.* Philadelphia.
Journal Royal Inst. of Brit. Arch. *s-qr.* London.
Jour. Roy. United Service Inst. *m.* London.
Journal of Sanitary Institute. *qr.* London.
Jour. of South African Assn. of Engineers. *m.* Johannesburg, S. A.
Journal of the Society of Arts. *w.* London.
Jour. Transvaal Inst. of Mech. Engrs., Johannesburg, S. A.
Jour. of U. S. Artillery. *b-m.* Fort Monroe, U. S. A.
Jour. W. of Scot. Iron & Steel Inst. *m.* Glasgow.
Journal Western Soc. of Eng. *b-m.* Chicago.
Journal of Worcester Poly. Inst., Worcester, U. S. A.
Locomotive. *m.* Hartford, U. S. A.
Machinery. *m.* New York.
Madrid Cientifico. *t-m.* Madrid.
Manufacturer's Record. *w.* Baltimore.
Marine Review. *w.* Cleveland, U. S. A.
Men. de la Soc. des Ing. Civils de France. *m.* Paris.
Métallurgie. *w.* Paris.
Minero Mexicano. *w.* City of Mexico.

Mines and Minerals. *m.* Scranton, U. S. A.
Mining and Sci. Press. *w.* San Francisco.
Mining Journal. *w.* London.
Mining Reporter. *w.* Denver, U. S. A.
Mittheilungen des Vereines für die Förderung des Local und Strassenbahnwesens. *m.* Vienna.
Municipal Engineering. *m.* Indianapolis, U. S. A.
Municipal Journal and Engineer. *w.* New York.
Nature. *w.* London.
Nautical Gazette. *w.* New York.
New Zealand Mines Record. *m.* Wellington.
Oest. Wochenschr. f. d. Oeff. Baudienst. *w.* Vienna.
Oest. Zeitschr. Berg & Hüttenwesen. *w.* Vienna.
Plumber and Decorator. *m.* London.
Popular Science Monthly. *m.* New York.
Power. *m.* New York.
Practical Engineer. *w.* London.
Pro. Am. Soc. Civil Engineers. *m.* New York.
Pro. Canadian Soc. Civ. Engrs. *m.* Montreal.
Proceedings Engineers' Club. *qr.* Philadelphia.
Pro. St. Louis R'way Club. *m.* St. Louis, U. S. A.
Pro. U. S. Naval Inst. *qr.* Annapolis, Md.
Quarry *m.* London.
Queensland Gov. Mining Jour. *m.* Brisbane, Australia.
Railroad Gazette. *w.* New York.
Railway Age. *w.* Chicago.
Railway & Engineering Review. *w.* Chicago.
Railway and Loc. Engng. *m.* New York.
Railway Master Mechanic. *m.* Chicago.
Revista d Obras. Pub. *w.* Madrid.
Revista Tech. Ind. *m.* Barcelona.
Revue de Mécanique. *m.* Paris.
Revue de Métallurgie. *m.* Paris.
Revue Gén. des Chemins de Fer. *m.-* Paris.
Revue Gén. des Sciences. *w.* Paris.
Revue Industrielle. *w.* Paris.
Revue Technique. *b-m.* Paris.
Rivista Gen. d Ferrovie. *w.* Florence.
Rivista Marittima. *m.* Rome.
Schiffbau. *s-m.* Berlin.
Schweizerische Bauzeitung. *w.* Zürich.
Scientific American. *w.* New York.
Scientific Am. Supplement. *w.* New York.
Sibley Jour. of Mech. Eng. *m.* Ithaca, N. Y.
Soc. Belge des Elect'ns. *m.* Brussels.
Stahl und Eisen. *w.* Düsseldorf.
Stevens Institute Indicator. *qr.* Hoboken, U. S. A.
Street Railway Journal. *w.* New York.
Technograph. *yr.* Urbana, Ill.
Technology Quarterly. *qr.* Boston, U. S. A.
Tijds v h Kljk. Inst. v Ing. *qr.* Hague.
Tramway & Railway World. *m.* London.
Trans. Am. Ins. Electrical Eng. *m.* New York.
Trans. Am. Ins. of Mining Eng. New York.
Trans. Am. Soc. Mech. Engineers. New York.
Trans. Inst. of Engrs. & Shipbuilders in Scotland, Glasgow.
Transport. *w.* London.
Verkehrstechnische Woche und Eisenbahntechnische Zeitschrift. *w.* Berlin.
Wood Craft. *m.* Cleveland, U. S. A.
Yacht. *w.* Paris.
Zeitschr. f. d. Gesamte Turbinenwesen. *w.* Munich.
Zeitschr. d. Mitteleurop. Motorwagon Ver. *s-m.* Berlin.
Zeitschr. d. Oest. Ing. u. Arch. Ver. *w.* Vienna.
Zeitschr. d. Ver. Deutscher Ing. *w.* Berlin.
Zeitschrift für Elektrochemie. *w.* Halle a S.
Zeitschr. f. Werkzeugmaschinen. *b-w.* Berlin

NOTE—*Our readers may order through us any book here mentioned, remitting the publisher's price as given in each notice. Checks, Drafts, and Post Office Orders, home and foreign, should be made payable to* THE ENGINEERING MAGAZINE.

Blacksmithing.

The Blacksmith's Guide. By J. F. Sallows. Size, 7 by 4½ in.; pp. 160. Ills., 165. Price, cloth, $1.50; leather, $2. Brattleboro, Vt.: The Technical Press.

A book for blacksmiths, machinists and toolmakers, containing instructions on forging, welding, hardening, tempering, treatment of high-speed steel, case hardening, annealing, coloring, brazing and general blacksmithing. The processes outlined are those used by the author during his twenty-seven years' experience as a blacksmith in nearly all kinds of shops, and they are of proved value as time and money savers. A great deal of valuable practical information is contained in a small compass and the book can be confidently recommended both to foremen blacksmiths and to the young mechanic anxious to improve himself in his trade.

Electrical Equipment.

Electrical Installations of the United States Navy. By Commander Burns T. Walling, U. S. N., and Julius Martin, E. E. Size, 9 by 6 in.; pp., 648. Ills. Price, $6. Annapolis, Md.: The United States Naval Institute.

A manual of the latest approved material, including its use, operation, inspection, care and management, and method of installation on board ship, originally published in the Journal of the United States Naval Institute and now issued in book form. It discusses exhaustively incandescent and arc lamps and search lights, wire and wiring appliances, generator and motor installation, operation and inspection, the various applications of electric motors aboard ship, and facilities for interior and exterior communication. The book by no means confines itself to mere description. While machines, appliances and instruments are described in detail, the main object is to give an idea of the practical details of their installation, operation and maintenance and to this end attention is paid to constructional features. The book is interesting to engineers generally as showing the immense importance of electricity on shipboard, and will prove of great value as a handbook for those whose duties require an intimate knowledge of the applications of electricity in the modern warship.

Gas Manufacture.

The Chemistry of Gas Manufacture. By Harold M. Royle. Size, 9 by 6 in.; pp., xv, 328. Ills. Price, $4.50. New York: The Norman W. Henley Publishing Company.

A practical treatise for the use of gas engineers, gas managers and students dealing entirely with the chemistry of gas manufacture to the exclusion of details of manufacturing. A knowledge of the elementary truths and processes is presupposed and the different subjects are treated in as concise a manner as possible. The book is intended to be a manual covering questions and points likely to arise in the ordinary course of the duties of the engineer or manager of a gas works not large enough to necessitate the employment of a separate chemical staff. The subjects treated are: preparation of standard solutions, coal analysis, testing and regulation of furnaces, products of carbonisation, analysis of crude coal gas, analysis of lime, ammonia, analysis of oxide of iron, naphthalene, analyses of fire-bricks and fire-clay, weldon, and spent oxide, photometry and gas testing, and carburetted water-gas. A large amount of useful information is given in the three appendices, consisting of selected extracts from matter previously published, including examples of official regulations and instructions for testing coal gas for illuminating power, calorific value and impurities, tables, data, statistics, etc.

Grinding.

Grinding and Lapping. By Joseph V. Woodworth. Size, 9 by 6 in.; pp., ix, 162. Ills. Price, $2. New York and London: Hill Publishing Company.

A comprehensive work on the design, construction and operation of grinding and lapping tools and devices written in a very practical manner and intended for the use of the mechanic. The book is divided into five sections. The first deals with grinding operations, describing conditions, rules, methods, processes, machines and attachments for accurate grinding, and the use and preparation of abrasives. In like manner the second section discusses lapping practice, describing the construction and use of tools and processes for finishing gages, tools, dies and machine parts to accurate dimensions.

In the third section grinding fixtures and jigs are considered, particularly their construction, use and operation for finishing repetition work and articles of metal, small hardened and tempered steel parts and special work. Sections 4 and 5 treat of the heat treatment of steel, the former describing the hardening and tempering of interchangeable tool steel parts of delicate structure which require to be ground and lapped afterwards, and the latter discussing the percentage of carbon crucible-steel parts and tools should contain, temper colors to which they should be drawn, and degrees of heat for giving them proper tempers. A valuable feature of the book is that all devices and processes mentioned are illustrated in an unusually clear manner.

Mathematics.

A Course in Mathematics. By Frederick S. Woods and Frederick H. Bailey. Volume I, containing algebraic equations, functions of one variable, analytic geometry, differential calculus. Size, 8½ by 6 in.; pp., 385. Diagrams. Price, $2.40. New York and London: Ginn and Company.

The first of two volumes of a course in mathematics designed for the use of students in the first and second years at engineering schools, the first volume containing the work of the first year. The course is designed to present in a consecutive and homogeneous manner an amount of material generally given in distinct courses under the various names of algebra, analytic geometry, differential and integral calculus, and differential equations. The authors' claim to novelty lies in the fact that the traditional divisions of mathematics into distinct subjects are ignored and the principles of each subject are introduced as needed and the subjects developed together. It is hoped that this arrangement will give the student a better grasp of mathematics as a whole, and of the interdependence of its various parts, and to accustom him to use, in later applications, the method best adapted to the problem in hand.

Pumps.

Notes on the Construction and Working of Pumps. By Edward C. R. Marks. Size, 7 by 5 in.; pp., viii, 259. Ills. Price, $1.50, 4/6. London: The Technical Publishing Company; New York: D. Van Nostrand Company.

A second edition of a work first published in 1902. The information given is intended for users of pumping machinery rather than for makers. The leading types of pumps are considered in detail and their construction and capabilities described with a view to enabling the buyer to choose the type best suited to his own particular requirements. In the second edition the material has been brought up to date by the addition of new illustrations and particularly by the addition of several chapters on the high lift turbine pump.

Rand.

The Commercial Aspect of Rand "Profits." By George A. Denny. Size, 10¼ by 5½ in.; pp., 40. Price, 1/. London: The Mining Journal.

A reprint of a series of articles which appeared originally in *The Mining Journal*, where they attracted a great deal of attention both in England and abroad. The situation on the Rand is a subject of absorbing interest both technically and economically. The mining interests in the field recognize the imminent danger of the exhaustion of the high-grade ore deposits and every possible means of reducing working costs is being taken in the effort to save the industry. Mr. Denny's well-informed review shows that a radical change in methods of management is necessary before the gold-mining industry can be placed on a firm foundation and that given rational financial methods, the exhaustion of the high-grade deposits becomes a phenomenon of secondary importance so far as the position of the Rand as the greatest gold field of the world is concerned.

Steam Plant Chemistry.

Engine-Room Chemistry, a Compend for the Engineer and Engineman. By Augustus H. Gill. Size, 7 by 4½ in.; pp., 200. Ills. Price, $1. New York and London: Hill Publishing Company.

This is a very useful little book for the engineer in charge of steam plants as it gives in clear and concise form the chemical information necessary for the intelligent testing and analysis of boiler fuels and waters, flue gases and lubricants. Such tests are being more and more recognized as an essential factor in the economical management of boilers and engines and the clearness and thoroughness with which this handbook is written should give it a wide usefulness. It gives in simple language only the essential principles of chemistry necessary for the understanding of the practical directions for conducting tests, to which the greater part of the book is devoted. By the aid of the latter he can make his own analysis of the lubricants best suited to his purpose; he can make his own analysis of the flue gases; he will know the quality of the fuel he buys and the qualities of the feed water and how to neutralize the scale forming propensities of the latter. The subject of the regulation of combustion is also covered in a thorough and interesting manner.

Steam Traps.

Modern Steam Traps (English and American): their Construction and Working. By Gordon Stewart. Size, 7 by 5 in.; pp., viii, 104. Ills. Price, 3/, $1.25. London: The Technical Publishing Co., Limited; New York: D. Van Nostrand Co.

A reprint of a series of articles which appeared originally in *The Practical Engineer.* The book illustrates and describes all the leading types of steam traps and while it does not pretend to deal exhaustively with the subject, it places before the user in an interesting and practical manner a large amount of information concerning the points most vital to the proper and efficient working of such apparatus.

Strength of Materials.

Problems in Strength of Materials. By William Kent Shepard. Size, 9½ by 6 in.; pp., 70. Diagrams. Price, $1.30. New York and London: Ginn and Company.

A very useful little book, intended to supplement text-books on the theory of structures and strength of materials which usually lack a sufficient number of practical examples fully to impress upon the mind of the student the applications of the theory. The book contains a collection of over five hundred graded practical problems on Tension, Compression and Shear; Elastic Deformation; Thin Cylinders and Spheres; Riveted Joints; Cantilever and Simple Beams; Overhanging, Fixed and Continuous Beams; Columns and Struts; Torsion; Combined Stresses; Compound Columns and Beams; Thick Cylinders and Guns; and Flat Plates. Tables necessary for the solution of the problems are given in the appendix. The use of this book will greatly lessen the labor of class-room teaching.

Water Power.

How to Use Water Power. By Herbert Chatley. Size, 7 by 5 in.; pp., 92. Price, $1. London: The Technical Publishing Co., Limited; New York: D. Van Nostrand Co.

The author has not attempted to produce an exhaustive treatise but a clear account of the methods and principles of hydraulic principles as at present practiced in a form that can be easily grasped by the craftsman or student with limited knowledge of mechanics and mathematics. Among the subjects treated are: the sources and transmission of power, the hydraulic press, water wheels, turbines, pumps, hydraulic engines, tidal power, and water supply problems.

Water Supply.

Clean Water and How to Get it. By Allen Hazen. Size, 8 by 5½ in.; pp., vi, 178. Ills, 14. Price, $1.50, 6/6. New York: John Wiley & Sons; London: Chapman & Hall, Limited.

This little volume deals with the means adopted in the best American practice to secure clean water for cities, and with the applications of these means to new problems. It is not intended to serve as an exhaustive treatise on water-supply problems, but merely to help beginners to understand something of first principles. The author fully attains his purpose to present a clear and concise discussion of the broad general principles of water-supply engineering and the seeker after this sort of information will find the work invaluable. Illustrations are freely used and typical plants are described wherever the application of principles to practice can be best shown by concrete examples. Besides the purely technical information the book contains a great deal to interest those connected with the administration of water-supply systems. Matters of general policy, the sale of water, the financial management of water works, etc., are considered in broad outline. Taken as a whole the book should prove of great value to city officials who are charged with the responsibility for procuring pure water and who have no previous technical or financial knowledge of water-supply engineering.

Wood Distillation.

The Utilisation of Wood Waste by Distillation. By Walter B. Harper. Size, 12 by 9 in.; pp., 156. Ills., 74. Price, $3. St. Louis, Mo.: Journal of Commerce Company.

A general consideration of the industry of wood distilling, including a description of the apparatus used and the principles involved, also methods of chemical control and disposal of the products. The material contained in the book was first published in a series of articles in *The Lumberman* which were among the most valuable contributions to the literature of the problem of extracting chemical products from the waste wood of saw mills. The book emphasizes the growing importance of the utilization of waste products, gives a complete discussion of the products which may be derived from pine wood by distillation, describes the various methods and processes used, and gives many hints as to the technical and economic management of plants for this purpose. It should prove of value to those connected with the industries in which turpentine, resin, tar, alcohol, etc., play an important part as well as to the lumberman and paper maker in whose interests the book is primarily written.

BOOKS RECEIVED.

Water Supply. By Frederick E. Turneaure. Size, 9½ by 6½ in.; pp., 143. Ills. Price, $1. Chicago, Ill.: American School of Correspondence.

Mechanical Drawing. By Ervin Kenison. Size, 9½ by 6½ in.; pp., 142. Ills. Price, $1. Chicago: The American School of Correspondence.

Plane Surveying. By Alfred E. Phillips. Size, 9½ by 6½ in.; pp., 207. Ills. Price, $1.50. Chicago, Ill.: American School of Correspondence.

Report of the Chief of the Weather Bureau for 1907. By Willis L. Moore. Size, 9 by 6 in.; pp., 56. Ills. Washington: Department of Agriculture.

Proceedings of the Master Car Builders' Association, Vol. 41, 1907. Size, 9 by 6 in.; pp., 642. Ills. Plates. Chicago, Ill.: Published by the Association.

The "Mechanical World" Pocket Diary and Year Book for 1908. Size, 6½ by 4 in.; pp., 392. Ills. Price, 6d. Manchester: Smmott & Company, Limited.

Machine Shop Work. By Frederick W. Turner. Size, 9½ by 6½ in.; pp., 190. Ills. Price, $1.50. Chicago: The American School of Correspondence.

Annual Report of the Smithsonian Institution, 1906. Size, 9 by 6 in.; pp., li, 546. Ills. Maps. Washington, D. C.: Published by the Smithsonian Institution.

The Steam Engine. By Walter B. Snow and Walter S. Leland. Size, 9½ by 6½ in.; pp., 156. Ills. Price, $1. Chicago: The American School of Correspondence.

Valve Gears and Indicators. By Walter S. Leland and Carl S. Dow. Size, 9½ by 6½ in.; pp., 111. Ills. Price, $1. Chicago: The American School of Correspondence.

Highway Construction. By Austin T. Byrne and Alfred E. Phillips. Size, 9½ by 6½ in.; pp., 136. Ills. Price, $1. Chicago, Ill.: American School of Correspondence.

State of Ohio, Highway Department; Bulletin No. 9: Proposed Amended State Aid Law. By Sam Huston. Size, 9½ by 6½ in.; pp., 67. Columbus, O.: State Highway Department.

Report of the Proceedings of the Fortieth Annual Convention of the American Railway Master Mechanics' Association. Size, 9 by 6 in.; pp., 453. Ills. Chicago, Ill.: Published by the Association.

Annual Report of the Minister of Mines of British Columbia for the Year Ending December 31, 1906. Size, 10 by 7 in.; pp., 276. Ills. Maps. Victoria, B. C.: Department of Mines of British Columbia.

Mechanical Engineering Materials, their Properties and Treatment in Construction. By Edward C. R. Marks. Size, 7 by 5 in.; pp., viii, 98. Price, 2/6. London: The Technical Publishing Company, Limited.

La Construction d'une Locomotive Moderne. By Dr. Robert Grimshaw. Translated from the second German edition by P. Poinsignon. Size, 9 by 6¼ in.; pp., xiv, 64. Ills., 42. Price, 3 fr., 75 c. Paris: Gauthier-Villars.

British Standard Specifications for Steel Castings for Marine Purposes. Prepared by the Engineering Standards Committee and edited by the Secretary, Leslie S. Robertson. Size, 13 by 8 in.; pp., 11. Price, 2/6. London: Crosby Lockwood and Son.

Contributions to Economic Geology, 1906; Part I; Metals and Non-Metals, except Fuels. Published by the United States Geological Survey under the direction of S. F. Emmons and E. C. Eckel. Size, 9 by 6 in.; pp., 505, iv. Ills. Washington, D. C.: Department of the Interior.

A Study of Four Hundred Steaming Tests made at the Fuel-Testing Plant, St. Louis, Mo., in 1904, 1905 and 1906. By Lester P. Breckenridge. Bulletin No. 325 of the United States Geological Survey. Size, 9 by 6 in.; pp., 196. Ills. Washington, D. C.: Department of the Interior.

Kahn System Standards; a Hand Book of Practical Calculation and Application of Reinforced Concrete. Compiled by the Engineering Department of the Trussed Concrete Steel Company. Size, 8 by 5 in.; pp., 106. Ills. Price, $1.50. London, Detroit, and Toronto: Trussed Concrete Steel Company.

The Mechanics of Hoisting Machinery. By Dr. Julius Weisbach and Professor Gustav Herrmann. Translated from the second German edition by Karl P. Dahlstrom. Size, 9 by 6 in.; pp., 177. Price, $3. New York: The Macmillan Company; London: Macmillan and Company, Limited.

Report upon Smoke Abatement. An Impartial Investigation of the Ways and Means of Abating Smoke; Results attained in other Cities; Merits of Patented Devices, together with Practical Suggestions to the Department of Smoke Abatement, the Steam Plant Owner and the Private Citizen. Size, 9 by 6 in.; pp., 42. Syracuse, N. Y.: Chamber of Commerce.

VOL. XXXIV.   FEBRUARY, 1908.   No. 5.

THE WASTE OF LIFE IN AMERICAN COAL MINING

By Clarence Hall and Walter O. Snelling.

Published by permission of the Director of the United States Geological Survey.

It is not necessary to emphasize the urgent importance of this theme, nor indeed possible to point its significance more impressively than the authors do in the opening sentence. We are glad to be able to present so authoritative a discussion of the situation as it exists and of the possibilities of betterment—a discussion based upon the careful studies and the exhaustive data assembled during many months work by the explosives expert and the explosives chemist of the United States Geological Survey. To them and to the Director our acknowledgements are made for this special presentation of their conclusions in our pages.—THE EDITORS.

THE four recent mine disasters in the United States, with the loss of nearly one thousand lives, have given terrible point to the bulletin just issued by the Technologic Branch of the United States Geological Survey on "Coal Mine Accidents, Their Causes and Prevention." The figures given in this report indicate that during the year 1906 nearly seven thousand men were killed or injured in the coal mines of the United States, and that the number of these accidents caused directly or indirectly by mine explosions has been steadily increasing. It is also indicated that this increase has been due in part to the lack of proper and enforceable mine regulations; in part to the lack of reliable information concerning the explosives used in mining, and the conditions under which they can be used safely in the presence of the gas and dust encountered in the mines; and in part to the fact that in the development of coal mining not only is the number of miners increasing, but many areas from which coal is being taken are either deeper or farther from the entrance, where good ventilation is more difficult and the dangerous accumulations of explosive gas are more frequent.

The increase both in the number and the seriousness of mine explosions in the United States during past years may be expected to
continue unless the country adopts means that have proved successful in European countries, where the proportionate death rate in the
mines has been materially reduced.   Information is needed at once
concerning the explosives used in the mines, the conditions under
which they may be used safely in the presence of coal dust and gas,
and the general conditions which make for health and safety in coal-
mine operations.   In 1906 the total number of men killed in the mines
was 2,061, and the injured 4,800.   In 17 years 22,840 men have been
killed in the coal mines of the United States.   In 1890 the number
killed was 701.   The number of killed has steadily increased until
1905, when 2,097 met with violent deaths.   The great increase in the
production of coal during the last decade and the related increase in
the number of men employed in the industry do not account altogether for the increase in the number of fatal accidents, although this
may seem to be true.   In 1895 2.67 persons were killed in the coal
mines for each 1,000 men employed.   This ratio has increased until
in 1906 it reached 3.40, and this was exceeded in 1905 and 1902.
In all the European coal-producing countries, the output of coal has
greatly increased during the last 10 years, but the number of deaths
per 1,000 miners, instead of increasing as in America, has undergone
a marked and decided decrease.   This has been due to the effect of
mining legislation in those countries for the safeguarding and protection of the lives of the workmen, and has been made possible by
Government action in establishing testing stations for the study of
problems relative to safety in mining, including the use of explosives.

In Belgium between 1831-40 before the testing station was built
the average number of men killed per each 1,000 men employed
was 3.19.   From 1901-1906 this ratio had been reduced to 1.02.
This is one-third of the ratio now existing in the United States.   The
beneficial result of this testing station is seen in the fact that the number of lives lost in the Belgian mines for each 1,000 men employed
in 1906 was 0.94, while in 1895 the ratio was 1.40.   Great Britain
has reduced its ratio of 1.50 in 1891 to 1.29 in 1906.   In Prussia the
ratio of deaths per 1,000 men employed has decreased from 2.94 in
1880 to 1.80 in 1904.   France's ratio has dropped in five years from
1.03 per 1,000 men employed to 0.84.

In the following tables giving the average ratio for the last
period of five years for which statistics are at hand, the position of

the United States may be compared with that of the principal European coal-producing countries.

NUMBER OF MEN KILLED FOR EACH 1,000 MEN EMPLOYED—AVERAGES FOR FIVE YEARS.

France (1901-1905) 0.91
Belgium (1902-1906) 1.00
Great Britain (1902-1906)............................... 1.28
Prussia (1900-1904) 2.06
United States (1902-1906)............................... 3.39

CAUSES OF MINE ACCIDENTS.

Fifty per cent of all the fatal accidents and 39 per cent of all the non-fatal accidents in the United States were the result of falls of roof and coal. The following statement of deaths from this cause per 1,000 men employed shows how the United States compares with several of the European coal-producing countries in this respect.

DEATHS FROM FALLS OF ROOF AND COAL PER 1,000 MEN EMPLOYED.

Belgium 0.40
France 0.47
Great Britain 0.64
Germany 0.92
United States 1.70

In France, taking five-year periods, the number of men killed from falls of roof and coal for each 1,000 men employed has been reduced from a ratio of 1.26 in 1875 to 0.47 in 1900. In Belgium the number of men killed from falls of roof and coal for each 1,000 employed has been reduced from 0.998 in 1860 to 0.406 in 1904. In Great Britain and Germany the results are similar to those in France and Belgium. In all the European coal-producing countries the use of excessive charges of explosives is prohibited by law and definite limits are set as to the amount of any explosive that may be used. Although these regulations were framed with the object of preventing gas explosions, it is believed that they have been of marked effect in preventing accidents from falls of roof and coal, as the very great disturbing and jarring effect exerted by the discharge of large amounts of explosives in a mine is believed to be one of the most important causes of fall of roof. The actual fall of the rock or coal may not occur at the time of firing the charge, but the heavy shots weaken the wall and roof and start cracks that impair the support of the roof, so that months after the blast, without warning, it falls.

It is also to be noted that explosions of fire damp and coal dust have a similar effect in jarring and weakening the walls and roof, and in gassy regions, where small explosions of fire damp are of frequent occurrence, falls from this cause may make up a considerable percentage of the total number. But even in such regions the use of explosives and of unsatisfactory forms of lamps is generally the indirect cause of the fall, for nearly all explosions of fire damp are caused by ignition from the shots used in breaking out the coal or from naked lights used by the miners.

In all European countries from which statistics are available restrictions in the maximum amount of explosives allowed to be used have not only increased the safety from explosions of gas, but have also materially diminished the loss of life due to falls of roof.

In the United States, during 1906, 11 per cent of the deaths in coal mines were due to gas and dust explosions. In Belgium the number of men killed in explosions of fire damp for each 1,000 men employed has been reduced from 0.965 in 1840 to 0.039 in 1904. The present death rate from these causes is about one-tenth of that of 30 years ago. This has been due to systematic testing of safety lamps, only those forms being allowed to be used which are capable of withstanding rapidly moving currents of fire damp under all conditions likely to be encountered in mines, and to a thorough testing of all explosives to determine the amount of each which can be fired without danger of explosions of fire damp or coal dust.

When the total number of fatal accidents in the coal mines of the United States is considered, with reference to the number of tons produced, it is seen that the number of men killed for each 1,000,000 tons of coal produced has not changed materially in the last fifteen years. Considered in periods the average results obtained are as follows:—

NUMBER OF MEN KILLED IN THE COAL MINES OF THE UNITED STATES FOR EACH MILLION TONS OF COAL PRODUCED, BY PERIODS.

General average, 1890-1906	5.93
1890-1895	5.97
1895-1900	5.77
1901-1906	6.04

In Great Britain in the decade 1874-1883 the number of men killed in the coal mines for each million tons of coal produced was 7.42. This has been reduced to the ratio of 4.31 in 1906. In France in 1900 the number of men killed for each million tons of coal produced was 5.55. In 1905 this was reduced to 4.17, but probably the 1906 ratio was larger. In Belgium in 1895 the number of men killed per each million tons of coal produced was 7.70. In 1906 this had been reduced to 4.96.

These figures show in regard to deaths per million tons of coal that the United States not only occupies a position worse than most

European countries, but also shows a general increase in the rate, whereas every other country has shown a decrease. This situation is still worse when it is considered that the natural conditions in America for getting out coal with the minimum amount of danger to the workmen employed are as favorable as in any other country in the world. With the depletion of the thicker and more favorably mined seams of coal, thinner and less regular seams must be worked. This will undoubtedly be a most important change within a comparatively few years, and the natural result would be to increase the death rate greatly. With the mining of the small beds of coal and the gradual development of properties worked with more difficulty, bringing the mining conditions in the United States more nearly to a position of equality with those elsewhere, a great increase in accidents must be expected, unless proper steps are taken to remedy the conditions which have brought about the remarkably high death rates in the coal mines of the United States. An important feature which must be considered in the mine accidents of the United States is the nationality of the miners. Most of them are foreign-born. A large proportion of them are unable to understand English freely, and a still larger number are unable to read or write that language. Some of them are inexperienced and do not take proper precautions for their own safety or for the safety of others. This becomes a most serious menace unless they are restrained by carefully enforced regulations.

Prevention of Mine Accidents.

In a number of European countries the Government is maintaining testing stations at which all the investigations into the use of explosives are conducted. These testing stations have already arrived at certain definite conclusions. The compulsory use of safety lamps in mines having a large amount of fire damp is general in all European coal-producing countries. The keeping of at least two safety lamps in a mine (whether thought to be gaseous or not) is also required in some countries; laws regulating the locking of safety lamps and requiring that all lamps should be capable of being so locked by magnetic devices, compressed air, or other means, as not to be opened by any other person, are also in effect. The employment of relighting stations, at which locked safety lamps may be relighted in a mine when they are accidentally extinguished, is permitted in several countries. In Germany the lamps in use are equipped with relighting devices of approved design and are known to be practically free from danger of igniting fire damp, so that relighting stations are unnecessary. In the United States in many mines the safety-lamp provision is violated.

In all European coal-producing countries regulations have been made forbidding the use of such explosives as are known to be liable to ignite fire damp, in all mines in which fire damp is represented to be present to a dangerous extent. In the United States in a few States the mine inspectors have control over the amount of explosives used.

In England a list of "permitted explosives" is kept, including all those explosives that answer the required tests; and all explosives used in mines that are dangerous because of gas or coal dust must be selected from the permitted list, which now includes more than fifty explosives. Control of like kind is exercised in France, Belgium and Germany. In all these countries, the restriction of unsafe explosives has caused the development of "safety" explosives which have proved greatly superior to those previously in use. In Belgium, owing to the greater amount of fire damp present in the mines of that country, a further restriction known as the "charge limit" has also been put in force. Through this restriction the maximum amount of any explosive which is allowed to be used in any one single charge is defined. The actual amount allowed varies according to the liability of the explosive to ignite fire damp, but with all the better class of explosives it is well above the amount needed for use in a single charge under proper mining conditions. These restrictions in the kind and amount of explosives have yielded additional benefit in the fact that the reduction in the amount used not only reduces the number of accidents, but also to a large degree prevents coal waste. In the United States there is no limit set as to the amount of explosives to be used.

Regulations governing the storage of explosives for use in coal mines are in general effect in European coal-producing countries; they provide that no explosives shall be taken into a mine except in canisters containing not more than five pounds, and that no unused powder shall be left in a mine over night. The amount of powder in the possession of a workman at any one time is also limited, and the amount which can be taken into a mine is usually restricted to that which will be used in one shift. Regulations enforced in Great Britain provide that in all mines in which inflammable gas is present, or has been found in dangerous quantities within the previous three months, all charges of explosives shall be fired by a competent person termed a "shot firer." The object is to place the responsibility in the hands of men better qualified for the work than the average miner. The shot firer begins work after the miners have left for the day, each miner on leaving having prepared the proper number of holes in the places where he wishes charges to be fired. The shot firer charges the holes with explosive and tamps and fires the charges. It will be seen that by this method the charges are fired at a time when but few men are in the mine, and that accordingly if a fire-damp explosion occurs fewer lives will be lost than if the miner fired his shots at a time when other miners were in the workings.

Still another factor of greater importance is that the shot firer, through constant experience, soon becomes used to the proper use of explosives, and also learns to charge the holes and fire them in the safest way possible. The shot firer is also free from the temptation to use excessive charges in the desire to get out a great amount of coal with the least possible effort. It is well-known to all persons experienced in coal mining that overcharging of holes and mining by means of large shots fired in the solid mass of coal bring about many fire-damp and coal-dust explosions.

At the present time several electrical shot-firing devices are being tested with promise of effecting a considerable saving of life. The main principle of these devices is the firing of the charges one after another, at proper time intervals. As already stated, the firing of the charges takes place after the miners have left for the day, and the entire operation of the device is automatic. It is of course necessary, after the charges are fired, to have some one examine each room in which a shot has been fired to see that conditions are normal, and that no feeders of gas have been ignited which might start a mine fire. The use of shot-firing devices or some other means equally safe is required by regulation in Great Britain.

The watering of the sides and roof of coal-mine passageways has been found to be of material benefit in preventing local ignitions of coal dust from becoming general. The watering prevents the coal dust from being stirred up by small explosions, and the reduction in the explosive properties of the wet coal dust has the general result of preventing an explosion from traveling into other workings. The watering is effected by sprinkling or spraying devices of various kinds.

The mining regulations of Germany, France, or Belgium, require that mines shall be supplied with means for administering to the comfort of injured workmen and that proper appliances shall be provided for first aid. Rescue devices equipped with small tanks of

compressed air, or compressed oxygen, or with some chemical appliance for generating a supply of oxygen so as to enable a person to
enter a mine working in which an irrespirable atmosphere is present,
are required in some countries. By the aid of such apparatus rescue
parties may enter the mine working immediately after the fire-damp
disaster and rescue injured miners who would soon fall victims to the
poisonous atmosphere surrounding them.

Enforcement of Regulations.

In all European countries, the enforcement of regulations in regard to coal-mining operations is placed in the hands of a specially
appointed force of mine inspectors, familiar with the industry and
able to examine mine conditions intelligently and to see that the
mining laws are obeyed. In the United States the mines are within
the jurisdiction of the States which actually produce the coal, each
State having a code of its own, the laws being different in different
States. In Germany mines are under the control of the district
officers, the superior mine offices, and the Minister of Commerce and
Trade. The district officers are the first instance in all those affairs
that belong to the mine authorities and which are not expressly
turned over to the superior mine offices. The management of the
extensive mining police is their most important charge. The district
officers have under them the inspectors. Each inspector controls
several mines which he has to inspect, reporting to the Department
of Mines within a certain time. Especially dangerous mines are to be
visited oftener than others. The inspection takes place at any time,
night or day, usually without previous notice. Inspectors are authorized to call the attention of the employees of the mining company to
certain offences against the mining police regulations and to other
bad conditions, and to talk over improvements with them. Several
functions are imposed upon the district officers by the local police
authorities, the lower administration authorities, and the industrial
chambers of mines, which give them a great influence upon the affairs of the workmen. Superior to the district officers are the superior mine offices, which hold the rights of the government within the
territory. They are tribunals of first instance in all matters and
also for appeals and plans. The most important right given to the
superior mine offices as a mine-policing authority is the right of
issuing independently police orders and police directions on the basis
of the common mining law. To carry on the business of the third
instance (being chiefly the instance of appeals) there has been estab-

lished in the Department of Commerce and Trade a Department of Mines, Smelting, and Saline Works, which is presided over by the superior mine captain. In order to ensure the successful control according to the Government mining law, the law imposes upon the owner the following duties as being an important foundation :—at least four weeks' notice to the mining authorities regarding the starting of work order or stopping of mine; and the operation of the mine only under the superintendence of persons who have been recognized by the authorities to be capable. The mine owner is bound to have made, by a recognized surveyor, a plan of the mine in duplicate, and to have made in due time all additional entries in the same. One of the plans is filed by the mining authorities for their use. The mining authorities may stop the work, carry out their directions by a third person at the expense of the mine owners, inflict fines, and withdraw the right to conduct and control the mine. Stopping of work can be ordered by a district officer either for a whole mine or for parts of the same for the following reasons :—when the mine owner is working without a plan of a mine or against it; when the work is conducted by a person who is not in possession of a license of capacity; when a mining police direction given in the face of menacing danger is either not at all or hesitatingly carried out.

In Belgium there is a very complete law for the inspection and regulation of mines. Legal provisions are also made for the establishment of relief societies, of which there seem to be a large number established. The head of the administration of mines is the Minister of Industry and Labor, who has a corps of mining engineers under him. It is their duty to watch the execution of the mining law; to examine all gas and steam apparatus; the work of the miners, especially women and children; and the control of explosives. The engineers of the first, second, and third classes attached to the districts, inspect as often as their chief thinks necessary the mines, etc., in their respective districts, watch for the execution of the law and the regulations, etc. When inspecting the mines they have to pay special attention to the safety of the workmen, conservation and constructions, the waters on the surface, the lighting apparatus, the maintenance, descent of workmen, life-saving devices, and the storage and use of explosives.

In France the inspection and regulation of mines is under the Minister of Public Works and the prefects. The Minister has under him a number of mining engineers who have extensive powers of inspection with a view to preventing danger and providing for the preservation of buildings and the safety of individuals. The prefects have also power at the request of the engineers to require mine owners at their own cost to execute the works necessary to protect buildings and to secure the safety of the public, the mines, the workmen, and of ways of communication, mineral waters, and the water supply for towns, villages or public establishments, at the cost of the *concessionaires*. Mine owners are required to keep on their premises, in proportion to the number of workmen and the extent of the workings, such remedies and aid appliances as may be required by the Minister of Public Works, and to conform to the regulations approved by him in that behalf. The mining law provides that in cases of accidents notice must at once be given to the authorities, who shall have power to take all necessary measures either to put an end to the danger or to prevent the consequences; and the owners and managers of neighboring mines are required to furnish all the assistance they can, any expenses of assistance to wounded men or repair of works being borne by the owners of the mine where the accident occurred. The mining engineers or other authorities must report upon the accidents, and in cases of negligence the mine owners and managers may be punished by imprisonment and fine besides being liable in damages. There is nothing approaching compulsory insurance of workmen in France, but by the law of July 11, 1868, any workmen may make provision for the case of death or accident, the insurance fund being administered by the Government officials, and by this law workmen may make similar provision for the case of old age. In addition to the public insurance institutions, there are numerous private associations or mutual-aid societies, the objects of which are usually to make provision for sickness and burial, but occasionally also to provide pensions for old age or incapacity for work. Some of the mining companies make special terms with hospitals for the benefit of their workmen. Others have established hospitals; others have established co-operative shops; others erect houses and establish schools for the benefit of their workmen, while certain companies make free distribution of coal among the workmen.

At the head of the mining administration is the Minister of Public Works, while in each Department the prefect is the head of the service under the authority of the Minister of Public Works. Under the immediate orders of the Minister of Public Works is the Council General of Mines. This is composed of the inspectors-general of the first and second class, presided over by the Minister of Public Works, or in his absence by a vice-president appointed for a year by the

Minister out of the inspectors-general. Below the inspectors-general of the first and second class come the engineers-in-chief, and under these are the ordinary engineers. Engineers-in-chief have to render account to the prefects of mining works, and to receive and execute their orders relative to inspection; to report to the Director-General and the prefects all breaches of the law and regulations as to mines, and to inspect the mines in their Departments in turn under orders of the Director-General. The ordinary engineers must visit at least once a year all the mines being worked in their several departments, and in the case of a breach of the law or any danger or accident, being notified at the working, they must attend at the place and prepare a report for the engineer-in-chief with the view of effecting a remedy. There is a head office of mines in Paris, and the rest of the country is divided into various districts and sub-districts for the purposes of mining administration. No child is employed under the age of 12. The duration of effective work of children of the male sex from 12 to 16 years old in the subterranean galleries of mines, etc., shall not exceed 8 hours in 24, being interrupted by a rest of one hour at least.

The mines of Great Britain are controlled by the Metalliferous Mines Regulation Act and the Coal Mines Regulations Act. England is divided into twelve districts, controlled by Royal Inspectors of Mines. Each inspector is charged with surveying a district. It is the duty of the inspectors especially to visit the mines of their district in order to assure themselves that all prescriptions of the law and particular regulations are duly attended to. In cases of infractions they may advise the mine owners thereof, or provoke a prosecution before the judicial tribunal, which can punish the faulty parties by fines up to £20, or by detentions of three months in serious cases. The visits of the inspectors are made either officially or in consequence of complaints filed, even by anonymous letters. The inspectors have the right to take any measure not provided for in the law or regulations should there be a situation creating a danger. If the mine owner does not agree with the inspector's opinion, an arbitration is created consisting of two arbitrators, one appointed by the inspector and the other by the owner of the mine. If they do not agree a third arbitrator is appointed. The arbitration sentence is a definitive one, without appeal. The value of this sentence for the mine owner is that it is a legislative prescription; every infraction would come under the law. Inspectors are appointed by the Secretary of State. Every mine in which more than 30 persons are employed must be under the control, management, and direction of a manager or under-manager,

who may be the owner or agent or some other person nominated by
the owner or agent, and must be registered as the holder of a first-
class certificate. Daily personal supervision of every such mine must
be exercised by the manager or by the under-manager. The inspectors must be notified in writing within 24 hours of any accident. In
case of loss of life or serious personal injury, the place where the
accident occurred shall be left as it was immediately after the accident
for a period of three days until the visit of the inspector, unless this
would tend to increase or continue danger, or would impede the
work of the mine.

The Explosives Order contains the list of explosives which have
passed the government test, called "permitted explosives." The order
applies especially to two classes of seams in coal mines:—seams in
which inflammable gas has been found within a previous three months
in such quantity as to be indicative of danger, and seams which are
not naturally wet throughout. The use of permitted explosives in
main haulage roads and main intakes is subject to the further condition that every part of the roof, floor, and sides within a distance of
20 yards from the place where the shot is fired must, unless naturally
wet, be thoroughly watered at the time of firing. If explosives not
in the permitted list are used, the workmen, with certain exceptions,
must be withdrawn from the mine when the shots are fired, but if
permitted explosives are used, the workmen need not be withdrawn
from the mine provided the roof, floor and sides within 20 yards from
the place where the shot is fired be thoroughly watered at the time of
firing. The list of permitted explosives may be seen in the copies of
the order which are supplied by the Home Office to the mine owners. To each explosive further conditions are attached which must
be carefully observed, and the provisions of the order are not in
substitution for those in the Coal Mines Regulation Acts or Rules,
or of any special rules of the mine, but are in addition to them.

THE REMEDY.

However necessary legislation may be on this subject, the problem
in the United States is so complex that those most familiar with
mining are convinced that scientific investigation must precede legislation. The coal operators of West Virginia recently adopted a resolution to the effect that:

"The United States Government should take the necessary steps to
determine the causes before any attempt is made to apply legislative

remedies, and when the causes have been ascertained and the remedies suggested, we pledge ourselves to co-operate with the National Congress and State legislatures in the framing and passing of any proper and effective legislation for the protection of life or property which may result advantageously to the National Government, States, labor and capital."

It is the general opinion that extensive scientific investigations should be carried on in the United States to determine the causes of the gas and dust explosions, and also those of other accidents in the mines. It is suggested that an experiment station similar to those in Belgium and Germany be erected at some point within the coal region, at which these experiments may be made. Tentative plans for such a station have been drawn, and the erection of the station simply awaits the action of Congress in the matter of a sufficient appropriation to carry on this work. The station as planned would provide for an explosives gallery, observation house, lamp-testing rooms, and explosives laboratory. The gallery would be made of boiler plate and in the form of a cylinder 100 feet long and 6 feet in diameter. A series of safety valves would be arranged on hinges along the top to allow the escape of gas following an explosion. Port holes along the sides, covered with one-half-inch plate glass, would allow those in the observation house to see whether an explosion had taken place in the gallery during the tests. The cylinder would be filled with fire damp and air, or coal dust and air, and the explosives would be hurled into the gallery by means of a steel mortar fire by electricity from the observation house sixty feet away.

The fire damp and air would be thoroughly mixed by an electric fan. Tests would be made with various explosives, and the maximum quantity of each explosive that can be safely used in mines would be published under the heading "permissible explosives."

Explosives known as "safety explosives" in which the temperature at the point of detonation is low and the flame of short duration, would have a higher "limit charge" than the less safe explosives.

Until these experiments are made and the results known, it is generally conceded that the State legislatures can do but little. West Virginia and Pennsylvania have struggled with the problem for many years and have some excellent statutes on their books tending to guard the lives of the miners. Nevertheless accidents are occurring and recurring with greater frequency. It is hoped that the findings will be of such a nature that when the recommendations are made they will be incorporated into the statutes of the various coal-producing States.

Some of the unsettled problems in connection with the coal-mine explosions in the United States are as follows:—

1. EXPLOSIVES.—The influence of the quantity and quality of each explosive and the methods of using the same, upon the risk of igniting either fire damp or dust, or a mixture of fire damp and dust, and thus causing an explosion in coal mines.

Standardizing explosives used in coal mines in the United States in such manner as to prevent unnecessary variations in the composition of the explosive for which the miner or shot firers would not be prepared.

2. FIRE DAMP.—Variations in the composition of this gas in different coal mines, and the influence of these variations, together with the variation of atmospheric conditions (pressure, etc.) within mines, in lessening or increasing the risk of fire-damp explosion.

Influence of the depth of the mine and distance from the outcrop and of the composition of the coal on the quantity and quality of gas in the coal.

Origin of reported unusual outbursts of gas and how these may be anticipated and counteracted.

3. DUST.—General conditions under which a dust explosion in coal mines may or may not be possible or probable from "windy" or "blown-out" shot, from electric sparks, or other means.

The influence of varying admixtures of fire damp in increasing the risk of or intensifying dust explosions.

The influence of the character and composition of dust from different coals in increasing or decreasing the risk of or force of gas explosions or dust explosions and upon the health of the miners.

The most practicable methods of maintaining the proper degree of humidity of the atmosphere in coal mines.

4. ELECTRICITY.—The conditions under which electricity may be safely used in coal mines for haulage and other purposes—whether an electric spark may cause a coal-dust explosion, or if gas is also necessary, what percentage, etc.

5. ATMOSPHERIC CONDITIONS.—(barometric pressure, temperature, etc); their influence, if any, in inducing mine explosions or conditions favorable or unfavorable to the same.

FIRE PREVENTION IN HIGH BUILDINGS. THE NEED OF AUXILIARY EQUIPMENT.

By J. K. Freitag, C. E.

THE burning of the twelve-story fireproof Parker Building at Fourth Avenue and 19th Street, New York City, on the night of January 10, 1908, is of peculiar interest to architects, engineers, and fire protectionists, in that the destruction by fire of individual fire-resisting buildings of great height has been comparatively rare. The behavior of such buildings under fire test is ever an interesting and vital speculation, and each trial of modern fire-resisting methods in high buildings, while deplorable from the standpoint of loss of life and property, is nevertheless of great value in helping to determine the worth or worthlessness of so-called fireproof construction. The present instance is not without its valuable lessons.

Widespread conflagrations such as those experienced by Paterson, N. J., Baltimore, and San Francisco, have pointed many morals which we are only too slow to follow, but the ruin in such calamities is so widespread, and the lessons presented are so multitudinous, that the full force of the application to individual buildings is apt to be overlooked for the more general, but quite as important, moral regarding the necessity of *uniform fire-resisting construction* within the congested areas of large cities. Again, past conflagrations have shown the almost utter uselessness of attempting to cope with fire when once it has reached the proportions of what is usually termed a conflagration. No structural materials of which we have present knowledge are equal to the task of resisting successfully such severe test conditions, at least to a point which would justify any reliance upon their use; while the points of attack in conflagration are so numerous and all conditions are so severe that the standards by which it is reasonable to judge fire-resisting construction are overthrown.

Fires in individual buildings with which the fire department is called upon to cope, prove, therefore, more instructive, and of such examples of fires in fireproof buildings exceeding ten stories in height, where the structure has been subjected to great or total loss, the Parker building furnishes but the fourth instance of prominence. It is only the second noteworthy example of practically complete loss to a high fire-resisting building in which the fire originated on the premises.

This fire serves to illustrate vividly certain prevalent weaknesses in fire-resisting design and construction; the value of such construction, notwithstanding faults of omission and commission; and, especially, the limitations of fire-department efficiency where very high buildings are concerned, and the imperative necessity of auxiliary equipment which shall aid the department in both the discovery and the fighting of fire.

The complete destruction of this building and contents, a loss now estimated at approximately two millions of dollars, may be directly attributed to two serious defects.

The first was a very prevalent fault in design and planning, *viz,* the open stair well and elevator shaft. Much has been said and written regarding the hazard of vertical openings in mercantile and other buildings, but no fire has ever more clearly demonstrated the widespread ruin which may result from failing to consider or failing to provide against the communication of fire from floor to floor by these means. I happened to be at the site of the fire in question when the alarm was turned in, and, before a single engine had arrived, and when flames had broken through only a few of the fifth-story windows on 19th Street, the fire could be plainly seen through the closed iron shutters on the east wall, working up rapidly from story to story through the open stairwell at that location, until, as the upper floor was reached, the flames spread out umbrella-like, enveloping the upper story, and then worked downwards again through other shafts.

Here was a "fireproof" building, provided with steel frame, adequate masonry walls, fire-resisting floor arches and column protection, and withal, open stairways and elevator shafts. Is it not a pertinent question to ask why this or any other similar building should be provided with fire-resisting floors and partitions, whose offices are obviously to *prevent* the communication of fire from room to room or from floor to floor, if the entire scheme or plan of fire-resistance is to be rendered null and void by the introduction of vertical flues which, in time of fire, present the surest and most effective means of communicating flame from floor to floor, and from basement to attic?

It is, perhaps, difficult to reconcile the theoretical advantages of isolated stair and elevator well-holes with the practical disadvantages which sometimes result from such isolation; for when the requirements of adequate fire protection seriously interfere with the conventional architectural treatment of the interior of the building, the introduction of any methods of fulfilling such requirements is most difficult to obtain.

In the latest and best examples of fire-resisting construction the problem of an architectural and yet efficient elevator enclosure has been solved through the use of wire glass, which, while still allowing the radiation of great heat, nevertheless confines the flames to the flue or shaft so protected, thus forming a temporary barrier to the horizontal spread of fire at each floor, efficient at least until the fire department work is well organized; and lateral spread is prevented. But it is still indisputably true that stairways should also be isolated, preferably to serve as safe fire-escapes within masonry partitions, with fire-resisting doors at each floor; or as has been the case in a few commendable buildings, they may be placed within a shaft adjacent to the elevator well, with a stairway enclosure opening onto each floor of the same construction and architectural treatment as the elevator front. Doors in such partitions may have counterweights and fusible links, thus closing automatically at a certain heat.

The second serious defect in the Parker building instance, was the absence of thermostats or automatic fire-alarm system. This fire had evidently gained great headway before it was discovered by the watchman, and, although the building was equipped with special building signal fire-alarm box, it is asserted that the watchman turned in the first alarm from a street box. The building was also provided with standpipe and basement open sprinklers, but these auxiliary appliances are required by law, and the provision of such protection is too often perfunctory, to cover some requirement in the local building laws or regulation of the fire department, rather than to provide and *suitably maintain* a system of fire-protection auxiliaries which may very likely prove of the utmost importance. For fire protection, viewed in a comprehensive light, must include not only those *passive* qualities of fire-resistance which result from the employment of proper materials in suitable design and construction, but also those *active* means of fire-detection and fire-fighting appliances which do so much to supplement and make effective the purely passive elements in the scheme of fire protection. *Fire protection must be aggressive as well as purely resistant.* Fire protection must consider the contents of a building as well as the structure itself. An incombustible building does not eliminate damage or destruction to the contents. Many a heavy loss has occurred to stock or contents through a quick sweep of flame which has worked comparatively little damage to the building itself. Such a result, while possibly affording a most gratifying object lesson as to the structural efficiency of the building, might be most satisfactory to the owner of the structure, and yet spell ruin to lessees of the premises. Hence adequate fire protection must consider the contents as well as the surrounding building.

Furthermore, fire protection worthy the name means not only the salvage of the main structural elements of the building after trial by fire; but considerations of uninterrupted business, and insurance interests, both require that the expense and time consumed in making repairs and reconstruction shall be reduced to a minimum. Without auxiliary equipment it is problematical how far incombustible construction *per se* will accomplish such a result. The neglect of those automatic or supplementary safeguards which even the best construction needs, may readily permit a fire to attain such area and intensity as to be beyond the power of the fire department.

The fire in the Parker building illustrates these truths most forcibly. Had the building been equipped with automatic fire alarms, notification of the fire would have been instantly and automatically sent to the central station, which, in turn, would have transmitted the alarm to the fire department. Any authoritative statement as to the result is out of the question, but it may safely be said that the probability is strong that the fire would never have attained sufficient magnitude before the arrival of the department to have completely gutted the building. Had automatic sprinklers with central-station notification system been installed, the result would probably have been even more satisfactory, for, in addition to the alarm which would have been instantly sounded on the premises (and to the supervising central station as well, if installed upon latest and most approved principles), the discharge of water from the sprinkler heads would have immediately begun its work of extinguishing, or at least checking, the rapid spread of the fire. In such a building as the present example, devoted to various mercantile interests, automatic alarms or sprinklers, or both, would not only have been advisable adjuncts in any thorough and comprehensive scheme of fire protection, but should have been considered indispensable, especially in view of the limitations of fire-department work, which will now be considered.

Since the advent of the many-storied office building in most American cities, it has been plain to those actively employed in the business of fighting fire, as well as to those who have thoughtfully followed the course of high-building construction, that the days of the ladder and hose in the hands of courageous and intrepid firemen were gradually passing from their hitherto unquestioned fire-fighting sufficiency, to partial if not complete impotence where the high building was concerned. This fact had been realized partially in the early nineties, while office buildings were rapidly developing from the modest ten stories of the Home Insurance Company's building in Chicago—the pioneer of skeleton construction methods—to the twenty-storied Masonic Temple in the same city, and the Manhattan Life building in New York City; but it was not until 1898, when the Home Life building in New York City was almost completely destroyed by fire above the eighth story (an "exposure" fire), that it became clear to those most intimately engaged in the fire problem—*viz,* the municipal fire department—that new fire-fighting facilities would have to be relied upon more and more in the successful coping with fire above the ordinary range of the department's efficient efforts.

This burning of the Home Life building—a sixteen-story office building which was considered thoroughly fire-resisting according to the standards of its day—demonstrated the fact that, notwithstanding the most modern apparatus of the New York City Fire Department, it was impossible to combat the flames successfully above the eighth story, at least after the fire had assumed serious proportions. In other words, modern *portable* fire apparatus is not and cannot be made efficient for fire-fighting purposes above the limit of about 125 feet above the street level. This is practically equivalent to saying that fire once established above the tenth floor of modern high buildings must be left to burn itself out, as far as portable department apparatus is concerned.

The difficulty experienced in fighting fire in high buildings may be said to increase about as the square of the height. Buildings 300 feet high are no longer marked curiosities in New York City, where no building regulations limit the height of structures (except tenements), and a census of the high buildings recently made by the New York City Building Department shows a total of 538 structures which are 10 stories or more in height;—*viz,* one building (uncompleted) of 48 stories, one of 41, two of 26, three of 25, two of 23, four of 22, nine of 20, two of 19, nine of 18, two of 17, nineteen of 16, nineteen of 15, eighteen of 14, thirteen of 13, one hundred and sixty-nine of 12, one hundred and one of 11, and one hundred and sixty-four of 10 stories. In other words, there are these 538 buildings in New York City which, as far as efficient fire protection is concerned, are above the effective fire-fighting range of the fire department, when working without auxiliary aids; and when it is remembered that the 48-storied tower of the Metropolitan Life building—658 feet high—and the 41-storied tower of the Singer building—611 feet high—are both to contain offices to a height greater than the Washington Monument, the necessity for every possible form of auxiliary equipment looking to the discovery and extinguishment of fire, as well as the need of adequate provisions against smoke and panic, are sufficiently apparent.

Such added facilities for fire protection could not be looked for in any possible improvement of the personnel of the departments, as that had already reached a high degree of excellence; nor in added pumping capacity in fire engines sufficient to meet the new demands of excessive height, although the new high-pressure service soon to be installed in New York will prove of immense value in high-building fires; nor in heavier or stronger hose to withstand the bursting pressures induced by great heights, even though no criticism might be made regarding the bursting of poor hose, as was the case at the Parker building fire; not, in fact, through new portable apparatus of any kind which would solve the problem. The only possible remedies lie, therefore, in the *improved design and construction* of high buildings, and in the employment of *auxiliary fire-detecting and fire-fighting appliances* which should especially be incorporated in the inherent design of each and every building exceeding the height limit of fire-department efficiency.

The matter of design includes such considerations of planning as limiting areas, enclosed vertical openings, exposure hazards, and the isolation of dangerous risks of any and all kinds.

Construction concerns the carrying out of such design through the employment of suitable fire-resisting materials, and with a minimum use of combustible trim, etc. But, beyond all of these elements of fire-resisting design, there are still required those auxiliary aids for detecting and extinguishing incipient fire, such as auxiliary fire alarms and automatic sprinklers, and also those auxiliary appliances which go so far to aid the fire department work under severe conditions. Such appliances include the open or basement sprinklers for flooding underground areas, standpipes and hose connections and reels at all floors, and monitor roof nozzles. It is only through the installation and use of such auxiliary equipment that buildings over ten stories high may be considered in any sense fire-resistive.

It is true that the Parker building was destroyed in spite of a standpipe equipment which was used by the firemen until they were either overwhelmed by falling floors or surrounded by flames which had worked around and below them. It is also true that there was much criticism regarding poor water pressure and bursting hose. But had the building been equipped with automatic fire alarms or sprinklers, and had all vertical openings been thoroughly protected, the loss might easily have been confined to inconsiderable proportions.

MODERN SYSTEMS FOR THE VENTILATION AND TEMPERING OF BUILDINGS.

By Percival Robert Moses.

Mr. Moses treats, with the authority of a specialist, one of the modern sub-divisions of mechanical engineering in which American practice is far in the lead of any other. He presents a general review of the accepted principles and available type of apparatus.—THE EDITORS.

WITH the development of greater scientific knowledge of the requirements of the human system, a demand has arisen for more perfect regulation and moderation of the temperature and quality of the air in our dwellings and workshops. This demand has led to the development of systems of heating and cooling far more elaborate than the old systems, and more complete in the results accomplished. Coincident with the development of improved heating and ventilating arrangements, has come the development of illuminants which do not use up the oxygen and which do not produce excessive heat. Necessarily the installation of these complete equipments is confined to large buildings used for public or semi-public purposes, or to houses owned by wealthy people who can afford to pay for conveniences and comforts.

A modern heating, ventilating and cooling equipment, or as it may be termed, a "Tempering Equipment," should give the following results: an abundant supply of fresh tempered air, free from dust or other impurities, and containing a regulated supply of moisture.

Apparatus for obtaining these results includes a more or less extensive arrangement of screens to take out the dust from incoming air, or washing apparatus to wash out the particles, fans or blowers designed to create a difference in pressure which will cause the air from out doors to flow through the screens into the fan or blower and be discharged at suitable velocities to the rooms requiring tempered air, and similar fans to exhaust the vitiated air and discharge it out doors; and heating or cooling devices to bring the air to a temperature and a degree of humidity suited to the requirements of comfort and hygiene. The tempering of the air proper may be carried on either before or after entering the fan.

The two operations of heating and cooling are performed in similar ways, with the same or similar appliances. For heating the air,

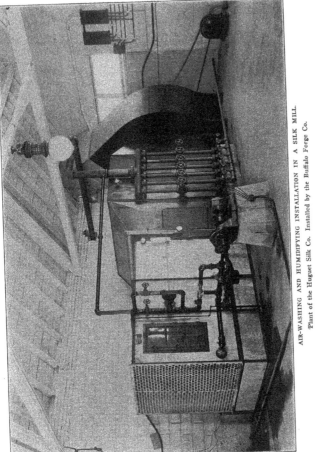

AIR-WASHING AND HUMIDIFYING INSTALLATION IN A SILK MILL.
Plant of the Huguet Silk Co. Installed by the Buffalo Forge Co.

coils containing steam are placed on either the intake or discharge; while for cooling, coils containing cold brine or expanding ammonia gas are used. The admission of steam or brine or ammonia to these coils is controlled automatically by sensitive thermostatic mechanisms located in the rooms. This automatic control is not only a convenience but an economy, as it tends to prevent waste. When the air is heated its percentage of saturation is decreased, and its ability to take on moisture increased. This dryness is objectionable in cold weather, and to overcome it the humidifier is employed, consisting of a coil of pipe containing steam immersed in a bath of water. Steam is automatically turned on, evaporating the water, or shut-off from this coil, as the percentage of humidity in air is to be increased or decreased.

HEATING INSTALLATION, ANGUS SHOPS, CANADIAN PACIFIC RAILWAY.
The B. F. Sturtevant Co.

On the other hand, when the air is cooled, the percentage of saturation is increased, and the tendency to deposit moisture on cold surfaces is increased. This may result in injury to woods or fabrics, and where the amount of cooling is considerable, means for drying the air must be provided. The method of obtaining comparatively dry air is to cool the air far below the temperature at which it is to be delivered to the rooms, causing it to deposit some of its moisture on the pipes by which it is cooled, and then to allow it to heat up again by contact with other air or by passing over steam coils, thus reducing the percentage of saturation. This operation is also

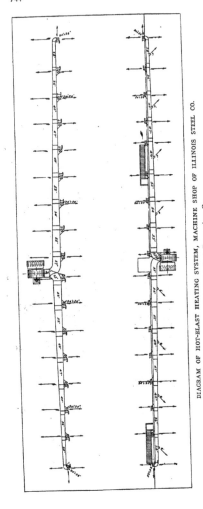

DIAGRAM OF HOT-BLAST HEATING SYSTEM, MACHINE SHOP OF ILLINOIS STEEL CO.
The Green Fuel Economizer Co.

automatically controlled, and the heating is accomplished by a mixing damper or by opening or closing the supply of steam to the coil.

Recently methods have been developed for cooling and heating air and regulating its humidity, by driving it through a finely divided shower of water, passing over heating or cooling coils. The water drops to a pan and is re-circulated. It would seem that such a process would result in the delivery of air completely saturated with moisture and, therefore, liable to deposit a part of its charge as soon as it was cooled. In practice, however, the plan works well. When the outside air is cold its percentage of humidity is low, and when it is heated by merely passing over steam coils its moisture per cubic foot—i. e., its humidity—is still further reduced.

The objectionable

features of very dry air are well-known. The membranes of the throat and nose, and the skin itself, are robbed of their moisture by the thirsty air, and a parched, dried-up feeling results, while the membranes become subject of inflammation familiar as colds and sore throats. By regulating the temperature of the water, and the rapidity of its flow, the humidity of the air delivered can be regulated and the excessive dryness prevented. On the other

HEATING AND VENTILATING INSTALLATION, DOYER PLANT OF THE CHICAGO
PNEUMATIC TOOL CO.
The American Blower Co.

hand, when the air is cooled the object is usually to reduce the humidity, which is generally excessive, and this result is also accomplished by the direct contact of the air with the falling particles of very cold water or brine. These cold particles apparently reduce the temperature of the portions of the air with which they come in contact below the dew point, if the air is at all near saturation, and instead of water being evaporated from the shower, water is actually squeezed out of the air, which is thus dried and cooled at the same time. This result is clearly shown in the plant of the New York Hippodrome, where instead of having to replenish the supply of circulated cooling water, it is necessary to draw off water condensed from the air.

There are important mechanical advantages inherent in such systems of heating and cooling due to the small amount of pipe radiating surface required on account of the greater ratio of heat transmission between heating or cooling surface and water as compared with that between the same surfaces and air. This results in smaller installations, which will necessarily be less costly, and in a much simpler equipment, as filtering screens are eliminated and the number of coils, with valves and traps, greatly reduced.

OUTLETS OF HEATING DUCTS, LOUISVILLE & NASHVILLE R. R. MACHINE SHOP.
The B. F. Sturtevant Co.

After being thus tempered, the air must be discharged into the rooms in such direction and at such velocity as will not cause appreciable draft on the people in the room and, at the same time, will insure a perfect diffusion and distribution throughout the entire space. As cold air is heavier than warm air, the natural impulse would be to discharge the heated air at the bottom, allowing it to rise, and the cold air at the top, allowing it to fall. Such an arrangement is used for large auditoriums such as theatres where the space is very high, but a contrary arrangement has proved necessary for smaller rooms, or rooms having a great area compared to their height, in order to secure distribution of the fresh supply throughout the space. The warm air enters such rooms at a point about two-thirds of the height from the floor, and as it meets the cooler air of the room it becomes cooled and falls. The cold air entering the room near the floor, sufficiently high above the floor to avoid dust, becomes heated by contact with the air of the room and rises. Where it is possible to do so, the air is usually discharged from the inside wall toward an outside wall or exposed surface, and the vitiated air removed from the wall containing the inlet. For warm air this discharge is taken out near the bottom; for cold air it should be taken out near the top.

The same ducts may be used for cool and heated air, but for a perfect system of distribution there should be two outlets in each of the supply and exhaust ducts, the top supply and the bottom exhaust outlets being used for heating, and the others for cooling.

The vitiated air of the rooms is perfectly removed mechanically by a fan, similar to the fan used for the supply, through ducts and outlets as stated.

HEATING INSTALLATION, FIRST NATIONAL BANK, CHICAGO.
The B. F. Sturtevant Co.

This is briefly the method of accomplishing the results, and no attempt will be made to consider all the methods and modifications used in practice, but there are a number of points which may be worth noting. Commencing with the entrance of the air, it is important that the air obtained shall be as clean as possible; hence a

HEATING AND VENTILATING INSTALLATION, MOLINE SHOPS, C., R. I. & P. R. R.

The B. F. Sturtevant Co.

supply from the level of the street or from any location liable to contain impure air should be avoided. On the other hand, it is not necessary to go up to the top of the building to obtain pure air, as the air obtained from a court or from a point sufficiently high above the ground to avoid the direct entrance of impurities meets the needs of the case. The entering velocity of the air should be low, hence the area of the intake must be large. The usual velocity for such incoming air is ten feet a second; where it is higher than this, the tendency is to draw dust and dirt in with it. It may be lower without any bad effects, provided the space is available.

HEATER, SHOWING ARRANGEMENT OF COILS.
The American Blower Co.

In designing screening apparatus it is of the utmost importance that the screens shall be readily cleaned, and the cleaning done away from the intake of the fan. This means removable screens, and the placing of the screens in a position where they can be very readily taken down; and further the use of a suction blower or similar arrangement which will draw the dirt from the screen and discharge it where it will do no harm. More than half the screens in use, it is safe to say, are not properly cleaned or looked after, and when thus neglected they become a menace to health and a hindrance to the proper and efficient operation of the fan. The screens are almost always arranged in zig-zags so as to reduce the space re-

THE GREEN AIR HEATER.

Utilizes the waste heat of flue gases from boiler or other furnaces, applying it to the heating of air for ventilating or drying purposes. The air passes through the tubes, and the flue gases circulate around them. The Green Fuel Economizer Co.

AN AIR WASHER, COMPLETE, SHOWING THE SPRAY SYSTEM.
The Buffalo Forge Co.

quired and, at the same time, give the necessary surface. The amount of surface required varies with the location of the air intake, but an average figure is one square foot of surface for every thirty-five cubic feet of air supplied per minute. These screens are made of cheese cloth framed in galvanized iron mesh. Where vacuum cleaning systems are in use a suction outlet with flexible hose and collector is arranged for cleaning the screens.

The coils for heating or cooling the air, as has been stated, are located alternatively on the intake or discharge. The most efficient and economical location for the cooling coils is on the intake, and for the heating coils is on the discharge; such an arrangement gives the smallest size of fan for a given quantity of air moved per minute, as the air is at its maximum density in passing through the fan. There are a number of reasons, however, militating against this arrangement, and the most usual plan is to place the coils on the intake;

THE GREEN HEATER.
Straight tubes, expanded into the headers.
Easily cleaned, plugged, or replaced.
Positive circulation. Used with live
or exhaust steam or with water.
Green Fuel Economizer Co.

because where space is limited, there will be a more perfect distribution of air over the coils where the flow is caused by atmospheric pressure discharging into the neck or reduced-pressure end of the fan intake, than where the flow is caused by a positive directive force such as a centrifugal fan exerts. Where the space is not limited, it is better to put the heating coils on the discharge, as the capacity of the fan will be from 20 to 25 per cent greater than with the other arrangement, but the coils must be located sufficiently far from the fan to allow the velocity of flow to decrease gradually before entering the coils, and to increase after leaving the coils.

There is no reason why the same coils that are used for heating should not be used for cooling, if brine made up of calcium chloride is circulated through them.

This brine is cooled by one of the commercial types of refrigerating machine in the ordinary manner, by the expansion of ammonia in cooling pipes surrounded by the brine which it is desired to cool. It would seem cheaper and simpler to use the ammonia directly in the cooling coils, and this was the first method adopted; but it was not found entirely practicable, as the expansion of the ammonia from liquid into a gas was so rapid because of the rapid passage of the incoming air over the coils that the moisture in the air froze on the pipes, making it difficult to force the air through. The approved practice, with coil systems therefore, is to expand the ammonia in coils surrounded by brine and to circulate the brine through the cooling coils of the ventilating system. As the brine is necessarily of a higher temperature than the ammonia, this involves more surface for brine circulation than for ammonia expansion, but it is found that the amount of surface required for heating is far in excess of that required for cooling, even with brine, so that where the some coils are used for both purposes there is but little extra expense for the brine system. There is a further advantage in using the brine circulation; that is, the reduction in danger of ammonia in cooling pipes surrounded by the brine which it is no possibility of this with the brine system, but with the ammonia system, if a leak of any magnitude should develop in the coils, a considerable quantity of ammonia might be carried into the rooms before the feed could be shut off.

The method of controlling automatically the supply of steam or brine to the coils, as before stated, is a combination of thermostats and valves controlled by these thermostats. The thermostatic control accomplishes usually two results: First, to prevent the delivery of untempered air to the rooms under any circumstances; and second, to maintain the temperature of the rooms at a fixed amount. To accomplish the first result, it is necessary to put an operating thermostat in the duct leading to the room or space to be applied with tempered air. This thermostat is arranged to open the valve in the supply whenever the temperature of the air in the duct falls or rises

A VENTILATING AND HEATING OUTFIT.
The American Blower Co.

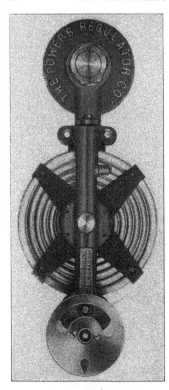

MECHANISM OF THE POWERS THERMOSTAT.

A liquid of low boiling point within the expansible disk exerts in it a pressure varying directly with ordinary temperatures. The resultant movements of the disk, through an elastic diaphragm behind, connected with an air valve, control the flow of compressed air to an air motor operating against a coil-spring resistance, and hence moving the steam valve or mixing dampers to a position corresponding always with the temperature of the thermostat disk. The upper circular portion is the "accelerating valve" for converting gradual action into positive action.

above a stated point. To accomplish the second result, the operating thermostat is hung in the room at some point free from drafts or air currents and at about a man's height from the floor. The thermostat in the duct usually controls a few sections of the heating coil, the remaining sections being thrown on by the room thermostats. There are several such thermostatic systems on the market operating under patent rights, such as the Johnson system and the Powers system, the difference being mainly in the construction of the thermostats.

Where the space to be supplied with tempered air is large and undivided, the thermostat in the room and the thermostat in the duct suffice with a single coil or set of coils. Where the spaces are subdivided and separated, it is necessary to have a thermostat in each space separately controlling its air supply. There are two ways of regulating such separated supply: The first is to have a reheating coil at the base of the duct leading to the space in question, this reheating coil being controlled by the thermostat in the room. The other arrangement is to have a double duct system, one duct containing cold air, and the other tempered air. These ducts are joined together at the base of the flue leading to the space to be supplied, and the relative amounts of cool and warm air admitted are adjusted by a damper, which damper is controlled by the thermostat in the room. With either of these methods the required amount of air for ventilation can be supplied and the temperature maintained at about a given point; while the first method is the simpler, the regulation of temperature is not as quick on account of the heat retained in the reheating coil, even after the steam is shut

off. In order to avoid possible misunderstanding it may be well to state that the thermostats do not directly operate the valves or dampers. Their function is similar to that of a relay in telegraphy— that is, to open or close a circuit, allowing compressed air, electricity, or other energy transmitter to flow and operate the valves through unbalanced pressure on the underside of the diaphragm or similar arrangement. The regulation of humidity is accomplished in a similar manner, wet and dry bulb measuring apparatus being used to open and close the circuits, instead of the ordinary measuring devices.

The ducts conveying the air from the fans to the rooms are

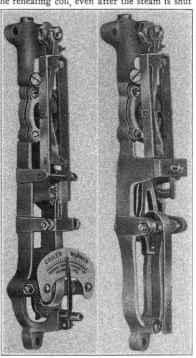

MECHANISM OF THE JOHNSON THERMOSTAT.
The metal strip at the bottom is the actuating member, operating by the different expansion of two metals under heat changes. Its movement opens and closes the small air valve at the top, which in turn governs the air-motors regulating the hot water or steam valves or the hot and cold air dampers. The Johnson Service Co.

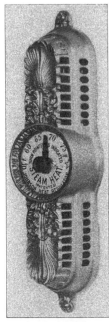

JOHNSON THERMOSTAT.

usually of galvanized iron varying in thickness from No. 20 gauge for large ducts to No. 24 and even No. 26 for vertical flues. The sizes are fixed by allowable velocity which in horizontal ducts is from 15 to 20 feet per second and from 12 to 15 feet per second in vertical flues. Where the air enters a room its velocity should not exceed 6 feet a second, except in extreme cases where the air has to be discharged a long distance and a question of drafts is not important. The design of the ducts has a great bearing on the success or failure of the system. Air is readily and rapidly moved by very small differences in pressure, and such small differences in pressure are liable to exist from all sorts of causes, such as open dampers and registers, etc. It is important, therefore, that a system of ducts should be designed to insure a proper division of air between the different supply ducts to rooms without having to maintain a greater pressure than would be necessary to insure the required velocity with a natural division of the flow of air. If the air has to turn sharp corners, and thread its way through numerous bends on one duct, it becomes necessary to throttle down the supply to all the other ducts and increase the pressure in the main supply in order to get the required amount through the crooked duct. Such an adjustment can not give satisfactory results, as the increased pressure causes increased veloc-

WALL PLATE OF THE POWERS THERMOSTAT.

Showing the ports for the compressed air pipes.

ity, and although the quantity supplied through a certain branch may be reduced by shifting its deflector, there will probably be a whistling sound and the most direct branches will receive the most air. A properly designed system should provide for easy and gradual change in direction of the flow of the air with as few bends as possible, and the main ducts should be gradually reduced, as branches are taken off, so that the natural tendency will be for the air to divide and go partly through the small branch and partly through the large duct.

Every such system of ducts should have movable deflectors at each branch, as no system can be planned before hand to take care of variations in pressure in rooms on different sides of the building.

It is usual to cover the main ducts with some insulating material, such as air-cell covering, in order to prevent the unpleasant radiation of heat. As the temperature of the air in the ducts is frequently 120 degrees, unless they were covered, radiation would have a considerable effect on the surrounding rooms and the air would be cold before delivery. For the same reason, it is usual to cover the housings of the fans, where the heater is on the intake, and the housing of the coils. This housing is also advisable to prevent loss through radiation and leaks through sudden changes of temperature of the heating surface as steam is turned on and off automatically.

A fruitful source of trouble in the design of heating and ventilating systems

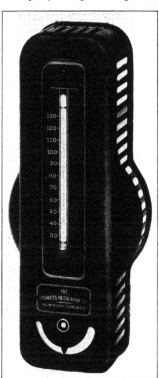

POWERS THERMOSTAT, COMPLETE.
The Powers Regulator Co.

DOUBLE SQUARE FLUE DAMPER.

The hot and cold-air flues are superposed, and the dampers are at right-angles and on the same spindle. As one opens, the other closes. The air motor controlled by the thermostat is rigidly attached below. The Johnson Service Co.

is in the exhaust system. Some positive method of exhausting the air is necessary. It is not safe to rely upon natural ventilation to get rid of the heated air or cool air, as variations in the direction of the wind, opening and closing of windows, opening and closing of doors, all affect the operation, and may make a system a complete failure unless there is some more positive means of exhausting the air from the room than can be obtained by natural circulation. A great many of the school houses of New York City were designed with merely a forcing apparatus for discharging the air into the rooms, but no means were provided for drawing the air out. Large flues were provided to allow the air to come out naturally, but it was found in many instances that instead of the air going out cold air came in, and it was necessary to board up the openings in order to prevent the inrush of cold air.

Unless the air to be exhausted is confined to one large space, it is not possible to put a fan at each space from which the air is to be taken. The question then becomes one as to whether it is better to put the fan at the end of the series of ducts, or to put it somewhere between the end of the series of ducts and the discharge to the atmosphere. It is my experience that the latter is the best. If the fan is located at the end of the series of ducts just before the air is discharged to the atmosphere, there are liable to be all sorts of inleakages, and all the friction of the ducts is pulling on the air, preventing it from being freely fed to the fan. It is my practice to locate the fan used to exhaust the air at some point between the long discharge flue and the various flues leading in from the rooms. In this way the work on the fan is divided equally between the discharge and the suction side.

It is of the utmost importance in the design of the exhaust system that the area of the ducts be maintained throughout their length. Any contraction in area, or any sudden sharp bends will materially affect the operation of the system. The amount of air flowing from an outlet to the fan, depends solely upon the frictional resistance of the ducts and inlets. The vacuum or minus pressure exerted by the fan suction is a constant for a given speed, so that it does not matter how many ducts there may be leading to a fan suction, if the total quantity of air discharged by the fan is not greater than its rated capacity; the shutting off of other ducts will not benefit the exhaust from any one duct through which the proper supply of air may not be coming, due to improper design and excessive friction. In this way the exhaust system differs materially from the supply system. In the supply system if one of the ducts is getting more than its share, it may be cut down by the use of deflectors, but in the exhaust system, the use of deflectors shutting off one, two or three ducts would not increase materially, the amounts drawn through the other ducts. In some cases the smokestack is enclosed by a ventilating flue, the heat of the stack being used to heat the air in the surrounding flue, causing this to rise and draw in the air from below. Such a system is all right where the space to be ventilated is at or near the base of the stack, but it is of no use at all where the space to be ventilated is near the top of the stack. In one instance, rooms on the fourth and fifth floors of a seven-story building were to be ventilated in this manner, but instead of the air from the rooms going out around the stack, the air from the space around the stack poured into the

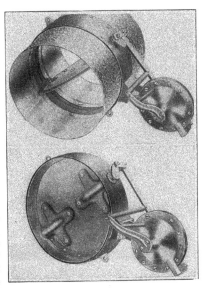

SINGLE ROUND-FLUE DAMPERS, CLOSED AND OPEN.

Operated by thermostatically controlled air-motor. The Johnson Service Co.

rooms, heating them up to such a degree as to make them un-
bearable. The difficulty in this case was that whenever there was
the slightest wind pressure on the top of the stack, the resistance to
egress around the stack of the heated air was much greater than it
was into the room of the building, and of course the heated air chose
the path of least resistance.

There is one feature in connection with the cooling of rooms to
which attention should be directed, as it has been the cause of a great
deal of trouble and the failure of cooling systems. It is a matter
which is generally overlooked, because it does not enter into the
consideration of heating buildings. This point is the amount of heat
given out by the condensation of moisture in the atmosphere. As
before stated, as the air is cooled, the moisture contained in it is condensed and deposited; and each pound of this moisture so condensed
gives out approximately 1,000 B. t. u. which must be taken up by the
refrigerating medium. The amount of refrigeration required to overcome this may be much greater than the amount necessary to over-

A LARGE DIAPHRAGM VALVE.
The Johnson Service Co.

come the heat radi-
ated through walls
and windows into the
building and from the
persons in the build-
ing. In one instance
that I have in mind,
where the cooling of
a large store floor
was planned, t h e
number of B. t. u. re-
quired for cooling the
air needed for ventila-
tion and for replacing
the heat l o s s e s
through walls a n d
windows was about
800,000 per h o u r,
while the number of
B. t. u. required be-
cause of the conden-
sation of moisture

was 1,300,000. In this particular instance, there were 1,200,000 cubic
feet of air per hour which it was supposed would enter at 92 degrees
F. at 95 per cent saturation, and leave at 82 degrees F. and 65 per cent

A COMPLETE HEATING AND VENTILATING OUTFIT.
The Green Fuel Economizer Co.

saturation; the amount of moisture condensed would be 1,384 pounds. These are extreme cases, but they serve to show the importance of the consideration of the humidity factor.

If a water tower is installed for cooling the circulating water, the total cost per annum, taking into account the lesser number of days in which cooling is required, compared to the number in which heating is required, would be about the same for cooling as for heating, including fixed charges on the equipment. This estimate is based on the supposition that cooling would be required for only 45 days a year, as compared with 180 days for heating. The cost of operation is, therefore, not excessive, and it is only a question of a short time before the cooling of large buildings will be as much a matter of course as their heating, particularly in climates subject to such extremes in temperatures as our own in New York.

If exhaust steam is available from an electrical plant or pumping machinery, the costs of both heating and cooling are greatly reduced through the reduction of both the fuel and labor items. The use of exhaust steam for heating is old, but its use for cooling is novel and not so widely known. It is, however, entirely practical and successful. This steam is used in the absorption system of refrigeration to evaporate and separate ammonia gas from a mixture of ammonia and water. The ammonia is then liquefied by cooling water and allowed to return to its gaseous state in the cooler and be re-absorbed in the

HEATING APPARATUS IN PAINT SHOP, N. Y., N. H. & H. R. R., READVILLE.
The B. F. Sturtevant Co.

mixture of water and ammonia in the absorber in the usual way. No high-pressure steam is required, except for a very small pump. With such a system, the cost of cooling is reduced to the fixed charges, repairs, water cost, and some labor, and this cost may be further reduced, where the number of days operation and size of plant warrant, by the installation of a cooling tower.

In every branch of science and mechanics, development has been wonderfully rapid in the past thirty years, and when a system of cooling, heating, moistening, and drying the air as described is compared with the old-time methods of isolated stoves, it is evident that the art of heating and ventilating has not stood still.

There is no reason at the present time why a building, if properly designed for the purpose, should not be maintained all the year at a comfortable, healthy temperature with air containing the proper percentage of moisture and at the same time free from dust or other impurities. The great difficulty is to persuade the owner to spend the requisite amount of money for a complete system, and to persuade the architect to adapt his building plans to allow the system to be installed perfectly. This involves usually some sacrifices in decorative effect and in available space, but from the point of view of the man who is to live or work in the completed building, these matters are of no importance when compared with the results to be obtained.

THE NEVADA COPPER FIELDS.

By A. Selwyn-Brown.

Various extensions in the use of copper as a material of construction—especially the vast and progressive expansion of the electrical industry and the application of electric power —have altogether combined to place copper in the position of a rival to iron as a "barometer of trade." It is expected by many, at least, to give a more prompt and more sensitive indica-tion of progress in the resumption of a normal stage of industrial activity. The attention thus centered on the metal heightens interest in Mr. Selwyn-Brown's excellent review of one of the newest and most important copper-bearing regions. His clear definition of the geology will be found particularly interesting and valuable.—THE EDITORS.

IN the early mining days in the West, Nevada became famous as a great silver-producing State. The mines on the Comstock and at Eureka, Tenabo, Austin, Belmont, Ely, Golconda, Tuscarora, Candelaria, and a dozen other fields, gave employment to many thou-sands of men and produced immense quantities of silver ore. The decline in the value of silver early in the eighties, however, caused silver-mining operations to become unprofitable and, by degrees, the silver camps became deserted, or were continued by a few miners who were able to lead a precarious life by mining gold and other ores on a small scale. A revival of mining interest in this State was induced in 1905-7 by the rich gold discoveries in the Tonopah, Gold-field, Bullfrog and other gold-mining districts in the southern por-tions of the State, and the returns from these districts promise to force Nevada to the head of the gold-producing States within the next few years. But the mineral wealth of Nevada is diversified, and when the great copper plants now in course of construction are completed and run to their full capacities, Nevada will also take rank among the fore-most copper-producing States. At a later date, when investors turn their attention to Nevada's wealth in zinc ores, her zinc production also will become extensive.

In the early days of mining in Nevada smelting practice was not well-known and the miners were compelled to carry out expensive furnace experiments. All over the State the remains of their plants are to be met with, and many are historically celebrated. It is gen-erally conceded that the birth-place of American lead smelting is at Eureka, Nevada. It was there that Kustel, Stetefeldt, and other re-

nowned metallurgists developed the smelting processes that have since
placed American smelting practice in advance of that of other countries. Experiments with new methods of concentrating copper, lead,
and zinc ores, preparatory to smelting operations, are being carried
on in various parts of Nevada which promise to work another revolution in metallurgical practice, and thus add another interesting chapter
to the State's mining history.

It is proposed in this article to survey briefly the principal copper
centers of the State and to describe several of the most important
copper-reduction plants now in course of construction.

TOPOGRAPHY.

Nevada consists of an immense tableland suspended on the Sierra
Nevada on the west and the outlying ranges of the Rockies on the
east. The elevation varies between 3,000 and 7,000 feet above sea
level. The State is traversed throughout by innumerable mountain
ranges having a general trend in a northeasterly direction. Arid valleys occur between the parallel ranges. There is very little natural
vegetation beyond sagebrush, and stunted trees such as juniper, *piñon,*
cottonwood, and pogonip or white pine, which grow in patches on the
mountains a little below the snow belt. Large areas of the desert
valleys, however, are very fertile when irrigated.

GEOLOGY.

The geological features of Nevada are in many respects remarkable and have inspired some of the finest geological monographs
hitherto published in America. The pioneering work was done by
Major Powell, Clarence King, Hague, Gilbert, Turner, Wheeler,
Walcott, and Curtis. Messrs. Spurr, Rowe, Weeks, and Ball of the
United States Geological Survey and Professor Lawson of the University of California, have recently published valuable geological
monographs dealing with specific areas.

The most remarkable feature of the geology of Nevada is the
evidence of volcanic activity everywhere visible. In places it is seen
in immense sheets of volcanic rocks, intrusive dikes, or gigantic ruins
of volcanic craters. In other places it is shown by gigantic fissures,
the result of earthquakes, marvelous metamorphic phenomena, and
immense beds of sinter and volcanic ashes. Often steaming springs
and fumeroles are met with, showing that volcanic activities are still
in operation, and unmistakably indicating that Nevada is still within
the historic "Circle of Fire," the great volcanic belt that rims the
Pacific Coasts of the Americas, as well as those of south-eastern Asia

and Australia, and continues in an undeviating course through many of the islands in the southern seas.

Investigations made by Mr. Spurr in 1900 in various districts in Nevada led him to believe that the whole of the State, as well as a large part of the Mojave desert, in California, is underlain by a single body of molten magma, which has supplied, at different periods, lavas · of similar composition to all parts of the overlying surface. This great territory thus forms a single petrographic province occupied by similar rock formations. More recent 'researches have increased the limits of this great province and many of the best authorities on the geology of the West Coast of America are agreed that one vast petrographic province stretches from Alaska away through British Columbia, Idaho, Nevada, Arizona and Mexico, down south into the Andean regions of South America.

The general sequence of lavas in Nevada is in the following order : I. Rhyolite, (Eocene.) 2. Andesite, (Miocene.) 3. Rhyolite and basalt, (Miocene-Pliocene.) 4. Andesite, (Late Pliocene-early Pleistocene.) 5. Basalts and occasional rhyolités (Pleistocene). This most remarkable petrographical province is accompanied by a coextensive metallographic province carrying some of the richest ore deposits hitherto discovered, and it is furthermore worthy of remark that a large proportion of these ore deposits are similar in age and in the manner of their occurrence, and have an almost identical geological history. For example :—the ore deposits of the Comstock, Tonopah, and De Lamar are, as regards their mineralization and geological history, almost identical with the ore deposits at Guanajuato, Pachuca, and other great mining districts in Mexico.

The above description of. the Pacific petrographic and metallographic province will explain the similarity of the geological features of most of the mining districts in the State. If we were considering the. gold-mining districts no other geological formations need be referred to than the volcanic strata named above; for it is in the rhyolite and andesite rocks that the principal gold veins are found. The great copper deposits, however, are found mostly in association with certain sedimentary rocks that require to be briefly mentioned. In White Pine County, a series of limestone formations occur in association with quartzites and siliceous and argillaceous rocks in which copper, silver, lead, zinc and other ores are present. The Nevada limestone, extensively exposed at Eureka, is over 6,000 feet in thickness and is of Devonian age. The Lone Mountain limestone formation, also well seen at Eureka, is of Silurian age and is from 1,800 to 2,000

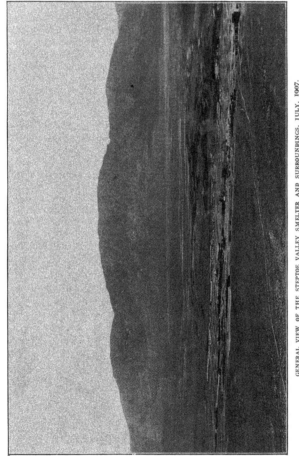

GENERAL VIEW OF THE STEPTOE VALLEY SMELTER AND SURROUNDINGS, JULY, 1907.

Situated at McGill, 14 miles north of Ely.

STATION OF THE NEVADA NORTHERN RAILWAY AT ELY, NEVADA.

feet in thickness. The Pogonip, Prospect Mountain, Red Wall, and Wasatch limestones occupy large areas in Central Nevada. The Ogden quartzite covering the Paleozoic area west of the Wasatch Range, in Utah, occurs in many places in White Pine and Humboldt counties. Less extensive sedimentary formations will be referred to when describing the geological features of particular copper mining localities.

THE ELY COPPER FIELD.

The most promising copper-mining district in Nevada at present is at Ely, in White Pine County. It is the scene of great activity and numerous important copper-mining companies with a total capital of over $200,000,000 are developing large areas of copper-bearing land, and erecting extensive concentrating and smelting plants.

Ely is situated in the southern portion of White Pine County, in North-eastern Nevada, about 40 miles due west of the Utah boundary line. It is connected by rail with the Southern Pacific main line at Cobre. The distance by rail between Salt Lake City and Ely is about 317 miles.

The town possesses a population of about 4,000 and is nicely situated at the mouth of a cañon in the Egan range known as the

Robinson cañon. It is one of the oldest mining centers in Nevada, and has had a very checkered history. It first became prominent in 1865, shortly after the discovery, in the autumn of that year, of some very rich silver veins on the western side of Mount Pogonip, a prominent landmark in north-eastern Nevada, which rises to an elevation of 10,792 feet above sea level, and is the highest peak in the White Pine Range, running parallel to the Egan Range on the west. The main road to the White Pine mining district passed through Robinson cañon, and thus Ely became an important base of supplies for the prospectors who led to the district by stories of the fabulous richness of the Pogonip silver veins, set to work and prospected the whole of the north-eastern districts in the State. As the result of their labors, deposits of silver, gold, zinc, molybdenum, manganese, copper, opal, turquoise, sapphire, chiastolite, of more or less value, have been discovered in and around the Ely district and throughout the whole of White Pine County.

The Egan Range begins at a point a little north of the fortieth parallel, which was made celebrated by Clarence King's geological surveys for the United States Geological Exploration of the Fortieth Parallel Reports, and runs southward for a distance of 150 miles. This range is composed chiefly of a strongly defined central ridge, which, in the vicinity of Ely, is considerably disturbed by intrusive volcanic rocks. The range is cut through by several cañons, of which the Egan and Robinson are the best known. Granite of Archean age, overlain by quartzite and quartzitic schists of Cambrian age, is exposed near the eastern end of the Egan cañon. From Egan cañon south to Ely the Egan Range is composed chiefly of stratified rocks. A well-formed dike of hornblende-tonalite-porphyry runs through the strata in Robinson cañon and runs eastward with a number of similar dikes toward Howell's Peak. It has, doubtless, played an important part in the mineralization of the district.

The Ely copper deposits occur in a belt of intrusive quartz-porphyry, or monzonite. Two well-defined areas are occupied by the porphyry. One commences at Copper Flat and runs eastward for a distance of three miles to a hill called Rusty Ridge. This belt is about a mile in width. The other porphyritic area stretches from Ocher valley eastwards to the Chainman mine. It is three miles long and about three-quarters of a mile in width. The Ruth limestone and White Pine shales are the principal Palaeozoic sedimentary formations intruded by the porphyry laccolites. It was in these sedimentaries that the early gold and silver mines were operated. Roughly speaking, the gold-bearing veins are distributed in the sedimentary formations lying to the south of the copper-bearing porphyry, while the silver-lead deposits are met with chiefly in the limestone formations on the north of the porphyry. The silver-lead-zinc-bearing Ruth limestones are of later geological age than the porphyry; the gold-bearing sedimentaries south of the porphyry are older than the copper formation.

PLANT OF THE CHAINMAN CONSOLIDATED COPPER COMPANY, ELY, NEVADA.
Taken in July, 1907. The light spot at the left is a dust whirlwind.

The porphyry is mineralized in parts only. It is a light, yellowish quartz-feldspar rock. Through its contact in a molten state with the shale and limestone formations, large bodies of quartzites and garnet rocks have been formed, which also carry chalcopyrite as a product of contact metamorphism. The porphyry has been greatly disturbed in places by faults and dikes. One of the dikes cut in the workings of the Ruth mine is found to be composed of minette carrying sapphires like a similar rock in which sapphires are mined in Montana.

THE MINES AT ELY.

The present impetus to mining at Ely is due to the efforts of Messrs. Bradley and Requa of San Francisco, who on learning several years ago of the similarity of the copper-bearing porphyry to the Bingham porphyry in Utah, secured numerous leases and developed the ground. In 1904 they employed Professor Lawson, of the Uni-

versity of California, to survey the field geologically, and on information gained by a study of Lawson's maps and monograph very large
tracts of country were leased. Shortly afterwards an era of promotion set in and the Nevada Consolidated, Giroux Consolidated, Cumberland-Ely, Ely Central, and other companies were formed to operate copper mines on the field.

OPEN CUT, EUREKA MINE OF THE NEVADA CONSOLIDATED COPPER COMPANY.
The railroad track rests on solid ore, ready for the steam shovel.

The porphyry is usually cut at a depth of from 80 to 300 feet from
the surface. The ore-bearing portions have been proved to a depth of
1,600 feet and for an equal width. The average value of the ore is a
little over 2½ per cent copper, or say 50 pounds of copper per ton.
The cost of production is estimated at 7.03 cents per pound net, delivered in New York.

Each of the above mentioned companies has developed immense
ore bodies. The ore blocked out in the Eureka and Ruth mines of the
Nevada Consolidated Company alone is estimated at more than 15,-
000,000 tons with an average copper content of 40 pounds of copper
per ton, and a little gold and silver.

BELOW, THE GIROUX CONCENTRATOR. ABOVE, THE BUTTE-ELY MINES IN THE FORE-
GROUND WITH THE GIROUX CONCENTRATOR IN THE DISTANCE.

Taken Oct. 29, 1907.

VETERAN SHAFT OF THE CUMBERLAND-ELY COPPER COMPANY.
The beginning of a four-compartment shaft; taken Oct. 31, 1907.

772

The Steptoe Valley Smelting and Mining Company, in which the Cumberland-Ely and Nevada Consolidated companies are interested, is building a concentrating and smelting plant at McGill, 14 miles north of Ely. An area of eight square miles has been secured for a smelter site, and the water rights over a larger area have been purchased. The directors of the company, being experienced copper smelters, secured all this land with a view of minimizing the damage done by the smelters and thus avoiding the expensive litigation which has constantly harassed the Montana and Utah smelters. The smelters will have a daily capacity from 2,000 to 5,000 tons of ore. The power plant, to furnish power for the smelter, concentrator, and mine plants, will have a capacity of 6,500 horse power. The concentrates will be treated in the smelter, which consists of roasters, reverberatories, blast furnaces and converters.

STAR POINTER SHAFT, NEVADA CONSOLIDATED COPPER COMPANY.

The claim area of the Nevada Consolidated Copper Company amounts to about 850 acres, consisting of 63 claims. It is divided into two general sections, the Eureka group, comprising 27 claims, and the Ruth group of 36 claims.

GENERAL VIEW OF THE STEPTOE PLANT (OWNED JOINTLY BY THE NEVADA CONSOLIDATED AND CUMBERLAND-ELY COMPANIES) TAKEN IN THE LATTER PART OF NOVEMBER.

774

THE THIRD AND THE UPPER WILFLEY FLOORS OF THE STEPTOE CONCENTRATOR.
Steptoe Valley Mining & Smelting Co.

ALPHA SHAFT, GIROUX CONSOLIDATED MINES CO. ONE OF THE FIVE MAIN WORKING SHAFTS.

The deepest in the district; nearing the 1,100 ft. level. Has ample ore and is equipped with a upright plant of large bins of high-grade ore, and is equipped with a power plant of 250,000 gals. per diem capacity for supplying water for the concentrator.

The copper ore occurs mainly as chalcocite as a secondary e n r i c h - ment; this chalcocite, in the form of s p e c k s , veinlets and bunches, being in the porphyry, of w h i c h there are large intrusive masses in both the Eureka and the R u t h groups. Besides copper in the form of chalcocite, there is also some chalcopyrite. Iron pyrites is associated with both minerals.

The G i r o u x Consolidated Company owns 52 claims comprising an area of 1,050 acres. This mine is about two miles in length and from 2,000 to 5,000 feet in width. It lies at an elevation of 7,300 feet above sea level. The mine has been developed through five shafts and the porphyry h a s been found ore-bearing at a depth of 1,200 feet. The company has erected a concentrating and smelting plant that will shortly be working.

YERINGTON.

The mines in Douglas and Lyon Counties in the vicinity of Yerington are the next in importance to the Ely mines at present. Yerington is situated on the Walker River, about 35 miles east of the California boundary, and about 12 miles southwest of Wabuska, a station on the Carson & Colorado Railroad. The altitude of the mines is about 4,300 feet above sea level. The principal workings are situated on a portion of the Mason Valley range, which is composed of a core of granitic rocks saddled by later volcanic formations and by sedimentaries. A remarkable feature of the geology is the occurrence of a large body of garnet rock, as at Ely, which lies between the intrusive granite and the limestone. It is evidently a product of contact metamorphism. Large masses of iron are associated with the garnet rock. Close to the garnet rock, the limestone has been changed into wollastonite, fluorite, and marble in places. There is an intimate connection between the garnet rock and ore bodies. The copper ore is not found in regular lodes, but in vughs and pockets filling irregular chambers and fissures in the limestone, garnet rock and granite. It would seem that the molten mass of intrusive lava, now forming the granite, melted portions of the limestone, and on cooling, the lighter upper portion crystallized out as garnet rock. The prevailing garnet is grossularite composed of lime and alumina. This in places is changed into andralite, the iron garnet, by iron replacing the alumina. This crystallization of the garnets took place simultaneously with the mineralization of the ore bodies.

The geological features of this field very closely resemble those at Ely. But up to date the granite has not been found to carry large low-grade bodies of disseminated copper like the Ely porphyry. The copper in the Yerington mines is found in large chambers in the overlying limestone or in the granite, and in shear planes and other fractures in the garnet rock, limestone, and granite. Some immense quantities of copper ore have been blocked out for stoping in the Nevada-Douglas, Ludwig, and other leading mines and important extensions to the metallurgical plants are being planned.

In addition to the deposits at Yerington, a large amount of prospecting for copper ore is being carried on right along the Walker River, Smith Valley, Excelsior, Candelaria, White, Silver Peak and other ranges running along the boundary between Nevada and California. Copper deposits have been located in numerous places

THE STEPTOE CONCENTRATOR AND RAILROAD TRESTLE. LOOKING NORTH.

in all the above named mountains, and numerous promising copper mines are being developed. The geological features of the ranges and of the copper deposits are the same from north to south, for they lie within the petrographic and metallographic province described above. These ranges will in time become the seats of great copper-mining industries.

BELMONT.

This is a famous old mining center situated at the south-eastern end of the Toquima Mountains, in Nye County, and about 50 miles north of Tonopah. The Toquima Range is composed of volcanic rock, chiefly; but in places limestone and schists occur. Near Belmont the stratified rocks, which dip at high angles, have an estimated thickness of 5,000 feet. There has been a large amount of ore deposition in the mountains, and a large amount of mining has been carried on during the past 30 years. Copper ore occurs in well-defined quartz veins traversing the sedimentary formations, and in the granite. Some of the most important mining properties in the vicinity of Belmont have been acquired by powerful financial interests lately, and it is proposed to ship the Toquima copper ores to Ely, when the proposed

Tonopah and Salt Lake Railroad, which is to pass through both Belmont and Ely, is completed.

TOYABE.

The Toyabe Mountains run parallel with the Toquima Range on its western side. They are about 100 miles in length. The mountains are composed chiefly of volcanic rock. The peaks are high and sharp, and the intervening cañons are deeply cut. The range has had a varied mining history. A large number of valuable mines have been discovered and worked at various times during the past with more or less success. At present there is little activity in this section, except in the vicinity of Austin, the oldest mining town in Nevada, where several old mines are being opened up again. Important lead and zinc deposits as well as copper deposits occur in the Toyabe Mountains.

ELLSWORTH.

This is an old copper-mining center, now almost abandoned, situated at the southern end of the Ellsworth Mountains, sometimes called the Desatoya Range. This is a range composed of ancient volcanic rocks and derived tuffs with numerous limestone cappings, running parallel with the Toyabe and Reese River Ranges in Central Nevada. The geological features of the district are favorable for mineral deposits, and a large number of mines have been located and prospected. At present the district is too remote from the railroad to enable the mines to be operated with profit.

REVEILLE.

The Reveille Mountains are in the western portion of Nye County, directly east of Tonopah. The range is composed of igneous and palaeozoic sedimentary formations, and contains valuable deposits of lead and copper. Several large mines are being developed near the town of Reveille by the Nevada Smelting and Mines Corporation, which is likewise developing valuable properties at Tybo, in the Kawich Range north-west of Reveille.

CONCLUSION.

Some of the most important of the copper-mining centers in Nevada have now been briefly reviewed; but many of the copper districts have not been mentioned owing to the lack of activity of the mine owners. There is hardly an important mountain range in Nevada that does not contain copper deposits of more or less economic

THE GIROUX CONCENTRATOR.
A 500-ton unit of the 1,000-ton plant, designed to treat low-grade porphyry ores.

importance. Owing, however, to the aridity of a large portion of the State, and the bad and costly transportation systems, slow progress is necessarily made in mining developments. It may, however, be confidently asserted that the immense ore reserves developed in the copper mines at Ely and Yerington alone, will assure Nevada a prominent position among the copper-producing States. It will, possibly, be two or three years before the mines are in full operation, but in that time Nevada will, doubtless, equal Utah in copper production. Then the success of her gold and copper-mining industries will draw attention to her other metallic wealth and consequently enhance the importance of the Cinderella State. Thus past neglect will be atoned for.

A SIMPLE SYSTEM OF RECORDING SHOP COSTS.

By C. J. Redding.

In the following article Mr. Redding outlines a system that has proved successful in an English works employing three thousand men. The most noteworthy feature of the system is the use made of a preliminary "shop summary." These shop summaries are in reality a specification of processes by which the course of the work is mapped out in advance. They represent a growing practice in some of the most advanced shops and will be adopted sooner or later as the proper starting point for all shop costs. Mr. Redding demonstrates their value from more than one point of view.—THE EDITORS.

THE success of every cost system depends on the accuracy with which the workman's time is recorded. The significance of this truth is not properly appreciated by most managers. Many works are equipped with elaborate systems for collecting the costs of material, but that of the wages is often relegated to the background, as being of no great importance. How great a mistake this is soon appears when these costs are examined by a practical person, the wages charged for some parts proving too high, for others far too low. The result of such an examination is a general condemnation of the whole system and no confidence is placed in any future costs. The wages side of cost-keeping is the most important, and this fact must be realised by any man who thinks about it at all. It seems to be the general opinion that any one is capable of taking a man's time. This idea is quite right if the system is a good one, but if it is not, then only the very best men obtainable are good enough to be time takers.

The object of a cost system is to obtain the cost of an order; but how much more valuable is the system when the cost of each detail of the order is also obtained. The more detailed manner in which the men's time is recorded, the better are the results. With this object in view I propose to explain a system which I have worked successfully for some time in a large engineering works employing about 3,000 hands.

In many works time takers are employed to record the men's time; in others the men record their own, on time sheets, etc. Both systems are good, as long as the time takers and the men each do their part faithfully, but experience has demonstrated that neither is good enough for an up-to-date works, in which facts alone are required.

Many managers and foremen point to the time wasted by the workmen recording their time, and are convinced of the economy of time takers. Is the workman's time saved by employing time takers? When the time taker arrives the man does not go on working, but stops immediately and proceeds to tell him all he has done, and probably dilates on the weather or some other subject as well. From long observation of time-taking methods I have come to the conclusion that more mistakes are made when time takers are employed than when the men record their own time.

Another point against time takers is that the men bear no responsibility for the accuracy of the time recorded. They know quite well that, in case any time is questioned, they can say the time taker has made a mistake and put down the time wrong. There can be no appeal for the time takers, as it is only a case of one man's word against another's. After all, when one time taker has to take a number of men's time, and he happens to be behind-hand in doing it, it is only human nature to economise work as much as possible. Take any works where time takers are employed, and examine the time books. See how many jobs each man is engaged on each day for the first three days of the week; then see how many there are for the last three days, but more especially the last one. I have often looked through these time books, and always thought it a remarkable coincidence that on the last day of the week nearly every man was engaged on one job only, for the entire day. This result speaks for itself, and in my opinion deals a fatal blow at any system which employs time takers alone to record the men's time.

The old method of making the men make out their own time sheet was costly, and the foremen were justified in drawing attention to the expense. One man often made up the time sheets for five or six, with the result that a lot of time was wasted, but no means were given to the men to lessen the time taken in making out the time sheets. These are some of the arguments against the men recording their own time. Give the men every facility for recording their time, and the result is all that can be desired.*

---

* An interesting mode recently advanced in the United States substitutes an annunciator system for the time taker. The workman rings up his starting and stopping times on the annunciator, the job ticket being stamped at the signal by a cost clerk at the central office and the elapsed time noted with the aid of the calculagraph.

These facilities are practically the combination of workmen and time takers, and the system evolved is undoubtedly the best from every point of view. The results obtained soon furnish the proof of this statement and are summed up in the three essentials for every business—accuracy, speed, and economy.

When this combination is worked the number of jobs a man is engaged upon on the last day of the week is just about the same as on the first or any of the other days. This system is as follows :—

Each department of the works is supplied with cards called "shop summary cards" ruled and printed as shewn :—

		SHOP SUMMARY CARD						
		FOR ORDER_____						
	Tracing No._____			Received in Shop_____				
Refer-ence	No. Off	Description	Operation	Man's Name	Man's No.	Date Begun	Date Fin-ished	Total Hours

The Engineering Magazine

THE SHOP SUMMARY CARD.

As soon as an order or drawing is received in the shop, the foreman or some capable person examines it and then proceeds to fill up a shop summary card for it. He writes down the order number, tracing number, and date the drawing was received in the shop, and copies each reference letter with its description on the card, leaving spaces after each reference in which to place the various operations each has to undergo in its course through that department. Consequently each shop summary card will contain the order number, tracing number, and all the details on it, with the reference letter of each and all the operations shewn on the tracing. By this means the manager is assured that each tracing, on its appearance in the shop, has had the close attention of the foreman. The foreman can thus arrange his work to the best advantage and small details are not overlooked as otherwise so often happens to great annoyance.

In the case of an order for a standard part which does not require a drawing, the person responsible must deal with this in the same manner as if one had been sent into the shop for it. He must shew each detail with the necessary operations to each, on the shop summary card. Many managers may say that they cannot have their foreman's time taken up by this work, but if they go carefully into the question and examine the various advantages to be obtained they will soon come to the conclusion that it is time well spent. If each tracing is examined properly on its receipt in the shop and the foreman makes out the shop summary card at once, it is really surprising how little time is required. It is not advisable to let the tracings wait till a number have to be done, as this leads to a superficial inspection only. They should be done immediately they are received in the shop. If it is found inconvenient for the foreman to do this work, some other capable person, who knows the work and the shop, can do it equally as well, but it is better in all respects for the foreman to do it.

Strict instructions must now be given that no workman may start a job without a work card, obtained from the shop office. To this end the foremen are provided with small note books, in which all the leaves are perforated. These books must be of a convenient size so that they may be carried in the pocket. They are ruled and printed as shewn :—

.... Shop.

Give Man's Name............................ Shop No............

Work Card for Order No.........................

Tracing No....................... Reference......................

Operation to be done: Bore, Turn, Face, Screw, Grind, Plane, Mill, Shape, Slot, Drill, Polish, Fit, Erect, Test, Disconnect.

Part to be done..

....

Date........................ Signed ....
                                                    Foreman.

FOREMAN'S NOTE BOOK FOR JOB ORDERS.

When the foreman gives the workman a job, he must also give him a note from this pocketbook. This note must have all the blanks shewn on the specimen filled in. It is only necessary to put a pencil mark through or under the operation he requires the workman to do. After the man has received this note he must send it by a laborer or messenger boy to the shop clerk. The workmen should be notified that they are not allowed to go to the clerk for these work cards. Any excuse of this nature for being away from their machine should never be accepted.

On receiving the note from the workman the clerk finds the shop summary card for that particular tracing for order number shewn on the note. He then places on the shop summary card, against the operation called for on the note, the man's name and number, also the date. He then makes out a work card for the man, giving the necessary information to enable the part to be identified. These work cards are ruled and printed as shewn :—

.... Shop.

Work Card for Order No..................... Man's No............

Tracing No.................... Reference.........................

Date begun ....

Work ordered.

Piece-Work Price £......................

Day work.

On completion of work this card must be defaced with an ✕ placed across face.

Signed ....

Date	Hours	Date	Hours	Date	Hours	Date	Hours
		Brought Forward		Brought Forward		Brought Forward	
						Total—	

THE WORK CARD.

The shop summary card contains all the information required by the clerk to enable him to fill up all the blanks (except the hours) on these work cards. No work cards may be given out by the clerk without having all these requirements fulfilled. He must also be clearly instructed that no work cards may be made out until he has filled in the man's name and number and date on the shop summary card against the work to be ordered on the work card, as shewn on the note received from the workman. This is absolutely necessary in order that the shop summary cards shall be complete in every detail, and that every work card may be traced in case it has not been returned to the office. It must also be clearly understood by the workmen that no time will be paid for unless a work card has been obtained for the work entailing the expenditure of that time. This is of vital importance and must be rigidly enforced. These work cards may be sent to the workmen by the messenger boys, periodically through the day, but in every case, all cards must be delivered to the men at least half an hour before leaving off time. A separate work card must be given to the man for each job he is engaged on. Each workman now has a work card giving the order number and all the necessary information required to identify and do the work ordered. Each day before leaving off work he must record on these cards the date and the hours he has spent on that particular date doing the work ordered on each card.

These cards are then placed in suitable boxes fixed in a convenient place in the shop so that no time may be lost by the men in putting their cards in them.

All cards for work which has been completed must be defaced by the workman with an X placed across the whole face of the cards, so that the time taker can see at a glance that the order is completed. This cross must be placed only on completed work cards.

Every morning on starting work the time takers collect the cards from the boxes and place them in rotation according to the men's shop numbers. They then check the total day's time charged by the men on the cards, with that recorded by the timekeeper, making any necessary corrections on the cards. They then copy into the time books (which are ruled and printed as shewn), the time of each man, giving the order number, tracing number, reference, description of the work, with the operation, all as recorded on the work cards.

After this has been done the time books are called over with the work cards thus checking the accuracy of the entries. At the end of each week these time books are signed by the person calling over as having been examined and found correct.

Order No.	Tracing & Reference	Description of Work	Daily Time								Total Time h. h. r.w.	Amount of Wages	Remarks
			Wed. Night	Th.	Fr.	Sat.	Mon.	Tue.	Wed.				
										Total			

FORM OF RULING FOR TIME BOOK.

Man's No. _____

Name _____ Rate _____

The Engineering Magazine

All work c a r d s which are not defaced w i t h the completed cross, are then returned to the men before the midday break. Compliance with this rule must be strictly enforced. In most factories it is quite possible for a man to ring his key on entering the shop and then go out by another way and return ringing another man's key, thus recording time for a man who is actually absent. If the work cards are returned to each man personally before the midday break it must prevent a man being paid for time when he is actually absent from the works. The delivery of the day's work into the boxes at leaving off time insures the attendance of the workman after t h e midday break.

It will be seen at once from the above how great a check this system affords in connection with mechanical time recorders, and how it prevents fraud on the part of the workmen. Most men look on mechanical time recorders as machines to be "done," if possible, much in the same way as some people regard the automatic machines at the railway stations.

All completed cards are now cast by the time takers and the total hours placed in the total column. These cards are then returned to the shop-office clerk who enters these totals, together with the date the work was completed, on the shop summary cards, covering the work ordered on the work cards. This work must be done immediately by the clerk and must never be allowed to get in arrears on any consideration.

At the end of the week, in order that no time may be lost in the cost department, the time takers collect the cards after the men leave off work, and stop overtime that night doing the work which they usually do the next morning. They also cast the hours of each man in the time books and place the total hours spent on each job in the columns provided, and the grand total of hours for each man for the week. These time books are then rated and the total hours on each job extended at the rate of pay of each man. The total wages of each man are then obtained and agreed with the pay roll of the shop.

To insure proper attention one time taker should be employed for every 300 men. This is a reasonable ratio and should ensure accuracy. In the case of any shop having a number of men engaged on small work, or in which a number of orders are in progress at the same time, this ratio of course must be lowered, and vice versa. Experience will soon fix the proper amount of work for each time taker. As this work is only copying, smart lads may with advantage be employed to do it. It should be the stepping stone to a better position for them. As vacancies occur in the office, those lads showing most promise should be promoted to fill them. The training as time takers would come in very useful for them on their promotion and would be advantageous to the firm because of their familiarity with the technical names and of the processes each part undergoes during the course of its construction in the shops. This is a great item to be considered with new clerks, who as a rule are of very little use in a works for some months.

To ensure that the time books are checked daily with the work cards, it is advisable that an independent person should periodically see that this has been done. This will tend to make the time takers careful in their daily checking of the time books.

The shop clerk may give only one work card to a man for any one operation on any one part of an order. When the order calls for a number of parts off the same pattern, then any one operation on all these parts may be ordered on one work card. For instance:— supposing an order called for twelve connecting rods off tracing number —; then a work card should be made out for turning the twelve rods, provided one man was going to do the whole lot; if more than one man would be engaged on the turning of the twelve rods, then as many work cards must be made out for the turning as there are men engaged on it. It must be clearly shewn on the work cards, and the shop summary cards, how many rods each man has orders to do. The planing, slotting, milling, drilling, etc., of these rods must be carried out in the same manner as the turning.

Another instance of what is required in the matter of work cards is this:— An order is received for three stop valves; separate work cards would have to be made out for boring and turning the three bodies, boring and turning three covers, boring and turning three hand wheels, turning and screwing three spindles, and so on. One work card should not be made out for all the lathe work for all the parts of the stop valves, but separate cards for each operation on each of the parts comprising the valves.

In the case of men engaged on the "marking out" slab it is only necessary to give a work card for all the "marking out" on each order number. It is almost impossible for these men's time to be charged directly to each detail of the order. In any case, it would cost far more to record this time than it would be worth. The same applies in some cases to drillers on very small work such as pin holes, etc.

These exceptions must be thoroughly studied and rules laid down to cover the extent of these "omnibus" cards.

Practice will soon determine into what detail it is advisable to divide the work. It should, however, not be forgotten that the greater the detail the better the result from a cost standpoint. Increases in costs can be more readily discovered if the detail is great and thus be more easily dealt with.

As each operation on a part is finished, the time charged by the workman is entered every day by the shop clerk on the shop summary card. Under these circumstances only one card can be given by the clerk for any one operation on a part, because he has instructions that a work card must not be made out until he has entered on the shop summary card the man's number and the date, and if he were asked for a work card for an operation on a part against which the columns for the man's number and date in the shop summary card were filled up, he would see at once that something was wrong as he could not fill in these columns again. In this case he would see the foreman at once and get an explanation. It would probably be caused by the part having to be replaced owing to bad work or material. If this were so, the foreman would then add, to the shop summary card for that order, the part which was replaced, entering all the operations to be performed on it, and marking it "spoiled work."

If the operation is piece work, it will be seen from the above that it is an impossibility, unless there be gross carelessness on the part of the clerk, for a man to be paid twice for the same operation on a part. It is not altogether an unheard of occurrence for a man to be paid more than once for the same work. Again, if an order calls for fifty articles and the operators on these are piece workers, it is also impossible for the men to be paid for more articles than are actually ordered, because the number of articles to be done is shewn on the work cards and consequently on the shop summary cards. If the clerk were asked for work cards for more than the fifty articles he must of necessity see that something was wrong and would then get an explanation. In how many factories can such a statement be made with confidence? How often are the men paid for more articles than are actually sent into the shop? Many cases have happened in my own experience where such payments have been made.

With this system the foreman has a record every day of every operation that is finished on each order. He can easily see by examining the shop summary cards how each part of any order is progressing, and by noticing the blanks in the hours columns he can see which operations are "hanging fire." If he went at once to the men, to whom the orders for these operations were given, he could tell if there were good reasons for the delay. Idle men would thus be soon found out and brought into line or discharged, as the case might be.

It would be an assistance if the clerk made a return daily or weekly, as might be determined, of the blanks in the hours columns of all shop summary cards. The foreman would thus have his attention drawn to those operations and could arrange accordingly.

Again, when an order is approaching completion, the clerk should make a return to the foreman of all tracing numbers with the reference letters and description of any parts, shewn on the shop summary cards, against which the number and date columns are not filled up. By this means no details ordered on a tracing are likely to be missed, because the foreman with this return in front of him can inform the necessary people so that these details may be sent into his shop without further delay. This would stop any short shipments of orders, owing to parts being overlooked in other departments.

When an order is completed, the clerk refers to the shop summary card and if there be any blanks he must immediately go to the workmen who had the orders for the operations against which the columns are filled up, and enquire why the cards have not been returned. It is very probable first that several blanks will be seen on the summary work cards after an order is completed. In nearly every case it will be found they are due to the fact that the workmen omitted to deface the work card when the work was completed; consequently they were returned to the men, instead of the clerk. If the foreman takes this matter up, strictly, with these men it will not occur very often. It should however be explained to the men that severe notice will be taken of any men making a practice of such carelessness.

Should a workman lose his work card, the clerk must obtain from the cost department the total hours charged on that operation prior to the loss and place it on the new work card before giving it to the man. This is necessary in order that the total hours shewn on the shop summary card shall be the actual time paid for.

If an order be completed and there are blanks in the shop summary cards which are not filled up (owing to the men losing or destroying the undefaced completed work cards which had been returned to them), the clerk must obtain, from the cost department, the total hours charged by these men on the operations and place them in the proper columns of the summary cards. When this system is in proper working order the foreman has a record of the time spent on each operation on every part shewn on the tracings of all orders. How important this is, in the case of repeat orders, every manager must admit. It is a great incentive to the foreman to beat his previous record and thus reduce costs.

The foreman can see every day how each order is progressing and average his work accordingly. He can easily "get in touch" with every man who is "hanging on to his job" and thus stop such behaviour. He can be sure he is not paying twice for the same thing, nor yet for more work than is actually done, and thus he has no occasion to do detective's work.

By having to examine each tracing on its receipt in the shop, he becomes familiar with the work required and thus has no occasion to be repeatedly turning up the tracing to see what is wanted.

He knows that all details will come along in the usual course, consequently has no need to bother about them. If they don't he knows he will be advised in good time.

Any questions which the manager may ask as to the condition of any order in the shop may be answered correctly by the clerk, who has all the necessary information in front of him. Under these circumstances the foreman has no need to draw on his imagination to try and answer the manager satisfactorily and thus avoid any unpleasantness.

The manager can see for himself how every order is progressing, and at the same time, if not satisfied with the time charged by the men, he can take up the matter in a sensible manner with the foreman because he has all the facts in front of him.

The men will see that it is not advisable to "hang on to a job," consequently the work will be done more quickly.

As the foreman will have all these shop summary cards in his possession, he can, when repeat orders come along, produce them with all the time records. Therefore when the hours charged by a man on any part are not satisfactory and the matter is taken up with him, the workman can see for himself that he is not being asked to do any more than has already been done. The men see they are not being imposed on and a good tone will soon prevail in that shop. This alone would more than pay for any expense the system might cause, without taking into consideration any of the advantages mentioned above.

In conclusion, for this system to be a success no deviation from the following general rules must be allowed on any consideration:—

1.—The foreman must make out the shop summary card as soon as the tracing or order is received in the shop.

2.—The clerk must not make out a work card before consulting and filling in the columns of the shop summary card.

3.—No workman shall be allowed to do any work without a work card.

4.—A workman shall be paid only for the time charged on his work card after it has been agreed with the timekeeper's records.

The above are the fundamental rules of the system, and if they be carried out in the manner explained in this article, the manager can rely on having a thoroughly practical time-recording system, from which the advantages obtained, and the satisfaction he will feel, will more than repay him for the trifling expenses he may be put to in instituting it.

WATER FOR ECONOMICAL STEAM GENERATION.

By J. C. Wm. Greth.

Mr. Greth treats his topic from the point of view most important to the power-plant owner and manager—the point also from which it is generally most difficult to get a true perspective; that is, the actual saving, in pounds of coal and in money, which can be counted upon in ordinary practice. In the following pages and yet more fully in a second article to be published next month, he gives the figures from a number of working stations, taken under regular operating conditions and covering periods sufficiently long to make the results a true average.—THE EDITORS.

WATER is the most abundant of all substances and the one most used—the greatest solvent known, and plays an important part in all the process of life and industry.

Natural waters are rarely, if ever, pure; the impurities may be divided into two classes—insoluble and soluble. The insoluble impurities can be removed by sedimentation or filtration. The soluble impurities cannot be removed unless they can first be thrown into suspension, and then removed by sedimentation or filtration.

The impurities in a water used for manufacturing purposes affect all the processes in which it is used by decreasing the output, contaminating it, and increasing the cost of production; such instances occur in the tanning processes on account of the loss of tannic acid due to lime and magnesia salts, and in all scouring and washing processes where soap is destroyed by the soluble salts of lime and magnesia. In nearly every industrial or domestic use the impurities in water cause enormous waste and expense. The losses due to the incrustation and corrosion of steam-generating apparatus amount to many millions of dollars annually.

Natural waters fed into steam boilers are taken from all sources. The relative degrees of purity of natural water supplies may be stated as follows:—

Rain water or melted snow.

The waters of creeks, rivers and lakes.

Well water.

Water of mineral springs.

Sea water.

The amount and kind of impurities in a water play an important part in the economical operation of a steam boiler. A steam boiler

793

fed with a clear water low in scale-forming substances and free from corroding substances will last almost indefinitely without cleaning or repairs. On the other hand, a boiler fed with a water containing incrusting solids must frequently be cleaned in order that steam can be made at all, as the scale formed is constantly accumulating, reducing the efficiency of the boiler and increasing the repairs, due to the higher temperature to which the metal must be raised to generate the same quantity of steam as with a clean boiler. Corrosive substances, also, increase repairs and materially shorten the life of the boiler. When both corroding and incrusting substances are present in a water, the scale formed on the surface conceals the corrosive action going on underneath it, and the disintegration of the boiler is not detected until the damage is done.

To obtain a natural water supply practically free from incrusting solids and corrosive substances is next to impossible. Very few natural supplies can be fed into a boiler without loss, from the following causes:—

First: Loss by waste of fuel, due to scale reducing the efficiency of the boiler. It is almost impossible to determine this loss accurately since it varies with the thickness and composition of the scale. All sorts of tables (some of them on good authority) have been published, but the only way to determine the actual fuel loss is to operate the same boiler under exactly the same conditions with and without scale.

Second: Loss due to labor required for cleaning heaters, piping, and boilers. In the average steam plant the money expended for this purpose alone is more than would be the interest on an investment for obtaining a good water supply.

Third: Loss due to the investment in spare boilers to be put in commission when it is necessary to take boilers out of service for cleaning or repairs.

Fourth: Loss due to cost of boiler compounds. Boiler compounds, if efficient, are irrational in application, and at best only increase the density of the water by throwing into suspension the scale-forming substances together with the compound introduced, making it necessary to generate steam from a water having in suspension a muddy, insoluble mass, which has no place in a boiler.

Fifth: Loss due to repairs to boilers necessitated by their being subjected to overheating from the insulating effect of scale covering the heating surface, or to the action of corrosive substances pitting and corroding the tubes and shell.

Sixth: Loss of fuel due to heat lost in cooling a boiler for cleaning or repairs and that required to bring it again to steaming heat.

Seventh: Loss due to reduced efficiency of heaters and econo-mizers, resulting in reduced temperature of feed water, which ma-terially increases fuel consumption.

Eighth: Loss due to cost of city water over cost of pumping and softening an available supply.

Ninth: Cost of tube-cleaning machines, repairs to them, interest and depreciation on the money invested, and labor and power re-quired for operating them.

Tenth: Loss of efficiency and earning power of improved fur-naces and stokers installed to increase evaporation, which correspend-ingly increases the concentration of impurities, thus forming a greater deposit of scale or producing more active corrosion, and hence a great reduction in the efficiency or life of the boilers.

Eleventh: Loss due to cost of skimmers, surface blow-offs, and other mechanical devices for preventing scale in boilers, as well as loss of water and heat.

The formation of scale in a boiler is due to two causes:—

First: the water at the increased temperature loses its power of holding in solution the scale-forming substances.

Second: the evaporation of the pure water into steam concen-trates the impurities in the water remaining until it eventually becomes supersaturated, when the impurities in excess crystallize or otherwise deposit.

Impurities in solution, although small in amount (even in a bad water) as compared with the volume of water holding them in solu-tion, will form scale because the concentration taking place in the boiler cannot be prevented no matter how frequently a boiler is blown off. For instance a 100 horse-power return-tubular boiler (say 66 inches in diameter by 16 feet long, with 98 3-inch tubes) which holds about 1,600 gallons of water, if evaporating 400 gallons of water per hour (basis of 33 1-3 pounds, or 4 gallons per horse-power per hour) having in solution 30 grains of scale-forming substances per United States gallon, if all of it remains in solution, will have at the end of 4 hours 60 grains, and at the end of 8 hours 90 grains, and so on.

The table on the following page will give some idea of the scale deposited in a boiler by waters containing different amounts of scale-forming substances in solution. From the figures any one can readily see why so large an amount of scale is deposited in boilers. The thick-ness of scale deposited in a boiler in a given time depends, first, on the amount of scale-forming substances contained in the feed water; sec-ond, on the ratio between the surfaces of the boiler and the volume of water contained therein. The modern water-tube boilers contain a considerably smaller water space than fire-tube boilers, hence greater concentration occurs in a given period of time and a thicker scale is the result on account of the smaller surface which receives the deposit.

In a 100 Horse-Power Boiler Evaporating 400 Gallons of Water per Hour.

Grains per U. S. Gallon.	24 hours.	6 days, 24 hrs. per day.	Month, 25 days.	Year, 300 days.
15 grains	20.6 lb.	123.6 lb.	515 lb.	6180 lb.
20 "	27.4 "	164.4 "	685 "	8220 "
30 "	41.2 "	247.2 "	1030 "	12360 "
40 "	54.8 "	328.8 "	1370 "	16440 "
50 "	68.6 "	411.6 "	1715 "	20580 "

Among the methods employed for the purification of water for boiler feed are exhaust- and live-steam heaters, which have for their object the precipitation of the scale-forming substances in the heater instead of in the boiler. Practice has proved, however, that these devices cannot do more than partially accomplish the removal of incrusting substances, and that neither one will prevent the accumulation of scale in a boiler; therefore, when they are used, the cleaning of the boiler as well as the heater becomes a necessity. This type of apparatus is rarely, if ever, built large enough to allow for complete precipitation to take place. It is impossible to obtain complete precipitation by means of heat alone in any apparatus, inasmuch as complete precipitation cannot be accomplished by heat without concentration, which cannot be effected in any type of feed-water heater.

A feed-water heater in its function of imparting heat from the steam to the water is essential in any power plant, but the maximum of temperature can be obtained only when the heater is kept perfectly clean; therefore, when it is used as a purifier for precipitating and collecting impurities, its efficiency is reduced in direct proportion to the amount of impurities deposited. Every degree of temperature lost through inefficiency of the heater must be compensated for by heat under the boiler, which means an increased consumption of fuel.

Surface condensers are frequently installed to obtain a good boiler-feed water by using the condensed steam for that purpose, but this is accomplished only with steam from turbine engines, and when used in connection with other types of engines is open to several objections. It is practically impossible to remove all the oil from the exhaust steam or condensed water by mechanical devices, as the oil forms an emulsion with the water which can be separated only by means of chemical treatment or continued boiling and skimming. If the oil is allowed to enter the boilers together with the scale-forming substances in the make-up water (the addition of which is necessary because it is impossible to condense all the steam to return it) it is apt to cause a dangerous condition, due to the non-conducting film of oil likely to be deposited and also to the scale-forming substances held by it. In many power plants operating with surface condensers the condensed steam is thrown away on account of the oil it contains. Another objection is that of keeping of the condensing apparatus in efficient service, especially when the water used for condensing contains carbonates of lime and magnesia in solution.

A surface condenser has its own office to perform, and unless the conditions as to oil-removal and make-up water are properly met, the use of this apparatus does not afford a satisfactory means of obtaining a good boiler-feed water.

The method which gives ideal results in purifying boiler feed water is that which brings about the removal of the scale-forming and corrodng substances by chemical treatment with exact quantities of predetermined reagents for their precipitation; the precipitates are then removed by sedimentation and filtration. The apparatus used is called a water-softening and purifying system.

There are two distinct types of such systems—the "intermittent" and the "continuous." There are on the market numerous intermittent and continuous systems. In the strictly intermittent system, known quantities of water are treated with the exact amount of reagents, time is allowed for sedimentation, and perfect clarification is effected by filtration. In other intermittent systems the reagents are introduced proportionally (as in a continuous system) while the water flows into settling tanks which are used alternately. In the continuous systems the reagents are introduced by proportional feed devices actuated by the water entering the system; the water enters either at the top or bottom and flows out by displacement at the same rate at which it enters. In the intermittent system definite quantities of water are treated, so that no matter how variable a source of supply may be, accurate treatment can always be made. The intermittent system, too, is independent of the rate at which the water is used. In the continuous system, the reagents being introduced by proportional feed devices, with a variable water supply and wide variations in the quantity used, it is difficult to get accurate results, and for the best results accurate treatment is essential. Either system, however, when properly designed and operated will give a soft, clear, boiler-feed water.

A water-softening system in order to give the best results must be designed to meet the specific conditions under which it is to be operated, as well as the kind of water to be treated; when properly designed and operated the reduction of scale-forming substances will be such that no scale can accumulate, all corroding substances will be removed, and the purified water will be perfectly clear. The reduction of the scale-forming substances to 5 grains will usually make a satisfactory boiler water. There are, however, many cases where it is necessary to reduce the scale-forming substances to less than 3 grains to prevent the accumulation of scale; and others where good results are obtained with 6 grains. It is impossible to remove from a water all the scale-forming substances, as calcium carbonate and magnesium hydrate are slightly soluble, nor is it possible to remove from a water any sodium salts; these cannot be removed except by distillation. All sodium salts are soluble and no chemical treatment will throw them into suspension, consequently they cannot be removed by sedimentation or filtration. Sodium salts in a properly treated water do not work any harm to the boiler, as they are neutral, non-corrosive and non-scaling.

As a general rule, anyone investigating the subject or desiring to install a system should consult a reliable manufacturer of this class of apparatus, whose business it is to install these systems and whose reputation depends on the results obtained with them.

A water-softening system to give the best results must comply with the following requirements. (A properly designed intermittent system meets them all and a properly designed continuous system meets them approximately.)

1.—Accurate chemical treatment:—the introduction of the proper reagents in exact quantities required to react with the impurities in the water.

2.—Thorough mixture of reagents with the water:—in order that chemical reaction will take place the reagents must be thoroughly mixed with the water.

3.—Accelerated chemical reaction:—this is brought about by thorough mixture and by mixing the sludge of previous softening with the new finely divided precipitate, and by heating the water if the nature of the water is such that heat is required to complete the chemical reaction.

4.—Complete chemical reaction:—which is brought about by a thorough mixture of the reagents with the water and by having the apparatus large enough to allow sufficient time for all the reactions to take place, and the apparatus so designed that every part of it is effective.

5.—Rapid sedimentation:—this takes place when the new finely divided precipitate is weighted by the sludge of previous precipitaton, causing it to settle more rapidly and perfectly.

6.—Perfect clarification:—this is accomplished by allowing time for sedimentation and final clarification by perfect filtration.

Possbly the best way to show the various effects of different water supplies would be to make use of some figures which I have at hand concerning two plants, one operating with a natural supply almost ideally adapted as a boiler-feed water, the water in question being a city water taken from a mountain stream, and the other plant operating with a well water to save the cost of city water. Later, the plant using the well water installed a water-softening system. Both plants operate 500 horse-power boilers, each consisting of four 125 horsepower return-tubular boilers of practically the same design. For convenience, the plant using city water, which is a clear soft water, will be indicated by "A," the plant using the well water by "B," and the same plant later using a water softened before being fed into the boilers by "C." The statement on the next page shows the cost of operation of the two boiler plants exclusive of fuel used per year.

It will be readily apparent to anyone from the statements given that the ideal condition of affairs is a natural soft, clear water. However, on account of the cost of the city water, plant "C" operating with the softened well water is by far the best. The plants "A" and "B" of course are not operating under exactly the same conditions. Plant "B" was used as a standard, the same prices for labor and material applied throughout, so as to get a comparison of expenditure on the same basis. Fuel saving was not taken into consideration, except that used for heating to steam boilers cooled for cleaning. It was impossible to obtain a record of the fuel used inasmuch as in plant "B" fuel was used for other purposes than for the boilers. It is reasonable to suppose that there was a decided saving in fuel effected under "A" and "C" over "B," as a clean boiler will certainly evaporate more water per pound of fuel than one that is scaled.

From the above it is evident that, from the standpoint of economy, the best water that can be obtained for boiler-feed use is the cheapest, and if water supply "A" could have been obtained for the cost of pumping it would of course have been the cheapest and best; but when the cost of city water is taken into consideration, then of the two raw supplies "B" is somewhat the cheaper. The best supply by far, from the standpoint of economy, is the well water using the softening system, "C."

A.

500 horse-power boilers evaporating 33⅓ pounds, or 4 gallons of water
 per horse-power per hour, 2,000 gallons per hour, 24 hours per day,
 300 days per year, = 14,400,000 gallons at 8 cents per 1,000 gallons $1,152.00
Washing out each boiler once every six months, 8 washings a year,
 requiring 1 man 1 day at each washing, at $1.50 per day.........   12.00
Heating to steam boilers cooled for cleaning, 8 boilers 125 horse-power
 each, requiring 1 ton of coal per 100 horse-power (estimated),
 = 10 tons of coal at $1.90 per ton..............................   19.00
                   ──────
                   $1,183.00

B.

500 horse-power boilers evaporating 33⅓ lbs. or 4 gallons of water per
 horse-power per hour, 2,000 gallons per hour, 24 hours per day,
 300 days per year, = 14,400,000 gallons, pumping at ¾ cent per
 1,000 gallons (estimated)...................................... $108.00
Cleaning each boiler once every two weeks, 104 cleanings, 1 man 2 days
 per cleaning, 208 days' labor at $1.50 per day.................. 312.00
Heating to steam boilers cooled for cleaning, 104 boilers 125 horse-
 power each, requiring 1 ton of coal per 100 horse-power (estimated) = 130 tons of coal at $1.90 per ton..................... 247.00
Cleaning open heater once each month, 12 cleanings a year, 1 day for
 each cleaning, at $1.50 per day.................................   18.00
Boiler compounds, 4,800 lbs. at 6 cents per lb......................... 288.00
Boiler repairs 121.70
                   ──────
                   $1,094.70

C.

500 horse-power boilers evaporating 33⅓ pounds or 4 gallons of water
 per horse-power per hour, 2,000 gallons per hour, 24 hours per day,
 300 days per year, = 14,400,000 gallons, pumping at ¾ cent per
 1,000 gallons (estimated)...................................... $108.00
Softening 14,400,000 gallons of water per year, at 1½ cents per 1,000
 gallons 216.00
Washing out each boiler once every 4 months, 12 washings a year,
 requiring 1 man 1 day at each washing, at $1.50 per day..........   18.00
Heating to steam boilers cooled for cleaning, 12 boilers 125 horse-
 power each, requiring 1 ton of coal per 100 horse-power (estimated) = 15 tons of coal at $1.90 per ton.....................   28.50
Depreciation at 10 per cent on investment of $2,000 for water-soften-
 ing system 200.00
Interest at 6 per cent on investment of $2,000 for water-softening sys-
 tem 120.00
                   ──────
                   $690.50

The following table taken from the records of a small electric
light plant operating a 175 horse-power water-tube boiler, shows the
saving in fuel that can be effected by supplying boilers with clear
water free from scale-forming substances. The load at which the
boiler is operated is practically constant, varying only with the length
of the day or the necessity for light on dark days. The figures cited
are taken from a daily record of every pound of coal burned and cover
a period of five months from January 1, 1905 to May 31, 1905 (the

boiler was cleaned every two weeks and compound was used), also from January 1, 1906 to May 31, 1906, with a water-softening plant.

COAL CONSUMPTION.

Without Water-Softening Plant.		With Water-Softening Plant.		SAVING.
Jan. 1905	307,070 lb.	Jan. 1906	248,430 lb.	58,640 lb.
Feb. 1905	280,455 "	Feb. 1906	221,470 "	58,985 "
Mar. 1905	300,352 "	Mar. 1906	241,380 "	58,972 "
Apr. 1905	286,307 "	Apr. 1906	210,600 "	75,707 "
May 1905	281,748 "	May 1906	212,860 "	68,888 "

Analyses of the water before and after treatment at this particular plant are as follows:—

WELL WATER.

Raw.	Grs. per U. S. Gallon.	Treated.	Grs. per U. S. Gallon.
Volatile and Organic Matter.	3.60	Volatile and Organic Matter.	.40
Silica	.30	Silica	.17
Oxides of Iron and Alumina.	trace	Oxides of Iron and Alumina.	trace
Calcium Carbonate	11.84	Calcium Carbonate	1.45
Calcium Sulphate	8.43	Calcium Sulphate	.15
Magnesium Carbonate	5.25	Magnesium Hydrate	.49
Magnesium Chloride	1.53	Sodium Sulphate	8.90
Magnesium Nitrate	4.73	Sodium Chloride	4.21
Sodium Chloride	2.37	Sodium Nitrate	5.34
TOTAL SOLIDS	38.05	TOTAL SOLIDS	21.11
Suspended Matter	.05		
Free Carbonic Acid	2.37		
Incrusting Substances	35.68	Incrusting Substances	2.66

The labor required for the operation of a water-softening plant need not be taken into consideration, because ninety per cent of the water softening plants in successful operation today are operated by the engine-room employee without interference with his regular work; and a wise engineer or boiler-room employee will give a water-softening plant proper attention to avoid the work of cleaning boilers, which at best is a disagreeable task.

A water-softening apparatus is a good investment for any steam plant wherever there is any trouble or expense in connection with the operation of the boiler due to scale or corrosion. The boiler is designed as a steam generator and not as a chemical retort for the formation of precipitates. Any work thrown on the boiler other than generating steam reduces its efficiency.

(To be continued.)

A DECADE OF AMERICAN RAILROAD HISTORY IN GRAPHIC FORM.

By Harold Vinton Coes.

IT is intended to set forth in this article, in graphic form, the various facts showing the growth of American railroads during the period from 1895 to 1905. To the average busy man tables and statistics mean but little; it takes too long and is too wearisome to get at the facts through them. But when the data are cast into graphic form, you have a picture readily appreciated by both eye and brain.

The total·number of roads designated as operating roads on June 30, 1905, was 1,169. Of these 907 are independent so far as contractual relations are concerned.

Chart 1, Plate I, shows the increase in single-track, second-track, third-track, and fourth-track mileage, as well ·in the total mileage. The total mileage in operation in 1895 was 233,000 miles; in 1905 it was 307,000 miles, an increase of 31¼ per cent. Chart 2 gives the density of mileage on different bases.* In 1895 the number of miles of line per 10,000 inhabitants was 26.0. This gradually decreased until in 1899 the minimum 25.3 was reached. The curve then rises, showing 26.4 in 1905. The number of miles of line per 100 square miles of territory, unlike the former curve, shows a yearly increase. This probably means that the population in certain sections of the country increased faster than railroad construction, for the latter curve shows an increase of 1.3 miles per 100 square miles per ten years, or 0.13 miles per 100 square miles per year. This last figure, therefore, can be used as a unit.

Charts 3 and 4, Plate II, pertain to railroad equipment. Chart 3 deals with cars only. The negative or decreasing direction of the curve marked "Fast Freight Service" can probably be explained by the fact that large corporations dealing in perishable goods had been gradually buying their own cars.†

The classification and sizes of both simple and compound locomotives are given in Table I. Locomotives have been classified ac-.cording to White's System. (See footnote, page 804.)

---

* The bases are 82,494,575 population and 2,970,038 square miles area.

† NOTE.—The curves take no account of privately operated or privately owned cars.

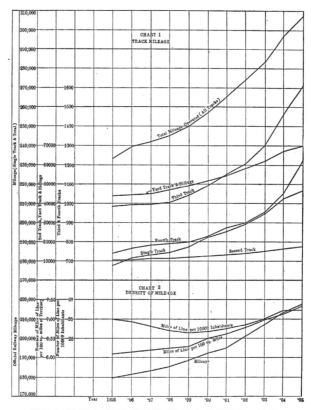

PLATE I. CHART I, TRACK MILEAGE. CHART 2, DENSITY OF MILEAGE.

A comparison of the earlier but incomplete records pertaining to locomotives with Chart 8, Plate IV, develops the fact that while American locomotives have been increasing in size at the rate of 2½ per cent a year, the size of the train to be handled has been increasing at the rate of 10 per cent a year.

A comparison of Table II (page 806) with figures for the previous

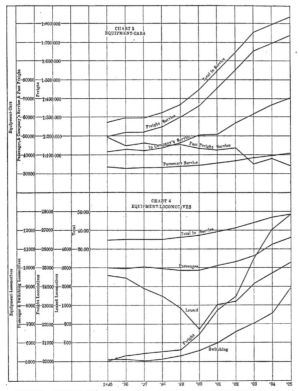

PLATE II.   CHART 3, EQUIPMENT, CARS.   CHART 4, EQUIPMENT, LOCOMOTIVES.

year discloses some significant facts. First, single-expansion locomotives increased in number from 43,246 to 45,033, or at the rate of 4.12 per cent, while compound locomotives decreased from 2,900 to

NOTE.—White's locomotive classification is based on the representation by numerals of the number and arrangement of the wheels, commencing at the front. Thus 260 means a Mogul and 460 a ten-wheel engine, the cipher denoting that no trailing truck is used. The total weight is expressed in 1,000 pounds. Thus an Atlantic locomotive weighing 176,000 pounds, would be classified as a 442-176 type. If the engine is compound the letter C is used; thus, 442C176. If tanks are used in place of a separate tender, the letter T is used similarly to the letter C.

TABLE I. DETAILED CLASSIFICATION OF LOCOMOTIVES.

	2	3	4	5	6	7
				AVERAGE.		
Class.	Number.	Tractive Power.	Grate Surface.	Heating Surface.	Weight, exclusive of Tender.	Weight on Drivers.
		Pounds.	Sq. Ft.	Sq. Ft.	Tons.	Tons.
040—33	1117	14087	13	812	33	33
040 C₄ 13..........	1	4900	7	450	13	13
060—53	5135	22650	23	1406	53	53
060 C₂ 64..........	30	29972	28	1577	64	64
080—53	149	23530	31	1432	53	53
050—82	14	37132	41	2498	82	82
0660—167	1	70185	72	5600	167	167
240—24	42	10860	14	683	30	24
260—51	5188	22337	27	1609	60	51
260 C₂ 61..........	136	25976	30	1838	71	61
260 C₄ 70..........	158	28199	38	2347	82	70
280—69	11578	32906	42	2367	78	69
280 C₂ 79..........	350	36718	43	2901	90	79
280 C₄ 84..........	663	38133	51	2837	96	84
2100—87	10	39556	67	2948	98	87
2100 C₄ 106.......	5	51502	61	4328	119	106
420—14	7	5427	12	510	24	14
440—31	10732	14591	21	1291	46	31
440 C₂ 17..........	4	7843	14	635	26	17
440 C₄ 38..........	17	12631	40	1534	59	38
460—50	8680	21912	28	1842	64	50
460 C₂ 62..........	273	27029	35	2615	83	62
460 C₄ 65..........	443	26622	40	2730	87	65
480—68	496	33064	42	2398	84	68
480 C₂ 70..........	50	29942	37	2427	84	70
042—24	15	9939	11	595	29	24
042 C₄ 37..........	1	12582	76	1429	70	37
062—55	13	20008	22	1549	66	55
222—6	2	1767	12	177	14	6
242—43	57	21369	56	2407	78	43
262—66	326	26976	46	2948	91	66
262 C₂ 68..........	26	31550	49	3225	95	68
262 C₄ 70..........	168	28889	52	3618	102	70
282—97	85	45098	57	3903	138	97
282 C₄ 79..........	37	39060	45	3773	102	79
2102—112	1	61840	58	4796	138	112
2102 C₄ 117.......	85	62729	58	4796	149	117
442—50	917	23231	50	2788	87	50
442 C₄ 48..........	202	20598	46	2940	92	48
462—67	302	29539	47	3284	103	67
462 C₄ 72..........	3	23418	49	3053	115	72
044—24	76	10196	14	660	31	24
044 C₂ 20..........	1	8207	17	528	28	20
044 C₄ 24..........	10	7709	18	502	36	24
244—29	44	12244	15	836	52	29
264—53	19	22748	45	1532	87	53
444—33	2	10710	18	1043	65	33
464—33	7	18743	21	1213	72	33
046—28	5	8396	15	707	50	28
246—36	10	16373	23	1251	83	36
266—53	3	19425	36	1843	64	53

C₂ = 2 Cylinder Compound.
C₄ = 4 " "

2,663, or at the rate of 8.9 per cent. Second, the comparison shows
an increase in power and size of the average single-expansion locomotive. The average tractive power per locomotive of this class increased from 22,206 pounds to 23,178 pounds. The grate surface
increased from an average of 29 square feet to 30 square feet, and
the heating surface from 1,681 square feet to 1,759 square feet. The
average weight, exclusive of tender, increased from 60 to 62 tons,
and the average weight on drivers increased from 49 to 51 tons per
locomotive.

TABLE II. CONDENSED CLASSIFICATION OF LOCOMOTIVES.*

	Single Expansion.	Four-cylinder Compound.	Two-cylinder Compound.
Number	45033	1793	870
Tractive Power—Pounds	23178 (Average)	32326 (Average)	31056 (Average)
Grate Surface—Sq. ft...	30 "	47 "	38 "
Heating Surface—Sq. ft.	1759 "	2941 "	2569 "
Weight, exclusive of Tender—Tons	62	95 "	83 "
Weight on Drivers—Tons.	51 "	73	69 ..

Plate III shows the assignment of equipment per 1,000 miles of
line, but as it is the passenger mileage and the freight mileage rather
than the distribution that gives the true measure of equipment efficiency, we may pass on to Chart 8, Plate IV. The increase in passenger-miles per passenger locomotive for the year 1905 over the
year 1904 was 100,174, and the increase in ton-miles per freight
locomotive was 233,854. The total increase for the ten years in pas-
senger-mile per passenger locomotive was 829,591, or 82,959.1 passenger-miles per passenger locomotive per year. The increase in ton-
miles per freight locomotive was 2,431,879, or 243,187.9 ton-miles per
freight locomotive per year. The curve marked "Freight Cars per
1,000,000 Tons of Freight Carried" shows a rapid decrease. This
does not mean, as might appear at first sight, that freight traffic had
fallen off, but is due rather to the increase in the use of cars of higher
capacity. The increase in density of loading for both passenger and
freight cars has consequently fallen off, due to the cause mentioned.
In the year 1905 one less passenger car was used per 1,000,000
passengers carried than was the case during the preceding year, and
79 fewer freight cars per 1,000,000 tons of freight carried. It is
interesting to note that of the various products carried by the railroads—products of agriculture, products of animals, products of
mines, products of forests, manufactures, merchandise, miscellaneous
—50 per cent of the tonnage was made up by the products of mines.

* Excludes 661 unclassified locomotives.

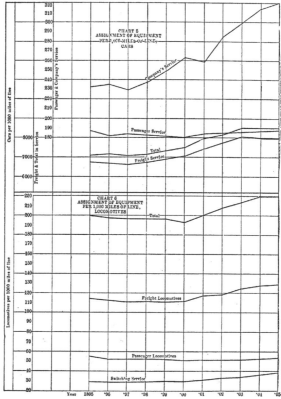

PLATE III.   CHART 5, ASSIGNMENT OF CAR EQUIPMENT.   CHART 6, ASSIGNMENT
OF LOCOMOTIVE EQUIPMENT.

Chart 8, Plate IV, and Charts 9 and 10, Plate V, and the figures
for 1905 derived therefrom indicate that railway transportation in the
United States, despite criticism to the contrary, has attained a high
degree of efficiency as compared with 1904 and the previous years.
These curves show also a marked increase in the amount of business
and density of traffic. The amount of passenger traffic during the

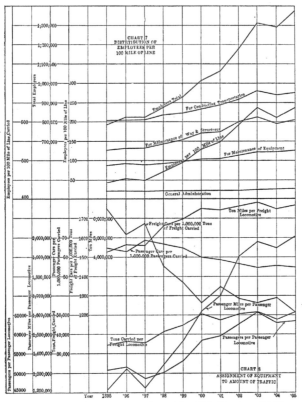

PLATE IV. CHART 7, DISTRIBUTION OF EMPLOYEES FOR 100 MILES OF LINE.
CHART 8, ASSIGNMENT OF EQUIPMENT TO AMOUNT OF TRAFFIC.

year 1905 is especially noticeable. The number of passengers reported carried during this fiscal year (June 30, 1904 to June 30, 1905) was 738,800,000, an increase of 23,400,000 as compared with the previous year. The increase in density of traffic—that is, passenger-miles per miles of line—is also marked, this density having increased from 104,200 passenger-miles per mile of line in 1904 to 109,900 in 1905. The increase in freight traffic, while not relatively so great, is nevertheless very noticeable. The ton-mileage accomplished during the fiscal year 1905 was 186,500,000,000, being an increase of 12,000,000,000 ton-miles. The density of freight traffic is shown by the fact that 861,000 ton-miles per mile of line were car-

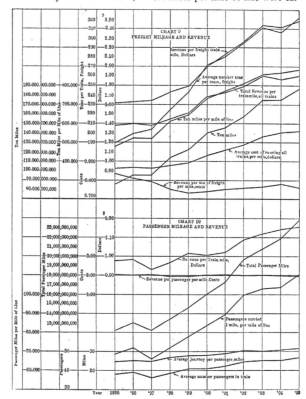

PLATE V. CHART 9, FREIGHT MILEAGE AND REVENUE. CHART 10, PASSENGER MILEAGE AND REVENUE.

ried. In this connection it may be interesting to note that the average haul in miles is less for the year 1905 than for 1904, the figure for 1905 being 237.56 miles against 244.30 miles for 1904. In 1899 (last authentic figure), it was 246.58 miles.

In Charts 9 and 10, Plate V, the cost of carrying freight and passengers is compared. The revenue per freight train-mile has increased

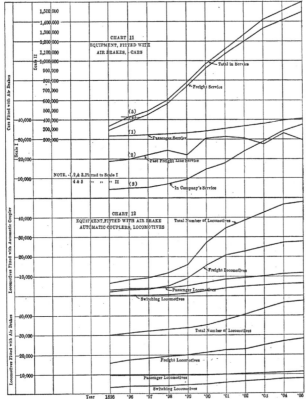

PLATE VI. CHART 11, AIR-BRAKE EQUIPMENT, CARS. CHART 12, DITTO, LOCOMOTIVES.

in the last ten years from \$1.61 to \$2.50 and from \$.97 to \$1.15 per passenger train-mile. The revenue per ton of freight per mile, as well as the revenue per passenger per mile, has decreased. Both of these curves are largely influenced by the policies of the various railroads, the action of the several State legislatures, and the attitude of the Federal Government. It is well to note that the average revenue per ton of freight per mile covers all kinds and classes of freight, and may be quite as much influenced by the change in the relative quantity of the different classes of freight as by a change in the rate itself.

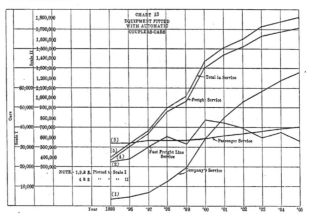

PLATE VII.   CHART 13, AUTOMATIC COUPLER EQUIPMENT, CARS.

Plates VI and VII are self-explanatory and need no comment except to note that the increase in the use of the automatic coupler is paralleled by the use of the air brake.

In Plate VIII are differentiated the various items of expense on a percentage basis, to show by comparison where and what they are. It is well to state in regard to the relative percentages of operating expenses to operating income, that the changes in this ratio from year to year are largely influenced by the policy of the railroads in charging improvements to operating accounts. On June 30, 1905, the payrolls of the railroads showed 1,382,196 persons in railway employment earning \$839,944,680. The operating expenses per annum at this time were \$1,390,602,152, from which it appears that the compensation paid to railway employees amounts to about 60 per cent of the total expense of operation.

Plate IX shows the Gross Earnings, Operating Expenses, Total Income, Income from Operation, Dividends, Net Income, and Surplus, all per mile of line operated, and the official mileage upon which the figures are based.

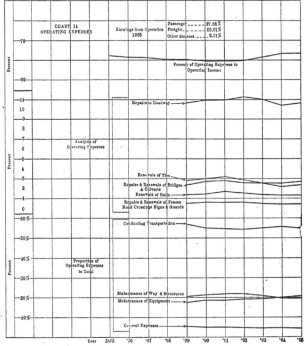

PLATE VIII. CHART 14, OPERATING EXPENSES ON A PERCENTAGE BASIS.

The last set of curves on Chart 16, Plate X deal exclusively with the par value of railway securities. Railway stocks are classified as common and preferred, and the funded debt as bonds, miscellaneous obligations, income bonds, and equipment trust obligations. The

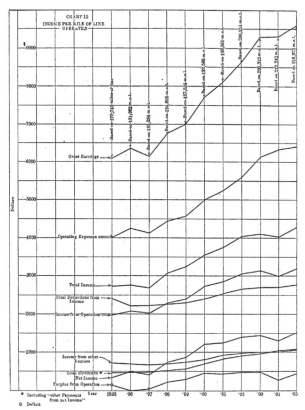

PLATE IX. CHART 15, EARNINGS, OPERATING EXPENSES AND INCOME.

aggregate of these securities makes up the total of the par value of railway capital. The most interesting curve of this set is the one marked "Per Cent of Stocks paying Dividends", which shows that only 29.9 per cent of the stocks paid dividends in the year 1895, while 62.8 per cent paid dividends in the year 1905; and this despite the periodic financial upheavals. It will be observed that the average

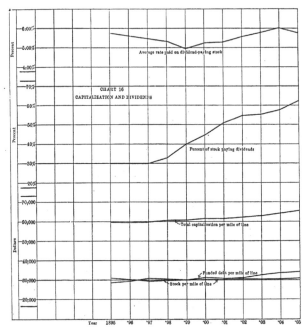

PLATE X.   CHART 16, CAPITALIZATION AND DIVIDENDS.

rate paid on dividend-paying stocks during the year 1905 is 5.78
per cent, which is a slight decrease from the previous year (6.02
per cent). The explanation of this decrease may be found in the
increase in the amount of stocks paying dividends (57.45 per cent
in 1904 and 62.84 per cent in 1905). If so it does not indicate a
reduction in the dividend-paying ability of the railroads, as might appear at first sight.

The Panama Canal.

ACCORDING to Washington dispatches, Lieutenant-Colonel Goethals has stated to a Senate committee that he can not tell if the cost of the Panama Canal will reach $250,000, or $300,000, or $500,000 ; indeed, that he can not fix any estimate at all of the total cost. The admission is evidence of the colonel's frankness, and of course does not suggest the slightest disparagement to his ability. Accurate estimates are peculiarly difficult on the Isthmus. Uncertainties of geology, of climate, of labor, of appropriations, and of legislation operate cumulatively to disturb an exact solution of the problem.

The elements which make determination of costs finally impossible, however, are inherent in the system under which the chairman of the Commission must work. But try to conceive of the chief engineer for a great contractor making a similar admission of uncertainty!

Politics and Engineering.

A MORE sinister aspect of political interference with economical and orderly management of public work was displayed in the indecent attempt to discredit the Commissioners of the Board of Water Supply of New York in the matter of the Ashokan dam. Not only public interest but private reputation was contemptuously trampled in the attempt to seize, for political and personal profit, appointments which had been pledged to conscientious and impartial selection and which were filled with ability and honor. It has been pointed out by some of our contemporaries that the result of such attacks would be that service would be-

come intolerable to men of high standing in their profession and in the community—that they would hold aloof from all public enterprises and leave them to the soiled hands of the political place hunters. Apparently this was exactly the result which the "investigation" sought to accomplish.

It failed completely, and recoiled upon the heads of its promoters, for two principal reasons. First, the strength of the position of the Commissioners, secure in absolute sincerity and single-minded loyalty to the interests committed to them ; and second, the great wave of determined support which swelled up from the public press, technical and non-technical. The Corporation Counsel's advice to the Mayor is that which he might have taken from his own better judgment at the outset. He now has to drop a matter he should have never have taken up.

A Mistaken Public Enterprise.

IF the temper of retrenchment and economy should so far affect New York State as to lead to serious reconsideration of the barge-canal project, the pains of the present business disturbance would not be without their reward. Superintendent Stevens is far within the mark in saying that the project now under construction will be obsolete when it is completed. It is obsolete already. It is ither of the dimensions nor on the 1.. the waterway which would be . serviceable to transportation a. mmerce. The waste of money in ashing it so far is deplorable, but it would be better to charge that off as loss and save what is left of the original estimate than to sink more in the ho. of making any practical success of t existing plans.

Train Ferry for the English Channel.

AMONG great engineering proposals which appeal to popular fancy for a sort of perennial vitality, the project for a tunnel under the English Channel has long been prominent. In its most recent revival, its advocates urged not merely the possible increase in the comfort, and therefore the volume, of passenger travel between England and the Continent, but the supposed great advantages to be secured by a through freight service, without transhipment at Channel ports. But if any such benefits as alleged would accrue to British trade by mere passage of "goods wagons" from British to Continental railway lines without breaking bulk, the ferry system would appear to afford all the necessary facilities with a fraction only of the uncertainty, cost, and inconvenience.

The transport of trains across the sea is no theoretical or untried proposal. Both on Lake Michigan and amongst the Danish Islands this method of transshipment has been in use for a considerable time. Nor is the mechanism of such transport either complex or expensive. Given a good harbor at either end and specially constructed vessels capable of taking made-up trains upon their decks, nothing more remains than to provide tidal platforms which enable the transfer to or from the boat to be made at any state of the tide. There is nothing particularly new nor particularly difficult in the provision of these tidal platforms. They may be seen in considerable number on the Mersey, where, it is true, only road vehicles are dealt with, and the enormous Prince's Landing Stage at Liverpool is a very fair example of a large floating structure of the kind which has worked for a great many years without the least trouble or difficulty.

To construct a tidal ferry in the Straits of Dover would not, as estimated, cost more than a twentieth of the capital outlay involved in piercing a tunnel. It would have many advantages. Under the stimulating effects of a large increase of traffic its facilities could be multiplied and extended quickly, cheaply and without any international complications whatever. It would avoid the menace—for a menace it seems to many British military strategists—of opening a possible line for sudden entrance by the forces of a Continental enemy. It would not weaken Britain's insular position, so strongly secured by the Channel fleet. No military problem is involved. The line of communication could be interrupted at a moment's notice by the smallest gunboat. It is not on the other hand a grandiose scheme, nor does it offer immense rewards to clever financiers, nor deeply stir the imagination like the great engineering feat of tunneling the ocean, but it is at least a practicable scheme. It secures the one thing worth securing, viz., greater advantages to commerce. It does this cheaply and therefore efficiently. More than this seems unnecessary.

Concerning Radio-active Medication.

THE efficacy of the "radio-energizing" of medicinal substances, mentioned in M. Boyer's article in our November issue, is doubtless believed in good faith by many persons; but the best authorities on the subject are at least sceptical, except possibly as to the cure of some forms of cancer. The emanation, even if therapeutically of benefit, dies away very quickly. To quote Mr. A. S. Eve, "in a month only a millionth of the activity remains, though polonium appears later." We must also correct a typographical error in the description of the mode of measuring radio-activity; the piezo-electric quartz is a plate, not a "sphere."

THE PREVENTION OF RAILWAY ACCIDENTS.

A PLEA FOR THE MORE GENERAL ADOPTION OF THE AUTOMATIC PRINCIPLE IN RAILWAY
SIGNALLING AND TRAIN CONTROL.

The Engineer.

WHILE fatal railway accidents are much less frequent in Great Britain than in the United States, several collisions and derailments on British railways during the last few months, due to the breakdown of the "man" and not the "machine," have proved conclusively that there is urgent need for further protection of the lives of passengers than can be afforded by even the highest excellence of materials and construction in permanent way and rolling stock. With the report of nearly every accident it becomes more and more apparent that the neglect of duty on the part of the signalman or engine driver is the chief contributory cause of railway fatalities of these two classes. In extension of Mr. Coes' demonstration of this fact in connection with collisions on the railroads of the United States, presented in THE ENGINEERING MAGAZINE last month, the following forceful plea for the extension of the automatic principle in signalling and train control, taken from *The Engineer,* London, for November 22, 1907, is of great and timely interest.

"If we examine the early and middle history of accidents, we shall find that they were to a great extent the result of structural failures, such as broken axles, broken bridges, broken rails, and broken tires. There were serious collisions also; but few derailments. Speeds were comparatively speaking, low. The lines were not crowded. We do not hesitate to say that if machinery, workmanship, and materials had been as good thirty years ago as they are now, everything else remaining the same, accidents would scarcely have occurred at all. It is well to keep this constructive or structural excellence in view, or what follows will not be fully intelligible. Now, a reference to the reports of the Railway Department of the Board of Trade for the last few years will show that the main causes of deaths and disablement on our railways have been collisions and derailments; and again, if we search a little further, we shall find that in the majority of cases the derailments have been due to a neglect of signals—trains running at high speed round curves, when caution or stop signals were shown, and plain instructions were supplied to the drivers. Collisions have been brought about by defective signalling, or by want of understanding on the part of the driver or fireman, or guards of a train. In other words, it may be said that railway mechanism is no longer the cause of disaster. Accidents are the result of the breakdown of the man, not of the machine. Formerly the machinery was more stressed than the man. In the present day the position is reversed, the man is more stressed than the machine. The logical deduction is, that in the future the abolition of death on the railway is not to be sought in further improvements in material or workmanship, or design of permanent way or rolling stock, but in the relief of stress in the human machine. This desirable result cannot be obtained by shortening hours. There is no reason to assume that under modern conditions 'brain fag'—to use a modern phrase which has no true physiological meaning—in so far as the time of service is concerned, has much to do with the matter. Signalmen have made mistakes after a couple of hours work. The only direction in which there is any promise of improvement—and by many that is still held to be problematical —lies in the extension of the automatic principle. Almost from the first automatic signals were suggested. More than half a century ago, in the Exhibition of 1851, there was shown a device by which a train passing a signal-post raised the arm to danger by running over a sort length of bar at the side of the rail. The arm was controlled by a clock in the signal-post, and would not fall for a period fixed according to the exigencies of traffic, but, generally speaking, of five minutes. There was, of course, no block system, but this invention served to maintain a time interval between two trains. From that date forth invention has followed invention; but the signalman still controls the traffic. The reason is simply that the man is held to be more trustworthy than automatic mechanism. There is, however, an apparent inconsistency. Although in certain respects, and within limits, implicit confidence is placed in him, he is not trusted with the control of signals connected with points. These are invariably interlocked, with the result that responsibility is transferred not to the mechanism, as would appear at first sight, but to the engine driver. A signalman cannot leave the siding points open and show main line clear for an express; but this does not eliminate all risks, because the driver may run through his signals and wreck his train. To secure absolute perfection the system should be such that the driver cannot, if he would, run through his signals.

"It is quite beyond our purpose to deal here with automatic mechanism. It is in use in train brakes; it controls signals from station to station, or cabin to cabin. In the present day of electrical development it is admitted generally that there is no physical reason why a driver should not have it forced upon his sight and dinned into his ears that he must stop. But there remains

behind the certain faith—it is not knowledge—that automatic signalling is not nearly as trustworthy as human signalling. We are assured that automatic signals break down and cannot be trusted. If, however, we look below the surface, we shall find that this is in the main the result of expecting too much. The inventor provides all that he considers needful, and assures the head of the signal department that his mechanism is so complete in every detail that it will work for six months without attention. The superintendent, after making allowances for the utterances of a sanguine inventor, himself lays down an impossible standard of excellence. The thing is automatic. If it cannot be trusted without supervision, then it cannot be trusted at all. But he does not trust the existing automatic locking of points and signals. The telegraph instruments and live wires are kept under constant observation. In precisely the same way, or rather to a much greater extent, all the mechanism of automatic signalling should be kept under the most careful observation. Brains must be devoted to this work instead of to looking out for trains.

"It is not to be supposed that the signalman can be entirely dispensed with. At important stations he will always be needed. Without him it would be very difficult to retain the necessary flexibility in working the traffic. But even while retaining his services, the extension of the automatic system beyond its present limits could easily be employed as a check on the signalman, which would prevent him from making mistakes. To take a very common case. Collisions have frequently occurred simply because the signalman has forgotten that a train is standing on the line close to his cabin. Now there is, of course, no reason why a bell out of his reach and beyond his control, should not ring, say, every minute while the train remains still. Or, if an audible signal of the kind might be supposed to worry him and prevent him from attending to other trains, then a visual indicator might easily be provided. There are a dozen ways in which, having due regard for the influence of climate, the principle of making the train set its own signals may be used. But no real progress will be made in this direction so long as those controlling the traffic hold to the theory that a man must be more trustworthy than, say, an electro-magnet. At one time this theory had, certainly, a strong basis of fact, the stronger that too much was expected from the mechanism employed. No doubt the man was then more trustworthy than the magnet. But this is no longer true. It may be now taken as proved that the utmost has been got out of the man. Nothing now remains but to supplement him. Even the most sceptical will, we venture to think, admit that, given efficient supervision, an automatic system will avoid the blunders and mistakes and errors of judgment which have of late years been mainly responsible for collisions and derailments on the railways of this country. The United States supply a valuable object lesson. The last Report of the Inter-States Commission tells the world that, in round numbers, 5,000 persons lost their lives on American railways last year. After deducting all other causes of death, we have a fearful list of derailments caused by bad

permanent way, and of collisions due to the failure of drivers and signalmen to do their duty. We have, in short, a combination of ancient conditions of bad materials and crude construction, and modern conditions of heavy traffic. The first we have in this country eliminated, the latter remains. Who can doubt that the extended adoption of the automatic system would be productive of good in the United States? It has formed no part of our purpose to name any particular invention, or to show how, had it been installed, a particular railway accident could not have taken place. Various discussions—notably at the recent Conference of the Institution of Civil Engineers—prove that automatic signalling is quite compatible with the highest standard of safety. It remains to be seen if any argument based on modern information can be urged against it in principle. Let it once be conceded that it may be more trustworthy than the man, and its adoption with all the resultant benefits will follow in good time."

THE COMMERCIAL ASPECT OF ALPINE RAILWAY TUNNELS.

A REVIEW OF PAST AND PRESENT PROJECTS FOR THE IMPROVEMENT OF OVERLAND COMMUNICATION BETWEEN NORTHERN AND SOUTHERN EUROPE.

Engineering News.

IN connection with a description of the Lötschberg tunnel between Brieg and Frutigen in Switzerland, *Engineering News* for December 5, 1907, published a long editorial on the commercial aspects of present and proposed Alpine railway tunnels, of great interest in making clear the importance and exact status of these projects which of recent years have become so numerous and perplexingly complicated. In the following extracts the main features of this extensive discussion are given.

"In the first place it should be explained that there is a great 'trade route' running southeast across Europe from the west coast to the Mediterranean, and this is obstructed by the mountain ranges of the Alps, extending northward between France and Italy, and eastward between Switzerland and Italy; further eastward, where it separates Italy from Austria, the barrier is less serious. The most formidable part of the range lies across the 'trade route' above

mentioned. The development of the existing and proposed routes across the Alps has been undertaken to provide improved facilities for existing traffic, rather than to develop new traffic. It has been governed largely by national commercial rivalry, and to an important extent also by municipal rivalry (especially as regards the Italian connections), and by railway rivalry in France. It has also been governed by two traffic conditions: (1) The freight traffic between northern European ports and industrial centers and the Mediterranean ports. (2) The passenger traffic, including the great tourist traffic to Italy and also that of what is known as the 'overland route' to India and the East. For many years the steamships on the oriental routes have made Brindisi (in Italy), near the southern end of the Adriatic Sea, a port of call for passengers and mails. Here the outbound steamers from England, after passing around France and Spain and down the Mediterranean, receive passengers and mail leaving London some days later than the steamers. In the same way, the homeward bound steamers land passengers and mails for British and continental cities, which are thus saved considerable time as compared with the longer and slower route by sea. Thus, very costly works may be warranted by a resultant saving of a few hours in time of transit.

"The trade and travel between Italy and its ports and Switzerland and northern Europe was, up to 1870, carried by the three great mountain highways crossing the Alps at the Simplon Pass (built by Napoleon in 1806), the Splügen Pass (1822), and the St. Gothard Pass (1824). In 1867, the Brenner Ry. (with an open summit in the Brenner Pass, in Austria) was opened, connecting the Austrian railways with those of Italy. At about the same time the Mt. Cenis railway was opened, crossing the Alps by a switchback line having steep grades, operated by the Fell grip-rail system. The traffic capacity of this was limited, but four years later, in 1871, the Mt. Cenis tunnel was opened, giving a direct line westward into France. It became evident that these new trade routes would absorb so large a proportion of the business that in order to regain a due share of this business and to protect Italian and Swiss interests it would be necessary to build a railway to connect Italy with the countries lying north of the Alps. The three highway routes were all examined with a view to the possibility of railways on approximately the same lines. The Simplon was rejected as being tributary to the interests of France rather than of Switzerland and Germany. The Splügen was rejected on military grounds, as being too near the Austrian frontier, and liable to seizure in case of war. The St. Gothard route was adopted, and in 1869 a treaty providing for the construction of the railway was made between Italy and Switzerland. In 1871, Germany also became a party to the treaty. The great summit tunnel was commenced in 1872 and completed in 1881. While this tunnel is the special engineering feature, it should be understood that the St. Gothard Ry. involved the construction of some 53 tunnels, aggregating about 25 miles in length.

"The Arlberg tunnel, on a railway connecting Austria-Hungary with Switzerland and western Europe, was completed in 1883. This line was 80 miles long, from Innsbruck to Bludentz, and had maximum grades of 2.5 per cent. South from Innsbruck runs the line to the Brenner Pass, already mentioned."

As an offset to the loss of traffic to the French railways by the opening of the St. Gothard line, the Simplon tunnel between Italy and Switzerland was proposed as early as 1870, but for various reasons the actual beginning of construction was delayed until 1895. The object of this tunnel was to give France communication with Italy by a shorter and easier route than that offered by either the Mt. Cenis or St. Gothard lines. Both these objects have been attained, the Simplon giving a route from Paris to Milan 55 miles shorter than the St. Gothard and 59 miles shorter than the Mt. Cenis, but this by no means put an end to projects for other lines. What the French railways are striving after is a direct route into Italy equally favorable with that of Germany and northern Europe. At present the only direct connections are by the Mt. Cenis tunnel and a coast line from Marseilles to Genoa, but this route is too long to enter into competition for international traffic.

The project now receiving the greatest attention for this purpose is that for the Mt. Blanc tunnel which has been proposed ever since 1836. The route now advocated is essentially the same as that proposed in 1878 from Chamounix to Courmayeur connecting with the Italian railways at Aosta. This route would shorten the line from Paris to Genoa over the Mt. Cenis by 52 miles, over the Simplon by 57 miles, and over the St. Gothard by 110 miles. It would have the disadvantage of high elevation and heavy grades and considerable auxiliary construction would be necessary before the line could be really competitive, but these considerations do not seem to carry much weight with the French promoters who are anxious for a direct mountain crossing at any cost.

As an alternative to the Mt. Blanc route, another route is proposed under the little St. Bernard, while Italian promoters advocate a line through the Great St. Bernard into Switzerland as a rival to both these routes. Other tunnel projects as auxiliary to the lines already in service are also very numerous, the most important being the Splügen Pass, the Lötschberg, now actually under construction, and the Faucille in France. The Lötschberg tunnel is to give Germany a direct connection with the Simplon line, while the Faucille project is generally regarded as a preliminary to the Mt. Blanc route. The smaller projects for the improvement of approaches to one or other of the mountain tunnels are too numerous for description.

"It has been suggested that the enormous cost of these great mountain tunnels might have been avoided, or postponed, by the construction of steep-grade switchback lines through the passes, as has been done in several cases in this country. There were two great objections to such a method, however: (1) the deep snows and heavy storms of the upper Alps; (2) (but most important) the necessity for speed and traffic capacity on railways competing for the business which already existed on a great 'trade route' and which would increase with better facilities for transportation. As a matter of fact, however, the first railway to cross the Alps was an open line, following closely the route of the highway or post road over the Mt. Cenis. This was operated by the Fell grip-rail system, in which pairs of horizontal driving wheels on the engines were powerfully gripped against a double-headed rail laid on its side on supports between the running rails. This line was opened some years before the completion of the tunnel, but was abandoned.

"With all subsequent projects, however, the problem has been much more difficult than that of simply getting a line across the mountains which has been the problem in the construction of railways across the Rocky Mountains. The Alpine tunnels are incidental parts of great traffic routes, and with each new project it has been necessary to locate a line which would be superior to preceding lines in regard to distance, time and operating conditions. Thus, with tunnel routes existing by the Mt. Cenis and St. Gothard routes, it would have been waste of time to even consider a steep-grade switchback line for the rival

Simplon route. The actual length of tunnel was a secondary consideration. What was needed was a line that would attract traffic by its more favorable conditions. It may be noted, for example, that the through traffic of the Simplon tunnel was estimated at 140,000 tons per year, of which 70 per cent. would be taken from the St. Gothard and 30 per cent. from the Mt. Cenis routes. The same principles and conditions govern the selection of the later routes now being proposed. It may be noted, however, that an open-air line operated on the Fell system has recently been proposed, on a route from Oulx, on the Italian side of the Mt. Cenis railway to Briancon, in France. This, however, is not on the main traffic route to the northwest, but is for an improved connection between the south of France and the great industrial centers of Turin and northern Italy. It would shorten the distance between Turin and Marseilles by 62 miles as compared with the present route by Genoa and Ventimiglia. The line would be 24 miles long and would cost $3,000,000.

"Summarizing the above review of Alpine tunnels and tunnel projects, it may be said that while some years of discussion and preliminaries must necessarily pass before any new Alpine line is definitely adopted and commenced, the indications are that the next international Alpine railway enterprise to be carried into effect will be that for a direct route between Italy and France (crossing only a small corner of Switzerland), as an offset to the Simplon route. It is significant to note again in this connection that the former routes have been built successively in rivalry of their predecessors and for competitive purposes. When the Mt. Cenis tunnel had connected France and Italy, the St. Gothard was planned to give Switzerland and Germany similar (but even more favorable) communication. This in turn was followed by the Simplon tunnel in the interests of France. Hardly had this been completed when work was started on the Lötschberg tunnel to give Germany a connection with the Simplon. And at the present time the greatest activity centers about a more direct French-Italian route in order to restore to France the possession of a route superior to all its predecessors."

ELECTRICAL ENGINEERING EDUCATION

A DISCUSSION OF THE TEACHING AND CURRICULUM BEST SUITED TO THE EDUCATION OF THE
ELECTRICAL ENGINEER.

Chas. F. Scott, Chas. P. Steinmetz—American Institute of Electrical Engineers.

THE great interest in electrical engineering education manifested by the attendants at the last annual convention of the American Institute of Electrical Engineers resulted in November, 1907, in the appointment of a committee to organize the work of promoting discussion on this important subject. On January 24 a special meeting was held at New York under the auspices of this committee, and judging from the interest and breadth of the discussion, the work of the committee is likely to be productive of much good. At this meeting, as an introduction to the discussion, a digest of all the papers on the subject presented before the Institute since 1892 was introduced by Mr. Chas. F. Scott and this was followed by a paper by Dr. Chas. P. Steinmetz, both of which are reviewed below.

Mr. Scott's short introductory paper emphasized the general agreement as to the importance of the subject, but pointed out that while the technical schools are willing and anxious so to shape their courses as to produce the type of engineer best suited to the demands of industry, there is no general agreement as to what is really desirable in the young graduate. Some demand skilled artisans and some trained engineers. Some expect technical specialists and some men of all-round knowledge and capabilities. In some cases the young graduate is expected to be immediately productive, while a number of the larger manufacturers provide practical courses of training. Mr. Scott believes that the best engineering education is that which fits the individual student for his largest development and usefulness *in the long run* and that eventually this sort of training will be accepted as the solution of the present much debated problems. In the meantime, however, the discussion centres around the following points, on which widely divergent opinions have been expressed in papers previously presented before the Institute, a digest of which follows Mr. Scott's paper:

"1. The desirable characteristics of the acceptable graduate with respect to:

a. Practical familiarity with electrical apparatus which will enable him to be immediately useful, versus a less practical and more general training, which is to be supplemented by an apprenticeship course or its equivalent.

b. Specialized technical training and technical knowledge, versus a broader education aimed to develop intellectual power rather than the acquisition of technical knowledge.

"2. The arrangement of subjects and courses which will best produce the desired results. The following questions arise:

a. The relative attention to be given to the practical or industrial, the scientific, and the cultural.

b. The relative proportion between subjects which are valuable for imparting technical knowledge and those affording training in scientific and logical methods.

c. The relation between school instruction and practical work; whether one should precede the other, and if so which one should come first, or whether they should alternate once a day, once a year, or at some other rate.

d. The importance of current engineering practice; of lectures by practising engineers; of discussion of current topics in local meetings of the Institute.

e. The degree of desirable differentiation in courses or methods on account of differences in the characteristics of individual students or in the fields of work they expect to enter.

f. The sequence of subjects—whether the theoretical ground-work should be laid during the first few years and the practical subjects reserved until the latter part of the course, or whether an intermingling of the two in accordance with the concentric method outlined by Professor Karapetoff is to be preferred."

Dr. Steinmetz's paper was in thorough agreement with Mr. Scott's view that training in the fundamental principles rather than education in a mass of memorized facts is the only satisfactory solution of present difficulties. The conditions for good electrical engineering education are more favorable in the United States than anywhere else. The great magnitude of the development of the electrical engineering industries, the general demand for technically trained men, the close co-operation between the industrial and educational branches of the profession are all circumstances favoring excellence in the results of technical instruction. There are, however, glaring defects in methods which prevent the attainment of the success in technical education which should be secured by the ample means at the disposal of most technical schools.

"The great defect of the engineering college is the insufficient remuneration of the teaching staff: the salaries paid are far below those which the same class of men command in industrial work, and as a result the college cannot compete with the industry for its men, but most of the very best men are out of reach for the colleges. The teaching forces of the colleges, therefore, consist of: 1. A few of the very best men, who are specially interested in educational work to such an extent that they are willing to sacrifice financial returns for it. These men have made the engineering college what it is; but even many of these men are ultimately forced by considerations of family, etc., to leave college work and enter industrial employment. 2. Many younger men interested in teaching, enter college work to give it a trial. Some of these remain, but many return to industrial work, when they are forced to realize the small prospect of financial return offered by the college. 3. First-class men who devote a part of their time to the college and a part to industrial work, usually consulting engineering. This arrangement is probably the best for the college, handicapped as it is by the policy of salaries which may have appeared sufficient in branches in which no industry competes, but which are suicidal in the engineering department. Still it would be far preferable if the colleges could get the benefit of the whole time and the undivided interest of these men." A vast improvement

could be made if a large part of the sums now expended on magnificent buildings and extensive laboratory equipment were used in procuring the highest grade of men as teachers.

Grave objections may also be made to the curricula and methods of teaching of the engineering schools in that quantity seems to be sought before quality. Students are forced to memorize for examination purposes a vast amount of detail more than they can possibly digest.

"It stands to reason that with the limited time at his disposal, it is inadvisable for a student to waste time on anything which he forgets in a year or two; only that which is necessary to know should be taught, and then it should be taught so that at least the better student understands it so thoroughly as never to forget it. That is to say, far better results would be obtained if half or more of the mass of details which the college now attempts to teach, were dropped; if there were taught only the most important subjects—the fundamental principles and their applications —in short, all that is vitally necessary to an intelligent understanding of engineering, but this taught thoroughly, so as not to be forgotten. This, however, requires a far higher grade of teachers than are needed if the mere memorizing of text-book matters, reciting them, at the end of the term passing an examination on the subject and then dropping it. The salaries offered by the colleges are not such as to attract such men. When the student enters college he is not receptive to an intelligent understanding, for after a four years' dose in the high school of the same vicious method of memorizing a large mass of half and even less understood matters, the student finds it far easier to memorize the contents of his text-books than to use his intelligence to understand the subject-matter. After graduation, years of practice do for the better class of students what the college should have done—teach them to understand things. It is, however, significant that even now young graduates of foreign universities, in spite of the inferior facilities afforded abroad, do some of the most important electrical development work of this country. Men who never had a college education rise ahead of college

graduates. This would be impossible if
our college training gave what it should,
an intelligent understanding of electrical
engineering subjects."

Dr. Steinmetz makes a strong plea for
the general education of the electrical engineer. The development of the electrical
industry is so rapid that the graduate of
the school which has dropped from its curriculum everything not required in electrical engineering soon finds himself out
of the race for success when subjects become of importance which were not considered as part of the trade of electrical
engineering at the time when he received
his education. Such an one is utterly help-
less when any occasion arises which requires general knowledge outside of his
trade as he learned it. In a good many
technical schools this fact is recognized
and an attempt is made to teach or review
general subjects, both cultural and technical. Such attempts, however, usually fail
to be of any real service owing to the
brevity of the courses or wrong methods of
teaching. In the same way instruction in
chemistry, civil engineering, etc., is often
of little real service, being totally unsuited
to the requirements of the electrical engineer. Dr. Steinmetz believes that an important factor in the inferiority of the

product of American technical schools to
that of those abroad is due in large measure to the inferior teaching of the sciences
allied to electrical engineering.

"In conclusion," he says, "the main defects in the present electrical engineering
training in some of our colleges appear to
me as follows:

1. The insufficient remuneration of the
teachers, which makes most of the best
men unavailable for educational purposes
and is, therefore, largely responsible for
the other defects.

2. The competition between colleges,
which leads to a curriculum marked more
by the quantity of the subjects taught than
by the thoroughness of the teaching. The
graduates are sent out with a mass of half
understood and undigested subjects, quickly forgotten, and deficient in understanding of the fundamental principles and in
the ability to think.

3. The tendency of some colleges to
teach the trade of electrical engineering
rather than educate intelligent and re-
sourceful electrical engineers.

4. The unsatisfactory state of the teaching of allied sciences, which gives instead
of general view and understanding of the
science, a fragmentary knowledge of some
details."

THE ECONOMIC MANAGEMENT OF STEAM PLANTS.

A SUMMARY OF THE FACTORS INFLUENCING THE EFFICIENCY AND ECONOMY OF OPERATION
OF STEAM PLANTS.

William H. Patchell—Association of Engineers-in-Charge.

THERE are few problems in modern
industry of more importance than
that of the economical production
of power. Economy and efficiency in power
production is especially important in
steam plants on account of the steady rise
in the price of coal, and the factors entering into the securing of these conditions are of the utmost interest to the engineer. An interesting summary of some
of the more important means of securing
economical operation is given in the recent presidential address of Mr. William
H. Patchell before the Association of Engineers-in-Charge, of which the following
is a brief abstract.

Dealing first with boiler feed water, Mr.

Patchell believes that in a great many
cases it is more economical to take water
from a well on the premises than from
city mains. The bored well has proved its
reliability and by the use of geared pumps
or, preferably, when the water supply is
plentiful, the air lift system a low priced
and very efficient water supply system can
be obtained. Of course in making the
change from a supply from mains to a supply from a well the quality of the water
must be considered carefully, for it is
never economical to feed impure water
into the boiler. If the water has scale-
forming propensities the loss in heat is
very serious, while if it forms only sludge
a great amount of time is consumed in

blowing off. Feed water should always be purified as much as possible and the open heater is one of the simplest and most efficient types of purifying apparatus.

The rising price of coal makes the problem of securing the most efficient combustion possible one of the utmost importance. In many cases economy will lie in the burning of smaller and cheaper grades, not only in connection with mechanical stokers but also with hand firing. In burning small sizes the stoker must be made to keep his fires thin to permit of an ample circulation of air through the bed of fuel. Steam jets under the grate form a simple means of assisting the draught, but unless carefully watched the amount of steam blown away assumes most uneconomical dimensions. It is much better to use a fan whenever possible. Forced draught requires a smaller fan, but the induced draught system has certain advantages when the chimney draught has to be assisted only on peak loads.

"Gas analysis has of late years received attention, though for many years it was ignored alike by experts when testing steam boilers and by the users of boilers; it is only of late years that the percentage of CO_2 in the flue gases has been recorded on boiler tests. Thousands of tons of coal are wasted yearly by the neglect of the subject. The average amount of CO_2 in the gases may be taken at about seven per cent, which means a loss of twenty-six per cent. in the calorific value of the fuel. Greater care would increase the percentage of CO_2 to, say, twelve per cent, which means a loss of fifteen per cent only, and would represent a saving of ten per cent, which would quickly pay for the apparatus involved. The benefit of recording the CO_2 in the flue gases was first realized in Germany, whence the earlier types of instruments were developed. Since their introduction into England many of the faults in the earlier types have been eliminated, and some of those now on the market merit a much larger recognition than has yet been accorded to them. In many German works it has long been the custom to pay the stokers a premium on the amount of CO_2 recorded, with beneficial results.

"Purchase of fuel by analysis is very im-

portant, but when we consider the ease with which coal merchants can sell their stock, it is not surprising that they do not generally view the system with favor, and resent the incorporation of such clauses in their contracts. More united effort will be needed to get them educated up to this equitable method of buying fuel. Experience gained by several years' trial has proved to me that sampling and analysis may be done quite commercially under a contract which states the amount of volatile hydrocarbons and ash allowable, with a penalty if the amount of ash is exceeded."

In selecting mechnical stokers care must be taken to choose a type which will burn different classes and sizes of fuel, and it must also be remembered that stokers which give excellent service under a certain type of boiler may be useless under others. Where only one or two boilers are in use no saving in labor can be made by installation of mechanical stokers and the saving effected in the class of fuel used must be very definite to warrant the additional capital outlay.

"Steam superheating, although increasing rapidly in favor, is still conspicuous by its absence in many plants where it could very readily be adopted with a distinct gain in economy. No matter whether the steam is to be used in an engine or for heating purposes, the gain due to superheating is very marked. The ill effect of condensation in steam pipes is cumulative, and the application of steam traps to remove the water offers a fine field for further waste. Pipe covering is frequently only literally applied, all the flanges being left bare. Although gills and flanges are recognized as necessary adjuncts on radiator pipes, it is strange that when an attempt is made to prevent radiation from steam pipes, these are the very parts which are usually left bare. The excuse is sometimes offered that leakage and the necessity for attention to joints prevent the adoption of covers to flanges. If the joints are properly made and the flanges stiff they may be safely covered up and their existence forgotten. The water formed by condensation at a flange will often cause a leak that would never have occurred if steam only had been present."

When the heat in the gases can be spared water heating by economizers is a great source of economy, increasing the evaporative capacity of the boiler and preventing local cooling and consequent straining. Water heaters of both the open and closed types are most valuable in the case of non-condensing plants.

"Air heating has been tried again and again, but it is very doubtful if it pays. Perhaps the greatest advantage it possesses is in the suppression of smoke, and in that field it certainly has an important place. Smoke prevention has been run hard both by fanatics and by those who are really anxious to see trade carried on upon a commercial basis with the least possible inconvenience either to the traders or to their neighbors. All who have considered the matter will admit that, though black smoke means waste, it is not equally true that a smokeless chimney means economy. On a series of tests, the lowest efficiency was recorded with a minimum amount of air, and the highest with a moderate amount of air, when some smoke was made. Every effort should be made to prevent the emission of smoke, the underlying principle of which is the admission of the proper quantity of air at the proper time, and the removal of cooling surfaces so far as is possible from the gases until combustion is complete, as if the temperature is lowered below the critical temperature before combustion is effected no amount of care in air regulation will prevent smoke. . . .

"It is an unfortunate fact that in many cases an engineer in charge does not really know what his plant is doing, and all for the want of tests which he could carry out himself with very slight expense or inconvenience. Coal may be weighed without appreciably increasing the cost of handling it. Water may be measured by positive meters, which do not need much attention so long as they are not run at a high speed or cut by grit. An interesting and novel type of water-meter is Lea's notched weir, which is very simple and not likely to be put out of action by the causes that affect other water-meters. It is a very ingenious application of an indicating and recording attachment to measure the flow over a V-notch, the value of which was first investigated by James Thompson about fifty years ago. These two measurements, coal and water, give sufficient data for checking the cost of evaporation, which is the most important factor in the works cost of private plants.

"Engine tests appear to have had more fascination for experimenters than boiler tests, but comparison of records goes to show that there is not much difference in the efficiency of different types of boilers which vary largely in design; certainly the figures are not so far part as are the results of tests of similar boilers worked under different conditions. This puts all the stronger emphasis on the necessity of an engineer in charge knowing what his boilers are doing. The more a man knows of his plant the keener will he be to keep its performance up to the highest level obtainable, and nothing shows a staff the capabilities of a plant and secures confidence in it better than a few tests, the educational value of which on all concerned cannot be overestimated."

THE WEATHERING OF COAL.

THE RESULT OF A SERIES OF INVESTIGATIONS TO DETERMINE THE CONDITIONS WHICH PROMOTE OR RETARD THE DETERIORATION OF STORED COAL.

S. W. Parr and N. D. Hamilton—Economic Geology.

A SERIES of investigations on the weathering of coal, the final results of which seem likely to be of considerable practical and economic interest in many industries, is now being carried out by the Chemical Department of the University of Illinois, in co-operation with the Illinois State Geological Survey and the Engineering Experiment Station. In connection with the rapid extension of the practice of storing large amounts of coal in many industries, the question of deterioration is of very vital importance. Storage plants with a capacity of 100,000 tons are by no means uncommon and, although a deterioration of one per cent. in value in this amount of coal means a loss of 1,000 tons, there is very little exact information as to the conditions which promote or retard the loss of heat value by weathering. It is information of this sort that the investigations now in progress are designed to furnish. The results of a series of preliminary tests made during the past year are given in the following abstract of a report by S. W. Parr and N. D. Hamilton in *Economic Geology* for November, 1907.

"Richter, after extended experiments in 1868, formulated an explanation for the weathering of coal, which does not seem to be disproven by more recent experimenters, to the effect that the weathering of coal is due to the absorption of oxygen, a part of which goes to the oxidation of carbon and hydrogen in the coal, and part is taken into the composition of the coal itself. We certainly need more definite information in order to formulate a final and satisfactory explanation of all the phenomena involved, but Richter's theory conforms to many of the known conditions and indicates the close relationship between the matter of deterioration and spontaneous combustion."

The first series of tests was made on small amounts of coal, in lots of from ten to twenty pounds, and it may be questioned whether deterioration would occur in large heaps at the same rate. It was desired, however to determine only the conditions attending the process of weathering as a preliminary to more extensive quantitative tests which are to follow.

The coal was sampled as soon as possible after it was mined. It was found during the investigation that the early examination of samples to determine initial conditions is a most important point, as deterioration of a certain sort takes place even under the most ideal laboratory conditions. The coal used was of the small lump, or nut size. Each sample of about one hundred pounds was subdivided to subject the same kind of coal to the four conditions of the tests, viz.: (a) outdoor exposure; (b) exposure to a dry atmosphere at a somewhat elevated temperature, from 85 to 120 degrees F.; (c) under the same temperature conditions as (b) but drenched with water two or three times per week; (d) submerged in ordinary water at

a temperature approximately 70 degrees F. The periods of examination of these samples as nearly as the work would permit were: (1) an initial analysis of the fresh coal; (2) after exposure for five months; (3) after exposure for seven months, (4) after exposure for nine months, at the expiration of which time the tests were ended. At these periods the calorific values of the samples were determined under uniform conditions by the Parr calorimeter and the results calculated to the ash- and water-free basis. The results of the determinations are given in a series of diagrams which show graphically the changes which took place in each sample.

An examination of the diagrams shows a distinct difference between the submerged coal and the samples exposed to the air. With two exceptions, the samples showed practically no deterioration in the submerged coal, and the values found did not vary more than would be expected with inevitable modifications of conditions of sampling, temperature and manipulation.

The samples subjected to outdoor exposure uniformly showed marked deterioration but of varying amount. The treatment of the samples was identical, the coal remaining in shallow boxes exposed to the various temperature and moisture changes from October to July. The variations in heat loss, therefore, ranging from two to ten per cent, must be ascribed to inherent properties of the coals themselves. All showed a tendency to disintegrate, but they varied distinctly with regard to the ease with which they crumbled under pressure.

The results of the tests on the coals subjected to a dry atmosphere and a slightly elevated temperature were rather unexpected in that, with one exception in which the deterioration was practically the same, they showed a greater deterioration than in the case of outdoor exposure. This would seem to contradict the popular idea that a roof over coal in storage is supposed to be preferable to open exposure. The samples subjected to high temperature with frequent wetting down behaved in general like those exposed to outdoor influences, though in some cases a greater deterioration was observed in the former samples. Here the results are undoubtedly variable in accordance with the variation of structure and composition of the coals themselves. In general, a greater persistence of value might be expected in the dense and less friable coals and in those with less of iron pyrites throughout their texture.

In conclusion the authors summarize the results as follows:

"(a) Submerged coal does not lose appreciably in heat value.

(b) Outdoor exposure results in a loss of heating value varying from 2 to 10 per cent.

(c) Dry storage has no advantage over storage in the open except with high-sulphur coals, where the disintegrating effect of sulphur in the process of oxidation facil-

itates the escape of hydrocarbons or the oxidation of the same.

(d) In most cases the losses in storage appear to be practically complete at the end of five months. From the seventh to the ninth month the loss is inappreciable.

(e) The results obtained in small samples are to be considered as an index of the changes affecting large masses in kind rather than in degree, but, since the losses here shown are not beyond what seems to conform in a general way to the experience of users of coal from large storage heaps, they may be not without value as an indication of weathering effects in actual practice."

LARGE GAS ENGINES IN GERMANY.

INTERESTING DATA ON THE EXTENT AND RESULTS OF THE USE OF GAS ENGINES IN
METALLURGICAL WORKS.

R. E. Mathot—Revue de Mécanique.

IN Germany the rapid development of the large gas engine has been largely due to its introduction for the utilization of blast-furnace or coke-oven gas in the iron and steel industries. In a review in these columns in October, 1906, it was noted that at that time, out of 49 iron works in Germany, 32 had already installed gas engines and 9 had ordered them. In the following extracts from an article in the *Revue de Mécanique* for November, 1907, M. R. E. Mathot reports further extension of their use in metallurgical works and gives some interesting data of the results obtained.

Of the 50 large iron works in Germany 45 have already installed gas engines for the utilization of the waste gases of the blast furnaces and coke ovens or are preparing for their introduction. These represent altogether 350 units, designed to develop a total of about 400,000 horse power. The largest of these installations is of 35,000 horse power and there are 15 which vary between 10,000 and 15,000 horse power. In some few cases the plants have been laid out so as to permit the installation of gas producers in case they are needed to keep the engines working at their full capacity.

In collieries and coking plants the strife between the internal-combustion motor and

the steam engine has been less marked on account of the large number of old coke ovens in use, from which the gas available can be used only for raising steam under boilers. Nevertheless in such plants there are now about 25 to 30 motors either working or about to be installed, developing a total of from 40,000 to 45,000 horse power. The greater part of the gas engines employed in blast-furnace plants and in collieries are of the double-acting type, some of them being two-cycle and some four-cycle. The latter are more generally used, however, on account of their high efficiency.

The increasing favor which the gas engine is finding in the metallurgical industries is easily explained by the following figures: an ordinary blast furnace with an output of 100 tons of pig per 24 hours produces almost 9,000 cubic metres per hour of gas available for power generation with a calorific value of about 1,000 calories per cubic metre. This amount of gas if used for raising steam could develop about 2,500 horse power while, used in gas engines, it develops 4,200 horse power, an increase over steam engines of 1,700 horse power or 70 per cent.

Such excellent results as these cannot be obtained, however, unless the plant is provided with all modern improvements, among which the most important have to do with systems of cleaning the gas. Cleaning devices have of late been given a great deal of attention by both constructors and users of gas engines. To clean the gas of the impurities it contains, such as dust, tar and chemical products prejudicial to the satisfactory working of the gas engine, and to reduce the temperature of the gas before its admission to the cylinder, a complete washing, a purification, and a cooling process are necessary. These operations are performed by means of fans, rotating washers or similar devices which use 1.5 to 2 litres of water per cubic metre of gas cleaned. It is possible by this means to reduce the dust content of the gas from 3 grammes to 0.2 grammes per cubic metre.

The power required to drive the fans and washers depends both on the system adopted and on the amount of impurities to be eliminated and varies between 2.5 and 8 horse power or on the average about 3 per cent. of the power contained in the gas. The cooling water necessary for the motors themselves, per horse-power hour, varies between 8 and 12 litres for the piston and piston rod, and 30 to 40 litres for the cylinder casing, valve seat, etc. The oil necessary for the lubrication of a good engine should not exceed 1 to 1.5 grammes per horse-power hour.

In order to realize as completely as possible one of the indubitable advantages of the steam engine with which the internal-combustion motor has to compete, the latter are designed to give a large margin of power. They are usually designed for a pressure of about 5 kilogrammes per square centimetre for normal and constant working although the excellent construction of the motors permits of the use of a pressure of from 6 to 6.5 kilogrammes per square centimetre. So far as fuel economy is concerned a thermal efficiency of 30 per cent. is attained. This almost corresponds to the production of an effective horse-power hour with 2,100 calories, as well in the small single-acting motors as in the large double-acting machines, working on a normal charge.

The regulation of the motors is effected by variation of the mixture admitted to the cylinder in each cycle, either by changing the volume with constant composition, varying the composition of a fixed volume of gas, or by combining these two methods. The first method implies a variable compression and, as a consequence, some loss due to the partial vacuum produced in the cylinder with small charges, but this is offset by the high power output of all the charges. The second method, although it appears less economical, possesses certain mechanical advantages. The third method, according to its advocates, possesses all the advantages of the two former methods with none of their disadvantages. The two systems combined, however, lead to a complication of mechanical details and it is a question whether, on this account, the operation of the engine is more satisfactory. The large gas engines are, in general, of the double-acting, tandem, horizontal type.

THE TESTING OF HEAT-INSULATING MATERIALS.

A DESCRIPTION OF SMALL AND CONVENIENT APPARATUS WHICH GIVES RELIABLE RESULTS IN SMALL-SCALE TESTS.

Charles R. Darling—Engineering.

IN *Engineering* for December 6, 1907, Mr. Charles R. Darling describes a simple and efficient piece of apparatus for the testing of heat-insulating materials under working conditions, which should prove of great interest and value to managers of steam plants. As Mr. Darling points out, tests such as are conducted on a small scale in the works laboratory are usually devoted to the determination of the absolute conductivity of the material and, while this is of importance, it can serve as only a partial guide and may be very misleading as to the behavior of the material under working conditions. While there are other methods of testing which will give accurate results, for example, those of coating a length of pipe with the insulating material and heating either by passing steam through the pipe or by an electrical resistance, the apparatus is too bulky for use in the ordinary plant. In the apparatus devised by Mr. Darling, described in the following extracts, reliable tests can be made on small quantities of material with an accuracy equal to that obtained with the larger and more costly methods of testing.

"The testing chamber is an air-tight copper vessel, cylindrical in shape, the height being 8 inches and the diameter 6 inches. This vessel is completely covered with the material under test of uniform thickness. The source of heat is a 32-candle-power incandescent lamp, without capping, the leads from which pass through a stuffing-gland, which may be removed to permit of the entrance of the lamp into the vessel. A voltmeter is placed across the lamp terminals, and an ammeter and rheostat are inserted in one of the wires leading to the mains. A brass tube communicates with the interior of the cylinder, and is connected by means of a glass tube and a piece of flexible tubing to a mercury cistern. This arrangement enables the internal temperature of the cylinder to be determined by the pressure of the enclosed air, on the well-known principle of the constant-volume air-thermometer; and as the average pressure is registered, the true average temperature will be obtained. As the temperature rises the cistern is raised, so as to keep the mercury at a fixed mark on the glass tube, the increase in pressure being measured by reading on a scale the difference between the starting and final level of the mercury in the cistern. A cup, furnished with a tap, and communicating with the interior of the glass tube, contains a drying material, which prevents the entrance of moisture; and on opening the tap the enclosed air attains the pressure of the atmosphere and the mercury stands at the same level in the tube and cistern. The whole arrangement will be recognised as a gas-pressure thermometer, of which the copper cylinder constitutes the bulb.

"In commencing a series of readings, the tap is opened and the height of the barometer noted. The temperature of the cylinder is obtained by means of an accurate thermometer, placed in contact with the lid beneath the lagging, and allowed to remain until stationary. The cistern is then moved until the mercury in the tube is opposite the fixed mark, and the scale reading opposite to the level of the mercury in the cistern is noted. The difference of level which must subsist between the starting and final positions of the cistern may then be obtained for any given temperature by calculation, or reference to a suitable table. For example, if a temperature of 180 degrees Cent., corresponding to 146 pounds absolute steam pressure, were required, the barometer standing at 765 millimetres, and the temperature of the cylinder at starting being 10 degrees Cent., the required difference of level would be 45.4 centimetres, or 17.8 inches. The tap is now closed and the current turned on full—the resistance being cut out—and the temperature allowed to rise until the pressure exceeds that required. by 2 or 3 centimetres; the cistern having been raised at intervals, so as to keep the mercury at the mark in the glass tube. A portion of the resistance is now brought into the circuit, and the temperature allowed to fall gradually until the correct value is indicated by the mercury level in the cistern. If the resistance has been carefully adjusted, the temperature will remain steady; and when stationary for 10 or 15 minutes the readings of the ammeter and voltmeter are taken. This process may be repeated at intervals of 10 degrees by carefully increasing the resistance, and the heat passing through the lagging obtained in each case by reading the ammeter and voltmeter, and making use of the expression

Volts $\times$ amperes $\times$ 0.24 $=$ calories per second,

or

Volts $\times$ amperes $\times$ 0.057 $=$ B.Th.U. per minute.

An equal thickness of a second material may now be substituted, and the observations repeated. The experiments should be carried out in a place not liable to large fluctuations of atmospheric temperature.

"The author has subjected the apparatus described to an extended trial, and has found it accurate and easy of manipulation. With any good form of lagging, ¾ inch thick, a range of temperature from 220 degrees Cent. downwards may be achieved, thus ensuring that the test may be performed at working steam temperatures. Higher temperatures still may be obtained by the use of a greater thickness of the material. For lower temperatures, such as 130 degrees Cent. and downwards, a 16-candle-power lamp is used in place of the 32 candle-power requisite for the higher range. The readings may, by the use of proper instruments, be made very sensitive; the change in temperature occasioned by an alteration of 1/100 ampere being detectable. For direct-current circuits, working from 100 to 110-volt mains, a moving-coil ammeter—range, 0 to 2 amperes, and capable of being read to 1/100 ampere—will be found suitable; for 210-volt mains a range of 0 to 1 ampere suffices. Moving coil voltmeters having a range of 60 to 110 volts, or 120 to 220 volts, according to supply, are also well suited to the purposes of the test, and should permit of being read to 0.5 volt. For alternating circuits hot-wire instruments of the same range and delicacy of reading may be substituted. A continuous wire resistance of range 0 to 250 ohms will permit of the necessary adjustment on any of the above circuits.

"The author is of opinion that this test will prove of great service, not only to the makers of boiler and steam-pipe coverings, but also to engineers in general, in enabling them to decide on the best type of cover for any given purpose. It is proposed to publish, at a later date, the results of tests made with various commercial types of lagging, together with the effects of thickness, atmospheric temperature, and nature of surface on the efficiency. No reliable information on these points exists at present, as it has not hitherto been found possible to obtain satisfactory data without the use of apparatus too large and elaborate to permit of the carrying out of the observations involved in a reasonable time. The contradictory assertions in published reports of tests are sufficient to indicate that a standard method of procedure is urgently needed, and the author ventures to hope that the apparatus described will help to place our knowledge of the subject on a better scientific basis."

THE ZIEGLER SYSTEM OF PEAT UTILIZATION.

A DESCRIPTION OF THE LATEST PLANT FOR THE BY-PRODUCT COKING OF PEAT RECENTLY ESTABLISHED AT BEUERBERG, BAVARIA.

Engineering.

T O the problem of the utlization of peat, which is now attracting so large an amount of attention in the United States, Continental practice offers two solutions. The latest, that of gasifying the peat in the gas producer with recovery of by-products, the results of which were presented in these columns in THE ENGINEERING MAGAZINE for December, 1907, is particularly adapted to cases where the demand for power outweighs all other considerations. The Ziegler system of by-product coking has been brought to such a stage of perfection, however, that the excellence of the coke and by-products obtained may cause it to be preferred in many localities in the United States and Great Britain. The following description of the latest development of this system in a large plant at Beuerberg in Bavaria is taken from *Engineering* of November 15, 1907.

The by-products obtainable are of great variety and excellence, and find a ready sale. "The farmers readily buy the ammonium sulphate, while chemists are quite aware of the value of the peat tar, because it resembles the lignite tar and yields, like it, tar proper, pitch, phenols, paraffin, acetic acid, and methylated spirits, and the peat coke can replace charcoal as a preserving and absorbing medium. These statements may not be of particular interest to the engineer, but what will directly appeal to him is the fact that peat-coke proves itself quite equal, if not superior, to charcoal in the metallurgy of iron and steel, and in the hardening and welding operations of armor-plate and engineering works, and also that it gives smokeless briquettes for marine boilers. There is undoubtedly a great opening for peat coke, which many districts will be able to produce at low cost in substitution for the now scarce charcoal. There is, further, the peat power-gas, and in addition still another side to the problem. It has been found in several localities—and, for in-

stance, at Beuerberg—that the clay underlying the peat bogs will, when mixed with
the sand also occurring in those strata,
yield an excellent farming soil. Thus we
may hope to convert the 1860 square miles
of bog of Ireland into productive ground,
and to raise and nurse industries at the
same time."

Ziegler developed his process from that
of the distillation of lignites and his first
plant was built at Oldenburg in 1897.
Though the financial results were not satisfactory the technical success of the process attracted so much attention as to be
made the subject of a special investigation
and report by a Commission appointed by
the Prussian Government. The next plant
built by Ziegler was at Redkino, in Russia,
but the projected auxiliary chemical works
were never completed. In 1905, however, all
the Ziegler patents were taken up by a
company and the large works at Beuerberg, about 24 miles south of Munich,
which embody the latest improvements in
the system, were opened in 1906. The plant
is situated on the edge of a large peat
moor which at the present rate of working
will furnish raw material for the next 40
years.

A large factor in the success of this plant
has been the adoption of electric power
throughout. Portable elevator huts, each
containing a Dolberg peat process run by a
small motor, follow on light tracks the
progress of the cutters. The peat as cut
with the spade is thrown on an elevator
which conveys to the press. From the
two nozzles of the latter, two strings of
pressed peat emerge, each about 4 by 4¼
inches in section. These are caught on
wooden pallets and are cut into blocks
about 16 inches long. The pallets are then
conveyed on trucks to the drying ground
where the peat is left until the moisture
content is reduced by air drying to below
25 per cent, the blocks being piled in
heaps as they become hard enough to
handle. When fully dried the blocks are
trucked to the furnace house, a four-story
structure, where they are elevated to the
charging floor in the top story. Two types
of furnace are installed which are described as follows:

"The two furnaces of type I are vertical
kilns, about 40 feet high. Each kiln rep-

resents a pair of retorts approximately
semi-circular in section. The lower portion of the kiln, from which the coked
peat is withdrawn by means of a helix, is
common to both the retorts. But the chief
part of the kiln is subdivided by a vertical
partition, which forms a system of flues,
and the whole kiln is jacketed with flues.
The hot gases which circulate in these
flues have a temperature of 1000 degrees C.
(1832 degrees F.), while the temperature
in the retort itself does not rise above 600
degrees C. (1000 degrees F.). The retort
is thus completely surrounded by flues, and
the construction of the furnaces is not simple. About 10 feet above the lower level
are the auxiliary grates, which are charged
with dry peat at the commencement of
operations, to start the furnace. Next to
these grates are the three auxiliary lower
gas furnaces, which are fired with the flue
gases which are escaping above. About 6
feet above we find two more of these gas
furnaces. In the upper portion of the kiln
there are two pipes—one for each retort—
which establish connection between the
flues and the interior of the retort by
means of dampers. With the aid of these
dampers or other devices, the operations
can be so regulated that the peat will be
dried in the one retort of a pair, while
coking goes on in the other retort, and the
different furnaces, inlets, and flues permit
of regulating the temperature. While drying is going on, the vapours are discharged
into the chimney; afterwards the gases
are sent into a system of piping for the recovery of the chemical products, with purification of the power-gas.

"The charging takes place on the top
floor. Each retort has its fill-hole, an
iron pipe luted with clay and alabaster;
the fill-holes are usually kept closed, and
they are covered by a big cowl, from which
the gases and vapours are withdrawn to
join the other flue gases. Steam can be
injected to keep out the air during charging. The peat is put in with the shovel
at intervals, and steam is injected into the
furnace to keep the air out during this operation, while the admission of the flue
gases is stopped. The charging arrangements of the furnaces of type II. are improved. The hot coke is drawn off below.
The air-dry peat sods are quite hard, and

not likely to crumble in the kiln. The charge remains in the kiln for about 18 hours, and from 8 to 10 tons of peat-coke are gained per day of 24 hours. The coke falls into trucks, which are closed by lids. The trucks are run out into the yard, where they cool sufficiently in six or eight hours to be taken into the sorting-shed.

"The furnaces or kilns of type II. serve for the rapid partial coking of peat, which cannot be sufficiently air-dried for use in the retorts of type I. They are round shaft kilns of the same height as the other furnaces (40 feet), and differ from them chiefly in so far as the flue gases from type I., or from some independent source, enter directly into the interior of the kiln, to dry and to gasify the peat. The shafts are not jacketed with flues, but are subdivided to a certain extent by two or more iron plate-rings fitted into the wall; these rings may be likened to the necks of bottles, widening out above. At Beuerberg the type II. furnace is worked in conjuncton with type I., as already explained; special coils for producing superheated steam are embedded in the jacketed walls of the furnace of type I. The flue gases (together with superheated steam) are admitted at two spots, beneath the iron rings mentioned. At first the valve under the upper ring is opened, while the outlet on the top of the furnace is also kept open. When the moisture is expelled from the peat, and tarry vapours make their appearance, the flue gases are made to enter under the lower plate, and the escaping vapours and gases are collected. The resulting power gas is not so rich as that of the retorts of type I., but it can be used in gas-engines. From 12 to 14 tons of semi-coke are produced in 24 hours." This semi-coke is a good · substitute for coal and, while it burns with a flame, it produces hardly any smoke.

A description is given of a third type of furnace invented by Ziegler and others for the production of power gas and chemicals from inferior peat, which has recently been tried with satisfactory results at the Fleiss Iron and Steel Works at Shelecken. The recovery of the by-products from the flue gases is then outlined. From a diagram it is learned that the amounts and values of the various products of the process from 1,000 tons of raw peat are as follows: 350 tons of coke at £2 per ton; 4 tons sulphate of ammonia at £12 per ton; 6 tons acetate of lime at £6 to £8 per ton; 2 tons methyl alcohol at £35 to £45 per ton; 8 tons crude paraffin at £22.10.0 per ton; 4 tons fine paraffin oil at £10 per ton; 18 tons gas oil at £5 per ton; 2 tons pitch at £2.10.0 per ton; and 2 tons creasote oil at £3 per ton.

The demand for peat coke in Germany far exceeds the supply. A number of the largest steel makers have adopted it as a substitute for charcoal in case-hardening and other processes.

"Of the ashes of the peat-coke nearly half is silica, bound together chiefly by lime and alumina; carbonates, sulphates, and phosphates make up together a small percentage, and in the other bases are iron oxide, magnesia, and alkali. Calculations for the coke show that the sulphur contents ranged from 0.101 to 0.117 per cent., and the phosphorus contents from 0.058 to 0.009 per cent. The composition of the ashes as well as of the peat, varies, of course, with the local conditions; but, so far, the peat-coke ashes have proved suitable for metallurgical applications, as they yield a slag of low melting-point, and the percentages of sulphur and phosphorus are small.

"The peat coke is produced in blocks 6 by 4 centimeters, and from 10 to 25 centimeters in length (2½ inches by 1¼ inches and 4 inches to 10 inches.) It is a hard, dense material, not betraying its origin in appearance, dull black in color, and hardly porous. It may be obtained as firm as, and ringing like, coal-coke, particularly if the peat is well dried before entering the furnace. The peat-coke retains its heat better and burns much less rapidly than charcoal, and does not spark or burst; it can therefore replace anthracite, when applied in small pieces, and is recommended for welding, soldering, and brazing. The peat-coke from the first Oldenburg plant answered very well as a substitute for charcoal in the iron works at Schmalkalden, and this experience has been confirmed in the Ural and in other places. The excellent reputation of Swedish iron is largely due to the use of charcoal in the furnaces; and the peat-coke is claimed to be at least equal to charcoal in this respect, and more economical in use, apart from its lower price. A new combined peat-coke and blast-furnace plant is now in course of erection near Oldenburg, where bog ore and Swedish ore are to be smelted. The blast-fur-

nace is to be 21 meters high and 6 meters maximum width (69 feet and 20 feet), and will produce 150 tons of pig daily. The peat-coke will be made in nine double-retort furnaces, each furnace to yield 9 tons of coke daily."

A CENTRAL RESCUE ORGANIZATION.

A DESCRIPTION OF THE EQUIPMENT AND TRAINING FACILITIES FOR RESCUE WORK IN THE AACHEN DISTRICT IN PRUSSIA.

Professor Stegemann—Glückauf.

F OUR coal-mine explosions within as many months, resulting in the loss of nearly a thousand lives, have brought forcibly before the people of the United States the necessity for strict inspection and regulation of mining methods and conditions if this shocking and unnecessary waste of life is not to continue. On another page of this issue Messrs. Hall and Snelling of the United States Geological Survey draw a striking parallel between conditions in European countries, where the loss of life has been decreasing steadily under efficient inspection and careful investigation of mining conditions, and the United States in which fatalities have become more numerous year by year. They indicate clearly the lines along which improvement of methods and regulations is necessary. It is hardly to be expected, however, that, even with the most perfect precautionary and preventative measures, accidents can be entirely eliminated and it must not be forgotten that the provision of adequate equipment for rescue work is of almost equal importance with the establishment of improved methods and efficient inspection. In view of the extreme difficulty under which rescue work was carried on after the recent disasters, the following description of a central rescue station lately established at Aachen, which shows the extent to which rescue work has been developed in Prussia, should amply demonstrate the great benefit which might be expected from the introduction of such equipment and methods of training in the coalfields of the United States.

The system of mine inspection boards which exists in Germany is described by Messrs. Hall and Snelling. Prof. Stegemann points out that these boards are in-

clining more and more to the establishment of more stringent rules for the provision of rescue appliances and their use is now common in many districts. For a long time only the Breslau district board prescribed the provision of rescue appliances at all the mines in the district, but within the past year a large number of other commissions have followed this example, one of the most recent being the Bonn district board, whose ordinance went into effect October 1, 1907. It is provided here that "at least two rescue apparatus, rendering possible entrance into noxious gases, must be kept at hand or easily procurable at each hoisting shaft and that it is the duty of the manager to see that these apparatus are always in working order and that a sufficient number of superintendents and miners are instructed in their use."

In the Aachen district mine fires are extremely rare, as the coal has no tendency to spontaneous combustion, and explosions of gas and dust are proportionately infrequent. Notwithstanding these favorable conditions, however, the Aachen Coal-Mining Association has been considering the subject of rescue appliances for some time, impelled thereto by the disasters at Courrières and Borussia, which proved conclusively the practical utility of rescue appliances. The scheme finally agreed upon was the building of a central rescue station to serve the whole district which covers a comparatively small area. The district has really two divisions, the mines on the west, on account of the hard, almost anthracitic quality of their coal, being much less hazardous than those on the east, with their open-burning and fat coals. A site was given for the station near the shaft house of the Maria, the central mine of the eastern division, where the Association already has an explosive testing station.

The scheme of the building provides for a large storage room for the apparatus which also serves as a cleaning and repair room, a large two-story drill hall, a dressing and bath room, a room for the apparatus wagon, and an office for the official in charge of the station. It is connected to the drainage, water, steam and lighting systems of the mine.

The drill room, which is, of course, the most important feature of the station, is 20 metres long and 7 metres wide. The practice galleries in two stories are built in a horse-shoe shape around a central observation room, and the width of the building permitted the building of a third gallery along one side of the building. All the passages are connected by shafts and inclines and give a total practice length of 120 metres. The whole interior of the drill hall is designed to give the impression of a mine on a small scale and, to make the drills interesting, the interior arrangements are altered from time to time. Wood and iron were used exclusively for its construction, the only brickwork being in the smoke producer. On the floors of the galleries are placed haulage ways with turntables; ventilation, drainage and compressed air pipes are installed, air doors are erected and a fall of the roof indicated, etc. For the production of smoke, hair refuse, cotton waste, etc., are used, these materials being stored in a small room near the producer. As a precautionary measure all the galleries on both floors are equipped with safety doors. As in other rescue stations work-measuring instruments are installed, in this case the devices being self-recording. In the wagon room the apparatus wagon will always stand ready for immediate service. Whether the wagon will be drawn by horses or will be driven by a motor is as yet undecided.

The rescue appliances consist of two hose and twelve oxygen apparatus, namely, one each of the Original König Nr. III and Westfalia devices of the hose type, each for two persons, and six each of the Westfalia and Draeger oxygen-generating appliances arranged for both mouth and helmet breathing. The Pneumatogen device has been entirely abandoned, its peculiar disadvantages outweighing the advantage of light weight.

In all, nine rescue divisions have been trained for the mines of the district, the number of trained men now amounting to one per cent. of the total number of miners. When the rescue corps was first organized 110 men were admitted. In choosing them only trustworthy and enthusiastic men were considered, the first choice being given to the mine superintendents as being most familiar with the layout of the mines underground, and after them the mine foremen, fire bosses, and lastly the ordinary miners and laborers. All the men were subjected to a physical examination, and it was taken into consideration whether they lived near the mine, were easy to reach in case of need, and worked distributed in various parts of the workings. Each rescue division is placed under the general leadership of the oldest superintendent, but they are divided into groups of five, each group being under the command of a group leader. The direction of the training as a whole .lies in the hands of a mine inspector especially appointed for this purpose and trained in the theory and practice of rescue work at the Bochum school. Herr Stegemann, as manager of the mines, has a general oversight over the rescue work.

The training begins with a general course to the whole corps on the design, construction and use of the rescue appliances. Following this the practical training takes place in groups and consists of ten two-hour drills. The aim of the whole course of training is to get the men so used to the apparatus as to have perfect confidence in them and willing to make the fullest possible use of them in case of need. The drills are graduated as follows:

(1) and (2) with the hose apparatus.

(1) in the open; putting on and carrying apparatus (½ hour); breathing under air pressure (½ hour); the performance of light work under air pressure (1 hour).

(2) in the drill room; exploration of the galleries without smoke (½ hour); with smoke (½ hour); the performance of light work (1 hour).

(3) to (10) with the oxygen apparatus; (3) to (5) in the open; (6) to (10) in the drill room.

(3) Putting on and carrying the Draeger and Westfalia apparatus with and without helmet, natural breathing, each ½ hour.

(4) the same as (3) under artificial respiration.

(5) the same performing work.

(6) Carrying the apparatus and working in the drill room without smoke and with natural breathing, in other respects like (3).

(7) Same as (6) with smoke and artificial respiration.

(8) Performance of work using the Draeger apparatus, 1 hour each for mouth and helmet types.

(9) Same with Westfalia apparatus.

(10) Two hours' continuous drill in the smoke-filled chambers, the men being required to make five tours of the galleries and do 10,000 kilogramme-metres of work, apparatus and breathing methods being left to their individual choice.

When this course of training is finished the men will be kept in condtion by means of quarterly drills which will extend to other work, such as the rescuing of victims, etc. This part of the training will take place partly in the drill room, but for the most part with the use of the apparatus wagon at the various mines of the district.

THE VOLUMETRIC COMPOSITION OF CAST IRON

A DISCUSSION OF THE COMPOSITION OF CAST IRON WITH REFERENCE TO THE PROPORTION BY VOLUME, INSTEAD OF BY WEIGHT OF THE IMPURITIES PRESENT.

Henry M. Lane—American Society of Mechanical Engineers.

AN interesting theory to account for many of the seeming contradictions in the behavior of alloys was put forward by Mr. Henry M. Lane at the December, 1907, meeting of the American Society of Mechanical Engineers. Mr. Lane pointed out that engineers are accustomed to speak of the percentage of impurities present in alloys with reference to the weight of the entire mass, to the exclusion of all consideration of the relative volumes occupied by the pure metal and the impurities. Taking the particular case of cast iron, Mr. Lane suggests that this volume relation offers a satisfactory explanation of many otherwise inexplicable phenomena, as, for example, the fact that the microscope cannot be applied to the study of cast iron with the gratifying results which have followed its application to the metallurgy of steel. The following abstract gives an outline of Mr. Lane's argument:

"Cast iron in its various forms may be considered as an alloy. By weight it rarely contains more than 94 per cent. metallic iron or pure iron, and in some cases scarcely more than 90 per cent. Many attempts have been made to account for the apparently disproportionate effect produced by the additon of comparatively small amounts of impurities. If, however, we consider the relative volumes occupied by these impurities by themselves, and also the

volumes probably occupied when in combination with the cast iron, we obtain an entirely new light upon the subject.

"The ordinary No. 2 foundry iron contains 2 per cent. of silicon, 0.04 per cent. of sulphur, 0.70 per cent. phosphorus, 0.70 per cent. manganese, and 3.50 per cent. carbon; or a total of 6.94 per cent. of the principal so-called impurities. . . . Manganese is a metal very closely related to iron and having nearly the same specific gravity, and hence when it is alloyed with iron it changes the volume of the mass but little. Some of the other elements, however, behave very differently. Pure silicon has a specific gravity of less than 2, while iron has a specific gravity variously reported from 7.77 to 8.0. This means that a given weight of pure silicon occupies four times the volume of a given weight of iron; therefore we may call the volume factor for silicon, 4. In like manner we find that sulphur occupies more than four times the volume of iron, and its volume factor would then be 4.07. The volume factor for phosphorus would be 4.44, and the volume factor for manganese 1.01. Carbon in the graphitic form, which is practically the condition in which it occurs in grey cast iron, occupies more space than any other of the constituent elements, and its volume factor would be 6.79. Using these factors, we may make some very interesting calculations concerning some of the more common iron mixtures, and these will lead us to an understanding of a number of the peculiarities of castings."

A table is given of the percentages by weight and by volume of the elements other than iron, viz., silicon, sulphur, phosphorus, manganese and carbon, contained in No. 2 foundry iron, stove plate, machinery iron and malleable iron, of ordinary average composition. The analyses on which the table is based were made on standard brands in an endeavor to show average conditions as far as possible. The percentages of impurities found in these classes of iron, by weight and by volume, the volumes being calculated from the volume factors given above, were as follows:

	By weight.	By volume.
No. 2 foundry iron	6.94	35.73
Stove plate	7.52	41.22
Machinery iron	5.90	34.01
Malleable iron	4.09	24.19

These volumetric relations are also shown graphically and Mr. Lane then proceeds to give practical proof of the fact that the impurities in cast iron actually do occupy so large a part of the volume of the total mass.

"This has been very nicely proved by an actual demonstration in which as much of the iron as possible was removed from the castings by a long pickling in dilute hydrochloric acid. The pickle was a wash used for dipping the castings into previous to nickel-plating. Some castings fell by accident to the bottom of the vat and remained there several months. When they were removed it was found that they were exceedingly light; and while they looked exactly like iron castings, they were brittle and very difficult to handle without breaking. An analysis showed that by far the largest portion of the iron had been removed by the pickle, and hence the remaining constituents formed simply a skeleton or sponge."

Several other examples are given of the results of pickling on castings. In all cases careful measurements show that the dissolving out of the iron made practically no change in the size of the casting. It is also interesting to note that the resulting spongy castings show the character of the original castings remarkably well. In them the dividing plane along the center of the casting due to its solidificaton from the outside in, which is not noticeable in the solid iron, can be clearly traced.

Among the phenomena which Mr. Lane mentions as explained by this theory is the fact that a long annealing of a malleable iron casting in the presence of an oxidizing medium results in a greatly weakened casting. This is because by long annealing practically all the graphitic carbon is removed from the casting and about 20 per cent. of the bulk of the casting is taken up by the resulting voids. Annealing would also result in the oxidation of a part of the iron and this would be another source of weakness. Mr. Lane offers this as the true explanation of the fact that the strongest malleable castings are produced by heating just above the recalescence point and holding there merely long enough to convert all of the carbon from the combined to the temper form, without any loss of this element.

The paper brought out a very interesting discussion, the following report of which is taken from *The Foundry* for January, 1908:

"The ensuing discussion brought out a very general disinclination to accept the theories set forth by Mr. Lane. Mr. Johnson, the first speaker, said that both in theory and practice the paper was absolutely in error so far as malleable castings were concerned. Carbon can penetrate through iron, he continued. The only part of the malleable casting which is worth anything is the outside skin, which is decarbonized. The casting otherwise might almost as well be hollow, except for the opposition offered to collapse. Volumetric study, he concluded, was without basis, except so far as it concerned the graphitic carbon in cast iron.

"Mr. Lane was questioned in regard to the nature of the pickling bath in which several small castings entering largely into his discussion had remained a number of months. He said it was very dilute hydrochloric acid, about a 1 per cent. bath. Mr. Keep said this acid was used to take the silicon off the casting. A sulphuric acid bath is also employed, but in time hydrochloric acid will disintegrate iron. Mr. Outerbridge offered a written discussion, describing the action of the weak acids in the bilge water from the hold of a freight steamer in producing an effect on a cast iron strainer similar to that described by Mr. Lane, and cited another case where oil refiners' acids had so changed the character of a cast iron pipe that picks penetrated it without difficulty.

"Prof. Stoughton said that Mr. Lane's theory was interesting, whether correct or not. It would hold, in his opinion, for graphitic carbon, but not for the other elements because they are in a combined state. Graphitic carbon does occupy space and therefore is of interest. He told of the theory advanced by Sir William Austen dealing with the molecular volume of impurities and stated that this bore many points of resemblance to that offered by Mr. Lane. The former was put forth many years ago. Prof. Stoughton also took issue with the author of the paper over the statement made in its presentation that the microscope was of little value to the foundryman. It has already helped the foundryman greatly, he declared, and can be of as much or even more assistance in the found-

ry as in steel works, all of which, provided they are of any considerable size, are now equipped with a complete microscopic department. The problem of the iron foundryman is much more complex than that of the steel founder and therefore the microscope can properly be applied and will prove of much importance. He cited the case of a concern operating a number of gray iron foundries having taken up within a year the microscopic study of cast iron. In refutation of the statement made by Mr. Lane that the microscope is of little value in the foundry, Prof. Bradley Stoughton supported his contention that the contrary is true with a large number of lantern slide illustrations of micrographs of different grades of iron and steel. Another speaker, in closing the discussion for the morning, said that Mr. Lane had given a few facts, but that the further value of his paper was entirely negatived by his failure to take into consideration the fact of the elements combining."

THE ELECTRIFICATION OF THE SWEDISH STATE RAILWAYS

THE LEADING FEATURES OF THE PROJECT FOR THE ELECTRIFICATION OF THIRTEEN HUNDRED MILES OF RAILWAY IN SOUTHERN SWEDEN.

Engineering.

A PLAN for the electrification of steam railways on a more extended scale than in any similar project hitherto attempted has been prepared under the official direction of the Swedish government and there is every indication of its being carried into effect. For the last two years a series of very thorough investigations on the practicability and economy of trunk-line electric traction has been carried out and the results have amply demonstrated the suitability of the system to the demands of the Swedish railways and have afforded a basis for the exact and exhaustive calculations on which the favorable report mentioned above is based. *Engineering* for December 13, 1907, publishes an abstract of this report of which the following extracts give the leading features.

"The decision in question affects all the railways south of Bollnäs (some 150 miles north of Stockholm), with the exception of the three following lines: Laxa-Char-

lottenburg, Orebro-Svarta, and Gothenburg-Stromstad, which sections for sundry reasons have not been included in the report. The large system of railways that comes under the scheme will, it is proposed, be supplied with electricity from five power stations at the Karsefors, the Trollhättan (both on or near the West Coast), the Motala, the Hammarby (having a central position), and the Alfkarleby Fall (near the East Coast and more northerly than the rest). The number of transformer stations is about thirty-five. It is distinctly pointed out that as the electric working is more economical with a large traffic, the calculations based upon the present available statistics do not compare favorably with what may be the state of affairs some years hence, when the electrification may have been brought about.

"The traffic mileage of the railways included in the present scheme amounts in the aggregate to some 1,300 miles, of which there are 1,230 miles single line, seventy miles double line, besides 440 miles of sidings. Owing to the increase of the traffic it will, no doubt, be necessary to build more double lines and to increase the mileage of the sidings. This latter increase is put at forty-five per cent, and it is taken that the lines on which an increase of sixty per cent in the traffic (which, according to recent experiences, may be expected by the year 1920) makes the increase necessary, carry an annual traffic of 300,000 carriage-axle kilometres per kilometre of rail, and would have been transformed into double lines by the year 1920. On this basis the lines Stockholm - Gothenburg, Stockholm - Bollnäs, Katrineholm - Malmö and Hallsberg-Frövi will have been made double-tracked at that time. This surmise, it is pointed out, makes the calculations look less flattering, inasmuch as the cost of the electric conductors is only slightly increased through increased traffic, while the second road means an increase of cost under that head of forty per cent. A larger calculated increase in the traffic of, say, 100 per cent or more would also make the calculations look better than one of sixty per cent."

Single-phase alternating current will be used and the conductors will carry a pressure of 50,000 volts in the feeders from the power stations and 15,000 volts in the contact wires. Both power and contact wires will be carried on the same poles, the former some distance above the latter to facilitate repair work. The poles will be spaced 83 feet apart and will be placed alternately on the right and left side of the railway. The contact wires, 65 square millimetres in section on main lines and 50 square millimetres on sidings, will be placed 20 feet above the surface of the rail except in tunnels and under bridges. The estimates include the raising of tunnels and bridges to a height of 18 feet. Between the water falls and the railways the power will be transmitted on two lines of wires, independent of each other, carried on iron supports with concrete foundations spaced 366 feet apart.

"All the transmission lines are calculated to be constructed with as complete reserve arrangements as possible. From all the power-stations except the one at Hammarby the transmission lines will proceed in two directions, and, as already stated, each with two independent sets of wires. From the point where these wires reach the railway double power lines will be installed, one on each side of the railway. These power-lines will feed the transformer stations, of which there are to be thirty-seven, lying along the railways at a distance of some 30 miles from each other. These transformer stations in their turn will feed the contact-wires in two directions, so that each point, under ordinary circumstances, receives current from two sides. The contact-wires on a few shorter sections, which only receive current from one side, will have instead two feeding-wires, coupled parallel with the contact-line, one on each side of the railway. Each of the four power-transmitting wires running along the railway will be able to be automatically switched off between each two transformer stations. These four 15-millimetre square wires will also go to those parts of the railway which are furthest away from the power-station, and which consequently adjoin the area of the next power-station in order to enable the one power-station to help the other by taking over the working of the nearest portion.

"The wires along the railway are so calculated that the normal maximum decrease of pressure, even with the most unfavourable train arrangements, shall not exceed 15 per cent, inclusive of 4 per cent. drop in the transformers of the sub-stations. These calculations are based upon the highest possible number of trains and of weight per train, and the most unfavourable position of the trains. . . . . .

"Water-power stations are proposed, built at the Karsefors Fall, in the River Lagan; Trollhättan Fall, in the Göta river; Motala river, between Norrby Lake and Roxen; Hammarby Fall, in the River Jäslean, and Alfkarleby Fall, in the Dalar river. These falls now all belong to the state, and the power calculations are subject to the requisite water regulations. The different power stations are intended to supply the following railway sections with power:

"The Karsefors station is to provide power for all state railways south of Falkenberg and Stockaryd. The scheme also embraces a peat power station in the province of Smaland, although the latter is not required for the present-day traffic. From the Karsefors station it is proposed to carry power lines to Laholm and Ousby.

"The Trollhättan station will supply power to the Falkenberg-Gothenburg-Laxa line, the Falköping-Nässjö and the Sköfde-Karlsborg line; power wires will be carried to Gothenburg and Falköping.

"The Motala station will provide power for the Stockaryd-Gnesta and the Mjölby-Hallsberg line; power lines are to be carried to Linköping and Karlsby.

"The Hammarby Fall station will supply power to the Laxa-Katrineholm and the Hallsberg-Frövi line, with a power line to Ervalla. It is proposed to construct an auxiliary peat power station for the last two stations in the vicinity of Katrineholm, but this will not be required until 1920.

"The Alfkarleby Fall is intended for the Gnesta-Stockholm-Bollnäs section, and the Krylbo-Frövi line, besides the Kilafors-Stugsund line, by means of power lines to Storvik and Upsala. A special power line will be carried between Potebro and Saltskog, so as to obtain a direct high-tension connection without having to take it through Stockholm."

The report goes into particulars as to the bases on which the estimates of cost of installation and operation were made. All possible factors were taken into account and full allowance was made for depreciation and unavoidable waste. The results of the elaborate computations are given in the following tables:

"The aggregate cost, based upon the traffic of the year 1905 amounts to

	£
Electric lines, transformer stations, &c.	2,070,000
Power-stations	1,330,000
Total	3,400,000

"With 60 per cent. larger traffic, which is expected by the year 1920, the figures will be:—

	£
Electric lines, transformer stations, &c.	2,610,000
Power-stations	1,540,000
Total	4,150,000

"The expenditure, based on the traffic of the year 1905, amounts to:—

	£
On electric lines, transformer stations, staff of men, &c	152,000
At power-stations	126,000
Management	5,500
Total	283,500

"The annual expenditure, based on an increase of 60 per cent. in traffic, as expected in 1920, amounts to:—

	£
On electric lines, transformer stations, staff men, &c.	191,000
At the power-stations	157,000
Management	6,600
Total	355,500

In 1905, 480 steam locomotives of all classes were required to handle the traffic. It is expected that the number of electric locomotives required to handle the same traffic will be 385 and these can be built at an additional cost of £127,000. It is expected, however, that the existing steam locomotives will serve their time owing to the gradual progress of the electrification and the increased locomotive cost is put at 8 per cent. of the total difference of cost, or about £10,000. The following tables give the comparative figures:

"The calculated expenditure for electric locomotives embraces (for present traffic):

	£
45 express locomotives	212,000
260 passenger and goods locomotives	1,300,000
80 smaller locomotives	222,000
Total	1,734,000

"The cost of steam locomotives of the same actual capacity amounts to:—

	£
57 express locomotives	171,000
335 goods and passenger locomotives	1,247,000
88 smaller locomotives	189,000
Total	1,607,000

The probable savings of electric over steam traction are then discussed and the method of making the estimates explained. On the basis of the traffic for 1905, "the aggregate savings by the introduction of electric traction are assumed to be as under:—

	£
Fuel, cost	204,000
" transport	16,000
Interest on coal in stock	6,600
Lubricant, &c.	5,700
Repairs	41,000
Pay to locomotive staff	24,000
" pumpers, &c.	28,800
Pumping-stations, &c.	5,000
Total	331,100

Under the conditions expected in 1920, the annual saving is put at £82,000.

"The calculations and particulars set forth above only apply to the State Railways of Southern Sweden, but as northern Sweden is singularly rich in water power, and as the traffic also in those parts is likely materially to increase, it is probable that —if the soundness of the calculations be confirmed—electric traction will, in due course, be adopted for the greater portion of the Swedish State Railways. The line most likely to be the first transformed from steam to electric traction is the important Stockholm-Gutenburg line."

THE ENGINEERING INDEX

THE KEYSTONE
IN THE ARCH OF APPLIED SCIENCE

The following pages form a descriptive index to the important articles of permanent value published currently in about two hundred of the leading engineering journals of the world—in English, French, German, Dutch, Italian, and Spanish, together with the published transactions of important engineering societies in the principal countries. It will be observed that each index note gives the following essential information about every publication:

(1) The title of each article, (4) Its length in words,
(3) A descriptive abstract, (5) Where published,
(2) The name of its author, (6) When published,

(7) *We supply the articles themselves, if desired.*

The Index is conveniently classified into the larger divisions of engineering science, to the end that the busy engineer, superintendent or works manager may quickly turn to what concerns himself and his special branches of work. By this means it is possible within a few minutes' time each month to learn promptly of every important article, published anywhere in the world, upon the subjects claiming one's special interest.

The full text of every article referred to in the Index, together with all illustrations, can usually be supplied by us. See the "Explanatory Note" at the end, where also the full title of the principal journals indexed are given.

CIVIL ENGINEERING

BRIDGES.

Accidents.

Blackfriars Bridge Accident. Gives the summing up of Dr. F. J. Waldo, at the inquest held. 3500 w. Engng—Dec. 20, 1907. No. 89259 A.

Bascule.

The Ringeplaat Bridge (De Ringeplaatbrug). A detailed description of an electrically - operated railway bascule bridge in Delft harbor. Ills. 3500 w. Ingenieur—Nov. 23, 1907. No. 88885 D.

Blackwell's Island.

The Erection of the Anchor Arms of the Blackwell's Island Bridge. Brief illustrated description. 2500 w. Eng Rec —Dec. 21, 1907. No. 89102.

Erection of the Manhattan Approach of the Blackwell's Island Bridge. Illustrated detailed description. 1600 w. Eng Rec— Dec. 28, 1907. No. 89205.

The Queen's Approach to the Blackwell's Island Bridge, New York. Illustrated detailed description of an approach

We supply copies of these articles. See page 878.

consisting chiefly of a steel viaduct with spans from about 56 to 166 ft. 4000 w. Eng Rec—Dec. 7, 1907. No. 88772.

Cantilever.

The New Bridge over the Rhine between Ruhrort and Homberg (Die neue Rheinbrücke zwischen Ruhrort und Homberg). An illustrated description of the design and construction. 1700 w. Serial. 1st part. Deutsche Bau—Nov. 9, 1907. No. 88841 B.

The Bridge over the Rhine between Ruhrort and Homberg (Die Strassenbrücke über den Rhein zwischen Ruhrort und Homberg). Julius Stanek. Illustrated description giving costs of construction and materials. 2500 w. Oest Wochenschr f d Oeffent Baudienst—Nov. 23, 1907. No. 88846 D.

See also Blackwell's Island, under BRIDGES.

Compression Members.

A Series of Failure Tests of Full-Size Compression Members, Made for the Pennsylvania Lines West of Pittsburg. C. P. Buchanan. A detailed report of compression tests on full size bridge members. Also editorial. Ills. 6500 w. Eng News—Dec. 26, 1907. No. 89150.

Concrete.

The Design and Construction of Concrete Arch Bridges of Moderate Span (Ein Vorschlag zur Ausführung und Berechnung von Betonbogenbrücken mittlerer Spannweite). Emil Heidecker. A mathematical discussion. Ills. 4400 w. Serial. 1st part. Oest Wochenschr f d Oeffent Baudienst—Nov. 9, 1907. No. 88845 D.

Hudson Memorial.

The Design for the Hudson Memorial Bridge; New York City. Editorial criticism of the design prepared by the Dept. of Bridges of New York City. 2200 w. Eng News—Dec. 19, 1907. No. 89093.

The Engineering Features of the Proposed Henry Hudson Memorial Bridge. Leon S. Moisseiff. A report to the Dept. of Bridges, New York. Detailed description. 4000 w. Eng Rec—Dec. 28, 1907. No. 89204.

Pontoon.

Pontoon-Bridges for Road-Traffic Over Rivers in the Darbhangah District, Bengal. Edward Golding Barton. Describes the construction of five pontoon-bridges. Ills. 2000 w. Inst of Civ Engrs—No. 3656. No. 89084 N.

Reinforced Concrete.

A Reinforced Concrete Arch. Howard C. Ford. Illustrates and describes the replacing of a steel highway bridge across Boulder Creek, Colo., by a 70-foot reinforced concrete arch, giving cost-data. 1200 w. Jour of Engng, Univ of Colo—No. 3. No. 89220 N.

See also Concrete, Hudson Memorial, and Viaducts, under BRIDGES.

Steel.

Vauxhall Bridge, 1906. William Charles Copperthwaite. Describes the old bridge and the new steel-arch structure, the construction work, etc. Ills. 10000 w. Inst of Civ Engrs—No. 3670. No. 89085 N.

The Marien Bridge over the Danube Canal at Vienna (Die Marienbrücke über den Wiener Donaukanal). Dr. Karl Rosenberg. Illustrated description of the design and construction of this steel-arch structure. 2000 w. Serial. 1st part. Zeitschr d Oest Ing u Arch Ver—Nov. 22, 1907. No. 88853 D.

See also Bascule, Blackwell's Island, Cantilever, Compression Members, and Viaducts, under BRIDGES; and Steel Stresses, under CONSTRUCTION.

Viaducts.

The Woodbury Viaduct. Brief illustrated description of a double-track steel viaduct of plate-girder spans on the Erie & Jersey R. R., crossing Bonney Brook. 1000 w. Eng Rec—Dec. 14, 1907. No. 89011.

Reinforced-Concrete Viaduct on the Richmond and Chesapeake Bay Railway, Richmond, Va. Illustrated detailed description of an unusual structure. 1200 w. Eng News—Dec. 12, 1907. No. 88944.

The Reinforced-Concrete Viaduct of the South Holland Electrical Railway Company at Rotterdam (De Spoorwegviaduct van Gewapend Beton der Zuid-Hollandsche Elektrische Spoorweg-Maatschappij te Rotterdam). A. J. A. Braak. Illustrated description of the design and construction. 5000 w. Ingenieur—Nov. 9, 1907. No. 88884 D.

CONSTRUCTION.

Beams.

New Special Structural Shapes. Concerning new steel beam sections, known as the Bethlehem beams. 1200 w. Ry Age—Nov. 29, 1907. No. 88673.

Tests of Reinforced Concrete Beams, Series of 1906. Arthur N. Talbot. A report of tests made on rectangular beams, methods, materials, etc. Ills. 6500 w. Univ of Ill—Bul. 14. No. 89224 N.

A New Method of Determining the Position of the Neutral Axis in Reinforced-Concrete Beams (Ein neues Verfahren zur Bestimmung der Lage der Neutralachse bei armierten Betonkörpern). Leo Blondek. A mathematical discussion. Ills. 2500 w. Zeitsch d Oest Ing u Arch Ver—Nov. 29, 1907. No. 88854 D.

Buildings.

See Reinforced Concrete and Steel Buildings, under CONSTRUCTION; and Lighthouses, under WATERWAYS AND HARBORS.

Caissons.

See Breakwaters, under WATERWAYS AND HARBORS.

Columns.

The Design of Latticed Metallic Columns (Sur le Calcul des Pièces Métalliques Chargées de Bout dont les Ames sont à Treillis). M. F. Chaudy. A mathematical paper. Ills. 2000 w. Mem Soc Ing Civ de France—Aug., 1907. No. 88806 G.

Concrete.

The Finish of Concrete Surfaces. M. C. Tuttle. Read before the Boston Soc. of Civ. Engrs. Gives methods of producing a variety of surface textures. 1600 w. Eng Rec—Dec. 28, 1907. No. 89210.

See also Tunnel Lining, under CONSTRUCTION; Cement, under MATERIALS; and Locks, under WATERWAYS AND HARBORS.

Drainage.

See Reclamation and Tunnels, under CONSTRUCTION; and Ditches and Reservoir Lining, under WATER SUPPLY.

Earth Pressures.

The Bracing of Trenches and Tunnels, with Practical Formulas for Earth Pressures. Discussion of J. C. Meem's paper. 10800 w. Pro Am Soc of Civ Engrs—Nov., 1907. No. 88757 E.

The Bracing of Trenches and Tunnels, with Practical Formulas for Earth Pressures. Continued discussion of paper by J. C. Meem. Ills. 7000 w. Pro Am Soc of Civ Engrs—Dec., 1907. No. 89227 E.

Excavation.

Method and Cost of Loading Dump Wagons from an Ingeniously Designed Hopper or Table. J. C. Black. Illustrates and describes methods and outfit used at Portland, Ore., in excavating for a business block. 1500 w. Engng-Con—Dec. 11, 1907. No. 88998.

Piles.

See same title, under WATERWAYS AND HARBORS.

Reclamation.

The Method and Cost of Surveying and Reclaiming Wet Farm Land. L. G. Hicks. Paper read before the Illinois Soc. of Engrs. & Survs. describing a drainage system used in Nebraska. 3000 w. Engng-Con—Dec. 4, 1907. No. 88795.

Regulations.

New Reinforced Concrete Regulations in Philadelphia. A copy of the regulations of the Bureau of Building Inspection, approved Oct. 8, 1907. 3000 w. Cement—Nov., 1907. No. 89191 C.

Official Instructions Current in France for Reinforced - Concrete Construction (Instrucciones Oficiales Vigentes en Francia para el Empleo del Hormigón Armado). Ills. 10000 w. Revista Tech Indus—Oct., 1907. No. 88803 D.

Reinforced Concrete.

A Reinforced-Concrete Addition to a Brick Building. L. J. Mensch. Illustrated description of work in Salt Lake City, Utah. 1200 w. Eng News—Dec. 5, 1907. No. 88787.

A Reinforced Concrete Hotel Building in Oakland, Cal. Edw. L. Soule and John B. Leonard. Illustrated detailed description. 1500 w. Eng Rec—Dec. 21, 1907. No. 89107.

Personal Experiences with Reinforced Concrete. Observations, by E. P. Wells, on the report by the Committee of the Royal Institute of British Architects. 2800 w. Am Archt—Dec. 7, 1907. No. 88727.

Concrete Construction at Jamestown. Percy H. Wilson. Illustrated descriptions of the Government Pier and History Building, built mainly of reinforced concrete. 2000 w. Cement Age—Dec., 1907. No. 89190.

See also Beams, Regulations, Retaining Walls, Roofs and Stacks, under CONSTRUCTION; Reinforced Concrete, under MATERIALS OF CONSTRUCTION; Sewers, under MUNICIPAL; Lighthouses, under WATERWAYS AND HARBORS; and Car Houses and Subway Stations, under STREET AND ELECTRIC RAILWAYS.

Retaining Walls.

Retaining Walls on the Delaware, Lackawanna & Western R. R. at Buffalo. Brief illustrated description of reinforced concrete walls for track elevation work. 1200 w. Eng Rec—Dec. 21, 1907. No. 89103.

Rivets.

See same title, under MECHANICAL ENGINEERING, MEASUREMENT.

Roofs.

A Light Roof Girder System in a Reinforced Concrete Garage. Critical discussion of the design used for a garage at Sea Gate, Coney Island, in New York City. Ills. 3000 w. Eng News—Dec. 12, 1907. No. 88946.

The Design of 75-ft. Reinforced-Concrete Girders for the Mammoth Garage, White Plains, N. Y. C. E. Tirrell. Illustrated description of the most important points in the design of girders for a long-span roof. 4000 w. Eng News—Dec. 12, 1907. No. 88947.

Stacks.

The Design of Power Plant Chimneys. Frank Kingsley. Discusses methods of stack design, and reports investigations made, and effects of stated conditions. 5500 w. Eng Rec—Dec. 21, 1907. No. 89106.

Reinforced Concrete Chimneys. Abstract of a report of investigation made for the Assn. of Am. Portland Cement Mfrs. by Sanford E. Thompson. Gives

We supply copies of these articles. See page 878.

general conclusions, requirements of design and construction, etc. 3300 w. Eng Rec—Dec. 28, 1907. No. 89208.

Steel Buildings.
Two New Record-Breaking Office Buildings in New York City. Illustrated detailed descriptions of the Singer Building, and of the City Investing Building. 3500 w. Eng News—Dec. 5, 1907. No. 88784.

See also Beams, Columns, and Steel Stresses, under CONSTRUCTION; Lighthouses, under WATERWAYS AND HARBORS; and Rivets, under MECHANICAL ENGINEERING, MEASUREMENT.

Steel Stresses.
Working-Stresses in Steel Construction. C. A. P. Turner. Discusses points of general specifications when applied to structures of unusual size. 4000 w. Eng News—Dec. 12, 1907. No. 88945.

Tunnel Lining.
Method of Mixing and Placing Concrete for a Tunnel Lining. Illustrated account of work on the Burton tunnel of the Southern Ry., in Indiana. 1500 w. Engng-Con—Dec. 11, 1907. No. 88997.

Tunnels.
The Tunnel and River Shaft of the Detroit Water Works. James Ritchie. Descriptive account of the work. 2500 w. Jour Assn of Engng Socs—Nov., 1907. No. 88926 C.

Three Colorado Trans-Continental Tunnel Projects. George J. Bancroft. Describes mining, transportation and water diversion features. Ills. 1400 w. Min Sci—Dec. 19, 1907. No. 89111.

A New Highway Tunnel Under the Thames River at London, England. E. H. Tabor. Introductory review of the tunnels and bridges providing communication across the Thames, with illustrated detailed description of the Rotherhithe tunnel. 4000 w. Eng News—Dec. 19, 1907. No. 89089.

The Loetschberg Tunnel and Its Relation to Alpine Railway Routes. An account of a tunnel under construction which will give a new route to the Simplon tunnel. Also long editorial discussion on "The Commercial Aspects of Present and Proposed Alpine Railway Tunnels." Maps. 8500 w. Eng News—Dec. 5, 1907. No. 88786.

Repair Work in Tunnel (Travaux de Réparation de Tunnels). M. Siegler. Describes work on various tunnels of the Eastern Railway of France and valuable lessons which may be learned from the faults disclosed. Ills. 8000 w. Rev Gen d Chemins de Fer—Nov., 1907. No. 88877 G.

See also Earth Pressures, and Tunnel Lining, under CONSTRUCTION.

MATERIALS OF CONSTRUCTION.

Brick.
Frost Resistance of Brick. George E. Walsh. On the importance of securing quality in bricks for certain uses, the tests made and the results. 2500 w. Clay Rec—Nov. 15, 1907. No. 88641.

Cement.
Hydraulic Properties of Reground Cement Mortars. Henry S. Speckmand and Robert W. Lesley. Read before the Assn. of Am. Portland Cement Mfrs. Gives results of investigations. 2800 w. Eng Rec—Dec. 21, 1907. No. 89114.

Efficiency of Cement Joints in Joining Old Concrete to New. Concerning experiments reported by M. Mesnager, in *Ann. des Ponts et Chaus.*, showing the value of a cement wash in joining new concrete to old. 800 w. Eng News—Dec. 12, 1907. No. 88949.

Paints.
Preservation of Iron and Steel by Paint. Briefly considers means of protecting from exposures of various kinds. 2500 w. Met Work—Dec. 28, 1907. No. 89202.

Reinforced Concrete.
A Few Tests and Experiments with Reinforced Concrete. R. T. Surtees. An illustrated account of experiments carried out to determine the cause of the great difference found in work with this material. 3000 w. Eng Rec—Dec. 14, 1907. No. 89012.

Timber Decay.
Dry Rot in Timber. A discussion of the origin and cause, the prevention and cure. 1600 w. Builder—Nov. 23, 1907. No. 88695 A.

Timber Preservation.
Preservative Treatment of Poles by the Open-Tank Process. D. A. Rockwell. Describes the single- and double-tank methods, and plant requirements for the treatment. 2000 w. Elec Wld—Dec. 14, 1907. No. 89021.

See also Timber Decay, under MATERIALS OF CONSTRUCTION.

MEASUREMENT.

Mapping.
Sketch-Mapping, with Special Reference to Southern Nigeria. Lindow Hereward Leofric Huddart. Gives methods that have been found suitable under unfavorable conditions of ground and transport. 5500 w. Inst of Civ Engrs—No. 3634. No. 89083 N.

Surveying.
See Reclamation, under CONSTRUCTION; and Mapping, under MEASUREMENT.

Testing Materials.
See Compression Members, under BRIDGES.

We supply copies of these articles. See page 878.

MUNICIPAL.

Garbage Disposal.

Municipal Refuse Disposal: An Investigation. J. T. Fetherston. Gives result of an investigation of local household refuse, with a series of tests in burning mixed wastes; and facts concerning a number of mixed refuse destructors. Ills. 14500 w. Pro Am Soc of Civ Engrs—Nov., 1907. No. 88755 E.

Pavements.

Improving Asphalt Pavements in Kansas City. E. A. Harper. Read before the Am. Soc. of Munic. Imp. An account of recent methods and tests made. 1200 w. Munic Engng—Dec., 1907. No. 89194 C.

Materials for Filling Joints of Brick Pavements. W. A. Howell. Read before the Am. Soc. of Munic. Imp. An account of methods in use in Newark, N. J. 1600 w. Munic Engng—Dec., 1907. No. 89193 C.

Repair Department.

The Construction and Repair Division of the City of Chicago. William D. Barber. Shows how this department has been organized and put upon a business basis. 4500 w. Eng News—Dec. 5, 1907. No. 88785.

Sewage Disposal.

Experimental Purification of Boston Sewage. Abstract of a paper by C. E. A. Winslow and E. B. Phelps, presented to the Boston Society of Civil Engineers. A study of local problems. 2200 w. Eng Rec—Nov. 30, 1907. No. 88662.

Sewage Disposal for Institutions and Small Communities. Theodore Horton. Paper before the Conference of N. Y. San. Officers. Considers domestic sewage and an installation that should obtain a sterile effluent suitable to be discharged into any stream. 3500 w. Munic Engng—Dec., 1907. No. 89192 C.

The Disposal of Domestic and Industrial Sewage and Rain Water at Gera (Häusliche Abwässer, Fabrikabwässer und Regenwässer in Gera). Herr Geissler. Describes the sewerage system of this Russian city of 47,000 inhabitants. 7000 w. Gesundheits-Ing—Nov. 2, 1907. No. 88862 D.

Sewers.

A Large Reinforced Concrete Sewer in the Borough of Queens, New York City. Illustrated detailed description of a circular trunk sewer under construction. 3000 w. Eng Rec—Nov. 30, 1907. No. 88659.

The Reconstruction of Large Sewers in St. Louis, Mo. Illustrates and describes methods used to increase the capacity of sewers. 1800 w. Eng Rec—Dec. 7, 1907. No. 88775.

The Harlem Creek Sewer, St. Louis,

Mo. Information in regard to the construction of a reinforced-concrete sewer over a mile in length, and of large section. 1400 w. Eng Rec—Dec. 14, 1907. No. 89013.

Smoke Prevention.

Smoke Prevention in Cities. A late study of the problem at Syracuse, N. Y. 3000 w. Ir Age—Dec. 5, 1907. No. 88781.

The Solution of the Smoke Problem (Die Lösung der Rauchschaden-Frage). John H. Mehrtens. The first part of the serial is devoted to a review of past agitations and attempts for the abatement of the smoke nuisance in German cities. 5200 w. Serial. 1st part. Glasers Ann —Nov. 1, 1907. No. 88860 D.

Street Design.

Street Engineering. Rutger B. Green. Remarks on the need of supervision of street design, locations of pipes, sewers, etc., and suggestions of a system. 1800 w. Jour Assn of Engng Socs—Nov., 1907. No. 88927 C.

WATER SUPPLY.

Artesian Wells.

See Anticlinal Structure, under MINING AND METALLURGY, MISCELLANY.

Brisbane, Australia.

Brisbane and Its Water Supply. Allen Hazen. Describes conditions in Queensland, and the water supply of Brisbane, the intermittent filtration, and other features. Ills. 2500 w. Pro Am W-Wks Assn—1907. No. 88741 N.

Dams.

Raising a Dam at Lennep, Germany. Brief illustrated description of the raising of the crest of a dam to increase the capacity. 1600 w. Eng Rec—Dec. 28, 1907. No. 89209.

The Toxaway Dam. Milo S. Ketchum. Illustrated description of the construction of an earth dam for the creation of an artificial lake in North Carolina. 1500 w. Jour of Engng, Univ of Colo—No. 3. No. 89222 N.

The Munich Water-Power Plant at Moosburg on the Isar (Die Wasserkraftanlage der Stadt München bei Moosburg an der Isar). K. Dantscher. The first part of the serial describes the regimen of the river and the reinforced-concrete weir. Ills. Serial. 1st part. 2200 w. Beton u Eisen—Nov., 1907. No. 88865 F.

See also Reservoirs, under WATER SUPPLY; and Hydro-Electric, under ELECTRICAL ENGINEERING, GENERATING STATIONS.

Ditches.

Ditch Construction on the Seward Peninsula. Notes from a paper by James A. Kelly, before the Pacific N.-W. Soc.

We supply copies of these articles. See page 878.

of Engrs, describing the ditch of the Taylor Creek Ditch Co. 2000 w. Eng Rec—Dec. 7, 1907. No. 88773.

See also Reclamation, under CONSTRUCTION; and Reservoir Lining, under WATER SUPPLY.

Filtration.

Water Filter for South Norwalk, Conn. Illustrated detailed description of the plant. 1800 w. Munic Jour & Engr—Dec. 4, 1907. No. 88900.

The Care of a Mechanical Filter Plant. J. M. Diven. On the construction, coagulant feed, washing, sterilization, etc. Discussion. 4500 w. Pro Am W-Wks Assn—1907. No. 88734 N.

See also Brisbane, Australia, under WATER SUPPLY.

Fire Protection.

Fire Protection System of the American Dock Stores. Frank Sutton. Describes a dry pipe sprinkler system installed on Staten Island, N. Y. 2000 w. Eng Rec—Dec. 7. 1907. No. 88774.

Connections from Public Water Supply for Private Fire Protection Service. A. W. Hardy. Also Rates and Regulations at Atlanta, Ga., and at Elmira, N. Y. From papers, discussed together. 12000 w. Pro Am W-Wks Assn—1907. No. 88752 N.

Friction.

See Pipe Friction, under WATER SUPPLY.

Goldfield, Nev.

The Water Supply of Goldfield, Nevada. An illustrated account of the supply for this mining district in the center of a desert region. 2500 w. Eng Rec—Dec. 7, 1907. No. 88770.

Greeley, Colo.

A Gravity Water Supply System at Greeley, Col. An illustrated description of a system having various special features of arrangement and design. 6000 w. Eng Rec—Dec. 14, 1907. No. 89007.

Interchange.

Plan for Interchange Between Low and High Service Systems. Henry L. Lyon. Illustrated description of a "Manifold Valve System," designed by the writer to meet conditions at Buffalo, N. Y. Short discussion. 1200 w. Pro Am W-Wks Assn—1907. No. 88747 N.

Irrigation.

See Pumping Plants, under MECHANICAL ENGINEERING, HYDRAULIC MACHINERY.

Legislation.

Some notes on Rules, Ordinances and Court Rulings Governing the Operation of Water Plants. S. J. Rosamond. Concerning contracts, rules, free water, etc. 5000 w. Pro Am W-Wks Assn—1907. No. 88733 N.

London.

The Future Water Supply of London. Detailed discussion of the proposals which have been put forward. Also editorial. 3300 w. Engr, Lond—Nov. 29, 1907. Serial, 1st part. No. 88965 A.

Pipe Corrosion.

Note as to the Action of Water Upon Lead and Zinc. Dr. W. P. Mason. A discussion of the amount of metallic salts in solution that may be allowed with safety. 2200 w. Pro Am W-Wks Assn—1907. No. 88743 N.

Pipe Friction.

Loss of Pressure in Water Flowing Through Straight and Curved Pipes. Arthur William Brightmore. A record of experimental investigations carried out at the Royal Indian Engineering College, Cooper's Hill, which has now ceased to exist. 5500 w. Inst of Civ Engrs—.o. 3670. No. 89082 N.

Friction Coefficients for Water (Reibungscoefficienten für Wasser). Discusses the subject with reference to pipes, canals, etc. 3000 w. Serial, 1st part. Electrotech u Polytech Rundschau—Nov. 28, 1907. No. 88852 D.

See also Pipe Line Cleaning, under WATER SUPPLY.

Pipe Line Cleaning.

Tuberculation and the Flow of Water in Pipes. Nicholas S. Hill, Jr. Considers the deposits that lessen the carrying capacity of water pipes, the method of removing them, formulae for determining flow, etc. Also short paper on "Cleaning Water Mains," by Park Woodward, and discussion. Ills. 20000 w. Pro Am W-Wks Assn—1907. No. 88746 N.

Pipe Line Repairs.

Repairing a Broken Force Main. C. W. Wiles. Describes work at Delaware, Ohio. 800 w. Pro Am W-Wks Assn—1907. No. 88749 N.

Repairing Broken Submerged Pipe Line. Harry A. Lord. Brief account of work at Ogdensburg, N. Y. 1200 w. Pro Am W-Wks Assn—1907. No. 88750 N.

Pipe Lines.

Rifled Pipe Line for Conveying Oil on the Southern Pacific. Illustrated description of a line under construction in California for conveying crude oil. 2000 w. Ry Age—Dec. 13, 1907. No. 89002.

Remodeling of Discharge Pipes, Springfield Ave. Pumping Station, Chicago, Illinois. W. A. Levering and H. S. Baker. Illustrated description of the work, giving cost. 700 w. Pro Am W-Wks Assn—1907. No. 88748 N.

Pipe Lines for Hydraulic Power Plants. Abstract of a paper by Arthur Jobson, presented at meeting of the Telluride Institute. Discusses the methods of determining the most economical diameter of penstocks, loss of head in circular pipes, etc. 3000 w. Eng Rec—Dec. 21, 1907. No. 89104.

The High-Pressure Pipe Line and Power Station at Nordhausen (Die Hochdruckwasserleitung und das angeschlossene Kraftwerk der Stadt Nordhausen). Herr Michael. A pipe line 11 kilometers long with a total fall of 222 meters, and a hydro-electric station is illustrated and described. 4500 w. Zeitschr d Ver Deutscher Ing—Nov. 30, 1907. No. 88876 D.

Pitometer.

The Pitometer. Edward S. Cole. Explains the use of the portable pitometer for measuring the flow of water and facilitating the detection of waste. Ills. 4000 w. Jour Fr Inst—Dec., 1907. No. 88923 D.

The Pitometer and Water Works Losses. Edward S. Cole. Shows the extent of water waste and considers the remedy to be the use of a recording Pitot tube, which is illustrated and described. 3800 w. Pro Am W-Wks Assn—1907. No. 88735 N.

Pollution.

Peculiar Instance of Contamination of a Well Water. Dr. W. P. Mason. A brief account of a well at Wellingford, Conn., contaminated by gasoline. 800 w. Pro ·Am W-Wks Assn—1907. No. 88740 N.

Protection.

The Sanitary Protection of Surface Water Supplies. George A. Soper. Discusses measures available and rules. 2500 w. Pro Am W-Wks Assn—1907. No. 88753 N.

Pumping.

See under MECHANICAL ENGINEERING, HYDRAULIC MACHINERY.

Purification.

Laboratory Experiments in Water Treatment. Edward Bartow and J. M. Lindgren. An account of investigations made for remedying stated troubles. 3000 w. Pro Am W-Wks Assn—1907. No. 88754 N.

Notes on Water Purification at St. Louis. W. F. Monfort. A statement of the chemical and bacterial results for the past year, at one of the larger plants, using lime and iron sulphate as coagulants, without the use of filters. General discussion. 3000 w. Pro Am W-Wks Assn—1907. No. 88744 N.

The Best Method of Purifying Ground Water from Iron and Manganese (Ueber die besten Verfahren der Reinigung des Grundwassers von Eisen und Mangan). Heinrich Többen. 9000 w. Serial, 1st part. Gesundheits-Ing—Nov. 16, 1907. No. 88863 D.

See also Filtration, under WATER SUPPLY.

Reservoir Lining.

Lining Ditches and Reservoirs to Prevent Seepage Losses. Information from a bulletin (No. 188) issued by the University of California, concerning investigations and experiments with ditch linings, etc. 3500 w. Eng News—Dec. 5, 1907. No. 88789.

Reservoirs.

The Urft Dam and Hydro-Electric Power Distribution. Illustrated detailed description of this artificial lake, created in Rhenish, Prussia, to raise the low water level of the district, to prevent floods, and to supply electric light and power. 5500 w. Engng—Nov. 29, 1907. Serial, 1st part. No. 88961 A.

Stream Flow.

Flow of Water in Open Conduits. A. P. Merrill. Gives an outline of a thesis investigation, discussing the results. 3000 w. Eng Rec—Dec. 28, 1907. No. 89207.

Tunnels.

See same title, under CONSTRUCTION.

Water Meters.

Meters and Meter Systems. W. Volkhardt. Discusses the advantages of the meter system. 2200 w. Pro Am W-Wks Assn—1907. No. 88729 N.

The Cost of Meters at Rochester, N. Y. George W. Rafter. Reviews statistics of metered water in Rochester, comparing with other cities, especially in Europe. 5500 w. Pro Am W-Wks Assn —1907. No. 88730 N.

See also Pitometer, and Water Rates, under WATER SUPPLY.

Water Rates.

Rates for Water Service. Dabney H. Maury. President's address, outlining the principles which should govern the determination of a fair rate. 2000 w. Pro Am W-Wks Assn—1907. No. 88728 N.

Water Works Plants and the Proper Rates for Domestic and Public Service. John Ericson. Discusses methods of adjusting rates. 2000 w. Pro Am W-Wks Assn—1907. No. 88745 N.

Water Consumption, Waste and Meter Rates. James L. Tighe. Gives information concerning the city of Holyoke, Mass., as an example of the benefits of the meter system. 4000 w. Pro Am W-Wks Assn—1907. No. 88731 N.

Discussion of papers by Messrs. W. Volkhardt, George W. Rafter, and James L. Tighe. 6500 w. Pro Am W-Wks Assn—1907. No. 88732 N.

See also Water Meters, under WATER SUPPLY.

Water Works.

Some Notes on Oriental Water Works. George A. Johnson. Brief account of features of water-works in Japan, India, China, and other Oriental countries. 5000 w. Pro Am W-Wks Assn—1907. No. 88742 N.

See also Pumping Engines and Pumping Plants, under MECHANICAL ENGINEERING, HYDRAULIC MACHINERY.

Water-Works Management.

Some Experiences in the Management of a Small Water Works. Howard Williams. 2500 w. Pro Am W-Wks Assn—1907. No. 88751 N.

WATERWAYS AND HARBORS.

Breakwaters.

The Port of Rosario (Puerto del Rosario). Domingo Carrique. An illustrated description of the rubble-stone masonry breakwater, its design and construction and particularly the building of the foundations in compressed air caissons. Serial, 1st part. 4000 w. Ingenieria—Oct. 30, 1907. No. 88805 D.

British Canals.

The Canal Inquiry. Discussing the evidence collected by the Royal Commission appointed to report on inland navigation of the United Kingdom, with the writer's views on this question. 3300 w. Elec Rev, Lond—Dec. 13, 1907. Serial, 1st part. No. 89131 A.

Coast Erosion.

The Royal Commission on Coast Erosion. A summary of the most important evidence from an engineering point of view. 4500 w. Engr, Lond—Dec. 20, 1907. Serial, 1st part. No. 89262 A.

Docks.

Notes on the Arrangements for Receiving Importations of Timber at Portishead Dock, Bristol. John William Hitchen. Discusses the considerations which led to the adoption of a particular method of dealing with timber, giving notes on the works constructed. 5000 w. Inst of Civ Engrs—No. 3672. No. 89081 N.

Dredges.

Sand-Pump Dredges for the Thames. Illustrated description of the twin-screw dredger, Lord.Desborough. 500 w. Engr, Lond—Nov. 22, 1907. No. 88718 A.

Hydraulic Dredge Used on the New York State Barge Canal. Emile Low. Illustrated detailed description of the special dredge "Oneida." 1200 w. Eng News—Dec. 5, 1907. No. 88788.

Dredging.

The Method of Operating a Lobnitz Rock Cutter in Canal and Harbor Work. Lindon Bates, Jr. Illustrates and describes a machine to shatter submerged rock to a size capable of being handled by dipper dredges. 2500 w. Engng-Con—Dec 18, 1907. No. 89068.

Dry Docks.

Floating Dock at Rotterdam. Illustrated detailed description. 1400 w. Engr, Lond—Dec. 6, 1907. No. 89055 A.

Lighthouses.

The Construction of the Sanganeb Light House (Construction du Phare de Sanganeb). M. M. Charvaut. A description of the design and construction of this steel lighthouse on reinforced concrete foundations, recently built in the Red Sea. Ills. 6000 w. Mem Soc Ing Civ de France—Aug., 1907. No. 88807 G.

Locks.

Methods of Constructing Concrete Lock Walls by Means of Traveling Forms. Describes methods used near Boston for the lock through the Charles River dam. 700 w. Engng-Con—Dec. 4, 1907. No. 88794.

See also Mechanical Locks, under WATERWAYS AND HARBORS.

Mechanical Locks.

The Ship Lift on the Dortmund-Ems Canal. Illustrated description of the apparatus near Henrichenburg. 800 w. R R Gaz—Dec. 6, 1907. No. 88902.

The Oelhafen-Loehle Type of Lift Lock (Des Schiffshebewerk System Oelhafen-Loehle). K. E. Hilgard. Illustrated description of the mechanism. 2200 w. Schweiz Bau — Nov. 16, 1907. No. 88843 B.

Panama Canal.

Conditions Along the Panama Canal. Information from the recent annual report of the Commission, and the "Canal Record." Ills. 7800 w. Eng Rec—Nov. 30, 1907. No. 88657.

Panama Canal Progress. An illustrated account of the organization and division of work, accommodations, etc. 4500 w. Sci Am Sup—Dec. 7, 1907. No. 88791.

Isthmian Geology and the Panama Canal. Ernest Howe. An account based upon an official report of five months of field work. Examines dam and lock sites, materials of construction, etc. Ills. 7000 w. Ec-Geol—Oct., 1907. No. 89216 D.

Piers.

See Reinforced Concrete, under Construction.

Piles.

Diagrams to Determine the Bearing Power of Piles. G. F. Stickney. Gives instructions for facilitating work on the New York Barge Canal. By their use the safe load for any pile may be instantly determined. 600 w. Eng Rec—Dec. 28, 1907. No. 89211.

We supply copies of these articles. See page 878.

Rio Grande.

Aggraded Terraces of the Rio Grande. Charles R. Keyes. Describes the remarkable terraced drainage-way. 2200 w. Am Jour of Sci—Dec., 1907. No. 89160 D.

Rock Dredging.

See Dredging, under WATERWAYS AND HARBORS.

Tiber.

Direct Communication Between Rome and the Sea and the Exposition of 1911 (La Comunicazione diretta fra Roma e il Mare e l'Esposizione del 1911). A. Brunelli. Discusses the importance of Italian commerce and the necessity of a navigable waterway from Rome to the sea, the cost and feasibility of the project, etc. 5500 w. Ann d Soc d Ing e d Arch Ital—Oct. 15, 1907. No. 88800 F.

U. S. Waterways.

Fifty Millions a Year for Waterway Improvements: Where Should It Be Spent? Editorial discussion of this question. 2300 w. Eng News—Dec. 12, 1907. No. 88948.

Wharf.

A Large Coal-Storage Wharf at Superior, Wisconsin. Illustrated detailed description of the construction and equipment of the wharf of the Berwind Fuel Co. 3000 w. Eng Rec—Nov. 30, 1907. No. 88661.

MISCELLANY.

Alaska.

Natural Conditions Affecting Engineering Work in Alaska. George B. Pillsbury. A brief account of the conditions to be expected on the Pacific Coast and in the interior. 1000 w. Eng News—Dec. 26, 1907. No. 89151.

Caisson Disease.

The Cause, Treatment and Prevention of the "Bends" as Observed in Caisson Disease. Prof. J. J. R. Macleod. Briefly describes the chief symptoms, discussing the cause, treatment, etc. 7000 w. Jour Assn of Engng Socs—Nov., 1907. No. 88925 C.

Earthquakes.

Construction to Withstand Earthquake Shocks (Edificacion contra Temblores). Domingo Selva. Discusses the stresses arising from earthquake shocks and the best means of building to withstand them, referring principally to small and medium-sized dwelling houses. Ills. 16000 w. Anal d l Soc Cien Argentina—July, 1907. No. 88804 N.

Slide Rule Computing.

Accuracy of Slide Rule Computations. John Berg. Gives results obtained and explains the convenience of its use. 1800 w. Eng Rec—Dec. 21, 1907. No. 89105.

ELECTRICAL ENGINEERING
.

COMMUNICATION.

Cable Laying.

See Cable Drawing, under TRANSMISSION; and Cable Ship, under MARINE AND NAVAL ENGINEERING.

Composite System.

A Composite Telegraph and Telephone System for Interurban Railways. Illustrates and describes a system on trial at Des Moines, Iowa. 2000 w. Elec Wld—Dec. 28, 1907. No. 89243.

Radio-Telegraphy.

The Poulsen Wireless Telegraph Station at Cullercoats. Describes a station where the Poulsen and spark methods are working side by side. Ills. 2500 w. Elect'n, Lond—Dec. 20, 1907. No. 89251 A.

A Directive System of Wireless Telegraphy. E. Bellini and A. Tosi. Describes some experiments carried out between three coast stations of the French Government. 3500 w. Elec Wld—Dec. 21, 1907. No. 89120.

Dr. Branly's Apparatus for Control of Distant Mechanical Effects. Illustrates and describes apparatus for controlling the lighting of lamps, radio-telegraphy

explosion of mines, sending up rockets, and various mechanical movements. 1200 w. Sci Am—Dec. 28, 1907. No. 89154.

Telautograph.

The Reproduction of Handwriting by Electricity. William Yale. Reviews the development of the telautograph, describing the instrument of George T. Tiffany. 1700 w. Yale Sci M—Dec., 1907. No. 89164 C.

Telegraph Lines.

Telegraph Construction in the German Colonies (Telegraphenbau in den deutschen Kolonien). Herr Hartung. Describes various features, poles, insulators, protective devices, etc., of telegraph line construction in the German colonies in Africa. Ills. 2500 w. Elektrotech Zeitschr—Nov. 14, 1907. No. 88880 D.

DISTRIBUTION.

See also under TRANSMISSION.

Current Rectifier.

Carborundum as a Rectifier for Electric Currents and Electric Oscillations. George W. Pierce. Abstract from *Phys. Rev.* Carborundum is shown to be unila-

We supply copies of these articles. See page 878.

terally conductive. 2000 w. Elect'n, Lond —Dec. 20, 1907. No. 89255 A.

Insulation.

Rubber Insulation for Conductors. Fred J. Hall. Gives general facts regarding rubber and information based on experience. 2500 w. Elec Wld—Nov. 23, 1907. Serial, 1st part. No. 88724.

Wiring.·

Wiring for Direct Current and Alternating Current Motors. Louis J. Auerbacher. Directions with diagrams and information. 2000 w. Elec Wld—Dec. 7, 1907. No. 88910.

See also Illumination, under LIGHTING.

DYNAMOS AND MOTORS.

Air Gap.

The Reluctance of the Air-Gap in Dynamo Machines. T. F. Wall. Abstract of paper read before the Manchester Sec. of the Inst. of Elec. Engnrs. Describes the results of measurements made to ascertain the increase of the reluctance of the air-gap of a dynamo due to slotting the armature, and the distribution of the flux. 2500 w. Elect'n, Lond—Dec. 20, 1907. No. 89254 A.

Characteristic Curves.

A Simple Method of Deriving the Characteristic Curves of an Alternating-Current Machine (Eine einfache Herleitung der Betriebkurven einer Wechselstrommaschine). Hermann Zipp. A mathematical explanation of the method. Ills. 4000 w. Zeitschr d Ver Deutscher Ing—Nov. 2, 1907. No. 88868 D.

Coil Insulation.

Coil Insulation in Electrical Apparatus. Reviews the paper of J. A. Jacobs, presented at meeting of the Ohio Soc. of Mech., Elec., and Steam Engrs., which considers the three classes of insulation used in coils. 2000 w. Elec Wld—Dec. 7, 1907. No. 88908.

D. C. Dynamos.

The Influence of the Rated Speed and Output on the Design of Continuous-Current Generators. H. M. Hobart and A. G. Ellis. Gives data for 250 and 1000 kw. generators, with a number of designs for the same rated outputs at low and high speeds. 1500 w. Elec Rev., Lond—Dec. 20, 1907. Serial, 1st part. No. 89250 A.

D. C. Turbo-Generators.

The Development of Turbo-Generators. Dr. Robert Pohl. Discusses the difficulties in direct-current turbo-generator design, and describes a modified winding for direct-current machines which offers great advantages as regards ·increased output. 4000 w. Inst of Elec Engrs—Sept. 17, 1907. No. 88958 N.

Enclosed Motors.

The Heating of Enclosed Motors. A.

G. Wessling. Argues that the enclosed type should only be used when absolutely necessary. 700 w. Sib Jour of Engng—Dec., 1907. No. 89201 C.

Induction Motors.

Leakage Coefficient of Induction Motors. R. E. Hellmund. Gives a formula possessing the requisite accuracy and giving good results. 1200 w. Elec Wld—Nov. 23, 1907. No. 88721.

The Leakage of Induction Motors. Dr. Rud. Goldschmidt. Aims to develop methods for the calculations of the no-load current and the leakage of three-phase, two-phase and single-phase motors, taking account of all influences as far as possible. 3000 w. Elect'n, Lond—Nov. 29, 1907. Serial, 1st part. No. 88977 A.

Single Phase Induction Motor Diagrams. Charles F. Smith. Gives the approximate construction of the Heyland diagram which is usually employed for commercial coil purposes, and also a diagram from which the performance of the motor may be studied under all conditions of loading. 2500 w. Mech Engr—Dec. 21, 1907. No. 89247 A.

See also Railway Motors, under DYNAMOS AND MOTORS.

Leakage.

See Induction Motors, under DYNAMOS AND MOTORS.

Railway Motors.

The Interpole Railway Motor. A Graphic Explanation. Norman G. Meade. 700 w. Elec Ry Rev—Dec. 28, 1907. No. 89231.

Single-Phase versus Continuous-Current Motors for Interurban Railway Service. Discusses the principal features of the single-phase railway motor, and indicates the lines along which future development of the continuous-current traction motor will proceed. 3500 w. Mech Engr—Dec. 7, 1907. No. 89035 A.

General Considerations of the Development and Possibilities of Single-Phase Motors for Railway Work. E. Heyland. Abstract of a lecture before the "Verband Deutscher Elektrotechniker." Describes simple single-phase induction motors supplied with slip-ring rotors, but without commutators, claiming superior results over commutator motors. 1500 w. Elect'n, Lond—Nov. 22, 1907. No. 88703 A.

See also Induction Motors, and Single-Phase, under DYNAMOS AND MOTORS.•

Reluctance.

See Air Gap, under DYNAMOS AND MOTORS.

Repulsion Motors.

The Regulation of Repulsion Motors Through Lead of Brushes (Regelung von Repulsionsmotoren durch Bürstenver-

We supply copies of these articles. See page 878.

schiebung). Karl Schnetzler. A theoretical and practical consideration of the method. Ills. Serial, 1st part. 2000 w. Elektrotech Zeitschr—Nov. 14, 1907. No. 88879 D.

Single Phase.

Recent Developments in Single-Phase Work for Power and Traction. V. E. Walters. Read before the Birmingham and District Elec. Club. Describes some of the recent developments in the single-phase motor, and its modern applications. 3500 w. Elec Engr, Lond—Dec. 20, 1907. No. 89249 A.

See also Induction Motors and Railway Motors, under DYNAMOS AND MOTORS.

Synchronous Motors.

Representation of Armature Reaction of the Synchronous Motor as an Equivalent Reactance. A. S. Langsdorf. Gives a proof of the accuracy of the simplifying assumptions, upon which all practical mathematical treatments of the phenomena of synchronous alternators have been based. 700 w. Elec Wld—Nov. 30, 1907. No. 88650.

Turbo-Generators.

See D. C. Turbo-Generators, under DYNAMOS AND MOTORS.

ELECTRO-CHEMISTRY.

Electro-Metallurgy.

The Manufacture of Glass by Electricity (Fabbricazione Elettrica del Vetro). Discusses the composition of various kinds of glass and the application of the electric furnace to their production. 3000 w. Serial, 1st part. Elettricita—Nov. 22, 1907. No. 88801 D.

See also same title, under MINING AND METALLURGY, IRON AND STEEL.

Electroplating.

A Simple Method of Depositing Metal Upon Glass, Wood, or Other Non-Metallic Substances. A description of the process. 1800 w. Brass Wld—Dec., 1907. No. 89185.

Nitric Acid.

Manufacture of Nitric Acid from the Air. Brief illustrated description of the Thomas and Barry process for the fixation of atmospheric nitrogen. 1000 w. Sci Am Sup—Dec. 21, 1907. No. 89063.

Nitric Acid from the Air. I. Moscicki, in *Elektrotechnische Zeit.* An account of the work of L Moscicki and I. von Kowalski on the fixation of atmospheric nitrogen. Ills. 2500 w. Electro-Chem and Met Ind—Dec., 1907. No. 89059 C.

ELECTRO-PHYSICS.

Cells.

See Standard Cells, under ELECTRO-PHYSICS.

Coil Windings.

Multiple Wire Windings. Charles R Underhill. Discusses some conditions affecting the turns. 500 w. Elec Wld—Dec. 21, 1907. No. 89121.

Resistance.

The Variation of Resistances with Atmospheric Humidity. E. B. Rosa and H. D. Babcock. Discusses the variations in the values of manganin resistances, the causes of the changes, etc. 5000 w. Bul Bureau of Stand—Dec., 1907. No. 88931 N.

Resonance.

On Relative Resonance in Alternating-Current-Circuits (Ueber relative Resonanz im Wechselstrom-Kreis). F. Grünbaum. Discusses methematically and theoretically the influence of an iron core and the resistance in self-Induction coils. Ills. Serial, 1st part. 4000 w. Elektrotech Zeitschr—Nov. 21, 1907. No. 88881 D.

Solenoids.

Characteristics of the Solenoid. Charles R. Underhill. A mathematical study, giving curves of several solenoids of various lengths. 800 w. Elec Wld—Nov. 23, 1907. No. 88723.

The Influence of Frequency on the Resistance and Inductance of Solenoidal Coils. Louis Cohen. A report of investigations. 3000 w. Bul Bureau of Stand—Dec., 1907. No. 88932 N.

Standard Cells.

Clark and Weston Standard Cells. F. A. Wolf and C. E. Waters. Details and results of work done at the Bureau of Standards on their constancy and reproducibility. Ills. 20800 w. Bul Bureau of Stand—Dec., 1907. No. 88928 N.

The Electrode Equilibrium of the Standard Cell. F. A. Wolf and C. E. Waters. Describes results of a study of Clark and Weston cells. 1800 w. Bul Bureau of Stand—Dec., 1907. No. 88929 N.

GENERATING STATIONS.

Alternators.

See Parallel Operation, under GENERATING STATIONS.

Auxiliary Plants.

Steam Auxiliary to Hydro-Electric Station. William Lee Church. Aims to show that a steam auxiliary may, in certain cases, prove the largest earning factor. 1500 w. Elec Wld—Dec. 7, 1907. No. 88911.

Central Stations.

The Reconstruction of the Power System of the New Orleans Railway & Light Company. An illustrated account of consolidation and reconstruction. 7000 w. St Ry Jour—Dec. 7, 1907. No. 88762.

We supply copies of these articles. See page 878.

Power Station of the Northern Colorado Power Company. Paul Affolter. Illustrated description of an electrical generating station at Lafayette, Colo. Ills. 2000 w. Jour of Engng, Univ of Colo—No. 3. No. 89223 N.

New Installations of the Electricity Supply Service of Brussels (Nouvelles Installations du Service de l'Electricité de la Ville de Bruxelles). Fernand Loewenthal. Detailed description of a new central station and transmission lines. Ills. 9000 w. Serial, 1st part. Soc Belge d Elec'ns—Nov., 1907. No. 88812 E.

See also Boiler Management, under MECHANICAL ENGINEERING, STEAM ENGINEERING.

Construction.

Cost of Constructing Steam-Driven Electric Power Plants. Frank Koester. Gives costs representing an average arrived at by a comparison of costs of various plants. 1700 w. Eng News—Dec. 19, 1907. No. 89090.

Design.

An Unwise Power Plant Layout and Some of the Results Entailed. H. R. Mason. Describes a trying case, the results, and the changes made. 1500 w. Power—Dec., 1907. No. 88994 C.

Hydro-Electric.

The Cost of Hydro-Electric Power Development in the Province of Ontario. A review of the reports of a study of present and future power conditions. 3500 w. Eng News—Dec. 19, 1907. No. 89091.

Rotating Screen of Power Canal, Salt River Project. F. Teichman. Illustration, with brief description. 600 w. Pro Am Soc of Civ Engrs—Nov., 1907. No. 88756 E.

The Tusciano Hydraulic Plant. C. L. Durand. Illustrated description of the principal features of a turbine station in Italy. 2000 w. Elec Rev, N Y—Nov. 30, 1907. No. 88648.

Hydro-Electric Generating Station on the Waipori River, in New Zealand. Illustrated description of the only high-tension transmission plant in Australasia. 5000 w. Elec Wld—Nov. 23, 1907. No. 88720.

Munaar Valley Electrical Power Scheme. Richard Fenwick Thorp. Detailed description of a plant to furnish power for the tea estates in India. 4500 w. Inst of Civ Engrs. No. 89078 N.

Electric Power Plants on Upper Missouri River. A. Floyd Bushnell. Illustrations, with brief description of a plant, the water for which is supplied by a steel dam at Hauser Lake, supplying current at 70,000 volts to Butte, Helena and Anaconda. 1000 w. Eng & Min Jour—Dec. 28, 1907. No. 89213.

The Water-Power Development of the Great Northern Power Co., near Duluth, Minn. Illustrated description of interesting development on the St. Louis River; especially the pressure regulating system at the Thomson, Minn., plant. 5000 w. Eng News—Dec. 26, 1907. No. 89152.

The Great Falls Power Plant of the Southern Power Company. L. T. Peck. Illustrated description of the power development on the Catawba River, which will furnish current for both Carolinas. 3500 w. Elec Jour—Dec., 1907. No. 89186.

Southern Water Power Developments. An account of development in the "Piedmont region," N. C., to furnish electric power for the driving of cotton mills. When completed it will be one of the largest power schemes in the world. Ills. 1600 w. Elec Wld—Dec. 28, 1907. No. 89240.

See also Auxiliary Plants, under GENERATING STATIONS; and Dams, Pipe Lines, and Reservoirs, under CIVIL ENGINEERING, WATER SUPPLY.

Isolated Plants.

See also Mechanical Plants. and Newspaper Plant, under MECHANICAL ENGINEERING, POWER AND TRANSMISSION.

Management.

Operating a Small Electric Plant. W. H. Wakeman. Suggestions for engineers of such plants. Ills. 2000 w. Elec Wld—Dec. 7, 1907. No. 88909.

Parallel Operation.

Notes on the Parallel Operation of Alternators. Waldo V. Lyon. Discusses the cause and effect of the interchange component of the armature current. 4500 w. Elec Wld—Dec. 28, 1907. No. 89241.

LIGHTING.

Arc Lamps.

Recent Progress in Flame Arc Lamps. A. Blondel. Discusses from a theoretical point of view the conditions necessary and those occurring in flame arcs. Also gives a comparison of the candle-power and prices of various systems of lighting. 3800 w. Elect'n, Lond—Dec. 20, 1907. No. 89253 A.

Electric Arc Lamps without Regulating Devices with Especial Reference to the Beck Lamp (Ueber elektrische Bogenlampen ohne Regulierwerk mit besonderer Berücksichtigung der Beck-Bogenlampe). Oskar Arendt. A general discussion. Ills. 4500 w. Elektrotech u Polytech Rundschau—Nov. 6, 1907. No. 88851 D.

See also Magnetite Arc, under LIGHTING.

Illumination.

Lighting of a Large Retail Store. Frederick J. Pearson. Illustrates and describes the lighting adopted by Marshall Field & Co. of Chicago, reporting tests made. 3500 w. Elec Rev, N Y—Dec. 7, 1907. No. 88906.

The Electric Illumination and Wiring of the New York Public Library. Charles Jenkins Spencer. Illustrated description of an important system now being installed in an imposing structure. 3000 w. Elec Age—Dec., 1907. No. 89118.

The Uniform Illumination of Horizontal Planes. Alfred A. Wohlauer. A theoretical discussion of the possibilities, with report of tests made. 2500 w. Elec Wld—Dec. 21, 1907. No. 89122.

Experimental Data on Illuminating Values. Clayton H. Sharp and Preston S. Millar. Condensed from paper read before the Assn. of Edison Ill. Cos. Deals with the various distributions given by a number of lighting arrangements. 3300 w. Elec Rev, N Y—Dec. 21, 1907. No. 89098.

See also Theatres, under LIGHTING.

Incandescent Lamps.

On the Determination of the Mean Horizontal Intensity of Incandescent Lamps. E. P. Hyde and F. E. Cady. Abstract from the Bul. of the Bureau of Standards. A report of investigations. 1800 w. Elect'n, Lond—Nov. 22, 1907. No. 88705 A.

A Comparative Study of Plain and Frosted Lamps. Edward P. Hyde and F. E. Cady. Describes the apparatus used and methods of study, discussing the results. 8800 w. Bul Bureau of Stand—Dec., 1907. No. 88930 N.

Metallic-Filament Lamps—How Will They Affect Electricity Undertakings. Argues the need of alterations in systems of connection and charging to meet the new conditions. 5000 w. Elec Engr, Lond—Dec. 6, 1907. No. 89038 A.

See also Tantalum Lamp, under LIGHTING.

Magnetite Arc.

Characteristics of the Magnetite Arc. G. M. Dyott. Gives results of some investigations with metallic arcs. 600 w. Elec Wld—Dec. 7, 1907. No. 88907.

Mercury Vapor.

A New Form of Cooper Hewitt Mercury Vapor Lamp. F. H. von Keller. Illustrated description of the new type, with other information. 3300 w. Jour Fr Inst—Dec., 1907. No. 88920 D.

Photometry.

The Problem of Color Photometry. J. S. Dow. A valuable résumé of this subject. Also editorial. 5500 w. Elec Wld—Nov. 30, 1907. No. 88651.

The International Photometric Commission (La Commission Internationale de Photométrie). M. F. Laporte. A report of the meeting held at Zurich in July, 1907. 4200 w. Bul Soc Inter d Elec'ns—Nov., 1907. No. 88813 F.

Street.

The City Lighting. Haydn Harrison. Reports results of a series of tests made in London to ascertain the value of the various means adopted in different parts of the city. 700 w. Elec Rev, Lond—Dec. 6, 1907. No. 89041 A.

Tantalum Lamp.

The Tantalum Lamp with High Resistance Filament. L. H. Walter. A brief account of investigations of the physical properties. 1200 w. Elect'n, Lond—Nov. 22, 1907. No. 88702 A.

Theatres.

The Electric Lighting of the Stuyvesant Theatre, New York City. Illustrated description of the elaborate equipment of a new theatre in West 44th St. 2000 w. Elec Rev, N Y—Dec. 21, 1907. No. 89097.

MEASUREMENT.

Alternator Testing.

The Determination of Load Resistance for Large Alternating-Current Generators (Die Berechnung der Belastungswiderstände für grosse Wechselstromgeneratoren). Carl Richter. A mathematical discussion and description of the water resistance method of testing. Ills. 4500 w. Elektrotech u Maschinenbau—Nov. 17, 1907. No. 88856 D.

Dynamo Testing.

Method of Testing Direct-Current Dynamos. E. S. Lincoln. Explains how to measure insulation resistance, locate connections and determine the rating and characteristics of a dynamo. 3000 w. Power—Dec., 1907. No. 88996 C.

Inductance.

On a Standard of Mutual Inductance. Albert Campbell. Abstract of a paper read before the Royal Soc. Considers the most suitable design for such a standard. 1400 w. Elect'n, Lond—Nov. 22, 1907. No. 88704 A.

Instruments.

Torque Analysis of Induction Meters. A. R. Dennington. A study of the production of the torque. 1200 w. Elec Wld—Dec. 21, 1907. No. 89124.

The Most Economical Shape of Winding for Electrical Measuring Instruments. A. P. Young. Gives a solution of a problem in the design proportion of electrical measuring instruments of the moving-coil type. 700 w. Elec Wld—Dec. 28, 1907. No. 89242.

Iron Losses.

Measurement of Iron Losses. Dr. J.

We supply copies of these articles. See page 878.

Sahulka. Describes a method of rapidly determining the relation connecting the core loss per cycle with the frequency, giving results showing that loss increases with the frequency. 1000 w. Elect'n, Lond—Dec. 15, 1907. No. 89132 A.

Meters.
Some Recent Improvements in the Wright Mercury Electrolytic Meter. H. Stafford Hatfield. A short account of the later developments of this house service meter and the method of manufacture. 2800 w. Elect'n, Lond—Dec. 6, 1907. Serial. 1st part. No. 89043 A.

Meter Testing.
Notes on the Testing of Multi-Range Ammeters and Voltmeters and Low-Reading Voltmeters. A. E. Moore. Considers the best method of recording the readings, of expressing the accuracy or error of the instrument, etc. 2000 w. Elec Engr, Lond—Nov. 29, 1907. No. 88972 A.

Units.
Energy Transformations from the Electrical Engineers' Standpoint. H. M. Hobart. Advocates the general adoption of the kilowatt-hour as the unit of energy. The present number discusses the relation between heat and temperature. Also editorial. 2500 w. Elec Rev, N Y—Dec. 14, 1907. Serial. 1st part. No. 89019.

Wire Testing.
Commercial Testing of Wire and Cable. F. M. Farmer. Reviews the various tests made, describing methods. Ills. 5500 w. Elec Age—Dec., 1907. No. 89119.

TRANSMISSION.

See also under DISTRIBUTION and GENERATING STATIONS.

Alternating Current.
The Theory of Alternate Current Transmission in Cables. Explains the graphical method of dealing with circuits having resistance, capacity, and leakage; and gives the simplest possible mathematical treatment, leading to the employment of Dr. A. E. Kennelly's methods of calculation. 2500 w. Elect'n, Lond—Dec. 6, 1907. Serial. 1st part. No. 89042 A.

Balancing.
The Balancing of Multiple Wire Systems of Electrical Distribution. Prof. J. H. Dobson. A report of a study, made by the writer, of balancing in its various aspects. Chiefly directed to three-wire systems. 4500 w. Jour S African Assn of Engrs—Oct., 1907. No. 88696 F.

Cable Drawing.
Notes on Drawing-In Cables. W. Pleasance. Discusses cables drawn into pipes or ducts. 2200 w. Elec Rev, Lond—Nov. 29, 1907. No. 88976 A.

A Portable, Electrically-Driven Cable Reel (Eine fahrbare Kabelwinde mit elektrischen Antrieb). Herm. Schultz. An illustrated description of a wagon carrying a reel driven by an electric motor, the power for which is generated by a dynamo driven by a 6 horse-power benzine engine, used by the State Telegraph Department in Germany. 2000 w. Elektrotech Zeitschr—Nov. 28, 1907. No. 88-882 D.

Cable Testing.
See Wire Testing, under MEASUREMENT.

Circuit Breakers.
Electrical Apparatus for High and Low Tension (L'Appareillage Electrique à Haute et Basse Tension). M. Vedovelli. Discusses principally circuit breakers and switchboards, noting the points to be observed in choosing and installing apparatus, and illustrating and describing types. Bul Soc Inter d Elec'ns—Nov., 1907. No. 88814 F.

Grounded Neutral.
Discussion on "The Grounded Neutral," at New York, October 11, 1907. 13500 w. Pro Am Inst of Elec Engrs—Dec., 1907. No. 88933 D.

Insulation.
See same title, under DISTRIBUTION.

Line Construction.
The Construction of Overhead Electric Transmission-Lines. Alexander Pelham Trotter. Deals mainly with the mechanical considerations of construction. Discussion. 26000 w. Inst of Civ Engrs—No. 3703. No. 89087 N.

Overhead Construction for High-Tension Electric Traction or Transmission. R. D. Coombs. Discusses the causes of interruption of service and how to guard against them; allowance for ice and wind loads; catenary stresses; and the requirements of construction. Ills. 7700 w. Pro Am Soc of Civ Engrs—Dec., 1907. No. 89225 E.

See also Telegraph Lines, under COMMUNICATION; and Wire Suspension, under STREET AND ELECTRIC RAILWAYS.

Line Design.
Long - Distance Power - Transmission Lines. Eric Adolf Löf. Shows the different factors to be taken into consideration for the proper design of the transmission line. 1000 w. Elec Rev, N Y—Nov. 30, 1907. Serial. 1st part. No. 88649.

Lines.
Joint Construction Pole Lines. H. B. Gear. Explains conditions in Chicago, and presents the advantages and disadvantages of joint lines. 2000 w. Cent Sta—Dec., 1907. No. 88799.

We supply copies of these articles. See page 878.

See also Central Stations and Hydro-Electric, under GENERATING STATIONS.

Poles.
See Timber Preservation, under CIVIL ENGINEERING, MATERIALS OF CONSTRUCTION.

Regulations.
Governmental Regulations for the Distribution of Electrical Energy (Règlement d'Administration Publique sur les Distributions d'Energie). The provisions of the French decree which went into effect Oct. 26, 1907. 2500 w. Electricien —Nov. 9, 1907. No. 88818 D.

Transformers.
Abnormal Primary Current and Secondary Voltage on Placing a Transformer in Circuit. J. Murray Weed. An explanation of the results obtained by Messrs. Jensen and Andree, as recorded in the *Elec. Wld.* of Sept. 14. 2000 w. Elec Wld—Nov. 23, 1907. No. 88722.

Wire Suspension.
See same title, under STREET AND ELECTRIC RAILWAYS.

MISCELLANY.

History.
The Development of Electrical Technology (Der Werdegang der Elektrotechnik). F. Niethammer. A general review of the historical development of all branches of electrical science. 6500 w. Elektrotech u Maschinenbau—Nov. 3, 1907. No. 88855 D.

Kelvin.
The Late Lord Kelvin. An appreciative editorial review of his life and work. 9500 w. Engng—Dec. 20, 1907. No. 89-261 A.

Lightning.
The Rolling of Thunder. D. S. Carpenter. Tends to show that rolling in thunder is due to the primary sounds of successive discharges. 1500 w. Elec Wld —Dec. 21, 1907. No. 89123.

Seismograph.
An Electrically Registering Seismograph (Séismograph à Enregistrement Électrique). R. Goldschmidt. Illustrated description. 2400 w. Soc Belge d Elec'ns —Nov., 1907. No. 88811 E.

• INDUSTRIAL ECONOMY

Cost Systems.
A Uniform Cost System for Jobbing Foundries as Recommended by the Jobbing Founders' Association. James S. Stirling. Presented before the Philadelphia Foundrymen's Association, Dec. 4, 1907. 4500 w. Ir Age—Dec. 12, 1907. No. 88919.

Education.
Rational Trades Instruction. O. M. Becker. An illustrated article discussing the work of industrial schools. 5000 w. Cassier's Mag—Dec., 1907. No. 88979 B.
The Trade School Problems of Skilled Labor. C. Alfred Smith. Calls attention to facts and conditions in England. 2000 w. Elec Rev, Lond—Nov. 29, 1907. No. 88974 A.
The Royal Naval College at Greenwich, and the Training of Engineer Officers for the Royal Navy. J. W. W. Waghorn. An illustrated description of the buildings, and account of the training. 4000 w. Int Marine Engng—Jan., 1908. No. 89144 C.
Secondary Technical Education Applied to Mining. L. R. Young. Abstract of paper read at meeting of the Am. Min. Cong. 2000 w. Min Rept—Dec. 5, 1907. No. 88766.
The Mining Engineer (L'Ingénieur des Mines). M. Clément. An address on the work and education of mining engineers delivered at the Technical High School at Delft, Oct. 4, 1907. 3500 w. Ingenieur —Nov. 2, 1907. No. 88883 D.

Engineering Societies.
The Mechanical Engineer and the Function of the Engineering Society. F. R. Hutton. Presidential address. 16000 w. Pro Am Soc of Mech Engrs—Dec., 1907. No. 88759 C.

Exhibitions.
The Franco-British Exhibition of 1908. An illustrated account of structural features in progress for this great exhibition dedicated to science, art, and the industries of the two nations. 1400 w. Engng—Nov. 29, 1907. Serial. 1st part. No. 88960 A.

Gold Stock.
The Mechanical Management of the World's Stock of Gold. Alex. Del Mar. Gives a plan, suggested for increasing its mobility and service to commerce. 2500 w. Engineering Magazine—Jan., 1908. No. 89174 B.

Labor.
Why Good Workmen Are Scarce. Egbert P. Watson. Discusses some of the causes that tend to restrict the supply of skilled workmen. 2000 w. Cassier's Mag—Dec., 1907. No. 88980 B.
See also Education, under INDUSTRIAL ECONOMY.

Patents.
Reform of the Patent Law (Reform des Patentrechtes). Wolfgang Koch. Discusses changes necessary in the German

Patent Law. 2500 w. Serial. 1st part. Zeitschr f Werkzeug—Nov. 15, 1907. No. 88850 D.

Salesmen.
The Functions of the Engineer as a Salesman. Sterling H. Bunnell. Shows that the fields of mechanics and business must overlap in engineering salesmen. 1600 w. Engineering Magazine—Jan., 1908. No. 89179 B.

Shop Management.
See Management, under MECHANI-

CAL ENGINEERING, MACHINE WORKS AND FOUNDRIES.

Wages.
The Bonus System of Wages (Les Salaires à Primes). M. Paul Lecler. An elaborate discussion. Compares the bonus system with other systems of paying labor, describes the various developments of the system, the methods of operation, results, applications, etc. Ills. 20000 w. Mem Soc Ing Civ de France—Aug., 1907. No. 88808 G.

MARINE AND NAVAL ENGINEERING

Artillery.
Modern French Field-Guns. Illustrated detailed descriptions of the Schneider-Canet guns of the P. D. and of the P. R. patterns. Plate. 3500 w. Engng—Dec. 6, 1907. No. 89050 A.

Rifled Artillery. A. G. Greenhill. A theoretical investigation of the least amount of spin requisite for stability in flight of an elongated body through the air. 2500 w. Engr, Lond—Nov. 22, 1907. Serial. 1st part. No. 88716 A.

Some Comparisons Between French and English Artillery. Inaugural address of the President, M. Gustave Canet, delivered before the Junior Inst. of Engrs. 8500 w. Engng—Nov. 22, 1907. No. 88715 A.

Cable Ships.
A New Cable Ship. Frederick C. Coleman. Illustrated description of the "Guardian' and its machinery for laying submarine cables. 1200 w. Sci Am Sup—Dec. 14, 1907. No. 88953.

Destroyer.
H. M. Torpedo-Boat-Destroyer "Mohawk." Illustrated description of the vessel and its equipment with report of trial. 2500 w. Engng—Nov. 22, 1907. No. 88714 A.

Education.
See same title, under INDUSTRIAL ECONOMY.

Electric Power.
Modern Marine Transportation. William T. Donnelly. An illustrated article discussing the electric unit system of marine transportation. 2000 w. Int Marine Engng—Jan., 1908. Serial. 1st part. No. 89143 C.

Fishing Steamers.
The Principles of the Construction of Fishing Steamers (Die Grundlagen der Konstruktionsgleichungen für Fischdampfer). P. Knipping. A discussion of their design according to the kind of

service for which they are intended. Ills. 2400 w. Serial. 1st part. Schiffbau—Nov. 13, 1907. No. 88844 D.

Mechanical Stoking.
See Locomotive Stokers, under RAILWAY ENGINEERING, MOTIVE POWER AND EQUIPMENT.

Model Basin.
Simple Explanation of Model Basin Methods. An explanation of the work of an experimental model basin. 4000 w. Sci Am—Dec. 7, 1907. No. 88790.

Motor Boats.
The German Institution of Naval Architects. This first part gives a review of Dr. M. Bauer's paper on "High-speed Motor Boats." 1700 w. Engr, Lond—Nov. 29, 1907. Serial. 1st part. No. 88963 A.

The Motor-Yacht "Swietlana." Illustrated description of a vessel built in England for use on rivers in Central Russia. 800 w. Engng—Dec. 13, 1907. No. 89134 A.

See also Petroleum, under MECHANICAL ENGINEERING, COMBUSTION MOTORS.

Resistance.
A Note on the Theory of Ship's Resistance (Beitrag zur Theorie des Schiffswiderstandes). H. Lorenz. A mathematical and theoretical discussion. Ills. 4500 w. Zeitschr d Ver Deutscher Ing—Nov. 16, 1907. No. 88872 D.

Ship Heating.
The Heating and Ventilating of Ships. Sydney F. Walker. The present article discusses the special requirements on board ship and the difficulties peculiar to ship work. 3500 w. Int Marine Engng—Jan., 1908. Serial. 1st part. No. 89142 C.

Steamboats.
Who Built the First Steamboat? C. Seymour Bullock. Gives a review of the work of a number of early mechanics and inventors, showing that the idea of ap-

We supply copies of these articles. See page 878.

plying steam to the propulsion of vessels was in many minds about the same time. Ills. 6500 w. Cassier's Mag—Dec., 1907. No. 88984 B.

Steamships.

New Egyptian Mail Turbine Steamship Helopolis. Allan McPherson. Illustrated detailed description. 2000 w. Int Marine Engng—Jan., 1908. No. 89141 C.

Submarine Signalling.

Sound as a Means of Controlling Operations at a Distance. John Gardner. Describes a new method of automatically governing actions at a distance without solid connections. 2000 w. Elec Rev, Lond—Dec. 13, 1907. No. 89130 A.

Train Ferry.

The Proposed Channel Train-Ferry. Sir Nathaniel Barnaby. Briefly reviews the projects that have been suggested for railway connection across the English Channel, describing and discussing the train-ferry scheme. 2200 w. Westminster Rev—Dec., 1907. No. 89064 D.

Turbines.

See Steamships, under MARINE AND CANAL ENGINEERING.

U. S. Navy.

The American Fleet from an English Point of View. Archibald S. Hurd. Reviews the latest additions to the fleet and the programme for naval construction. Ills. 4000 w. Cassier's Mag—Dec., 1907. No. 88982 B.

MECHANICAL ENGINEERING

Ariel.

The 1908 Ariel Car. Illustrated description of this British vehicle. 500 w. Auto Jour—Dec. 7, 1907. Serial. 1st part. No. 89030 A.

Brakes.

Notes on Brakes. Edward T. Humphries. Briefly discusses foot, hand, and mechanical brakes. 1600 w. Autocar—Dec. 7, 1907. No. 89032 A.

Brake Testers.

An Electrical Brake Tester. Illustrated description of the Morris & Lister device. 900 w. Autocar—Dec. 21, 1907. No. 89245 A.

Buick.

The 18-H.P. and 24-H.P. Buick Petrol Cars. Illustrates and describes cars of American manufacture. 1800 w. Auto Jour—Dec. 14, 1907. Serial. 1st part. No. 89126 A.

Carburetting.

The Principles of Carburetting, as Determined by Exhaust Gas Analysis. Dugald Clerk. Read before the Incor. Inst. of Auto. Engrs. Describes the results obtained by the application of a new method to the study of perfect carburation. 5000 w. Autocar—Dec. 14, 1907. No. 89127 A.

Gas Analysis and Carburation. W. R. Ormandy and J. H. Lester. Shows that gas tests are as important as brake and consumption tests and how gas analysis may be used in connection with carburetter design. 4000 w. Autocar—Dec. 7, 1907. No. 89033 A.

Design.

Modern Motor Vehicles. Lt. Col. Rookes Evelyn Bell Crompton. Describes existing vehicles as examples of typical systems. General discussion. Ills. 41800

w. Inst of Civ Engrs—No. 3669. No. 89088 N.

Exhibitions.

Notes on the Recent Motor Show. A general review of the exhibition at Olympia. 4000 w. Engng—Dec. 6, 1907. No. 89049 A.

The Tenth Automobile and Cycle Exhibition (Le 10 Salon de l'Automobile et du Cycle). J. de Vergnes. A description of the exhibits at the recent show held under the auspices of the Automobile Club of France. Ills. Serial. 1st part. 3000 w. La Métallurgie—Nov. 20, 1907. No. 88823 D.

Farm Motor.

A Successful Farm Motor. C. M. Eason. An illustrated account of automobiles adapted to agriculture. 1500 w. Sci Am Sup—Dec. 7, 1907. No. 88793.

France.

The Automobile Situation in France. Jacques Boyer. Discusses the cause of the crash in automobile securities, and reviews the present status of automobile construction in France, with especial reference to the propelling machinery. Ills. 3500 w. Cassier's Mag—Dec., 1907. No. 88983 B.

The Automobile in France (Les Progrés de l'Automobilisme en France). P. Rimauce. Discusses the development of the industry, the growing use of the automobile, description of cars, costs, etc. Tables and curves. Ills. Serial. 1st part. Génie Civil—Nov. 23, 1906. No. 88827 D.

Fuels.

The Fuel System of Automobiles. Thomas J. Fay. Discusses details of the fuel question of interest to the user. 3500

We supply copies of these articles. See page 878.

w. Automobile—Dec. 26, 1907. Serial.
1st part. No. 89168.

Garages.
The Electromobile Co.'s New Garage.
Illustrated detailed description of a ga-
rage in London. 2500 w. Elec Rev,
Lond—Nov. 29, 1907. No. 88973 A.

Modern Garage Plants (Moderne Ga-
ragen-Anlagen). Otto Rambuscheck.
Gives suggestions for building small
garages and illustrates and describes the
arrangement of several large plants in
Berlin. 2700 w. Zeitschr d Mit Motorwagen Ver—Nov. 30, 1907. No. 88886 D.

Gears.
The Jackson Epicyclic Gear for Cars,
Cabs, Vans, and 'Buses. Illustrated description of a change-speed mechanism.
1800 w. Auto Jour—Dec. 14, 1907. Serial.
1st part. No. 89125 A.

Humphris System of Gear-Wheel Design. Illustrates and describes this mechanism. Also abstract of paper on "Gears
and Gearing," by F. Humphris, read before the Royal A. C. 3800 w. Auto Jour
—Dec. 21, 1907. Serial. 1st part. No.
89244 A.

Ignition.
The Relation Between Power and
Spark. W. Watson. A comparison between accumulator and high tension magneto ignition. 2500 w. Autocar—Nov.
30, 1907. No. 88957 A.

See also Spark Coils, under AUTOMO-
BILES.

Lamps.
Automobile Lamps and Their Lenses.
Victor Lougheed. Discusses the efficiency
of a typical auto headlight, and matters
related. Ills. 4000 w. Automobile—
Dec. 12, 1907. No. 88917.

Motors.
Gasoline Engines for Automobiles. Da-
vid Fergusson. A comparison of the four-
cylinder and six-cylinder vertical types.
Ills. 1700 w. Cassier's Mag—Dec., 1907.
No. 88985 B.

It Costs More to Produce a Six Than
a Four. F. B. Stearns. Discusses the
relative merits. Also several short articles by different writers. 3700 w. Automobile—Dec. 19, 1907. No. 89096.

Omnibuses.
The Electrobus. An illustrated account
of these vehicles, running in London, with
details of the method of handling the accumulators, and the cost of running. 2000
w. Elect'n, Lond—Dec. 6, 1907. Serial.
1st part. No. 89044 A.

Pilgrim.
The "Pilgrim" Motor-Car. Illustrated
detailed description. 1200 w. Engng—
Dec. 6, 1907. No. 89051 A.

Progress.
Learned in the Process of Evolution.

Charles B. Hayward. An illustrated description of poorly designed or poorly
made parts and their improvement. 2500
w. Automobile—Dec. 5, 1907. No. 88796.

Six-Cylinder.
The Standard 20-H.P. Six-Cylinder
Car. Illustrated detailed description.
1200 w. Autocar—Nov. 23, 1907. No.
88698 A.

Spark Coils.
The Design and Operation of Spark
Coils. F. W. Springer. Discusses in detail the methods and effects of varying
the various design factors of different
coils, and the constructive and electrical
characteristics of each kind of coil. 7000
w. Elec Wld—Dec. 14, 1907. No. 89020.

Speed Recorders.
Principles of Speed and Distance Recorders. Charles B. Hayward. An illustrated explanation of the working of
these instruments. 3500 w. Automobile
—Dec. 19, 1907. No. 89095.

Talbot.
The New 35-H.P. Talbot Car. Illustrated description. 700 w. Autocar—Nov.
30, 1907. Serial. 1st part. No. 88955 A.

Timing Device.
The Holden-Elphinstone Electric Timing Apparatus. Illustrated description
of apparatus for timing cars. 1500 w.
Auto Jour—Dec. 7, 1907. No. 89029 A.

White.
The 1908 White Steam Cars. Illustrates the new external thermostat and
other improvements. 3000 w. Auto Jour
—Nov. 23, 1907. No. 88697 A.

Wind Screens.
A Study of Wind Screens. Illustrates
and describes various types. 1200 w.
Autocar—Nov. 30, 1907. No. 88956 A.

COMBUSTION MOTORS.

Air Cooling.
Light Air-Cooled Internal-Combustion
Motors (Les Moteurs Légers à Explosions avec Refroidissement par Circulation d'Air). M. J. Ambroise Farcot. Discusses various types but particularly the
hot-air engine. Ills. 6000 w. Mem Soc
Ing Civ de France—Sept., 1907. No.
88809 G.

By-Product Producers.
See Coking By-Products and Peat, un-
der MINING AND METALLURGY,
COAL AND COKE.

Exhaust Gas Analysis.
See Carburetting, under AUTOMOBILES;
and Flue Gas Analysis, under STEAM EN-
GINEERING.

Gas Cleaning.
See Blast Furnace Gas, under MINING
AND METALLURGY, IRON AND STEEL.

Gas Engines.
The Gas-Engine Cycle. W. H. Booth.

Discusses the two-stroke and the four-stroke engine. 1600 w. Power—Dec., 1907. No. 88993 C.

The Cost of Depreciation. A. S. Atkinson. Discusses the question as related to gas engines. 2000 w. Engr, U S A—Dec. 2, 1907. No. 88688 C.

600-Horse-Power Two-Cycle Gas Engine. Illustrated description of an engine for driving a cotton spinning mill. 1800 w. Engr, Lond—Dec. 13, 1907. No. 89140 A.

Gas Power Plant at an Engineering Works. Illustrated description of a plant installed recently by Messrs. David Rowan & Co., Glasgow, for driving Marine Engine Works. 3300 w. Engr, Lond—Dec. 20, 1907. No. 89264 A.

The Evolution of the Internal Combustion Engine. S. A. Reeve. States the problem of the internal combustion engine in the light of past experience, and its solution. 16700 w. Pro Am Soc of Mech Engrs—Dec., 1907. No. 88758 C.

Novelties in Large Gas Engines (Nouveautés dans les Grosses Machines à Gaz). Herr von Handorff. Abstract translation of a paper in the *Zeitschr. d. Ver. Deutscher Ing.* for Aug. 14, 1907. Ills. 4000 w. Rev d Métal—Nov., 1907. No. 88817 E + F.

The Employment of Large Gas Engines in Germany (Notes sur L'Emploi des Grands Moteurs à Gaz en Allemagne). R.-E. Mathot. The extent of their use in Germany is discussed and some test results are given. 3000 w. Rev de Mécan—Nov. 30, 1907. No. 88890 E + F.

See also Gas Plant Testing, Ignition, and Thermal Efficiency, under COMBUSTION MOTORS; Refrigeration, under HEATING AND COOLING; Pumping Plants, under HYDRAULIC MACHINERY; Newspaper Plant, under POWER AND TRANSMISSION; Blast-Furnace Gas, under MINING AND METALLURGY, IRON AND STEEL; and Interurban, under STREET AND ELECTRIC RAILWAYS.

Gas Engine Wreck.
A Peculiar Gas Engine Wreck. A brief illustrated description of a wreck in a factory at Newark, N. J. 700 w. Power—Dec., 1907. No. 88991 C.

Gasoline.
See Motors, under AUTOMOBILES.

Gas Plant Testing.
Efficiency Tests of a Lignite-Gas Engine Generating Set (Abnahmeversuche an Braunkohlen-Grossgasdynamos). Report of a 1400 horse-power plant driven by two Nürnberg gas engines fed with gas from lignite briquettes. Ills. 1600 w. Elektrotech Zeitschr—Nov. 7, 1907. No. 88878 D.

See also Pumping Plants, under HYDRAULIC MACHINERY.

Gas Producers.
See also Gas Engines and Gas Plant Testing, under COMBUSTION MOTORS; Pumping Plants, under HYDRAULIC MACHINERY; and Coking By-Products and Peat, under MINING AND METALLURGY, COAL AND COKE.

Hot-Air Engine.
See Air Cooling, under COMBUSTION MOTORS.

Ignition.
A New Method of Working Internal-Combustion Engines. Describes a method recently devised by Messrs. Gebrüder Sulzer. 1200 w. Mech Engr—Dec. 21, 1907. No. 89248 A.

Oil Engines.
Oil-Motor for Agricultural Purposes. Illustrates and describes an agricultural oil-engine built in England. 2000 w. Engng—Nov. 29, 1907. No. 88962 A.

Petroleum Motors at the Kiel Motor-Boat Show (Petroleummotoren auf der Kieler-Motorbootausstellung). Herr Pflug. Illustrated description of the principal exhibits. 2200 w. Zeitschr d Mit Motorwagen Ver—Nov. 15, 1907. No. 88858 D.

Thermal Efficiency.
On the Limits of Thermal Efficiency in Internal Combustion Motors. Dugald Clerk. A report of investigations carried out by the Institution Committee. Also discussion. 17000 w. Inst of Civ Engrs—No. 3696. No. 89086 N.

HEATING AND COOLING.

Air Washers.
Air Washing and Humidifying and Some of Its Applications to Industrial Purposes. W. A. Rowe. Read before the Ohio Soc. of Mech., Elec., & Steam Engrs. The requirements and design of air washers are considered and their applications. 3500 w. Engr, U S A—Dec. 2, 1907. No. 88686 C.

Cooling Towers.
Cooling-Tower Practice. Dr. Jos. H. Hart. An illustrated review of the kinds of towers in use and some of the applications, working principles, and operation. 2000 w. Engineering Magazine—Jan., 1908. No. 89177 B.

Cooling Towers in Refrigerating Plants. B. Franklin Hart, Jr. Discusses types of cooling apparatus, and shows when cooling towers are available. 2500 w. Ice & Refrig—Dec., 1907. No. 89031 C.

Ice Making.
The Vacuum System of Ice Making. John Patten. A process for freezing water by its own vaporization in an almost complete vacuum. 3000 w. Ice & Refrig—Dec., 1907. No. 89025 C.

We supply copies of these articles. See page 878.

Industrial Buildings.

Modern Methods of Heating Industrial Buildings. E. Newton. A study of the conditions that make for workshop warmth and comfort. 2200 w. Wood Craft—Dec., 1907. No. 88640.

Plumbing.

Plumbing Work in the Washington Y. M. C. A. Building. A. R. McGonegal. Illustrated detailed description. 4500 w. Dom Engng—Dec. 28, 1907. No. 89228.

Radiation.

Data for a Heating Engineer's Handbook. Compiled by Gerard W. Stanton. Tables and explanatory notes. 1000 w. Heat & Vent Mag—Dec., 1907. Serial. 1st part. No. 89200.

Refrigeration.

Refrigeration and Power-Plant Practice. Joseph H. Hart. Shows the interrelations between the two. 2000 w. Elec Rev, N Y—Dec. 28, 1907. No. 89239.

Mechanical Production of Low Temperature. Sydney F. Walker. Explains why power is required and why heat is employed in refrigeration. 2500 w. Power —Dec., 1907. No. 88989 C.

Producer Gas in Refrigeration. Ellis L. Phillips. Some facts concerning the economy and reliability of gas power for use in refrigeration plants. Ills. 1700 w. Ice & Refrig—Dec., 1907. No. 89024 C.

Heat Transfer in Coolers and Condensers of the Double Pipe Type. R. L. Shipman. Gives results of recent theoretical and experimental investigations. 1800 w. Ice & Refrig—Dec. 6, 1907. No. 89036 C.

The Fish Freezing and Storage Plant of the Consolidated Weir Co., Provincetown, Mass. Howard S. Knowlton. Illustrated detailed description of a large fireproof fish-freezing warehouse and its equipment. 1600 w. Eng Rec—Dec. 28, 1907. No. 89206.

See also Cooling Towers, under HEATING AND COOLING; and Refrigeration, under RAILWAY ENGINEERING, TRAFFIC.

School House.

Heating and Ventilation of a School House. Charles L. Hubbard. Considers matters of detail, and treatment of special rooms in a typical city high school. Ills. 1800 w. Met Work—Nov. 30, 1907. No. 88642.

Valves.

See same title, under STEAM ENGINEERING.

Ventilation

Heating and Ventilating. George W. Nistle. A general review of present practice. 2200 w. Dom Engng—Dec. 21, 1907. No. 89094.

Important Differences Between Direct and Indirect Heating., R. S. Thompson. Discusses the proper ventilation of rooms heated by the two systems, showing that what is correct with one system is incorrect with the other. 2000 w. Met Work —Dec. 14, 1907. No. 88954.

Constructive Provisions for Preventing the Reversal of the Air Current in the Ground Floor, with Central Ventilating Plants (Konstructive Vorkehrungen zur Vermeidung der Umkehr des Luftstromes im Erdgeschoss bei zentralen Lüftungsanlagen). G. Recknagel. The first part of the serial gives a mathematical discussion of the subject. Ills. Serial. 1st part. 4500 w. Gesundheits-Ing—Nov. 30, 1907. No. 88864 D.

HYDRAULIC MACHINERY.

Accumulators.

The Design of Hydraulic Accumulators. N. S. Trustrum. Considers the design of the ram and cylinder, and other features of importance. 1300 w. Prac Engr—Nov. 29, 1907. No. 88968 A.

Centrifugal Pumps.

Centrifugal Pumps. E. F. Doty. Discusses the use of centrifugal vs. reciprocating pumps for domestic water supply. 2500 w. Engr, U S A—Dec. 2, 1907. No. 88689 C.

Centrifugal Pumps. E. F. Doty. Illustrates and describes their application for fire protection and condenser service. 2500 w. Engr, U S A—Dec. 16, 1907. No. 89062 C.

The Theory of Centrifugal Pumps and Fans (Zur Theorie der Zentrifugalpumpen und Ventilatoren). V. Blaess. A theoretical and mathematical paper on the principles of their action. Ills. 3000 w. Serial. 1st part. Zeitschr f d Gesamte Turbinenwesen—Nov. 9, 1907. No. 88848 D.

General Equations for Centrifugal Pumps (Allgemeine Gleichung für Zentrifugalpumpen). E. Busse. Discusses the general expression of the relation between head, speed of revolution, and quantity of water and the plotting of characteristic curves. Ills. 3000 w. Zeitschr f d Gesamte Turbinenwesen— Nov. 20, 1907. No. 88849 D.

See also Electric Pumping, under HYDRAULIC MACHINERY.

Dredges.

See same title, under CIVIL ENGINEERING, WATERWAYS AND HARBORS.

Electric Pumping.

Electrically-Operated Pumping Plants. T. L. Kolkin. Considers the turbine pump, and the ram pump for this use. 1500 w. Elec Rev, Lond—Dec. 6, 1907. No. 89040 A.

Electrically Driven Centrifugal Pump, Buffalo, N. Y. Henry L. Lyon. Illustrated description of the electric pump, with comparison of cost with steam. Discussion. 4000 w. Pro Am W-Wks Assn—1907. No. 88738 N.

Pipe Friction.
The Friction of Oil in Pipes (Oelreibung in Röhren). R. Camerer. A record of elaborate experiments to determine the efficiency of automatic oil-pressure governors for water turbines. Ills. 3500 w.. Zeitschr f d Gesamte Turbinenwesen—Nov. 9, 1907. No. 88847 D.
See also same title, under CIVIL ENGINEERING, Water Supply.

Pumping Engines.
Pumping-Engines and Pumps for the Kidderminster Water Supply. Illustrated detailed description. 800 w. Engng—Nov. 22, 1907. No. 88713 A.
High Duty vs. Low Duty Pumping Engines. Irving H. Reynolds. Calls attention to points of importance in making a decision between the two types. Discussion. 9500 w. Pro Am W-Wks Assn—1907. No. 88739 N.

Pumping Plants.
Pumping Water by Producer Gas Plant at St. Stephen, N. B. F. A. Barbour. States the conditions and describes in detail the gas producer plant adopted. 3800 w. Pro Am W-Wks Assn—1907. No. 88737 N.
Test of a Producer Gas Pumping Unit. C. H. Johnson and A. L. Sparrow. How tests were made of a producer-gas power plant and the results obtained. 1700 w. Engr, U S A—Dec. 2, 1907. No. 88687 C.
The Test of Irrigation Pumping Plants. A report of field work in California to determine the actual cost of pumping water for irrigation by various methods and under various conditions. 3500 w. Eng News—Dec. 19, 1907. No. 89092.
Greater Economy in Small Pumping Plants. H. G. H. Tarr. Considers the coal consumption of steam pumps in relation to water-works plants, with suggestions for substituting the gas engine and producer gas. Discussion. Ills. 4500 w. Pro Am W-Wks Assn—1907. No. 88736 N.

Turbine Governing.
See Pipe Friction, under Hydraulic Machinery.

Water Wheels.
Tests of a 12-Inch Doble Water-Wheel. A. L. Westcott. Describes tests made by students of the University of Missouri. 700 w. Penn—Dec., 1907. No. 88995 C.

MACHINE ELEMENTS AND DESIGN.

Bearings.
Improvements and Novelties in Machine Parts (Maschinenteile. Fortschritte

und Neuerungen). C. Volk. Illustrates and describes new types of bearings and couplings. 2200 w. Zeitschr d Ver Deutscher Ing—Nov. 9, 1907. No. 88869 D.

Cams.
Laying Out Automatic Screw-Machine Cams. F. E. Anthony. Illustrates and describes how the cams are laid out for operating the turret and cross slides. 3000 w. Am Mach—Vol. 30. No. 49. No. 88779.

Couplings.
See Bearings, under Machine Elements and Design.

Fly Wheels.
On Determining Size of Fly-Wheels for Motor-Driven Planers. W. Owen. Explains method of ascertaining the correct dimensions. 1000 w. Mach, N Y—Dec., 1907. No. 88693 C.

Gears.
Finding Change Gears by Prime-Factor Tables. John Parker. Gives a factor table from 1 to 10000, showing how change gears are calculated. 1500 w. Am Mach—Vol. 30. No. 49. No. 88777.
Tumbler Gear Design. John Edgar. Considers the faults of these gears, their design, etc. 2500 w. Mach, N. Y.—Dec., 1907. No. 88691 C.
See also same title, under Automobiles.

Rings.
Strength of Rings. An investigation showing the method of calculating the safe load. 1500 w. Engr, Lond—Dec. 20, 1907. No. 89263 A.

Roller Bearings.
Salient Principles of Roller Bearings. J. F. Springer. Discusses the practical differences between straight and tapered types, alinement of rollers, etc. Ills. 3000 w. Power—Dec., 1907. No. 88987 C.
Test of Bearings—Hyatt Roller vs. Babbitted. L. P. Alford. A comparative test which showed a saving in friction for the former. 1500 w. Am Mach—Vol. 30. No. 50. No. 88938.

MACHINE WORKS AND FOUNDRIES.

Boiler Making.
See Rivets, under Measurement; and Boiler Design and Boiler Repairs, under Steam Engineering.

Boring Machine.
Horizontal Turning and Boring Machine. Illustration with brief description, of a machine built in England for the Japanese Government. 500 w. Engng—Dec. 13, 1907. No. 89137 A.

Brass Founding.
The Brass Foundry. W. S. Quigley. Explains brass foundry conditions and discusses modern methods of melting and handling metal. Also discussion. 7500 w.

Pro N Y R R Club—Nov. 15, 1907. No. 89183.

Case Hardening.
See same title, under MINING AND METALLURGY, COPPER.

Castings.
Making Castings Off Broken Castings. Walter J. May. Directions for duplicating a broken casting where parts are missing. 800 w. Prac Engr—Dec. 6, 1907. No. 89034 A.

A Machine for the Production of Very Thin Castings. Illustrates and describes a new process for the casting of metals and alloys, invented by P. Bohin. 1400 w. Mech Engr—Nov. 30, 1907. No. 88970 A.

See also Phosphor-Bronze, under MATERIALS OF CONSTRUCTION.

Cost Keeping.
See Cost System, under INDUSTRIAL ECONOMY.

Cupolas.
See Foundry Fans, under MACHINE WORKS AND FOUNDRIES.

Cutting Tools.
See Tool Steels, under MATERIALS OF CONSTRUCTION.

Drilling.
Drill Heads for Closely Spaced Holes. Illustrated description of the mechanism of the Langelier Mfg. Co. to drive any number of drills very close together. 1200 w. Am Mach—Vol. 30. No. 50. No. 88937.

Drilling Machine.
A Precision Drilling and Reaming Machine. Illustrates and describes a machine for laying out and drilling work on a drill press so each piece will be an exact fit with many others. 1700 w. Am Mach—Vol. 30. No. 52. No. 89146.

Forging.
See Rolling, under MINING AND METALLURGY, IRON AND STEEL.

Foundry Costs.
See Cost System, under INDUS-TRIAL ECONOMY.

Foundry Fans.
Blast for Cupolas. E. L. Rhead. Discusses fans, blowers and blowing engines, and air furnaces. Ills. 3000 w. Mech Engr—Nov. 30, 1907. No. 88969 A.

Foundry Practice.
See Brass Founding, under MACHINE WORKS AND FOUNDRIES.

Gear-Hubbing Machine.
A New British Gear-Hubbing Machine. I. W. Chubb. Illustrated detailed description. 1000 w. Am Mach—Vol. 30. No. 52. No. 89147.

Grinding Machines.
An English Universal Grinding Machine. J. E. Storey. Illustrates and de-

scribes a tool having interesting features of construction. 1800 w. Am Mach—Vol. 30. No. 52. No. 89148.

Machine Tools.
Interesting Operations in a Large Shop. Illustrates and describes machining with special tools, at the Westinghouse works. 1500 w. Am Mach—Vol. 30. No. 49. No. 88776.

Management.
General Arrangement of Manufacturing Buildings. Oscar E. Perrigo. Third of a series of articles on shop management and cost keeping. Ills. 2000 w. Ir Trd Rev—Dec. 5, 1907. No. 88798.

The Management of Production in a Great Factory. George F. Stratton. The first of a group of articles explaining the production methods of leading manufacturing plants in the United States. The present number deals with the General Electric Co.'s shops. 2500 w. Engineering Magazine—Jan., 1908. No. 89175 B.

See also Wages, under INDUSTRIAL ECONOMY.

Master Plates.
Master Plates and How They Are Made. E. M. King. An illustrated article describing accurate results obtained by using buttons and plugs and setting the plate by means of a sensitive test indicator. 3500 w. Am Mach—Vol 30. No. 51. No. 89069.

Microscope.
The Microscope in the Tool Room. F. A. Stanley. An illustrated article showing how a great variety of operations in connection with tool work may be facilitated by the use of the microscope. 4000 w. Am Mach—Vol. 30. No. 50. No. 88934.

Milling.
Locating Holes Accurately in Special Tools. J. G. Vincent. Describes exact work on the milling machine accomplished by verniers to check the distances. 1500 w. Am Mach—Vol. 30. No. 52. No. 89149.

Milling Machines.
Milling Machines. Illustrated detailed descriptions of types. 20000 w. Engr, Lond—Nov. 22, 1907. No. 88717 A.

Molding Machines.
Machine Molding for Cast-Iron Bathtubs (Moulage Mécanique des Baignoires en Fonte par le Système P. Dupont). H. Mamy. An illustrated description of the machine and process. 2500 w. Génie Civil—Nov. 2, 1907. No. 88824 D.

Pipe Manufacture.
The Manufacture of Cast Iron Pipe. W. B. Robinson. Illustrated description of the continuous process of making pipe, at Scottdale, Pa. 2000 w. Ir Trd Rev—Dec. 5, 1907. No. 88797.

We supply copies of these articles. See page 878.

Pneumatic Tools.
Note on Pneumatic Striking Tools (Note sur les Frappeurs Pneumatiques). A. Baril. Illustrates and describes in detail several types of pneumatic chipping drills, riveters, etc., with a mathematical discussion of their operation. 9000 w. Rev de Mécan—Nov. 30, 1907. No. 88889 E +F.

Shop Heating.
See Industrial Buildings, under HEATING AND COOLING.

Shops.
The Avonside Engineering Works. Illustrated description of a new plant in Bristol, Eng. 1300 w. Engr, Lond—Dec. 20, 1907. No. 89268 A.

See also same title, under RAILWAY ENGINEERING, MOTIVE POWER AND EQUIPMENT.

Soldering.
Some Soldering Appliances. Trans. from Edmund Schlosser's Das Löten. Illustrates and describes blowpipes, furnaces and other appliances. 4000 w. Sci Am Sup—Dec. 14, 1907. No. 88951.

Spindles.
A Method of Broaching Tang Slots in Spindles. John Edgar. Illustrated description of an ingenious method of obtaining the tang slot. 1000 w. Am Mach —Vol. 30. No. 51. No. 89071.

Woodworking Machines.
The Bearings of Wood-working Machines. W. J. Blackmar. Considers both upright spindle and horizontal bearings. 1800 w. Am Mach—Vol. 30. No. 51. No. 89072.

MATERIALS OF CONSTRUCTION.

Alloy Steels.
Vanadium Steel. J. Kent Smith. States some of the attributes given steel by vanadium. General discussion. 5000 w. Pro Ry Club of Pittsburgh—Sept., 1907. No. 89161 C.

Heat Insulation.
A Laboratory Test for Heat-Insulating Materials. Charles R. Darling. Describes an apparatus for conducting reliable tests on small quantities of material. 1500 w. Engng—Dec. 6, 1907. No. 89053 A.

Phosphor-Bronze.
Some Notes on the Manufacture and Casting of Phosphor-Bronze. Erwin S. Sperry. Explains what phosphor-bronze is, its making, etc. 2500 w. Brass Wld —Dec., 1907. No. 89184.

Steel.
Hand Bending Tests. Capt. H. Riall Sankey. An account of tests made on eight types of steel experimented upon by the Alloys Research Committee, comparing the results with the tensile and other tests. 2000 w. Engng—Dec. 20, 1907. No. 89256 A.

Tool Steels.
The Theory of High Speed Tool Steel. George Auchy. Gives conclusions drawn from a comparison of authorities. 6000 w. Ir Age—Dec. 26, 1907. No. 89167.

A Comparison of the Productive Values of High-Speed Tool Steels and Carbon Steels. George Bilham. Examines cases illustrating the difference in productive values. 2500 w. Am Mach—Vol. 30. No. 50. No. 88935.

MEASUREMENT.

Extensometer.
A Method of Measuring the Extension of Test-Pieces under Tensile Stress. Wesley John Lambert. Describes an optical method of measuring the extensions. 700 w. Inst of Civ Engrs—No. 3694. No. 89080 N.

File Testing.
Testing the Cutting Quality of Files. Edward G. Herbert. Illustrates and describes a testing machine, giving some results of tests made with ordinary and special steels and correctly shaped teeth. 2000 w. Am Mach—Vol. 30. No. 51. No. 89073.

Pyrometry.
Measuring Industrial Temperatures. Thomas T. Read. Illustrates and describes types of pyrometers. 3500 w. Min & Sci Pr—Dec. 7, 1907. No. 89005.

Rivets.
An Investigation of the Influence of Some of the Commoner Defects in Riveted Work Upon the Strength of Joints. J. C. Black. Reports tests made to demonstrate the effect of certain features of riveted work commonly considered defects. 3000 w. Cal Jour of Tech—Nov., 1907. No. 89189.

Speed Indicators.
Some Speed Indicators (Quelques Indicateurs de Vitesse). A large number of types are described and their mechanism illustrated in detail. 7500 w. Rev de Mécan—Nov. 30, 1907. No. 88891 E + F.

POWER AND TRANSMISSION.

Belts.
The Proper Care of Belts in the Shop. William H. Taylor. Suggestions with illustrated description of appliances to insure equal tension. 2500 w. Am Mach— Vol. 30. No. 49. No. 88778.

Compressed Air.
See same title, under MINING AND METALLURGY, MINING.

Electric Driving.
Electric Drive in Iron and Steel Mills. W. Edgar Reed. Discusses methods of equalizing loads. 2000 w. Elec Jour— Dec., 1907. No. 89187.

Characteristics of Electric Motors. Edwin H. Anderson. How the work to be

We supply copies of these articles. See page 878.

done determines a motor's continuous and intermittent capacity and its design and rating. 2500 w. Am Mach—Vol. 30. No. 50. No. 88936.

Electrical Driving in a Cloth Factory. Details concerning a practical comparison between group driving and individual machine driving carried out in a German cloth factory. 900 w. Elec Rev, Lond—Nov. 29, 1907. No. 88975 A.

The Electrical Equipment of the Workshops of the Buenos Ayres Western Railway Company at Liniers. Illustrated description of an installation for electric driving on the three-phase system at 440 volts. 5800 w. Elec Engr, Lond—Dec. 13, 1907. No. 89129 A.

Gas Power.
See Gas Engines and Gas Producers, under COMBUSTION MOTORS.

Lubrication.
Experiments with Graphite Lubrication. C. H. Benjamin. Gives results of tests showing that graphite improves the lubricating qualities of most oils. 2200 w. Am Mach—Vol. 30. No. 51. No. 89070.

Friction and Lubrication. Dr. J. T. Nicholson. Aims to explain the phenomena of the resistance offered to the relative motion of lubricated surfaces, in particular of journals and bearings. Describes modern methods of automatic and forced lubrication. 3000 w. Mech Engr—Nov. 30, 1907. Serial. 1st part. No. 88971 A.

Mechanical Plants.
New Mechanical Equipment of the Enlarged Tribune Building, New York. Illustrated description of the new layout for this 20-story office building. 4500 w. Eng Rec—Nov. 30, 1907. No. 88660.

Mechanical Plant of the Brooklyn Institute Building. Outlines the large extensions recently completed and gives an illustrated description of the power plant. 4500 w. Eng Rec—Dec. 14, 1907. Serial. 1st part. No. 89009.

The Power Plant of the Commercial National Bank Building. Illustrated description of a Chicago building plant with fire-tube boilers, vertical compound generating units and Corliss pumping engines. 2500 w. Engr, U S A—Dec. 2, 1907. No. 88685 C.

See also Newspaper Plant, under POWER AND TRANSMISSION.

Newspaper Plant.
The Mechanical Plant of the Boston Herald. Howard S. Knowlton. Illustrated description of the gas-engine driven generating station and electric driving of machines. 2500 w. Eng Rec—Nov. 30, 1907. No. 88658.

STEAM ENGINEERING.

Boiler Design.
Design of Riveted Joints for High-Pressure Boilers. Vernon Smith. Discusses the best and cheapest type of joint for any boiler. Ills. 1800 w. Prac Engr —Dec. 13, 1907. No. 89128 A.

Boiler Explosions.
The Greenwich Boiler Explosion. A report of evidence brought forward in a Board of Trade inquiry. 2000 w. Engr, Lond—Dec. 20, 1907. No. 89266 A.

Boiler Management.
Decree Regulating the Management of Steam Boilers (Décret Régulating l'Emploi des Générateurs à Vapeur). The text of the new French law. 3500 w. Serial. 1st part. Electricien—Nov. 30, 1907. No. 88821 D.

Fuel Economy in Central Stations (Les Economies de Combustible dans les Stations Centrales). J. Izart. Discusses means of attaining greatest economy in fuel consumption, as a commentary on a table showing the coal consumption per kilowatt in several large generating stations in England. 3500 w. Serial. 1st part. Electricien—Nov. 9, 1907. No. 88819 D.

See also Plant Management, under STEAM ENGINEERING.

Boiler Repairs.
Renewing Tubes in a Horizontal Tabular Boiler. J. E. Sexton. Describes method employed and the tools required. 800 w. Power—Dec., 1907. No. 88988 C.

Boilers.
Grille Water-Tube Boilers at the Bordeaux Exhibition. Illustrates and describes a Grille boiler of the new type, giving an account of the working. 2000 w. Engng—Dec. 13, 1907. No. 89135 A.

Cooling Towers.
See same title, under HEATING AND COOLING.

Engine Foundations.
Erecting Machinery Foundations. J. A. Pratt. The characteristics of materials used are discussed and the construction methods. Ills. 2500 w. Mach, N Y—Dec., 1907. No. 88694 C.

Engine Governing.
An Investigation of Indirect Steam Engine Governing (Untersuchung einer Mitelbaren Dampfmaschinenregelung). W. Gensecke. Discusses hydraulic governing, the first part of the serial from a theoretical standpoint. Ills. Serial. 1st part. 3500 w. Zeitschr d Ver Deutscher Ing—Nov. 16, 1907. No. 88871 D.

See also Hoisting Engines, under MINING AND METALLURGY, MINING.

Engines.
A Portable Engine with Superheater. Illustrated description of a single-cylinder engine developing about 25 h.p. at 200 revolutions. 1500 w. Engr, Lond—Dec. 6, 1907. No. 89054 A.

We supply copies of these articles. See page 878.

See also Pumping Engines, under HY-
DRAULIC MACHINERY; Auxiliary Plants
and Central Stations, under ELECTRI-
CAL ENGINEERING, GENERATING STA-
TIONS; and Hoisting Engines, under
MINING AND METALLURGY, MIN-
ING.

Entropy Diagram.
Application of the Entropy Diagram.
Prof. Sidney A. Reeve. Its use made
plain in computing any kind of action in
or upon steam. 4000 w. Power—Dec.,
1907. No. 88992 C.

Flue-Gas Analysis.
The Analysis of Flue and Exhaust
Gases. Arnold Hartley Gibson. Shows
how the accuracy of the analysis may be
tested, and what interpretation should be
put on the results obtained. 3000 w.
Inst of Civ Engrs—No. 3648. No.
89079 N.

Fuels.
Pulverized Coal and Its Industrial Applications. William D. Ennis. Second
instalment of two articles reviewing this
subject. The present number considers
modes of firing and construction of the
furnace. Ills. 2500 w. Engineering
Magazine—Jan., 1908. No. 89176 B.

See also Fuel, under RAILWAY EN-
GINEERING, MOTIVE POWER AND EQUIP-
MENT.

Heat Insulation.
See same title, under MATERIALS OF
CONSTRUCTION.

Mechanical Stoking.
See Locomotive Stokers, under RAIL-
WAY ENGINEERING, MOTIVE POWER
AND EQUIPMENT.

Plant Cost.
See Construction, under ELECTRI-
CAL ENGINEERING, GENERATING STA-
TIONS.

Plant Management.
Economc Considerations on the Management of Plant. William H. Patchell.
Address of the President before the Association of Engineers-in-Charge. 4500
w. Engng—Nov. 22, 1907. No. 88712 A.

Plants.
See Pumping Plants, under HYDRAULIC
MACHINERY; Mechanical Plants, under
POWER AND TRANSMISSION; and Auxiliary
Plants and Central Stations, under ELEC-
TRICAL ENGINEERING, GENERATING
STATIONS.

Separators.
Value of Separators. T. E. O'Donnell.
Explains why they are necessary for the
economical operation of a steam plant.
Ills. 2500 w. Engr, U S A—Dec. 16,
1907. No. 89061 C.

Smoke Prevention.
See same title, under CIVIL ENGI-
NEERING, MUNICIPAL.

Speed Indicators.
See same title, under MEASUREMENT.

Stacks.
See same title, under CIVIL ENGI-
NEERING, CONSTRUCTION.

Steam vs. Electricity.
See Centrifugal Pumps, under HY-
DRAULICS.

Superheating.
The Specific Heat of Superheated
Steam. Prof. Carl C. Thomas. Gives results showing both true and mean specific
heats. Describes apparatus and experiments. 7000 w. Pro Am Soc of Mech
Engrs—Dec., 1907. No. 88760 C.

See also same title, under RAILWAY
ENGINEERING, MOTIVE POWER AND
EQUIPMENT.

Thermodynamics.
A Theory of Heat Influence. John
Batey. Abstract of a paper read before
the Birmingham Assn. of Mech. Engrs.
An explanation of heat effect, with arguments. 4500 w. Prac Engr—Dec. 20,
1907. No. 89246 A.

Turbines.
Practical Side of Steam-Turbine Operation. W. J. Kennedy. An illustrated
article giving results of three years experience. 2000 w. Power—Dec., 1907.
No. 88986 C.

Steam Turbines at the A. S. R. A.
Convention. Reviews papers by Messrs.
Chilton, Kruesi, and Bibbins on the latest
turbine construction and performance.
9500 w. Power—Dec., 1907. No. 88990 C.

Design of a 400-Kilowatt Reaction Turbine. Henry F. Schmidt. Illustrates and
describes a practical method utilizing
simple mathematics. 2000 w. Engr, U S
A—Dec. 16, 1907. No. 89060 C.

The Practical Proportioning of the Reaction Steam-Turbine. Examines the empirical rules used by many builders for
determining the proportions necessary
3300 w. Engng—Dec. 13, 1907. No.
89133 A.

The Zvonicek Steam Turbine (Dampf-
turbine System J. Zvonicek). J. Zvonicek. An illustrated description. 2000 w.
Elektrotech u Maschinenbau—Nov. 24,
1907. No. 88857 D.

The Operation and Efficiency of Steam
Turbines (Fonctionnement et Rendement
des Turbines à Vapeur). M. Hanocq.
Discusses particularly the de Laval turbine. A mathematical paper. Ills. 5500
w. Bul Scien d l'Assn d Élèves d Ecoles
Spéc—June, 1907. No. 88810 D.

See also Steamships, under MARINE
AND NAVAL ENGINEERING; and
Blowing Engines, under MINING AND
METALLURGY, IRON AND STEEL.

Valves.
Advantages of the Double Disk Design

We supply copies of these articles. See page 878.

of Gate Valve. Carlisle Mason. Types of gate valves are described and their merits and defects discussed. 2500 w. Engr, U S A—Dec. 2, 1907. No. 88690 C.

TRANSPORTING AND CONVEYING.

Cableways.
Cableways. F. T. Rubridge. Illustrated detailed description of various types, discussing their advantages and disadvantages, and related information. 3500 w. Jour of Engng, Univ of Colo—No. 3. No. 89221 N.

Car Dumping.
Car Dumping Elevators (Elévateurs-Basculeurs pour Wagons). A short illustrated description of the construction and mechanism of a type built by the Société des Ateliers d'Augsbourg et de Nuremberg. 1400 w. Génie Civil—Nov. 30, 1907. No. 88828 D.

Coal Handling.
The Coal-Handling Apparatus of a Large Coke Oven Plant. Illustrated description of an installation of unusual magnitude and arrangement, at Solvay, Ill. 3500 w. Eng Rec—Dec. 28, 1907. No. 89203.
See also Wharf, under CIVIL ENGINEERING, WATERWAYS AND HARBORS; and Coke Drawing, under MINING AND METALLURGY, COAL AND COKE.

Cranes.
Design of Light Structural Jib Cranes. W. H. Butz. The methods of computing the stresses in the various members are considered. 2500 w. Mach, N Y—Dec., 1907. No. 88692 C.
See also Docks, under CIVIL ENGINEERING, WATERWAYS AND HARBORS.

Elevators.
Discussion—A High-Speed Elevator. Discussion of Charles R. Pratt's paper. Ills. 6000 w. Pro Am Soc of Mech Engrs—Dec., 1907. No. 88761 C.

Haulage.
See same title, under MINING AND METALLURGY, MINING.

Hoisting.
See Electric Hoisting, under MINING AND METALLURGY, MINING.

Ore Handling.
See same title, under MINING AND METALLURGY, MINING; and Nickel, under MINING AND METALLURGY, MINOR MINERALS.

MISCELLANY.

Aeronautics.
Santos Dumont's Latest Aeroplane. Illustrated description. 1200 w. Sci Am—Dec. 14, 1907. No. 88950.
Military Airships. J. L. Bagg. Discusses the present development of these airships. 2000 w. Yale Sci M—Dec., 1907. No. 89165 C.

Die Making.
Die-Sinking and Engraving-Machines. Illustrated description of London built machines and their operation. 1000 w. Engng—Nov. 22, 1907. No. 88711 A.

MINING AND METALLURGY

COAL AND COKE.

Accidents.
Coal Mine Accidents; Their Causes and Prevention. Clarence Hall and Walter Snelling. Extract from Bul. No. 333 (1907) of the U. S. Geol. Surv. Reviews the accidents of 1906 in the United States and their causes, discussing means of prevention. 4000 w. Min Wld—Dec. 28, 1907. No. 89234.

Belgium.
See Electric Plants, under MINING.

Briquettes.
See Russia, under COAL AND COKE; and Gas Plant Testing, under MECHANICAL ENGINEERING, COMBUSTION MOTORS.

Coke Drawing.
The Marmac Coke Drawing and Loading Machine. Illustrated description of an underdrawing machine, consisting of an extractor and conveyor. 1800 w. Ir Age—Dec. 26, 1907. No. 89166.

Coke Ovens.
The Koppers Coke Oven. Illustrated description of a by-product system introduced from Europe. 2500 w. Ir Age—Dec. 12, 1907. No. 88918.

Coking.
See also Colorado, under COAL AND COKE.

Coking By-Products.
Dr. Ostwald's Process for the Production of Nitric Acid and Nitrate of Ammonia from Ammoniacal Liquor, as Applied to the Gas and Coke-Oven Industries. F. D. Marshall. An introductory review of other methods, and a description of the process named. Ills. 3300 w. Ir & Coal Trds Rev—Dec. 20, 1907. No. 89269 A.

Colorado.
Coal Mining and Coke Making in the Trinidad, Colorado, District. Describes this district and its development. Ills. 4800 w. Eng Rec—Dec. 14, 1907. No. 89010.

We supply copies of these articles. See page 878.

Eastern United States.

Were the Appalachian and Eastern Interior Coal Fields Ever Connected? George H. Ashley. A study of this problem with statement of conclusions. 3000 w. Ec-Geol—Oct., 1907. No. 89217 D.

Explosion.

Disaster at Monongah Coal Mines Nos. 6 and 8. Floyd W. Parsons. An illustrated account of an explosion in W. Va., probably caused by a loaded trip of cars breaking away and severing a trolley wire. 1800 w. Eng & Min Jour—Dec. 14, 1907. No. 89017.

Handling.

See Coke Drawing, under COAL AND COKE; and Coal Handling, under MECHANICAL ENGINEERING, TRANSPORTING AND CONVEYING.

Japan.

The Coal Mines of Kyushu, Japan. Report by W. B. Cunningham, student interpreter of the British Embassy at Tokio. Map. 2000 w. Min Jour—Nov. 23, 1907. Serial. 1st part. No. 88706 A.

Michigan.

Mining the Coal Measures of Michigan. Lee Fraser. Describes the deposits, which are worked chiefly by pick and shovel. Ills. 900 w. Eng & Min Jour—Nov. 30, 1907. No. 88655.

Mine Dust-Removal.

A Coal Mine Vacuum Cleaner Plant. Illustrated description. 1200 w. Ir & Coal Trds Rev—Nov. 20, 1907. No. 88966 A.

Montana.

The Operation of Coal Mines in Montana. Floyd W. Parsons. Illustrates and describes the coal operations, methods of preparation, mine equipment, etc. 2800 w. Eng & Min Jour—Dec. 7, 1907. No. 88915.

Peat.

The Utilization of Peat for Heating and the Direct Production of Power. Abstract of an address by Prof. Adolf Frank to the Society for the Development of German Moors, with remarks by Dr. N. Caro. 2000 w. Sci Am Sup—Dec. 7, 1907. No. 88792.

The Production of Power from Peat. T. Tomlinson. Opening address before the Dublin Loc. Sec. of the Inst. of Elec. Engrs. Considers a scheme of utilization in the gas producer with recovery of by-products. 2300 w. Elec Engr, Lond—Dec. 6, 1907. Serial. 1st part. No. 89039 A.

Pennsylvania.

The Anthracite Mines at Alden, Penn. M. S. Hachita. Illustrated description of plant, methods, average output, haulage cost, etc. 2500 w. Eng & Min Jour—Dec. 28, 1907. No. 89215.

Rescue Appliances.

Rescue Methods in the Wurm and Inde District (Das Rettungswesen im Wurm-und Inde-Revier). Prof. Stegemann. Describes appliances and methods in a central district organization. Ills. 2000 w. Glückauf—Nov. 16, 1907. No. 88839 D.

Russia.

Tests of Russian Coal and Briquettes. Gives results of tests made by the Russian Marine Department. 1000 w. Col Guard—Nov. 22, 1907. No. 88709 A.

Safety Lamps.

The Dangers of Fulminating Caps for Lighting Safety Lamps. G. Chesneau, in *Ann. des Mines.* Detailed account of experiments with fulmnating caps, and also with white phosphorus igniters. 2700 w. Col Guard—Dec. 6, 1907. No. 89048 A.

Weathering.

The Weathering of Coal. S. W. Parr, and N. D. Hamilton. Reports results of a series of tests, stating conclusions. 1300 w. Ec-Geol—Oct., 1907. No. 89219 D.

Wyoming.

Mining Anthracite Coal in the Wyoming Valley. M. S. Hachita. Describes methods used. Steel beams support the roof. Ills. 1800 w. Eng & Min Jour—Dec. 21, 1907. No. 89101.

COPPER.

Alloys.

Alloys of Cobalt and Copper (Sur les Alliages de Cobalt et de Cuivre). N. Konstantinow. A note on some recent researches. Ills. 2000 w. Rev. de Métal—Nov., 1907. No. 88815 E + F.

Blowing Engines.

See same title, under IRON AND STEEL.

British Columbia.

Mining in Western Canada. H. Mortimer Lamb. Describes conditions in the Yukon and coast districts, and other mining regions of British Columbia. Ills. 3500 w. Mines & Min—Dec., 1907. No. 88680 C.

Mining in the Rossland District, British Columbia. Ralph Stokes. The first of a series of illustrated articles describing the mines of this district, and the yield of copper, gold and silver. 2200 w. Min Wld—Dec. 21, 1907. Serial. 1st part. No. 89115.

California.

Recent Developments at Furnace Creek Copper Mines. Francis C. Nicholas. An account of present conditions at these mines in the Greenwater district of California. Ills. 1800 w. Min Wld—Dec. 21, 1907. No. 89117.

Case Hardening.

Face-Hardening Copper and Bronze. Walter J. May. Describes methods used by the writer. 700 w. Prac Engr—Nov. 22, 1907. No. 88699 A.

We supply copies of these articles. See page 878.

Chile.
The Braden Copper Mines in Chile.
William Braden. Illustrated description
of two mines in the Andes, the geological
occurrence, system of mining, etc. 2500
w. Eng & Min Jour—Dec. 7, 1907. No.
88913.

The Paderosa Copper Mine of Calla-
huasi, Tarapacá, Chile. Robert Hawx-
hurst. Abstract translation from *El Na-
cional.* Descriptive. 1200 w. Min Jour—
Nov. 23, 1907. No. 88707 A.

Lake Superior.
Mines of the Lake Superior Copper Region. Dwight E. Woodbridge. Describes
geologic and mining features, mechanical
innovations, and the present status of operations. 2500 w. Min Sci—Dec. 12, 1907.
No. 89057.

North Carolina.
The Gold Hill Copper Mine, and Its
Development. Francis C. Nicholas. An
illustrated account of this gold-copper re-
gion in North Carolina. 1200 w. Min
Wld—Dec. 7, 1907. No. 88916.

Shetland.
See same title, under IRON AND STEEL.

GOLD AND SILVER.

Assay Furnaces.
Construction and Manipulation of a
Gasoline Assay-Furnace. Wilton E. Dar-
row. Gives drawing and description of
a home-made furnace, with an account of
the making of a control assay on sulphides. 1800 w. Min & Sci Pr—Dec. 14,
1907. No. 89066.

British Columbia.
See same title, under LEAD AND ZINC.

Cyanidation.
Cyanidation with the Brown Vat.
Francisco Narvaez. Diagram and description. 900 w. Min & Sci Pr—Nov. 30,
1907. No. 88768.

Cyanide Filter.
The Burt Rapid Cyanide Filter. E.
Burt. Illustrated description of a plant
and its operation. 1500 w. Min & Sci
Pr—Dec. 7, 1907. No. 89006.

Cyanide Tank.
The B. and M. Circulating-tank. F. C.
Brown. General description of apparatus,
method of working, etc., with diagrams.
3000 w. N Z Mines Rec—Oct. 16, 1907.
No. 89157 B.

Dredging.
Modern Gold Dredging Practice and
Equipment. Horace J. Clark. An illustrated discussion of the dredging of
shallow bars and beds of rivers as a
commercial possibility. 1000 w. Min
Wld—Nov. 30, 1907. No. 88665.

Goldfield, Nev.
The Association of Alunite with Gold
in the Goldfield District, Nevada. Freder-

ick Leslie Ransome. Describes the close
association of gold and alunite, giving
a microscopic study of the rocks and discussing the origin of the deposits. Ills.
9000 w. Ec-Geol—Oct., 1907. No. 89-
218 D.

See also same title, under CIVIL EN-
GINEERING, WATER SUPPLY.

Gold Production.
The Increased Gold Production and Its
Effect Upon the Cost of Living. F. Lynwood Garrison. An analysis of the present production and its economic effects.
2000 w. Jour Fr Inst—Dec., 1907. No.
88922 D.

Gold Refining.
Gold Refining. Donald Clark. Deals
with the processes for producing refined
metal. 2500 w. Aust Min Stand—Oct.
30, 1907. Serial. 1st part. No. 89028 B.

History.
The History of Gold and Silver. James
W. Malcolmson. A summary of the ancient and modern uses of the precious
metals, the sources of supply, and the effect of supply on value. 3500 w. Eng
& Min Jour—Nov. 30, 1907. No. 88654.

Mexico.
The Mines of Balanos, Old Mexico.
Ernest E. Behr. History and description
of these old silver mines and their pro-
duction. 2500 w. Cal Jour of Tech—
Nov., 1907. No. 89188.

Topographical and Other Notes on the
Choix-Guadalupe y Calvo Mining Districts. A. W. Warwick. Map and description of these districts in Mexico.
3000 w. Min & Sci Pr—Nov. 30, 1907.
No. 88767.

Nevada.
Mining Developments in Nevada. A.
Selwyn-Brown. An estimate of conditions and prospects of gold production.
3800 w. Engineering Magazine—Jan.,
1908. No. 89182 B.

See also Goldfield, Nev., under GOLD
AND SILVER; Nevada, under LEAD AND
ZINC; and Goldfield, Nev., under CIVIL
ENGINEERING, WATER SUPPLY.

Placers.
See Drilling, under MINING.

Rand.
See Ore Handling, under MINING.

Roasting.
See Telluride Ores, under GOLD AND
SILVER.

Telluride Ores.
The Roasting of Telluride Ores. R. L.
Mack and G. H. Scibibd, with an introduction by T. T. Read. States the difficulties presented in the treatment of
these gold-bearing ores, and describes experiments in roasting. 4000 w. Min &
Sci Pr—Dec. 14, 1907. Serial. 1st part.
No. 89067.

IRON AND STEEL.

Assaying.

Assay of Iron. Symbol Fe. Atomic Weight 56. Evans W. Buskett. Gives the permanganate and the bichromate methods. 1200 w. Mines & Min—Dec., 1907. No. 88681 C.

Metals and Alloys. Abstract translation from *Rassegna Mineraria.* Describes the chief methods of analysis of alloys and ferro-alloys. 3300 w. Min Jour—Nov. 23, 1907. No. 88708 A.

See also Nickel Determination, under MINOR MINERALS.

Blast-Furnace Gas.

Blast Furnace Gas Cleaning Plants on the Zschocke System (Installations pour l'Epuration des Gaz de Hauts Fourneaux, Système Zschocke). M. Wolf. Illustrates and describes this system of gas cleaning and several of the more recent installations. 3500 w. Génie Civil—Nov. 16, 1907. No. 88825 D.

Blowing Engines.

A 750 Horse-Power Brown-Boveri-Rateau Turbo-Fan (Turbogebläse, Bauart Brown-Boveri-Rateau, von 750 PS). K. Rummel. Illustrated description of the design, giving results of operation, of this steam-turbine driven fan. 5000 w. Zeitschr d Ver Deutscher Ing—Nov. 23, 1907. No. 88873 D.

Electro-Metallurgy.

The Electro-Thermic Production of Iron and Steel. Joseph W. Richards. A review of the progress made in the electro-thermic metallurgy of iron, describing furnaces and methods, etc. Ills. 3000 w. Jour Fr Inst—Dec., 1907. Serial. 1st part. No. 88924 D.

The Electro-Metallurgy of Iron and Steel (L'Electrométallurgie du Fer et de l'Acier). A general review of methods, processes, products, etc. Serial. 1st part. 4000 w. Revue d'Electrochim et d'Electrométal—Oct., 1907. No. 88887 F.

The Desulphurization of Iron in the Kjellin Induction Furnace (Beitrag zur Entschweflung des Eisens im Kjellinschen Induktionsofen). A. Schmid. Gives the results of tests. 2000 w. Stahl u Eisen —Nov. 6, 1907. No. 88831 D.

The Röchling-Roderhauser Electric Induction Furnace (Der elektrische Induktionsofen nach dem System Röchling-Rodenhausen). H. Wedding. Illustrated detailed description of a new furnace at the Röchling steel works at Völklingen, Germany, with details of its operation. 4000 w. Stahl u Eisen—Nov. 6, 1907. No. 88830 D.

See also Steel Making, under IRON AND STEEL.

Forging.

See Rolling, under IRON AND STEEL.

History.

The Iron Making of Antiquity (Das Eisenhütten-wesen im Altertum). F. Freise. A discussion of the raw materials, smelting arrangements, blowing devices, processes and results in ancient iron smelting. Ills. 3500 w. Serial. 1st part. Stahl u Eisen—Nov. 6, 1907. No. 88832 D.

Rolling.

Forging and Rolling Mild Steel. Suggestions for successful treatment. 1200 w. Mech Engr—Nov. 23, 1907. No. 88700 A.

Rolling Mills.

Cartridge-Metal Rolling Mill. Illustrated description. 600 w. Engng—Dec. 20, 1907. No. 89260 A.

Segregation.

A Further Study of Segregation in Ingots. Henry M. Howe. Discusses the influence of ingot size; and the rate of cooling, in the degree of segregation in steel ingots. Also other questions bearing on the subject. 5000 w. Eng & Min Jour—Nov. 30, 1907. No. 88653.

Shetland.

Iron and Copper Mining in Shetland. Robert W. Dron. Read before the Geol. Soc. of Glasgow. An account of the mines and their limited development. 1500 w. Ir & Coal Trds Rev—Nov. 22, 1907. No. 88719 A.

Steel Making.

Qualitative Work in the Production of Steel and in Electric Smelting (Qualitative Arbeit in der Stahlerzeugung und elektrisches Schmelzverfahren). O. Thallner. A criticism of the methods of steel production with reference to their economy and especially considering electric smelting. 6300 w. Serial. 1st part. Stahl u Eisen—Nov. 20, 1907. No. 88834 D.

Steel Works.

The Friedrich-Alfred Iron and Steel Works, Rheinhausen. Illustrated detailed description of this plant of the Krupp firm. 3500 w. Ir & Coal Trds Rev—Nov. 29, 1907. No. 88967 A.

The Witkowitz Company, Moravia, Austria-Hungary. G. B. Waterhouse. History and illustrated description of these iron and steel works. 2300 w. Ir Age —Dec. 5, 1907. No. 88780.

Recent Austrian Steel Works (Neues in Oesterreichischen Eisenhüttenwerken). Theodor Naske. The first part of the serial is devoted to the Witkowitz works. Ills. 4000 w. Serial. 1st part. Stahl u Eisen—Nov. 13, 1907. No. 88833 D.

See also Electric Driving, under MECHANICAL ENGINEERING, POWER AND TRANSMISSION.

Wyoming.

The Iron Ores and System of Mining at Sunrise Mine, Wyoming. B. W. Val-

We supply copies of these articles. See page 878.

lat. From *Pro. Colo. Sci Soc.* Describes the deposits and their development. 2500 w. Min Rept—Nov. 28, 1907. No. 88663.

LEAD AND ZINC.

British Columbia.
The Duncan Section of the Lardeau Country, North Kootenay. E. Jacobs. Illustrated article describing the topography, geology, physiography, etc. 6000 w. B C Min Rec—Oct., 1907. No. 88676 B.

The St. Eugene Silver-Lead Mine, British Columbia. Ralph Stokes. Describes the deposits and methods of mining and milling. 2500 w. Min Wld—Nov. 30, 1907. No. 88667.

Gangue.
Gangue and Associated Minerals of Lead and Zinc. Otto Ruhl. Special reference to the deposits of the Missouri-Kansas fields. 1600 w. Min Sci—Dec. 19, 1907. No. 89112.

Joplin District.
Costs of Mining Operations in Joplin District. Presented at meeting of Am. Min. Cong. Gives estimates of costs with explanatory notes. 1200 w. Min Rept—Dec. 5, 1907. No. 88764.

Mining Sheet Ground in the Joplin District. Doss Brittain. Describes the drifting and stoping, breaking the ore, handling, etc. 1400 w. Eng & Min Jour —Dec. 14, 1907. No. 89016.

The Relation of Ores to Mining in the Joplin District. Otto Ruhl. Outlines the conditions governing classification. 2000 w. Min Rept—Nov. 28, 1907. No. 88664.

Lead Smelting.
Lead Smelting in Utah. Robert B. Brinsmade. A description of the methods used at the plants at Bingham Junction and at Murray. Ills. 5000 w. Mines & Min—Dec., 1907. No. 88678 C.

Leadville Geology.
Structural Geology at Leadville. F. L. Barker. A study of the formations and the structural problems. Ills. 3000 w. Mines & Min—Dec., 1907. No. 88679 C.

Milling.
See Zinc Milling, under ORE DRESSING AND CONCENTRATION.

Missouri.
Lead and Zinc Deposits of the Ozark Region. E. R. Buckley. Extract from the report of Directors of Missouri Bureau of Geology and Mines. 2200 w. Min Wld—Nov. 30, 1907. No. 88668.
See also Gangue and Joplin District, under LEAD AND ZINC.

Nevada.
The Silver-Lead Mines of Eureka, Nevada. Walter Renton Ingalls. Illustrated account of an old mining district now being reopened. 5000 w. Eng & Min Jour —Dec. 7, 1907. No. 88912.

MINOR MINERALS.

Aluminium.
The Actual State of the Aluminium Industry (Etat Actuel de l'Industrie de l'Aluminium). Robert Pitaval. The first part of the serial gives a historical review of the industry and begins a description of the process of manufacture. Ills. 5000 w. Serial 1st part. Génie Civil—Nov. 16, 1907. No. 88826 D.

Antimony.
Antimony: Its Use and Treatment. F. T. Havard. Classifies the different kinds of ores, and describes the various processes used for the extraction of the metal. 1500 w. N Z Mines Rec—Oct. 16, 1907. No. 89158 B.

Cadmium.
Metallic Cadmium (Ueber das Metallische Cadmium). Paul Speier. Production, uses, statistics, prices, etc. 2000 w. Oest Zeitschr f Berg u Hüttenwesen— Nov. 30, 1907. No. 88836 D.

Cobalt.
See Alloys, under COPPER.

Diamonds.
Diamond Mining. Henry Leffmann. Abstract of a lecture giving a review of methods of mining and the cutting of the stones. 2000 w. Jour Fr Inst—Dec., 1907. No. 88921 D.

Emery.
Turkish Emery Ore. E. L. Harris. Brief description of the deposits. 600 w. Min Wld—Dec. 28, 1907. No. 89235.

Gypsum.
Extent and Importance of Oklahoma Gypsum Deposits. Charles N. Gould. Describes the geology, mode of occurrence and quality of the deposits, discussing present and possible development. 1500 w. Min Sci—Dec. 12, 1907. Serial. 1st part. No. 89058.
See also Plaster of Paris, under MINOR MINERALS.

Monazite.
Mining and Cleaning Monazite. Douglas B. Sterrett. Describes methods of working the gravel deposits, the cleaning by electrical machinery, etc. 1400 w. Min Wld—Dec. 28, 1907. No. 89233.

Natural Gas.
See Anticlinal Structure, under MISCELLANY.

Nickel.
The Development of the Nickel Deposits of New Caledonia (Die Aufschliesung der Nickelerzlagerstätten in Neukaledonien). G. Dietrich. Geology, mining methods, transportation, etc. Ills. Serial. 1st part. 8200 w. Zeitschr d Ver Deutscher Ing—Nov. 16, 1907. No. 88-870 D.

Nickel Determination.
A New Method for the Determination

of Nickel and Its Application to the Separation of Nickel from Iron, Aluminium, Zinc and Cobalt (Eine Neue Bestimmungs-methode des Nickels und ihre Anwendung in der Analyse zur Trennung des Nickels vom Eisen, Aluminium, Zink und Kobalt). Hermann Grossmann and Bernhard Schück. 2000 w. Oest Zeitschr f Berg u Hüttenwesen—Nov. 23, 1907. No. 88835 D.

Oil.
See Anticlinal Structure, under MISCELLANY.

Plaster of Paris.
Investigations on Plaster of Paris (Essais sur le Plâtre). E. Leduc and Maurice Pellet. A description of tests on the influence of various factors, fineness of material, water content, etc., on the quality of plaster of Paris, with tables of results. Ills. 7000 w. Bul du Lab d' Essais—Nov. 11, 1907. No. 88888 F.

Silica Sand.
The Silica Sand Industry. Beverley S. Randolph. An account of the sources of supply, preparation, etc. 1500 w. Eng & Min Jour—Dec. 28, 1907. No. 89214.

Sulphur.
An Improved Method of Mining Sulphur. Herman Frasch. Illustrated description of an invention relating to the obtaining of sulphur from an underground deposit by melting the sulphur and removing it in the melted condition. 3000 w. Min Wld—Dec. 14, 1907. No. 89004.

MINING.

Cableways.
See same title, under MECHANICAL ENGINEERING, TRANSPORTING AND CONVEYING.

Compressed Air.
Applications of Compressed Air to Mining. Jos. H. Hart. On the importance of a good and efficient air compression plant, and some of the problems connected with its use. 2000 w. Min Wld—Dec. 21, 1907. No. 89116.

Drilling.
Testing Placer Ground with the Keystone Drill. John Power Hutchins. Illustrates and describes methods explaining some of the difficulties. 5000 w. Eng & Min Jour—Dec. 21, 1907. No. 89099.

A-B-C of Steam Percussion Drill Practice. John Power Hutchins. Suggestions for unloading, moving and setting up Keystone drills used in testing placer ground in California. Ills. 3500 w. Eng & Min Jour—Dec. 14, 1907. No. 89015.

Notes on Churn Drill Placer Prospecting. John Power Hutchins. An illustrated discussion of methods of handling core material, recording results, care of tools and equipment, etc. 4500 w. Eng & Min Jour—Dec. 28, 1907. No. 89212.

Education.
See same title, under INDUSTRIAL ECONOMY.

Electric Hoisting.
Electric Hoisting at Grangesberg, Sweden. J. B. Van Brussell. Illustrated description of plant which raises 1200 tons in eight hours. 1600 w. Eng & Min Jour—Dec. 21, 1907. No. 89100.

Electric Plants.
Some Electric Mining Operations in Belgium. Frank C. Perkins. Illustrates and describes electric plants used in coal mines. 7000 w. Min Wld—Nov. 30, 1907. No. 88666.

Explosives.
The Manufacture of High Explosives. Illustrates and describes the methods of work at the Nobel factory in Scotland. 3000 w. Sci Am—Dec. 28, 1907. No. 89153.

Haulage.
Underground Haulage. John Bell. Read before the British Soc. of Min. Students. Considers the main and tail rope, and the endless rope systems. 5000 w. Can Min Jour—Dec. 1, 1907. Serial. 1st part. No. 88765.

Hoisting Engines.
Stopping and Governing Devices on Steam Hoisting Engines (Stau- und Regelvorrichtungen bei Dampffördermaschinen). Herr Grunewald. A general discussion of the subject is contained in the first part of the serial. Ills. Serial. 1st part. 4500 w. Zeitschr d Ver Deutscher Ing—Nov. 2, 1907. No. 88867.

See also Electric Hoisting, under MINING.

Mine Dam.
An Emergency Water Supply for a Coal Breaker. John H. Haertter. Describes a successful experiment to avert suspension for need of water, by constructing a mine dam. 1500 w. Eng & Min Jour—Dec. 14, 1907. No. 89018.

Mine Temperature.
The Influence of High Temperatures on the Output of the Workers in Saltpeter Mines (Ueber den Einfluss höherer Temperaturen in Kalisalzgruben auf die Leistung der Arbeiter). Herr Barnitzke. A record of extensive observations. Ills. 5500 w. Glückauf—Nov. 9, 1907. No. 88838 D.

Ore Handling.
Labor-Saving Appliances at the Mines of the New Kleinfontein Co., Transvaal. Edward J. Way. Illustrated description of appliances installed, and a report of the cost reduction effected. 7000 w. Inst of Mech Engrs—Nov. 15, 1907. No. 88-710 N.

See also Nickel, under MINOR MINERALS.

We supply copies of these articles. See page 878.

Quarrying.
Granite Quarrying. T. Nelson Dale. Brief discussion of methods used, utilization of waste, etc. 1400 w. Min Wld— Dec. 28, 1907. No. 89232.

Quarry Plant.
A Modern Quarry Plant. Illustrated description of a plant in England. 1700 w. Engr, Lond—Dec. 13, 1907. No. 89139 A.

• **Stoping.**
Methods of Stoping at Cripple Creek. G. E. Walcott. Describes the methods and gives a comparison of costs. 2500 w. Eng & Min Jour—Nov. 30, 1907. No. 88652.

Stowing.
Waste Stowing Plant at the Deutscher Kaiser Mine at Hamborn (Spülversatzanlagen auf Zeche Deutscher Kaiser bei Hamborn). Illustrated description of a plant for stowing the mine with sand from blast-furnace slag, with details of operation. 3200 w. Glückauf—Nov. 2, 1907. No. 88837 D.

Timbering.
Substitution of Steel for Timber in Mines. R. B. Woodworth. Abstract of a report made to the Carnegie Steel Co. An investigation of the timber conditions in the anthracite region. Ills. 4500 w. Mines & Min—Dec., 1907. No. 88677 C.
See also Wyoming, under COAL AND COKE.

Tunnels.
See same title, under CIVIL ENGINEERING, CONSTRUCTION.

Ventilation.
Notes on the Ventilation of Deep Mines (Einige Betrachtungen über die Erzeugung des Wetterstromes in tiefen Gruben). K. Kegel. Discusses theoretically the influence of changes of temperature, humidity, etc. 2000 w. Glückauf— Nov. 16, 1907. No. 88840 D.

ORE DRESSING AND CONCENTRATION.

Concentration.
See Monazite, under MINOR MINERALS.

Copper.
The Utah Copper Company's Garfield Plant. Illustrated description of a 6000-ton capacity concentrating mill and accessory power house. 1200 w. Min Sci— Dec. 12, 1907. No. 89056.

Elmore Process.
The Elmore Vacuum Process at Dolcoath. Edward Walker. An illustrated account of the successful treatment of the complex tin-copper tungsten ores of Cornwall. 1500 w. Eng & Min Jour—Dec. 14, 1907. No. 89014.

Magnetic Separation.
See Monazite, under MINOR MINERALS.

Zinc Milling.
Milling Practice at Gundling No. 5,

Joplin District. Doss Brittain. An illustrated description of method of treating ore which is mainly zinc sulphide. 2200 w. Min Wld—Dec. 14, 1907. No. 89003.

MISCELLANY.

Alloys.
The Actual State of the Theories on the Equilibrium of the Iron-Carbon System (Etat Actuel des Théories sur l'Equilibre du Système Fer-Carbone). M. A. Portevin. A comparative review of the theories, drawing conclusions. Ills. 5500 w. Rev. de Métal—Nov., 1907. No. 88816 E + F.

Anticlinal Structure.
Relation of Anticlinal Structures to Gas, Oil and Water. Arthur Lakes. An illustrated study. 1000 w. Min Sci—Dec. 19, 1907. No. 89113.

Faulting.
Faulting in the Red Cloud Mine. H. W. Turner. Describes conditions found in these lead-silver mines in Idaho. 1500 w. Min & Sci Pr—Dec. 14, 1907. No. 89065.

Klondike Geology.
The Geology of the Klondike. Dr. Willis Eugene Everette. Describes geological investigations of the rocks of the Yukon River. 4000 w. Sci Am Sup— Dec. 28, 1907. Serial, 1st part. No. 89156.

New South Wales.
Minerals and Metals in New South Wales. F. Danvers Power. Gives a list of minerals found and the uses to which they may be applied. 6500 w. Aust Min Stand—Oct. 30, and Nov. 6, 1907. Serial. 2 parts. No. 89027 each B.

Norway.
Mining in Norway. Joseph Ralph. An account of the mining outlook, labor conditions, etc., in the present number. 3000 w. Min Jour—Nov. 30, 1907. Serial, 1st part. No. 88959 A.

Oklahoma.
Mineral Resources of Oklahoma. Charles N. Gould. Information concerning deposits of coal, oil, gas, rock asphalt, salt, gypsum, etc. 2000 w. Mfrs Rec— Dec. 5, 1907. No. 88726.

Ore Deposits.
The Problem of the Metalliferous Veins. James Furman Kemp. From a presidential address before the New York Academy of Science. Discusses the Geological formation. 2200 w. Sci Am Sup—Dec. 28, 1907. Serial, 1st part. No. 89155.

Ore Genesis.
Genesis of Ores in the Light of Modern Theory. Horace V. Winchell. From a commencement address before the Montana Sch of Mines. Shows the importance of mass, time, temperature, climate, etc., as factors in enrichment. 3500 w. Eng & Min Jour—Dec. 7, 1907. No. 88914.

RAILWAY ENGINEERING

CONDUCTING TRANSPORTATION.

Accidents.

The Prevention of Railway Accidents. Editorial, discussing the need of the extended adoption of the automatic engine-stop system of signalling. 1500 .w. Engr, Lond—Nov. 22, 1907. No. 89270 A.

Collisions.

Fatal Railway Collisions. Description of three collisions in England, prepared from Board of Trade reports. 2000 w. Engr, Lond—Dec. 20, 1907. No. 89267 A.

Can Railroad Collisions be Reduced to a Theoretical Minimum? Harold Vinton Coes. An argument for an automatic mechanical engine-stop system. 3300 w. Engineering Magazine—Jan., 1908. No. 89181 B.

Signalling.

Block Signals in 1907. Reviews the history of the applications of signals to American railways, and the recent action by the Interstate Commerce Commission. 1600 w. Ry Age—Dec. 27, 1907. No. 89236.

See also Cab Signals, under MOTIVE POWER AND EQUIPMENT.

Trains.

The "Black Diamond" on the Lehigh. An illustrated description of this well-known train, and the country through which it passes. 1000 w. Ry & Loc Engng—Dec., 1907. No. 88682 C.

MOTIVE POWER AND EQUIPMENT.

Adhesion System.

The Hanscotte System of Adhesion Traction (Zugförderung mit mittlerer Reibschiene, Bauart Hanscotte). R. Bonnin. An illustrated description of the mechanism and its application to electric and steam railroads. 2500 w. Zeitschr d Ver Deutscher Ing—Nov. 23, 1907. No. 88874 D.

Cab Signals.

Automatic Cab-Signalling on Locomotives. J. Pigg. Illustrated detailed description of Raven's Electrical System of Cab-Signalling, with a statement of the requirements it was designed to meet. 9500 w. Inst of Elec Engrs—Oct. 17, 1907. No. 89037 N.

The Bounevialle System of Electric Signalling on Locomotives. Illustrates and describes a system that gives both visual and audible signals in the cab of the locomotive. 2000 w. Elect'n, Lond—Dec. 20, 1907. No. 89252 A.

Couplings.

The Strength of a Steam Locomotive Coupling (Die Beanspruchung der Kupplung einer Dampflokomotive). Herr Strahl. A mathematical and general discussion of the stresses to which the coupling between tender and locomotive is subjected. Ills. 6500 w. Glasers Ann—Nov. 1, 1907. No. 88859 D.

Electrification.

The Electrification of the Swedish State Railways. Gives an abstract of the report, prepared by Robert Dahlander, as the basis of a scheme for the whole country, with two exceptions. 4200 w. Engng—Dec. 13, 1907. No. 89136 A.

An Investigation of Steam Railroad Electrification, with Particular Reference to the Suburban Lines at Melbourne, Victoria. Thomas Tait. Abstract of a recent report showing the progress and results. Also editorial, map. 9000 w. R R Gaz—Dec. 13, 1907. No. 88940.

Fuel.

The Influence of Heat Value and Distribution on Railway Fuel Cost. J. G. Crawford. Deals with the economical purchase and supply of coal. Also discussion. 10500 w. Pro W Ry Club—Nov. 19, 1907. No. 89163 C.

Locomotive Chimneys.

Trials, with Chimneys and Blast-Pipes on Locomotives. E. Höhn. Brief report of trials. 1000 w. Bul Int Ry Cong—Nov., 1907. No. 89196 E.

Locomotive Repairs.

See Shops, under MOTIVE POWER AND EQUIPMENT.

Locomotives.

Four-Wheel Shunting Tender Locomotive. Brief illustrated description. 400 w. Engng—Dec. 20, 1907. No. 89257 A.

Tandem Compound Engine with Bollinckx Valve-Gear. Plate and description. 1000 w. Engng—Dec. 20, 1907. No. 89258 A.

Tank Locomotive for the Midland Railway. Illustrated detailed description of an English engine for long distance running. 800 w. R R Gaz—Nov. 29, 1907. No. 88644.

Ten-Wheel Locomotive for General Service. Illustrated description of class D-3-B locomotive for the Delaware & Hudson Co. 800 w. Am Engr & R R Jour—Dec., 1907. No. 88783 C.

Four-Cylinder de Glehn Compound Locomotive for the Paris-Orleans Railway of France. Illustrated detailed description of American built engines. 900 w. Ry Age—Nov. 29, 1907. No. 88671.

We supply copies of these articles. See page 878.

Atlantic Type Passenger Locomotive for the North British Railway Company. Illustrated detailed description of engines for heavy express service. Plate. 900 w. Engng—Dec. 6, 1907. No. 89052 A.

Locomotive for the G. I. P. Railway. Plate and description of one of the latest types of locomotives introduced into India for fast mail and passenger traffic. 400 w. Engr, Lond—Dec. 13, 1907. No. 89138 A.

Locomotives for South Manchurian Railroads. Illustrates and describes a type of which twenty consolidation locomotives for freight service have been built in America. Also two steam inspection cars. 500 w. R R Gaz—Dec. 6, 1907. No. 88903.

4-5 Coupled, Compound Freight Locomotive of the Italian State Railway (4-5 gekuppelte Verbund Güterzugloko-motive der Italienischen Stattsbahn). G. Heise. Illustrated detailed description. 1600 w. Zeitschr d Ver Deutscher Ing—Nov. 30, 1907. No. 88875 D.

The Four-Cylinder Compound Locomotive, Series C 4-5, with Four Coupled Axles, of the Gothard Railway (Die Vier-zylinder-Verbund-Lokomotive met vier gekupperten Achsen, Series C 4-5, der Gotthardbahn). Illustrated description. 2500 w. Schweiz Bau—Nov. 9, 1907. No. 88842 B.

Locomotive Stokers.

Mechanical Stoking on Locomotives and Marine Boilers. C. S. Vesey-Brown. Discusses some of the latest methods of stoking by mechanical appliances. Ills. 3000 w. Cassier's Mag—Dec., 1907. No. 88981 B.

Shops.

The Readville Locomotive Repair Shops. Illustrated description of a N Y, N H, & H R R improvement. 4000 w. Ir Age—Dec. 5, 1907. No. 88782.

The Harmon Shops for the New York Central Electric Zone. Illustrated detailed description. 2500 w. Ry Age—Dec. 13, 1907. No. 89001.

Some Engineering Features of the Parsons Shops of the Missouri, Kansas & Texas Ry. Illustrates and describes features of the locomotive and repair shops at Parsons, Kans. 4500 w. Eng Rec—Dec. 7, 1907. No. 88769.

The Beech Grove Shops of the Big Four. Reviews the conditions affecting the planning of this Indiana plant, and describes the grouping and operating arrangements. Ills. 3500 w. R R Gaz—Nov. 29, 1907. No. 88646.

Smoke Consuming.

The Smoke Consuming Question Forty-eight Years Ago. C. H. Caruthers. An illustrated account of a series of trials made on the Pennsylvania R R in 1859.

6000 w. R R Gaz—Dec. 13, 1907. No. 88942.

Steam vs. Electricity.

See Interurban, under STEAM AND ELECTRIC RAILWAYS.

Superheating.

Test of Vauclain Superheater on the Rock Island. Illustrations, with account of tests. 1800 w. Ry Age—Dec. 13, 1907. No. 89000.

Train Lighting.

The Evolution of the Lighting of Passenger Carriages on the Prussian and Hessian State Railways. Mr. Wedler. Gives a short sketch of lighting methods formerly used and describes the new system introduced. Ills. 3500 w. Bul Int Ry Cong—Nov., 1907. No. 89195 E.

Wheel Contacts.

Areas of Contact Between Wheels and Rails. George L. Fowler. Reprinted from a volume of reports made to the Schoen Steel Co. describing an extensive series of tests. Ills. 2000 w. R R Gaz—Dec. 20, 1907. No. 89077.

Wheel Friction.

Co-efficients of Friction Between Wheels and Rails. George L. Fowler. From a volume of reports made to the Schoen Steel Wheel Co. Gives results of experimental investigations of the causes of slipping. 2500 w. R R Gaz—Dec. 27, 1907. No. 89171.

Wrecking Outfits.

The Organization and Working of Wrecking Outfits. A committee report presented to the Chicago convention of the Roadmasters' and Maintenance of Way Assn. 1000 w. R R Gaz—Dec. 20, 1907. No. 89076.

NEW PROJECTS.

Congo.

The Congo Railroad. Brief account of its construction and operation. 1200 w. R R Gaz—Dec. 13, 1907. No. 88939.

PERMANENT WAY AND BUILDINGS.

Car Dumping.

See same title, under MECHANICAL ENGINEERING, TRANSPORTING AND CONVEYING.

Clearing.

Methods and Cost of Clearing and Grubbing Land. Describes work in connection with railroad construction. 1800 w. Engng-Con—Dec. 25, 1907. No. 89145.

Crossing Gates.

An Electrically-Operated Automatic Railway Gate (Una Nuova Barriera Ferroviaria Automatica a Comando Elettrico). Emilio Gerli. Illustrated description. 2000 w. Ing Ferroviaria—Nov. 16, 1907. No. 88802 D.

Crossings.

Grade Crossing Abolition at Newton

Highlands and Newton Centre, Mass. Walter C. Whitney. Illustrated description of recent work on the B & A R R. 5000 w. Eng Rec—Nov. 30, 1907. No. 88656.

Earth Slides.

Controlling Earth Slides. H. Rohwer. Reprint from Bul No 90 of the Am Ry Engng & Main of Way Assn. Describes work showing the importance of thorough drainage in maintaining a good roadbed. Ills. 2000 w. R R Gaz—Dec. 13, 1907. No. 88943.

Improvements.

Southern Pacific Improvements. Brief illustrated description of important changes to better traffic conditions. 1200 w. Eng Rec—Dec. 14, 1907. No. 89008.

Roadbed and Terminal Improvements, Baltimore & Ohio R R. Describes the recent revision of the main line between Hollofields and Davis, Md. Ills. 2500 w. Ry & Engng Rev—Dec. 28, 1907. No. 89238.

Vandalia Track Elevation and Improvement Work at Indianapolis. Map and illustrated description of serious and puzzling problems solved in work made necessary by track elevation ordinance requirements. 2000 w. R R Gaz—Dec. 27, 1907. No. 89169.

Rails.

The Sandberg Rail. An account of these high silicon rails, with report of tests. 2500 w. Tram & Ry Wld—Dec. 5, 1907. No. 89046 B.

Rail Specifications. Gives the full majority report of the American Ry Assn, with extracts from other specifications, Hollofields and Davis, Md. Ills. 2500 w. R R Gaz—Dec. 20, 1907. No. 89074.

Steel Rails. Franklin E. Abbott. Gives outline description of the process of manufacture and discusses their use from the utilitarian point of view. 11500 w. Pro Cent Ry Club—Nov. 8, 1907. No. 89162 C.

Reconstruction.

Double Tracking Through Eagle River Canyon on the Denver & Rio Grande. Illustrated description of work costing more than $100,000 a mile. 1000 w. R R Gaz—Dec. 13, 1907. No. 88941.

Shops.

See same title, under MOTIVE POWER AND EQUIPMENT.

Stations.

The New Union Station at Washington, D C. Illustrated detailed description. 3000 w. Engr, Lond—Nov. 29, 1907. No. 88964 A.

Passenger Stations and Train Yards in the United States. E. Giese and Dr. Blum. Plans and descriptions of recently designed stations and yards. 16000 w.

Bul Int Ry Cong—Nov., 1907. No. 89198 E.

Terminals.

The New Locomotive Terminal of the Chicago Junction. Illustrated description. 1000 w. R R Gaz—Dec. 20, 1907. No. 89075.

The Bush Terminal Company. Describes the work accomplished at this deep water terminal in South Brooklyn, and the plant and methods. 1800 w. R R Gaz—Dec. 27, 1907. No. 89172.

Track Construction.

Prussian Railroad Tests. William Mayner. A synopsis of Prof. Frahm's report of the official experiments on the trial track at Oranienburg. 1200 w. Sci Am Sup—Dec. 14, 1907. No. 88952.

Tunnels.

See same title, under CIVIL ENGINEERING, CONSTRUCTION.

Yards.

The New Shunting Yard at Mannheim. Plan and description. 1200 w. Bul Int Ry Cong—Nov., 1907. No. 89199 E.

TRAFFIC.

Agreement.

The Great Northern and Great Central Agreement. Discusses the proposed working agreement. 2000 w. Engr, Lond—Dec. 20, 1907. No. 89265 A.

Car Hire.

"Per Diem" Suspended in New England. An account of the controversy over the rate per diem for car hire between the principal roads in New England and the rest of the country. 2800 w. Ry Age—Dec. 13, 1907. No. 88999.

Demurrage.

Digest of Reciprocal Demurrage Laws. Compiled by the American Railway Clearing House. 2000 w. Ry Age—Nov. 29, 1907. No. 88672.

M. H. Smith on Excess of Traffic Over Facilities in Alabama. From a statement to the Committee of the Alabama Legislature on reciprocal demurrage. 1000 w. R R Gaz—Dec. 27, 1907. No. 89173.

Locomotive Pooling.

Pooling of Locomotives in General. D. R. McBain. Read at Trav Engrs' Con. Gives reasons why pooling is necessary and how to make it a success. 2000 w. Ry & Loc Engng—Dec., 1907. No. 88684 C.

Refrigeration.

Recent Investigations in the Handling of Perishable Products for Transportation. S. J. Dennis. Discusses results of experiments made by the Dept. of Agriculture to determine the bearing of different methods of handling and shipping. 3000 w. Ice & Refrig—Dec., 1907. No. 89026 C.

Wastes.

Wastes of Fuel, Power and Time in Railroad Operation. D. C. Buell. From a paper presented to the Trav Engrs Assn. Calls attention to channels through which energy is wasted. Also causes of waste of time. 2200 w. R R Gaz—Dec. 6, 1907. No. 88904.

MISCELLANY.

Africa.

Railway Progress in the Dark Continent. J. Hartley Knight. An illustrated account of recent projects and their progress. 3000 w. Engineering Magazine—Jan., 1908. No. 89180 B.

England.

Transportation in England (Die Engländer und ihr Verkehrswesen). Herr

Frahm. A general review of railway conditions and organization in Great Britain. Ills. 7500 w. Glasers Ann—Nov. 15, 1907. No. 88861 D.

Railway Problems.

Symposium on Railway Conditions and Problems. Gives responses received to a letter sent out asking for views on present conditions. 22000 w. Ry Age—Dec. 27, 1907. No. 89237.

St. Gothard.

Silver Wedding of the St. Gothard Railway. W. Berdrow. A review of the history of the road for twenty-five years. 5500 w. Bul Int Ry Cong—Nov., 1907. No. 89197 E.

Train Ferry.

See same title, under MARINE AND NAVAL ENGINEERING.

STREET AND ELECTRIC RAILWAYS

Adhesion System.

See same title, under RAILWAY ENGINEERING, MOTIVE POWER AND EQUIPMENT.

Axles.

Broken Axles. W. Park. Discusses the principal causes of axle breakage. 2000 w. Tram & Ry Wld—Dec. 5, 1907. No. 89047 B.

Brake Tests.

Electro Mechanical Brake Tests at Leeds. An illustrated account of tests made of A. W. Maley's invention. 2500 w. Elect'n, Lond—Nov. 29, 1907. No. 88978 A.

Braking Tests in Leeds, England. Reports a series of track brake tests carried out recently. Ills. 1600 w. St Ry Jour—Jan. 4, 1908. No. 89352.

Buenos Aires.

The Lacroze Tramway System and the Buenos Aires Central Railway. Illustrated detailed description of a combined steam and electric service. 3000 w. Tram & Ry Wld—Dec. 5, 1907. No. 89045 B.

Car House.

The New Fourteenth Street Concrete Storage Car House of the Capital Traction Company, Washington. Illustrated detailed description. 3800 w. St Ry Jour Dec. 21, 1907. No. 89108.

Communication Systems.

See Composite System, under ELECTRICAL ENGINEERING, COMMUNICATION.

Controllers.

The Automatic Controller. W. B. Kou-

wenhoven. Illustrated description of a device which gives a uniform rate of acceleration. 3000 w. Ry & Loc Engng—Jan., 1908. No. 89339 C.

Electrification.

See same title, under RAILWAY ENGINEERING, MOTIVE POWER AND EQUIPMENT.

Funeral Trains.

Electric Funeral Trains in Milan. Illustrated description of the electric car service for funerals. 1200 w. St Ry Jour—Jan. 4, 1908. No. 89351.

Instruction Sheets.

Shop Prints of the Public Service Corporation of New Jersey. Gives a series of instruction sheets on wiring, brush setting, etc., prepared for distribution among the employes. 1000 w. St Ry Jour—Dec. 14, 1907. No. 89023.

Interurban.

The Fort Dodge, Des Moines & Southern Railway. Illustrated description of this electric line and its operation. 600 w. R R Gaz—Dec. 6, 1907. No. 88901.

The Atlantic Shore Line. Illustrated description of the last link in a system in New England, with information relating to the route and operation. 4000 w. St Ry Jour—Dec. 14, 1907. No. 89022.

The Milwaukee Northern Railway. Illustrated detailed description of a line in process of construction which will ultimately connect Milwaukee and Fond du Lac. Its equipment, gas engine power, etc. 5800 w. Elec Ry Rev—Dec. 7, 1907. No. 88905.

The Easton & Washington Traction Company. Illustrated description of a new line in the Delaware Valley, eventually to reach Lake Hopatcong. 2000 w. St Ry Jour—Dec. 28, 1907. No. 89229.

The Indianapolis & Louisville 1200-Volt Direct-Current Line. John R. Hewett. Illustrated detailed description of the scheme of electrification on the portion of the line completed. 3500 w. St Ry Jour —Jan. 4, 1908. No. 89350.

The Development of Electric Transportation Service and Its Effect on St. Louis. B. R. Stephens. Discusses the comparative efficiency of steam and electric roads, the growth of business, etc. 3500 w. Elec Ry Rev—Dec. 21, 1907. No. 89110.

Electric Railways in Sparsely Settled Communities. E. P. Roberts. Abstract of a paper read before the Am St & Int Ry Assn. Considers the requirements for success for suburban and interurban electric lines. 3000 w. R R Gaz—Dec. 27, 1907. No. 89170.

See also Single Phase, under STREET AND ELECTRIC RAILWAYS.

Locomotives.

New Locomotives for the Illinois Traction System. Illustrated detailed description. 800 w. Elec Ry Rev—Dec. 28, 1907. No. 89230.

Large Electric Locomotives for Heavy Service. Bela Valatin. A comparison of 15-cycle, single-phase locomotives and 15-cycle three-phase locomotives in weight and performance. 1500 w. St Ry Jour— Jan. 4, 1908. No. 89353.

Lotschberg.

The Lötschberg Railway (Chemin de Fer du Lötschberg). The first part of the serial discusses the various projected routes and the line finally chosen, the costs, etc. Ills. 3000 w. Bul Tech d l Suisse Rom—Nov. 25, 1907. No. 88822 D.

Motors.

See Railway Motors, under ELECTRICAL ENGINEERING, DYNAMOS AND MOTORS.

New York Subway.

See Subways, under STREET AND ELECTRIC RAILWAYS.

Single Phase.

Windsor, Essex, and Lake Shore Rapid Railway. Illustrates and describes the first single-phase electric railway to be built and operated in Canada. 2500 w. Elec Ry Rev—Dec. 21, 1907. No. 89109.

The Westinghouse System of Single-Phase Electric Traction on European Railways (La Traction Electrique par Courant Altérnatif Simple sur les Chemins de Fer en Europe, Système Westinghouse). M. Henry. Begins a complete detailed description. Ills. 2300 w. Serial, 1st part. Electricien—Nov. 16, 1907. No. 88820 D.

The New Haven System of Single-Phase Distribution with Special Reference to Sectionalization. W. S. Murray. Reviews briefly alternative systems that were considered, and discusses the methods and lengths involved in sectionalizing the single-phase distribution. 4000 w. Pro Am Inst of Elec Engrs—Jan., 1908. No. 89327 D.

Brembana Valley Single-Phase Railway. Illustrated description of a recently completed line, one of the first in Europe employing the Westinghouse single-phase system. 2200 w. Elect'n, Lond—Dec. 27, 1907. No. 89378 A.

See same title, under ELECTRICAL ENGINEERING, DYNAMOS AND MOTORS.

Subways.

Methods of Increasing the Capacity of the New York Subway. Extracts from the preliminary report of B. J. Arnold, suggesting methods of improving the service. 2500 w. Eng Rec—Dec. 7, 1907. No. 88771.

The Great Northern, Piccadilly and Brompton Electric Tube Railway in London (Die elektrisch betriebene Great Northern, Piccadilly and Brompton-Röhrenbahn in London). E. A. Ziffer. A general description of the tubes, stations, power stations, rolling stock, etc. Ills. 3500 w. Mit d Ver f d Förd d Lokal u Strassenbahnwesens—Oct., 1907. No. 88829 F.

Subway Stations.

An arched Station of Reinforced and Hooped Concrete on the Considère System (Station voutée en Béton armé et fretté du Système Considère). Henry Lossier. An illustrated detailed description of the design of one of the stations of the Paris Subway. 2500 w. Beton u Eisen—Nov. 1907. No. 88866 F.

Third Rail.

The Third-Rail Problem. A. D. Williams. Jr. A discussion of the physical difficulties in installation and maintenance. 3800 w. Engineering Magazine—Jan., 1907. No. 89178 B.

Tokio.

The Tokio Electric Street Railway Company. Henry K. Brent. Map and account of street railway progress in this city of Japan. 3000 w. St Ry Jour— Dec. 7, 1907. No. 88763.

Wire Suspension.

A New Suspension for the Contact Wires of Electric Railways Using Sliding Bows. Joseph Mayer. Discusses the economical construction of a safe contact wire, giving formulas for calculating stresses, and describing a strain adjuster and explaining the theory. Ills. 10000 w. Pro Am Soc of Civ Engrs—Dec., 1907. No. 89226 E.

We supply copies of these articles. See page 878.

EXPLANATORY NOTE—THE ENGINEERING INDEX.

We hold ourselves ready to supply—usually by return of post—the full text of every article indexed in the preceding pages, *in the original language*, together with all accompanying illustrations; and our charge in each case is regulated by the cost of a single copy of the journal in which the article is published. The price of each article is indicated by the letter following the number. When no letter appears, the price of the article is 20 cts. The letter A, B, or C denotes a price of 40 cts.; D, of 60 cts.; E, of 80 cts.; F, of $1.00; G, of $1.20; H, of $1.60. When the letter N is used it indicates that copies are not readily obtainable and that particulars as to price will be supplied on application. Certain journals, however, make large extra charges for back numbers. In such cases we may have to increase proportionately the normal charge given in the Index. In ordering, care should be taken to *give the number* of the article desired, not the title alone.

Serial publications are indexed on the appearance of the first installment.

SPECIAL NOTICE.—To avoid the inconvenience of letter-writing and small remittances, especially from foreign countries, and to cheapen the cost of articles to those who order frequently, we sell coupons at the following prices:—20 cts. each or a book of twelve for $2.00; three books for $5.00.

Each coupon will be received by us in payment for any 20-cent article catalogued in the Index. For articles of a higher price, one of these coupons will be received for each 20 cents; thus, a 40-cent article will require two coupons; a 60-cent article, three coupons; and so on. The use of these coupons is strongly commended to our readers. They not only reduce the cost of articles 25 per cent. (from 20c. to 15c.), but they need only a trial to demonstrate their very great convenience—especially to engineers in foreign countries, or away from libraries and technical club facilities.

Write for a sample coupon—free to any part of the world.

CARD INDEX.—These pages are issued separately from the Magazine, printed on one side of the paper only, and in this form they meet the exact requirements of those who desire to clip the items for card-index purposes. Thus printed they are supplied to regular subscribers of THE ENGINEERING MAGAZINE at 10 cents per month, or $1.00 a year; to non-subscribers, 25 cts. per month, or $3.00 a year.

THE PUBLICATIONS REGULARLY REVIEWED AND INDEXED.

The titles and addresses of the journals regularly reviewed are given here in full, but only abbreviated titles are used in the Index. In the list below, *w* indicates a weekly publication, *b-w*, a bi-weekly, *s-w*, a semi-weekly, *m*, a monthly, *b-m*, a bi-monthly, *t-m*, a tri-monthly, *qr*, a quarterly, *s-q*, semi-quarterly, etc. Other abbreviations used in the index are: Ill—Illustrated: W—Words; Anon—Anonymous.

Alliance Industrielle. *m.* Brussels.
American Architect. *w.* New York.
Am. Engineer and R. R. Journal. *m.* New York.
American Jl. of Science. *m.* New Haven, U. S. A.
American Machinist. *w.* New York.
Anales de la Soc. Cien. Argentina. *m.* Buenos Aires.
Annales des Ponts et Chaussées. *m.* Paris.
Ann. d Soc. Ing. e d Arch. Ital. *w.* Rome.
Architect. *w.* London.
Architectural Record. *m.* New York.
Architectural Review. *s-q.* Boston.
Architect's and Builder's Magazine. *m.* New York.
Australian Mining Standard. *w.* Melbourne.
Autocar. *w.* Coventry, England.
Automobile. *w.* New York.
Automotor Journal. *w.* London.
Beton und Eisen. *qr.* Vienna.
Boiler Maker. *m.* New York.
Brass World. *m.* Bridgeport, Conn.
Brit. Columbia Mining Rec. *m.* Victoria, B. C.
Builder. *w.* London.
Bull. Bur. of Standards. *qr.* Washington.
Bulletin de la Société d'Encouragement. *m.* Paris.

Bulletin du Lab. d'Essais. *m.* Paris.
Bulletin of Dept. of Labor. *b-m.* Washington.
Bull. Soc. Int. d'Electriciens. *m.* Paris.
Bulletin of the Univ. of Wis., Madison, U. S. A.
Bulletin Univ. of Kansas. *b-m.* Lawrence.
Bull. Int. Railway Congress. *m.* Brussels.
Bull. Scien. de l'Assn. des Elèves des Ecoles Spéc. *m.* Liége.
Bull. Tech. de la Suisse Romande. *s-m.* Lausanne.
California Jour. of Tech. *m.* Berkeley, Cal.
Canadian Architect. *m.* Toronto.
Canadian Electrical News. *m.* Toronto.
Canadian. Engineer. *m.* Toronto and Montreal.
Canadian Mining Journal. *b-w.* Toronto.
Cassier's Magazine. *m.* New York and London.
Cement. *m.* New York.
Cement Age. *m.* New York.
Central Station. *m.* New York.
Chem. Met. Soc. of S. Africa. *m.* Johannesburg.
Colliery Guardian. *w.* London.
Compressed Air. *m.* New York.
Comptes Rendus de l'Acad. des Sciences. *w.* Paris.
Consular Reports. *m.* Washington.

878

Deutsche Bauzeitung. *b-w.* Berlin.
Domestic Engineering. *w.* Chicago.
Economic Geology. *m.* New Haven, Conn.
Electrical Age. *m.* New York.
Electrical Engineer. *w.* London.
Electrical Review. *w.* London.
Electrical Review. *w.* New York.
Electric Journal. *m.* Pittsburg, Pa.
Electric Railway Review. *w.* Chicago.
Electrical World. *w.* New York.
Electrician. *w.* London.
Electricien. *w.* Paris.
Electrochemical and Met. Industry. *m.* N. Y.
Elektrochemische Zeitschrift. *m.* Berlin.
Elektrotechnik u Maschinenbau. *w.* Vienna.
Elektrotechnische Rundschau. *w.* Potsdam.
Elektrotechnische Zeitschrift. *w.* Berlin.
Elettricità. *w.* Milan.
Engineer. *w.* London.
Engineer. *s-m.* Chicago.
Engineering. *w.* London.
Engineering-Contracting. *w.* New York.
Engineering Magazine. *m.* New York and London.
Engineering and Mining Journal. *w.* New York.
Engineering News. *w.* New York.
Engineering Record. *w.* New York.
Eng. Soc. of Western Penna. *m.* Pittsburg, U. S. A.
Foundry. *m.* Cleveland, U. S. A.
Génie Civil. *w.* Paris.
Gesundheits-Ingenieur. *s-m.* München.
Giorn. dei Lav. Pubb. e d Str. Ferr. *w.* Rome.
Glaser's Ann. f Gewerbe & Bauwesen. *s-m.* Berlin.
Heating and Ventilating Mag. *m.* New York.
Ice and Refrigeration. *m.* New York.
Il Cemento. *m.* Milan.
Industrial World. *w.* Pittsburg.
Ingegneria Ferroviaria. *s-m.* Rome.
Ingenieria. *b-m.* Buenos Ayres.
Ingenieur. *w.* Hague.
Insurance Engineering. *m.* New York.
Int. Marine Engineering. *m.* New York.
Iron Age. *w.* New York.
Iron and Coal Trades Review. *w.* London.
Iron Trade Review. *w.* Cleveland, U. S. A.
Jour. of Accountancy. *m.* N. Y.
Journal Asso. Eng. Societies. *m.* Philadelphia.
Journal Franklin Institute. *m.* Philadelphia.
Journal Royal Inst. of Brit. Arch. *s-qr.* London.
Jour. Roy. United Service Inst. *m.* London.
Journal of Sanitary Institute. *qr.* London.
Jour. of South African Assn. of Engineers. *m.* Johannesburg, S. A.
Journal of the Society of Arts. *w.* London.
Jour. Transvaal Inst. of Mech. Engrs., Johannesburg, S. A.
Jour. of U. S. Artillery. *b-m.* Fort Monroe, U. S. A.
Jour. W. of Scot. Iron & Steel Inst. *m.* Glasgow.
Journal Western Soc. of Eng. *b-m.* Chicago.
Journal of Worcester Poly. Inst., Worcester, U. S. A.
Locomotive. *m.* Hartford, U. S. A.
Machinery. *m.* New York.
Manufacturer's Record. *w.* Baltimore.
Marine Review. *w.* Cleveland, U. S. A.
Men. de la Soc. des Ing. Civils de France. *m.* Paris.
Métallurgie. *w.* Paris.
Mines and Minerals. *m.* Scranton, U. S. A.
Mining and Sci. Press. *w.* San Francisco.
Mining Journal. *w.* London.

Mining Science. *w.* Denver, U. S. A.
Mining World. *w.* Chicago.
Mitthcilungen des Vereines für die Förderung des Local und Strassenbahnwesens. *m.* Vienna.
Municipal Engineering. *m.* Indianapolis, U. S. A.
Municipal Journal and Engineer. *w.* New York.
Nature. *w.* London.
Nautical Gazette. *w.* New York.
New Zealand Mines Record. *m.* Wellington.
Oest. Wochenschr. f. d. Oeff. Baudienst. *w.* Vienna.
Oest. Zeitschr. Berg & Hüttenwesen. *w.* Vienna.
Plumber and Decorator. *m.* London.
Popular Science Monthly. *m.* New York.
Power. *w.* New York.
Practical Engineer. *w.* London.
Pro. Am. Soc. Civil Engineers. *m.* New York.
Pro. Canadian Soc. Civ. Engrs. *m.* Montreal.
Proceedings Engineers' Club. *qr.* Philadelphia.
Pro. St. Louis R'way Club. *m.* St. Louis, U. S. A.
Pro. U. S. Naval Inst. *qr.* Annapolis, Md.
Quarry *m.* London.
Queensland Gov. Mining Jour. *m.* Brisbane, Australia.
Railroad Gazette. *w.* New York.
Railway Age. *w.* Chicago.
Railway & Engineering Review. *w.* Chicago.
Railway and Loc. Engng. *m.* New York.
Railway Master Mechanic. *m.* Chicago.
Revista d Obras. Pub. *w.* Madrid.
Revista Tech. Ind. *m.* Barcelona.
Revue d'Electrochimie et d'Electrométallurgie. *m.* Paris.
Revue de Mécanique. *m.* Paris.
Revue de Métallurgie. *m.* Paris.
Revue Gén. des Chemins de Fer. *m.-* Paris.
Revue Gén. des Sciences. *w.* Paris.
Revue Technique. *b-m.* Paris.
Rivista Gen. d Ferrovie. *w.* Florence.
Rivista Marittima. *m.* Rome.
Schiffbau. *s-m.* Berlin.
School of Mines Quarterly. *q.* New York.
Schweizerische Bauzeitung. *w.* Zürich.
Scientific American. *w.* New York.
Scientific Am. Supplement. *w.* New York.
Sibley Jour. of Mech. Eng. *m.* Ithaca, N. Y.
Soc. Belge des Elect'ns. *m.* Brussels.
Stahl und Eisen. *w.* Düsseldorf.
Stevens Institute Indicator. *qr.* Hoboken, U. S. A.
Street Railway Journal. *w.* New York.
Technology Quarterly. *qr.* Boston, U. S. A.
Tramway & Railway World. *m.* London.
Trans. Am. Ins. Electrical Eng. *m.* New York.
Trans. Am. Ins. of Mining Eng. New York.
Trans. Am. Soc. Mech. Engineers. New York.
Trans. Inst. of Engrs. & Shipbuilders in Scotland, Glasgow.
Transport. *w.* London.
Verkehrstechnische Woche und Eisenbahntechnische Zeitschrift. *w.* Berlin.
Wood Craft. *m.* Cleveland, U. S. A.
Yacht. *w.* Paris.
Zeitschr. f. d. Gesamte Turbinenwesen. *w.* Munich.
Zeitschr. d. Mitteleurop. Motorwagon Ver. *s-m.* Berlin.
Zeitschr. d. Oest. Ing. u. Arch. Ver. *w.* Vienna.
Zeitschr. d. Ver. Deutscher Ing. *w.* Berlin.
Zeitschrift für Elektrochemie. *w.* Halle a S.
Zeitschr. f. Werkzeugmaschinen. *b-w.* Berlin.

CURRENT RECORD of NEW BOOKS

NOTE—*Our readers may order through us any book here mentioned, remitting the publisher's price as given in each notice. Checks, Drafts, and Post Office Orders, home and foreign, should be made payable to* THE ENGINEERING MAGAZINE.

Electric Lighting.

Pocket Book of Electric Lighting and Heating. By Sydney F. Walker. Size, 6½ by 4 in.; pp., 438. Ills. Price, $3. New York: The Norman W. Henley Publishing Company.

The author of this handbook makes no claim to originality of matter. The aim has been to place in handy form for the central-station engineer and contracting engineer information as to the appliances under their control or which they may be required to furnish in an emergency which is usually contained in trade catalogues or is otherwise difficult to obtain. While it relates principally to materials of British manufacture, it should prove of value to engineers in all parts of the world. The author has included as much information as possible as to sizes, weights, efficiencies, dimensions, etc., and in this departure from the regular field of the electrical handbook has produced a very practical guide for the lighting and heating engineer. The eight sections into which the book is divided include an Introduction, on the fundamental principles of electrical engineering, and sections on Generators, Accumulators, Switch-boards, Circuit Breakers, Cables, Measuring Instruments, Lamps and Accessories, and Heating Appliances.

Management.

Works' Organization, Accounting, and Cost Systems of the firm of Ludw. Loewe & Co., Berlin. (Fabrikorganisation, Fabrikbuchführung und Selbstkostenberechnung der Firma Ludw. Loewe & Co.). By J. Lilienthal. Size, 10½ by 7½ in.; pp., 220. Ills. Price, M. 10. Berlin: Julius Springer.

A complete exposition of the management system of this important German firm, thoroughly illustrating the forms and methods used in accounting and cost keeping. As an exemplification of the best current Continental practice in works' management and organization, the book can be recommended to the attention of anyone interested in the problems of systematizing and cheapening production in industrial operations.

Specifications.

Specifications and Contracts. By J. A. L. Waddell and John C. Wait. Size, 9 by 6 in.; pp., 169. Price, $1. New York: The Engineering News Publishing Company.

This excellent little book contains reprints of Dr. Waddell's lectures on "Specifications" and "Engineering Contracts," delivered before the engineering students of the Rensselaer Polytechnic Institute, supplemented by a large amount of new matter giving examples for practice in the drawing of specifications and contracts and a "Note on the Law of Contracts" by Mr. John C. Wait. *Engineering News* has done students of engineering a distinct service in publishing in such convenient form so interesting and authoritative a discussion of these important subjects.

BOOKS RECEIVED.

The "Mechanical World" Electrical Pocketbook for 1908. Size, 6 by 4 in.; pp., 247. Ills. Price, 6d. Manchester: Emmett & Co., Limited.

Moving Loads on Railway Underbridges. By H. Bamford. 8 by 6½ in.; pp., 78. Ills. Price, $1.25. New York: The Macmillan Company. London: Whittaker & Co.

Annuaire pour l'An 1908 publié par le Bureau des Longitudes. Avec des Notices Scientifiques. Size, 6 by 4 in.; pp., 958. Ills. Price, 1 fr. 50 c. Paris: Gauthier-Villars.

Coal-Mine Accidents: their Causes and Prevention. By Clarence Hall and Walter O. Snelling. Size, 9 by 6 in.; pp., 21. Washington, D. C.: United States Geological Survey.

Annual Report of the Chief of the Bureau of Steam Engineering to the Secretary of the Navy for the Fiscal Year 1907. Size, 9 by 6 in.; pp., 56. Washington, D. C.: Navy Department.

The State of Wyoming. A Book of Reliable Information Published by Authority of the Ninth Legislature. Size, 9 by 6 in.; pp., 141. Ills. Cheyenne, Wyo.: Commissioner of Public Lands.

The Hardening Process of Hydraulic Cements. By Dr. W. Michaelis, Sr. Translated by Dr. W. Michaelis, Jr. Size, 8 by 5½ in.; pp., 29. Price, 50 cents. Chicago, Ill.: Cement & Engineering News.

THE ENGINEERING MAGAZINE

THE WORLD IS ITS FIELD

VOL. XXXIV. MARCH, 1908. No. 6.

CONDITIONS AND PROSPECTS IN THE AMERICAN IRON INDUSTRY.

By Edwin C. Eckel.

Late geologist in charge of investigations of iron ores and structural materials, U. S. Geological Survey.

THE year which recently closed opened with superficially brilliant prospects, but with underlying conditions of serious financial stress. For a time these threatening conditions were masked by the continuation of previous industrial activity, and it seemed as though the crisis might be postponed if not entirely avoided. By the middle of the year, however, the tension had greatly increased, and it needed little to bring about a crash. This was afforded in October by the uncovering of some remarkable innovations in the business methods of certain New York banks, and by the end of that month financial and industrial activity was almost completely suspended.

Though dragged down in the fall of the other industries, American iron and steel producers can be acquitted of any share of the blame. The expansion of iron and steel manufacture during the past few years has been rapid, and the prices of the products have at times escaped effective control. But in general the management of the iron industry has been conservative and few weak spots in the iron trade were revealed when the panic arrived. In these respects it affords a marked and pleasant contrast to the copper trade, where a danger spot had been created in the American industrial situation.

The course of the iron trade during 1907 was so directly affected by the financial events of the year, that it would be difficult to discuss them intelligently without briefly reviewing this financial history.

Financial History of the Year.

It has been said that 1907 opened with apparently brilliant prospects in all lines of trade. Mills and factories were running overtime in a vain attempt to catch up with accumulated orders, merchants reported unprecedented business, and the farming population
had just received record prices for record crops. Underlying all this,
however, was serious financial tension, international in its scope.
During the years immediately preceding, excessive demands had been
made upon the world's capital. Funds in unexampled amounts had
been withdrawn for new railway construction in the United States.
Two costly wars had been financed, while an immense amount of
fixed capital had been destroyed by fire and earthquake. Real-estate
and mining speculation had exacted heavy toll, particularly from the
smaller investors.

With such conditions as these existing, it was evident that extreme conservatism and general confidence, would be required if serious disaster was to be avoided. Unfortunately conservatism was
lacking in some high financial quarters, while public confidence had
been gradually undermined. There had been unpleasant revelations
as to the business ethics of prominent financiers, there had been story
after story—some true but mostly false—of graft and rottenness in
American business institutions. This sort of thing had gone so far
that a large and vociferous section of the people could see no remedy
but the wholesale nationalization of the more important industries.
All this combined to make capital timorous as to further investments, and apprehensive as to the safety of those in hand.

It is perfectly true that reform was urgently needed in some of
the methods of high finance, and that the unforeseen increase of corporate power demanded means of regulation not provided by earlier
statutes. Few would dispute these statements, and no honest man
could have the slightest objection to such reform or regulation if
carried out in a fair and reasonable manner. Early in 1907, however, radical legislative and judicial action, particularly in the southern and western States, gave rise to the general feeling that the
new movement would assume a punitive rather than a corrective
character, and that the status of existing investments was endangered. It became impossible to float new issues of railroad bonds
or notes except at prohibitive rates of interest, and even municipal
securities of undoubted standing could not be satisfactorily disposed
of. The result was that expansion of business and new construction
were effectively checked by a combination of scarcity of money and
lack of confidence.

It was at a time like this, when all conditions suggested that extreme conservatism was desirable everywhere, that the industrial world witnessed with pained interest an attempt to market an important product at prices far above its real value. The general business activity of 1905 and 1906 had resulted in a steady and increasing demand for copper. Eighteen cents a pound for copper meant profits for all the copper mines in the country, and large profits for many of them, and at this price there was no lack of demand. Certain important copper interests, however, had come to the conclusion that consumers could be forced to pay any price for the metal. Disregarding the strained financial situation, and forgetting an earlier painful experience in the same line, the nominal price of copper was advanced steadily until it reached twenty-six cents a pound. In spite of the fact that buying stopped early in the year, the high-level prices were maintained until July, and mines and smelters were pushed to their fullest capacity. The result was the accumulation of an enormous stock of unsold copper, and this finally led to a series of sharp breaks in copper prices which did not terminate until, in October, copper sold at less than twelve cents a pound. Then, and not till then, it was decided to restrict production, a step which might better have been taken in April.

From the very beginning of the year, securities of all kinds had sagged steadily in value, their downward course being marked by three particularly sudden breaks in March, August, and October respectively. The two earlier collapses in stocks were not accompanied or followed by serious industrial changes. The October collapse, however, with its banking failures and money stringency, caused even the most optimistic to realize that prosperity had passed for a time, and that a period of severe industrial depression was at hand. Business of all kinds fell off with remarkable suddenness, and the great manufacturing industries were as usual affected first and most markedly.

THE IRON TRADE IN 1907.

The iron industry shared with other lines of business both the earlier prosperity and the later depression. For the first nine months of the year the furnaces and mills were operated at their fullest capacity, and the earnings of the iron and steel companies were record-breaking. Prices softened somewhat about the middle of the year, and new orders were less pressing, but at the close of September the prospects seemed good that the iron output of the year would be very largely in excess of the production of 1906, itself a record year. In the last quarter of the year, however, the collapse came.

New orders did not lessen gradually—they simply stopped suddenly
and completely; and the only thing that held prices fairly steady was
an equally sudden and complete shut-down of furnaces and mills.

The tonnage of pig iron produced in the first nine months or so of
1907 had been so great that even the shut-downs of the latter part
of the year were not sufficient to carry the year's total below the 1906
record. In consequence, the final figures for the year, just published by Mr. James M. Swank, of the American Iron and Steel Association, show that in 1907 the furnaces of the United States produced
25,781,361 tons of pig iron, as compared with 25,307,191 tons in
1906. A more detailed statement of this production by States, as
compared with the output of the two preceding years, is presented in
the following table:

TOTAL PRODUCTION OF PIG IRON, 1905-1907, BY STATES.

	Production, including spiegeleisen and ferromanganese, in long tons.		
	1905.	1906.	1907.
Pennsylvania	10,579,127	11,247,869	11,348,549
Ohio	4,586,110	5,327,133	5,250,687
Illinois	2,034,483	2,156,866	2,457,768
Alabama	1,604,062	1,674,848	1,686,674
New York	1,198,068	1,552,659	1,659,752
Virginia	510,210	483,525	478,771
Tennessee	372,692	426,874	393,106
Colorado Missouri }	407,774	413,040	b 468,486
Maryland	332,096	386,709	411,833
New Jersey	311,039	379,390	373,189
Wisconsin Minnesota }	351,415	373,323	322,083
Michigan	288,704	369,456	a 436,507
West Virginia	298,179	304,534	291,066
Kentucky	63,735	98,127	127,946
Georgia Texas }	38,699	92,599	55,825
Connecticut Massachusetts }	15,987	20,239	19,119
Total	22,992,380	25,307,191	25,781,361

b. Including also small output in Pacific Coast States.
a. Including Indiana.

The relation of the year 1907 to the recent past of the American
iron and steel trade is, however, brought out more clearly in the following diagram, than could be done in tables. The curves here
shown—page 885—represent the production of iron ore, pig iron
and steel respectively in the years 1870 to 1907 inclusive. All the
data used in preparing the diagram are official except the production
of iron ores and steel in 1907, which are believed to be close estimates.
On examining this figure it will be seen that 1905 and 1906 showed

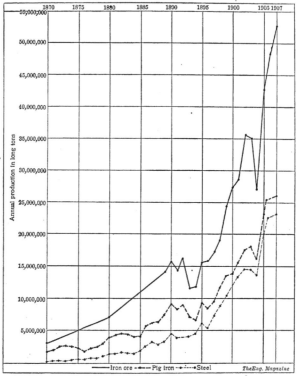

PRODUCTION OF IRON ORE, PIG IRON, AND STEEL, 1870-1897.

a most remarkable growth in the output of all three products. In 1907 this growth was sharply checked in the case of pig iron and steel, but the iron-ore production could not be slowed down in time. The result is that the stocks of iron ore now on hand are probably larger than ever before at the same season of the year.

So far as can be foretold now, the production figures of 1907 will remain as record figures until 1909 and possibly longer. It is very unlikely that the recovery from the present depression will come soon enough to give 1908 a chance to surpass or even approach the figures of 1907. The prospects are briefly discussed later.

Price-Cutting versus Shut-Downs.

An examination of the respective courses followed by the copper and iron industries during the past year discloses interesting differences of policy. At the beginning of every period of severe commercial depression, owners of mines and manufacturing plants are usually offered two alternative ways of adjusting their business to the changed conditions. They may keep on with practically no reduction of output, trusting to reduction of price to bring about a strong consumptive demand. This course of action may finally lead to selling their product below cost, for a greater or less time. On the other hand, the situation may be met by reducing the output, and waiting until demand reasserts itself naturally. This may lead to an absolute shut-down of the works if the period of depression is prolonged.

In some manufacturing businesses, particularly in those making a special or distinctive line of goods, there is a possibility of reasonable choice between these two alternatives. In this case the manufacturer may argue that the fixed charges of his investment are great, relative to his operating costs; that his trade may be secured by rivals in case of a shut-down; and that his office and mill force, especially if it includes many highly trained men, will be hopelessly scattered and disorganized.

With a mine, or a mill manufacturing any of the great staples, these arguments have less weight. A mine which sells its product at less than cost, or even at an unfairly low profit, is simply using up its supply of raw material without being properly recompensed for it. It is drawing on a fixed and limited reserve of ore which is becoming naturally scarcer and more valuable each year. It is doing this, moreover, in order to supply a demand which if not satisfied at the moment would simply accumulate until it became strong enough to justify fair prices. Its product is in no way different from that of other producers of the same metal, so that it has no individual "trade" to lose by a shut-down. Its working force is relatively simple and easily reorganized after a stoppage. The same statements which apply to a mine are true, though to a somewhat less extent, of industries whose product is made directly from mineral raw materials, and whose output is substantially the same in grade throughout the country. Iron furnaces and cement plants are manufactories of this type. In industries of this class it is better to try to maintain reasonable prices by curtailment of output or by total shut-down, than to try to encourage sales by serious price reductions.

There is, it is true, a third possible mode of action, but it demands such enormous working capital and such an unfailing trust in the future that it is rarely possible to carry it out in any large industry, except to bridge over a brief and temporary period of depression. It is to run the mine or mill as usual, and stock the product, not forcing it on the market.

In reviewing the industrial history of the year it has been seen that the leading American copper producers tried all three of these plans in turn, and tried them in an order which inevitably led to disaster. The principal iron interests are trying the second method named—reduction of output. The October panic had scarcely ended when, warned by decreases in new buying and by attempts to cancel orders already booked, the leading iron and steel producers took active steps to prevent entire demoralization in the industry. Acting on the initiative of the United States Steel Corporation, meetings were arranged at which leading interests decided upon substantial unity of action in these matters. It was argued that as producers had not forced consumers to pay unreasonably high prices for steel products during the boom, there was now no reason to market products at unprofitably low prices during the depression. It is probable enough that if the present depression is unexpectedly prolonged, prices will be gradually lowered to some extent, but it is very unlikely that we shall witness the extreme demoralization in prices which has accompanied other periods of poor business. And the lesson of the panic will not have been entirely wasted if buyers are taught that a contract is not a one-sided affair, to be enforced or cancelled at the buyer's option.

THE STRENGTH OF THE INDEPENDENTS.

The fact that concerted action is possible is due largely to a factor often overlooked by those who are interested only in the stock-selling side of the steel industry. That factor is the existence at present of a number of strong and well-equipped independent companies. In the extreme west the Colorado Fuel & Iron Co. is in a strong technical position, through its control of coal and iron lands, though its financial results have been far from brilliant. Jones & Laughlin and the Republic Iron and Steel Company are strongly established in the Pittsburg-Ohio district, while the latter company has also important Southern holdings. The Lackawanna Steel Company at Buffalo lost the principal benefits of the past period of prosperity, but should hereafter be a noteworthy competitor. In central Pennsylvania and eastward are the Pennsylvania, Cambria, Midvale, and Bethlehem steel companies, and several strong furnace interests— Wharton, Thomas, Empire, etc. In the South, now that the Tennessee company is a subsidiary of the Steel Corporation, the leading independents are Sloss-Sheffield, Republic, and Woodward, none of which has as yet taken up steel manufacture.

Some of the independent iron and steel producers, of course, exist simply on sufferance, and will find it hard to keep alive through any prolonged period of depression. The majority of those noted above, however, seem to be at least as strongly based as the Steel Corporation itself.

A number of the independent producers are particularly strong in their possession of large supplies of raw materials, and in their relatively low fixed charges. These factors are important even during a period of prosperity, but they become vital during business reactions. Small bond-issues, cheap coke, and cheap ore will enable even a small plant to meet the competition of a much larger concern heavily loaded with fixed charges or deficient in supplies of raw material. Some very remarkable differences are revealed when the capitalization of different companies is compared, using as a basis their relative output of pig iron or of steel per year.

THE STEEL CORPORATION IN THE SOUTH.

In the latter part of 1906 the event which gave promise of having the greatest influence on the future of the American iron industry was the acquisition of the so-called Hill or Great Northern ore lands by the United States Steel Corporation. This lease removed from the market the only large remaining tract of undeveloped Lake Superior ore lands, strengthened the Steel Corporation's ore reserves immensely, and made it obvious that future competition could not effectively be based on the Lake district ores.

During the latter part of 1907 another step, of still greater importance, was taken by the Steel Corporation, in entering the Southern field. This advance was an indirect and rather unexpected result of the October break in the stock market. A few years ago interests closely allied to those already in control of the Republic Iron & Steel Co. secured control of the Tennessee Coal, Iron & Railroad Co. Extensive improvements were inaugurated, and it seemed as if the Republic-Tennessee combination would soon be a serious rival to the larger Steel Corporation. The financial crisis of 1907, however, delayed the renovation of the Tennessee properties, and at the time of the October panic it developed that the stock necessary for the control of the Tennessee company was being carried by a New York bank which was itself in urgent need of assistance, and which could not readily realize on this stock in the state of the market. The difficulty was solved by the acquisition of the Tennessee company by the Steel Corporation, Steel Corporation bonds being exchanged for Tennessee stock. As the bonds were saleable enough the banking difficulty was relieved, and the Steel Corporation absorbed a company which might have proven troublesome in the future.

This acquisition is a matter of importance, both to the Steel Corporation and to the Southern iron industry. To the Steel Corporation it means that very large reserves of coal and iron ore have been secured on a ridiculously low valuation, and that with one step it becomes the leading iron and steel producer of the South. Of greatest immediate value is the fact that it secures an open-hearth steel plant and a rail mill full of orders. To the Southern iron industry in general the advent of the Steel Corporation promises to bring greater steadiness and a more rapid advance in development, particularly along the line of finished products.

SEABOARD PLANTS AND FOREIGN ORES.

One feature of American industry to which relatively little attention has been paid is the necessity for providing for the handling of water-borne trade—coastal and export—in an efficient and economical manner. This lack of foresight is particularly noticeable in the iron trade, now that we have reached the stage when exports are of interest. I recently had occasion to examine and report upon this situation in connection with a closely related industry, and the general conclusions then reached may be summarized as follows:

The best steam and coking coals in the United States are those of the Appalachian coal field, which reaches from Pennsylvania to Alabama in a belt trending almost parallel to the Atlantic Coast, and distant 150 to 350 miles from it. These Appalachian coals are, in general, far superior to those of other portions of the United States. For most of its extent this coal field is paralleled, on its eastern or coastal side, by a series of extensive iron-ore deposits. These iron ores, however, though enormous in quantity, are usually of relatively low grade; but they can be placed on the seaboard at a very low price per unit of iron. A plant located at the coastal end of a coal-carrying railroad could therefore depend on securing a cheap supply of high-grade fuel, and a cheap suply of low-grade domestic ore. But it would also be in a position to avail itself of still cheaper supplies of high-grade foreign ores, imported from Canada, the West Indies, South America and Spain. The duty on such ores is rebated in case the products made from them are exported, but even without this provision the ores used by a seaboard plant would not cost more than Lake ores at Pittsburg.

In view of these facts it is remarkable that only one plant—that of the Maryland Steel Co. near Baltimore—is located so as to take advantage of this situation, though several eastern Pennsylvania plants depend partly at least on foreign ores. The principal points which offer the combination of factors necessary for the success of a seaboard steel plant are Baltimore, Norfolk, Brunswick, Pensacola and Mobile. Of these, Norfolk has such distinct advantages that it is improbable that it will long remain entirely undeveloped.

THE PROSPECTS FOR THE FUTURE.

At the date of writing it cannot be said that any improvement in industrial conditions has set in since October, or even that there is any immediate prospect of such improvement. Money conditions have eased, as is usual after such a crisis, but railroad earnings are showing decreases and general trade is stagnant. In the American iron industry, for example, furnaces with a total maximum capacity of about 30,000,000 tons of pig iron per year are now being operated at a rate of not much over 12,000,000 tons. Few expect much improvement until the early summer, the majority probably believe that improvement will be delayed until after election, while an extreme minority hold that we are on the verge of a long drawn out industrial agony such as that of 1893-1896.

As an offset to purely temporary financial embarrassment and to political agitation which, though less temporary, will not always be with us, there are many factors which are even now at work to bring about better conditions in the iron trade. Basal conditions are sound, thanks to conservative action during our time of prosperity. There is hardly an iron producer, large or small, which is not in infinitely better technical and financial condition now than in 1903. During the past four years most companies have spent more money on improvements, or on decreasing fixed charges, than they have disbursed in dividends. The saving feature of the situation is the rapidity with which production was curtailed, when once the change in conditions had become obvious. Our reserves of iron ore and of good coking coals, though still large, are being decreased every year. These raw materials in consequence, are gradually increasing in value, and there is little justification for drawing heavily on them during a period of depression. The natural increase in value of raw materials, taken in combination with the sound financial condition of most producers, is a guarantee against demoralization in the industry. Stability in prices is causing what may be called an accumulated demand for iron and steel products, and when once consumers feel free to commence buying, it is probable that the resumption in activity will be rapid. This is particularly true of the railroad trade, for large quantities of rails and material are urgently needed by American railroads.

THE PRODUCTION SYSTEM OF THE WESTING-
HOUSE ELECTRIC AND MFG. COMPANY.

By H. M. Wharton.

Mr. Wharton was engaged in various capacities in the Westinghouse Electric & Manufacturing Company's works for twelve years; for four years he was assistant superintendent
in charge of production, and for three years engineer of works. He therefore writes from
intimate pesonal knowledge of the production system of this establishment—one of the most
interesting and effective in the world—and his faculty for exact and descriptive expression
makes the organization and its practical workings appear before the reader in complete
outline and in correct perspective.—THE EDITORS.

THE ultimate purpose of any manufacturing organization being
to produce certain finished articles that are to be sold, it may
be said that, from the purchaser's point of view, the whole
organization is engaged in production and, therefore, that a description of its production system would mean a description of the methods used throughout the entire organization.

For the purpose of this article, however, a more restricted definition has been assumed for the word "production," and only those
methods are considered which are used by the departments directly
concerned in creating the product. Even with this limitation, an
accurate description of the method of regulating production in a
large and complex organization would be possible only by exhibiting
the multiplicity of forms used, necessitating an article of great
length and of questionable interest. It has been deemed best, therefore, to outline only the general method of procedure, omitting, as
far as possible, all confusing details.

Under the plan of organization of the Westinghouse Electric
and Manufacturing Company there are six departments directly
concerned in production, viz:—1. The Correspondence Department, 2. The Engineering Department, 3. The Purchasing Department, 4. The Manufacturing Department, 5. The Raw and
Finished Material Stores, 6. The Shipping Department. The heads
of the first three report directly to one of the vice-presidents while
the last three are in charge of the manager of works who reports
also to a vice-president.

Although the selling force cannot be properly classified with
those departments engaged directly in production, yet, in order to
give a proper understanding of the general system, mention must be

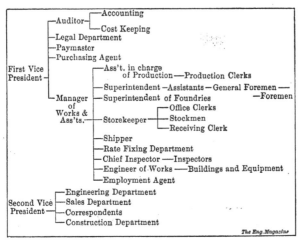

GENERAL OUTLINE OF ORGANIZATION, THE WESTINGHOUSE ELECTRIC & MFG. CO.

made of the fact that district sales offices are maintained in most of the larger cities and that practically all negotiations with either real or prospective customers are conducted by the managers of these offices as agents for the company. The district office, therefore, is an important element in the line of communication between the customer and the works, since all correspondence with the customer is transmitted through the nearest district office.

THE CORRESPONDENCE DEPARTMENT.

This department handles all correspondence between the works and the district sales offices relative to preliminary inquiries and shipping orders, the district offices holding no communication upon these subjects with any of the other works departments. Although for convenience it is so arranged that each correspondent devotes his attention to some particular class of apparatus, it is not to be supposed that he can answer unaided the variety of questions constantly being asked, many of which are highly technical. It is usually necessary to obtain from some of the other works departments the information needed to answer an inquiry, but the official answer is dictated by the correspondent and the letters are filed in his office.

In case a preliminary inquiry from a customer finally leads to an actual order, this order is also transmitted by the district agent to the correspondent and the latter, having at hand all the papers referring to the subject, is then in a position to enter the official order covering the apparatus that is to be shipped to the customer. Such an official order is called a "general" order and, as usually made out, bears a serial number and contains a complete list of all the apparatus wanted, together with the directions for shipping and billing to the customer. No attempt is made to segregate different types and sizes of apparatus for convenience in manufacture by issuing separate general orders for each, because, as will be explained further on, the general order is not a manufacturing order.

After a general order has been entered the correspondent must still keep closely in touch with it during its progress through the works, to inform the customer of the probable date of shipment.

Raw and Finished-Material Stores.

Theoretically, all shipments are supposed to be made directly from the stock of finished product carried in the warehouse, manufacturing orders being issued from time to time to replenish that stock or to maintain it at certain prescribed limits; and as the storekeeper is the custodian of all finished product, as well as of the raw-material stores, all shipping orders, or *general orders,* are issued to him and not to the manufacturing department. Whatever manufacturing orders are necessary for apparatus to be supplied to maintain the warehouse stock are issued by the storekeeper and are known as "stock orders."

In the actual carrying on of a business which includes the building of a great deal of large and costly special machinery, it would not be possible, of course, to keep such special machinery in stock. Orders of this kind, however, are handled in the same way as orders for standard apparatus regularly kept in stock—a *general order* is issued to the storekeeper, who in turn issues *stock orders* for the apparatus to be manufactured—the only difference being that a notation is made upon the stock order of the number of the general order for which the particular apparatus is wanted.

Although a general order, as previously explained, may call for a variety of types and sizes of apparatus, a stock order, on the other hand, is limited to only one type and size although it may call for any convenient number of completed units. In other words, as a stock order is purely a manufacturing order, every complete unit made for any given stock order must be identical with every other unit made for the same order. Copies of all stock orders as they are issued are sent to the engineering department, to the manufacturing department, and to the cost-keeping department.

Upon receipt of a general order in the storekeeper's office provision for the shipment must be made at once by ascertaining from the stock ledgers which of the items called for are in stock and which items must be provided for by means of stock orders, these orders being issued immediately. A record is then made upon the original order sheet, and also upon the carbon copies attached, opposite each item, stating either that it is in stock or else giving the number of the stock order upon which it is to be manufactured. The original general order sheet is retained by the storekeeper for his use and that of the shipper, while the carbon copies are returned to the correspondent who is thereby informed as to the numbers of the stock orders which have been entered to provide for the apparatus wanted for that general order. These numbers are necessary in dealing with the manufacturing department as that department keeps no file of general orders. Orders for repairing apparatus that has been shipped back for that purpose and is to be returned to the customer after being repaired are not considered to be orders for stock, and are, therefore, handled according to a somewhat different method. Such orders are issued by the correspondent directly to the manufacturing department, a copy going to the storekeeper and to the shipper.

HEADINGS AND RULINGS OF STOCK LEDGER. ORIGINAL PAGE IS 14 BY 17 INCHES, ADAPTED FOR LOOSE-LEAF BINDING.

The Eng. Magazine

The whole stores system being built upon the theory that a stock of both finished product and raw material will always be maintained sufficient to meet the expected demands, it may be of interest to note the system of accounting used to accomplish this result without an excessive investment in material.

As will be shown further on, the engineering department determines what material shall be used for manufacturing purposes and issues to the manufacturing department bills or lists of the material required for every stock order. Copies of these bills are sent to the storekeeper at the same time so that he may know in advance what he will be required to provide.

The material accounts are all kept in the storekeeper's office, but in order to reduce the expense of handling to a minimum, the raw-material storerooms are so distributed throughout the works that each class of material may be kept as near as possible to the section of the works in which it will be used. Each of these storerooms is in charge of a stockman, and material is issued from them only upon requisitions signed by the shop foremen. As the stockman is responsible only for the stock itself and not for the accounts, all requisitions he receives are forwarded to the storekeeper's office where they are used to make the proper entries in the stock ledgers.

The sample page of one of the stock ledgers shown somewhat reduced in size will give a very fair idea of the storeroom accounting system and will serve to show that the intention is to make its operation as nearly automatic as possible and not dependent upon the judgment of the men in charge of the ledgers. The form shown is used for both classes of accounts—finished-stock and raw-material. The first of the upper row of columns is used for raw-material accounts to show the minimum amount of material liable to be drawn for any one order. The next column is used for finished-stock accounts to show the minimum number of pieces that can be manufactured economically. The supply to be carried in stock is determined from time to time by the storekeeper after consultation with the heads of other departments interested. The lower columns headed "Cost" are for inventory purposes. The amount "In Stock" is of course the difference between "Receipts" and "Issues." In the column headed "Required by Specification" are posted the amounts required for each order as shown by the bills of material from the engineering department. As these bills are received the amounts required are entered in this column and the total subtracted from the sum of the amounts given under "Ordered" and "In Stock," leaving the "Balance Available" which is the key to the successful operation of the system. Expressed in the form of an equation:

$$\text{"Balance Available"} = \text{"Ordered"} + \text{"In Stock"} - \text{"Required by Specification."}$$

Abstracts from the stock ledgers showing the condition of the stock of certain classes of apparatus are issued weekly for the information of the selling force and the manufacturing department.

Requisitions for new material are sent to the purchasing agent as the accounts show such purchases to be necessary. These are approved by the storekeeper and duly entered in the ledgers, a copy being sent to the receiving clerk stating to which storeroom the material is to be delivered when it arrives. Receipts of material are entered in the ledgers from the reports furnished by the receiving clerk who is required to make out a report of everything received in the works.

THE ENGINEERING DEPARTMENT.

The engineering department is responsible for all matters of design, both electrical and mechanical, and furnishes the specifications, drawings, bills of material, and all other engineering information necessary to enable the manufacturing department to make the apparatus called for by stock orders.

For every stock order a list is furnished by the engineering department of all the drawings that will be needed for that order, together with the information as to which sections of the shop are to make the different parts—or in other words, the "routing" of the order through the shop. This routing is established, of course, only after consultation with the mechanical experts of the manufacturing department.

All drawings according to which parts are to be manufactured are accompanied by bills of material giving every item of material required to make such parts. These bills are sent not only to the shop but to the storekeeper and to the cost department as well. The storekeeper's copies serve as advance information of the drafts that are to be made upon his stock, and enable him to place his requisitions for material as far as possible ahead of the actual requirements. In the cost department they are used in determining the material cost of the apparatus manufactured.

When new drawings do not have to be made the engineering information can easily be issued without serious delay and the programme for the work in the shop can be laid out in accordance with the time of delivery specified on the order. In case new drawings are required, however, a schedule, giving the dates set for completion of the drawings for each part, is made up as soon as possible and sent to the shop so that the shop schedule can be prepared without waiting for the receipt of the drawings. A copy of this schedule is kept in the drawing office until the drawings are finished.

For apparatus that has been made standard and given a style number permanent drawing lists are retained in the shop. Orders for apparatus of this kind, therefore, require no attention on the part of the engineering department.

THE MANUFACTURING DEPARTMENT.

The shop is under the direct charge of the manager of works, who with his assistants is responsible for the equipment and maintenance of buildings, tools, and machinery, for the manufacturing methods pursued and methods of paying the operatives, for the discipline maintained, and for the quantity, quality and cost of the output.

Under the plan of shop organization adopted the works are divided into a number of smaller shops, known as "sections," each equipped for the manufacture of a particular class of apparatus. To avoid duplication of equipment, certain operations, such as coil winding and insulating, punching sheet steel, pattern making, and foundry work, which are common to a number of different classes of apparatus, are performed in separate sections devoted exclusively to such work.

Each shop section is in charge of a general foreman and a production clerk, and for convenience is subdivided according to the nature of the different operations to be performed, each group of operations being directed by a foreman who may have one or more assistants.

Copies of every stock order as it is issued by the storekeeper are sent to each shop section that will have any work to do in connection with that order, and for the information of the production clerks the order sheet gives the required date of delivery and states whether the apparatus is for stock, or gives the number of the general order for which it is wanted and the name of the correspondent who entered the general order. The delivery date, of course, is of most importance to the production clerk of the section in which the apparatus is completed or assembled, and as it is usually necessary to establish the dates for completion of parts according to the requirements of the assembling section, the production clerk of that section assumes the burden of setting all dates for parts furnished by other sections, as well as for the finished apparatus.

When a stock order calls for apparatus the design of which has been standardized and given a style number, the clerk, after receiving the order, does not have to wait for further instructions but has only to turn to the permanent information issued by the engineering department for that style number to get the list of drawings and the bills of material necessary. He then makes out a card for each foreman in his section, giving him the number of the stock order, a brief statement of what it calls for, the name of the part upon which work is to be done, the numbers of the drawings upon which the part is detailed, the date set for the delivery of parts made by other sections, and the date upon which the foreman must have his work finished. At the same time he sends notices to the production clerks of the other sections of the shop from which parts are to be supplied, advising them when such parts must be delivered to the assembling section. The clerks of these sections thereupon issue similar cards to the foremen in their sections.

Upon receiving a date card from the production clerk the foreman first ascertains whether he has blue prints of the drawings specified. If not, they are ordered from the blue-printing room. He then estimates the length of time required to do the work, and from the date of completion given on his card he sets a date for starting his work and files his card under that date. In case it is impossible, on account of other work, to finish his work by the time specified on his card, the foreman immediately notifies the production clerk and either a new date is set or else arrangements are made, after consultation with the general foreman, for extending the time set for work of less importance. In case a new date is set on orders for stock, the storekeeper, of course, must be notified; but should the apparatus be wanted for shipment on a general order, a date for completion later than that specified on the order itself cannot be set without notifying the correspondent whose name appears on the order.

When the date arrives for starting work the foreman is reminded of that fact by the card in his file, and he makes sure that the work is actually started before transferring the card in the file to the date the work is to be finished. As the storekeeper receives copies of all bills of material issued by the engineering department, it can safely be assumed that material for standard apparatus will always be kept in stock and that the storekeeper will need no advance information as to the time the material will be used by the shop.

Should any unforeseen circumstances arise during the time the work on any part of an order is in progress that will cause a delay in its completion, the foreman immediately notifies the production clerk of his section, who in turn notifies the production clerk of the section finishing the order and to which the part is to be delivered. It then becomes the duty of the latter clerk to make whatever re-arrangement of the programme may be necessary, or even possible. If shipment will be delayed the correspondent must be notified. Neglect on the part of any foreman to deliver work to another section of the shop at the time specified cannot escape attention, as each production clerk keeps a record of what his section should receive each day and reports such neglect immediately to the clerk of the section in which it occurs.

The method of keeping track of orders for apparatus for which new drawings have to be made and for which special material has to be bought is, in general, the same as that used in the case of orders for standard apparatus, but as the number of possible complications is increased, greater watchfulness is required on the part of the production clerks. For orders of this kind the production clerk of the assembling section makes out a regular schedule for each element, including the tools and patterns and the principal items of raw material. This schedule being based upon the promise received from the engineering office for completion of the drawings, the production clerk must see to it that the drawings are finished as promised, or, if they are not, he must promptly notify the correspondent, who will take the matter up with the engineering department before a serious delay can occur.

As the drawings with their accompanying bills of material are issued, the storekeeper must be advised as to when the material should be brought in, castings must be ordered and the foundry notified when the patterns will be finished, and all other sections must be instructed to have their parts ready by certain dates. As a record of these transactions is made on the schedule, the production clerk or the general foreman, by glancing over the schedule, can ascertain, at any time, what parts have been finished and what the expectations are for the remaining parts.

Upon completion of the order the apparatus is delivered to the warehouse properly tagged, a receipt taken for it, and as far as the manufacturing department is concerned, the order is closed and a report to that effect is sent to the cost-keeping department.

THE PURCHASING DEPARTMENT.

As its name indicates, this department makes all purchases of material and supplies, and assumes all responsibility for prices; but, as it is not responsible for the maintenance of a sufficient supply of raw material, purchases of such material are made only upon requisition from the storekeeper.

Purchases of tools, machinery, fuel, office furniture, and other supplies which are not classified as raw material for manufacturing purposes and which are, therefore, not kept in the storekeeper's stock, are purchased upon requisitions issued by the department using such supplies. Requisitions for machine tools, for instance, are issued by the manufacturing department and sent directly to the purchasing agent. In all such cases a copy of the requisition is sent to the receiver to enable him to identify the shipment when it arrives and to deliver it to the proper department.

Purchase orders are sent out by the purchasing agent for material wanted to fill the various requisitions as they are received, and as soon as promises of shipment can be obtained, the storekeeper (or other department issuing the requisition) is notified of the probable date of shipment. When material, such as copper, for instance, is used in large quantity and the market price is subject to wide variations, contracts are usually made at periods of low prices for an amount sufficient to supply the needs of the works for some time in the future. The actual orders, however, are placed only as these needs are indicated by the storekeeper's requisitions.

When material is ordered in accordance with patterns, dies, or models which are to be furnished to the dealer, instructions for the shipment of these must be given to the shipper by the purchasing department. Shipping instructions for the return of rejected material are given in the same way.

THE SHIPPING DEPARTMENT.

This attends to the packing and arranges for the transportation of all shipments that are to leave the works. Such shipments can be made only upon general orders or upon instructions issued by the purchasing agent. All general orders, after the storekeeper has made provision to fill them, are passed to the shipper to enable him to make whatever preparations may be necessary, and as each shipment is made, whether on a general order or on instructions from the purchasing agent, a report is sent to the accounting department.

The methods used by any organization will depend largely upon the plan of the organization itself, and upon the character and volume of the work to be done; and the value of any system adopted can be measured by its success in meeting the demands of a rapidly increasing business. Except for some changes of minor importance, the system herein outlined has been used for about fifteen years by a company engaged in the manufacture of practically every type and size of electrical machinery and apparatus, the value of the yearly output having increased during that time about 1,000 per cent.

POWER EQUIPMENT FOR THE SMALL FACTORY.

By Percival Robert Moses.

In the following pages Mr. Moses presents the problems confronting the designer of factory equipment, and shows the more important factors which should receive consideration in determining the character of the plant so that it may do its work in "the easiest and cheapest way—cheapest meaning all things considered, and not lowest in installation cost" only. The author speaks his own conclusions, based upon large experience as a consulting engineer in this special line of practice.—THE EDITORS.

THIS article is limited to the discussion of the "small-factory equipment," by which is meant a factory depending upon one, two or three units with total capacity between 100 and 300 horse power, and with the work confined either to a single building or to a small group of buildings. The conclusions reached, and the methods advised, will apply equally to large factories; but the requirements in such installations are so diverse and the character of the equipment generally so different as to make the consideration of each type of large plant worthy of separate discussion.

The problem presented in the ordinary small factory is the generation and delivery of from 100 to 300 horse power, and its distribution for the purpose of operating various small tools or machines located at several points. Besides the operation of machinery, there is generally a requirement of low-pressure steam for heating, and steam of other pressures for purposes of drying, boiling, evaporation, etc. Power is also required for lighting, and possibly for electric heating; frequently water must be pumped, and fire protection must be maintained. In special classes of work, refrigeration is required for maintaining stated temperatures and humidities, and ventilation, either forced or exhaust, or both, for removing impurities in the air or waste products of manufacturing processes. Each instance, therefore, requires individual consideration in determining the type of plant required and the character of the apparatus needed to secure the best results for the least money. But all such decisions require basic facts of costs of installation and operation, and it is the intention to give these as far as it is possible to do so in a single paper.

In the choice of equipments we are limited practically to five methods of obtaining power:—first, the steam engine; second, the gas engine; third, the oil engine; fourth, water power; fifth, street service.

TABLE.

Cost of Producing 124,000 K.W. Hrs. per Year.

Conditions—Plant Operates 10 Hrs. per Day 310 Days per Year.

Character of Plant.	Amount of Fuel.			Rate of Fuel.	Cost of Fuel.		Cost of Labor.	Cost of Water.	Cost of Oil, etc.
	Without heating.	With heating.	Amount of Labor.		Without heating.	With heating.			
Steam Plant Non-condensing Coal Fuel	434	551	Eng. Fire One Man	2	868	1102	1200	110	500
	377			3	1302	1653	1200	110	500
		494		4	1508	1976	1200	110	500
				5	1885	2470	1200	110	500
Steam Plant Non-condensing Oil Fuel	2000	2468	Eng.	.50	1000	1234	900	110	500
				.75	1500	1851	900	110	500
				1.00	2000	2468	900	110	500
				1.20	2400	2962	900	110	500
				1.50	3000	3702	900	110	500
Gas Engine and Producer	124	357	Eng.	2	248	714	900 1200	200	500
				3	372	1071	900 1200	200	500
				4	496	1428	900 1200	200	500
				5	620	1785	900 1200	200	500
Oil Engine	600	1532	Eng.	.50	300	766	900 900	200	500
				.75	450	1149	900 900	200	500
				1.00	600	1532	900 900	200	500
				1.20	720	1838	900 900	200	500
				1.50	900	2298	900 900	200	500

Heating is figured for building 80 by 125 x 70 ft. high. New York City conditions.

Fixed charges at 10 per cent include interest and depreciation charges.

The first problem presented to the engineer is the choice between these ways. This may be said to be the most important and the most difficult part of the engineer's work. If he makes a mistake in advising as to the best method of the five mentioned to be employed, and his report is accepted, he may handicap the plant for the rest of its existence; as, after the investment is once made, it is no longer possible to choose on an equal basis between the various methods, the method chosen having on its side the cost of the investment and the cost of making the change, in which latter is included the inconvenience and delay. These are generally sufficient to prevent the change being made, and consequently the plant is handicapped during the rest of its working existence.

There is only one safe way to decide upon this question of power. The probable requirements of the plant being designed should be plotted carefully, either from data from similar plants or, when this

TABLE.

COST OF PRODUCING 124,000 K.W. HRS. PER YEAR.

Plant 75 to 90 K.W. Average Load for 10 Hrs. 40 K.W.

Cost of Repairs.		Total Mfg. Cost, Inc. Heat.	Value of Heating.	Net Mfg. Cost.	Gross Mfg. Cost per K.W. Hr.	Net Mfg. Cost per K.W. Hr.	Fixed Charge.		Total Cost per K.W. Hr.	
							No Allow. for Heat.	Allow. for Heat.	No Allow. for Heat.	Allow. for Heat.
200		3112	936	2176	2.51	1.75			3.32	2.48
200		3663	1170	2493	2.95	2.01			3.76	2.74
200		3986	1402	2584	3.21	2.08	.81	.73	4.02	2.81
200		4480	1635	2845	3.61	2.29			4.42	3.02
250		2994	936	2058	2.42	1.66			3.23	2.39
250		3611	1169	2442	2.92	1.97			3.73	2.70
250		4228	1402	2826	3.42	2.28	.81	.73	4.23	3.01
250		4722	1588	3134	3.82	2.53			4.63	3.26
250		5462	1862	3600	4.42	2.91			5.23	3.64
300	350	2964	936	2028	2.39	1.63			3.24	2.56
300	350	3321	1170	2151	2.68	1.73			3.53	2.66
300	350	3678	1402	2276	2.96	1.83	.85	.93	3.81	2.76
300	350	4035	1635	2400	3.26	1.93			4.11	2.86
200	350	2716	936	1780	2.19	1.43			3.04	2.36
200	350	3099	1169	1930	2.49	1.55			3.34	2.48
200	350	3482	1402	2080	2.81	1.68	.85	.93	3.66	2.61
200	350	3788	1588	2200	3.05	1.78			3.90	2.71
200	350	4248	1862	2386	3.42	1.93			4.27	2.86

The quantity of coal is expressed in tons and that of oil in barrels, both per annum.

All costs are given in dollars, per unit or per annum.

is impossible, from a careful study of the conditions that will exist in the plant when completed.

In making an estimate of the amount of power required from the prime mover, the mistake is made frequently of assuming the average and maximum loads of motors at far too high an amount, and the lighting load also. The ratio of maximum load to connected capacity —i. e., capacity of motors connected—varies from 80 per cent with high-speed continuous machinery, such as wood-working, to 33 1/3 per cent with machine-shop work. This ratio will also be greatly affected by the extent to which subdivisions of power is carried; e. g., the ratio will be much higher if there is only one motor than if there are half a dozen to the floor. In one building, used for sugar manufacture, in which motors were used for cranes, centrifugals, crystallizers, conveyors, pumps, etc., the ratio of maximum load to connected capacity was 33 1/3 per cent; i. e., with 360 kilowatts of motors, the maximum power load never exceeded 120 kilowatts. In a factory building containing wood-working machinery driven by one motor to each floor for about half the building, and book-binding machinery, sewing machines for garments, and rolling machines for the remainder, also motor-driven, the ratio of maximum load to connected capacity is 70 per cent. In a third, used entirely for printing, with motors partly on each press and partly used for driving groups of presses, the ratio was 30 per cent.

After the probable power and steam requirements have been settled, it is possible to determine which type of power producer is advisable and how much more the purchaser is warranted in paying for one type of producer than for another.

If water power is immediately available in sufficient quantity for the use of the factory, it is almost always the cheapest form of power to use. By this is meant, water power directly convertible into power for operation of machinery—not water power used to drive electrical machinery from a distance perhaps, of fifty or sixty miles. In plants where water wheels can be obtained to develop the power in the mill, the cost is practically limited to the investment charges, the cost of oil and repairs forming a very small fraction of the operating cost. I have figures from a number of water-power plants varying in size from a few hundred horse power up to several thousand, and the repairs on these plants are extremely small. On one 600 horse-power plant, the labor for engineers and switchboard attendant amounted to $1,220 a year, the cost of oil being less than $135 a year, and repairs $500 a year. At another plant, in the southern States, the cost was $1.24 a year per horse power installed; this included the maintenance of the electric equipment, and the total cost of operation in this latter case was less than $5.00 per horse power installed per annum. In plants where logs and ice interfere with maintenance of rack and water way, the cost of repairs, etc., may run up to $3.00 per horse power installed, but this is exceptional.

However, most factories are not located directly on a water power, and the choice is limited to a gas engine, an oil engine, a steam engine, or no engine at all, the service being supplied from a central-station company.

The table of operating costs on pages 902-903 is based upon the following assumed conditions of size and heating requirements :—

The building is 80 by 125, 70 feet high, with weather conditions those of New York and vicinity. In allowing for value of exhaust steam for heating, it is estimated that only one-half the heating re-

GAS-ENGINE AND HIGH-SPEED STEAM-ENGINE INSTALLATION.

That above is a gas-engine plant installed for the Firth-Stirling Steel Co., Washington, D. C.
It consists of two 16 by 18 producer-gas engines, 190 horse power each, supplied by
Westinghouse anthracite pressure producers. That below is a high-speed steam-
engine installation at the Pittsburg & Lake Erie Terminal, Pittsburg. Three
14 and 24 by 14 compound steam engines of 250 horse power each at 150
lb. Both by the Westinghouse Machine Co.

quirements will be supplied by exhaust steam. With oil fuel, four barrels of oil are taken as equivalent to one ton of No. 1 buckwheat coal, and sixty-two kilowatt hours are supposed to be obtained from burning of one barrel of fuel oil under boilers. The latter rate is somewhat low, but probably a fair average for small non-condensing factory plants. The first cost forming the basis of estimate of fixed charges, includes boilers, engines, dynamos, switchboard, steam and wiring connections, steam auxiliary apparatus such as feed heater, separator, pump, stack and foundations. In making allowance for heating, the fixed charges with the steam plant are reduced, because part of the apparatus installed would be required for heating whether a plant were installed or not. With the gas-engine and producer plant, and the oil engine, the first cost is increased (to provide for the heating) by the cost of boiler with setting, stack and connection. The fixed charges do not cover the heating or electric system of the building—merely the provision in the plant to supply these systems. Cost of labor will vary with conditions, but the comparative costs should be about correct. Water is charged at $1.00 per 1,000 cubic feet. Oil, etc., includes waste, packing, gauge glasses, sundries, tools, but no incandescent-lamp renewals or motor repairs.

The table shows the cost of producing 124,000 kilowatt hours a year, with a total connected capacity of between 150 and 200 horse power, and a maximum load of between 75 and 90 kilowatts, the average load between 40 and 50 kilowatts, the operation of the plant 10 hours a day. These are usual factory conditions where the load is a varied one. If the work done is in the nature of wood-working or electric heating, the average will be much higher, but this table will serve to show the relation between the cost with various types of power. It is evident that with the water-power plant omitted, the gas-producer and the gas-engine plant is more economical than steam power where coal costs more than $3.00 a ton, and more economical than oil fuel where coal costs less than $4.00 a ton and oil fuel more than 3 cents per gallon. It seems strange that under these conditions, the gas engine and gas producer have not had a wider application. There are good reasons for this, and the first and most important lies in the requirement of low-pressure steam for various purposes before enumerated. The effect of this is clearly shown in the columns of the table where allowance is made for value of heating and cost of supplying this factor with different types of plant. The second reason is the comparative novelty of the gas engine and gas producer and the consequent lack of standardization and confidence.

KOERTING ENGINE AND SUCTION GAS-PRODUCER PLANT, RUSHMORE DYNAMO WORKS.
Four-cycle 100 horse-power engine direct-connected to 125-Volt generator. De La Vergne
Machine Co. and C & C Electric Co.

A 65 HORSE-POWER TWO-CYLINDER HORNSBY-AKROYD OIL ENGINE OPERATING ALTER-
NATING-CURRENT GENERATOR, KUTZTOWN, PA., ELECTRIC LIGHT AND POWER
STATION.

One of the few municipally owned plants which is commercially successful; 125 horse-power
has been added, of the same type of engine, since the photograph was taken.
The De La Vergne Machine Co.

I think there is little question that the gas engine and gas producer
will find a very much wider application than they do at present, and
the heat now wasted in exhaust gases and in jacket water will be used
in the same manner as exhaust steam in the steam engine. There has
been a third reason why the gas engine and gas producer have not
been more widely used, though this reason has now practically disappeared; I refer to the excessive cost of the gas engine. A complete
gas-engine and gas-producer equipment can now be installed for about
the same price as steam boilers, steam engine, smoke stack, and auxiliary apparatus required in a steam plant. Heretofore, or to within
.the last couple of years, the first cost of the steam equipment was so
much less than the cost of the gas-engine equipment that with any low
load factor— that is, a low average load as compared to the maximum
load—the gas engine was not economically advisable, and this condition still holds true in many instances where the steam requirements
are a large factor, in which case the installation of a gas-engine and
gas-producer plant does not preclude the necessity of a steam-boiler
installation and smoke stack. In some installations where the amount
of power used is comparatively small, and the amount of steam very

large, the size of boiler plant may not be in any way increased by the amount of steam used for power purposes, and in such instances the comparison of cost must be made between the cost of the gas engine and gas producer and the cost of the steam engine alone. In such instances the gas-producer plant is rarely advisable.

The following table gives prices per kilowatt capacity of boilers, engines, dynamos, switchboard, steam and auxiliary apparatus required for plant operation, and similar figures for gas engine, gas producer, etc. :—

TABLE OF COST PER KILOWATT CAPACITY.
(Based on personal experience in New York and vicinity.)

	Per Kilowatt Plant Capacity.
BOILERS (erected and set in masonry) :	
Horizontal-Tubular	$14—$18
Water-Tube	16— 20
STEAM ENGINES :	
High Speed, Simple Direct-Connected........................	20— 25
Medium Speed, Compound Non-Condensing, Direct-Connected.	28— 35
Low Speed, Compound Condensing, Belted....................	20— 25
" " Simple Slow-Speed, Belted	25— 30
GAS ENGINES	50— 60
OIL ENGINES	75— 85
GAS PRODUCERS	15— 20
DYNAMOS :	
Direct-Connected to High-Speed Engine.....................	13— 16
Belt-Connected to Engine..................................	12— 15
Direct-Connected to Corliss Engine.........................	16— 20
SWITCHBOARD	5— 10
FOUNDATIONS	5— 10
STEAMFITTING—including auxiliary apparatus—such as feed heater, grease separator, exhaust head, tanks, covering, etc............	20— 30

In the table of comparative costs of operation, costs are shown per kilowatt hour with fuel at $2, $3, $4, and at $5 a ton, and oil at 50 cents, 75 cents, $1.20, and $1.50 a barrel. It is, of course, evident that the cost of fuel and its quality has a very important bearing on the choice of the plant. It may absolutely decide the character of the plant. If wood is plentiful and cheap, wood will be the fuel used, and the plant would be a steam-operated plant. If coal is plentiful and cheap, the plant would still be a steam-operated plant. If coal is expensive, or if wood is scarce, but the only available fuel, the gas-producer plant would be indicated; and if fuel oil is obtainable in large quantities and low price, the choice is restricted to the oil engine and to the oil burner under boilers operating a steam engine.

As between these two propositions—that is, the oil engine and the oil fuel under the boiler—there is little to choose as regards the first cost of the installation. Where steam is required for heating and other purposes, the steam plant has the preference. Where it is not required, there is no objection to the oil engine which, if properly designed and built, is an entirely reliable piece of apparatus, only requiring intelligent care and a little hard work. It has the great advantage of simplicity and small stand-by losses, when compared with a steam-boiler and engine plant.

If there is a supply of natural gas available, the conditions are again different, and the question then becomes one of the cost of natural gas and the requirements of steam for other purposes.

If service from a central plant is available, a direct comparison may be made between the cost per kilowatt hour shown in the table and the price obtainable from the central service. If the size of building and its heating requirements are similar to those noted in the table, the figures may be used directly; otherwise, the value of heating must be computed with the actual requirements in mind. For New York conditions, from 1/3 to 2/3 ton per 1,000 cubic feet per year will be found to be the amount of fuel required for heating, depending upon dimensions of building, amount of heat required, exposure, and construction. To this must be added the cost of operating boilers, including labor, repairs, water, removal of ashes, sundries, and tools. The sum of these items constitutes the value of heating as shown in the table for the typical building.

After the question of the character of the power to be used has been settled—that is whether it shall be steam, gas or oil, or water power, or none at all—the question of the type of steam engine or steam boiler to be used, or gas engine or producer, is next to be settled. There are already well defined limits in steam practice for the different types of apparatus. For small powers, that is for 150 horse power or less and steam pressures of 125 pounds or less, the horizontal return-tubular boiler, brick-set, has the preference. Where the capacity of the individual boiler runs up over 150 horse power or over 200 horse power, and where the pressures exceed 125 pounds, it is usual for purposes of safety to divide the heating surface and make the boiler a water-tube boiler.

As between high-speed, medium-speed, and low-speed engines, for rapidly varying loads, and total loads of 150 kilowatts or less, the direct-connected high-speed engine with its automatic oiling system is extensively used. Where space permits, however, the slow or moderate-speed engine is really preferable on account of its heavy weight, the large size of its parts and ample bearing surfaces, and ease of maintenance in its first condition. It is a matter of simple arithmetic to show that if a piston travels over a certain point in the cylinder

FLEMING ENGINES IN INDUSTRIAL POWER-PLANT INSTALLATIONS.

The upper one is a Fleming medium-speed four-valve engine, 16 by 16, direct-connected to Crocker-Wheeler 100-kilowatt generator, plant of the McClintic-Marshall Construction Co. That below shows the Fleming piston-valve side-crank type, with all reciprocating parts enclosed, driving Bullock alternators. The nearer large unit is 300-kilowatt, with exciter driven by silent chain. The farther unit is 200-kilowatt. The Harrisburg Foundry & Machine Works.

RICE & SARGENT AND FOUR-VALVE NON-RELEASING GEAR CORLISS-ENGINE INSTALLA-
TIONS.

The upper figure shows a releasing-gear Rice & Sargent type Corliss of 150 kilowatts, speed
150 revolutions. The lower is a four-valve non-releasing gear high-speed Corliss
engine running a textile mill, size 200 kilowatts, speed 200 revolutions;
with belted exciter. Both built by the Providence Engineering Works.

FOUR VERTICAL CENTER-CRANK CROSS-COMPOUND SINGLE-VALVE AUTOMATIC ENGINES,
POWER PLANT OF THE FIRST NATIONAL BANK BUILDING, CHICAGO.
Ball engines, each direct-connected to a 150-kilowatt Crocker-Wheeler direct-current gener-
ator operating at 225 revolutions.   The Ball Engine Co.

300 times a minute, it will wear that cylinder much faster than a
piston will traveling over a long-stroke cylinder 100 times a minute,
and the same condition, more or less, applies to all parts of the
machinery.

In plants where but little steam for heating and other purposes
is required, the engines will run condensing; and if the water is poor,
the condenser will be of the surface type; if it is good, and the water
is plentiful, it will probably be of the jet type.   There is only one way
to determine whether it will pay to run condensing or non-condensing,
and that is to estimate both ways the total cost of running the plant

with a condenser and without a condenser, doing the heating and supplying the other steam requirements.

Steam turbines are now making a strong bid for public favor, but they are not sufficiently standard and perfect, in my opinion, to be adopted as yet in small plants. Where they are run non-condensing, their main advantage is in the saving of oil, and this advantage is far more than counterbalanced by the excessive use of steam operating under these conditions. Where they are run condensing, they offer several advantages; viz., the absence of any cylinder oil, tending to foul the boilers, and an economy equal to the best compound-condensing engine, together with small floor space, low cost of foundations, and a very compact installation. The possibility of their use is worth carefully considering where operation condensing is feasible and advisable.

As between vertical and horizontal steam engines, the verdict has been in favor of horizontal engines for 500 horse power or less, on account of the ease of adjustment of the moving parts, and carrying the weight of these moving parts on a solid bed rather than on the moving parts themselves.

As to gas engines and gas producers, the two main varieties are the pressure producer and the suction producer. There are special producers arranged for the recovery of by-products, such as ammonium-chloride, but these are not to be considered in small factory equipments; they become valuable only where the use of coal runs into the hundreds of tons per day. In the usual type of suction producer, anthracite coal must be used, and this system has the merit of simplicity and low first cost. For a single-unit plant, where anthracite is economically available, a suction producer is the best because it is the simplest and least expensive. Where more than one unit is in use, and the units are small in size, the suction-producer plant, if properly piped, is entirely satisfactory; but for large installations, or where bituminous coal must be used, the pressure producer, with gas tanks to receive the gas before it reaches the engine, is preferable.

Where bituminous coal is available and anthracite coal is not, on account of its price, the bituminous producer is entirely satisfactory. It requires a great deal more attention than the suction producer and it is harder to maintain, but provided proper arrangements are made for taking out the tar and scrubbing the gas, no especial trouble with the engines need be feared.

Gas engines designed for use with illuminating or natural gas are not suitable for operation on producer gas, unless the power to be

MODERN GAS-PRODUCER PLANT FOR POWER AND FUEL.
A 300 horse-power automatic pressure-producer generator (below) and wet scrubbers
(above), installed for the Penn Hardware Co., Reading, Pa., by the Wile
Power-Gas Co. Started in February, 1907.

WESTINGHOUSE STEAM-TURBINE INSTALLATIONS IN FACTORY POWER PLANTS.
The machine above is in the factory of the Griffin Wheel Co. That below is installed for
the Suburban Manufacturing Co., Waltham, Mass. Both by the Westinghouse
Machine Co.

CURTIS TURBINE INSTALLATIONS IN INDUSTRIAL PLANTS.

The upper one is a 75-kilowatt 250-Volt machine installed for the Ferracute Machine Co.
Bridgeton, N. J. The lower is a 150-kilowatt 250-Volt turbine at the Wilton, Ky.,
plant of the North Jellico Coal Co. The General Electric Co.

derived from them is very greatly reduced, because producer gas has a value of from 125 to 150 British thermal units only per cubic foot; while natural gas has a value of 1,000 British thermal units, and illuminating gas of 700 British thermal units per cubic foot.

A 50 HORSE-POWER TWO-CYLINDER HORNSBY-AKROYD ENGINE OPERATING GENERAL MACHINERY IN A FOUNDRY.

The De La Vergne Machine Co.

Oil engines of the Diesel or Hornsby-Akroyd type are properly gas engines, the gas being formed by the vaporization of oil. In the Diesel engine vaporization takes place in connection with compressed air cooled before being mixed with the oil. The combined charge of oil and air is then injected into the engine cylinder where it meets the air of this cylinder compressed by the returning piston. By this compression the air is heated to such a point as to ignite the mixture of fuel oil and air, and this burns, propelling the piston forward. With the Hornsby-Akroyd engine, the oil is injected directly into a vaporizer connected to the cylinder of the engine, and an explosion takes place when the returning piston compresses the air mixing it with the fuel oil, the mixture being heated by the heat of the cylinder walls together with the heat of compression. The Diesel engine uses at full load approximately 1/11 of a gallon of fuel oil per brake horse-power per hour. The Hornsby-Akroyd engine uses approximately 1/8 to 1/9

TYPICAL INSTALLATIONS OF THE DIESEL ENGINE.

The upper view shows two units of 170 brake horse power each, coupled to 120-kilowatt
G E. direct-current generators, 200 revolutions, in the Traction Terminal Building,
Indianapolis. The lower shows the Kimberley & Clark Paper Mills, auxiliary to
water power; one unit of 450 horse power with 300-kilowatt generator.
two of 225 with 150-kilowatt generators, and one of 120 driving
engine compressors. The American Diesel Engine Co.

919

of a gallon of fuel oil per hour. As to which of these types of engines be used, the question of cost of fuel and cost of apparatus must be considered in each case in order to arrive at a decision. The Hornsby-Akroyd engine is somewhat simpler and less expensive than the Diesel engine, but its guaranteed efficiency is also less.

After the question of the type of power-generating apparatus has been decided, the next important question is that of its application. The use of electricity for transmitting the power from the engine to the tool is gradually superseding other methods of transmission, except where large amounts of steady power are to be delivered to long continuous shafts, such as is the case with looms, yarn mills, etc. Wherever the apparatus driven works intermittently, or where it is spread out over considerable area, the electric drive offers so many advantages as to make its choice almost a foregone conclusion.

The question whether to use individual drive or group drive from a single motor is one that is not so definitely settled. The individual drive, if it were not for its expense, is preferable in all cases, as it allows adjustment of the operation of each machine to the need of the work and to the operator, and allows its location in any desired position; and when the apparatus is shut down no power is being used and none wasted in friction losses. Where there are a number of small machines or pieces of apparatus using but little power for each one, it seldom pays, however, to use a separate motor for each machine, and it is customary and advantageous to group a number of small machines around a single countershaft and drive this shaft from a motor. The electric drive offers the further advantage of ready metering, and this is particularly important where the building is occupied by more than one tenant and it is desirable to sell the current used either at cost of so much per horse-power per year or at the usual meter rates, so much per kilowatt per hour. With the old system of belt drive it was difficult to tell how much power a tenant (or a department) was using, an dthe results were neither fair to the user nor to the supplier of power.

If electric drive is used the question immediately arises as to whether the current supplied shall be alternating or direct, and, if direct, whether at 120 volts or 240 volts. For most factory work the direct-current motor with its easily variable speed, its large starting power, and its small starting current, is preferable to the alternating-current motor. The special field of the alternating-current motor is for continuous service and for locations where possible sparking at the brushes may be dangerous, or where dust and dirt are liable to accumulate and damage the motor.

ALTERNATING CURRENT ELECTRIC-GENERATOR INSTALLATIONS FOR FACTORY POWER
PLANT.

The unit above is a 200-kilowatt 240-volt 3-phase alternator at the Eagle Tanning Co., Grand
Haven, Mich.; that below is an alternator in the plant of the Hecla Portland Cement
Co., Bay City, Mich. Both installed by the Western Electric Co.

ELECTRIC GENERATORS IN STEAM-DRIVEN MANUFACTURING PLANTS.
The upper picture shows Crocker-Wheeler 200-kilowatt 250-volt generator driven by Hewes &
Phillips engine at the Patton paint works; the lower one shows two 75-kilowatt 125-volt
Crocker-Wheeler generators driven by Ball engines at 275 revolutions, Sheriff Street
Market & Storage Co., Cleveland, Ohio. The Crocker-Wheeler Co.

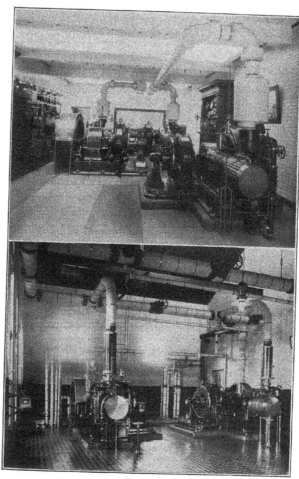

MODERN ELECTRIC GENERATING SETS.

Above, 50 and 100-kilowatt units in the store of John C. Haynes, Boston, Mass. Below, two 75-kilowatt units at Dartmouth College, Hanover, N. H. Both are McEwen engines and Thompson-Ryan dynamos, installed by the Ridgway Dynamo & Engine Co.

If there is much lighting to be done in the factory, the voltage to be used should be 120 volts. If there is not much lighting, 240 volts is preferable, as it reduces the cost of feeders, and the switches are not so expensive. The 120-volt system means more substantial motors and switches, and all parts are heavier than for the 240-volt system, so that unless the cost of feeders is sufficiently greater on the lower voltage system to make the saving an object, the 120-volt motors are preferable, and they will be found to give much less trouble in operation than the higher voltage motors. This is due partly to the fact that there is a smaller difference in pressure existing between adjacent commutator segments on the motors, and also because the size of the wire used in the winding of the fields and in the winding of the armature on the lower voltage motors is four times as heavy as the wire used on the higher voltage motors.

In general, therefore, the equipment for a small factory would be designed for 120 volts, unless special conditions should make the use of alternating-current motors advisable, in which case the voltage would be 200.

With any type of plant described in this article the operation is comparatively simple, but it is important that some record should be kept of the operation, so that wastes may be checked before they have gone too far and (what is perhaps more important) to have the engineer in charge of the plant familiar with the quantities of electricity supplied, steam used, and the amount of fuel required to produce these quantities of steam and electricity. The records should include some hourly system of records, similar to a ship's log, to insure that someone is attending to business at least once an hour, and to prevent any continuation of wrong conditions, which might otherwise pass unnoticed until they had resulted in damage to the equipment.

On page 925 is shown a simple log blank for a small factory building containing one unit, and others for more complex plants will be found in an article entitled "Systematized Operation of Isolated Plants" which appeared in THE ENGINEERING MAGAZINE of December, 1903. These give all the essential information, and if summarized weekly or monthly would prove a valuable aid in maintaining the plant at its best efficiency. Such records should be supplemented from time to time by indicator cards of the engines, and by special reports on electric lamps and other supplies used in the plant. On page 926 a blank showing the method adopted in some plants to keep track of the electric lamps used and allow a comparison of the life of various makes is shown. Similar forms can be gotten up for packing and other supplies, but no attempt has been made in this

DAILY LOG OF POWER PLANT

Time	Amp's	Volts	STEAM PRES.	BACK PRES.	METERS	EDISON LIGHT	POWER	SUMMARY 11-28-06
7 A.M.	60	230	90	2½	Pres. Reading			Avg. Temp. Day
8	210	"	"	"	Prev. Reading			Kind of Day
9	250	"	"	"	Diff.			
10	230	"	"	"	Units			Cans Ashes 8
11	200	"	"	"	Rate			Qual. Coal
12 M	170	"	"	"	Cost			
1 P.M.	200	"	"	"		Quan.	Rate	Cost	Coal on Hand
2	230	"	"	"	Cyl. Oil			
3	220	"	"	"	Eng. Oil			
4	240	"	"	"	Greases			
5	260	"	"	"	Coal	5000	5.62½	
6	70				Gas			
7	2340				Lamps			
8					Sundries			
9					H&T 72		635	
10	2270						$11.97½	
11								

TIME SHEET

REMARKS 538 K W H

Employee	Arr.	Due
Engineer		
Fireman		
Elev.		
Machine	Start	Stop
Boiler 1		
Boiler 2		
Generator	655	6

The Eng. Magazine

LOG BLANK FOR POWER-PLANT DAILY RECORDS.

paper to take up the minor details which have so much to do with the success or failure of the plant, as such a discussion would involve a book rather than an article. The intention has been to present the problem confronting the designer of a factory equipment, and to show the more important factors which should receive consideration in determining the character of the plant. After the general design has been determined, the question of the comparative merits of different types of furnaces, feed-water heaters, oil burners, pumps, steam and grease separators, engines, dynamos and motors, etc., is one to be settled by practical experience governed and directed by a knowledge based on the experience of others, commonly called theoretical knowledge. While not strictly germane to the subject of power-plant design, as a consulting engineer I may be permitted to use the engineering privilege of looking at all possible views of a matter, and a few words on the subject of so-called practical and theoretical knowledge may be in order.

— Lamp Record —

Lamp No.	C. P.	Make	Plain or Frosted	Watts	Put in	Re-placed	Re-placed	Re-placed	Re-placed	Re-placed	Re-placed	Re-placed	Re-placed

Lamp Voltage _____　　　Plant _____

The Eng. Magazine

FORM FOR LAMP RECORDS.

By practical knowledge is meant the knowledge gained by personal experience, such as the knowledge the operating engineer gains of boilers and engines, or the electrician of wiring or motors, or the machinist of machine tools. Such knowledge is necessarily limited to the type and kind of apparatus which the engineer, electrician, or machinist has used personally, or has personally seen used. The highest grade mechanic is usually the one too who will hold a position the longest with one firm, being promoted from grade to grade, and it follows from this that his knowledge, exclusive of his theoretical knowledge gained from books, while perfect in respect to the apparatus used in his own plant, may be restricted as to apparatus used in other plants, particularly if these be working along different lines from his own; unless such practical knowledge is combined with theoretical engineering knowledge based on results obtained elsewhere, or on theoretical knowledge of the fundamental laws of mechanics and properties of materials, progress in design will be along very narrow lines.

STURTEVANT GENERATING SETS IN FACTORY INSTALLATIONS.
Above is a 150-kilowatt set supplied to the power house of the Thomas G. Plant Co. Below
is a 250-kilowatt set with vertical cross-compound engine and 8-pole generator, at the
Sturtevant works. Both by the B. F. Sturtevant Co.

AMERICAN-BALL ANGLE-COMPOUND STEAM ENGINE INSTALLED IN THE MANUFAC-
TURERS' POWER PLANT.

The unit shown is of 160 horse-power, 300 revolutions, and is direct-connected to a 100-
kilowatt American-Ball generator. Dimensions of the set: 112½ in. long, 112 in. wide,
117 in. high with wheel above the floor. Weight 21,000 lb.

Practical knowledge and theoretical knowledge may be compared
to the horse and driver. A horse without a driver will find his way
back to the stable and may be able to travel roads he has traveled
before, but if new ways are to be gone over and new destinations
arrived at, a driver with a knowledge of the conditions, obstacles, and
comparative advantages of available roads must be looked to for
guidance. We must look to the practical mechanic for the work of
erection and manufacture, but the theoretical engineer is needed to
plan the way to do the work to gain the desired end in the easiest and
cheapest way—cheapest meaning "all things considered," and not
"lowest in installation cost."

WATER FOR ECONOMICAL STEAM GENERATION.

By J. C. Wm. Greth.

In an article published in the preceding issue of this Magazine, Mr. Greth presented a clear analysis and discussion of the relative purity of water supplies from various natural sources, and of the losses due to their direct use for boiler feed. The character and effect of the most conspicuous impurities were described, and an outline was given of the methods commonly followed to meet the difficulties in plant operation thus caused. The two types of water-softening system were indicated, and the results which an efficient system should secure were succinctly tabulated. Examples from the records of typical stations were shown to point out in the concrete the effect of the system advocated by the author.

In the pages following, this practical aspect of the matter is given precedence and a full showing is made of the actual savings effected in a large number of cases, the figures being proven wherever possible by complete records from the books, covering a length of time sufficient to assure true averages.—THE EDITORS.

A 3-INCH BOILER-FEED LINE REDUCED BY SCALE TO
I INCH.

To supply the boilers working at 110 lb., the pump gauge
registered from 160 to 175 lb. An open heater
was in use.

SINCE our natural water supplies are such that boilers fed with them cannot be operated as economically as they should be, it is absolutely essential that means be taken to purify the water by removing from it such substances as interfere with the boilers being operated at the highest efficiency. Preventing the accumulation of scale by throwing the scale-forming substances into suspension in the boiler itself is only a make-shift. The offending material is still there and, by concentration, in an ever-increasing quantity, which cannot be prevented or removed by blowing off; even if no scale accumulates in the boilers they must frequently be opened to remove the collected sludge.

PART OF 1¼-INCH WATER-COLUMN LINE, OPENING REDUCED BY SCALE TO ½ INCH.
The water was supplied to the boiler through the line shown on the preceding page.

Since there is a method which will successfully purify water before it is fed into a boiler to prevent the accumulation of scale and corrosion, it will be apparent to anyone who has given any consideration to the economy of steam-boiler operation, or who has ever examined the bills in connection with it, that a water-softening system is an important factor in the economical generation of steam, where the water supply is not soft and clear and all that it should be. To determine whether a water-softening system would be a paying investment, let anyone operating a steam boiler compile figures showing the following:—

1.—Cost of city water.
2.—Cost of cleaning boilers.
3.—Cost of cleaning heaters, piping, and economizers.
4.—Cost of repairs.
5.—Cost of compounds.
6.—Cost of tube cleaners.
7.—Cost of fuel required to heat the boilers cooled for cleaning and repairs. (One ton of coal per 100 horse power may be used as an average figure for this calculation.)
8.—Interest and depreciation on spare or idle boilers. Against this should be figured for operation with a water-softening system the following:—

1.—Cost of pumping water (if city water is used in figures under item 1 above), which will vary from 1 mill to about 2 cents per 1,000 gallons, depending on the amount, type of pump, and distance pumped.

2.—Cost of softening water (can easily be determined from the analysis of the water).

3.—Cost of washing out boilers (once in six months with an efficient water-softening system).

4.—Cost of heating to steam boilers cooled for washing.

5.—Depreciation of 10 per cent; the investment may be taken roughly at $5.00 per horse power for plants up to 500 horse power; at $4.00 per horse power for plants from 500 to 800 horse power; $3.50 per horse power for plants from 800 to 1,250 horse power; and above this from $3.00 to $1.50 per horse power for plants of 6,000 horse power or greater. The above figures include foundations and ordinary connections.

6.—Interest on investment at 6 per cent.

The above does not take into consideration any saving in fuel. That a saving in fuel will be effected when using a clear, soft water as compared with the use of a water which forms scale, is beyond question, since with the clear, soft water the boilers will always be clean, while with an impure water, even though the boilers may be regularly cleaned, the amount of scale will vary from cleaning to cleaning, depending upon how thoroughly the cleaning is done; it is, however, practically impossible to clean thoroughly by mechanical means many of the various types of boilers now on the market, and if the prevention of scale formation is accomplished with boiler compounds there will not be obtained the same economy in fuel consumption as when the boilers are fed with a clear, soft water.

An absolutely safe figure for use in calculating the possible saving in fuel due to increased evaporation may be taken at 1 per cent. In other words, a boiler under average operating conditions which is regularly cleaned with tube-cleaning apparatus and with the use of a compound, evaporates 10 pounds of water per pound of coal; the boiler being clean all the time, the evaporation is increased to 10.1 pounds. This saving as compared with the figures already given is conservative, and perfectly safe for anyone to use who is investigating the subject to determine whether a water-softening system would be a paying investment. One of the largest steam users in the United States, who has gone into the subject of water softening to probably a greater extent than any other steam user, makes the

SCALE DEPOSITED IN A WOODEN TROUGH CARRYING HOT WELL WATER TO THE SEWER.
The sample is 3 in. thick and was 22 ft. long. The water had been raised to 110 degrees in ammonia and steam condensers.

statement that in one of his plants having an output of 40,000 tons of steel annually, he is saving 200 pounds of fuel per ton of product. This plant has 1,300 horse power boiler capacity. On the basis of 30 pounds of water per horse power per hour, when the boilers were regularly cleaned and compound used the evaporation. was 6 pounds of water per pound of coal; with the water-softening system the evaporation was increased to 7½ pounds of water per pound of coal, or a saving of 25 per cent in fuel. The analyses before and after treatment are as follows :—

RIVER WATER.

Raw.	Grs. per U. S. Gallon.	Treated.	Grs. per U. S. Gallon.
Volatile and Organic Matter.	1.10	Volatile and Organic Matter.	.75
Silica	.25	Silica	.20
Oxides of Iron and Alumina.	.15	Oxides of Iron and Alumina.	.05
Calcium Carbonate	6.55	Calcium Carbonate	1.09
Calcium Sulphate	5.68	Calcium Sulphate	.22
Calcium Chloride	7.50	Magnesium Carbonate	1.07
Magnesium Chloride	3.21	Sodium Sulphate	5.97
Sodium Chloride	21.45	Sodium Chloride	30.50
TOTAL SOLIDS	45.89	TOTAL SOLIDS	39.85
Suspended Matter	2.35		
Free Carbonic Acid	.11		
Incrusting Substances	24.44	Incrusting Substances	3.38

To determine whether a water supply will prove to be a satisfactory boiler feed it is advisable not to wait for effects in the boilers, but to have a complete analysis made in order to determine the effect on the boilers and the purifying reagents to be employed in case treatment is necessary. The usual reagents are lime and soda ash; these are the cheapest known as well as the best adapted for the treatment of ordinary water supplies. . The cost may vary in different localities, but lime at ½ cent per pound and soda ash at 1¼ cents per pound would be safe figures for nearly all parts of the United States.

The following analyses represent typical supplies taken from various sources through the United States (together with the analyses after treatment), with the cost of reagents (based on above figures) required for treatment to reduce the scale-forming substances to a point where experience has shown that the water will neither scale nor corrode boilers.

WELL—INDIANAPOLIS, IND.

Raw.		Treated.	
	Grs. per U. S. Gallon.		Grs. per U. S. Gallon.
Volatile and Organic Matter.	1.60	Volatile and Organic Matter.	.40
Silica	1.10	Silica	.27
Oxides of Iron and Alumina.	trace	Oxides of Iron and Alumina.	trace
Calcium Carbonate	12.96	Calcium Carbonate	1.39
Calcium Sulphate	10.54	Magnesium Hydrate	.65
Magnesium Carbonate	8.04	Sodium Carbonate	.11
Magnesium Chloride	2.41	Sodium Sulphate	11.21
Sodium Chloride	1.63	Sodium Chloride	4.65
		Sodium Hydrate	.30
TOTAL SOLIDS	38.28	TOTAL SOLIDS	18.98
Suspended Matter	.75		
Free Carbonic Acid.........	1.10		
Incrusting Substances	36.65	Incrusting Substances	2.71

Cost of treatment, 3.7 cents per 1,000 gallons.

WHITE RIVER—INDIANAPOLIS, IND.

Raw.		Treated.	
	Grs. per U. S. Gallon.		Grs. per U. S. Gallon.
Volatile and Organic Matter..	.80	Volatile and Organic Matter..	.11
Silica	.35	Silica	.17
Oxides of Iron and Alumina..	trace	Oxides of Iron and Alumina..	trace
Calcium Carbonate	10.37	Calcium Carbonate	1.63
Calcium Sulphate	6.80	Calcium Hydrate	.15
Magnesium Carbonate	3.51	Magnesium Hydrate	.27
Magnesium Chloride	.87	Sodium Sulphate	7.18
Sodium Chloride	6.68	Sodium Chloride	7.78
TOTAL SOLIDS	29.38	TOTAL SOLIDS	17.29
Suspended Matter	3.65		
Free Carbonic Acid.........	.44		
Incrusting Substances	22.70	Incrusting Substances	2.33

Cost of treatment, 2 cents per 1,000 gallons.

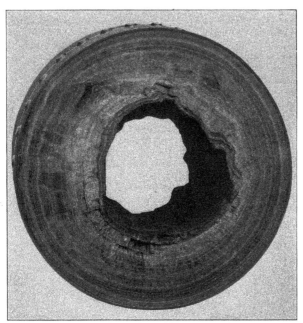

SECTION OF 10-INCH SPIRAL RIVETED EXHAUST OUTLET FROM OPEN HEATER.
Opening at heater reduced by scale to 2 in., tapering to full opening in 8 ft. Back pressure
on engine increased from about ½ lb. to 8 lb.

MONONGAHELA RIVER—McKEESPORT, PA.

Raw.	Grs. per U. S. Gallon.	Treated.	Grs. per U. S. Gallon.
Volatile and Organic Matter..	.25	Volatile and Organic Matter..	.05
Silica	.55	Silica	.25
Oxides of Iron and Alumina.	.80	Oxides of Iron and Alumina..	trace
Calcium Sulphate	6.07	Calcium Carbonate	1.39
Magnesium Sulphate	2.88	Magnesium Carbonate	.74
Magnesium Chloride	.20	Sodium Sulphate	13.14
Sodium Chloride	.58	Sodium Chloride	.82
TOTAL SOLIDS	11.33	TOTAL SOLIDS	16.39
Suspended Matter	1.75		
Free Carbonic Acid..........	.88		
Free Sulphuric Acid.........	2.22		
Incrusting Substances	10.75	Incrusting Substances	2.43

Cost of treatment, 2 cents per 1,000 gallons.

NIAGARA RIVER—BUFFALO, N. Y.

Raw.		Treated.	
	Grs. per U. S. Gallon.		Grs. per U. S. Gallon.
Volatile and Organic Matter.	trace	Volatile and Organic Matter.	trace
Silica	1.85	Silica	.15
Oxides of Iron and Alumina..	trace	Oxides of Iron and Alumina..	trace
Calcium Carbonate	2.20	Calcium Carbonate	1.25
Calcium Sulphate	2.11	Magnesium Hydrate	.25
Magnesium Carbonate	.48	Sodium Sulphate	2.21
Magnesium Chloride	.05	Sodium Chloride	.80
Magnesium Nitrate	1.16	Sodium Nitrate	1.31
Sodium Chloride	.76		
TOTAL SOLIDS	8.61	TOTAL SOLIDS	5.97
Suspended Matter	.10		
Free Carbonic Acid.........	1.43		
Incrusting Substances	7.85	Incrusting Substances	1.65

Cost of treatment, .8 cent per 1,000 gallons.

BAD WELL WATER.

Raw.

	Grs. per U. S. Gallon.
Volatile and Organic Matter............................	6.10
Silica	1.05
Oxides of Iron and Alumina............................	trace
Calcium Sulphate	92.28
Magnesium Carbonate	10.31
Magnesium Sulphate	28.38
Magnesium Chloride	1.55
Sodium Chloride	.23
TOTAL SOLIDS	139.90
Suspended Matter	2.90
Free Carbonic Acid...................................	1.00
Incrusting Substances	139.67

Treated with Lime and Soda Ash.		Treated with Barium Carbonate, Lime and Soda Ash.	
	Grs. per U. S. Gallon.		Grs. per U. S. Gallon.
Volatile and Organic Matter..	.75	Volatile and Organic Matter..	.35
Silica	.20	Silica	.15
Oxides of Iron and Alumina..	trace	Oxides of Iron and Alumina..	trace
Calcium Carbonate	1.85	Calcium Carbonate	1.96
Magnesium Hydrate	.45	Magnesium Hydrate	.55
Sodium Carbonate	.11	Sodium Sulphate	7.19
Sodium Sulphate 131.06		Sodium Chloride	2.14
Sodium Chloride	2.10		
Sodium Hydrate	.07		
TOTAL SOLIDS	136.59	TOTAL SOLIDS	12.34
Incrusting Substances	3.25	Incrusting Substances	3.01
Non-Incrusting Substances ...133.34		Non-Incrusting Substances ...	9.33

Cost of treatment, 20.2 cents per thousand gallons.   Cost of treatment, 42.3 cents per thousand gallons.

For the last analysis of raw water two treatments are shown. The water treated with lime and soda ash did not make a satisfactory boiler water on account of foaming, so barium carbonate, lime, and soda ash were used with good results. This is the only available supply unless a pipe line were run some eight miles for a better one. It was found impossible to operate with the raw water on account of high cost of fuel and repairs.

ANOTHER BAD WELL WATER.

	Raw. Grs. per U. S. Gallon.		Treated. Grs. per U. S. Gallon.
Volatile and Organic Matter..	1.05	Volatile and Organic Matter..	.35
Silica	2.10	Silica	.55
Oxides of Iron and Alumina..	.35	Oxides of Iron and Alumina..	trace
Calcium Carbonate	14.37	Calcium Carbonate	.75
Calcium Sulphate	62.90	Magnesium Hydrate	.52
Magnesium Carbonate	4.54	Sodium Carbonate	.48
Sodium Sulphate	31.95	Sodium Sulphate	97.70
Sodium Chloride	15.00	Sodium Chloride	15.00
Sodium Nitrate	5.59	Sodium Nitrate	5.67
		Sodium Hydrate	.09
TOTAL SOLIDS	137.85	TOTAL SOLIDS	121.11
Suspended Matter	1.55		
Free Carbonic Acid..........	1.65		
Incrusting Substances	85.31	Incrusting Substances	2.17
Non-Incrusting Substances ...	52.54	Non-Incrusting Substances ...	118.94

This water after treatment compares very closely in the total non-incrusting solids with the preceding water treated with lime and soda ash, which in that case could not be used on account of priming and foaming.

A BOILER TUBE ENTIRELY FILLED WITH SCALE.
No trouble from scale had been suspected until it was found necessary to remove more than 100 tubes like this.

Priming and foaming are brought about by conditions not thoroughly understood, and usually are attributed entirely to the water used. Experience, however, has shown that foaming is often the result of surface conditions in the boiler, and in nearly all cases can be directly traced to the presence of suspended matter, coupled with the condition of the boiler, the pressure and rate at which the boiler is operated, the type of boiler, and the arrangement of steam piping. Clean boilers fed with a clear water practically free f r o m scale-forming substances are not likely to foam. Foaming very often is attributed to the sodium salts which a water contains, and which cannot be removed by treatment. This may be true to a certain extent, but in the above case experience has shown that a water containing almost 119 grains of sodium salts per U. S. gallon, when fed into a clean boiler, did not cause priming or foaming. It will be noted that the scale-forming substances in the analysis shown above have been reduced to a low point, that the water is entirely free from suspended hatter, and that even with the concentration of the sodium salts no trouble from foaming or priming is experienced. This treated water is used in water-tube boilers, rated at 300 horse power each, operated at about 10 per cent above their rated capacity.

SCALE LOOSENED FROM TUBE BY BOILER COMPOUND, BUT CEMENTED AGAIN BY SCALE-FORMING SUBSTANCES IN THE FEED WATER.

Some soft clear natural waters corrode boilers on account of containing such substances as magnesium chloride and magnesium nitrate, as for instance the following taken from a well in Mississippi; this water, aided by its carbolic acid, causes corrosion due to the dissociation of the magnesium chloride into magnesium hydrate and hydrochloric acid, the alkalinity of the water being too low to fix the acid set free.

	Grs. per U. S. Gallon.
Volatile and Organic Matter	.25
Silica	.85
Oxides of Iron and Alumina	trace
Calcium Carbonate	.55
Calcium Sulphate	.27
Calcium Nitrate	.23
Magnesium Carbonate	.17
Magnesium Chloride	.23
Sodium Chloride	.30
TOTAL SOLIDS	2.85
Suspended Matter	.05
Free Carbonic Acid	1.32
Incrusting Substances	2.55

Cost of treatment, ⅓ cent per 1,000 gallons.

In some natural waters the scale-forming substances in solution, such as calcium carbonate, calcium sulphate, or magnesium carbonate, are not present to such an extent that they of themselves would form scale, but the presence of suspended matter in conjunction with them causes the formation of scale. This is due to the fact that the action of heat and the concentration of the soluble impurities form a precipitate that acts as a cement. The analysis of such a water, taken from one of the large American rivers, is as follows: Perfect clarification by sedimentation and filtration would make this an ideal supply.

RIVER·WATER.

	Grs. per U. S. Gallon.
Volatile and Organic Matter	.75
Silica	.35
Oxides of Iron and Alumina	.25
Calcium Carbonate	.52
Calcium Sulphate	1.85
Magnesium Carbonate	.50
Sodium Sulphate	.11
Sodium Chloride	.33
Sodium Nitrate	.16
Total Solids	4.82
Suspended Matter	18.05
Free Carbonic Acid	.33
Incrusting Substances	4.22

The following analysis shows a supply which is quite common in some sections of the United States; it contains a considerable amount of scale-forming substances, but in actual practice boilers are operated from two weeks to a month without cleaning; no scale accumulates but a great deal of sludge is thrown down, which is removed by frequent blowing off and washing out. The formation of scale is probably prevented by the sodium carbonate. While this water does not form scale, better results would be obtained if the scale-forming substances were removed from it before being fed into boilers.

WELL WATER.

	Grs. per U. S. Gallon.
Volatile and Organic Matter	.85
Silica	.15
Oxides of Iron and Alumina	.15
Calcium Carbonate	9.29
Magnesium Carbonate	3.70
Sodium Carbonate	5.01
Sodium Sulphate	2.13
Sodium Chloride	3.96
Total Solids	25.24
Suspended Matter	.15
Free Carbonic Acid	.22
Incrusting Substances	14.14

I have in my possession a large number of statements of savings effected by different concerns by obtaining a proper water supply for boiler feed. On account of the wide difference in the methods of bookkeeping as well as in the records kept at various points, it is impossible to obtain a complete statement of the savings effected; but the savings of which record could be obtained in each case are sufficient to have warranted the investment, and to bring a return of the money invested in a comparatively short period of time. If all the savings that are actually effected are taken into consideration, in every case the investment in a proper apparatus for supplying the boilers with a clear soft water will, under average conditions, be returned within two or three years.

SECTION OF BOILER TUBE ORIGINALLY FULL OF SCALE, SHOWING DISINTEGRATION
OF THE SCALE BY SOFTENED WATER.

The following statements are based on actual working conditions, and wherever estimates of figures were made it is so stated. In each case the analyses of the water are given before and after treatment. Of course all water supplies vary more or less and there will be some variation from the analyses given both in the raw and treated waters.

CASE I.

A 3,000 horse-power boiler plant located·on the bank of a creek and buying city water. The city water on account of the quantity used is obtained at 4 cents per thousand gallons, which is an exceptionally low rate. The city water contains free sulphuric acid at times, in addition to some scale-forming substances. The creek water is contaminated with mine drainage and is considerably harder than the city water. The analyses before and after treatment are as follows:—

CREEK WATER.

Raw.	Grs. per U. S. Gallon.	Treated.	Grs. per U. S. Gallon.
Volatile and Organic Matter..	.80	Volatile and Organic Matter...	.35
Silica	.55	Silica	.15
Oxides of Iron and Alumina..	.10	Oxides of Iron and Alumina..	trace
Calcium Carbonate	1.75	Calcium Carbonate	1.63
Calcium Sulphate	10.25	Magnesium Carbonate	.21
Magnesium Carbonate	3.40	Magnesium Hydrate	.38
Sodium Sulphate	1.44	Sodium Carbonate	.09
Sodium· Chloride ....:..	4.79	Sodium Sulphate	12.17
Sodium Nitrate	.10	Sodium Chloride	4.91
		Sodium Nitrate	.11

Total Solids	23.18		
Suspended Matter	4.35		
Free Carbonic Acid..........	.44	Total Solids	20.00
Incrusting Substances	16.85	Incrusting Substances	2.72

OPERATING WITH CITY WATER.

88,720,000 gallons of water (taken from meter readings) per year, at 4 cents per 1,000 gallons..................................	$3,548.80
Soda ash used as a compound, 60,000 lb. per year at 1 cent per lb...	600.00
Two men cleaning boilers throughout the year, at $45.00 per month each	1,080.00
Repairs, new tubes, patching and other incidental expenses due to bad water	1,790.00
	$7,018.80

OPERATING WITH WATER-SOFTENING SYSTEM TREATING CREEK WATER.

Softening 88,720,000 gallons of water per year (based on cost of reagents purchased during year)............	$1,552.60	
Cost of pumping 88,720,000 gallons of water at ¼ cent. per 1,000 gallons (estimated)........................	221.80	
One man washing out boilers and looking after water-softening plant, at $45.00 per month (water tender operating water-softening system without interfering with regular work at night)........................	540.00	
Interest at 6 per cent on investment of $5,700.00 in water-softening system	342.00	
Depreciation at 10 per cent on investment of $5,700.00 in water-softening system	570.00	3,226.40
Savings effected..		$3,792.40
Or over 66 per cent on an investment of $5,700.00.		

In the above statement it will be noted that the fuel saving is not taken into consideration because some of the boilers were operated with the waste heat from the furnaces and fuel was used for other purposes than for steam generation; therefore it was impossible to obtain the cost of the fuel used for steam-making purposes only. The cost of pumping the water from the creek was estimated at ¼ cent per thousand gallons, which is believed to be ample.

CASE 2.

A 1,000 horse-power boiler plant operating 10 hours per day at about 900 horse power, 300 days a year, using city water (a natural supply of average quality).

CITY WATER.

Raw.		Treated.	
	Grs. per U. S. Gallon.		Grs. per U. S. Gallon.
Volatile and Organic Matter..	1.55	Volatile and Organic Matter..	.29
Silica	.35	Silica	.20
Oxides of Iron and Alumina..	trace	Oxides of Iron and Alumina..	trace
Calcium Carbonate	13.12	Calcium Carbonate	1.90
Calcium Sulphate	.61	Magnesium Hydrate	.60
Magnesium Carbonate	3.90	Sodium Sulphate	.71
Magnesium Chloride	1.30	Sodium Chloride	1.87
Sodium Chloride	.21	Sodium Hydrate	.08
TOTAL SOLIDS	21.04	TOTAL SOLIDS	5.65
Suspended Matter	.75		
Free Carbonic Acid..........	.10		
Incrusting Substances	20.83	Incrusting Substances	2.99

OPERATING WITH CITY WATER.

5371.2 tons coal at $1.87 per ton.................................		$10,044.14
Boiler compounds per year, 3 kinds, 4 months, each kind:		
1,800 lb. at 9 cents lb.............................	$162.00	
2,000 " " 6 " "	120.00	
1,200 " " 12 " "	144.00	426.00
Repairs, new tubes, etc...........................:		219.00
		$10,689.14

OPERATING WITH WATER-SOFTENING SYSTEM TREATING CITY WATER.

4526.7 tons coal at $1.87 per ton........................	$8,464.93	
Total cost of reagents for softening water..............	251.00	
Depreciation at 10 per cent. on investment of $2,900.00..	290.00	
Interest at 6 per cent. on investment of $2,900..........	174.00	9,179.93
Savings effected..		$1,509.21
Or over 52 per cent. on an investment of $2,900.00.		

At this point I was able to obtain from the books of the concern a complete statement of the fuel used. The year previous to installing the water-softening system at this plant, contracts had been made with several companies to furnish boiler compounds. Three different compounds were tried during the year. No labor charge is included in this statement because the boilers were cleaned by the firemen, who were paid by the month.

<h2 style="text-align:center">CASE 3.</h2>

A 4,800 horse-power plant using a river water of about average quality, pumped by the plant. In this case again it was impossible to obtain a statement of fuel saving. One ton of coal per 100 horse power was used as a basis in estimating the amount of fuel required for heating to steam boilers cooled for cleaning. This, of course, also takes into consideration the heat lost due to cooling down the boilers.

RIVER WATER.

Raw.	Grs. per U. S. Gallon.	Treated.	Grs. per U. S. Gallon.
Volatile and Organic Matter..	2.16	Volatile and Organic Matter..	.60
Silica	.58	Silica	.25
Oxides of Iron and Alumina..	.30	Oxides of Iron and Alumina..	trace
Calcium Carbonate	4.67	Calcium Carbonate	1.25
Calcium Sulphate	5.31	Magnesium Hydrate	.55
Magnesium Carbonate	1.26	Sodium Carbonate	.25
Magnesium Chloride	1.20	Sodium Sulphate	5.55
Sodium Chloride	2.01	Sodium Chloride	3.50
TOTAL SOLIDS	17.49	TOTAL SOLIDS	11.95
Suspended Matter	2.11		
Free Carbonic Acid	.22		
Incrusting Substances	15.48	Incrusting Substances	2.65

OPERATING WITH RIVER WATER.

Cleaning twelve 400 horse-power Stirling boilers, one each week, 52 cleanings a year, two men 300 days a year at $1.37½ per day each	$825.00
Heating to steam 52 boilers 400 horse-power each, cooled for cleaning and repairs, 4 tons of coal per boiler, 208 tons at $1.40 per ton (estimated)	291.20
Repairs, new tubes, and incidental expenses, tube cleaners, etc., due to bad water	1,121.00
Boiler compounds, 48,000 lb. at 6¼ cents per lb.	3,000.00
	$5,237.20

OPERATING WITH WATER-SOFTENING SYSTEM TREATING RIVER WATER.

Reagents for softening water (about .9 cent per thousand gallons)	$1,296.00	
Washing out one boiler every two weeks, 1 day for one man, 26 washings a year, 26 days at $1.37½ per day..	35.75	
Heating to steam 26 boilers cooled for washing out, 4 tons coal per boiler, 104 tons at $1.40 per ton (estimated)	145.60	
One-sixth man's time looking after softening plant at $40.00 per month	80.00	
Repairs, incidental expenses on boilers	219.00	
Depreciation at 10 per cent on investment of $7,200	720.00	
Interest at 6 per cent on investment of $7,200	432.00	2,928.35
Savings effected		$2,308.85

Or over 32 per cent on an investment of $7,200.00, without allowance for savings of fuel.

CASE 4.

A 2,000 horse-power plant shows the following expense for six months operation without water-softening system and for six months operation with water-softening system.

WELL WATER

Raw.		Treated.	
	Grs. per U. S. Gallon.		Grs. per U. S. Gallon.
Volatile and Organic Matter..	.70	Volatile and Organic Matter..	.35
Silica	.85	Silica	.41
Oxides of Iron and Alumina..	trace	Oxides of Iron and Alumina..	trace
Calcium Carbonate	9.05	Calcium Carbonate	2.05
Calcium Sulphate	5.17	Magnesium Carbonate	.35
Magnesium Carbonate	7.31	Magnesium Hydrate	.57
Magnesium Chloride	.71	Sodium Carbonate	.11
Sodium Chloride	1.10	Sodium Sulphate	5.29
		Sodium Chloride	2.00
TOTAL SOLIDS	24.89		
Suspended Matter	.30	TOTAL SOLIDS	11.13
Free Carbonic Acid.........	.44		
Incrusting Substances.......	23.79	Incrusting Substances	3.73

OPERATING WITH WELL WATER.

Fuel	$14,718.40
Cleaning	864.44
Repairs	188.50
Incidentals	200.58
Labor	1,080.00
	$17,051.92

OPERATING WITH WATER-SOFTENING SYSTEM TREATING WELL WATER.

Fuel	$13,078.75	
Cleaning	40.00	
Repairs	94.00	
Incidentals	392.00	
Labor	1,080.00	
Interest at 6 per cent on investment (6 months)........	112.50	
Depreciation at 10 per cent on investment.............	187.50	14,984.75
Savings effected......................................		$2,067.17

Or over 55 per cent on an investment of $3,750.00.

CASE 5.

The statement given below is taken from a 450 horse-power boiler plant consisting of four 150 horse-power units, three of them being in operation all the time and the fourth being either cleaned or repaired From the cost of repairs that were necessary it is apparent that one of the boilers is in process of repair all the time. This plant is in operation at a coal mine, hence coal was not taken into consideration, as the plant owner does not realize that coal wasted under his boilers is

SECTION OF TUBE FROM A BADLY SCALED BOILER.

The hard scale has been loosened by the action of softened water.

worth to him that for which it could be sold. His principal reason for installing a plant was the cost of repairs. Each boiler was practically re-tubed every year.

WELL WATER.

Raw.	Grs. per U. S. Gallon.	Treated.	Grs. per U. S. Gallon.
Volatile and Organic Matter..	1.85	Volatile and Organic Matter..	.55
Silica	1.05	Silica	.12
Oxides of Iron and Alumina..	trace	Oxides of Iron and Alumina..	trace
Calcium Sulphate	30.60	Calcium Carbonate	2.11
Magnesium Carbonate	11.34	Magnesium Carbonate	.51
Magnesium Sulphate	2.15	Magnesium Hydrate	.70
Sodium Sulphate	4.39	Sodium Carbonate	.13
Sodium Chloride	1.81	Sodium Sulphate	38.78
		Sodium Chloride	1.80
TOTAL SOLIDS	53.19	TOTAL SOLIDS	44.70
Suspended Matter	.10		
Free Carbonic Acid.........	.55		
Incrusting Substances	46.99	Incrusting Substances	3.99

OPERATING WITH WELL WATER.

Boiler repairs, new tubes, and labor...............................	$1,450.00
Cleaning boilers, two days each week, 52 cleanings a year, 104 days at $1.60 each..	166.40
Cleaning heater twice a week (soda ash used in heater), 5 hours each cleaning, 52 days' labor at $1.60..........................	83.20
Soda ash, 30,000 lb. at 1 cent per lb...............................	300.00
	$1,999.60

OPERATING WITH WATER-SOFTENING SYSTEM
TREATING WELL WATER.

Boiler repairs (three new tubes).........................	$111.00	
Cleaning each boiler, 4 boilers, 4 times a year, 2 days each cleaning, 32 days at $1.60 per day....................	51.20	
Cost of treating water (about 4 cents per thousand gallons)	531.36	
Depreciation at 10 per cent on investment of $1,900........	190.00	
Interest at 6 per cent on investment of $1,900............	114.00	997.56
Savings effected..		$1,002.04

Almost 53 per cent on an investment of $1,900.00.

CASE 6.

A 5,000 horse-power boiler plant located on the bank of a river and buying city water. The plant is a power station of a large street-railway system. The statement is taken from the power-house record for the month of March, 1905, and for the same month of the year 1906. The engines in this plant are run surface-condensing. When the boiler was fed with this condensed steam and city water for make-up, it was found necessary to use a high-grade mineral oil for the cylinders, because the removal of oil by skimming devices from the condensed steam (in order to fit the latter for boiler feed) is more completely effected with a pure mineral oil than with an emulsifying mixture containing animal or vegetable oil. This will explain the decided drop in the cost of oil per month. (See next page.)

RIVER WATER.

Raw.	Grs. per U. S. Gallon.	Treated.	Grs. per U. S. Gallon.
Volatile and Organic Matter..	1.10	Volatile and Organic Matter..	.55
Silica	.50	Silica	.50
Oxides of Iron and Alumina..	.15	Oxides of Iron and Alumina..	.15
Calcium Carbonate	5.70	Calcium Carbonate	1.07
Calcium Sulphate	1.73	Magnesium Hydrate	.36
Magnesium Carbonate	.88	Sodium Carbonate	.29
Sodium Sulphate	.32	Sodium Sulphate	2.09
Sodium Chloride	1.32	Sodium Chloride	1.35
TOTAL SOLIDS	11.70	TOTAL SOLIDS	6.36
Suspended Matter	3.35		
Free Carbonic Acid.........	.66		
Incrusting Substances	10.06	Incrusting Substances	2.63

	March. 1905.	March, 1906.	Increase.	Decrease.
Labor—				
Operating Expenses:				
Boiler Room	$500.30	$491.00	......	$9.30
Boiler Repairs	61.71	37.50	......	24.21
Material—				
Oil and Grease.........	388.66	183.79	......	204.87
Water	126.14	66.68	......	59.46
Purifying Water	......	35.25	$35.25	.....
Total K.W.H.	1,156,975	1,305,897	148,922	.....
Total tons coal............	2,385	2,231	......	154
Lb. coal per K.W.H........	4.60	3.82	......	.78
Cost coal per K.W.H........	$0.00339	$0.00305	......	$0.00034
Expense per K.W.H........	$0.006	$0.00507	......	$0.00093
Avg. tons coal per day.....	77	71.97	......	5.03
Cost coal per ton...........	$1.47	$1.59	$0.12	.....
Cost coal per car per day...	$1.92	$1.75	......	$0.17
Cost oil per 1,000 K.W.H...	$0.336	$0.157	......	$0.179

SAVINGS EFFECTED OPERATING WITH WATER-SOFTENING SYSTEM.

Boiler-Room Labor Saving		$9.30
Boiler Repairs, saving...		24.21
Oil and Grease..		204.87
Water		59.46
Fuel, 154 tons at $1.59 per ton....................................		244.86
		$542.70
Cost of purifying water......................................	$35.25	
Depreciation charge per month at rate of 10 per cent on $7,000.00	58.33	
Interest charge per month at rate of 6 per cent on $7,000.00.	35.00	128.58
Savings effected per month...............................		$414.12
Savings effected per year...................................		$4,969.44
Or almost 71 per cent on an investment of $7,000.00.		

The above statements represent only average conditions and not by any means the worst, for often an entire power plant is shut down on account of the condition of the boilers, due either to the accumulation of scale or to the action of corrosive acids or salts.

It will be noted in the statements that 10 per cent for depreciation and 6 per cent for interest on the investment in the softening system have been allowed. As nearly as possible the actual cost of installing the system was obtained, which included the system itself, necessary foundations and connections. The life of a system of this character should be at least twenty years, so that the difference in the amount deducted for depreciation to take care of the cost of replacing the system will easily take care of the cost of repairs which may be necessary and for keeping it in proper condition.

It is a well-known fact that in changing the feed of a boiler from a hard to a soft water the scale becomes loosened, and that by the continued use of the soft water the boiler will in time become clean. The rate at which the scale will be loosened depends on its character and thickness and the conditions under which the boiler operates. The removal of the scale usually occasions some expense on account of exposing defects in the boiler which had been covered by it; and further, the removal of the scale by the change of water is sometimes accompanied by foaming or priming due to the loosened scale being whipped fine by the disintegrating effect of the circulation of the water. These effects are not chargeable against the soft water, but are after-effects of the use of bad water, and with care and the continued use of soft clear water will entirely disappear.

The ideal supply for boiler feed is a clear, soft, natural water which will not cause an accumulation of sludge or scale, or corrode the boiler. In the absence of such a supply, the best thing to do is to change by chemical treatment an impure water into one which is clear and soft before being fed into the boiler, and not to resort to make-shift methods for the prevention of scale.

Our city supplies, as a rule, are as hard as supplies which can usually be obtained more cheaply by pumping. It is better, then, to soften and purify the supply which can be pumped. When a water supply is not clear, soft, and free from corroding substances, a water-softening and purifying system will always earn large dividends on the investment, allowing liberally for cost of maintenance, depreciation and interest.

A steam-plant owner will install boilers of the most improved design, with mechanical stokers, damper, and feed-water regulators, install feed-water heaters and economizers and the most economical engine obtainable; but the purity of the water—the most essential factor which enters into the cost of producing power—is often totally neglected, and as a result of using an impure feed water, only a partial economy can be effected in the operation of the entire plant. The steam boiler is the heart of the steam plant; from it extend the arteries conducting the vital fluid which sustains the activities of the whole plant, and upon its successful operation depends to a great extent the economy of the entire mechanical system. The limit of possible economy is much more distant in boiler than in engine practice, and the engineers of the future are likely to be employed more with the problem of conveying a larger number of heat units from the fuel to the water than with the problem of making a pound of steam develop more foot pounds of work. The first essential in economy of boiler operation is that the water, the medium for transforming the energy of the fuel into power, should be of such a quality that it will not interfere with utilizing the greatest number of heat units in every pound of fuel, at a minimum cost of boiler maintenance.

FEED-WATER HEATING FOR THE POWER PLANT.

By Sidney A. Reeve.

Professor Reeve may always be counted upon to be interesting, whether his argument leads along well established lines or whether, as in this case, he strikes across the boundaries of ordinary practice to open up new suggestions. The interest here for those who may not agree with the ideas advanced—or at least not without modification—is heightened by Mr. Bolton's appended criticism and Professor Reeve's rejoinder, giving to the whole the fulness of a professional discussion.—THE EDITORS.

IN power plants making any pretense to high efficiency, the method of preheating the boiler-feed is a factor of considerable importance. This is true not only because it bears upon the heat which the fuel must supply, per pound of water passing through the boiler itself, but also because it affects the temperature at which the gases leave the economizer and enter the stack.

Such power plants may be divided into three classes, viz:

1.—Factory power plants run non-condensing, using their exhaust for heating purposes in winter.

2.—Factory power plants run condensing, except to such a degree as receiver steam is used, in whole or part, for heating.

3.—Power plants run condensing at all times.

In plants of the first and second classes the prime feed-water problem lies in the oil in the radiator returns. This, of course, can be and is minimized by the use of oil separators, etc. But the danger of oil being carried into the boilers in quantity is always there, while there is the certainty at all times that the boilers are subject to a certain degree of fouling, with its sequel of correctives and repairs.

In plants of the first class the best treatment of the question is to use only fresh, clean feed water; but to pass it, before use, through a closed heater through which the condensed returns pass in the opposite direction. In some cases, of course, the cost of the fresh water may be prohibitive of this plant, but not often. As to the extra cost of the heater, that will usually be much less than the cost involved in caring for the oil and making the repairs which it entails. And finally, as to the lessened temperature of water-supply, that is slight; and if the heater be arranged to utilize the appreciable quantity of steam which the returns bring with them, and which commonly goes to waste, it will be still less.

948

In plants of the second class there can be no question as to the advisability of using only fresh clean water. In these plants the primary heater constitutes a free supply of feed water of from 100 degrees to 125 degrees temperature. The primary heater is in reality a little preliminary surface-condenser. If the warm oily returns from the heating system be thrown to waste, and fresh cold water be run through the primary heater to the full amount needed by the boilers, the only penalty incurred is the purchase of a somewhat larger primary heater. The fresh supply of hot water is had without cost in the shape of fuel.

Incidentally, too, there is some reduction of load upon the condenser. Indeed, in plants working under a highly variable load, this latter feature may be of considerable value, in helping to take care of temporary overloads without spoiling the vacuum.

The same remarks are of course even more applicable to plants of the third class. No hot-well water should be used, except when the cost of a fresh supply of clean water is prohibitive.

From the primary heater the feed water should always go direct to the economizer. If it be replied to this that the economizer is not a fit piece of apparatus, constructively, to handle water of this temperature, it still remains true that the water from the primary heater "should" go there. In many cases economizers are handling cold water successfully. In every case they ought to do so. It is the prime object of the economizer's existence to cool the chimney-gases; and this it cannot do effectively if supplied with hot water.

From the economizer, or from the primary heater if there be no economizer, or from the original supply if there be no primary heater, the feed water should enter a closed heater and be heated by exhaust-steam from the pumps *under heavy back pressure.* The pumps should never exhaust into and through a heater under atmospheric pressure.

The policy of using direct-acting, non-expansive types of pumps is established in the United States beyond question. This policy needs no criticism, *provided* the pumps be used under the right conditions.

The correct conditions for efficiency in non-expansive working are a very slight range of pressure and temperature within the pump cylinder. For instance, engineers are now familiar with the idea that reciprocating engines, even when working expansively and compounded, cannot profitably lower their exhaust pressure below 24 inches to 26 inches of vacuum, corresponding to temperatures of 125 to 140 degrees Fahrenheit. Yet the turbine can go profitably far below this region.

Similarly, the direct-acting steam pump is a type which improves its efficiency only slightly as its back pressure is lowered. Indeed, it soon reaches a point where a lower back pressure is a positive harm. This point varies with the boiler pressure, of course, and with the design of pumps; but assuming an efficiency of action within the pump of 50 per cent, the following table gives the most profitable back pressure for a pump fed under several different boiler pressures:—

Boiler-pressure	120		150	180	250 lb. by gauge.
Best back-pressure..	7 inches vacuum		0	4	13 " " "

This, of course, is based upon a consideration of merely the efficiency of the pump itself. But the heat in the pump exhaust furnishes temperature to a much larger quantity of water than merely that finding entrance, as steam, to its own cylinders, and thus determining its own efficiency. The pump exhaust is useful for feed heating, and the hotter this feed can be gotten the better.

There is no limit to this last statement. The boiler will work at its best efficiency when the entering feed is at steam temperature. The pump will work, in one sense, at its best thermodynamic efficiency when its exhaust temperature is close to its admission temperature. The entire plant will work at its best efficiency when no heat escapes unused.

All these considerations point the wisdom of placing a heavy back pressure upon the heater into which the pumps exhaust, as near to boiler pressure as may be practicable; and then of running the pump on a virtually wide-open throttle, releasing the heater drip if more speed be desired.

The usual practice in pump operation is to throttle the available boiler pressure to a small fraction of itself at the pump admission. If the pump is to work under this greatly reduced effective pressure it is far better that the throttling be done on the exhaust, giving the plant the benefit, in its feed water, of the temperature range ordinarily lost in the throttling, than that it should be done on the incoming steam.

The limit to this practice, of course, is the ability of the feed water to condense the pump exhaust in the heater; and this in turn depends upon the temperature of the water entering the heater. If the back pressure be placed too high (and the pump steam-cylinder be made big enough still to provide sufficient power) the pump will exhaust more steam than the feed water can condense. If it be placed too low, the pump will be too efficient; it will pass so little steam that the feed water will enter the boiler colder than it might, with the same con-sumption of fuel.

Assuming a temperature of discharge from the economizer, or of entrance to the secondary heater, rather, of 250 degrees F., and allow-ing a temperature-difference between steam and water in the heater of 20 degrees, the following table gives the figures suited to several dif-ferent boiler pressures:—

1,000-Horse Power Plant.   18,000 Lb. Feed-Water per Hour.

Boiler pressure, by gauge..................	120	150	180	250 lb.
Proper dimensions for duplex steam pump, the same for all boiler pressures, no extras being allowed for..............		16 — 4¼ x 12 inches		
Back pressure on pump..................	102	130	157	222 "
Pressure difference	18	20	23	28 "
Pressure difference required to balance boiler pressure on plunger............	8½	10½	13	18 "
Temperature of boiler steam............	349	365	379	406 deg. F.
" " pump exhaust	339	355	369	396 " "
" " feed water entering boiler.	319	335	349	376 " "
Probable weight of pump steam per 1 lb. of feed water, pump being in good or-dinary condition	0.083	0.103	0.120	0.156 lb.

It is one of the happy coincidences of this plan that the varying pressure- and volume-effects, as the boiler pressure varies, work out to such a nice balance that the same proportion of pump cylinders will lead to the same temperature drop from boiler steam to pump exhaust, namely 10 degrees, and so to the same discrepancy between boiler temperature and feed temperature, namely 30 degrees, for all boiler pressures.

The natural arrangement of apparatus is shown diagrammatically on the next page. The system is controlled primarily by the set of the discharge trap T from the secondary heater, which should be adjust-able so as to maintain automatically different back pressures in this heater. It is also controlled by the valve A. Since the effective pres-sure on the pump piston needed to produce motion is only some 10 to 20 pounds, while the difference between boiler and back pressures is to be not over 20 or 30 pounds, comparatively little throttling will suffice to control the pump over all ranges of speed ordinarily needed. Nor-mally the valve A is expected to be left wide open.

For sudden and wide changes of load these methods of control are too sluggish. This objection is met by the adaptation of the usual auxiliary or relay pump, provided for emergency, to the needs of regulation. This pump is of ordinary dimensions. Should the main feed pump go out of commission, the relay pump may be supplied with boiler steam through the valve C, whereupon it throws cold feed

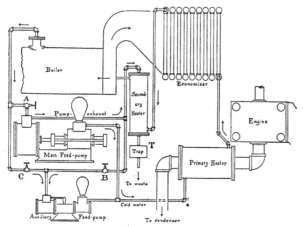

DIAGRAMMATIC ILLUSTRATION OF THE PROPOSED ARRANGEMENT FOR FEED HEATING.

water into the same circuit as the main feed pump. But this connection is for use only when the main feed pump is out of commission.

For purposes of regulation this auxiliary pump is also piped to
receive back-pressure steam from the exhaust of the main pump,
through the valve B. Should a sudden demand for feed exceed the
ability of valve A and trap T to pick up the speed of the main pump
sufficiently, the opening of valve B both accelerates the main pump
and augments its water capacity by the addition of that of the auxiliary pump.

The degree of saving which such a system should make over the
ordinary arrangement is impossible of computation. The only source
of saving directly visible is that due to the lowered temperature of
chimney gases, due to colder water in the economizer. The second
saving is that much discussed and incomprehensible, but real and un-
questionable, one due to feeding the boiler with water of its own
temperature, even when the feed heating is done with live steam.

The third saving is the still more elusive, but none the less important, one involved in lessened wear upon the boilers, due to the
absence of cold water from their interiors. It is such practical considerations as these, rather than refinements in fuel efficiency, which
support the advocacy of this system.

THE ARGUMENT FOR THE OPEN FEED HEATER.

A DISCUSSION OF PROFESSOR REEVE'S PROPOSALS.

By Reginald Pelham Bolton.

As explained in the introduction to Professor Reeve's article, Mr. Bolton's comment was prepared at the invitation of the Editors.

THE author of the foregoing paper appears to have an exaggerated conception of difficulty in eliminating oil from the exhaust of engines. I have found no such difficulty, where a properly proportioned and properly drained oil separator is provided; and where loads are very variable, I have secured complete freedom by placing the separator in series with an open feed-water heater.

The only case of a failure of boiler tubes by reason of oil from the returns of radiators, in plants with which I have had to deal, was one on Broadway, which was provided with a closed feed heater and oil separator improperly drained.

Fresh water has in it more sources of danger to boilers than oil in the returns. The open feed heater receives and deals with the make-up water, and even where fresh water can be had for nothing, the arrangement of returning part of the exhaust in the form of direct condensation has the advantage. The author must have some unusually defective systems in mind when he speaks of the cost of "making the repairs" which oil entails. I have plants which have been running twelve years with open feed heaters without a cent of expense for such a purpose having been entailed.

As to plants of the second class, that is run condensing, using receiver steam for heating, there is no less advantage in the open feed heater; in fact, the combination is precisely that of marine practice, when the open feed heater is standard and successful.

As to class three, I would refer to my paper before the American Society of Heating and Ventilating Engineers, 1905, "Steam Heating in Connection with Condensing Engines," in which I gave the results obtained from a primary heater on the exhaust of a triple-expansion high-pressure cylinder in my Milburn test of 1904.

The increase of temperature is very limited with these heaters, as the most that can be secured is to catch part of the difference in

953

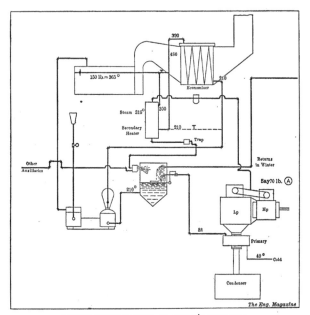

DIAGRAM ILLUSTRATIVE OF MR. BOLTON'S ARGUMENT.

Showing that the second heater is unnecessary with the economizer.

temperature between the exhaust at the opening of the valve, and
the average of the condenser temperature. Under the circumstances
noted the increase did not exceed 60 degrees F. and averaged only
48 degrees.

The plan proposed and shown in diagram on page 952 does not
appear to me to be practical. In the first place, the primary heater
does not drain. The only relief for the secondary heater is the trap,
and it would require an extraordinarily nice balance of steam cylinder to heating surface and feed to prevent its backing up the condensation to the pump.

A direct-acting pump will short-stroke severely with an excessive back pressure such as proposed, and the pump would waver
about its center of motion while its cylinder losses would be excessive.

All the heat in the trap discharge of superheated water goes to waste, as well as the exhaust of the auxiliary pump whenever that should be used.

Better results would be attained by the well-known Scotch marine system of taking the steam supply for the feed pumps from the first receiver and discharging and condensing its exhaust in the open feed heater.

The feed supply can usually be taken by gravity through a primary heater to the open heater.

The arrangements shown in the sketch on page 954 indicate that, where a temperature of 210 degrees can be reached by the exhaust of auxiliary machinery, there is little to be gained by the primary heater.

Further, where an economizer is desirable and is fed from an open (or closed) heater with water at or near 210-212 degrees, the output temperature is so near the boiler temperature that a secondary heater would effect little if any increase. If a secondary heater, rather than an economizer, be desirable, then the steam for its work may best be borrowed from the intermediate receiver or steam chest, which steam has already done its major work in expansion.

There is no great advantage in, or need for, the introduction of feed into boilers at the steam temperature; for first, such a temperature cannot be gained without economic loss either in flue gases or in steam used for the purpose, and, second, a moderate difference in temperature increases the effectiveness of heat transfer and promotes circulation.

Professor Reeve, being invited by the Editors of THE ENGINEERING MAGAZINE to make rejoinder to Mr. Bolton's comment, closes the discussion thus:—

"Mr. Bolton and myself are obviously writing of different things. He is writing of a machine, of which the illustration is taken as a working drawing. I am advocating a policy, of which the illustration is a diagram. The arrangement as presented may quite likely develop some of the objections he mentions. But it is supposed to be the office of the engineer, if the pump kicks or the heater will not drain, to correct those incidents by a modification of previous practice, rather than to abandon his basic plan of design because of these insignificant features.

"The crux of the whole article—the question whether it were better to feed our boilers with steam-hot water, and whether it were better to do it by live-steam heating or by the plan suggested—Mr. Bolton quite ignores. To him, as I understand him, 212-degree feed from an open heater is hot enough. The author finds in Mr. Bolton's remarks no reason for modifying his previous dissent from this view."

LABOR-COST DISTRIBUTION AT THE GENERAL ELECTRIC SHOPS.

By George Frederic Stratton.

WHEN eleven-thousand workpeople are paid $150,000 for their previous week's exertions upon a great variety of operations resulting in a great variety of products, it seems almost incredible to one not familiar with the highly systematized methods of great industrial corporations that the manager should receive, within a day or two, a detailed statement of exactly what every dollar of that money was paid out for. Upon that statement he can see, to a cent, how much has gone into production, construction, and expense. And these three classes are subdivided sometimes to an extraordinary degree. From that statement he knows the amount expended upon a single piece of apparatus, or a total expenditure upon some one great class of manufacture. He knows how many dollars and cents the handling of materials and supplies about the plant has absorbed, the cost of sweeping up the shops or any particular shop, or of clearing away a snow-fall.

To divide and charge up a heavy pay roll properly requires a system which shall be simple without sacrificing one iota of comprehensive detail; shall be economical without neglecting the keenest supervision and correct distribution of the various items. The system in use by the General Electric Company is most interesting, both for the large sums involved and the minute subdivision of those sums into their proper debits. At its Lynn (Mass.) plant there are eleven-thousand employees drawing a total of $150,000 weekly, including salaries. About one-third of these hands are on day work, the others on piece work. As this plant manufactures over five-thousand distinct varieties of articles, large and small, the number of operations is enormous, over twenty-thousand piece-work prices being in operation. The foundation upon which the entire system rests is the shop order. No service performed in these works is commenced without an order being issued first and a number assigned. Outside of the clerical force, every man, whether he be engineer or lumper, draughtsman or furnace man, piece worker or day worker, works under some distinct, numbered, written order.

The system of numbering appears complex in description, but is really simple when understood. The figure which stands for hundreds of thousands also stands for one of six great classifications. Thus, orders numbered 100,000 always relate to production; 200,000 to construction; 300,000 to repairs; 400,000 to expense; 500,000 to experimentation, and 600,000 to engineering, and draughting. But there is much subdivision of these classes, and to meet this the numbers are further systematized. On production orders the first figure indicates the class (production) the second and third the subdivision, and the last three the particular variety of apparatus or the special machine required.

For example: in the order 127,436 the figure 1 shows at a glance that it is a production order; the figures 27 show that it belongs to the division of railway motors, and the last three figures are the distinguishing numbers by which some peculiar type or size of motor is required. The subdivisions of production all relate to some distinct class of apparatus, as dynamos, railway motors, stationary motors, arc lamps, transformers, and so on. There are about fifty such subdivisions.

The construction orders relate to the erection of new buildings, the making of special tools, and everything which in any way increases the value of the plant or the equipment. The orders of this description always commence with the figure 2, thus showing instantly that construction is meant. The last five figures refer to the detailed instructions or plans of whatever construction is desired.

Repair orders refer entirely to repairs made by the company upon outside apparatus, or apparatus sent in for that purpose. As new parts are often used on such jobs, and as they always produce income, they are really "productions," and are treated as such in the yearly summary; but in the distribution of labor they are treated as a separate class. The first or indicating number is 3—the following numbers being, of course, an index to the particular job.

Expense is indicated by the number 4. Such orders are issued for all operations which are clearly operating expense—as running the steam plants, unloading cars and vessels, cleaning windows, etc. —all of which operations are again subdivided on the distribution of labor sheets, by being charged to the particular shop or department where the expense is incurred.

The number 5 relates to all work of an experimental nature— the making of new apparatus and the investigation and testing of appliances which may be purchased.

With number 6 we reach the engineering and draughting departments. As in the other classes, the five figures following refer to the special machine or instrument which is being designed or improved.

All these shop orders contain only such simple instructions as will enable foremen and men clearly to identify the work done with the number assigned. Form 1 shows this plainly. Detailed instructions are given separately to the foremen, accompanied by blueprints and specifications—all, however, bearing the same number.

	Shop Order	No. *103432*

Mr. _ _ _ _ _ _ *Lincoln* _ _ _ _ _ _ _ _ _ _ _ _ _

Date _ _ _ _ _ _ _ _ _ _ _ $^{8}/_{7} - 07$ _ _ _

Make 100 D. M. Railway Motors.

Duplicates to

_ _

The Eng. Magazine

FORM I. SHOP ORDER.

The use of these numbers becomes apparent upon a consideration of the time cards and piece-work slips in use. The time cards are filled up daily by the workman. They must specify the times at which he commenced, or quit, upon any one job and must give the order number of the job, whether it be work on a lathe or firing a boiler—cleaning windows or melting steel. No matter what the time books (which are kept separately) say, no day-worker is credited on the pay roll with ten-hours work unless time cards showing ten-hours work, and upon what order it was spent, are turned over to the paymaster. These cards are O K'd by the foreman every night and sent to the paymaster each morning, and it will thus be seen that he receives, each day, vouchers or invoices (which they really are) for every hour of the previous day's labor.

Thompson–Houston Electric Co.

Workman's No. *4100*   *8/9*   190 *7*

Name   *John Smith*

Order No. *133024*   Part

Assemble Motors

Commenced *9.15*

Finished *3.30*

Hrs. *6*   Amt. *1 05*

Rate *20 ¢*   *The Eng. Magazine*

FORM 2. TIME CARD.

The piece-work slips (Form 3) are filled up by the workman in a similar manner, every day. They are O K'd by a regular inspector and sent to the paymaster on the following morning.

The story of the assembling of all these vouchers, or invoices, and the making up of the pay roll, does not properly belong to this

COMPLETED WORK SLIP

Thompson–Houston Electric Co.

West Lynn, Mass.,   *8/7*   190 *7*

No. *7380*   Name *H. Jones*

List   No.   Order No. *133024*

Part No. *7*   No. Completed *9*

     each *11*

Oper. No. *14*   Price { per 100

Operation *Drilling Bases*   Amount

     Grand Total   *99*

The Eng. Magazine

FORM 3. PIECE-WORK SLIP.

article: It is sufficient to explain that the clerks figure the time upon each time card and write the amount due upon that card (as shown) whether the time is for ten hours or for only one-quarter of an hour. The clerks in this department also check the piece-work prices upon the piece-work slips and figure up the amount due upon each slip.

It will thus be seen that, for every minute's service—either of day worker or piece worker—there comes into the office a written voucher which gives the number of the order under which the work was done, and the dollars or fractions of dollars claimed for such work.

After the pay roll is made up and all these vouchers are turned over from the paymaster's department to the distribution of labor department, they are found to have been arranged in the order of the workmen's numbers, so that credit for the amounts could be entered against each name on the pay roll. Now, however, a re-arrangement is necessary and is at once made by young clerks. The vouchers are sorted out according to the order numbers, and it is at this point that the peculiar significance of those numbers becomes apparent. First, sorting into the six great classifications by observing the first figure on each, the sorting into subdivisions is continued by means of the second and third figures. The work, which is of enormous proportions, is thus simplified and condensed.

The number of these little vouchers turned in weekly is very great. Frequently day workers will turn in five or six time cards in one day (every separate operation must be entered on a separate rate card). And on piece-work operations, although many men, working only on one operation, can enter their entire day's work upon one slip, there are so many minor operations in this plant— so many detail parts—that the average number of piece-work slips turned in by the workers is nine or ten each week. In all a total of over two-hundred thousand vouchers or invoices reach the Distribution of Labor department each week.

Sorted into neat bundles, with all slips bearing like order numbers assembled together, they reach the entry clerks, who add together all the amounts paid upon each order number, and enter the total in its proper place upon the "Labor Sheet."

A glance at Form 4 will show this method. The order number is at the left; the entries of the amounts paid to day workers and piece workers are made separately, and the total brought out to the right. Thus the manager can see, at a glance, the total amount spent upon Type C motors for the week, or he can find, by means

Distribution of Labor

Week ending   *August* 10' 07.

Class	Order Number	Day Work	Hrs.	Piece Wk.	Hrs.	Total	Total Hrs.	Grand Total
Type C. Motor	133924	321 17		1046 20		1367 37		
	133410	42 00		304 19		346 25		
	133623	86		208 17		209 03		1922 65
Arc Lamps	171303	178 24		394 20		572 44		
	171042	399 17		876 19		1275 36		
	171493	22 09		43 02		65 11		1912 91
Tool Expense								
Factory A.	410170	43 02				43 02		
" B.	410193	17 01				17 01		
" C.	410214	56 56				56 56		
" D.	410762	63 07				63 07		179 66
Helpers								
Factory A.	412203	33 06				33 06		
B.	412606	20 17				20 17		
C.	412327	170 43				170 43		
D.	412410	130 60				130 60		354 26
)		1497 51		2871 97		4369 48		4369 48

FORM 4. DISTRIBUTION-OF-LABOR SHEET.

of the order numbers and reference to the original order, the amount spent upon any one variety of that Type C motor.

It is so all through the sheets. The subdivision of production has been mentioned. On expense it is carried out still more fully. Under "sweepers," for instance, there will be an entry for every separate shop in the plant where an individual sweeper is employed.

Under "tool expense" the same subdivision occurs, and so through all items of shop expense—they are separated by means of those order numbers and charged against the individual departments.

In all of the six great classes there are about one-hundred subdivisions, and generally one-thousand separate entries under all those subdivisions. That means that the entire force has been working under one-thousand different orders during the week. (It will be understood that these are not commercial orders. They are purely shop orders, many of them being "standing orders.")

. In addition to the shop wages, the salaries of officials, clerks, and foremen are properly entered upon these labor sheets, with the result that the grand total balances the total on the pay rolls for the corresponding week.

In describing this system it may appear to be intricate and cumbersome, but it must be remembered that the amount of detail finally reduced to a few significant totals is very great. In considering it fairly one must even put aside the number of the hands and the amount of money paid out, and consider instead the vast quantity of items which must be handled to bring any result of value out of great confusion of record. For each one of those two-hundred thousand time cards and piece-work slips is really an invoice—complete in date, name of creditor, details of service rendered, and amount due and paid; and they must be so treated—unless the manager should be content simply to note the total amount of wages paid each week. And the value of this system, or one producing similar results, can hardly be estimated by any but those who have been connected with large aggregations of workers and have thus seen the continual utility and value of such statistics. Within a week following the payment of the wages there is placed before the manager a small slip of paper (Form 5) with seven totals written upon it. They are of Production, Construction, Repairs, Expense, Experimental, Engineering and the grand total of all these. Down in one corner is another little group of figures something like the following:—

Production	65 per cent
Construction	6 per cent
Repairs	3 per cent
Expense	15 per cent
Experimental	4 per cent
Engineering	7 per cent
	100 per cent

Those are the figures the manager's eyes will fall upon first. He knows what the normal percentages should be and he notes any change, instantly. Taking from a pigeon hole a similar slip for the preceding week, he compares the seven totals. Suddenly he calls for the "Labor Sheets," turns hastily to the expense section and scans it. Still unsatisfied, he demands the sheets for the preceding week and runs his finger down the expense subdivisions, comparing the two weeks.

Summary of Labor Sheets		
Week ending Aug. 10/07		
Production		97576
Construction		8941
Repairs		4516
Expense		22487
Experimental		6114
Engineers		10432
	$	150066
Production	65¢	
Construction	6%	
Repairs	3%	
Expense	15%	
Experimental	4%	
Engineering	7%	
	100	

The Eng. Magazine.

FORM 5. SUMMARY OF LABOR DISTRIBUTION.

The finger stops at an item. The charge for helpers in some one department—(factory D, we will say) has increased from $120 to $130 since the preceding week. That means an extra man. But the total of production has not increased by any material percentage. A telephone call brings the foreman to the office and he explains why that additional helper has been necessary.

That is the marvel of the well-designed and correctly entered distribution of labor sheet—that it enables the manager of eleven-thousand hands to detect even so small a leak in expense as the engagement of one extra man. No foreman of a department, however large that department may be, can employ even an extra errand boy without its appearing as a distinct increase in the charge under that heading, and it is little likely to escape the keen eye of any careful manager. No clerical force of any department can be increased by even a $6-a-week junior without that increase showing in the sub-divided expense entries.

Of course such minute changes as these can be noted only in the expense columns, as the expense employees of a factory are usually a very stable factor. Production may fluctuate materially and the causes be perfectly natural. But it is the expense which managers watch, and need to watch, so closely. And it is here that the labor sheet is at once an indicator and a safety valve.

Its use in production is not far behind in value. In a great manufacturing plant, where foremen of large departments, with several hundreds of hands, are often more or less bothered in getting parts and supplies through in order to keep their crews thoroughly busy, it is not unusual that some classes of work are pushed ahead unduly while others get behind. But the manager studying the labor. sheets can detect this tendency, at once, by noting the abnormally large amount paid out for wages on any one class of apparatus, and can at once take steps to avoid a possible congestion at one point and a shortage at others.

These labor sheets are also the means by which the cost department obtains the labor cost on any special machine, and also checks the estimates on standard apparatus. In obtaining the cost, say of a special size of Type C motor, order No. 133,262—all that is necessary is to examine the labor sheets under Type C heading, from the date of the issue of the order until its completion. The time occupied may be one week, or ten, but the addition of all charges against that order number gives, to a cent, the amount the company has paid in wages on that machine. And it is quickly done because the headings on the sheets are in a regular, fixed order which is never changed, and can be found as readily as a numbered page of a book.

The distribution of labor sheet easily becomes, in the hands of the manager, the greatest help he possesses in the control of his factory operations. It is, to his keen and observant eye, what the pulse of the patient is to the delicate and experienced perception of the physician. The most trifling changes in the conditions of output and expense may be easily noted by one who understands and studies the summary and its component sheets.

The warning of disproportion in production, or an insiduous increase of expense, comes quickly. The change, or trouble, which happens this week is shown up in unerring, indisputable, and locating figures next week.

A COMPRESSED-AIR LOCOMOTIVE FOR MINE HAULAGE.

Gauge 2 ft. 0 in.; weight 12,600 lb., all on the drivers. Cylinders 7 by 14 in.; storage pressure 700 lb., working pressure 125 lb. Built by the American Locomotive Co.

ECONOMICAL MATERIAL-HANDLING EQUIP-
MENTS FOR INDUSTRIAL PLANTS.

By Werner Boecklin.

"On the other side, true dispatch is a rich thing. For time is the measure of business, as money is of wares; and business is bought at a dear hand where there is small dispatch."—*Francis Bacon.*

THE question how to handle materials in industrial plants is one continually presenting itself to owners and engineers for proper solution. It is a problem entering into nearly all lines of business, in some assuming an important place in the general economy. As the demands upon the manufacturing plant have increased and a keener competition has necessitated radical changes, methods to meet the new conditions have been devised. Mechanical handling machinery has kept pace with the general mechanical development of the past decade and is now recognized as a distinct branch of manufacturing.

The ever-present question of cost, which the manager has daily to consider, is the germ from which have sprung the various systems designed to help the manufacturer to cut his operating expenses and to enable him to place his product, whether it be power, light, machinery, or one of a thousand different articles, upon the market at a lower price than was possible before the introduction of labor-saving devices. The engineer or owner is interested in securing that equipment which, being best adapted for his needs, will first materially

ELECTRIC LOCOMOTIVES FOR INDUSTRIAL HAULAGE.

Above, 9,500-lb. locomotive for Carpenter Steel Co. Two motors, 250-volts; gauge 2 ft.
Below, 10,000-lb. locomotive for the Commissioners of Water Works, Cincinnati; two
motors, 80 volts, gauge 1 ft. 9½ in. The Westinghouse Electric & Manufacturing Co.

reduce certain costs entering into the total cost of product; and second,
will be safe and easy to operate. Cases often arise where any one of
many different systems may be installed for doing the work. They
are probably not all equally well adapted for the situation, and it is
of vital importance, especially in large installations, that the condi-

tions be carefully studied with the view to securing the most economical equipment. In considering an investment, it should be borne in mind that the unit cost for handling material with a given equipment is not a true measure of the efficiency of that equipment unless it is worked to its full capacity the year round, and such a condition prevails only in a limited degree. Such unit costs, too, may have little or no significance in certain cases, the effectiveness of the equipment being measured rather by its general all-round usefulness, which it is impossible to render into dollars and cents.

STORAGE-BATTERY LOCOMOTIVE HAULING GREEN LUMBER, POTLATCH LUMBER COMPANY.

Seven-ton locomotive built by the Jeffrey Manufacturing Co.

The conditions which are presented vary to such an extent that each case must of necessity be taken up as an independent problem. The facts for any specific case under consideration may be marshaled in some such form as follows:—

1.—Material to be handled.

2.—Its nature or condition; whether heavy or light, in small or large pieces, wet or dry, hot or cold, sticky, having active chemical properties, in packages or in bulk.

3.—The amount in number or weight to be moved per hour.

4.—Whether material is to be moved horizontally or vertically only, or both.

5.—The power available.

6.—General information as to local conditions which will affect the installation.

KOPPEL INDUSTRIAL RAILWAY TRACKS AND CARS ON A BIG IRRIGATION JOB IN THE
WESTERN STATES.

Showing the heavy loads which can be hauled by one mule: this train is made up of nine
cars, each holding 36 cu. ft. The ease with which the track can be shifted is also
apparent. The Arthur Koppel Co.

There are two general methods of handling materials; viz, the intermittent and the continuous. In the first instance the material is transported at intervals in a condensed form, as exemplified in a crane carrying a bundle of steel or a box of castings. In the second instance the material is transported in a more or less continuous stream, as for example a conveyor taking a supply of coal from an overhead storage pile. These two methods may for the purpose of this article be subdivided as follows:—

INTERMITTENT HANDLING.

 1.—Industrial Railways
 2.—Automatic "
 3.—Cable "
 4.—Traveling Hoists
 5.—Aerial Cableways
 6.—Cranes and Derricks
 7.—Elevators or Hoists

CONTINUOUS HANDLING.

 8.—Bucket Elevators or Conveyors
 9.—Flight Conveyors
 10.—Spiral or Screw Conveyors
 11.—Trough or Pan Conveyors
 12.—Belt Conveyors
 13.—Reciprocating Conveyors

Of these general types there are various combinations and modifications.

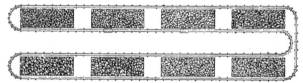

ARRANGEMENT OF INDUSTRIAL RAILWAY TRACKS, ROANE IRON WORKS.
The ore is dumped into the ore bins from standard-gauge railway cars. It is spouted thence into the industrial cable-railway cars and from these dumped to the furnace hoists; 1,000 cable cars are handled every ¾ hours. Length of track 1,300 ft. Installed by the C. W. Hunt Co.

INTERMITTENT HANDLING MACHINES.

Industrial Railways are generally understood to mean systems having gauges of 20 to 36 inches and cars having capacities from 1 to 10 tons. The track may be built in sections, the rails being securely attached to the steel cross ties by means of clips or rivets and shipped in this form ready for placing. Curves, frogs, switches, crossings and turntables are made standard and usually kept in stock ready for shipment. Cars are made in many forms and sizes suitable for various conditions. The trucks are frequently fitted with roller bearings, thereby reducing the axle friction from one-fifth to one-third. For inter-shop transportation the industrial railway is often indispensable and the charges for maintenance are insignificant.

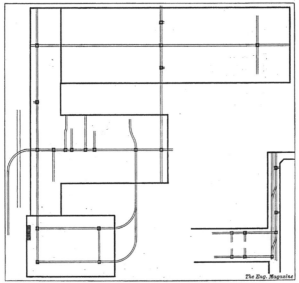

The Eng. Magazine

KOPPEL INDUSTRIAL RAILWAY SYSTEM, INSTALLED AT THE WAGNER ELECTRIC PLANT.
Scale 1 in. equals 72½ ft.   Gauge of track 23½ in.

Such a system can be operated either by hand or by using some form of power. The kind of power to be employed will naturally be an important consideration. The two agents generally recognized as best adapted for factory and yard systems are the storage-battery locomotive and the trolley locomotive.

The advantages to be secured in the use of the storage battery are convenience, safety, and economy. The machine is always ready, night and day; no high-grade labor is required to operate it; it may be run into any part of the works where either permanent or temporary tracks can be laid. Compared with steam or trolley, its safety is easily recognized as there are no sparks or overhead wires to contend with. The claim of economy, although not so easily proven, is borne out by an investigation of a number of installations now in operation. Considering the matter of economy it may be well to bear the following points in mind: the power is used only when the locomotive is in service; only enough power is taken to do the work required; the locomotive may be recharged at intervals during the day when it is not in use; with such a locomotive the material in an ordinary factory may be easily handled by one man.

Aside from the question of available room for tracks, curves, switches, etc., an important factor in selecting a locomotive will be the maximum grade to overcome. Storage-battery locomotives may be used economically on grades below 5 per cent. For higher grades they are not suitable, nor are they fitted for long hauls, or speeds exceeding about 350 feet per minute.

THE CABLE ROAD AND ORE BINS, ROANE IRON WORKS.
The C. W. Hunt Co.

The other form of electric traction, the trolley locomotive, is applicable to longer hauls, steeper grades, and heavier loads than the storage locomotive. As with the latter type, no skilled labor is required for operation and it consumes no energy when idle. For factory purposes, however, the trolley locomotive is not suitable on account of the overhead wire being in the way. In yards this objection may not be a serious one. By the introduction of the electromagnetic system the difficulty with trolley wires or third rails may be overcome, and one company at least is building industrial locomotives on this system. All contact buttons through which the current is supplied excepting those under the locomotive are "dead." Grades may run up to 10 or 15 per cent, but the average working grades for economy should not exceed 3 per cent for long hauls or 5 per cent for short ones.

INDUSTRIAL RAILWAY TRACKS, THE WESTINGHOUSE ELECTRIC & MANUFACTURING COMPANY.
Installed by the C. W. Hunt Co.

The average cost for hauling coal by this method in the case of seven different installations, taking into account labor, depreciation, and supplies, interest not being figured, was less than 3 cents per ton. The amount hauled ranged from 300 to 900 tons per day.

The Automatic Railway is one of the oldest methods in use for eliminating labor from the problem of handling materials. The essential parts are the track, having a gauge of about 22 inches, properly supported on trestle, an automatic dumping car, and the "triangle," or counterweight, used as a storage of energy, with cable and necessary sheaves. The system is extensively used for handling

INSTALLATIONS FOR THE HANDLING OF COAL AND STONE.
Above, is an automatic road for conveying sand and stone at the yards of the N. Ryan Co.,
N. Y.; it shows the "triangles" or counterweights. Below is a 5-ton storage-battery
locomotive hauling coal on a 3 per-cent grade, Bridgeport Malleable Iron Co.
plant. Both installations made by the Mead-Morrison Manufacturing Co.

SHUTTLE CABLE ROAD FOR COAL HANDLING, NEWTOWN DOCK, BROOKLYN RAPID TRANSIT RAILROAD.

The road is 350 ft. long, and uses one 3-ton car; capacity 600-800 tons per day. Installed by Mead-Morrison Mfg. Co.

coal, or other material in bulk, along the water fronts for delivering to storage piles.

To operate successfully the track should have a grade of about 3 per cent and the length should not exceed 400 feet. The car acquires sufficient momentum to elevate the counterweight which by its fall returns the empty car to the loading point. It is made to discharge its load at any point by means of a suitable tripping device.

INDUSTRIAL-RAILWAY INSTALLATION IN THE KOPPEL PLANT.
Cars are loaded by traveling crane, and carry the material from machine to machine until the work on the individual piece is completed. Then they take it to the assembling shop. The car does not have to turn back, nor does it encounter another in its travel. All the buildings are connected by this narrow-gauge track.
The Arthur Koppel Co.

The cost of such an installation is comparatively low, and cost of operation is reduced to a minimum. An automatic road 250 feet long, with an average run of 100 feet, and operating a 1-ton car, will handle 60 to 70 tons of coal per hour. Such roads are installed in many cases with mast and gaff rig and electric or steam hoists for handling tubs where the tonnage is too light to warrant larger plants. In one case where 20,000 to 30,000 tons of coal are handled yearly, using tubs and an electric mast and gaff equipment, it costs, including labor and all fixed charges, about 7 cents per ton. This figure would be cut to less than half in the case of large plants. The current in this case must be purchased, which, at the high local rate, adds materially to the cost.

A Cable Railway consists of the driver, endless steel cable with

APPLICATIONS OF THE PORTABLE ELECTRIC HOIST IN SHOP WORK.

On the left, a jib crane with 2-ton portable hoist serving boring mills and controlled from the floor. Each hoist serves from two to four mills. On the right, monorail equipped with 2-ton hoist for handling boiler plate. It is brought from stores on the trolley, transferred to chain blocks on the jib crane, swung to the punches, and returned the same way. The single-hook suspension enables the hoist to work at an angle as great as 45 degrees. Built by the Yale & Towne Co.

guide, curve, and supporting sheaves and take-up, one or more cars with cable grips, and necessary track. Two systems of railways of this class are in use, the shuttle, or reversible, and the continuous. The first is used for small capacities, (40 tons an hour and under,) the second for capacities of 60 tons and over. Two cars are usually employed on the shuttle system and two tracks. While one car is being carried loaded to the end of the line, the other one is returning empty to the loading point. The cable driver must therefore be a

reversing machine. On a continuous cable road the cable runs in one direction, the cars being gripped to it at intervals. The speed of cable will vary from 200 to 500 feet per minute, and the cars carry from 2 to 5 tons of material. Cars may be spaced a mini-

A MONORAIL ELECTRIC HOIST, WITH CAGE.
Manufactured by the General Electric Tool Co.

mum distance apart of 50 feet. Grades of 7 to 10 per cent may be used without difficulty.

Aside from minor repairs and renewals the chief items of operating cost are labor, power, and renewal of cable. A cable with ordinary attention may last from 9 months to a year, and, with special care, much longer than this.

A Traveling Hoist is one of the commonest and most economical methods for handling material in the factory. Such a system consists in its simplest form of a properly supported beam, or track, carrying a trolley carriage from which is suspended a sling for handling loads, or preferably, a chain block attached to the trolley for hoisting the load. An electric or pneumatic hoist may be used with suitable trolley carriage, and, by adding a small motor to the trolley carriage in case of the electric hoist, both movements may be accomplished by power. By the further addition of a second trolley carriage and a suitable frame and cage for the operator, a very effective machine may be produced, and one which can be employed in an endless number of

ELECTRIC MONORAIL TRAVELING HOIST WITH CAGE, FOR HANDLING SHEET STEEL.
Manufactured by the Sprague Electric Co.

TELPHER TRAIN WITH THREE TRUCKS, WORKS OF THE WALTER BAKER CHOCOLATE
COMPANY.
Each truck carries 2,000 lb. of cocoa in bags.   Installed by the Dodge Coal Storage Co.

situations.   Such a system may be laid out with suitable switches and curves to cover a large portion of a given floor space, or it may be used to transport over long distances and difficult ground. Hoisting speeds vary from 20 to 50 feet per minute, and trolley speeds range from 50 to 250 feet per minute.   By means of transfer beams a machine of this type may be used for carrying loads from one part of the shop to another, crossing from bay to bay.   As stated, compressed air hoists are used in connection with a monorail system, but they are not so flexible as electric hoists although highly practicable under special conditions.

SHEPARD ELECTRIC HOIST UNLOADING CARS, RENSSELAER MANUFACTURING COMPANY.
Hoist is of 10-tons capacity. Installed by General Pneumatic Tool Co.

The cost of handling material by means of such a system is always difficult and in most cases impossible to determine. We know that, taking into account the various shiftings and hoistings of materials throughout a plant, a vast amount of time and much money can be saved by such an overhead method, and that the installation in some cases will pay for itself many times over in a year.

Aerial Cableways consist of a main cable on which the trolley carrying the load runs, stretched between elevated towers, a traversing cable for moving the trolley, and a hoisting cable for the vertical movement of the load. In some cases a third cable is used for dumping the load. The operating cables are controlled by one man at one end of the line. Cableways are practicable up to a length of 2,500 feet. When the length exceeds about 250 feet, fall-rope carriers are employed. These are small hangers running on the main cable and spaced at regular intervals to prevent the hoisting rope from sagging. If not used the weight of the rope will be sufficient to lift the bucket or hook off the ground and prevent loading. Equipments of this class are used extensively by contractors and also find a limited application in industrial plants. They are specially applicable for transporting material over difficult ground, or where existing structures or other obstacles preclude the use of any other system. Steam or electric power is commonly used for operating. In the case of a steam equipment either a separate boiler is installed, or, as is often possible in an industrial plant, steam is drawn from the main power station. In this latter case and with the use of an electric hoist, power is used only when needed. A skilled operator is absolutely necessary. To operate a steam equipment of this kind will cost, including the wages of engineer, fuel, and supplies, from 5 to 7 dollars a day.

DOUBLE-DRUM ELECTRIC AERIAL CABLEWAY HANDLING SAND.
Length 700 ft.; capacity of bucket 1½ yds.; a trip every 3 minutes. Installed for the Pittsburg Plate Glass Co. by the Mead-Morrison Manufacturing Co.

STATIONARY HAND BRIDGE CRANE, NEW YORK, NEW HAVEN & HARTFORD RAILROAD.
It is of 30-tons capacity. Built by the Brown Hoisting Machinery Co.

Cranes may be conveniently sub-divided as follows :—

1.—Overhead Traveling Cranes.
2.—Gantry Cranes.
3.—Cantilever Cranes.
4.—Stationary Bridge Cranes.
5.—Pillar Cranes.
6.—Jib Cranes.
7.—Locomotive Cranes.
8.—Derricks.

One or more overhead traveling cranes will be found in any large machine shop or foundry of the present day. Depending upon the importance of the location and the character of the service, these cranes are operated either by hand or by power, usually electric. In its application to the shop or yard one of the chief advantages of this type is that the full ground area between rails or runways can be covered by the hook of the hoist. For quick action, long travels, and heavy loads, the power crane is preferred to the hand-operated machine. The main hoist, supported on a suitable trolley carriage which travels from end to end of the bridge, has a hoisting speed under full load between 10 and 25 feet per minute. The trolley

UNLOADING YARD CRANE, SHOPS OF THE GENERAL ELECTRIC COMPANY.

Built by the Sprague Electric Co.

travels at about 100 feet per minute under full load, and the bridge at about 250 feet per minute. Much higher speeds for bridge travel have been advocated, but there is a chance for an undue amount of swinging of the load at very high speeds.

The nature of the work may be such that an auxiliary hoist for light loads and high speeds may be advantageously installed. The saving in operating expense in a case where, with a 50-ton crane for example, 75 per cent of the hoisting can be handled by an auxiliary,

EXAMPLES OF HEAVY CRANES IN SERVICE.

Above, 40-ton crane and two Morgan electric floor-type open-hearth charging machines, open-hearth department of the Lackawanna Steel Co. Below, a 120-ton double-trolley crane, each trolley of 60-tons capacity with 5-ton auxiliary hoist; span 65 ft.; operated by seven electric motors. Coal & Coke Railway Co., Elkins, W. Va. All built by the Morgan Engineering Co.

GANTRY CRANE, WITKOWITZER BERGBAU UND EISENHUTTEN GEWERKSCHAFT, MORAVIA, AUSTRIA.
Of 4-tons capacity. Built by the Brown Hoisting Machinery Co.

will be at once apparent. When the time and money saved by the installation of an overhead-crane system are considered, the charge against operation becomes a negligible quantity.

Gantry Cranes consist essentially of a bridge supported at either end upon movable piers. There are cases where one pier is eliminated and one end of the bridge runs on an elevated track. In some machines of this class the piers are rigidly attached to the bridge, with the frequent result that the irregularities in the track produce a severe racking throughout the structure. In other designs the bridge is not rigidly connected to the piers, but is pivoted at one end, and suspended from a shear at the other. If one end runs ahead of the other, or if the track is out of level or alignment, the crane still operates without

CRANE INSTALLATIONS IN MACHINE SHOP AND FOUNDRY.
Above is a 2-ton transfer crane. Below, a pair of heavy electric traveling cranes, 60 and 40 tons, 75-ft. span, the former with 10-ton auxiliary hoist, supplemented by traveling wall cranes of 10-tons capacity, 25-ft. arms. Installed for the Reading Iron Co. by the Niles-Bement-Pond Co.

TRAVELING CANTILEVER DREDGE CRANE, WITH CLAM-SHELL BUCKET.

Installed at Garfield plant of the Utah Copper Co., for handling wet copper concentrates from settling basins to cars. Length of bridge over all, 41 ft. 6 in.; wheel base 19 ft. Capacity of trolley 3½ tons; hoisting speed 60 ft., trolley travel 150 ft., gantry travel 30 ft. per minute. Pawling & Harnischfeger.

SPECIAL ELECTRIC GANTRY CRANE FOR HANDLING STRUCTURAL MATERIAL.

It is of 3-tons capacity, 50-ft. span. A self-contained hoisting machine travels on the lower flange of a bridge girder which overhangs the supporting frame 10 ft. This enables the crane hook to reach a car or other point that far beyond the lot line, and also facilitates the swinging of steel angles and beams. The trolley can run off the crane onto a fixed I-beam track. Built by the Northern Engineering Works.

any undue strain being imposed upon any part. Such cranes are built with spans up to 300 feet, and capacities of 5 and 10 tons are common. The hoisting speed under full load ranges between 200 and 300 feet per minute, trolley speed 800 to 1,200 feet per minute, and bridge speed 300 to 600 feet per minute. Either steam or electric power may be employed for operating and the machinery may be conveniently housed on the double-track pier. There are three ways commonly used of applying the power to the bridge wheels:—power may be transmitted across the bridge through a universal-joint shaft, and thence to the wheels by gears and vertical shafting; the vertical shafting may be replaced by sprocket chain; or independent motors may be placed on the piers and the wheels driven as in case of street-car motors. The three motions are under the control of one operator. Gantries are extensively used in structural yards, areas for storage of materials such as coal and ores, ship-building yards, and similar localities, and are built with various modifications to suit requirements.

Cantilever Cranes may be classed as modified gantry, the bridge of the latter being extended in either direction and the piers brought closer together. Cranes of this type are built either to travel, to revolve, or to have the two motions combined. They have been constructed with spans exceeding 350 feet. The general principle of operation is the same as for gantries. A counterweight is employed which balances the load by automatically assuming a position on its arm corresponding to that held by the load on the opposite arm. All movements are at comparatively high speeds, and capacities of 10 or 15 tons are common. As generally built, the gauge of the runway track is about 20 feet. These machines are largely used in ship-building yards, coal-storage and rehandling plants, and steel-plant yards. The principal advantage in the use of such a crane lies in the fact that heavy loads can be handled with ease and dispatch over very large areas, without interfering in any way with other operations going on in the same area, and irrespective of any obstacle in that area.

Stationary Bridge Cranes are rather limited in application, being used principally for railroad stations and freight yards. Under these conditions they are seldom in commission, and are not as a general thing equipped with power. The crane consists of a main girder, or bridge, sufficiently long to span the required space, and rigidly supported at either end by suitable columns, or piers. In the case of hand machines the hoisting and trolleying mechanism is attached to the side of one of the piers at a convenient height. This crane is sometimes installed in industrial plants for loading and unloading cars, and is built in sizes from 1 to 2 tons to about 30-tons capacity.

Pillar Cranes have, in addition to the operating mechanism, three structural parts, namely; the pillar, which is usually made of cast iron; the inclined boom, made of structural shapes; and the ties, made of round rods, or eye bars, to support the outer end of the boom. Cranes of this type are installed where, on account of local conditions, the mast can not have a support for its upper end. This necessitates a mast of heavy construction to withstand the bending strains put upon it. Either hand or power equipment is employed, depending upon the importance of the crane. These cranes are built with lifting capacities from 1 to 30 tons, and with radii from 10 to 40 feet, and are suitable for use in freight yards, industrial plants, at docks, etc.

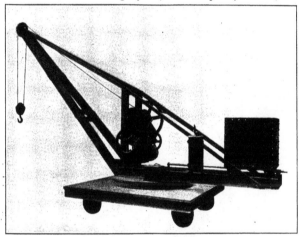

ELECTRIC TRUCK CRANE, FIXED RADIUS, 5-TONS CAPACITY.
Radius about 15 ft., hoist operated by alternating-current motor; truck propelled by hand.
Northern Engineering Works.

Jib Cranes are a type belonging to the early history of cranes. Of the smaller machines for handling materials in the industrial plant, the jib crane in its various forms is probably more used than any other class. Its application is primarily local, with certain exceptions to be mentioned later, being limited in scope by its radius, which is small. The designs are various, but the essential parts are the hoisting mechanism, the mast, strut or tie, and boom. Upper and lower

JIB CRANE IN BURNSIDE SHOPS OF THE ILLINOIS CENTRAL RAILROAD.

Equipped with the Western Electric Company's crane motors.

LOCOMOTIVE CRANE AT THE INGERSOLL-RAND WORKS.
Of 10-tons capacity, with 52-in. Electric Controller & Supply lifting magnet. The Brown
Hoisting Machinery Co.

bearings are provided for the mast to swing in. These cranes are built in sizes ranging from less than 1 ton capacity to 10 and 20 tons, with radii to 20 and 25 feet. In the smaller capacity cranes hand power is commonly used for all movements, although with the modern development of light electric hoists they are frequently equipped with plain trolleys carrying electric hoists. Local conditions determine the kind of power, which may be hand, pneumatic, hydraulic, steam, or electric. Electric equipment for this class of work is monopolizing the field, principally on account of the great flexibility of the agent. The jib crane when converted into a traveling wall or floor crane loses in part its local characteristic and assumes some of the functions of the overhead crane, and in fact the wall crane is generally used as auxiliary to the main shop crane. For handling plates and castings over tools, for general foundry and machine-shop service, and for similar work, jib cranes have become indispensable, and pay for themselves many times over during the year.

Locomotive Cranes are built in most instances to travel on standard-gauge tracks, in order that they may be utilized on the existing track system throughout the shops and yards where they are to be used. Cases arise, especially when greater stability is required, where broader gauges are employed. Three movements are generally sought; rotating, traveling, and hoisting, and these are accomplished through the medium of either steam or electric power. The operating machinery is mounted on a rotating bed, which revolves about

A LOCOMOTIVE CRANE TRANSFERRING COAL FROM CAR TO HOPPER.

The crane shown hauled two loaded coal cars containing 40 tons each about ¼ mile and unloaded them in 1½ hours. The Browning Engineering Co.

a central pin, the whole being attached to the truck frame. Vertical engines are preferable to horizontal, causing less vibration of the machine, and two coupled at right angles insure prompt starting. To travel the crane along its track requires more power than the other operations, and all the wheels in a four-wheeled crane should have power applied to them, thus insuring the use of all available traction. This is important on curves and wet or bad tracks, as the overhanging load may relieve one pair of truck wheels of a portion or of the whole of the weight which would come upon them. Capacities run from about 2 to 20 tons and radii from about 10 to 40 feet. When handling coal such a crane can easily make 50 trips per hour, and, figuring on a basis of a 1-ton grab bucket, the crane can handle 500 tons if working continuously for 10 hours.

Machines of this class are seldom called upon to do continuous service. An actual case may be cited where a locomotive crane averaged 170 tons per day at a cost, exclusive of interest and depreciation, of 2.14 cents per ton. Interest and depreciation will add about 1.75 cents per ton, giving a total of 3.89 cents per ton. The actual cost per ton is not so important as a comparison of costs between old and new methods of doing the work. In the case cited the reduction in cost per ton due to the use of the crane amounted to over 12 cents.

Derricks, as labor-saving devices, are probably more widely used than any other class of machines. A description is scarcely needed. There are however some recent developments which are worthy of mention. Manufacturers are equipping contractors' derricks with electric machinery making them self-contained. For swinging, two methods are in use: in the first, a bull wheel is attached to a frame under the derrick and is fixed, the derrick and platform being revolved by a steel rope actuated from a specially designed winch on the end of one of the drum shafts; in the second, a spur wheel replaces the bull wheel, and a vertical shaft to which a pinion is keyed

meshing with the wheel has power transmitted through it from a horizontal shaft. Derricks of some such design may be used in industrial plants where the current is available at a low price and they offer in such a position certain advantages over the steam-actuated derrick.

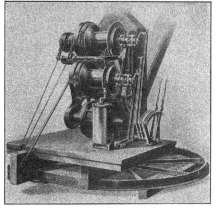

HOIST FOR REVOLVING DERRICK.
Electrically operated. Revolved by rope and clutch, operating directly on the bull wheel, with no gears. The Mead-Morrison Manufacturing Co.

Skip Hoists are used for handling such materials as ores, coal, limestone, sand, or ashes. The essential parts of such a plant consist of the hoisting mechanism, operated by steam or electricity; the cable and necessary sheaves; the skip, or bucket; and guides, or

COAL AND ASH-HANDLING INSTALLATIONS FOR LARGE BOILER-HOUSE WORK.
Above, a motor-driven monobar flight conveyor, 90 tons an hour capacity, boiler house
of the Allegheny City Water Works; below, ash hoppers feeding a Peck overlapping
bucket conveyor in tunnel of the Pullman Company boiler house. Both
installed by the Link Belt Co.

track. The electric hoisting apparatus may be either non-automatic or automatic. In the first, and this applies also to steam, the services of an operator are needed, whereas in the second, the throwing of a switch is all the service needed to start the hoist and return it ready for reloading. The skip is usually lowered into a pit below the floor level, so that material may be easily emptied into it, and is then automatically discharged by upsetting at the upper end of the runway. For charging blast furnaces this system is frequently used, the skip operating on an inclined track. In power houses such a plant is often well adapted for handling ashes, and here the track, or runway, is commonly built vertical. The serious trouble resulting from handling wet or hot ashes is practically overcome by the skip method as, with proper design, the moving parts are kept entirely free from water and ashes.

In one plant equipped with an automatic electric hoist, and handling 36 tons of ashes per day, the charge for power, the company generating its own current, amounts to $1.75 per day. The operation of a steam equipment, taking steam from the main power plant, would cost $4.00 to $5.00 per day.

CONTINUOUS HANDLING MACHINES.

Bucket Elevators and Conveyors comprise, in the continuous handling class, a large percentage of the various systems on the market. The reason for such universal use of the bucket type is primarily its adaptability to varying conditions. Such machines are designed with two objects in view, viz, to elevate the material vertically or to convey it horizontally. These two movements are accomplished either by two independent machines or by one machine designed for the purpose. These elevators and conveyors are made up of a line of buckets, built in an endless variety of shapes to suit varying conditions, attached to or suspended from a continuous belt or one or more continuous chains. In a very common design of bucket elevator either a canvas or so-called rubber belt is used, having the buckets secured to it by means of bolts. This type of elevator is particularly applicable to the handling of broken stone, sand, cement, or any sharp, gritty material, the belt withstanding the cutting action of such materials better in a great many cases than does a chain-type machine. The buckets are of sheet steel or malleable iron. In the handling of wet materials, and particularly where there is a chance for any chemical action, as in ashes or coal, malleable buckets are preferable to sheet-steel ones. So-called "link belts" are extensively used for conveyors and elevators, and the many standard designs afford possibilities for handling almost any class of material.

The Gravity Bucket Conveyor, which may be classed by itself among the bucket machines, is made up of a double strand of chain between which the buckets are suspended on trunnions. The links as made by the different manufacturers vary from 12 inches to about 36 inches in length. The links and buckets are carried by single flange wheels upon a suitable track. There are two types of gravity bucket conveyors, the contact and the overlapping. In the first, the ends of adjacent buckets touch or swing slightly free when the conveyor is on a horizontal run. ·In the second case, the ends of adjacent buckets overlap about an inch or an inch and a half. A filler, or mechanical device for loading the buckets is commonly used with the contact machine to prevent spilling; the object of overlapping in the other machine is to dispense with the filler. The speed of gravity bucket conveyors ranges from 40 to 60 feet per minute, and they can handle from 35 to 140 tons of coal per hour, depending upon size and speed. The type is well adapted for handling coal, ashes, hot cement clinker, and other bulk materials, and offers certain advantages, chief among which may be mentioned: material carried free

GRAVITY BUCKET CONVEYOR FOR CRUSHED ROCK, NAZARETH CEMENT COMPANY.
The run is 80 ft. horizontal and 60 ft. vertical; buckets 18 by 24 in.; capacity 50 tons per hour. Installed by the Webster Manufacturing Co.

FLIGHT CONVEYOR FOR THE PHILADELPHIA RAPID-TRANSIT COMPANY.
Capacity 100 tons an hour; 360 ft. long. Built by the Webster Manufacturing Co.

from contact with running parts, no dragging of material required, material may be carried both vertically and horizontally by one machine, discharging mechanism is simple, repairs are comparatively low. In carefully operated plants repairs and depreciation should not exceed 10 per cent, being lower than this in some cases.

Flight Conveyors are used for transporting bulk material either horizontally or on an incline. Such a machine consists of endless chain or chains having attached at regular intervals plates, or flights, the whole moving along a continuous trough in which the material is

OVERLAPPING BUCKET CARRIER FOR HOT CLINKERS, CHICAGO PORTLAND CEMENT
COMPANY.
Capacity 40 tons per hour. Installed by the Link Belt Co.

held, the flights pushing the material forward. The flights are supported upon runways by means of wearing shoes, rollers, as in case of roller chains, and wheels. Conditions determine which method to employ, but all things considered the latter method gives the best service, as the principles of design are more nearly correct. The scraping action is necessarily hard on the trough plate, and particularly so with gritty materials, such as sand, or stone. The usual speeds range from 50 to 80 feet per minute. The capacities given in the table, being calculated for a speed of 100 feet per minute, afford a quick means of determining capacities for any given speed.

CAPACITIES OF FLIGHT CONVEYORS.

Material—Coal, weight 50 lb. per cubic foot. Speed—100 feet per minute.

Size of Flight, Inches.	HORIZONTAL CONVEYORS.			INCLINED CONVEYORS.		
	Flights 16 Inches apart. Tons per Hour.	Flights 18 Inches apart. Tons per Hour.	Flights 24 Inches apart. Tons per Hour.	10 Degrees. Tons per Hour.	20 Degrees. Tons per Hour.	30 Degrees. Tons per Hour.
				Flights 24 Inches apart.		
4 x 10	33½	30	22½	18	14	10½
4 x 12	42½	38	28½	24	18	13½
5 x 12	51½	46	34½	28½	22½	16½
5 x 15	69½	62	46½	40½	31½	22½
6 x 18	....	80	60	49½	40½	31½
8 x 18	....	120	90	72	57	48
8 x 20	....	...	105	84	66½	56
8 x 24	....	...	135	120	96	72
10 x 24	....	...	172	150	120	90

EXAMPLES OF APRON CONVEYOR INSTALLATIONS.

The upper one is a steel-apron conveyor handling bottles at the Frank Fehr brewery, Louisville; the lower is a wood-apron conveyor in the Lamb Wire-Fence Co. plant, Adrian, Mich. Both built by the Jeffrey Manufacturing Co.

Spiral or Screw Conveyors date back to the time of Archimedes who seems to have been the first to make a practical application of the principle. The use is limited to the conveying of the finer grades of coal, sand, cement, grain, or similar materials. The design is simple, consisting of a spiral plate attached to a central shaft, the whole being supported in a box. By revolving the shaft the material resting in the box is pushed forward by the spiral plate which is really a continuous flight. There is the same dragging action in this as in the flight conveyor, and the box lining, usually made of steel curved to proper radius, receives the principal wear. Screw Conveyors are well adapted to confined places, as they occupy the minimum amount of space and require the simplest sort of a drive. They can be operated in either direction, which offers advantages in

WATER-JACKETED SCREW CONVEYOR.
For moist sugar. The Link Belt Co.

certain cases. The table gives sizes commonly used with maximum capacities for coal and recommended revolutions.

CAPACITIES OF SPIRAL CONVEYORS.
Handling Fine Coal.

Diameter, inches.	Maximum capacity per hour. Cu. ft.	Recommended revolutions per minute.
4	94	220
6	280	200
9	940	175
12	1860	150
16	4700	130
18	5600	120

The Trough or Pan Conveyor is used principally for carrying hot materials, like spikes, rivets, castings, etc. The successive pans, attached to the conveyor-chain links, overlap, forming a continuous trough. The whole is supported by wheels running on a suitable track. The conveyor is also used for large material like run-of-mine coal. It can not be classed among the higher grades of machines but is suitable under certain circumstances.

Belt Conveyors for the economical handling of various classes of materials have come into considerable prominence during the past few years. They are extensively used for transporting coal, lime, cement, sand, ores, bags, boxes, etc. The first systems of this type consisted of an endless cotton belt running flat on a series of wooden idlers, and were employed for carrying grain. Concentrators, or dish-shaped rollers, were then introduced, as a precaution against the spilling of materials over the edge. These rollers tilt the edge of the belt slightly, giving it a trough shape. The next step in the development of the belt conveyor, was the introduction of the trough-

BELT-CONVEYOR ON A STEEP INCLINE, NATIONAL LEAD COMPANY'S ATLANTIC WORKS.

Installed by the Robins Conveying Belt Co. to carry coal; 18-in. belt, about 150 ft. centers. Lower part on an incline of 20 degrees with horizontal, upper part level, with tripper; capacity 90 tons coal per hour.

ing idler, a refinement upon the concentrator principle. The belts are made of cotton duck, usually 4 to 6 ply, protected against wear by the application of coatings of various kinds, rubber and certain oil compounds being commonly used for this purpose.

BELT CONVEYOR FOR HANDLING BOXES, PLANT OF THE SINGER MANUFACTURING
COMPANY.
Installed by the C. W. Hunt Co.

Conveyors of this type are used to transport material both horizontally and on an incline. The angle of inclination depends upon the
nature of the material handled, but experience has shown that the
maximum inclination under favorable conditions is about 23 degrees.
The driving pulley and accessories may be located at the head end, at
the tail end, or at any intermediate point. This flexibility offers
certain advantages as the position of the drive may thus be controlled by the location of the source of power, by structural considerations, and by the fact that more than one conveyor and possibly other
machinery are to be driven from a central point.

The belt may have a speed of from 200 to 700 feet per minute, and
capacities depend upon width and speed. The table gives capacities
of conveyors usually employed.

DRIVING END OF RECIPROCATING CONVEYOR, WESTINGHOUSE AIR BRAKE COMPANY.

Installed by Heyl & Patterson.

CAPACITIES OF TROUGHED BELT CONVEYORS.

Material—Coal, weight 50 lb. per cubic foot.

Width of conveyor belt.	Speed 200 feet per minute.		Speed 400 feet per minute.		Speed 600 feet per minute.	
	Largest size of cube which can be carried.	Tons per hour.	Largest size of cube which can be carried.	Tons per hour.	Largest size of cube which can be carried.	Tons per hour.
12	2 inches	6	½ inch	16	¼ inch	22
16	3 inches	16	1 inch	34	¾ inch	50
18	4 inches	20	1½ inches	45	1 inch	70
20	5 inches	30	2 inches	60	1 inch	100
24	6 inches	50	3 inches	100	1 inch	190
30	7 inches	100	4 inches	200	2 inches	360
36	9 inches	180	6 inches	340	2 inches	600

The main item of expense in the up-keep of this system is that of renewal of the belt, and all possible means should be employed, by improvement in design or care in operation, to increase its life. With ordinary care and attention a good grade of conveyor belt should last from two to five years.

Reciprocating Conveyors are used in foundries, glass works, and similar places where perfect mixing of several ingredients into batches is desired. Such a machine consists of a rectangular trough, built of

RECIPROCATING SAND CONVEYOR, FOUNDRY OF BENJAMIN ATHA & COMPANY.
The sand is fed to conveyor at the rate of 37 tons an hour by a Link Belt conveyor outside
the building. Installed by the Link Belt Co.

wood, or steel, with bottom gates at required points for the discharge of material. The flights, which push the material forward, are attached to a rigid member, consisting usually of a wrought-iron pipe extending the full length of conveyor. This pipe is supported at intervals by guide frames, or by trucks which keep it in position. It is given a reciprocating motion by means of a crank and pitman attached to one end, and driven by suitable mechanism. On the forward motion the flights are held vertical, and push the material ahead, but on the return, being hinged, they ride freely up over the material.

There are many machines and combinations of machines which may be installed for doing any special sort of work and it is the province of the engineer to analyze the various schemes which the manufacturers submit as best suited for the situation. The depreciation of most material-handling equipments is necessarily high—that is, high for example when contrasted with the same item in the operating cost of steam engines, and I believe that in the majority of cases, an increase in the first cost which will materially decrease this charge is warranted. Careful attention on the part of operating engineers also will often prevent needless and expensive breakdowns, the immediate effect of which, aside from the expense to the owner, is a condemnation of the manufacturer and of the machine, including the meting out of a share of blame to the consulting engineer.

HOISTING MACHINERY FOR THE HANDLING OF MATERIALS.

By T. Kennard Thomson.

If any justification be needed for the space here given to the discussion of machinery for hoisting and handling, it is amply supplied by these papers themselves. To some extent, the field of hoisting appliances must overlap that of conveying devices already reviewed; but Mr. Thompson invests the description and analysis with additional interest by treating it in connection chiefly with a distinctive range of work—that of construction, rather than of production. And he brings out impressively that here as elsewhere in the domain of modern enterprise, economy in unit costs and maximum of output can be secured only where intelligent and well-advised use is made of the mechanical facilities afforded for the handling of materials. Mr. Thomson's review will be continued through two succeeding installments.—THE EDITORS.

THE last fifteen or twenty years have witnessed so vast an improvement in hoisting machines and their development as labor-saving devices in so great variety that many books could be filled with illustrations of the various kinds of machines. All that is possible in a magazine article is to refer to the general types and give a few examples of each, showing their range of work and cost.

A discussion of hoisting machines, by convention, must begin with the pyramids of Egypt, and yet no one has been able to decide how the enormous stones were handled, for, even allowing for the vast army of men, some sort of mechanical contrivance must have been used. One theory is that as each course of stone was laid, a sand embankment was built around it with long easy slopes so that the stones for the next course could be pushed up on rollers and slid into place without any actual lifting, and so on, the pyramid being in fact buried as fast as it was built, until the top was reached, when the stupendous job of removing the sand embankment was commenced.

This might account for the vast army of laborers used; for instance, it is said that in building the great pyramid of Cheops 100,000 men were employed for 30 years, although the quarry from which the stone was obtained was only 3,000 feet from the pyramid.

The City Investing building, the largest sky scraper in New York, happens to be nearly the same height as the pyramid. If we required the same number of days' labor for this building, which takes a year

SHORES USED IN BUILDING OPERATIONS, CORNER OF MAIDEN LANE AND WILLIAM
STREET, NEW YORK.

Some of the strongest shores ever put up to sustain a building. At the bottom may be
seen the screw-jacks by which they were adjusted. The Foundation Co., contractors.

to erect, as the Egyptians took, we would have 30 multiplied by 100,-
000, or 3,000,000 men working every day for one year on this one
building. Imagine a number of men equal to almost the entire population (including men, women and children) of New York City, trying
to reach the corner of Broadway and Cortlandt Street every morning.
Even if they got there, there would remain the impossibility of finding

standing room, let alone working room for them, or of providing a commissariat or hygienic department. Against this 3,000,000 men working every day for one year, the average number on the City Investing building would not exceed 2,000 men a day, while the maximum would probably never reach 3,000. While in some ways it is not fair to compare a stone structure with a steel building, still there is no doubt that the chief difference is due to the improved machinery, the greatest strides in which have been made within the last twenty years—many kinds of work now being handled in from one-half to one-third the time that was required two decades ago.

As we have seen, it is conceivable that the pyramids might have been built with no other mechanical aid than an ordinary crowbar or lever and, perhaps, a screw or hydraulic jack. These still remain the fundamental mechanisms whose principles, once thoroughly understood, make perfectly clear the otherwise puzzling theory of bending moments, balanced forces, etc. While the principle of the lever was being worked out and applied and many complicated forms were being evolved, the principle of the screw was probably not far behind. The screw combined with the lever is one of the most powerful and at the same time slowest of hoisting implements—for, as the nut revolves it is forced up the incline of the screw, about 60 degrees, and if the diameter of the screw is $\frac{1}{4}$ of an inch the nut has to be turned around 20 times to move one inch vertically. If the diameter is 6 inches, $2\frac{1}{4}$ revolutions are required; if the power be applied by a man at the end of a 5-foot wrench by hand, to raise the nut of a 6-inch screw 1 inch he would have to walk $2\frac{1}{4}$ times around a circle of which the diameter is 10 feet, which is equivalent to a travel of more than 70 feet. With a $1\frac{3}{4}$ inch screw he would have to go around 5 times, or cover about 157 feet to raise the nut 1 inch, and, as the power he applies to the screw jack so slowly is multiplied by the radius (length of the lever), in this case 5 feet or 60 inches, and divided by the radius of the screw, it will easily be seen that a force of 100 pounds would raise over 3 tons, less the friction. As this movement is so slow, it is possible (by placing an immense number of jacks under a building), to have one man go from jack to jack, giving each a slight turn each time, and thus unaided to lift the entire building gradually from its foundations. As a matter of fact, we all have probably seen buildings thus raised, sometimes only a fraction of an inch or hair line, until new foundations are placed under, and at other times enough to allow the building to be moved to another site.

One of the most interesting examples of this was the raising and moving of the Brighton Beach Hotel, a building 460 feet long, 120 feet wide and from two to five stories high. After it had been jacked up by one thousand two hundred screw jacks, twenty-four tracks were laid under it and one hundred and twelve flat cars were distributed under the building. A system of beams transferred the weight of the hotel to these cars, which were so connected by means of block and tackle to four or six railroad locomotives that the entire building was moved nearly 600 feet without injury, although at one time the engines pulled at the rate of 8 miles an hour; the block and tackle were made to act like the whiffle tree of a carriage, so that when the locomotives pulled the strain was distributed evenly over the entire area of that big frame structure, which was thus saved from the encroachment of the ocean.

NEEDLE BEAMS FOR UNDERPINNING THE NEW BUILDING AT 166 BROADWAY, NEW YORK.
Showing the screw-jacks for taking up the strain.   Clinton & Russell, architects; the Foundation Co., contractors.

More powerful than the screw and at the same time much quicker is the hydraulic jack—also long known. The principle on which it works might be explained by taking a box full of water, having its top arranged as a piston, and a small pipe containing another piston connected with the side or bottom of the box. Now we know that if the area of the big piston is one hundred times the area of the small piston, then the weight lifted by the big piston will be one hundred times that applied to the small piston; but, as this lifting can only be accomplished by forcing water or other fluid from the pipe into the box, it is obvious that at each stroke the little piston must move one hundred times as far as the big one. Here again we see that a very small force can raise an enormous load, but only by moving through a proportionately greater distance.

By using pumps, levers, etc., this principle is used for many purposes, such as operating elevators, launching vessels, forcing water through fire hydrants, operating cranes, etc. In launching ships, enormous pressures are sometimes used and water has even been forced to ooze through cast-iron plates 6 inches thick. Not only can heavy objects be thus raised slowly, but the reverse must also be true, and a great force can be used slowly to raise a small weight very quickly.

In underpinning a building (an everyday occurrence in New York now, owing to the new foundations being lower than the old—not to mention tunnels, etc.), a multitude of screw jacks are sometimes used, or in other cases a much smaller number of hydraulic jacks, the most common commercial sizes now used in New York running from 4-tons to 200-tons capacity. They are so designed that one 150-pound man can raise 10 tons 1 foot in about 1½ minutes, or 100 tons in 15 minutes. In forcing underpinning caissons 3 feet in diameter under the walls of a building, we usually use one or two jacks of from 60-tons to 100-tons capacity each; but I once stopped a foreman who was trying to shove a caisson down with four jacks aggregating 320 tons, for fear that he would injure the superimposed building, the cylinder or caisson having got jammed on the way down.

Screw jacks cost from $10 apiece up, while a good 60-ton hydraulic jack is quoted at about $200 and a 200-ton jack at from $500 to $600.

As the ancients probably understood levers and jacks, they were able by taking infinite pains and time and plenty of cheap labor to move almost anything. They were no doubt content; but in this age of rush, labor unions, and extreme pressure of time and cost, even a fraction of a minute on a single operation means a great deal of time and money saved or lost. This is probably more obvious in handling and printing the daily papers than in contract work, but it is none the less vital to the contractor.

Did the genius who invented the cart wheel also invent the pulley block, and which came first? Of course the first pulley consisted of a single wheel, and merely overcame friction without any increase in

CRABS USED FOR PULLING DOWN THE CAISSONS IN FOUNDATION WORK.

At the Commercial Cable building, New York; Arthur McMullen & Co., contractors.

power. By using two blocks with one sheave each the power is
doubled, and it can be doubled again as often as the sheaves or blocks
are doubled; but as before, the applied force has to travel twice as far
when the lines are doubled up, and so on each time, with slight allowance for friction, the blocks and sheaves being regulated according
to the work. For instance, in caisson work, when we have to lift
heavy parts of machinery, say from 15 to 20 tons, the blocks are
doubled up and the ponderous weights are carefully hoisted; but when

SUPERIMPOSED ENGINE FOR HOISTING. THE POWERFUL AND EFFICIENT SUCCESSOR OF MULTITUDES OF WORKERS EXERTING HAND POWER.

Double-drum double-cylinder direct-connected hoisting engine and single-drum double-cylinder traversing engine. Built by the C. W. Hunt Co.

we want to take a half-yard bucket of material out of a caisson the hoisting engine does not need the assistance of any pulley block, but great speed is required, so the rope is simply run over a single sheave and attached to the drum of the engine and by this means a bucket of materials can be taken out of a caisson air chamber 80 feet below ground, passed through the lock (only one door of which can be open at a time), lifted 20 or 30 feet above ground, emptied by being turned upside down, and returned through the lock to the air chamber twenty times an hour, as long as the men will fill the bucket quickly enough. On caisson work, as on most important contract work, the speed with which the material can be handled means thousands of dollars to the contractor or owner, for there is always an item called "general expense" covering the wages of superintendents, foremen, fuel, etc., amounting from one hundred to many hundreds of dollars a day. This expense runs right along practically unchanged whether a hundred or several thousands of yards of material are moved.

DERRICK BOATS USED IN BUILDING THE NORFOLK & WESTERN RAILROAD BRIDGE OVER THE OHIO RIVER AT KENOVA, W. VA.
A small one is moored in the foreground, and a larger one may be seen in the stream. This was used for building a pier 103 ft. high.

Probably not long after the invention of blocks with one or more sheaves each, the uncertainty and labor of continually pulling on the rope, whether the load was being lifted or simply held suspended, became apparent; and some wide-awake chap thought of passing the end of the rope over a drum, giving the rope two or three turns around (thus obtaining a better purchase) and then fastening the end of the rope to the drum and adding a crank or handle, thus saving several men. Continual holding of the handle, however, was still required, until the ratchet wheel and pawl were added, similar to that used for brakes of a railroad train. The next step was to place the drum on a frame and call it a "crab." This is still used on small jobs, especially in building construction, in which case the crab is usually attached to a two-legged frame called a house derrick. A much more advanced "hand" machine is the triplex block, which is geared to raise heavy weights by hand power and automati- cally checks it- self; though a little slow it is al- ways in place ready for use. A further step in the application of this principle has produced t h e compressed - a i r hoists, which are built of various sizes to lift from ½ ton to 5 tons, the small size costing $135 and the 5-ton $275. They are high- duty pneumatic motors geared to hoisting drums and generally operated under

AN AIR-MOTOR HOIST.

The "Imperial," made by the Ingersoll-Rand Co.

80-pounds air pressure. The range of speed is quite wide and the motion is controlled by worm gears instead of brakes.

It is stated that a 1-ton hoist with 80-pounds gauge pressure can lift 30 feet a minute with a consumption of 30 cubic feet of free air, while the 5-ton hoist can lift about 7 feet per minute using 45 cubic feet of free air; so where they are in constant use or the compression plant is used for other purposes, the cost of running the hoist is trifling. Dr. Frank Richards some time ago published an estimate that at 100-pounds gauge pressure, compressed air costs 5 cents per 1,000 cubic feet of free air. Triplex blocks, pneumatic hoists and other similar devices usually run on overhead trolleys and are more often used in shops than as shown in the cut below lifting a heavy girder in the yard.

AN AIR-MOTOR HOIST LIFTING GIRDERS IN A BRIDGE-BUILDING YARD.
The hoist shown is a "No. 10 Imperial," made by the Ingersoll-Rand Co. The air-supply and the mode of control from the ground are clearly indicated.

Before going on with the description of up-to-date steel hoisting machines which are rapidly supplanting the old-style wooden derricks, we might briefly refer to the latter, which have been so familiar on building and other construction.

ERECTION OF THE ELEVATED RAILROAD STRUCTURE ON FLATBUSH AVENUE, BROOKLYN.
Showing the derrick traveler used by the American Bridge Co.

Out in the country where space is not limited, the cheapest to erect and easiest to handle is probably the well-known guy derrick, consisting of a vertical mast on a simple foundation with a ball and socket contrivance at the base to permit the mast to revolve; but while this contrivance is simple, it makes all the difference in the world whether it is well designed or not, and many are not. The top of the mast carries a gudgeon pin some 4 inches in diameter and 4 or 5 feet long for a big derrick, over which a thick circular plate with five or more holes for the guy wires is dropped. A 2-ton derrick usually has about five ¾-inch diameter galvanized-iron guy wires,

A STIFF-LEG DERRICK ON PILES, FOR CONSTRUCTION OF THE OHIO RIVER BRIDGE AT MINGO JUNCTION, OHIO.

Arthur McMullen & Co., Contractors.

at least three times as long as the mast; if the locality permits they can be made much longer to advantage, while in some recent sky scrapers very short guys have been used, 25 to 50 feet from the mast; in this case considerably more than five guys have been used, and of course the range is very limited, as the boom could not be swung under the wires except when raised as high as possible. The boom usually fits into the socket of a casting attached to the foot of the mast, and a pin connection permits the boom to be raised or lowered while the mast revolves, which latter motion is usually obtained by means of a bull wheel 8 to 12 feet in diameter, around which a wire or hemp rope is run, the ends being passed over the drums of the hoisting engine.

The derrick irons for a 2-ton guy cost about $90, and a good engine with a 3,000-pound pull, 6¼ by 8-inch double cylinders, friction drum, pumping gear, and boiler, cost about $1,200; and, of course, less complete engines can be had for less money.

The cost of the timber and framing and hoisting wires (½-inch diameter) would vary greatly according to locality.

STEEL-DERRICK HIGH-SPEED SAND-HANDLING PLANT AT TULSA, OKLAHOMA.

For transferring sand from barges to cars. Steel derrick; bucket 2 cu. yd. capacity. Handles
about 1,000 cu. yd. a day at a cost of about 1 ct. per cu. yd. under continuous
operation. Designed and erected by the National Equipment Co.

A guy derrick for 20 tons generally has seven 1⅛-inch galvanized wire guys, derrick irons costing some $360; a double cylinder, friction-drum engine with boiler and pumping gear, the cylinder being 8½ by 12 inches, and the engine having a pull of 10,000 pounds and costing complete about $2,400. The hoisting rope is usually ¾-inch diameter.

A DERRICK WITH RIGID ARM AND TROLLEY, AT WORK BUILDING A WALL AT OSSINING, N. Y.

Mast 10 by 10 in., 30 ft.; trolley beam, 10-in. I beam 20 ft. long, 20 ft. above foot block. Crab had two drums, one for hoisting and one for trolley, with reversible rotary engine. Made by Dake Engine Co., Grand Haven, Mich. Derrick cost $500. Ford & Waldo, contractors.

Where the confined space does not permit a guy derrick, the next best thing on temporary work is a stiff-leg derrick. For ordinary contract work where 2 cubic-yard buckets are used for handling material, these are generally built with masts from 12 by 12 inch to 18 by 18 inch timbers, 20 to 45 feet high, and two stiff legs of 12 by 12 timber placed at right angles to each other and inclined vertically about 45 degrees, or very often 10 inches horizontally for each foot vertical. The top connection is made with a goose neck—a heavy steel bent plate some 10 inches wide and 2 inches thick, with about 6 feet bolted to the stiff leg, and a little less than 2 feet of the plate bent so as to fit horizontally over the gudgeon pin.

The goose neck was usually the weak point of a stiff-leg derrick, and in consequence they were made very heavy and of the most expensive iron, steel being too unreliable for such heavy forgings; many have now discarded that form of construction entirely in favor of built up steel caps for mast and stiff legs. From under the mast a 12 by 12 sill is usually run to the foot of each stiff leg, to which it is rigidly connected, as well as to a platform carrying a dead weight of stone, pig iron, or other weight, necessary to prevent the derrick from overturning. In places where the derrick rests on some stable structure, it is bolted to this instead.

The irons. for a 2-ton stiff leg cost about $130 and a high-grade engine complete with boiler, etc., as for the guy derrick, about $1,200; and a 20-ton derrick would also require the same engine, boiler, etc., as a guy—$2,400—and have iron fixtures costing about $570. These are commercial sizes, sold by regular manufacturers; but many contractors have learned by sad experience that a poorly designed derrick and improving for years, and which, in a few cases, may cost less than the above, though usually they cost much more. Nearly all contractors have learned by sad experience that a poorly designed derrick is a very expensive investment, as a single accident often costs a good many times as much as the derrick itself.

In many cases neither stiff-leg nor guy derrick can be used, and some combination has to be designed to fit the conditions; an example is shown opposite, being a combination of both with a gantry crane which was used by Ford & Waldo on their contract for the New York Central Railroad at Ossining. A boom swinging across railroad tracks should never be allowed, as it is always dangerous and has been the cause of many accidents, and this is probably why the contractors omitted the boom in this case, putting in instead a 10-inch I-beam trolley runway 20 feet long and 20 feet above the foot block. The crab had two drums, one for hoisting and one for moving the trolley, and was operated by a Dake reversible rotary engine occupying no more space than an ordinary crab handle, the whole outfit costing about $500.

Some contractors who still stick to the "big stick" or wooden derricks get all their pine or fir masts and booms by the car loads direct from Oregon, the forests there supplying the best material obtainable.

Ordinarily stiff-leg derricks are sometimes mounted on trucks and called travelers; but as a rule travelers are more elaborately and compactly designed and have from one to four booms. A heavy four-boom traveler designed to lift 20 tons per boom has been built in New York

A SPECIAL DEVELOPMENT OF THE TRAVELING CRANE. GANTRY CRANE WITH CANTILEVER EXTENSION, ILLINOIS STEEL CO., CHICAGO. Center, to center of rails 56 ft.; extensions 28¾ ft. on one end and 12¾ ft. on the other; 15-tons capacity. Uprights are so constructed as to permit the trolley to travel the full length of the girders with its suspended load. Nils-Bement-Pond Co.

for $3,000; this did not include the engines, boilers, and hoisting tackle, which were complete for each boom. These four-masted travelers are now frequently used in New York city, the latest improvement being the substitution of steel for wood throughout.

FOLDING-BOOM STATIONARY-JIB 30-TON GANTRY CRANE FOR CHESAPEAKE & OHIO RAILWAY, NEWPORT NEWS, VA.
Width at bottom 40 ft.; height from top of rail to bottom of trolley-track girders, 44 ft. 10 in.; total height 78 ft.; length of boom 51 ft.; total length of trolley-track girders 105 ft. Would cost about $35,000. Installed by the Wellman-Seaver-Morgan Co.

A very neat steel derrick was recently used for building the concrete steel warehouse for Montgomery & Ward in Chicago. It had a steel frame mast 135 feet high, revolving inside of a 10-foot square steel-frame tower 80 feet high, above which an 85-foot steel boom was attached to the mast; it was capable of lifting 12 tons anywhere within a complete circle the diameter of which was 170 feet. It had a vertical range of 150 feet, so that it could lift an object from the ground to 15 feet above the roof, and was so arranged that the 10-foot square tower fitted into a single floor panel, the guy wires being moved as required. As the entire building is 750 feet long and 270 wide, several of these derricks were provided for, saving an immense amount of time, shifting derricks, etc.

The introduction of steel cranes marked the greatest advance

LOCOMOTIVE CRANE AND TRESTLE FOR UNLOADING COAL FROM BOATS, LEHIGH COAL & NAVIGATION CO., PHILADELPHIA.

The crane unloads the boats and discharges direct into bunker or into an electric hopper car depositing on ground storage piles under the trestle. The car travels between the girders and the crane on top of them. Coal is reclaimed from the ground by the crane operated from the trestle, the car carrying the coal to bunker. Capacity 60 tons per hour, operated by two men. Crane furnished by R. H. Beaumont & Co.

in labor and time-saving devices, in the shop, field, and on the wharves, etc. Their value was appreciated in England earlier than it was in the United States—especially that of the three-motor electric traveling cranes. At first American contractors imported some of the British machinery, but it was soon found to be not adapted for the rough and heavy work in the United States. The imported cranes were strengthened up where they were weak and lightened where unnecessary material had been used, and while they have been used, and while they have been improved so that often little resemblance to the old remains, the improvement is still going on every day. There is still much room for improvement, not only in the design of the machines but also in the selection of the proper machine in the proper place, in nearly all classes of work, and in many cases this is so obvious that it seems a criminal waste of time and money to continue using manual labor or inadequate or antiquated tools. Very few wharves or freight depots, for example, are adequately fitted up, and hence not only is the direct cost much greater than it should be, but also the indirect cost due to demurrage, etc. A case in point:—an enterprising manufacturer had convinced the chief engineer of a railroad that it would pay him to spend $300,000 for plant at a certain terminal, but the chief engineer could not persuade the Board to raise the money; so the manufacturer offered to furnish the money himself and to take a royalty of something like 2 cents a cubic yard on the material handled, which eventually returned to him a far greater profit than he could have made by selling the machines outright.

Contractors for extensive excavations, masonry and such work, are probably the most enterprising in seeking after new methods, for not only do they save several hundred dollars general expense for every day saved, but also the interest on their money; sometimes (I have seen this happen) a single day's delay will cause the completion of the work to be put off for months, on account of sudden floods, ice, etc.

As an example of how cheaply material can be handled:—coal has been taken out of ships and placed on the wharf at the rate of 3 cents per ton, and even this included a sinking fund which would pay for the plant itself in 20 years. Simple as all this may seem, it is astonishing how little the general public and even many engineers and contractors appreciate it. A single case in point:—I have seen a contractor for a whole year work six or seven men on an old-fashioned pulley and crab, and keep five or six men waiting, while he accomplished work that one man could have done with a good pulley block in much less time, or a stationary hoisting engine could have done even more cheaply, considering the number of men kept idle, etc. The improved methods would have allowed him to make a good profit on his contract, in place of the actual loss he sustained.

A HAND PILLAR CRANE AT THE U. S. NAVAL STATION, PENSACOLA, FLA.
35-ft. radius, 30-ton capacity.   The Brown Hoisting Machinery Co.

The pillar crane is an improvement or substitute for the old guy and stiff-leg derricks. It is more permanent, and being anchored to the ground the necessity of guys and stiff legs is done away with, economizing room and giving more freedom of work. It is generally made of cast iron or steel, or both, and so pivoted that it can revolve in a complete circle while the boom is raising, lowering, or holding its load. Pillar cranes are made of many capacities, as a rule from 2 to 30 tons, the only limit, in this as in all other cases, being the cost as compared with the cost of some other machine for doing the same work. The cost of a 2-ton hand-power pillar crane with a boom radius of 15 feet is about $450, and for one of the same radius to lift 30 tons, $3,600. A 20-ton electrical-power pillar crane with a 15-foot boom radius would cost about $2,700.

A LOCOMOTIVE CRANE SETTING A 60-FOOT ONE-PIECE STACK.
The Browning Engineering Co.

Pillar cranes are often mounted on flat cars for wrecking purposes, etc. Where the swinging of a boom would interfere with the work the pillar jib crane is used, a horizontal arm carrying a small trolley taking the place of the boom.

In shops, and elsewhere when possible, a plain jib crane is used instead of a pillar jib crane, the top of the mast or pillar being held as well as the bottom—a much more economical arrangement in many places, as the framework of the shop or building provides easy facilities for the top connection; here, however, the range of the swing is often limited to half a circle, especially when set against a wall. In

THREE-TROLLEY GANTRY CRANE FOR LACKAWANNA STEEL CO.

Length of girders 160 ft.; capacity 10 tons. Equipped with lifting magnets for unloading. Such a machine would cost about $30,000. Built by the Wellman-Seaver-Morgan Co.

jib and pillar jib cranes the length of the horizontal arm naturally has much to do with the design and cost of the machine, as has also the load to be lifted; for instance, 2 tons lifted 20 feet away from the mast would cause as great cantilever strains as 4 tons with a 10-foot radius would.

FLOATING CANTILEVER CRANE FOR THE U. S. NAVY YARD, NEW YORK.

100-tons capacity. Pontoon 60 by 100 ft., 11 ft. deep; the rigid legs are far enough apart to pass a load 40 ft. wide from end to end of the cantilever. In addition to water ballast, a 250-ton counterweight traVes so as to keep the keel always horizontal. There is no swaying when heaVy turrets (for example) are placed. Adjustment can be made to 1/16 in. The crane weighs about 1,200 tons and cost about $225,000. Built by the Brown Hoisting Machinery Co.

The next step from a jib crane is a gantry having two vertical legs or frames carrying an overhead bridge, the legs being arranged to run on parallel tracks, the gauge of which varies from 5 to 200 feet or more; and as the height also varies to clear local objects, it seems almost impossible to give any idea of the cost. Each individual case is a study by itself, so the cost is given only under special photographs.

One of the early very large cranes in America is that at Sparrow's Point, Maryland, built by the Pennsylvania Steel Company to lift 150 tons, the steel used in the construction weighing 775 tons.

Another of these imense hoisters was built to lift the same weight for the Imperial Dock at Bremerhaven, with a clear height of 118 feet and a horizontal reach of 72 feet, allowing it to pick a 150-ton load off the ships and clear the smokestacks, etc. This contains only 412 tons of steel as compared with the 775 tons of the Maryland shears.

Stationary cranes are operated by hand, steam, electric or hydraulic power. The first is desirable only where the loads are to be moved at considerable intervals and time is a secondary object. Steam is very useful where only one crane is to be run and that steadily, but is inferior to electric or hydraulic power when there are a large number of machines to be worked from the same central station; for it has been found that in such a case the maximum working load is often three or four times as great as the average, and if the engine has to be kept fired to take the maximum load at any moment the waste of fuel is very great.

Steam is slower in its action than electric or hydraulic power; but on the other hand, where initial cost is a prime consideration, a steam plant has the advantage. Local conditions must always determine which to use, although electricity is fast supplanting everything else on account of being so quickly applied and always ready for use, without being wasted when the machines are idle or partly idle.

THE FUNDAMENTAL PRINCIPLES OF WORKS ORGANIZATION AND MANAGEMENT.

By P. J. Darlington.

In this article and a second paper to follow next month, Mr. Darlington undertakes the very interesting task of reducing some of the greatest problems of works management to their fundamental elements. The maze of special methods and systems, employed in a multitude of shops, he believes can be separated, sorted, and classified so that the apparent confusion will prove to be but a mingling of a very few, simple, clearly defined and easily understood methods and systems. Each of these systems is then tersely explained, with a statement of the results it secures and the conditions under which it is best and those under which it fails. If the often mysterious subjects of stock keeping and cost keeping can be thus simplified into a scant dozen easily understood elements, Mr. Darlington believes that the manager can readily combine from them such organization, routine, process, or records as will best serve the purposes of his works, and may be certain in advance what he can expect his system to accomplish and what difficulty or cost will attend its use.—THE EDITORS.

ENTERPRISE, ingenuity and good judgment in matters commercial and technical are the real forces of successful manufacturing. The plan of organization and management, with its rules, routines, and records is merely mechanism for utilizing those forces.

This mechanism may fail either from lack of essential features, or from too great complication by dissipating effort in red-tape. We learn by experience that unnecessary complication discourages effort and invention and reduces quality and quantity of product. Our problem, therefore, is not to evolve an elaborate and logical system of organization and management, but rather to build the very simplest and most direct plan that will substantially meet our needs.

Manufacturing works differ widely in the character and variety of their product, the material used, and the labor and intellect employed. They differ likewise as to commercial circumstances and conditions. For example, one works may be producing under fixed conditions while another is choosing new lines of manufacture, for sale in close competition, where a knowledge of the normal cost of widely different articles is essential for guiding the growth of business into most profitable channels.

We therefore may not hope to find a ready-made plan of organization and management suited to our own conditions. Study of successful works may suggest new features and details, but usually not in

practical proportion for our needs. Our problem is similar to that of
a designer who must produce a machine for specified work by applying his knowledge of elementary mechanical forces, movements, and
materials. He adopts the fewest and simplest parts for minimum
cost, internal friction and risk of breakdown.

The direct application of the simplest means available to each
clearly defined purpose is the secret of building up a vigorous and
sturdy organization and management. "System" beyond this is idle
mechanism and a source of loss and risk. What must be accomplished
can be determined only by those well acquainted with the business
and conditions. They alone can define the needs and purposes in correct proportion of importance. There appears, however, to be no
good reason for the present almost total lack of comparative data
and discussion of the "means available," or what we may call "elements of organization and management."

The purpose of this article is to describe and discuss a few of
these elements, in an effort to place the subject on a scientific
basis so that organizations, routines, clerical processes, and records
may be chosen for any given purpose with a fair knowledge of what
they may be expected to accomplish and of the difficulties and costs
of using them.

We put value into our manufacturing business for interest on
investment, for operation and maintenance, for purchase of materials,
and for employment of labor and intellect. We combine them into a
product saleable at a profit over these constituent costs. This profit
depends not only on cheap manufacture but also on wise choice of
product to be made and marketed.

The operation of our business is divided and subdivided into such
transactions or undertakings; for example, the addition of a new
product, the investment in special tools for it, the setting of stock limits on parts, etc., even down to the choice of operations in its manufacture.

Our annual profits depend not only on the successful forwarding
of each undertaking, but also on our increased experience from it
and our improved judgment thereafter. Evidently organization and
management must not only solve each problem and carry through each
transaction, but also must make the experience available for the
future. The sum total of such experience may become a large asset
of the business, and is of no less actual value because it cannot be
estimated in dollars.

With good organization and management each employee understands the purpose of his efforts. He gains respect for his own work

and that of his associates, and he enters into *esprit de corps* far more lasting and valuable than aimless enthusiasm.

Good organization and management considers human nature. Men of mechanical education and experience are not annoyed by clerical duties; every employee feels a normal share of responsibility and is at the same time confident that results will be appreciated at their true values; every man, from the inventor to the mechanic, is given the incentive and working data for the best individual effort and results as well as for the best co-operation of individuals and departments.

An organization and management based on accurate knowledge and results weeds out incompetent employees and increases the average efficiency of the working force. Clearly defined purposes and definite means allow each duty to sift down to the cheapest grade of intellect that can handle it, and give each employee an opportunity to study ahead of his position and advance to the limit of his capabilities.

With prompt and practical information in regard to orders, stock, men, and equipment we may balance our departments and operate our works at highest efficiency and maximum output without danger of overloading and breaking down some part of the industrial mechanism, thereby throwing the works into confusion and nonproductiveness. Accurate and comparative knowledge of costs enables us to develop the more profitable lines of manufacture and so increase the profit factor of our business ahead of less scientific competitors. In many ways good organization and management increases efficiency and adds annual profits not calculable in dollars.

These as well as more measurable benefits are very largely dependent upon our correct understanding and wise application of elementary principles which are capable of scientific study. A good understanding of human nature, a high sense of justice, and other personal good qualities lose most of their force in the confusion of a mismanaged industry where fact is smothered under form and noise mistaken for results. On the other hand, we must remember that clerical system is *not a prime force*. For example, if our product be unnecessarily complicated in design, or our stock scattered about without needed facilities for handling and storing it, clerical system will add to the trouble and may greatly reduce output.

STOCK.

Fundamentally, stock may be classified as:—

1.—Accidental.
2.—Standard.

1.—Accidental stock is an accumulation of materials, parts, or products which it is not intended to continue to carry in stock. It may be extra parts which are normally made on an assembling order, or it may be unusual, special, worn, or superseded articles. That it has not gone to scrap implies a hope of using it. All that we can do is to advertise it to those departments that can suggest a use for it. To assure its receiving attention at the right time we record it in connection with the standard article on which it may be used, or for which it may be substituted, or we may cross-index it to several such.

When accidental stock is to be used in manufacture it must be clearly called for on the using order, with a statement of the standard stock, if any, for which it is to be substituted. This records its use, avoids duplication with new material, and prevents the using order receiving credit for labor not expended.

Accidental stock must not be confused with standard stock. There is no connection between them and they are very different in principle. There will be no further reference to accidental stock and hereafter stock will mean standard stock.

2.—Standard stock may be raw material, worked parts, or saleable product maintained in stock according to limits or on other authorized basis. When an article becomes standard stock it must, therefore, be made always as such on its own stock orders. Otherwise using orders would not get credit for all labor expended and the cost and stock records would be in error.

Our purpose is to fill orders promptly, cheaply, with lowest investment, and with least depreciation and loss from changing designs; also at least expense for handling and storing. When and on what limits to establish an article as standard stock is therefore determined in consideration of:

a. The nature, number and variety of its uses.

b. How far we can anticipate demand and how soon replenish stock.

c. The importance of large quantity in reducing cost and improving quality.

d. The amount of money invested in the stock and the value of the store space occupied.

e. Depreciation of the stock and possible lack of confidence in the stability of the design.

If a machine part is assembled into several different sizes or designs of product, at different times and on different orders, then by making that part standard stock and thus concentrating its manufacture on one large order for a long-time supply, we may often gain enormously in cost, quality, shop capacity, investment, and clerical expense. Otherwise if the part be costly in material, special to that one product, and if it can be obtained in time to take its place in manufacture and assembling, then it may be better to make it on the general or assembling order for the machine. To establish such an article as standard stock might result in useless and expensive shop orders, stock transactions, records, investment, and cost for storage.

Evidently an article should be established in standard stock only with a clearly defined purpose and very definite gains in view. Divided orders for divided cost record are usually costly and useless as will be shown later. The cost office should serve manufacture, rather than the reverse.

When stock falls to the "low limit" an order is issued for either a "standard order" quantity or for a quantity to bring the total to a "high limit." The standard-order plan makes quantities uniform, costs comparable, and shortage evident. It is useful with bin stock systems, as described hereafter. The high-limit plan is more flexible to meet accumulated or extraordinary demand, and is therefore essential to the "order balance" stock systems. It may be modified by writing the orders to the nearest tens or hundreds to show shortages.

Stock limits are determined as the estimated demand for an assumed period of time. This demand and period may be recorded as a basis for adjusting limits to new conditions without a resolution of the entire stock problem for that article. If demand or design be uncertain the high limit may be omitted, thus assuring special consideration of and authority for each new order.

Stock Keeping.

We must have goods available when wanted and yet maintain economy of investment, storage, and handling with security against theft and loss. Reduced to elementary principles, stock-keeping systems are definable as:—

1.—Bin Stock.
    a. Bin View.
    b. Bin Record.
    c. Bin Record from Advance Tickets.
    d. Bin Record with Tickets Audited.
2.—Order Balance Stock.

1.—Bin-stock plans show the quantity of the article in the bin or rack. They take no account of quantities due to or from the bin on orders in progress. Therefore, they are applicable where demand is frequent and supply quick, and especially where supply and demand are continuous, as in a works producing a small standard article in large quantity with one department making the part and another assembling it into the product. The bin is the reservoir between the making and using departments. By watching the rise and fall in the bin the supply is regulated to the demand.

1a. Bin View is a bin-stock plan in which the low limit is marked on the bin or rack and stock is kept up by observation and without records. It is applicable to articles that can be cheaply stored in plain view and are easier to identify and describe by sight than in writing; also to articles difficult or costly to measure, weigh, or count, and not worth stealing. In short, this plan is useful where demand is so frequent and for so many purposes, and the goods are so cheap, that any other system would cost, in clerical labor and delays to production, more than the results would be worth.

1b. Bin Record is a bin-stock plan in which the material stock and its movements are duplicated by clerical transactions. Every delivery to or from the bin is parallcled by a material ticket from the foreman to the stock keeper, and charged or credited to a bin record whose balance should be the quantity in the bin or rack.

The advantage of this plan is in concentrating the information and allowing stock to be more compactly and cheaply stored, out of sight, at a distance or divided among branch stock rooms, the bin record indexing the location of stock. In using this plan it is often a great help to designate each stock article by catalogue number, which reduces the writing and makes stock keeping more automatic.

In this plan there is usually not much safe-guard against loss or theft, as anyone can issue a ticket or raise the quantity on it.

1c. Bin Record from Advance Tickets is a bin-record plan in which the material tickets are limited to the quantities authorized by the material list of the shop order and are issued in advance from a clerical department to the foreman, filed on his desk behind guides, and surrendered by him to the stock keeper when the goods are delivered.

The tickets may thus be more cheaply and accurately made out, but they suggest to the stock keeper what the clerical department expects the goods to weigh or measure, and this reduces accuracy. There is not much safe-guard, as any department can raise the ticket to cover theft. The plan is inflexible and slow for partial delivery and in case of substituting a different material. It puts an extra clerical department ahead of shop production and in position to delay and reduce output.

This plan evidently requires that design, manufacture, and clerical work be very settled and stable; also that the business be large enough to keep a sufficient force to avoid delay to production from the absence of a small part of the clerical force.

1d.   Bin Record with Tickets Audited:—With this plan when a shop order is complete its material tickets are sorted, totaled for each article, compared with the material list, and corrections made in the bin record.   To identify a ticket with its entry in the bin record either both must be serially numbered or both must quote order and date.   A discrepancy between bin and bin record is still not proof of theft or loss, for there may be many errors on uncompleted orders which cannot be checked up otherwise than by inventory and comparison with their material tickets—often a difficult task.

The following defects are evident in these bin-record plans.

First.   Great clerical expense for tickets, entries, and comparisons.

Second.   Possible delay and reduction of output from productive men and equipment waiting while tickets are being issued, goods weighed, and receipts exchanged.

Third.   The many clerical errors, omissions, and losses that may be only a normal proportion of the very many scattered documents and entries.

Fourth.   The costly investigations of and arguments over these old errors, etc.

Fifth.   The very doubtful or long delayed discovery of loss or theft.

In the material list we have a checked and proven statement of materials needed for the shop order, condensed, classified, and convenient.   It may not be wise to scatter these data about on hundreds of material tickets to be later collected, sorted, and tabulated for comparison with the original.

2.—Order Balance Stock Plan takes account of orders in progress. It shows the balance of the article in stock if all orders to make, buy, use and sell it were completed.   It shows the balance in and due to bin less stock due from bin.

This plan is useful wherever stock goods must be ordered in advance of and on account of a known demand (such as for an assembling order).   This includes nearly all general machine-making business, using stock materials or parts, and especially those building large machinery.   This plan does not show what quantity is in a bin and it takes no account of actual deliveries across the counter between stock room and shop.   It therefore does not require material tickets.

In an order register a page is assigned to each article and all orders to make, buy, use or sell it are registered as issued. In the column "To Stock" are entered quantities ordered to make or buy, and in the column "From Stock" are entered quantities ordered to ship or use. The third column is reserved for "Balance." As soon as a shipping or an assembling order is issued its material list is written up in the From Stock columns of the different articles. If any balance is brought below the low limit, a stock order is at once issued to make or purchase stock up to the high limit.

If we desire to carry a large stock of a certain finished product and only a small repair stock of a given part together with a little special material from which the part is made, this is accomplished by simply placing the high limit as low as we wish on the part and on its raw material. Shop orders will combine the small limits with the large immediate demands. With any bin-stock plan we would have carried large stock each of the product, the part, and the material, with an immense waste of investment and storage space.

To detect shortage (indicating theft or loss) it is necessary to count the quantity in the bin, add that due to the bin, and deduct that due from the bin. For this purpose the order register will give the numbers of the unclosed orders. Orders to ship, buy, or make the article will usually show quantities completed, and still due, but orders to use the article will not show this unless bin deliveries are recorded. Lacking such record the using orders in progress must be inventoried for the article in question. The best time for such a balancing is when a replenishing order is requested, for there will usually then be no orders in progress to buy or to make.

If the balance shows a discrepancy between the actual stock and the order register, we may investigate for theft or loss and make an entry in red ink "unaccounted for" in the proper column to balance the register with stock.

This order-balance plan furnishes the best clerical safe-guard against theft or loss, for all entries are opposite their order numbers, and are therefore, easily checked with their orders. Suppose that to cover theft the entry from stock had been raised in the order register; then under that same order number on another page, it could be seen that the assembled machines delivered were less than should result from the quantity of parts shown as used. Moreover, if the register shows considerable divergence from quantity ordered, either in the assembled machines delivered or in the parts used, an irregularity is indicated and scrap reports may be consulted to account for parts spoiled.

Orders cannot always be filled to exact quantities called for, on account of accidents. To maintain accuracy of the stock registers it may be desirable to enter the quantity in pencil when the order is issued, and to confirm or correct it in ink when the order is finished. This cannot be done on the orders to use unless a delivery record is kept, but these "from stock" entries may be cut down in proportion to the shortage of complete machines delivered. This keeps the books on the safe side and causes any loss to appear in the balance.

A copy of the material list may be used as delivery record between stock and shop. When so used it furnishes data for this correcting of quantities used and also for balancing order registers (instead of inventory). Such delivery record completes the order-balance plan, but is not essential to its practical success and is often very costly.

It will be noted that the order-balance plan is a record of the progress of the article from its purchase as raw material to its sale as part of a finished product. The article is safe-guarded from the moment it is paid for until it is billed to a customer, assuming. of course, the customary precautions to insure actual receipt of all that we pay for on both purchase and making orders. This safe-guard is of such nature that it is not considerably reduced by keeping the order registers in the stock room where they may be useful as an index to order numbers and stock location.

It is characteristic of this order-balance stock plan that the few cheap entries from orders as issued form the basis of a very practical safe-guard against wholesale or accumulating theft. The refinements are only for greater accuracy. A register of orders to make or buy an article is the "Reverse Arrangement" or index from article to order number and is usually found necessary for reference and record. The order-balance plan should not therefore be charged with the entire cost of the order registers.

SUMMARY OF STOCK PLANS.

The above covers all the elements of stock keeping. The bin plans and the order-balance plans are entirely different and for different purposes. Both may be applied to the same article. They may be combined on the same sheet. By adding two sets of vertical columns to the order register, a bin record may be kept thereon, tabulated to order number. This makes an absolutely complete and self-contained record, but often a very large document.

The user may choose the one or several of these stock plans best suited to his needs. For example, if his stock be made up of a few valuable and many cheap articles, he may put brass and worked parts on the order-balance plan, and the remainder on the bin-view with orders to buy or to make registered. In practice it is usually found that the great majority of small, scattering transactions are in cheap steel and iron articles, such as bolts, screws, nuts, and bar steel, for which an order-balance plan would be relatively costly and comparatively unnecessary.

When product is serially numbered we may state on the assembling order the block of machine numbers assigned for use. In an order register we may immediately enter a brief description, the order number, and all the machine numbers (a line for each). As shipping orders are received we enter in pencil, opposite the machine, the shipping-order number and the name of the customer to whom it is assigned. At the time of shipment we enter the dates and confirm shipment in ink.

We may add a column showing date of receipt of each machine as it comes from the shop to the stock room.

It will be noted that this is merely a combination of order-balance and bin-record detailed to individual machine numbers. It shows the balance available as well as stock in hand. Its practical value is in the ease and certainty with which errors and loss can be traced and in its usefulness as a record of the manufacture and disposition of each numbered machine.

A thorough understanding of these stock plans will help to simplify stock management. Every plan described has its own profitable uses, but misunderstanding and misapplication of them have probably caused more loss of production and profit than any other form of mismanagement.

(To be continued.)

The Cause of Good Roads.

THE creation of a State Department of Highways, as urged in a report of the Joint Committee of the Senate and Assembly of the State of New York, seems to be the best means yet proposed for effective and economical work in the improvement of the public roads. Of course, as in every similar case, everything depends upon the personnel of the department; but the conditions favor the selection of men of broad and sound views and the removal of their effort and undertakings from the most disturbing effects of frequent political change. Further, while the proposed "constructive legislation" establishing the department carefully preserves the principle of home rule in county and town affairs, a strong and sane policy consistently carried out on the State roads would have a powerful educational and moral influence upon the local officers and their work.

Neglect of the roads is one of our notorious economic sins in the United States, and preaching from the text of Continental example is of little power to work reform. The creation of a system of good State roads, penetrating and connecting the principal cities and villages, would establish everywhere a demonstration from experience of the enormous saving possible in haulage to and from the farm, and inevitably would result in the gradual extension of betterment in the work done and the result secured on market and town roads.

At present we are suffering from total want of any consistent policy, and almost universal prevalence of shortsighted local favoritism and jealousy. The commissioner "favors" his section of the town; the contractor "favors" his end of the district. Scarce one seems to realize that it is the road that must be traveled, not that immediately in front of his fence, that saves or wastes time, money, horseflesh and temper. Improvement miles away is no less important than improvement at the gate.

The need for this breadth of view is one great reason for committing the general management of the highway improvement fund to a non-partisan and a continuing commission, rather than to the change and chance of the State engineer's office—excellently filled now, perhaps less so after another election. And the movement is too important economically to be merged in the multitudinous detail of an already heavily burdened office. Mr. W. Pierrepont White, in an address recently delivered, aptly contrasted the possibilities of the highway system in New York State with those of the Erie Canal. In total volume and in breadth of distribution of benefits, the former is surely the greater.

Forest Preservation.

ANOTHER direction in which State administration must be extended and stimulated is that of forestry. We have often urged the point, and we are sincerely glad to see that so able an influence as that of our great engineering societies is active in the cause. A resolution lately adopted by the Board of Directors of the American Institute of Electrical Engineers declares that as "the value of water powers is determined in great measure by regularity of flow of streams, which regularity is seriously impaired by the removal of forest cover at the headwaters . . . this impairment frequently being permanent because of the impossibility of reforestation, owing to the destruction of essential elements of the soil by fire and its loss by erosion . . . the attention of the National and State Governments should be called to the importance of taking such *immediate* action as may be necessary to protect the headwaters of important streams from deforestation and to secure . . . the introduction of scientific forestry and the elimination of forest fires . . ."

The italics are ours. It is not a remote danger, but a present and progressive damage with which we have to deal. The Ohio valley is having a bitter demonstration, and about the upper watersheds of the Delaware, the Hudson, and the Connecticut, destruction productive of like disaster is already under way. The difficulty of the problem, solvable only by State intervention, as we have already shown, is that the cutter of timber—or worse, of pulp wood or "acid" wood—feels no concern as to the aftermath; and those who do have no power of control and no (or but very remote) redress. The creation of Eastern forest reserves, as proposed in the Lever, Currier, Gallinger and Bradley bills, now in the House Committee on Agriculture, would do much; but beyond this Federal legislation and protection there would seem to be crying need for a strong State forest law which would prohibit and prevent the absolute denudation now carried on, and enforce the preservation of sufficient cover, if not actually provide for reforestation. It would beyond doubt soon prove its wisdom and its interest even for those now engaged solely in destruction and removal; but this education apparently must be at first compulsory. May it not be long delayed!

The Safeguarding of Workmen.

IT gives us much pleasure to afford space for the following notice, issued by the American Institute of Social Service.

The movement looking to the reduction of waste of life and limb in industrial occupations has always received our strong sympathy, and in furtherance of it The Engineering Magazine some time since secured from M. Jacques Boyer two articles exhibiting the latest advance in this effort in France, where both manufacturers and legislators offer many models for valuable study. One of these papers will deal with the museum of the Conservatoire des Arts et Métiers at Paris, where the latest safety devices and ideas in industrial hygiene are assembled, and the other will be concerned with Continental practice in the promotion of safety in mining. Their appearance will be happily coincident with the exposition. The announcement is as follows:—

"An Exposition of two months will be held early in April in New York, under the auspices of the American Museum of Safety Devices and Industrial Hygiene, for showing the best methods of safeguarding workmen and protecting the general public. The exhibits will consist of safety devices, protected machinery in actual operation, models and photographs. During the Exposition, illustrated lectures by engineers will explain industrial conditions and hazardous occupations and the most approved methods of safety.

"Believing that many accidents are preventable and to stimulate further invention, three solid gold medals are offered for the best safety devices in the field of transportation, mining, motor vehicles and motor boats. Two prizes of $100 each, one for the best essay on The Economic Waste due to Accidents, the other on The Economic Waste due to Occupational Diseases, are offered.

"The Chairman of the Committee of Direction is Charles Kirchhoff, and of the Committee of Exhibits, Prof. F. R. Hutton. There will be no charge for space. All inquiries and applications for space should be made to Dr. W. H. Tolman, at the Museum, 231 West 39th St., New York City."

TECHNICAL EDUCATION AND INDUSTRY.

A PLEA FOR THE RECOGNITION OF COMMERCIAL LIMITATIONS AND INDUSTRIAL ENGINEERING
IN ENGINEERING SCHOOLS.

Charles B. Going—The Columbia University Quarterly.

THE discussion which centers about the defects and improvement of methods of technical education is a most complicated one. It is generally admitted that defects exist and that improvements are necessary, but there is no settled agreement as to what the defects really are and the possibility of improvement is precluded by this uncertainty. It would seem that the only firm foundation, on which those earnest workers who have the interests of technical education at heart can base their labors for improvement, lies in the oft-repeated assertion that the technical graduate must be trained for some time in practical work before he can be trusted to work out results which will be economically valuable. On this premise, Mr. Charles B. Going, in *The Columbia University Quarterly* for December, 1907, makes an interesting and thoughtful plea for the recognition in technical courses of the economic aspects of engineering.

The inherent tendency of all human systems is to crystallize and the history of thought in every domain has been a recurrent cycle of hardening creeds broken through from time to time by an outpouring of protesting revolution. The rise of the schools of applied science in the last century was itself a revolution in existing educational ideals. The schools have done a splendid work in the development of engineering, but they in their turn are growing academic and away from the close parallel with life which made their past achievements possible. The prevalent belief that the technical graduate is of little value for at least two years after graduation indicates that his training has failed to include recognition of some important elements in the problems he is supposed to have been taught to solve; in other words, that the subjects of his engineering course are not taught as they must be practiced. The cause of this defect is the fear of commercialism which has always haunted the university. In engineering anything within the laws of nature is feasible, but whether it is advisable is a question of dividends. In engineering projects the dividing line between commercial success and failure is usually narrow and the engineer whose function it is to carry works of construction or production to practical success must be able clearly to recognize the point at which technical refinement ceases to be economical. The fault of the engineering school is that they give their graduates too academic a devotion to the ideal of technical perfection and an inadequate idea of the imperative character of commercial laws.

"It is not for a moment desirable that the vision of the goal set by pure science be obscured, but only that the understanding be increased of the approximations which it is expedient to accept in actual performance, and the conditions by which they are determined. For illustration: the development of the steam engine from Savery's 'Miner's friend' to the modern Corliss, or (say) the engines of the Inchmarlo, is a wonderful exhibition of improvement in efficiency in which the data of thermodynamics have been the inspiring and directing influence. Everyone must wish the student of mechanical engineering to possess sufficient knowledge of the science to understand what has been done and to take an intelligent part, if necessary, in continuing it. But the choice of power plant in any given case depends on a careful comparison of the net total of interest on investment, expense of maintenance, depreciation, wages of skilled attendance, and fuel costs. It may easily prove that—in a Pennsylvania coal mine, for example—a plain slide-valve machine, wasteful of coal but cheap and easily run, will be most serviceable to its owners. The average graduate is many times more likely to meet the need of advising as to such an installation than he is to find opportunity for further betterments in steam-engine design; but is he one-fourth as well-informed in the conditions of *commercial* efficiency as he is in thermodynamics?

"Or again: the design of a piece of machinery may require a sound, scientific knowledge of mechanics and kinematics, a careful mathematical calculation of stresses and weights, and a high degree of skill at the drawing board. But if cost of construction is important—and it usually is very essential—it requires also attention to economy in the use of the material and especially of the labor involved in manufacture; the selection of stock sizes; the standardization of gears, arbors, bores, tapers, bolts and screws; avoidance of difficult problems in pattern making and moulding, and minimization of labor in machining and assembling of parts. Does the average course impress the student with these common-sense considerations in the handling of materials as strongly as it does with theoretical mechanics or the technique of draftsmanship? Can it possibly do so, until the economics of engineering are given an entity in the curriculum, and made to occupy an arc of the educational horizon as large as that which they will find in the graduate's discovery of his life work?

"More curious, however, than this incomplete recognition of the industrial element in the generally accepted engineering courses is the absence of any adequate attempt to supply systematic training for the newest of the engineering occupations—manufacturing. This is the great exception, already referred to, in the sufficiency of the technical school to follow the specializations of modern applied science. The factory is surely no less characteristic and no less fertile a department of the industrial era than the railroad or the steamship. It employs, probably, a larger number of technical graduates than stationary, locomotive, or marine engineering, and the proportion is increasing. But while institutions so conservative even as the Lords of the Admiralty and the Navy Department have realized that the modern warship is a huge, coherent and interlocking, sea-going and power-driven fighting machine, and that it must be officered by engineers trained in seamanship and navigation, steam engineering and hydraulics, electricity, ordnance, sanitation, discipline, language and international law, our technical faculties seem to have been less responsive in their sphere. The civil engineering college has been very slow to recognize the factory as a vast, coordinated manufacturing machine, demanding a personnel instructed not merely in dynamic engineering, but in executive organization and management, economical works design and construction, transportation and handling appliances, tool-room practice, stores-keeping, wage systems, time records, shop discipline and sanitation, and by no means least, cost analysis and study. Hardly a course has been founded by any technical school which will ground the student thoroughly in this concept—that the manufacturing plant is, or should be, a harmoniously coordinated if intricate device, in which, by skilled organization, the factors of power, machinery, and labor are made to cooperate in the manipulation of material, so as to use the priceless and irreplaceable element of *time* for the maximum of production with the minimum of expenditure or depreciation in any part. It has remained, so far, largely for outsiders to preach this new doctrine of the times. It is not surprising that a new profession should arise outside of the schools; many of us can remember when all the instruction given in electricity was a part of the physics course. But while electrical engineering was promptly adopted and has grown to be one of the noblest children of its *alma mater,* industrial engineering still stands unacknowledged on the doorstep knocking vainly for admission. . . . . .

"I have been tempted to expand this point of current criticism upon the technical curriculum because just here the difference is most discernible between the preparation which the school gives and the equipment which the work of the graduate demands. The difference is not so much quantitative as qualitative. For if in theoretical training the courses of instruction lean too as qualitative. For if in theoretical training they lean too far toward the handicrafts. The aim (to illustrate by an intentional exaggeration) seems to be to turn out a composite of the natural philosopher and the mechanic—a sort of scientist in overalls. Now it is not manual training, however glorified, that is needed, but a trained observation of, and a capacity to measure, the forces and factors entering into production or construction, under competitive commercial conditions.

"Practicing the future mechanical superintendent in chipping and filing, or setting him at elementary work on the planer or the lathe, is about as fruitful of good results as teaching an architect to lay brick or a mining engineer to slog a drill. At best, with the hours available in a four-years' course, he can scarcely acquire the dexterity of a junior apprentice. What he needs to know is *not* how to do filing and fitting, but how to avoid them so far as possible, as enormously wasteful and expensive processes which the genius of modern manufacturing endeavors to the utmost to eliminate. . . . . .

"With much diffidence, and some trepidation, the suggestion is advanced that (without any change in the full courses laid out for the specializing student) the general educational scheme might be brought nearer to the plane of practical professional work —nearer to the life the graduate must follow—if a more careful scheme of minor courses not *elemental,* but *fundamental,* were worked out, proportioning more usefully the *Nebensachen* to the *Hauptsache* for the candidate for any given degree, and thus affording room for the desirable larger number of these coordinate studies.

"Esoteric knowledge is demanded only of the adept. The technical graduate ordinarily does not need to head a rivet with the skill of an expert boilermaker; neither does he need the mathematics of an astronomer. Profound acquaintance with electro-physics probably will never be exacted of him (after his examination for his degree), but he ought to be amply qualified to select proper equipment for a lighting station, or to dissect the representations of an impetuous salesman of dynamos. He might even design an excellent steam plant while he still remained a little vague on the subject of entropy; and he might direct and check the work of draftsmen who have many times his manipulative skill (and command a fraction only of his salary), though his own lettering be far inferior to copper-plate.

"Superficial? No; he may be absolutely thorough in everything, so far as he goes; and except in his specialty (in which the technical student must go very far indeed) thoroughness of training as a whole depends quite as much upon amplitude as upon altitude. If the industrial effort works a little too close to the ground, the academic tendency is a little too urgent toward the empyrean; but with closer cooperation their resultant might be the definition of a body of education of ideal proportions. We need, especially in our schools of applied science, a closer fellowship between town and gown. The jealousy of many universities, which discourages the teacher from maintaining close touch with active professional work, is one of the regrettable influences forcing our practical science courses out of parallel with the progress of industry. A more weighty one still, perhaps, is the small opportunity given the alumnus to be heard in university councils and to bring back within university bounds the fresh spirit and suggestion of the constantly changing world without. The result is an inevitable set toward formalism and scholasticism on one side, and toward apathy on the other. Through freer, wider, and more frequent intercourse between faculty and alumni, probably more effectually than in any other way, will the creative forces of the system of technical education be kept fresh, vigorous, and instantly responsive to the demands of industrial life."

FLOATING DRY DOCKS.

A DESCRIPTION OF THE VARIOUS STRUCTURAL TYPES OF THE FLOATING DRY DOCK AND ITS ADVANTAGES OVER THE STATIONARY DOCK.

Harry R. Jarvis—North-East Coast Institution of Engineers and Shipbuilders.

THE rapid development of naval engineering and marine transport has presented to the civil engineer many problems of ever increasing magnitude in connection with the building and extension of harbor works. One of the most important of these is the provision of adequate dry-docking facilities and in this field the introduction of the floating dock is the most interesting development. Recently floating docks were made the subject of an interesting paper by Mr. Harry R. Jarvis before the North-East Coast Institution of Engineers and Shipbuilders. The following extracts from this discussion outline the main structural features of the types now in use and the advantages of the floating over the stationary dock.

"It will be as well, before proceeding with a description, to classify the various types of floating docks. It has been found convenient to designate various types of non-'selfdocking' and 'selfdocking' floating docks as follows:—The 'box' type of floating dock is non-selfdocking. The 'selfdocking' types include the 'depositing,' 'offshore,' 'sectional' pontoon, 'Havana,' and 'bolted sectional' docks.

"The non-selfdocking 'box' dock, which does not need any elaborate description, consists simply of a pontoon with two side walls forming one homogeneous whole. Being solidly built, this type is naturally the strongest form of floating dock which can be constructed. Whether 'selfdocking' facilities in a floating dock are absolutely essential depends entirely upon its situation. It is generally found that marine growths, which are so detrimental to a ship on account of loss of speed, do not in any way interfere with the life of a floating dock, but rather tend to preserve the material. This statement is not based on theory but on actual experience over many years. It is not, however, the portion of the dock always entirely submerged which suffers most from corrosion but that portion between wind and water, about the normal waterline of the dock, which constantly va-

ries with the amount of freeboard given to the floor of the pontoon. This part is very easy of access when the dock is not in use, for by pumping out the water and heeling the dock sideways or endways, it is always possible to clean and paint that portion as often as may be found convenient. Naturally, this operation is not so satisfactory as selfdocking the dock in a proper manner.

"To enable the underwater parts of a floating dock to be examined, cleaned, painted, and if necessary renewed, with perfect safety and expedition, there was designed for the Russian Government in 1877 a single-sided outrigger dock which was designated a 'depositing' dock. This dock which, in end elevation, is shaped like the letter L, is of peculiar construction, has a single side wall which contains the pumping machinery, and the pontoon is made up of a series of 'fingers' with spaces between them. The necessary stability to enable the pontoon to be lowered and raised is obtained by an outrigger which is connected to the side wall of the dock by a series of parallel booms hinged at each end. As the wall of the dock is divided into two equal lengths, it will be readily seen that to selfdock either portion of the dock it is merely necessary to separate the two halves and disconnect the booms from the portion to be dealt with; the remaining portion of the dock may then be docked in the usual manner. This type of dock can not only dock vessels but, owing to the peculiar construction of the pontoons, is enabled to 'deposit' them on a grid, formed by groups of piles. It is obvious, therefore, that in certain situations, such as non-tidal harbors or rivers, comparatively cheap docking accommodation need only be limited by the number of grids provided.

"The 'off-shore' dock, which was designed and patented in 1884, is a modification of the 'depositing' dock. The pontoon is continuous and L-shaped in end view. Unlike the 'depositing' dock, however, the 'off-shore' dock is, as its name implies, attached to a series of shore girders or triangulations, by parallel booms, hinged at each extremity. The inner end of the top boom is not rigidly attached to the shore girder, but is suspended on a crank which allows the dock to take a small inclination in either direction. This movement is indicated in a simple manner to the man in charge of the dock, and enables him to correct, by manipulation of the valves, any tendency of the dock to go out of trim, and so avoid any undue strains on the shore girders or other parts of the structure. The 'off-shore' dock is very simple in construction and is easily handled. It can be very profitably employed in suitable situations. When submerged to receive a vessel it is merely necessary to sheer the latter in sideways, or the vessel can enter the dock from either end. It is then rapidly centred by telescopic mechanical side shores. When properly centred the vessel is lifted in the usual way, and, without stopping the pumps, it is supported by mechanical bilge shores, which consist of strong timber booms hinged at their inner ends to the pontoon dock, and supported at their outer ends by a cast steel rack, which is raised or lowered by worm and pinion gear, driven by suitable gearing from the deck of the side wall. The sound of a hammer need never be heard when docking a ship on an 'off-shore' dock. Many examples of this type of dock have been constructed, all of which have given the greatest satisfaction.

"The 'sectional' pontoon dock is made up of a number of separate tanks or pontoons, which are held together by two side walls. This type is very suitable for vessels of medium dimensions, and owing to its form can be very easily constructed in places where a dock of any other description would be practically impossible to complete. The separate sections of the pontoon can be easily erected and launched, after which the walls can be erected in place.

"The 'Havana' type of floating dock consists of two continuous side walls, between which are attached three or more (usually three) separate pontoons, two end sections and a central section. At intervals along the walls two tiers of lugs project from the face of the wall; corresponding lugs are built on to the side of the pontoons, the lower set of lugs being just above the light

waterline and the top set in line with the top of the pontoon. The actual joint between the pontoon and wall is made by pairs of fishplates, secured by large tapered screw bolts. Several examples of this type of dock have been constructed during recent years, but although the 'Havana' type is a very strong and useful form of dock, the latest and undoubtedly the best and strongest form of selfdocking floating dock is that known as the 'bolted sectional' type, two examples of which have lately been constructed, and a third, destined to be towed to Callao in Peru, is now in an advanced stage of construction at Wallsend.

"The 'bolted sectional' dock is, generally speaking, a modified form of 'box' dock, having the longitudinal strength of the latter combined with the selfdocking facilities of the former. This type is built in three sections of approximately equal lengths, the two end sections being symmetrical. The ends of the walls are stepped down, the lower step forming a docking land, when selfdocking the centre section. The joint between the three sections is usually a combination of butting angles bolted together for the underwater portion, and butt straps for the remainder. The end bulkheads of the various sections of the dock occur about two feet from the section ends. It therefore follows that when the sections are brought into position for coupling up, a chamber about four feet long is formed, from which the water may be removed and so allow the bolts in the joint angles across the bottom and up the outer sides of the pontoon to be inserted. The ends of the centre section are recessed, and a compressible compound inserted to enable a tight joint to be made, so that water may be removed from the joint chamber when making or breaking the connection between sections.

"With regard to the facilities for self-docking possessed by the various types of floating docks as enumerated and described, the 'bolted sectional,' from actual experience, is as simple as any of the other, excepting, perhaps, the 'off-shore.' The 'Havana' type certainly calls for very careful manipulation to enable all the tapered bolts to be withdrawn, and, moreover, the dock has to be heeled over sideways twice in each direction when selfdocking the centre section, or the two end sections, which are usually taken together. In the 'sectional' type of dock, the pontoons are easily self-docked. The bolts connecting the bottom flanges of the walls to the pontoon having been removed, water is let into the section to be selfdocked, to overcome its surplus buoyancy; it is then removed, the remainder of the dock sunk to the necessary draught and the free section docked between the walls, in the usual manner. However, as long as proper precautions are taken to preserve the interior of the dock and the outer shell above light waterline by suitable protective composition, there is no necessity to selfdock at more frequent intervals than, say, from 10 to 15 years."

There is no class of repair work which cannot be satisfactorily completed on a floating dock. Extensive bottom repairs and important ship lengthening contracts have been expeditiously carried out. It is especially useful in the case of vessels which require to be sighted after ground-

ing, as the docking and undocking is frequently accomplished in well under three hours. For cleaning and painting the floating dock is pre-eminent. The free circulation of the air during painting is of immense advantage.

Ease and rapidity of construction are important points to the advantage of the floating dock. There are numerous examples of very rapid and satisfactory construction in cases where time was of importance in providing docking facilities for specific purposes. The "Havana" dock, with a lifting capacity of 10,000 tons, was built, launched and towed a distance of 6,500 miles in a little under eleven months and was ready for immediate service on arrival at Havana. This performance also illustrates the mobility of the floating dock which is another of its great advantages. The transport of the Havana dock is only illustrates the mobility of the floating dock, have been towed long distances with perfect safety and success.

THE DESIGN OF STEAM-ELECTRIC POWER PLANTS.

THE BEARING OF THE MANAGEMENT AND EFFICIENCY OF THE DESIGNING DEPARTMENT ON FIRST COST AND ECONOMY OF OPERATION.

Frank Koester—Electrical Review.

IN the design of power plants the aim is to secure the highest economy and efficiency of operation compatible with reasonably low first cost. While careful superintendence is necessary to get the most out of a plant, the possibility of securing thoroughly satisfactory operation depends primarily on the ability of the designer and his staff. The following abstract of an article by Mr. Frank Koester in the *Electrical Review* for January 18, 1908, takes up a novel aspect of power-plant design, namely, the management and efficiency of the designing department and their bearing on first cost and economy of operation of the finished plant.

In the design of steam-electric power plants, "the designer is frequently in the employ of the company; contracts are sometimes made with a firm of consulting and contracting engineers to furnish the plan only, or in some instances the entire plant for a fixed sum. Sometimes a contract is made between capitalist and contracting en-

gineer under which the latter is paid for the design and erection of the plant by a fixed percentage of the total actual cost of the plant. Another method is to pay the consulting engineer, for his services in designing and supervising the erection of the plant, a fixed fee plus a percentage on the total cost of the plant."

In the first case the first cost is not likely to be excessive but the capitalist runs the risk that his designer may fall very far short of economy and efficiency of operation unless his experience is known to have lain along the lines of power-plant design. In the second case efficiency of operation is usually attained on account of the wide experience of the designer but there is a tendency for him to raise the first cost as much as possible in order to swell his own profit. The third form of contract is perhaps the most satisfactory as the percentage is necessarily small and the designer has no very great incentive to increase the first cost.

In letting contracts for the design and construction of power plants the capitalist should not only investigate the financial standing and business reputation of the contractor, but should satisfy himself of the ability of the contractor's designing staff and particularly of the designer in charge. This is especially true in the case of those contracting firms who for one reason or another make frequent changes in their staff. Changes in the designing staff, particularly of the chief designer or his principal assistants, are likely both to increase the first cost of the plant and to interfere seriously with the securing of efficient operation. Primarily the satisfactory operation of the plant depends upon the designer, and as this fact is more and more recognized it may be expected that guarantees of plant efficiency will be required of contractors in the same way that manufacturers are required to guarantee the efficiency of individual machines.

"As the scope of power-plant design is broad, embodying mechanical, electrical, structural, architectural and often civil engineering work, it is necessary, particularly in the design of large plants, to employ able assistant designers and draftsmen in each of these branches.

"For instance, in designing a plant of 25,000-kilowatt capacity the staff may consist of the chief, one assistant (all-around man), who is familiar with the various branches of engineering required in connection with the plant, and four draftsmen (assistant designers), who are experienced in the four branches previously mentioned. The mechanical designer, who is always most familiar with the operation of steam-power plants, should arrange the general scheme and particularly the system of piping, as well as all foundations for machinery. The electrical designer, of course, will handle the wiring and switchboard equipment, and the structural man will have charge of the design of the structural work of the building, as well as such structural steel work as is necessary for supporting coal-handling machinery, boilers, etc. The architect is not usually called upon, as is evident from the character of many power plants, the architectural features having been sadly neglected. In plants of smaller capacity, the employment of a special archi-tect is not justified, but an architect also qualified as a civil engineer might be employed to handle the architectural features and others, such as retaining walls, condenser-water tunnels, etc.

"In order to bring about economy in the drawing room, for the less important work, such as tracing, lettering, etc., a few tracers must be employed, who may be shifted from one branch to another as necessity requires.

"While checkers are necessary for every branch of designing, it is not feasible to employ one for each branch. It is customary, however, in the structural steel branch, to have a special checker in connection with the design of very large plants. In the design of such plants where the employment of a special checker for each branch is not warranted, the checking should be done by turning over the drawings to one of the other assistant designers, who will check it in conjunction with the original draftsman. This applies purely to the checking of dimensions, etc., as the work of designing has been carried on under the supervision of the principal designer and the all-around man. Any attempt to make it a part of the checker's duty to check the design would be a repetition of work, and the checker employed may not be competent to check the designs pertaining to each and every branch. It will, therefore, be seen that the employment of a general checker for checking the dimensions and the designs pertaining to all branches would not be justifiable, and a man with such general knowledge, including also thorough knowledge of every detail of the design, construction and operation of the plant, could not be found, because engineering, which now covers such an extensive field, has become highly specialized. To employ designers and able draftsmen for each individual branch, for a small, or even medium-sized plant, can not always be done, as the item of engineering expense would be too high in proportion to the cost of the plant. As the character of work done by a firm of consulting engineers covers not only power-plant design, but also general engineering work, and the chief designer frequently is not an expert in the particular work of power-plant design, an all-around power-plant engineer is required, to act in a supervising capacity in connection with these various branches, whose duty may also consist in checking all drawings which pertain to the general design. Having employed an all around designer in this capacity the lower grade of draftsmen may be shifted from one line of work to another, thus not only cutting down the running expenses of the department, but also training these draftsmen for the duties of a general power-plant designer.

"When capable draftsmen or assistant designers have an idea or scheme in connection with the work in their particular line, they should be encouraged in developing it and should not, as is too often the case, be overruled by the chief simply because he, as the head, has the mistaken idea that all good suggestions should emanate from him. If these subordinates receive such encouragement the efficiency of the department will be materially increased, and, naturally, the running expenses of the department will be accordingly decreased. Furthermore, if such ideas or schemes are good this may be seen at a glance by a capable chief, and the subordinate should be given an opportunity to develop them, which usually will not absorb much time,

because of the interest which he takes in them. Credit should also be given the subordinates for such ideas, which should not be appropriated by the chief and put forth as his own scheme, nor should the chief throw the blame for all bad features of his own schemes upon his subordinates. This is one of the principal reasons why rising designers do not stick to the drawing-board, but seek opportunities elsewhere, and as a consequence, capable draftsmen and designers are scarce, such men devoting their energies to other branches where greater rewards are in store for them. For this very reason some of the universities discourage the students in the pursuit of occupation at the drawing-board.

"With proper management on the part of the executive head of the firm the conditions described would not exist in connection with its staff of engineers and assistants; and capitalists should be most careful in the selection of a firm of consulting engineers which is to take charge of their work, that such executive heads, and others concerned in the business management, have been chosen with proper regard to their ability to select and control the engineering staff, and that such managers do not owe their preferment to favoritism."

STATE CONTROL OF ELECTRICITY SUPPLY IN GERMANY.

A DISCUSSION OF THE RECENT PROPOSAL FOR THE NATIONALIZATION OF ALL ELECTRICITY-SUPPLY UNDERTAKINGS.

Engineering.

THE proposal that the State should take over all the electricity-supply undertakings in Germany as a means of augmenting the Imperial revenues has not been taken very seriously by the technical press, but from latest reports it seems that the proposal was serious in intent and is likely to be brought up for legislative action. The effects of such State control of electricity supply were discussed in an interesting editorial in *Engineering* for January 10, which we reprint in full.

"Considerable stir has been created in Germany by the statement that the Imperial Treasury contemplated making the supply of electricity an imperial monopoly. When the news was first launched some weeks ago, people refused to credit what was ap-

parently merely a rumor born out of the alarm over the serious state of the imperial finances. But the statement has been repeated; certain members of the Reichstag have declared that they regard the suggestion without any antipathy, and some newspapers and journals have discussed the proposal with an astonishing *naïveté*, as if the taxation of electricity could in a few years relieve the treasurer of all his troubles, without detriment to other interested parties. Nothing is definitely known as to the intention of the Government; but it is believed that the supply of electricity is to be made a monopoly, while the manufacture of electric machinery and apparatus is to be left free to industrial enterprise. The Continent is more accustomed to Government monopolies than we are, and the unsatisfactory state of the German Imperial finances, which have steadily deteriorated of late, may be thought to demand a drastic cure. We cannot help feeling that the Bundesrat—the Government delegates of the different States constituting the German Empire—would do well, if not definitely committed, to prevent serious disturbance of the prosperous German electric industry by such alarming rumors. The Bundesrat will have to approve of any such bill before it can be placed before the Reichstag.

"Broadly speaking, the expenditure of the German Empire, as distinct from the expenditures of the States, is covered by the indirect taxes—the Customs dues—a certain surplus of which flows back into the treasury of the respective States—and stamp duties. Some articles of consumption—spirits, beer, tobacco, sugar, and salt—are also taxed in the imperial interest. The direct taxes go to the States, the rates to the communities. If the electricity supply is to become an imperial monopoly, the municipalities will suffer. The public electric power stations are either owned and managed by the municipalities, or they will become their property at the expiration of the concessions granted to companies. Most of these electric-supply undertakings are doing well; a smaller number do not pay their way, and would hardly be more prosperous under Government administration, which does not, as a rule, work more economically and successfully than private enterprise. If the empire is to buy these undertakings out, it will saddle itself with a heavy debt, and the municipalities will have to impose further rates to compensate themselves for the loss of revenue when they are no longer able to trade in electric light and power. That the State would succeed in increasing its income by raising the rates for the electricity supply is not at all probable, though it has been suggested. If the electric light ress made by gas illumination and gas light of twenty years ago, electricity might disregard its competitors. But it is too often forgotten that, astounding as the development of electricity has been, the progress made by gas illumination and gas power supply is still more astounding, and that the gas progress-curves continue to rise more steeply than the electricity curves.

Any attempt to raise the electricity rates would benefit the gas industry and cripple the electric industry.

"In addition to the public electricity plants, Germany has a considerable number of private electricity works. We need not trouble about the small, strictly private installations. The industrial electricity works, which are private only in so far as they do not sell electricity, but generate it for their own use, can, however, not be dismissed as they have been by some journals. Those writers who placidly suggest that it cannot much matter to big works whether they generate their own electricity or take it from the State, betray an amazing ignorance. Twenty years ago it would have been possible without great risk to pass such a law. The State supply of electricity might then have turned out as economic and as rational as the network of streets and tramways of a newly-planned city. At the present time it is as reasonable to prescribe to the owners of works where to take their power from, as it would be to dictate to them where to buy their raw materials. Huge industrial establishments, metallurgical and chemical plants, mines, mills, etc., are absolutely dependent upon electricity. They generate their own electricity because they find it profitable. Whether steam, gas, pneumatic, or hydraulic power is to be adopted for a particular purpose is a question as to which opinions may change as the plant is extended. Central electric power-stations have certainly proved a success in some districts, but they cannot answer everywhere; and if the engineer is to-day to be commanded that he must buy his electricity of the State, he may to-morrow be dictated that he must obtain his coal, steam, and gas from the State. If he is not satisfied with the electricity and the steam, he may appeal to Parliament, where he is not likely to obtain much satisfaction.

"The whole idea seems to us totally unsound. All the troubles and quarrels that are now fought out in the town councils would become national parliamentary questions, and the legislative machinery still more clumsy, and it is difficult to see how the scheme can benefit the Imperial Treasury without at the same time impoverishing municipalities and industrials. The majority of the German electricians may, however, regard the threatened change with equanimity. We hear complaints over here that a few high officials are, as a rule, overpaid, while the staff are underpaid. These complaints are much louder in Germany, where we find in the big electric and other works large numbers of academically-trained and probably competent engineers working for poor pay, without hope of promotion. These engineers might welcome a State monopoly. They would become civil servants, entitled to a pension, with an increased certainty of tenure of office, and an easier time probably.

"We should be sorry to interfere in a German political question. We are only concerned with the general features of the matter. We have State monopolies in the postal and telegraph services, and we do not wish to abolish these monopolies. In federal Germany the railways are practically a Government monopoly, and the United States may adopt that policy. We do not want more centralization than is absolutely necessary in Great Britain, as we are not citizens of a Federation of States, and we are not in favor of monopolies on the whole. Yet we can understand those whom the fear of the abuses of trusts converts into believers in State monopolies, although they see in private enterprise better guarantees of healthy and steady progress. The electric railways would naturally fall under State control in a country of State railways. But the supply of light and power are local problems, and we cannot see how they can be improved by being made the object of a State monopoly."

THE CARELESSNESS OF WORKMEN.

A DISCUSSION OF THE LOSSES OF LIFE AND PROPERTY DUE TO THE GROSS CARELESSNESS OF THE AMERICAN WORKMAN AND THEIR BEARING ON EMPLOYERS' LIABILITY.

Thomas D. West—Philadelphia Foundrymen's Association.

SINCE its inception THE ENGINEERING MAGAZINE has always been one of the strongest supporters of that phase of the industrial betterment movement which has to do with safeguarding the workman from accident. In its pages many writers of prominence have shown that it is not only common humanity but sound economy for an employer to use every means in his power to secure his employees against the manifold dangers to life and limb that are the accompaniment of modern industrial conditions. In the following review, however, is presented an element in the question which is new to the pages of this magazine, but which has a profound bearing not only on the special case of safety in manufacturing pursuits but also on the wider problem of security in all fields of industrial activity and the employer's liability in case of accident. This is the carelessness of the typical American workman, the effects of which were discussed in a paper recently read by Mr. Thomas D. West before the Philadelphia Foundrymen's Association, and reprinted in *The Iron Age* for January 9, 1908, from which the following extracts are taken.

Mr. West first notes the popular habit of placing all the blame for accidents upon the employer and the demand for an Employers' Liability law which shall make the employer legally responsible for every accident regardless of its real cause, and points to the ill-advised proposals of President Roosevelt in his Jamestown speech and his recent message to Congress, and to articles in the popular magazines, as prominent examples of the popular ignorance as to actual conditions. "It is right and just to hold an employer responsible when he does not provide the best safeguards to life and limb that can practically be devised, but unless we enact laws that provide for fining and imprisoning a person for carelessness, we will never be guaranteed the protection that can and should exist to decrease greatly our loss of life and property through this cause. What can be conducted in Europe as to the liability of employers cannot be done as justly in America.

"It is safe to say that where there is one casualty in Europe caused by the downright carelessness of the employees, there are nearly half a dozen in America. Why? Simply because the American employee is less careful, more flippant and indifferent to his duties. In fact, as an example, if in a railroad crew or a shop's gang, one of their number exhibits any great interest in being careful or economical, he is generally derided and in time often expelled from among them, and made to seek another situation. The 'don't-care' spirit is so prevalent in America that the foreigner, who at the start exhibits much seriousness in aiming to be careful, is in a short time often worse than the native, for the reason that many know no bounds when they discover the flippancy that is tolerated and so often common to many workers, a defect that is imbibed in their youth, chiefly through the American spirit of independence, buoyancy and pleasure seeking, which tends to create a nation of 'don't cares' and 'sports' instead of earnest, careful workers. In passing, it is to be said that the qualities generating carelessness in workers are more pronounced in prosperous than dull times, which accounts for the increase of casualties in the past five years."

To prove his contention that the American workman is much more careless than the European, Mr. West instances the comparative records of the United States and European countries in railroad and coal-mining accidents. In the former case, out of each 100,000,000 passengers carried on the railroads of the United States during the years 1901-1905, 31.64 were killed and 72.23 injured, against 1.27 and 4.88 respectively in the United Kingdom; while the number of coal-mining fatalities in the United States is three times as great per 1,000 men employed as in most of the coal-producing countries of Europe. Mr. West considers this unsatisfactory showing to be due to no other cause than the recklessness and gross carelessness of the American temperament. And the effects of this carelessness are not confined to the destruction of life but are almost equally to be deplored in connection with wholesale waste and destruction of property. A striking example of this is the fact that the loss in New York City by fires having their origin in the careless throwing away of burning matches and cigarettes amounts to $2,500,-000 annually, or on the same basis, about $50,000,000 for the whole of the United States.

"Considering the subject of losses by manufacturers through carelessness, it is safe to figure that breakage and other losses caused by operatives' carelessness are, on an average, fully $20 per employee each year. Allotting one-sixth of the population of the United States as being employed in manufacturing, we have a total loss of about $28,000,000 from this class of employment.* Add to manufacturing the work of the farmer and domestic labor, and the grand total can easily reach $150,000,000 that could have been saved to the people and added yearly to our nation's wealth, had there been due care exercised in the labors or actions of the people."

Mr. West considers that the foundryman suffers more from the carelessness and inattention of his employees than he does from their ignorance and lack of skill, and that it is to the interests of the various foundrymen's associations throughout the country to prosecute a vigorous educational campaign against these evils.

"As the first organized step to make the subject of carelessness one of national issue along educational lines few measures, if any, could be more beneficial than to have our various foundrymen's associations take up the work of collecting statistics of shop accidents and losses caused by individual carelessness, so that those outside of the actual control of manufacturers could have facts to influence moral support in portraying the chief cause of casualties, and thereby aid other societies and organizations to influence legislative bodies in enacting laws and devising other agencies that would be very helpful in greatly decreasing present day casualties and insure a greater lease of life and preservation of limbs to all civilians. The above is merely a suggestion given with the belief that actions not words are what we must chiefly look to in order to achieve the reform in carelessness that should and can be brought about.

"To reduce the present great loss of life and limb in connection with that of property, some may say, use the power of discharge to discipline workers. This would help in dull times, but in brisk periods, like the past five years, when a man had to walk but a few streets to find a dozen po-

---

* Apparently an error. One sixth of the population is 14,000,000 persons, giving an aggregate of $280,000,000.—ED. E. M.

sitions awaiting him, this would have little restraint.

"It would seem that with the 600,000 persons who are maimed or killed, in connection with the $150,000,000 worth of property lost yearly, and chiefly through carelessness, there would be a universal awakening to apply needed remedies. There could be better discipline in our homes and schools and societies, for the development of carefulness could be formed, but the most effective remedy that can be had would be to enact laws whereby any guilty party, whether employer or employee, can be summoned to answer charges for carelessness, and be liable to fine and imprisonment for any accidents of which they may be the cause. To support such laws we will have to educate popular opinion, as by reading such papers as this, and by the presentation of actual facts demonstrating that the great majority of accidents are due to sheer carelessness of individuals.

"Another phase of this subject is the effect that the placing of the whole responsibility and cost of employees' injuries on employers would have in debarring many with small capital from starting in business for themselves. It would kill most of such prospects for many, and few movements would do more to help make this strictly a rich man's country and take from us present advantages which permit many employees to become proprietors, all for the reason that no one of small means would dare to risk his hard earned savings to help pay for an unfortunate man's life or limbs lost through some chance accident that was beyond his employer's control.

"Some will claim that employers' liability insurance companies can be relied upon to protect employers in paying their costs of accidents. Make the new Pennsylvania law for holding proprietors wholly universal in our country, instead of the old law which excused an employer, if negligence on the part of an employee could be proved, and these insurance companies would in a few years be a thing of the past, or their rates would be so high that a small beginner could not spare the capital necessary to insure himself. Whichever way we turn, the plan of loading the whole responsibility and cost upon employers for accidents is an ill advised one, and, like a boomerang, will turn to injure also those that believed themselves furthest from harm. The courts can heavily fine the employer and those who can carry the load will endeavor, if possible, to make the consumer pay, thereby still further increasing the cost of living. How much more sensible for all to labor for every possible removal of the chief cause of accidents, which is individual carelessness, and hereby save our country's great loss of life and limb, as well as the many millions of the people's money.

"The endeavor to place the whole responsibility and cost due to individual carelessness upon the employer will have little weight in this country in preventing accidents under the present régime. It suggests the question if the public is satisfied to continue enduring its loss of millions and further lessen opportunities for small capitalists to start business, as well as taking the hazardous risk of life and limb that they must under the present unpunishable dire carelessness of so many of our employees and unresponsible citizens."

THE LABORATORY IN GOLD MINING

A DISCUSSION OF THE VARIOUS WAYS IN WHICH THE LABORATORY SERVES THE ENGINEER IN THE ECONOMICAL MANAGEMENT OF THE RAND MINES.

A. McArthur Johnston—Chemical, Metallurgical and Mining Society of South Africa.

CHEMISTRY has always been intimately associated with mining and the assayer has followed the engineer closely as a necessary factor in the successful development of mining enterprises. In most cases, however, the application of chemical science to mining has been limited to the one branch of assaying or metallurgical chemistry. But there is no doubt that the extension of the usefulness of the laboratory to the testing of the various materials used in mining work, along the same lines in which it has proved its value in large railway and manufacturing enterprises, would be of immense value, particularly in the case of companies operating on a large scale on low-grade deposits. This is the view taken on the Rand, where, within the last few years, the chemist has come to be regarded as the most important ally of the engineer in securing economy and efficiency of plant operation. The various ways in which the services of the chemist are of assistance to the engineer in the particular case of the Consolidated Gold Fields Company, one of the largest operators on the Rand, were outlined in a recent paper by Mr. A. McArthur Johnston before the Chemical, Metallurgical and Mining Society of South Africa, of which the following is a brief abstract.

To be successful the laboratory must be conceived as a sound financial concern, and the ultimate object should be the testing of all supplies used in the mine. In Britain the works chemist is scarcely tolerated, but other countries are much further advanced in this respect. Germany's attention to the work of the chemist has enabled her to outstrip her rivals in certain of the arts while the rapid progress of the United States in all branches of metallurgy is undoubtedly due to the research work done by chemists and metallurgists.

"Unfortunately the laboratory has in the past on the Rand been looked upon as an expensive luxury. The financiers naturally expect some tangible return, from an outside source, of the value expended in the initial cost and running expenses, and as it very often happens that such is not forthcoming, its importance is not appreciated. They forget that it is there that most of the initial work in experimenting and in building up data on which to base changes in running operations, is carried out. Many of the results obtained undoubtedly are negative, but if occasionally one or other idea turns up trumps then the industry is the gainer. Looked at in this way, I may safely affirm that these initial experiments cost but a mere trifle as compared with what they would cost were they attempted in actual practice and, as often happens, found wanting."

To the manager of the reduction works the laboratory can be of immense assistance. As a striking instance of the value of laboratory experiments may be mentioned the discovery that the loss of gold

when working with new amalgamating plates can be greatly reduced by treating the plates with silver amalgam. To the cyanide manager tests of lime and cyanide are of importance. On the Rand these materials are bought on a guaranteed percentage basis, the former on the basis of the caustic lime content and the latter on the percentage of cyanide, calculated as KCN, and the absence of sulphides. There is little difficulty in buying cyanide of the requisite strength and purity, but the caustic lime test is a sore point with the sellers though it is the only equitable basis of purchase. Occasional tests of the working cyanide solution have to be carried out and experiments of this kind, made necessary by alterations in the ore or in crushing methods, are much better carried out in the laboratory than by the cyanide manager who usually has but little time to devote to research work.

To the engineer the assistance of the laboratory in making coal tests is of the utmost importance, and more especially so since colliery managers have combined to raise the price of coal. The ideal basis for the purchase of fuel would be that of a guaranteed minimum of steam raised per unit of weight, but so many variables enter into this question that the system is impossible of attainment. The simplest and most efficient plan is to purchase coal by proximate analysis and determination of calorific value. As the value of laboratory tests of this kind are more and more recognized, engineers place more reliance on them and less on steam raising trials. In coal testing one of the most important elements is the sample. This should always be taken in the same way and the sample taken during the discharge into the bins is undoubtedly the most reliable. Colliery samples and car samples can never be reckoned as absolute. In the Consolidated Gold Fields laboratory the evaporative factor is determined by the Mahler bomb method which is found to give extremely steadfast results. The estimation of volatile matter, fixed carbon and ash in the proximate analysis is carried out in the assay muffle. Since the calorific value of coal depends largely on the amount of ash present, coal bought under contract should be sampled daily and the ash determined.

The complete proximate analysis and calorific value test should be carried out on a weekly sample made up from the daily samples. In this way a rough daily check can be had on all the coal supplied.

Second in importance only to fuel analysis is the testing of boiler feed waters. Most of the Rand mines are obliged to use the water taken from the mines and some degree of purification or softening is absolutely essential. In installing purifying plants the services of the chemist are of the utmost importance. Many such plants have been erected in the Rand fields which are absolutely useless. The mine waters change month by month and regular analyses of these and the purified water are necessary if the plant is to give any satisfaction.

In the testing of lubricating oils the most important points to be considered are the flash point, the percentage of fatty oils present, the acidity and the gumming. The first three usually give little trouble, but the gumming test is more important, oils failing in this regard giving very unsatisfactory results in service. The film formed by gumming oils is due to the presence of resin, to oxidation, or to the presence of soaps formed by the chemical action of the free acids on the metal of the bearings. The presence of resins can be determined only by fractioning. Viscosity tests are not of much importance since most engineers are able to estimate the viscidity necessary for their special requirements. Rope greases call for some attention. Acidity must always be looked for and also the presence of grit likely to set up irritation or to interact chemically with the bearing surfaces.

In the examination of candles the aim is to determine by practical tests their value in actual working conditions. The intensity of the light, the height of the flame and the melting point are of course determined, but most attention is paid to the actual burning, the cost per hour consumed in still air and in draughts of about 60 feet per minute, noting in the latter the guttering of the wick and the hardness or softness of the candle.

The testing of Portland cements, of which large quantities are used, are carried out on an extended scale and also the testing of rubber which is of considerable

importance to the engineer. Explosives as supplied on the Rand are usually of excellent quality, but laboratory tests on the blasting power of different explosives with the aid of the lead cylinder have an important application. Laboratory tests on fuse are also of value, being much quicker and less likely to lead to bad results than testing by actual experience. The microscopic analysis of metals has been introduced recently with good results.

"One of the most important of the many questions to be decided in the laboratory is that of keeping our metallurgical ideas up to date. All of you know how often, on reading of some new development on these or other fields, the desire to test this on your own account, or to prove its applicability to some of your own problems, has assailed you. Necessary apparatus for conducting these trials on a small scale is usually wanting on the mine, and a request for trying them under working conditions generally results in a curt demand to keep down costs. Setbacks in this direction are certainly not conducive to individual effort, and the heads of our industry recognize that more concerted action in this direction would be advantageous, but hitherto the sum total of this recognition has been its recognition. An 'experiments fund' was inaugurated some two or three years ago by our largest group here and, quoting a local weekly, it seems that already £15,000 has been expended in trying new devices, and that in one case an unsuitable device cost £4,500 prior to its being abandoned.

"Now, I have no hesitation in saying that quite a considerable portion of this could have been saved had the initial tests been carried out on a laboratory scale, meaning thereby a mechanical as well as metallurgical laboratory. I may be permitted to draw the attention of members, however, to one source of false conclusions which are only too liable to crop up in initial experimental work, that is, the test must represent the absolute duplication of the working conditions. That many failures on the large scale are undoubtedly due to neglecting this is only too true, and an impartial test by a thorough practical man should always be insisted on prior to money being spent in large tests. We know how prone the inventor is to overlook small details, which in themselves may latterly cause the failure of the trial. From the point of view also of the originator we find that his grievance is, that on a mine sufficient attention is not given to the carrying out of his ideas to the best advantage.

As against this, the engineer, mill manager and cyanide manager usually have sufficient daily worries of their own and are sometimes inclined to afford scant courtesy to the man who increases these worries, more especially if he be a non-practical worker."

SMOKELESS COMBUSTION OF COAL.

THE ESSENTIAL CONDITIONS OF BOILER CONSTRUCTION AND MANAGEMENT FOR THE SMOKELESS COMBUSTION OF HIGH-VOLATILE FUELS.

L. P. Breckenridge—Bulletin of the University of Illinois.

THE Engineering Experiment Station of the University of Illinois has recently added to its series of valuable bulletins one on the problem of the smokeless combustion of coals of a high volatile content, particularly those of the State of Illinois. While the writer, Mr. L. P. Breckenridge, does not claim to present any new information beyond the results of careful and extensive boiler trials, the bulletin gives an extremely interesting discussion of the smoke-abatement problem, of which the following abstract is a brief outline.

Mr. Breckenridge prefaces a review of the chemical principles of combustion thus: "The problem of smoke prevention is the problem of perfect combustion. There is no such thing as smoke consumption and this term should never be used. There is such a thing as perfect combustion and this means smokeless combustion." Continuing he says:

"It may perhaps be profitable to picture an ideal perfect combustion, and then inquire in what ways actual combustion falls short of the ideal. The given fuel, composed of carbon, various volatile hydrocarbon gases, and perhaps sulphur, is to be burned in air. Theoretically, each atom of the fuel finds and seizes upon the number of oxygen atoms with which it will combine. Each atom will meet with two oxygen atoms at a temperature sufficiently high for ignition. They will combine, and the resulting CO_2 will pass out of the furnace, carrying with it the heat arising from the combustion; likewise with the hydrogen and sulphur atoms. No more air will be delivered than is just sufficient to furish the exact number of oxygen atoms, and no carbon or hydrogen atoms will pass out of the furnace without finding oxygen atoms with which they can combine.

"Actual combustion deviates from ideal conditions in many respects. If only the theoretical amount of air is supplied, on account of the difficulty of properly mingling the fuel and air, some of the fuel atoms will not find oxygen atoms, and will escape uncombined. Or some of the carbon may burn to carbon monoxide instead of to carbon dioxide, and the CO will escape without further combustion. It is found in practice that to insure complete combustion, an excess of air must be furnished. This excess is usually 50 per cent., and may reach 100 per cent.; i.e., while only 11.3 pound of air are required for the complete combustion of 1 pound of carbon, it is usually necessary to furnish 18 to 24 pounds. Since the heat of combustion is distributed throughout the excess of air introduced into the furnace as well as the products of combustion, the furnace temperature is lowered by the presence of the extra air.

"In another important particular, the actual state of affairs is likely to be quite different from the ideal combustion outlined above. Carbon and oxygen atoms will not unite unless a certain temperature, the ignition temperature, is reached. In parts of the furnace, the temperature may fall below the ignition point because of the inrush of an excess of air, or because of cold bounding surfaces. As a result, carbon particles, even in the presence of plenty of oxygen, will refuse to burn."

The products of combustion, carbon dioxide, steam and sulphur dioxide, are colorless gases. It is only when combustion is imperfect that visible smoke is produced and its appearance is due to the driving off of volatile hydrocarbons from the heated coal. The percentage of volatile constituents in coal varies from 3 per cent. in the eastern anthracites to 50 per cent. in some of the western lignites, and the difficulty of smoke prevention increases with the amount of volatile matter. When driven off from the coal, if they encounter a sufficiently high temperature, the volatile hydrocarbons decompose into carbon and hydrogen. If sufficient air is present the hydrogen immediately combines with the oxygen to form steam and the carbon burns to CO_2. If, however, sufficient air is not admitted, the carbon, which requires favorable conditions under which to combine with oxygen, will not burn but will be carried along with the products of combustion as soot. The mere explanation of the formation of black smoke, therefore, immediately suggests the means for its prevention, namely, the introduction into the furnace of air in sufficient volume and at a sufficiently high temperature to burn the carbon particles liberated from the volatile gases. To insure proper combustion, after the gases are driven from the coal, they should intimately mingle with sufficient air in a chamber in which a high temperature can be maintained and should not come in contact with the heating surface of the boiler until combustion has been completed.

The absence of smoke does not necessarily indicate perfect combustion. It may mean simply excessive air supply. But the appearance of black smoke is the signal of incomplete combustion and uneconomical operation. The fuel loss in soot, even under the worst conditions of smoke production, is so small that it would not be worth while to provide elaborate devices to prevent this loss were it the only one. It is the prevention of the escape of the unconsumed gases indicated by the presence of smoke which is of importance hygienically and economically. Smokelessness is a fairly safe indication that the total heat of the fuel has been liberated, but it gives no indication of the degree of efficiency of heat utilization. The highest efficiency demands smokelessness with a minimum air supply.

Mr. Breckenridge described the method used for the observance and recording of smoke production and then proceeds to a discussion of the results of the extensive tests carried out during the past two years on the boiler plant of the University of Illinois. This plant consists of nine units with a total capacity of 2,000 horse power, four of the boilers being Babcock & Wilcox, three Stirling, and one each National water tube and Heine. Two of the B. & W. boilers were equipped with the B. & W. chain grate stoker and the other two with Roney stokers; two of the Stirling boilers were equipped with the Stirling bar grate stoker; the remaining units were equipped with the Green chain grate stoker. In all, over 200 separate boiler tests were made in this plant to determine furnace conditions, many changes in the constructive features of the furnaces and boiler baffling being made to admit of a thorough examination of the smoke problem. The results of these tests reveals the fact that any one of the four well-known types of boilers may be set over at least three well-known types of automatic stokers and be operated without objectionable smoke. The University of Illinois plant was operated without objectionable smoke fully 90 per cent. of the time, and Mr. Breckenridge is convinced that "any fuel may be burned economically and without smoke if it is mixed with the proper amount of air at a proper temperature." The boiler has very little to do with the smoke problem except that some types lend themselves more easily to the necessary furnace construction. The arrangements of construction and baffling which gave satisfactory results are fully illustrated and described in detail. With the Heine boiler, an exact duplicate of the one used in the government fuel tests at St. Louis except that it was mechanically instead of hand fired, over 100 10-hour tests were made on a large number of different coals and in each case absolutely no smoke was produced. Mr. Breckenridge considers that with this setting it is impossible to make smoke. He describes the process of combustion thus:

"The fresh coal, fairly uniform in size, advances slowly from the hopper along on the grate toward the furnace where the temperature is very high. The combustible volatile matter is continually being distilled from the coal, more and more rapidly, but with much uniformity, while it is passing under the combustion arch. Some of the necessary air flows in through the coal in the hopper, more through the grate under the arch, but by far the most flows through the redhot coals on that part of the grate beyond the arch. This air is thus heated and made ready for combining with the volatile products flowing from beneath the arch, and all together mix and roll along on the bottom of the tiles forming the roof of the furnace.

"The bottom row of boiler tubes is covered with suitably formed tiles, which prevent the still actively mingling gases from being cooled by coming in contact with the tubes, and so the combustion processes go on until completed before reaching the point where the gases pass in among the cooling tubes. The tiles in the adjoining rows touch each other so that no gases pass between them."

There is an increasing tendency to drive boilers above their rated capacity, but there is a limit beyond which smoke will be produced. This limit varies for different types of furnaces and methods of baffling and for different kinds of coal, but in general it may be taken that when boilers are forced much over 130 per cent. of their rated capacity, the probability of smoke rapidly increases. It would seem necessary that when a plant has reached say 140 per cent. of its rated capacity, additional boilers should be installed, in the interests of smoke abatement, until further advances in furnace construction and operation have been made.

AERIAL NAVIGATION.

A REVIEW OF PROGRESS AND AN ESTIMATE OF FUTURE POSSIBILITIES.

Engineering News.

TO the popular mind, the attention which is now being paid to experiments in aeronautics by the military departments of a number of Governments seems to indicate that the day of successful aerial navigation is near at hand. *Engineering News,* however, in an editorial review of progress in this field in the issue of January 16, 1908, which is abstracted below, regards the complete removal of the difficulties and dangers inherent in any form of air navigation as impossible of accomplishment and doubts that the air ship will ever be developed sufficiently to have any practical application outside the sporting field and a limited use in military operations.

"Aeronautical practice of the present time easily divides itself into three lines of work with distinct types of apparatus, common balloons, dirigible gas bags and 'heavier-than-air' flying machines. No little publicity has been given to recent balloon work and while greater skill and knowledge in handling these great gas bubbles is evident than ever before, yet the apparatus is much the same as that devised and used 125 years ago.

"With balloons, far greater altitudes can be reached than it seems probable any true airship could possibly attain. The simple balloon has greater inherent safety and stability than any dirigible or airship and greater ease of control in so far as it can be controlled at all.

"The desire to make an airship which can go where its navigator desires and not simply float with the winds has led to the various dirigibles with their peculiar elongated supporting body and comparatively small propelling and steering apparatus." Among the early experimenters Giffard (1852), Dupoy de Lome (1872), and Tissandier (1883), built dirigibles and operated them with some success. The immediate prototype of the successful dirigibles of the present day was, however, the "La France" of Renard and Krebs, whose feats have not been greatly exceeded.

"Among the most notable experiments with dirigibles should be noted those of Alberto Santos-Dumont which resulted in his winning the Henri Deutsch prize of 100,000 francs at Paris in 1901. His designs were evidently founded on the work of Renard and Krebs; and while it is often stated that his gas bag itself was in many ways inferior to that of 'La France,' yet with a superior source of power their record was greatly exceeded. Santos-Dumont travelled from St. Cloud to the Eiffel Tower with the wind in 8 minutes 45 seconds, circled the tower and returned against the wind in 20 minutes 30 seconds and descended in 1 minute 40 seconds. His gas bag was smaller than that of 'La France,' being 111.5 feet long by 20 feet maximum diameter. The propelling screw was 13 feet in diameter and was driven by a 16-horse-power gasoline motor at 200 revolutions per minute. Space can hardly be given to an enumeration of all the subsequent dirigible balloons which have been wholly or partially successful. However, the names of Von Parsifal and Count Von Zeppelin, both in Germany, and of Henri Deutsch, Count de La Vaulx, the Lebaudy brothers and their engineer Julliot, all in France, should be mentioned as noteworthy experimenters. . . . . .

"Sir Hiram S. Maxim's big aeroplane is credited as being the first true flying machine to leave the ground. The lifting power developed overcame the restraint put upon its ascent and it was wrecked in consequence of its overdoing. When Maxim first started his experiments little was known of the properties of aeroplanes and aerial propellers and the modern internal combustion engine was entirely undeveloped. A steam boiler and engine were designed, remarkable for lightness in comparison to the boiler and engines of that time. This apparatus was the forerunner of the steam-automobile boiler and engine as they exist now. By 1890 Maxim's investigations were well under way and his power apparatus was all designed. In 1894 Maxim gave an exhibition of a huge machine weighing, when loaded with three men, 200 pounds of naphtha fuel and 600 pounds of water, a total of about 8,000 pounds. A wing area of some 4,000 feet was designed to develop a lifting force of 10,000 pounds. It will be remembered that the machine broke through the inverted track, intended to limit its flight for these experiments, turned crosswise and was so badly wrecked that little further has been done with it. The results of this pioneer work, however, became available to later experimenters and have served as a basis of later developments.

"The next important contributions to the science were in the efforts from 1890 to 1895, of the late Professor S. P. Langley, Secretary of the Smithsonian Institution at Washington. The experiments with model 'aerodromes' were entirely successful, but the manipulation of the full sized machine brought practical difficulties which were not surmounted through lack of funds when the death of the inventor occurred. Langley's large machine was 12 feet long and 8 feet wide with four stationary wings and a steering pair about half the size of the larger ones. Steam power was used to drive two small wooden propellers. Mr. Chas. M. Manley, for seven years associated with Professor Langley on this work, is authority for the statement that the failures were of the launching apparatus rather than of the 'Aerodrome' itself and that the work with this type has not been abandoned but only suspended. Langley's experiments are admittedly the most thorough and scientific ever made in this field and, while no practical commercial machine was developed, the design of lifting planes was placed on a more exact and well defined basis."

About this time also Lilienthal came into notice with his "artificial birds," huge, soaring kites which were intended to imitate the flights of certain birds. His experiments were not very successful, but he laid the foundation on which his successors in this line of work, Chanute, Herring and especially the Wright brothers, have brought the sailing type of air ship much nearer practical success.

"Santos Dumont not satisfied with his success with dirigible balloons turned his attention to aeroplane flyers. His arrangement seems to be two lateral sections like double-decked wings with a box-kite-like rudder attached to a long lever in front of the machine. A 50-horse-power eight-cylinder gasoline engine drives a small screw propeller in the rear. The greatest flight obtained with this machine is stated to have been 300 yards.

"Occasional press notices for a few months back have chronicled the experiments of Henry Farman in Paris. The latest report is that on Jan. 13 he made a circuit of about 1,300 meters and won the Deutsch-Archdeacon prize which was offered for the first 'heavier-than-air' machine to cover a circular kilometer. It is stated that this aeroplane weighs about 300 pounds. A frame of ash and piano wire seems to be arranged much in the form of the Wright flyer in this country. A forward rudder like Santos' Dumont's scheme is provided. An eight-cylinder, 50-horse-power gasoline motor, weighing about 175 pounds, is used for propulsion. The final trial for the Deutsch-Archdeacon prize was held a few miles outside Paris in the presence of members of the French Aero Club, several noted aeronauts and about 500 spectators. It is stated that the machine sailed gracefully, about 30 feet above the ground, at a speed of some 25 miles per hour."

Passing mention should also be made of Dr. Alexander Graham Bell's experiments with kites made up of tetrahedral cells. Recent trials with a large model carrying a man are reported to have shown this type to be very stable in the air and experiments with gasoline motor propulsion are to be undertaken at once.

"There is a good deal of misconception as to the probable field for successful aerial navigation. To the lay mind anything is possible, and the 'navies battling in the blue' appears a reasonable probability of the future. The engineer, however, realizes the inherent dangers and difficulties of any form of air navigation. No invention or ingenuity can neutralize the danger to a body poised in mid air and dependent only upon the supporting power of the air to save it from a disastrous fall to the earth. No invention or ingenuity can make safe the navigation of a medium subject to such commotions as hurricanes and tornadoes. By no possibility can carriage of freight or passengers through mid air compete with their carriage on the earth's surface.

"The field, then, for aerial navigation is limited to military use and for sporting purposes. The former is doubtful; the latter is fairly certain. As regards the former, it will be recalled that military balloons have been in use for more than half a century; and that modern high-power, long-range guns have greatly limited the usefulness of such methods of observation. There is every reason to believe that the same thing will limit the usefulness of the military dirigible balloon or flying machine. As for the use of any form of aerial navigation for carrying messages, the development of wireless telegraphy and telephony appears to have blocked all prospect of success there.

"The sporting field is therefore the real field at which the flying machine inventor is aiming. Already ballooning and driving dirigibles are among the most popular fads of the sporting set in Paris. Aside from this the imitation of the birds has fascinated thousands and millions of people to an extent that makes certain that the search for a successful flying machine will go steadily on, even though there be no prospect of great commercial returns to the inventor."

ELECTRIC WELDING.

A DESCRIPTION OF THE BENARDOS PROCESS AS APPLIED TO THE REPAIR OF DEFECTIVE STEEL CASTINGS.

C. B. Auel—The Electric Journal.

WITHIN recent years the steel casting has replaced the iron casting in the construction of machinery of all kinds on account of its comparatively light weight for a given service, but it still fails to equal the iron casting in reliability. There are certain defects of steel castings, such as blow holes and mis-running at the high points of the mould, that no refinements of moulding and pouring seem to be able effectually to guard against. These defects are the more troublesome to the purchaser in that they rarely appear on the surface and are disclosed only on machining, and the loss of time and money in machining and the delay in replacing defective castings have been the cause of much inconvenience and dissatisfaction. In *The Electric Journal* for January, 1908, Mr. C. B. Auel describes a novel application of the Benardos process of electric welding which should remove a great deal of this inconvenience by permitting the purchaser to make his own repairs rapidly and cheaply. The apparatus and processes are described in the following abstract of Mr. Auel's paper.

In the Benardos process the metal to be welded forms one terminal of an electric circuit and an arc is drawn between it and a carbon electrode which forms the other terminal. The requirements for the welding of steel castings are a direct current source of supply, a rheostat, a carbon electrode and fire clay or carbon blocks for moulding purposes. Welding operations are best carried on in an isolated part of the shop owing to the intense glare of the arc, and the body of the operator should be protected against the rays of the arc. the effects of which are somewhat similar to sunburn. The hands should be covered with heavy gloves and the head with a canvas hood with a window of colored glass.

Current up to 220 volts may be used, but this voltage is very wasteful of power. Most economical results are obtained with a 110 to 125 volt current. Where there is enough welding to be done to keep one man steadily employed, the current should be taken from an independently driven dynamo of 75 to 100 kilowatt capacity at 100 to 125 volts instead of from the supply mains. With the dynamo should be provided a small switchboard with the necessary instruments.

"The rheostat may be of the grid type, though a very satisfactory one is easily constructed by using two water-tight barrels placed side by side. The positive cable of the circuit is carried from the dynamo to the switchboard and from the switchboard to the water rheostat. At the rheostat this cable divides into two smaller ones, these being fastened to separate triangular steel plates not less than one-fourth inch thick, suspended above the barrels by means of pulleys and counterweights, so that the plates may be readily lowered into or withdrawn from the barrels as occasion requires the adjusting of the water resistance. Similar cables are run down the inside of each barrel and one end likewise fastened to a heavy plate of steel, which lies on the bottom. The other end of each of these cables is attached to the casting to be welded or the cables may be fastened to a metal cable and the casting simply laid upon it, always providing good contact is made. Discarded steel castings may be substituted for the steel plates in the bottom of the barrels. They should weigh about 20 or 30 pounds each and not occupy too much room. The negative cable of the circuit is carried from the dynamo

to the switchboard and from the switchboard to the vicinity of the casting to be welded, where it is provided with a metal terminal and clamp into which the carbon electrode is tightly fitted. In order to manipulate the carbon electrode during welding, the negative terminal is held in a wood insulating handle, to which is attached a shield of asbestos or other fireproof insulating material. The exact form of the terminal and clamp, the insulating handle and shield or the terminal plates of the water rheostat is immaterial, as is the method of attaching the cables to their respective terminals as long as good and sufficient contact is made, thus preventing undue heating at the joints."

The selection of the proper carbon requires some care. If the carbon flakes and crumbles under the hard treatment to which it is subjected the pieces falling into the weld produce a high-carbon steel which is very difficult to machine. The best results are obtained with a hard solid carbon one or one and one-half inches in diameter and six to twelve inches long, which will wear to a round blunt end instead of a long pencil point.

For the repairing of steel castings it is preferable to use for filling Norway iron rods, about ⅜ inch in diameter, though for small welds small pellets from steel scrap may be used.

"The positive terminal of the circuit may be clamped directly to the casting to be welded or it may simply be laid upon a metal table and the terminal clamped to the latter. The positive terminal is thus connected instead of the negative terminal so as to direct the flow of current from the casting to the carbon electrode, and in this way prevent carbon, when the electrode is vaporized, from entering the weld. The steel plates of the water rheostat are lowered into the barrels which have been previously filled with water, the circuit breaker and the switch closed, when the actual welding is ready to be undertaken.

"The operator places himself directly in front of the casting, holding the negative terminal with its carbon electrode in one hand by means of the wood insulating handle, and having within reach of the other several pieces of iron rod. He then pulls the canvas cap well down over his head, touches the carbon to the casting, thereby closing the circuit and thus producing an arc. As soon as the arc is sprung, the carbon is withdrawn to a distance of two inches or more (too short an arc will tend to produce a hard weld), and the arc allowed to play upon the casting until the metal commences to boil. It is advisable not to concentrate the arc on any one spot, but to give it a circular movement so as to heat the casting very thoroughly within the immediate vicinity of the proposed weld. This will tend to prevent too rapid cooling of the metal with its consequent chilling and hardening effect. The end of one of the iron rods is now placed directly in the midst of the boiling metal, where it gradually melts and mixes with it, the arc

meanwhile being continued. As the rod melts away it is fed into the weld and this process is continued with one or more additional pieces of rod until the weld has been completed. The surface of the weld may be hammered as it cools off to produce a closer grain or to make it conform to some particular shape."

Besides the repairing of steel castings there are many other applications in which the Benardos process will give thoroughly satisfactory results commercially. Among these may be mentioned the removal of surplus metal, the boring of large holes in castings or plates and the welding of flanges, elbows and couplings to pipes. It can be learned by any workman of average ability in a few weeks time.

THE RESISTANCE OF THE AIR

THE METHODS AND RESULTS OF M. EIFFEL'S EXPERIMENTS AT THE EIFFEL TOWER IN PARIS.

Revue de Mécanique.

A N exact knowledge of the value of the resistance offered by the air to moving bodies is of importance in many branches of engineering. In the propulsion of vehicles of all kinds, the resistance of the air is one of the most important forces to be overcome, as, for example, in aerial navigation and railway traction, to mention two of the most obvious fields in which such knowledge is necessary. Many experiments have been made to establish exact data on the subject, but the results have given such widely divergent values that none of them can be taken as finally authoritative. During the last three years, however, M. Eiffel has carried out at the Eiffel tower in Paris an extensive series of very careful investigations which, from the uniformity of the results obtained, may be fairly considered as the most important contribution to the subject. The following details of these experiments are taken from the *Revue de Mécanique* for December, 1907, after M. Eiffel's recent publication of his result in book form.

The experiments were carried out from the second story of the Eiffel tower at a height of 120 metres from the ground. The body dropped from this height weighed 120 kilogrammes. The fore part of the weight

consisted of the surface under test and behind this was placed a conical box containing the delicate measuring instruments. The most difficult problem in connection with the experiments was to devise a means of stopping the weight without damage to the apparatus. In the arrangement finally adopted the apparatus slid along a smooth cable held perfectly vertical. About 20 metres above the ground a progressive enlargement of the cable began and this enlargement by opening a series of jaws connected to powerful springs carried by the apparatus, brought the weight to a stop by a perfectly regular braking action in a distance of ten metres.

During the fall the apparatus was connected to the surface under test by carefully calibrated springs which acted as a dynamometer. The action of the springs being antagonistic to the resistance of the air on the surface, the resistance could be deduced from the displacement of the surface with respect to the rest of the apparatus. This displacement was indicated by a stylus attached to the surface under test and resting lightly on a vertical blackened cylinder carried by the apparatus. The cylinder rotated about its axis, the rotation being produced by a very fine toothed roller held against the cable by a spring. The rotation of the cylinder was in each case proportional to the height of the drop and the line traced upon it showed the tension of the dynamometer springs at each instant.

The aim of the investigations, however, was not to ascertain the resistance of the air for every point of the drop but for each variation in speed. Hence the stylus was mounted on a tuning-fork making 100 vibrations per second, so that in addition to registering the tension of the springs and the distance traversed, it indicated also the elapsed time from the beginning of the fall. The fine sinusoidal curve traced on the cylinder showed, therefore, the three elements of the problem, space, time and the tension of the springs; the speed was deduced from the first two and the resistance of the air from the third. The calculations were made by an extremely simple yet accurate graphical method.

All diagrams which showed the slightest indication of inaccuracy were rejected and in making the experiments the utmost possible precautions were taken to secure accurate results, particular attention being paid to the protection of the falling weight from air currents. Repeated tests were carried out on each surface at different times in order to eliminate the possibility of accidental errors in the apparatus. The results in each case were reduced to their equivalent at a temperature of 15 degrees

C. and a pressure of 760 millimetres. The experiments were carried out between the speeds of 18 and 40 metres per second, or 65 to 144 kilometres per hour.

M. Eiffel summarizes the results as follows:

"We have established that, within the limits of the investigations, that is, for speeds between 18 and 40 metres per second, the resistance of the air is approximately proportional to the square of the velocity. In reality, the exponent of the speed, for plates, seems to increase continuously and to reach the value 2 for a speed of about 33 metres per second; but in all cases the value is so close to 2 that for all practical purposes this figure can be taken as correct and the resistance of the air in kilogrammes can be represented by the product KSV^2, where K is a coefficient depending only on the surface, S the surface of the plate in square metres and V the speed in metres per second.

"We have not found for the coefficients K the high values, reaching 0.13, proposed by some experimenters and used by many engineers. Our investigations, repeated under many varying conditions and giving results of remarkable uniformity, seem to us to establish the value of the coefficients as between 0.07 and 0.08, at a temperature of 15 degrees C. and a pressure of 760 millimetres. The latter value appears to be a maximum only attained in plates of very large dimensions."

COMPRESSION AND EFFICIENCY IN GAS ENGINE OPERATION.

A SUMMARY OF THE RESULTS OF TESTS CARRIED OUT BY THE GAS ENGINE RESEARCH COMMITTEE OF THE INSTITUTION OF MECHANICAL ENGINEERS.

Prof. F. W. Burstall—Institution of Mechanical Engineers.

THE third report of the Gas Engine Research Committee of the Institution of Mechanical Engineers, presented to the Institution by Prof. F. W. Burstall at the meeting of January 17, deals with the relation between compression and thermal efficiency. As shown by the following extracts, which summarize the conclusions reached by the Committee, the results of the investigations do not support the common belief that a high compression is necessary for economy.

A Premier scavenging engine was used in the investigations. It developed 150 horse power at a speed of 170 revolutions per minute, the size of the cylinder being 20 inches in diameter by 24 inches stroke. In order to enable the engine to run at a compression pressure of 200 pounds per square inch, with charges which were estimated to give an initial pressure of 600 pounds, the diameter of the cylinder was reduced to 16 inches. The compression was varied by altering the length of the connecting rod, or by bolting packing pieces or junk pieces at the back of the piston.

"The working of the engine is as follows:—Starting with the suction stroke, the combined air-and-gas valve is opened to a predetermined point by a pivoted lever under the control of the governor and a positively driven pecker block, actuated by the half-speed shaft, the governor thus controlling the opening of the air-and-gas valve. The mixture, after passing through this valve, enters through the breech end into an annular casing, which contains the inlet valve, and then into the cylinder itself. After shutting the inlet valve the usual sequence of compression, explosion, expansion, and exhaust follows, but about half-way along the exhaust stroke a second valve, scalled the scavenger valve, lying alongside the mixing valve, is opened from the lay shaft, and allows a current of cold air from the differential piston to enter into the motor cylinder. This serves the double purpose of clearing out the exhaust products, and at the same time cooling the inner surfaces. During the idle stroke of the engine this scavenging charge is simply compressed and expanded in the passages leading up to the mixing and scavenger valves. In order to prevent, as far as possible, any possibility of pre-ignitions occurring through hot surfaces, every part of the engine exposed to the flame is water-jacketed, and in order to estimate the amount of heat rejected through each of these surfaces, the water services are taken from separate measuring tanks, and discharged without a mixture from water from any other surface. The temperatures of discharge were in each case measured by thermometers placed in the outlet pipes.

"The tests were undertaken to determine in the first place the thermal efficiencies based on the indicated horse-power, at various compressions, having regard to the richness of mixture, and in the second place to formulate if possible the law connecting efficiency and compression. Thus at each compression it was proposed to run a series of trials with different mixtures, which was done by using a number of different mixing valves in which the ratio of the air and gas ports varied. Had the composition of the gas throughout the tests been uniform this would have been a simple matter, but as the producer plant was in general worked at a fairly light load, it was impossible to ensure beforehand that the composition of the gas should be exactly what was required for the particular valve employed. The calorific value of the gas aimed at throughout the tests was 160 B.Th.U. per cubic foot (lower value).

"The heating values were obtained not only from the analysis of a sample drawn continuously throughout the whole of the tests, but also from the Junker calorimeter, which also ran continuously. During the preliminary tests it was noticed that the values from analysis and from the calorimeter did not agree, and that they varied capriciously. As the lower value deduced from the calorimeter necessitated the determination of the amount of water condensed, and this quantity varied according to the hygrometric state of the atmosphere, it was considered essential that the calorimeter should either work with dry air and dry gas or saturated air and saturated gas, and the latter condition was chosen. In this particular calorimeter, both air and gas are led through coke-packed towers, over which a stream of water is steadily trickling, so that they issue from these towers saturated. From these the air and gas are led through a coil of pipe placed in the outlet of the calorimeter, so that incoming air, incoming gas, and outgoing products of combustion are automatically reduced to the same temperature.

"In all other respects the calorimeter was used as was customary. This arrangement had the effect of giving the amount of the condensed water in the calorimeter a closer value than usual to that obtained from analysis, and also produced a remarkably close agreement between the calorimeter values calculated from analysis and those observed.

"Considering the thermal efficiencies shown in the tabulated results it is apparent that for each compression there is a particular mean pressure which gives the highest economy for that compression. This pressure appears to range between 85 lb. and 95 lb. for all the compressions, with the tendency to increase as the compression goes up. Higher mean pressures than these caused the efficiency to fall off. This result does not accord with the usual belief that economy increases with compression, when a suitable mixture is used. The cooling action of the walls, however, affects the result materially. Consider the contents of the cylinder at the end of compression. The gas is confined in a space 16 inches in diameter, at the highest compression about 3¼ inches long, and at the lowest compressions about 6½ inches long, the gas being entirely surrounded by water-cooled surfaces. This being the case, the leakage of heat during compression will be greater proportionately at the high than the low compressions, because the higher compression is accompanied by a higher density and by a temperature difference between the walls and the charge, and this more than compensates for the reduction of the area of surface exposed to the gases. Hence, after some definite compression is reached, further compression will result in a loss of economy and not a gain. For this particular engine the most economical compression pressure is apparently 175 pounds per square inch; but, of course, the particular compression that will give the highest economy will vary according to the design of the clearance spaces, but it does not seem to be probable to get a design which will give better results than in the engine experimented upon.

"In order to obtain higher thermal efficiencies by the aid of higher compressions, it would be necessary to increase the stroke of the engine in proportion to its diameter. In the particular engine experimented upon the stroke is one and a-half times the diameter. If the stroke were twice the diameter, it might be possible to employ a higher compression pressure. In this way the disc of hot gas might still be kept fairly thick, but, of course, such a method would mean slower speeds of rotation for a given piston speed, and thus it is quite probable that the lower speed of rotation might produce prejudicial effects, which would more than counterbalance the gain due to heat losses. Very high mean pressures, extending to some 114 pounds per square inch, were proved to be very decidedly uneconomical, the economy falling from 39 per cent. to nearly 32 per cent. in some cases.

"In the actual tests the maximum pressure that was allowed was 550 pounds, and this was only rarely reached. The heat rejected to the cooling water does not represent the whole of the heat lost to the walls, because the scavenger charge carries some portion of heat from the interior walls of the cylinder, and that heat is thrown into the exhaust. Hence, the values found for the heat rejected into the jacket water are lower than those which are generally obtained for non-scavenger engines.

"The whole of the experiments appear to point conclusively to the fact that the most economical mean pressure is very considerably below the maximum which can be obtained, and that the highest economies are obtained with a comparatively low maximum temperature. Both these results imply that the engine should not only be subjected to lower pressure, but to lower temperatures as well, and thus many of the difficulties which arise in large engines from rich charges might be avoided, and the maximum pressures kept down to quite reasonable limits.

"This, of course, only applies to the indicated power, and the conclusions as to the brake horse-power would be widely different. If, however, the engine is constructed to work only with these moderate pressures and temperatures, the whole of the working parts might be very much lightened, and thus a good mechanical efficiency obtained with the very moderate mean pressures.

"The question of the liability to premature ignitions, of gas containing larger or smaller percentages of hydrogen, was borne in mind throughout these experiments, but in every case of premature ignition which occurred—and many such cases occurred—with compressions higher than 160 pounds per square inch, it was traced to dirt or carbonised oil in the cylinder, or to some part having got overheated, and such premature ignitions took place equally with a weak as with a rich gas.

"The reporter is of opinion that, as far as premature ignition goes, the compression might be made a great deal higher than any which have been used during these experiments, but in view of the fact that the economy falls off after a certain point, there does not seem to be any useful object gained in going to any higher compression."

The following pages form a descriptive index to the important articles of permanent value published currently in about two hundred of the leading engineering journals of the world—in English, French, German, Dutch, Italian, and Spanish, together with the published transactions of important engineering societies in the principal countries. It will be observed that each index note gives the following essential information about every publication:

(1) The title of each article, (4) Its length in words,
(3) A descriptive abstract,   (5) Where published,
(2) The name of its author,   (6) When published,

(7) *We supply the articles themselves, if desired.*

The Index is conveniently classified into the larger divisions of engineering science, to the end that the busy engineer, superintendent or works manager may quickly turn to what concerns himself and his special branches of work. By this means it is possible within a few minutes' time each month to learn promptly of every important article, published anywhere in the world, upon the subjects claiming one's special interest.

The full text of every article referred to in the Index, together with all illustrations, can usually be supplied by us. See the "Explanatory Note" at the end, where also the full title of the principal journals indexed are given.

DIVISIONS OF THE ENGINEERING INDEX.

CIVIL ENGINEERING

BRIDGES.

Arches.
See Reinforced Concrete, under BRIDGES.

Bascule.
The Ohio St. Bascule Bridge at Buffalo, N. Y. Illustrated detailed description of a bridge built after the patented design of Thomas E. Brown. 2800 w. Eng News—Jan. 16, 1908. No. 89565.

DoubleTrack Trunnion Bascule Bridge Over Bodine Creek, Staten Island Rapid Transit Railway. Illustrated detailed description of a counterweighted bascule bridge of the Strauss pattern. 1500 w. Eng News—Jan. 16, 1908. No. 89567.

Blackwell's Island.
The Construction of the Queens Approach to the Blackwell's Island Bridge. Describes methods of work in the construction of a steel viaduct structure, carrying a roadway, sidewalks, and 6 railroad tracks. Ills. 1500 w. Eng Rec—Jan. 11, 1908. No. 89465.

We supply copies of these articles. See page 1103.

Columns.

Safe Stresses in Steel Columns. J. R. Worcester. Brief review of the history of the development of the column formulæ used in bridge specifications, and presents a new formula. 2000 w. Pro Am Soc of Civ Engrs—Jan., 1908. No. 89824 E.

Twelve Tests of Carbon-Steel and Nickel-Steel Columns. J. A. L. Waddell. Gives data relatng to tests recently made of full-size compression members to determine the strengths of nickel-steel and carbon-steel. Ills. 4000 w. Eng News—Jan. 16, 1908. No. 89569.

Compression Members.

See Columns, under BRIDGES.

Drawbridges.

Temporary Jack - Knife Drawbridge Over Bronx River on the New York, New Haven & Hartford R. R. Illustrated description. 3000 w. Eng Rec—Jan. 4, 1908. No. 89356.

Minimum End-Lift Device of a German Swing-Bridge. Illustrates and describes an arrangement used on a bridge at Oldenburg, Germany. 1000 w. Eng News—Jan. 9, 1908. No. 89422.

Erection.

See Steel, under BRIDGES; and Cranes, under MECHANICAL ENGINEERING, TRANSPORTING AND CONVEYING.

Foundations.

Foundation Work for Six Lift Bridges. C. M. Ripley. Illustrates and describes work made necessary by the construction of the new East Chicago Canal. 1200 w. Eng Rec—Jan. 11, 1908. No. 89466.

Latticing.

Proportioning of Lattice Bars. Discussion, opened by H. S. Prichard. 5000 w. Pro Engrs' Soc of W Penn—Dec., 1906. No. 89501 D.

Riveted Lattice for Railroad Bridges of Maximum Span; A Plea for a Return to Rational Design. George Huntington Thomson. Also editorial. Aims to give the consensus of opinions representative of the "lattice school" of engineers. 4800 w. Eng News—Jan. 23, 1908. No. 89807.

Lift Bridges.

See Foundations, under BRIDGES.

Plate Girders.

Experimental Determination of Stresses in Web Plates and Stiffeners of Plate Girders. F. E. Turneaure. Presents results of experiments on plate girder webs and stiffeners, with theoretical discussion. Ills. 6000 w. Jour W Soc of Engrs—Dec., 1907. No. 89551 D.

Quebec.

Summary of the Quebec Bridge. Frank W. Skinner. Describes the design, construction, and failure. Ills. 2200 w. Cornell Civ Engr—Dec., 1907. No. 89687 C.

Reinforced Concrete.

Method and Cost of Constructing a Concrete Ribbed Arch Bridge at Grand Rapids, Mich. A bridge consisting of 7 parabolic arch ribs of 75 ft. clear space and 14 ft. rise. 1000 w. Engng-Con—Jan. 8, 1908. No. 89431.

See also Steel and Viaduct, under BRIDGES.

Steel.

The Proportioning of Steel Railway Bridge Members. Discussion of paper by Henry S. Prichard. 11000 w. Pro Engrs' Soc of W Penn—Dec., 1907. No. 89500 D.

Erection of the Long Plate Girders of the Towanda Bridge. Illustrated description of the 14-span plate girder bridge across the Susquehanna River. 500 w. Eng News—Jan. 30, 1908. No. 89922.

New Bridges Over the Nile at Cairo. States the requirements of the designs and gives an illustrated description of the steel-girder bridge and the tests imposed by the Government. 3300 w. Engng—Jan. 10, 1908. No. 89653 A.

The Development of the Soudan. An account of the steel-truss bridge to be built over the Blue Nile at Khartoum, and an illustrated description of the design. 1200 w. Engr, Lond—Jan. 3, 1908. No. 89530 A.

The Construction of the Marien Bridge in Vienna (Die Bauausführung der Marienbrücke in Wien). Karl Brenner. Details and historical sketch of the construction of this steel arch highway bridge. Ills. 3300 w. Zeitschr d Oest Ing u Arch Ver—Dec. 20, 1907. No. 89751 D.

The Steel and Concrete Bridge over the Mouth of the Guindy at Tréguier (Pont en Acier et Béton sur le Guindy Maritime, à Tréguier). M. Harel de la Noé. A steel, three-pin arch, resting on steel and concrete supports, and floored with reinforced concrete. Ills. 9000 w. Ann des Ponts et Chauss—1907—IV. No. 89710 E + F.

See also Bascule, Blackwell's Island, Columns, Drawbridges, Latticing, Plate Girders, Quebec, Struts, Suspension, and Viaducts, under BRIDGES.

Struts.

The Design of Struts. W. E. Lilly. Considers the problems involved in the design of a strut, pointing out in what way the usually-applied formulæ fail to give correct values when estimating its strength; also examines the causes to which the failure of the Quebec bridge may be attributed. Ills. 3300 w. Engng—Jan. 10, 1908. No. 89652 A.

Suspension.

Erection of the Manhattan Bridge Across the East River. Illustrated de-

We supply copies of these articles. See page 1103.

scription of features, and methods of erection of this bridge, which is said to be the heaviest and strongest ever built. 2000 w. Sci Am—Feb. 1, 1908. No. 89928.

Viaducts.

Reinforced Concrete Railroad Viaducts at Seville, Spain. E. Ommelange. Illustrates and describes viaducts and wharves of reinforced concrete on a private line for the Spanish Cala Iron Co. 1000 w. R R Gaz—Jan. 17, 1908. No. 89587.

Curved Viaduct Leading to the Austerlitz Bridge (Viaduc Hélicoidal d'Accés au Pont d'Austerlitz). L. Biette. An illustrated description of this steel-truss viaduct in Paris. 3500 w. Génie Civil—Dec. 7, 1907. No. 89723 D.

CONSTRUCTION.

Brickwork.

Brickwork Details. Halsey Wainwright Parker. An illustrated series of articles on the ornamental possibilities of brickwork. 1800 w. Br Build—Dec., 1907. Serial. 1st part. No. 89285 D.

Buildings.

See Concrete, and Steel Buildings, under CONSTRUCTION.

Columns.

See same title, under BRIDGES.

Concrete.

An All-Concrete Laundry Building. Illustrated description of a 4-story structure in Salem, Mass. 2000 w. Eng Rec—Jan. 11, 1908. No. 89464.

Concrete Building Stone. C. A. P. Turner. Illustrated description of a new material, its process of manufacture and some of the uses. 3000 w. Cement Age—Jan., 1908. No. 89684 C.

Concrete of Exposed Selected Aggregates. Albert Moyer. Illustrates and commends a house in South Orange, N. J., describing its construction. 2200 w. Munic Engng—Jan., 1908. No. 89485 C.

On the Influence of Temperature on Masonry, Cement and Reinforced Concrete Construction (Sull Influenza della Temperature nelle Costruzioni in Muratura, Cemento e Cemento Armato). Ills. 3000 w. Il Cemento—Nov., 1907. No. 89703 D.

See also Concrete Testing, under MEASUREMENT; and Locks, under WATERWAYS AND HARBORS.

Contracts.

Substantial Performance of a Building Contract. George Doan Russell. Considers some of the cases which arise in courts, and discusses the legal side of building operation. 3500 w. Columbia Engr—1906. No. 89498 N.

Earthquakes.

Effect of Earthquake Shock on High Buildings. R. S. Chew. A study and statement of conclusions as to the type of construction for the vicinity of San Francisco. 1500 w. Pro Am Soc of Civ Engrs—Jan., 1908. No. 89825 E.

Earthquakes and Their Relation to Engineering Construction (Die Erdbeben in ihren Beziehungen zur Technik und Baukunst). Dr. Franz Ritter. The first part of the serial discusses the vast amount of damage that has been done by earthquakes and reviews literature on the subject. Serial. 1st part. 6000 w. Zeitschr d Oest Ing u Arch. Ver—Dec. 6, 1907. No. 89750 D.

Excavation.

Comments on the Use of the Mattock. Deals with the use and misuse of this earth-working tool. 1800 w. Engng-Con—Jan. 15, 1908. No. 89581.

Factories.

Doubling a Factory's Floor Space. C. M. Ripley. Describes an interesting piece of work just completed at Passaic, N. J. A new structure was built over a wing without interruption of the operation of the old plant. 1000 w. Eng Rec—Jan. 18, 1908. No. 89614.

See also Shops, under MECHANICAL ENGINEERING, MACHINE WORKS AND FOUNDRIES.

Failures.

Failures of Reinforced Concrete. H. F. Porter. Aims to clearly set forth the causes of the failure of the Kodak Building in Rochester, N. Y., and probably of other structures. 2800 w. Cornell Civ Engr—Dec., 1907. No. 89688 C.

Fireproof.

An Analysis of the Fire in the Parker Building, New York City. Peter Joseph McKeon. Describes the construction of the building and course of the fire, stating the conclusions. 1500 w. Eng News—Jan. 16, 1908. No. 89573.

See also Fire Protection, under WATER SUPPLY.

Regulations.

Regulations of the Bureau of Building Inspection in Regard to the Use of Reinforced Concrete. Emile G. Perrot. Gives the text of the new regulations now in force, discussing some of the important features. General discussion. Ills. 9500 w. Pro Engrs' Club of Phil—Oct., 1907. No. 89548 D.

Reinforced Concrete.

Notes on Reinforced Concrete. Discusses concrete in compression, grooped columns, allowable stresses, etc. 3000 w. Jour Worcester Poly Inst—Jan., 1908. No. 89695 C.

Something About Reinforced Concrete. E. Horton Jones. Discusses its qualities, application, and uses. 1800 w. Min Sci—Jan. 16, 1908. No. 89634.

We supply copies of these articles. See page 1103.

Reinforced Concrete from the Contractor's Standpoint. H. H. Fox. Read before the Nat. Assn. of Cement Users. How to make the best possible reinforced concrete. 2500 w. Eng News—Jan. 30, 1908. No. 89926.

A Reinforced Concrete Observation Tower. Illustrated description of a tower recently erected at Vicksburg, Miss. 900 w. Eng Rec—Jan. 25, 1908. No. 89831.

The Stadium of Syracuse University. Illustrated description of the recently completed reinforced concrete stadium for athletic games. 4000 w. Eng Rec—Jan. 18, 1908. No. 89617.

The Influence of Bond on Size of Reinforcement Bars for Concrete. William P. Creager. Aims to provide a rational method of determining the maximum size of the bar that can be used in any given span and system of loading, consistent with the adopted working intensity of adhesion or bond stress in the bar. 1200 w. Eng Rec—Jan. 25, 1908. No. 89830.

The Choice of Diameter for Round-Iron Reinforcement (Beitrag zur Berechnung der Haftspannungen und Ermittlung der entsprechenden Rundeisendurchmesser). Rich. Wuczkowski. A mathematical discussion on the design of reinforced-concrete beams with round-iron reinforcement. Ills. 2500 w. Beton u Eisen—Dec., 1907. No. 89775 F.

Recent Researches in Reinforced-Concrete Construction (Neuere Forschungen im Gebiete des Eisenbetonbaues). M. Foerster. Discusses principally the Considére system of reinforcement and new types of reinforcing metal. Ills. 4500 w. Stahl u Eisen—Dec. 4, 1907. No. 89729 D.

The Tensile Strength of Concrete in Reinforced-Concrete Structural Design (Berechnung der Eisenbetonbauten unter Berücksichtigung des Zugspannungen des Betons.) Chr. Vlachos. A mathematical paper, reviewing the researches of Wayss and Freytag, and Schüle on the ratio $n = E_e/F_b$. Ills. 3500 w. Beton u Eisen—Dec., 1907. No. 89777 F.

See also Failures and Regulations, under CONSTRUCTION; Reinforced Concrete, under MATERIALS OF CONSTRUCTION; and Conduits and Reservoirs, under WATER SUPPLY.

Stacks.

A Tall Brick Chimney with Acid-Proof Lining. Illustrates and describes the 366-ft. brick chimney of the Eastman Kodak Co., Rochester, N. Y., for carrying off strong acid fumes. 1500 w. Eng Rec—Jan. 4, 1908. No. 89357.

Steel Buildings.

The New Hearst Building, San Francisco, Cal. Illustrated description of the engineering features of a steel-cage build-

ing to be built in an earthquake district. Plates. 2500 w. Am Archt—Jan. 18, 1908. No. 89664.

Structural Features of the Warwick Shops of the Lehigh and Hudson River Ry. A steel and concrete building at Warwick, N. Y., is illustrated and described. 2500 w. Eng Rec—Jan. 11, 1908. No. 89468.

See also Columns, Latticing and Plate Girders, under BRIDGES.

Tunnels.

Method of Drilling and Mucking in a Rock Tunnel and a Comparison in the Tunneling on the Rand. Abstract of articles by W. P. J. Dinsmoor. 2500 w. Engng-Con—Jan. 1, 1908. No. 89314.

The Bernese Alpine Tunnel (Lötschbergbahn). Fritz Hromatka. Illustrated description of a line to be constructed and operated electrically. 3800 w. Bul Int Ry Cong—Dec., 1907. No. 89509 E.

The Detroit River Tunnel. James C. Mills. Illustrated description of the construction of this tunnel under the Detroit River. 3500 w. Cassier's Mag—Jan., 1908. No. 89454 B.

The Flushing Tunnel for the Gowanus Canal in Brooklyn, N. Y. Illustrated detailed description of the construction of this tunnel, built to correct the unsanitary condition of the canal caused by sewerage and a variety of wastes. 3500 w. Eng Rec—Jan. 11, 1908. No. 89462.

Air Compressors on New York Tunnel Work. Frank Richards. An illustrated article giving information in regard to the great amount of tunnel work completed and in prospect in the vicinity of New York City, with details about the use of compressed air. 4000 w. Compressed Air —Jan., 1908. No. 89484.

Waterproofing.

Waterproof Engineering. Edward W. De Knight. Treats of methods and materials and their application. Ills. 6000 w. Jour Assn of Engng Socs—Dec., 1907. No. 89676 C.

MATERIALS OF CONSTRUCTION.

Brick.

The Sand-Lime Industry (Conference sur l'Industrie Silico-Calcaire). E. Leduc. A discussion of sand-lime brick, their composition, chemical reactions, manufacture, etc., describing leading plants for their manufacture in France. Ills. 8700 w. Bul Soc d'Encour—Nov., 1907. No. 89718 G.

Cement.

The Manufacture of Commercial Portland Cement. Richard K. Meade. Briefly describes the burning of raw material, the fuel used, the grinding of the clinker, etc. Ills. 2500 w. Min Sci—Jan. 23, 1908. No. 89851.

Reinforced Concrete.

Tests of Bond Between Plain Bars and Concrete. L. R. Viterbo. A report of tests made at Washington University laboratory. 800 w. Engng-Con—Jan. 1, 1908. No. 89313.

See also same title, under Construction.

Steel.

See Columns, Latticing, and Plate Girders, under Bridges; and Steel, under MECHANICAL ENGINEERING, Materials of Construction.

Timber Preservation.

The Galesburg Timber Preserving Plant of the Burlington. Illustrated description of this large new plant embodying recent ideas. 1000 w. R R Gaz—Jan. 10, 1908. No. 89443.

The Seasoning and Preservative Treatment of Hemlock and Tamarack Cross-Ties. W. F. Sherfesee. Detailed account of experiments made to devise methods for bettering the treatment of these close-grained woods. Ills. 7000 w. U S Dept of Agri, Forest Serv—Circular 132.· No. 89690 N.

The Estimation of Moisture in Creosoted Wood. Arthur L. Dean. Discusses the value of J. Marcusson's method for estimating the water and acid contents of lubricants, when applied to creosoted wood. Ills. 1200 w. U S Dept of Agri, Forest Serv—Circular 134. No. 89691 N.

See also Poles, under ELECTRICAL ENGINEERING, Transmission.

Waterproofing.

See same title, under Construction.

MEASUREMENT.

Concrete Testing.

The Analysis of Concrete. Royal L. Wales. Outlines work showing the writer's method of determining the ratio of cement to sand in mortar, and the ratio of mortar to rock, and the importance of such analyses. 1600 w. Eng News—Jan. 9, 1908. No. 89421.

Precipitation.

The Automatic Registration of Precipitation. James L. Bartlett. Illustrates and describes the instruments used by the United States Weather Bureau. 1200 w. Wis Engr—Dec., 1907. No. 89511 D.

Standardizing.

How Our Measures of Length Are Tested. Herbert T. Wade. Illustrates and describes the instruments employed and the methods. 2000 w. Sci Am—Feb. 1, 1908. No. 89930.

Stream Flow.

Theoretical Considerations on the Gauging of Streams with Changing Bottoms (Considérations Théoriques sur les Jaugeages des Cours d'Eau à Fond Mo-

bile). M. R. Tavernier. Draws practical conclusions from theoretical considerations. Ills. 11000 w. Ann des Ponts et Chauss—1907—IV. No. 89711 E + F.

Surveying.

The Panoramic Camera Applied to Photo-Topographic Work. Charles Will Wright. An illustrated description of this method, discussing its advantages and accuracy. 4000 w. Bul Am Inst of Min Engrs—Jan., 1908. No. 89894 C.

The Alaskan Boundary Survey. Prof. O. M. Leland. Reviews the difficulties arising in regard to this boundary line, and gives an account of the controversy and its settlement, and the methods used in the final survey and marking of the line. Ills. 3500 w. Cornell Civ Engr—Jan., 1908. No. 89902 C.

MUNICIPAL.

Experiment Station.

Lawrence Experiment Station. A brief statement of the origin of the station and of the work done by the Mass. State Board of Health. Ills. 1500 w. Munic Jour and Engr—Jan. 15, 1908. No. 89537.

Garbage Disposal.

Destructor Plant for Chiswick Urban District Council. Illustrated description of a plant dealing not only with house-refuse, but with solids and sludge from sewage works. 900 w. Engng—Jan. 3, 1908. No. 89524 A.

Municipal Refuse Disposal: An Investigation. Discussion of the paper by J. T. Fetherston. 10000 w. Pro Am Soc of Civ Engrs—Jan., 1908. No. 89826 E.

Report on Garbage and Refuse Disposal, Milwaukee, Wis. A review of the report of Mr. Rudolph Hering, recently submitted to the city authorities. 3500 w. Eng News—Jan. 16, 1908. No. 89566.

The Importance of Garbage and Sewage Burning to the Electrical Industry (Ueber die Bedeutung der Müll- und Canalisationsschlammverbrennung für die Elektrotechnik). Discusses garbage and sewage destructors from a hygienic point of view and also their importance as sources of electric power. 3000 w. Elektrotech Rundschau—Dec. 24, 1907. No. 89746 D.

Germany.

Municipal Engineering in Germany. C. F. Wike. A summary of information including almost every branch of municipal work connected with sanitation. 6500 w. Surveyor—Jan. 10, 1908. No. 89638 A.

Pavements.

The Municipal Asphalt Pavement Repair-Plant at New Orleans. Information from a recent report by W. J. Hardee concerning the first year's work of this plant. Ills. 1500 w. Eng News—Jan. 2, 1908. No. 89272.

A Municipal Repair Plant for Pavements, New Orleans, La. Plan and description of works and equipment, and reports of results. 2200 w. Eng Rec—Jan. 11, 1908. No. 89463.

Roads.

The Digging of Macadam Roads by Machinery (Note sur le Piochage Mécanique des Empierrements). M. Bret. Describes machines for breaking up macadam roads, giving costs. Ills. 10000 w. Ann des Ponts et Chauss—1907—IV. No. 89712 E + F.

The Destructive Effects of High-Speed Automobiles on Macadamized Roads (Effets Destructeurs des Grandes Vitesses Automobiles sur les Empierrements). M. Salle. Conclusions drawn after the recent *Grand Prix* trials of of the Automobile Club of France. Ills. 11000 w. Ann des Ponts et Chauss—1907—IV. No. 89709 E + F.

Septic Tanks.

The Cameron Septic Tank Patent Sustained by the Court of Appeals. Gives the full text of the decision and opinion of the U. S. Circuit Court of Appeals declaring the *process claims* of the Cameron septic tank patent are valid. Also editorial. 6500 w. Eng News. Jan. 23, 1908. No. 89806.

Sewage Disposal.

Notes on Sludge Disposal. George W. Fuller. Emphasizes the importance of good management in sewage-disposal plants, and of due attention to sludge disposal. 2500 w. Eng Rec—Jan. 18, 1908. No. 89610.

A Review of the Sewerage Problem of the City of Baltimore. Ezra B. Whitman. A brief account of the work being carried on to give the city a complete system of sanitary sewers and sewage disposal works. 4000 w. Cornell Civ Engr—Jan., 1908. No. 89904 C.

A Small Sewage Disposal Plant in Central Iowa. A. Marston. States the conditions often found in small towns of the middle west in the United States, and discusses how best to meet the difficulties. 1200 w. Cornell Civ Engr—Dec., 1907. No. 89689 C.

The Sewage Disposal System of Wilmersdorf (Die Entwässerungsanlagen der Gemeinde Wilmersdorf). Herr Müller. An illustrated description of the sewers, pumping plants, purification beds, etc., of this extensive system. Serial, 1st part. 5000 w. Zeitschr d Ver Deutscher Ing—Dec. 14, 1907. No. 89779 D.

See also Garbage Disposal and Septic Tanks, under MUNICIPAL.

Sewers.

The Completion of the Los Angeles, Cal., Outfall Sewer. Explains the conditions which made necessary a new outfall, and illustrates and describes its construction. 4000 w. Eng Rec—Jan. 25, 1908. No. 89829.

A Simple Formula for the Design of Drainage Canals (Eine einfache Formel für die Berechnung von Entwässerungskanälen). Fr. V. Meyer. A simplification of the Kutter formula. Serial, 1st part. 1100 w. Gesundheits-Ing—Dec. 21, 1907. No. 89768 D.

See also Sewage Disposal, under MUNICIPAL.

Smoke Prevention.

Smoke Prevention in Newark, N. J. Gives a new ordinance recently adopted, and describes improved methods of preventing smoke, especially the Thomas device. 1200 w. Eng Rec—Jan. 18, 1908. No. 89615.

Street Cleaning.

See San Francisco, under WATER SUPPLY.

WATER SUPPLY

Aqueducts.

See Edinburgh, under WATER SUPPLY.

Artesian Wells.

See Air Lift and Pumping, under MECHANICAL ENGINEERING, HYDRAULIC MACHINERY.

Conduits.

Conduit of Special Design in Ogden, Utah. Describes a reinforced-concrete conduit, designed to meet special conditions. 1600 w. Eng Rec—Jan. 18, 1908. No. 89611.

Dams.

Final Report on the Award of the Ashokan Reservoir Dam Contract, New York City. Gives the final conclusions of the Commissioners of Accounts, with editorial comment. 3500 w. Eng News—Jan. 2, 1908. No. 89276.

A New Permanent and Movable Lath Dam (Nuova Diga Permanente Instabile a Panconcelli). Describes a movable dam of pine lath for use in emergencies at hydro-electric stations in Italy. 4000 w. Elettricita—Dec. 20, 1907. No. 89702 D.

See also Sluice Gates, under WATER SUPPLY; and Flood Protection, under WATERWAYS AND HARBORS.

Edinburgh.

The Talla Water Supply, Edinburgh, Scotland. Describes the reservoir and aqueduct which gives an additional supply of about 14,500,000 imp. gal. per day. 1700 w. Eng Rec—Jan. 4, 1908. No. 89359.

Faucets.

Faucets for Limiting Flow (Les Limitateurs de Débit). P. A. Bergès. Describes and illustrates faucets of German design which prevent waste by automatically stopping unnecessary flow. Serial, 1st part. 3200 w. Génie Civil—Dec. 7, 1907. No. 89724 D.

Filtration.

Experiences in the Practical Operations of a Mechanical Filter Plant. C. H. Cobb. Abstract of a paper read before the Illinois Soc. of Engrs. & Survrs. 1000 w. Eng News—Jan. 30, 1908. No. 89924.

Fire Protection.

The New York City Fire-Protection Water System. Describes the high-pressure system designed to protect the drygoods district. Ills. 3000 w. Eng Rec—Jan. 4, 1908. No. 89360.

Fire Prevention in High Buildings. The Need of Auxiliary Equipment. J. K. Freitag. A discussion of the Parker Building disaster in New York City, and its lessons. 3000 w. Engineering Magazine. Feb., 1908. No. 89878 B.

See also Hydrants and San Francisco, under WATER SUPPLY; and Fireproof, under CONSTRUCTION.

Great Britain.

Water Supply in 1907. Mainly a review of British progress and improvements. 3000 w. Engr, Lond—Jan. 3, 1908. No. 89529 A.

Hydrants.

Experiments on Various Types of Fire Hydrants. Charles L. Newcomb. Abridged from paper presented to the A. S. M. E. in 1899. An illustrated report of tests on types commonly used. Discussion. 17000 w. Jour N Eng W-Wks Assn—Dec., 1907. No. 89596 F.

Irrigation.

Irrigation in Egypt. Illustrates and describes the steam pumping plants which have been built during the last fifteen years for the purpose of irrigation in Egypt. 1500 w. Engr, Lond—Jan. 10, 1908. Serial. 1st part. No. 89658 A.

See also Nevada, under MINING AND METALLURGY, GOLD AND SILVER.

Orifices.

Discharge of Water from Minute Orifices. W. R. Baldwin-Wiseman. Describes experiments to determine the discharge at different pressures, from circular orifices of small diameter piercing pipes of moderate thickness. 1500 w. Surveyor—Jan. 10, 1908. No. 89639 A.

Pipe Corrosion.

See same title, under MECHANICAL ENGINEERING, HEATING AND COOLING.

Pipe Flow.

Experiments with Submerged Tubes 4.0 Feet Square at the University of Wisconsin. Clinton B. Stewart. Abstract of a forthcoming bulletin on "An Investigation of Flow Through Large Submerged Orifices and Tubes." Ills. 6500 w. Wis Engr—Dec., 1907. No. 89510 D.

Research on the Discharge of Partly Filled Pipe Lines, Canals, Streams and Rivers (Ermittelung der Abflussmengen in Teilweise gefüllten Rohrleitungen, Kanälen, Bächen und Flüssen). Adolf Städing. A mathematical discussion of experimental results. Ills. Serial. 1st part. 5000 w. Gesundheits-Ing—Dec. 21, 1907. No. 89767 D.

Pipe Lines.

Note on the Determination of the Most Efficient and Satisfactory Pipe Diameter for Water-Works Pipe Lines (Beitrag zur Berechnung der wirtschaftlichsten Rohrdurchmesser bei Pumpwerks-Wasserleitungen). Joh. Pelinka. A mathematical discussion. Ills. 4000 w. Zeitschr d Oest Ing u Arch Ver—Dec. 20, 1907. No. 89752 D.

Pipe Specifications.

Cast-Iron Pipe Specifications. William R. Conard. Aims to show what may be done. General discussion. 3000 w. Jour N Eng W-Wks Assn—Dec., 1907. No. 89595 F.

Pipe Testing.

The Testing of Pipes by Outward Pressure (Die Prüfung von Rohren auf äusseren Ueberdruck). E. Preuss. A review of the methods used, and formulae developed by various investigators. Ills. 2800 w. Stahl u Eisen—Dec. 18, 1907. No. 89735 D.

Purification.

The Sterilization of Water by Electricity (Esterilizacion Eléctrica de las Aguas). Eduardo Gallego. Illustrates and describes small household devices in the first instalment. Serial. 1st part. 2000 w. Energia Elec—Dec. 10, 1907. No. 89705 D.

See also Filtration and Reservoirs, under WATER SUPPLY; and Boiler Waters, under MECHANICAL ENGINEERING, STEAM ENGINEERING.

Reservoirs.

The Croton Falls Reservoir, Croton Water System, New York. Illustrated description of one of a series of reservoirs now being completed to increase the water supply. 3000 w. Eng Rec—Jan. 18, 1908. No. 89608.

Recent Improvements to the Water-Works at Peabody, Mass., Including Pumping Plant and Distributing Reservoir. Frank A. Barbour. Gives a brief history of the works and describes the new pumping plant and reservoir. Ills. 4400 w. Jour N Eng W-Wks Assn—Dec., 1907. No. 89594 F.

The Waltham Reservoir. Bertram Brewer. Gives an illustrated detailed description of the construction of this large reinforced-concrete standpipe. Also discussion. 7500 w. Jour N Eng W-Wks Assn—Dec., 1907. No. 89593 F.

We supply copies of these articles. See page 1103.

The Construction of a Masonry Reservoir with a Capacity of 22,000 Cubic Meters for the Water Supply System of Dinan (Notice sur la Construction d'un Réservoir en Maçonnerie de 22,000 Mètres Cubes destiné à l'Alimentation de la Ville de Dinan). M. Daubert. Ills. 4000 w. Ann des Ponts et Chauss—1907—IV. No. 89713 E + F.

The Reservoir with Iron-Removal Plant of the Plauen Curtain Manufacturing Company at Plauen (Das Hochreservoir mit Enteisenungsanlage der Gardinenfabrik Plauen Act.-Ges. in Plauen im Vogtlande). Max Sieb. A reservoir for the purification of bleaching water is illustrated and described, the first part of the serial giving a mathematical discussion of the design. Serial. 1st part. 6000 w. Beton u Eisen—Dec., 1907. No. 89776 F.

See also Edinburgh, under WATER SUPPLY.

San Francisco.
Proposed Auxiliary Water System for Fire Protection and Flushing Purposes, San Francisco, Cal. Gives an outline of a proposed scheme at an estimated cost of $5,500,000. 900 w. Eng News—Jan. 16, 1908. No. 89574.

Sluice Gates.
Electrically Operated Sluice Gates for the Shoshone and Pathfinder Dams. F. W. Hanna. Illustrated description of high-pressure gates for the control of dams remarkable for the high heads under which they are to be operated. 2700 w. Eng News—Jan. 2, 1908. No. 89273.

Stream Flow.
Stream Flow Data from a Water-Power Standpoint. Charles E. Chandler. Gives tabulated data with methods of analyzing river discharges. General discussion. 6500 w. Jour N Eng W-Wks Assn—Dec., 1907. No. 89597 F.

Stream Gauging.
See Stream Flow, under MEASUREMENT.

Valuation.
The Appraisal and Depreciation of Water-Works and Similar Properties. William H. Bryan. States eleven methods that may be used, briefly considering their merits; outlines methods of computing depreciation, etc. General discussion. 16000 w. Jour Assn of Engng Socs—Dec., 1907. No. 89677 C.

WATERWAYS AND HARBORS.

Barge Canal.
Report on the New York Barge Canal Work by the Superintendent of Public Works. Abstract of the annual report of Hon. Frederick C. Stevens favoring the Federal Deep Waterway. 2500 w. Eng News—Jan. 23, 1908. No. 89808.

British Harbors.
Harbors and Waterways in 1907. Prin-

cipally a review of improvements at British ports, but briefly considering United States waterways and foreign ports. 6500 w. Engr, Lond—Jan. 3, 1908. No. 89527 A.

Canals.
The Teltow Canal. Map and illustrated description of this German canal, constructed chiefly to overcome difficulties in draining Berlin suburban districts. 1800 w. Engr, Lond—Dec. 27, 1907. No. 89392 A.

See also Barge Canal and Panama Canal, under WATERWAYS AND HARBORS.

Chicago.
Chicago Harbor and Water-Front Park Improvement Question. Gives Mayor Busse's message to the city council protesting against devoting the whole of the water front to parks, urging improvements in harbor facilities. Map and editorial. 3300 w. Eng News—Jan. 23, 1908. No. 89809.

Docks.
The Steel Ore Dock at Narvik, Norway. John Birkinbine. Explains the terms of the Swedish Government ore contract, and illustrates and describes the facilities at Narvik for exporting the iron ore. 2500 w. Ir Age—Jan. 9, 1908. No. 89412.

See also Harbors, under WATERWAYS AND HARBORS.

Dredging.
Dredging Costs on the St. Lawrence River and in Other Parts of Canada. Emile Low. Brief account of work with tabulated statement of cost during fiscal year 1905-06. 1700 w. Eng News—Jan. 30, 1908. No. 89921.

See also Dredges, under MARINE AND NAVAL ENGINEERING.

Flood Protection.
The Building of the Friedrichswald Valley Barrier and Its First Year's Operation (Ueber den Bau der Friedrichswalder Talsperre und Bericht über das erste Betriebsjahr). Viktor Czehak. A description of a large dam and other flood protection works in this district in Austria. Ills. Serial, 1st part. 2700 w. Zeitschr d Oest Ing u Arch Ver—Dec. 6, 1907. No. 89749 D.

Florida.
The Coast and Inland Waters of Florida. M. B. Claussen. Illustrated description of the picturesque inland route from one end of Florida to the other, through rivers, bayous, sounds and connecting canals. 4500 w. Rudder—Jan., 1908. No. 89394 C.

Harbors.
The Physical Difficulties of Modern Harbor and Dock Extension. H. C. M. Austen. Discusses some of the financial and physical difficulties in connection with

preparing docks and harbors for vessels of ever-increasing size. 1600 w. Engng—Jan. 10, 1908. No. 89654 A.

See also British Harbors, Chicago, Italian Harbors, Rangoon, and St. John, N. B., under WATERWAYS AND HARBORS.

Italian Harbors.

Navigation in Italian Harbors in the Years 1904-1905 (La Navigazione nei Porti Italiani negli Anni 1904-1905). Antonio Tess. A statistical review of increasing commerce. Ills. 14000 w. Rivista Marittima—Dec., 1907. No. 89700 E + F.

Locks.

Construction and Unit Costs of Concrete Lock, Rough River, Kentucky. Compiled from the Annual Report of the Chief of Engineers, U. S. Army. Ills. 2000 w. Eng News—Jan. 9, 1908. No. 89420.

See also Mechanical Locks, under WATERWAYS AND HARBORS.

Mechanical Locks.

The Oelhafen-Löhle Ship Lift (Das Schiffshebewerk, System "Oelhafen-Löhle"). K. E. Hilgard. Illustrated description. 2500 w. Deutsche Bau—Dec. 14, 1907. No. 89745 B.

The Efficiency of Ship Lifts (Die Wirtschaftlichkeit von Schiffshebewerken). Hermann Bertschinger. A thorough discussion of their design and the advantages and disadvantages of their operating features. Ills. Serial. 1st part. 4000 w. Zeitschr d Ver Deutscher Ing—Dec. 1, 1907. No. 89778 D.

Panama Canal.

At Panama. Fullerton L. Waldo. Illustrates and describes the work as viewed by the writer, giving information concerning life on the isthmus, progress of work, etc. 4500 w. Jour Fr Inst—Jan., 1908. No. 89554 D.

Statement of Col. George W. Goethals, Chairman of the Isthmian Canal Commission, before the Senate Committee on Interoceanic Canals. Interesting extracts from the testimony. 6500 w. Eng News—Jan. 30, 1908. No. 89923.

The Culebra Cut of the Panama Canal. A. S. Zinn. Briefly outlines the history of the project, and gives an illustrated description of the nature of the country and general conditions, the work that has been accomplished, the equipment, etc. 5500 w. Jour W Soc of Engrs—Dec., 1907. No. 89552 D.

Rangoon.

New Port Works at Rangoon. Describes previous conditions at this port of India, and describes the new works. 1200 w. Engr, Lond—Jan. 17, 1908. No. 89873 A.

St. John, N. B.

Schemes Showing the Possibilities of St. John, N. B., as a Great Port; and How the Interior of New Brunswick Can Be Opened Up to Ocean Traffic. J. S. Armstrong. Also notes and amendments. Maps. 8500 w. Can Soc of Civ Engrs—1907. No. 89602 N.

Wharves.

See Viaducts, under BRIDGES.

MISCELLANY.

Review of 1907.

Engineering in the United States in 1907. Railways, tramways, bridges, tunnels, canals, and harbors are considered in the present number. 4000 w. Engr, Lond—Jan. 10, 1908. Serial. 1st part. No. 89656 A.

ELECTRICAL ENGINEERING

COMMUNICATION.

Radio-Telegraphy.

Land Wireless Telegraphy. Major Edgar Russell. Reviews the work of the U. S. Signal Corps in this field, describes stations established and their equipment, modern field telegraphs, etc. Ills. 8000 w. Jour U S Art—Nov., 1907. No. 89502 D.

Electrical Oscillations on Helices. C. J. Watson. Brief explanation of methods used in investigating and the results. 300 w. Elect'n, Lond—Jan. 3, 1908. No. 89517 A.

On Magnetic Oscillators as Radiators in Wireless Telegraphy. J. A. Fleming. Considers the closed, or magnetic, type of oscillator. 2800 w. Elect'n. Lond—Dec. 27, 1907. Serial. 1st part. No. 89379 A.

Telephone Tolls.

The Determination of Telephone Rates for Large Exchanges. William H. Crumb. Points out some of the questions that must be given consideration in order to establish equitable rates for large telephone exchanges. Discussion. 10500 w. Jour W Soc of Engrs—Dec., 1907. No. 89550 D.

DISTRIBUTION.

Potential Regulator.

The Tirrill Automatic Potential Regulator (Ueber den selbsttätigen Spannungsregler System Tirrill). Dr. Gustav Grossmann. Illustrated description with details of operation. Serial. 1st part. 2000 w. Elektrotech Zeitschr—Dec. 12, 1907. No. 89786 D.

We supply copies of these articles. See page 1103.

Regulations.
See same title, under TRANSMISSION.

Wiring.
Interior Wiring Progress in the Rocky Mountain Districts. W. J. Canada. Gives a summary of the conditions in the region as viewed from the insurance inspection standpoint. 2500 w. Elec Wld—Feb. 1, 1908. No. 89913.

Wiring of Residences at Small Cost. Thomas W. Poppe. Explains how a dwelling can be wired at small cost without damage to walls, decorations or floors. 2000 w. Elec Wld—Jan. 4, 1908. No. 89418.

DYNAMOS AND MOTORS.

A. C. Dynamos.
Test of a Low-Voltage Alternator for Calcium Carbide Furnaces. A report of tests made of an alternator for calcium carbide works in Ireland. 1000 w. Elec Rev, Lond—Jan. 17, 1908. No. 89867 A.

The Non-Synchronous Generator in Central Station and Other Work. W. L. Waters. Considers the general characteristics of the non-synchronous generator, its excitation; showing it to be superior to the synchronous generator for supplying power to motor-generators and synchronous converters through an underground cable system, and other applications, especially in connection with turbine-driven generators. 8500 w. Pro Am Inst of Elec Engrs—Feb., 1908. No. 89896 D.

Theoretical and Practical Notes on the Parallel Operation of Alternating-Current Machines (Theoretisches und Praktisches über den Parallelbetrieb von Wechselstrommaschinen). Gustav Benischke. Ills. 2800 w. Elektrotech u Maschinenbau—Dec. 29, 1907. No. 89748 D.

A. C. Motors.
The Torque Conditions in Alternate Current Motors. Val. A. Fynn. Deals with torque conditions in types of single and polyphase commutator and squirrelcage or slip-ring motors, comparing their relative weight efficiencies. Diagrams. 13500 w. Inst of Elec Engrs—1907. No. 89670 N.

Ball-Bearings.
See same title, under MECHANICAL ENGINEERING, MACHINE ELEMENTS AND DESIGN.

D. C. Dynamos.
The Unipolar Dynamo. E. J. Noe. A brief description of the electrical operation of these machines. Ills. 2000 w. Wis Engr—Dec., 1907. No. 89512 D.

The Early Days of the Dynamo. Sydney F. Walker. Gives a résumé of the early history of the development of this machine. 3000 w. Elec Rev, N Y—Jan. 11, 1908. No. 89471.

Induction Motors.
Magnetic Leakage in Induction Motors. R. E. Hellmund. A discussion of the factors which influence the leakage coefficient without causing an appreciable change in the main flux of the motor. 3000 w. Elec Wld—Jan. 25, 1908. No. 89822.

Insulation.
Cellulose Acetate as Wire Insulation. R. Fleming. Extracts from an article in the Gen. Elec. Rev. Gives a comparison with silk and cotton insulation. 700 w. Eng News—Jan. 16, 1908. No. 89568.

Railway Motors.
A Single-Phase Railway Motor. E. F. Alexanderson. Describes a "series repulsion motor," and its operation, stating the advantages claimed. Diagrams. 4500 w. Pro Am Inst of Elec Engrs—Jan., 1908. No. 89326 D.

Repulsion Motors.
Control of Repulsion Motors by Brush Displacement. K. Schnetzler. From Elektrotech. Zeitschr. Considers how far this method of regulation may be regarded as satisfactory. 1700 w. Elect'n, Lond—Jan. 3, 1908. No. 89518 A.

Single Phase.
See Railway Motors, under DYNAMOS AND MOTORS.

ELECTRO-CHEMISTRY.

Electrolytic Assaying.
See Assaying, under MINING AND METALLURGY, COPPER.

Electro-Metallurgy.
See same title, under MINING AND METALLURGY, IRON AND STEEL; and Lead Refining, under MINING AND METALLURGY, LEAD AND ZINC.

Electro-Plating.
A Study of the Causes of Impure Nickel Plate with Special Reference to the Iron. D. F. Calhane and A. L. Gammage. Describes experiments made to ascertain the amount of iron present, and to study the conditions that influence its deposition. 2500 w. Jour Worcester Poly Inst—Jan., 1908. No. 89697 C.

ELECTRO-PHYSICS.

Alternating Currents.
A Plea for the Physical Treatment of Alternating-Current Phenomena. Lamar Lyndon. Shows how alternating-current phenomena may be explained and made as simple as continuous-Current. Also editorial. 2500 w. Elec Rev, N Y—Jan. 11, 1908. No. 89472.

Demagnetization.
On the Best Method of Demagnetizing Iron in Magnetic Testing. Charles W. Burrows. Outlines a plan whereby a specimen may be freed from all effects of previous magnetic treatment and so be

brought to a standard condition. 12500 w. Bul Bureau of Stand—Jan., 1908. No. 89692 N.

Magnetic Alloy.

Heusler's Magnetic Alloy. A. D. Ross. Slightly abridged paper read before the Roy. Soc. of Edinburgh. An account of tests carried out. 1000 w. Elect'n, Lond —Dec. 27, 1907. No. 89380 A.

Solenoids.

Solenoids in Series with Resistance. Charles R. Underhill. Considers the resistance in the circuit external to the resistance of the winding. 1500 w. Elec Wld—Jan. 18, 1908. No. 89625.

GENERATING STATIONS.

Central Stations.

New Turbine Station of the Fall River Electric Light Company. Illustrated detailed description. 1800 w. Elec Wld— Jan. 25, 1908. No. 89821.

The Cos Cob Power Plant of the New York, New Haven & Hartford Railroad. E. H. Coster. Illustrated detailed description of the plant supplying power for the division between New York and Stamford. 5000 w. Elec Jour—Jan., 1908. No. 89678.

The Northern Colorado Power Company. · N. A. Carle. Brief illustrated description of a steam-power plant located in the coal fields at Lafayette, and distributing current by 80 miles of transmission lines. 1500 w. Min Sci—Jan. 9, 1908. No. 89492.

Plant of the Terre Haute (Ind.) Traction & Light Co. Illustrated description of a new turbine plant with high-tension transmission for interurban railway service and mercury arc rectifier for street lighting. 4000 w. Engr, U S A—Jan. 15, 1908. No. 89590 C.

The Condition and Development of the Mainz City Electric Plant (Die Beschaffenheit und Entwicklung des städtischen Elektrizitätswerkes Mainz). Herr Furkel. An illustrated description of this municipal plant. Serial. 1st part. 1200 w. Elektrotech Zeitschr—Dec. 5, 1907. No. 89783 D.

See also Gas Engines, under MECHANICAL ENGINEERING, COMBUSTION MOTORS.

Compensators.

See Voltage Drop, under TRANSMISSION.

Design.

Management and Efficiency of the Designing Department, and Its Bearing Upon the First Cost and Economy in Operation of Steam-Electric Power Plants. Frank Koester. 2200 w. Elec Rev, N Y—Jan. 18, 1908. No. 89931.

Hydro-Electric.

Hydro-Electric Power at Duluth. Dwight E. Woodbridge. Illustrates and describes the Great Northern Power Company's developments. 4000 w. Ir Age— Jan. 2, 1908. No. 89290.

Power Development of the Kootenay River for the West Kootenay Power and Light Company, Limited. Robert A. Ross and Henry Holgate. Illustrated description of power development at Upper Bonnington Falls. 3000 w. B C Min Rec— Nov, 1907. No. 89698 B.

Recent Power Developments at Montpelier, Vt. Illustrated description of the high-tension systems of two companies furnishing current for lighting and power. Gives data of a test made upon motors used in a granite shed, at Barre, Vt. 3000 w. Elec Wld—Jan. 4, 1908. No. 89423.

The Augst-Wylen Water-Power Plant (Die Wasserkraftanlage Augst-Wylen). Illustrated description of this plant on the Rhine near Basle, Switzerland. 2500 w. Schweiz Bau—Dec. 14, 1907. No. 89744 B.

The Hohenfurth Electric Plant (Das Elektrizitätswerk Hohenfurth). P. Ehrlich. An illustrated description of this plant in Austria on the Moldau. 2300 w. Zeitschr d Oest Ing u Arch Ver—Dec. 27, 1907. No. 89753 D.

See also High Tension, under TRANSMISSION; and Turbine Plants, under MECHANICAL ENGINEERING, HYDRAULIC MACHINERY.

Isolated Plants.

The Turbine Plant for the Shops of the Technical High School at Charlottenburg (Die Dampfturbinenanlage des Maschinenbaulaboratoriums der kgl. Technischen Hochschule Charlottenburg). E. Josse. Illustrated detailed description of a plant for generating electric power for use in the extensive laboratories and shops of this institution. Serial. 1st part. 2500 w. Zeitschr f d Gesamte Turbinenwesen —Dec. 10, 1907. No. 89773 D.

See also Mechanical Plants, under MECHANICAL ENGINEERING, POWER AND TRANSMISSION.

Operation.

The Operation of a Small Electric Plant. W. H. Wakeman. Suggestions in regard to motors and lamps. Ills. 2500 w. Elec Wld—Feb. 1, 1908. No. 89917.

Noisy Operation of Dynamos. Nathaniel R. Craighill. Shows how to determine whether a machine will run noisy, and how to remedy the trouble. 700 w. Elec Wld—Feb. 1, 1908. No. 89916.

Convenient Tests for Central Station Operators. W. M. Hollis. Outlines methods for measuring insulation resistance for locating grounds, etc. 1200 w. Elec Wld—Feb. 1, 1908. No. 89915.

We supply copies of these articles. See page 1103.

Rates.
Rates and Systems of Charging. Discusses systems of charging in England and gives the writer's idea of an ideal rate. 1800 w. Elec Rev, Lond—Dec. 27, 1907. No. 89375 A.

Switchboards.
See same title, under TRANSMISSION.

LIGHTING.

Arc Lamps.
The New Type of Arc Lamp (Nouveau Type de Lampe à Arc). H. Guérin. Discusses the Bremer, Siemens and Blondel types of metallic electrode arc lamps giving photometric results. Ills. 3000 w. Génie Civil—Dec. 7, 1907. No. 89725 D.

Illumination.
Distribution of Light. Otto Foell. Read before the Ill. Engng. Soc. Shows what an ideal distribution curve should be, and what can be accomplished with a proper light distribution. 2200 w. Elec Rev, N Y—Jan. 4, 1908. No. 89338.

General Office Building of the Philadelphia Electric Company. Illustrated description of the building and its equipment, especially the lighting. 2800 w. Elec Wld—Jan. 11, 1908. No. 89481.

Lighting of the New Plaza Hotel. H. Thurston Owens. An illustrated detailed description of the lighting installation of a hotel costing nearly $13,000,000. 2000 w. Elec Wld—Jan. 4, 1908. No. 89416.

Artistic Illumination—Murray's Restaurant, New York City. H. Thurston Owens. Illustrated description of an elaborately lighted restaurant. 1400 w. Elec Wld—Feb. 1, 1908. No. 89912.

Incandescent Lamps.
Manufacture of Incandescent Lamps. George Loring. Read before the Ohio Soc. of Mech., Elec., and Steam Engrs. Describes the process in detail. 1800 w. Engr, U S A—Jan. 15, 1908. No. 89592 C.

Some Facts Regarding Metal-Filament and Carbon-Filament Lamps. George Loring. Shows comparative results of carbon, tantalum, and tungsten filament lamps, and advantage of substituting metal-filament for carbon-filament lamps. 1200 w. Elec Wld—Jan. 4, 1908. No. 89417.

Photometry.
The Utilization of Photometric Data. William E. Wickenden. An explanation of the principles, methods and tables by which such data may be made useful. 1500 w. Wis Engr—Dec., 1907. No. 89513 D.

A New Universal Photometer. Clayton H. Sharp and Preston S. Millar. Illustrates and describes an instrument designed by the writers, and its performance. 2200 w. Elec Wld—Jan. 25, 1908. No. 89823.

Street.
Street Lighting in 1907. Alton D. Adams. A review of progress in England and America. 3000 w. Munic Jour & Engr—Jan. 1, 1907. No. 89298.

Street Lighting and New Illuminants. Dr. Louis Bell. Remarks on the gain in efficiency and related subjects. 1700 w. Elec Wld—Jan. 11, 1908. No. 89483.

MEASUREMENT.

Cable Testing.
See Cables, under TRANSMISSION.

Galvanometer.
Direct-Reading Galvanometer Scales. J. Rymer-Jones. Gives a method which saves time and calculation. 1000 w. Elec Rev, Lond—Jan. 10, 1908. No. 89645 A.

Instruments.
A New Recording Electrical Voltmeter and Ammeter of Great Sensitiveness and Accuracy. Illustrated description of a milli-voltmeter recently devised by Prof. Wm. H. Bristol. 600 w. Eng News—Jan. 2, 1908. No. 89274.

Instrument Testing.
See Potentiometer, under MEASUREMENT.

Potentiometer.
A Deflection Potentiometer for Voltmeter Testing. Shows what features should be incorporated in an instrument of this kind for such work as the testing of voltmeters, and describes a type of instrument recently constructed. 6500 w. Bul Bureau of Stand—Jan., 1908. No. 89693 N.

Reactance.
See same title, under TRANSMISSION.

Resistance.
On the Comparison of Resistances. F. E. Smith. Critical discussion of the universal slide wire bridge devised by Dr. Drysdale, and the corrections necessary where great accuracy is required. Also brief communication from Dr. C. V. Drysdale. 3000 w. Elect'n, Lond—Jan. 10, 1908. No. 89647 A.

Wattmeters.
The Design of Prepayment Watt-Hour Meters. Arthur Pestel. Considers the more important points involved, indicating the way the difficulties may be overcome. Diagrams. 2500 w. Elec Wld—Jan. 18, 1908. No. 89624.

TRANSMISSION.

Cable Heating.
See Wire Heating, under TRANSMISSION.

Cables.
High-Tension Cables and Their Testing (Ueber Hochspannungs-Kabel und ihre Prüfung). C. Feldmann and J. Herzog. Discusses cable sections, dielectric materials, testing methods, etc. Ills.

2000 w. Elektrotech Zeitschr—Dec. 5, 1907. No. 89782 D.

See also Insulation, under TRANSMISSION.

Conduits.

Municipal Conduit System of the City of Baltimore, Md. Illustrated description. 2000 w. Elec Wld—Jan. 4, 1908. No. 89415.

Fault Finding.

The Earth-Potential Method of Fault Finding. W. A. Toppin. Shows how this method may be applied on three-wire networks, two-wire networks, D. C. or A. C., and traction feeders, provided that current can be passed through the faulty cable to earth. 500 w. Elec Rev, Lond—Jan. 3, 1908. No. 89516 A.

Four-Wire.

Four-Wire Distribution. Harold H. Jowers. Brief explanation of the advantages claimed for this system. 600 w. Elec Rev, Lond—Dec. 27, 1907. No. 89-376 A.

High-Tension.

High-Tension Energy Transmission in Peru. An illustrated description of the stations and their equipment, the lines and of some of the difficulties. 1800 w. Elec Wld—Feb. 1, 1908. No. 89911.

Discussion on High-Tension Transmission Papers, at Niagara Falls, N. Y., June 26, 1907. 5000 w. Pro Am Inst of Elec Engrs—Feb., 1908. No. 89900 D.

Inductance.

The Self and Mutual Inductances of Linear Conductors. Edward B. Rosa. A discussion of the formulæ used, with the derivation of new expressions, and illustrated by examples. 6500 w. Bul Bureau of Stand—Jan., 1908. No. 89694 N.

Insulation.

The Dielectric Strength of Insulating Materials and the Grading of Cables. Alexander Russell. Discusses the laws of disruptive discharge, the methods of measuring the dielectric strengths of gases, liquids, and solids, and the electric stresses on the insulating materials of a single-core cable, with special reference to the grading of cables. 7800 w. Inst of Elec Engrs—Nov. 14, 1908. No. 89859 N.

Insulators.

High Voltage Insulator Manufacture. Walter T. Goddard. An illustrated article describing the process of manufacture, testing, etc., and discussing materials, design, and related matters. 5500 w. Can Soc of Civ Engrs—Dec. 19, 1907. No. 89603 N.

High Tension Insulators from an Engineering and Commercial Standpoint. C. E. Delafield. Read before the Can. Elec. Assn. Briefly considers the present requirements of high tension transmission, discussing the design of porcelain insulators, their manufacture and testing. 3300 w. Can Elec News—Jan., 1908. No. 89605.

Lines.

See High-Tension, under TRANSMISSION.

Poles.

The Life and Efficiency of Wooden Poles in Austria (Lebensdauer und Gebrauchswert hölzerner Leitungsmaste in Oesterreich). Robert Nowotny. The experience of the State Telegraph Department with natural and treated poles. Ills. 3500 w. Elektrotech u Maschinenbau—Dec. 1, 1907. No. 89747 D.

Protective Devices.

The Application of Protective Devices to Trolley Lines (Application des Appareils de Protection aux Lignes de Trolley). M. Mariage. A discussion of trolley accidents and a description and comparison of the leading types of protective devices. Ills. 4200 w. Bul Soc Int d'Elec'ns—Dec., 1907. No. 89708 F.

Appliances for Guarding against the Consequences of Ruptures of Overhead Wires (Appareils de Protection contre les Conséquences des Ruptures des Conducteurs Aériens). M. Barré. An illustrated description of various types of suspension devices which automatically break the circuit in case of accident. 4000 w. Bul Soc Int d'Elec'ns—Dec., 1907. No. 89707 F.

Protective Relays.

Protective Relays. M. C. Rypinski. Considers the various classes and the conditions under which they operate. Ills. 2500 w. Elec Jour—Jan., 1908. Serial. 1st part. No. 89581.

Reactance.

The Slide Rule in the Calculation of Magnetic Reactance. M. S. Howard. Gives a slide-rule method for calculating the reactance of alternating-current transmission lines. 1000 w. Elec Wld—Jan. 18, 1908. No. 89626.

Regulations.

Rules Issued by the Verband Deutscher Elektrotechniker for Overhead Wires, Cables, and Electric Wiring. Gives the revised rules that come into force Jan. 1, 1908. 2800 w. Elect'n, Lond—Dec. 27, 1907. No. 89377 A.

Rotary Converters.

Some Developments in Synchronous Converters. Charles W. Stone. Points out some of the more important improvements made in the last few years. Ills. 2500 w. Pro Am Inst of Elec Engrs—Feb., 1908. No. 89897 D.

Some Features of Railway Converter Design and Operation. J. E. Woodbridge. Considers six-phase versus three-phase converters; the alternating-current starting of synchronous converters; and

We supply copies of these articles. See page 1103.

the compounding of converters. Ills. 6500 w. Pro Am Inst of Elec Engrs—Feb., 1908. No. 89898 D.

See also A. C. Dynamos, under DYNAMOS AND MOTORS.

Switchboards.

Electrically Operated Switchboards. S. Q. Hayes. Deals with the panels, desks, pedestals, etc., illustrating and describing both recent and older equipments, to show the progress. 3000 w. Elec Wld—Feb. 1, 1908. No. 89914.

Three-Phase.

See Voltage Drop, under TRANSMISSION.

Voltage Drop.

Voltmeter Compensation for Drop in Alternating-Current Feeder Circuits. William Nesbit. Considers methods, describing types of compensators and discussing their adjustment. 3500 w. Elec Jour—Jan., 1908. No. 89680.

Inductive Voltage Drop on Three-Phase Transmission, with the Conductors Lying in the Same Plane. Alfred Still. Explains a practical method of calculating the inductive effects when the conductors are arranged in line at a stated distance. 1600 w. Elec Engr, Lond—Jan. 10, 1908. No. 89644 A.

Wire Heating.

Temperature Rise of Conductors. C. C. Badeau. Gives a chart showing the relation between temperature rise and current. Also editorial. 1200 w. Elec Wld—Jan. 11, 1908. No. 89482.

Wire Suspension.

Pringle's Tangential System of Supporting Overhead Trolley Wires. Illustrated description of a system in which each suspension is made by a wire attached to two ears a considerable distance apart. 3500 w. Elect'n, Lond—Jan. 10, 1908. No. 89648 A.

See also Protective Devices, under TRANSMISSION.

MISCELLANY.

Government Ownership.

State Control of Electricity Supply in Germany. Editorial discussion of the proposal to make electricity supply in Germany a monopoly of the Imperial treasury. 1000 w. Engng—Jan. 10, 1908. No. 89798 A.

Review of 1907.

The Electrical Industries in 1907. A brief review of the important developments of the past year. 3500 w. Elec Rev, N Y—Jan. 11, 1908. No. 89470.

Varnishes.

Heat-Dissipating Varnishes. A. R. Warnes. Brief notes on their composition and heat-dissipating properties. 1200 w. Elec Rev, Lond—Jan. 17, 1908. No. 89866 A.

INDUSTRIAL ECONOMY

Apprenticeship.

Grand Trunk Apprentice System. Information concerning the entrance requirements, training, wages, etc. Ills. 2500 w. Am Engr & R R Jour—Jan., 1908. No. 89318 C.

The Training of Engineering Apprentices. Commences a general review of the methods of training employed by various firms. Ills. 3500 w. Engr, Lond—Jan. 17, 1908. Serial, 1st part. No. 89871 A.

Copper Trade.

See Trade, under MINING AND METALLURGY, COPPER.

Cost Keeping.

A Simple System of Recording Shop Costs. C. J. Redding. Outlines a system used successfully in an English works employing three thousand men. The use of a "shop summary" card is explained. 4500 w. Engineering Magazine—Feb., 1908. No. 89881 B.

Depreciation.

See Valuation, under CIVIL ENGINEERING, WATER SUPPLY.

Education.

A Plea for the Broader Education of the Chemical Engineer. Address by Clifford Richardson before the American Assn. for the Advancement of Science. 5500 w. Eng Rec—Jan. 4, 1908. No. 89358.

The Education of Young Men for Mechanical Pursuits. R. T. Crane, Sr. An explanation of the writer's views on manual training, technical education, trade schools, and training of apprentices. General discussion. 18700 w. Pro W Ry Club—Dec. 17, 1907. No. 89546 C.

The Relations of Technical Education to Industry. Charles Buxton Going. A criticism of the preparation given in technical schools, with suggestions for bringing it more in harmony with practical industry. 3000 w. Columbia Univ Qr—Dec., 1907. No. 89543 D.

The Best Engineering Education. Charles F. Scott. Introduction to a digest of previous papers on engineering education, with a view to renewing discussion. 4500 w. Pro Am Inst of Elec Engrs—Jan., 1908. No. 89328 D.

. We supply copies of these articles. See page 1103.

Electrical Engineering Education. Charles P. Steinmetz. A criticism of some defects. 2500 w. Pro Am Inst of Elec Engrs—Jan., 1908. No. 89329 D.

The Place of the Laboratory in the Training of Engineers. A. L. Mellanby. Discusses the general course of training for engineers, particularly laboratory work in the colleges of the United Kingdom. Ills. 3000 w. Inst of Engrs & Shipbldrs in Scotland—Nov. 19, 1907. No. 89861 N.

Laboratory Instruction in Testing Materials. Edgar Marburg. A summary of the methods adopted in the Engineering Department of the University of Pennsylvania. 2000 w. Cement Age—Jan., 1908. No. 89685 C.

Higher Education in South America. William R. Shepherd. An account of the universities and institutions for higher education and matters relating to them. 2200 w. Columbia Univ Qr—Dec., 1907. No. 89545 D.

History of the Engineering Schools of Columbia University. A series of articles by C. F. Chandler, William H. Burr, Ira H. Woolson, Francis B. Crocker, and Charles Edward Lucke, reviewing the history of the various departments. Ills. 11000 w. Columbia Univ Qr—Dec., 1907. No. 89544 D.

See also Apprenticeship, under INDUSTRIAL ECONOMY.

Eight-Hour Day.

The Reduction of the Working Day to Eight Hours (La Réduction de la Journée de Travail à Huit Heures). Maurice Alfassa. A long discussion of the experiences with the system in French industries and also in other parts of the world. 25000 w. Bul Soc d'Encour—Nov., 1907. No. 89719 G.

Employer's Liability.

The Dire Cost and Best Remedy for Carelessness. Thomas D. West. Read before the Philadelphia Found. Assn. Discusses the subject of accidents due to carelessness, and the best remedy. 3000 w. Ir Age—Jan. 9, 1908. No. 89414.

Industrial Museums.

A New Engineering Museum in Berlin, Germany. Bruno Braunsburger. An illustrated account of this recent addition to Germany's educational institutions, and the exhibits. 1800 w. Eng News—Jan. 9, 1908. No. 89419.

Inventory.

Taking an Inventory. Sterling H. Bunnell. Describes a method of taking a careful shop inventory. 2000 w. Eng News—Jan. 2, 1908. No. 89275.

Iron Trade.

See Trade, under MINING AND METALLURGY, IRON AND STEEL.

Labor.

Trades Unions, Trusts, Employers' Associations (Arbeitsnehmerverbände, Kartelle, Arbeitgeberverbände). Leo Vossen. A discussion of the labor question in German. 2800 w. Stahl u Eisen—Dec. 4, 1907. No. 89733 D.

Pension System.

A Modern System of Pensioning and Insuring Employes. Gives an outline of the system of benefits put into effect by Deere & Co., Moline, Ill. 1500 w. Ir Trd Rev—Jan. 23, 1908. No. 89820.

Purchasing.

A Systematized Purchasing Department. Explains a method for handling buying records, devised by T. P. Artaud, purchasing agent for the Hudson Companies of New York. 2500 w. Ir Age—Jan. 23, 1908. No. 89699.

Shop Management.

See Management, under MECHANICAL ENGINEERING, MACHINE WORKS AND FOUNDRIES.

Stores Keeping.

See Inventory, under INDUSTRIAL ECONOMY; and Locomotive Maintenance, under RAILWAY ENGINEERING, MOTIVE POWER AND EQUIPMENT.

Wages.

See same title, under MINING.

MARINE AND NAVAL ENGINEERING

British Navy.

The Naval Situation. Editorial review of the present condition of the Navy, the vessels built in 1907, and the performance of the vessels. Gives results of steam trials of armored ships during the year. 3000 w. Engng—Jan. 3, 1908. No. 89533 A.

Cable Ship.

A New Cable Ship. Illustrated detailed description of the "Guardian," built in England for an American Company. 1500 w. Engr, Lond—Jan. 3, 1908. No. 89528 A.

Destroyer.

Boilers of H. M. Torpedo-Boat Destroyer "Tartar." Illustrates the vessel and gives illustrated description of the six Thornycroft water-tube boilers. 700 w. Engng—Dec. 27, 1907. No. 89385 A.

Dredges.

An Electrically Operated Sea-Going Steam Dredge. Illustrated description of

We supply copies of these articles. See page 1103.

the sea-going dredge "Thor," designed for electrical operation of all winches and mechanisms. 1500 w. Int-Marine Engng —Feb., 1908. No. 89813 C.

Dry Docks.
Floating Docks. Harry R. Jarvis. Read before the N.-E. Coast Inst. of Engrs. & Shipbldrs. Classifies the various types, giving descriptions, especially of the details of self-docking types. 3500 w. Mech Engr—Jan. 18, 1908. No. 89863 A.

Heating.
The Heating and Ventilating of Ships. Sydney F. Walker. The present number discusses methods of heating available. Diagrams. 3500 w. Int Marine Engng— Feb., 1908. Serial, 1st part. No. 89814 C.

Lusitania.
The Cunard Liner Lusitania. Illustrated description of the construction of the vessel and its equipment, and report of performance. 15600 w. Jour Am Soc of Nav Engrs—Nov., 1907. No. 89600 H.

Marine Transport.
Presidential Address of Mr. John Ward. Deals especially with the evolution of marine transport. 13500 w. Trans Inst of Engrs & Shipbldrs in Scotland— Oct. 22, 1907. No. 89686 N.

The Ocean Carrier. J. Russell Smith. Shows how certain types of carriers have been developed while other types have been discarded, and discusses the conditions that have affected modern ocean commerce. 5400 w. R R Gaz—Jan. 10, 1908. Serial, 1st part. No. 89446.

Mechanical Draft.
Mechanical Draft in Marine Practice. Walter B. Snow. Briefly reviews early applications, the process of development, American naval practice, and related matters. 3000 w. Int Marine Engng—Feb., 1908. Serial, 1st part. No. 89815 C.

Refrigeration.
Heat Insulation on Shipboard (Isolierungen an Bord). H. Schoeneich. A discussion of cold-storage for provisions on shipboard. Ills. Serial, 1st part. 2500 w. Schiffbau—Dec. 25, 1907. No. 89765 D.

Resistance.
The Coefficient k in the Formula for the Resistance of Flat and Prismatic Moving Bodies (Der Beiwert k in der Formel für den Wasserwiderstand bewegter plattenförmiger und prismatischer Körper). H. Engels and Fr. Gebers. A record of tests to establish the value of the coefficient k in the formula $W = k \gamma F v^2/2g$. Ills. Serial, 1st part. 5500 w. Schiffbau—Dec. 25, 1907. No. 89764 D.

Review of 1907.
Shipbuilding and Engineering in 1907. A review showing a decrease in tonnage, diminution in profits, and dismissal of workers. 9000 w. Engng—Jan. 3, 1908. No. 89526 A.

Speed.
The Fastest Ships in the World. Brief review of the growth in speed since the beginning of the century. 1500 w. Int Marine Engng—Feb., 1908. No. 89812 C.

Steam Boilers.
The Water-Tube Boiler with Schulz Superheater (Wasserkessel mit Ueberhitzer nach Patent Schulz). D. Dietrich. An installation suited to turbine steamers is described. Ills. 3500 w. Schiffbau— Dec. 11, 1907. No. 89763 D.

Steam Engines.
Engine Efficiency and Effective Turning Moments and Their Experimental Determination. Carl A. Richter. Trans. from the German. Discusses methods of measurement and apparatus used. Ills. 12000 w. Jour Am Soc of Nav Engrs— Nov., 1907. No. 89598 H.

Inertia and Torsional Stresses and Pressures on Bearings, Together with an Investigation of the Lubrication Problem, of the Port Main Engine U. S. S. Tennessee. Lewis Hobart Kenney. Gives computations and investigations made to determine the causes of the heating of the crank-pin and main bearings. 4000 w. Jour Am Soc of Nav Engrs—Nov., 1907. No. 89601 H.

See also British Navy and Steamships, under MARINE AND NAVAL ENGINEERING.

Steamships.
The Twin-Screw Allan Liner "Corsican." Drawings and illustrations of the machinery of this large merchant steamer, with brief notes. 300 w. Engng—Jan. 10, 1908. No. 89655 A.

See also Lusitania, under MARINE AND NAVAL ENGINEERING.

Steam Turbines.
The Turbine Plant of the "Mauretania." Illustrated description of the essential features of these large steam turbines. 3000 w. Power—Jan. 7, 1908. No. 89396.

Description and Test of 27-Inch Curtis Turbine for 50-Foot U. S. Navy Cutter. Paul E. Dampman. Ills. 1200 w. Jour Am Soc of Nav Engrs—Nov., 1907. No. 89599 H.

See also British Navy, and Lusitania, under MARINE AND NAVAL ENGINEERING.

Torpedo Boats.
The Development of Torpedo Craft. Archibald S. Hurd. Discusses the efficiency of the torpedo boat as revealed during the late war in the Far East, and describes recent vessels, especially those of the British Navy. Ills. 5000 w. Cassier's Mag—Jan., 1908. No. 89452 B.

Ventilation.
See Heating, under MARINE AND NAVAL ENGINEERING.

MECHANICAL ENGINEERING

AUTOMOBILES.

Berliet-Mixte.

The Six-Cylinder Berliet Marche Mixte. Illustrates and describes this engine, which is primarily a six-cylinder petrol motor, with its cylinders cast in pairs and valve chambers on the left-hand side of the engine. 1500 w. Autocar—Jan. 18, 1908. No. 89857 A.

Cabs.

The Development of Motor Cab Traffic in 1907 in Paris, London and Berlin (Die Weiterentwickelung des Motordroschken-verkehrs im Jahre 1907 in Paris, London und Berlin). Herbert Bauer. A review of experiences and financial returns. Ills. 5500 w. Zeitschr d Mit Motorwagen-Ver—Dec. 31, 1907. No. 89758 D.

Certus.

The "Certus" Friction-Driven Cars. Illustrated detailed description of gearless vehicles having a gradually variable speed mechanism. 1500 w. Auto Jour—Jan. 18, 1908. Serial, 1st part. No. 89854 A.

Commercial Vehicles.

See Cabs, Farm Motors, and Omnibuses, under AUTOMOBILES.

Control.

The Positive Driving System for Automobiles. M. C. Krarup. Discusses the principal causes of automobile accidents and presents the advantages of the system named. 4000 w. Ir Age—Jan. 2, 1908. No. 89291.

Electric.

Progress in Electric Automobile Construction (Ueber Fortschritte im Bau von Elektromobilen). W. A. Th. Müller. The first part of the serial discusses arrangement of motor and batteries, driving by front or rear axle, etc. Ills. Serial, 1st part. 5000 w. Elektrotech Zeitschr—Dec. 12, 1907. No. 89785 D.

See also Omnibuses, under AUTOMOBILES.

Farm Motors.

A New Agricultural Oil Motor. An illustrated article showing the possibilities of the kerosene traction engine on the farm. 1500 w. Sci Am Sup—Jan. 11, 1908. No. 89450.

Flywheels.

The Gyroscopic Action of a Fly-wheel. Roger B. Whitman. Explains gyroscopic action and its application to the flywheel of automobile engines as a cause of accidents. 1800 w. Automobile—Jan. 16, 1908. No. 89578.

Germany.

Automobiles in Germany (Automobile-

verkehr in Deutschland). Dr. R. Bürner. Gives statistics of the use of automobiles by provinces, the uses to which they are put, power, etc. 4000 w. Zeitschr d Mit Motorwagen-Ver—Dec. 31, 1907. No. 89-759 D.

Ignition.

A New Step in Magneto Development. Ernest Coler. Illustrated description of the new Miller magneto. 1600 w. Automobile—Jan. 2, 1908. No. 89312.

Morriss.

The Morriss Steam Car. Illustrated detailed description. 2000 w. Auto Jour—Dec. 28, 1907. No. 89372 A.

Motors.

An Interesting Chapter from the Past. Pol Ravigneaux. Trans. from *La Vie Automobile*. A short résumé of Daimler's career, describing his internal-combustion motors. 1700 w. Automobile—Jan. 16, 1908. No. 89579.

Omnibuses.

Petrol-Electric Motor Omnibuses. Describes the Hart-Durtnall polyphase alternating-current system. 2000 w. Elec Engr, Lond—Jan. 3, 1908. No. 89515 A.

The Hallford Petrol Electric 'Bus. Illustrated detailed description of a vehicle embodying the Stevens system and the "S. B. & S." drive. 2200 w. Auto Jour—Jan. 18, 1908. Serial, 1st part. No. 89-855 A.

The Petrol-Electric Automobile (Le Automobile Petroleo - Elettriche). Describes the leading vehicles of this type, referring principally to omnibuses. Ills. 5000 w. Ing Ferroviaria—Dec. 16, 1907. No. 89704 D.

Reo.

The 18-H.P. Reo Car. Illustrated detailed description of a moderate priced car. 2200 w. Auto Jour—Jan. 4, 1908. Serial, 1st part. No. 89514 A.

Testing.

A Critical Examination of Some Methods of Measuring the Tractive Power of Automobiles (Examen Critique de quelques Méthodes de Mesure de la Puissance Utile des Voitures Automobiles). J. Auclair. Ills. 9500 w. Bul du Lab d'Essais—No. 12, 1907. No. 89714 F.

Weigel.

The 25 H.P. Weigel Car. Illustrated detailed description. 2300 w. Autocar—Jan. 11, 1908. No. 89640 A.

COMBUSTION MOTORS.

Crankshafts.

Crankshafts for Internal-Combustion Engines. George F. Fuller. Considers

We supply copies of these articles. See page 1103.

improvements in design and manufacture. Ills. 1200 w. Cassier's Mag—Jan., 1908. No. 89458 B.

Fuels.

The Use of Water and Steam in Internal-Combustion Engines. Henry Henderson. Shows the ideas and methods adopted by past and present inventors and manufacturers. 5000 w. Cassier's Mag —Jan., 1908. No. 89459 B.

Gas and Oil Engine Diagrams and Fuel Data. Peter Eyermann. Diagrams showing how various fuels act in internal combustion engines, with tables giving information on these fuels. 600 w. Power— Jan. 14, 1908. Serial, 1st part. No. 89541.

Gas-Engine Regulation.

Gas-Engine Regulation for Direct-Connected Units. Charles E. Lucke. Reviews some important papers on steam engine regulation, some of the phases being common to gas engines, giving a summary of the variations of conditions entering into the regulation problem and the conclusions of the study. 8500 w. Pro Am Inst of Elec Engrs—Feb., 1908. No. 89895 D.

Gas Engines.

The Gas Engine. Cecil P. Poole. Part first gives an illustrated explanation of the elementary principles, pressures and temperatures, cooling and heat loss, valves and valve-gear. 6500 w. Power— Jan. 14, 1908. Serial, 1st part. No. 89542.

The Explosion of Gases. John Batey. A discussion of the mechanical and thermodynamic effects of the explosions of gas in the cylinder of the gas engine. 2500 w. Cassier's Mag—Jan., 1908. No. 89453 B.

On the Measurement of Temperatures in the Cylinder of a Gas Engine. Prof. H. L. Callendar, and Prof. W. E. Dalby. An illustrated account of investigations and statement of conclusions. 6000 w. Engng—Dec. 27, 1907. No. 89387 A.

Largest Gas Engines for Electrical Work. Cecil P. Poole. Illustrates and describes construction details and operating data of the gas engines at the Martin station of San Mateo (Cal.) Power Co. 3000 w. Power—Jan. 14, 1908. Serial, 1st part. No. 89538.

300 Horse-Power Suction Gas Plant. Illustrates and describes a late design of twin-cylinder producer-gas engine. 1000 w. Engr, Lond—Dec. 27, 1907. No. 89393 A.

Four 4,000 H.P. Gas Engines. James Cooke Mills. Illustrated description of the installation for the California Gas and Electric Corporations, San Francisco. 700 w. Mach, N Y—Jan., 1908. No. 89345 C.

Gas-Engine Equipment of a Rolling Mill. Illustrated description of gas engines successfully operated on blast-fur-

nace gas, in parallel with steam-driven units, in a large tube mill. 1000 w. Power—Jan. 21, 1908. No. 89665.

See also Fuels, under COMBUSTION MOTORS; Blast-Furnace Gas, under MINING AND METALLURGY, IRON AND STEEL; and Electric Power, under MINING AND METALLURGY, MINING.

Gasoline Engines.

Gasoline Motors for Aeronautical Work. Illustrates and describes recent light weight motors of French design. 2800 w. Sci Am Sup—Jan. 11, 1908. No. 89451.

Gas Producers.

The Jahns System of Transforming Solid Fuel into Gas. Oskar Nagel. Illustrated description of the Jahns producer plant which is used in German and Belgian coal mines. 800 w. Elec-Chem & Met Ind—Jan., 1908. No. 89490 C.

Gas Producers (Les Gazogènes). An illustrated detailed description of the latest types of producers, charging devices, gas meters, and calorimeters. 7500 w. Rev de Mécan—Dec. 31, 1907. No. 89732 E + F.

Governing.

See Gas-Engine Regulation, under COMBUSTION MOTORS.

Oil Engines.

New Type of Internal Combustion Motor. H. Addison Johnston. Illustration, with brief description of an improved type of kerosene or crude oil engine of 60 h.p., fitted with a compressed air starting device. 1500 w. Sci Am Sup—Jan. 25, 1908. No. 89810.

See also Farm Motors, under AUTOMOBILES; and Fuels, under COMBUSTION MOTORS.

Troubles.

Cause and Effect in Engine Troubles. Herbert L. Towle. Considers the possible mechanical and electrical defects of internal-combustion engines and the effects which they produce. Ills. 3000 w. Rudder—Jan., 1908. Serial, 1st part. No. 89395 C.

HEATING AND COOLING.

Cooling Towers.

Cooling Towers. Charles L. Hubbard. Discusses some of the advantages, and describes the principle upon which a cooling tower operates, giving illustrated description of types. 2500 w. Elec Rev, N Y—Jan. 25, 1908. No. 89828.

Fan Blowers.

Variation in Fan Blower Efficiency. Walter B. Snow. Discusses the efficiency of operation in connection with heating and ventilating service. 2500 w. Met Work—Jan. 18, 1908. No. 89604.

We supply copies of these articles. See page 1103.

Hot-Air Heating.

Furnace Heating of a Hotel. Plan and description of a building of eighty-three rooms served by two furnaces with trunk air mains. 1200 w. Met Work—Jan. 4. 1908. No. 89331.

Present Practice in Fan-Blast Heating. A report read before the Am. Soc. of Heat. & Vent. Engrs., giving questions and results of a canvass of hot-blast heating practice. 5800 w. Heat & Vent Mag —Jan., 1908. No. 89844.

Hot-Water Heating.

Data for the Design of Hot-Water Heating Systems. A revised report of the Committee on Hot-Water Heating, presented at the New York meeting of the Am. Soc. of Heat. & Vent. Engrs. 2000 w. Eng News—Jan. 30, 1908. No. 89927.

Industrial Buildings.

See Steam Heating, under HEATING AND COOLING.

Pipe Corrosion.

Relative Corrosion of Wrought Iron and Soft Steel Pipes. T. N. Thomson. Read before the Am. Soc. of Heat. & Vent. Engrs. An illustrated account of investigations with a summary of the conclusions. 3000 w. Heat & Vent Mag— Jan., 1908. No. 89843.

Refrigeration.

Modern Commercial Uses of Refrigeration. The present article deals with brewery refrigeration, illustrating machines used. 6000 w. Ice & Cold Storage—Jan., 1908. Serial, 1st part. No. 89535 A.

Pipe Line Refrigeration. R. H. Tait. Read at Jamestown Exposition Convention. Calls attention to the possibilities of extending the operation of refrigerating plants and so increasing the revenue. 1800 w. Ice & Cold Storage—Jan., 1908. No. 89536 A.

See also same title, under MARINE AND NAVAL ENGINEERING.

Steam Heating.

The Determination of Pipe Diameters in Steam Heating Plants (Bestimmung der Rohrdurchmesser bei Dampfheizungsanlagen). Max Hottinger. Illustrating by curves a mathematical discussion. Serial, 1st part. 1500 w. Gesundheits-Ing— Dec. 21, 1907. No. 89769 D.

Heating and Ventilating in the New Building of L. Löwe & Co., A.-G., Berlin (Heizung und Lüftung im Fabrikneubau der L. Löwe & Co. A.-G., Berlin). Theodor Fröhlich. Illustrates and describes the machinery and installations for heating and ventilating these extensive shops. 3000 w. Gesundheits-Ing—Dec. 28, 1907. No. 89770 D.

Ventilating.

Modern Systems for the Ventilation and Tempering of Buildings. Percival Robert Moses. A general review of the principles and available type of apparatus for furnishing an abundant supply of fresh tempered air. Ills. 5000 w. Engineering Magazine—Feb., 1908. No. 89879 B.

The Unhealthfulness of the Air of Living Rooms and its Improvement by Ozone (Ueber die Gesundheitsschädlichkeit der Luft bewohnter Räume und ihre Verbesserung durch Ozon). Dr. A. Lübbert. Discusses the vitiation of the air in dwellings and an ozone-generating apparatus for its regeneration. Ills. 4400 w. Gesundheits-Ing—Dec. 7, 1907. No. 89766 D.

See also Fan Blower, and Steam Heating, under HEATING AND COOLING; and Heating, under MARINE AND NAVAL ENGINEERING.

HYDRAULIC MACHINERY.

Air Lift.

Artesian Well Pumping by Compressed Air. H. Tipper. Explains the advantages of the air-lift system of delivering artesian water. 1700 w. Eng News— Jan. 16, 1908. No. 89571.

Air-Lift Pumps for Slimes. Illustrates and describes air lifts in use at the East Rand Proprietary Mines, Ltd., reporting tests. 3500 w. Engr, Lond—Jan. 10, 1908. No. 89657 A.

Centrifugal Pumps.

Investigation of Centrifugal Pumps. Clinton Brown Stewart. A discussion of the theory of the centrifugal pump and tests of a six-inch vertical centrifugal pump. Ills. 20000 w. Bul Univ of Wis, No. 173—Part 1. No. 89424 N.

Pumping.

Methods of Pumping Deep Ground Waters. Charles B. Burdick. Reviews the means now in use, their applicability and economy; referring particularly to supplies below suction depth. Discussion. Ills. 13000 w. Jour W Soc of Engrs— Dec., 1907. No. 89549 D.

Pumping Plants.

See Sewage Disposal, under CIVIL ENGINEERING, MUNICIPAL: and Irrigation and Reservoirs, under CIVIL ENGINEERING, WATER SUPPLY.

Pumps.

Hinged Cylinder Head for a Vertical Sinking Pump. A. H. Hale. Illustrated description. 600 w. Power—Jan. 28, 1908. No. 89886.

Turbine Plants.

Safety Devices for Pressure Conduits in Turbine Plants (Sicherheitsvorrichtungen für die Turbinenleitung). Discusses various devices for preventing excessive pressure, particularly the indirect-acting hydraulic safety valve. Ills. 2300

w. Zeitschr f d Gesamte Turbinenwesen —Dec. 30, 1907. No. 80774 D.

Water Hoist.
The Loss of Power in Hoisting Water from Bore Holes (Die bei der Schöpfbewegung in Bohrlöchern entstehende Druckverminderung). Richard Sorge. A record of investigations to determine friction and other losses in this method of raising water. Ills. 3500 w. Glückauf— Dec. 14, 1907. No. 89737 D.

MACHINE ELEMENTS AND DESIGN.

Ball Bearings.
Manufacture and Tests of Double Ball Bearings. Illustrates and describes how Chapman bearings are produced by specialized tools, reporting comparative tests. 2500 w. Am Mach—Vol. 31, No. 4. No. 89804.

The Use of Ball Bearings on Electric Motors. States the principal advantages gained by their use, and considers also the disadvantages, illustrating types. 2500 w. Elec Rev, Lond—Jan. 10, 1908. No. 89646 A.

Ball and Roller Bearings in Practical Operation. Samuel S. Eveland. Discusses their design, giving tests and results obtained, illustrating and describing types, and the methods of manufacture, etc. General discussion. 5800 w. Pro Engrs' Club of Phil—Oct., 1907. No. 89547 D.

Ball Bearings (Kugellager - Konstruktionen). Reviews the use of ball bearings on automobiles illustrating a large number of types of construction. 2700 w. Zeitschr d Mit Motorwagen-Ver—Dec. 15, 1907. No. 89757 D.

Chains.
The Strength of Chain Links (Beanspruchung der Kettenglieder). J. Stieghorst. A mathematical discussion. Ills. 2500 w. Schiffbau—Dec. 11, 1907. No. 89762 D.

Crane Hooks.
A Diagram for Designing Hoisting Hooks. Axel Pedersen. Gives diagram plotted from Prof. Bach's formula, explaining how hooks may be properly proportioned without using elaborate mathematics or tests. 1500 w. Am Mach—Vol. 31, No. 5. No. 89910.

Crankshafts.
See same title, under COMBUSTION MOTORS.

Dies.
A Press Die for Magnet-Head Punchings for Inclosed Arc Lamp. Julius F. A. Vogt. Illustrates and describes a method of laying out multiple follow dies. 1200 w. Am Mach—Vol. 31, No. 5. No. 89909.

Drying Drums.
Design of Rotating Drums. Discusses

the principal points, considered in calculating the strength and proportion of rotating cylinders. Ills. 2500 w. Mach, N Y—Jan., 1908. No. 89344 C.

Gears.
The Safe Working Loads for Gear Teeth. Charles H. Logue. On the importance of proper design, giving formulæ. 3000 w. Am Mach—Vol. 31, No. 3. No. 89564.

The Variation of the Strength of Gear Teeth with the Velocity. Ralph E. Flanders. Some suggestions in regard to proposed tests of the influence of velocity on the safe stress. 2200 w. Mach, N Y— Jan., 1908. No. 89346 C.

The Design of Bevel Gears for Precision Machines (Berechnung von Kegelrädern für Präzisionsmaschinen). C. E. Berck. An illustrated mathematical paper giving formulæ. 1500 w. Zeitschr f Werkzeug—Dec. 25, 1907. No. 89756 D.

Roller Bearings.
Requisites of Practical Roller Bearings. J. F. Springer. How they should be classified; the constructive features which insure serviceability, etc. Ills. 1000 w. Power—Jan. 14, 1908. Serial. 1st part. No. 89539.

See also Ball Bearings, under MACHINE ELEMENTS AND DESIGN.

MACHINE WORKS AND FOUNDRIES.

Brass Castings.
A Troublesome Driving Brass Casting. Louis Luhrsen. An account of difficulties experienced in making a locomotive bearing. 1200 w. Foundry—Jan., 1908. No. 89409.

Bronze Foundry.
An Ornamental Bronze Foundry. Illustrated description of a new plant at Mt. Vernon, N. Y. 1800 w. Foundry— Jan., 1908. No. 89408.

Case-Hardening.
Quick Case - Hardening. Walter J. May. Brief description of process. 600 w. Prac Engr—Jan. 17, 1908. No. 89860 A.

Castings.
Steel Castings by the McHaffie Process. Illustrated description of the methods in use at Chester, Pa. 1000 w. Ir Trd Rev —Jan. 2, 1908. No. 89308.

Development of the Steel Casting Industry. W.- M. Carr. Briefly considers the defects of early attempts, and the reasons for recent advances. 1500 w. Ir Trd Rev—Jan. 2, 1908. No. 89307.

The Szèkely Process of Casting in Metallic Moulds. Illustrated description. 400 w. Engng—Jan. 3, 1908. No. 89525 A.

See also Brass Casting, under MACHINE WORKS AND FOUNDRIES.

Chain Making.

Chain-Stud Recessing Machine. Illustrated description of an interesting machine for chain work. 1100 w. Am Mach—Vol. 31, No. 1. No. 89283.

Cores.

Foundry Cores, Core Sand, and Core-Making Machinery. George H. Wadsworth. Abstract of an address before the Chicago Foundry Foremen. 2000 w. Am Mach—Vol. 31, No. 2. No. 89441.

Cupola Practice.

Melting Iron in the Foundry Cupola. Prof. H. McCormack. Describes work at the Armour Inst. of Technology undertaken to secure data, of economic value, on cupola practice. 1000 w. Elec-Chem & Met Ind—Jan., 1908. No. 89489 C.

Cylinder Grinding.

Machining Gasolene Engine Cylinders. Illustrates and describes the different operations of machining the cylinder and its valve chamber. 1800 w. Am Mach—Vol. 31, No. 4. No. 89800.

Extrusion Process.

Description of the Extrusion or "Squirting" Process for Making Brass Rods and Special Shapes. History with description of the machine and its operation. Ills. 2000 w. Brass Wld—Jan., 1908. No. 89682.

Forging.

See Crankshafts, under COMBUSTION MOTORS.

Foundries.

The Goulds Manufacturing Company's New Foundry. Illustrated description of a plant at Seneca Falls, N. Y., for the production of pump castings. 3500 w. Ir Age—Jan. 2, 1908. No. 89289.

See also Bronze Foundry, under MACHINE WORKS AND FOUNDRIES.

Foundry Design.

Design of the Iron Foundry. Oscar E. Perrigo. Discusses its location and the arrangement of various departments. 2200 w. Foundry—Jan., 1908. No. 89411.

Foundry Furnaces.

Electric Furnaces for the Iron and Brass Foundry. John B. C. Kershaw. Deals with the use of electricity for melting, showing that under certain conditions it may prove economic. Ills. 2500 w. Ir Trd Rev—Jan. 2, 1908. No. 89306.

Foundry Practice.

The Situation of Modern Foundry Practice (Stand des modernen Eisengiessereiwesens). O. Leyde. A review of progress. 3500 w. Stahl u Eisen—Dec. 4, 1907. No. 89730 D.

See also Cupola Practice, under MACHINE WORKS AND FOUNDRIES.

Furnaces.

Fuel Oil for General Shop-Furnace Use. Holden A. Evans. Illustrated description of the construction and operation of furnaces successfully and economically fired with oil at the Mare Island Navy Yard. 5500 w. Am Mach—Vol. 31, No. 5. No. 89907.

See also Foundry Furnaces and Tempering Ovens, under MACHINE WORKS AND FOUNDRIES.

Gear Cutting.

Gear-Cutting Machinery. Ralph E. Flanders. An illustrated article dealing with methods of cutting gear teeth, the machinery and its operation. 6500 w. Mach, N Y—Jan., 1908. Serial, 1st part. No. 89347 C.

A Hobbing Machine and Dividing-Head Tests. Illustrated description of how accurate worm wheels for dividing heads are hobbed in a special machine and the heads afterwards tested. 2000 w. Am Mach—Vol. 31, No. 1. No. 89282.

Gun Making.

Interesting Work at Watervliet. Illustrates and describes backing off hardened taps with a grinder, shaping octagon holes, rifling a gun slope, and other work. 2000 w. Am Mach—Vol. 31, No. 2. No. 89439.

Hardening.

Hardening Methods of a Large Shop. C. R. Clover. Describes methods of work. 1800 w. Am Mach—Vol. 31, No. 4. No. 89803.

Lapping.

Accurate Lapping Operations and Small Tools. Illustrated description of how the members of the mold used in the monotype casting machine are finished accurately by grinding and lapping. 2000 w. Am Mach—Vol. 31, No. 4. No. 89801.

Lathes.

The Gridley Multiple Spindle Automatic. Illustrates and describes a new turret lathe for simultaneous operations on four bars of stock. 2000 w. Ir Age—Jan. 30, 1908. No. 89906.

New Lathes (Neuerungen an Drehbänken). The first part of the serial is devoted to an illustrated description of new lathes recently brought out by the Gebrüder Böhringer, Göppingen. Serial, 1st part. 2000 w. Zeitschr f Werkzeug—Dec. 5, 1907. No. 89754 D.

Machine Tools.

Machine Tools for Gas Engine Construction. Illustrates and describes recent British machine tools in the present number. 2000 w. Cassier's Mag—Jan., 1908. Serial, 1st part. No. 89460 B.

Management.

System in Sheet Metal Work. Illustrated account of a system applied in a shop at Cleveland, O. 2200 w. Met Work—Jan. 11, 1908. No. 89469.

Theory and Practice of Shop and Factory Management. Oscar E. Perrigo.

Fourth of a series of articles on cost keeping and shop management. 3000 w. Ir Trd Rev—Jan. 2, 1908. No. 89309.

See also Tool Rooms, under MACHINE WORKS AND FOUNDRIES; and Inventory, Purchasing, and Stores Keeping, under INDUSTRIAL ECONOMY.

Metallic Molds.
See Casting, under MACHINE WORKS AND FOUNDRIES.

Microscope.
The Microscope in the Manufacturing Plant. F. A. Stanley. Illustrates and describes some of the advantages of the microscope in connection with manufacturing processes. 1200 w. Am Mach—Vol. 31, No. 3. No. 89560.

Milling.
Machining a Turret-Lathe Turning Tool. A. J. Baker. Illustrates and describes some jigs and fixtures that are indexed to set the work at the proper angles while milling and drilling. 1500 w. Am Mach—Vol. 31, No. 1. No. 89284.

Milling Operations on Vise Parts. John Edgar. Illustrates and describes how the different phases of machining are done on the milling machine. 2000 w. Mach, N Y—Jan., 1908. No. 89349 C.

Milling Cutters.
Notes on the Manufacture and Upkeep of Milling Cutters. Dr. H. T. Ashton. Describes a system of manufacture developed particularly to meet the difficulties of high-speed steel. Ills. 3000 w. Inst of Mech Engrs—Dec. 20, 1907. No. 89-382 N.

Molding.
Method of Molding Large Bed Plate. W. W. McCarter. Illustrated description of the method employed in making the bed plate for an anti-friction pulverizer. 1500 w. Foundry—Jan., 1908. No. 89410.

Molding Machines.
Molding with Machinery. Joseph H. Hart. Describes some applications of mechanical devices to molding. 1800 w. Am Mach—Vol. 31, No. 3. No. 89563.

Molds.
See Castings, under MACHINE WORKS AND FOUNDRIES.

Planer.
The Largest Planer in the World. Illustrated description of a machine weighing over 420 tons and requiring more than 200-h.p. to drive it. 3800 w. Am Mach—Vol. 31, No. 1. No. 89281.

Pneumatic Tools.
Pneumatc Power Hammer Tests. Illustrates and describes interesting tests made to show the amount of power consumed in driving three different sizes. 500 w. Engr, Lond—Jan. 17, 1908. No. 89875 A.

Rake Making.
Tools for Making Garden Rakes. Julius F. A. Vogt. Sketches and description. 1000 w. Am Mach—Vol. 31, No. 3. No. 89561.

Shops.
The Creation of a Manufacturing Plant. Walter B. Snow. Illustrated account of the building of the new works of the B. F. Sturtevant Co., Hyde Park, Mass. 2500 w. Columbia Engr—1906. No. 89496 N.

The Establishments of Henschel and Son at Cassel, Rothenditmold and Hattingen am Ruhr (Etablissements Henschel und Sohn, Cassel—Rothenditmold Hattingen am Ruhr). Robert Le Chatelier. A description of these locomotive and heavy machinery shops. Ills. 5500 w. Rev d Métall—Dec., 1907. No. 89716 E + F.

See also same title, under RAILWAY ENGINEERING, MOTIVE POWER AND EQUIPMENT; and under STREET AND ELECTRIC RAILWAYS.

Steam Hammers.
A New Steam Hammer (Ein neuer Dampfhammer). B. Simmersback. Illustrated description. 1600 w. Zeitschr f Werkzeug—Dec. 15, 1907. No. 89755 D.

Tempering Ovens.
Electric Heating and Tempering Ovens (Elektrische Glüh- und Harteöfen). Illustrates and describes types and compares the cost of operation with that of gas-fired ovens. 2000 w. Elektrochem Zeitschr—Dec., 1907. No. 89728 D.

Tool Making.
See Milling, under MACHINE WORKS AND FOUNDRIES.

Tool Rooms.
Tool-room Arrangement and System. William H. Taylor. Discusses the laying out, fitting with racks and boxes, and the general management. 2800 w. Am Mach—Vol. 31, No. 4. No. 89802.

Welding.
Electric Welding. C. B. Auel. Detailed description of the Benardos process as applied in connection with steel castings, pipes and plates. Ills. 2200 w. Elec Jour—Jan., 1908. No. 89679.

MATERIALS OF CONSTRUCTION.

Alloy Steels.
See Columns, under CIVIL ENGINEERING, CONSTRUCTION.

Metallography.
The Microscopic Examination of Metals, Alloys, and Other Opaque Material. William Campbell. Explains the preparation of specimens and the methods used, and what the microscope tells. Ills. 4500 w. Columbia Engr—1906. No. 89497 N.

See also Wheels, under RAILWAY ENGINEERING, MOTIVE POWER AND EQUIPMENT.

Specifications.

Specifications for Metallic Materials (Les Cahiers des Charges pour la Réception des Matières Métalliques). M. G. Charpy. A plea for more careful drawing and more rigid enforcement of specifications for materials of construction. 10800 w. Rev de Métall—Dec., 1907. No. 89715 E + F.

Steel.

Hand Bending Tests. Capt. H. Riall Sankey. Gives a comparison of some hand-bending tests. 900 w. Engng—Dec. 27, 1907. No. 89386 A.

Cold-Rolled and Cold-Drawn Steel Bars. Arthur J. Wood. Reports test results and conclusions. 2000 w. Eng News—Jan. 16, 1908. No. 89570.

The Variability of Steel. Alexander Jude. Abstract of paper read before the Staffordshire Iron & Steel Inst. Points out the great range of variation in test results, and considers the reasons for the variability. Ills. 600 w. Ir & Coal Trds Rev—Dec. 27, 1908. No. 89447 A.

MEASUREMENT.

Dynamometers.

A Magnetic Absorption Dynamometer. C. M. Garland. Illustrated description of a type where the power is transformed into heat in an armature revolving in a magnetic field, and the heat is absorbed by water. 1200 w. Am Mach—Vol. 31, No. 5. No. 89908.

Elastic Limit.

The Elastic Limit and the Testing of Materials. H. Gansslen. Considers the term "elastic limit" and methods of testing. 2000 w. Mach, N Y—Jan., 1908. No. 89348 C.

Gages.

Testing the Lead of Taps and Screws. Eric Oberg. Describes testing the lead by gages; and comparators for the lead of taps and screws. Ills. 1200 w. Mach, N Y—Jan., 1908. No. 89799 C.

Pyrometry.

The Chauvin and Arnoux Thermo-Electric Industrial Pyrometers (Sur les Pyromètres Thermo-Electriques Industriels de MM. Chauvin et Arnoux). M. J. Violle. An illustrated description of the principle and working of this instrument. 2500 w. Bul Soc d'Encour—Nov., 1907. No. 89717 G.

Testing Machines.

Hydraulic Diaphragm Testing-Machines. Briefly reviews the use of these hydraulic gauges, and especially considers the researches of Prof. A. Martens. Ills. 2000 w. Engng—Dec. 27, 1907. No. 89383 A.

Testing Materials.

Testing Materials by Impact on Nicked Bars (Die Kerbschlagprobe im Material-

prüfungswesen). Dr. Ehrensberger. Records in curves and tables the results of extensive tests comparing this with other methods of testing and drawing conclusions as to its efficiency. Ills. Serial, 1st part. 3500 w. Stahl u Eisen—Dec. 11, 1907. No. 89734 D.

See also Elastic Limit, under MEASUREMENT; and Education, under INDUSTRIAL ECONOMY.

Thermometry.

See Gas Engines, under COMBUSTION MOTORS.

POWER AND TRANSMISSION.

Air Compressors.

See Turbo-Compressors, under POWER AND TRANSMISSION.

Compressed Air.

See Tidal Power, under POWER AND TRANSMISSION; Tunnels, under CIVIL ENGINEERING, CONSTRUCTION; and Compressed Air, under MINING AND METALLURGY, MINING.

Electric Driving.

The Electric Driving of Rolling Mills. William T. Dean. An illustrated study of the conditions to be met in the application of electricity, favoring the steam turbine as a generating unit. 4500 w. Ir Trd Rev—Jan. 2, 1908. No. 89305.

The Electrically-Driven 30-Inch Universal Plate Mill of the Illinois Steel Co. Illustrated description of the electric drive for this heavy reversing mill. 5000 w. Ir Trd Rev—Jan. 9, 1908. No. 89430.

See also Armor Plates, under MINING AND METALLURGY, IRON AND STEEL.

Electric Power.

Cost of Electrical Power for Industrial Purposes. John F. C. Snell. Compares the costs of supply from town installations and power companies, and their relation to the cost of production from plant of various kinds installed in factories, discussing the economical possibilities. 9500 w. Inst of Elec Engrs—1907. No. 89671 N.

Lubricants.

See Boiler Waters, under STEAM ENGINEERING.

Lubrication.

See Steam Engines, under MARINE AND NAVAL ENGINEERING.

Mechanical Plants.

Mechanical Plant of the Stuyvesant High School New York. Describes a building to be devoted to instruction in manual training and the mechanical plant to serve it. Ills. 3500 w. Eng Rec—Jan. 18, 1908. Serial, 1st part. No. 89613.

Tidal Power.

Compressing Air by Tidal Power. William A. Webber. Illustrates and describes

We supply copies of these articles. See page 1103.

details of the plant at Rockland, Me. 1200 w. Am Mach—Vol. 31, No. 2. No. 89440.

Turbo-Compressors.
A Review of the Theories of the Turbo-Compressor (Zusammenstellung der Theorien von Turbokompressoren). Rudolf Mewes. A comparison of the aerodynamic and thermodynamic principles of design. Ills. 3000 w. Die Turbine—Dec. 20, 1907. No. 89771 D.

STEAM ENGINEERING.

Boiler Design.
Estimating the Cost of a Return Tube Boiler. F. C. Douglas Wilkes. Shows how the cost is estimated. 4500 w. Boiler Maker—Jan., 1908. No. 89286.

Estimating the Horsepower of Boilers and Engines. A. S. Atkinson. A discussion of some things that affect the Boiler's capacity. 2000 w. Boiler Maker—Jan., 1908. No. 89287.

Boiler Furnaces.
A Boiler Furnace for Burning High-Volatile Coals. Describes a furnace with water-tube boilers, used in the Rocky Mt. region, for burning lignites and similar fuel, which embodies an application of the gas-producer method. 1700 w. Eng Rec—Jan. 11, 1908. No. 89467.

See also Smoke Prevention, under STEAM ENGINEERING.

Boiler Management.
Fuel Economy. L. J. Wing. Read before the Am. Soc. of Heat. & Vent. Engrs. Brief suggestions on the subject of grates and draft. 1200 w. Heat & Vent Engrs—Jan., 1908. No. 89845.

See also Fuels, under STEAM ENGINEERING.

Boiler Plates.
The Formation of Cracks in Boiler Plates (Der heutige Stand der Frage der Rissbildung in Kesselblechen). R. Baumann. Discusses the effects of construction, materials and operation on the question of the cracking of boiler plates. 7700 w. Zeitschr d Ver Deutscher Ing—Dec. 14, 1907. No. 89780 D.

Boilers.
See Destroyer and Steam Boilers, under MARINE AND NAVAL ENGINEERING.

Boiler Waters.
Water for Economical Steam Generation. J. C. William Greth. Gives facts relating to savings in actual water-softening plants. 3500 w. Engineering Magazine—Feb., 1908. Serial, 1st part. No. 89882 B.

Simple Methods of Testing Feed-Water and Lubricants. James E. Noble. Brief description of methods and treatment. 1700 w. Power—Jan. 14, 1908. No. 89540.

The Purification of Feed-Water. Charles L. Hubbard. Considers methods of purifying the water before it is fed to the boilers. Ills. 2500 w. Elec Rev, N Y—Jan. 4, 1908. Serial, 1st part. No. 89337.

Some Notes on the Electrical Purification of Feed Water. Refers to the working of a purifying plant on the Davis-Perrett electrical system, capable of dealing with 8,000 gallons of water per hour. 1500 w. Elec Engr, Lond—Jan. 17, 1908. No. 89864 A.

Condensers.
See Engine Efficiency, under STEAM ENGINEERING.

Cooling Towers.
See same title, under HEATING AND COOLING.

Cylinder Condensation.
Cylinder Condensation and Preventives. Discusses steam jacketing, superheating, and compounding. 2000 w. Engr, U S A—Jan. 1, 1908. No. 89280 C.

Engine Bearings.
See Steam Engines, under MARINE AND NAVAL ENGINEERING.

Engine Design.
See Boiler Design and Thermodynamics, under STEAM ENGINEERING.

Engine Efficiency.
The Influence of Back Pressure on the Steam Consumption of Reciprocating Engines (Der Einfluss des Gegendruckes und der Zwischendampfentnahme auf den Dampfverbrauch von Kolbendampfmaschinen). Chr. Eberle. Gives in curves and tables the results of extensive tests. Ills. Serial, 1st part. 7000 w. Zeitschr d Ver Deutscher Ing—Dec. 21, 1907. No. 89781 D.

Steam Engine Economies. Thomas Hall. Discusses the economy of reducing the pressure for light loads. 1000 w. Sib Jour of Engng—Jan., 1908. No. 89887 C.

Engine Erection.
Lining Up a Horizontal Engine. T. E. O'Donnell. Simple and complete directions. Ills. 3300 w. Engr, U S A—Jan. 1, 1908. No. 89279 C.

Engine Foundations.
Foundations for the Steam Engine. Considers the materials used and the requirements for a good foundation. Ills. 3500 w. Engr, U S A—Jan. 1, 1908. No. 89278 C.

Engine Governing.
See Valve Setting, under STEAM ENGINEERING.

Engine Lubrication.
See Boiler Waters, under STEAM ENGINEERING; and Steam Engines, under MARINE AND NAVAL ENGINEERING.

Engines.
Historical Sketch of the Steam Engine.

An illustrated review of its development and its inventors. 2000 w. Engr, U S A —Jan. 1, 1908. No. 89277 C.

The Large Billet Mill Engine at the Donora Steel Works. Illustrated description of the horizontal girder frame engine to drive the 30-inch billet mill of the Carnegie Steel Co. 1400 w. Ir Trd Rev—Jan. 23, 1908. No. 89818.

A Novel High-Speed Engine. Description and illustrations of an engine designed by C. Lee Cook. The special features are the governor and a collapsible slide valve. 1800 w. Engr, U S A—Jan. 15, 1908. No. 89591 C.

See also Steam Engines and Steamships, under MARINE AND NAVAL ENGINEERING.

Entropy Diagram.
Use of the Entropy Diagram in Engine Tests. Sidney A. Reeve. Directions for constructing an entropy diagram from a steam engine indicator diagram. 2000 w. Power—Jan. 21, 1908. Serial, 1st part. No. 89668.

Exhaust Steam.
See Turbines, under STEAM ENGINEERING.

Flue-Gas Analysis.
A New CO_2 Recorder. C. O. Mailloux. Illustrated description of an improved apparatus for flue-gas analysis, with remarks on the utility of such records, etc. 5400 w. Pro Am Inst of Elec Engrs—Jan., 1908. No. 89324 D.

Fuels.
A Comparison of the Results of 400 Boiler Tests with Coals from Various Parts of the United States. Gives the general conclusions drawn from tests made by the U. S. Geol. Survey at St. Louis. 600 w. Eng News—Jan. 30, 1908. No. 89920.

See Comparison, under MINING AND METALLURGY, COAL AND COKE.

Heat Transmission.
A Study of the Movement of Heat in the Walls of Steam Engine Cylinders (Etude sur le Mouvement de la Chaleur dans les Parois des Cylindres de Machines à Vapeur). F. Thonet. A mathematical discussion. Ills. 7500 w. Rev de Mécan—Dec. 31, 1907. No. 89731 E + F.

Mean Pressure.
Mean Effective Pressure. Fred R. Low. Explains what it means, and how it is computed with and without clearance. 2500 w. Power—Jan. 21, 1908. No. 89666.

Mechanical Draft.
See same. title, under MARINE AND NAVAL ENGINEERING.

Packings.
Labyrinth Packings. Describes a packing adopted by C. A. Parsons to diminish leakage losses in his steam turbine, and gives formula for calculating the discharge through the packing. 2000 w. Engng—Jan. 10, 1908. No. 89651 A.

Plants.
See Central Stations and Isolated Plants, under ELECTRICAL ENGINEERING, GENERATING STATIONS.

Saturated Steam.
Methods of Determination of the Amount of Liquid Water Mechanically Mixed with the Steam of Boilers (Méthodes de Détermination de l'Eau Liquide Mécaniquement Entrainée par la Vapeur des Chaudières). Ills. 3500 w. Génie Civil—Dec. 21, 1907. No. 89727 D.

Smoke Prevention.
How to Burn Illinois Coal Without Smoke. L. P. Breckenridge. Discusses the principles that apply to smokeless furnace construction and operation, illustrating by units in actual operation. Ills. 12000 w. Univ of Ills—Bul. No. 15. No. 89429 N.

Specific Heats.
Specific Heats of Nitrogen, Carbon Dioxide, and Steam. Gives particulars of recent investigations conducted by Prof. L. Holborn and Prof. F. Henning in the Reichsanstalt. 2000 w. Engng—Jan. 3, 1908. No. 89521 A.

Superheating.
Fittings for Superheated Steam. Information from a paper by A. H. Kruesi, presented at the last meeting of the Assn. of Edison Ill. Cos. 1800 w. Elec Wld —Feb. 1, 1908. No. 89918.

See also Cylinder Condensation, under STEAM ENGINEERING: and Steam Boilers, under MARINE AND NAVAL ENGINEERING.

Thermodynamics.
The Practical Significance of the Carnot Cycle. Joseph H. Hart. A simple discussion of the fundamental principles of engine design. 3000 w. Cassier's Mag —Jan., 1908. No. 89455 B.

Turbines.
Mechanical Points in Connection with Steam Turbines. Discusses steam pipe, couplings, bearings, oil service, governing, etc. 4800 w. Engr, Lond—Dec. 27, 1907. No. 89388 A.

The Willans - Parsons Steam - Turbine. Plates and detailed description of these turbines and their construction. 6000 w. Engng—Jan. 3, 1908. No. 89522 A.

An Exhaust Steam Turbine Plant. Henry H. Wait. Illustrated detailed description of a low-pressure turbine equipment at the Wisconsin Steel Company's mill at South Chicago, with report of results. 6500 w. Pro Am Inst of Elec Engrs—Jan., 1908. No. 89323 D.

Low-Pressure Turbines for Exhaust

We supply copies of these articles. See page 1103.

Steam (Niederdruckturbinen für Auspuff-dampf). W. Heym. Discusses their design and operation and describes several leading installations. 3000 w. Die Turbine—Dec. 20, 1907. No. 89772 D.

See also Packings, under STEAM ENGINEERING; Electric Driving, under POWER AND TRANSMISSION; Central Stations and Isolated Plants, under ELECTRICAL ENGINEERING, GENERATING STATIONS; and Steam Turbines, under MARINE AND NAVAL ENGINEERING.

Valve Setting.
Setting the Valves of the Fitchburg Engine. Herbert E. Collins. An illustrated, simple explanation of the valve motion and governor, with clear instructions for setting and adjusting the valves. 2500 w. Power—Jan. 28, 1908. No. 89884.

TRANSPORTING AND CONVEYING.

Cableways.
Aerial Ropeway at Loch Leven. Illustrated description of the ropeway employed in the construction of the dam in this valley. 1200 w. Engr, Lond—Jan. 10, 1908. No. 89659 A.

Cableways (Kabelbahnen). J. André de la Porte and A. Snethlage. Discusses cableways for various types of service and illustrates and describes several of the more important European installations. 9000 w. De Ingenieur—Dec. 14, 1907. No. 89788 D.

The Cableway from Oettingen to Differdingen of the Deutsch - Luxemburg Mining and Smelting Company (Die Drahtseilbahn Oettingen-Differdingen der Deutsch - Luxemburgischen Bergwerks-und Hütten Akt.-Ges.). A. Pietrkowski. Illustrates and describes a cableway for ore transport nearly eight miles long. 2000 w. Glückauf—Dec. 14, 1907. No. 89738 D.

See also Cranes, under TRANSPORTING AND CONVEYING.

Coal Handling.
See Conveyors, under TRANSPORTING AND CONVEYING.

Conveyors.
The Coaling Machinery at Savona Harbor. F. Collischonn. Trans. from *Elek. Kraftbetriebe und Bahnen.* Illustrated description of a station in Italy equipped with the Hunt system of movable elevators and with a cable railway on the same system. 2000 w. Elect'n, Lond—Jan. 17, 1908. No. 89868 A.

See also Cranes, under TRANSPORTING AND CONVEYING.

Cranes.
Crane Design. Alton L. Smith. Briefly outlines the development of lifting devices. 2500 w. Jour Worcester Poly Inst—Jan, 1908. No. 89696 C.

Modern Systems of Hoisting and Conveying. Walter G. Stephan. An illustrated outline of the systems available and the special field of each. 5000 w. Ir Trd Rev—Jan. 2, 1908. No. 89304.

Electrical Power in Railway Goods Warehouses. H. Henderson. A discussion of its advantages. Ills. 3500 w. Inst of Elec Engrs—Jan. 13, 1908. No. 89858 N.

Erection Traveler for the Genesee River Viaduct. Illustrated description of an overhead traveler used for viaduct erection on the Erie System. 1500 w. Eng Rec—Jan. 18, 1908. No. 89609.

See also Crane Hooks, under MACHINE ELEMENTS AND DESIGN.

Elevators.
Boiler Power for Elevators. Charles L. Hubbard. Explains method of computing the power for hydraulic and electric elevators. 1500 w. Power—Jan. 28, 1908. No. 89885.

Ore Handling.
See Cableways, under TRANSPORTING AND CONVEYING; and Docks, under CIVIL ENGINEERING, WATERWAYS AND HARBORS.

MISCELLANY.

Aeronautics.
A Review of Progress in Aerial Navigation. Editorial review of the more important efforts to solve the problems of aërial navigation. 2500 w. Eng News—Jan. 16, 1908. No. 89572.

The Roe Aeroplane. Photographs and descriptions of a full-sized man-carrying machine. 1000 w. Auto Jour—Dec. 28, 1907. No. 89373 A.

New European Aeroplanes and Airships. Illustrates and describes the latest models, and motors. 1500 w. Sci Am—Jan. 18, 1908. No. 89575.

Henry Farman's Record Flight. An illustrated account of the flight that won the Deutsch-Archdeacon prize. 2000 w. Auto Jour—Jan. 18, 1908. No. 89856 A.

Farman Wins $10,000 Prize for Kilometer Flight. Brief illustrated account of the test of the heavier-than-air machine and its success. 1000 w. Automobile—Jan. 16, 1908. No. 89577.

The Farman Aeroplane. A résumé of the flights made with the machine that won the Deutsch-Archdeacon prize. 2000 w. Sci Am Sup—Feb. 1, 1908. No. 89932.

See also Gasoline Engines, under COMBUSTION MOTORS.

Shipping Weights.
Graphic Estimation of Shipping Weights. C. F. Cukor. Gives a method of determining approximate shipping weights of hoisting engines which can be applied to any power-generating outfit. 600 w. Power—Jan. 21, 1908. No. 89669.

We supply copies of these articles. See page 1103.

MINING AND METALLURGY

COAL AND COKE.

Accidents.

Fatal Accidents in Coal Mines. Frederick L. Hoffman. Reports the number of men killed in coal mines of North America in 1906, discussing causes. 2800 w. Eng & Min Jour—Jan. 4, 1908. No. 89403.

The Waste of Life in American Coal Mining. Clarence Hall and Walter O. Snelling. Discusses the loss of life in the mines of the United States as compared with Europe, the causes and remedy. 6500 w. Engineering Magazine—Feb., 1908. No. 89877 B.

See also Explosions, under COAL AND COKE.

Briquetting.

Briquetting of Fuels in British Columbia. Description of Pacific Coal Company's Plant at Bankhead, Alberta. G. J. Mashek. Ills. 1200 w. Can Min Jour—Jan. 15, 1908. Serial. 1st part. No. 89631.

Cape Breton.

Coal Mining in Cape Breton During 1907. A review of the year, predicting a steady increase in production. 2500 w. Can Min Jour—Jan. 15, 1908. No. 89630.

Coke-Oven Gas.

See Blast-Furnace Gas, under IRON AND STEEL.

Coke Ovens.

Retort versus Beehive Coke-Ovens. Alfred Ernst. A comparison of costs and economies in operation. 2000 w. Ir Age—Jan. 16, 1908. No. 89559.

The Sheldon Retort Coke Oven and Process. S. B. Sheldon. Illustrates and describes this system, the special features being a chamber for precoking and the coal under compression during the coking operation. 4500 w. Ir Age—Jan. 16, 1908. No. 89557.

The Armstrong By-Product Coke Oven. Illustrates and describes this verticircular oven for manufacturing high-class coke and saving all the by-products. 1600 w. Col Guard—Jan. 17, 1908. No. 89869 A.

Coking Plants.

A Recent Plant for the Utilization of Small Coal. E. M. Hann. From Pro. S. Wales Inst. of Engrs. Illustrates and describes the by-product coke oven and coal washing plant at Bargold, Wales, and its operation. 4500 w. Ir & Coal Trds Rev—Jan. 17, 1908. No. 89876 A.

Comparison.

Pure Coal as a Basis for the Comparison of Bituminous Coals. W. F. Wheeler. A discussion based upon work in the laboratories of the State University of Illinois. 3500 w. Bul Am Inst of Min Engrs—Jan., 1908. No. 89890 C.

Electric Power.

See same title, under MINING.

England.

Coal Mining in Northumberland, England. George Raylton Dixon. Describes in detail the methods adopted in working out the coal so as to take advantage of the seams' cleavage. Ills. 3000 w. Eng & Min Jour—Jan. 25, 1908. No. 89839.

Explosions.

Monongah Mine Disaster. H. H. Stoek. Illustrated description of the methods of working these mines in W. Va., and the conditions before and after the explosion. 2500 w. Mines & Min—Jan., 1908. No. 89335 C.

Coal Dust as a Cause of Mine Explosions. Day Allen Willey. A discussion of recent disasters which seem to indicate that coal dust was the cause. 1800 w. Sci Am—Feb. 1, 1908. No. 89929.

See also Accidents and Explosives, under COAL AND COKE.

Explosives.

Miss-Fires and the Ignition of Firedamp with Permitted Explosives. From *Annales des Mines*. Report of investigations in France. 2000 w. Col Guard—Dec 27, 1907. No. 89381 A.

Safety Explosives (Sicherheitssprengstoffe). Karl Scholze. Results of tests on various explosives relating to force, products, temperature of detonation, etc. Ills. Serial. 1st part. 2500 w. Oest Zeitsch f Berg- u Hüttenwesen—Dec. 21, 1907. No. 89743 D.

Kentucky.

The Middleboro Coal Fields, Kentucky. John Howard. An illustrated account of a district mined entirely by drifts and slopes. 2400 w. Eng & Min Jour—Jan. 18, 1908. No. 89623.

Leasing System.

Leasing the Federal Coal Lands. H. Foster Bain. Present arguments for and against the proposed leasing system for United States coal lands. 2000 w. Min & Sci Pr—Jan. 11, 1908. No. 89584.

Mine-Gas Detection.

An Apparatus for Indicating Fire Damp in Coal Mines. Illustrates and describes the invention of Henry Guy Carleton. 1600 w. Sci Am—Jan. 11, 1908. No. 89448.

Mining.

See Economics, under MINING.

We supply copies of these articles. See page 1103.

Production.
The Coal Production of the World, 1864-1907.. A review of the production in the principal coal-producing countries. 2200 w. Min Wld—Jan. 25, 1908. No. 89848.

Railway Properties.
See Coal Lands, under RAILWAY ENGINEERING, MISCELLANY.

Rescue Appliances.
Breathing Apparatus in Mines. Considers the requirements and describes the principal types that have been experimented with in Europe. Ills. 5000 w. Mines & Min—Jan., 1908. No. 89332 C.

Reviews of 1907.
The Coal and Coke Trades of the United Kingdom in 1907. General, and district reviews. 12000 w. Ir & Coal Trds Rev—Jan. 3, 1908. No. 89531 A.

Coal Mining in the United States in 1907. Reviews, by different authors, of the general mining conditions of important anthracite and bituminous centers.. 15000 w. Eng & Min Jour—Jan. 4, 1908. No. 89406.

Shaft Stations.
Pit Bottom Arches and General Arrangements for Modern Collieries. J. S. Barnes. Considers details of arrangement and design and the method of procedure in determining the various factors. 4500 w. Ir & Coal Trds Rev—Jan. 10, 1908. Serial. 1st part. No. 89662 A.

Tipples.
See same title, under MINING.

Wages.
See same title, under MINING AND METALLURGY, MINING.

Washing.
See Coking Plants, under COAL AND COKE.

Wyoming.
The Diamondville Coalfield, Wyoming. A. T. Shurick. Describes the systems of mining. Ills. 2200 w. Eng & Min Jour—Jan. 11, 1908. No. 89478.

The Coal Mines of Southern Wyoming. Floyd W. Parsons. Information concerning the deposits and their development Ills. 1400 w. Eng & Min Jour—Jan. 11, 1908. No. 89479.

COPPER.

Alaska.
The Copper River District, Alaska. W. M. Brewer. Brief discussion of the projected lines of railway is given in the present number. 2500 w. Min & Sci Pr —Jan. 11, 1908. Serial. 1st part. No. 89583.

Assaying.
Electrochemical Analysis with Rotating Anodes in the Industrial Laboratory. Andrew M. Fairlie and Albert J. Bone. Illustrated description of a system of rotating anodes evolved in the laboratory of the Tennessee Copper Co. for the estimation of copper in slags and other products. 1000 w. Elec-Chem & Met Ind— Jan., 1908. Serial. 1st part. No. 89488 C.

Australia.
The Production of Copper in South Australia. John Plummer. Describes the mines, their development and production. 1200 w. Min Wld—Jan. 4, 1908. No. 89367.

British Columbia.
Notes on the Tyee Copper Mine. Walter Harvey Weed. Describes the peculiar characteristics of the ore bodies in the coast region of Vancouver Island, B. C. 2000 w. Eng & Min Jour—Jan. 25, 1908. No. 89834.

Colorado.
The Evergreen Copper-Deposit, Colorado. Etienne A. Ritter. Illustrates and describes a very peculiar and interesting deposit, in which both bornite and chalcopyrite occur as rock-minerals. 2500 w. Bul Am Inst of Min Engrs—Jan., 1908. No. 89889 C.

Congo Free State.
The Copper Mines of Katanza. Substance of a recent report by Robert Williams, giving information of the ores and their treatment. 1800 w. Eng & Min Jour—Jan. 25, 1908. No. 89835.

Converter Hoods.
Movable Converter Hoods. A. H. Wethey. Detail drawings with description of hoods in use at the Butte Reduction Works. 700 w. Eng & Min Jour— Jan. 11, 1908. No. 89474.

Flue Dust.
The Deposition of Flue Dust. Charles F. Shelby. Gives results of a study made at Cananea, of flue dust deposition and its sizes at different distances from the furnaces. 800 w. Eng & Min Jour—Jan. 25, 1908. No. 89836.

Germany.
See Electric Power, under MINING.

Hungary.
See same title, under GOLD AND SILVER.

Mexico.
See same title, under GOLD AND SILVER.

Montana.
Methods of Mining and Handling Ore in Butte. Edwin Higgins. Data concerning mines where the bulk of the ore comes from below the 1000-ft. level, and where electric power is extensively used. 1200 w. Eng & Min Jour—Jan. 11, 1908. No. 89473.

Nevada.
The Nevada Copper Fields. A. Selwyn-Brown. Gives a brief survey of the principal copper centers of the State and describes several of the more important cop-

We supply copies of these articles. See page 1103.

per-reduction plants. Ills. 4000 w. Engineering Magazine—Feb., 1908. No. 89880 B.

See also same title, under GOLD AND SILVER.

Peru.

The Cerro de Pasco Mining District, Peru. Clarence C. Sample. Map and illustrated description of the region with an account of an ancient silver camp, now the scene of extensive copper mining and smelting. 3000 w. Eng & Min Jour —Jan. 18, 1908. No. 89620.

Mining and Smelting at Cerro de Pasco, Peru. Clarence C. Sample. An illustrated account of methods old and new, used at these mines. 4000 w. Eng & Min Jour—Jan. 25, 1908. No. 89838.

Precipitation.

Precipitation of Copper from Cupriferous Waters. Frank H. Probert. Illustrated description of practice in handling cupriferous waters, with remarks and suggestions. 2200 w. Min & Sci Pr—Jan. 4, 1908. No. 89435.

Production.

The Copper Production of North America. Report of the production for 1907 in United States, Mexico, and Canada. 7000 w. Eng & Min Jour—Jan. 4, 1908. No. 89399.

The Copper Mining Industry, 1845-1907. Interesting information relating to the production in the United States and throughout the world. 2800 w. Min Wld —Jan. 25, 1908. No. 89847.

Pyritic Smelting.

See Gold Milling, under ORE DRESSING AND CONCENTRATION.

Smelter Contracts.

Relation Between the Assay-Value of Mill Products and Smelter Contracts. Gelasio Caetani. Shows how any contract may be expressed mathematically by a formula, and be analyzed. 1700 w. Min & Sci Pr—Jan. 4, 1908. No. 89434.

Smelters.

The Smelter of the Mammoth Copper Mining Company, at Kennett, California. Donald F. Campbell. Illustrates and describes this plant which has been recently much enlarged. 1800 w. Min & Sci Pr— Jan. 4, 1908. No. 89436.

Engineering Features of the Southwest Smelting and Refining Works, Jarilla Junction, N. Mex. Illustrated detailed destruction of features of design and construction for new copper smelting works. 4000 w. Eng Rec—Jan. 4, 1908. No. 89354.

See also Converter Hoods, under COPPER; and Electric Power, under MINING.

Smelting.

Copper Smelting in Tennessee. J. Parke Channing. Information concerning the working and progress during the past

year: 10000 w. Min & Sci Pr—Jan. 18, 1908. No. 89816.

The Production of Converter-Matte from Copper-Concentrates by Pot-Roasting and Smelting. George A. Packard. Describes experiments made at the Missouri School of Mines. 1500 w. Bul Am Inst of Min Engrs—Jan., 1908. No. 89892 C.

The Metallurgy of Copper in 1907. Walter Renton Ingalls. Briefly considers the practice at different works, and the improvements introduced. 1300 w. Eng & Min Jour—Jan. 4, 1908. No. 89428.

See also Flue Dust and Peru, under COPPER; and Sulphur, under MINOR MINERALS.

Trade.

The Year in Copper. H. M. Cole. A review of the price movements for the year and the tremendous shrinkages. 2500 w. Ir Age—Jan. 2, 1908. No. 89294.

Utah.

See Mills, under ORE DRESSING AND CONCENTRATION.

GOLD AND SILVER.

Assaying.

Notes on Cobalt Ores. W. K. McNeil. On the methods and apparatus employed in the assay of these complex ores. 900 w. Can Min Jour—Jan. 1, 1908. No. 89370.

The Assay of Telluride Ores. George T. Halloway and Leonard E. B. Pearse. An investigation of the causes of losses in treating telluride ores on a large scale and in assaying them, and how to avoid them. 9500 w. Inst of Min & Met, Bul. 39—Dec. 12, 1907. No. 89672 N.

See also Laboratory, under MINING.

Australia.

Mining Practice at Kalgoorlie, West Australia. Gerard W. Williams. An illustrated account of the gold mining. The telluride ores occur in lenses and are extracted by methods insuring good ventilation. 2000 w. Eng & Min Jour—Jan. 25, 1908. No. 89833.

Cobalt.

See Assaying, under GOLD AND SILVER.

Colorado.

See Placers, under GOLD AND SILVER.

Cyaniding.

Use of Compressed Air in Cyanidation. A. Grothe. Describes the method employed in New Zealand, known as the Brown process. 1500 w. Min Wld—Jan. 11, 1908. No. 89493.

See also Nevada, under GOLD AND SILVER.

Dredging.

Gold Dredging in 1907. John Power Hutchins. Discusses the attempts made

We supply copies of these articles. See page 1103.

to dredge frozen ground, variation in practice, steam-shovel dredging, etc. 2200 w. Eng & Min Jour—Jan. 4, 1908. No. 89425.

Gold Refining.
The Clean-Up, Melting and Refining of Gold Bullion. Gerard W. Williams. Describes methods used in South Africa. 2500 w. Min Wld—Jan. 4, 1908. No. 89368.

Homestake.
See Mine Fire, under MINING.

Hungary.
Nagybanya, Hungary. Edward Skewes. A descriptive account of this district of eastern Hungary, its mines of gold, silver, copper, etc., the methods of mining and milling, the management of the Government mines, and other subjects of interest. Ills. 5000 w. Min & Sci Pr—Jan. 11, 1908. No. 89582.

Hydraulic Mining.
Hydraulic Mining in British Columbia. Howard W. Dubois. Considers the conditions necessary for successful hydraulic operations, and shows what has been accomplished. 2500 w. Columbia Engr—1906. No. 89499 N.
See also Yukon, under GOLD AND SILVER.

Larder Lake.
The Larder Lake District. R. W. Brock. A report of this region, based on field work. Ills. 3000 w. Can Min Jour—Jan. 1, 1908. Serial. 1st part. No. 89371.

Mexico.
Mexico. Progress in 1907. C. A. Bohn. A review of the year. Map & Ills. 3500 w. Min & Sci Pr—Jan. 4, 1908. No. 89438.
The Mineral Resources of Sonora. F. J. H. Merrill. The chief metals are copper, gold and silver, though other metals are mined. The mines are discussed by districts. Map. 7000 w. Min & Sci Pr—Jan. 4, 1908. No. 89437.
The Promontorio Silver-Mine, Durango, Mexico. Francis Church Lincoln. Illustrated description of the vein, the ore, the relations of the primary vein-minerals, the mining, milling, etc. 4000 w. Bul Am Inst of Min Engrs—Jan., 1908. No. 89893 C.

Milling.
See Gold Milling, under ORE DRESSING AND CONCENTRATION.

Nevada.
Some Bullfrog Mines. W. H. Spaulding. Information concerning the Bullfrog reduction mill, the cyanide treatment, the Mayflower, Tramps, and Gold Bar mines. 2400 w. Eng & Min Jour—Jan. 18, 1908. No. 89621.
Nevada, the Silver State, and Government Irrigation in Nevada. The Truckee-

Carson Project. Prof. Oscar C. S. Carter. Reviews the history of Nevada, its ore deposits, topography, climate, and the irrigation projects. Ills. & Map. 7000 w. Jour Fr Inst—Jan., 1908. No. 89553 D.
Genesis of the Formation and Deposition of the Nevada Desert Gold, Silver, and Copper Mines. Dr. Willis E. Everette. Thinks there was an upward radial deposition and percolation from the upthrust of a series of "laccoliths" which exist far below the surface. 7000 w. Sci Am Sup—Jan. 25, 1908. No. 89811.

New Mexico.
Genesis of the Lake Valley Silver-Deposits. Charles R. Keyes. Describes in detail the geology of this region in New Mexico, the ores and their mineralogic character, discussing their origin. 10000 w. Bul Am Inst of Min Engrs—Jan., 1908. No. 89888 C.

New Zealand.
The Waihi Gold Mine in New Zealand. Ralph Stokes. First of a series of illustrated articles describing the mines and methods. 2500 w. Min Wld—Jan. 11, 1908. Serial. 1st part. No. 89494.

Peru.
See same title, under COPPER.

Placers.
The Breckenridge Gold Placers, Colorado. Arthur Lakes. Illustrates and describes this region. 1100 w. Min Wld—Jan. 4, 1908. No. 89366.

Production.
The World's Supply of Gold and Silver. Charles C. Schnatterbeck. Interesting statistics showing the progress made in the production. 6000 w. Min Wld—Jan. 25, 1908. No. 89846.
Gold, Silver, and Platinum in 1907. Frederick Hobart. Review of the influences and commercial conditions affecting the production of the precious metals. 3500 w. Eng & Min Jour—Jan. 4, 1908. No. 89398.
The Great Gold Mines. T. A. Rickard. A review of the production and profit of the world's greatest gold mines. 3000 w. Min & Sci Pr—Jan. 4, 1908. Serial. 1st part. No. 89432.

Rand.
The Origin of the Gold in the Rand Banket. Discussion of J. W. Gregory's paper on this subject. 13800 w. Inst. of Min & Met, Bul. 38—Nov. 14, 1907. No. 89675 N.
The Economics of the Deep-Deep Level Mines of the Rand. B. J. Collings. Discusses the serious financial side of such developments and other difficulties in working. 1000 w. Min Jour—Jan. 11, 1908. No. 89649 A.
See also Transvaal, under GOLD AND SILVER.

We supply copies of these articles. See page 1103.

Transvaal.

The Transvaal Mines—1907 Results. Gives comparison with 1906, and reviews the results of the year. 2500 w. Min Jour—Jan. 11, 1908. No. 89650 A.

Utah.

Mines of Tintic District, Utah. Robert B. Brinsmade. An illustrated description of the region and methods employed. 6000 w. Mines & Min—Jan., 1908. No. 89336 C.

Yukon.

Development of the Bonanza Creek Gold Mines. Francis C. Nicholas. An illustrated description of the region, with review of the past history, and account of present conditions. 1800 w. Min Wld —Jan. 8, 1908. No. 89632.

IRON AND STEEL.

Armor Plate.

The Use of Electricity in the Manufacture of Armor Plates. J. W. Warr. Describes the process of manufacture and comments on the applications made of electricity for power. Ills. 3500 w. Elec Rev, Lond—Dec. 27, 1907. No. 89374 A.

The Development of Armor-Plate Manufacture (Die Entwicklung der Panzer-Fabrikation). Max Kraleipper. A review of improvements in materials and processes of manufacture. 4000 w. Oest Zeitschr f Berg u Hüttenwesen—Dec. 21, 1908. No. 89742 D.

Blast-Furnace Gas.

The Use of Blast Furnace and Coke Oven Gas. F. E. Junge. Gives results of an investigation of the latest developments along this line in Germany. 5700 w. Ir Trd Rev—Jan. 2, 1908. No. 89302.

See also Gas Engines, under MECHANICAL ENGINEERING, COMBUSTION MOTORS.

Blast Furnaces.

The New Blast Furnace of the Shenango Furnace Co. at Sharpsville, Pa. Illustrated detailed description. 1500 w. Ir Trd Rev—Jan. 2, 1908. No. 89301.

The Inland Steel Company's New Furnace. Illustrated description of the new plant at Indiana Harbor, Ind. 2000 w. Ir Age—Jan. 16, 1908. No. 89558.

Recent Progress and Present Problems in the Blast-Furnace Industry. John J. Porter. A lecture before the Cincinnati Sec. of the Am. Chem. Soc. Discusses engineering and metallurgical problems connected with the operation of the blast furnace. 8800 w. Ir Trd Rev—Jan. 2, 1908. No. 89303.

Chute Process.

The Chute Steel-Making Process. Illustrated description of a process claiming to accomplish the same results attained by the duplex process, but in a simpler and much cheaper manner. 2000 w. Ir Trd Review—Jan. 23, 1908. No. 89819.

Electro-Metallurgy.

The Manufacture of High-Grade Steel in the Electric Furnace. Reviews articles by O. Thallner, in *Stahl und Eisen,* dealing with steel refining in general, and the Héroult furnace in particular. 2000 w. Elec-Chem & Met Ind—Jan., 1908. No. 89491 C.

A Modification of the Induction Furnace for Steel Refining. H. Wedding. Trans. from *Stahl und Eisen.* Illustrated description of a new development in the design of the induction furnace for steel refining at the Roechling works. 1400 w. Elec-Chem & Met Ind—Jan., 1908. No. 89486 C.

See also Foundry Furnaces, under MECHANICAL ENGINEERING, MACHINE WORKS AND FOUNDRIES.

Lake Superior.

Greatest Year in the Ore Ranges' History. Oliver J. Abell. Review of the record at the mines and on railroads and lakes. 5500 w. Ir Trd Rev—Jan. 2, 1908. No. 89310.

Lake Superior Iron Mining in 1907. Dwight E. Woodbridge. A review of the production, improved methods, developments, etc. 3500 w. Ir Age—Jan. 9, 1908. No. 89413.

Towns Built on Iron Ore Foundation. George E. Edwards. Describes the ore deposits and changes in methods of mining which will make necessary the removal of the village of McKinley and other towns in the Mesabi iron range. 2000 w. Min Wld—Jan. 11, 1908. No. 89495.

Operating Changes in the Lake Superior Iron Mines. Dwight E. Woodbridge. An account of recent and prospective changes in methods of operating the mines and handling the ores. 1000 w. Eng & Min Jour—Jan. 25, 1908. No. 89837.

Production.

Iron Ores Outside the Lake Superior Region. John Birkinbine. Gives a résumé of the sources from which iron ore is now derived in quantity, with suggestions. 2500 w. Ir Trd Rev—Jan. 2, 1908. No. 89311.

Rhode Island.

Contributions to the Geology of Rhode Island: I. Notes on the History and Geology of Iron Mine Hill, Cumberland. B. L. Johnson. II. The Petrography and Mineralogy of Iron Mine Hill, Cumberland. C. H. Warren. 11500 w. Am Jour of Sci—Jan., 1908. No. 89321 D.

Rolling Mills.

See also Gas Engines, under MECHANICAL ENGINEERING, COMBUSTION MOTORS; Electric Driving, under MECHANICAL ENGINEERING, POWER AND TRANSMISSION; and Engines, un-

We supply copies of these articles. See page 1103.

der MECHANICAL ENGINEERING, STEAM ENGINEERING.

Steel Making.

See Chute Process, under IRON AND STEEL.

Steel Works.

The Grey Structural Mill at South Bethlehem. Illustrated description of the Grey mill, which forms a part of the new Saucon plant. 2500 w. Ir Age—Jan. 2, 1908. No. 89288.

The Krupp Plant at Rheinhausen, Prussia (L'Asine Métallurgique de la Maison Krupp à Rheinhausen, Prusse). Abstracted from *Stahl u Eisen*. Plate. Ills. 3000 w. Génie Civil—Dec. 14, 1907. No. 89726 D.

See also Blast Furnaces, under IRON AND STEEL.

Tin Plate.

Making Tin Plate at Griffiths Iron Mills. Illustrated description of the process used in these charcoal iron mills. 1500 w. Met Work—Jan. 4, 1908. No. 89330.

The Sheet and Tin Plate Trades in 1907. B. E. V. Luty. A general review of the industry. 3800 w. Ir Age—Jan. 2, 1908. No. 89297.

Trade.

The Chicago Iron Trade in 1907. T. J. Wright. A review of the year, giving a summary of mill and furnace construction, and discussing the outlook. 3500 w. Ir Age—Jan. 2, 1908. No. 89295.

The Cincinnati Pig Iron Trade in 1907. James A. Green. A review of the year. 2800 w. Ir Age—Jan. 2, 1908. No. 89292.

The Philadelphia Iron Trade in 1907. A. A. Miller. A review of the year, the prices, material, etc. 4800 w. Ir Age—Jan. 2, 1908. No. 89296.

The Pittsburgh Iron Trade in 1907. Robert A. Walker. A general review. 5000 w. Ir Age—Jan. 2, 1908. No. 89293.

Ten Months' Prosperity in the Pittsburg District. W. B. Robinson. The story of the past year. 3500 w. Ir Trd Rev—Jan. 2, 1908. No. 89300.

The Iron and Steel Trades in 1907. General and district reviews of the trade in the United Kingdom. 1300 w. Ir & Coal Trds Rev—Jan. 3, 1908. No. 89532 A.

The Iron and Steel Industry in 1907. Reviews by various authors showing immense production, followed by partial collapse. 11000 w. Eng & Min Jour—Jan. 4, 1908. No. 89402.

The Iron Trade of the World with Special Reference to the Conditions of the United States. J. Stephen Jeans. A discussion of present conditions and the further outlook. 5500 w. Mfrs' Rec—Jan. 2, 1908. No. 89299.

The Future of the Iron and Steel Industry from a British Point of View. T. Good. 2500 w. Cassier's Mag—Jan., 1908. No. 89457 B.

LEAD AND ZINC.

Lead Refining.

The Direct Extraction and Refining of Lead by Electrolysis (Extraction Directe et Affinage du Plomb par Electrolyse). Jean Escard. A general review of various processes. Ills. 3000 w. Electricien—Dec. 21, 1907. No. 89722 D.

Lead Smelting.

The Metallurgy of Lead in 1907. Walter Renton Ingalls. Discusses the results of the processes in use. 2700 w. Eng & Min Jour—Jan. 4, 1908. No. 89427.

Leadville Geology.

Present Views of Genesis of Leadville Limestone Ores. S. F. Emmons. From Bull. 320, U. S. Geol. Surv. 2500 w. Eng & Min Jour—Jan. 11, 1908. No. 89476.

Milling.

See Zinc Milling, under ORE DRESSING AND CONCENTRATION.

Production.

The Production of Lead and Spelter in 1907. Review of mining conditions, giving statistics of production and price. 8000 w. Eng & Min Jour—Jan. 4, 1908. No. 89400.

The Production of Zinc Ores (Das Vorkommen der Zinkerze). Franz Peters. Gives a review of zinc ore production in all the leading countries for the last three years. 3500 w. Glückauf—Dec. 21, 1907. No. 89739 D.

Zinc Smelting.

Zinc Smelting in the United States in 1907. Walter Renton Ingalls. Reports the zinc smelting capacity, and describes the modifications introduced at different works. 3500 w. Eng & Min Jour—Jan. 4, 1908. No. 89426.

MINOR MINERALS.

California.

California Minerals. Arthur S. Eakle. A brief description of the new minerals found, and the date of their discovery. 1800 w. Min & Sci Pr—Jan. 18, 1908. No. 89817.

Diamonds.

Diamond Mining in South Africa. William Taylor. Describes the geological formation and methods of mining. Ills. 2200 w. Mines & Min—Jan., 1908. No. 89334 C.

Lime Burning.

Modern Lime Kilns; the Plant of the Knickerbocker Lime Co., Mill Lane, Pa. Rex C. Wilson. Illustrated detailed description of a large modern lime-producing plant, and its operation. 800 w. Eng News—Jan. 30, 1908. No. 89919.

Manganese.

Manganese Ores: Their Uses, Occurrence and Production. Gives a résumé of information now available. 6500 w. Bul Imp Inst—Vol. 5. No. 3. No. 89322 N.

Oil.

The Petroleum Industry of the United States. Articles reporting the production in various fields. 6500 w. Eng & Min Jour—Jan. 4, 1908. No. 89407.

Extent and Importance of Oklahoma Oil Fields. Dr. Charles N. Gould. Describes geological features, giving the present production and discussing the possible development. 2000 w. Min Sci—Jan. 16, 1908. No. 89635.

The Location of Bore Holes in Oil Fields (Zur Wahl der Bohrpunkte in den Erdölgebieten). H. Höfer. A discussion of the geological features to be taken into account. Ills. 2000 w. Oest Zeitschr f Berg- u Hüttenwesen—Dec. 14, 1907. No. 89741 D.

Phosphate.

Phosphate Mining in Tennessee. H. D. Ruhm. A report of this industry during 1907, and the present outlook. 1200 w. Eng & Min Jour—Jan. 18, 1908. No. 89619.

Platinum.

See Production, under GOLD AND SILVER.

Sapphires.

Anakie Sapphire Fields, Central Queensland. Information from a recent report of C. F. V. Jackson. Ills. 1200 w. Queens Gov Min Jour—Dec. 14, 1907. No. 89832 B.

Slate.

Slate Mining in Wales and Cause of Its Decline. Discusses the causes of the decline of this industry and gives an illustrated description of the Oakeley mine and its methods. 2000 w. Eng & Min Jour—Jan. 18, 1908. No. 89618.

Sulphur.

Apparatus for Obtaining Sulphur from Furnace Gases. Franklin R. Carpenter. Illustrated description of the writer's sulphur recovery apparatus from the smelting of sulphide ores. 2500 w. Min Wld —Jan. 4, 1908. No. 89365.

Tin.

Notes on Tin. Prof. A. Humboldt Sexton. The present number considers its physical and chemical properties, uses, value; tin minerals and ores, their distribution, etc. 3500 w. Mech Engr—Jan. 11, 1908. Serial. 1st part. No. 89641 A.

Government Tin Prospecting in Transvaal. An encouraging report, taken from an article by U. P. Swinburne. 1200 w. Min Jour—Jan. 4, 1908. No. 89520 A.

The Bolivian Tin Mining Industry and Railways. G. Preumont. An account of the deposits and mines and the railways proposed which will affect their develop-

ment. 3000 w. Min Jour—Jan. 14, 1908. Serial. 1st part. No. 89519 A.

The Utilization of Tin Scrap (Aprovechamiento de los Recortes de Hojalata). Reviews the various methods for the recovery of tin from scrap and refuse. Serial. 1st part. 3000 w. Ingenieria—Nov. 30, 1907. No. 89706 D.

See also Magnetic Separation, under ORE DRESSING AND CONCENTRATION.

Tungsten.

The Dragoon, Arizona, Tungsten Deposits. Ralph W. Richards. Notes on the occurrence and origin of rubnerite and scheelite ores, the production and development. 1200 w. Min Sci—Jan. 23, 1908. No. 89852.

MINING.

Bore-holes.

The Deviation of Rand Bore-Holes from the Vertical. Discussion of Joseph Kitchin's paper on this subject. 10500 w. Inst of Min & Met, Bul 39—Dec. 12, 1907. No. 89673 N.

Compressed Air.

High vs. Low Pressure for Compound Air in Mines. Robert B. Brinsmade. Discusses the economical limit of pressure, the theory of the intercooler, preheaters, etc. 1500 w. Eng & Min Jour—Jan. 18, 1908. No. 89622.

Costs.

Variation in Mining Costs. J. R. Finlay. Gives a comparative table of working costs, and discusses the considerations that give rise to the wide difference in estimates. 3000 w. Min & Sci Pr—Jan. 4, 1908. No. 89433.

Development Work.

Some Experiences with Exploration Tunnels. Arthur Lakes. Reasons why exploration work should be by sinking and cross-cutting, rather than by cross-cutting at depth and raising. 2500 w. Min Sci—Jan. 2, 1908. No. 89369.

Drills.

Notes on Small Stope Drills. E. M. Weston. Considers some of the reasons for past failures of small drills, describes and illustrates types now used, and their operation. 10000 w. Jour Chem, Met, & Min Soc of S Africa—Oct., 1907. No. 89643 E.

Economics.

False Economy in Mining. M. Mownley. Read before the Nat. Assn. of Col Mgrs. Gives cases from the writer's experience, and discusses details. General discussion. 7000 w. Ir & Coal Trds Rev —Jan. 10, 1908. No. 89663 A.

Electric Hoisting.

The Highest Speed and Time of Acceleration and Retardation in Electric Hoisting with Constant Current (Ermittlung der höchsten Geschwindigkeit, der Be-

We supply copies of these articles. See page 1103.

schleunigungs- und Verzögerungsdauer
elektrisch betriebener Fördermaschinen
bei Anfahren und Stillsetzen mit kon-
stantem Strome). Eugen Kulka. Ills.
1600 w. Elektrotech Zeitschr—Dec. 12,
1907. No. 89784 D.

Electric Power.
Electric Power in Coal Mining. An
illustrated article discussing its adaptation
to pumping, winding, coal cutters, etc.
2500 w. Cassier's Mag—Jan., 1908. No.
89456 B.
Gas-Electric Power at the Mansfield
Copper Mines and Smelting Works. Summary of articles by H. R. Speyer, on the
present electric equipment of these mining
and smelting works. Ills. 2500 w. Elec-
Chem & Met Ind—Jan., 1908. No. 89487 C.
See also Montana, under COPPER.

Explosives.
See same title, under COAL AND COKE.

Haulage.
See Tipples, under MINING.

Hoisting.
The Foby Safety Appliance for the
Prevention of Overwinding. Illustrated
description, translated from *Comptes Ren-
dus.* 500 w. Col Guard—Jan. 17, 1908.
No. 89870 A.

Laboratory.
The Laboratory: Its Economic Value.
A. McArthur Johnston. Outlines the
basis on which laboratory work should be
conducted and the methods in use in the
Consolidated Gold Fields laboratory. 8000
w. Jour Chem, Met, & Min Soc of S
Africa—Oct, 1907. No. 89642 E.

Mine Fire.
The Homestake Mine Fire: Unusual
Methods Employed in Fighting It, and the
Lessons That It Taught. Bruce C. Yates.
Gives a brief description of the ore bodies
and method of extracting the ore, with il-
lustrated description of the flooding of
this South Dakota mine, its unwatering,
etc. 8000 w. Eng News—Jan. 2, 1908.
No. 89271.

Quarrying.
Stone: Quarrying and Preparation for
Sale. Allan Greenwell, and J. V. Elsden.
The first of a series of articles containing
information upon matters connected with
quarrying. 3000 w. Quarry—Jan., 1908.
Serial. 1st part. No. 89534 A.

Reviews of 1907.
Mineral and Metal Production in 1907.
Gives statistics of the output of the more
important substances. 1200 w. Eng &
Min Jour—Jan. 4, 1908. No. 89397.
Review of Mining in Foreign Countries.
Gives details of production during 1907 in
various parts of the world. 22500 w. Eng
& Min Jour—Jan. 4, 1908. No. 89405.
Mining in the United States During
1907. Different authors give reports of

progress from various camps. 25000 w.
Eng & Min Jour—Jan. 4, 1908. No. 89404.

Shaft Sinking.
Sinking the Clonan Shaft at Mineville,
N. Y. Guy C. Stoltz. Describes the conditions and the methods adapted. 1200
w. Eng & Min Jour—Jan. 11, 1908. No.
89477.

Shaft Stations.
See same title, under COAL AND COKE.

Sinking Pumps.
See Pumps, under MECHANICAL
ENGINEERING, HYDRAULIC MACHIN-
ERY.

Stoping.
The Method of Breast Stoping at Cripple Creek. G. E. Wolcott. Illustrates and
describes the methods employed at the
Portland mine, which require much timber, permit complete extraction, but do
not increase the safety. 1400 w. Eng &
Min Jour—Jan. 11, 1908. No. 89475.

Tipples.
The Green Self-Dumping Car Haul. Illustrated description of a new system of
handling and dumping mine cars. 2500
w. Ir Age—Jan. 30, 1908. No. 89905.

Unwatering.
See Mine Fire, under MINING.

Valuation.
Calculation of Mine-Values. R. B.
Brinsmade. An attempt to form a formula by which a mine can be quickly
evaluated, after the physical data has been
collected. 1500 w. Bul Am Inst of Min
Engrs—Jan., 1908. No. 89891 C.

Wages.
The Difficulty of the Wage Question in
Mining (Schwierigkeiten des Lohnwesens
im Bergbau). Dr. Herbig. A discussion of
the various factors to be taken into account in establishing scales of wages. 3500
w. Glückauf—Dec. 28, 1907. No. 89740 D.
Wage Scales in the British and Rhine-
Westphalian Coal Fields (Ueber Lohnta-
rife im britschen und rheinisch-westfä-
lischen Steinkohlenbergbau). Herr Hilgen-
stock. An elaborate discussion of wages,
costs, labor conditions and labor troubles
in these districts. Ills. Serial. 1st part.
10000 w. Glückauf—Dec. 7, 1907. No.
89736 D.

ORE DRESSING AND CONCENTRATION.

Concentrates.
See Smelter Contracts, under COPPER.

Costs.
See same title, under MINING.

Gold Milling.
Progress in Gold-Ore Treatment During 1907. Alfred James. Fine sliming
and the treatment of slimes, improved apparatus, and progress in cyanidation are
considered. 2500 w. Eng & Min Jour—
Jan. 4, 1908. No. 89401.

We supply copies of these articles. See page 1103.

The Question of Ore Concentration. H. P. Dickinson. . Considers the limitations of water concentration as compared with fire treatment. 1200 w. Min Sci—Jan. 11, 1908. No. 89636.

See also Hungary and Nevada, under GOLD AND SILVER; and Air Lift, under MECHANICAL ENGINEERING, HYDRAULIC MACHINERY.

Magnetic Separation.

The Separation of Tin Oxide from Wolfram. Discussion of the paper by Amos Treloar, and Gurth Johnson. 3000 w. Inst of Min & Met, Bul 39—Dec. 12, 1907. No. 89674 N.

Mills.

The Six-Thousand Ton Concentrator of the Utah Copper Co., Garfield, Utah. R. L. Herrick. Illustrates and describes the machinery equipment and methods of milling. 4000 w. Mines & Min—Jan, 1908. No. 89333 C.

Silver Milling.

See Mexico, under GOLD AND SILVER.

Slimes Treatment.

See Air Lift, under MECHANICAL ENGINEERING, HYDRAULIC MACHINERY.

Zinc Milling.

Sludge Concentration in the Joplin District. Doss Brittain. An illustrated explanation of methods used to recover the zinc and lead wastes. 3500 w. Min Wld—Jan. 4, 1908. No. 89364.

MISCELLANY.

Ontario.

Ontario's Mining Progress in the Last Decade. Phillips Thompson. A review of the increase in production, the development work, etc. 2000 w. Can Min Jour—Jan. 15, 1908. No. 89629.

Temperature Gradients.

The Internal Temperature Gradient of Metals. Schuyler B. Serviss. Detailed description of apparatus used, and observations and results of experiments made, comparing with work by Dr. C. B. Thwing. 5000 w. Am Jour of Sci—Dec., 1907. No. 89159 D.

Washington.

Reconnaissance into Okanogan Mountains, Washington. Horace F. Evans. The present article reviews briefly the early explorers of this region, and begins a description of the geology. 2500 w. Min Wld—Jan. 18, 1908. Serial, 1st part. No. 89633.

RAILWAY ENGINEERING

CONDUCTING TRANSPORTATION.

Signalling.

Overrunning Signals. Mr. Scholkmann. Concerning the tests made on the Prussian State Railway, the devices proposed, etc. 3700 w. Bul Int Ry Cong—Dec., 1907. No. 89507 E.

See also Switches, under PERMANENT WAY AND BUILDINGS.

Signals.

Alternating Current Automatic Block Signals on the Highland Division of the New York, New Haven & Hartford. Illustrated description of the system installed. 1800 w. R R Gaz—Jan. 17, 1908. No. 89588.

See also same title, under STREET AND ELECTRIC RAILWAYS.

MOTIVE POWER AND EQUIPMENT.

Adhesion System.

The Puy de Dome Adhesion Railway. Illustrated detailed description of the special system used on very steep grades, using steam locomotives with supplementary adhesion gear. 3000 w. Engr, Lond—Dec. 27, 1907. No. 89389 A.

Brakes.

Some Defects of the E. T. Brake. G. W. Kiehm. A discussion. 2000 w. Ry & Loc Engng—Jan., 1908. No. 89340 C.

Continuous Brakes and Goods Trains. A. Huberti and J. Doyen. A contribution to the question of drawing up a programme for making goods-train brake trials. 4800 w. Bul Int Ry Cong—Dec., 1907. No. 89503 E.

Car Repairing.

Repairing Steel Freight Cars. Illustrates and describes methods of maintaining and repairing these cars at the McKees Rocks shops of the Pittsburgh & Lake Erie. 2000 w. Am Engr & R R Jour—Jan., 1908. No. 89315 C.

Cycle.

Railway Cycle. A. Honemann. Describes a cycle devised by O. Grasshoff which has given excellent results. Ills. 1700 w. Bul Int Ry Cong—Dec., 1907. No. 89508 E.

Driving Wheels.

The Influence of the Diameter of Driving Wheels. Editorial discussion of Prof. Goss's investigations and the conclusions. 1700 w. Engr, Lond—Dec. 27, 1907. No. 89391 A.

Electrification.

Practical Aspects of Steam Railroad Electrification. W. N. Smith. Discusses the electrification project, and railroad operation. 6000 w. Pro Am Inst of Elec Engrs—Jan., 1908. No. 89325 D.

We supply copies of these articles. See page 1103.

Locomotive Efficiency.

The Efficiency of Locomotives in Relation to Their Main Constructional Features and Their Speed (Die Leistungsfähigkeit der Lokomotiven in Abhängigkeit von ihren baulichen Hauptverhältnissen und der Fahrgeschwindigkeit). Albert Frank. Serial. 1st part. 5000 w. Glasers Ann—Dec. 15, 1907. No. 89761 D.

Locomotive Maintenance.

Handling Locomotive Supplies. E. Fish Ensie. Considers the practical care, upkeep, supervision, and economy in the handling of engine equipments. Ills. 1700 w. Am Engr & R R Jour—Jan., 1908. No. 89316 C.

Locomotives.

Four-Wheel Switch Locomotive for the Burden Iron Company. Illustrated description. 400 w. Ry Age—Jan. 3, 1908. No, 89362.

Shay Locomotive for the Southern Railway. Illustrated description of engines used on heavy grades. 300 w. Ry Age—Jan. 3, 1908. No. 89363.

Locomotive with Water-Tube Fire-Box. Illustrated description of engine fitted with a Brotan fire-box. 500 w. Engr, Lond—Jan. 10, 1908. No. 89661 A.

Types of Compound Locomotives. Explains the action of the compounds most commonly used. 1200 w. Ry & Loc Engng—Jan., 1908. No. 89341 C.

Pacific Locomotive for the Lake Shore. Illustrated description of two types of heavy passenger engines. 1000 w. Ry Age—Jan. 24, 1908. No. 89849.

Simple Prairie Type Locomotives. Illustrated description of engines for freight service on the Wabash R. R. 600 w. Am Engr & R R Jour—Jan., 1908. No. 89320 C.

Simple Consolidation Locomotive. Illustrated description of Class D, heavy locomotives built for the C., N. O. & T. P. Ry. 700 w. Am Engr & R R Jour—Jan., 1908. No. 89319 C.

The British Locomotive. A. W. S. Graeme. Read before the Rugby Engng Soc. Explains conditions in Great Britain which influence the design of locomotives, describes types used, and briefly explains why compounding is not economical under present conditions. 5500 w. Mech Engr—Jan. 18, 1908. No. 89862 A.

Italian 4-Cylinder Compound Double-Ended Locomotive. Illustrated detailed description of engines adapted for express and freight service upon the steepest grades in mountainous regions. 1500 w. Ry Age—Jan. 17, 1908. No. 89627.

Recent Improvements in the Rolling Stock of Italian Railways (Le Recenti Migliorie nel Materiale Rotabile delle Ferrovie Italiane). Luigi Greppi. De-

scribes principally new types of locomotives, illustrating with detailed plates. Ills. 10500 w. Ann d Soc d Ing e d Arch Ital. —Nov.-Dec., 1907. No. 89701 F.

A Comparison of the Two- and Three-Coupled High-Speed Locomotives of the Prussia-Hesse State Railways from a Theoretical Point of View (Ein Vergleich der zwei- und dreigekuppelten Schnellzug-Lokomotiven der preussisch-hessischen Staatsbahnen auf Theoretischer Grundlage). J. Zillgen. Ills. Serial, 1st part. 4500 w. Glasers Ann—Dec. 15, 1907. No. 89760 D.

See also Steam vs. Electricity, under MOTIVE POWER AND EQUIPMENT.

Motor Cars.

Gasoline-Electric Motor Car for Delaware & Hudson Railroad. Illustrated description of the car and its operation and equipment. 2500 w. St Ry Jour—Jan. 25, 1908. No. 89827.

Shops.

A Few Locomotive Tools and Fixtures. L. E. Salmon. Illustrates and describes special appliances used on railroad work which save time and money. 2300 w. Am Mach—Vol. 31. No. 3. No. 89562.

Arrangement of Railroad Shops. George A. Damon. Read before the Canadian Railway Club. Considers the design, arrangement, and equipment of repair shops from a financial standpoint. Plans. 5000 w. R R Gaz—Jan. 10, 1908. No. 89445.

New Freight Car Repair Shops of the Santa Fé. Illustrations, with description of new shops at Topeka, Kan. 1500 w. R R Gaz—Jan. 17, 1908. No. 89586.

See also Car Repairing, under MOTIVE POWER AND EQUIPMENT; and Steel Buildings, under CIVIL ENGINEERING, CONSTRUCTION.

Steam vs. Electricity.

Electric Working on Standard-Gauge Lines with Steep Gradients. Mr. Mühlmann. Shows that electric working cannot yet compete financially with steam-working. 4500 w. Bul Int Ry Cong—Dec., 1907. No. 89506 E.

Discussion on "Comparative Performance of Steam and Electric Locomotives," at New York, Nov. 8, 1907. Paper of Albert H. Armstrong. 6500 w. Pro Am Inst of Elec Engrs—Jan., 1908. No. 89899 D.

Valve Gears.

Walschaert Valve Gear. Illustrated description of a design being applied by the Canadian Pacific Ry. 1000 w. Am Engr & R R Jour—Jan., 1908. No. 89317 C.

Wheels.

The Car Wheel. George L. Fowler. Gives an illustrated account of investigations made under the auspices of the Schoen Steel Wheel Co. Also general discussion. Ills. 11500 w. Pro Ry Club of Pittsburgh—Nov., 1907. No. 89683 C.

Comparative Physical Tests of Car Wheels and Tires. George L. Fowler. Report of tests and results, reprinted from report made to the Schoen Steel Wheel Co. 2000 w. R R Gaz—Jan. 3, 1908. No. 89343.

Microphotographs of Steel Wheels and Tires. George L. Fowler. Illustrations and report of examinations with the mi-- croscope of sample wheels tested by the Schoen Steel Wheel Co. 2000 w. R R Gaz—Jan. 24, 1908. No. 89840.

NEW PROJECTS.

Alaska.

The Most Wonderful Railroad of the North. Katherine Louise Smith. Illustrated description of the White Pass and Yukon Railway, and its construction. 1500 w. Sci Am—Jan. 11, 1908. No. 89449.

Austria.

The New Austrian Alpine Railways. Map, plate, illustrations, and description of new lines to give more direct communication with the port of Triest. Many interesting engineering features. 4000 w. Engr, Lond—Jan. 17, 1908. No. 89872 A.

Bolivia.

See Tin, under MINING AND METALLURGY, MINOR MINERALS.

Central America.

A Central American Railway. Theodore Paschke. Urging the construction of an Intercontinental Railway from Mexico to Panama, discussing the line and method of procedure. 2800 w. Eng Rec —Jan. 18, 1908. No. 89612.

Yosemite Valley.

The New Yosemite Valley Railroad. L. E. Danley. An illustrated description of a line through this famous region, and the interests it will serve. 1000 w. Ry Age—Jan. 24, 1908. No. 89850.

PERMANENT WAY AND BUILDINGS.

Construction.

Railroad Construction in Western Montana. John C. Breedlove. Mainly a description of extensive work on the Northern Pacific Ry. 1800 w. Cornell Civ Engr—Jan., 1908. No. 89903 C.

Construction of the Guelph & Goderich Railway. J. Grant MacGregor. Illustrated description. 1300 w. Eng Rec—Jan. 18, 1908. No. 89616.

Grade Reduction.

Grade Revision on the Canadian Pacific Railway in British Columbia. Brief account of important work in progress on the steepest part of the road. Ills. 1000 w. Eng News—Jan. 23, 1908. No. 89805.

Improvements.

The Pennsylvania Railroad Improvements at Blairsville, Pa. G. W. Phillips,

and J. Gwilliam. Illustrated description of a change of line about 9 miles long from Blairsville to Tunnelton, Pa., costing about $2,500,000. 2800 w. Eng Rec—Jan. 4, 1908. No. 89355.

Rails.

The Present Status of Rail Specifications. A review of the action taken by the American Railway Assn. 2000 w. Ry Age—Jan. 17, 1908. No. 89628.

Stations.

New Station at Sydney, N. S. W. Illustrates and describes new station buildings and arrangements for handling heavy traffic. 2500 w. Engr, Lond—Dec. 27, 1907. No. 89390 A.

The New Station at Stuttgart. Plans and description of proposed alterations. 1700 w. Bul Int Ry Cong—Dec., 1907. No. 89505 E.

Switches.

New Interlocking Plant, Hoboken Terminal Yard, Delaware, Lackawanna & Western Ry. Illustrated detailed description of a plant possessing novel features tending to greater safety and dispatch. 4500 w. Eng News—Jan. 30, 1908. No. 89925.

Ties.

See Timber Preservation, under CIVIL ENGINEERING, MATERIALS OF CONSTRUCTION.

Yards.

The New Shunting Yard at Wath, in England. Mr. Frahm. Plan and description. 1100 w. Bul Int Ry Cong—Dec., 1907. No. 89504 E.

TRAFFIC.

Europe.

Observations on European Railroads. Including Freight House and Freight Office Work. Albert T. Perkins. Illustrates and describes typical team-yard and freight-house arrangements, methods of work, etc. 2000 w. Pro St Louis Ry Club—Dec., 1907. No. 89461.

Freight Sheds.

See Cranes, under MECHANICAL ENGINEERING, TRANSPORTING AND CONVEYING.

MISCELLANY.

Canada.

The Railroads of Canada. J. L. Payne. Information concerning the growth of the railroad system, earnings, betterment, operation, etc. 4000 w. R R Gaz—Jan. 24, 1908. No. 89841.

Coal Lands.

Railway Coal Properties Under the Hepburn Act. Shows some of the problems arising from the Act to be solved by lawyers and the law department of the government. 4500 w. Ry Age—Jan. 3, 1908. No. 89361.

We supply copies of these articles. See page 1103.

STREET AND ELECTRIC RAILWAYS

Adhesion System.

A Mountain Electric Railway. Frank C. Perkins. Illustrated description of a railway to the summit of the Altenberg. A combined adhesion and rack railway. 700 w. Sci Am Sup—Jan. 18, 1908. No. 89576.

Cars.

Novel Car Used in Troy, N. Y. Illustrated description of a car practically without platforms, which cannot be boarded after the doors are closed. 1000 w. St Ry Jour—Jan. 18, 1908. No. 89607.

The Best Design of Car for Suburban Traffic at Melbourne. Thomas Tait. Presents the disadvantages of the type of cars in use and suggests desirable changes. 1700 w. R R Gaz—Jan. 17, 1908. No. 89589.

Conduit System.

The Conduit System of Electric Tramway Construction and Recent Improvements. Fitz Roy Roose. Read before the Jun. Inst. of Engrs. Compares the original method of construction of the conduit system with that at present in vogue. Ills. 1400 w. Elec Engr, Lond—Jan. 17, 1908. Serial. 1st part. No. 89865 A.

Double Traction.

The Colonna Double Traction System for Tramways (La Double Traction Electrique des Tramways Système Colonna). J. A. Montpelier. Illustrates and describes a simple double traction system in use at Naples which avoids the expense necessary to equip cars with multiple-unit control apparatus. 2300 w. Electricien—Dec. 14, 1907. No. 89721 D.

Electrification.

See same title and Steam vs. Electricity, under RAILWAY ENGINEERING, MOTIVE POWER AND EQUIPMENT.

Interurban.

See Single Phase, under STREET AND ELECTRIC RAILWAYS.

Locomotives.

Electric Locomotive for Passenger and Freight Service; Metropolitan Railway, of London. Illustrated detailed description. 1500 w. R R Gaz—Jan. 17, 1908. No. 89585.

Motors.

See Railway Motors, under ELECTRICAL ENGINEERING, DYNAMOS AND MOTORS.

Power Plants.

See Central Stations and Hydro-Electric, under ELECTRICAL ENGINEERING, GENERATING STATIONS.

Rack Railways.

See Adhesion System, under STREET AND ELECTRIC RAILWAYS.

Shops.

Anderson Shops of the Indiana Union Traction Company. R. C. Taylor. Illustrated detailed description. 2000 w. Elec Ry Rev—Jan. 18, 1908. No. 89637.

Signals.

Signals for Trolley Car Lines. Illustrates and describes the signal systems for trolley car lines in America and England. 2500 w. Tram & Ry Wld—Jan. 12, 1908. No. 89555 B.

Single Phase.

Single-Phase Equipment of the Windsor, Essex & Lake Shore Rapid Railway. S. C. DeWitt. An illustrated description of the construction and operation of this line in Canada. 3000 w. St Ry Jour—Jan. 11, 1908. No. 89480.

Subways.

Fires on Underground Electric Railways. Briefly discusses this matter especially as regards London underground railways, been made to guard against danger. 2000 w. Engr, Lond—Jan. 17, 1908. No. 89874 A.

The Project for Underground Railways as a Solution for the Traffic Problems of Berlin (Der Entwurf unterirdischer Strassenbahnen in Berlin und die Berliner Verkehrsfrage). E. C. Zehme. An illustrated description of the plan. 2300 w. Elektrotech Zeitschr—Dec. 19, 1907. No. 89787 D.

Systems.

A Critical Study of the Various Systems of Electric Traction (Etude Critique des Différents Systèmes de Traction Electrique). Ills. 9000 w. Soc Belge d'Elec'ns Dec., 1907. No. 89708 E.

Track Bonding.

Bonding. E. Goolding. Discussing the effect of bonding on the resistance of tracks used for return current, giving diagrams for measuring the resistance. 700 w. Tram & Ry Wld—Jan. 2, 1908. No. 89556 B.

Track Construction.

A Comparison of Substructures for Tracks in Streets. H. L. Weber. Illustrates and suggests track schemes for lessening cost and lengthening the life of good tracks. 1200 w. Elec Ry Rev—Jan. 25, 1908. No. 89853.

Track Reconstruction in San Francisco. Illustrates and describes interesting track work carried out since the earthquake. 1200 w. St Ry Jour—Jan. 18, 1908. No. 89606.

Trolley Wires.

See Protective Devices and Wire Suspension, under ELECTRICAL ENGINEERING, TRANSMISSION.

We supply copies of these articles. See page 1103.

EXPLANATORY NOTE—THE ENGINEERING INDEX.

We hold ourselves ready to supply—usually by return of post—the full text of every article indexed in the preceding pages, *in the original language*, together with all accompanying illustrations; and our charge in each case is regulated by the cost of a single copy of the journal in which the article is published. The price of each article is indicated by the letter following the number. When no letter appears, the price of the article is 20 cts. The letter A, B, or C denotes a price of 40 cts.; D, of 60 cts.; E, of 80 cts.; F, of $1.00; G, of $1.20; H, of $1.60. When the letter N is used it indicates that copies are not readily obtainable and that particulars as to price will be supplied on application. Certain journals, however, make large extra charges for back numbers. In such cases we may have to increase proportionately the normal charge given in the Index. In ordering, care should be taken to *give the number* of the article desired, not the title alone.

Serial publications are indexed on the appearance of the first installment.

SPECIAL NOTICE.—To avoid the inconvenience of letter-writing and small remittances, especially from foreign countries, and to cheapen the cost of articles to those who order frequently, we sell coupons at the following prices:—20 cts. each or a book of twelve for $2.00; three books for $5.00.

Each coupon will be received by us in payment for any 20-cent article catalogued in the Index. For articles of a higher price, one of these coupons will be received for each 20 cents; thus, a 40-cent article will require two cou,ons; a 60-cent article, three coupons; and so on. The use of these coupons is strongly commended to our readers. They not only reduce the cost of articles 25 per cent. (from 20c. to 15c.), but they need only a trial to demonstrate their very great convenience—especially to engineers in foreign countries, or away from libraries and technical club facilities.

Write for a sample coupon—free to any part of the world.

CARD INDEX.—These pages are issued separately from the Magazine, printed on one side of the paper only, and in this form they meet the exact requirements of those who desire to clip the items for card-index purposes. Thus printed they are supplied to regular subscribers of THE ENGINEERING MAGAZINE at 10 cents per month, or $1.00 a year; to non-subscribers, 25 cts. per month, or $3.00 a year.

THE PUBLICATIONS REGULARLY REVIEWED AND INDEXED.

The titles and addresses of the journals regularly reviewed are given here in full, but only abbreviated titles are used in the Index. In the list below, *w* indicates a weekly publication, *b-w*, a bi-weekly, *s-w*, a semi-weekly, *m*, a monthly, *b-m*, a bi-monthly, *t-m*, a tri-monthly, *qr*, a quarterly, *s-q*, semi-quarterly, etc. Other abbreviations used in the index are: Ill—Illustrated: W—Words; Anon—Anonymous.

Alliance Industrielle. *m.* Brussels.
American Architect. *w.* New York.
Am. Engineer and R. R. Journal. *m.* New York.
American Jl. of Science. *m.* New Haven, U. S. A.
American Machinist. *w.* New York.
Anales de la Soc. Cien. Argentina. *m.* Buenos Aires.
Annales des Ponts et Chaussées. *m.* Paris.
Ann. d Soc. Ing. e d Arch. Ital. *m.* Rome.
Architect. *w.* London.
Architectural Record. *m.* New York.
Architectural Review. *s-q.* Boston.
Architect's and Builder's Magazine. *m.* New York.
Australian Mining Standard. *w.* Melbourne.
Autocar. *w.* Coventry, England.
Automobile. *w.* New York.
Automotor Journal. *w.* London.
Beton und Eisen. *qr.* Vienna.
Boiler Maker. *m.* New York.
Brass World. *m.* Bridgeport, Conn.
Brit. Columbia Mining Rec. *m.* Victoria, B. C.
Builder. *w.* London.
Bull. Bur. of Standards. *qr.* Washington.
Bulletin de la Société d'Encouragement. *m.* Paris.

Bulletin du Lab. d'Essais. *m.* Paris.
Bulletin of Dept. of Labor. *b-m.* Washington.
Bull. Soc. Int. d'Electriciens. *m.* Paris.
Bulletin of the Univ. of Wis., Madison, U. S. A.
Bulletin Univ. of Kansas. *b-m.* Lawrence.
Bull. Int. Railway Congress. *m.* Brussels.
Bull. Scien. de l'Assn. des Elèves des Ecoles Spéc. *m.* Liège.
Bull. Tech. de la Suisse Romande. *s-m.* Lausanne.
California Jour. of Tech. *m.* Berkeley, Cal.
Canadian Architect. *m.* Toronto.
Canadian Electrical News. *m.* Toronto.
Canadian Engineer. *m.* Toronto and Montreal.
Canadian Mining Journal. *b-w.* Toronto.
Cassier's Magazine. *m.* New York and London.
Cement. *m.* New York.
Cement Age. *m.* New York.
Central Station. *m.* New York.
Chem. Met. Soc. of S. Africa. *m.* Johannesburg.
Clay Record. *s-m.* Chicago.
Colliery Guardian. *w.* London.
Compressed Air. *m.* New York.
Comptes Rendus de l'Acad. des Sciences. *w.* Paris.

Consular Reports. *m.* Washington.
Deutsche Bauzeitung. *b-w.* Berlin.
Die Turbine. *s-m.* Berlin.
Domestic Engineering. *w.* Chicago.
Economic Geology. *m.* New Haven, Conn.
Electrical Age. *m.* New York.
Electrical Engineer. *w.* London.
Electrical Engineering. *w.* London.
Electrical Review. *w.* London.
Electrical Review. *w.* New York.
Electric Journal. *m.* Pittsburg, Pa.
Electric Railway Review. *w.* Chicago.
Electrical World. *w.* New York.
Electrician. *w.* London.
Electricien. *w.* Paris.
Electrochemical and Met. Industry. *m.* N. Y.
Elektrochemische Zeitschrift. *m.* Berlin.
Elektrotechnik u Maschinenbau. *w.* Vienna.
Elektrotechnische Rundschau. *w.* Potsdam.
Elektrotechnische Zeitschrift. *w.* Berlin.
Elettricità. *w.* Milan.
Engineer. *w.* London.
Engineer. *s-m.* Chicago.
Engineering. *w.* London.
Engineering-Contracting. *w.* New York.
Engineering Magazine. *m.* New York and London.
Engineering and Mining Journal. *w.* New York.
Engineering News. *w.* New York.
Engineering Record. *w.* New York.
Eng. Soc. of Western Penna. *m.* Pittsburg, U. S. A.
Foundry. *m.* Cleveland, U. S. A.
Génie Civil. *w.* Paris.
Gesundheits-Ingenieur. *s-m.* München.
Giorn. dei Lav. Pubb. e d Str. Ferr. *w.* Rome.
Glaser's Ann. f Gewerbe & Bauwesen. *s-m.* Berlin.
Heating and Ventilating Mag. *m.* New York.
Ice and Cold Storage. *m.* London.
Ice and Refrigeration. *m.* New York.
Il Cemento. *m.* Milan.
Industrial World. *w.* Pittsburg.
Ingegneria Ferroviaria. *s-m.* Rome.
Ingenieria. *b-m.* Buenos Ayres.
Ingenieur. *w.* Hague.
Insurance Engineering. *m.* New York.
Int. Marine Engineering. *m.* New York.
Iron Age. *w.* New York.
Iron and Coal Trades Review. *w.* London.
Iron Trade Review. *w.* Cleveland, U. S. A.
Jour. of Accountancy. *m.* N. Y.
Journal Asso. Eng. Societies. *m.* Philadelphia.
Journal Franklin Institute. *m.* Philadelphia.
Journal Royal Inst. of Brit. Arch. *s-qr.* London.
Jour. Roy. United Service Inst. *m.* London.
Journal of Sanitary Institute. *qr.* London.
Jour. of South African Assn. of Engineers. *m.* Johannesburg, S. A.
Journal of the Society of Arts. *w.* London.
Jour. Transvaal Inst. of Mech. Engrs., Johannesburg, S. A.
Jour. of U. S. Artillery. *b-m.* Fort Monroe, U. S. A.
Jour. W. of Scot. Iron & Steel Inst. *m.* Glasgow.
Journal Western Soc. of Eng. *b-m.* Chicago.
Journal of Worcester Poly. Inst., Worcester, U. S. A.
Locomotive. *m.* Hartford, U. S. A.
Machinery. *m.* New York.
Manufacturer's Record. *w.* Baltimore.
Marine Review. *w.* Cleveland, U. S. A.
Men. de la Soc. des Ing. Civils de France. *m.* Paris.

Métallurgie. *w.* Paris.
Mines and Minerals. *m.* Scranton, U. S. A.
Mining and Sci. Press. *w.* San Francisco.
Mining Journal. *w.* London.
Mining Science. *w.* Denver, U. S. A.
Mining World. *w.* Chicago.
Mittheilungen des Vereines für die Förderung des Local und Strassenbahnwesens. *m.* Vienna.
Municipal Engineering. *m.* Indianapolis, U. S. A.
Municipal Journal and Engineer. *w.* New York.
Nautical Gazette. *w.* New York.
New Zealand Mines Record. *m.* Wellington.
Oest. Wochenschr. f. d. Oeff. Baudienst. *w.* Vienna.
Oest. Zeitschr. Berg & Hüttenwesen. *w.* Vienna.
Plumber and Decorator. *m.* London.
Power. *w.* New York.
Practical Engineer. *w.* London.
Pro. Am. Soc. Civil Engineers. *m.* New York.
Pro. Canadian Soc. Civ. Engrs. *m.* Montreal.
Proceedings Engineers' Club. *qr.* Philadelphia.
Pro. St. Louis R'way Club. *m.* St. Louis, U. S. A.
Pro. U. S. Naval Inst. *qr.* Annapolis, Md.
Quarry *m.* London.
Queensland Gov. Mining Jour. *m.* Brisbane, Australia.
Railroad Gazette. *w.* New York.
Railway Age. *w.* Chicago.
Railway & Engineering Review. *w.* Chicago.
Railway and Loc. Engng. *m.* New York.
Railway Master Mechanic. *m.* Chicago.
Revista d Obras. Pub. *w.* Madrid.
Revista Tech. Ind. *m.* Barcelona.
Revue d'Electrochimie et d'Electrométallurgie. *m.* Paris.
Revue de Mécanique. *m.* Paris.
Revue de Métallurgie. *m.* Paris.
Revue Gén. des Chemins de Fer. *m.-* Paris.
Revue Gén. des Sciences. *w.* Paris.
Revue Technique. *b-m.* Paris.
Rivista Gen. d Ferrovie. *w.* Florence.
Rivista Marittima. *m.* Rome.
Schiffbau. *w.* Berlin.
School of Mines Quarterly. *q.* New York.
Schweizerische Bauzeitung. *w.* Zürich.
Scientific American. *w.* New York.
Scientific Am. Supplement. *w.* New York.
Sibley Jour. of Mech. Eng. *m.* Ithaca, N. Y.
Soc. Belge des Elect'ns. *m.* Brussels.
Stahl und Eisen. *w.* Düsseldorf.
Stevens Institute Indicator. *qr.* Hoboken, U. S. A.
Street Railway Journal. *w.* New York.
Surveyor. *w.* London.
Technology Quarterly. *qr.* Boston, U. S. A.
Tramway & Railway World. *m.* London.
Trans. Am. Ins. Electrical Eng. *m.* New York.
Trans. Am. Ins. of Mining Eng. New York.
Trans. Am. Soc. Mech. Engineers. New York.
Trans. Inst. of Engrs. & Shipbuilders in Scotland, Glasgow.
Wood Craft. *m.* Cleveland, U. S. A.
Yacht. *w.* Paris.
Zeitschr. f. d. Gesamte Turbinenwesen. *w.* Munich.
Zeitschr. d. Mitteleurop. Motorwagen Ver. *s-m.* Berlin.
Zeitschr. d. Oest. Ing. u. Arch. Ver. *w.* Vienna.
Zeitschr. d. Ver. Deutscher Ing. *w.* Berlin.
Zeitschrift für Elektrochemie. *w.* Halle a S.
Zeitschr. f. Werkzeugmaschinen. *b-w.* Berlin.

Lightning Source UK Ltd.
Milton Keynes UK
UKHW032011020219
336576UK00009BA/143/P